Hematopathology

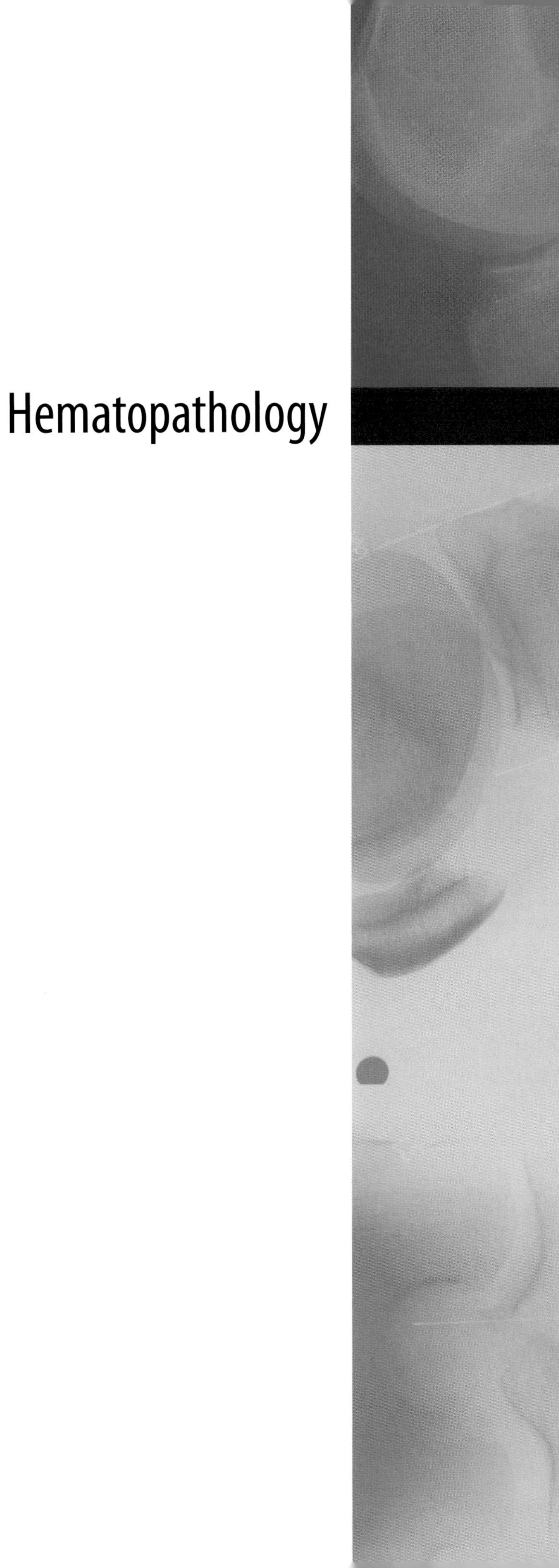

Faramarz Naeim M.D., FASCP

Professor Emeritus, David Geffen School of Medicine, UCLA
Director of Hematopathology
Department of Pathology and Laboratory Medicine
VA Greater Los Angeles Healthcare System
Los Angeles, California

P. Nagesh Rao Ph.D., FACMG

Professor, Pathology and Lab Medicine and Pediatrics
Director of Clinical and Molecular Cytogenetics Laboratories
David Geffen School of Medicine, UCLA
Los Angeles, California

Wayne W. Grody M.D., Ph.D., FACMG, FCAP, FASCP

Professor, Departments of Pathology and Laboratory Medicine,
Pediatrics, and Human Genetics
Director of Molecular Pathology Laboratories
David Geffen School of Medicine, UCLA
Los Angeles, California

Hematopathology

Morphology, Immunophenotype, Cytogenetics and Molecular Approaches

AMSTERDAM • BOSTON • HEIDELBERG • LONDON
NEW YORK • OXFORD • PARIS • SAN DIEGO
SAN FRANCISCO • SINGAPORE • SYDNEY • TOKYO

Academic Press is an imprint of Elsevier

Academic Press is an imprint of Elsevier
525 B Street, Suite 1900, San Diego, CA 92101-4495, USA
30 Corporate Drive, Suite 400, Burlington, MA 01803, USA
84 Theobald's Road, London WC1X 8RR, UK
Radarweg 29, PO Box 211, 1000 AE Amsterdam, The Netherlands

First edition 2008

Notice

Medicine is an ever-changing field. Standard safety precautions must be followed, but as new research and clinical experience broaden our knowledge, changes in treatment and drug therapy may become necessary or appropriate. Readers are advised to check the most current product information provided by the manufacturer of each drug to be administered to verify the recommended dose, the method and duration of administrations, and contraindications. It is the responsibility of the treating physician, relying on experience and knowledge of the patient, to determine dosages and the best treatment for each individual patient. Neither the publisher nor the authors assume any liability for any injury and/or damage to persons or property arising from this publication.

British Library Cataloguing-in-Publication Data
A catalogue record for this book is available from the British Library

Library of Congress Cataloging-in-Publication Data
A catalog record for this book is available from the Library of Congress

ISBN: 978-0-12-370607-2

For information on all Academic Press publications
visit our website at elsevierdirect.com

Typeset by Charon Tec Ltd., A Macmillan Company.
(www.macmillansolutions.com)

Printed and bound in China
09 10 11 12 10 9 8 7 6 5 4 3 2 1

To my Grandchildren Arya and Shayan
Faramarz Naeim

To my Father
P. Nagesh Rao

To my Father
Wayne W. Grody

Contents

Preface ix
Acknowledgements xi
List of Contributors xiii

1. Structure and Function of Hematopoietic Tissues 1
Faramarz Naeim

2. Principles of Immunophenotyping 27
Faramarz Naeim

3. Principles of Cytogenetics 57
P. Nagesh Rao

4. Principles of Molecular Techniques 65
Wayne W. Grody, P. Nagesh Rao, Faramarz Naeim

5. Morphology of Abnormal Bone Marrow 81
Faramarz Naeim

6. Reactive Lymphadenopathies 101
Faramarz Naeim

7. Bone Marrow Aplasia 115
Faramarz Naeim, P. Nagesh Rao, Wayne W. Grody

8. Myelodysplastic Syndromes 129
Faramarz Naeim, P. Nagesh Rao, Wayne W. Grody

9. Chronic Myeloproliferative Diseases 155
Faramarz Naeim, P. Nagesh Rao, Wayne W. Grody

10. Myelodysplastic/Myeloproliferative Diseases 191
Faramarz Naeim, P. Nagesh Rao

11. Acute Myeloid Leukemia 207
Faramarz Naeim, P. Nagesh Rao

12. The Neoplasms of Precursor Lymphoblasts 257
Faramarz Naeim, P. Nagesh Rao, Wayne W. Grody

13. Acute Leukemias of Ambiguous Lineage 279
Faramarz Naeim

14. Lymphoid Malignancies of Non-precursor Cells: General Considerations 287
Faramarz Naeim, P. Nagesh Rao, Wayne W. Grody

15. Mature B-Cell Neoplasms 297
Faramarz Naeim, P. Nagesh Rao, Wayne W. Grody

16. Plasma Cell Myeloma and Related Disorders 373
Faramarz Naeim, P. Nagesh Rao, Wayne W. Grody

17. Mature T-cell and NK-Cell Neoplasms 397
Faramarz Naeim, P. Nagesh Rao, Sophie Song, Wayne W. Grody

18. Hodgkin Lymphoma 441
Sophie Song, Wayne W. Grody, Faramarz Naeim

19. Non-neoplastic and Borderline Lymphocytic Disorders 455
Faramarz Naeim, P. Nagesh Rao, Wayne W. Grody

20. Mastocytosis 477
Faramarz Naeim

21. Histiocytic and Dendritic Cell Disorders 489
Faramarz Naeim

22. Granulocytic Disorders 513
Faramarz Naeim

23. Disorder of Red Blood Cells: Anemias 529
Faramarz Naeim

24. Disorders of Megakaryocytes and Platelets 567
Tom E. Howard, Faramarz Naeim

Index 583

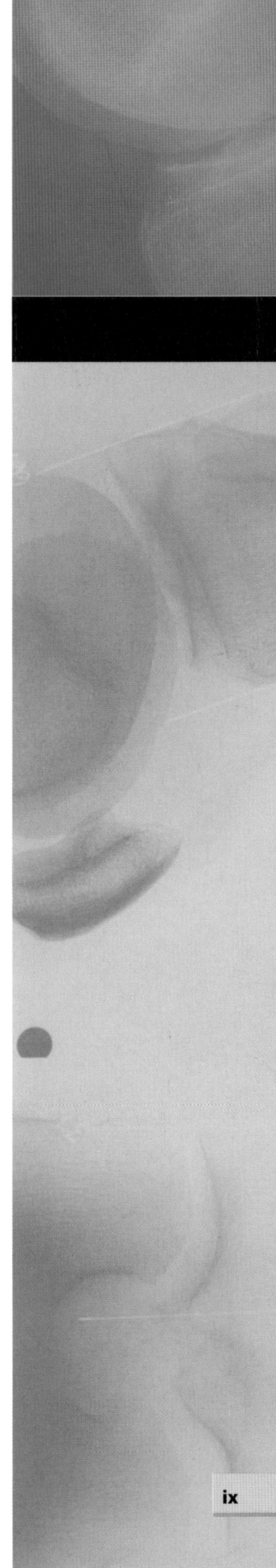

Preface

The main purpose of this book is to provide the reader a relatively comprehensive and concise source of information on the topic of hematopathology. The authors have selected a multidisciplinary approach by correlating morphology with biochemistry, immunophenotyping, cytogenetics and fluorescence *in-situ* hybridization, molecular studies, and clinical aspects. This book offers important information to practicing physicians and those in pathology and hematology training, which will help them to better understand the nature of the hematologic disorders and improve their diagnostic skills along the way. It also functions as a valuable referral book for researchers who work in hematology-related areas.

The book consists of 24 chapters. The first 6 chapters are devoted to normal structure and function of hematopoietic tissues, principles of immunophenotyping, cytogenetics and molecular genetics, an overview of abnormal bone marrow morphology, and reactive lymphadenopathies. Chapters 7 through 24 deal with various types of clonal and non-clonal hematopoietic disorders. These disorders are classified according to the currently published classification by the World Health Organization (WHO).

In planning this book, the authors considered that while there are some excellent hematopathology texts and atlases available, no single volume combines the two in a way that reflects the thought processes a pathologist goes through in evaluating a real-world case. The explosion of new molecular, cytogenetic, and proteomic techniques applicable to pathology has not rendered histologic examination obsolete but rather offers powerful ancillary information to facilitate differential diagnosis, predict prognostic behavior, and help in the selection of targeted molecular therapies. With this in mind, we have constructed most chapters along the general format for each disease category to encompass etiology and pathogenesis, pathologic features (morphology, immunophenotype, cytogenetics, and molecular studies), clinical aspects, and differential diagnosis. To facilitate a better understanding of the text material and the grasp of the information, numerous tables and images are provided throughout the book. Images represent and illustrate the characteristic features of morphology, immunophenotype, cytogenetics, and molecular studies in the majority of the disorders discussed in the book.

Faramarz Naeim, M.D.
P. Nagesh Rao, Ph.D.
Wayne W. Grody, M.D., Ph.D.

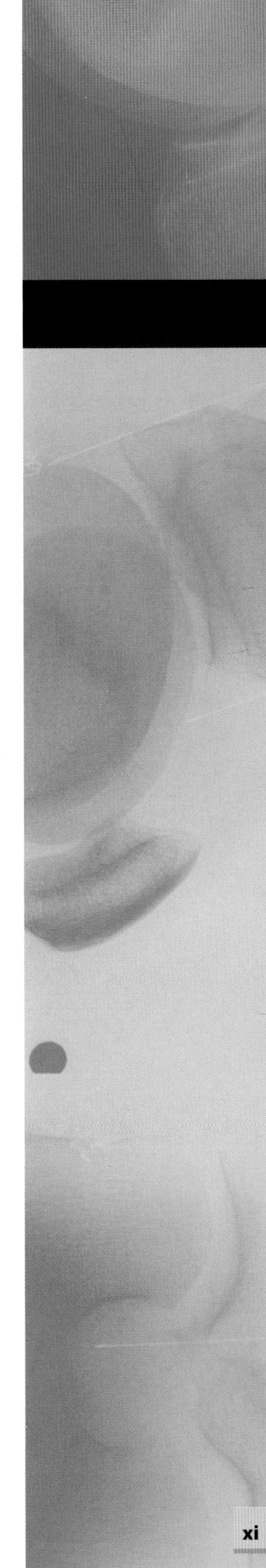

Acknowledgements

We are grateful for the support and encouragement of Farhad Moatamed, M.D., Department of Pathology and Laboratory Medicine, Department of Veterans Affairs, Greater Los Angeles Healthcare System. He generously facilitated our work and offered valuable advice throughout the process of this production.

The authors are thankful to Drs. Sophie Song and Tom Howard for their contributions in writing Chapters 17 and 18, and Chapter 24, respectively.

Preparation and completion of this book would not have been possible without the technical and laboratory assistance of Myrna Fisher and Cecille Repinsky in flow cytometry, Diana Tanaka-Mukai and Eva Archuleta in hematology, and Ivanna Klisak and Audry Teng in cytogenetics. The authors also acknowledge all the Residents, Fellows, and Staff at VA Greater Los Angeles Healthcare System and UCLA for providing the opportunity to teach, learn, and exchange ideas that are reflected in this book.

We would also like to thank Mara E. Conner, the Publishing Editor, and Megan Wickline, the Developmental Editor of Academic Press, for their assistance in the production of this book.

List of Contributors

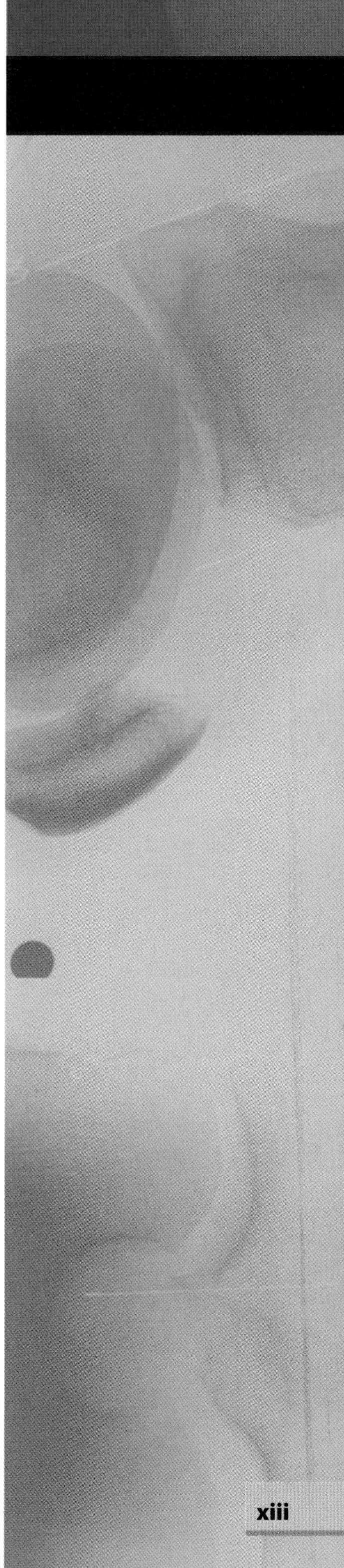

Wayne W. Grody M.D., Ph.D., FACMG, FCAP, FASCP
Professor
Departments of Pathology and Laboratory Medicine, Pediatrics, and Human Genetics
Director of Molecular Pathology Laboratories
David Geffen School of Medicine
UCLA, Los Angeles, California

Tom E. Howard, M.D., Ph.D.
Director
Hemostasis and Pharmacogenetics Laboratories
VA Greater Los Angeles Healthcare System
Associate Professor, Pathology and Laboratory Medicine
Keck School of Medicine
University of Southern California
Los Angeles, California

Faramarz Naeim M.D., FASCP
Professor Emeritus
David Geffen School of Medicine, UCLA
Director of Hematopathology
Department of Pathology and Laboratory Medicine
VA Greater Los Angeles Healthcare System
Los Angeles, California

P. Nagesh Rao Ph.D., FACMG
Professor
Pathology and Lab Medicine and Pediatrics
Director of Clinical and Molecular Cytogenetics Laboratories
David Geffen School of Medicine, UCLA
Los Angeles, California

Sophie Song, M.D., Ph.D.
Assistant Clinical Professor
Director, Clinical Flow Cytometry/ Bone Marrow
Department of Pathology and Laboratory Medicine
David Geffen School of Medicine at UCLA/UCLA Medical Center
Los Angeles, California

1 CHAPTER

Structure and Function of Hematopoietic Tissues

Faramarz Naeim

BONE MARROW STRUCTURE AND FUNCTION

Bone marrow is a mesenchymal-derived complex structure consisting of hematopoietic precursors and a complex microenvironment that facilitates the maintenance of hematopoietic stem cells (HSCs) and supports the differentiation and maturation of the progenitors. All differentiated hematopoietic cells including lymphocytes, erythrocytes, granulocytes, macrophages, and platelets are derived from HSCs.

In the early embryonic life, HSCs first appear in yolk sac and mesodermal tissue of the aorta-gonad-mesonephros region [1–3]. These stem cells then migrate and colonize in a series of early hematopoietic sites including liver, thymus, spleen, and omentum [1–3]. They eventually reside in bone marrow as their permanent home, where they give rise to sequential generations of blood cells throughout adult life. Stem cells have highly specific homing properties, demonstrate very high self-renewal potential, and are capable of differentiation. They share morphologic features of blast cells but are distinguished by their functional properties, such as various colony-forming units (CFUs) and expression of certain differentiation-associated macromolecules. The most primitive (pluripotent) HSCs express CD34 and are negative for CD38 and HLA-DR [4–6]. These primitive cells, which include long-term repopulating stem cells, are also characterized by low level expression of c-kit receptor (CD117) and absence of lineage specific maturation markers. There is a spectrum of heterogeneity in the bone marrow stem cell pool: a continuum of cells with decreasing capacity for self-renewal and increasing potential for differentiation. This trend is also associated with changes in immunophenotypic features. For example, the committed stem cells (short-term repopulating cells), in addition to CD34, appear to express CD38 and/or HLA-DR. The pluripotent HSCs comprise about 1 per 20,000 of bone marrow cells, and only a small fraction of them are active, whereas the remaining majority are in a "resting" phase, on call for action when it is necessary [5–7]. Based on the "clonal succession" hypothesis, a series of stem cells successively contribute to the clonal expansion to maintain a balanced hematopoiesis throughout life [8–10].

The choice of the bone marrow stem cells between self-renewal and differentiation appears to be stochastic, meaning that the commitment of a stem cell to self-renewal or to a particular pair of progeny of given differential potential is a random event and follows the probability rules of statistics [11–13]. In this random process, activation of certain complex nuclear transcription factors appears to play an important role.

Similar to the hematopietic cells, bone marrow stromal cells are derived from pluripotent stem cells [14–16]. In other words, two separate and distinct pluripotent stem cells are simultaneously at work in bone marrow: hematopoietic and stromal. These two systems not only co-exist but closely interact with each other. Stromal cells are composed of a heterogeneous cell population including adipocytes, fibroblast-like cells, endothelial cells, and osteoblasts [17–20]. They produce a number of cytokines and a group of proteins that are involved in facilitating cell–cell interactions and presenting the cytokines and growth factors to the hematopoietic progenitor cells (Table 1.1). Stromal cells with their extracellular matrix make a mesh of fibrovascular environment to home and support the hematopoietic precursors [20–25]. The thin-walled

TABLE 1.1 The main adhesion molecule families.

Adhesion molecule families	Major distribution	Ligand/matrix
(a) Leukocyte cell adhesion molecules (Leu CAM)		
CD11a (LFA-1α)	Leukocytes	ICAM-1
CD11b (MAC-1)	Neutrophils, monocytes	C3bi, ICAM-1
CD11c (gp150/95)	Granulocytes, monocytes	C3bi
CD18 (LFA-1β)	Widespread	CD11a, b, c
(b) Immunoglobulin superfamily		
CD2 (LFA-2)	T lymphocytes	LFA-3
CD50 (ICAM-3)	Leukocytes	LFA-1
CD58 (LFA-3)	Widespread	CD2
CD54 (ICAM-1)	Widespread	LFA-1
CD102 (ICAM-2)	Endothelial cells	LFA-1
CD106 (VCAM-1)	Dendritic cells, endothelial cells	VLA-4
ICAM-4	Erythroid	AlphaVbeta3
(c) Selectins		
CD62E (E-selectin)	Activated endothelial cells	Sialylated Lewis-X
CD62L (L-selectin)	Leukocytes	CD34
CD62P (P-selectin)	Megakaryocytes, endothelial	Sialylated Lewis-X
(d) Sialomucins		
CD34	Stem cells, blasts	L-selectin
CD45	Hematopoietic cells	Heparan sulfate
CD162 (PSGL-1)	Leukocytes	Selectins
(e) Very late activation antigens (VLA)		
CD49a (VAL-1)	Lymphocytes, fibroblasts	Laminin
CD49b (VAL-2)	Fibroblasts, megakaryocytes	Collagen
CD49c (VAL-3)	Stromal cells	Fibronectin, laminin, collagen
CD49d (VAL-4)	Widespread	VCAM-1, fibronectin
CD49e (VAL-5)	Erythroid precursors	Fibronectin
CD49f (VAL-6)	Widespread	Laminin

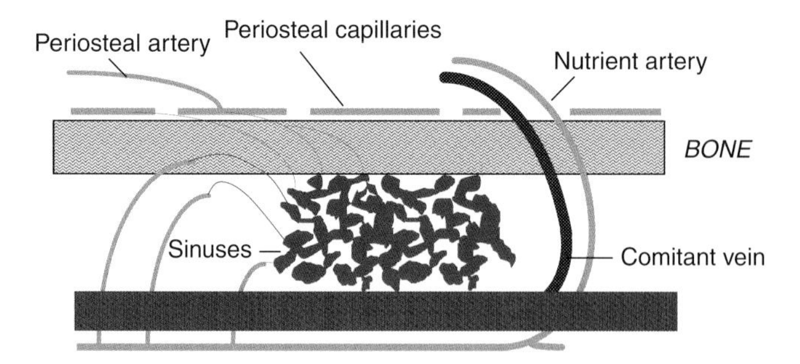

FIGURE 1.1 Schematic of microvascular circulation in the bone marrow. Adapted from De Bruyn PPH. (1981). Structural substrates of bone marrow function. *Semin Hematol* **18**, 179.

TABLE 1.2 Major components of extracellular matrix in bone marrow.

Type	Comments
Collagen (reticulin)	Consisting of various subtypes. Erythroid and myeloid precursors adhere to collagen types I and VI.
Fibronectin	Attaches to early erythroid precursors and other hematopoietic and stromal cells.
Hemonectin	Myeloid precursors adhere to laminin. Regulates leukocyte chemotaxis.
Proteoglycans	Components containing heparin sulfate, chondroitin sulfate, and hyaluronic acid. Interact with laminin and type IV collagen and play a role in cytokine presentation and cell differentiation.
Thrombospondin	Interacts with collagen, fibronectin, and CD36.

venous sinuses are the most prominent vascular spaces in the bone marrow. They consist of an inside layer of endothelial cells supported by an outer layer of fibroblast-like (parasinal, adventitial) stromal cells. They receive blood from the branches of the nutrient artery and periosteal capillary network. The nutrient artery penetrates the bony shaft, branches into the bone marrow cavity, and forms capillary–venous sinus junctions [26, 27]. The periosteal capillary network connects with the sinuses at the bone marrow junction through the Haversian canals. The smaller venous sinuses drain into larger centrally located sinuses, which connect together to form the comitant vein. The comitant vein and the nutrient artery run through the bone marrow cavity adjacent to one another in the same vascular canal (Figure 1.1).

The HSCs reside in microenvironmental niches. These niches, which are composed of stromal cells, accessory cells (such as T lymphocytes and macrophages), components of extracellular matrix (Table 1.2), and various regulatory cytokines (Table 1.3), play an important role in the regulation of hematopoiesis and proliferation of the committed stem cells [28–30]. These niches create topographical patterns in the bone marrow [25, 31–32]. For example, granulopoiesis primarily takes place in periosteal areas where the concentration of hemonectin is higher. Recent studies suggest

TABLE 1.3 Regulatory cytokines.

Cytokine	Primary effect
GM-CSF[1]	Granulocyte and macrophage colony formation, functional enhancement of mature forms
G-CSF[2]	Granulocyte colony formation, functional enhancement of granulocytes
M-CSF (CSF-1)[3]	Macrophage colony formation, functional enhancement of monocytes and macrophages
Erythropoietin (EPO)	Erythropoiesis, possible enhancement of megakaryocyte proliferation
Thrombopoietin (TPO)	Megakaryocyte proliferation, platelet production
Steel factor (c-kit ligand)	Stem cell and mast cell proliferations
Interleukin (IL)-1	Promoter of hematopoiesis, inducer of other factors, B- and T-cell regulators, endogenous pyogen
IL-2	T-cell growth factor, may inhibit G/M colony formation and erythropoiesis
IL-3 (multi-CSF)	G/M colony formation, syngeneic effects on EPO, eosinophil, mast cell, and megakaryocyte colony formation
IL-4	B-cell proliferation, IgE production
IL-5	Eosinophil growth and B-cell differentiation
IL-6	B-cell differentiation, synergestic effects on IL-1
IL-7	Development of B- and T-cell precursors
IL-8	Granulocyte chemotactic factor
IL-9	Growth of mast cells and T-cells
IL-10	Inhibitor of inflammatory and immune responses
IL-11	Synergestic effects on growth of stem cells and megakaryocytes
IL-12	Promoter of Th1 and suppressor of Th2 functions
IL-13	B-cell proliferation, IgE production
IL-15	Activates T-cells, neutrophils and macrophages
IL-16	Chemotactic factor for helper T-cells
IL-17	Promotes T-cell proliferation, pro-inflammatory activities
IL-18	Activates T-cells, neutrophils, and fibroblasts
IL-19	Member of IL-10 family, transcriptional activator of IL-10
IL-20	Member of IL-10 family with epidermal function
IL-21	Improves proliferation of T- and B-cells, and enhances NK cytotoxic activities
IL-22	Member of IL-10 family; induces inflammatory responses
IL-23	Activates autoimmune responses
IL-24	Member of IL-10 family, tumor suppressor molecule
IL-25	Capable of amplifying allergic inflammation
IL-26	Member of IL-10 family; plays a role in mucosal and cutaneous immunity
TGF-β[4]	Suppresses BFU-E, CFU-S, and HPP-CFC
Interferons	Suppress BFU-E, CFU-GEMM, and CFU-GM
TNF[5]-α and -β	Suppress BFU-E, CFU-GEMM, and CFU-GM
PGE[6]-1 and -2	Suppress GFU-GM, GFU-G, and GFU-M
Lactoferrin	Suppresses release of IL-1

[1] Granulocyte and macrophage colony-stimulating factor.
[2] Granulocyte colony-stimulating factor.
[3] Macrophage colony-stimulating factor.
[4] Transforming growth factor.
[5] Tumor necrosis factor.
[6] Prostaglandin E.

that endosteal osteoblasts and their precursors play a role in the creation of stem cell niches. Osteoblasts release regulatory components that influence stem cell function, such as G-CSF, M-CSF, GM-CSF, IL-1, and IL-6, and others [14]. Erythropoiesis takes place in distinct anatomical foci referred to as "erythroblastic islands." These islands are rich in fibronectin and consist of erythroid precursors surrounding a central macrophage [33, 34]. They are usually away from bone trabeculae and are located subjacent to the vascular structures. Megakaryocytes are mostly located in proximity of the sinuses. In this location a portion of their cytoplasm enters into the sinusoidal space to release platelets. The bone marrow niches support and regulate hematopoiesis, leading to the production of huge numbers of progenitor cells and differentiated mature blood cells (Figure 1.2).

Every day, an estimated 2.5 billion red cells, 2.5 billion platelets, and 1.0 billion granulocytes are produced per kilogram body weight in normal conditions.

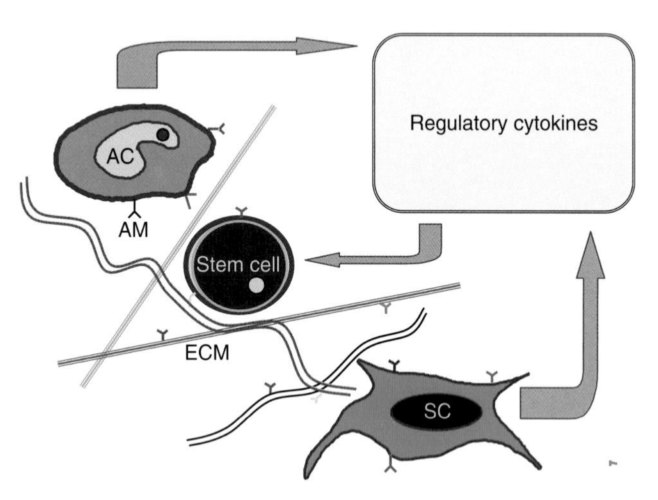

FIGURE 1.2 Cytokines released from accessory cells (AC) (e.g. macrophages, T-cells) and stromal cells have a regulatory effect on stem cells. The extracellular matrix (ECM) and adhesion molecules (AM) support cell–cell, cell–matrix, and cell–cytokine interactions.

Erythropoiesis apparently begins with the commitment of a small pool of pluripotent stem cells to a primitive committed cell to non-lymphoid lineages referred to as CFU-GEMM. GEMM stands for granulocytes, erythrocytes, macrophages, and megakaryocytes. The most primitive committed erythroid progenitor in humans is the erythroid burst-forming unit (BFU-E), which divides and forms subpopulations of erythroid colonies known as colony-forming units (CFU-E). This process requires a combination of cytokines, such as erythropoietin (EPO), Steel factor (SF; c-kit ligand), and interleukin-3 (IL-3) (Figure 1.3) [33, 35, 36]. EPO is necessary for the CFU-E formation and terminal differentiation of erythroid progenitors. SF has no erythroid colony-forming ability but has marked synergistic effects on BFU-E formation in the presence of EPO [36]. Similarly, IL-3 and granulocyte-macrophage colony stimulating factor (GM-CSF) enhance EPO-dependent erythropoiesis (Table 1.3). Proliferation and maturation of CFU-E leads to the formation of erythroblasts, more mature erythroid precursors, and eventually enucleated reticulocytes and erythrocytes. The entire process requires approximately 2 weeks. Except for the newborns, only erythrocytes and polychromatic erythrocytes (reticulocytes) are released into the blood circulation.

Myelopoiesis begins with the differentiation of a small population of pluripotent stem cells to CFU-GEMM and then to the committed primitive myeloid precursors, granulocyte/macrophage colony-forming units (CFU-GM) [37–40]. This process requires GM-CSF, SF, IL-3, and IL-6 (Figure 1.3). CFU-GM give rise to more mature colony-forming units CFU-G, CFU-M, CFU-Eo, and CFU-Baso which in turn differentiate into neutrophils, macrophages, eosinophils, and basophils, respectively (Figure 1.3) [39, 41].

The neutrophilic precursors in bone marrow consist of two major compartments: mitotic and post-mitotic. The mitotic compartment consists of cells that are able to proliferate, such as myeloblasts, promyelocytes, and myelocytes. The post-mitotic compartment includes metamyelocytes, bands, and segmented neutrophis, representing more differentiated cells with no proliferating capacity. The released

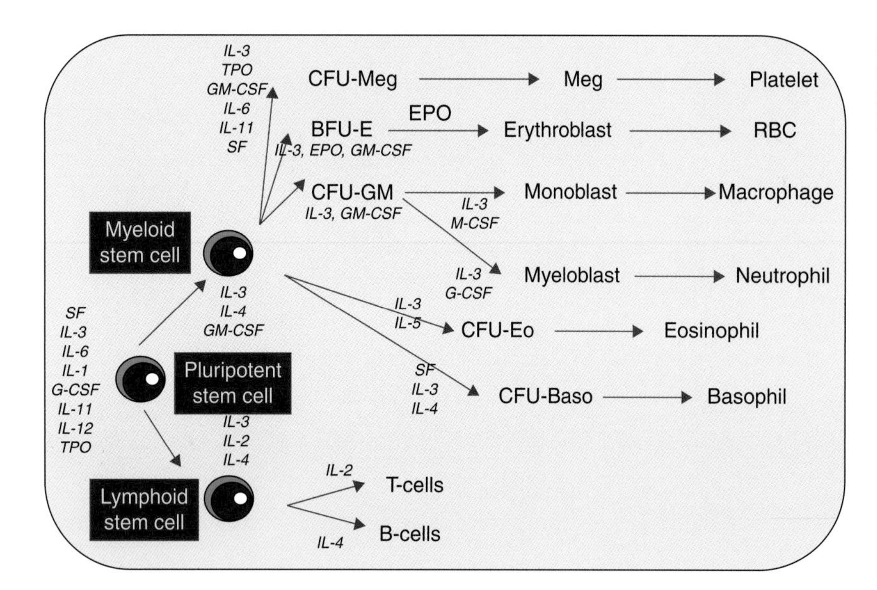

FIGURE 1.3 Current scheme of hematopoiesis demonstrating the differentiation of the multipotent stem cell to hematopoietic precursors and various levels of cytokine interaction.

granulocytes from bone marrow into the circulation consist of two components: the marginating pool and the circulating pool. These two components are in equilibrium. Granulocytes reside in the blood for an average of 10h and leave the circulation toward the inflammation sites in various tissues.

Eosinophil and basophil maturation appears to be analogous to neutrophil maturation, though the involved regulatory cytokines are different. GM-CSF and IL-5 play a major role in the development of the eosinophilic lineage, whereas SF, IL-3, IL-4, IL-9, and IL-10 regulate the basophilic development (Figure 1.3) [39, 41].

Monocytic maturation, assisted by GM-CSF and M-CSF, begins with the formation of monoblasts followed by promonocytes and monocytes. Monocytes are released into the circulation, and from there, they travel into various tissues and become various types of tissue macrophages (histiocytes) and dendritic cells. Pulmonary alveolar macrophages, hepatic Kupffer cells, pleural and peritoneal macrophages, osteoclasts, Langerhans and interdigitating dendritic cells in various tissues, and perhaps microglial cells in the central nervous system are all examples of cells derived from a monocytic lineage [42, 43].

Thrombopoiesis begins with maturation of CFU-GEMM into a colony-forming unit with a high proliferating response to cytokines (e.g. IL-1, IL-3, and IL-6) referred to as the high proliferative potential–colony-forming unit–megakaryocyte (HPP-CFU-MK) [44, 45]. The next step is the formation of a burst forming unit (BFU-MK), which is capable of producing numerous megakaryocytic colony-forming units (CFU-Meg). This process is regulated by SF, IL-3, GM-CSF plus thrombopoietin (TPO), and IL-11 (Figure 1.3) [45, 46]. Maturation of CFU-Meg leads to the formation of megakaryoblasts, megakaryocytes, and, eventually, platelets [45,47].

Lymphopoiesis begins in the bone marrow with the committed lymphoid stem cells [25, 48–50]. The precise mechanism involved in the differentiation of HSCs into lymphoid precursors is not well understood. However, it is clear that the first step is the separation of B progenitor cells from non-B progenitor cells (T-cells and natural killer cells) [51–53]. This progression does not need interaction with exogenous antigen and therefore is considered the "antigen-independent" phase.

The development of B progenitor cells is influenced by a number of regulatory cytokines including IL-1, IL-2, IL-4, IL-10, and interferon gamma (Figure 1.3) [51, 54, 55]. The B-cell precursors in bone marrow are known as hematogones [56, 57]. These cells are found in small numbers in normal bone marrow (usually between 5% and 10% in young children and <5% in adults) but may increase in regenerating marrows.

T lymphocytes derive from the precursor lymphoid cells in the marrow (pre-thymic phase) under the influence of several cytokines, such as IL-1, IL-2, and IL-9, and then migrate to the thymus for further maturation [50–53, 58]. A subclass of large granular lymphocytes, natural killer (NK) cells, appears to share a common progenitor cell with T-cells in the marrow [59–62]. The NK-cells demonstrate HLA-nonrestricted cytotoxicity and release various regulatory cytokines, such as IL-1, IL-2, IL-4, and interferons.

Bone Marrow Examination

Bone marrow samples are obtained and prepared for pathologic evaluation in different ways, such as biopsies, clotted aspirated marrow particles, marrow smears, and touch preparations (Figures 1.4 and 1.5) [26, 63–67].

Bone marrow biopsy sections are evaluated for the estimation of bone marrow cellularity and for the identification of pathological processes, such as primary hematologic disorders, granulomatosis, amyloidosis, fibrosis, osteosclerosis, and metastasis. Biopsy sections are routinely stained with hematoxylin and eosin (H&E stain) (Figures 1.5 and 1.6). In addition, in some laboratories, sections are stained with periodic acid Schiff (PAS) technique. Bone marrow cellularity is defined as the percentage of the bone marrow areas occupied by cells (% cellularity = 100% area occupied by fat). Cellularity of the bone marrow varies depending on the location of the marrow sample and the age of the individual. For example, bone marrow cellularity is higher in the vertebrae than in the pelvic bone, and higher in the children than in the elderly. Bone marrow cellularity approaches 100% at birth and continues to decline approximately 10% for each decade of life. In a 50-year-old healthy person, the average bone marrow cellularity is about 50% (Figures 2.5 and 2.6).

Bone marrow clot sections (particle sections) are prepared from aspirated bone marrow and therefore are devoid of bone trabeculae. They only represent cells and lesions that are released by aspiration (Figures 1.4 and 1.5). Clot sections are routinely stained with H&E. Some laboratories may also use PAS stain.

Bone marrow smears are prepared by smearing the aspirated marrow over the glass slides. Marrow smears are usually stained with Wright's (or in some laboratories with Geimsa) stains (Figures 1.5–1.7). They are used primarily for the cytological evaluations, cellular details, maturation steps, differential count, and assessment of the myeloid:erythroid (M:E) ratio (normal range = 2–3). Differential counts reflect the percent of different hematopoietic cells in bone marrow smears. At least 200 cells are counted by randomly selected areas of a properly stained and adequately cellular marrow smear to calculate the differential count (Table 1.4). Marrow smears are also useful for special cytochemical stains and evaluation of the bone marrow iron stores.

Bone marrow touch (imprint) preparations are made by gently touching (pressing) the glass slides over the biopsy sample and are routinely stained with Wright's and/or Geimsa stains (Figures 1.4–1.6). Touch preparations most often are not optimal for morphologic evaluations, because their preparation creates significant artifacts. However, they are the only source of cytologic evaluation when bone marrow aspiration fails to yield (dry tap).

Morphologic Characteristics of Hematopoietic Cells

Granulocytic Series

Granulocytic series include neutrophilic, eosinophilic, basophilic, and mast cell lineages. The morphologic steps

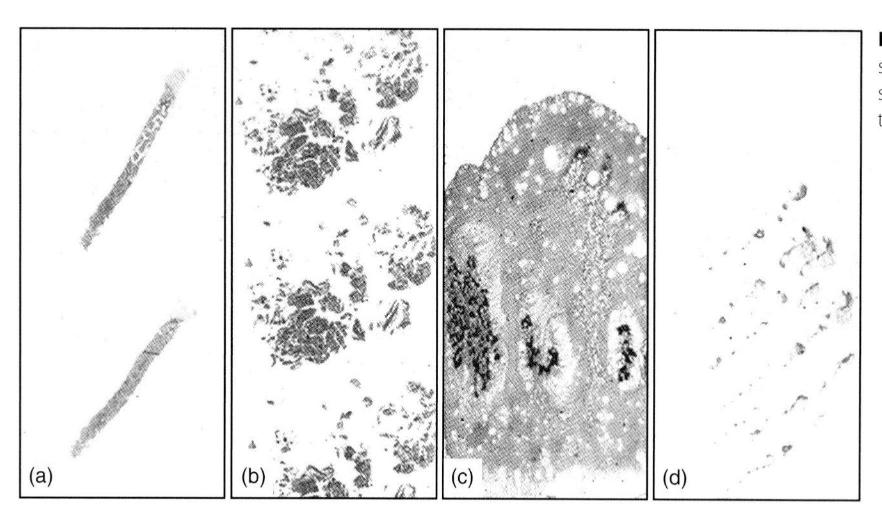

FIGURE 1.4 Representative examples of glass slide preparations of bone marrow biopsy (a), clot sections (b), bone marrow aspirate smear (c), and touch preparation (d).

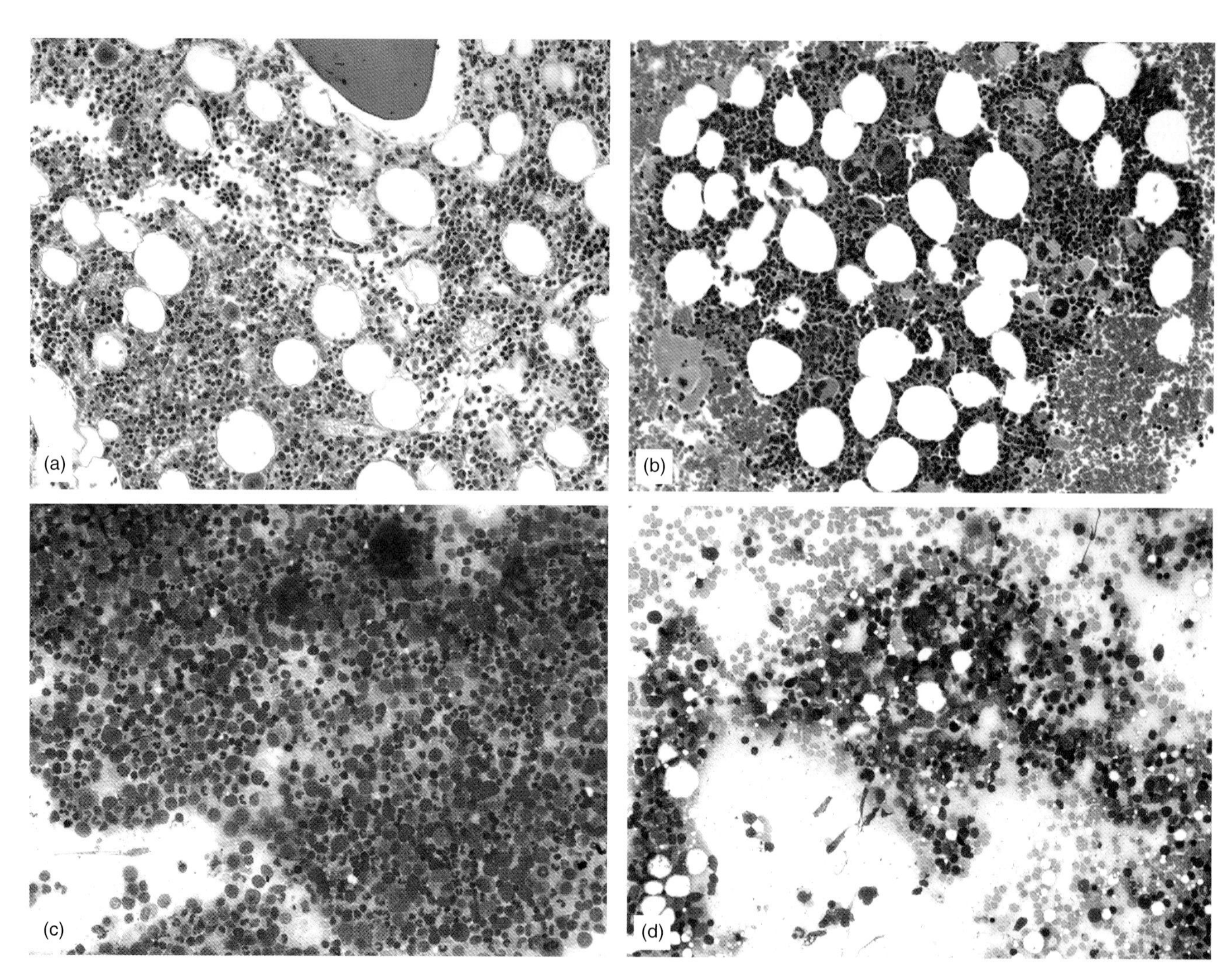

FIGURE 1.5 Bone marrow preparations: Biopsy section (a), clot section (b), aspirate smear (c), and touch preparation (d).

in the maturation process of the granulocytic series include myeloblast, promyelocyte, metamyelocyte, band, and segmented cell (Figures 1.7–1.9). During this process, the cytoplasmic:nuclear ratio increases, cytoplasm accumulates lyzosomal granules that are non-specific at first (primary granules, azurophilic granules) and become specific (secondary granules) later. The nuclear chromatin becomes coarser and denser, and the nucleoli appear less prominent

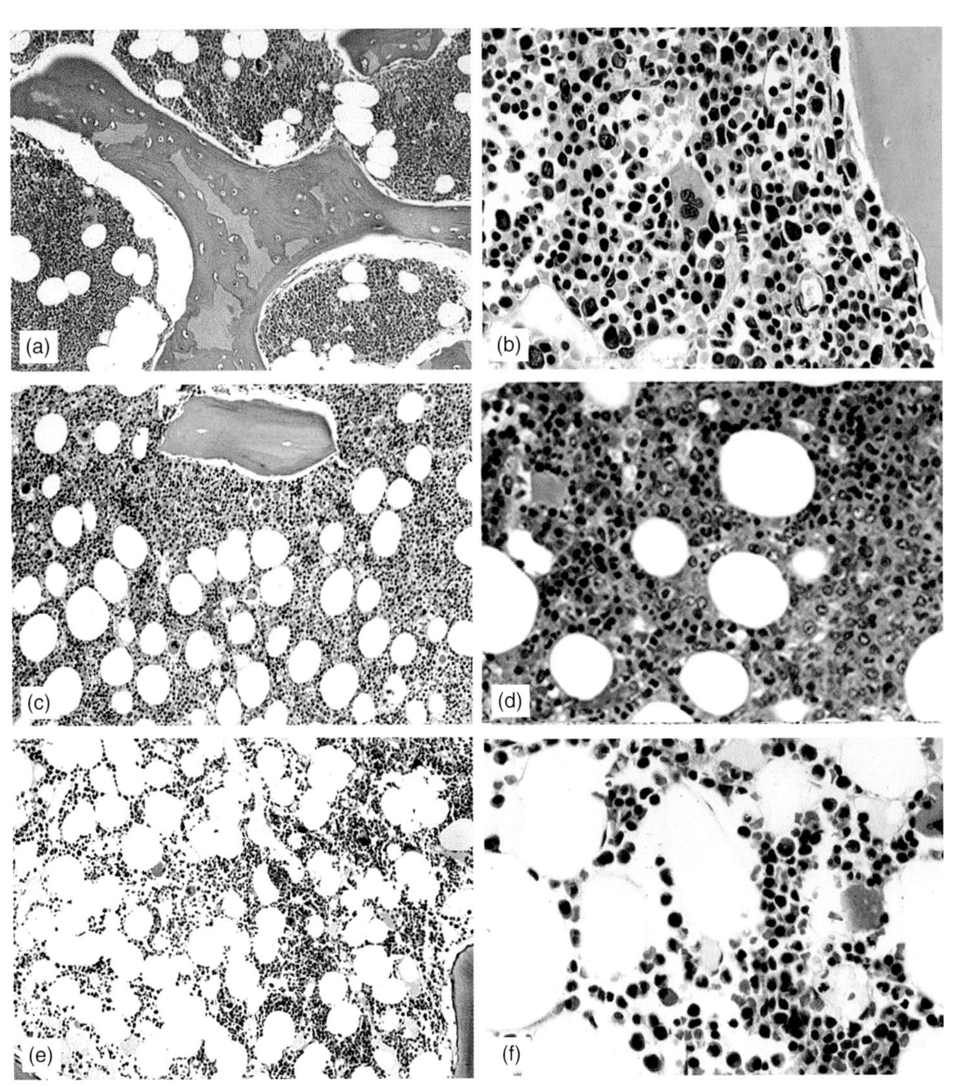

FIGURE 1.6 Bone marrow cellularity declines by age. (a) and (b); (c) and (d); and (e) and (f) are biopsy sections from 2-year, 55-year, and 75-year-old individuals, respectively. From Ref. [63] by permission.

and indistinct. The nuclear shape gradually changes from round/oval to kidney-shaped and segmented forms.

Myeloblasts are the earliest granulocytic precursors identified by morphologic evaluations. They range in size from 10 to 20 μm and are characterized by a high nuclear:cytoplasmic (N:C) ratio, a centrally located round or oval nucleus, finely dispersed chromatin and several nucleoli. Based on their cytoplasmic granules, myeloblasts are divided into three types (Figure 1.8) [68]:

Type I myeloblasts contain no cytoplasmic granules.

Type II myeloblasts contain <20 cytoplasmic azurophilic granules.

Type III myeloblasts contain >20 cytoplasmic azurophilic granules.

Myeloblasts are positive for CD13, CD33, and HLA-DR and may express CD117, CD34, and myeloperoxidase (MPO) [69, 70].

Promyelocytes are overall larger than myeloblasts, ranging from 13 to 25 μm in diameter. They carry more cytoplasm and contain larger quantities of azurophilic granules than myeloblasts. They depict a perinuclear pale area (a well-developed Golgi system) and a round or oval nucleus, which is often eccentric. Type III myeloblasts and promyelocytes share overlapping morphologic features, and therefore their distinction at times is difficult (Figures 1.7 and 1.8). Myeloblasts are HLA-DR-positive and may express CD34, whereas promyelocytes are negative for HLA-DR and CD34. Promyelocytes are positive for CD13, CD33, and MPO and may express CD117 [69, 70].

Myelocytes are smaller than promyelocytes and are characterized by a reduced N:C ratio with ample granular cytoplasm containing both primary and secondary granules. At the myelocytic stage the production of primary granules stops and the synthesis of specific granules enhances. Myelocytes depict a round or oval nucleus with coarse chromatin and often lack distinct nucleoli (Figures 1.7 and 1.9). Myelocytes are positive for CD13, CD15, and MPO, and may express CD33. They are negative for CD34, CD117, and HLA-DR.

Metamyelocytes are slightly smaller than myelocytes and are characterized by abundant granular cytoplasm with predominance of specific granules, kidney-shaped or indented nucleus, coarser chromatin, and lack of distinct

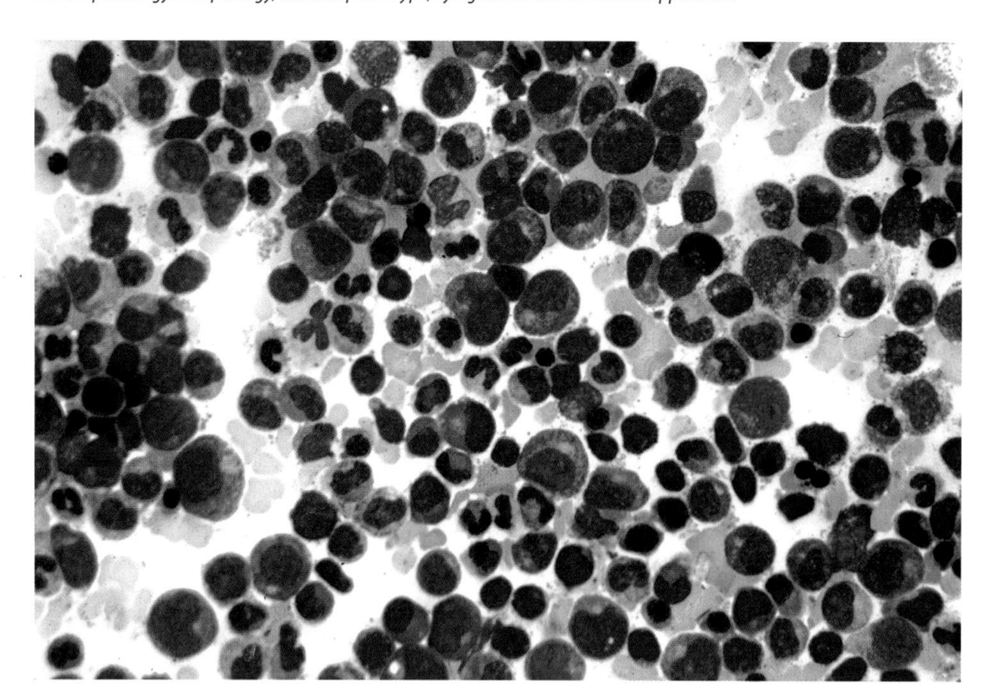

FIGURE 1.7 Bone marrow smears demonstrating myeloid cells in various stages of maturation. Scattered erythroid precursors and lymphocytes are also present.

TABLE 1.4 Approximate ranges of bone marrow differential counts in healthy persons.*

	Age	
Cell type	**18 months**	**Adult**
Myeloid		
Myeloblast	N/A	1–5
Promyelocyte	1–2	1–8
Neutrophilic series		
Myelocytes	2–4	5–19
Metamyelocyte	8–16	13–22
Band/Segs	14–25	21–40
Eosinophilic series	1.5–3.5	0.5–3
Basophilic series	<1	<1
Monocytic series	1–3	1–4
Erythroid		
Rubriblast	<1	0.5–2
Prorubricyte	0.5–1	1.5–6
Rubricyte	4–10	5–25
Metarubricyte	<1	2–20
Megakaryocyte	<1	0.5–2
Lymphocyte	40–42	3–20
Plasma cell	<1	0.5–2
M:E ratio	4–5:1	3–3.5:1

*References: Williams (1990), Brunning (1994), Bain (1996), Naeim (1998), and Greer (1999).

nucleoli (Figure 1.9). Metamyelocytes express CD13, CD15, and MPO and are negative for CD33, CD34, CD117, and HLA-DR.

Bands and **segmented cells** are the end stage cells in the granulocytic series and are distinguished by abundant cytoplasm with specific granules, lack of sparse primary granules, condensed nuclear chromatin with indistinct nucleolus, and nuclear lobulation or segmentation. Neutrophilic bands (stabs) are cells with bilobed nuclei with no filament formation, and neutrophilic segmented cells (Segs) demonstrate up to five distinct nuclear lobules (segments) connected to one another by filaments (Figure 1.9). Neutophilic bands and segmented cells demonstrate alkaline phosphatase activity and are positive for CD11c, CD15, CD16, and MPO and may express CD13.

Other granulocytic lineages, such as ***eosinophils*** and ***basophils***, undergo more or less similar differentiation steps. Mature eosinophils, unlike segmented neutrophils, usually have bilobed nuclei and are loaded with eosinophilc granules [71, 72]. Eosinophilic granules are larger than the neutrophilic granules (Figure 1.10). Mature basophils contain a large number of coarse basophilic granules and show less nuclear segmentation than the neutrophils (Figure 1.10) [73–75]. ***Mast cells*** appear to be closely related to the basophils by sharing certain characteristics, such as basophilic granules, IgE receptor, and histamine content [73, 75, 76]. However, mast cells live longer, are larger, have more abundant cytoplasm than basophils, and their nucleus is round, oval, or spindle-shaped without segmentation. Mast cell cytoplasmic granules are MPO negative and are more numerous and more variable in appearance than the granules in basophils (Figure 1.11). Mast cells express CD117 and tryptase [77, 78].

Monocytes and macrophages are derived from the same committed stem cells (CFU-GM) as the granulocytic cells [42, 43]. The maturation process in this lineage starts from *monoblast*, and then goes through *promonocyte*, *monocyte*, *macrophage (histiocyte)*, and *multinucleated giant cell* (such

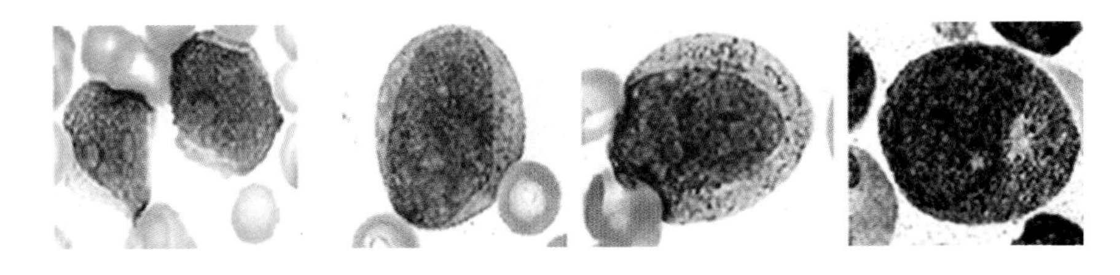

FIGURE 1.8 Left to right: myleoblast types I, II and III and a promyelocyte. From Ref. [63] by permission.

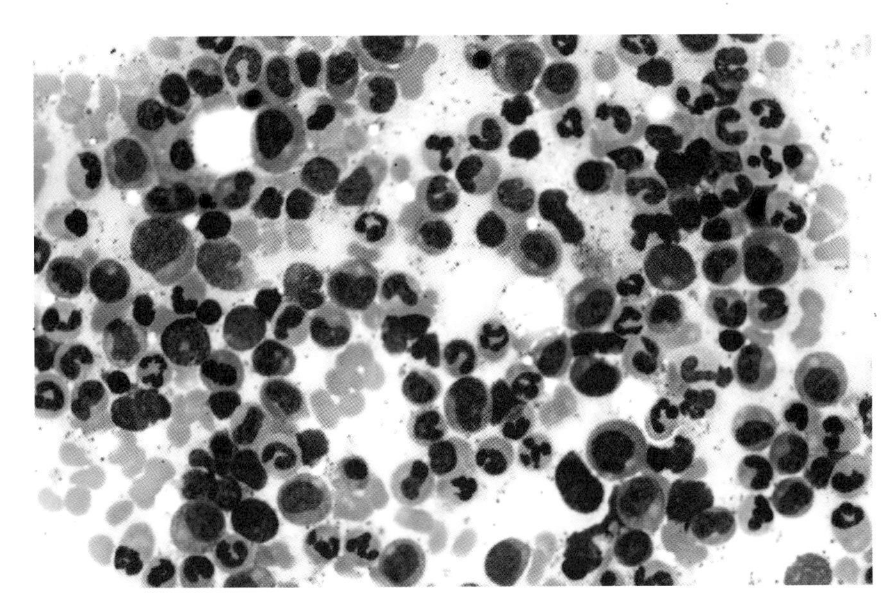

FIGURE 1.9 Bone marrow smears demonstrating myeloid cells in various stages of maturation.

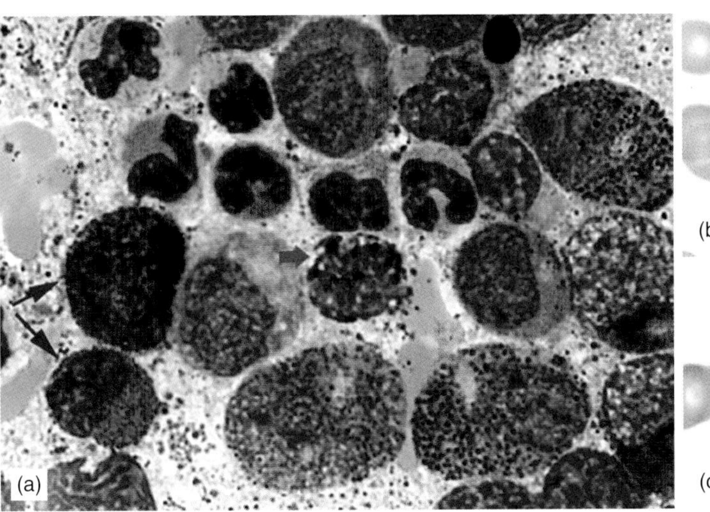

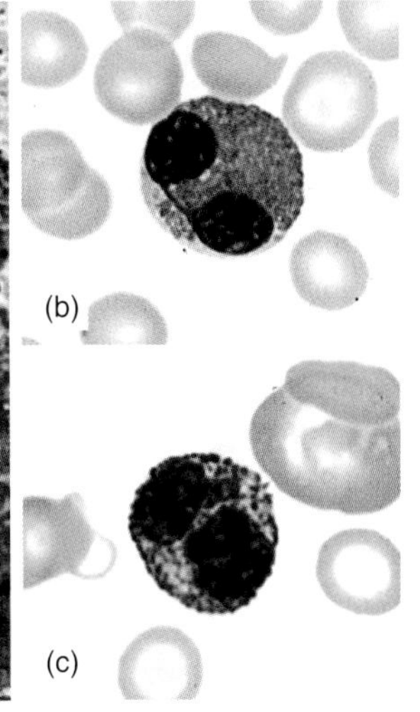

FIGURE 1.10 (a) A bone marrow smear showing granulocytic precursors including eosinophilic myelocytes (blue arrows) and a basophil (red arrow). An eosinophil (b) and a basophil (c) are demonstrated in blood smears. From Ref. [63] by permission.

as osteoclasts or giant cells in granulomas) (Figures 1.12 and 1.13). During the maturation process from monoblast to monocyte, nucleoli become folded, nuclear chromatin gets more condensed, nucleoli disappear, and cytoplasm acquires lysosomal granules (Figure 1.12). Monocytes are positive for CD13, CD14, CD15, CD64, CD11 (b and c), HLA-DR, and non-specific esterase, and may express CD33 and/or CD68 [79]. Monocytes are released from bone marrow into the blood circulation, and from there they migrate out into various tissues, and finally transform to soft tissue histiocytes (macrophages). Iron is stored in bone marrow macrophages as hemosiderin (insoluble aggregates) or less abundantly as ferritin (soluble). Prussian blue (potassium ferrocyanide) stains hemosiderin as dark blue cytoplasmic granules (Figure 1.13a).

Dendritic cells are considered a subclass of histiocytic lineage and are primarily involved in antigen presentation to lymphocytes [80–83]. Dendritic cells, except for the follicular dendritic cells (FDCs), are derived from bone marrow stem cells. These cells are divided into two groups: Langerhans cells (LCs) (Figure 1.13c) and interdigitating dendritic cells (IDCs). LCs are primarily located in the skin and are characterized by the ultrastructural Birbeck granules and the expression of CD1a, CD4, S100, HLA-DR, and Langerin (CD207) [82]. Unlike macrophages, LCs are usually negative for CD68, non-specific esterase,

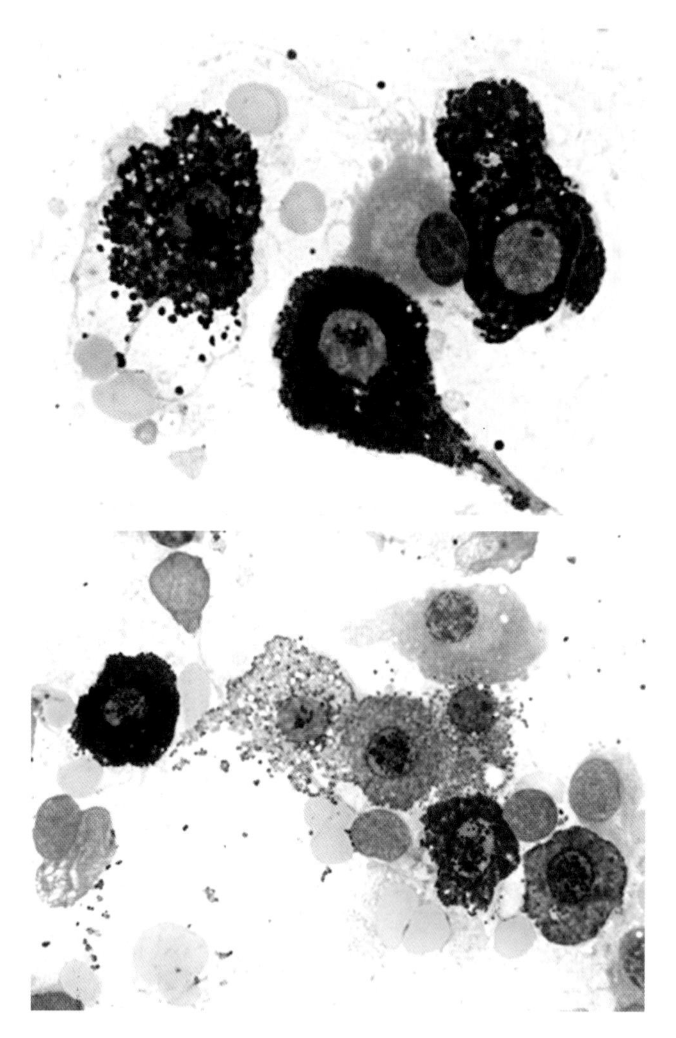

FIGURE 1.11 Bone marrow smears demonstrating mast cells with various amounts of cytoplasmic granules.

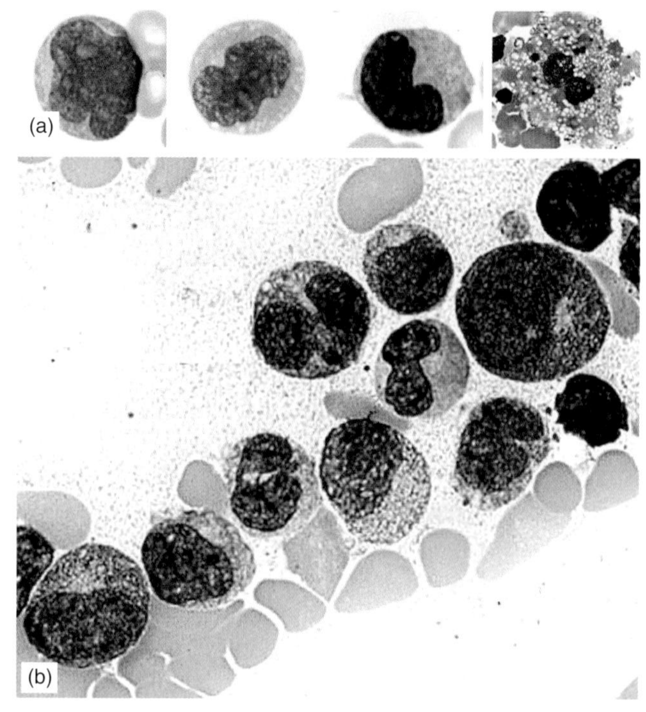

FIGURE 1.12 Left to right: (a) monocytic maturation from monoblast to promonocyte, monocyte, and macrophages. Bone marrow smear (b) showing several monocytic cells. From Ref. [63] by permission.

and lysozyme. IDCs are found in the lymphoid tissues and show immunophenotypic features similar to those of LCs, except for lack of CD1a and CD4 expression. The FDCs are derived from the mesenchymal cells in the follicular structures in the lymph nodes and express CD21 and CD35 [84, 85].

Erythroid Precursors

During the maturation process in erythropoiesis, cells gradually become smaller, the cytoplasmic:nuclear ratio increases, cytoplasm accumulates hemoglobin, the nuclear chromatin becomes denser and pyknotic, and the nucleoli appear less prominent and indistinct. The nucleus is eventually extruded from the cell, resulting in the development of polychromatophilic, and then mature red blood cells (RBCs) (Figures 1.14 and 1.15) [26, 64]. Erythroid precursors express several membrane-associated molecules, such as CD71, CD235 (glycophorin), CD238 (Kell blood group), CD240 (Rh blood group), and CD242.

Rubriblasts (erythroblast, pronormoblast) are the earliest morphologically distinguished eythroid precursors. They measure 15–30 μm in diameter and have a high

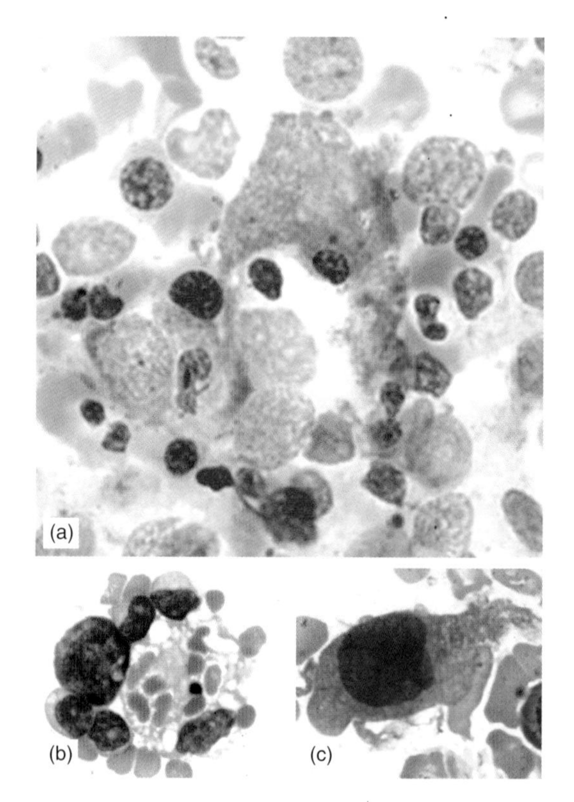

FIGURE 1.13 (a) A bone marrow clot section with iron stain showing iron-laden macrophages. (b) An erythrophagocytic macrophage surrounded by lymphoid cells and a rubriblast. (c) A Langerhans cell.

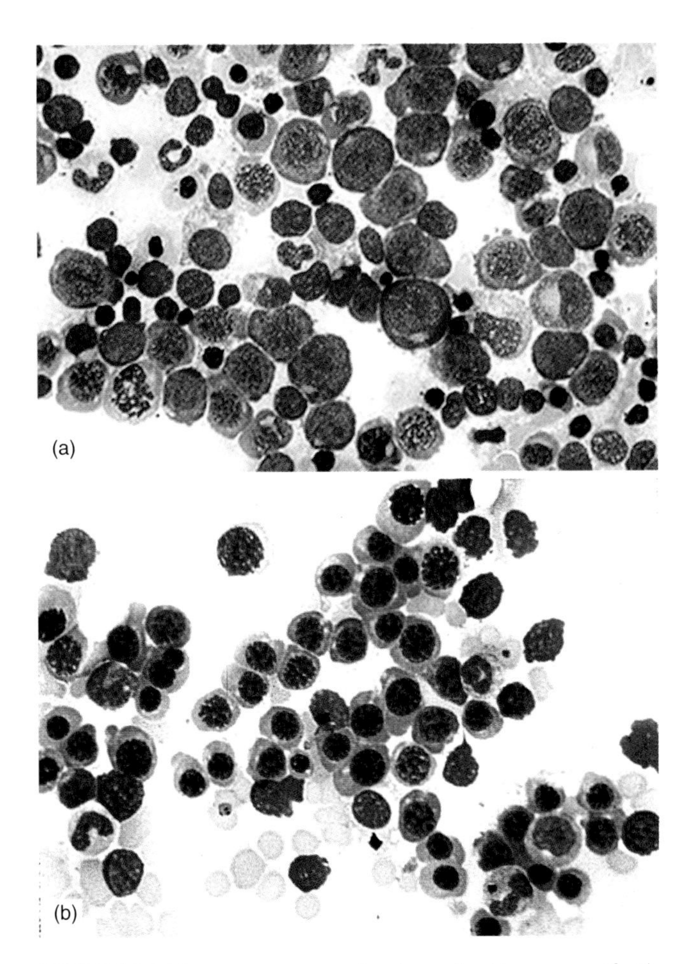

FIGURE 1.14 A bone marrow smear showing erythroid precursors of early (a) and intermediate (b) stages of maturation.

N:C ratio with a deep blue non-granular cytoplasm and a perinuclear pale area (the Golgi system). The nuclear chromatin is fine and one to two nucleoli are present. *Prorubricytes* (basophilic erythroblasts, basophilic normoblasts) are smaller and have more condensed nuclear chromatin than rubriblasts. They depict dark blue cytoplasm and indistinct nucleoli. Prorubricytes undergo three cell divisions and continue maturation to form *rubricytes* (polychromatophilic normoblasts), and subsequently *metarubricytes* (orthochromic normoblast) which are not able to divide but continue hemoglobin synthesis (Figure 1.15). Metarubricytes lose their nuclei and become polychromatophilic RBCs (reticulocytes). Reticulocytes gradually lose their ribosomes (in 1–2 days) and become mature RBCs.

Platelet Precursors

Megakaryoblasts (promegakaryoblasts, group 1 megakaryocytes) are the earliest morphologically identifiable platelet precursors (Figure 1.16) [86, 87]. Megakaryoblasts undergo endomitosis (nuclear division without cytoplasmic division) once or twice and become *promegakaryocytes* (group II megakaryocytes). Endomitosis continues, cells become larger, the nuclear lobulation and volume increase, and the end result is the formation of *granular megakaryocytes* (group III megakaryocytes) which are able to release platelets (Figure 1.16) [45, 47, 88, 89]. Granular megakaryocytes are the largest hematopoietic cells in the bone marrow. The duration from formation of megakaryoblats to platelet production is about 1 week [89, 90]. Platelets are released into the bloodstream with a proportion (approximately one-third) pooled in the spleen. Their average life span is about 8–10 days. Cells from megakaryocytic lineage express CD41, CD42, CD31, CD61, and factor VIII [90, 91].

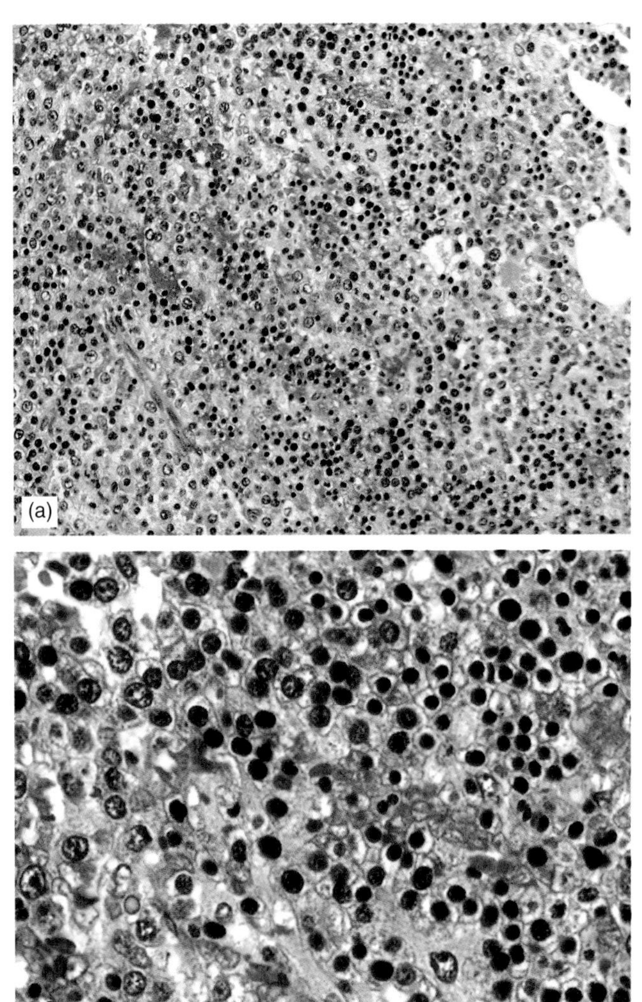

FIGURE 1.15 A bone marrow clot section (a, low power and b, high power) showing erythroid precursors at various stages of maturation.

Lymphoid Lineage

Lymphocytes, similar to the other hematopoietic cells, are derived from the multipotent stem cells [48, 49, 92, 93]. Lymphoblasts, the earliest morphologically identifiable lymphoid cells, have a high N:C ratio with a narrow rim of dark blue non-granular cytoplasm, a round or oval nucleus with fine chromatin and one to two nucleoli. Mature lymphocytes are slightly larger than erythrocytes and are characterized by scanty blue cytoplasm, round nucleus, coarse chromatin, and inconspicuous nucleolus (Figure 1.17a). They may be of B- or T-cell origin. A variable proportion of lymphocytes are larger with abundant cytoplasm and cytoplasmic azurophilic granules. These *large granular lymphocytes* (LGL) are more frequently identified in normal blood smears than bone marrow smears and often express CD16, CD56, and/or CD57 molecules (Figure 1.17b) [61, 95]. They are of two types: NK-cells and cytotoxic T-cells. NK-cells are negative for surface CD3 and show no T-cell receptor (TCR) gene rearrangement but may express CD8, and cytotoxic T-cells express CD3 and CD8 and show *TCR* gene rearrangement [94–96].

Prolymphocytes are larger than lymphocytes (more cytoplasm and a larger nucleus), display a coarse chromatin, and often show a prominent nucleolus. They are either of B- or T-cell origin (Figure 1.17c).

Activated lymphocytes are transformed B-, T-, or NK-cells. These are large cells with abundant cytoplasm and a highly polymorphic nuclear morphology (Figure 1.17d). They are more frequently identified in blood smears than marrow smears.

Hematogones represent normal bone marrow precursor B-cells. These cells consist of a heterogeneous population [56, 57]. The earlier forms (stage 1 hematogones) often express TdT, CD34, CD10, and CD19. The more mature

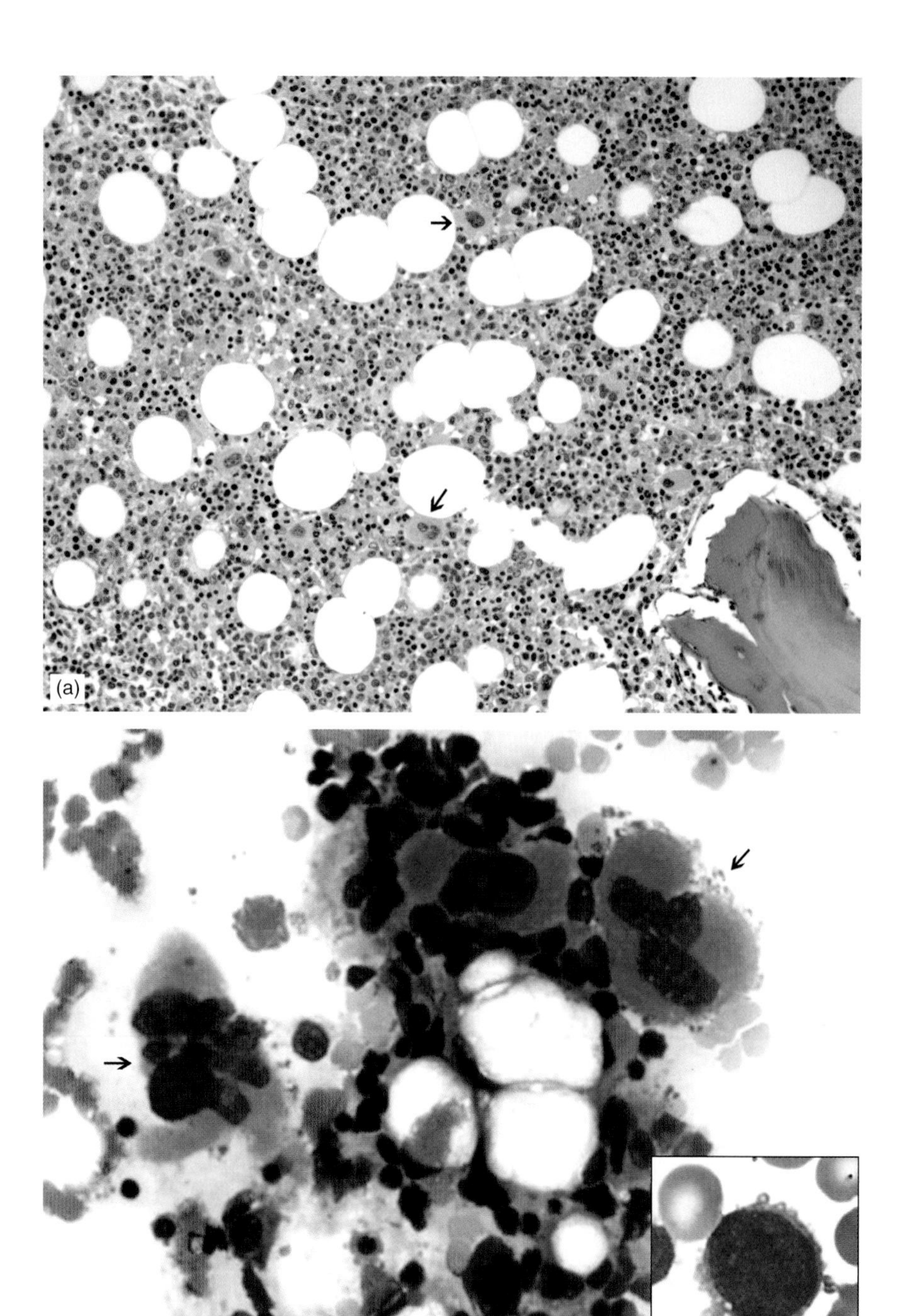

FIGURE 1.16 The megakaryocytic lineage: (a) bone marrow biopsy section demonstrating two megakaryocytes, several eosinophils, and numerous erythroid and neutrophilic precursors; (b) bone marrow smear demonstrating megakaryocytes. A megakaryoblast with cytoplasmic budding is demonstrated in the inset.

forms (stage 2 hematogones) lack the CD34 and TdT expression. These cells may morphologically resemble lymphoblasts but usually show somewhat denser chromatin and absent or inconspicuous nucleoli (Figures 1.18 and 1.19). Hematogones display a distinctive pattern in SSC versus CD45 studies by flow cytometry (Figure 1.20). They are $CD45^{dim}$ and appear as the tail of the mature lymphocytes ($CD45^{strong}$). Hematogones account for about 5–10% of the bone marrow cells in children and <5% of the bone marrow cells in adults.

Plasma cells are the end product of the B-cell lineage and are characterized by abundant dark blue cytoplasm, a perinuclear pale area (Golgi system), and an eccentric nucleus with coarse chromatin (cartwheel appearance) (Figure 1.21). Plasma cells may show small cytoplasmic vacuoles (Mott or morula cells), or large eosinophilic cytoplasmic inclusions (Russell bodies) or nuclear inclusions (Dutcher bodies) (Figure 1.22). Russell bodies and Dutcher bodies are more often seen in plasma cell disorders than in normal plasma cells. The vacuoles and inclusions contain immunoglobulin. Rarely, ovoid-, angular-, or rod-shaped immunoglobulin crystals are found in plasma cells. Cell-membrane-associated molecules CD19, CD38, CD138, CD79b, and sometimes CD117 are expressed by plasma cells [97–99].

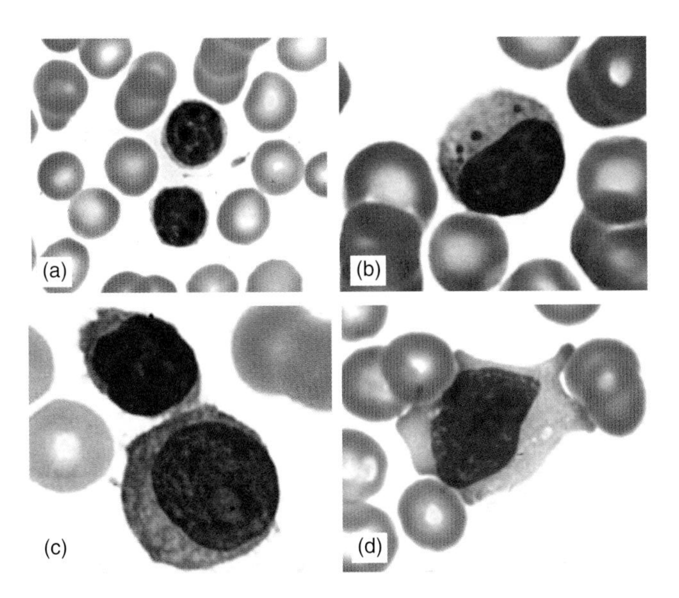

FIGURE 1.17 Examples of various lymphoid cells: (a) mature lymphocytes, (b) a large granular lymphocyte, (c) prolymphocytes, and (d) a reactive lymphocyte.

Lymphoid aggregates are relatively common findings in bone marrow sections, particularly in the elderly. They are well-defined round or oval structures that are randomly distributed in the marrow, usually in close association with small blood vessels and apart from bone trabeculae. They primarily consist of small mature lymphocytes comprising a mixture of B- and T-cells (Figure 1.23). Scattered macrophages, eosinophils, and plasma cells may also be present within or around the lymphoid aggregates.

Other Bone Marrow Cells

Osteoblasts are derived from a multipotent mesenchymal stem cell, a lineage different from that of the HSC [26, 64]. These cells are elongated or oval cells that contain an eccentric round or oval nucleus and one or more nucleoli. Osteoblasts may resemble plasma cells, except that they are larger, their Golgi are not as close to the nucleus, and their nuclear chromatin is finer than that of plasma cells. Osteoblasts in biopsy sections are located along the bone trabeculae, and in bone marrow smears usually appear as individual or small cluster of cells (Figure 1.24). Osteoblasts release a number of matrix molecules and cytokines, such as collagen type 1, proteoglycans, osteoid, GM-CSF, M-CSF, and IL-6 [14, 100].

Osteoclasts are derived from the monocytic lineage [101, 102]. They are multinucleated giant cells involved in

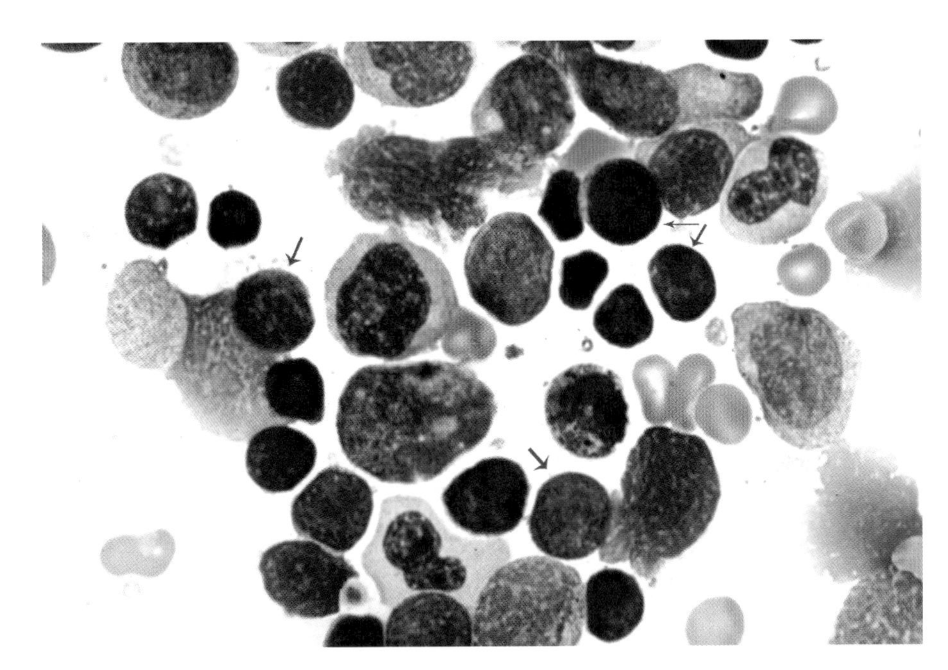

FIGURE 1.18 Several hematogones (arrows) are presented in this bone marrow smear.

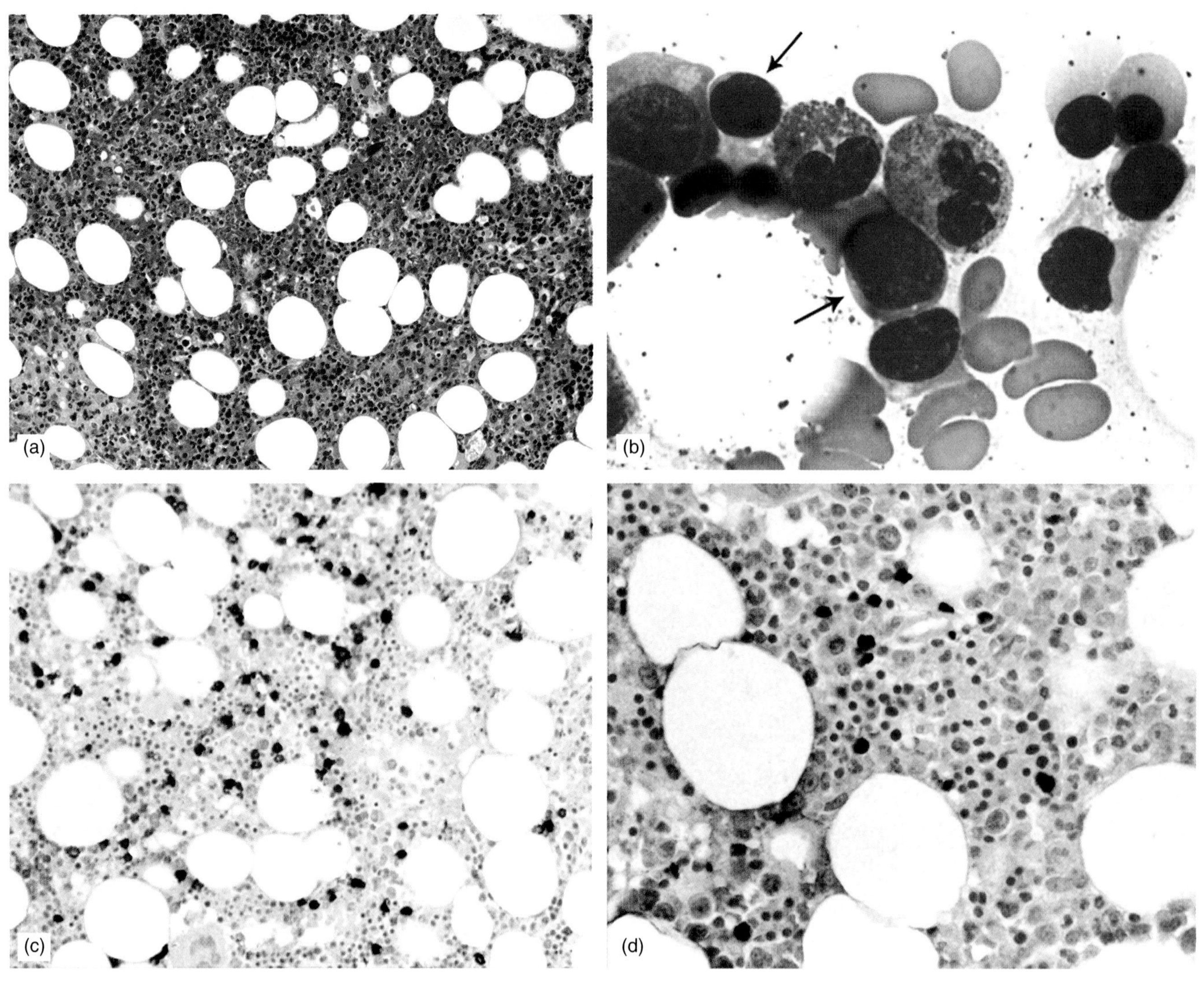

FIGURE 1.19 (a) An unremarkable bone marrow biopsy section showing progressive multilineage maturation. (b) A bone marrow smear demonstrating hematogones (arrows) and an eosinophil. (c) Scattered CD20-positive B-cells. (d) Rare TdT-positive cells which may represent early hematogones.

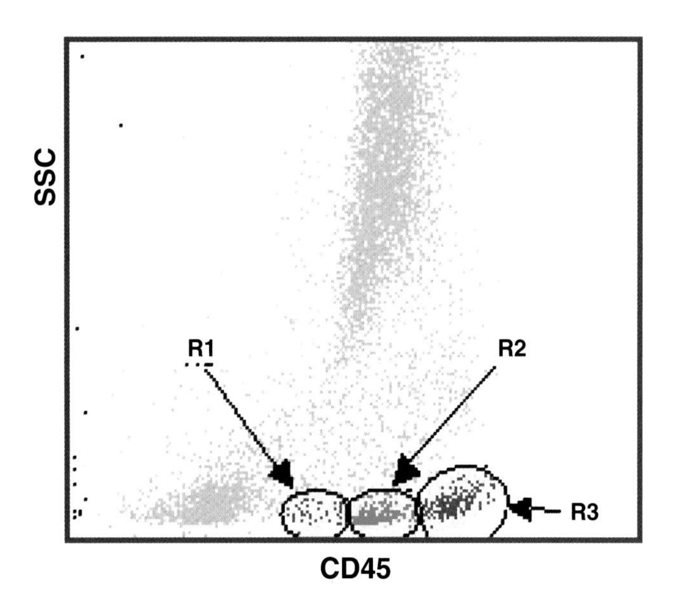

FIGURE 1.20 Flow cytometry. The sidescatter (SSC)/CD45 dot plot of a bone marrow sample demonstrates aggregates of early (R1) and late (R2) hematogones, and mature lymphocytes (R3).

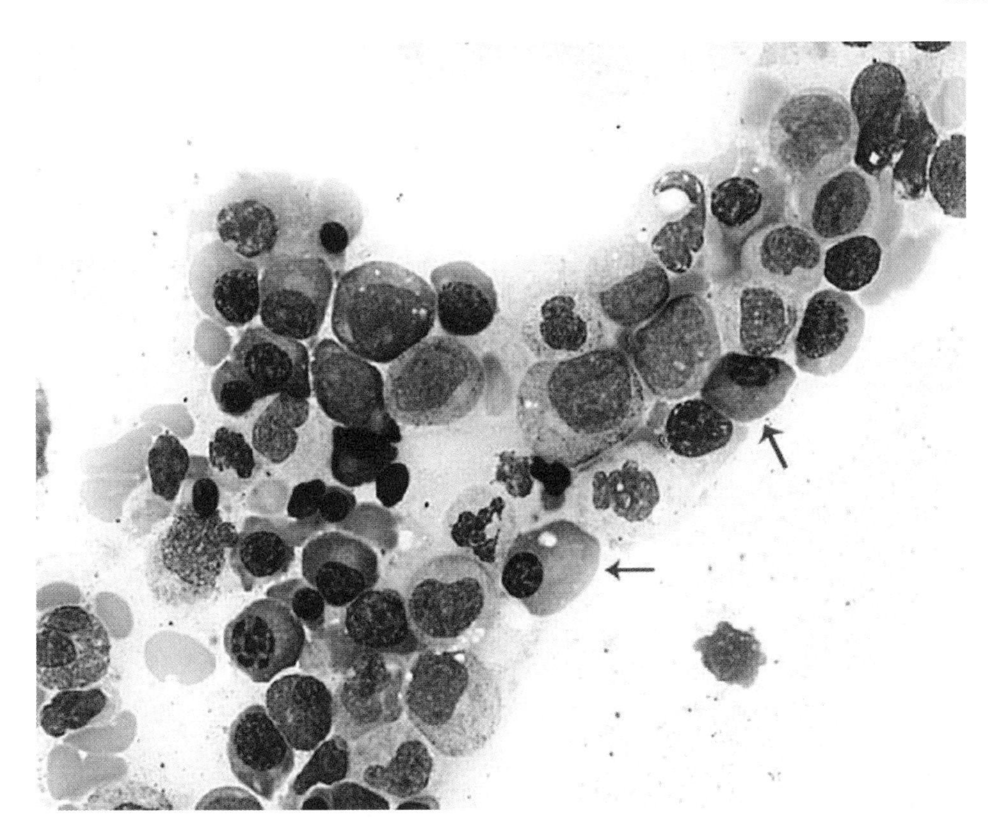

FIGURE 1.21 A bone marrow smear demonstrating several plasma cells (arrows).

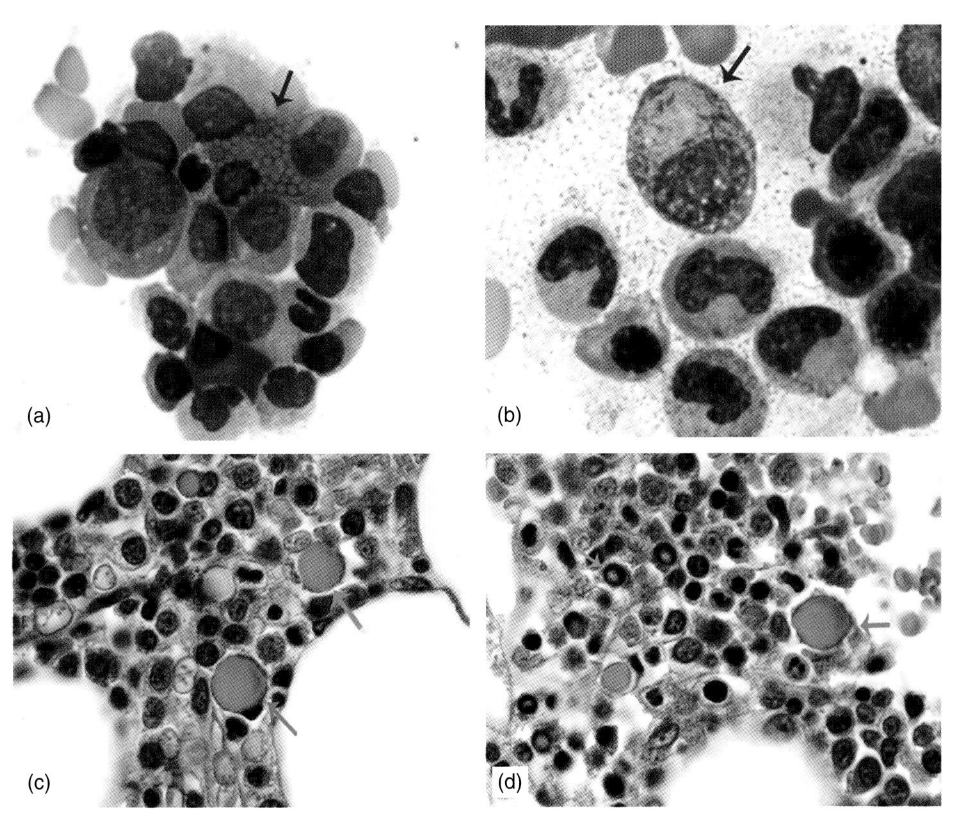

FIGURE 1.22 Plasma cells with inclusions: (a) a grape-like plasma cell (Mott cell), (b) plasma cells with cytoplasmic rod-like Ig crystals, (c) cytoplasmic Ig inclusions (Russell bodies), and (d) Russell bodies (green arrow) and nuclear Ig inclusions (Dutcher bodies) (blue arrows).

bone resorption and remodeling (Figure 1.25). They have abundant cytoplasm which contains numerous azurophilic granules. They are found along the bone trabeculae and may resemble megakaryocytes, except they have multiple separated nuclei that are uniform in size. Osteoclasts are frequently observed in bone marrow of patients with hyperparathyroidism, chronic renal failure, and Paget's disease.

Adipocytes (fat cells, lipocytes) are mesenchymal-derived cells with abundant fat-laden cytoplasm and a

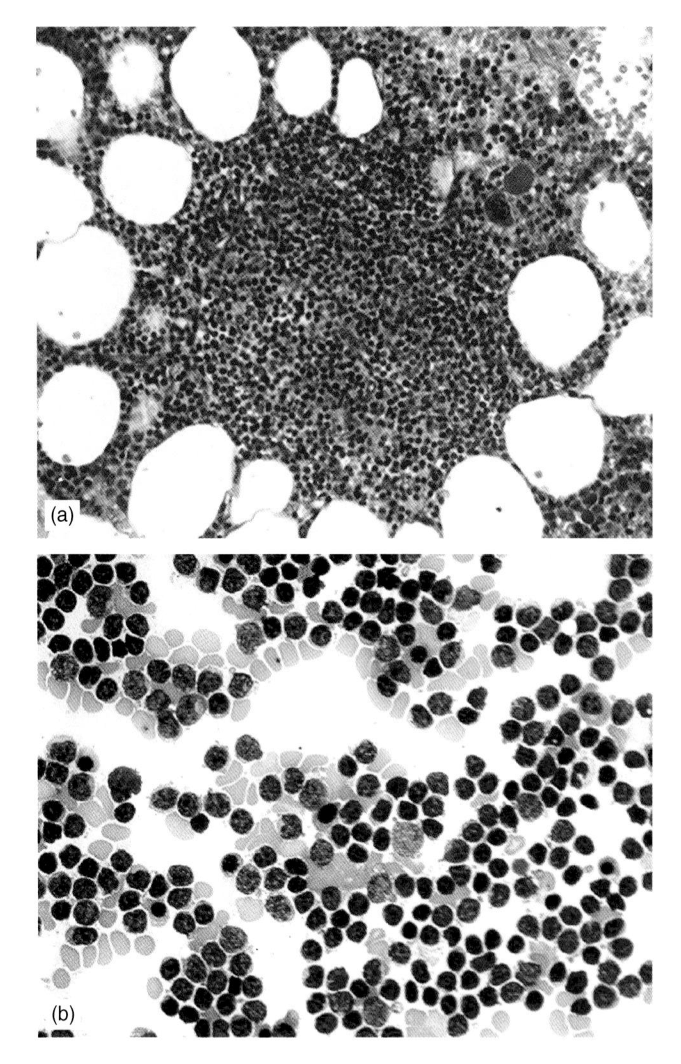

FIGURE 1.23 A well-defined lymphoid aggregate is shown from a bone marrow biopsy section (a) and numerous small mature lymphocytes are present in a bone marrow smear (b).

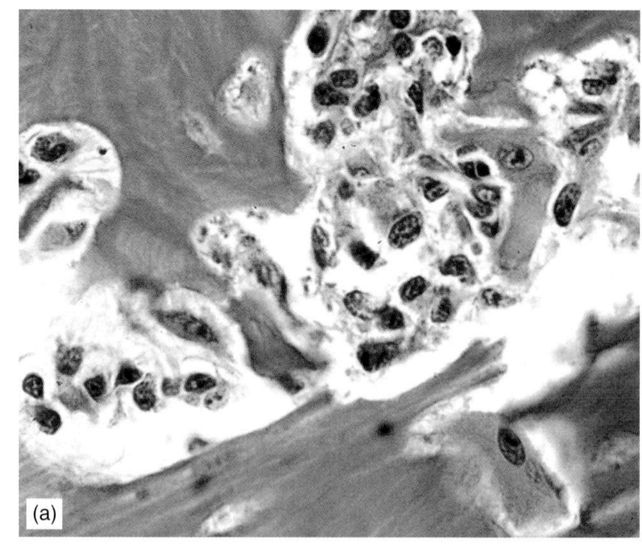

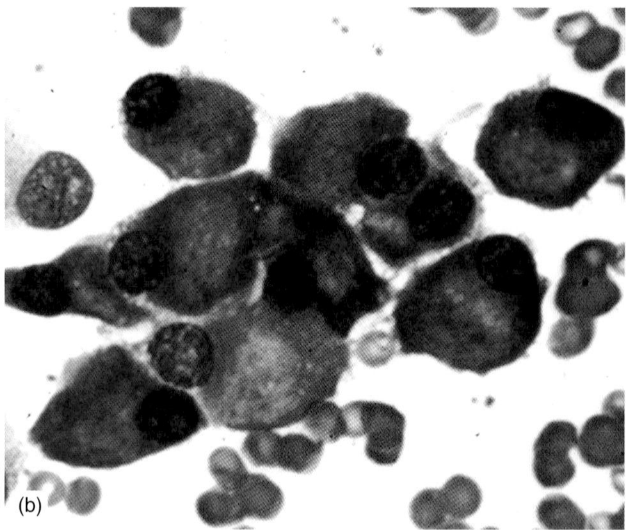

FIGURE 1.24 Numerous osteoblasts are demonstrated in a bone marrow biopsy section (a) and a bone marrow smear (b).

small nucleus often pushed toward the cell membrane (Figure 1.26) [103]. Bone marrow adipocytes can be rapidly replaced by hematopoietic tissue when there is a need for increased hematopoiesis.

Fibroblast-like cells and **endothelial cells** support the wall of the bone marrow sinuses and build the framework of the marrow stroma that supports the hematopoietic cells [104]. These are usually elongated or polygonal cells (15–30 μm) with variable amount of pale cytoplasm and round, oval, or folded nuclei. Their nuclear chromatin is fine and they may depict one or more nucleoli. Fibroblast-like cells support proliferation of myeloid and lymphoid progenitor cells. They are negative for CD33 and CD34 but may express CD10 [105]. Endothelial cells are involved in the regulation of homing and trafficking of the hematopoietic cells, as well as proliferation and differentiation of hematopoietic precursors (Figure 1.26). They express CD31, CD34, and CD146 and carry various receptors, such as receptors for IL-3, EPO, and SF [106,107].

Blood Smear Examination

Morphologic evaluation of blood smear is important in routine hematology work-up, because unremarkable CBC results by automated instruments may not necessarily reflect normal hematopoiesis [108, 109]. For example, in hereditary spherocytosis, lead poisoning, or malaria, the CBC may be within normal limits, but the peripheral blood smears show spherocytes, basophilic stippling, or RBC-containing parasites, respectively. Blood smears should be thin, evenly distributed over the glass slides and quickly air-dried and stained (Wright's stain is the most popular stain).

RBC Morphology

In normal conditions, red cells are relatively uniform in shape and size and contain no inclusions. They are normocytic (an average of 7–8 μm in diameter) and normochromic (the pale central area less than 1/2 of the RBC diameter) (Figures 1.27 and 1.28) [110]. One to two percent of erythrocytes are larger and polychromatophilic (bluish-red)

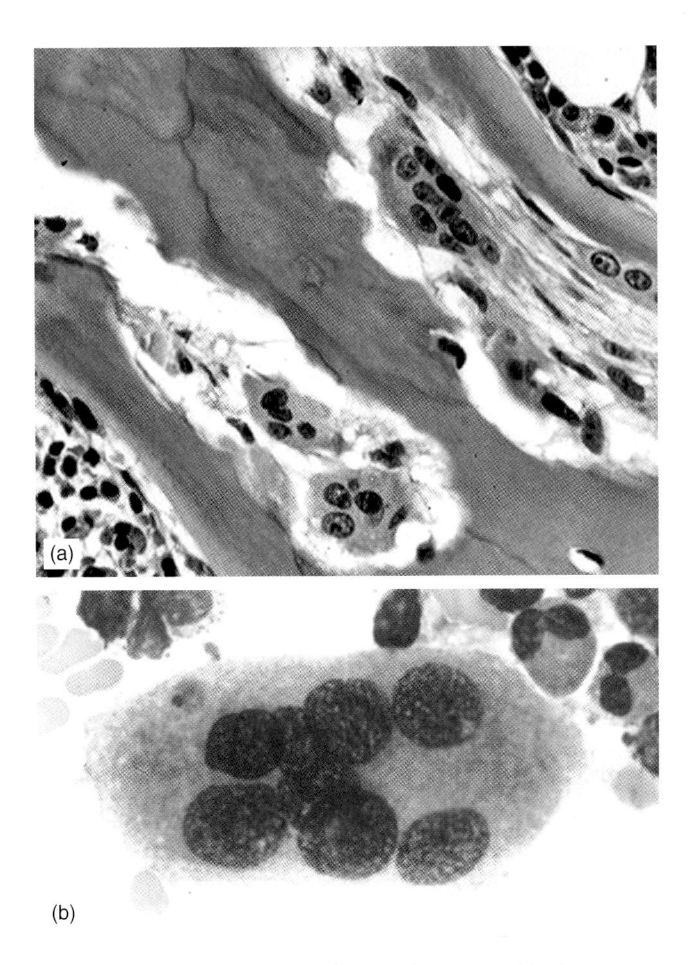

FIGURE 1.25 (a) Numerous osteoclasts are demonstrated in a bone marrow biopsy section. (b) An osteoclast with multiple separated nuclei and finely granular cytoplasm is shown.

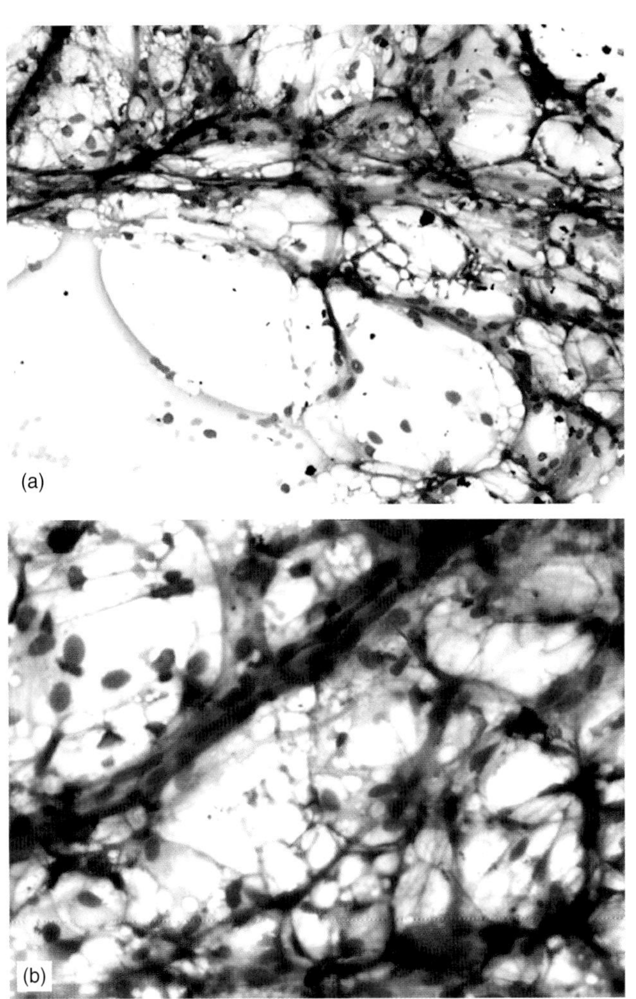

FIGURE 1.26 Bone marrow smears showing adipose tissue and stromal cells. A collapsed capillary (arrow), lined by endothelial cells, is present in (b).

(Figure 1.29). These represent reticulocytes [111]. Except in newborns, nucleated red cells are not normally found in peripheral blood.

Leukocyte Morphology

In normal conditions, peripheral blood smears show various proportions of neutrophilic segmented cells (Segs) and bands (stabs), lymphocytes, monocytes, eosinophils, and basophils (Figures 1.27–1.29). The white blood cell (WBC) count ranges from 3 to 10 $\times$ 10^3 cells/μL with a differential count shown in Table 1.5.

Certain conditions such as exercise, emotional disturbances, menstruation, anesthesia, convulsive seizures, and electric shock may be associated with a transient neutrophilic granulocytosis. This is due to the demargination of the neutrophilic granulocytes and their release into the circulating pool. The presence of immature leukocytes in the peripheral blood should be considered abnormal.

Platelet Morphology

Platelets are the end products of the megakaryocytic lineage and are released into the circulation as cytoplasmic fragments of granular megakaryocytes. They are the smallest hematopoietic elements (measuring 2–4μm in diameter), with a count ranging from 150,000 to 400,000/μL (Figures 1.27–1.29). A rough estimate of the platelet count is calculated in wedge smear preparations by the number of platelets per oil-immersion field $\times$ 20,000. Approximately 7–21 platelets are found per 100 $\times$ oil-immersion field in an evenly distributed normal blood smear. Anti-coagulants or agglutinins (IgM or IgG) which are found in patients with autoimmune disorders, chronic liver disease, or malignancy may cause platelet aggregation.

STRUCTURE AND FUNCTION OF THE SPLEEN

The spleen represents the largest filter of the blood circulation in our body. In normal conditions, it weighs between 75 and 200g and has a deep indentation (the hilum), where blood vessels enter and leave. The spleen is surrounded by a fibrous capsule with many trabeculae radiating from the

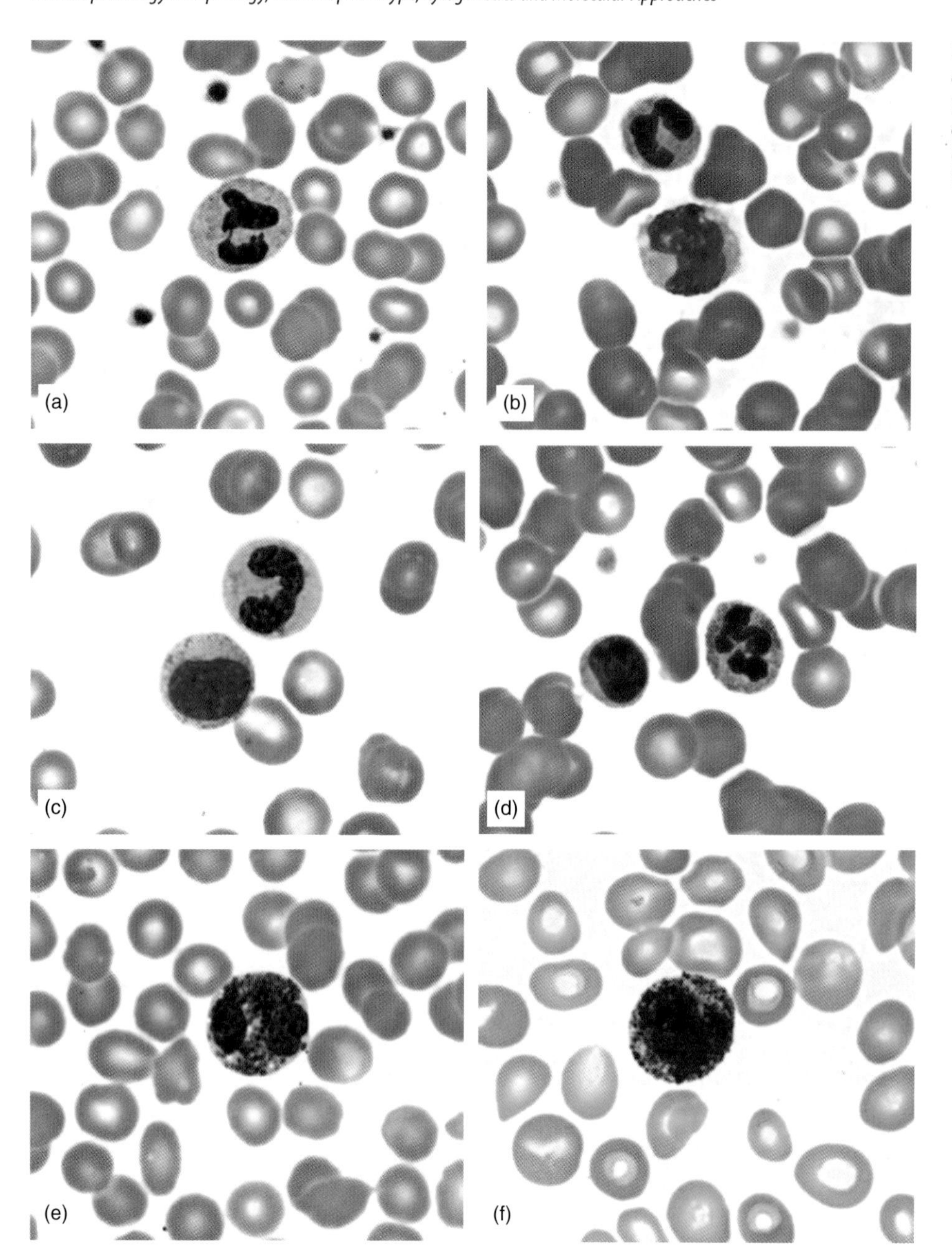

FIGURE 1.27 Blood smears demonstrating segmented neutrophils (a, b, and d), monocytes (b and d), a large granular lymphocyte (c), a lymphocyte (d), an eosinophil (e), and a basophil (f). Platelets are present in (a, b, and d).

hilum and from the internal surface of the capsule into the splenic parenchyma. The splenic artery branches into the trabecular arteries and these branches in turn give off smaller branches that leave the trabeculae and are called central arteries. Central arteries run through the splenic lymphoid tissue (white pulp) and extend to the marginal zone and the red pulp. Therefore, the splenic parenchyma consists of three distinct components: the white pulp, the marginal zone, and the red pulp (Figures 1.30 and 1.31) [112, 115].

The White Pulp

The white pulp consists of lymphoid structures organized in B- and T-cell zones (Figures 1.30 and 1.31) [116–118]. The T-cell zone is represented by the periarteriolar lymphoid sheath, primarily consisting of tightly packed lymphocytes and the presence of IDCs. The T-cells interact with the dendritic cells and passing B lymphocytes. The B-cell zone consists of follicles, which are structurally similar to the follicular structures in the lymph nodes (see lymph node structure later). The follicles are separated from the marginal zone by a densely packed mantle zone and frequently contain germinal centers consisting of large blast-like lymphocytes (centroblasts), smaller lymphocytes (centrocytes), FDCs, and scattered macrophages. Follicles are the place for clonal expansion of activated B-cells, leading to isotype switching and somatic hypermutation.

The Marginal Zone

The marginal zone is the transit area for cells that are leaving the bloodstream and entering the white pulp (Figures

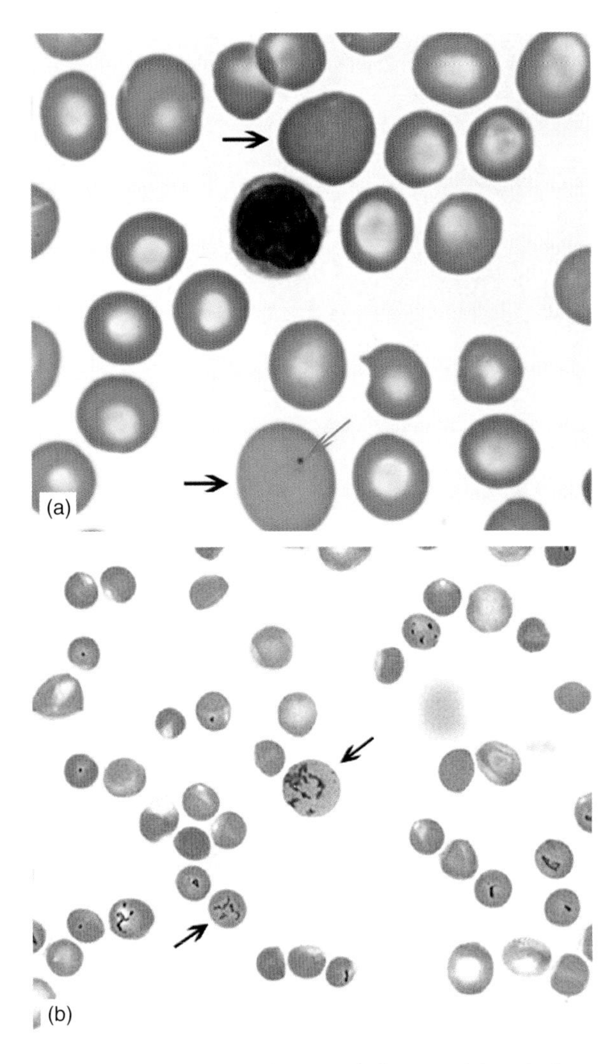

FIGURE 1.28 (a) Blood smear showing polychromatophilic red cells (black arrows) and a Howel-Jolly body (green arrow). (b) Reticulocytes are demonstrated by a supravital stain (arrows).

1.30 and 1.31) [116–120]. However, a large number of cells, such as macrophages, B-cells, and dendritic cells, reside in the marginal zones in order to regulate and facilitate the back and forth transit flow of cells between the blood and the white pulp. The marginal zone macrphages are of two subtypes: the outer ring macrophages, which are close to the blood vessels, and the inner ring macrophages, which are in the proximity of the white pulp. In between the inner and the outer rings, marginal zone B-cells and a subset of dendritic cells reside. Marginal zone B-cells are medium- to large-sized cells with pale cytoplasm and irregular nuclei, resembling monocytes. That is why they were originally called monocytoid B lymphocytes. Because of the presence of variable amounts of cytoplasm, they show nuclear spacing in sections and appear lighter in color and less dense compared to the cells present in the mantle zone. The marginal zone B-cells do not express CD5, CD10, and CD23. They have mutated Ig-V genes and express surface IgM and IgD.

TABLE 1.5 The range of WBC differential counts in normal adults.

Cell type	Range (%)
Granulocytes	
Segs	33–72
Bands	0–13
Eosinophils	0–6
Basophils	0–3
Lymphocytes	16–48
Monocytes	1–13

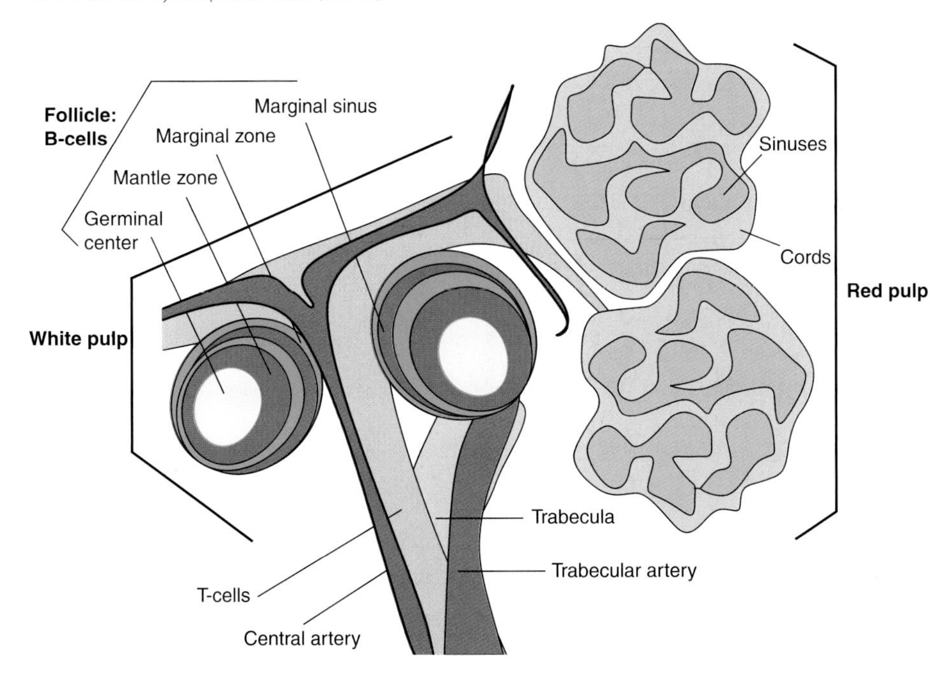

FIGURE 1.29 Schematic of a spleen demonstrating the white pulp, the red pulp, and the marginal zone. Adapted from Greer JP, *et al.* (2004). *Wintrob's Clinical Hematology*, 11th ed., Williams & Wilkins Lippincott.

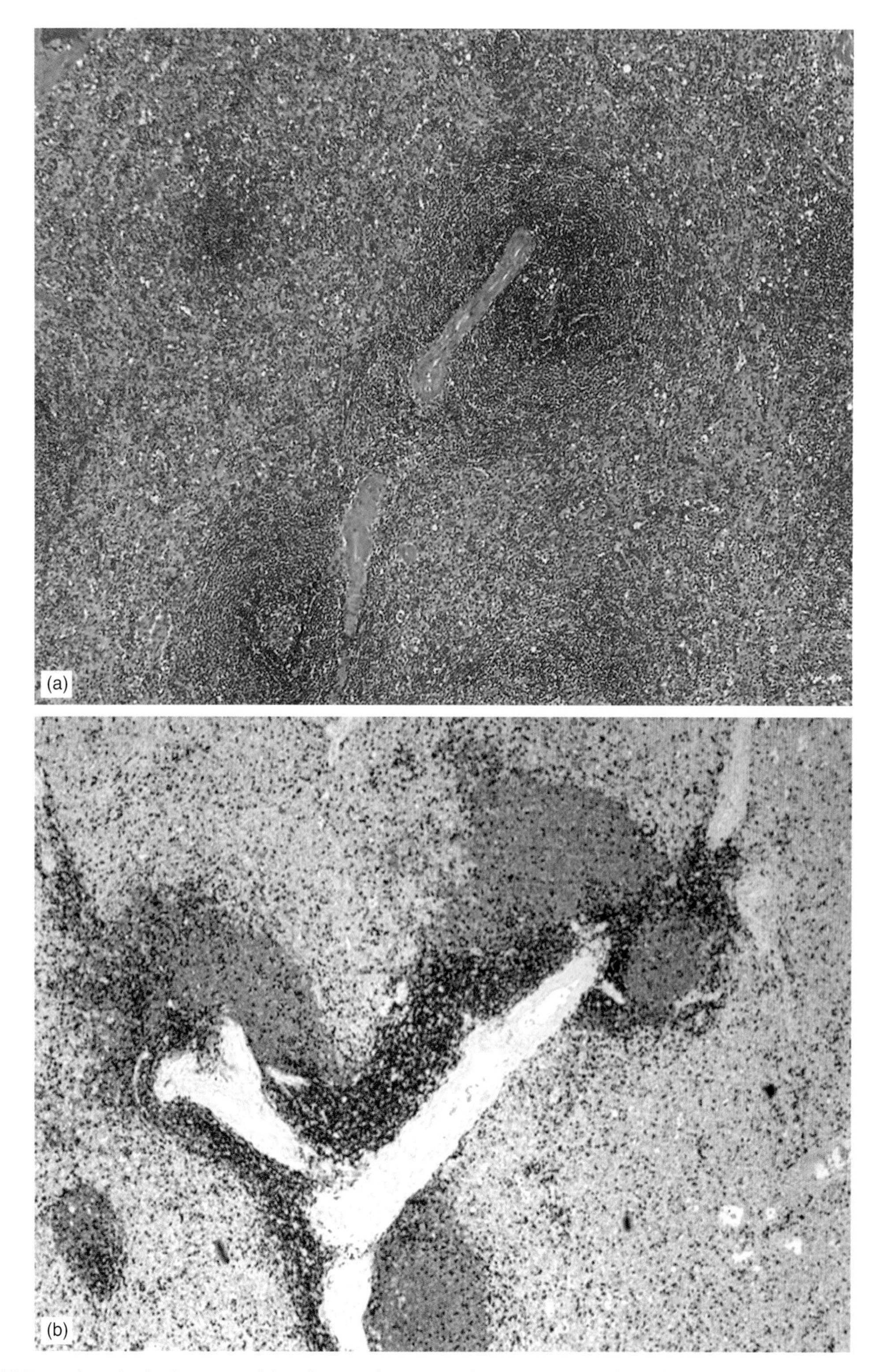

FIGURE 1.30 (a) White pulp and red pulp regions of the spleen are demonstrated in an H&E section. (b) Dual immunohistochemical staining for CD3 (brown) and CD20 (red) demonstrates T- and B-cell areas, respectively.

The marginal zone is a place where the blood-borne pathogens are challenged by the adaptive immune system. Numerous arteriolar branches are present in this region, some with funnel-shaped orifices. These funnel-shaped orifices facilitate the release of arterial content into the mantle zone. However, the marginal zone is devoid of sinuses. Macrophages with their specific pattern-recognition receptors can effectively take up the pathogens and also activate the marginal zone B-cells and dendritic cells. Entry of activated marginal zone B lymphocytes and dendritic cells

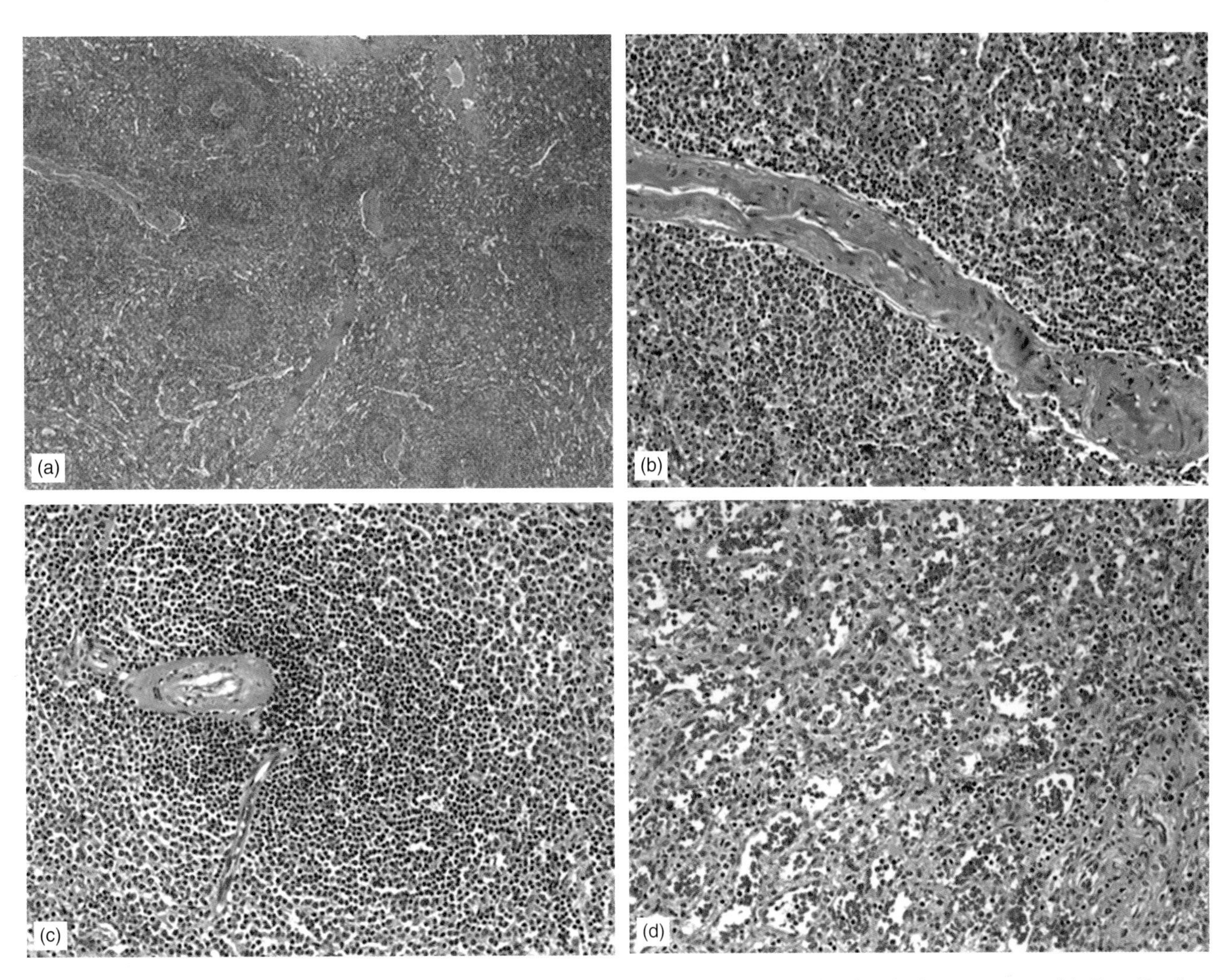

FIGURE 1.31 (a) A low power microscopic view of the spleen demonstrated white and red pulp. (b) and (c) show higher power views of the T- and B-cell regions, respectively. (d) A higher power view of the red pulp with numerous blood-containing sinuses.

into the white pulp initiates an adaptive immune response against the blood-borne pathogens.

The Red Pulp

The red pulp consists of splenic cords and the sinusoidal system (Figures 1.30 and 1.31) [116–118,121]. Cords are composed of a meshwork of fibroblast-like cells supported by extracelluar matrix and reticulin fibers. They form cavernous spaces with no endothelial lining and directly receive arterial blood from terminal arterioles and arterial capillaries. Numerous macrophages are present in the cords which are able to remove the damaged, abnormal, or aged blood cells, while the blood passes through into the venous sinuses. Unlike the cords, these sinuses are lined by the endothelial cells. There are slit-like gaps between the endothelial cells which allow blood cells to penetrate from the cordal space into the sinusoidal lumen. Abnormal RBCs, such as sickle cells, or cells with inclusions, such as Heinz bodies, might not be able to pass through these slits. The sinus basement membrane consists of a network of contractile thick and thin reticular fibers (stress fibers) running circumferentially and longitudinally, respectively. The network is connected to the extracellular matrix of the splenic cord and its contraction helps the blood to pass through the cords into the sinuses. Activity of stress fibers might also help to retain erythrocytes and platelets in the spleen, thereby forming a reservoir for these cells.

The spleen demonstrates three major functions:

1. *Phagocytosis*: Invading micro-organisms and pathogens are effectively removed by macrophages in the spleen. Also, abnormal, damaged, and dysfunctional blood cells are filtered and removed by macrophages when blood passes through the spleen.
2. The splenic white pulp is an important component of the cell-mediated and humoral immune systems.
3. The splenic sinus system serves as a big reservoir for blood cells.

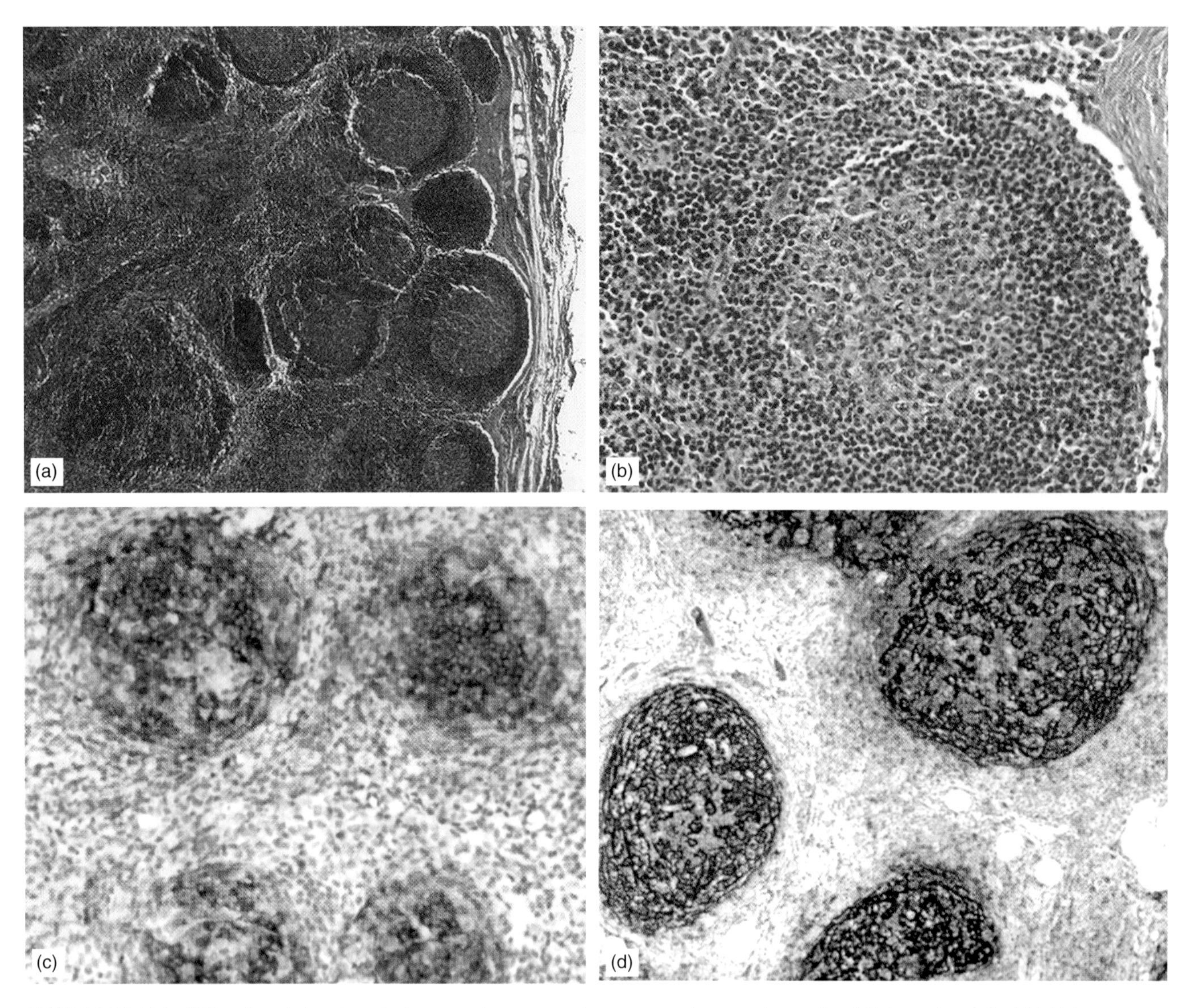

FIGURE 1.32 (a) An H&E section of a lymph node demonstrating cortex and paracortex, primary and secondary follicles, and medulla. (b) A higher power view demonstrating a secondary follicle with a germinal center surrounded by a mantle zone, (c) demonstrating CD10 expression on follicular B-cells, and (d) showing CD21 expression on follicular dendritic cells.

STRUCTURE AND FUNCTION OF THE LYMPH NODES

Lymph nodes are the major components of the lymphatic system and consist of round or oval structures located along the major blood vessels, in peritoneum and mediastinum, and at the base of the extremities. They measure from several millimeters to around 1 cm in diameter and are surrounded by a fibrous capsule. Incoming lymphatic vessels penetrate the capsule and release their content into the subcapsular sinuses. Blood vessels enter and leave the lymph nodes through the hilum. Several fibrous trabeculae extend from the inner part of the capsule into the lymph node parenchyma, forming a supporting meshwork and dividing the lymph node into many subsections. The lymph node parenchyma is divided into a peripheral zone, the *cortex*, and a deeper, centrally located zone, the *medulla* (Figure 1.32) [122–124]. The cortex consists of a superficial part, immediately located under the capsule, and a deeper part or *paracortex*. The following anatomical structures are recognized in lymph node sections (Figure 1.32).

Follicular Structures

Follicular structures are the primary home of the B lymphocytes [122, 124–127]. The ones that are not yet exposed to antigens are called primary follicles and consist of packed, uniform-looking small mature lymphocytes. Secondary follicles have been already exposed to antigenic stimulation. They have a pale central area, called *germinal*

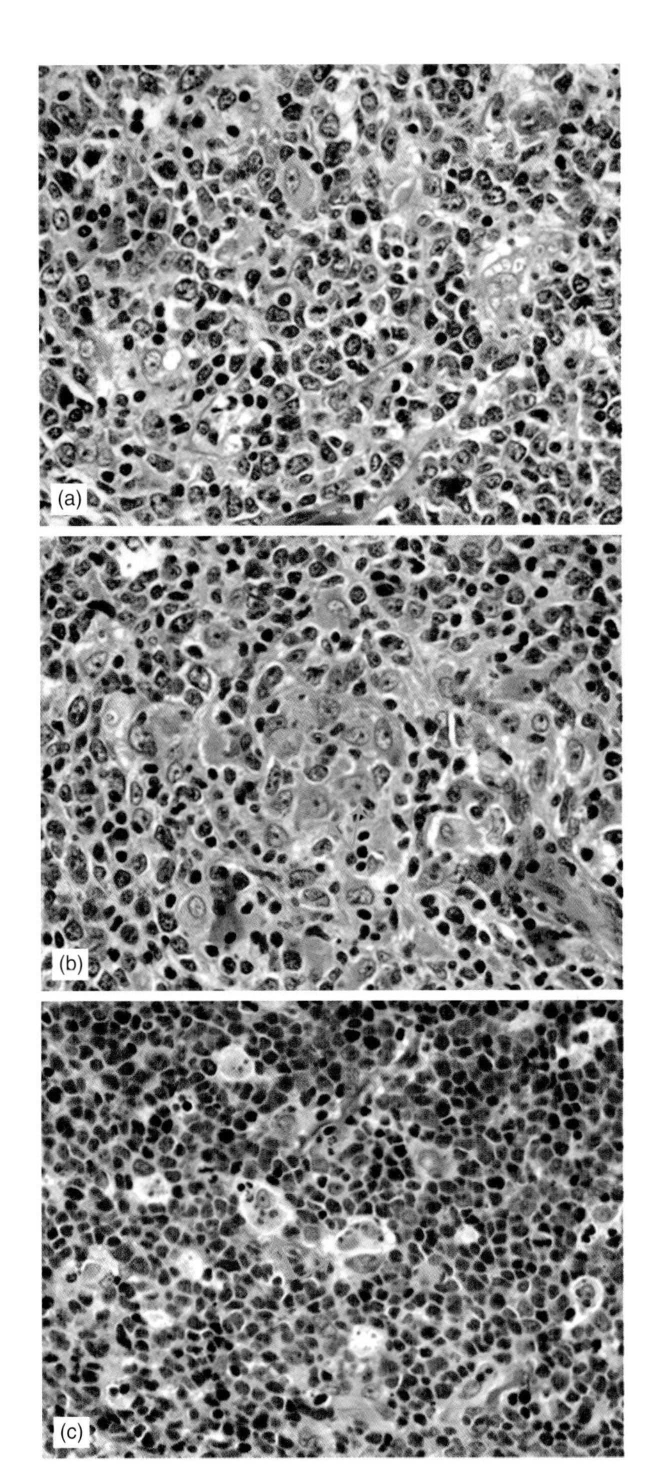

FIGURE 1.33 High power views of a germinal center in a lymph node section demonstrating centroblasts and centrocytes with the presence of mitotic figures (a, b, and c), follicular dendritic cells (b, arrows), and tingible body macrophages (c, arrows).

center, consisting of a mixture of large and small cells. The smaller uniform-looking lymphocytes surrounding the germinal center are packed as a darkly stained crescent known as *mantle zone* (Figures 1.32 and 1.33). Lymphocytes of the mantle zones (mantle cells) are B-cells, but functionally heterogeneous, consisting of bone marrow-derived naive cells and lymphocytes previously exposed to antigens. However, a significant proportion of the mantle cells express B-cell associated markers, such as CD19, CD20, CD79a, as well as CD5, FMC-7, and CD43. Mantle cells are negative for CD10 and CD23.

Within the germinal center, two types of lymphoid cells are present [124, 126, 127]:

1. *Centroblasts*, which are apparently derived from mantle cells and are large, non-cleaved cells located at the bottom of the germinal center and often show frequent mitotic figures. They have a vesicular nuclear chromatin, and multiple distinct nucleoli usually located close to the nuclear membrane (Figure 1.33).
2. *Centrocytes*, which are evolved from centroblasts and consist of small cleaved lymphoid cells located at the upper part of the germinal centers. These cells show scant cytoplasm, dispersed nuclear chromatin, and inconspicuous nucleoli. Centrocytes mature to marginal zone or monocytoid B-cells and leave the germinal center (Figure 1.33).

Centroblasts and certrocytes express B-cell-associated antigen, such as CD19, CD20, CD22, and CD79a. They are often positive for CD10 and bcl-6 and negative for CD5.

In addition to the B lymphocytes, follicles contain FDCs, which function as antigen presenting cells (Figure 1.32d) [124, 126, 127]. FDCs are derived from the mesenchymal cells in the follicular structures (not originated from bone marrow stem cells). They often appear in pairs, have round or irregular nuclei with dispersed chromatin, and often one small, centrally located nucleolus. FDCs are characterized by expressing CD21 and CD35. Also, scattered macrophages, some with tingible bodies, are present in the germinal centers (Figure 1.32c).

The Paracortex

The paracortical area is the primary home of T-cells. These cells slowly flow in the spaces provided by the paracortical cords. The cords consist of a centrally located venule lined by tall, cuboidal endothelial cells (high endothelial venules) surrounded by narrow corridors outlined by reticular fibers [124, 127, 128]. In these corridors T-cells interact with antigens presented to them by the stationary IDCs. IDCs, unlike FDCs, are derived from bone marrow stem cells and express HLA-DR and S-100 protein. The T-cells have passed through the thymic developmental processes (post-thymic T-cells) and are divided into *helper* and *suppressor* T-cells. Helper T-cells are CD4-positive and release regulatory cytokines to facilitate the immune responses and are divided into two major subtypes: (1) Th_1 cells which secrete IL-2 and interferon γ and provide help to other T-cells and macrophages and (2) Th_2 cells which secrete IL-4, IL-5, IL-6, and IL-10 and assist B-cells in their antibody production. Suppressor T-cells express CD8 and are primarily involved in cytotoxic reactions. There are more CD4-positive than CD8-positive T-cells in lymph nodes.

The Medulla

The medulla consists of medullary cords loaded with T- and B-cells, plasma cells, and macrophages [126, 127].

Vascular and Lymphatic Structures

The main artery, after entering the lymph node through the hilum, branches and gives rise to numerous arterioles that pass through the trabeculae and reach the cortex [126, 127, 129]. There, they make a capillary network. The capillaries empty into the high endothelial venules in the center of the cortical corridors. Venules join together and make larger branches, extend from the cortex to the medulla, and finally leave the hilum as veins.

Afferent lymphatic vessels penetrate the lymph node capsule and empty into the subcapsular sinuses, which are connected to the cortical sinuses. The sinuses are lined by endothelial cells which have no basement membrane. The sinusoidal lumen is subdivided into smaller interconnecting spaces by fibrous septa covered by endothelial cells. Sinuses guide the lymphatic flow from the capsule into the medulla and eventually terminate to the efferent lymphatics at the hilum and leave the lymph node. Sinusoidal spaces are loaded with macrophages.

References

1. Tavian M, Peault B. (2005). Embryonic development of the human hematopoietic system. *Int J Dev Biol* **49**, 243–50.
2. Peault B, Tavian M. (2003). Hematopoietic stem cell emergence in the human embryo and fetus. *Ann NY Acad Sci* **996**, 132–40.
3. McGrath KE, Palis J. (2005). Hematopoiesis in the yolk sac: More than meets the eye. *Exp Hematol* **33**, 1021–8.
4. Bonnet D. (2003). Biology of human bone marrow stem cells. *Clin Exp Med* **3**, 140–9.
5. Kucia M, Ratajczak J, Ratajczak MZ. (2005). Are bone marrow stem cells plastic or heterogenous – that is the question. *Exp Hematol* **33**, 613–23.
6. Shizuru JA, Negrin RS, Weissman IL. (2005). Hematopoietic stem and progenitor cells: Clinical and preclinical regeneration of the hematolymphoid system. *Annu Rev Med* **56**, 509–38.
7. Sharkis SJ, Neutzel S, Collector MI. (2001). Phenotype and function of hematopoietic stem cells. *Ann NY Acad Sci* **938**, 191–4.
8. Lanzkron SM, Collector MI, Sharkis SJ. (1999). Hematopoietic stem cell tracking *in vivo*: A comparison of short-term and long-term repopulating cells. *Blood* **93**, 1916–21.
9. Chertkov LD, Deryugina EI, Drize NJ, Sadovnikova EY. (1989). Limited proliferative potential of primitive hematopoietic stem cells: Hematopoiesis by clonal succession. *Haematol Blood Transfus* **32**, 178–82.
10. Brecher G, Beal SL, Schneiderman M. (1986). Renewal and release of hemopoietic stem cells: Does clonal succession exist? *Blood Cells* **12**, 103–27.
11. Quesenberry P, Abedi M, Dooner M, Colvin GS, anchez-Guijo FM, Aliotta J, Pimentel J, Dooner G, Greer D, Demers D, Keaney P, Peterson A, Luo L, Foster B. (2005). The marrow cell continuum: Stochastic determinism. *Folia Histochem Cytobiol* **43**, 187–90.
12. Ogawa M. (1999). Stochastic model revisited. *Int J Hematol* **69**, 2–5.
13. Hume DA. (2000). Probability in transcriptional regulation and its implications for leukocyte diff erentiation and inducible gene expression. *Blood* **96**, 2323–8.
14. Taichman RS. (2005). Blood and bone: Two tissues whose fates are intertwined to create the hematopoietic stem-cell niche. *Blood* **105**, 2631–9.
15. Clark BR, Keating A. (1995). Biology of bone marrow stroma. *Ann NY Acad Sci* **770**, 70–8.
16. Janowska-Wieczorek A, Matsuzaki A, Marquez LA. (2000). The hematopoietic microenvironment: Matrix metalloproteinases in the hematopoietic microenvironment. *Hematology* **4**, 515–27.
17. Conget PA, Minguell JJ. (1999). Phenotypical and functional properties of human bone marrow mesenchymal progenitor cells. *J Cell Physiol* **181**, 67–73.
18. Nardi NB, Alfonso ZZ. (1999). The hematopoietic stroma. *Braz J Med Biol Res* **32**, 601–9.
19. Boggs SS. (1999). The hematopoietic microenvironment: Phylogeny and ontogeny of the hematopoietic microenvironment. *Hematology* **4**, 31–44.
20. Verfaillie CM, Gupta P, Prosper F, Hurley R, Lundell B, Bhatia R. (1999). The hematopoietic microenvironment: Stromal extracellular matrix components as growth regulators for human hematopoietic progenitors. *Hematology* **4**, 321–33.
21. Tocci A, Forte L. (2003). Mesenchymal stem cell: Use and perspectives. *Hematol J* **4**, 92–6.
22. Kassem M. (2004). Mesenchymal stem cells: Biological characteristics and potential clinical applications. *Cloning Stem Cells* **6**, 369–74.
23. Whetton AD, Spooncer E. (1998). Role of cytokines and extracellular matrix in the regulation of haematopoietic stem cells. *Curr Opin Cell Biol* **10**, 721–6.
24. Evans CA, Ariffin S, Pierce A, Whetton AD. (2002). Identification of primary structural features that define the differential actions of IL-3 and GM-CSF receptors. *Blood* **100**, 3164–74.
25. Parreira L, Neves H, Simoes S. (2003). Notch and lymphopoiesis: A view from the microenvironment. *Semin Immunol* **15**, 81–9.
26. Naeim F. (1997). *Pathology of Bone Marrow*. Williams & Wilkins, Baltimore.
27. Gulati GL, Ashton JK, Hyun BH. (1988). Structure and function of the bone marrow and hematopoiesis. *Hematol Oncol Clin North Am* **2**, 495–511.
28. Han W, Yu Y, Liu XY. (2006). Local signals in stem cell-based bone marrow regeneration. *Cell Res* **16**, 189–95.
29. Heissig B, Ohki Y, Sato Y, Rafii S, Werb Z, Hattori K. (2005). A role for niches in hematopoietic cell development. *Hematology* **10**, 47–53.
30. Quesenberry PJ, Colvin GA, Abedi M, Dooner G, Dooner M, Aliotta J, Keaney P, Luo L, Demers D, Peterson A, Foster B, Greer D. (2005). The stem cell continuum. *Ann NY Acad Sci* **1044**, 228–35.
31. Naeim F. (1995). Topobiology in hematopoiesis. *Hematol Pathol* **9**, 107–19.
32. Haylock DN, Nilsson SK. (2005). Stem cell regulation by the hematopoietic stem cell niche. *Cell Cycle* **4**, 1353–5.
33. Ingley E, Tilbrook PA, Klinken SP. (2004). New insights into the regulation of erythroid cells. *IUBMB Life* **56**, 177–84.
34. Goltry KL, Patel VP. (1997). Specific domains of fibronectin mediate adhesion and migration of early murine erythroid progenitors. *Blood* **90**, 138–47.
35. Ratajczak J, Kijowski J, Majka M, Jankowski K, Reca R, Ratajczak MZ. (2003). Biological significance of the different erythropoietic factors secreted by normal human early erythroid cells. *Leuk Lymphoma* **44**, 767–74.
36. Jelkmann W. (2004). Molecular biology of erythropoietin. *Intern Med* **43**, 649–59.
37. Barreda DR, Hanington PC, Belosevic M. (2004). Regulation of myeloid development and function by colony stimulating factors. *Dev Comp Immunol* **28**, 509–54.
38. Rosenthal J, Cairo MS. (1995). The role of cytokines in modulating neonatal myelopoiesis and host defense. *Cytokines Mol Ther* **1**, 165–76.
39. Herrmann F, Mertelsmann R. (1992). Regulatory effects of cytokines on myelopoiesis. *Immunol Ser* **57**, 339–49.

40. Nelson DA. (1990). The biology of myelopoiesis. *Clin Lab Med* **10**, 649–59.
41. Kawamoto H, Minato N. (2004). Myeloid cells. *Int J Biochem Cell Biol* **36**, 1374–9.
42. Gordon S, Taylor PR. (2005). Monocyte and macrophage heterogeneity. *Nat Rev Immunol* **5**, 953–64.
43. Hume DA, Ross IL, Himes SR, Sasmono RT, Wells CA, Ravasi T. (2002). The mononuclear phagocyte system revisited. *J Leukoc Biol* **72**, 621–7.
44. Abraham R, Basser RL. (1997). Megakaryocyte growth and development factor: A review of early clinical studies. *Oncologist* **2**, 311–18.
45. Dolzhanskiy A, Basch RS, Karpatkin S. (1996). Development of human megakaryocytes: I. Hematopoietic progenitors (CD34+ bone marrow cells) are enriched with megakaryocytes expressing CD4. *Blood* **87**, 1353–60.
46. Burmester H, Wolber EM, Freitag P, Fandrey J, Jelkmann W. (2005). Thrombopoietin production in wild-type and interleukin-6 knockout mice with acute *inflammation. J Interferon Cytokine Res* **25**, 407–13.
47. Long MW. (1998). Megakaryocyte differentiation events. *Semin Hematol* **35**, 192–9.
48. Maillard I, Fang T, Pear WS. (2005). Regulation of lymphoid development, differentiation, and function by the Notch pathway. *Annu Rev Immunol* **23**, 945–74.
49. Baba Y, Pelayo R, Kincade PW. (2004). Relationships between hematopoietic stem cells and lymphocyte progenitors. *Trends Immunol* **25**, 645–9.
50. Hirose J, Kouro T, Igarashi H, Yokota T, Sakaguchi N, Kincade PW. (2002). A developing picture of lymphopoiesis in bone marrow. *Immunol Rev* **189**, 28–40.
51. Radtke F, Wilson A, MacDonald HR. (2004). Notch signaling in T- and B-cell development. *Curr Opin Immunol* **16**, 174–9.
52. Radtke F, Wilson A, Mancini SJ, MacDonald HR. (2004). Notch regulation of lymphocyte development and function. *Nat Immunol* **5**, 247–53.
53. Akashi K, Reya T, Dalma-Weiszhausz D, Weissman IL. (2000). Lymphoid precursors. *Curr Opin Immunol* **12**, 144–50.
54. Milne CD, Fleming HE, Zhang Y, Paige CJ. (2004). Mechanisms of selection mediated by interleukin-7, the preBCR, and hemokinin-1 during B-cell development. *Immunol Rev* **197**, 75–88.
55. Kincade PW, Yamashita Y, Borghesi L, Medina K, Oritani K. (1998). Blood cell precursors in context. Composition of the bone marrow microenvironment that supports B lymphopoiesis. *Vox Sang* **74**(2), 265–8. (Suppl).
56. McKenna RW, Asplund SL, Kroft SH. (2004). Immunophenotypic analysis of hematogones (B-lymphocyte precursors) and neoplastic lymphoblasts by 4-color flow cytometry. *Leuk Lymphoma* **45**, 277–85.
57. Babusikova O, Zeleznikova T, Mlcakova A, Kusenda J, Stevulova L. (2005). The knowledge on the 3rd type hematogones could contribute to more precise detection of small numbers of precursor B-acute lymphoblastic leukemia. *Neoplasma* **52**, 502–9.
58. Zuniga-Pflucker JC. (2004). T-cell development made simple. *Nat Rev Immunol J* **4**, 67–72.
59. O'Connor GM, Hart OM, Gardiner CM. (2006). Putting the natural killer cell in its place. *Immunology* **117**, 1–10.
60. Hallett WH, Murphy WJ. (2004). Natural killer cells: Biology and clinical use in cancer therapy. *Cell Mol Immunol* **1**, 12–21.
61. Lanier LL. (2005). NK cell recognition. *Annu Rev Immunol* **23**, 225–74.
62. Zingoni A, Sornasse T, Cocks BG, Tanaka Y, Santoni A, Lanier LL. (2005). NK cell regulation of T cell-mediated responses. *Mol Immunol* **42**, 451–4.
63. Naeim F. (2001). *Atlas of Bone Marrow and Blood Pathology*. W.B. Saunders, Philadelphia.
64. Foucar K. (2001). *Bone Marrow Pathology*, 2nd ed. ASCP Press, Chicago.
65. Krenacs T, Bagdi E, Stelkovics E, Bereczki L, Krenacs L. (2005). How we process trephine biopsy specimens: Epoxy resin embedded bone marrow biopsies. *J Clin Pathol* **58**, 897–903.
66. Kim S, Zehnder JL. (2007). Bone marrow aspiration and biopsy: Indications and technique. *UpToDate*.
67. Graf BL, Korte W, Schmid L, Schmid U, Cogliatti SB. (2005). Impact of aspirate smears and trephine biopsies in routine bone marrow diagnostics: A comparative study of 141 cases. *Swiss Med Wkly* **135**, 151–9.
68. Goasguen JE, Bennett JM. (1992). Classification and morphologic features of the myelodysplastic syndromes. *Semin Oncol* **19**, 4–13.
69. Pirruccello SJ, Young KH, Aoun P. (2006). Myeloblast phenotypic changes in myelodysplasia. CD34 and CD117 expression abnormalities are common. *Am J Clin Pathol* **125**, 884–94.
70. Auewarakul CU, Promsuwicha O, U-Pratya Y, Pattanapanyasat K, Issaragrisil S. (2003). Immunophenotypic profile of adult acute myeloid leukemia (AML): Analysis of 267 cases in Thailand. *Asian Pac J Allergy Immunol* **21**, 153–60.
71. Walsh GM. (2001). Eosinophil granule proteins and their role in disease. *Curr Opin Hematol* **8**, 28–33.
72. Rothenberg ME. (2007). Eosinophils in the new millennium. *J Allergy Clin Immunol* **19**, 1321–2.
73. Gessner A, Mohrs K, Mohrs M. (2005). Mast cells, basophils, and eosinophils acquire constitutive IL-4 and IL-13 transcripts during lineage differentiation that are sufficient for rapid cytokine production. *J Immunol* **174**, 1063–72.
74. Uston PI, Lee CM. (2003). Characterization and function of the multifaceted peripheral blood basophil. *Cell Mol Biol (Noisy-le-grand)* **49**, 1125–35.
75. Bochner BS, Schleimer RP. (2001). Mast cells, basophils, and eosinophils: Distinct but overlapping pathways for recruitment. *Immunol Rev* **179**, 5–15.
76. Shiohara M, Koike K. (2005). Regulation of mast cell development. *Chem Immunol Allergy* **87**, 1–21.
77. Horny HP, Sotlar K, Valent P. (2007). Mastocytosis: State of the art. *Pathobiology* **74**, 121–32.
78. Hauswirth AW, Florian S, Schernthaner GH, Krauth MT, Sonneck K, Sperr WR, Valent P. (2006). Expression of cell surface antigens on mast cells: Mast cell phenotyping. *Meth Mol Biol* **315**, 77–90.
79. Steinbach F, Thiele B. (1994). Phenotypic investigation of mononuclear phagocytes by flow cytometry. *J Immunol Meth* **174**, 109–22.
80. Pieri L, Domenici L, Romagnoli P. (2001). Langerhans cells differentiation: A three-act play. *Ital J Anat Embryol* **106**, 47–69.
81. Robinson SP, Saraya K, Reid CD. (1998). Developmental aspects of dendritic cells *in vitro* and *in vivo*. *Leuk Lymphoma* **29**, 477–90.
82. Nunez R, Garay N, Bruno A, Villafane C, Bruno E, Filgueira L. (2004). Functional and structural characterization of two populations of human monocyte-derived dendritic cells. *Exp Mol Pathol* **77**, 104–15.
83. Call ME, Wucherpfenning KW. (2007). Antigen presenting cells. *UpToDate*.
84. Munoz-Fernandez R, Blanco FJ, Frecha C, Martin F, Kimatrai M, Abadia-Molina AC, Garcia-Pacheco JM, Olivares EG. (2006). Follicular dendritic cells are related to bone marrow stromal cell progenitors and to myofibroblasts. *J Immunol* **177**, 280–9.
85. Wallet MA, Sen P, Tisch R. (2005). Immunoregulation of dendritic cells. *Clin Med Res* **3**, 166–75.
86. Bruno E, Hoffman R. (1998). Human megakaryocyte progenitor cells. *Semin Hematol* **35**, 183–91.
87. Hoffman R, Murrav LJ, Young JC, Luens KM, Bruno E. (1996). Hierarchical structure of human megakaryocyte progenitor cells. *Stem Cells* **1**(Suppl), 75–81.
88. Cramer EM. (1999). Megakaryocyte structure and function. *Curr Opin Hematol* **6**, 354–61.
89. George JN. (2000). Platelets. *Lancet* **355**, 1531–9.
90. Law HK, Bol SJ, Palatsides M, Williams NT. (2000). Analysis of human megakaryocytic cells using dual-color immunofluorescence labeling. *Cytometry* **41**, 308–15.
91. Higuchi T, Koike K, Sawai N, Koike T. (1997). Proliferative and differentiative potential of thrombopoietin-responsive precursors: Expression of megakaryocytic and erythroid lineages. *Exp Hematol* **25**, 463–70.

92. Tsuji K, Feng MA, Wang D. (2002). Development of human lymphohematopoiesis defined by CD34 and CD81 expression. *Leuk Lymphoma* **43**, 2269–73.
93. Busslinger M, Nutt SL, Rolink AG. (2000). Lineage commitment in lymphopoiesis. *Curr Opin Immunol* **12**, 151–8.
94. O'Malley DP. (2007). T-cell large granular leukemia and related proliferations. *Am J Clin Pathol* **127**, 850–9.
95. Alekshun TJ, Sokol L. (2007). Diseases of large granular lymphocytes. *Cancer Control* **4**, 141–50.
96. Ahmad E, Kingma DW, Jaffe ES, Schrager JA, Janik J, Wilson W, Stetler-Stevenson M. (2005). Flow cytometric immunophenotypic profiles of mature gamma delta T-cell malignancies involving peripheral blood and bone marrow. *Cytometry B Clin Cytom* **67**, 6–12.
97. DiGiuseppe JA. (2007). Flow cytometric immunophenotyping of plasmacytic neoplasms. *Am J Clin Pathol* **127**, 172–4.
98. Seegmille AC. (2007). Immunophenotypic differentiation between neoplastic plasma cells in mature B-cell lymphoma vs plasma cell myeloma. *Am J Clin Pathol* **127**, 176–81.
99. Bataille R, Jego G, Robillard N, Barille-Nion S, Harousseau JL, Moreau P, Amiot M, Pellat-Deceunynck C. (2006). The phenotype of normal, reactive and malignant plasma cells. Identification of "many and multiple myelomas" and of new targets for myeloma therapy. *Haematologica* **91**, 1234–40.
100. Mackie EJ. (2003). Osteoblasts: Novel roles in orchestration of skeletal architecture. *Int J Biochem Cell Biol* **35**, 1301–5.
101. Horowitz MC, Lorenzo JA. (2004). The origins of osteoclasts. *Curr Opin Rheumatol* **16**, 464–8.
102. Quinn JM, Gillespie MT. (2005). Modulation of osteoclast formation. *Biochem Biophys Res Commun* **328**, 739–45.
103. Gimble JM, Nuttall ME. (2004). Bone and fat: Old questions, new insights. *Endocrine* **23**, 183–8.
104. Short B, Brouard N, Occhiodoro-Scott T, Ramakrishnan A, Simmons PJ. (2003). Mesenchymal stem cells. *Arch Med Res* **34**, 565–71.
105. Gregoretti MG, Gottardi D, Ghia P, Bergui L, Merico F, Marchisio PC, Caligaris-Cappio F. (1994). Characterization of bone marrow stromal cells from multiple myeloma. *Leuk Res* **18**, 675–82.
106. Strijbos MH, Kraan J, den Bakker MA, Lambrecht BN, Sleijfer S, Gratama JW. (2007). Cells meeting our immunophenotypic criteria of endothelial cells are large platelets. *Cytometry B Clin Cytom* **72**, 86–93.
107. Finney MR, Greco NJ, Haynesworth SE, Martin JM, Hedrick DP, Swan JZ, Winter DG, Kadereit S, Joseph ME, Fu P, Pompili VJ, Laughlin MJ. (2006). Direct comparison of umbilical cord blood versus bone marrow-derived endothelial precursor cells in mediating neovascularization in response to vascular ischemia. *Biol Blood Marrow Transplant* **12**, 585–93.
108. Pierre RV. (2002). Peripheral blood film review. The demise of the eyecount leukocyte differential. *Clin Lab Med Mar* **22**, 279–97.
109. Bessis M. (1977). *Blood Smears Reinterprited.* Springer-Verlag, Berlin.
110. Pierre RV. (2002). Red cell morphology and the peripheral blood film. *Clin Lab Med* **22**, 25–61. v–vi.
111. Pierre RV. (2002). Reticulocytes. Their usefulness and measurement in peripheral blood. *Clin Lab Med* **22**, 63–79.
112. Mebius RE, Kraal G. (2005). Structure and function of the spleen. *Nat Rev Immunol* **5**, 606–16.
113. Enriquez P, Neiman RS. (1976). *The pathology of the Spleen. A Functional Approach.* ASCP Press, Chicago.
114. Van Krieken JH, te Velde J. (1988). Normal histology of the human spleen. *Am J Surg Pathol* **12**, 777–8.
115. Weiss L. (1965). The structure of normal spleen. *Semin Haematol* **2**, 205–28.
116. Elmore SA. (2006). Enhanced histopathology of the spleen. *Toxicol Pathol* **34**, 648–55.
117. Suttie AW. (2006). Histopathology of the spleen. *Toxicol Pathol* **34**, 466–503.
118. Cesta MF. (2006). Normal structure, function, and histology of the spleen. *Toxicol Pathol* **34**, 455–65.
119. Mebius RE, Nolte MA, Kraal G. (2004). Development and function of the splenic marginal zone. *Crit Rev Immunol* **24**, 449–64.
120. Nolte MA, Arens R, Kraus M, van Oers MH, Kraal G, van Lier RA, Mebius RE. (2004). B cells are crucial for both development and maintenance of the splenic marginal zone. *J Immunol* **172**, 3620–7.
121. Weiss L. (1983). The red pulp of the spleen: Structural basis of blood flow. *Clin Haematol* **12**, 375–93.
122. Mebius RE. (2003). Organogenesis of lymphoid tissues. *Nat Rev Immunol* **3**, 292–303.
123. Schumacher HR, Rock WA, Stass SA. (2000). *Handbook of Hematologic Pathology.* Marcel Dekker, New York.
124. Jaffe EL, Harris NL, Stein H, Vardiman JW. (2001). *Pathology and Genetics: Tumors of Haematopoietic and Lymphoid Tissues. WHO Classification of Tumors.* IARC Press, Lyon.
125. Cupedo T, Mebius RE. (2005). Cellular interactions in lymph node development. *J Immunol* **174**, 21–5.
126. Allen CD, Okada T, Cyster JG. (2007). Germinal-center organization and cellular dynamics. *Immunity* **27**, 190–202.
127. Willard-Mack CL. (2006). Normal structure, function, and histology of lymph nodes. *Toxicol Pathol* **34**, 409–24.
128. Gretz JE, Anderson AO, Shaw S. (1997). Cords, channels, corridors and conduits: Critical architectural elements facilitating cell interactions in the lymph node cortex. *Immunol Rev* **156**, 11–24.
129. Weiss L. (1988). Lymphatic vessels and lymph nodes. In *Cell and Tissue Biology* (Weiss L, ed.). Urban and Schwarzenberg, Baltimore.

Principles of Immunophenotyping

CHAPTER 2

Faramarz Naeim

THE HUMAN CELL DIFFERENTIATION MOLECULES

"Human cell differentiation molecules" (HCDM) is a new terminology coined by the 8th International Workshop on Human Leukocyte Differentiation Antigen (HLDA) to describe surface molecules associated with human cell differentiation [1–3]. These molecules have been characterized in a series of international workshops studying a large number of monoclonal antibodies. The antibodies have been grouped according to their patterns of reactivity and are referred to as "clusters of differentiation" (CD). The 8th International Workshop on HLDA, held in Adelaide, Australia, December 2004, brought the total number of CD molecules to 339 (Table 2.1) [1–3]. These molecules characterize human leukocytes as well as other human cells such as endothelial and stromal cells. They are not only detected on the surface but also inside the cells [1–3].

Monoclonal antibodies are routinely used for the diagnosis and classification of hematopoietic malignancies and other hematologic disorders. However, it is important to remember the following facts:

1. If not all, by far, the vast majority of the available monoclonal antibodies raised against CD molecules are not tumor-specific and react with non-neoplastic hematopoietic cells.
2. These molecules are mostly differentiation associated and not lineage-specific.
3. They may react with non-hematopoietic human cells.

Because of these facts, the results of immunophenotypic studies, such as flow cytometry and immunohistochemical stains, should be always incorporated with morphology and other available data, such as cytogenetics and molecular studies. The following are examples of CD molecules most frequently used in diagnostic hematopathology at the present time [4–7].

B-CELL-ASSOCIATED CD MOLECULES

CD10

CD10, also known as *common acute lymphoblastic leukemia antigen* (CALLA), is a neutral endopeptidase, which cleaves peptides at the amino side of hydrophobic residues and inactivates several peptide hormones [5, 8]. It is expressed on the leukemic cells of the most common type of acute lymphoblastic leukemia (ALL), precursor B-cell ALL. CD10 is also present in hematogones (normal precursor B-cells in bone marrow), as well as cells in other B-cell lymphoid malignancies, such as follicular center cell lymphomas, Burkitt leukemia/lymphoma, and some cases of plasma cell myeloma, T-cell ALL, and acute myelogenous leukemia (AML) [4–6, 8]. This molecule is abundant in kidney, particularly on the brush border of proximal tubules and on glomerular epithelium [9, 10]. It is also present in granulocytes, fibroblasts, and a variety of normal and neoplastic epithelial cells [5, 9].

CD19

CD19 is a signal-transduction molecule that plays an important role in the regulation of development, activation, and differentiation of B-lymphocytes [5, 11]. It is the earliest lineage-restricted molecule expressed on B-cells throughout B-cell differentiation. Follicular dendritic cells also express

TABLE 2.1 The human cell differentiation molecules.

CD	Molecule	Main distribution
CD1a	T6/Leu-6, R4, HTA1	Cortical thymocyte, LC, IDC
CD1b	R1	Cortical thymocyte, LC, IDC
CD1c	M241, R7	Cortical thymocyte, LC, IDC
CD2	T11; Tp50; sheep red blood cell (SRBC) receptor; LFA-2	Thymocyte, T, NK, thymic B–cells
CD3	CD3 complex, T3, Leu4	Precursor T, thymocyte, T
CD4	OKT4, Leu 3a, T4	Helper T, thymocyte, M
CD5	Tp67; T1, Ly1, Leu-1	Thymocytes, T, B subset
CD6	T12	Thymocytes, T, B subset
CD7	Leu 9, 3A1, gp40, T-cell leukemia antigen	Precursor T, T, NK
CD8	OKT8, LeuT, LyT2, T8	Cytotoxic T, NK
CD9	Drap-27, MRP-1, p24, leukocyte antigen MIC3	Platelet, early B, Eo, Baso, endothelial
CD10	CALLA, membrane metallo-endopeptidase	Precursor B, B subset, G
CD11a	alphaL; LFA-1, gp180/95	All leukocytes
CD11b	alphaM; α-chain of C3bi receptor, gp155/95, Mac-1, Mo1	G, M, NK
CD11c	alphaX; α-chain of: complement receptor type 4 (CR4); gp150/95	G, M, NK
CDw12	P90-120	G, M, NK
CD13	Aminopeptidase N, APN, gp150, EC 3.4.11.2	G, M, endothelial, LGL subset
CD14	LPS receptor	M, DC subset
CD15	Lewis X, CD 15u: sulphated Lewis X. CD 15s: sialyl Lewis X	G, Reed-Sternberg cells
CD16	Fc gamma R IIIa	NK, G, M, macrophage
CDw17	LacCer, lactosylceramide	Platelet, G, M, B subset
CD18	β2-Integrin chain, macrophage antigen 1 (mac-1)	All leukocytes
CD19	Bgp95, B4	Precursor B, B
CD20	B1; membrane-spanning 4-domains, subfamily A, member 1	Precursor B subset, B
CD21	C3d receptor,CR2, gp140; EBV receptor	FDC, B subset, T subset
CD22	Bgp135; BL-CAM, Siglec2	Precursor B, B
CD23	Low affinity IgE receptor; FceRII; gp50-45; Blast-2	B, DC, M
CD24	Heat stable antigen homologue (HSA), BA-1	Precursor B, B, G
CD25	Interleukin (IL)-2 receptor α-chain; Tac-antigen	Activated T, B and M
CD26	Dipeptidylpeptidase IV; gp120; Ta1	Thymocyte, B, NK, macrophage, activated T
CD27	T14, S152	NK, thymocyte, B subset, T subset
CD28	Tp44	Thymocyte, T, PC
CD29	Integrin β1 chain; platelet GPIIa; VLA (CD49) β-chain	All leukocytes
CD30	Ki-1 antigen, Ber-H2 antigen	M, activated B, T, and NK
CD31	PECAM-1; platelet GPIIa'; endocam	Endothelial, platelet, leukocyte

(*Continued*)

TABLE 2.1 (Continued)

CD	Molecule	Main distribution
CD32	Fc gamma receptor type II (FcγRII), gp40	M, G, Eo, Baso, B, platelet
CD33	My9, gp67, p67	Precursor G, G, M
CD34	My10, gp105-120	Hematopoietic progenitor cells, endothelium
CD35	C3b/C4b receptor; complement receptor type 1 (CR1)	Erythroid, B, Eo, M, T subset
CD36	Platelet GPIV, GPIIIb, OKM-5 antigen	Platelet, M
CD37	Gp40-52	Mature B
CD38	T10; gp45, ADP-ribosyl cyclase	Early and activated hematopoietic cells, PC
CD39	Gp80, ectonucleoside triphosphate diphosphohydrolase 1	Leukocytes
CD40	Bp50, TNF Receptor 5	B, DC, macrophage, endothelial
CD41	Platelet glycoprotein GPIIb	Platelet
CD42a	Platelet glycoprotein GPIX	Platelet
CD42b	Platelet glycoprotein GPIb-a	Platelet
CD42c	Platelet glycoprotein GPIb-ß	Platelet
CD42d	Platelet glycoprotein GPV	Platelet
CD43	Leukosialin; gp95; sialophorin; leukocyte sialoglycoprotein	Leukocytes
CD44	Pgp-1; gp80-95, Hermes antigen, ECMR-III and HUTCH-I.	Leukocytes
CD45	LCA, B220, protein tyrosine phosphatase, receptor type, C	Leukocytes
CD45RA	Restricted T200; gp220; isoform of leukocyte common antigen	Naive T, B, M, NK
CD45RO	Restricted T200; gp180	Thymocyte, memory T, G, M
CD45RB	Restricted T200; isoform of leukocyte common antigen	T subset, B, G, M
CD46	Membrane cofactor protein (MCP)	Leukocytes
CD47	Integrin-associated protein (IAP), Ovarian carcinoma antigen OA3	Leukocytes
CD48	BLAST-1, Hulym3, OX45, BCM1	Leukocytes
CD49a	Integrin a1 chain, very late antigen, VLA 1a	Broad
CD49b	Integrin a2 chain, VLA-2-alpha chain, platelet gpla	Broad
CD49c	Integrin a3 chain, VLA-3-alpha chain	Broad
CD49d	Integrin a4 chain, VLA-4-alpha chain	Broad
CD49e	Integrin a5 chain, VLA-5-alpha chain	Broad
CD49f	Integrin a6 chain, VLA-6-alpha chain, platelet gplc	Broad
CD50	ICAM-3, intercellular adhesion molecule 3	Leukocytes
CD51	Integrin alpha chain, vitronectin receptor alpha chain	Platelet, endothelial cell
CD52	Campath-1, HE5	Thymocyte, B, T, NK, M
CD53	MRC OX-44	B, T, M,NK, G
CD54	ICAM-1, intercellular adhesion molecule 1	B, T, M, G, endothelial cell
CD55	DAF, decay accelerating factor	Broad

(Continued)

TABLE 2.1 (Continued)

CD	Molecule	Main distribution
CD56	NKHI, neural cell adhesion molecule (NCAM)	NK, T subset, neuroendodermal cells
CD57	HNK1	NK, T subset, neuroendodermal cells
CD58	LFA-3, lymphocyte function associated antigen-3	Broad
CD59	MACIF, MIRL, P-18, protectin	Broad
CD60	GD3 (CD60a), 9-0-acetyl GD3 (CD60b), 7-0-acetyl GD3 (CD60c)	Platelets, T subset
CD61	Glycoprotein IIIa, beta3 integrin	Platelets
CD62E	E-selectin, LECAM-2, ELAM-1	Endothelium
CD62L	L-selectin, LAM-1, Mel-14	B, T, M, NK subset, G
CD62P	P-selectin, granule membrane protein-140 (GMP-140)	Activated platelet, endothelium
CD63	LIMP, gp55, LAMP-3 neuroglandular antigen, granulophysin	Activated platelets, G, M, endothelium
CD64	FcgR1, FcgammaR1	Precursor G, G, M, DC subset
CD65	Ceramide dodecasaccharide 4c, VIM2	G, M
CD66a	BGP, carcinoembryonic antigen-related cell adhesion molecule 1	G, epithelium
CD66b	CGM6, NCA-95	G
CD66c	Nonspecific cross-reaction antigen, NCA-50/90	G, epithelium
CD66d	CGM1	G
CD66e	CEA	Epithelium
CD66f	PSG, Sp-1, pregnancy specific (b1) glycoprotein	Myeloid cell lines, placenta
CD68	gp110, macrosialin	M, G, DC subset, Baso, Mast cell
CD69	AIM, activation inducer molecule, MLR3, EA1, VEA	Activated leukocytes
CD70	CD27 ligand, KI-24 antigen	Activated B and T
CD71	Transferrin receptor	Erythroid precursors, proliferating cells
CD72	Lyb-2, Ly-19.2, Ly32.2	Precursor B, B
CD73	Ecto-5'-nucleotidase	B subset, T subset
CD74	MHC Class II associated invariant chain (Ii)	B, IDC, T subset
CD75	Lactosamines	B, activated T, macrophages, activated endothelium
CD75s	Since HLDA7, CDw76 has been renamed CD75s	B, T subset
CD77	Pk blood group antigen; Burkitt´s lymphoma associated antigen	Germinal center B
CD79a	MB-1; Iga	Precursor B, B, activated B
CD79b	B29; Igß	Precursor B, B, activated B
CD80	B7-1; BB1	Macrophages, activated T and B
CD81	Target of an antiproliferative antibody (TAPA-1); M38	Broad
CD82	R2; 4F9; C33; IA4, kangai 1	Broad
CD83	HB15	IDC, LC

(Continued)

TABLE 2.1 (Continued)

CD	Molecule	Main distribution
CD84	p75, GR6	CD84
CD85	ILT5; LIR3; HL9	B, thymocytes, M, macrophages, platelets
CD86	B7-2; B70	IDC, LC, B, and M subset
CD87	Urokinase plasminogen activator-receptor (uPA-R)	Subsets of T, NK, M and G
CD88	C5a-receptor	G, M, DC
CD89	Fca-receptor, IgA-receptor	Precursor myeloid, G, M
CD90	Thy-1	Hematopoietic stem cell
CD91	a2-macroglobulin receptor (ALPHA2M)	Broad
CDw92	p70	G, M
CDw93	GR11	G, M, myeloid blast, endothelium
CD94	kP43, killer cell lectin-like receptor subfamily D, member 1	NK, T subset
CD95	APO-1, Fas, TNFRSF6	Thymocytes, B and T subset
CD96	TACTILE (T-cell activation increased late expression)	Activated NK and T
CD97	BL-KDD/F12	DC, G, M activated B and T
CD98	4F2, FRP-1	Activated leukocytes
CD99	MIC2, E2	Broad
CD100	SEMA4D	Leukocytes, activated T, germinal center B
CD101	V7, P126	G, M, DC, activated T
CD102	ICAM-2	M, platelet, endothelium
CD103	Integrin alpha E subunit, HML-1	Intraepithelial lymphocytes, hairy cells
CD104	Integrin beta 4 subunit, TSP-1180	Epithelium
CD105	Endoglin	Endothelium, precursor B, activated M
CD106	VCAM-1 (vascular cell adhesion molecule-1), INCAM-110	DC, activated endothelium
CD107a	Lysosomal associated membrane protein (LAMP)-1	Degranulated platelet, activated T
CD107b	Lysosomal associated membrane protein (LAMP)-2	Degranulated platelet
CD108	GPI-gp80; John-Milton-Hagen (JMH) human blood group antigen	Erythroid
CD109	Platelet activation factor; 8A3, E123	Activated platelet, endothelium
CD110	Thrombopoietin receptor; c-mpl	Hematopoietic stem cells, platelets
CD111	PRR1, Nectin 1, Hve C1, poliovirus receptor related 1 protein	34^+ hematopoietic precursors
CD112	PRR2, Nectin 2, Hve B, poliovirus receptor related 2 protein	34^+ hematopoietic precursors
CDw113	PVRL3, Nectin3	Epithelium
CD114	G-CSFR, HG-CSFR, CSFR3	M, platelets
CD115	M-CSFR, CSF-1, C-fms	M, macrophages
CD116	GMCSF R alpha subunit	Myeloid cells

(Continued)

TABLE 2.1 (Continued)

CD	Molecule	Main distribution
CD117	SCFR, c-kit, stem cell factor receptor	Hematopoietic stem cells, mast cells, plasma cells, AML blasts
CD118	LIFR	Broad
CD119	IFN gamma receptor alpha chain	Broad
CD120a	TNFRI; TNFRp55	Broad
CD120b	TNFRII; TNFRp75	Broad
CD121a	Type I IL-1 receptor	Broad
CD121b	Type II IL-1 receptor	Broad
CD122	IL-2 receptor betachain, p75	B, T, NK, M
CD123	Interleukin-3 receptor alpha chain (IL-3Ra)	Hematopoietic precursors
CD124	IL-4 R alpha chain	Broad
CDw125	Interleukin-5 receptor alpha chain	Baso, Eo, activated B
CD126	IL-6 receptor alpha chain	T, M, activated B
CD127	IL-7 receptor alpha chain, p90	Precursor B, B, T
CD129	IL-9 receptor alpha chain	Hematopoietic cells
CD130	gp130	Broad
CD131	Common ß chain, low-affinity (granulocyte-macrophage)	Precursor myeloid, precursor B, M, G, Eo
CD132	Common gamma chain, interleukin 2 receptor, gamma	B, T, M, G, NK
CD133	AC133, PROML1, prominin 1	$CD34^+$ hematopoietic precursor
CD134	OX 40, TNFRSF4	Thymocyte, T
CD135	FLT3, STK-1, flk-2	Precursor B, precursor myleomonocytic
CDw136	Macrophage stimulating protein receptor, MSP-R, RON	Epithelium, M
CDw137	4-1BB, Induced by lymphocyte activation (ILA)	T, activated T
CD138	Syndecan-1, B-B4	Plasma cells, B subset, epithelium
CD139		B, M, G
CD140a	Alpha-platelet derived growth factor (PDGF) receptor	Mesenchymal cells
CD140b	Beta-platelet derived growth factor (PDGF) receptor	Mesenchymal cells, M, G
CD141	Thrombomodulin (TM), fetomodulin	Broad
CD142	Tissue factor, thromboplastin, coagulation factor III	Epithelium, M, endothelium
CD143	Angiotensin-converting enzyme (ACE), peptidyl dipeptidase A	Broad
CD144	VE-cadherin, cadherin-5	Endothelium
CDw145	None	Endothelium
CD146	Muc 18, MCAM, Mel-CAM, s-endo	Endotheilum, melanoma cells, activated T
CD147	Basigin, M6, extracellular metalloproteinase inducer (EMMPRIN)	Leukocyte, erythroid, platelet, endothelium
CD148	DEP-1, HPTP-n, protein tyrosine phosphatase, receptor type, J	G, M, T subset, DC, platelet

(Continued)

TABLE 2.1 (Continued)

CD	Molecule	Main distribution
CD150	SLAM, signalling lymphocyte activation molecule, IPO-3	Thymocyte, B, DC, T subset, endothelium
CD151	Platelet-endothelial tetra-span antigen (PETA)-3	Platelet, endothelium, epithelium
CD152	Cytotoxic T-lymphocyte antigen (CTLA)-4	Activated B and T
CD153	CD30 Ligand	Activated T and M
CD154	CD40 Ligand; TRAP (TNF-related activation protein)-1; T-BAM	Activated T
CD155	Polio virus receptor (PVR)	M, neurons
CD156a	ADAM-8, a disintegrin and metalloproteinase domain 8	G, M
CD156b	TACE, ADAM 17 snake venom like protease CSVP	Broad
CD157	BST-1 BP-3/IF7 Mo5	G, M, precursor B
CD158	Killer cell Ig-like receptor, three domains, long cytoplasmic tail, 1	NK, T subset
CD159a	Killer cell lectin-like receptor subfamily C, member 1	NK
CD160	BY55, NK1, NK28	NK, T subset
CD161	NKR-P1A, killer cell lectin-like receptor subfamily B, member 1	NK, T subset
CD162	P selectin glycoprotein ligand 1, PSGL-1	T, M, G, B subset
CD163	GHI/61, D11, RM3/1, M130	M, macrophage, activated T
CD164	MUC-24, MGC 24, multi-glycosylated core protein 24	M, epithelium, bone marrow stromal cells
CD165	AD2, gp 37	Thymocyte, T, platelet
CD166	ALCAM, KG-CAM, activated leukocyte cell adhesion molecule	Epithelium, activated T and M
CD167	Discoidin receptor DDR1 (CD 167a) and DDR2 (CD 167b)	Epithelium
CD168	RHAMM (receptor for hyaluronan involved in migration and motility)	Thymocyte
CD169	Sialodhesin, Siglec-1	Macrophage
CD170	Siglec 5 (sialic acid binding Ig-like lectin 5)	Myeloid cells
CD171	Neuronal adhesion molecule, LI	Neurons
CD172	SIRP, signal inhibitory regulatory protein family member	Leukocytes
CD173	Blood Group H2	Erythrocytes
CD174	Lewis Y blood group, LeY, fucosyltransferase 3	Erythrocytes
CD175	Tn Antigen (T-antigen novelle)	Carcinomas
CD176	Thomsen-Friedenreich (TF) antigen	Carcinomas
CD177	NB 1	
CD178	FAS ligand, CD95 ligand	T, NK
CD179a	V pre beta	Precursor B
CD179b	Lambda 5	Precursor B
CD180	RP105, Bgp95	Mantle zone and marginal zone B
CD181	CXCR1 (was CDw128A)	Leukocytes
CD182	CXCR2 (was CDw128B)	Leukocytes

(Continued)

TABLE 2.1 (Continued)

CD	Molecule	Main distribution
CD183	CXCR3 chemokine receptor, G protein-coupled receptor 9	T, $CD34^+$ hematopoietic cells, DC subset, Eo
CD184	CXCR4 chemokine receptor, Fusin	M, T subset
CD185	CXCR5; Chemokine (C-X-C motif) Receptor 5, Burkitt lymphoma receptor 1	Broad
CDw186	CXCR6; Chemokine (C-X-C motif) Receptor 6	T, epithelium
CD191	CCR1; Chemokine (C-C motif) Receptor 1, RANTES Receptor	T and NK subset
CD192	CCR2; Chemokine (C-C motif) Receptor 2, MCP-1 receptor	M
CD193	CCR3; Chemokine (C-C motif) Receptor 3, eosinophil eotaxin receptor	Eo, Baso, epithelium
CD195	CCR5; Chemokine receptor	T, M
CD196	CCR6; Chemokine (C-C motif) Receptor 6	DC and T subset
CD197	CCR7; (was CDw197) Chemokine (C-C motif) Receptor 7	DC and T subset
CDw198	CCR8; Chemokine (C-C motif) Receptor 8	Thymocyte, macrophage
CDw199	CCR9; Chemokine (C-C motif) Receptor 9	Intestinal T-cells
CD200	MRC OX 2	Broad
CD201	Endothelial protein C receptor (EPCR)	Endothelium
CD202b	TIE2, TEK	Endothelium, hematopoietic stem cell
CD203c	E-NPP3, PDNP3, PD-1beta	Mast cell, Baso
CD204	MSR, SRA, Macrophage scavenger receptor	Macrophage
CD205	DEC-205	DC
CD206	Macrophage mannose receptor (MMR)	M, macrophage, endothelium
CD207	Langerin	LC
CD208	DC-LAMP	IDC
CD209	DC-SIGN	DC subset
CDw210	IL-10 receptor	B, T, NK, M, macrophage
CD212	IL-12 receptor beta chain	Activated T and NK
CD213a1	IL-13 receptor alpha 1	Broad
CD213a2	IL-13 R alpha 2	B, M
CDw217	IL-17 receptor	Broad
CDw218	IL18Ralpha	
CD220	Insulin receptor	Broad
CD221	IGF I Receptor, type I IGF receptor	Broad
CD222	Mannose-6-phosphate receptor, insulin like growth factor II R	Broad
CD223	LAG-3 (Lymphocyte activation gene 3)	T and NK subset
CD224	Gamma-glutamyl transferase, GGT	Broad
CD225	Leu-13, interferon-induced transmembrane protein 1	Broad
CD226	DNAM-1, DTA-1	T, NK, M, platelet, B subset

(Continued)

TABLE 2.1 (Continued)

CD	Molecule	Main distribution
CD227	MUC1; episialin; PUM; PEM; EMA; DF3 antigen; H23 antigen	Broad
CD228	Melanotransferrin, p97	Melanoma cells, endothelium
CD229	Ly9	T, B
CD230	Prion protein, PrP(c), PrP(sc) abnormal form	Broad
CD231	TALLA-1, TM4SF2	Precursor T, neuroblastoma
CD232	VESPR	Broad
CD233	Band 3, AE1, anion exchanger 1, Diego blood group antigen	RBC
CD234	DARC, Fy-glycoprotein, Duffy blood group antigen	RBC
CD235a	Glycophorin A	RBC
CD235b	Glycophorin B	RBC
CD236	Glycophorin C/D	RBC, stem cell subset
CD236R	Glycophorin C	RBC, stem cell subset
CD238	Kell blood group antigen	RBC, stem cell subset
CD239	B-CAM, lutheran glycoprotein	RBC, stem cell subset
CD240CE	Rh blood group system, Rh30CE	RBC
CD240D	Rh blood group system, Rh30D	RBC
CD240DCE	Rh30D/CE crossreactive mabs	RBC
CD241	RhAg, Rh50, Rh associated antigen	RBC
CD242	LW blood group, Landsteiner-Wiener blood group antigens	RBC
CD243	MDR-1, P-glycoprotein, pgp 170, multidrug resistance protein I	Hematopoietic stem cell
CD244	2B4; NAIL; p38	NK, T subset
CD245	p220/240, DY12, DY35	T subset
CD246	Anaplastic lymphoma kinase (ALK)	Anaplastic large cell lymphoma
CD247	T-cell receptor zeta chain, CD3 zeta	T, NK
CD248	TEM1, Endosialin	Fibroblast, endothelium
CD249	Aminopeptidase A; APA, gp160	Epithelium
CD252	OX40L; TNF (ligand) superfamily member 4, CD134 ligand	T
CD253	TRAIL; TNF (ligand) superfamily member 10, APO2L	T
CD254	TRANCE; TNF (ligand) superfamily member 11, RANKL	T, M
CD256	APRIL; TNF (ligand) superfamily member 13, TALL2	Osteoclast, B subset
CD257	BLYS; TNF (ligand) superfamily, member 13b, TALL1, BAFF	B
CD258	LIGHT; TNF (ligand) superfamily, member 14	
CD261	TRAIL-R1; TNFR superfamily, member 10a, DR4, APO2	Broad
CD262	TRAIL-R2; TNFR superfamily, member 10b, DR5	Broad
CD263	TRAIL-R3; TNFR superfamily, member 10c, DCR1	Broad

(Continued)

TABLE 2.1 (Continued)

CD	Molecule	Main distribution
CD264	TRAIL-R4; TNFR superfamily, member 10d, DCR2	NK, T subset
CD265	TRANCE-R; TNFR superfamily, member 11a, RANK	M, DC
CD266	TWEAK-R; TNFR superfamily, member 12A, type I transmembrane protein Fn14	Broad
CD267	TACI; TNFR superfamily, member 13B, transmembrane activator and CAML interactor	Lymphocytes
CD268	BAFFR; TNFR superfamily, member 13C, B-cell-activating factor	B
CD269	BCMA; TNFR superfamily, member 17, B-cell maturation factor	B
CD271	NGFR (p75); nerve growth factor receptor (TNFR superfamily, member	Neurons
CD272	BTLA; B and T-lymphocyte attenuator	B, T subset
CD273	B7DC, PDL2; programmed cell death 1 ligand 2	Activated B and T
CD274	B7H1, PDL1; programmed cell death 1 ligand 1	Broad
CD275	B7H2, ICOSL; inducible T-cell co-stimulator ligand (ICOSL)	Broad
CD276	B7H3; B7 homolog 3	N/A
CD277	BT3.1; B7 family: butyrophilin, subfamily 3, member A1	
CD278	ICOS; inducible T-cell co-stimulator	Activated T
CD279	PD1; programmed cell death 1	Broad
CD280	ENDO180; uPARAP, mannose receptor, C type 2, TEM22	Macrophages
CD281	TLR1; TOLL-like receptor 1	Lymphocytes
CD282	TLR2; TOLL-like receptor 2	Lymphocytes
CD283	TLR3; TOLL-like receptor 3	Lymphocytes
CD284	TLR4; TOLL-like receptor 4	Lymphocytes
CD289	TLR9; TOLL-like receptor 9	Lymphocytes
CD292	BMPR1A; bone morphogenetic protein receptor, type IA	Broad
CDw293	BMPR1B; bone morphogenetic protein receptor, type IB	Broad
CD294	CRTH2; PGRD2; G protein-coupled receptor 44,	T subset
CD295	LEPR; leptin receptor	Platelets, G
CD296	ART1; ADP-ribosyltransferase 1	G
CD297	ART4; ADP-ribosyltransferase 4; Dombrock blood group glycoprotein	RBC
CD298	ATP1B3; Na^{+}/K^{+}-ATPase beta 3 subunit	Broad
CD299	DCSIGN-related; CD209 antigen-like, DC-SIGN2, L-SIGN	DC
CD300	CMRF35 FAMILY; CMRF-35H	M, G, B and T subsets
CD301	MGL; CLECSF14, macrophage galactose-type C-type lectin	Macrophages
CD302	DCL1; Type I transmembrane C-type lectin receptor DCL-1	Hodgkin lymphoma cell line
CD303	BDCA2; C-type lectin, superfamily member 11	DC subtype
CD304	BDCA4; Neuropilin 1	Broad
CD305	LAIR1; leukocyte-associated Ig-like receptor 1	B, T, NK

(Continued)

TABLE 2.1 (Continued)

CD	Molecule	Main distribution
CD306	LAIR2; leukocyte-associated Ig-like receptor 2	B, T, NK
CD307	IRTA2; immunoglobulin superfamily receptor translocation associated	B
CD309	VEGFR2; KDR (a type III receptor tyrosine kinase)	Endothelium
CD312	EMR2; EGF-like module containing, mucin-like, hormone receptor-like	Lymphocytes
CD314	NKG2D; killer cell lectin-like receptor subfamily K, member 1	NK
CD315	CD9P1; prostaglandin F2 receptor negative regulator	Lymphocytes
CD316	EWI2; immunoglobulin superfamily, member 8	Lymphocytes
CD317	BST2; bone marrow stromal cell antigen 2	Bone marrow stromal cells
CD318	CDCP1; CUB domain-containing protein 1	Hematopoietic stem cell subset
CD319	CRACC; SLAM family member 7	Activated T
CD320	8D6; 8D6 antigen; FDC	N/A
CD321	JAM1; F11 receptor	Epithelium, endothelium
CD322	JAM2; junctional adhesion molecule 2	Epithelium, endothelium
CD324	E-cadherin; cadherin 1, type 1, E-cadherin (epithelial)	Epithelium
CDw325	E-cadherin; cadherin 2, type 1, N-cadherin (neuronal)	Neurons
CD326	Ep-CAM; tumor-associated calcium signal transducer 1	Epithelium
CDw327	siglec6; sialic acid binding Ig-like lectin 6	Cell–cell adhesion
CDw328	siglec7; sialic acid binding Ig-like lectin 7	Cell–cell adhesion
CDw329	siglec9; sialic acid binding Ig-like lectin 9	Cell–cell adhesion
CD331	FGFR1; fibroblast growth factor receptor 1	Fibroblasts
CD332	FGFR2; fibroblast growth factor receptor 2 (keratinocyte growth factor receptor)	Fibroblasts
CD333	FGFR3; fibroblast growth factor receptor 3 (achondroplasia, thanatophoric dwarfism)	Fibroblasts
CD334	FGFR4; fibroblast growth factor receptor 4	Fibroblasts
CD335	NKp46; NCR1, (Ly94); natural cytotoxicity triggering receptor 1	NK
CD336	NKp44; NCR2, (Ly95); natural cytotoxicity triggering receptor 2	NK
CD337	NKp30; NCR3	NK
CDw338	ABCG2; ATP-binding cassette, sub-family G (WHITE), member 2	Epithelium
CD339	Jagged-1; Jagged 1 (Alagille syndrome)	Broad

From: http://mpr.nci.nih.gov/prow/

LC: Langerhans cell; IDC: interdigitating dendritic cell; NK: natural killer; M: monocyte; Eo: eosinophil; Baso: basophil; G: granulocyte; DC: dendritic cell; FDC: follicular dendritic cell; PC: plasma cell; AML: acute myelogenous leukemia; RBC: red blood cell.

CD19 [12]. This molecule is not expressed on immature and mature T-lymphocytes, monocytic and granulocytic series, or erythroid precursors. However, CD19 is occasionally expressed in patients with AML [4–6].

CD20

CD20 is a membrane-embedded surface molecule which plays a role in the development and differentiation of B-cells into plasma cells [5, 13]. It appears after HLA-DR, TdT, CD19, and CD10 expression and before cytoplasmic μ chain appearance in B-cell ontogeny. Similar to CD19, CD20 is a lineage-restricted molecule and is expressed on B-cells throughout B-cell differentiation prior to terminal differentiation of B-cells to plasma cell [13]. CD20 is expressed in the majority of B-cell lymphoid malignancies and some cases of plasma cell myeloma, T-cell leukemia/lymphomas, and AML [4–6].

CD21

CD21 is a receptor for EBV, C3d, C3dg, and iC3b [14]. Complement components may activate B-cells through CD21, which is part of a large signal-transduction complex that also involves CD19 and CD81 [5, 14]. This molecule is expressed on mature B-cells, from the stage when surface Ig is first expressed and then is lost upon activation. Follicular mantle zone B-cells and marginal zone B-cells express CD21. Also, follicular dendritic cells and subsets of thymocytes and T-cells express CD21 [4–6, 15].

CD22

CD22 is a single chain integral membrane molecule and a member of the immunoglobulin gene superfamily. It functions as an adhesion receptor and a signaling molecule, which appears to be involved in regulating the expression of surface IgM on peripheral B-cells and Ca^{++} flux in response to immunoglobulin signaling [16, 17]. Cytoplasmic CD22 is expressed at the earliest stages of B-cell differentiation, along with CD19 and prior to the expression of CD20. The majority of the TdT-positive precursor B-cells are also positive for cytoplasmic CD22. Expression of surface CD22 precedes or accompanies expression of surface IgM and/or IgD in mature B-lymphocytes, but it is lost in plasma cells. The neoplastic cells in various proportions of B-cell lymphoid malignancies, including precursor B-ALL, chronic lymphoid leukemias, and B-cell lymphomas, express CD22 [4–6, 18]. CD22 expression is particularly strong in hairy cell leukemia and prolymphocytic leukemia. T-cells and their malignant counterparts do not express CD22.

CD23

CD23 is an integral membrane glycoprotein involved in the regulation of IgE synthesis and pro-inflammatory activities, such as triggering the release of regulatory cytokines TNF, IL-1, IL-6, and GM-CSF by human monocytes [5, 19, 20]. It is expressed by activated B-cells, monocytes, follicular dendritic cells, and subsets of eosinophils and platelets [5]. Chronic lymphocytic leukemia (CLL) cells are positive for CD23, as are frequently follicular cell lymphoma cells [4, 5, 21]. The neoplastic cells in mantle cell and marginal zone B-cell lymphomas do not typically express CD23; neither do the neoplastic cells in plasma cell myeloma, ALL, T-cell, and myeloid malignancies [4–6].

CD24

CD24 is expressed on immature and mature B-cells except plasma cells. This molecule, however, is not lineage-restricted and is present on granulocytes and various benign and malignant epithelial cells [4–6, 22]. T-lymphocytes, monocytes, and erythroid precursors do not express CD24.

CD79

CD79 in association with surface Ig constitutes the B-cell antigen receptor complex on the surface of the B-lymphocytes [5, 23]. CD79 consists of α and β heterodimers and plays a critical role in B-cell maturation and activation. The pattern of CD79 expression on B-cells is closely similar to that of CD19. CD79a is expressed initially in the cytoplasm prior to cytoplasmic μ heavy chain expression, and later on, after the expression of surface Ig, appears on the cytoplasmic membrane. CD79a is usually negative in CLL cells and plasma cells, whereas CD79b may be expressed in plasma cells and a significant proportion of CLL patients [5, 24]. CD79 is an excellent B-cell marker, but some cases of T-cell ALL and AML may react positively with anti-CD79 monoclonal antibodies [5, 25, 26].

CD138

CD138 is a transmembrane sulfate proteoglycan, which functions as a receptor for cell–matrix interactions [5, 27]. This molecule appears to be involved in the cellular organization in various tissues. Plasma cells adhere to type 1 collagen through CD138 and are the only hematopoietic cells that express this molecule. CD138 is expressed by various mesenchymal and epithelial cells, such as fibroblasts, endothelial cells, and stratified epithelia [5, 6, 28, 29].

OTHER B-CELL-ASSOCIATED MARKERS

CD5

The description of CD5 is briefed in the section "T-Cell-Associated CD Molecules."

CD74

CD74 is an integral transmembrane molecule playing a role in intracellular sorting of MHC class II molecules [30, 31]. This molecule is expressed by most of the B-cells, particularly follicular center cells, mantle cells, and activated B-lymphocytes [5, 32]. A subpopulation of T-cells, macrophages, activated endothelial cells, and neoplastic plasma cells may also express CD74 [5, 32, 33].

CD103

CD103 is a membrane receptor which appears to play a role in the activation of intraepithelial lymphocytes [34]. It is expressed in >90% of intestinal intraepithelial lymphocytes and certain types of B- and T-cell lymphoid malignancies, such as *hairy cell leukemia* (B-cell), enteropathy-associated T-cell lymphoma, and adult T-cell leukemia/lymphoma [4, 6, 35, 36].

FMC7

FMC7 molecule binds to a particular conformation of the CD20 antigen, probably to a multimeric CD20 complex, and it is detected only when CD20 antigen is present in high densities and in the postulated multimeric complex formation [37]. FMC7 is weakly expressed or is negative in CLL cells, and strongly positive in hairy cell leukemia and prolymphocytic leukemia [38, 39]. Antibodies against FMC7 are routinely used in flow cytometric studies for the diagnosis and classification of B-cell lymphoproliferative disorders.

T-CELL-ASSOCIATED CD MOLECULES

CD1

CD1 is a member of the immunoglobulin supergene family consisting of MHC class 1-like glycoproteins [5]. So far, five distinct molecules of CD1 have been described: a, b, c, d, and e [5, 40]. The first three have been extensively used in diagnosis and classification of hematologic malignancies. CD1 molecules are expressed on thymocytes. They are absent on mature peripheral blood T-cells, but their cytoplasmic expression has been observed in activated T-lymphocytes. High levels of CD1a and less of CD1b and CD1c are present on Langerhans cells [41]. CD1c is expressed by the majority of cord blood and a subset of peripheral blood B-cells [5]. A subset of mantle zone and follicular center B-cells also express CD1c [4–6, 42, 43]. Follicular dendritic cells and monocytes/macrophages are CD1-negative.

CD2

CD2 is a transmembrane molecule and a member of the immunoglobulin supergene family and binds CD58, CD48, and CD59 [5, 44, 45]. The existence of this molecule was originally discovered by the ability of human T-cells to spontaneously bind sheep erythrocytes (E-rosette receptor) [5, 40]. CD2 plays an important role in T-cell activation, T- or NK-mediated cytolysis, apoptosis in activated peripheral T-cells, and the production of cytokines by T-cells [5, 44, 45]. It is expressed by thymic T-cells, peripheral T-cells, NK cells, and a subset of thymic B-cells [5, 44, 45]. CD2 is an excellent pan-T-cell marker and one of the earliest antigens which precedes CD1 but appears after CD7 on the T-cells [44–46]. However, some of the T-cell lymphoid malignancies, particularly peripheral T-cell lymphomas, may aberrantly lose CD2 expression. Also, some cases of AML, mainly promyelocytic type, may express CD2 [4–6].

CD3

CD3 is a complex structure composed of three different polypeptide dimmers: $\gamma\varepsilon$, $\delta\varepsilon$, and $\zeta\zeta$. CD3 in conjunction with T-cell receptor (TCR) makes the TCR complex (Figure 2.1). TCR molecules represent two different heterodimers: $\alpha\beta$ and $\gamma\delta$. The vast majority of T-cells bear TCR$\alpha\beta$ and only about 5% of T-cells express TCR-$\gamma\delta$ [5, 6, 47]. The $\alpha\beta$ T-cells divide into CD4$^+$ and CD8$^+$ cells and are widespread and found in all hematopoietic and lymphoid tissues, whereas

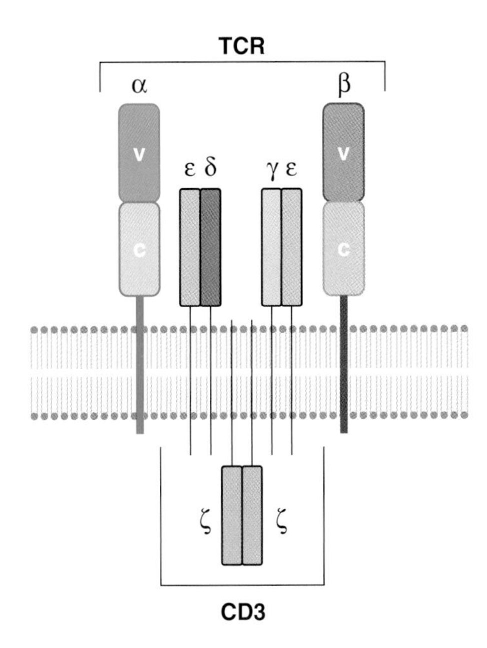

FIGURE 2.1 Schematic of TCR complex.

$\gamma\delta$ T-cells are negative for CD4 and CD8 and are primarily found in the spleen and intestinal epithelium [4–6]. NK cells do not express TCR complex but usually show cytoplasmic ε chain of CD3 [4–6].

Surface membrane CD3 is a pan-T-cell marker and is expressed by thymocytes and all mature T-cells of peripheral blood and lymphoid tissues. B-cells, granulocytic series, and monocytes/macrophages are all CD3-negative. NK cells are negative for membrane CD3 but may express cytoplasmic (ε chain) CD3 [5, 6]. Precursor T-lymphoblastic leukemias often show cytoplasmic CD3 expression.

CD4

CD4 is a membrane glycoprotein and a member of the immunoglobulin supergene family and a co-receptor in MHC class II-restricted T-cell activation [5, 48]. It also plays a role in the differentiation of thymocytes and the regulation of T-lymphocyte/B-lymphocyte adhesion [5, 48, 49]. CD4 is the primary receptor for HIV retroviruses [50]. CD4 is expressed in a large proportion of thymocytes (80–90%) and over 50% of the peripheral blood T-cells (helper/inducer subtype) [5, 48]. Most thymocytes coexpress CD4 and CD8. Monocytes, macrophages, and Langerhans cells (LC) express CD4. The majority of postthymic T-cell neoplasms are CD4-positive.

CD5

CD5 is a signal transducing molecule involved in tyrosine phosphorylation of intracellular proteins [5, 51]. It modulates signaling through the antigen-specific receptor complexes TCR and B-cell receptor (BCR) [5, 52]. CD5 is expressed at low density on thymocytes and at high density on all mature T-lymphocytes. It is also expressed at low density on a small subset of mature B-lymphocytes (B1a cells) which is expanded during fetal life and in several autoimmune disorders. Certain B-cell lymphoid malignancies, such as chronic lymphocytic leukemia/small lymphocytic lymphoma and mantle cell lymphoma, express CD5 [4–6, 53].

CD7

CD7 is a transmembrane glycoprotein and a member of the immunoglobulin supergene family. It appears to play an essential role in T-cell and T-cell/B-cell interactions during early lymphoid development [5, 54]. CD7 is the earliest T-cell-associated molecule to appear in stem cells and prethymic stages and extends its expression all the way to the mature stages. The pluripotent stem cells, capable of differentiating to T-cells as well as erythroid, megakaryocytic, and myeloid precursors, may express CD7. This molecule is also present on most NK cells. A subpopulation of AML, particularly those with monocytic differentiation, may express CD7 [4–6, 55]. CD7 is an excellent marker for the detection of T-cell lymphoproliferative disorders. However, it is usually absent or weakly expressed in mycosis fungoides/Sezary syndrome and adult T-cell leukemia/lymphoma [4–6].

CD8

CD8 is a cell surface glycoprotein and a member of the immunoglobulin supergene family that is involved in the mediation of cell–cell interactions within the immune system [5, 56, 57]. This molecule is found on cytotoxic/suppressor T-lymphocytes and the majority of the thymocytes. Approximately 80–90% of the thymocytes and 35–45% of the peripheral blood lymphocytes express CD8. Most thymocytes coexpress CD8 with CD4. A subpopulation of NK cells also express CD8 [4–6].

CD45RA AND CD45RO

These molecules represent two different isoforms of the CD45 cluster. CD45 is typically expressed in all hematopoietic cells (a pan-leukocyte marker) [4–6]. CD45RA is expressed on naive/resting T-cells and medullary thymocytes, whereas CD45RO is detected on memory/activated T-cells, cortical thymocytes, monocytes/macrophages, and granulocytes [5, 58–60].

T-CELL RECEPTOR MOLECULES

As mentioned earlier, TCR heterodimers, $\alpha\beta$ and $\gamma\delta$, in association with CD3 make the TCR complex (Figure 2.1). The vast majority of T-cells bear TCR$\alpha\beta$ and only about 5% of T-cells express TCR$\gamma\delta$ [5, 6, 61–63]. The $\alpha\beta$ T-cells are widespread and are found in all hematopoietic and lymphoid tissues, whereas $\gamma\delta$ T-cells are primarily found in the spleen and intestinal epithelium. NK cells do not express TCR; neither do the B-cells, monocytes/macrophages, or granulocytic cells.

OTHER T-CELL-ASSOCIATED MARKERS

CD26 is a T-cell co-stimulatory molecule with dipeptidyl peptidase activity and is considered as a T-cell activation molecule [64, 65]. CD26 expression is lost in the Sezary cells [66].

CD246 or anaplastic lymphoma kinase (ALK) is expressed by the neoplastic cells in anaplastic large cell lymphomas [67, 68].

CD247 is a component of the TCR complex and is expressed by T-cells [69].

CD CLUSTERS ASSOCIATED WITH LARGE GRANULAR LYMPHOCYTES

CD16

CD16 is a low affinity IgG receptor expressed on large granular lymphocytes (LGL) of both NK- and T-cell types [5, 70]. This molecule is involved in antibody-dependent cell-mediated cytotoxicity [70, 71]. Approximately 15–20% of the peripheral blood lymphocytes and a much smaller fraction (<5%) of bone marrow lymphocytes express CD16. CD16 is also expressed on granulocytes, tissue macrophages, and a subset of monocytes, eosinophils, and dendritic cells [5, 72, 73]. CD16 expression is reduced or lost in paroxysmal nocturnal hemoglobinuria (PNH) due to the structural abnormality of glycosyl-phosphatidyl-inositol (GPI) [74, 75].

CD56

CD56 is a transmembrane-anchored glycoprotein and a member of the immunoglobulin supergene family [5, 76–78]. It functions as an adhesion molecule on neural and NK cells, and a subset of T-cells. NK cells are divided into CD56^{bright} and CD56^{dim} [5, 79]. The CD56^{dim} subset represents about 90% of the NK cells, is CD16-positive, and contains higher levels of granzyme A and perforin, two molecules involved in exocytosis-mediated cytotoxicity [5, 78]. The CD56^{bright} NK cells are CD16^{dim} or negative. A subset of dendritic cells, known as plasmacytoid dendritic cells, coexpress CD56 and CD4 [80].

CD56 is an excellent marker for the detection of NK cells and T-LGL lymphoproliferative disorders, but it is also expressed in some cases of plasma cell myelomas, AML, and ALL [4–6]. Hematodermic neoplasms (blastic NK cell lymphomas) coexpress CD4 and CD56 and appear to be the plasmacytoid dendritic cell origin [80]. Neuroectodermal tumors, such as small cell carcinoma of lung, neuroblastoma, medulloblastoma, and astrocytoma, are CD56-positive [5, 81, 82].

CD57

The CD57 molecule is a glycoprotein expressed on NK cells, T-cell subsets, and some cells of neuroectodermal origin [5, 83]. The proportion and absolute number of CD57-positive cells in peripheral blood increases with age. In adults, CD57 is expressed by 10–25% of the peripheral blood mononuclear cells [5, 83–85]. The CD57-positive cells are proliferation incompetent and most of them coexpress CD8. A small fraction of $CD4^+$ T-cells express CD57 and appear to be associated with chronic inflammatory conditions, such as tuberculosis, malaria, and AIDS [5]. The $CD57^+CD4^+$ T-cells constitute a major subset of T-cells in the germinal center of the lymphoid tissues. Approximately 40% of the CD16-positive cells coexpress CD57 [5, 84]. The $CD16^+$ $CD57^+$ subset demonstrates strong cytotoxic activities.

CD57 is a good marker for the detection of LGL disorders. The $CD4^+CD57^+$ cells are increased in lymphocyte predominance Hodgkin lymphoma and chronic inflammatory conditions [5, 6]. CD57 is positive in a wide variety of tumors of neuroectodermal or mesenchymal origin [5, 86].

OTHER NK/LGL-ASSOCIATED MARKERS

The 8th International Workshop on HLDA in December 2004 designated 95 new CD clusters which include several NK-associated markers:

CD158 or killer cell inhibitory receptor is expressed by NK cells [87, 88].

CD161 is expressed on most NK cells and a subset of $CD4^+$ and $CD8^+$ T-cells [89, 90].

CD335 was previously known as NKp46, NCR1, (Ly94), or natural cytotoxicity triggering receptor 1 [91].

CD336 was previously referred to as NKp44, NCR2, (Ly95), or natural cytotoxicity triggering receptor 2 [92].

CD337 was previously known as NKp30, NCR3, or natural cytotoxicity triggering receptor 3 [93].

GRANULOCYTIC/MONOCYTIC-ASSOCIATED CD MOLECULES

CD13

CD13 is an integral membrane zinc-binding aminopeptidase which is expressed on the surface of about 40% of granulocytes/monocytes precursors and mature granulocytic/monocytic cells [5, 94]. This molecule is also expressed on endothelial cells, bone marrow stromal cells, osteoclasts, and a small proportion of LGL. This molecule is not expressed in other lymphocytes, erythroid cells, or platelets. CD13 is commonly used as a pan-myeloid marker for the diagnosis of AML [4–6, 95]. However, about 5–15% of acute lymphoid leukemias also express CD13 [4–6, 96]. The epithelia of renal proximal tubules and bile duct canaliculi may express CD13 [5, 94].

CD14

CD14 is a lipopolysaccharide-binding protein, which functions as an endotoxin receptor [5, 97]. It is anchored to the cell surface by linkage to GPI. CD14 is strongly expressed on the surface of monocytes and most tissue macrophages, and weakly expressed on the surface of granulocytes [4–6]. Myeloid precursors and monoblasts are negative for CD14. A small proportion of peripheral blood lymphocytes and mantle cells may weakly express CD14 [5]. T-cells, dendritic cells, and platelets are CD14-negative, though CD14 expression has been reported in non-myleoid cells [5, 98]. CD14 expression is reduced or lost in PNH due to the structural abnormality of GPI [99]. Anti-CD14 monoclonal antibodies are frequently used for the identification of leukemias with monocytic differentiation [4–6].

CD15

CD15 is a carbohydrate-based molecule expressed in the granulocytic series past the myeloblast stage [5]. A significant proportion of monocytes, a minority of macrophages/histiocytes, and a wide variety of epithelial cells and their malignant counterparts also express CD15 [4–6, 100, 101]. Erythroid precursors, B-cells, T-cells, and NK cells are CD15-negative. Reed–Sternberg cells and activated T-cells may express CD15 [4–6, 102].

CD33

CD33 is a sialoadhesin molecule and a member of the immunoglobulin supergene family [5, 103]. It is expressed by myeloid stem cells (CFU-GEMM, CFU-GM, CFU-G, and E-BFU), myeloblasts and monoblasts, monocytes/macrophages, granulocyte precursors (with decreasing expression with maturation), and mast cells [5, 103]. Mature granulocytes may show a very low level of CD33 expression. This molecule is not expressed in erythrocytes, platelets, B-cells, T-cells, and NK cells. CD33 is an excellent myeloid marker and is commonly used for the diagnosis of AML. However, approximately 10–20% of precursor B-ALL may express CD33 [4–6, 103, 104].

CD64

The CD64 molecule is a member of the immunoglobulin supergene family and functions as an FcIgG receptor [5, 105]. It is expressed by monocytes/macrophages, myeloid precursors, and follicular dendritic cells. CD64 and CD14 are considered good monocyte/macrophage-associated markers and are commonly used in flow cytometric studies to identify leukemias with myelomonocytic differentiation [4–8]. CD64 appears to be more sensitive but less specific monocytic marker than CD14 [106, 107]. Langerhans cells, interdigitating dendritic cells, B-cells, T-cells, NK cells, and erythroid and megakaryocytic lineages are CD64-negative.

CD68

CD68 is a sialomucin and a member of the scavenger receptor supergene family [5, 108]. This molecule is expressed by monocytes and macrophages as well as subsets of CD34-positive hematopoietic stem cells, dendritic cells, neutrophils, basophils, and mast cells [5, 109, 110]. Activated T-cells and a proportion of mature B-cells may also express CD68, which usually appears as a dot-like cytoplasmic or finely granular positivity by immunohistochemical techniques. Some non-hematopoietic cells, such as epithelium of renal tubules, may show CD68 positivity [5, 111].

OTHER GRANULOCYTIC/ MONOCYTIC-ASSOCIATED MARKERS

CD88 is a C5a receptor and is expressed by granulocytes, monocytes, mast cells, and subsets of dendritic cells [112, 113].

CD114 is the receptor for granulocyte colony-stimulating factor (G-CSF) [114–116]. It is expressed by cells of the granulocytic lineage in all stages of differentiation and is found in various proportions of monocytes, endothelial cells, and trophoblastic cells [116–118].

CD115 is the receptor for macrophage colony-stimulating factor (M-CSF) and is primarily expressed on cells of the monocyte/macrophage lineage [119].

CD116 is the α chain subunit of the GM-CSF receptor and is expressed by various myeloid cells including macrophages, neutrophils, eosinophils, and dendritic cells [120].

ERYTHROID-ASSOCIATED CD MOLECULES

CD235

CD235 molecules represent glycophorins A and B, the two major sialoglycoproteins of the human erythrocyte membrane [121, 122]. These molecules bear the antigenic determinants for the MN and Ss blood groups [122, 123]. Monoclonal antibodies against glycophorin A (GPA) are frequently used in immunophenotypic studies for the identification of erythroid precursors in hematologic disorders [4, 124].

CD238

The CD238 molecule is the Kell blood group transmembrane protein [125].

CD240

CD240 represents the CE and D antigens of the Rh blood group system, the second most clinically significant, and the most polymorphic of the human blood groups [126].

CD242

CD242 is an intercellular adhesion molecule (ICAM4) and represents the Landsteiner–Wiener (LW) blood group antigen(s) [127].

OTHER ERYTHROID-ASSOCIATED MARKERS

CD71 is the transferrin receptor and is expressed on all proliferating cells [128]. It is also expressed by erythroid precursors which need iron for the synthesis of heme molecules [128, 129]. CD71 in conjunction with glycophorin A (GLPA) is a helpful marker in the identification of erythroid precursor cells in hematologic disorders [4, 129].

Anti-hemoglobin antibodies are routinely used for immunophenotypic studies of erythroid precursors.

MEGAKARYOCYTE/PLATELET-ASSOCIATED CD MOLECULES

CD42

CD42 complex (a, b, c, and d) is restricted to the megakaryocytic lineage and platelets [130]. This complex facilitates adhesion of the platelets to the subendothelial matrices. Absence of the CD42 complex leads to the Bernard–Soulier syndrome [131]. Anti-CD42 monoclonal antibodies are routinely used for identification of megakaryoblasts and immature megakaryocytes in myeloproliferative disorders and myeloid leukemias [4, 132].

CD41 and CD61

CD41 (platelet glycoprotein IIb) and CD61 (platelet glycoproteins IIIa) form a calcium-dependent heterodimeric complex [133]. This glycoprotein complex (GPIIb-IIIa) binds plasma proteins, such as fibrinogen, fibronectin, von Willebrand factor, and vitronectin, and plays a critical role in platelet aggregation [134]. Hereditary defects of the GPIIb-IIIa receptor cause Glanzmann's thrombasthenia [131, 134]. Similar to CD42, anti-CD41 and CD61 monoclonal antibodies are frequently used for identification of megakaryocytic precursors in myeloproliferative disorders and myeloid leukemias [4].

OTHER MEGAKARYOCYTE/ PLATELET-ASSOCIATED MARKERS

CD110 or thrombopoietin receptor (TPO-R) is expressed on the megakaryocytic precursors and platelets, hematopoietic stem cells, and some of the hematopoietic precursors [135].

Factor VIII is another useful megakaryocytic marker used for identification of megakaryocytic precursors in myeloproliferative disorders and myeloid leukemias.

PRECURSOR-ASSOCIATED CD MOLECULES

CD34

CD34 is a transmembrane glycoprotein expressed on early lymphohematopoietic stem cells, progenitor cells, and endothelial cells [5, 136, 137]. Also, embryonic fibroblasts and some cells in fetal and adult nervous tissue are CD34-positive. Almost all pluripotent and committed stem cells in colony-forming assays express CD34 [5, 136]. The uncommitted progenitor cells are CD38-negative, and the committed ones are CD38-positive. In normal conditions, $CD34^+$ cells account for about 1–2% of the total bone marrow cells. The TdT^+ precursor B-cells (hematogones) are also positive for CD34. Approximately 40% of AML and over 50% of ALL express CD34 [4–6, 138, 139].

CD38

CD38 is a multifunctional ectoenzyme widely expressed in hematopoietic cells [5, 140, 141]. It plays a role in the regulation of cell activation and proliferation. It is expressed in committed hematopoietic stem cells and other hemopoietic precursors during early differentiation and activation [4, 141]. Very early erythroid and myeloid cells, precursor B-cells, thymocytes, and activated T-cells and NK cells express CD38 [4, 142]. It is also expressed at high levels on plasma cells [4, 143].

CD90

The CD90 molecule is a member of the immunoglobulin supergene family and is expressed by 10–40% of $CD34^+$ cells in bone marrow [144, 145]. The $CD34^+/CD90^+$ cells probably represent the most primitive hematopoietic progenitor cells [144, 145]. CD90 is also expressed in fibroblasts and other stromal cells.

CD99

CD99 is a transmembrane protein involved in homotypic cell adhesion, apoptosis, vesicular protein transport, and differentiation of T-cells [146, 147]. Its expression has been reported in acute lymphoid leukemias, germ cell tumors, and tumors of neuroectodermal origin [146–148].

CD117

CD117 (c-kit) is a tyrosine kinase receptor and a member of the immunoglobulin supergene family [149]. It is expressed in most of the hemopoietic stem and $CD34^+$ progenitor cells, and mast cells. The majority of AML cells are also CD117-positive. Plasma cells may also express CD117 [150]. CD117 is an excellent marker for the detection of mast cell disorders and identification of myeloblasts in acute leukemias [151–153].

OTHER PRECURSOR-ASSOCIATED MARKERS

TdT (terminal deoxynucleotidyl transferase) is a DNA polymerase present in precursor T- and B-cells and thymocytes. Anti-TdT antibodies are routinely used for the detection of precursor B- and T-acute lymphoid leukemias/lymphomas and lymphoid blast transformation in chronic myeloid leukemia (CML) [4–6]. A small proportion of AMLs are also TdT-positive [154].

OTHER MARKERS ROUTINELY USED IN HEMATOPATHOLOGY

CD11

CD11a, b, and c are components of heterodimer CD11/CD18 adhesion molecules [5, 155]. CD11a is a panleukocyte marker and is expressed by B- and T-lymphocytes, monocytes, macrophages, neutrophils, basophils, and eosinophils [5]. CD11b is strongly expressed by most of the granulocytes, monocytes/macrophages, and NK cells, and subsets of B- and T-cells [5]. CD11c expression is high in monocytes/macrophages, NK cells, and hairy cells, moderate in granulocytes, and weak in lymphocyte subsets [4–6].

CD30

The CD30 molecule is a member of the TNF receptor family and appears to be involved in TCR-mediated cell death [5, 156]. It is expressed by Reed–Sternberg cells and Hodgkin cells, cells of anaplastic large cell lymphoma (ALCL), and activated T-, NK-, and B-cells, and monocytes [4–6, 157, 158]. Some cases of embryonal carcinoma and mixed germ cell tumors also express CD30 [159, 160].

CD43

CD43 is a sialomucin transmembrane molecule expressed at high levels, on all leukocytes except most resting B lymphocytes [5, 161]. In hematopathology, CD43 is often considered as a T-cell-associated marker, because it is expressed by over 95% of thymocytes and peripheral blood T-cells. Interdigitating dendritic cells, Langerhans cells, epithelioid histiocytes, and multinucleated giant cells express CD43, whereas follicular dendritic cells and sinus histiocytes of the lymph nodes are usually CD43-negative [4–6].

CD43 may be expressed in mantle cell lymphoma, mastocytosis, and some cases of plasma cell disorders [4–6]. Loss or defect of CD43 has been reported in lymphocytes of patients with Wiskott–Aldrich syndrome [162, 163].

CD55

CD55 or decay accelerating factor (DAF) binds C3b and C4b to inhibit formation of the C3 convertases [164]. It is anchored to the GPI in the cell membrane, and, therefore, its expression is reduced or lost in patients with PNH [165, 166]. It is widely expressed on cells throughout the body, including hematopoietic cells.

CD59

CD59 is also a GPI-anchored molecule and inhibits formation of membrane attack complex (MAC), thus protecting cells from complement-mediated lysis [167]. Similar to CD55, CD59 expression is reduced or lost in patients with PNH [165, 166]. It is widely expressed on cells throughout the body, including hematopoietic cells.

Ki-67

Ki-67 is a proliferation-associated molecule [168]. Its expression is upregulated during the S phase of the cell cycle and is maximized during mitosis [169]. Anti-Ki-67 antibodies are used for the estimation of proliferating index in lymphoid malignancies.

PAX5

The *PAX5* gene encodes the B-cell lineage-specific activator protein (BSAP) which is a member of the highly conserved paired box (PAX)-domain family of transcription factors [170]. It plays a crucial role in B-cell development and commitment of the bone marrow multipotent progenitor cells to the B-lymphoid lineage [171]. Antibodies to PAX5 are used for the diagnosis of lymphoid malignancies, particularly precursor B-ALL [172, 173]. Neuroendocrine neoplasms and t(8;21)-AML may also express PAX5 [173].

IMMUNOGLOBULIN TRANSCRIPTION FACTORS

Oct1, Oct2, and BOB.1/OBF.1

Oct1 and Oct2 and their co-activator BOB.1/OBF.1 regulate immunoglobulin gene transcription [174]. Antibodies raised against these molecules are used for the characterization of certain types of B-cell lymphoid malignancies [174, 175]. They are also used for the identification of the L&H cells in the lymphocyte predominant Hodgkin lymphoma. Oct2 and BOB.1/OBF.1 are expressed in L&H cells but not in Reed–Sternberg cells of classical Hodgkin lymphomas [6].

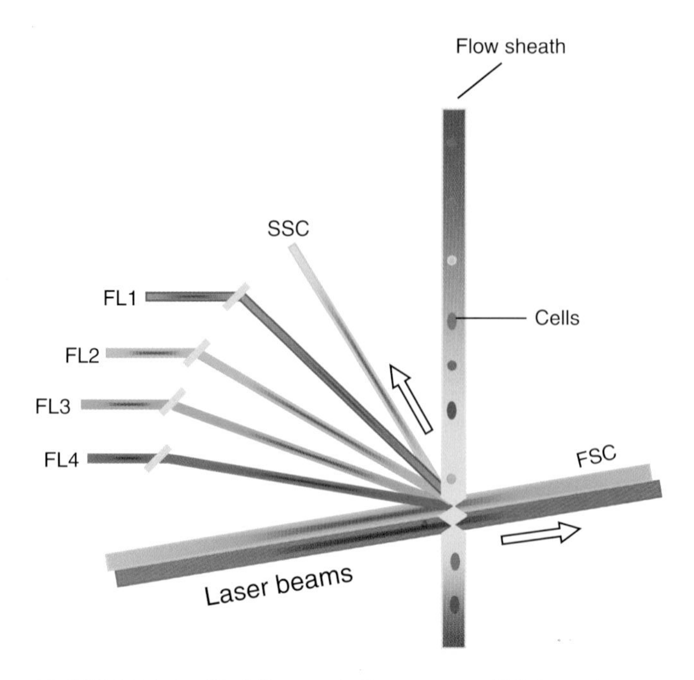

FIGURE 2.2 A simplified diagram of a flow cytometer. FSC: forward scatter, SSC: side scatter, FL1, -2, -3, and -4 represent various types of fluorochromes.

ZAP-70

ZAP-70 is a tyrosine kinase that plays a role in TCR-linked signal transduction [176,177]. This molecule is expressed in T-cells and NK cells, precursor B-acute lymphoblastic leukemia cells, and CLL cells, particularly in those cases with unmutated IgV_H genes [177, 178]. ZAP-70 expression in CLL appears to be associated with a worse prognosis in terms of progression and survival [178].

PRINCIPLES OF FLOW CYTOMETRY

Flow cytometry is now considered an integral component of immunophenotyping in hematopathology. Access to a huge number of antibodies against CD molecules and high quality and diverse fluorochromes, sophisticated and user-friendly flow cytometry instruments, advanced software, improved gating strategies, and multiparameter interpretation approaches have made flow cytometry a powerful method of immunophenotyping [179–182].

The flow cytometer is basically a particle analyzer. It measures cell properties when a stream of single cell (or particle) suspension passes through a laser beam (Figure 2.2). The cell size, texture (granularity), and membrane-associated, cytoplasmic or nuclear molecules that are marked by different fluorochromes are measured by a set of optical detectors and analyzed [179–182]. A fluorochrome is a chemical which can absorb energy from an excitation source (laser light) and emits photons at a longer wavelength (fluorescence). Most flow cytometry instruments are able to measure and analyze at least six parameters simultaneously: cell size

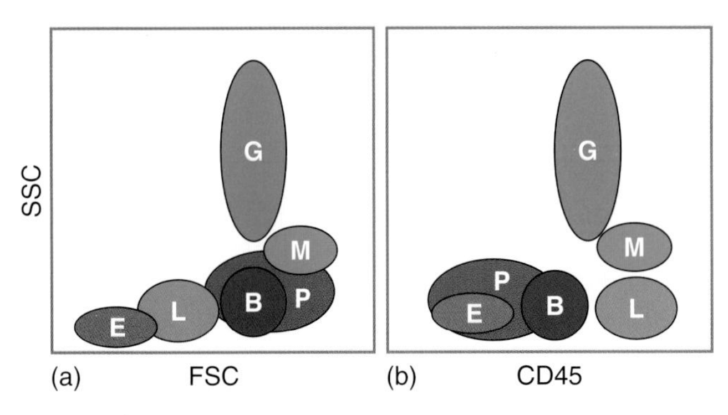

FIGURE 2.3 Patterns of cell aggregation in flow cytometry analysis of FSC/SSC (a) and CD45/SSC (b). E: erythroid precursors, L: lymphocytes, B: blasts, M: monocytes, G: granulocytes, and P: plasma cells.

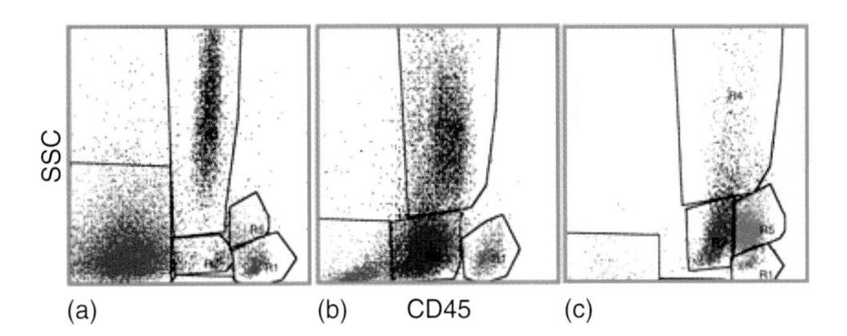

FIGURE 2.4 Multicolor cell plot presentations of flow cytometry demonstrating example patterns of a plasma cell myeloma (a, red cluster), an acute lymphoblastic leukemia (b, green cluster), and an acute myelomonocytic leukemia (c, green and red clusters).

depicted by forward scatter (FSC) laser light, cell granularity represented by side scatter (SSC) laser light, and at least four different fluorochromes defining four different molecular characteristics of the cells passing through the instrument (Figure 2.1). The basic principles of flow cytometry are briefly discussed later.

GATING

Gating refers to the selection of a population of cells in an electronic window. Cells sharing similar electronic signal tend to be aggregated together in the electronic windows. For example, in flow cytometric study of peripheral blood leukocytes, lymphocytes which are small and non-granular appear as an aggregate in the lower left section of the FSC versus SSC electronic window (Figure 2.3). Currently, the recommended gating strategy for hematopoietic tissues is a combination of FSC versus SSC and CD45 versus SCC (Figure 2.3). CD45 is strongly expressed by lymphocytes and monocytes while erythroid precursors are CD45-negative. Blast cells are CD45^{dim} and plasma cells are CD45-negative to CD45^{dim}. This gating strategy is extremely helpful in separating blast cells from non-blast cells, lymphoid cells from non-lymphoid cells, and monocytes from

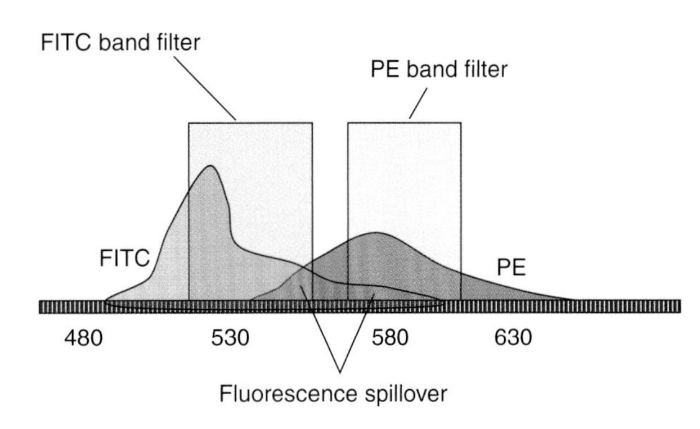

FIGURE 2.5 An example of fluorescence spillover between FITC and PE. Adapted from Wulff S. *Guide to Flow Cytometry*, DakoCytomation.

granulocytes (Figures 2.3 and 2.4). It also helps to distinguish the normal patterns of expression from the abnormal ones. Proper gating is a critical step in data analysis and interpretation of the results in flow cytometry.

In order to gate the right cell population for the flow cytometric analysis, we strongly recommend microscopic review of the samples and access to clinical information prior to the selection of monoclonal antibodies and gating processes.

COMPENSATION

When multiple fluorochromes are used, there is a possibility of fluorescence interference due to the overlapping emission spectra. For example, both fluorescein isothiocyanide (FITC) and phycoerythrin (PE) are excited at 488 nm, but their maximal emission peaks are at 520 and 576 nm, respectively [181, 183]. However, since the emission wavelength of these fluorochromes are relatively broad, there is an overlap between the emitted FITC and PE fluorochromes, even when the proper filters are used to limit this overlap (Figure 2.5). The currently available flow cytometry programs are able to correct (eliminate) the fluorescence overlap. This corrective measure is called compensation. Compensation is one of the most challenging technical aspects in multicolor flow cytometry. The use of new tandem fluorochromes, the emergence of newly developed compensation software, and publications providing guidelines for the standardization of compensation have significantly improved this process [183–186]. In practice, for monitoring compensation, we usually utilize two mutually exclusive markers such as CD4 and CD8 in peripheral blood samples. All peripheral blood T-cells express CD3, but they are positive for either CD4 or CD8 (Figure 2.6).

DATA ANALYSIS

Sophisticated flow cytometry instruments in conjunction with powerful and user-friendly software programs offer great

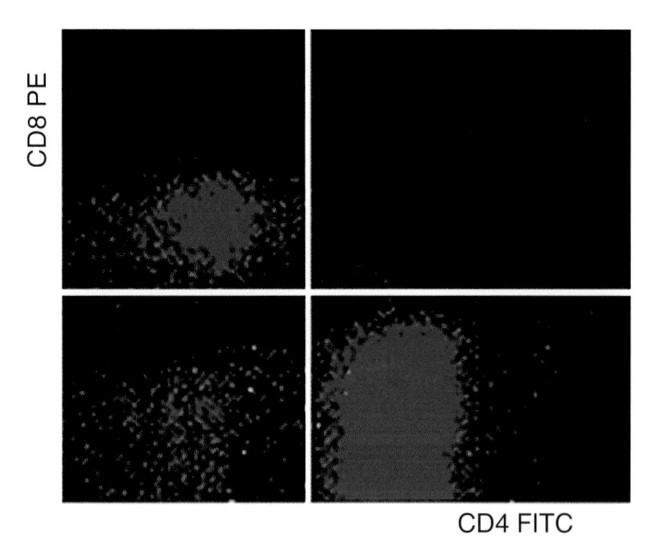

FIGURE 2.6 In practice, for monitoring compensation, two mutually exclusive markers such as CD4 and CD8 are utilized in a peripheral blood sample.

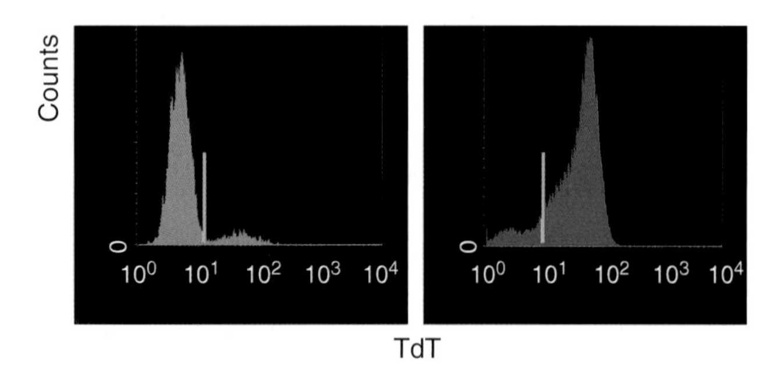

FIGURE 2.7 Histograms for a single parameter usually depict fluorescent intensity versus cell count. The TdT-negative and TdT-positive samples are demonstrated in left and right, respectively.

opportunities for hematopathologists, immunologists, and researchers to rapidly acquire data and analyze the results on large cell populations. Most available instruments are able to process >50,000 cells per second and detect at least six parameters (two light scatter and four fluorescent signals) simultaneously. Software programs provide a wide variety of options for the evaluation and analysis of the signals, including data collection on logarithmic or linear scales, and different options for histograms. The logarithmic scale is the preferred scale for most flow cytometric immunophenotypic studies. Histograms for a single parameter usually depict fluorescent intensity versus cell counts (Figure 2.7). Dot and density plots provide simultaneous information for two parameters. Two-parameter dot plot histograms depict four quadrants. The lower left quadrant represents negative cell cluster, the upper left quadrant shows the cell population positive for one parameter, the lower right quadrant depicts cells positive for the second parameter, and the upper right quadrant represents cells that coexpress both parameters

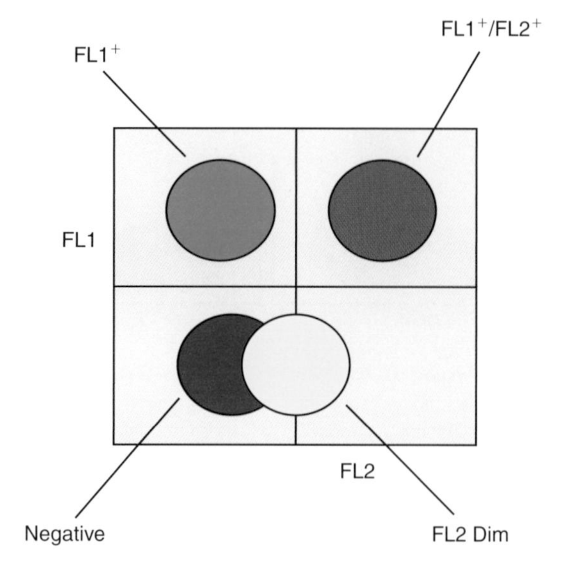

FIGURE 2.8 Two-parameter dot plot histograms depict four quadrants: Lower left: negative; upper left: positive for one parameter; lower right: positive for the second parameter; and upper right: positive for both parameters.

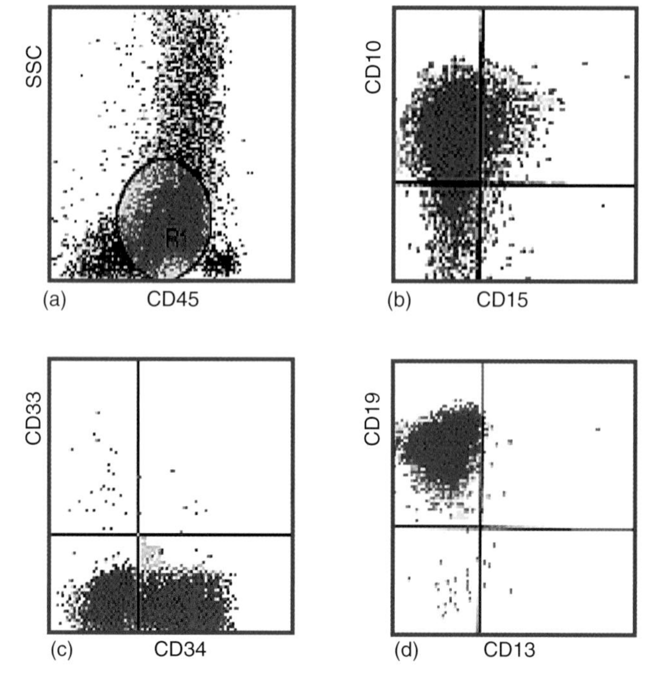

FIGURE 2.9 An example of two-parameter dot plot analysis showing a population of CD45^{dim} blast cells (a) expressing CD10 (b), CD34 (c), and CD19 (d).

(Figures 2.8 and 2.9). Contour histograms display the data as a series of encircling lines correlating with cellular density and distribution (Figure 2.10). Most programs also allow us to compare data from multiple samples by simultaneously overlaying their single parameter histograms on top of one another.

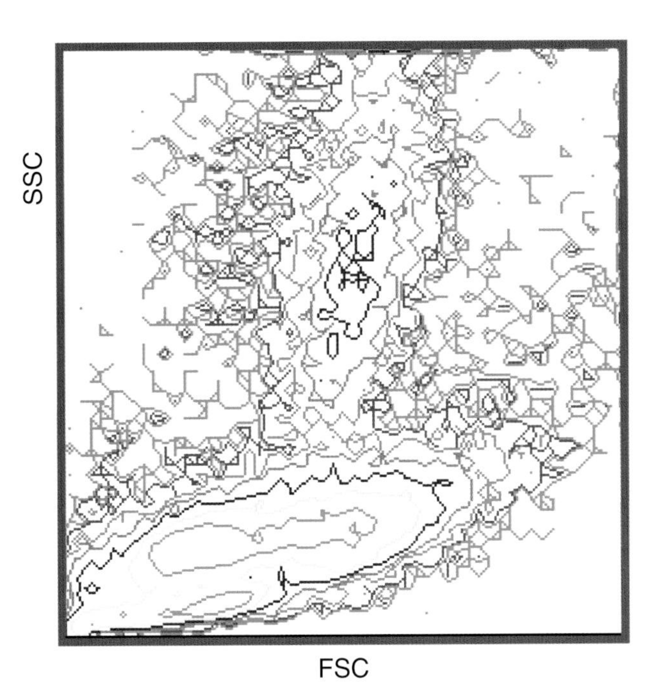

FIGURE 2.10 Counter histograms display the data as a series of encircling lines correlating with cellular density (higher in the center) and cell clusters.

QUALITY CONTROL AND QUALITY ASSURANCE ISSUES

Similar to all other instruments in the clinical laboratories, flow cytometry has its own quality control (QC) and quality assurance (QA) issues [183–187]. Many steps are involved in various aspects of flow cytometry, such as the optimization of instrument function, sample processing, acquiring and analyzing data, and reporting the results. Flow cytometers should be calibrated with samples consisting of a mixture of blank and predefined fluorescence-labeled microbeads. The performance of various components, such as fluidics, optical filters, multiplier tubes, and lasers, should be checked on a regular basis. Standardized protocols for each step of the process should be implemented to ensure reliable results, including verification of accuracy of the results with known samples.

PRINCIPLES OF IMMUNOHISTOCHEMISTRY

Immunohistochemistry has become a routine staining technique in most pathology laboratories. Enzyme-conjugated antibodies are used for the demonstration of antigens in tissue sections, smears, and cytospin preparations [188–190]. Horse radish peroxidase and/or alkaline phosphatase are the most frequently used enzymes for signal generation. Sections from frozen or fixed tissues are used [190–193].

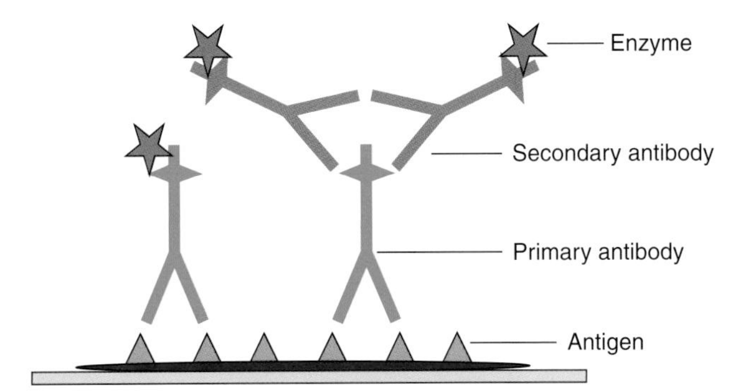

FIGURE 2.11 Schematic of immunohistochemical techniques demonstrating direct and indirect immunoenzyme methods. Adapted from Ref. [195].

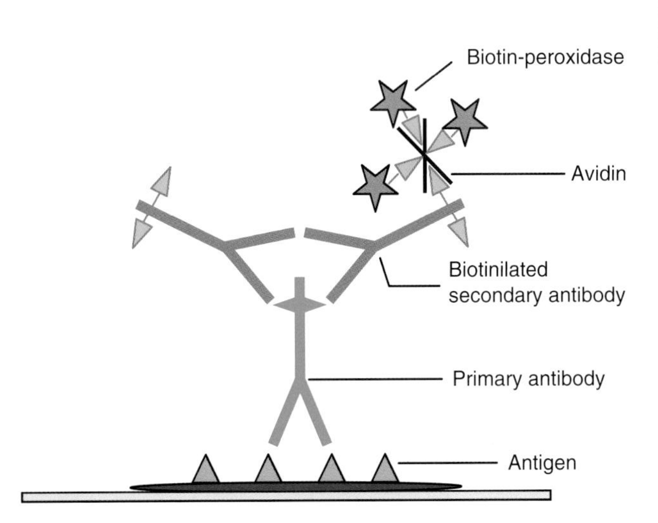

FIGURE 2.12 Schematic of immunohistochemical techniques demonstrating an indirect method using biotin–avidin–enzyme complexes. Adapted from Ref. [195].

Archival tissue blocks are sectioned and deparaffinized and then properly heated for epitope retrieval [188, 189, 194]. After blocking the endogenous peroxidase, the primary antibody (1–5 μg) is applied with proper incubation time (~30 min) and then the enzyme-conjugated or biotinylated secondary antibody is added (Figure 2.11) [195]. To amplify the signals, biotin–avidin–enzyme or biotin–streptavidin–enzyme complexes are used (Figure 2.12) [195]. Currently, there are automated machines available for performing single or dual immunohistochemical stains (Figure 2.13).

Immunohistochemistry provides information regarding pattern, intensity, and location of antigen(s) in tissues and cells. It is used in hematopathology for diagnosis and classification of leukemias and lymphomas, and differential diagnosis of primary hematopoietic neoplasms from non-neoplastic hematopoietic disorders and metastatic tumors [188, 189] (Figures 2.14 and 2.15). For example, most of the acute lymphoid leukemias are CD10, TdT, and HLA-DR positive, whereas these markers are not expressed in neuroblastoma, rhabdomyosarcoma, or Ewing's sarcoma. Immunohistochemical technique

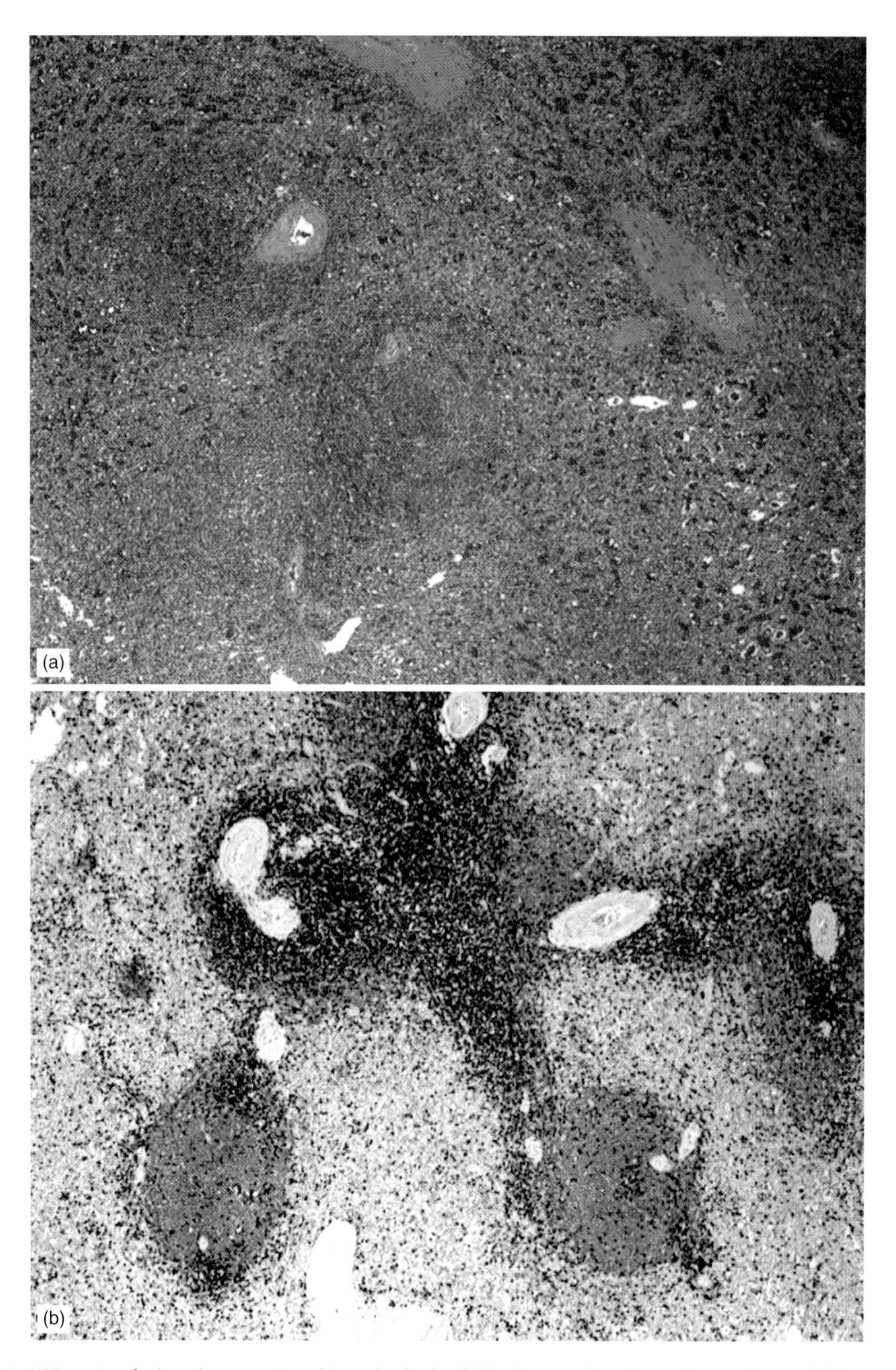

FIGURE 2.13 (a) An H&E section of spleen demonstrating white and red pulps. (b) Dual immunohistochemical stains showing T (CD3-positive, brown) and B (CD20-positive, pink) cells.

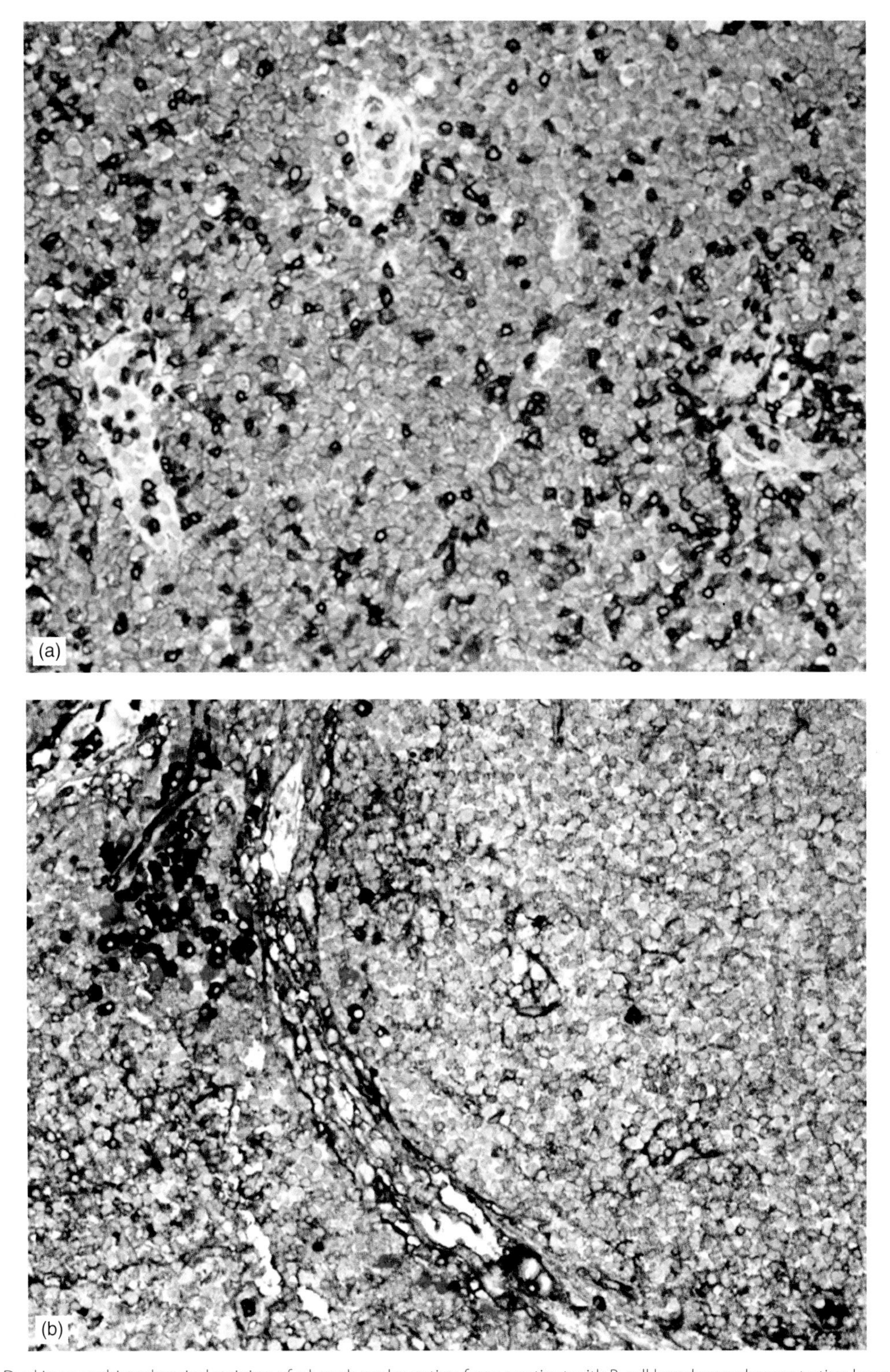

FIGURE 2.14 (a) Dual immunohistochemical staining of a lymph node section from a patient with B-cell lymphoma demonstrating large numbers of B-cells (CD20$^+$, pink) and scattered T-cells (CD3$^+$, brown). (b) Dual kappa (brown) and lambda (pink) staining shows a cluster of polyclonal plasma cells and sheets of kappa-positive lymphocytes.

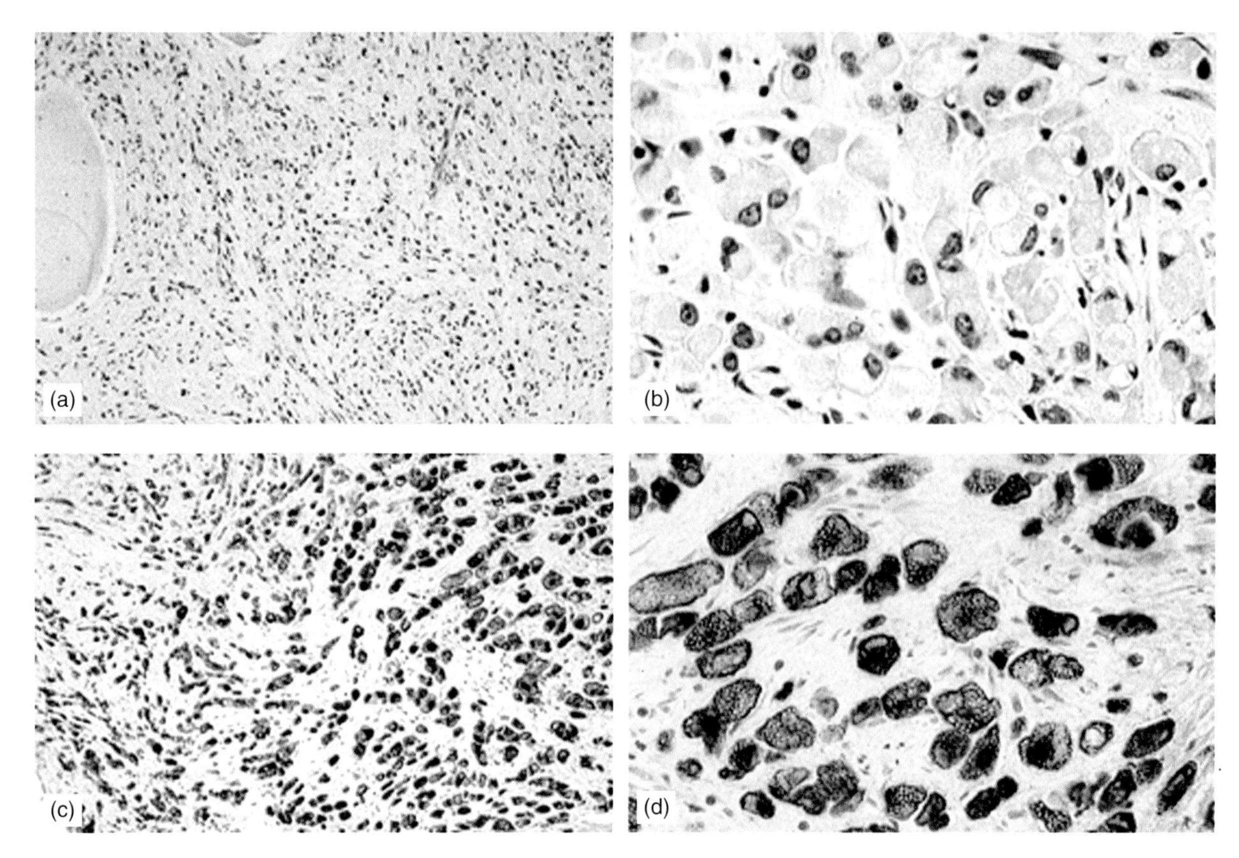

FIGURE 2.15 Metastatic adenocarcinoma simulating a histiocytic infiltrate (a and b: H&E; low and high power views). Immunohistochemical stain for cytokeratin shows numerous positive cells (c and d, low and high power views). From Naeim F. (2001). *Atlas of Bone Marrow and Blood Pathology*. W.B. Saunders, Philadelphia, by permission.

may help to detect occult metastatic lesions, such as metastatic breast carcinomas and neuroblastomas, or to identify their tissue of origin [196–198]. For example, metastatic prostatic carcinomas are positive for prostatic acid phosphatase and prostate-specific antigen (PSA), and metastatic rhabdomyosarcomas may demonstrate myosin, desmin, or myoglobin expression.

References

1. Zola H, Swart B. (2005). The human leucocyte differentiation antigens (HLDA) workshops: The evolving role of antibodies in research, diagnosis and therapy. *Cell Res* **15**, 691–4.
2. Swart B, Salganik MP, Wand MP, Tinckam K, Milford EL, Drbal K, Angelisova P, Horejsi V, Macardle P, Bailey S, Hollemweguer E, Hodge G, Nairn J, Millard D, Dagdeviren A, Dandie GW, Zola H. (2005). HLDA8 blind panel. The HLDA8 blind panel: Findings and conclusions. *J Immunol Methods* **305**, 75–83.
3. Zola H, Swart B, Nicholson I, Aasted B, Bensussan A, Boumsell L, Buckley C, Clark G, Drbal K, Engel P, Hart D, Horejsi V, Isacke C, Macardle P, Malavasi F, Mason D, Olive D, Saalmueller A, Schlossman SF, Schwartz-Albiez R, Simmons P, Tedder TF, Uguccioni M, Warren H. (2005). CD molecules 2005: Human cell differentiation molecules. *Blood* **106**, 3123–6.
4. Naeim F. (1997). *Pathology of Bone Marrow*, 2nd ed. Williams & Wilkins, Baltimore.
5. Knowles DM. (2000). *Neoplastic Hematopathology*, 2nd ed. Lippincott Williams & Wilkins, Philadelphia.
6. Jaffe ES, Harris NL, Stein H, Vardiman JW. (2001). *Pathology and Genetics. Tumors of Haematopoietic and Lymphoid Tissues*. IARC Press, Lyon.
7. Human Cell Differentiation Molecules. Web site: http://www.hlda8.org
8. Arber DA, Weiss LM. (1997). CD10: A review. *Appl Immunohistochem Mol Morphol* **5**, 125–40.
9. Avery AK, Beckstead J, Renshaw AA, Corless CL. (2000). Use of antibodies to RCC and CD10 in the differential diagnosis of renal neoplasms. *Am J Surg Pathol* **24**, 203–10.
10. Mazal PR, Stichenwirth M, Koller A, Blach S, Haitel A, Susani M. (2005). Expression of aquaporins and PAX-2 compared to CD10 and cytokeratin 7 in renal neoplasms: A tissue microarray study. *Mod Pathol* **18**, 535–40.
11. Carter RH, Tuveson DA, Park DJ, Rhee SG, Fearon DT. (1991). The CD19 complex of B lymphocytes. Activation of phospholipase C by a protein tyrosine kinase-dependent pathway that can be enhanced by the membrane IgM complex. *J Immunol* **147**, 3663–71.
12. Murakami T, Chen X, Hase K, Sakamoto A, Nishigaki C, Ohno H. (2007). Splenic CD19-CD35+ B220+ cells function as an inducer of follicular dendritic cell network formation. *Blood* **110**, 1215–24.
13. Macardle PJ, Nicholson IC. (2002). CD20. *J Biol Regul Homeost Agents* **16**, 136–8.
14. Prota AE, Sage DR, Stehle T, Fingeroth JD. (2002). The crystal structure of human CD21: Implications for Epstein-Barr virus and C3d binding. *Proc Natl Acad Sci U S A* **99**, 10641–6.
15. Liu YJ, Xu J, de Bouteiller O, Parham CL, Grouard G, Djossou O, de Saint-Vis B, Lebecque S, Banchereau J, Moore KW. (1997). Follicular dendritic cells specifically express the long CR2/CD21 isoform. *J Exp Med* **185**, 165–70.

16. Doody GM, Justement LB, Delibrias CC, Matthews RJ, Lin J, Thomas ML, Fearon DT. (1995). A role in B cell activation for CD22 and the protein tyrosine phosphatase SHP. *Science* **269**, 242–4.
17. Kelm S, Gerlach J, Brossmer R, Danzer CP, Nitschke L. (2002). The ligand-binding domain of CD22 is needed for inhibition of the B cell receptor signal, as demonstrated by a novel human CD22-specific inhibitor compound. *J Exp Med* **195**, 1207–13.
18. Huang J, Fan G, Zhong Y, Gatter K, Braziel R, Gross G, Bakke A. (2005). Diagnostic usefulness of aberrant CD22 expression in differentiating neoplastic cells of B-cell chronic lymphoproliferative disorders from admixed benign B cells in four-color multiparameter flow cytometry. *Am J Clin Pathol*, **127**, 826–32.
19. Aubry JP, Pochon S, Graber P, Jansen KU, Bonnefoy JY. (1992). CD21 is a ligand for CD23 and regulates IgE production. *Nature* **358**, 505–7.
20. Tu Y, Salim S, Bourgeois J, Di Leo V, Irvine EJ, Marshall JK, Perdue MH. (2005). CD23-mediated IgE transport across human intestinal epithelium: Inhibition by blocking sites of translation or binding. *Gastroenterology* **129**, 928–40.
21. DiRaimondo F, Albitar M, Huh Y, O'Brien S, Montillo M, Tedeschi A, Kantarjian H, Lerner S, Giustolisi R, Keating M. (2002). The clinical and diagnostic relevance of CD23 expression in the chronic lymphoproliferative disease. *Cancer* **94**, 1721–30.
22. Lim SC, Oh SH. (2005). The role of CD24 in various human epithelial neoplasias. *Pathol Res Pract* **201**, 479–86.
23. Chu PG, Arber DA. (2001). CD79: A review. *Appl Immunohistochem Mol Morphol* **9**, 97–106.
24. Schlette E, Medeiros LJ, Keating M, Lai R. (2003). CD79b expression in chronic lymphocytic leukemia. Association with trisomy 12 and atypical immunophenotype. *Arch Pathol Lab Med* **127**, 561–6.
25. Bhargava P, Kallakury BV, Ross JS, Azumi N, Bagg A. (2007). CD79a is heterogeneously expressed in neoplastic and normal myeloid precursors and megakaryocytes in an antibody clone-dependent manner. *Am J Clin Pathol* **128**, 306–13.
26. He G, Wu D, Sun A, Xue Y, Jin Z, Qiu H, Miao M, Tang X, Fu Z, Chen Z. (2007). CytCD79a expression in acute leukemia with t(8;21): Biphenotypic or myeloid leukemia? *Cancer Genet Cytogenet* **174**, 76–7.
27. Wijdenes J, Dore JM, Clement C, Vermot-Desroches C. (2002). CD138. *J Biol Regul Homeost Agents* **6**, 152–5.
28. Kambham N, Kong C, Longacre TA, Natkunam Y. (2005). Utility of syndecan-1 (CD138) expression in the diagnosis of undifferentiated malignant neoplasms: A tissue microarray study of 1,754 cases. *Appl Immunohistochem Mol Morphol* **13**, 304–10.
29. Braylan RC, Orfao A, Borowitz MJ, Davis BH. (2001). Optimal number of reagents required to evaluate hematolymphoid neoplasias: Results of an international consensus meeting. *Cytometry* **46**, 23–7.
30. Koch N, Moldenhauer G, Hofmann WJ, Moller P. (1991). Rapid intracellular pathway gives rise to cell surface expression of the MHC class II-associated invariant chain (CD74). *J Immunol* **147**, 2643–51.
31. Badve S, Deshpande C, Hua Z, Logdberg L. (2004). Expression of invariant chain (CD74) and major histocompatibility complex (MHC) class II antigens in the human fetus. *J Histochem Cytochem* **50**, 473–82.
32. Moller P, Moldenhauer G. (2000). CD74. *J Biol Regul Homeost Agents* **14**, 299–301.
33. Burton JD, Ely S, Reddy PK, Stein R, Gold DV, Cardillo TM, Goldenberg DM. (2004). CD74 is expressed by multiple myeloma and is a promising target for therapy. *Clin Cancer Res* **10**, 6606–11.
34. Oshitani N, Watanabe K, Maeda K, Fujiwara Y, Higuchi K, Matsumoto T, Arakawa T. (2003). Differential expression of homing receptor CD103 on lamina propria lymphocytes and association of CD103 with epithelial adhesion molecules in inflammatory bowel disease. *Int J Mol Med* **12**, 715–19.
35. Chen YH, Tallman MS, Goolsby C, Peterson L. (2006). Immunophenotypic variations in hairy cell leukemia. *Am J Clin Pathol* **125**, 1–9.
36. Babusikova O, Tomova A. (2003). Hairy cell leukemia: Early immunophenotypical detection and quantitative analysis by flow cytometry. *Neoplasma* **50**, 350–6.
37. Serke S, Schwaner I, Yordanova M, Szczepek A, Huhn D. (2001). Monoclonal antibody FMC7 detects a conformational epitope on the CD20 molecule: Evidence from phenotyping after rituxan therapy and transfectant cell analyses. *Cytometry* **46**, 98–104.
38. Ahmad E, Garcia D, Davis BH. (2002). Clinical utility of CD23 and FMC7 antigen coexistent expression in B-cell lymphoproliferative disorder subclassification. *Cytometry* **50**, 1–7.
39. Delgado J, Matutes E, Morilla AM, Morilla RM, Owusu-Ankomah KA, Rafiq-Mohammed F, del Giudice I, Catovsky D. (2003). Diagnostic significance of CD20 and FMC7 expression in B-cell disorders. *Am J Clin Pathol* **120**, 754–9.
40. Vincent MS, Xiong X, Grant EP, Peng W, Brenner MB. (2005). CD1a-, b-, and c-restricted TCRs recognize both self and foreign antigens. *J Immunol* **175**, 6344–51.
41. Hunger RE, Sieling PA, Ochoa MT, Sugaya M, Burdick AE, Rea TH, Brennan PJ, Belisle JT, Blauvelt A, Porcelli SA, Modlin RL. (2004). Langerhans cells utilize CD1a and langerin to efficiently present nonpeptide antigens to T cells. *J Clin Invest* **113**, 701–8.
42. Coventry B, Heinzel S. (2004). CD1a in human cancers: A new role for an old molecule. *Trends Immunol* **25**, 242–8.
43. Salamone MC, Roisman FR, Santiago J, Satz ML, Fainboim L. (1990). Analysis of CD1 molecules on haematological malignancies of myeloid and lymphoid origin. I. Cell surface antigen expression. *Dis Markers* **8**, 265–9.
44. Yang JJ, Ye Y, Carroll A, Yang W, Lee HW. (2001). Structural biology of the cell adhesion protein CD2: Alternatively folded states and structure-function relation. *Curr Protein Pept Sci* **2**, 1–17.
45. Kim M, Sun ZY, Byron O, Campbell G, Wagner G, Wang J, Reinherz EL. (2001). Molecular dissection of the CD2-CD58 counter-receptor interface identifies CD2 Tyr86 and CD58 Lys34 residues as the functional "hot spot". *J Mol Biol* **312**, 711–20.
46. Moingeon P, Chang HC, Sayre PH, Clayton LK, Alcover A, Gardner P, Reinherz EL. (1989). The structural biology of CD2. *Immunol Rev* **111**, 111–44.
47. Kastrup J, Pedersen LO, Dietrich J, Lauritsen JP, Menne C, Geisler C. (2003). *In vitro* production and characterization of partly assembled human CD3 complexes. *Scand J Immunol* **56**, 436–42.
48. Gaubin M, Autiero M, Houlgatte R, Basmaciogullari S, Auffray C, Piatier-Tonneau D. (1996). Molecular basis of T lymphocyte CD4 antigen functions. *Eur J Clin Chem Clin Biochem* **34**, 723–8.
49. Nakamura K, Yube K, Miyatake A, Cambier JC, Hirashima M. (2003). Involvement of CD4 D3-D4 membrane proximal extracellular domain for the inhibitory effect of oxidative stress on activation-induced CD4 down-regulation and its possible role for T cell activation. *Mol Immunol* **39**, 909–21.
50. Sattentau QJ, Weiss RA. (1988). The CD4 antigen: Physiological ligand and HIV receptor. *Cell* **52**, 631–3.
51. Raab M, Yamamoto M, Rudd CE. (1994). The T-cell antigen CD5 acts as a receptor and substrate for the protein-tyrosine kinase p56lck. *Mol Cell Biol* **14**, 2862–70.
52. Tarakhovsky A, Kanner SB, Hombach J, Ledbetter JA, Muller W, Killeen N, Rajewsky K. (1995). A role for CD5 in TCR-mediated signal transduction and thymocyte selection. *Science* **269**, 535–7.
53. Perez-Chacon G, Contreras-Martin B, Cuni S, Rosado S, Martin-Donaire T, Losada-Fernandez I, Vargas JA, Jorda J, Alvarez N, Garcia-Marco J, Perez-Aciego P. (2005). Polymorphism in the CD5 gene promoter in B-cell chronic lymphocytic leukemia and mantle cell lymphoma. *Am J Clin Pathol* **123**, 646–50.
54. Stillwell R, Bierer BE. (2001). T cell signal transduction and the role of CD7 in costimulation. *Immunol Res* **24**, 31–52.
55. Cruse JM, Lewis RE, Pierce S, Lam J, Tadros Y. (2005). Aberrant expression of CD7, CD56, and CD79a antigens in acute myeloid leukemias. *Exp Mol Pathol* **79**, 39–41.
56. Laky K, Fleischacker C, Fowlkes BJ. (2006). TCR and Notch signaling in CD4 and CD8 T-cell development. *Immunol Rev* **209**, 274–83.
57. Schepers K, Arens R, Schumacher TN. (2005). Dissection of cytotoxic and helper T cell responses. *Cell Mol Life Sci* **62**, 2695–710.

58. Pilarski LM, Deans JP. (1989). Selective expression of CD45 isoforms and of maturation antigens during human thymocyte differentiation: Observations and hypothesis. *Immunol Lett* **21**, 187–98.
59. Beverley PC. (1987). Human T cell subsets. *Immunol Lett* **14**, 263–7.
60. Bell EB. (1992). Function of CD4 T cell subsets *in vivo*: Expression of CD45R isoforms. *Semin Immunol* **4**, 43–50.
61. Mahajan VS, Leskov IB, Chen JZ. (2005). Homeostasis of T cell diversity. *Cell Mol Immunol* **2**, 1–10.
62. Carreno LJ, Gonzalez PA, Kalergis AM. (2006). Modulation of T cell function by TCR/pMHC binding kinetics. *Immunobiology* **211**, 47–64.
63. Girardi M. (2006). Immunosurveillance and immunoregulation by gamma delta T cells. *J Invest Dermatol* **126**, 25–31.
64. Ohnuma K, Inoue H, Uchiyama M, Yamochi T, Hosono O, Dang NH, Morimoto C. (2006). T-cell activation via CD26 and caveolin-1 in rheumatoid synovium. *Mod Rheumatol* **16**, 3–13.
65. Hegen M, Kameoka J, Dong RP, Morimoto C, Schlossman SF. (1997). Structure of CD26 (dipeptidyl peptidase IV) and function in human T cell activation. *Adv Exp Med Biol* **421**, 109–16.
66. Introcaso CE, Hess SD, Kamoun M, Ubriani R, Gelfand JM, Rook AH. (2005). Association of change in clinical status and change in the percentage of the CD4 + CD26- lymphocyte population in patients with Sezary syndrome. *J Am Acad Dermatol* **53**, 428–34.
67. Pulford K, Lamant L, Espinos E, Jiang Q, Xue L, Turturro F, Delsol G, Morris SW. (2004). The emerging normal and disease-related roles of anaplastic lymphoma kinase. *Cell Mol Life Sci* **61**, 2939–53.
68. Passoni L, Scardino A, Bertazzoli C, Gallo B, Coluccia AM, Lemonnier FA, Kosmatopoulos K, Gambacorti-Passerini C. (2002). ALK as a novel lymphoma-associated tumor antigen: Identification of 2 HLA-A2.1-restricted CD8+ T-cell epitopes. *Blood* **99**, 2100–6.
69. Housden HR, Skipp PJ, Crump MP, Broadbridge RJ, Crabbe T, Perry MJ, Gore MG. (2003). Investigation of the kinetics and order of tyrosine phosphorylation in the T-cell receptor zeta chain by the protein tyrosine kinase Lck. *Eur J Biochem* **270**, 2369–76.
70. Sun PD. (2003). Structure and function of natural-killer-cell receptors. *Immunol Res* **27**, 539–48.
71. Shibuya A. (2003). Development and functions of natural killer cells. *Int J Hematol* **78**, 1–6.
72. Stroncek D. (2002). Neutrophil alloantigens. *Transfus Med Rev* **16**, 67–75.
73. Edberg JC, Salmon JE, Kimberly RP. (1992). Functional capacity of Fc gamma receptor III (CD16) on human neutrophils. *Immunol Res* **11**, 239–51.
74. Thomason RW, Papiez J, Lee RV, Szczarkowski W. (2004). Identification of unsuspected PNH-type cells in flow cytometric immunophenotypic analysis of peripheral blood and bone marrow. *Am J Clin Pathol* **22**, 128–34.
75. Olteanu H, Karandikar NJ, McKenna RW, Xu Y. (2006). Differential usefulness of various markers in the flow cytometric detection of paroxysmal nocturnal hemoglobinuria in blood and bone marrow. *Am J Clin Pathol* **126**, 781–8.
76. Lanier LL, Cwirla S, Yu G, Testi R, Phillips JH. (1989). Membrane anchoring of a human IgG Fc receptor (CD16) determined by a single amino acid. *Science* **246**, 1611–13.
77. Lanier LL, Testi R, Bindl J, Phillips JH. (1989). Identity of Leu-19 (CD56) leukocyte differentiation antigen and neural cell adhesion molecule. *J Exp Med* **169**, 2233–8.
78. Lanier LL. (2007). Back to the future – defining NK cells and T cells. *Eur J Immunol* **37**, 1424–6.
79. Batoni G, Esin S, Favilli F, Pardini M, Bottai D, Maisetta G, Florio W, Campa M. (2005). Human CD56bright and CD56dim natural killer cell subsets respond differentially to direct stimulation with *Mycobacterium bovis* bacillus Calmette-Guerin. *Scand J Immunol* **62**, 498–506.
80. Petrella T, Bagot M, Willemze R, Beylot-Barry M, Vergier B, Delaunay M, Meijer CJ, Courville P, Joly P, Grange F, DeMuret A, Machet L, Dompmartin A, Bosq J, Durlach A, Bernard P, Dalac S, Dechelotte P, D'Incan M, Wechsler J, Teitell MA. (2005). Blastic NK-cell lymphomas (agranular CD4+ CD56+ hematodermic neoplasms): A review. *Am J Clin Pathol* **123**, 662–75.
81. Ravandi F, Cortes J, Estrov Z, Thomas D, Giles FJ, Huh YO, Pierce S, O'Brien S, Faderl S, Kantarjian HM. (2002). CD56 expression predicts occurrence of CNS disease in acute lymphoblastic leukemia. *Leuk Res* **26**, 643–9.
82. Kurokawa M, Nabeshima K, Akiyama Y, Maeda S, Nishida T, Nakayama F, Amano M, Ogata K, Setoyama M. (2003). CD56: A useful marker for diagnosing Merkel cell carcinoma. *J Dermatol Sci* **31**, 219–24.
83. Ibegbu CC, Xu YX, Harris W, Maggio D, Miller JD, Kourtis AP. (2005). Expression of killer cell lectin-like receptor G1 on antigen-specific human CD8+ T lymphocytes during active, latent, and resolved infection and its relation with CD57. *J Immunol* **174**, 6088–94.
84. Rose MG, Berliner N. (2004). T-cell large granular lymphocyte leukemia and related disorders. *Oncologist* **9**, 247–58.
85. Tarazona R, DelaRosa O, Alonso C, Ostos B, Espejo J, Peña J, Solana R. (2000). Increased expression of NK cell markers on T-lymphocytes in aging and chronic activation of the immune system reflects the accumulation of effector/senescent T-cells. *Mech Ageing Dev* **121**, 77–88.
86. Ng SB, Sirrampalam K, Chuah KL. (2002). Primitive neuroectodermal tumours of the uterus: A case report with cytological correlation and review of the literature. *Pathology* **34**, 455–61.
87. Lanier LL. (1998). NK cell receptors. *Annu Rev Immunol* **16**, 359–93.
88. Long EO, Barber DF, Burshtyn DN, Faure M, Peterson M, Rajagopalan S, Renard V, Sandusky M, Stebbins CC, Wagtmann N, Watzl C. (2001). Inhibition of natural killer cell activation signals by killer cell immunoglobulin-like receptors (CD158). *Immunol Rev* **181**, 223–33.
89. Metelitsa LS. (2004). Flow cytometry for natural killer T cells: Multiparameter methods for multifunctional cells. *Clin Immunol* **110**, 267–76.
90. McQueen KL, Parham P. (2002). Variable receptors controlling activation and inhibition of NK cells. *Curr Opin Immunol* **14**, 615–21.
91. Poggi A, Massaro AM, Negrini S, Contini P, Zocchi MR. (2005). Tumor-induced apoptosis of human IL-2-activated NK cells: Role of natural cytotoxicity receptors. *J Immunol* **174**, 2653–60.
92. Augugliaro R, Parolini S, Castriconi R, Marcenaro E, Cantoni C, Nanni M, Moretta L, Moretta A, Bottino C. (2003). Selective cross-talk among natural cytotoxicity receptors in human natural killer cells. *Eur J Immunol* **33**, 1235–41.
93. Poggi A, Prevosto C, Massaro AM, Negrini S, Urbani S, Pierri I, Saccardi R, Gobbi M, Zocchi MR. (2005). Interaction between human NK cells and bone marrow stromal cells induces NK cell triggering: Role of NKp30 and NKG2D receptors. *J Immunol* **175**, 6352–60.
94. Luan Y, Xu W. (2007). The structure and main functions of aminopeptidase N. *Curr Med Chem* **14**, 639–47.
95. Mason KD, Juneja SK, Szer J. (2006). The immunophenotype of acute myeloid leukemia: Is there a relationship with prognosis? *Blood Rev* **20**, 71–82.
96. Asahara SI, Saigo K, Hasuike N, Tamura M, Maeda Y, Tomofuji Y, Chinzei T, Tatsumi E. (2001). Acute lymphoblastic leukemia accompanied by chromosomal abnormality of translocation (12;17). *Haematologia (Budap)* **31**, 209–13.
97. Andra J, Gutsmann T, Garidel P, Brandenburg K. (2006). Mechanisms of endotoxin neutralization by synthetic cationic compounds. *J Endotoxin Res* **12**, 261–77.
98. Jersmann HP. (2005). Time to abandon dogma: CD14 is expressed by non-myeloid lineage cells. *Immunol Cell Biol* **83**, 462–7.
99. Sutherland DR, Kuek N, Davidson J, Barth D, Chang H, Yeo E, Bamford S, Chin-Yee I, Keeney M. (2007). Diagnosing PNH with FLAER and multiparameter flow cytometry. *Cytometry B Clin Cytom* **72**, 167–77.
100. Gocht A, Struckhoff G, Lhler J. (1996). CD15-containing glycoconjugates in the central nervous system. *Histol Histopathol* **11**, 1007–28.

101. Comin CE, Novelli L, Boddi V, Paglierani M, Dini S. (2001). Calretinin, thrombomodulin, CEA, and CD15: A useful combination of immunohistochemical markers for differentiating pleural epithelial mesothelioma from peripheral pulmonary adenocarcinoma. *Hum Pathol* **32**, 529–36.
102. Rudiger T, Ott G, Ott MM, Muller-Deubert SM, Muller-Hermelink HK. (1998). Differential diagnosis between classic Hodgkin's lymphoma, T-cell-rich B-cell lymphoma, and paragranuloma by paraffin immunohistochemistry. *Am J Surg Pathol* **22**, 1184–91.
103. Andrews RG, Torok-Storb B, Bernstein ID. (1983). Myeloid-associated differentiation antigens on stem cells and their progeny identified by monoclonal antibodies. *Blood* **62**, 124–32.
104. Wellhausen SR, Peiper SC. (2002). CD33: Biochemical and biological characterization and evaluation of clinical relevance. *J Biol Regul Homeost Agents* **16**, 139–43.
105. Cohen-Solal JF, Cassard L, Fridman WH, Sautes-Fridman C. (2004). Fc gamma receptors. *Immunol Lett* **92**, 199–205.
106. Davis BH, Bigelow NC. (2005). Comparison of neutrophil CD64 expression, manual myeloid immaturity counts, and automated hematology analyzer flags as indicators of infection or sepsis. *Lab Hematol* **11**, 137–47.
107. Fjaertoft G, Hakansson L, Foucard T, Ewald U, Venge P. (2005). CD64 (Fcgamma receptor I) cell surface expression on maturing neutrophils from preterm and term newborn infants. *Acta Paediatr* **94**, 295–302.
108. Holness CL, Simmons DL. (1993). Molecular cloning of CD68, a human macrophage marker related to lysosomal glycoproteins. *Blood* **81**, 1607–13.
109. Pulford KA, Sipos A, Cordell JL, Stross WP, Mason DY. (1990). Distribution of the CD68 macrophage/myeloid associated antigen. *Int Immuno* **12**, 973–80.
110. Re F, Arpinati M, Testoni N, Ricci P, Terragna C, Preda P, Ruggeri D, Senese B, Chirumbolo G, Martelli V, Urbini B, Baccarani M, Tura S, Rondelli D. (2002). Expression of CD86 in acute myelogenous leukemia is a marker of dendritic/monocytic lineage. *Exp Hematol* **30**, 126–34.
111. Svec A, Velenska Z. (2005). Renal epithelioid angiomyolipoma – a close mimic of renal cell carcinoma. Report of a case and review of the literature. *Pathol Res Pract* **200**, 851–6.
112. Wojta J, Kaun C, Zorn G, Ghannadan M, Hauswirth AW, Sperr WR, Fritsch G, Printz D, Binder BR, Schatzl G, Zwirner J, Maurer G, Huber K, Valent P. (2002). C5a stimulates production of plasminogen activator inhibitor-1 in human mast cells and basophils. *Blood* **100**, 517–23.
113. Hauswirth AW, Florian S, Schernthaner GH, Krauth MT, Sonneck K, Sperr WR, Valent P. (2006). Expression of cell surface antigens on mast cells: Mast cell phenotyping. *Methods Mol Biol* **315**, 77–90.
114. Sampson M, Zhu QS, Corey SJ. (2007). Src kinases in G-CSF receptor signaling. *Front Biosci* **12**, 1463–74.
115. Avalos BR. (1996). Molecular analysis of the granulocyte colony-stimulating factor receptor. *Blood* **88**, 761–77.
116. Tsuji K, Ebihara Y. (2001). Expression of G-CSF receptor on myeloid progenitors. *Leuk Lymphoma* **42**, 1351–7.
117. Touw IP, van de Geijn GJ. (2007). Granulocyte colony-stimulating factor and its receptor in normal myeloid cell development, leukemia and related blood cell disorders. *Front Biosci* **12**, 800–15.
118. Calhoun DA, Donnelly WH Jr., Du Y, Dame JB, Li Y, Christensen RD. (1999). Distribution of granulocyte colony-stimulating factor (G-CSF) and G-CSF-receptor mRNA and protein in the human fetus. *Pediatr Res* **46**, 333–8.
119. Pixley FJ, Stanley ER. (2004). CSF-1 regulation of the wandering macrophage: Complexity in action. *Trends Cell Biol* **14**, 628–38.
120. Dabusti M, Castagnari B, Moretti S, Ferrari L, Tieghi A, Lanza F. (2001). CD116 (granulocyte-macrophage colony stimulating factor receptor). *J Biol Regul Homeost Agents* **15**, 86–9.
121. Baum J, Ward RH, Conway DJ. (2002). Natural selection on the erythrocyte surface. *Mol Biol Evol* **19**, 223–9.
122. Lisowska E. (2001). Antigenic properties of human glycophorins – an update. *Adv Exp Med Biol* **491**, 155–69.
123. Palacajornsuk P. (2006). Review: Molecular basis of MNS blood group variants. *Immunohematology* **22**, 171–82.
124. Nakahata T, Okumura N. (1994). Cell surface antigen expression in human erythroid progenitors: Erythroid and megakaryocytic markers. *Leuk Lymphoma* **13**, 401–9.
125. Daniels G. (2005). The molecular genetics of blood group polymorphism. *Transpl Immunol* **14**, 143–53.
126. Wagner FF, Flegel WA. (2004). Review: The molecular basis of the Rh blood group phenotypes. *Immunohematology* **20**, 23–36.
127. Hermand P, Gane P, Huet M, Jallu V, Kaplan C, Sonneborn HH, Cartron JP, Bailly P. (2003). Red cell ICAM-4 is a novel ligand for platelet-activated alpha IIbbeta 3 integrin. *J Biol Chem* **278**, 4892–8.
128. Aisen P. (2004). Transferrin receptor 1. *Int J Biochem Cell Biol* **36**, 2137–43.
129. Miller JL. (2004). A genome-based approach for the study of erythroid biology and disease. *Blood Cells Mol Dis* **3**, 341–3.
130. Slupsky JR, Kamiguti AS, Rhodes NP, Cawley JC, Shaw AR, Zuzel M. (1997). The platelet antigens CD9, CD42 and integrin alpha IIb beta IIIa can be topographically associated and transduce functionally similar signals. *Eur J Biochem* **244**, 168–75.
131. Clemetson KJ, Clemetson JM. (1994). Molecular abnormalities in Glanzmann's thrombasthenia, Bernard–Soulier syndrome, and platelet-type von Willebrand's disease. *Curr Opin Hematol* **1**, 388–93.
132. Gassmann W, Loffler H. (1995). Acute megakaryoblastic leukemia. *Leuk Lymphoma* **18**(Suppl 1), 69–73.
133. Bennett JS. (2004). Structure and function of the platelet integrin alphaIIbbeta3. *J Clin Invest* **115**, 3363–9.
134. Fullard JF. (2004). The role of the platelet glycoprotein IIb/IIIa in thrombosis and haemostasis. *Curr Pharm Des* **10**, 1567–76.
135. Fishley B, Alexander WS. (2004). Thrombopoietin signalling in physiology and disease. *Growth Factors* **22**, 151–5.
136. Gangenahalli GU, Singh VK, Verma YK, Gupta P, Sharma RK, Chandra R, Luthra PM. (2006). Hematopoietic stem cell antigen CD34: Role in adhesion or homing. *Stem Cells Dev* **15**, 305–13.
137. Krause DS, Fackler MJ, Civin CI, May WS. (1996). CD34: Structure biology, and clinical utility. *Blood* **8**, 1–13.
138. Florian S, Sonneck K, Hauswirth AW, Krauth MT, Schernthaner GH, Sperr WR, Valent P. (2006). Detection of molecular targets on the surface of CD34+/CD38− stem cells in various myeloid malignancies. *Leuk Lymphoma* **47**, 207–22.
139. Cox CV, Evely RS, Oakhill A, Pamphilon DH, Goulden NJ, Blai A. (2004). Characterization of acute lymphoblastic leukemia progenitor cells. *Blood* **104**, 2919–25.
140. Deaglio S, Dianzani U, Horenstein AL, Fernandez JE, van Kooten C, Bragardo M, Funaro A, Garbarino G, Di Virgilio F, Banchereau J, Malavasi F. (1996). Human CD38 ligand. A 120-KDA protein predominantly expressed on endothelial cells. *J Immunol* **156**, 727–34.
141. Lee HC. (2000). Enzymatic functions and structures of CD38 and homologs. *Chem Immunol* **75**, 39–59.
142. Malavasi F, Funaro A, Alessio M, DeMonte LB, Ausiello CM, Dianzani U, Lanza F, Magrini E, Momo M, Roggero S. (1992). CD38: A multi-lineage cell activation molecule with a split personality. *Int J Clin Lab Res* **22**, 73–80.
143. Costello R, Sainty D, Bouabdallah R, Fermand JP, Delmer A, Divine M, Marolleau JP, Gastaut JA, Olive D, Rousselot P, Chaibi P. (2001). Primary plasma cell leukaemia: A report of 18 cases. *Leuk Res* **25**, 103–7.
144. Henniker AJ. (2001). CD90. *J Biol Regul Homeost Agents* **15**, 392–3.
145. Wetzel A, Chavakis T, Preissner KT, Sticherling M, Haustein UF, Anderegg U, Saalbach A. (2004). Human Thy-1 (CD90) on activated endothelial cells is a counterreceptor for the leukocyte integrin Mac-1 (CD11b/CD18). *J Immunol* **172**, 3850–9.
146. Kang LC, Dunphy CH. (2006). Immunoreactivity of MIC2 (CD99) and terminal deoxynucleotidyl transferase in bone marrow clot and

core specimens of acute myeloid leukemias and myelodysplastic syndromes. *Arch Pathol Lab Med* **130**, 153–7.
147. Dworzak MN, Froschl G, Printz D, Zen LD, Gaipa G, Ratei R, Basso G, Biondi A, Ludwig WD, Gadner H. (2004). CD99 expression in T-lineage ALL: Implications for flow cytometric detection of minimal residual disease. *Leukemia* **18**, 703–8.
148. Khoury JD. (2005). Ewing sarcoma family of tumors. *Adv Anat Pathol* **12**, 212–20.
149. Galli SJ, Zsebo KM, Geissler EN. (1994). The kit ligand, stem cell factor. *Adv Immunol* **55**, 1–96.
150. Lugli A, Went P, Khanlari B, Nikolova Z, Dirnhofer S. (2004). Rare KIT (CD117) expression in multiple myeloma abrogates the usefulness of imatinib mesylate treatment. *Virchows Arch* **444**, 264–8.
151. Wozniak J, Kopec-Szlezak J. (2004). c-Kit receptor (CD117) expression on myeloblasts and white blood cell counts in acute myeloid leukemia. *Cytometry B Clin Cytom* **58**, 9–16.
152. Scolnik MP, Morilla R, de Bracco MM, Catovsky D, Matutes E. (2002). CD34 and CD117 are overexpressed in AML and may be valuable to detect minimal residual disease. *Leuk Res* **26**, 615–19.
153. Orfao A, Garcia-Montero AC, Sanchez L, Escribano L, REMA. (2007). Recent advances in the understanding of mastocytosis: The role of KIT mutations. *Br J Haematol* **38**, 12–30.
154. Drexler HG, Sperling C, Ludwig WD. (1993). Terminal deoxynucleotidyl transferase (TdT) expression in acute myeloid leukemia. *Leukemia* **7**, 1142–50.
155. Ihanus E, Uotila L, Toivanen A, Stefanidakis M, Bailly P, Cartron JP, Gahmberg CG. (2003). Characterization of ICAM-4 binding to the I domains of the CD11a/CD18 and CD11b/CD18 leukocyte integrins. *Eur J Biochem* **270**, 1710–23.
156. Gruss HJ. (1996). Molecular, structural, and biological characteristics of the tumor necrosis factor ligand superfamily. *Int J Clin Lab Res* **26**, 143–59.
157. Schneider C, Hubinger G. (2002). Pleiotropic signal transduction mediated by human CD30: A member of the tumor necrosis factor receptor (TNFR) family. *Leuk Lymphoma* **43**, 1355–66.
158. Gruss HJ, Dower SK. (1995). Tumor necrosis factor ligand superfamily: Involvement in the pathology of malignant lymphomas. *Blood* **85**, 3378–404.
159. Cossu-Rocca P, Jones TD, Roth LM, Eble JN, Zheng W, Karim FW, Cheng L. (2006). Cytokeratin and CD30 expression in dysgerminoma. *Hum Pathol* **37**, 1015–21.
160. Rakheja D, Hoang MP, Sharma S, Albores-Saavedra J. (2002). Intratubular embryonal carcinoma. *Arch Pathol Lab Med* **126**, 487–90.
161. Rosenstein Y, Santana A, Pedraza-Alva G. (1999). CD43, a molecule with multiple functions. *Immunol Res* **20**, 89–99.
162. Remold-O'Donnell E, Rosen FS. (1990). Sialophorin (CD43) and the Wiskott–Aldrich syndrome. *Immunodefic Rev* **2**, 151–74.
163. Greer WL, Higgins E, Sutherland DR, Novogrodsky A, Brockhausen I, Peacocke M, Rubin LA, Baker M, Dennis JW, Siminovitch KA. (1989). Altered expression of leucocyte sialoglycoprotein in Wiskott–Aldrich syndrome is associated with a specific defect in O-glycosylation. *Biochem Cell Biol* **67**, 503–9.
164. Harris CL, Abbott RJ, Smith RA, Morgan BP, Lea SM. (2005). Molecular dissection of interactions between components of the alternative pathway of complement and decay accelerating factor (CD55). *J Biol Chem* **280**, 2569–78.
165. Parker C, Omine M, Richards S, Nishimura J, Bessler M, Ware R, Hillmen P, Luzzatto L, Young N, Kinoshita T, Rosse W, Socie G International PNH Interest Group (2005). Diagnosis and management of paroxysmal nocturnal hemoglobinuria. *Blood* **106**, 3699–709. Epub 2005 Jul 28.
166. Krauss JS. (2003). Laboratory diagnosis of paroxysmal nocturnal hemoglobinuria. *Ann Clin Lab Sci* **33**, 401–6. Review.
167. Maio M, Brasoveanu LI, Coral S, Sigalotti L, Lamaj E, Gasparollo A, Visintin A, Altomonte M, Fonsatti E. (1998). Structure, distribution, and functional role of protectin (CD59) in complement-susceptibility and in immunotherapy of human malignancies. *Int J Oncol* **13**, 305–18.
168. Tachibana KE, Gonzalez MA, Coleman N. (2005). Cell-cycle-dependent regulation of DNA replication and its relevance to cancer pathology. *J Pathol* **205**, 123–9.
169. Sheval EV, Churakova JV, Dudnik OA, Vorobjev IA. (2004). Examination of the proliferative activity of tumor cells in human lymphoid neoplasms using a morphometric approach. *Cancer* **102**, 174–85.
170. Dong HY, Liu W, Cohen P, Mahle CE, Zhang W. (2005). B-cell specific activation protein encoded by the PAX-5 gene is commonly expressed in merkel cell carcinoma and small cell carcinomas. *Am J Surg Pathol* **29**, 687–92.
171. Maier H, Hagman J. (2002). Roles of EBF and Pax-5 in B lineage commitment and development. *Semin Immunol* **14**, 415–22.
172. Poppe B, De Paepe P, Michaux L, Dastugue N, Bastard C, Herens C, Moreau E, Cavazzini F, Yigit N, Van Limbergen H, De Paepe A, Praet M, De Wolf-Peeters C, Wlodarska I, Speleman F. (2005). PAX5/IGH rearrangement is a recurrent finding in a subset of aggressive B-NHL with complex chromosomal rearrangements. *Genes Chromosomes Cancer* **44**, 218–23.
173. Tiacci E, Pileri S, Orleth A, Pacini R, Tabarrini A, Frenguelli F, Liso A, Diverio D, Lo-Coco F, Falini B. (2004). PAX5 expression in acute leukemias: Higher B-lineage specificity than CD79a and selective association with t(8;21)-acute myelogenous leukemia. *Cancer Res* **64**, 7399–404.
174. Browne P, Petrosyan K, Hernandez A, Chan JA. (2003). The B-cell transcription factors BSAP, Oct-2, and BOB.1 and the pan-B-cell markers CD20, CD22, and CD79a are useful in the differential diagnosis of classic Hodgkin lymphoma. *Am J Clin Pathol* **120**, 767–77.
175. Pileri SA, Gaidano G, Zinzani PL, Falini B, Gaulard P, Zucca E, Pieri F, Berra E, Sabattini E, Ascani S, Piccioli M, Johnson PW, Giardini R, Pescarmona E, Novero D, Piccaluga PP, Marafioti T, Alonso MA, Cavalli F. (2003). Primary mediastinal B-cell lymphoma: High frequency of BCL-6 mutations and consistent expression of the transcription factors OCT-2, BOB.1, and PU.1 in the absence of immunoglobulins. *Am J Pathol* **162**, 243–53.
176. Qian D, Weiss A. (1997). T cell antigen receptor signal transduction. *Curr Opin Cell Biol* **9**, 205–12.
177. Crespo M, Villamor N, Gine E, Muntanola A, Colomer D, Marafioti T, Jones M, Camos M, Campo E, Montserrat E, Bosch F. (2006). ZAP-70 Expression in normal Pro/Pre B Cells, mature B cells, and in B-cell acute lymphoblastic leukemia. *Clin Cancer Res* **12**, 726–34.
178. Vener C, Gianelli U, Cortelezzi A, Fracchiolla NS, Somalvico F, Savi F, Pasquini MC, Bosari S, Deliliers GL. (2006). ZAP-70 immunoreactivity is a prognostic marker of disease progression in chronic lymphocytic leukemia. *Leuk Lymphoma* **47**, 245–51.
179. Macey MG. (2007). *Flow Cytometry: Principles and Applications*. Humana Press, Totowa.
180. Song S, Naeim F. (2004). New applications of flow cytometry in cancer diagnosis. In *Cancer Diagnosis. Current and Future Trends* (Nakamura RM, Grody WW, Wu JT, Nagle RB, eds), pp. 199–232. Humana Press, Totowa.
181. Shapiro HM. (2003). *Practical Flow Cytometry*, 4th ed. Wiley, Hoboken.
182. Gorczyca W. (2006). *Flow Cytometry in Neoplastic Hematology*. Taylor & Francis, London.
183. Baumgarth N, Roederer M. (2000). A practical approach to multicolor flow cytometry for immunophenotyping. *J Immunol Methods* **243**, 77–97.
184. McCoy JP Jr. (2002). Basic principles of flow cytometry. *Hematol Oncol Clin North Am* **16**, 229–43.
185. Schwartz A, Marti GE, Poon R, Gratama JW, Fernandez-Repollet E. (1998). Standardizing flow cytometry: A classification system of fluorescence standards used for flow cytometry. *Cytometry* **33**, 106–14.
186. Gratama JW, Bolhuis RL, Van't Veer MB. (1999). Quality control of flow cytometric immunophenotyping of haematological malignancies. *Clin Lab Haematol* **21**, 155–60.

187. Stelzer GT, Marti G, Hurley A, McCoy P Jr., Lovett EJ, Schwartz A. (1997). U.S.-Canadian Consensus recommendations on the immunophenotypic analysis of hematologic neoplasia by flow cytometry: Standardization and validation of laboratory procedures. *Cytometry* **30**, 214–30.
188. Dabbs D. (2006). *Diagnostic Immunohistochemistry*, 2nd ed. Churchill Livingstone, London.
189. Polak JM, Van Noorden S. (2003). *Introduction to Immunocytochemistry*, 3rd ed. Academic Press, New York.
190. Jaffe ES. (1999). Hematopathology: Integration of morphologic features and biologic markers for diagnosis. *Mod Pathol* **12**, 109–15.
191. Myers JD. (1989). Development and application of immunocytochemical staining techniques: A review. *Diagn Cytopathol* **5**, 318–30.
192. Pettigrew NM. (1989). Techniques in immunocytochemistry. Application to diagnostic pathology. *Arch Pathol Lab Med* **113**, 641–4.
193. Swanson PE. (1988). Foundations of immunohistochemistry. A practical review. *Am J Clin Pathol* **90**, 333–9.
194. Shi SR, Cote RJ, Taylor CR. (2001). Antigen retrieval techniques: Current perspectives. *J Histochem Cytochem* **49**, 931–7.
195. Ramos-Vara JA. (2005). Technical aspects of immunohistochemistry. *Vet Pathol* **42**, 405–26.
196. Berger U, Bettelheim R, Mansi JL, Easton D, Coombes RC, Neville AM. (1988). The relationship between micrometastases in the bone marrow, histopathologic features of the primary tumor in breast cancer and prognosis. *Am J Clin Pathol* **90**, 1–6.
197. Thor A, Viglione MJ, Ohuchi N, Simpson J, Steis R, Cousar J, Lippman M, Kufe DW, Schlom J. (1988). Comparison of monoclonal antibodies for the detection of occult breast carcinoma metastases in bone marrow. *Breast Cancer Res Treat* **11**, 133–45.
198. Beck D, Maritaz O, Gross N, Favrot M, Vultier N, Bailly C, Villa I, Gentilhomme O, Philip T. (1988). Immunocytochemical detection of neuroblastoma cells infiltrating clinical bone marrow samples. *Eur J Pediatr* **147**, 609–12.

Principles of Cytogenetics

3

CHAPTER

P. Nagesh Rao

INTRODUCTION

Cancer is a genetic disease characterized by DNA changes at either the nucleotide or chromosomal level, or both. Malignancies can develop either from a genetic predisposition followed by acquired somatic mutations, or from an accumulation of somatic mutations that develop into a cancer phenotype. At the chromosome level, these mutations include changes in chromosome numbers (aneuploidy), loss of heterozygosity (LOH, whole chromosome or segmental region loss), chromosomal rearrangements (translocations and inversions), and gene duplications or amplifications. Many of these cytogenetically visible or cryptic (submicroscopic) aberrations are characteristic of a particular disease or disease subtype. Because specific chromosomal aberrations, especially in hematologic malignancies, provide diagnostic, prognostic, and/or treatment information for many cancers, they are, in many ways, true biomarkers for human cancer.

Clear insights into the genetic basis of cancer were obtained in the 1950s and 1960s when improved cell culture and slide preparation techniques made it possible to accurately enumerate the number of human chromosomes as 46 in 1956 [1]. Chromosome analysis is usually carried out on cells in mitosis (cell division) when the chromosomes become visible as distinct entities. After identifying each chromosome in a cell by its characteristic size, shape, and staining properties, a karyotype displaying the full chromosome complement of the cell can be prepared.

The first specific chromosome abnormality observed in a human tumor was seen in Philadelphia in 1960 by Nowell and Hungerford [2] who found an unusually small chromosome in the leukemic cells of patients with chronic myeloid leukemia (CML). This small chromosome was named the "Philadelphia" (*Ph*) chromosome. The discovery of the *Ph* chromosome aroused considerable interest in cancer cytogenetics as it gave the first direct evidence for a consistent DNA-associated change in a tumor. More than 30,000 cases of hematologic malignancies with chromosome aberrations have been reported, thereby making it the most thoroughly cytogenetically investigated group of all neoplastic disorders.

A second major breakthrough in cytogenetics was the development of microscopic staining techniques, generating a banding pattern along the length of the chromosomes [3]. With this banding pattern, all individual chromosomes could be identified and structural changes could be characterized in much greater detail. Consistent chromosome aberrations, which are uncommon or extremely rare in normal tissues, were found in different cells of a tumor and further karyotypic changes were shown to occur during tumor progression. Modifications of culture methods to improve yields of dividing cells and high-resolution banding of elongated chromosomes now allow for a more precise definition of rearrangements as well as the identification of previously undetected rearrangements. By using these techniques, most tumor cells can be shown to have some form of chromosomal defect. Possibly, the best correlation between the presence of highly specific chromosomal changes and a subtype of leukemia is the 15;17 translocation which is identified only in patients with acute promyelocytic leukemia (APL or AML-M3). Because the chromosomal aberrations determined at diagnosis is an independent prognosis, indicator in several leukemias and lymphomas, it is imperative that the karyotype analyses are completed. As a result of relative ease of obtaining bone marrow or peripheral blood specimens from leukemia patients, it is possible to do serial sampling, which allows for studying the cytogenetic patterns during the various stages of the clinical course, such as at diagnosis, remission, and relapse. This requires appropriate preparations and culturing of the bone marrow. A number of techniques are available to evaluate chromosomal aberrations in hematological malignancies.

CELL PREPARATION

Bone marrow cells from leukemia patients are an ideal source of tissue for cytogenetic studies of leukemias. The use of "direct" preparation to avoid selection during culturing has been considered advantageous for obtaining an accurate picture of the chromosomal constitution of the leukemic cells. However, it has also been suggested that culturing leukemic bone marrow cells may uncover both clonal abnormalities and a greater number of abnormal cells than the direct method of preparation. Chromosome analysis of leukemic cells is not always possible by the direct method because cases are often encountered in which only a few or no mitoses are present or whose chromosomes show blurred outlines and do not provide good banding patterns that are adequate for detailed analysis despite good mitotic index. It must also be noted that the 15;17 translocation characteristic of APL is usually not observed in the direct preparations but easily seen in 24-h cultured bone marrow cells. Thus, it is important that more than one kind of culture setup and harvest method be performed.

A prerequisite for chromosome analysis is dividing or mitotic cells. Spontaneously dividing cells suitable for direct chromosome preparations are found only in the rapidly proliferating tissues of the body such as the gonads and bone marrow, or in tissues with malignancies. Bone marrow is the tissue of choice for cytogenetic study of most hematological conditions. However, in chronic disorders where there is high white cell count, such as CML-blast crisis and chronic lymphocytic leukemia (CLL), a hypercellular peripheral blood sample is more appropriate. It is also important to recognize that even a very dilute sample can overgrow and subsequently fail. Spleen tissue, or more rarely ascitic and pleural effusions, is also amenable for cytogenetic study in some hematological disorders.

Chromosome studies of malignant lymphomas are usually based on studies from lymph node biopsies because bone marrow may not always be involved in the early stages of the disease. But occasionally, a bone marrow may be taken which can prove informative particularly if there is doubt as to whether or not the bone marrow involvement has occurred. The success of a cytogenetic analysis mostly depends on the quality of material investigated. Therefore, the key to successful cytogenetics is adequate sampling with a high viability. The sample must be drawn under sterile conditions with the aspiration syringe coated with preservative-free heparin to avoid clumping of blood components. Heparin can also be added to lymph node and spleen tissue after their surgical removal. Where there is a likelihood of a dry tap, especially in diseases such as primary or secondary myelofibrosis or due to faulty technique, a peripheral blood sample can be sent as an alternative. However, unless there are sufficient blast cells in these samples, the abnormal clone may go undetected. Care must also be taken in suspected APL cases where clotting is possible and may cause the sample to be unsuitable for culture and chromosome analysis.

A good quality bone marrow aspirate or bore core biopsy sample of 1–2 mL is adequate in most cases, although less than this could be accepted if the marrow is very cellular. The drawing of tissue must be done under sterile conditions because the chromosome analysis is, in most cases, preceded by short- or long-term cell culture (96h) which mandates a high degree of sterility. It is also advisable to determine if the initial aspirate has bone marrow spicules. If a previous aspirate has been obtained, aspiration of a second sample after repositioning is recommended for cytogenetic studies.

The shorter the duration of time (≤24h) between collection and culture setup in the cytogenetic laboratory, the greater the chances for a successful chromosome analysis with an accurate result. Every effort must be made to ensure that the bone marrow or lymph node biopsy samples are set up in cultures with a minimum delay. If a delay in the transportation of the sample is anticipated, the sample should preferably be collected and transferred to a tube containing transportation medium made up of a preservative-free heparin in an appropriate basal medium and supplemented with serum and antibiotics. The sample must never be frozen but can be stored at 4°C overnight or for up to 3 days. However, the cell viability is greatly reduced with time, yielding misleading or only normal results. Disorders such as acute lymphoblastic leukemia (ALL) and others with a high white cell count are particularly adversely affected by delays.

Same day or direct cultures are often recommended for ALL, and sometimes for CML. Studies suggest that erythropoietic cells divide rapidly in the first few hours of culture followed by granulopoietic cell divisions [4, 5]. On the basis of these observations, short cultures in erythroleukemia are more likely to yield good results. However, in a majority of cases a minimum of 16–24h unstimulated culture is appropriate. On the other hand, CLL and some ALL cases, which have B- or T-cell phenotype, need 3–5 day cultures with appropriate mitogens, as well as having some unstimulated cultures. Sometimes, due to poor response to the commonly used mitogens, (e.g. chronic B- and T-lymphoid leukemias), TPA (12-O-tetradecanoylphorbol-13-acetate), and EBV (Epstein–Barr virus), pokeweed or IL-2 are used as stimulants in the cultures. Several laboratories supplement the regular media with a condition medium derived from cultures of a human urinary bladder carcinoma or giant cell tumor (GCT) cell lines, which are capable of stimulating the proliferation and growth of human myeloid leukemia cells.

The chromosomes of patients with ALL are particularly difficult, fuzzy, and resistant to banding. Nevertheless, analysis of direct preparation of these marrows have shown that 50–78% of these patients have chromosomal abnormalities in their leukemic cells. But it is important to recognize that more than one technique is necessary to assess accurately the karyotypic constitution of the leukemic cells.

BANDING TECHNIQUES

The standard cytogenetics method consists of culturing a suspension of cells in mitogenic media for 24–72 h. Then, the dividing cells are arrested in metaphase by the addition of an inhibitor of the mitotic spindle, such as colchicine or vinblastine. The cells are submitted to a hypotonic solution (commonly 0.075M KCl) and stained with Giemsa (G-banding), which reveals characteristic banding patterns that are specific for each chromosome. These banding patterns allow the assignment of homologous chromosomes, the identification of extra or missing chromosomes as well as of structural aberrations. For G-banding patterns, pretreatment of chromosomes by enzymes such as trypsin or pancreatin is required. The mechanism of the G-banding is not fully understood yet but the chromosomes express dark and light G-bands. It is the most common and traditional banding technique used in the clinical setting. Other chromosome banding techniques are used to produce a reverse banding pattern (R-bands), a fluorescence banding technique using quinacrine derivatives (Q-bands), or centromeric staining (C-bands) to better define the chromosomal aberrations. The nucleolus organizer regions (NORs) located in the short arms of acrocentric chromosomes are visualized by staining with silver nitrate ($AgNO_3$).

ANALYSIS

Chromosome analysis is performed using a microscope commonly at 1000× magnification. With the development of image analysis hardware and software, computer-aided chromosome analysis systems are in use and have greatly reduced the turnaround times, and have also made the quality of the karyotyped images almost equal to that of photographed ones. Bone marrow karyotype analysis is often biased toward cells with poor morphology where there is a mixed population. Selection of only metaphases with good morphology can often lead to failure of detecting the cells from abnormal clones. A clone is defined as at least two cells with the same structural abnormality or gain of the same chromosome or at least three cells with the loss of the same chromosome. The karyotype results are interpreted using an International System for Human Cytogenetic Nomenclature (ISCN 2005, [6]). For example, a normal male and female karyotypes are designated as 46,XY and 46,XX, respectively. An abnormal karyotype such as 46,XY,t(9;22)(q34;q11.2) designates an abnormal male karyotype with a balanced translocation between chromosomes 9 and 22 at band 34 of the long arm of chromosome 9 and the long arm of chromosome 22 at band 11.2. In contrast, a karyotype 47,XY,t(9;22)(q34;q11.2), +der(22)t(9;22) delineates an abnormal male karyotype not only with a balanced 9;22 translocation as explained earlier, but also an additional chromosome that is derived from this translocation, i.e. an extra *Ph* chromosome. The number of cells observed for a particular clone is provided in brackets []. For example, a karyotype 47,XX,+8[12]/46,XX[8], indicates that of the total 20 metaphase cells analyzed from this female patient, 12 cells forming the abnormal clone, had trisomy 8 and the second clone of 8 cells had a normal karyotype.

There are three main types of cytogenetic aberrations in human cancer:

1. Balanced chromosomal rearrangements (translocations or inversions) (Figures 3.1 and 3.2).
2. Gain or loss of whole chromosomes (aneuploidy), or part of a chromosome (segmental aneuploidy) (Figures 3.3 and 3.4).
3. Loss of heterozygosity (LOH).

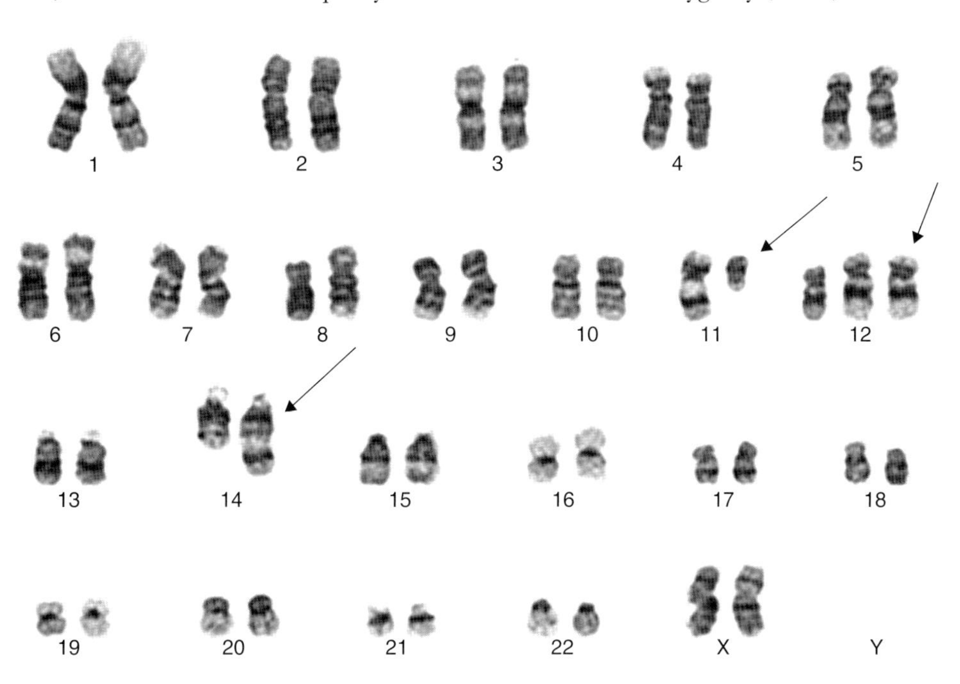

FIGURE 3.1 An abnormal female karyotype showing a balanced 11;14 translocation and trisomy 12. The karyotype is designated as 47,XX,t(11;14)(q13;q32), +12.

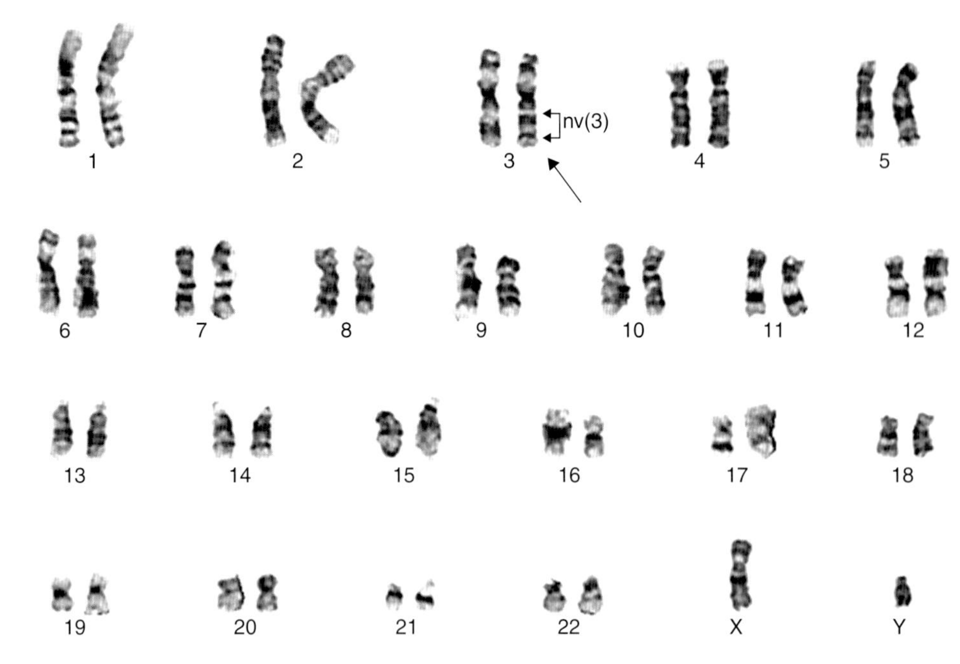

FIGURE 3.2 An abnormal male karyotype with a paracentric inversion in the long arm of one chromosome 3; the region of inversion is shown in brackets. The karyotype is 46,XY,inv(3)(q21q26).

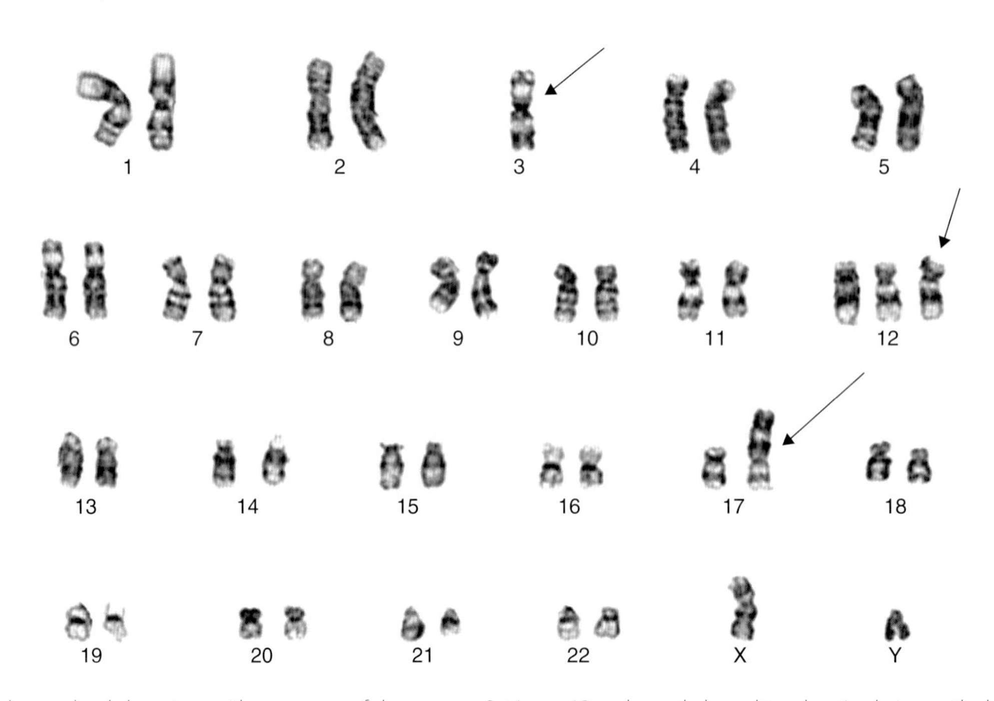

FIGURE 3.3 An abnormal male karyotype with monosomy of chromosome 3, trisomy 12, and an unbalanced translocation between the long arm of chromosome 3 and the short arm of chromosome 17 resulting in the deletion of short arm of chromosome 3 and distal 17p segment. The karyotype is designated as 46,XY, +12,der(17)t(3;17)(q13;p13).

These three types of chromosomal aberrations typically cause cell overgrowth through over-expression/activation of an oncogene, or by deletion of a tumor suppressor gene. Identification of recurrent chromosomal aberrations has become very important in the diagnosis of soft tissue and hematologic tumors. Especially in some hematologic malignancies, the identification of recurrent chromosomal aberrations is important for diagnosis, classification, prognosis, and therapy.

Balanced Rearrangements

Balanced rearrangements in cancer include translocations (exchange between two or more chromosomes) and inversions (orientation change relative to the centromere, within a single chromosome/arm). These rearrangements often result in chimeric cellular proteins that appear to disrupt the normal function of critical genes involved in normal cell growth or differentiation resulting in an abnormal process. More than

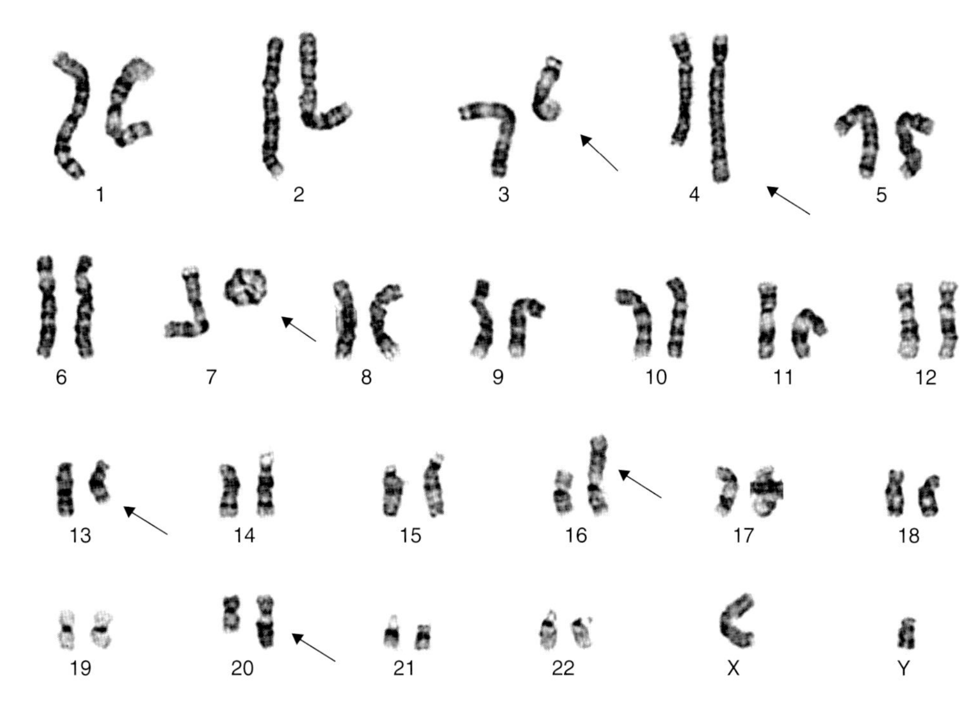

FIGURE 3.4 An abnormal complex male karyotype with several aberrations including a deletion of 3q, an interstitial deletion of 13q, and an unbalanced translocation with unidentifiable chromosomal segments of chromosomes 4q, 16p, and 20q, respectively, and ring chromosome 7. The karyotype is written as 46,XY,del(3)(q21),add(4)(q35), r(7),del(13)(q12q14),add(16)(p13),add(20)(q13).

600 neoplasia-related recurrent balanced cytogenetic aberrations have been reported to date [7]. Translocations and inversions usually cause cancer by fusing together two genes, resulting in aberrant expression. Currently, more than two hundred fusion genes responsible for human cancers have been reported in the literature [8, 9]. One classic example of an important translocation in human cancer is the t(9;22) in CML. The t(9;22) results in aberrant expression of a gene (*ABL1*) that normally functions in cellular proliferation by coming under control of a constitutively expressed gene (*BCR*). Approximately 50% of hematopoietic neoplasms acquire translocations somatically; most of these neoplasms are restricted to a single cell lineage (that in which the translocation originated) and are arrested in a particular stage of developmental maturation. Occasionally, more than one cell lineage is affected (e.g. *MLL* gene-related malignancies) suggesting that the involved genes were affected at the pluripotent stem cell stage. While balanced aberrations may be directly related to the etiology of the malignancy, the unbalanced translocations are often recognized as indicators of secondary tumor progression.

Chromosomal Aneuploidy

Chromosomal aneuploidy is extremely common in cancer, and can be either a primary or a secondary event [10]. Chromosomal gains (whole or partial) are designated in the karyotype with " + " or "add," and typically result in the over-expression of an oncogene. Despite the presence of a high frequency of aneuploidy in cancer, the exact role of aneuploidy in carcinogenesis is not very clear. Numerical aberrations as the sole karyotypic anomalies, including single or multiple losses and gains, are found in approximately 15% of all cytogenetically abnormal hematologic neoplasms (Figure 3.5) [8]. Although they are relatively frequent, numerical abnormalities have generally received less attention than the structural abnormalities, particularly the simple reciprocal translocations that are amenable to rigorous molecular analysis. The association of numerical aberrations with hematologic disorders, although well established, also appears less disease specific [11–14]. Trisomy 8, monosomy 7, and trisomy 21 have been found in different categories of leukemias both at initial presentation and as secondary cytogenetic events. Roughly, half of all numerical aberrations are trisomies. Other than trisomies for chromosomes 8, 9, 11, 12, and 21, autosomal trisomies are infrequent in hematologic disorders. Aneuploidy can be detected with the help of traditional metaphase cytogenetics (Figure 3.5), interphase cytogenetics (fluorescence *in situ* hybridization (FISH) (Figure 3.6), multicolor FISH, spectral karyotyping, comparative genomic hybridization techniques (CGH)), flow cytometry (FCM), and image cytometry (ICM). Flow cytometry and ICM can measure the relative DNA content of the cell with respect to reference diploid cells. Imbalances, i.e. aberrations that result in gain or loss of genetic material, are even more common than translocations and inversions in hematologic malignancies. These include amplifications, duplications, heterozygous or homozygous deletions, monosomies, and trisomies. Amplifications (several extra copies of a gene or chromosome region) may occur in the form of supernumerary marker chromosomes (SMCs), double minutes, and homologous staining regions (HSRs); and are a result of over-expression of one or more genes (Figure 3.6c).

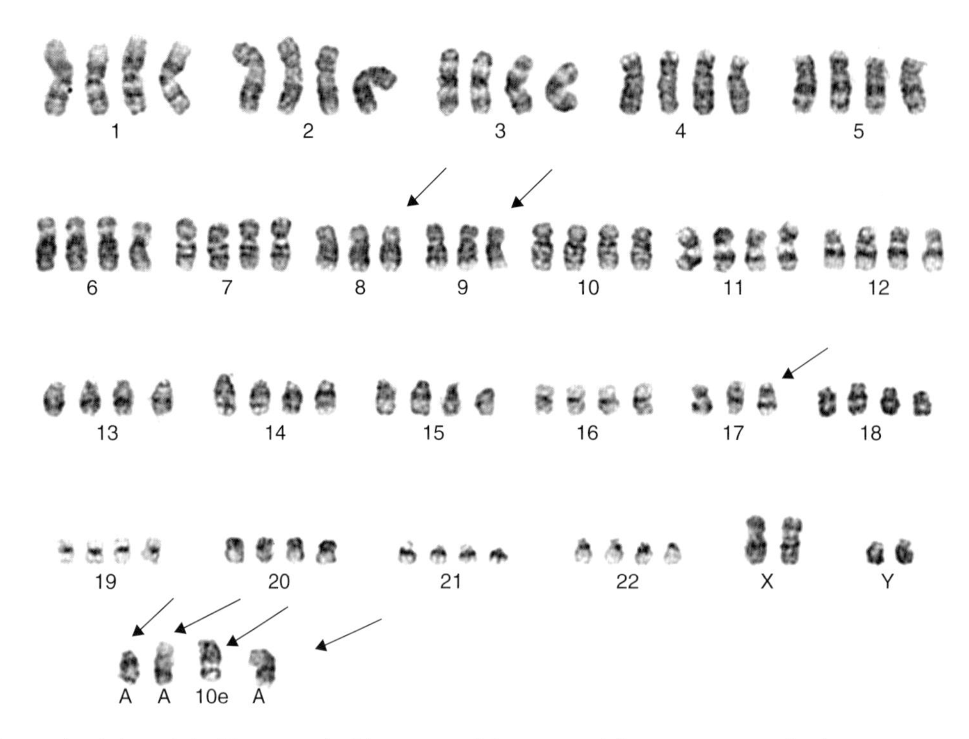

FIGURE 3.5 An abnormal male hyperdiploid (near-tetraploid) karyotype with four copies of all autosomes except for chromosomes 8, 9, 17, and extra copies of small marker chromosomes of unknown origin. This karyotype is written as 93,XY, +X, +Y, −8, −9, −17, +4mar.

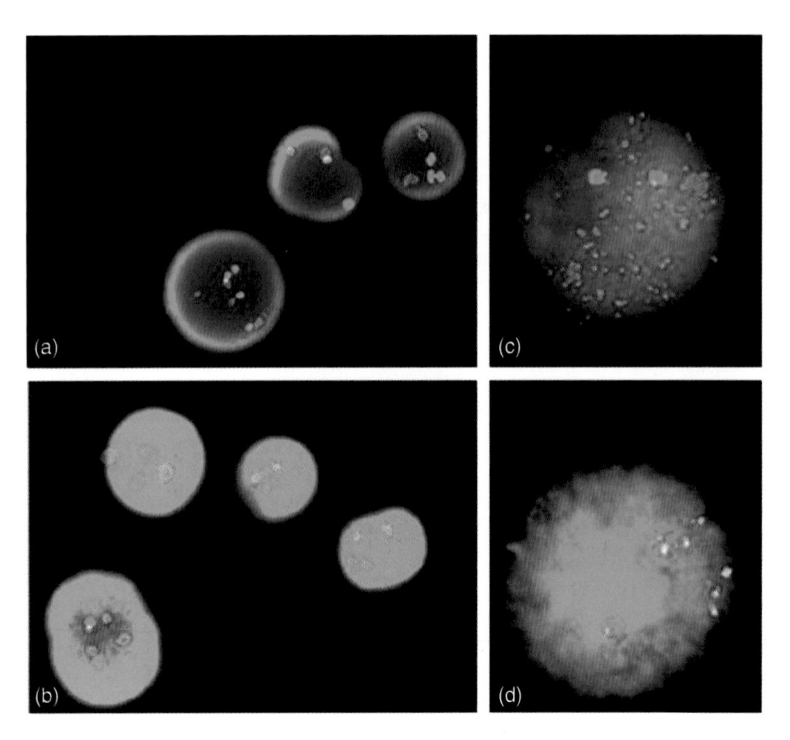

FIGURE 3.6 Panel showing identification of various chromosomal abnormalities by FISH. (a) two–four copies of the 11q13 (red) and 14q32 (green) loci, (b) four copies of 9q34 (red) and 22q11.2 (green) loci, (c) amplification of the *ABL* oncogene (red), and normal two copies of the BCR (green) locus, and (d) six copies of the *MLL* locus (yellow).

SMCs are small additional chromosomes whose origins are not readily identifiable by banding methodologies, and are designated as "+mar" in the karyotype (see Figure 3.5). Double minutes are specific types of SMCs that are characterized by a typical dumbbell shape and represent extra-chromosomal oncogene amplification. For example, *MYCN* gene amplification in the form of double minutes is commonly observed in neuroblastoma. The mixed lineage leukemia gene (*MLL*) is sometimes amplified in acute leukemia and can be easily visualized in interphase nuclei by FISH studies with MLL-specific probes (Figure 3.6d). HSRs are amplified oncogenes within the structure of a

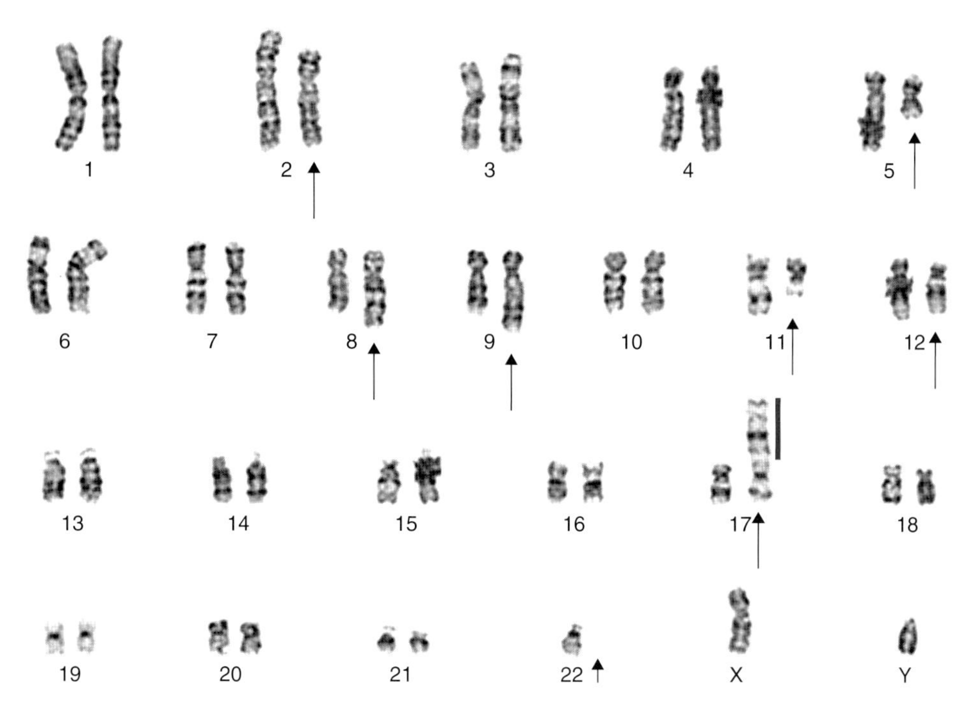

FIGURE 3.7 An abnormal male karyotype with an HSR of an unknown chromosomal region at the short arm of chromosome 17 (red line).

chromosome (Figure 3.7), and are designated as "HSR" in the karyotype. Usually *C-MYC* gene amplification on chromosome 8 in some acute myeloid leukemias (AMLs) is an example of an HSR that is typically detected with FISH. Chromosomal losses (whole or partial) are designated in the karyotype by "-" or "del," and are thought to result in deletion (or decreased activity) of tumor suppressor genes.

Loss of Heterozygosity

Loss of heterozygosity is defined as the loss of one parent's contribution to the cell and can be caused by deletion, gene conversion, mitotic recombination, or loss of a chromosome. LOH often occurs in cancer, where the second copy of a gene (typically a tumor suppressor gene) has been inactivated by other mechanisms, such as point mutation or hypermethylation. When a whole chromosome or a large segment of a chromosome is lost, the remaining chromosome or segment is often duplicated. With complete duplication of the remaining genetic material, the karyotype may appear normal, even though no normal genes are present. Though not easily detected by cytogenetic techniques, this duplication of the remaining chromosome or segment has been shown using molecular genetic techniques [5, 15]. At least in theory this type of LOH can be detected cytogenetically using chromosome heteromorphisms, though it is not often pursued.

Cytogenetic analyses can provide valuable and extremely relevant information to establish the presence of a malignant clone, determine the cell lineages in the disease process or clarify and confirm a diagnosis, provide prognostic predictive features, and monitor response to treatment and classification of neoplasms. The significance and usefulness of these will be discussed in the following chapters.

A close relationship between the pathologist and the cytogeneticist is essential if maximum useful information is to be produced from the cytogenetic studies of hematological disorders.

References

1. Tjio TH, Levan A. (1956). The chromosome number of man. *Hereditas* **42**, 1–6.
2. Nowell P, Hungerford D. (1960). A minute chromosome in human granulocytic leukemia. *Science* **132**, 1497.
3. Caspersson T, Zech L, Johansson C. (1970). Differential binding of alkylating fluorochromes in human chromosomes. *Exp Cell Res* **60**, 315–19.
4. Harrison CJ, Fitchett M, Potter AM, Swansbury GJ. (1987). A guide to cytogenetics studies in hematological disorders. *Eugenics Soc Occasional papers* **1**, 1–30.
5. Berger R, Bernheim A, Daniel MT, Valensi F, Flandrin G. (1983). Cytological types of mitoses and chromosome abnormalities in acute leukemia. *Leukemia Res* **7**, 221–35.
6. ISCN (2005). *An International System for Human Cytogenetic Nomenclature*, Shaffer LG, Tommerup N (eds); S. Karger, Basel.
7. Mitelman F. (2005). Cancer cytogenetics update. *Atlas Genet Cytogenet Oncol Haematol.*
8. Mitelman F. (2005). Database of Chromosome Aberrations in Cancer. Cancer Genome Anatomy Project (CGAP) http://cgap.nci.nih.gov/Chromosomes/Mitelman.
9. Mitelman F, Johansson B, Mertens F. (2007). The impact of translocations and gene fusions on cancer causation. *Nat Rev Cancer* **7**, 233–45.
10. Jallelapali PV, Lengauer C. (2001). Chromosome segregation and cancer: cutting through the mystery. *Nat Rev Cancer* **1**, 109–17.
11. Heim S, Mitelman F. (1986). Numerical chromosome aberrations in human leukemia. *Cancer Genet Cytogenet* **15**, 99–108.
12. United Kingdom Cancer Cytogenetics Group (UKCCG) (1992). Primary, single, autosomal trisomies associated with haematological disorders. *Br J Haematol* **16**, 841–51.

13. Ross FM, Stockdill G. (1987). Clonal chromosome abnormalities in chronic lymphocytic leukemia patients revealed by TPA stimulation of whole blood cultures. *Cancer Genet Cytogenet* **25**, 109–21.
14. Watson MS, Carroll AJ, Shuster JJ, Steuber CP, Borowitz MJ, Behm FG, Pullen DJ, Land VJ. (1993). Trisomy 21 in acute lymphoblastic leukemia: A pediatric oncology group study. *Blood* **82**, 3098–102.
15. Gorletta TA, Gasparini P, Elios MMD, Trubia M, Pelicci PG, Di Fiore PP. (2005). Frequent loss of heterozygosity without loss of genetic material in acute myeloid leukemia with a normal karyotype. *Genes Chromosomes Cancer* **44**, 334–7.

Principles of Molecular Techniques

4

CHAPTER

Wayne W. Grody,
P. Nagesh Rao
and Faramarz Naeim

FLUORESCENCE *IN SITU* HYBRIDIZATION

Karyotype analysis depends primarily on classical chromosomal banding techniques; and has the distinct advantage that the entire genome can be analyzed in a single experiment. In particular, it is useful for identifying whole chromosomes accurately and for identifying obvious chromosomal aberrations. However, karyotype studies are limited to actively dividing cells, and the resolution is limited to chromosomal rearrangements that are >3 Mb in size. In addition, though there is a history of several decades of clinical cytogenetic analysis of cancer cells, it has become apparent that suboptimal collection, transport, and culture of clinical specimens can lead to inappropriate (e.g. normal) results. Poorly spread or contracted metaphase chromosomes, low mitotic activity, and highly rearranged karyotypes with numerous marker chromosomes, common in neoplastic cell preparations, are often difficult to interpret unambiguously. Furthermore, chromosome preparations are labor-intensive and time-consuming and the interpretation of cytogenetic findings require extensive experience. Although automated karyotyping systems became available, analyzing metaphase spreads remains time-consuming. Techniques such as polymerase chain reaction (PCR) have the advantage to be more sensitive and to screen for a specific chromosome aberration without the need for dividing cells. However, such molecular analyses are limited to known fusion genes and do not allow for the screening of the whole genome for other (secondary) alterations. Thus molecular cytogenetic techniques have been developed to bridge the gap between classical cytogenetics and molecular DNA techniques. The limitations of classical chromosome studies have been overcome by the introduction of fluorescence *in situ* hybridization (FISH), which offers a molecular dimension to cytogenetic analysis. Different and new FISH technologies have emerged, each with their own particular advantages and applications, e.g. interphase FISH, comparative genomic hybridization (CGH), fiber-FISH, and multi-color FISH. These techniques are capable of detecting aberrations of an intermediate size (~10 kb to 5 Mb), and are commonly used in cancer cytogenetics laboratories today, for both diagnostic and research applications. These techniques are fast and provide an accurate but targeted analysis of whole tumor genomes in a single experiment. The FISH technologies provide increased resolution for the elucidation of structural chromosome abnormalities that cannot be resolved by more conventional cytogenetic analyses, including submicroscopic deletions, cryptic or subtle duplications and translocations, complex rearrangements involving many chromosomes, and marker chromosomes.

The FISH procedure has been developed for the tagging of DNA and RNA with labeled nucleic acid probes and is a process whereby chromosomes or portions of chromosomes are vividly painted with fluorescent molecules that anneal to specific regions. This technique has been used widely for the identification of chromosomal abnormalities. The method enables enumeration of multiple copies of chromosomes or detection of specific regions of DNA or RNA that represent associations with certain genetic characteristics and infectious disease.

The FISH methods widely employed in clinical laboratory studies involve hybridization of a fluorochrome-labeled DNA probe to an *in situ* chromosomal target and can be applied to a

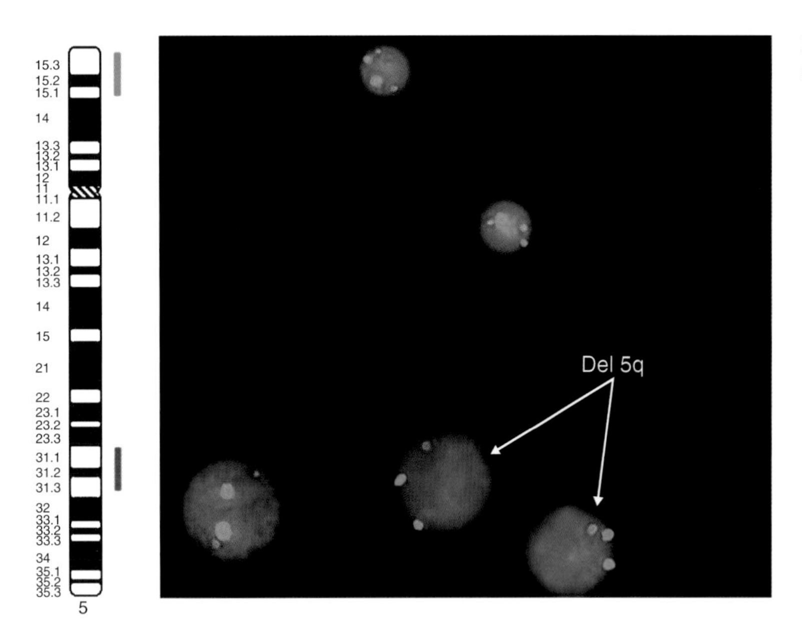

FIGURE 4.1 Deletion of 5q detected by 5p (green) (control) and 5q (red) specific FISH probes.

variety of specimen types and performed on nondividing interphase cells. Interphase nucleus assessment from uncultured preparations allows for a rapid screening for specific chromosome rearrangements or numerical abnormalities associated with hematologic malignancies. Interphase analysis may also be performed on fixed bone marrow cell suspensions, paraffin-embedded tissue sections or disaggregated cells from paraffin blocks, bone marrow or blood smears, and touch preparations of cells from lymph nodes or solid tumors. It is also commonly used when rapid or direct (i.e. without culturing) results are needed and can be performed on formalin-fixed paraffin-embedded (FFPE) tissue. FISH uses fluorescently labeled DNA probes (e.g. bacterial artificial chromosomes, or BACs) hybridized to either metaphase chromosomes or interphase nuclei, depending on the application. However analytically powerful and diagnostically useful interphase FISH might be, great care should be taken in the interpretation of interphase hybridization patterns. As a rule of thumb, FISH results should be interpreted in conjunction with the neoplastic karyotype.

Four different types of probes are commonly used, each with different ranges of applications:

1. Gene-specific probes target DNA sequences (Figure 4.1) present in only one copy per chromosome. They are used to identify chromosomal translocations, inversions and deletions, contiguous gene syndromes, and chromosomal amplifications in interphase and metaphase chromosomes. These probes are particularly useful for screening of specific chromosomal aberrations in metaphase spreads and interphase nuclei. For this purpose, the probes cover the chromosomal breakpoints and can specifically identify the genes involved in the chromosome alterations without the need for dividing cells. The same FISH experiments can subsequently be used to assess the efficacy of therapeutic regimens and to detect residual disease with a rather limited sensitivity of 0.5–5%.

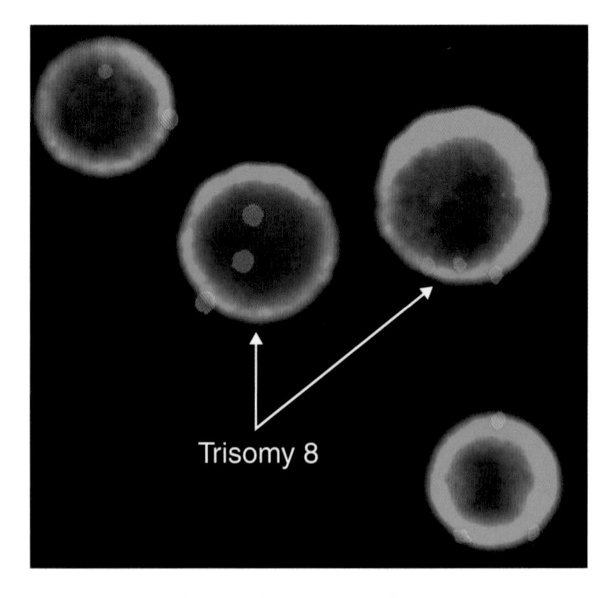

FIGURE 4.2 FISH studies reveal three copies of chromosome 8 with centromere specific probe in each cell (arrows), indicative of trisomy 8.

2. Repetitive sequence probes (Figure 4.2) (alpha-satellite sequences) bind to chromosomal regions that are represented by short repetitive base-pair sequences that are present in multiple copies (e.g. centromeric and telomeric probes). Centromeres are usually A–T rich, whereas telomeres are known to have repetitive TTAGGG sequences. Centromeric probes are extremely useful for identifying marker chromosomes and for detecting copy number chromosome abnormalities in interphase nuclei.
3. Subtelomeric probes (Figure 4.3) are frequently used to identify subtle or submicroscopic chromosomal rearrangements. The relative ease of performance and high resolution (0.5 Mb) of these unique sequence have made them popular to screen for chromosomal

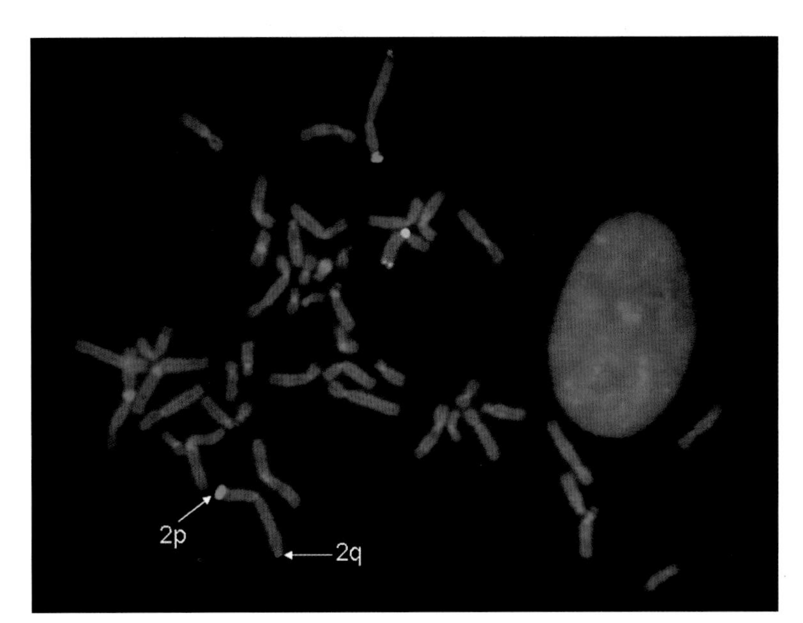

FIGURE 4.3 Subtelomere specific FISH probes are demonstrated in chromosomes 2 and Xp, respectively.

rearrangement probes, subtelomeric deletions, and marker chromosomes.

4. Whole-chromosome painting (WCP) probes (Figure 4.4) are complex DNA probes that are generated by degenerate oligonucleotide polymerase chain reaction or through flow sorting. WCP probes have high affinity for the whole chromosome along its entire length, with the exception of the centromeric and telomeric regions. These probes are most suitable for identifying genomic imbalances in metaphase chromosomes, especially the complex chromosomal arrangements observed in many cancers. Two variants of WCP probes have been developed, multi-color FISH (M-FISH) and spectral karyotyping (SKY). WCP probes have been developed that paint all human chromosomes in different colors (48 paints). And WCP probes can also offer simultaneous detection of each arm of all human chromosomes in a single hybridization. This technique is usually used in conjunction with chromosomal banding techniques for a more precise identification of chromosome aberrations. The two greatest limitations of WCP probes are: (1) extensive knowledge of genetic abnormality to enable the correct selection of the probes and (2) limited resolution to detect chromosomal inversions and very small deletions/amplifications and translocations due to its limited resolution of >2–3 Mb.

Three main probe strategies are utilized in FISH: (1) enumerating probes, (2) fusion probes, and (3) "break-apart" probes.

Enumerating Probe

The *enumerating*, or counting, probe strategy, as its name implies, is useful for counting the number of a particular locus or whole chromosomes within the cell. Counting probes are used to detect gains or losses of whole chromosomes (e.g. chromosomes 5, 7, 8, and 20 in myelodysplastic syndrome, MDS) (Figures 4.5 and 4.6) or deletions and duplications of genes involved in a disease (e.g. *TP53* and *RB1* gene probes in myeloma). These probes can be either BACs containing the gene or genes of interest, or alpha-satellite repeat sequences specific for the centromeric region of each chromosome. This strategy is also useful for detecting cryptic deletions that cannot be detected by classical metaphase chromosome analysis.

Fusion Probe

The fusion probe strategy is classically used to detect translocations or inversions [e.g. the t(9;22) in chronic myeloid leukemia, CML (Figures 4.7 and 4.9) and the t(15;17) in APL]. BAC probes complementary to chromosomal regions involved in the rearrangements are labeled with two different fluorophores (e.g. red and green) and analyzed under the microscope for signal overlap. Normal nuclei will have two red and two green signals, corresponding to the two normal (un-rearranged) chromosomes, while nuclei with rearrangements will have one or more yellow signals, corresponding to the overlap of the red and green signals and suggestive of refused chromosomes. Dual-fusion strategies are used to reduce false-positive signals produced by artifactual overlap caused by the three-dimensional structure of DNA compaction within the nucleus. Dual-fusion approaches utilize probes that overlap the two reciprocal translocation breakpoints and result in two yellow fusion signals corresponding to the two derivative chromosomes. This probe strategy is also useful to distinguish between variants, such as an extra Philadelphia chromosome in CML blast crisis.

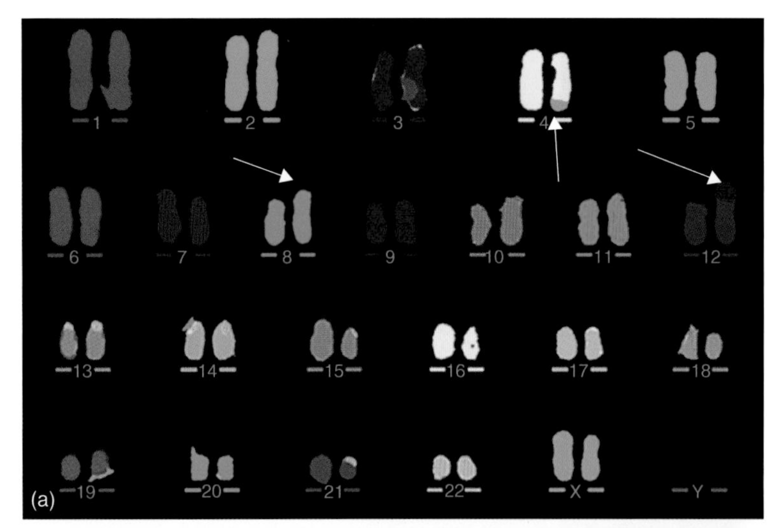

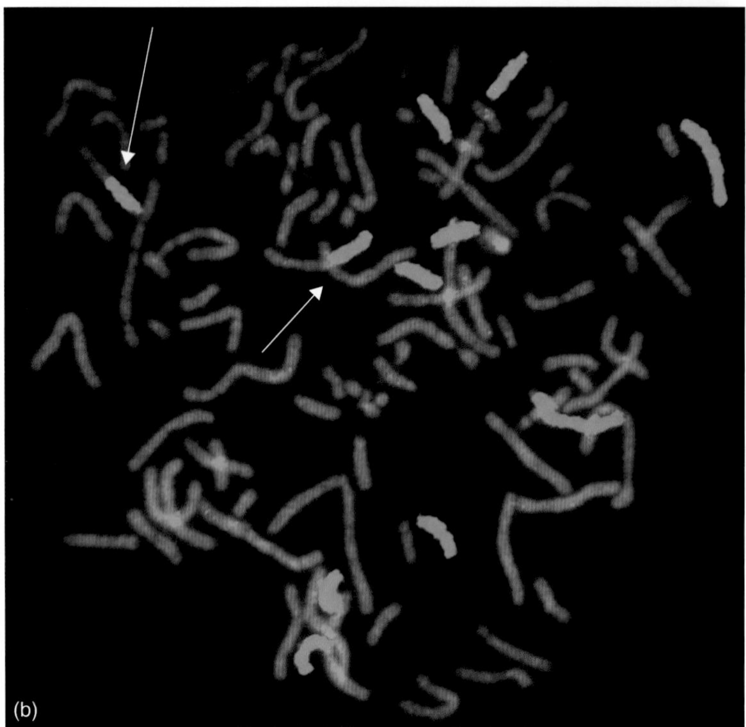

FIGURE 4.4 Chromosome painting. Translocations (arrows) are demonstrated by multi-color FISH technique (a) and chromosome 7 is highlighted with whole chromosome paint (b).

Break-Apart Probe

The break-apart probe strategy (Figure 4.8) is essentially the opposite of the fusion probe strategy and is most useful when a single locus is involved in several different rearrangements (translocations, inversions, deletions, etc.) involving multiple partners. For example, a dual-color FISH probe has been developed with probes on either side of the *MLL* gene breakpoint, resulting in separation of the normally co-localizing signals if the *MLL* gene is rearranged. The advantage of this system is that it can detect all recurrent and possibly novel *MLL* rearrangements in a single experiment. The *MLL* gene locus is involved in >70 recurrent translocations [1], all of which can be detected with the break-apart strategy. Two differently labeled BAC probes (e.g. red and green) normally bind to a single locus and produce the overlapped signal color (e.g. yellow). When the locus of interest is rearranged, the colors split apart. Normal nuclei will have two yellow (overlapped) signals, while nuclei with a rearrangement will have one yellow, one red, and one green signals.

While FISH can provide rapid results and is applicable for various sample types that are otherwise not amenable to classic cytogenetic analyses, it has limitations. FISH will only answer the particular question being asked regarding an exact probed locus. For example, cells probed with a

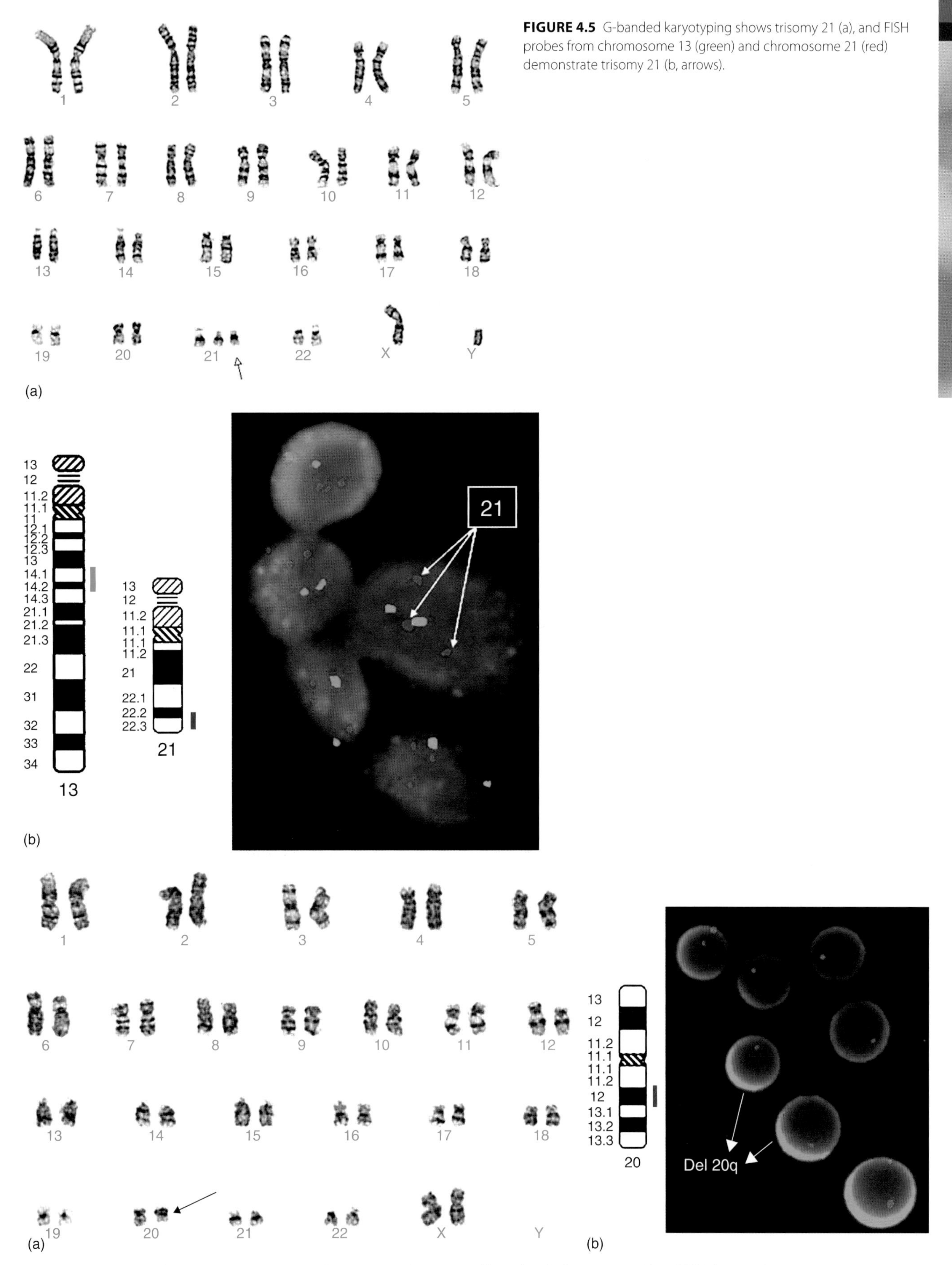

FIGURE 4.5 G-banded karyotyping shows trisomy 21 (a), and FISH probes from chromosome 13 (green) and chromosome 21 (red) demonstrate trisomy 21 (b, arrows).

FIGURE 4.6 Deletion of chromosome 20q, 46,XX,del(20)(q11.2) is demonstrated by G-banded karyotyping (a) and FISH (b, arrows).

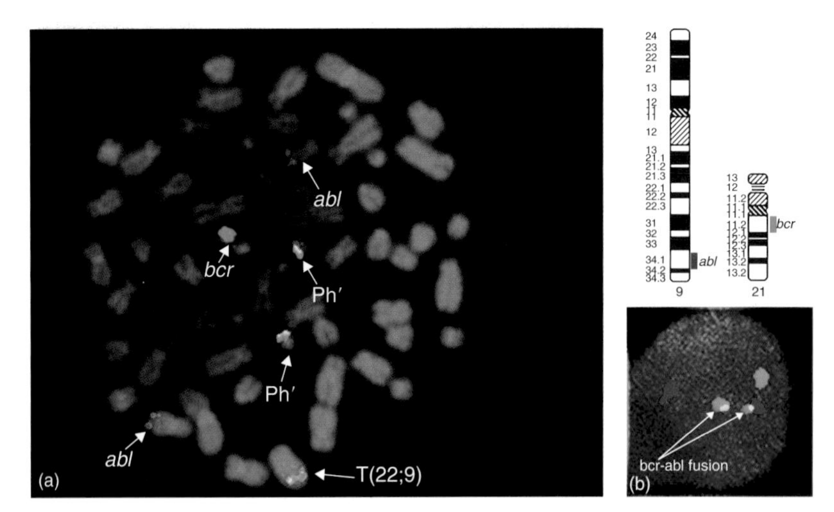

FIGURE 4.7 Metaphase (a) and interphase (b) FISH showing BCR–ABL fusion signals in CML.

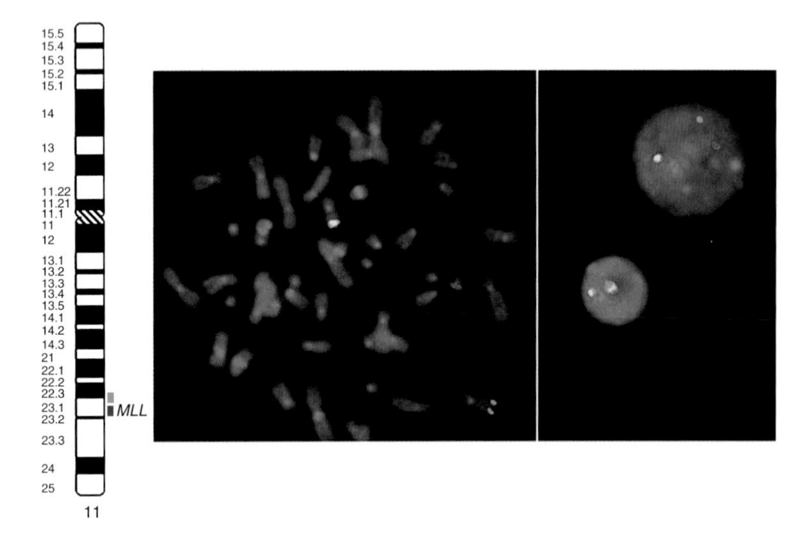

FIGURE 4.8 Break-apart probe signal patterns of MLL locus (yellow) in abnormal cells (b).

BCR–ABL1 fusion probe set may be positive for the t(9;22), but trisomy 8 cells within the sample (often seen in CML-blast crisis) would not be detected unless a chromosome 8 enumerating probe set is used in the probe mix. Similarly, an enumerating probe set consisting only of the alpha repeats from the chromosome 5 centromere will not detect a deletion of the long arm of chromosome 5 (5q−). Physicians ordering tests should be mindful of the questions they are trying to address and order FISH and/or karyotypes appropriately. FISH studies may also be an integral component of the diagnostic work-up if a specific genetic abnormality is suggested by histopathology, peripheral blood counts, or clinical parameters, or when cytogenetic analysis fails or provides a normal karyotypic result. For example, it is recommended that all cases of CML be studied at diagnosis by cytogenetic analysis and molecular cytogenetic methods to determine the initial clonal abnormalities and the FISH signal pattern both for prognostic information and for follow-up studies [2]. When questions arise regarding which FISH test to order, the laboratory should be consulted, as ordering a FISH test without specifying the probe set is inadequate. The application of FISH further provides increased sensitivity, in that chromosomal abnormalities have been detected in samples that appeared to be normal by conventional cytogenetic analysis, e.g. a chromosomally hidden or unidentifiable translocation has been evidenced for the first time in 1995 with the discovery of the t(12;21)(p13;q22), resulting in ETV6/AML1 fusion (Figure 4.9). Molecular cytogenetic investigations with probes specific for this gene rearrangement revealed the translocation to be present in 25% of all pediatric B-ALL [3].

The comparative genomic hybridization (CGH) technique is a relatively new molecular technique for identifying gains and losses in a test sample (e.g. a patient sample), relative to a control sample. DNA is extracted from both the test and control samples and digested with restriction enzymes or sonicated to break it into short (~500 bp) fragments. The test and control samples are differentially labeled with florescent dyes (e.g. red and green), denatured, and hybridized to metaphase chromosomes. Chromosomal regions that are equally represented in the test and control samples will hybridize equally to the chromosomes and produce an overlapped color (e.g. yellow). A loss (deletion) is detected when the control DNA fluoresces stronger within a region of the metaphase chromosomes, and a gain (duplication) is detected when the patient DNA fluoresces stronger.

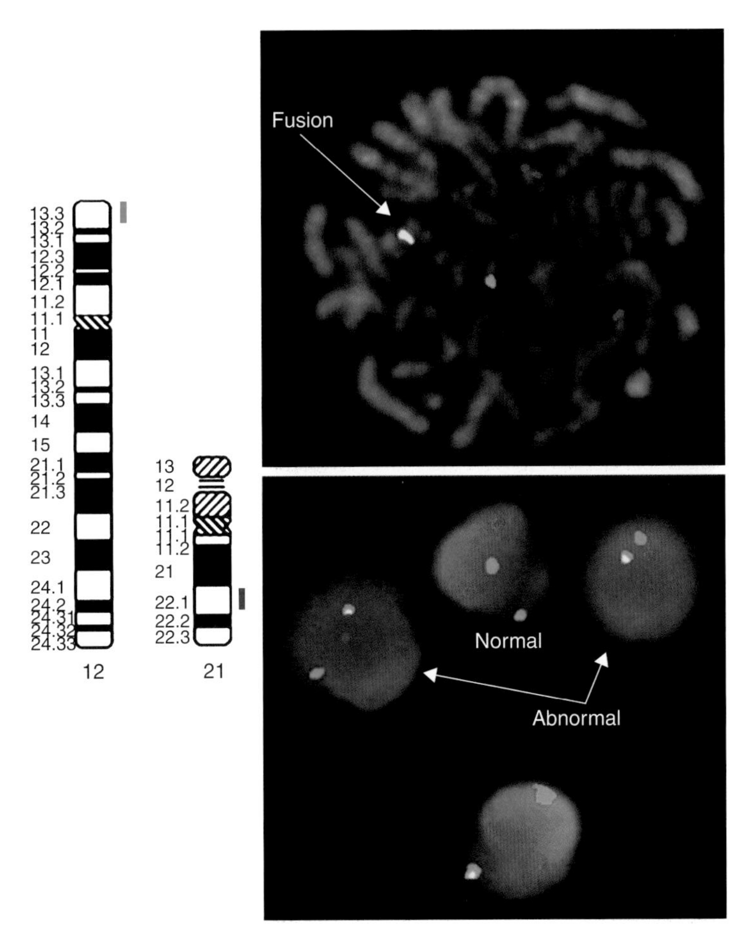

FIGURE 4.9 Demonstration of t(12;21): The fusion of *ETV6* (red) and *RUNX1* (green) genes demonstrates a yellow spot by dual-color FISH.

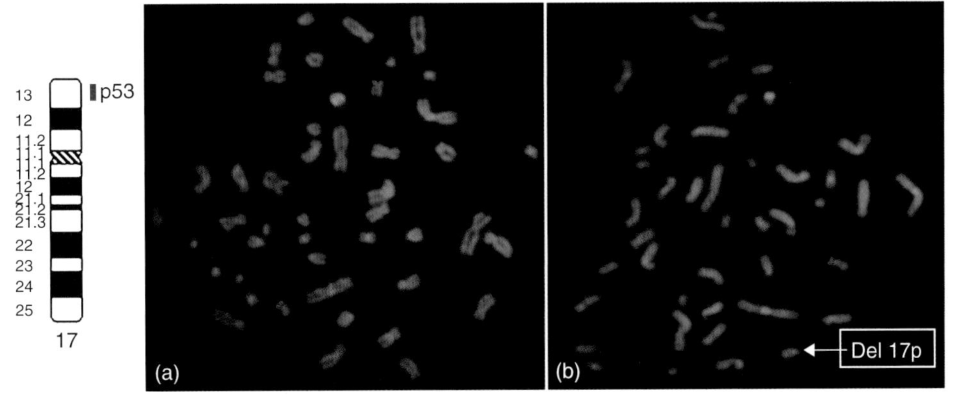

FIGURE 4.10 A normal metaphase with p53 FISH probe reveals normal 17p (two red signals (a)) and (b) a metaphase with deletion of 17p.

In array CGH (aCGH), the fluorescently labeled test and control samples are hybridized to an array of DNA sequences [e.g. BACs (Figure 4.11) or oligonucleotides (Figure 4.12)] rather than metaphase chromosomes. Array CGH has a much higher resolution than classic metaphase CGH. Although CGH and aCGH have been well established for use in detecting submicroscopic gains and losses in constitutional (inherited) disease [4], neither is currently appropriately established as a stand-alone technology for diagnosis in cancer. One reason is that the CGH technique cannot detect balanced chromosomal aberrations (translocations and inversions), which are very common in cancers, especially in hematological disorders. Also, because of tumor heterogeneity and general view of genomic instability, i.e. several clonal populations, CGH and aCGH do not necessarily provide a narrow and consistent genomic regions of interest that can be definitely implicated or identify previously unknown genomic regions of primary etiology. However, with improvements

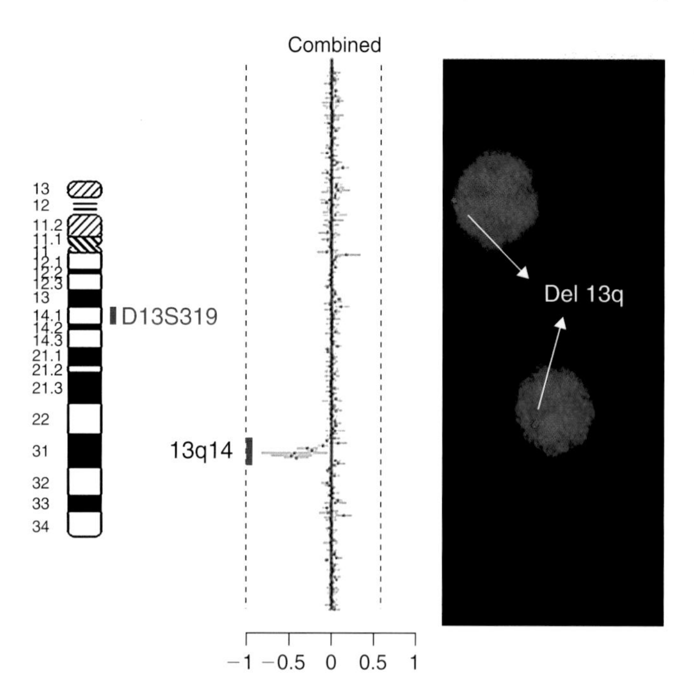

FIGURE 4.11 A whole-genome BAC-array CGH showing a deletion of 13q.

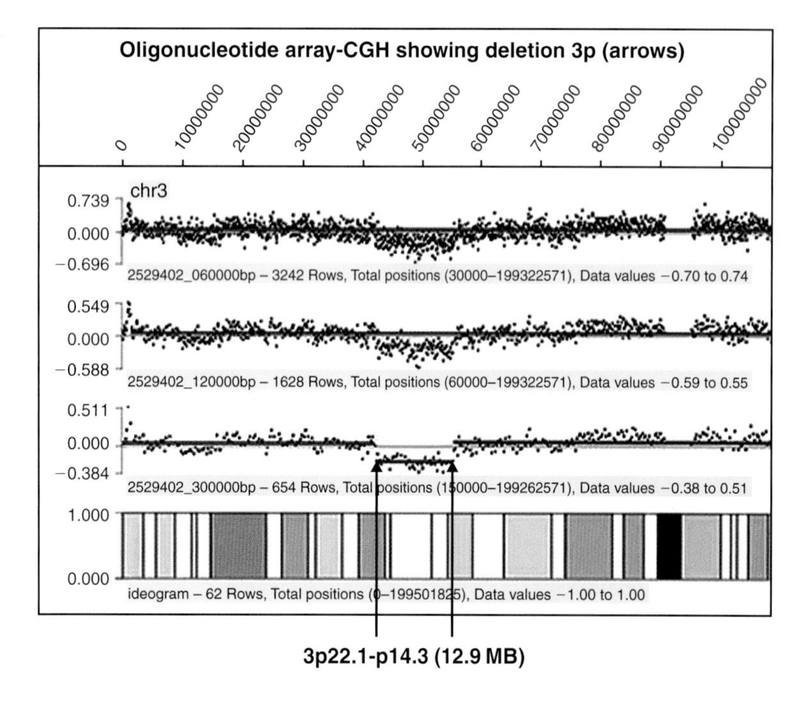

FIGURE 4.12 Shift at 42150000…55050000 bp; chromosome 3p22-p14.3 (12.9 MB).

in technology, software analyses and database collection, CGH will be one of the most useful techniques adopted by diagnostic laboratories [5]. Indeed, it would be interesting to see if aCGH enhances the efficiency of detecting subtle changes such as partial tandem duplication/amplification of the *MLL* gene observed in acute myeloid leukemia (AML). Furthermore, in chronic diseases such as MDS or CML, where a progressively evolving karyotype relates to worsening prognosis that may be used in therapeutic decisions, aCGH would provide very useful comparative genome-wide information in sequential clinical specimens.

Fluorescence *in situ* hybridization is also a useful tool to monitor remission status when clonal chromosome abnormalities have been identified at diagnosis and appropriate probes are available. For CML, sequential FISH studies are particularly useful to determine changes in clinical status in response to therapy and to assess for minimal residual disease. In patients with sex-mismatched bone marrow

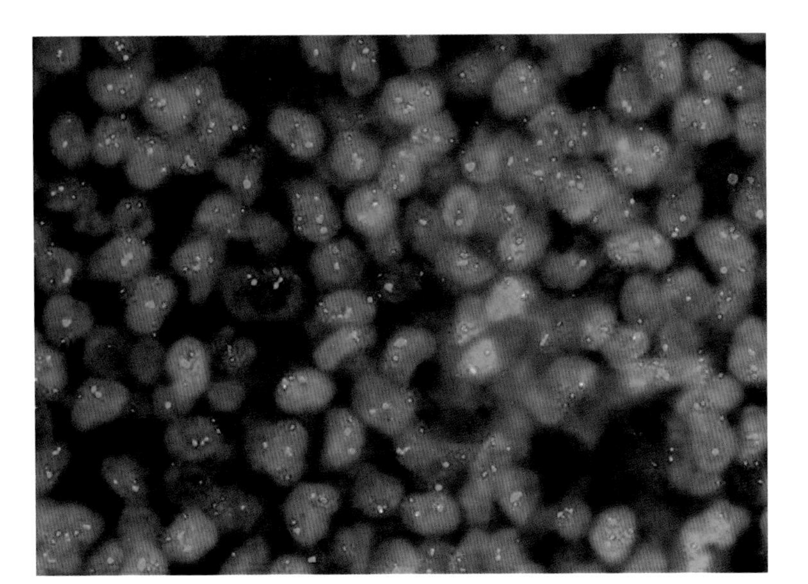

FIGURE 4.13 A typical image of FISH analysis on FFPE section showing overlapping truncated cells.

transplants for whom graft rejection, marrow suppression, or disease relapse is a clinical consideration, monitoring with a FISH assay that combines sex chromosome probes with or without probes to detect the patient's clonal abnormality can be valuable for graft assessment and for detecting residual or recurrent disease.

Fluorescence *in situ* hybridization may be performed on FFPE sections by hybridizing directly either to unstained thin sections (2–4 microns) of tissue that have been deparaffinized or to thick sections of tissue from which individual cell suspensions are made, to which standard FISH techniques may then be applied. Sections must be mounted on slides that will reduce the loss of tissue during FISH-pretreatment processes. Fixation in buffered formalin (pH 7.0) is the best fixative for FISH hybridization. However, fixation in solutions that contain a heavy metal or picric acid often results in unsuccessful FISH results. There are several limitations to FFPE FISH, including overlapping and truncated cells, making assessment of individual cells difficult (Figure 4.13). Also, improvements in the pretreatment, hybridization, optimal probe size, and postwash techniques will alleviate the commonly encountered problems such as background (nonspecific binding) or absence of probe signals. FISH on frozen-tissue sections will yield good results if the slides are first slowly thawed at room temperature and then fixed in 10% buffered formalin.

There are only a few commercially manufactured probe kits that have been approved by the Food and Drug Administration (FDA) for *in vitro* diagnostic testing. These FISH kits must meet the sensitivity and specificity parameters stated in package inserts provided by the manufacturer. The majority of probes used for clinical FISH testing are considered Analyte Specific Reagents (ASRs) that are exempt from FDA approval. When a new ASR probe is introduced in the laboratory, extensive validation is needed, including specific validation of the probe itself (*probe validation*) and validation of the procedures utilizing the probe (*analytical validation*) [6, 7]. Initially, it is important to become familiar with a probe's parameters including signal intensity and pattern and any cross-hybridization that is likely to confound test results. Probe sensitivity, defined as the percentage of metaphases with the expected signal pattern at the correct chromosomal location, should be established. Likewise, probe specificity, defined as the percentage of signals that hybridize to the correct locus and no other location, must also be assessed. Probes used for hematologic malignancy studies should have a high analytic sensitivity and specificity (>95%), particularly if they are to be used for minimal residual disease assessment.

Fluorescence *in situ* hybridization results should be interpreted within the broader context of probe and analytical validation [7]. The interpretation of FISH results should include consideration of the reason for referral for testing and, when available, additional laboratory findings including conventional cytogenetic analysis, hematopathology, and immunophenotyping [8]. When acute promyelocytic leukemia (APL) is the suspected diagnosis, FISH should be performed on a STAT basis with 24–48 h turnaround time to allow for timely treatment with all-trans-retinoic acid (ATRA) [9].

A system for FISH nomenclature, including both metaphase and interphase analysis, has been developed [10]. While the system may seem confusing to those not working directly with chromosomes, correct nomenclature designations are important to convey the precise nature of a result. For example, metaphase FISH ISCN (International Society of Chromosome Nomenclature) for a male patient with a 9;22 translocation resulting in fusion of the *BCR* and *ABL1* genes studied with conventional banding and with a dual-color, single-fusion *BCR/ABL1* probe set would be written: 46,XY,t(9;22)(q34;q11.2).ish t(9;22)(*ABL1*-;*BCR*+,*ABL1*+) indicating that the probe sequence from the *ABL1* locus is missing from the derivative chromosome 9 and is present on the derivative chromosome 22 distal to the *BCR* locus.

The same rearrangement expressed in interphase FISH nomenclature but using a dual-color, dual-fusion probe set would be: nuc ish 9q34(*ABL1* × 3), 22q11.2(*BCR* × 3)(*ABL1* con *BCR* × 2), indicating that each of the probes has been split apart and juxtaposed by the translocation. The use of such precise ISCN is valued by laboratories in the initial diagnostic work-up and continued monitoring of patients with a specific chromosome abnormality. The report must indicate any specific limitations of the assay, some of which may be described in the probe manufacture's package insert.

While interphase FISH analysis provides information only on specific probes used and generally does not substitute for complete karyotype analysis, it may, under some disease circumstances, be the preferred means of identifying an abnormal clone, e.g. FISH with the *ATM, CEP12, D13S319*, and *TP53* probe panel in B-cell chronic lymphocytic leukemia (CLL) [11], discrimination of the inversion 16 in AML (M4-Eo) [12], or FISH for the diagnostic abnormality in post-therapy patients who have hypocellular marrows.

Fluorescence *in situ* hybridization has now become an invaluable tool in defining and monitoring acquired chromosome abnormalities associated with hematologic and other neoplasias. The implementation of the technology into the routine diagnostic laboratory requires rigorous attention to when it is appropriate to apply the technology, a very systematic approach to the validation of probes and technical procedures involved in FISH and training of individuals who will perform the testing, and a comprehensive, but plain and simple, means of reporting out results. As the number of critical loci involved in neoplastic chromosome rearrangements or numeric abnormalities continues to expand, the diversity of FISH probes and unique probe sets will undoubtedly increase. FISH has become an important means both for the definition of the initial chromosome changes in a disease process and a reliable means for the ongoing monitoring of response to therapy and disease remission.

Apart from the diagnostic approaches, FISH can be used as a research tool to refine the breakpoint regions of novel chromosome abnormalities, which is often an essential step in the identification of new (partner) genes involved in leukemogenesis. Recent progress of the Human Genome Project has facilitated the characterization of the translocation breakpoints using FISH. PAC/BAC resources, covering the entire genome are available and can be easily found using the databases of the University of California, Santa Cruz and the National Center for Biotechnology Information (http://genome.ucsc.edu/, http://www.ensembl.org/, http://www.ncbi.nlm.nih.gov/genome/guide/human/). These clones can be used to determine more precisely the breakpoint regions and to search for genes involved in the translocations.

The brief overview of the genetic tools and strategies described earlier have been applied to cancer for fewer than 50 years, but they have quickly been recognized to be invaluable in the study and diagnosis of malignancy. The importance of cytogenetics in oncology is evidenced by the reclassification of certain hematological diseases by the World Health Organization, and the application of both classical and molecular cytogenetic methods to hematologic diseases will be presented throughout this book.

PCR AND RELATED TECHNIQUES

It is impossible to overstate the degree to which the advent of nucleic acid amplification techniques, especially the PCR, has revolutionized the molecular approaches to hematopathologic diagnosis and molecular diagnostics generally. PCR and reverse transcriptase PCR (RT-PCR) not only enable robust analysis of scant or degraded specimens, but also allow for precise quantitation of the analyte, fine dissection of a particular locus or sequence from among the over 3 billion nucleotides of the human genome, and access to otherwise difficult specimens such as FFPE tissue biopsies. These techniques have largely replaced the much more laborious and time-consuming Southern blot (discussed later) for most, but not all, applications in hematopathology. At a more basic level, they have also replaced difficult DNA-cloning procedures for many purposes, including the generation of DNA probes used in a variety of downstream applications such as the Southern blot. On the other hand, they have a number of limitations and technical pitfalls, one of which is that the sequence of portions of the target gene must be known, which is not necessarily the case for certain cloning experiments.

Basic Technique

Polymerase chain reaction is so simple in its conception that it is surprising no one in molecular biology happened to think of it before the seminal paper by Saiki *et al.* appeared in 1988 [13]. Fundamentally it merely mimics replication of DNA *in vivo*, using essentially the same enzymes (DNA polymerases), but in a highly specific and exponential fashion. The specificity results from the use of specific *primer* sequences that are constructed to be complementary only to the target gene or region of interest. The exponential amplification results from the use of a pair of primers which flank the target region and hybridize to opposite strands of the DNA (Figure 4.14). Serving as start-sites for the polymerase reaction, they promote a bidirectional synthesis of daughter strands that then become templates for the next round of replication. The replication cycles are controlled by alternate heating and cooling of the sample, denaturing the amplification products and then enabling the primers (which are present in great excess) to re-hybridize and begin the replication again. Since each replication cycle doubles the number of template molecules, the products accumulate in exponential, rather than linear, fashion. A typical experiment encompasses about 30 cycles, performed in a programmable heating/cooling instrument called a *thermal cycler*, which produces amplification of many millionfold. The precise timing and temperature settings will vary depending on the target sequence and application but generally fall in the range of about 94°C for denaturation and 50–60°C for renaturation. The elongation step, in which the actual DNA synthesis occurs, is run at 70–75°C. The reason it is not performed at physiologic temperature is that in modern PCR the DNA polymerases used are cloned from a variety of thermophilic microorganisms (e.g. *Thermus aquaticus*, source of so-called *Taq polymerase*) so that they do not degrade during each denaturation step [14].

Primer Design

For PCR to deliver the specificity desired, it is most important that primer sequences be chosen carefully so they do

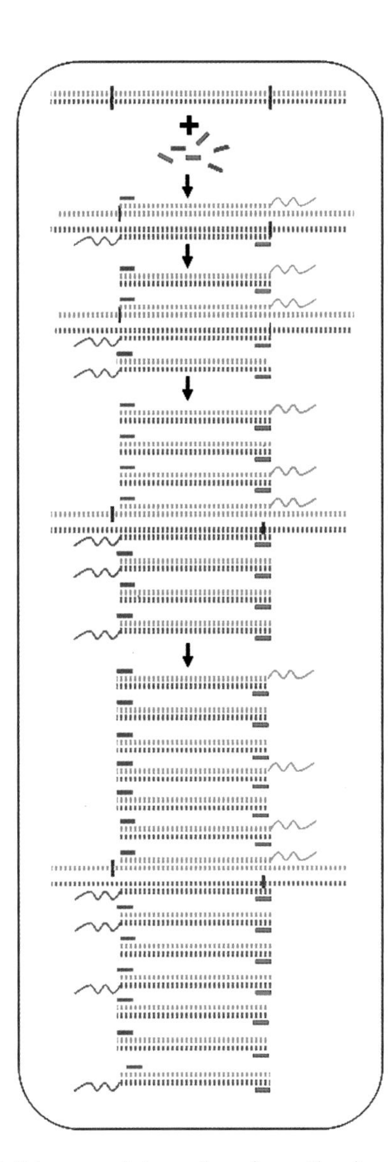

FIGURE 4.14 Polymerase chain reaction scheme. The oligonucleotide primers (rectangles) are designed to hybridize to opposite stands some distance apart on the target region of interest. When hybridized, they serve as priming sites for DNA polymerase replication of each strand. Alternate heating and cooling allows for multiple cycles of denaturation, rehybridization, and replication, increasing the amount of the target sequence exponentially.

not cross-hybridize with other regions of the genome at the renaturation temperatures used. Primers are typically about 18–25 nucleotides in length, and there are computer programs available on the Internet to work out optimal base composition such that the primers do not hybridize elsewhere or to themselves. For most clinical purposes, primers are designed to hybridize a few hundred nucleotides apart in the target sequence, since replication efficiency declines as target length increases. A number of long-range PCR protocols and commercial kits are available, allowing for amplification of targets many thousands of base pairs in length [15], but these are mostly reserved for research purposes.

Quality Control

Given the awesome sensitivity of PCR, theoretically down to a single-target DNA molecule [16], extreme care must be exercised to guard against the production of spurious amplification products by contaminant target molecules. In a clinical laboratory, these can come from other patient specimens, from the operator himself, or from residual amplification products of a previous assay. In practice, it is this last source that is the most concerning, since these products are present in infinitely greater excess than stray genomic contaminants from an individual specimen. A large number of precautionary and preventative procedures have been developed to guard against it, including one-way workflow from pre- to post-amplification areas, special dedicated pipets, re-gloving, and degradation of residual amplicons by ultraviolet light and other methods. Failing this, contamination is detected by the running of a "no DNA" or "no template" control tube in every PCR assay. This tube contains all the requisite enzymes and nucleotides for the reaction to proceed, but no target template to copy. If any amplicon is observed in this sample after PCR, it indicates a contamination problem with that run or with one of the reagents. In addition, known positive controls must be run with each assay, to demonstrate capability of the PCR conditions and reagents to detect and amplify the desired target sequence when present.

Product Analysis

For most applications, PCR is not an end in itself but rather a first step in a subsequent assay. Once the desired target has been amplified, it may be analyzed by hybridization with specific DNA probes in a dot blot format, by size fractionation in agarose gel or capillary electrophoresis, by digestion with sequence-specific restriction endonucleases, or by DNA sequencing. The choice of downstream technique will depend on the nature of the disease process being tested, as will become clear from the specific clinical examples presented throughout this book.

Reverse Transcriptases PCR

Certain disease applications, for example the detection of the *BCR–ABL* translocation in CML (see Chapter 7), involve detection and accurate quantification of an RNA, rather than a DNA, target. Since PCR uses *DNA* polymerases which do not replicate or transcribe RNA, amplification of such targets can only be accomplished by first converting them to DNA. Fortunately, there are viral-derived enzymes available that can do this, termed *reverse transcriptases* (RT). Thus, amplification and study of RNA targets utilizes the technique of RT-PCR, in which the first step is reverse transcription, followed then by conventional PCR of the resulting DNA products.

Real-Time PCR

As noted earlier, in most cases PCR amplification is the first step in an assay that then utilizes another method to analyze or quantify the products. This is because conventional thermal cyclers are closed systems in which amplification proceeds without observation until it is finished and the reaction tubes removed. In contrast, a newer generation of thermal cycler instruments can measure the accumulation of amplicon as it occurs (i.e. in real time) by measuring the incorporation of a fluorescent label into the elongating products. These instruments can provide very precise quantification of products and, by extrapolation, the amount of starting target material, in the sample. They are very useful for cancer applications in hematopathology designed to detect minimal residual disease after therapy.

Related Amplification Techniques

A large number of innovations to the basic PCR technique have been developed over the years to address particular applications or to circumvent certain pitfalls. Included are such techniques as nested PCR, whole-genome amplification, inverse PCR, hot-start PCR, and allele-specific PCR. For the most part they are beyond the scope of this chapter but will be mentioned in the context of particular disease applications where relevant elsewhere in the book. In addition, a number of non-PCR amplification techniques have been developed over the years, such as Qβ replicase, and ligate chain reaction, but for the most part they have fallen by the wayside in favor of PCR, at least for applications relevant to hematopathology (some are used in molecular microbiology and genetics testing).

BLOTTING TECHNIQUES

In blotting techniques, unique segments of nucleic acid sequences (DNA probes) are used to demonstrate the presence of complementary sequence of DNA or RNA in the sample. Since the complementary target is composed of hundreds or thousands of nucleotide bases, the reaction of the DNA probe to the target (hybridization) is the tightest and most specific intermolecular interaction [17].

Southern Blot

In this technique the DNA probes are labeled with a radioactive or nonradioactive signal moiety. The DNA target is treated with restriction enzymes (endonucleases), and the DNA fragments are separated by gel electrophoresis (usually an agarose gel). The DNA is then transferred to a membrane which is a sheet of special blotting paper. The blot is incubated with numerous copies of a single-stranded labeled DNA probe. The probe hybridizes with its complementary DNA sequence within the target sample to form a labeled double-stranded DNA molecule. The radioactively labeled copy of the DNA probe is then detected by autoradiography (Figure 4.15). In nonradioactive DNA labeling procedures, nucleotides (probes) are conjugated with biotin or other protein binders such as digoxigenin. Biotin binds specifically to the protein avidin with a very high affinity. Avidin is a polyvalent protein which can be linked to chromogenic enzymes, fluorescent compounds or electron-dense particles. The advantage of a radiolabeled probe over a nonradiolabeled probe is its higher sensitivity (5- to 10- folds), and its disadvantages are the elongated radiography step, which may extend to several days, radiation hazard, and requirement for special procedures for disposal of the radioactive contaminated wastes [17, 18].

Northern Blot

Northern blot is a technique basically similar to Southern blot, except that it is used to transfer RNA from a gel to a blot instead of DNA [19].

Dot Blot

Dot blot is similar to the other blotting techniques, except that it does not provide information regarding the size of

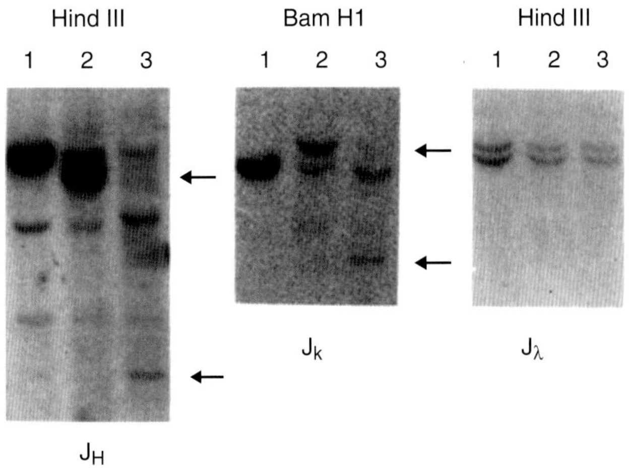

FIGURE 4.15 Southern blot analysis for Ig gene rearrangements demonstrating rearranged bands (arrows) for the joining region of the Ig heavy chain (J_H) and the joining region of the kappa light chain (J_k).

the hybridized fragment. With this technique, extracted DNA or RNA from the target specimen is spotted onto the filter without the prior electrophoresis and transfer steps.

MICROARRAY TECHNIQUES

The next generation of blotting techniques is a marriage of molecular biology and information technology: the microarray. In contrast to a dot blot or Southern blot in which target DNA is hybridized to one or a few probes, microarrays enable the hybridization of hundreds or hundreds of thousands of target sequences simultaneously. Because it is constructed at nanoscale using technology similar to that used to manufacture silicon-based computer chips [20], the vernacular term "DNA chip" is often applied. Essentially it is a reverse dot blot on a grand (and miniaturized) scale. The individual probes, in great numbers, are bound to the solid support, while the specimen DNA (or RNA) is hybridized to them after being labeled with a fluorescent marker (Figure 4.16). A chip-reader instrument interprets these signals and generates data output detailing which sequences hybridized and which did not, as well as the intensity of hybridization (which relates to the relative amount of each particular target sequence in the material being tested).

There are two basic kinds of microarrays: DNA sequence arrays and RNA expression arrays. The former are used to detect the presence of mutations and other sequence variants in the tested sample, whereas the latter are used to assay relative expression of hundreds or thousands of genes, as for example in comparing tumor mRNA to that in a corresponding normal tissue. Depending on how many probes are placed on the array, assays of varying comprehensiveness can be developed, even to the extreme of comprising the entire human genome. Microarrays can even be used for sequencing, if every possible target sequence is encompassed by the probes on the chip.

In a sense, microarrays represent one area where technology has outpaced biological knowledge, at least in the clinical setting. One would be hard pressed to come up with a disease state, in hematopathology or elsewhere, in which our knowledge is so extensive that we need to assay the sequence or expression of 20,000 genes, or even for that matter 200. For this reason, microarray technology remains, at time of this writing, largely in the research sphere, with few accepted clinical applications. However, most people recognize that this is indeed the technology of the future and will find many applications in the clinical laboratory as soon as our knowledge catches up to its technical capabilities. At the same time it must be kept in mind that such comprehensive whole-genome scanning raises a number of ethical issues related to defects and predictive risks that might unwittingly be revealed when applied to patients [21].

DNA SEQUENCING

In many ways DNA sequencing is the most definitive molecular biology technique because it gets directly at the genetic code of the specimen being analyzed, and that is what is at the core of everything else, both biologically and clinically. Other techniques, such as probe hybridization, restriction endonuclease digestion and even PCR, are surrogate assays whose outcome depends ultimately on the inherent sequence of the target material but which do not ascertain that sequence directly. For this reason DNA sequencing is often referred to as the "gold standard" for molecular testing in the clinical setting.

Like other techniques in molecular diagnostics, DNA sequencing arose in the research setting to satisfy research needs and resulted in Nobel Prizes for its inventors. Initially cumbersome, the techniques have been refined over the years, recently under the impetus of the Human Genome Project which demanded extremely-high-throughput, accurate, and inexpensive sequence analysis. The benefits of these

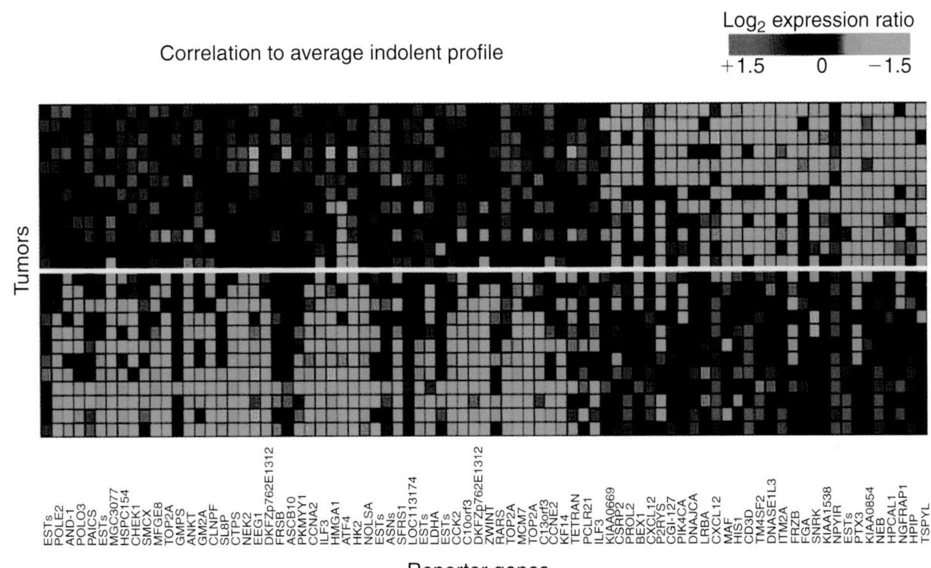

FIGURE 4.16 Gene expression profiling in follicular lymphomas using a microarray technique. Results demonstrate two populations of patients: those with aggressive disease are segregated above the solid yellow line and those with indolent clinical course are placed below it. This research was orginally published in *Blood*; Glas AM, Kersten MJ, Delahaye LJ, Witteveen AT, Kibbelaar RE, Velds A, Wessels LF, Joosten P, Kerkhoven RM, Bernards R, van Krieken JH, Kluin PM, van't Veer LJ, de Jong D. (2005). *Blood* **105**, 301–307, by permission.

innovations now spill over into the clinical arena where sequencing, previously considered a kind of "last resort" methodology for special cases, is now routine for a wide variety of applications.

Chemical Methods

Two basic approaches to sequencing were initially developed and used, but one of them has now assumed dominance for both manual and automated platforms. The chemical degradation method, developed by Maxam and Gilbert [22], uses various chemicals to cleave the double helix, based on their reaction with specific nucleotides. Sizing and aligning the resulting fragments by electrophoresis allows one to determine at which point along the length of the helix each nucleotide was positioned. This method has largely been abandoned for routine uses because of its technical difficulty and the noxious nature of the chemicals required.

It has been supplanted by the chain termination method, developed by Sanger *et al.* [23] at about the same time. This one, too, is based on fragment sizes determining the positions of each nucleotide, but the fragments are created not by degradation but by synthesis. Four DNA polymerase reactions are set up using the target sample as the template for replication. In each reaction tube, in addition to the four regular nucleotide substrates, a dideoxynucleotide derivative is also added. When one of these is incorporated into the growing daughter strand by the polymerase, replication is halted at that point because the dideoxynucleotide lacks the 3′-hydroxyl group required for chemical linkage to the next nucleotide that would otherwise be added. Since the derivative nucleotides are present in the minority, they are not inserted at every spot but rather at a random distribution whenever that nucleotide is required. This produces a complete spread of fragments whose size is based solely on the positions of that particular nucleotide in the original target material. If the fragments from each nucleotide reaction are separated based on size by electrophoresis, a "ladder" is created that reveals in order, from shortest to longest, the nucleotide sequence.

Sequence Detection and Analysis

The electrophoresis can be done in a flat-bed (usually vertical) gel set-up, typically a long polyacrylamide gel, using radioactive nucleotides that can be seen by autoradiography of the dried gel after the run. But more and more clinical laboratories have moved to automated sequencing instruments which use capillary electrophoresis of fluorescently labeled oligonucleotides to generate the sequence. The reactions are carried out in a single tube instead of four, and the various reactions are discriminated because each of the four dideoxynucleotides is labeled with a different-colored fluorophore. These instruments provide more precise sizing, much higher throughput, and sophisticated software for calling out and analyzing the sequence (Figure 4.17).

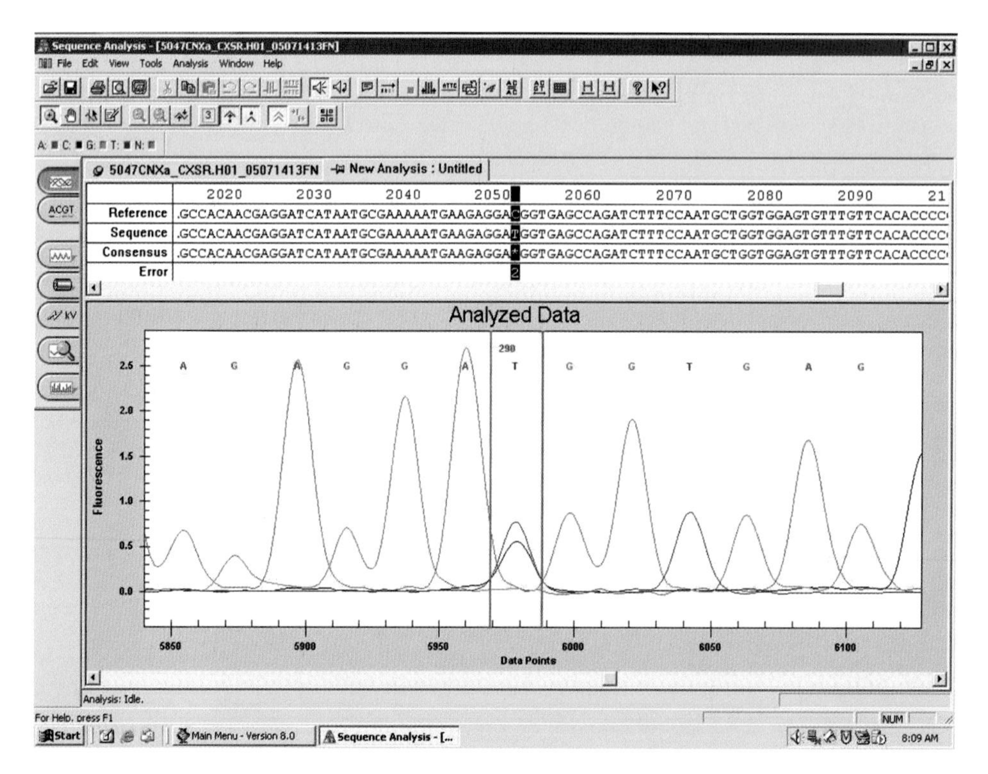

FIGURE 4.17 Example of the read-out from an automated capillary electrophoresis DNA sequencer. Each colored peak represents the capture of a fluorescently labeled DNA fragment by the laser detector of the instrument, in order of size fractionation. Each dideoxynucleotide in the reactions was labeled with a different-colored fluorophore, represented by the colors of the read-out and nucleotide calls by the instrument. Note the heterozygous single-nucleotide substitution (C→T) at position 290 (bracketed by the pink vertical lines).

Limitations of Sequencing

While it may be the "gold standard", DNA sequencing, like any other technique, has its limitations and pitfalls. For many applications in which only one particular mutation or size fragment is being probed, sequencing may represent "overkill" in that it yields a tremendous amount of extraneous data not relevant to the clinical question being asked. Some of that data, moreover, can be of questionable clinical relevance. Scattered throughout the genome of every human being are countless nonpathologic nucleotide substitutions, called *polymorphisms*. When they occur within protein coding regions (exons), they appear as missense mutations, causing the substitution of one amino acid for another in the gene product. If a sequencing test reveals one of these changes that has not been seen or reported in the literature before, it can be very difficult or impossible to decide whether it represents a pathologic missense mutation or merely a benign polymorphism. Certain physical attributes, such as its position within the protein and the biochemical nature of the amino acid substitution, may help one to deduce its impact, though there are many exceptions to these rules. Correlation with phenotype, presence or absence of the change in other affected or unaffected family members, or the population at large, can also be of help. Ultimately, functional studies of the altered gene *in vitro* may be needed, but these are not applicable to a clinical laboratory.

Aside from the clinical interpretation, even the detection of nucleotide substitutions in the heterozygous state may be problematic. Ironically, this seems to be more of an issue with automated sequencers than with manual systems which depend more on the eye of the operator. A heterozygous substitution will appear on the automated read-out as two differently colored peaks superimposed at the same nucleotide position (see Figure 4.17). For various technical reasons which are difficult to control, the two peaks may not be of the same intensity, even though they are supposedly present in equal amounts in the specimen. If one of the peaks is much smaller than the 20–30% level, it may be ignored by the instrument's software and called out as homozygous for the other nucleotide. One way to guard against missing a heterozygous change in this way is to perform the sequencing in both directions (i.e. using opposite strands of the same fragment as template) and compare the read-outs. It is unlikely that the instrument would miss the change both times.

References

1. Huret JL. (2005). MLL (myeloid/lymphoid or mixed lineage leukemia). *Atlas Genet Cytogenet Oncol Haematol2*.
2. Dewald G, Stallard R, Alsaadi A, Arnold S, Blough R, Ceperich TM, Rafael Elejalde B, Fiunk J JV, Higgins RR. (2000). A multicenter investigation with D-FISH BCR/ABL1 probes. *Cancer Genet Cytogenet* **116**, 97–104.
3. Bernard OA, Romana SP, Poirel H, Berger R. (1996). Molecular cytogenetics of t(12;21) (p13;q22). *Leuk Lymphoma* **23**, 459–65.
4. Shaffer LG, Bejjani BA. (2005). Medical applications of array CGH and the transformation of clinical cytogenetics. *Cytogenet Genome Res* **115**, 303–9.
5. Davies JJ, Wilson IM, Lam WL. (2005). Array CGH technologies and their applications to cancer genomes. *Chromosome Res* **13**, 237–48.
6. Test and Technology Transfer Committee (2000). Technical and clinical assessment of fluorescence *in situ* hybridization: An ACMG/ASHG position statement. I. Technical considerations. *Genet Med* **2**, 356–61.
7. American College of Medical Genetics. *Standards and Guidelines for Clinical Genetic Laboratories, Section E: Clinical Cytogenetics*. www.acmg.net.
8. Mrozek M, Heerema NA, Bloomfield CD. (2004). Cytogenetics in acute leukemia. *Blood Rev* **18**, 115–36.
9. Schad CR, Hanson CA, Paietta E, Casper J, Jalal SM, Dewald GW. (1994). Efficacy of fluorescence *in situ* hybridization for detecting PML/RARA gene fusion in treated and untreated acute promyelocytic leukemia. *Mayo Clin Proc* **69**, 1047–53.
10. International Society of Chromosome Nomenclature (ISCN) (2005). *An International System for Human Cytogenetic Nomenclature* (Shaffer LG, Tommerup N, eds). S Karger, Basel.
11. Shanafelt TD, Geyer SM, Kay NE. (2004). Prognosis at diagnosis: Integrating molecular, biologic insights into clinical practice for patients with CLL. *Blood* **103**, 1202–10.
12. Ravandi F, Kadkol SS, Ridgeway J, Bruno A, Dodge C, Lindgren V. (2003). Molecular identification of CBFbeta-MYH11 fusion transcripts in an AML M4 Eo patient in the absence of inv16 or other abnormality by cytogenetic and FISH analyses – a rare occurrence. *Leukemia* **17**, 1907–10.
13. Saiki RK, Gelfand DH, Stoffel S, Scharf SJ, Higuchi R, Horn GT, Mullis KB, Erlich HA. (1988). Primer-directed enzymatic amplification of DNA with a thermostable DNA polymerase. *Science* **239**, 487–91.
14. Cline J, Braman JC, Hogrefe HH. (1996). PCR fidelity of Pfu DNA polymerase and other thermostable DNA polymerases. *Nucleic Acids Res* **24**, 3546–51.
15. Cheng S, Fockler C, Barnes WM, Higuchi R. (1994). Effective amplification of long targets from cloned inserts and human genomic DNA. *Proc Natl Acad Sci USA* **91**, 5695–9.
16. Li HH, Gyllensten UB, Cui XF, Saiki RK, Erlich HA, Arnheim N. (1988). Amplification and analysis of DNA sequences in single human sperm and diploid cells. *Nature* **335**, 414–17.
17. Grody WW, Gatti RA, Naeim F. (1989). Diagnostic molecular pathology. *Mod Pathol* **2**, 553–68.
18. Southern EM. (1975). Detection of specific sequences among DNA fragments separated by gel electrophoresis. *J Mol Biol* **98**, 503–17.
19. Alwine JC, Kemp DJ, Parker BA, Reiser J, Renart J, Stark GR, Wahl GM. (1979). Detection of specific RNAs or specific fragments of DNA by fractionation in gels and transfer to diazobenzyloxymethyl paper. *Meth Enzymol* **68**, 220–42.
20. Cheung VG, Morley M, Aguilar F, Massimi A, Kucherlapati R, Childs G. (1999). Making and reading microarrays. *Nat Genet* **21**(Suppl), 15–19.
21. Grody WW. (2003). Ethical issues raised by genetic testing with oligonucleotide microarrays. *Mol Biotechnol* **23**, 127–38.
22. Maxam AM, Gilbert W. (1977). A new method for sequencing DNA. *Proc Natl Acad Sci USA* **74**, 560–4.
23. Sanger F, Nicklen S, Coulson AR. (1977). DNA sequencing with chain terminating inhibitors. *Proc Natl Acad Sci USA* **74**, 5463–7.

Morphology of Abnormal Bone Marrow

Faramarz Naeim

This chapter briefly describes morphologic features of bone marrow lesions that will not be covered in the following chapters. These lesions include gelatinous transformation of bone marrow, bone marrow necrosis, amyloidosis, granulomas, metastatic lesions, post-therapeutic changes, and, finally, bone and stromal changes.

GELATINOUS TRANSFORMATION

Gelatinous transformation of bone marrow is characterized by fat atrophy, accumulation of hyaluronic acid (gelatinous material), and bone marrow hypoplasia [1–3]. This condition has been observed in various chronic debilitating conditions such as anorexia nervosa and other forms of severe starvations, as well as in malignancies, tuberculosis, chronic renal failure ulcerative colitis, and AIDS [4–7].

The bone marrow involvement is often patchy, and the gelatinous amorphous, glassy deposit appears as a light blue substance by H&E stain and reacts positively with alcian blue and periodic acid Schiff (PAS) stains (Figure 5.1). Bone marrow smears stained with Wright's stain show a bluish-pink material mixed with adipocytes (Figure 5.1c).

The reversal of gelatinous transformation has been observed in patients with anorexia nervosa when their nutritional status has been improved [8].

BONE MARROW NECROSIS

Extensive bone marrow necrosis is infrequent and is usually associated with hematologic disorders, such as leukemias/lymphomas and sickle cell anemia, bone marrow metastasis, or infections [9–13]. The cause of bone marrow necrosis could be attributed to the following factors [14–19]:

1. Vascular occlusion caused by deformed red cells (sickle cell anemia), fibrin clot (disseminated intravascular coagulation, thrombotic thrombocytopenic purpura), or vascular occlusion by fungi (such as mucormycosis) or tumor emboli.
2. Inadequate blood supply, such as in severe anemia or hyperparathyroidism.
3. Infection, such as gram-positive and gram-negative bacteria, Q fever, typhoid fever, tuberculosis, diphtheria, and histoplasmosis.
4. Radiation and chemotherapy.
5. Release of tumor necrosis factor (TNF) in cancer patients.

Extensive bone marrow necrosis may be associated with extramedullary hematopoiesis or leukoerythroblastosis (presence of immature erythroid and myeloid cells in the peripheral blood) and pancytopenia. The bone marrow biopsy sections show coagulation or fibrinoid necrosis. The coagulation necrosis is more

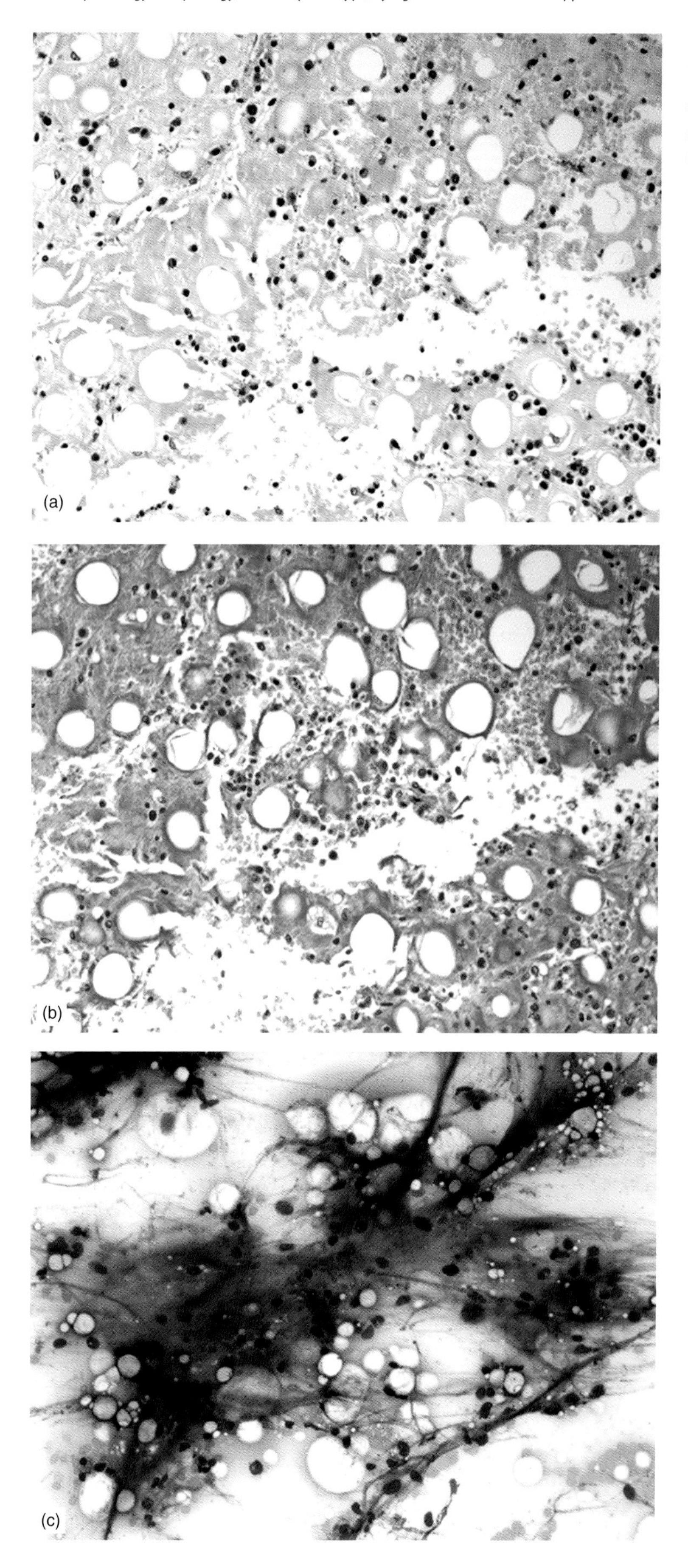

FIGURE 5.1 Gelatinous transformation of fatty tissue: (a) Bone marrow biopsy section demonstrating hypocellularity and partial replacement of the bone marrow fat by an eosinophilic amorphous substance. (b) The substance reacts positively with the alcian blue stain. (c) Bone marrow smear is hypocellular and shows increased cell debris.

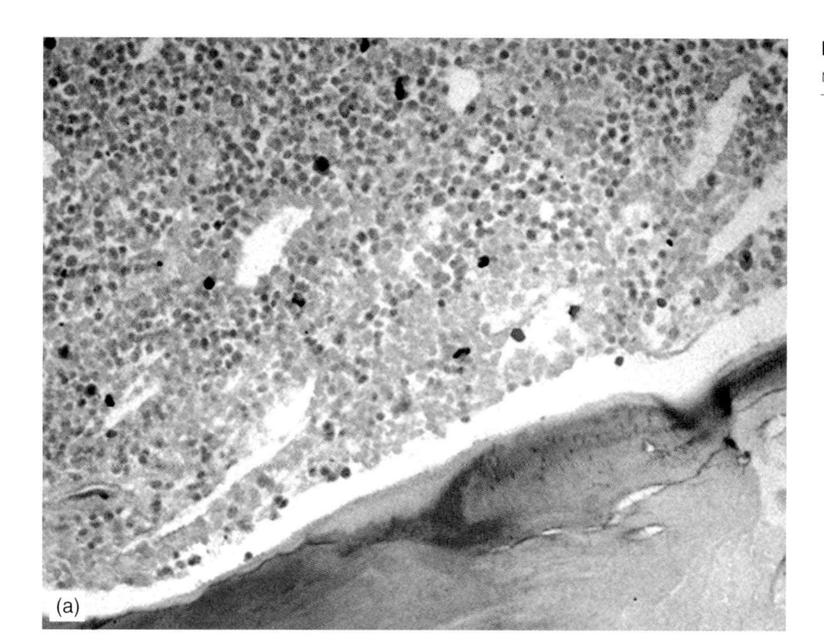

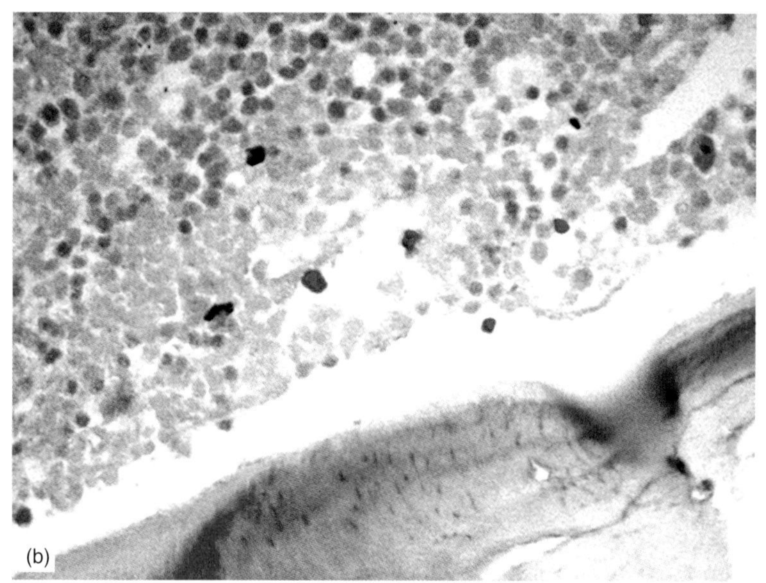

FIGURE 5.2 Low (a) and high (b) power views of a bone marrow biopsy section demonstrating coagulation necrosis. The shadow of the necrotic cells is visible.

frequently observed and is characterized by the preservation of the overall architectural framework of the tissues and the skeleton of the necrotic cells (Figure 5.2). Fibrinoid necrosis consists of the accumulation of an amorphous, granular cell debris and is more often associated with infections. Aspirated necrotic bone marrow contains pyknotic nuclei and blurred outlines of cells in a background of amorphous granular material.

Since hematologic malignancies and metastatic tumors are frequent causes of bone marrow necrosis, the presence of necrosis in bone marrow samples of patients with a history of malignant disease is highly suggestive of bone marrow involvement. In such cases, additional sections and/or samples are recommended.

AMYLOIDOSIS

Amyloid is an extracellular deposit which appears as a hyaline, eosinophilic, amorphous material on H&E sections (Figure 5.3). It predominantly consists of non-branching fibrils composed of polypeptide chains, which on X-ray crystallographic analysis yield a "cross-beta" pleated sheet [20, 21]. Amyloid is often recognized by methyl violet and Congo red histochemical stains. Amyloid deposits appear rose-pink by methyl violet stain and red-orange by Congo red stain. When Congo red stained tissue sections are viewed under a polarizing microscope, amyloid appears as an apple green birefringent deposit (Figure 5.4). It is also

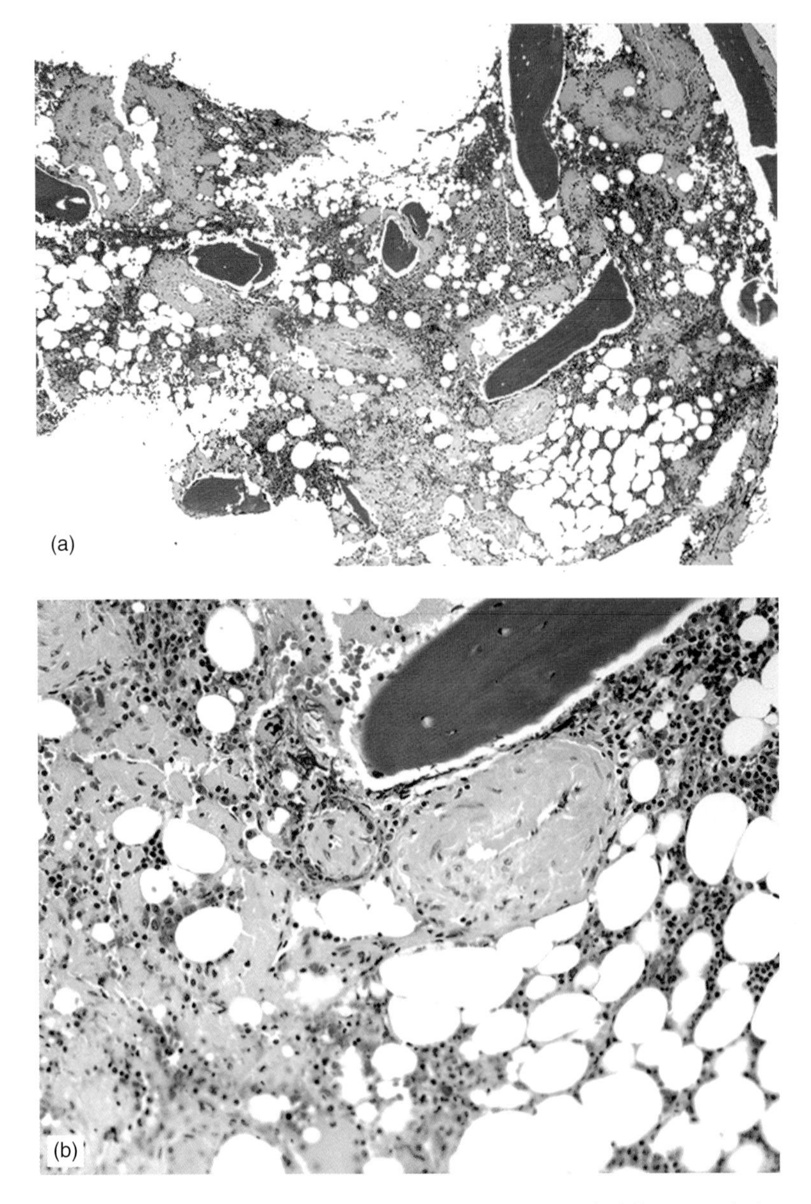

FIGURE 5.3 Low (a) and high (b) power views of a bone marrow biopsy section demonstrating amyloid deposits in the bone marrow space and vessel walls.

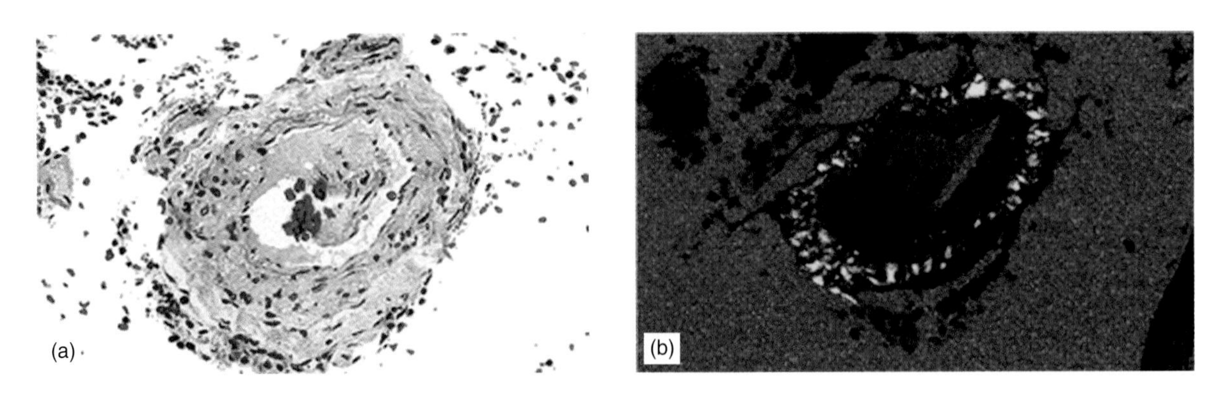

FIGURE 5.4 Amyloidosis. A sclerotic bone marrow blood vessel (a) displays an apple green birefringent deposit by Congo red stain (b).

possible to detect amyloid deposits by using antibodies raised against amyloid components (anti-AA, anti-AL) by immunofluorescence or immunohistochemical techniques.

Amyloid deposits in bone marrow are observed in primary systemic amyloidosis, associated with plasma cell dyscrasias (see Chapter 16), and secondary to chronic inflammatory disorders [22–26]. These deposits are usually multifocal and are found adjacent to or within the walls of small blood vessels. Vascular amyloidosis may lead to ischemia and focal bone marrow hypoplasia. Occasionally, bone marrow involvement with amyloid is extensive.

GRANULOMAS

Granulomas are aggregates of histiocytes which are frequently surrounded by lymphocytes, plasma cells, and eosinophils. Multinucleated giant cells, which are the result of fusion of closely packed histiocytes, may also be present. Granulomas may show focal areas of necrosis.

The most frequent causes of bone marrow granulomas are mycobacterial and fungal infections and sarcoidosis (Figures 5.5–5.8) [27–30]. However, other infectious and non-infectious conditions, such as viral infections (such as HIV and EBV), syphilis, Q fever, typhoid fever, Legionnaires' disease, Hodgkin and non-Hodgkin lymphomas, autoimmune disorders (Figure 5.9), and drug-induced inflammatory responses may lead to the information of granulomas (Table 5.1) [31–34]. Foreign body granulomas are occasionally seen in bone marrow biopsies which are usually the result of drug abuse or prosthetic pelvic operations (Figure 5.10).

The viral-associated granulomas are usually small, rarely show multinucleated giant cells, and are non-necrotizing [33–35]. Bone marrow granulomas in Q fever

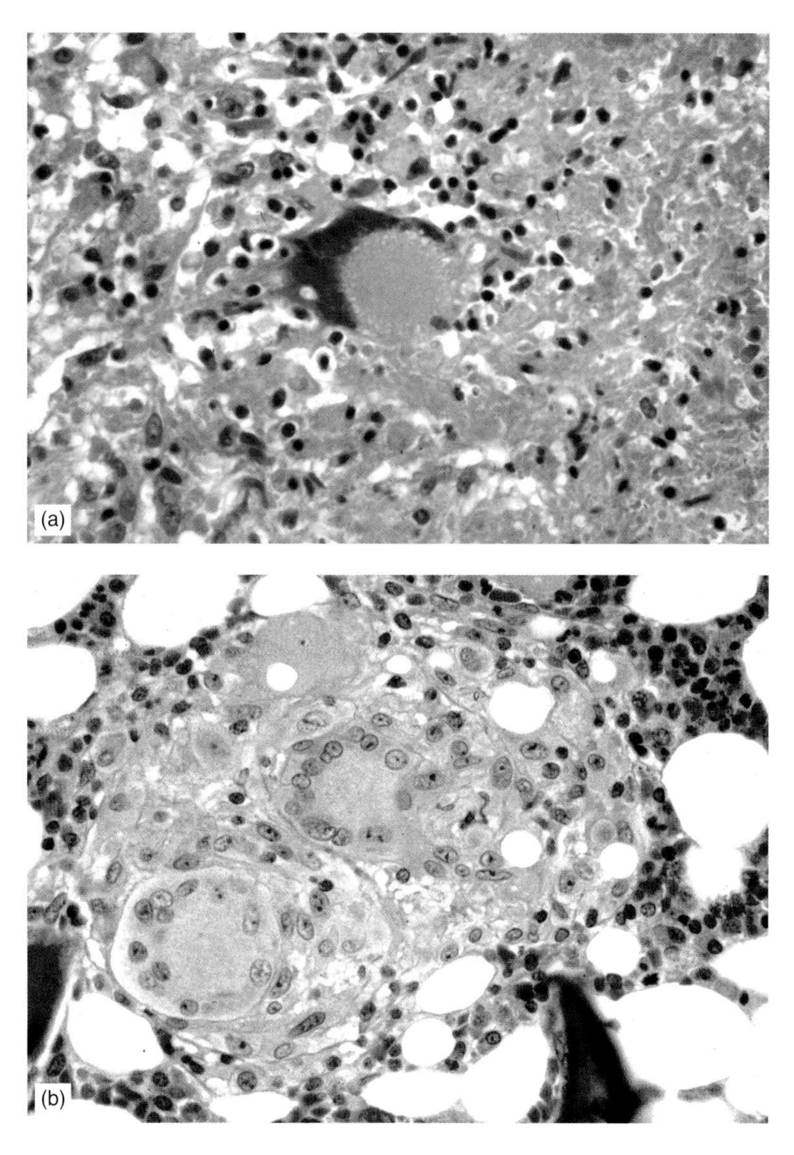

FIGURE 5.5 Bone marrow granulomas. (a) A necrotizing granuloma in a patient with tuberculosis. (b) A non-necrotizing granuloma in a patient with sarcoidosis.

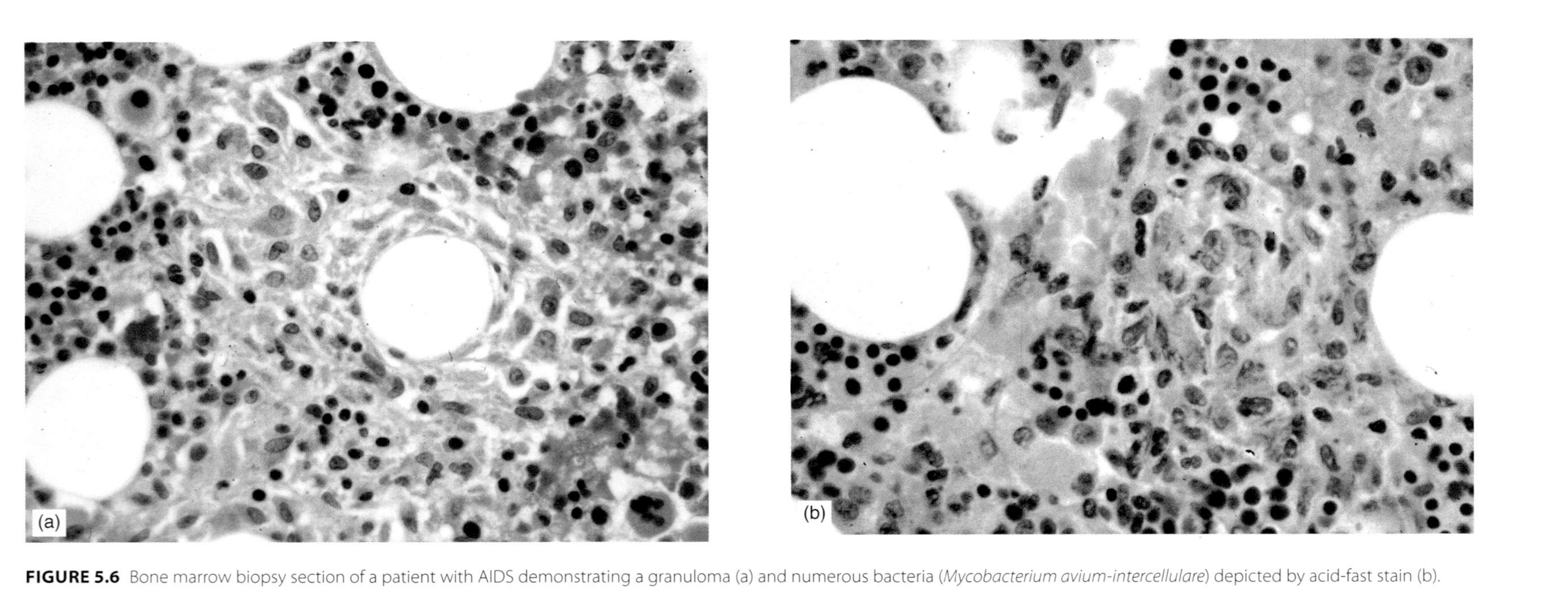

FIGURE 5.6 Bone marrow biopsy section of a patient with AIDS demonstrating a granuloma (a) and numerous bacteria (*Mycobacterium avium-intercellulare*) depicted by acid-fast stain (b).

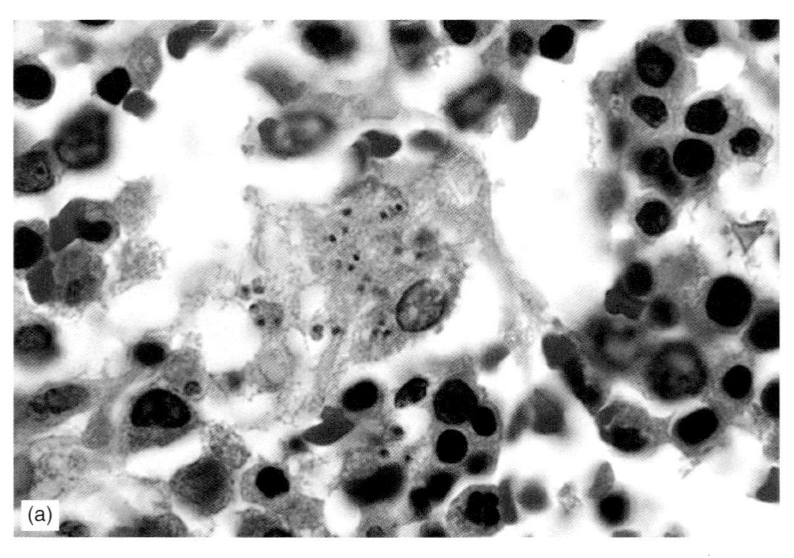

FIGURE 5.7 *Leishmania donovani*. Bone marrow biopsy section (a) and bone marrow smear (b) demonstrating histiocytes containing numerous organisms.

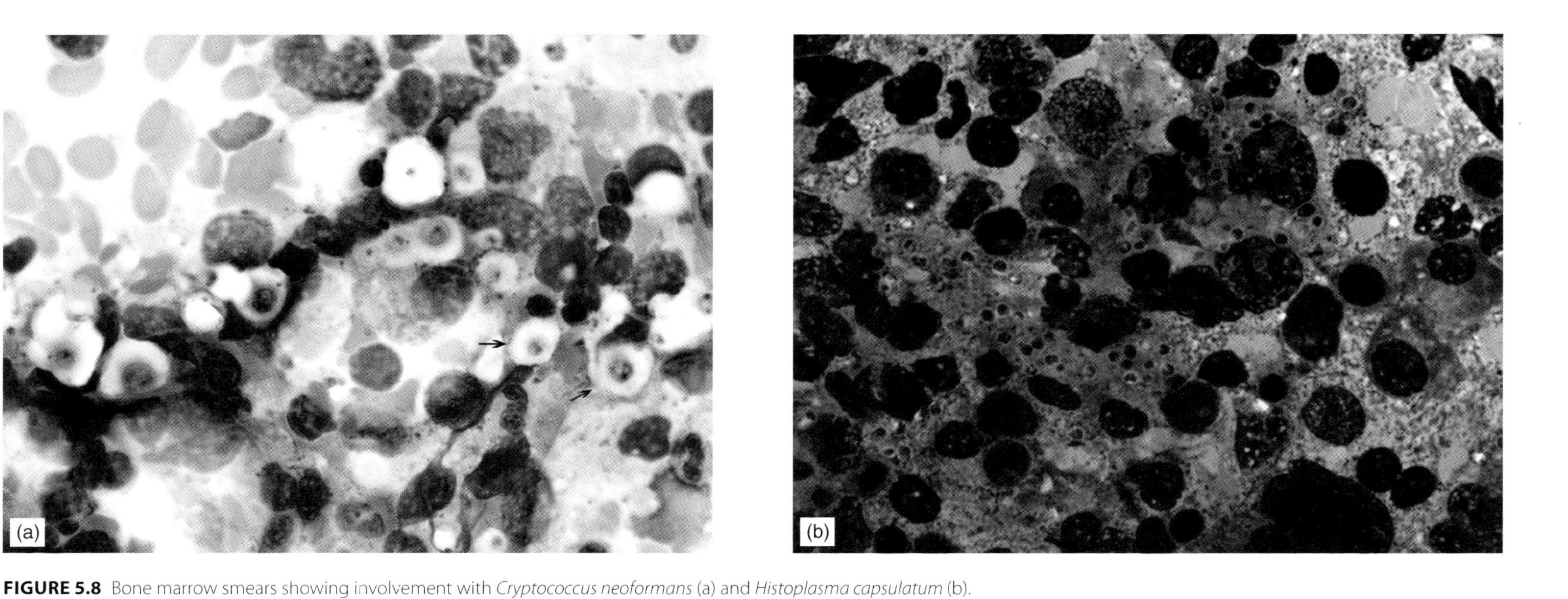

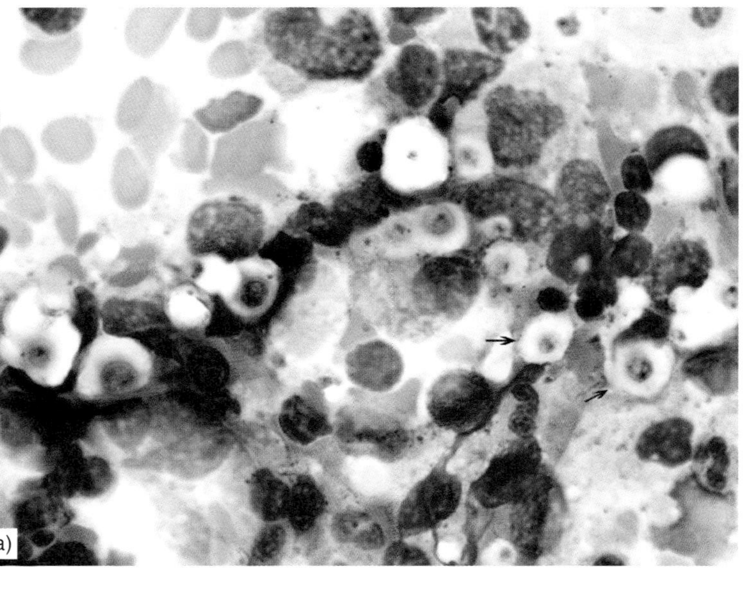

FIGURE 5.8 Bone marrow smears showing involvement with *Cryptococcus neoformans* (a) and *Histoplasma capsulatum* (b).

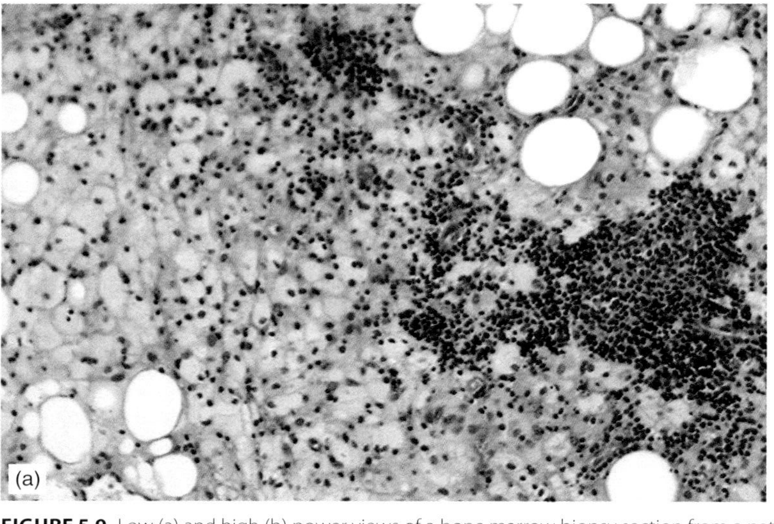

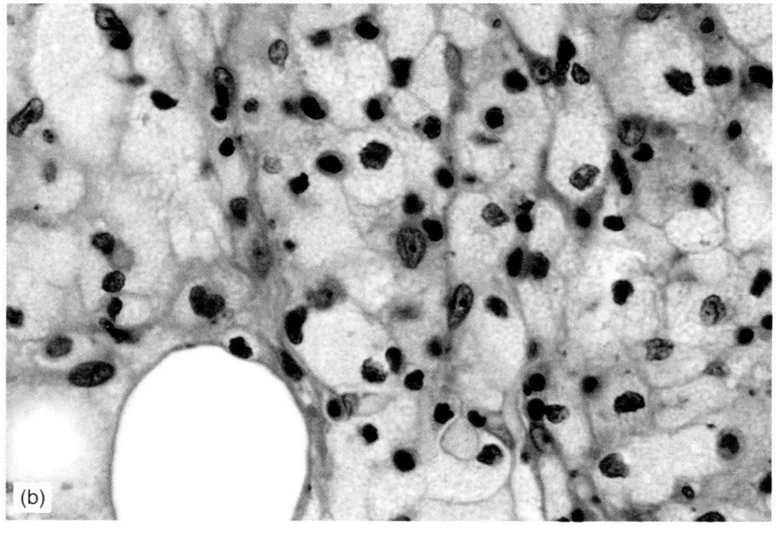

FIGURE 5.9 Low (a) and high (b) power views of a bone marrow biopsy section from a patient with Erdheim–Chester disease demonstrating aggregates of foamy histiocytes.

TABLE 5.1 Conditions associated with bone marrow granulomas.

Infections	Sarcoidosis
Tuberculosis	Hematologic malignancies
Mycobacterium avium-intracellulare	Hodgkin lymphoma
Leprosy	Non-Hodgkin lymphoma
Infectious mononucleosis	Plasma cell myeloma
Cytomegalovirus	
Herpes zoster	Autoimmune disorders
AIDS	
Mycoplasma	Drug-induced
Histoplasmosis	Phenytoin
Cryptococcosis	Procainamide
Brucellosis	Oxyphenbutazone
Typhoid fever	Chlorpropamide
Legionaire's disease	
Q fever	
Rocky Mountain spotted fever	

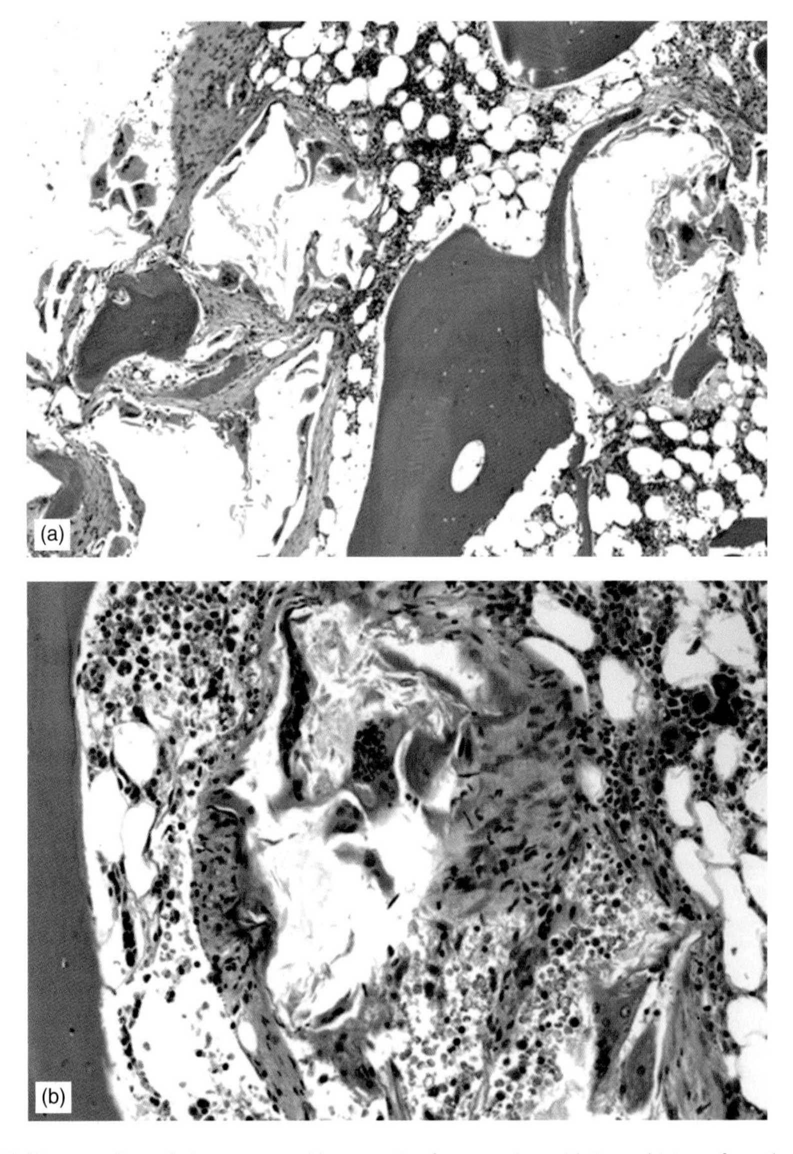

FIGURE 5.10 Low (a) and high (b) power views of a bone marrow biopsy section from a patient with 8-year history of prosthetic pelvic operation showing several cystic structures. These structures contain foreign material surrounded by multinucleated giant cells.

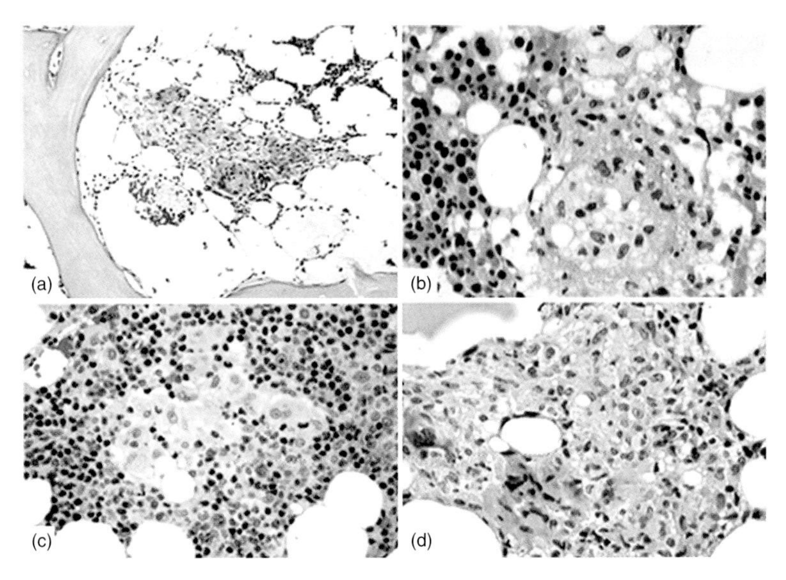

FIGURE 5.11 Small granulomas consisting of epithelioid histiocytes have been observed in the bone marrow of the patients with viral infections (such as HIV and EBV), syphilis, Q fever, typhoid fever, and Legionnaires' disease (a through d).

consist of small clusters of epithelioid histiocytes radially arranged around a fatty tissue containing fibrinoid necrosis (ring granulomas) (Figure 5.11) [36, 37]. The characteristic feature of bone marrow involvement in lepromatous leprosy is the presence of clusters of foamy histiocytes containing lepra bacilli (dirty histiocytes) [38]. Lipogranulomas consist of small aggregates of vacuolated, fat-containing histiocytes in bone marrow sections, frequently observed in association with lymphoid aggregates and plasmacytosis [39].

Granulomas are best detected in bone marrow biopsy sections. It is extremely difficult to detect them in bone marrow smears. A careful search for granulomas is recommended in patients with history of fever of unknown origin, mycobacterial or fungal infections, and in immunocompromised patients. Special stains for mycobacterial and fungal infections are recommended when a granuloma is detected in bone marrow samples.

Certain hematologic disorders, such as peripheral T-cell lymphomas with histiocytic component and metastatic carcinomas composed of cells with abundant cytoplasm (renal cell carcinoma, hepatoblastoma, pheochromocytoma), may resemble granulomas.

BONE MARROW METASTASIS

Bone marrow biopsy sections have a higher detection rate of metastasis than bone marrow clot sections or smears, because metastatic tumor cells trapped in dense fibrosis may not be aspirable. Also, intercellular organization of tumor cells such as glandular structures, rosette formations, and tumor-associated stromal alterations such as fibrosis are usually not detected in bone marrow smears (Figure 5.12). However, it is highly recommended that all various types of bone marrow slides, such as tissue sections, touch preparations, and smears, are thoroughly examined for the detection of metastatic lesions [1, 40–42].

Metastatic tumor cells tend to clump together, are usually larger than erythroid and myeloid precursors, and may show areas of necrosis (Figures 5.13 and 5.14). Reactive changes such as fibrosis, or the presence of macrophages, lymphocytes, plasma cells, and neutrophils, may be seen within or adjacent to the metastatic site [43, 44].

In bone marrow smears, metastatic tumor cells are usually well defined and in clusters, and are often detected at the periphery of the marrow particles. Small round cell tumors (oat cell carcinoma, neuroblastoma, retinoblastoma, rhabdomyosarcoma, Ewing's sarcoma) may resemble hematopoietic blast cells. However, they are usually larger than the hematopoietic blasts, often show significant nuclear pleomorphism, and may also show characteristic morphologic structures, such as organoid pattern (oat cell carcinoma) or rosette formation (neuroblastoma) (Figure 5.14c). In addition, small cell tumors may demonstrate specific features by electron microscopy, immunophenotypic studies, molecular and cytogenetic studies, or *in vitro* culture [45–50]. For example, the diagnosis of a metastatic neuroblastoma is facilitated by the existence of secretory granules by electron microscopy, the presence of cellular catecholamines and neurofilaments by immunohistochemistry, and by demonstrating the outgrowth of neurites in tissue culture.

The presence of bone marrow fibrosis in patients with a history of malignancy is strongly suspicious for metastasis. Metastatic carcinomas, particularly of breast origin, are frequently associated with bone marrow fibrosis. Clusters and strands of tumor cells are often trapped in a dense, highly collagenized, fibrous tissue. Sometimes, fibrosis may make the detection of small metastatic foci difficult. In such cases, examination of additional tissue sections may be helpful. As mentioned earlier, metastatic carcinomas composed of cells with abundant cytoplasm may resemble granulomas.

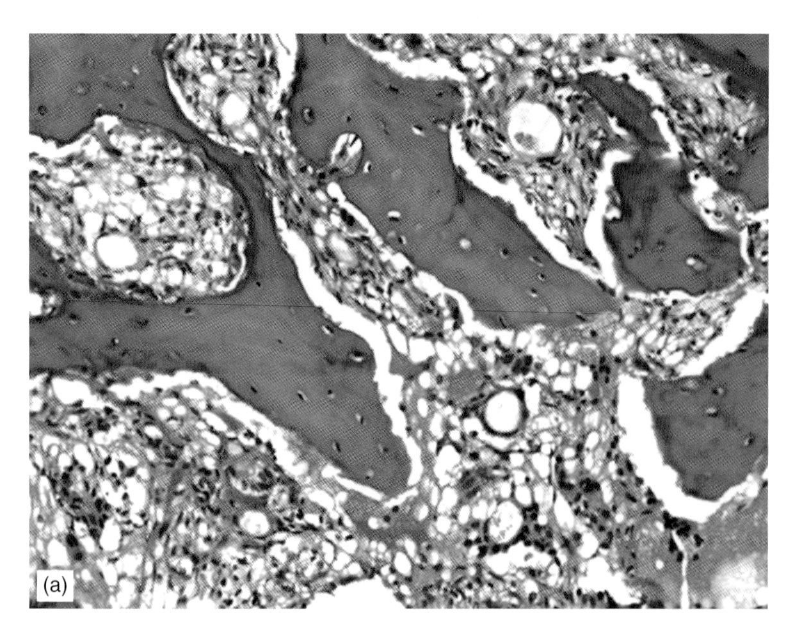

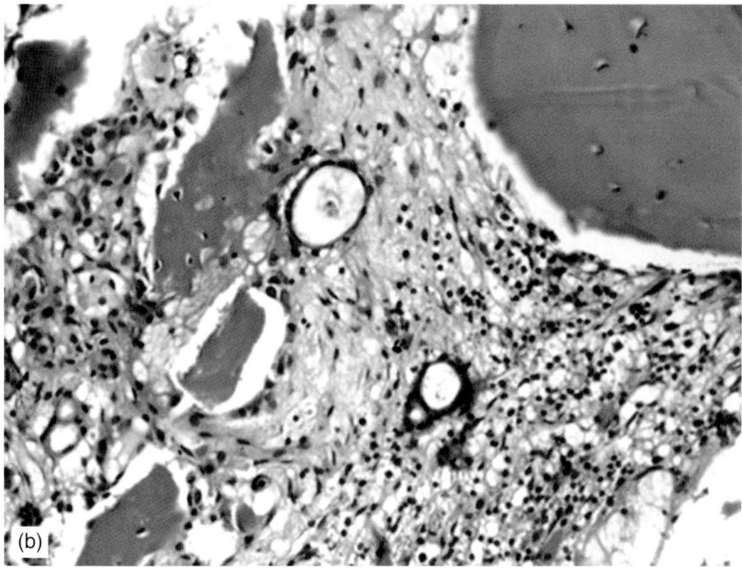

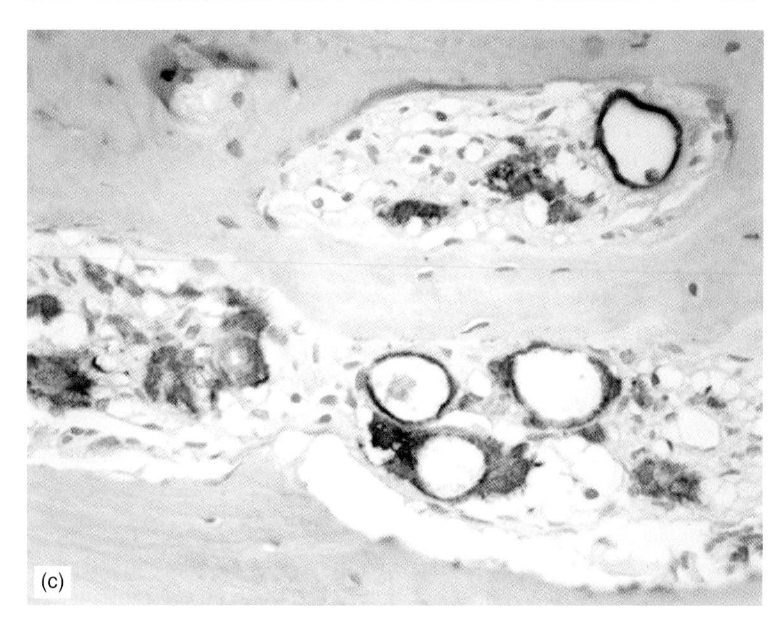

FIGURE 5.12 Bone marrow biopsy section, (a) low power and (b) high power, demonstrating sclerosis of the bone trabeculae and extensive fibrosis with scattered metastatic glandular structures expressing cytokeratin (c).

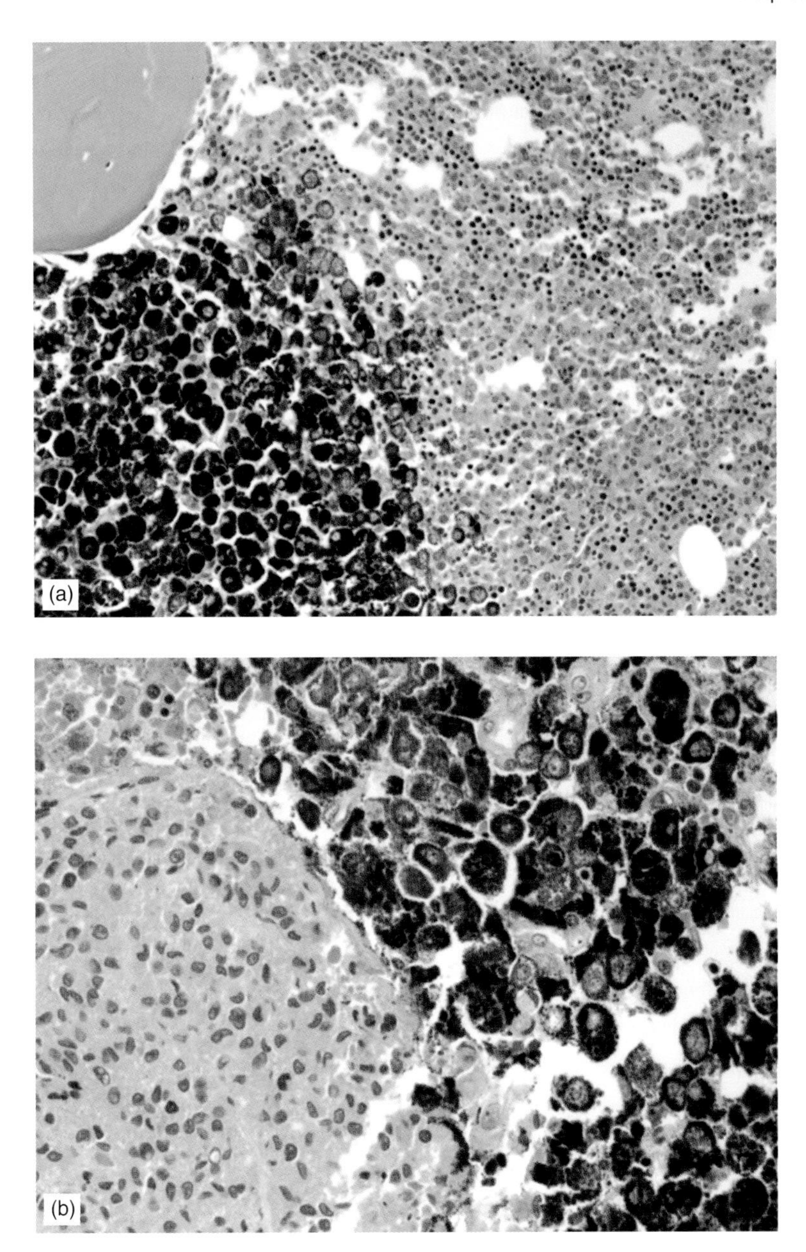

FIGURE 5.13 Bone marrow biopsy section depicting metastatic melanoma (a). The melanoma consists of pigmented and non-pigmented components (b).

Immunophenotypic studies are useful in (1) distinguishing metastatic neoplasms from primary bone marrow malignancies and reactive proliferations, (2) classification of metastatic tumors and their possible primary sites, and (3) detection of small, occult metastatic lesions [45, 47]. For example, tumors of epithelial origin express cytokeratin and are negative for CD45; metastatic prostate carcinoma is positive for prostate specific antigen (PSA) and prostatic acid phosphatase, and metastatic rhabdomyosarcoma may demonstrate myosin, desmin, and/or myoglobin.

POST-THERAPEUTIC CHANGES

The morphologic changes of the bone marrow after chemotherapy and/or irradiation are the result of rapidly progressive cellular death and a transient ineffective hematopoiesis. These changes include marked hypocellularity, fibrinoid necrosis, edema, dilated sinuses, multilobulated adipocytes, new bone formation, mild to moderate increase in reticulin fibers, and increased number of macrophages, frequently with phagocytic particles (Figures 5.15 and 5.16) [1, 51].

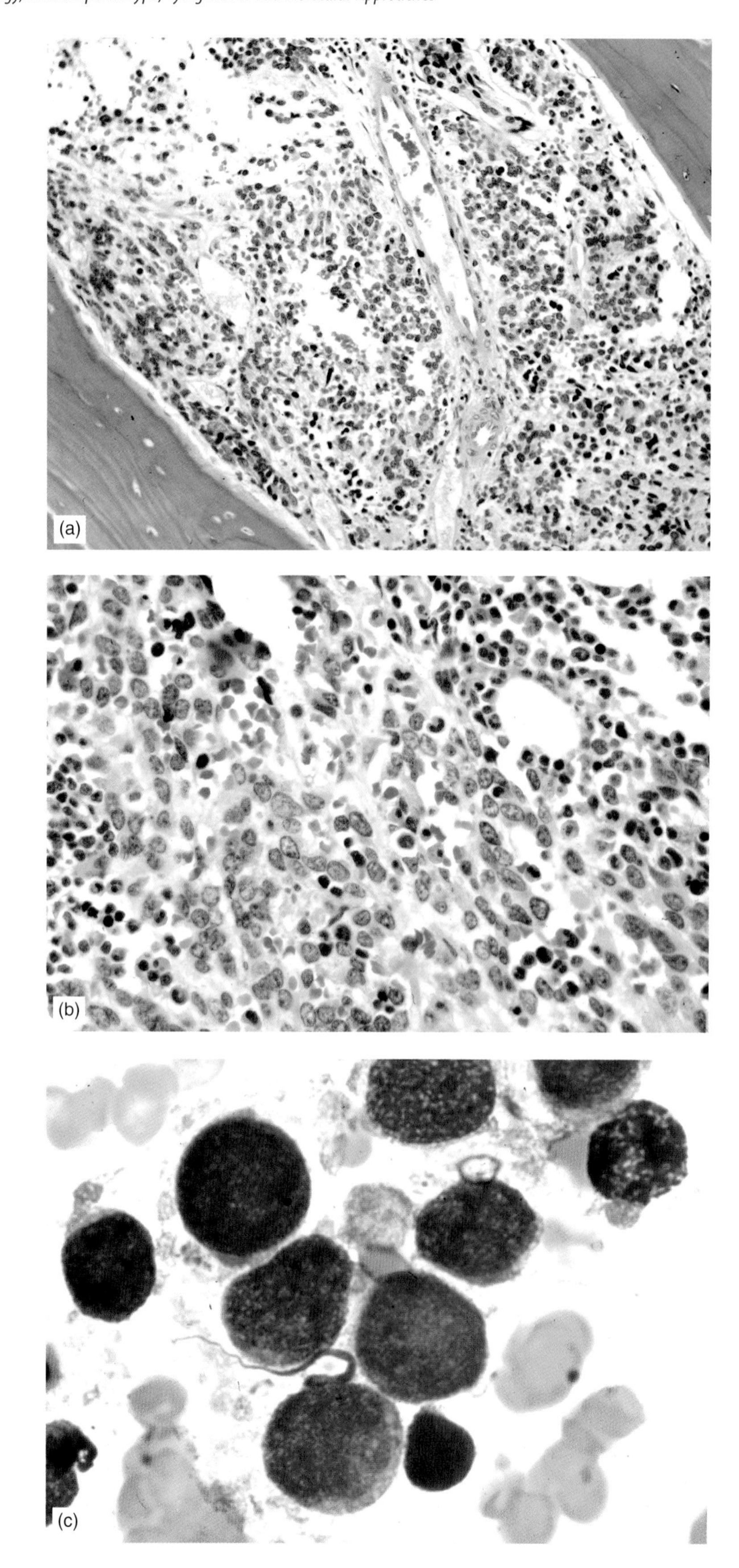

FIGURE 5.14 Metastatic neuroblastoma demonstrating clusters of small round tumor cells. Bone marrow biopsy section: (a) low power and (b) high power views. (c) Bone marrow smear.

Morphologic evidence of post-therapy bone marrow regeneration usually appears 1–2 weeks after therapy. Usually, erythroid and myeloid precursors appear sooner than megakaryocytes. Myeloid precursors are usually adjacent to bone, whereas erythroid clusters are far from bone trabeculae and are surrounded by fatty tissue. Rapid bone marrow regeneration is often associated with left-shifted hematopoiesis and increased hematogones [1].

Bone marrow changes following growth factor and/or interleukin therapy usually consist of increased cellularity, left shift, and may be associated with some degree of dysplastic changes. IL-3 therapy may occasionally cause marrow fibrosis, and GM-CSF and G-CSF cause myeloid preponderance and left shift, sometimes with marked eosinophilia [52–55].

BONE AND STROMAL CHANGES

Bone marrow stroma and bone play an important role in support and regulation of hematopoiesis [56–60]. However, pathological changes of hematopoiesis may significantly affect structure and function of the bone marrow stroma and the surrounding bone trabeculae.

Bone Changes

Hematopathologists and hematologists may overlook the bone changes when they review bone marrow biopsy sections. These changes are usually of two types: (1) associated with decreased bone formation (osteopenia) and (2) associated with increased bone formation (osteosclerosis).

Conditions Associated with Osteopenia

The receptor activator of nuclear factor-kappaB ligand (RANKL) is a pivotal regulator of osteoclast activity, and its inhibition may lead to osteopenia and hypercalcemia [61–65]. Osteopenia (osteoporosis, osteolysis) is caused by various conditions including circulatory disturbances, metabolic deficiencies, endocrine imbalance or deficiency, dialysis, inflammatory conditions, inactivity and immobilization, and the expansion of bone marrow space caused by metastasis, plasma cell myeloma, leukemia/lymphoma, and extensive bone marrow hypercellularity due to erythroid hyperplasia [1, 66]. Alteration of RANKL activity has been reported in many of these conditions, particularly plasma cell myeloma [67–69].

In osteoporosis, the bone marrow space is expanded at the expense of bone trabeculae. The bone trabeculae are thin and far apart. In osteomalacia the osteoid is partially decalcified.

Conditions Associated with Osteosclerosis

Osteosclerosis is demonstrated in bone marrow biopsy sections by increased thickness of the bone trabeculae. It is observed in chronic idiopathic myelofibrosis, leukemias, metastatic cancers, Erdheim–Chester disease, and mastocytosis (Figure 5.17) [1, 70, 71]. Osteopetrosis, an inherited disease which occurs in both autosomal dominant and recessive forms, is also characterized by increased bone density [72, 73].

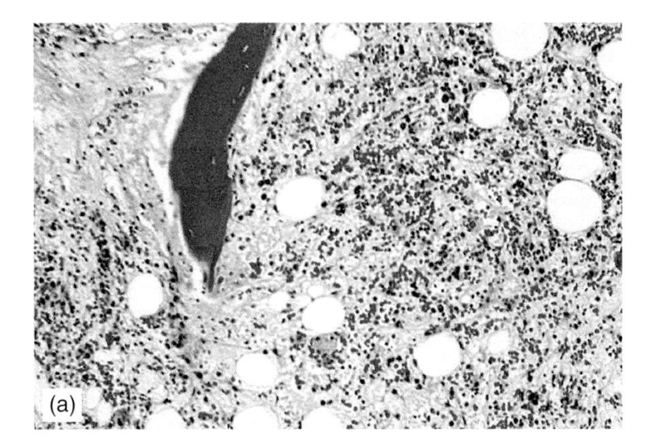

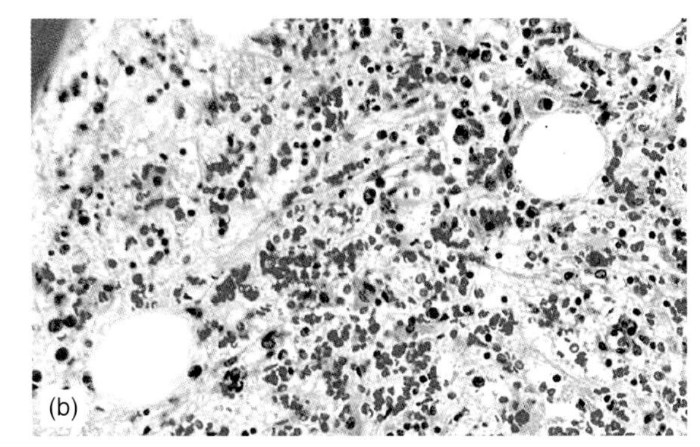

FIGURE 5.15 Post-chemotherapy bone marrow sections are hypocellular with areas of necrosis, scattered hematopoietic precursors, stromal tissue, and minimal fat: (a) low power and (b) high power.

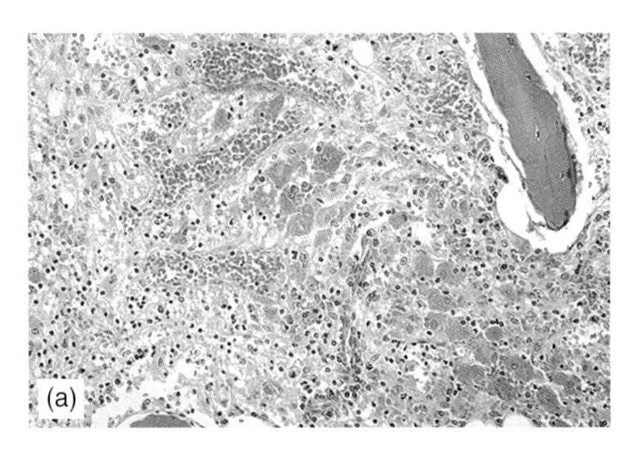

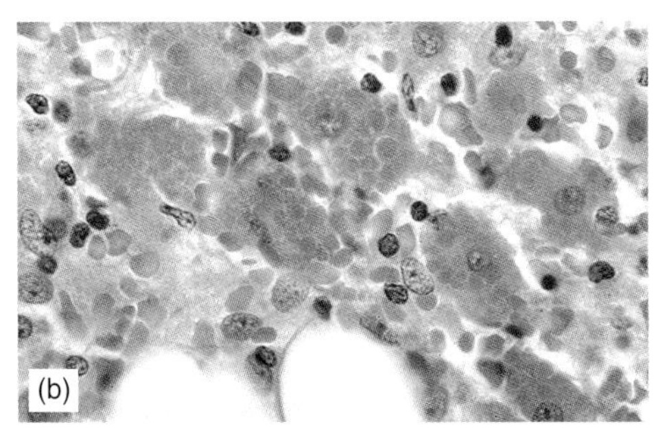

FIGURE 5.16 Bone marrow biopsy sections after chemotherapy or irradiation are hypocellular and may demonstrate necrosis, edema, and increased number of histiocytes (a). Higher power view (b) shows some hemophagocytic histiocytes.

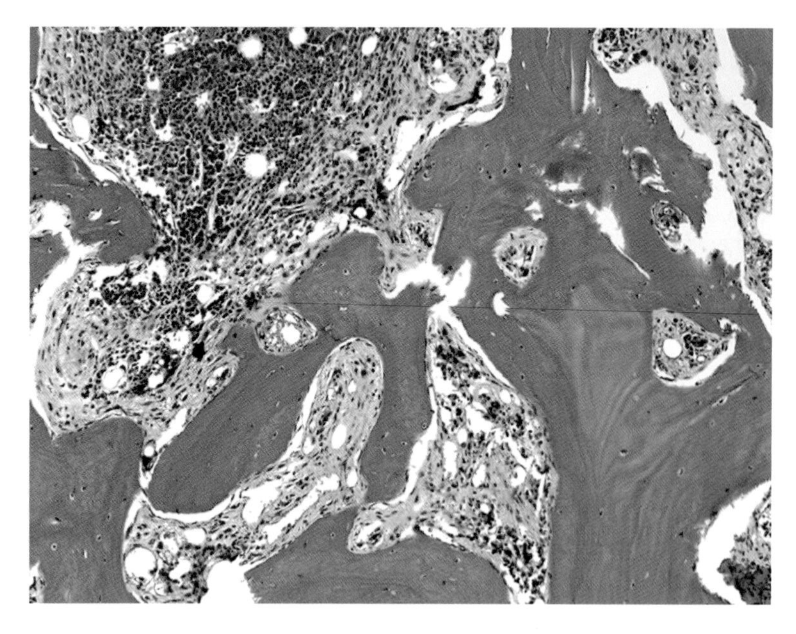

FIGURE 5.17 Bone marrow biopsy section from a patient with mastocytosis showing sclerotic bone trabeculae and areas of fibrosis.

Previous Biopsy Site (Repair)

The recovery to the damaged bone marrow following biopsy is relatively fast and complete. However, occasionally, when several biopsy attempts are made at short intervals, the biopsy may show features of tissue repair, such as neovascularization and proliferation of the fibroblasts (granulation tissue), edema, increased number of macrophages, and evidence of new bone formation (Figure 5.18). In addition, fragments of necrotic bone or bone marrow tissue may be still present. The reparative changes may mimic myelofibrosis, granuloma, or Paget's disease.

Bone Marrow Fibrosis

Bone marrow fibrosis is a common phenomenon and is observed in various pathological conditions such as myeloproliferative and myelodysplastic disorders, leukemias, lymphomas, mastocytosis, paroxysmal nocturnal hemoglobinuria, plasma cell myeloma, Gaucher's disease, granulomas, metastatic tumors, hyperparathyroidism, chronic renal failure, osteopetrosis, autoimmune disorders, and Paget's disease (Table 5.2) (Figures 5.19 and 5.20). Fibrosis may be focal (e.g. fibrosis associated with chronic renal failure or mastocytosis) or diffuse (e.g. fibrosis associated with myeloproliferative disorders).

Bone marrow fibrosis is a non-clonal reactive process and is caused by the release of fibroblastic growth factors by megakaryocytes, platelets, histiocytes, and other cells.

Vascular Changes

Vascular inflammatory changes such as arteritis, arteriolitis, and granulomatous vasculitis may involve bone marrow as part of a systemic process (Figure 5.21). Similarly, atherosclerotic and thrombotic lesions may be detected in bone marrow biopsy. Tumor emboli are the major sources of bone marrow metastasis, and when they are extensive may cause microangiopathic hemolytic anemia. As mentioned earlier, bone marrow amyloidosis may involve vascular structures.

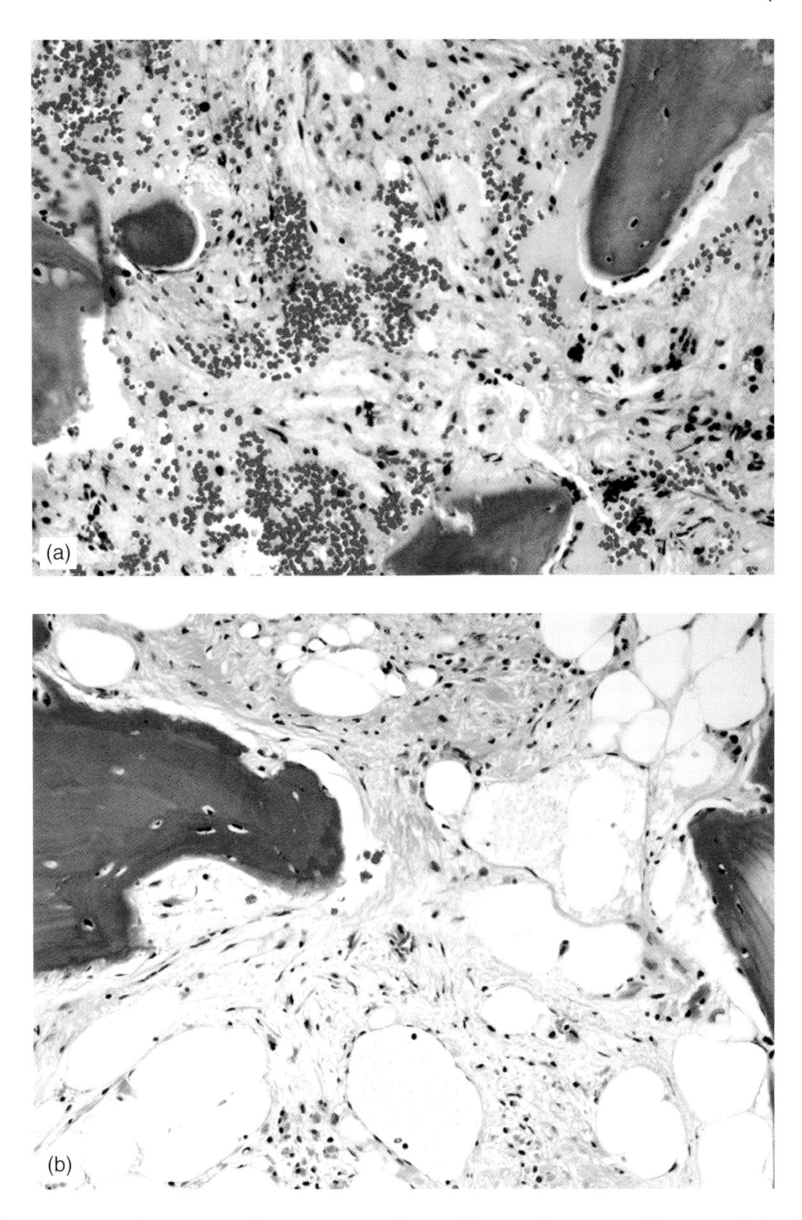

FIGURE 5.18 Bone marrow biopsy sections from previous biopsy sites may show evidence of bone remodeling, granulation tissue formation (a) and/or fibrosis (b).

TABLE 5.2 Conditions associated with bone marrow fibrosis.

Primary hematologic disorders
  Chronic myeloproliferative disorders
  Myelodysplastic syndromes
  Paroxysmal nocturnal hemoglobinuria
  Aplastic anemia
  Leukemias
  Lymphomas
  Plasma cell myeloma
  Mastocytosis
  Gray platelet syndrome

Metastatic lesions
Lysosomal storage diseases

Inflammatory and repair processes
  Granulomas
  Osteomyelitis
  Autoimmune disorders
  Following bone marrow necrosis
  Following bone marrow radiation
  Previous bone marrow biopsy site

Metabolic disorders
  Osteomalacia
  Osteopetrosis
  Chronic renal failure
  Primary hyperparathyroidism

Paget's disease

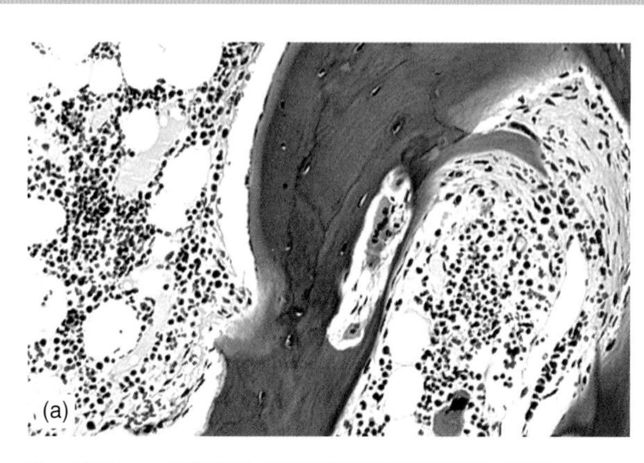

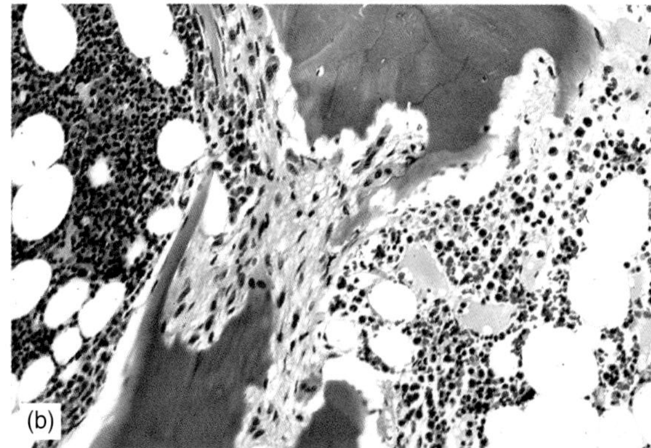

FIGURE 5.19 Bone marrow biopsy section from a patient with chronic renal failure demonstrating paratrabecular fibrosis, bone resorption and remodeling, and the presence of osteoclasts.

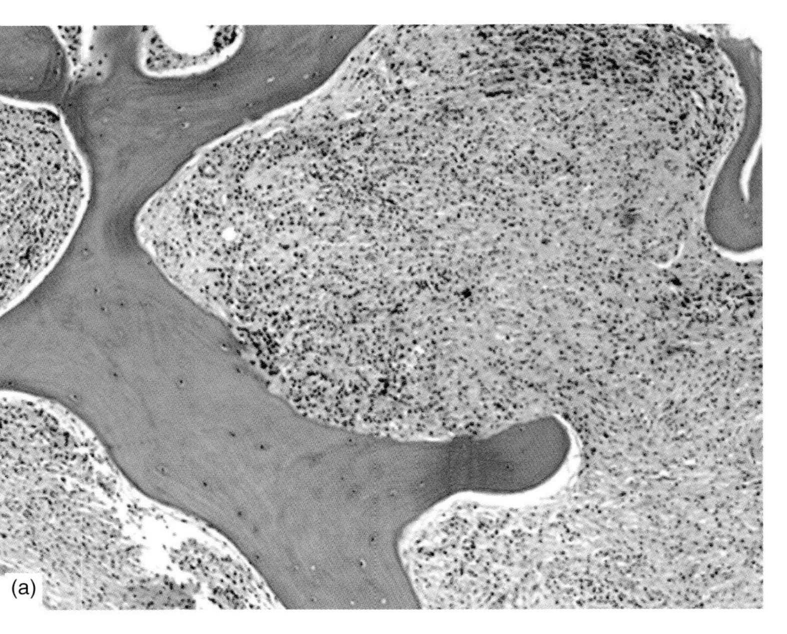

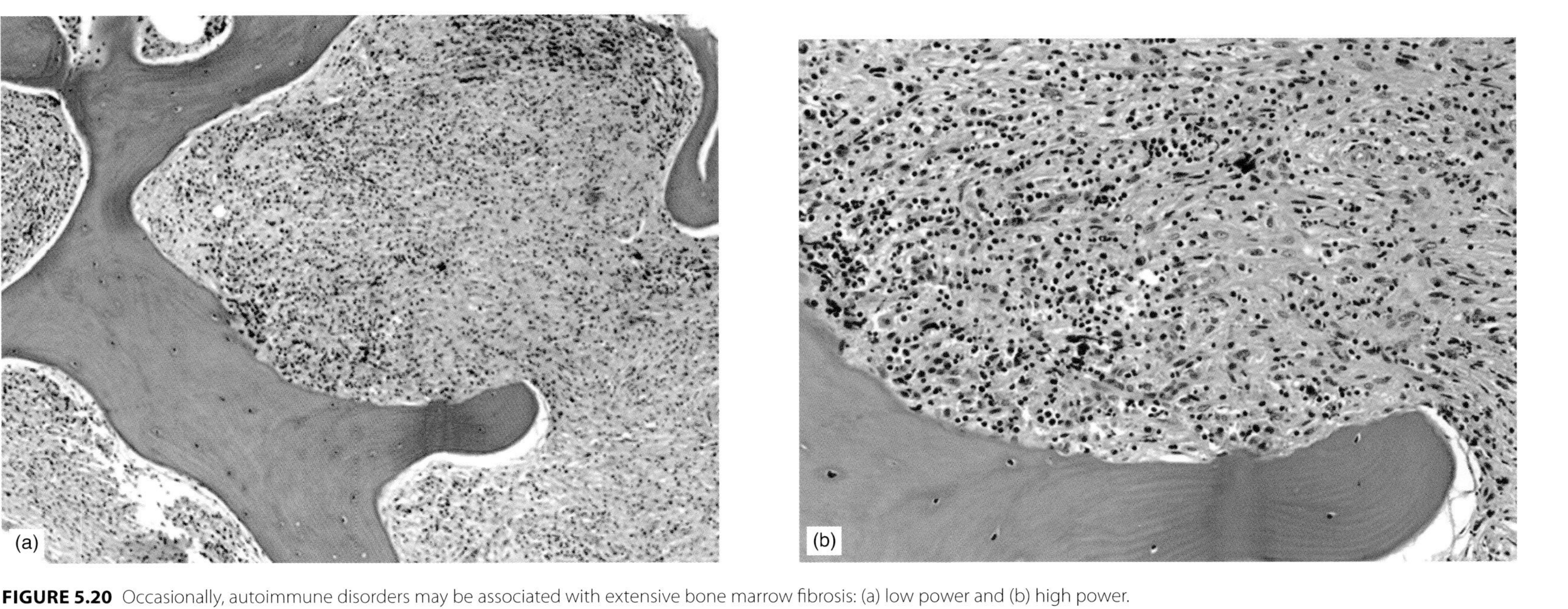

FIGURE 5.20 Occasionally, autoimmune disorders may be associated with extensive bone marrow fibrosis: (a) low power and (b) high power.

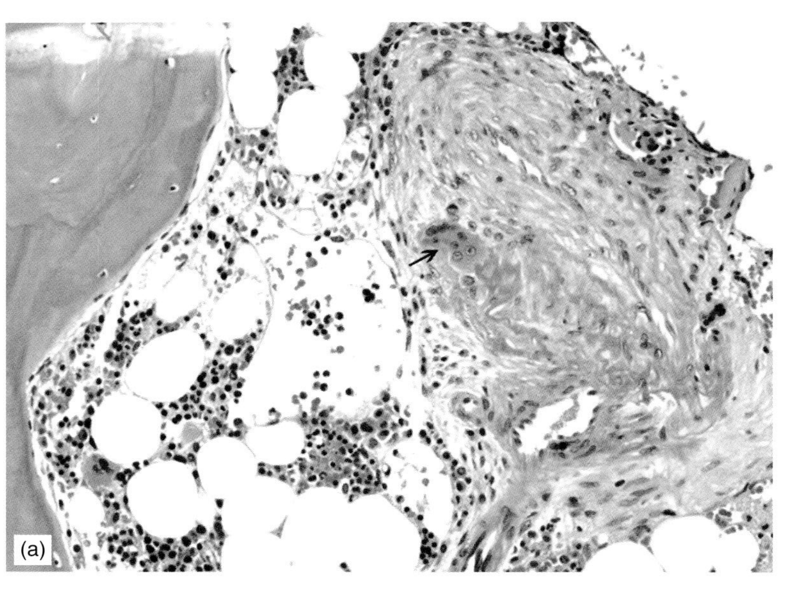

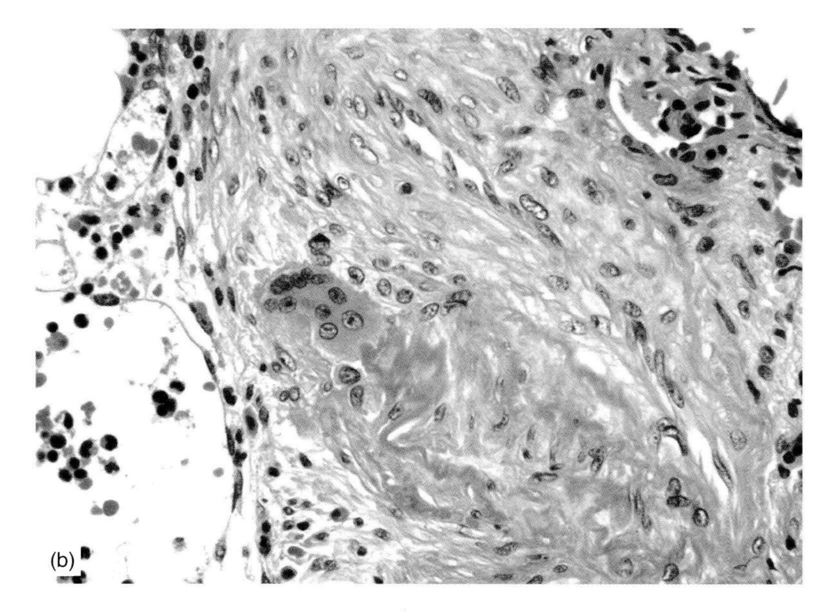

FIGURE 5.21 A vascular structure in a bone marrow biopsy section showing features of giant cell arteritis with a large multinucleated giant cell in the sclerotic wall: (a) low power and (b) high power.

References

1. Naeim F. (1997). *Pathology of Bone Marrow*, 2nd ed. Williams & Wilkins, Baltimore.
2. Seaman JP, Kjeldsberg CR, Linker A. (1978). Gelatinous transformation of the bone marrow. *Hum Pathol* **9**, 685–92.
3. Bohm J. (2000). Gelatinous transformation of the bone marrow: The spectrum of underlying diseases. *Am J Surg Pathol* **24**, 56–65.
4. Tavassoli M, Eastlund DT, Yam LT, Neiman RS, Finkel H. (1976). Gelatinous transformation of bone marrow in prolonged self-induced starvation. *Scand J Haematol* **16**, 311–19.
5. Jain R, Singh ZN, Khurana N, Singh T. (2005). Gelatinous transformation of bone marrow: A study of 43 cases. *Indian J Pathol Microbiol* **48**, 1–3.
6. Stroup JS, Stephens JR, Baker DL. (2007). Gelatinous bone marrow in an HIV-positive patient. *Proc (Bayl Univ Med Cent)* **20**, 254–6.
7. Murugan P, Chandrakumar S, Basu D, Hamide A. (2007). Gelatinous transformation of bone marrow in acquired immunodeficiency syndrome. *Pathology* **39**, 287–8.
8. Mant MJ, Faragher BS. (1972). The haematology of anorexia nervosa. *Br J Haematol* **23**, 737–49.
9. Brown CH III. (1972). Bone marrow necrosis. A study of seventy cases. *Johns Hopkins Med J* **131**, 189–203.
10. Norgard MJ, Carpenter JT Jr., Conrad ME. (1979). Bone marrow necrosis and degeneration. *Arch Intern Med* **139**, 905–11.
11. Kiraly III JF, Wheby MS. (1976). Bone marrow necrosis. *Am J Med* **60**, 361–8.
12. Maisel D, Lim JY, Pollock WJ, Yatani R, Liu PI. (1988). Bone marrow necrosis: An entity often overlooked. *Ann Clin Lab Sci* **18**, 109–15.
13. Paydas S, Ergin M, Baslamisli F, Yavuz S, Zorludemir S, Sahin B, Bolat FA. (2002). Bone marrow necrosis: Clinicopathologic analysis of 20 cases and review of the literature. *Am J Hematol* **70**, 300–5.
14. Janssens AM, Offner FC, Van Hove WZ. (2000). Bone marrow necrosis. *Cancer* **88**, 1769–80.
15. Noguchi M, Oshimi K. (2007). Extensive bone marrow necrosis and symptomatic hypercalcemia in B cell blastic transformation of chronic myeloid leukemia: Report of a case and review of the literature. *Acta Haematol* **118**, 111–16.
16. Dang NC, Johnson C, Eslami-Farsani M, Haywood LJ. (2005). Bone marrow embolism in sickle cell disease: A review. *Am J Hematol* **79**, 61–7.
17. Knupp C, Pekala PH, Cornelius P. (1988). Extensive bone marrow necrosis in patients with cancer and tumor necrosis factor activity in plasma. *Am J Hematol* **29**, 215–21.
18. Ataga KI, Orringer EP. (2000). Bone marrow necrosis in sickle cell disease: A description of three cases and a review of the literature. *Am J Med Sci* **320**, 342–7.
19. Forrest DL, Mack BJ, Nevill TJ, Couban SH, Zayed E, Foyle A. (2000). Bone marrow necrosis in adult acute leukemia and non-Hodgkin's lymphoma. *Leuk Lymphoma* **38**, 627–32.
20. Glenner GG. (1980). Amyloid deposits and amyloidosis. The beta-fibrilloses (first of two parts). *N Engl J Med* **302**, 1283–92.
21. Glenner GG. (1980). Amyloid deposits and amyloidosis: The beta-fibrilloses (second of two parts). *N Engl J Med* **302**, 1333–43.
22. Sirohi B, Powles R. (2006). Epidemiology and outcomes research for MGUS, myeloma and amyloidosis. *Eur J Cancer* **42**, 1671–83.
23. Rajkumar SV, Dispenzieri A, Kyle RA. (2006). Monoclonal gammopathy of undetermined significance, Waldenstrom macroglobulinemia, AL amyloidosis, and related plasma cell disorders: Diagnosis and treatment. *Mayo Clin Proc* **81**, 693–703.
24. Comenzo RL. (2000). Primary systemic amyloidosis. *Curr Treat Options Oncol* **1**, 83–9.
25. Friman C, Pettersson T. (1996). Amyloidosis. *Curr Opin Rheumatol* **8**, 62–71.
26. Kaloterakis A, Filiotou A, Koskinas J, Raptis I, Zouboulis C, Michelakakis H, Hadziyannis S. (1999). Systemic AL amyloidosis in Gaucher disease. A case report and review of the literature. *J Intern Med* **246**, 587–90.
27. Diebold J, Molina T, Camilleri-Broet S, Le Tourneau A, Audouin J. (2000). Bone marrow manifestations of infections and systemic diseases observed in bone marrow trephine biopsy review. *Histopathology* **37**, 199–211.
28. Bhargava V, Farhi DC. (1988). Bone marrow granulomas: Clinicopathologic findings in 72 cases and review of the literature. *Hematol Pathol* **2**, 43–50.
29. Bodem CR, Hamory BH, Taylor HM, Kleopfer L. (1983). Granulomatous bone marrow disease. A review of the literature and clinicopathologic analysis of 58 cases. *Medicine* (Baltimore) **62**, 372–83.
30. Browne PM, Sharma OP, Salkin D. (1978). Bone marrow sarcoidosis. *JAMA* **240**, 2654–5.
31. Chang KL, Gaal KK, Huang Q, Weiss LM. (2003). Histiocytic lesions involving the bone marrow. *Semin Diagn Pathol* **20**, 226–36.
32. Eid A, Carion W, Nystrom JS. (1996). Differential diagnoses of bone marrow granuloma. *West J Med* **164**, 510–15.
33. Sanal SM, Winiarski NB, Cardenas FJ. (1992). Crohn's disease with associated bone marrow granulomas. *South Med J* **85**, 646–7.
34. Karcher DS, Frost AR. (1991). The bone marrow in human immunodeficiency virus (HIV)-related disease. Morphology and clinical correlation. *Am J Clin Pathol* **95**, 63–71.
35. Nosanchuk JS. (1984). Bone marrow granulomas with acute cytomegalovirus infection. *Arch Pathol Lab Med* **108**, 93–4.
36. Okun DB, Sun NC, Tanaka KR. (1979). Bone marrow granulomas in Q fever. *Am J Clin Pathol* **71**, 117–21.
37. Delsol G, Pellegrin M, Familiades J, Auvergnat JC. (1978). Bone marrow lesions in Q fever. *Blood* **52**, 637–8.
38. Suster S, Cabello-Inchausti B, Robinson MJ. (1989). Nongranulomatous involvement of the bone marrow in lepromatous leprosy. *Am J Clin Pathol* **92**, 797–801.
39. Rywlin AM, Ortega R. (1972). Lipid granulomas of the bone marrow. *Am J Clin Pathol* **57**(4), 457–62.
40. Anner RM, Drewinko B. (1977). Frequency and significance of bone marrow involvement by metastatic solid tumors. *Cancer* **39**, 1337–44.
41. Ihde DC, Bilek FS, Cohen MH, Bunn PA, Eddy J, Minna JD. (1979). Bone marrow metastases in small cell carcinoma of the lung: Frequency, description, and influence on chemotherapeutic toxicity and prognosis. *Blood* **53**, 677–86.
42. Braun S, Vogl FD, Janni W, Marth C, Schlimok G, Pantel K. (2003). Evaluation of bone marrow in breast cancer patients: Prediction of clinical outcome and response to therapy. *Breast* **12**, 397–404.
43. Cotta CV, Konoplev S, Medeiros LJ, Bueso-Ramos CE. (2006). Metastatic tumors in bone marrow: Histopathology and advances in the biology of the tumor cells and bone marrow environment. *Ann Diagn Pathol* **10**, 169–92.
44. Moatamed F, Sahimi M, Naeim F. (1988). Fractal dimension of the bone marrow in metastatic lesions. *Hum Pathol* **29**, 1299–303.
45. Beck D, Maritaz O, Gross N, Favrot M, Vultier N, Bailly C, Villa I, Gentilhomme O, Philip T. (1988). Immunocytochemical detection of neuroblastoma cells infiltrating clinical bone marrow samples. *Eur J Pediatr* **147**, 609–12.
46. Athanassiadou P, Grapsa D. (2006). Recent advances in the detection of bone marrow micrometastases: A promising area for research or just another false hope? A review of the literature. *Cancer Metastasis Rev* **25**, 507–19.
47. Krishnan C, George TI, Arber DA. (2007). Bone marrow metastases: A survey of nonhematologic metastases with immunohistochemical study of metastatic carcinomas. *Appl Immunohistochem Mol Morphol* **15**, 1–7.
48. Zach O, Lutz D. (2006). Tumor cell detection in peripheral blood and bone marrow. *Curr Opin Oncol* **18**, 48–56.
49. Braun S, Harbeck N. (2001). Molecular markers of metastasis in breast cancer: Current understanding and prospects for novel diagnosis and prevention. *Expert Rev Mol Med* **3**, 1–14.
50. Baker M, Gillanders WE, Mikhitarian K, Mitas M, Cole DJ. (2003). The molecular detection of micrometastatic breast cancer. *Am J Surg* **186**, 351–8.

51. Wittels B. (1908). Bone marrow biopsy changes following chemotherapy for acute leukemia. *Am J Surg Pathol* **4**, 135–42.
52. Falk S, Seipelt G, Ganser A, Ottmann OG, Hoelzer D, Stutte HJ, Hubner K. (1991). Bone marrow findings after treatment with recombinant human interleukin-3. *Am J Clin Pathol* **95**, 355–62.
53. Naiem F, Champlin R, Nimer S. (1990). Bone marrow changes in patients with refractory aplastic anemia treated by recombinant GM-CSF. *Hematol Pathol* **4**, 79–85.
54. Schmitz LL, McClure JS, Litz CE, Dayton V, Weisdorf DJ, Parkin JL. Brunning, RD. (1994). Morphologic and quantitative changes in blood and marrow cells following growth factor therapy. *Am J Clin Pathol* **101**, 67–75.
55. Ryder JW, Lazarus HM, Farhi DC. (1992). Bone marrow and blood findings after marrow transplantation and rhGM-CSF therapy. *Am J Clin Pathol* **97**, 631–7.
56. Kacena MA, Gunberg CM, Horowitz MC. (2006). A reciprocal regulatory interaction between megakaryocytes, bone cells, and hematopoietic stem cells. *Bone* **39**, 978–84.
57. Calvi LM. (2006). Osteoblastic activation in the hematopoietic stem cell niche. *Ann NY Acad Sci* **1068**, 477–88.
58. Dazzi F, Ramasamy R, Glennie S, Jones SP, Roberts I. (2006). The role of mesenchymal stem cells in haemopoiesis. *Blood Rev* **20**, 161–71.
59. Aguila HL, Rowe DW. (2005). Skeletal development, bone remodeling, and hematopoiesis. *Immunol Rev* **208**, 7–18.
60. Blair JM, Zheng Y, Dunstan CR. (2007). RANK ligand. *Int J Biochem Cell Biol* **39**, 1077–81.
61. McClung M. (2007). Role of RANKL inhibition in osteoporosis. *Arthritis Res Ther* **9**(Suppl 1), S3.
62. McCormick RK. (2007). Osteoporosis: Integrating biomarkers and other diagnostic correlates into the management of bone fragility. *Altern Med Rev* **12**, 113–45.
63. Shinohara M, Takayanagi H. (2007). Novel osteoclast signaling mechanisms. *Curr Osteoporos Rep* **5**, 67–72.
64. Hadjidakis DJ, Androulakis II. (2006). Bone remodeling. *Ann NY Acad Sci* **1092**, 385–96.
65. Steindler A. (2006). The Classic: Osteoporosis. *Clin Orthop Relat Res*, **443**, 3–9.
66. Adams JE. (2002). Dialysis bone disease. *Semin Dial* **15**, 277–89.
67. Hjertner O, Standal T, Borset M, Sundan A, Waage A. (2006). Bone disease in multiple myeloma. *Med Oncol* **23**, 431–41.
68. Giuliani N, Rizzoli V, Roodman GD. (2006). Multiple myeloma bone disease: Pathophysiology of osteoblast inhibition. *Blood* **108**, 3992–6.
69. Oyajobi BO. (2007). Multiple myeloma/hypercalcemia. *Arthritis Res Ther* **9**(Suppl), S4.
70. Diamond T, Smith A, Schnier R, Manoharan A. (2002). Syndrome of myelofibrosis and osteosclerosis: A series of case reports and review of the literature. *Bone* **30**, 498–501.
71. Murray D, Marshall M, England E, Mander J, Chakera TM. (2001). Erdheim–Chester disease. *Clin Radiol* **56**, 481–4.
72. Balemans W, Van Wesenbeeck L, Van Hul W. (2005). A clinical and molecular overview of the human osteopetroses. *Calcif Tissue Int* **77**, 263–74.
73. Tolar J, Teitelbaum SL, Orchard PJ. (2004). Osteopetrosis. *N Engl J Med* **351**, 2839–49.

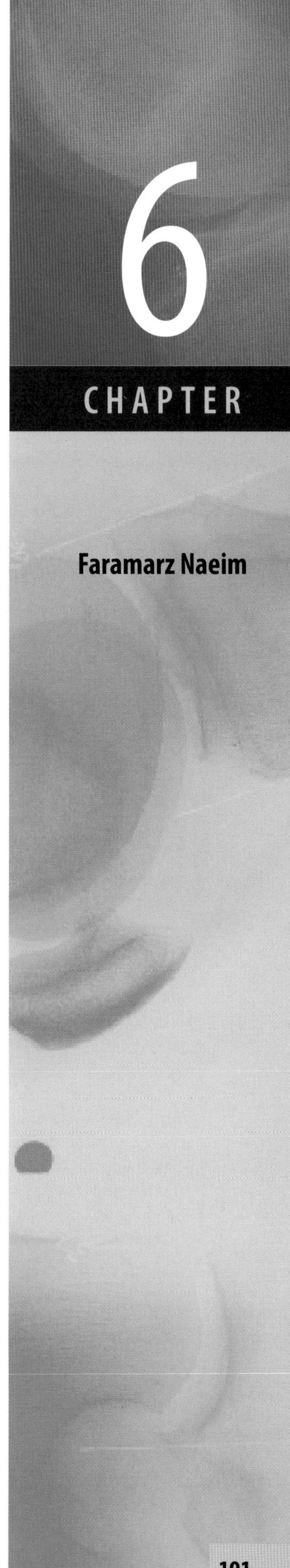

Reactive Lymphadenopathies

Faramarz Naeim

The reactive lymphadenopathies represent hyperplasia of different lymph node components and therefore demonstrate particular structural patterns such as follicular hyperplasia, paracortical hyperplasia, sinus histiocytosis, granulomas, or a mixed pattern (Table 6.1) [1–3].

FOLLICULAR HYPERPLASIA

In follicular hyperplasia, the lymph node enlargement is primarily due to the increased number and size of the follicles [1, 2]. Germinal centers

TABLE 6.1 Patterns of reactive lymphadenitis.

Patterns	Examples
Follicular hyperplasia Enlargement of germinal centers, follicles in variable size and shapes, intact mantle zones	Rheumatoid arthritis Castleman's disease Early HIV infection Bacterial infections
Paracortical hyperplasia Expansion of interfollicular areas, increased high endothelial venules, mostly T-cells	Infectious mononucleosis Postvaccinal lymphadenitis Drug-induced
Sinus pattern Expansion of sinusoidal spaces with proliferation of histiocytes or monocytoid B-lymphocytes	Sinus histiocytosis Rosai–Dorfman disease Whipple's disease Monocytoid B-cell hyperplasia
Granulomatous lymphadenitis Granulomas may be of epithelioid type, may show caseous necrosis, or may be suppurative	Tuberculosis Sarcoidosis Cat-scratch fever Fungal infections
Mixed pattern	Dermatopathic lymphadenitis Toxoplasmosis Kikuchi's disease Kimura's disease

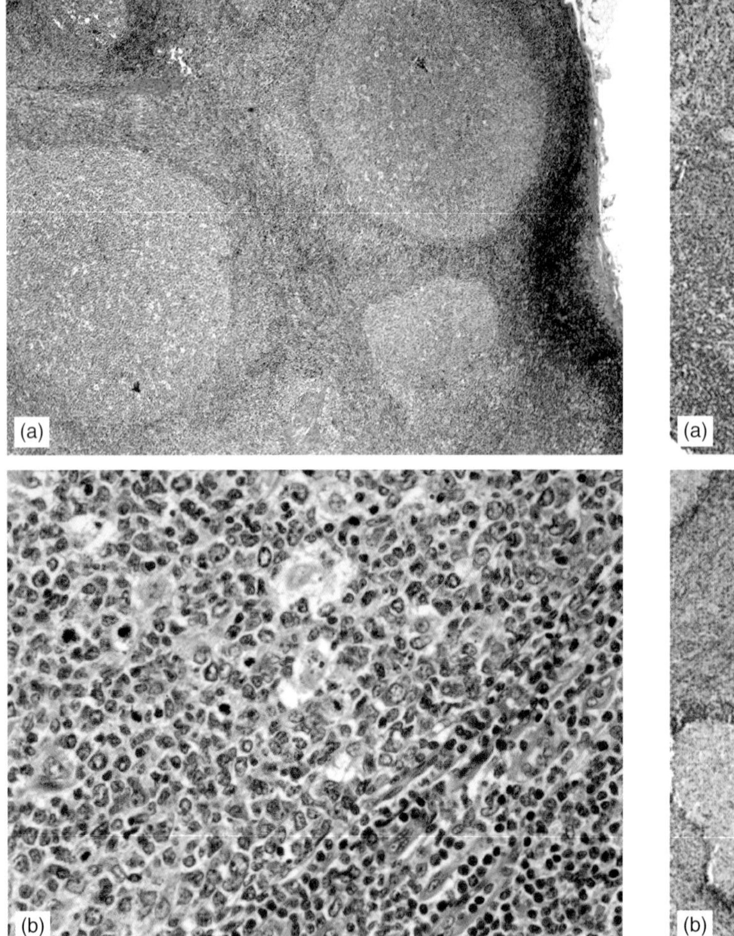

FIGURE 6.1 (a) Follicular hyperplasia with the expansion of germinal centers. (b) High power view of a germinal center showing centrocytes, centroblasts, tingible body macrophages, and mitotic figures.

are expanded and follicles appear in different sizes and shapes and are surrounded by well-defined mantle zones. Follicles occupy cortical, paracortical, and even sometimes medullary zones, but they are usually separated from one another by interfollicular lymphoid tissues. The expanded germinal centers show numerous tingible body macrophages intermixed with centrocytes and centroblasts and are characterized by the CD21-positive follicular dendritic cell network, and lack of the expression of BCL-2 (Figures 6.1 and 6.2) [1, 2, 4]. Mitotic figures and apoptotic bodies are frequent. Plasma cells may be present within the germinal centers, occasionally in large numbers. Interfollicular areas often show vascular proliferation and a mixture of small lymphocytes, plasma cells, and immunoblasts, sometimes with increased number of histiocytes and eosinophils. Medullary plasmacytosis is a frequent finding. As discussed in Chapter 14, differential diagnosis of follicular hyperplasia from follicular lymphoma is sometimes difficult [5, 6]. The major distinguishing features between these two categories are summarized in Table 6.2. Castleman's disease, rheumatoid arthritis (RA), and various infections, such as syphilis, are often associated with follicular hyperplasia. The morphologic features of RA and Castleman's disease are briefly discussed later.

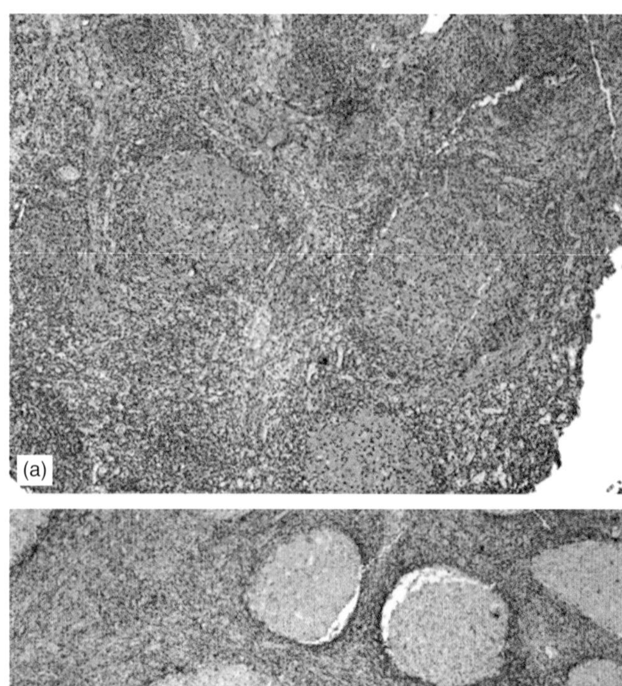

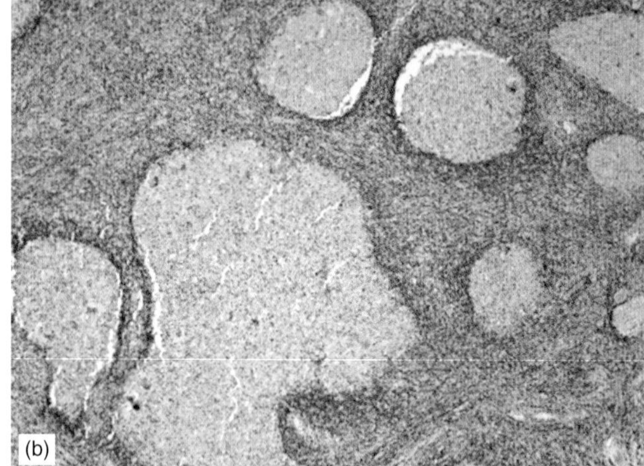

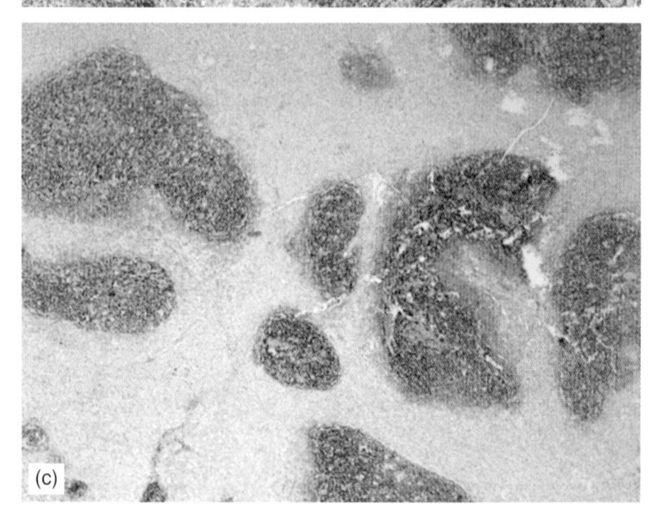

FIGURE 6.2 Immunohistochemical studies of hyperplastic follicles: (a) B-cells (CD20-positive, red) are the predominant cells in the germinal centers and T-cells (CD3-positive, brown) are the predominant cells in the interfollicular areas. (b) Reactive follicles do not express BCL-2 and (c) CD21 expression represents the network of follicular dendritic cells.

Rheumatoid Arthritis

Rheumatoid arthritis (RA) is an autoimmune, chronic systemic disorder affecting women more than men (male:female

TABLE 6.2 Morphologic features distinguishing follicular lymphoma (FL) from reactive follicular hyperplasia (RFH).

Features	FL	RFH
Follicles	High density per unit area	Low density per unit area
	Back to back or merging, often with loss of mantle zone	Separated, with preservation of mantle zone
	Lack of polarity	Presence of polarity
	Monoclonal	Polyclonal
	Commonly BCL-2 positive	BCL-2 negative
	Low Ki-67 fraction	High Ki-67 fraction
Interfollicular areas	Presence of CD10 positive cells	Absence of CD10 positive cells
	Presence of BCL-6 positive cells	Absence of BCL-6 positive cells

ratio of about 1:3) [7, 8]. Most patients are over 30 years of age. The juvenile form of RA is often associated with high spiking fevers and transient rashes (Still's disease) [9, 10]. The lymph node biopsy sections show prominent follicular hyperplasia with sinus histiocytosis (Figure 6.3) [2, 11, 12]. Increased plasma cells are frequently observed in the interfollicular regions. The lymphadenopathy in longstanding cases may be associated with the deposition of periodic acid Schiff (PAS)-positive hyaline material (negative for Congo red stain) with partial or complete replacement of the lymphoid tissue [2].

Castleman's Disease

Castleman's disease, also known as angiofollicular hyperplasia or giant lymph node hyperplasia [1, 2, 13], consists of two forms: (1) localized type and (2) multicentric type. The localized type of Castleman's disease is much more frequent than the multicentric type, usually involving mediastinum. Other sites such as axillary and cervical lymph nodes, skeletal muscles, and pulmonary parenchyma have been involved less frequently [1, 2]. There are two morphologic variants of Castleman's disease: (1) hyaline vascular type and (2) plasma cell type.

Hyaline Vascular Type

The hyaline vascular type of Castleman's disease is the common variant accounting for over 90% of cases in one large study [13–17]. The lesion is characterized by abnormal germinal centers, expanded mantle zones, and increased interfollicular vascularity. Germinal centers contain hyalinized vascular structure and show increased number of follicular dendritic cells (Figure 6.4). Follicular dendritic cells may appear large and bizarre. The mantle zone area is often expanded and concentrically arranged, creating an "onion skin" pattern (Figure 6.4b). The expanded mantle zones may sometimes completely obscure the germinal center [1]. Blood vessels are prominent in the interfollicular areas, some with prominent endothelial cells (Figure 6.5). Small lymphocytes and scattered plasma cells are found in between vascular structures.

The hyaline vascular type, because of the expansion of mantle zone areas, may resemble mantle cell lymphoma. Mantle cell lymphoma is monoclonal and demonstrates Ig light chain restriction, expresses BCL-1, and is usually devoid of hyalinized vessels.

Plasma Cell Type

The plasma cell type of Castleman's disease is less frequent (~10%) and is often associated with polyclonal gamma globulinemia, increased serum levels of IL-6, and elevated erythrocyte sedimentation rate. Anemia and elevated erythrocyte sedimentation rate are frequent findings. The most frequent site of involvement is the abdomen, particularly in the small bowel mesentery [1, 2, 18]. Lymph node sections show follicular hyperplasia with a well-defined mantle zone, surrounded by sheets of mature plasma cells and scattered immunoblasts. Vascular proliferation or hyalinization is usually absent. In approximately 40% of the cases the plasma cells are monotypic and express Ig lambda light chain [1, 19]. The plasma cell type resembles other follicular hyperplasias, such as those associated with RA, or other autoimmune disorders. Diagnosis of Castleman's disease of plasma cell type is a diagnosis of exclusion, when all other possible causes of follicular hyperplasia have been ruled out [1, 2].

PARACORTICAL (INTERFOLLICULAR) PATTERN

Lymphadenitis with paracortical pattern is characterized by the expansion of the interfollicular areas with a mixture of small lymphocytes, immunoblasts, and increased high endothelial venules. T-cells are the predominant cell type, but there are variable numbers of B-immunoblasts and plasma cells. Infectious mononucleosis, postvaccinal lymphadenitis, and drug-induced hypersensitivity reactions are classical examples of lymphadenitis with paracortical pattern.

SINUS PATTERN

Sinus Histiocytosis

Sinus histiocytosis is a relatively common phenomenon associated with a garden variety of infectious and inflammatory conditions. It is characterized by the expansion of sinuses due to the increased number of histiocytes (Figure 6.6). Pulmonary hilar lymph nodes in heavy smokers and elderly individuals contain a large number of sinus histiocytes

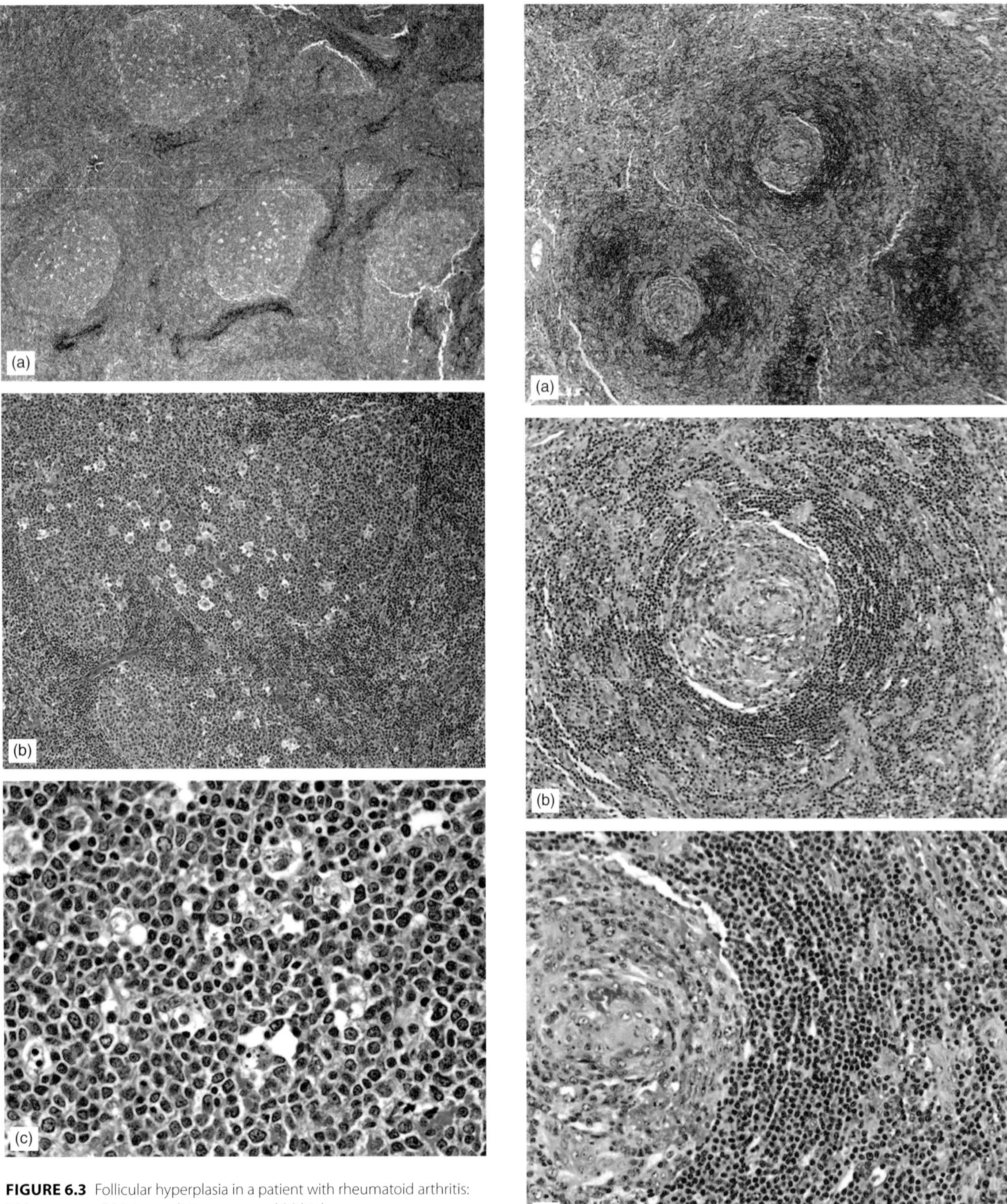

FIGURE 6.3 Follicular hyperplasia in a patient with rheumatoid arthritis: (a) low power, (b) intermediate power, and (c) high power.

FIGURE 6.4 Castleman's disease, hyaline vascular type, demonstrating follicular hyperplasia and expansion of mantle zones demonstrating an "onion skin" pattern: (a) low power, (b) intermediate power, and (c) high power.

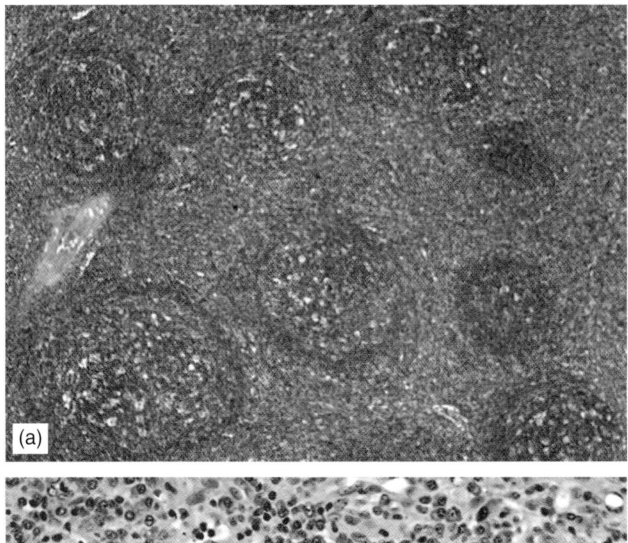

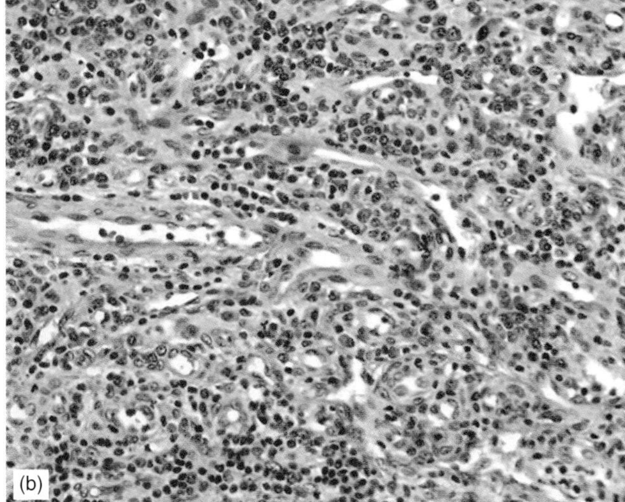

FIGURE 6.5 Lymph node biopsy section from a patient with Castleman's disease, hyaline vascular type, demonstrating follicular hyperplasia (a) and vascular proliferation in the interfollicular areas (b).

with carbon pigments. Sinus histiocytosis with pigments may be seen in the regional lymph nodes of tattooed areas. Melanin-containing histiocytes are seen in dermatopathic lymphadenitis (see later). Sinus histiocytosis is also noted in lysosomal storage diseases, hemophagocytosis (often with the presence of hemosiderin pigments), and post-lymphangiography [20, 21]. Histiocytes express CD68 and are strongly positive for lysozyme.

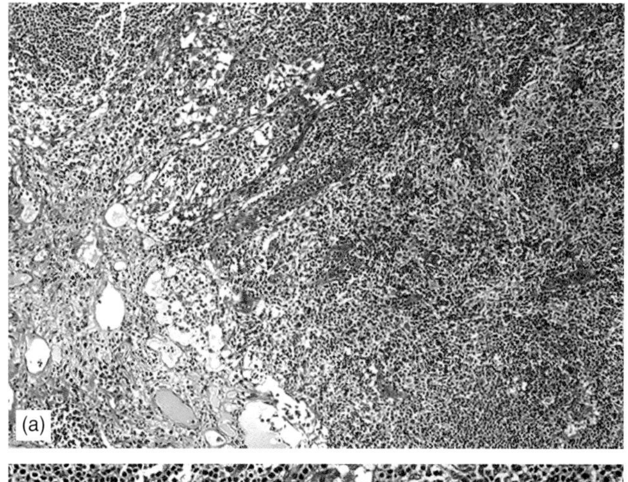

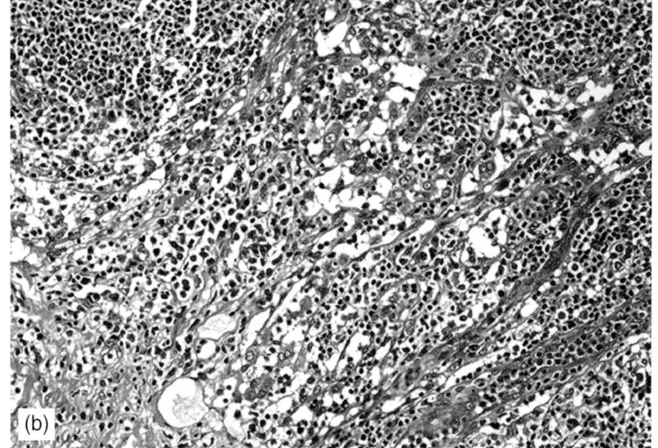

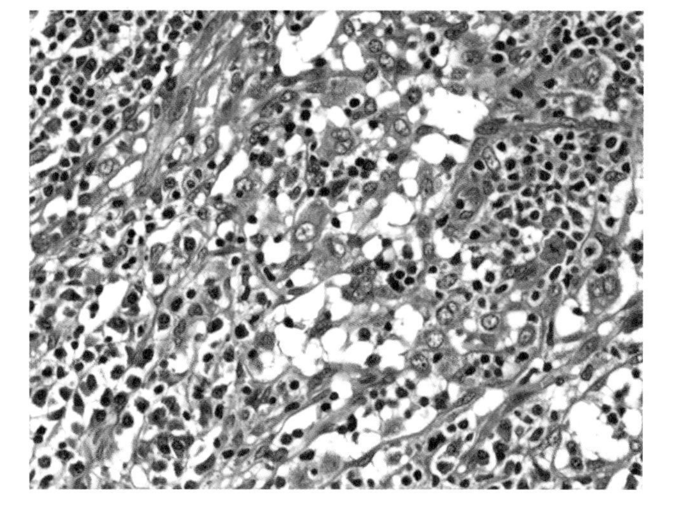

FIGURE 6.6 Sinus histiocytosis. Lymph node biopsy section demonstrating dilatation of the sinusoids and presence of numerous histiocytes: (a) low power, (b) intermediate power, and (c) high power.

Sinus Histiocytosis with Massive Lymphadenopathy

Sinus histiocytosis with massive lymphadenopathy (Rosai–Dorfman disease) is a bilateral, painless, massive cervical lymphadenopathy reported in patients under the age of 20 years [22–26]. An association with human herpes virus 6 has been reported in some cases [27]. The affected lymph nodes show a characteristic feature consisting of sheets of large sinus histiocytes with abundant clear or foamy cytoplasm containing numerous lymphocytes (or less frequently other hematopoietic cells) (Figure 6.7). These lymphocytes appear intact. The active penetration of cells into and through larger cells is called emperipolesis. Often, there is a marked thickening of the capsule with pericapsular fibrosis. Histiocytes express CD68 and are strongly positive for lysozyme [28].

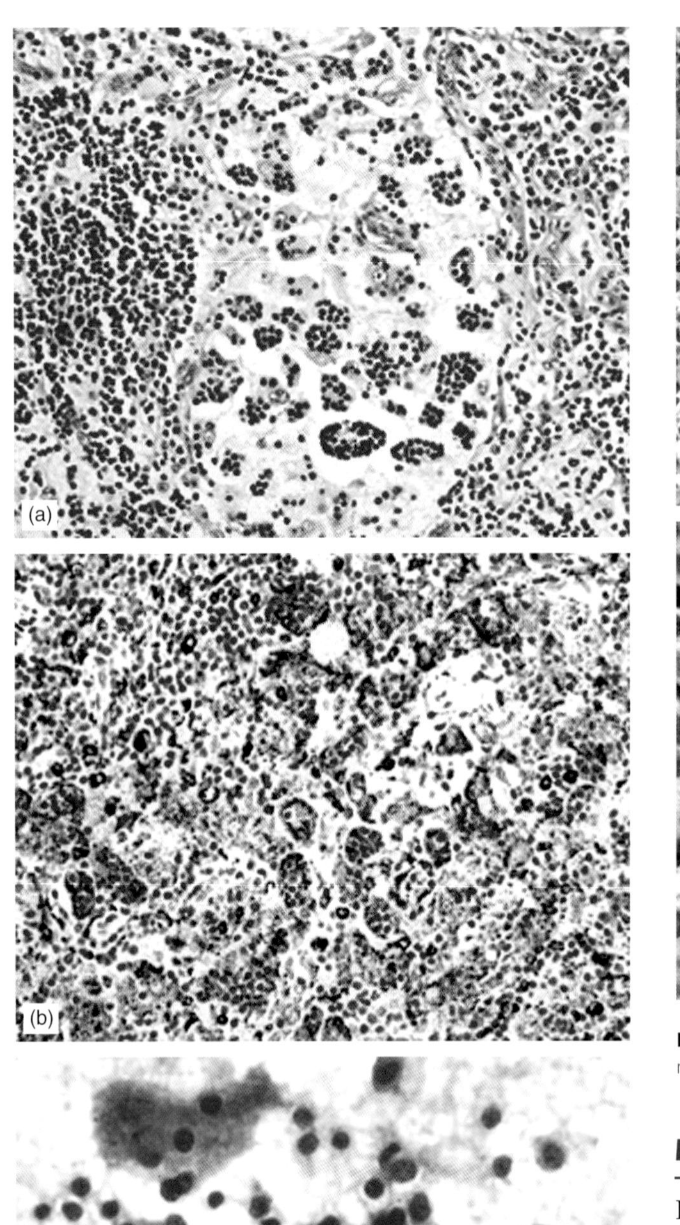

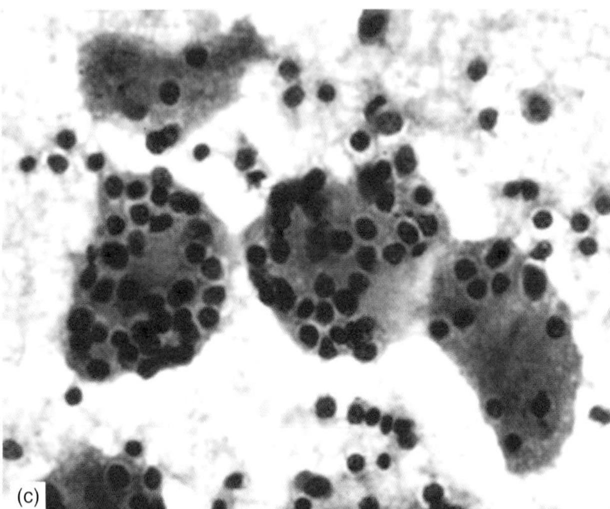

FIGURE 6.7 Sinus histiocytosis with massive lymphadenopathy (Rosai–Dorfman disease). Dilated sinuses contain numerous lymphocyte-containing histiocytes. Histiocytes demonstrate emperipolesis with the presence of numerous intact lymphocytes. (a) H&E section, (b) immunohistochemical stain for CD68, and (c) touch preparation. Courtesy of Sophie Song, M.D., Ph. D., Department of Pathology and Laboratory Medicine, UCLA Medical Center.

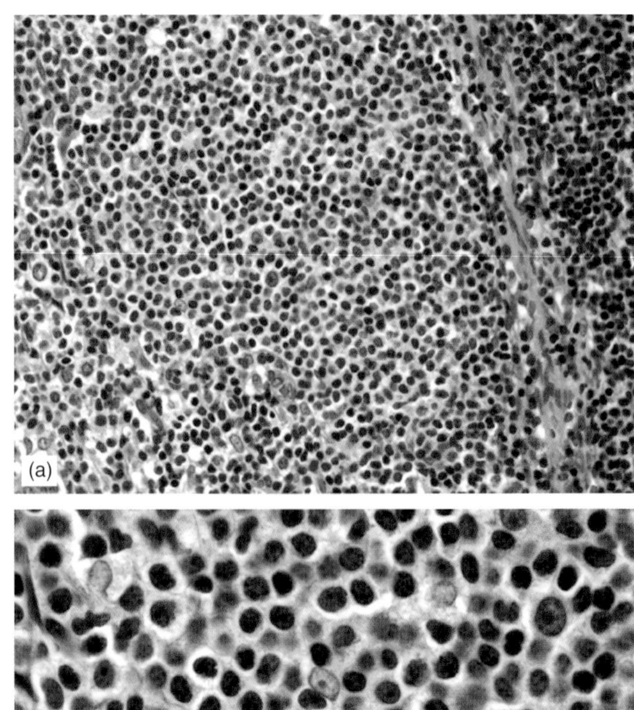

FIGURE 6.8 Lymph node section demonstrating hyperplasia of monocytoid B-cells: (a) low power and (b) high power.

Monocytoid B-Cell Hyperplasia

Hyperplasia of monocytoid B-cells (marginal zone B-cells) is predominantly sinusoidal, but it may also extend to the paracortical areas (Figure 6.8) [1, 29, 30]. Monocytoid B-cell hyperplasia has been observed in toxoplasmosis, viral infections, cat-scratch disease, and other non-specific lymphoid hyperplasias [1, 2, 31]. It is characterized by the presence of clusters of medium to large cells with abundant pale cytoplasm, round or convoluted nuclei, moderately condensed chromatin, and inconspicuous nucleoli. These cells express B-cell-associated antigens and lack monocytic markers. Monocytoid B-cell hyperplasia is often associated with follicular hyperplasia. The differential diagnosis includes monocytic/histiocytic disorders, marginal zone B-cell lymphoma, hairy cell leukemia, and mastocytosis.

Whipple's Disease

Whipple's disease is a multi-system bacterial infection caused by *Tropheryma whippeli*, most frequently involving the gastrointestinal tract and mesenteric lymph nodes

[1, 2, 32, 33]. The affected lymph nodes show sinus histiocytosis. Histiocytes show vacuolated cytoplasm and contain PAS-positive sickle-form particles. *Tropheryma whippeli* can be detected by the polymerase chain reaction (PCR) technique using species-specific sequences of the 16S ribosomal RNA [34].

Langerhans Cell Histiocytosis

Langerhans cell histiocytosis frequently affects lymph nodes in a sinus pattern (see Chapter 21 for more details) [35–37]. Langerhans cells are large mononuclear cells with abundant cytoplasm, folded or grooved nuclei, and inconspicuous nucleoli. They express CD1, S100, and Langerin (CD207), and demonstrate Birbeck granules in their cytoplasm by electron microscopy.

Hemophagocytic Syndromes

Hemophagocytic syndromes are discussed in Chapter 21 [38–40]. Affected lymph nodes show sinus histiocytosis with evidence of hemophagocytic activities.

GRANULOMATOUS LYMPHADENITIS

Granulomas may appear in different morphologic configuration, such as clusters of epithelioid histiocytes (e.g. sarcoidosis, fungal infections), with caseous necrosis (tuberculosis), or suppurative granulomas (e.g. cat-scratch disease, lymphogranuloma venereum). Sarcoidosis and cat-scratch disease are briefly discussed later as examples of this category.

Sarcoidosis

Sarcoidosis is a multicentric granulomatous disorder of unknown etiology characterized by well-defined epithelioid granulomas, often surrounded by lymphocytes and plasma cells, and less frequently by fibrosis (Figure 6.9) [1, 2, 41, 42]. Multinucleated giant cells may be present and may show asteroid bodies. Granulomas are usually non-necrotizing but occasionally may show small central fibrinoid necrosis. Sarcoid-like granulomas may be present in lymph nodes of patients with Hodgkin lymphoma, Crohn's disease, Whipple's disease, fungal infections, or tuberculosis. Therefore, morphologic diagnosis of sarcoidosis is based on the exclusion of other granulomatous lesions. Mediastinal and pulmonary hilar lymph nodes are the most frequent sites of involvements.

Cat-Scratch Disease

Cat-scratch disease is caused by a small, gram-negative bacterium, *Bartonella henselae*.

It is characterized by a variable-sized central fibrinoid necrosis, which in early stages is suppurative and contains variable numbers of neutrophils. The central necrosis is surrounded by palisading histiocytes, creating a stellate granuloma (Figure 6.10) [1, 2, 43–45]. The Warthin-Starry silver stain facilitates the detection of the bacteria. The involved lymph nodes often show follicular hyperplasia, sinus histiocytosis, proliferation of monocytoid B-cells, and increased number of immunoblasts.

Most of the cases are found in patients under the age of 18 years, and axillary lymph nodes are the most frequent sites of involvement. A history of exposure to cats is found in almost all cases [1, 2, 45].

MIXED PATTERN

Certain types of reactive lymphadenitis demonstrate a combination of follicular, paracortical, and/or sinus patterns. Examples are dermatopathic lymphadenitis, toxoplasmosis, Kikuchi's disease, and Kimura's disease.

Dermatopathic Lymphadenitis

Dermatopathic lymphadenitis is a reactive process often associated with inflammatory skin disorders [1, 2, 46]. The involved lymph nodes show nodular expansion of paracortical areas with pale-staining large cells consisting of histiocytes and Langerhans cells (Figures 6.11 and 6.12) [1, 2, 47]. Some of the histiocytes may contain melanin or hemosiderin pigments. Histiocytes are strongly CD68 and lysozyme positive and Langerhans cells express CD1 and CD100 (Figure 6.12) [48]. The nodular areas also contain scattered lymphocytes and plasma cells. The affected lymph nodes show sinus histiocytosis. Patients with mycosis fungoides may show regional enlarged lymph nodes with morphologic features very similar to dermatopathic lymphadenitis. Therefore, detection of nodal involvement in early stages of mycosis fungoides without significant architectural effacement is very difficult and may require gene rearrangement studies [49].

Toxoplasmosis

The involved lymph nodes in toxoplasmosis show follicular hyperplasia with expanded germinal centers. Multiple clusters of epithelioid histiocytes are present in the paracortical areas, adjacent to the follicles and within the germinal centers (Figure 6.13). Multinucleated giant cells are usually not present [50–52]. The subcapsular and trabecular sinuses are dilated and may contain numerous monocytoid B-cells. Similar morphologic features have been reported in enlarged lymph nodes of HIV-infected patients.

Kikuchi's Disease

Kikuchi's disease is a necrotizing lymphadenitis affecting young adults of particularly Asian descent. Women

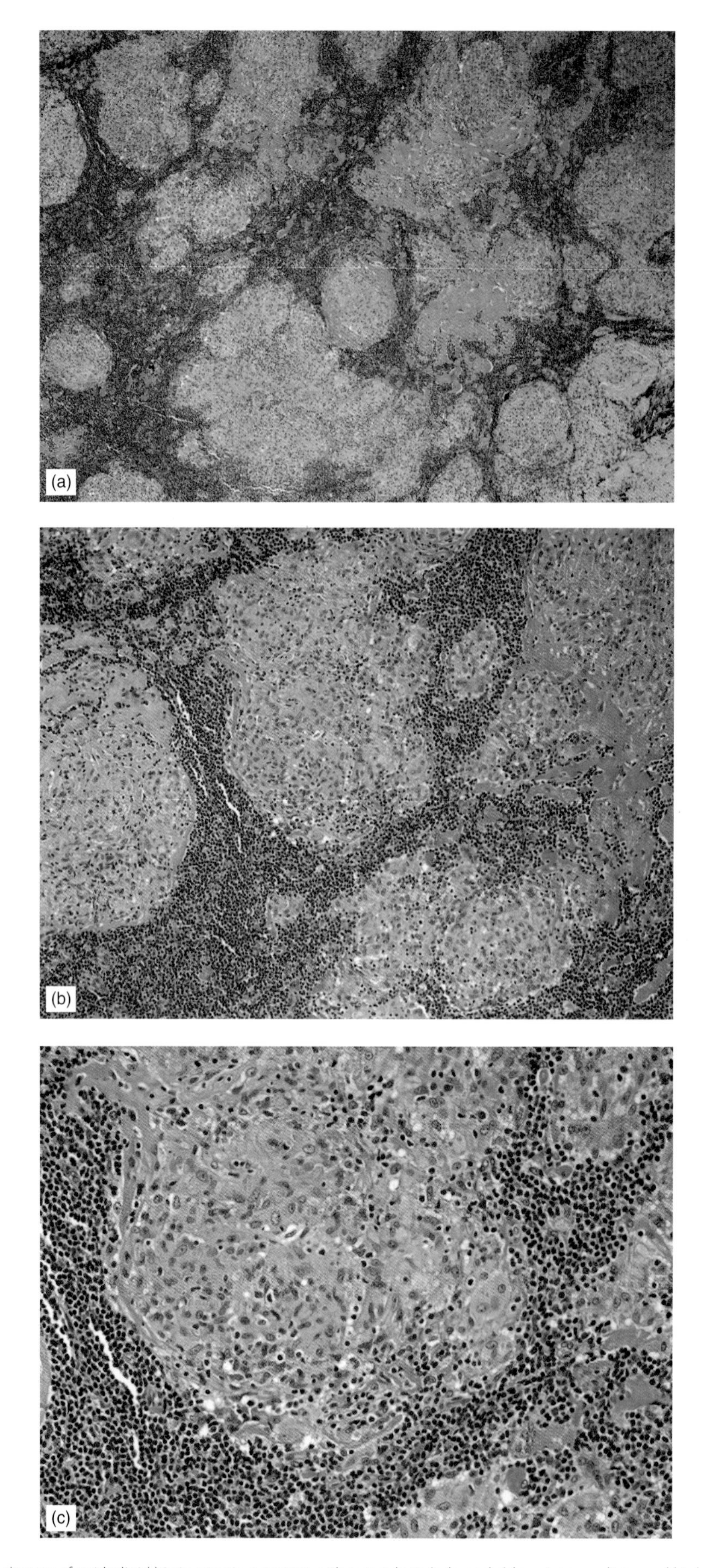

FIGURE 6.9 Lymph node: clusters of epithelioid histiocytes in a patient with sarcoidosis (a, b, and c) low, intermediate, and high power views, respectively.

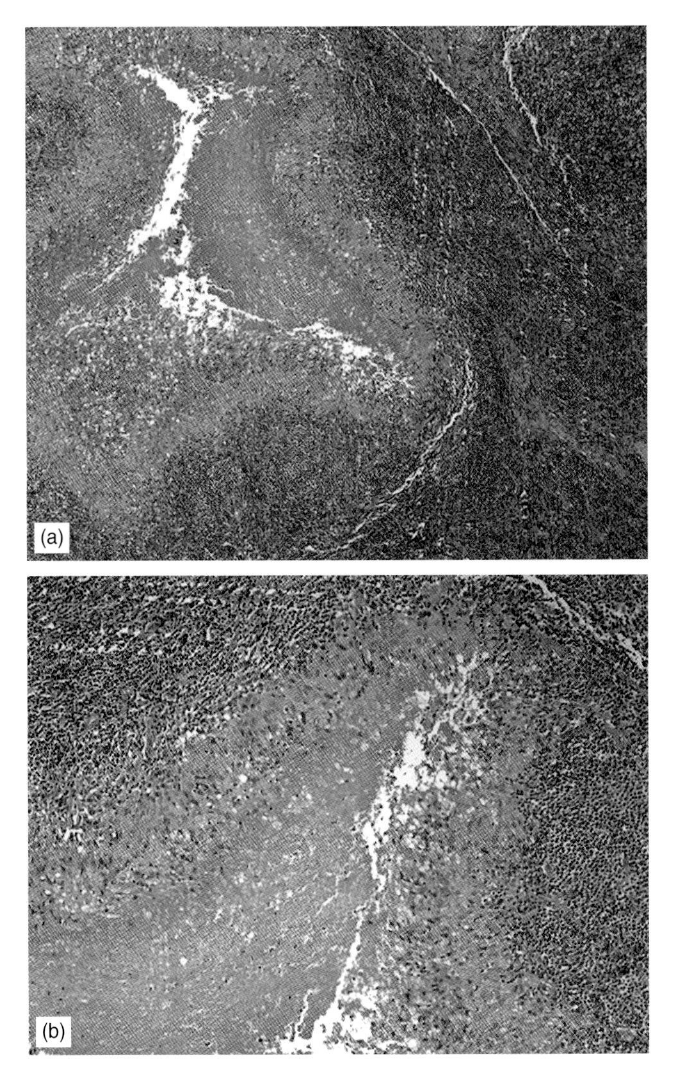

FIGURE 6.10 Lymphadenitis with irregular areas of necrosis surrounded by palisading histiocytes (stellate granulomas) are characteristic features of cat-scratch disease: (a) low power and (b) high power.

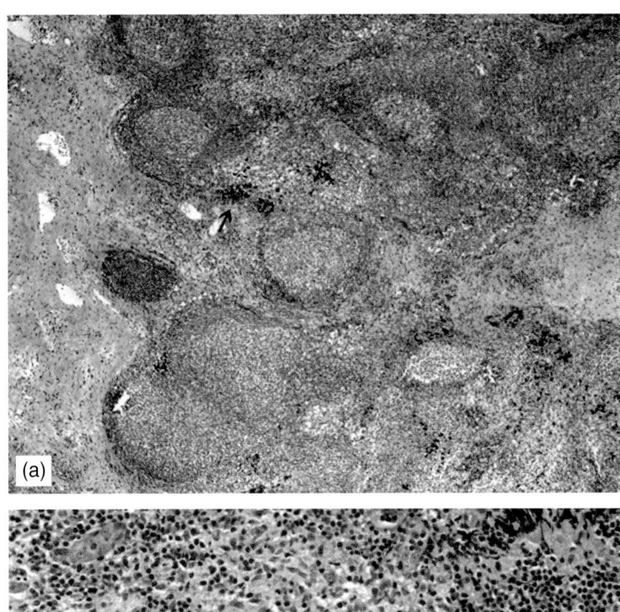

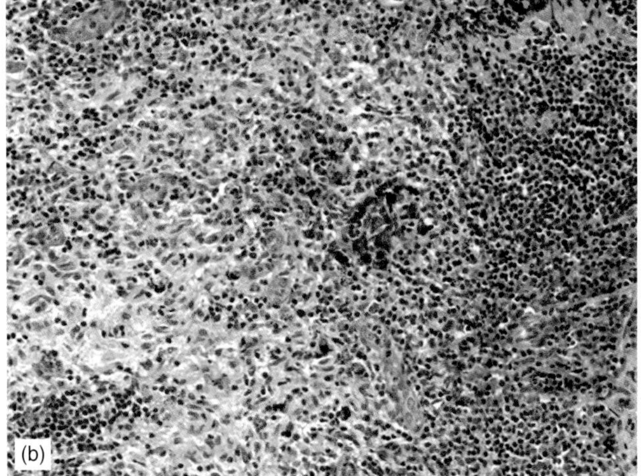

FIGURE 6.11 Dermatopathic lymphadenitis demonstrating follicular hyperplasia, sinus histiocytosis, pigmentation, and nodular expansion of paracortical areas with pale-staining large cells consisting of histiocytes and Langerhans cells: (a) low power and (b) high power.

are affected more than men [53–57]. It is a self-limited localized lymphadenitis, often involving cervical lymph nodes. The characteristic morphologic findings are patchy areas of necrosis with prominent karyorrhexis (apoptosis) and nuclear debris with or without coagulative necrosis. Neutrophils are rare or absent. The apoptotic foci are predominantly located in cortical and paracortical areas and are surrounded by large histiocytes and immunoblasts. These cells show some atypical features and may resemble Hodgkin cells. Some phagocytic histiocytes may show eccentric nuclei (signet ring histiocytes) and others may show foamy cytoplasms. Follicular hyperplasia may be present. Differential diagnosis of Kikuchi's disease includes Hodgkin lymphoma, systemic lupus erythematous, and herpes simplex lymphadenitis (Figure 6.14) [58].

Kimura's Disease

Kimura's disease (eosinophilic hyperplastic lymphogranuloma) is a rare form of chronic inflammatory disorder involving subcutaneous tissue, predominantly in the head and neck region. It is frequently associated with regional lymphadenopathy and/or salivary gland involvement [59]. This condition has a predilection for males of Asian descent [59–61]. The involved lymph nodes show follicular hyperplasia with the presence of numerous eosinophils in the follicle, paracortex, sinusoids, and perinodal soft tissues [2]. There is a distinct IgE-positive dendritic network in the germinal centers [2].

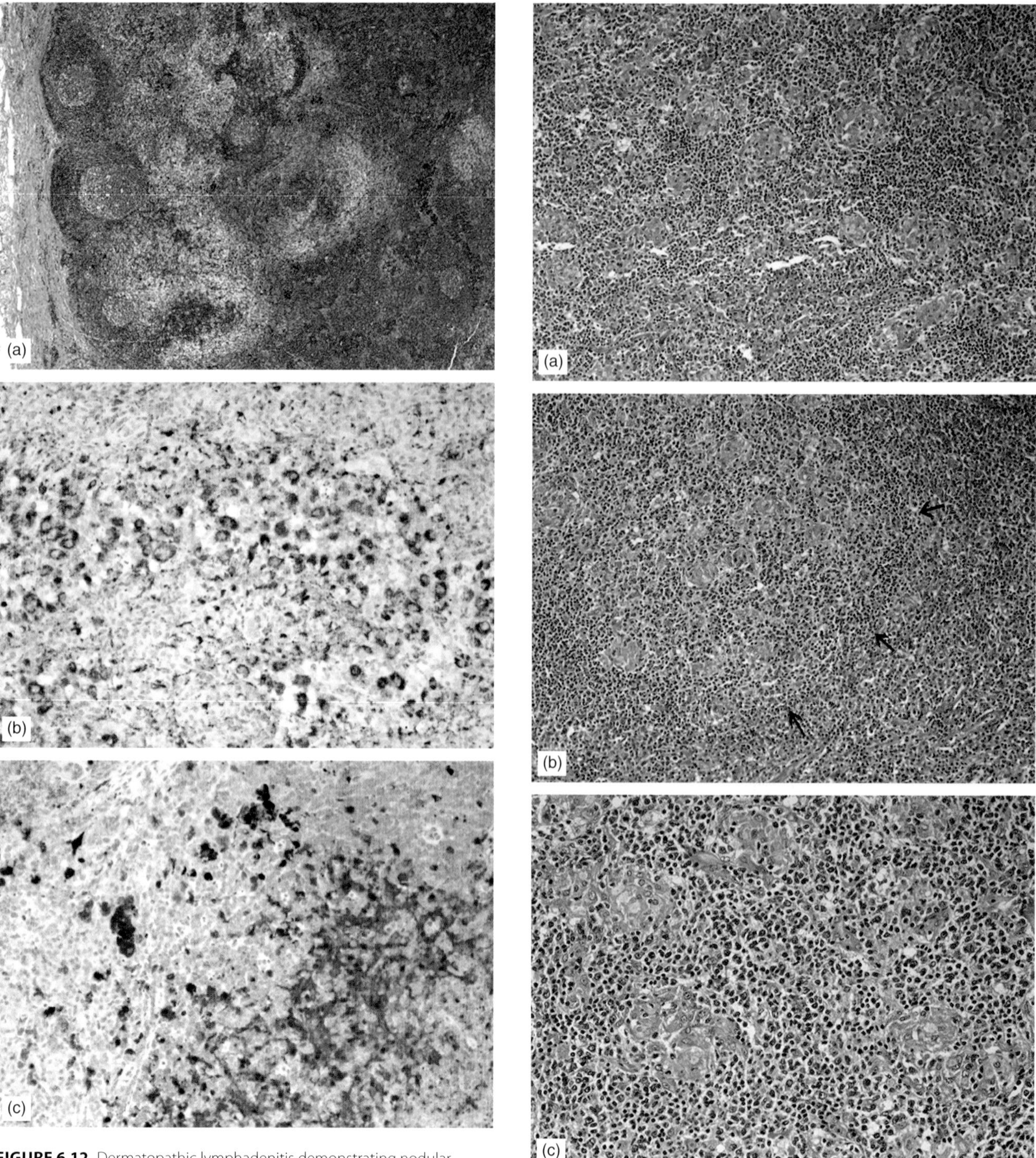

FIGURE 6.12 Dermatopathic lymphadenitis demonstrating nodular expansion of paracortical areas with pale-staining large cells (a). These cells consist of a mixture of Langerhans cells (b), S100-positive, and histiocytes (c), CD68-positive.

FIGURE 6.13 Epithelioid granulomas in toxoplasmosis. Clusters of epithelioid histiocytes are present in the paracortical areas (a) as well as within the follicles (b, arrows). A higher power view of the epithelioid granulomas is presented in (c).

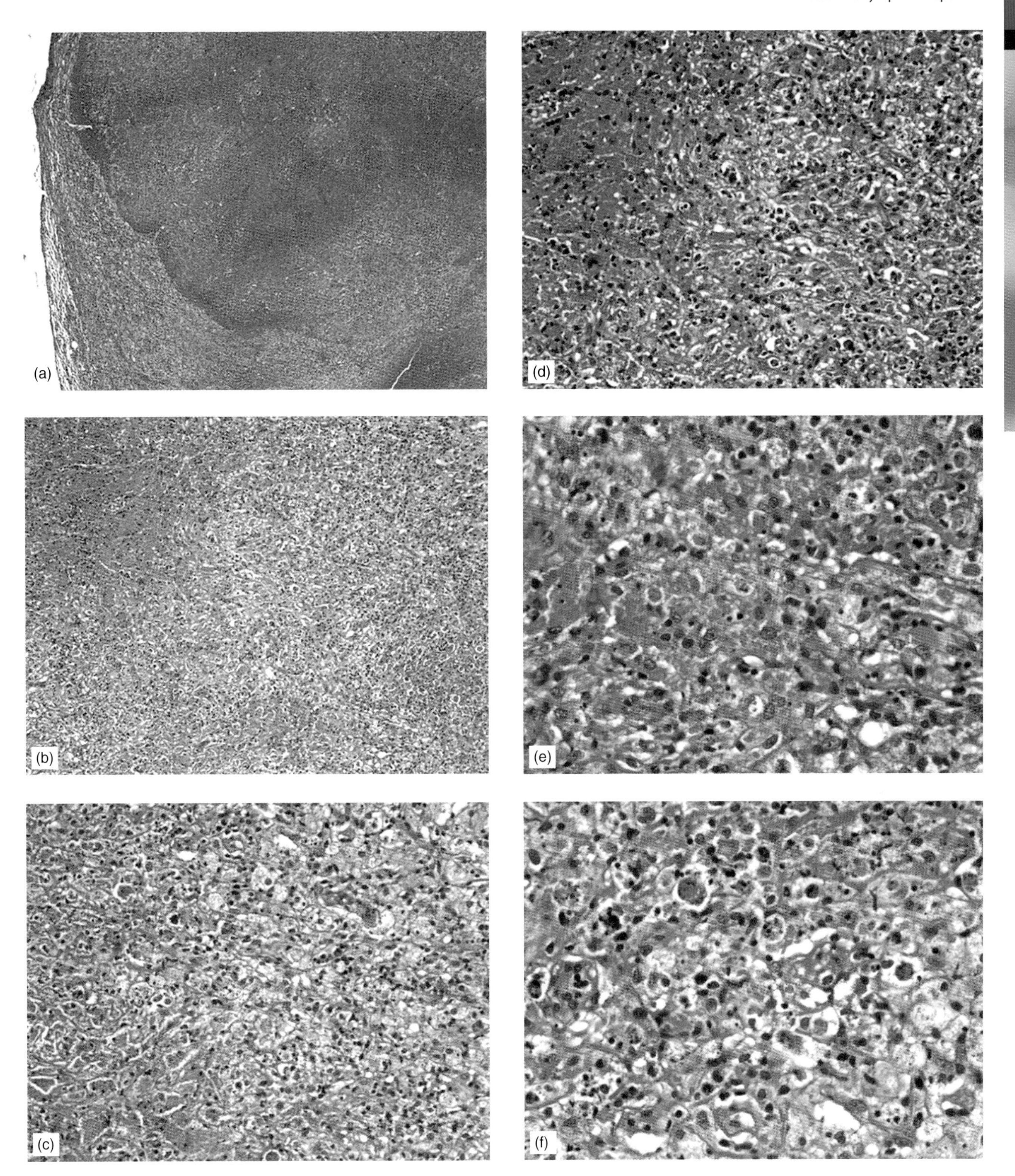

FIGURE 6.14 Kikuchi's disease. A large area of necrotic debris and histiocytes replacing a significant proportion of the lymph node: (a) low power, (b) intermediate power, and (c) high power views. The necrotic debris lack neutrophils (c and d). Numerous histiocytes show pale cytoplasm and contain nuclear debris (e and f). Courtesy of Sophie Song, M.D., Ph.D., Department of Pathology and Laboratory Medicine, UCLA Medical Center.

References

1. Knowels DM. (2001). *Neoplastic Hematopathology*, 2nd ed. Lippincott Williams & Wilkins, Philadelphia.
2. Ferry JA, Harris NL. (1997). *Atlas of Lymphoid Hyperplasia and Lymphoma*. W.B. Saunders, Philadelphia.
3. His ED. (2007). *Hematopathology*. Churchill Livingstone, Philadelphia.
4. Burke JS. (1988). Reactive lymphadenopathies. *Semin Diagn Pathol* **5**, 312.
5. Krishnan J, Danon AD, Frizzera G. (1993). Reactive lymphadenopathies and atypical lymphoproliferative disorders. *Am J Clin Pathol* **99**, 385–96.
6. Dorfman RF, Warnke R. (1974). Lymphadenopathy simulating the malignant lymphomas. *Hum Pathol* **5**, 519–50.
7. Grassi W, De Angelis R, Lamanna G, Cervini C. (1998). The clinical features of rheumatoid arthritis. *Eur J Radiol Suppl* **1**, S18–S24.
8. Horton MR. (2004). Rheumatoid arthritis associated interstitial lung disease. *Crit Rev Comput Tomogr* **45**, 429–40.
9. Haines KA. (2007). Juvenile idiopathic arthritis: Therapies in the 21st century. *Bull NYU Hosp Jt Dis* **65**, 205–11.
10. Ravelli A, Martini A. (2007). Juvenile idiopathic arthritis. *Lancet* **369**, 767–78.
11. Jackson CE, Puck JM. (1999). Autoimmune lymphoproliferative syndrome, a disorder of apoptosis. *Curr Opin Pediatr* **11**, 521–7.
12. Nosanchuk JS, Schnitzer B. (1969). Follicular hyperplasia in lymph nodes from patients with rheumatoid arthritis. A clinicopathologic study. *Cancer* **24**, 243–54.
13. Castleman B, Iverson L, Menendez VP. (1956). Localized mediastinal lymphnode hyperplasia resembling thymoma. *Cancer* **9**, 822–30.
14. Keller AR, Hochholzer L, Castleman B. (1972). Hyaline-vascular and plasma-cell types of giant lymph node hyperplasia of the mediastinum and other locations. *Cancer* **29**, 670–83.
15. Peterson BA, Frizzera G. (1993). Multicentric Castleman's disease. *Semin Oncol* **20**, 636–47.
16. Frizzera G. (1988). Castleman's disease and related disorders. *Semin Diagn Pathol* **5**, 346–64.
17. Danon AD, Krishnan J, Frizzera G. (1993). Morpho-immunophenotypic diversity of Castleman's disease, hyaline-vascular type: With emphasis on a stroma-rich variant and a new pathogenetic hypothesis. *Virchows Arch A Pathol Anat Histopathol* **423**, 369–82.
18. Dham A, Peterson BA. (2007). Castleman disease. *Curr Opin Hematol* **14**, 354–9.
19. Menke DM, Tiemann M, Camoriano JK, Chang SF, Madan A, Chow M, Habermann TM, Parwaresch R. (1996). Diagnosis of Castleman's disease by identification of an immunophenotypically aberrant population of mantle zone B lymphocytes in paraffin-embedded lymph node biopsies. *Am J Clin Pathol* **105**, 268–76.
20. Albores-Saavedra J, Vuitch F, Delgado R, Wiley E, Hagler H. (1994). Sinus histiocytosis of pelvic lymph nodes after hip replacement. A histiocytic proliferation induced by cobalt-chromium and titanium. *Am J Surg Pathol* **18**, 83–90.
21. Jaffe ES. (1988). Histiocytoses of lymph nodes: Biology and differential diagnosis. *Semin Diagn Pathol* **5**, 376–90.
22. Rosai J, Dorfman RF. (1969). Sinus histiocytosis with massive lymphadenopathy. A newly recognized benign clinicopathological entity. *Arch Pathol* **87**, 63–70.
23. Rosai J, Dorfman RF. (1972). Sinus histiocytosis with massive lymphadenopathy: A pseudolymphomatous benign disorder. Analysis of 34 cases. *Cancer* **30**, 1174–88.
24. Horneff G, Jürgens H, Hort W, Karitzky D, Göbel U. (1996). Sinus histiocytosis with massive lymphadenopathy (Rosai–Dorfman disease): Response to methotrexate and mercaptopurine. *Med Pediatr Oncol* **27**, 187–92.
25. Foucar E, Rosai J, Dorfman R. (1990). Sinus histiocytosis with massive lymphadenopathy (Rosai–Dorfman disease): Review of the entity. *Semin Diagn Pathol* **7**, 19–73.
26. Paulli M, Locatelli F, Kindl S, Boveri E, Facchetti F, Porta F, Rosso R, Nespoli L, Magrini U. (1992). Sinus histiocytosis with massive lymphoadenopathy (Rosai–Dorfman disease). Clinico-pathological analysis of a paediatric case. *Eur J Pediatr* **151**, 672–5.
27. Levine PH, Jahan N, Murari P, Manak M, Jaffe ES. (1992). Detection of human herpesvirus 6 in tissues involved by sinus histiocytosis with massive lymphadenopathy (Rosai–Dorfman disease). *J Infect Dis* **166**, 291–5.
28. Paulli M, Rosso R, Kindl S, Boveri E, Marocolo D, Chioda C, Agostini C, Magrini U, Facchetti F. (1992). Immunophenotypic characterization of the cell infiltrate in five cases of sinus histiocytosis with massive lymphadenopathy (Rosai–Dorfman disease). *Hum Pathol* **23**, 647–54.
29. Plank L, Hell K, Hansmann ML, Pringle JH, Lauder I, Fischer R. (1995). Reactive versus neoplastic monocytoid B-cell proliferations. *In situ* hybridization study of immunoglobulin light chain mRNA. *Am J Clin Pathol* **103**, 330–7.
30. Kojima M, Nakamura S, Itoh H, Yoshida K, Shimizu K, Motoori T, Yamane N, Joshita T, Suchi T. (1998). Occurrence of monocytoid B-cells in reactive lymph node lesions. *Pathol Res Pract* **194**, 559–65.
31. Hunt JP, Chan JA, Samoszuk M, Brynes RK, Hernandez AM, Bass R, Weisenburger DD, Müller-Hermelink K, Nathwani BN. (2001). Hyperplasia of mantle/marginal zone B cells with clear cytoplasm in peripheral lymph nodes. A clinicopathologic study of 35 cases. *Am J Clin Pathol* **116**, 550–9.
32. Sieracki JC. (1958). Whipple's disease-observation on systemic involvement. I: Cytologic observations. *AMA Arch Pathol* **66**, 464–7.
33. Sieracki JC, Fine G. (1959). Whipple's disease; observations on systemic involvement. II: Gross and histologic observations. *AMA Arch Pathol* **67**, 81–93.
34. Fenollar F, Fournier PE, Raoult D, Gerolami R, Lepidi H, Poyart C. (2002). Quantitative detection of *Tropheryma whipplei* DNA by real-time PCR. *J Clin Microbiol* **40**, 1119–20.
35. Willman CL, Busque L, Griffith BB, Favara BE, McClain KL, Duncan MH, Gilliland DG. (1994). Langerhans'-cell histiocytosis (histiocytosis X) – a clonal proliferative disease. *N Engl J Med* **331**, 154–60.
36. Kakkar S, Kapila K, Verma K. (2001). Langerhans cell histiocytosis in lymph nodes. Cytomorphologic diagnosis and pitfalls. *Acta Cytol* **45**, 327–32.
37. Lieberman PH, Jones CR, Steinman RM, Erlandson RA, Smith J, Gee T, Huvos A, Garin-Chesa P, Filippa DA, Urmacher C, Gangi MD, Sperber M. (1996). Langerhans cell (eosinophilic) granulomatosis. A clinicopathologic study encompassing 50 years. *Am J Surg Pathol* **20**, 519–52.
38. Risdall RJ, Mckenna RW, Nesbit ME, Krivit W, Balfour HH Jr., Simmons RL, Brunning RD. (1979). Virus-associated hemophagocytic syndrome: A benign histiocytic proliferation distinct from malignant histiocytosis. *Cancer* **44**, 993–1002.
39. Risdall RJ, Brunning RD, Hernandez JI, Gordon DH. (1984). Bacteria-associated hemophagocytic syndrome. *Cancer* **54**, 2968–72.
40. Yao M, Cheng AL, Su IJ, Lin MT, Uen WC, Tien HF, Wang CH, Chen YC. (1994). Clinicopathological spectrum of haemophagocytic syndrome in Epstein-Barr virus-associated peripheral T-cell lymphoma. *Br J Haematol* **87**, 535–43.
41. Mitchell DN, Scadding JG. (1974). Sarcoidosis. *Am Rev Respir Dis* **110**, 774–802.
42. Shaffer S. (1996). Benign lymphoproliferative disorders. *Semin Oncol Nurs* **12**, 28–37.
43. Caponetti G, Pantanowitz L. (2007). Cat-scratch disease lymphadenitis. *Ear Nose Throat J* **86**, 449–50.
44. Tamir E. (2007). Cat scratch disease. *Acta Dermatovenerol Croat* **15**, 47.
45. English R. (2006). Cat-scratch disease. *Pediatr Rev* **27**, 123–8.
46. Winter LK, Spiegel JH, King T. (2007). Dermatopathic lymphadenitis of the head and neck. *J Cutan Pathol* **34**, 195–7.
47. Gould E, Porto R, Albores-Saavedra J, Ibe MJ. (1988). Dermatopathic lymphadenitis. The spectrum and significance of its morphologic features. *Arch Pathol Lab Med* **112**, 1145–50.

48. Burke JS, Sheibani K, Rappaport H. (1986). Dermatopathic lymphadenopathy. An immunophenotypic comparison of cases associated and unassociated with mycosis fungoides. *Am J Pathol* **123**, 256–63.
49. Westhoff TH, Loddenkemper C, Hörl MP, Schmidt S, Anagnostopoulos I, Hummel M, Zidek W, van der Giet M. (2006). Dermatopathic lymphadenopathy: A differential diagnosis of enlarged lymph nodes in uremic pruritus. *Clin Nephrol* **66**, 472–5.
50. Stansfeld AG. (1961). The histological diagnosis of toxoplasmic lymphadenitis. *J Clin Pathol* **14**, 565–73.
51. McCabe RE, Brooks RG, Dorfman RF, Remington JS. (1987). Clinical spectrum in 107 cases of toxoplasmic lymphadenopathy. *Rev Infect Dis* **9**, 754–74.
52. Weiss LM, Chen YY, Berry GJ, Strickler JG, Dorfman RF, Warnke RA. (1992). Infrequent detection of *Toxoplasma gondii* genome in toxoplasmic lymphadenitis: A polymerase chain reaction study. *Hum Pathol* **23**, 154–8.
53. Park HS, Sung MJ, Park SE, Lim YT. (2007). Kikuchi-Fujimoto disease of 16 children in a single center of Korea. *Pediatr Allergy Immunol* **18**, 174–8.
54. Yilmaz M, Camci C, Sari I, Okan V, Sevinc A, Onat AM, Buyukhatipoglu H. (2006). Histiocytic necrotizing lymphadenitis (Kikuchi-Fujimoto's disease) mimicking systemic lupus erythematosus: A review of two cases. *Lupus* **15**, 384–7.
55. Boula AM, Bizakis JG, Tsirakis GE, Chimona TS, Stathopoulos EN, Alexandrakis MG. (2005). Kikuchi's disease: A benign cause of fever and cervical lymphadenopathy. *Eur J Intern Med* **16**, 356–8.
56. Famularo G, Giustiniani MC, Marasco A, Minisola G, Nicotra GC, De Simone C. (2003). Kikuchi Fujimoto lymphadenitis: Case report and literature review. *Am J Hematol* **74**, 60–3.
57. Lin HC, Su CY, Huang CC, Hwang CF, Chien CY. (2003). Kikuchi's disease: A review and analysis of 61 cases. *Otolaryngol Head Neck Surg* **128**, 650–3.
58. Dikov DI, Staikova ND, Solakov PT. (2000). Differential diagnosis of Kikuchi's disease and systemic lupus erythematosus lymphadenopathy: Clinicopathologic algorithm. *Folia Med (Plovdiv)* **42**, 34–6.
59. Chen H, Thompson LD, Aguilera NS, Abbondanzo SL. (2004). Kimura disease: A clinicopathologic study of 21 cases. *Am J Surg Pathol* **28**, 505–13.
60. Li TJ, Chen XM, Wang SZ, Fan MW, Semba I, Kitano M. (1996). Kimura's disease: A clinicopathologic study of 54 Chinese patients. *Oral Surg Oral Med Oral Pathol Oral Radiol Endod* **82**, 549–55.
61. Gao Y, Chen Y, Yu GY. (2006). Clinicopathologic study of parotid involvement in 21 cases of eosinophilic hyperplastic lymphogranuloma (Kimura's disease). *Oral Surg Oral Med Oral Pathol Oral Radiol Endod* **102**, 651–8.

7

CHAPTER

Bone Marrow Aplasia

Faramarz Naeim, P. Nagesh Rao and Wayne W. Grody

Bone marrow aplasia (aplastic anemia) refers to those hematologic conditions that are caused by a marked reduction and/or defect in the pluripotent or committed stem cells, or the failure of the bone marrow microenvironment to support hematopoiesis. The clinical outcome is anemia, leukopenia, and thrombocytopenia (pancytopenia) [1–4]. The term "aplastic anemia" (AA) is a misnomer, because the patients, in addition to anemia, also suffer from leukopenia and thrombocytopenia.

In this chapter, constitutional and acquired AA, dyskeratosis congenita, Shwachman–Diamond syndrome, Diamond–Blackfan anemia (DBA), amegakaryocytosis, and paroxysmal nocturnal hemoglobinuria are discussed (Table 7.1) [5–9]. Bone marrow failure due to myelodysplastic syndromes (MDS), leukemias, myelofibrosis, and other disorders are discussed in the following chapters.

TABLE 7.1 Classification of bone marrow aplasia.

1. Constitutional
   (a) Fanconi anemia
   (b) Dyskeratosis congenita
   (c) Shwachman–Diamond syndrome
   (d) Diamond–Blackfan anemia
   (e) Amegakaryocytosis
2. Acquired
   (a) Idiopathic aplastic anemia
   (b) Secondary aplastic anemia
      (i) Chemical and physical agents
         – Drugs and other chemicals
         – Radiation
      (ii) Infection
         – Viral: hepatitis, EBV, HIV
         – Others: tuberculosis, dengue fever
      (iii) Immunologic (humoral and/or cellular)
      (iv) Metabolic (pancreatitis, pregnancy)
3. Paroxysmal nocturnal hemoglobinuria
4. Others
   (a) Hypoplastic myelodysplastic syndromes
   (b) Bone marrow replacement
      (i) Malignant neoplasms
      (ii) Fibrosis
      (iii) Others

FANCONI ANEMIA

Fanconi anemia (FA) is the most common form of congenital bone marrow aplasia [10–12]. It is an autosomal recessive or X-linked disorder with a prevalence of about 1 in 300,000 in most populations, but with much higher frequencies in the Afrikaner population of South Africa and Ashkenazi Jews [10, 12, 13]. FA is associated with physical abnormalities and affects males more than females with a ratio of about 2:1 [10, 12]. The congenital AA without physical abnormalities is known as Eastern–Dameshek anemia.

Etiology and Pathogenesis

At least 11 complementation groups (FA-A to FA-L) have been reported, and so far, eight genes have been cloned [10, 12, 14–16]. The FA genes may contribute to genomic stability, DNA repair, and control of apoptosis [10, 12, 14–16]. The defects in DNA repair in FA patients may lead to spontaneous chromosomal breakage, which is significantly enhanced in homozygotes, particularly when the drugs mitomycin and diepoxybutane are used [10, 14–16]. The *FA-A* gene is responsible for 60–65% of the patients with FA and has been mapped to chromosome 16q24.3

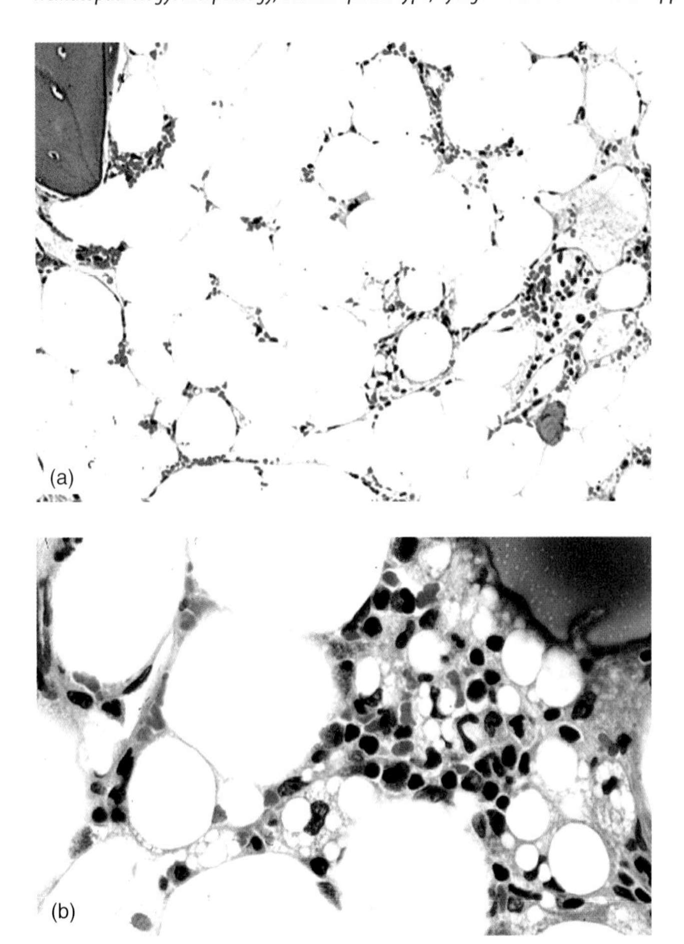

FIGURE 7.1 Bone marrow biopsy section from a patient with Fanconi anemia demonstrating marked hypocellularity with small foci of hematopoietic cells: (a) low power and (b) high power views.

[10, 14–16]. Interestingly, the *FAD1* gene turned out to be the same as the *BRCA2* gene involved in familial breast/ovarian cancer, a dominantly inherited cancer predisposition [11, 17, 18].

An abnormality in the interferon signaling mechanism has been suggested in FA which may lead to enhanced apoptosis. There is also some evidence of abnormal telomeres in FA patients [19]. Telomeres have been implicated in the control of both genomic stability and cell proliferation capacity [19].

Pathology

Morphology

Bone marrow biopsy sections in early stages of the disease may appear hyper- or normocellular with some megaloblastic changes but eventually become hypoplastic and depict marked hypocellularity with scattered foci of hematopoietic cells, predominantly erythroid (Figure 7.1) [1–3]. Often there is an increased proportion of plasma cells and lymphocytes. Bone marrow smears may show increased mast cells and evidence of hemophagocytosis, particularly in early stages of the disease. These morphologic features are not pathognomonic for FA and are also observed in patients with acquired AA [1–3].

Blood examination is usually normal at birth. Usually, microcytosis is the first detected abnormality, followed by elevated levels of fetal hemoglobin, thrombocytopenia, and neutropenia between the ages of 5 and 10 years [1–3, 12].

Molecular Genetics and Cytogenetics

Although the central diagnostic study for FA is the demonstration of the chromosome breakage defect, the rearrangements are so varied and inconsistent that the molecular study of the breakpoints has no practical diagnostic value [20]. However, the cloning of the genes associated with most of the complementation groups has made definitive molecular genetic diagnosis possible. DNA testing of several of the FA genes is now available, consisting of either gene sequencing or targeted mutation analysis such as the predominant IVS4 + 1A→T mutation in the Ashkenazi-Jewish population. Moreover, once the precise gene defect is discovered in a proband, the same approach can be used for prenatal diagnosis in future pregnancies within the family. In addition, population-based carrier screening (i.e. for those couples with no family history of the disorder) is being offered in some centers for Ashkenazi Jews of reproductive age, in whom the carrier frequency for mutations in the gene for FA type C is 1 in 90 [21].

It is worth noting that FA is the first disorder for which couples have availed themselves of the technique of preimplantation genetic diagnosis (involving single-cell biopsy and polymerase chain reaction (PCR) testing of early embryos conceived by *in vitro* fertilization) to select a baby having the identical HLA type to serve as a bone marrow donor for their living affected child [22]. Clearly, such efforts, although technically feasible, raise a number of troubling ethical issues [23, 24].

Clinical Aspects

Characteristic congenital malformations, such as generalized skin hyperpigmentation (café au lait spots) and areas of hypopigmentation, microcephaly, hypogonadism, abnormality of thumbs, and short stature, are present in up to 60–70% of the affected children [6, 10, 12]. Other abnormalities include microphthalmia, renal hypoplasia, horseshoe kidneys, or double urethras [6, 10, 12]. The hematologic findings evolve gradually and may take months to years to reach full-blown pancytopenia. Thrombocytopenia is among the most common initial findings [10, 12, 25].

FA patients have an increased risk of developing clonal bone marrow cytogenetic abnormalities, such as myelodysplastic syndrome (MDS) and/or acute myelogenous leukemia (AML) [26–28]. The actuarial risk of MDS and AML is over 50% by the age of 40. This risk is higher in patients with cytogenetic abnormalities. There is also an elevated risk of squamous carcinoma of head and neck and gynecologic system, and various other solid tumors, in patients with FA [26, 29].

OTHER CONGENITAL BONE MARROW APLASIAS

Dyskeratosis Congenita

Dyskeratosis congenita (DC) is an X-linked recessive trait which is characterized by bone marrow failure and a triad of mucosal leukoplakia, nail dystrophy, and abnormal skin pigmentation [12, 30–32]. Approximately 20% of the patients may also suffer pulmonary dysfunction characterized by reduced diffusion capacity [33].

The dyskeratin gene (*DKC1*) at chromosome Xq28 is mutated. This gene appears to play a role in ribosome synthesis and telomerase function [34]. Sequence analysis of the *DKC1* gene is available in a small number of laboratories. Detection of mutations can be used for diagnosis, carrier screening, and prenatal testing.

The approximate median ages for the demonstration of somatic abnormalities and bone marrow failure are 8 and 10 years, respectively [12, 33]. Bone marrow becomes markedly hypoplastic with morphologic features similar to those of FA.

Over 90% of the affected patients are male [12, 33, 35]. DC patients have a higher tendency to develop MDS, AML, and skin and oropharynx cancer [35, 36].

Shwachman–Diamond Syndrome

Shwachman–Diamond syndrome or Shwachman–Diamond–Oski syndrome is a rare autosomal disorder which presents its clinical symptoms during infancy [37, 38]. It is characterized by skeletal anomalies, short stature, pancreatic insufficiency, and progressive bone marrow failure [37, 38]. The pancreatic insufficiency in this syndrome is distinguished from cystic fibrosis by a normal sweat chloride test result. Mutations of a gene referred to as Shwachman–Bodian–Diamond syndrome (*SBDS*) have been reported; it is located on chromosome 7q11 and so far has no known function [39]. Sequencing of the *SBDS* gene is available in several reference laboratories [40]. Although exocrine pancreatic dysfunction is a feature of the disorder and the cystic fibrosis gene (*CFTR*) is located on chromosome 7 (distal at 7q31.2), the pathogenesis is different and these patients do not have cystic fibrosis, so *CFTR* mutation testing is of no value. Over-expression of *p53* has been observed in bone marrow [41], but this is a non-specific finding and not used for diagnostic purposes.

Neutropenia is the major hematologic characteristics of this disorder, which is often intermittent or cyclic (Figure 7.2). Patients are prone to infection, particularly caused by gram-negative organisms, *Hemophilus influenza* or *Staphylococcus aureus*. Elevated levels of fetal hemoglobin are detected in up to 80%, and AA is observed in 20–25% of patients with Shwachman–Diamond syndrome [38, 42].

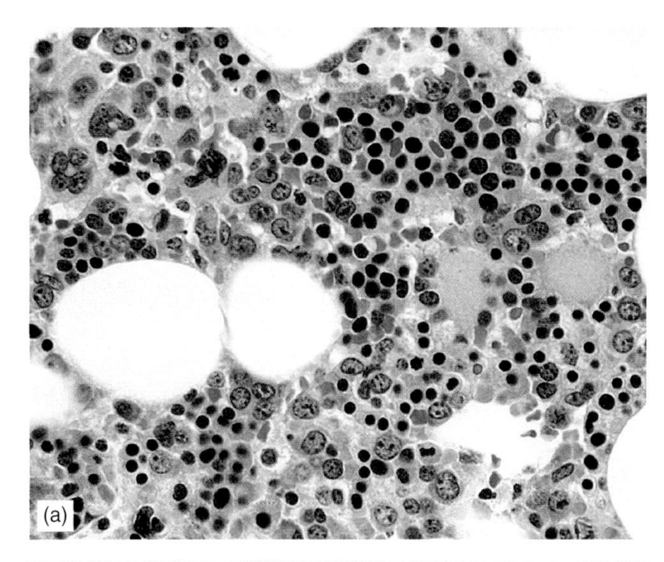

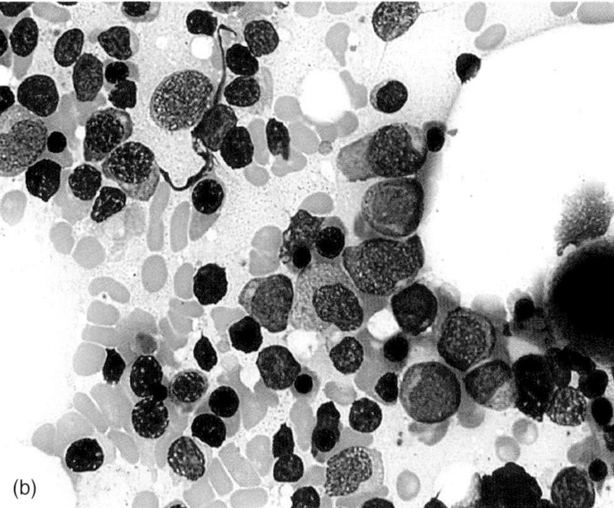

FIGURE 7.2 Bone marrow smears of patients with Shwachman–Diamond syndrome show reduced number of neutrophils and bands: (a) low power and (b) high power.

Diamond–Blackfan Anemia

Diamond–Blackfan anemia (DBA) is a pure red cell aplasia predominantly demonstrated in infancy and early childhood (Figure 7.3) [42–44]. DBA is about 45% familial and is often associated with physical anomalies, such as thumb malformations, growth retardation, and craniofacial deformities [42, 45]. Hematologic findings include macrocytic anemia, elevated fetal hemoglobin levels, and increased erythrocyte adenosine deaminase activity [42, 43]. DBA patients may eventually develop pancytopenia and aplastic bone marrow.

The first DBA gene, *RPS19*, located on chromosome 19q13, was found to be mutated in approximately 25% of patients. This gene encodes a protein which is a component of the 40S ribosomal subunit. Most patients are found to be heterozygous suggesting that the disease is caused by protein haploinsufficiency [46]. There are probably at least two other genes associated with some cases of DBA, but they have not been defined sufficiently for testing. The genetics of the disorder are complicated by the absence of family history in many cases, which could be due to either sporadic incidence or a dominant gene with low penetrance. Testing for parvovirus B19 by PCR of bone marrow samples may be performed as part of the differential diagnosis of red cell aplasia in an infant.

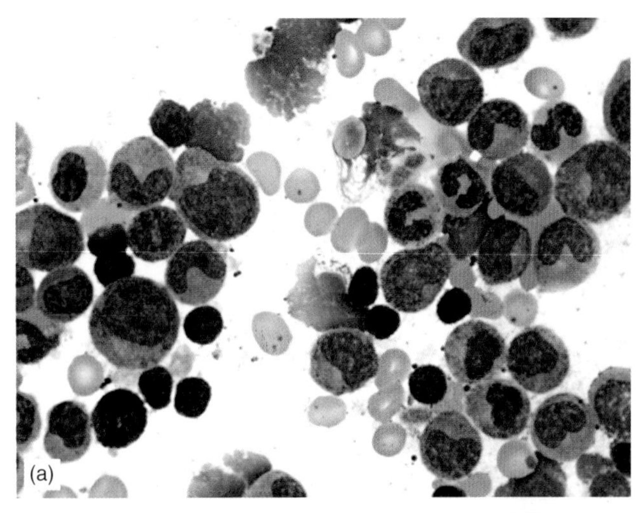

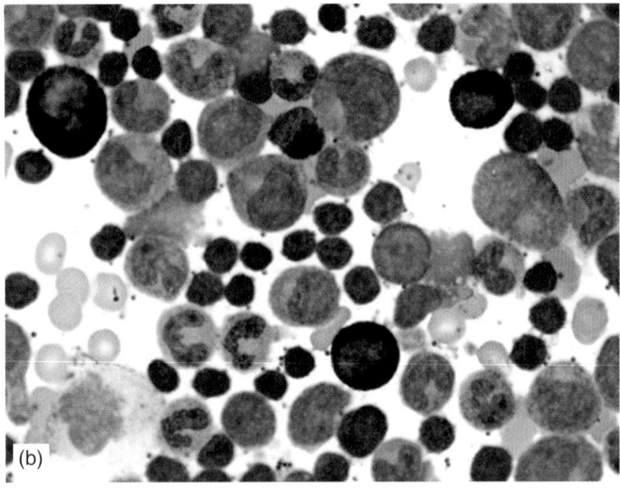

FIGURE 7.3 Bone marrow smears from a patient with pure red cell aplasia (Diamond–Blackfan anemia) demonstrating lack of erythroid precursors: (a) low power and (b) high power.

Amegakaryocytosis

Congenital amegakaryocytosis (amegakaryocytic thrombocytopenia) is a rare disorder of infancy with markedly reduced or absent megakaryocytes in bone marrow and therefore isolated thrombocytopenia [12, 47]. Approximately 50% of these patients may eventually develop AA and pancytopenia. The cause of this disorder in some children appears to be mutations of the thrombopoietin receptor gene, *MPL* (CD110) on chromosome 1p35 [48]. The autosomal recessive form of amegakaryocytosis is caused by mutations in the *MPL* gene [49], which block end-organ response to thrombopoietin (see Chapter 24). Because of the rarity of the disorder, testing for these mutations is available only on a research basis.

The serum concentration of thrombopoietin is elevated. Clinical symptoms include bleeding into the mucous membranes, gastrointestinal tract, and skin. Absence of radial bones is observed in the majority of the patients (thrombocytopenia with absent radius, TAR syndrome) [47, 50].

ACQUIRED APLASTIC ANEMIA

Acquired AA is characterized by severe bone marrow hypocellularity and pancytopenia. The term "acquired" refers to non-congenital causative mechanisms which could be immunologic, environmental, or unknown. Patients often have a history of exposure to a wide spectrum of chemical and physical agents and various diseases. However, since there is daily exposure to unlimited and widespread chemicals, such as insecticides, fertilizers, food additives, and herbal medicine, the exact causative factor(s) is not detected in about 50–75% of AA patients [51–53]. Therefore, acquired AA is divided into two major categories: (1) idiopathic AA (with no known etiology) and (2) secondary AA.

Etiology and Pathogenesis

The etiology of AA in most patients is still not clear (idiopathic). Several studies support the destruction or suppression of bone marrow stem cells by immune mechanisms as the major contributing factors [53–55]. These studies include:

1. Positive effects of various immunosuppressive agents such as antithymocyte globulin (ATG), cyclosporine, cyclophosphamide, or corticosteroids on the clinical outcome of the patients with AA [56–58].
2. Occurrence of AA in association with graft versus host disease following allogeneic bone marrow transplantation [51].
3. Association of AA with immunologic disorders such as eosinophilic fasciitis and Grave's disease [59, 60].
4. Evidence of clonal expansion of CD8-positive T-cells in patients with idiopathic AA [61, 62].
5. Evidence of a reduction in the bone marrow natural killer cells of patients with AA [63].
6. Lymphocyte activation by antigens, chemicals, or viruses may suppress hematopoiesis through the release of interferon gamma [64, 65].

In certain conditions damage to bone marrow stem cells and/or microenvironment is caused by agents that are considered relatively harmless. This could be due to the presence of several factors, such as severely depleted marrow stem cells, excessive acquired vulnerability of the hematopoietic precursors or stromal cells, or development of autoimmunity.

Many drugs, such as chloramphenicol, felbamate, nifedipine, gold, sulfonamides, and phenylbutazone, may play a role in the development of AA (Table 7.2) [51, 66–70]. The bone marrow suppression effect of these drugs is often reversible, meaning that by discontinuation of the medication the bone marrow activities eventually come back to normal. Of these drugs, chloramphenicol is perhaps the best documented one, as it was widely used in the United States between 1948 and 1967 [71]. The toxic effects of chloramphenicol are associated with vacuolization of the erythroid precursors, presence of ringed sideroblasts (accumulation of iron in mitochondria)

TABLE 7.2 Drugs associated with aplastic anemia.

1. Anti-inflammatory drugs and analgesics (a) Butazones (b) Diclofenac (c) Gold (d) Indomethacin (e) Piroxicam
2. Antibiotics (a) Chloramphenicol (b) Isoniazid (c) Penicillin (d) Streptomycin (e) Sulfonamides (f) Tetracycline
3. Anti-epileptic (a) Felbamate (b) Methionine (c) Methsuximide (d) Phenacemide (e) Troxidone
4. Anti-diabetic (a) Chlorpropamide (b) Tolbutamide
5. Anti-malarial (a) Mepacrine (b) Chloroquine (c) Piramethamine
6. Others (a) Allopurinol (b) Chlorpromazine (c) Nifedipine (d) Organic arsenicals

and increased serum iron. Occasionally, the toxic effects are irreversible, leading to a sustained AA [53].

Development of AA has been reported in association with the exposure to a garden variety of non-pharmacological chemicals, such as organic solvents, pesticides, and aniline dyes [72, 73]. There are reports of development of AA as a consequence of use of traditional herbal medications [74]. Benzene metabolites suppress DNA synthesis and inhibit proliferation of hematopoietic precursors leading to AA, MDS, and AML [75].

Radiation may induce AA by impairing hematopoietic stem cells (HSCs) and damaging the bone marrow microenvironment. Lethal or sub-lethal amounts of total body irradiation, long-term continuous exposure to small amounts of radiation, and high doses of local therapeutic radiation have all been considered as possible contributors to the development of AA [51, 74].

Certain viruses are able to damage stem cells and/or bone marrow microenvironment and cause AA [53, 76–78]. Hepatitis-associated AA is observed 2–3 months after the onset of acute hepatitis, though the responsible virus has not been identified yet. Between 2% and 5% of patients with AA show evidence of hepatitis [53, 76, 77]. Other viruses such as HIV, EBV, rubella virus, and parvovirus B19 may also cause AA [76, 78].

Reports of higher frequency of HLA-DR2 and HLA-B14 in patients with AA suggest a genetic predisposition [79, 80]. Also, there is a close relationship between AA and paroxysmal nocturnal hemoglobinuria (see later), which is considered a clonal stem cell disorder.

Pathology

Morphology and Laboratory Findings

The diagnosis of AA is established by bone marrow examination [1–3]. Bone marrow is markedly hypocellular with a very high proportion of fatty tissue and stromal cells (Figure 7.4). All hematopoietic elements are decreased but are morphologically normal. There is no evidence of a malignant infiltrate or diffuse fibrosis. Peripheral blood shows pancytopenia with reduced reticulocyte count. A severe AA is defined as [81]:

1. A bone marrow cellularity of <25% of normal cellularity for age in biopsy sections, or
2. A bone marrow cellularity of <50% of normal cellularity for age with <30% hematopoietic cells, plus at least two of the following:
   (a) Absolute erythrocyte count of <40,000/μL.
   (b) Absolute neutrophil count <500/μL.
   (c) Platelet count <20,000/μL.

When the criteria for severe AA are met and the absolute neutrophil count is <200/μL, the patient is considered to have a very severe AA [53].

The hypocellular bone marrow biopsy sections show scattered islands of hematopoietic cells randomly distributed throughout the marrow. These islands are predominantly erythroid and contain very few megakaryocytes. Bone marrow smears consist predominantly of adipocytes and stromal tissue with scattered hematopoietic cells. Occasionally, some of the aspirated smears may show cellular marrow particles, giving the wrong impression of a normocellular or even hypercellular marrow. For this reason, bone marrow biopsies are preferred for the establishment of the diagnosis of AA [1–3]. Some bone marrow smears may show increased proportion of lymphocytes, plasma cells, macrophages, and mast cells. These cells either appear as well-defined aggregates or are diffusely dispersed in the stroma. There may be evidence of hemophagocytosis, particularly in early stages of the disease.

Peripheral blood examination reveals pancytopenia. Anemia is usually normochromic and normocytic, but macrocytosis and anisocytosis may be present. The reticulocyte count is low, and platelets, neutrophils, monocytes, eosinophils, and basophils are reduced. Neutrophils may show toxic granulation. The lymphocyte count is normal or low.

Molecular Genetics and Cytogenetics

Owing to its heterogeneous and non-genetic etiology, there are no specific molecular tests for acquired AA. Mutation testing of genes associated with the hereditary disorders, and PCR-based detection of implicated viruses, may be

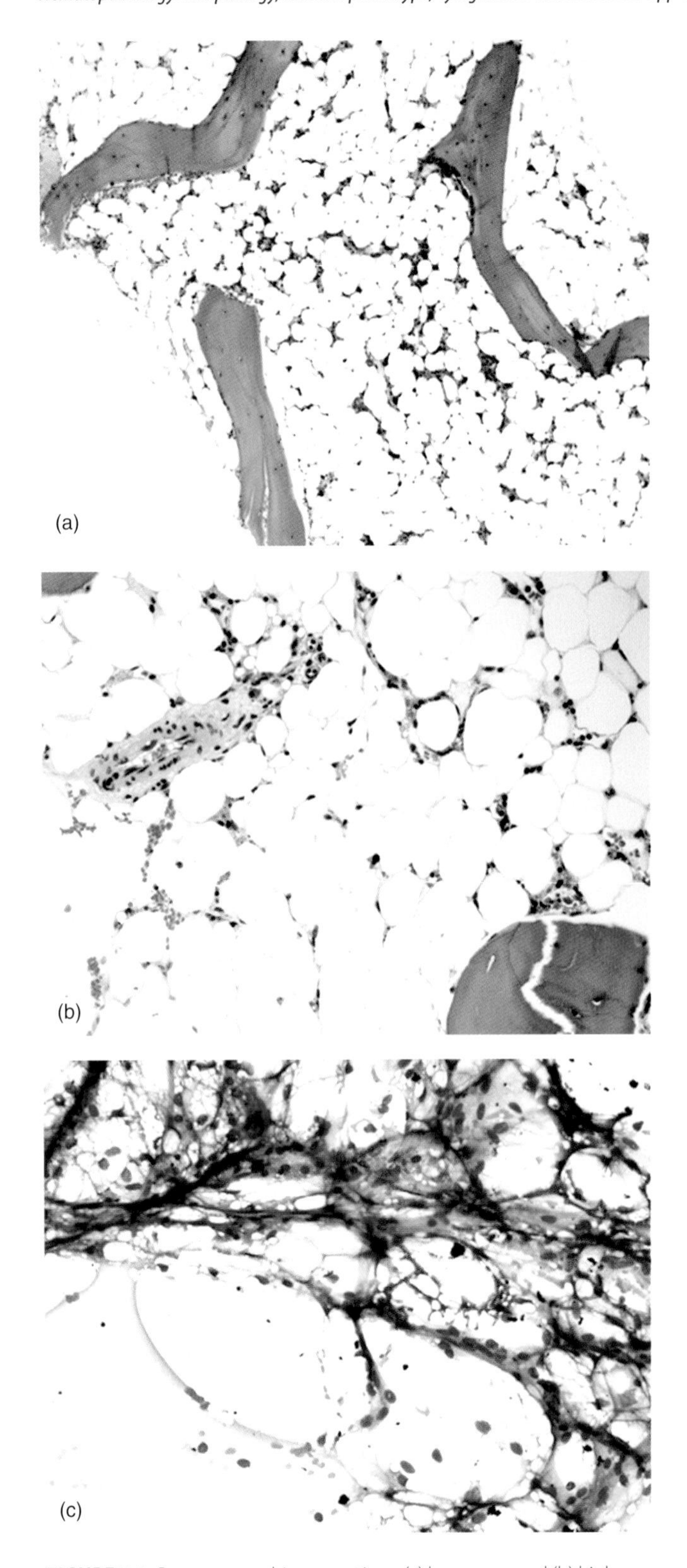

FIGURE 7.4 Bone marrow biopsy sections: (a) low power and (b) high power, from a patient with AA demonstrating marked hypocellularity. The bone marrow smear shows fatty tissue and stromal cells (c).

performed as part of the differential diagnostic work-up. Approximately 4% of patients with AA show cytogenetic abnormalities, such as 5q−, monosomy 7, and trisomy 6 or 8 (Figures 7.5 and 7.6) [82, 83].

Clinical Aspects

The incidence of AA is significantly higher (about fivefold) in the Far East than the West [84, 85]. Clinical manifestations of AA are non-specific and are usually related to pancytopenia. Pallor, fatigue, purpura and mucosal hemorrhage, and recurrent infections are common findings [51, 53]. Occasionally cardiopulmonary symptoms associated with severe anemia are the presenting clinical picture. The cause of death is usually infection, particularly disseminated fungal forms.

The outcome of untreated severe AA is very poor, with over 70% death rate within 1 year. Prognosis is also age-dependent, with better outcome in patients under 49 years than those over 60 years [51, 53, 86, 87]. The treatment of choice under the age of 45 is HSC transplantation [88, 89]. The major problem is lack of HLA-matched donors. Only 25–30% of AA patients find proper donors. Immunosuppressive therapy is recommended for patients over the age of 45. Immunosuppressive agents include ATG, corticosteroids, and cyclosporine [56, 90]. Hematopoietic growth factors, such as G-CSF, have been added to the immunosuppressive regimen with some beneficial effects.

PAROXYSMAL NOCTURNAL HEMOGLOBINURIA

Paroxysmal nocturnal hemoglobinuria (PNH) is an acquired stem cell disorder associated with a defect in cell membrane glycosyl phosphatidylinositol (GPI) anchor due to mutation of the *PIG-A* gene. This defect leads to partial or complete loss of certain GPI-linked membrane proteins, such as CD14, CD16, CD24, CD48, CD52, CD55, CD58, CD59, CD66, and CD73 (Table 7.3) [91–96]. Some of these proteins, such as CD55 and CD59, play an inhibitory role in the activation of the complement system, and therefore their absence leads to complement-induced lysis and hemolytic anemia [95–97]. CD55, also known as decay accelerating factor (DAF), is expressed by all hematopoietic cells and is an inhibitor of C3 and C5 convertases [95–97]. Similarly, CD59 is expressed by all hematopoietic cells. It is referred to as membrane inhibitor of reactive lysis (MIRL) and binds to C8 component of the complement system and prevents polymerization of the complement components [95–98]. PNH is characterized by hemolytic anemia (often with hemoglobinuria), venous thrombosis, and bone marrow failure [91–94].

Etiology and Pathogenesis

The fundamental pathogenic process in PNH is the defect in the production of GPI anchor [91, 94, 97, 98]. This defect is due to a somatic mutation in the phosphatidylinositol glycan complementation class A (*PIG-A*) gene at the level of HSC. So far, constitutional *PIG-A* defects have not been described. The *PIG-A* gene is located on the short arm of the X chromosome [91, 94, 99]. Patients with PNH have one or more stem cell GPI-deficient clones. More than one hundred different mutations have been reported

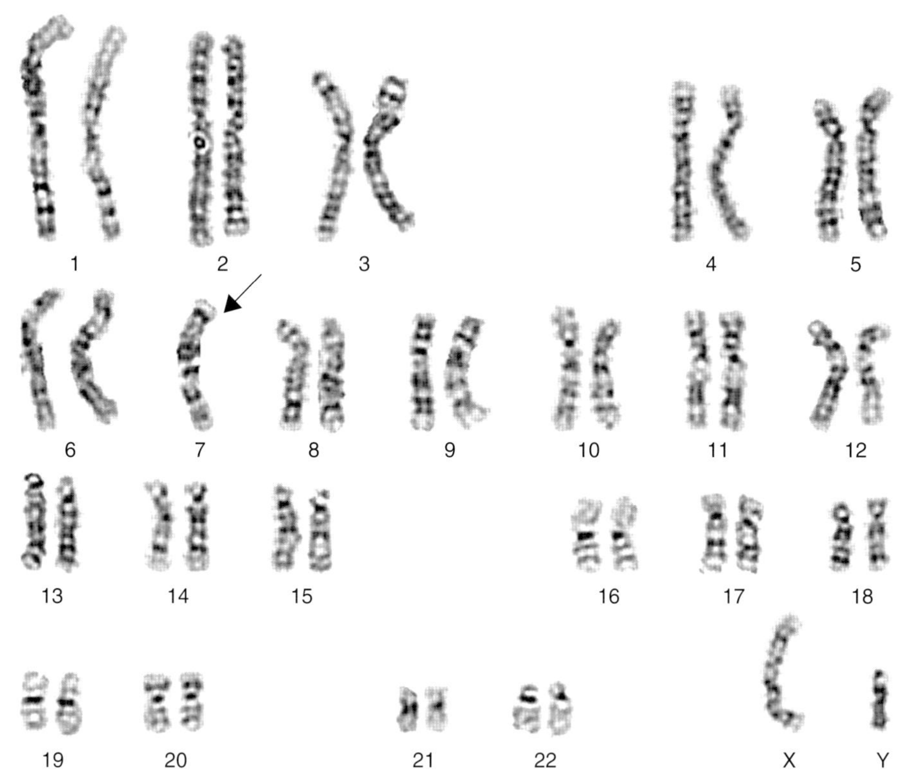

FIGURE 7.5 A G-banded karyotype showing monosomy 7 (arrow) in a patient with AA.

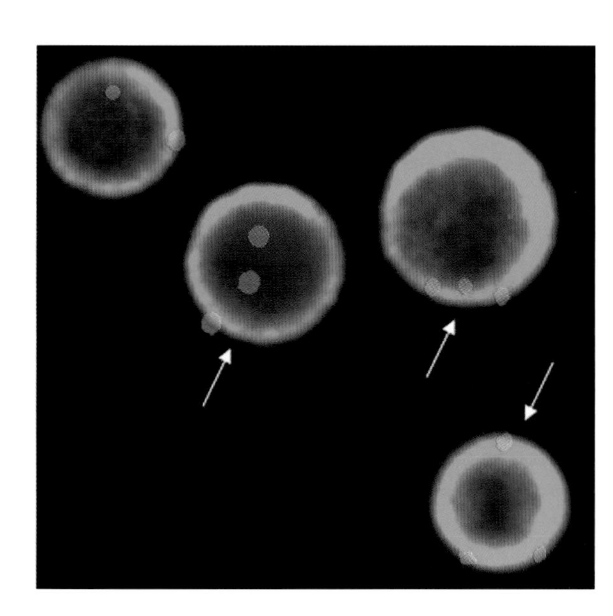

FIGURE 7.6 FISH studies for chromosome 8 in interphase. Cells demonstrating trisomy 8 (arrows).

[91, 99]. However, the defect in GPI production may not be sufficient enough for the manifestation of PNH, since *PIG-A* mutations have been described in the stem cells of normal individuals as well. It has also been shown that most patients with AA and refractory anemia who carry HLA-DRBI have an expanded population of GPI-deficient clones [85, 94, 95, 100, 101]. However, in most instances, the GPI-deficient clones found in AA and refractory anemia are generally small.

It seems that other hematopoietic alterations are necessary to allow the *PIG-A* deficient clone to dominate. For example, in some patients development of AA precedes PNH [91, 101, 102]. The dominant pathogenic theory for the development of PNH in patients with AA is that the development of AA suppresses the GPI-defective stem cells less than the normal stem cells, and therefore, when the degree of bone marrow suppression is reduced by therapy (such as ATG or cyclosporine), the GPI-defective clone has a better chance to emerge. An autoimmune process may play a role in the pathogenesis of PNH. The involved auto reactive T-cells may selectively damage the GPI-positive HSCs, whereas GPI-negative HSCs would escape damage and be able to proliferate [91, 101, 102].

Pathology

Morphology and Laboratory Findings

Bone marrow in most instances is markedly hypocellular and presents morphologic features similar to AA (Figure 7.7) [1–3]. However, some patients may show normo- or even hypercellular marrow. There is often erythroid preponderance. Stainable iron is usually absent, primarily due to loss of iron secondary to hemoglobinuria and hemosiderinuria [1–3].

Blood examination commonly reveals severe anemia with some degree of granulocytopenia and thrombocytopenia [1–3, 91, 92]. There is evidence of intravascular hemolysis by the presence of hemoglobinuria, hemosiderinuria and elevated reticulocyte count. There is a reduction in plasma haptoglobulin levels and an increase in plasma lactate

TABLE 7.3 Some of the GPI-linked proteins deficient in paroxysmal nocturnal hemoglobinuria*.

Molecule	CD	Comments
Complement Regulatory Molecules		
DAF	CD55	Decay accelerating factor
MIRL	CD59	Membrane inhibitor of reactive lysis
Enzymes		
Ecto-5'-nucleotidase	CD73	Lymphocytes
ADP-ribosyl transferase	CD157	T-cells and neutrophils
Adhesion Molecules		
Blast-1	CD48	Leukocytes; binds CD24
LFA-3	CD58	All hematopoietic cells
Adhesion molecule 1	CD66a	Granulocytes, epithelium
NCA-95	CD66b	Granulocytes
NCA-50/90	CD66c	Granulocytes, epithelium
Carcinoembryonic antigen	CD66e	Epithelium
Others		
NA1/NA2	CD16	Neutrophils and natural killer cells
Campath-1	CD52	Lymphocytes and monocytes
BA-1	CD24	B-cells and granulocytes
Thy-1	CD90	Stem cell subset, T-cell subset

* Adapted from Hall C, Richards SJ, Hillmen P. (2002). The glycosyphosphatidylinositol anchor and paroxysmal nocturnal haemoglobinuria/aplasia model. *Acta Haematol* **108**, 219–30.

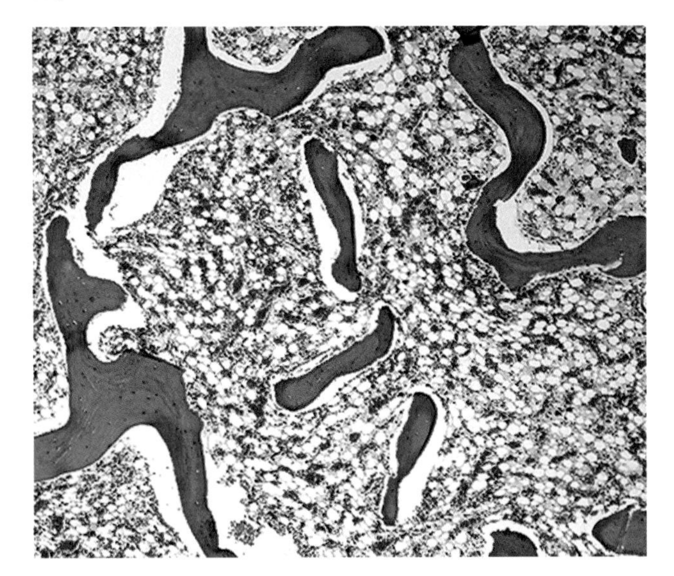

FIGURE 7.7 Bone marrow biopsy section from a patient with PNH demonstrating marked hypocellularity.

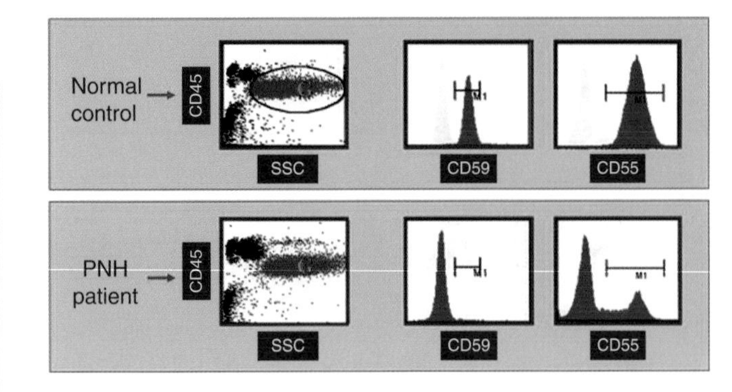

FIGURE 7.8 Flow cytometry comparison of CD55 and CD59 expressions on the granulocytes of a normal control and a patient with PNH. Granulocytes of the normal control show expression of CD55 and CD59, whereas in the PNH patient there is lack of expression of CD55 and CD59 expression is partially lost.

dehydrogenase (LDH) level [91–93]. The leukocyte alkaline phosphatase score is reduced.

For years, the diagnosis of PNH was based on the sensitivity of the red cells to lysis by complement. This was determined by the sucrose lysis screening test and the confirmatory Ham acid hemolysis test [91–93]. In the sucrose lysis test, the patient's red cells are incubated with serially diluted isotonic sucrose solutions. Under these conditions the complement system is activated and the test is considered positive if there is evidence of hemolysis. In the Ham test, the pH of serum is reduced to activate the complement system and to induce hemolysis in the PNH red cells. However, nowadays, immunophenotyping by flow cytometry (see the following section) is considered the standard procedure because of the higher sensitivity and specificity it provides.

Flow Cytometry

Study of the expression of GPI-linked proteins on the hematopoietic cells by flow cytometry is now the recommended approach in the diagnosis of PNH [103–106]. Complete or partial loss of these proteins is indicative of GPI deficiency. There is a garden variety of GPI-linked proteins to choose for different types of hematopoietic cells, and therefore there are a number of options regarding antibody selection, cell type, and gating strategies. Most laboratories include at least two GPI-linked proteins, often CD55 and CD59, for each cell type under study. Red cells and granulocytes are the most frequent cell types studied (Figure 7.8) [103–106]. A transmembrane antigen is also used as a positive control. Each cell type is divided into three categories according to the status of the protein expression:

1. Type I: cells with normal expression
2. Type II: cells with partial absence
3. Type III: cells with complete absence.

PNH patients show a significant proportion (usually over 10%) of the type II and/or type III cells.

Analysis of red cells: Monoclonal antibodies against CD55 and CD59 are used for the analysis of the red cells. The red cell population is gated by forward scatter (FSC) and side scatter (SSC) amplification in log mode, and the purity of red cells is checked by the percent glycophorin A-positive cells in the gated population. The red cell analysis is clearest in untransfused patients [103–106]. A marked variation in the distribution of types I, II, and III is observed from patient to patient, though sometimes the separation of subtypes is not clear-cut. The type III red cells appear to have a significantly shorter survival (about 17–60 days) than type I red cells. Red cells are suitable for testing up to 3 weeks if kept at 4°C.

Analysis of granulocytes: Anti-CD55 and CD59 monoclonal antibodies are often used for granulocyte analysis. The granulocyte population is gated by FSC and SSC characteristics, but some laboratories use a myeloid-associated marker, such as CD15 or CD33, to optimize the detection of granulocytes. Antibodies against other GPI-linked proteins, such as CD16 and CD66, could also be utilized [103–106]. However, since CD16 is not expressed by normal eosinophils, its use is not recommended for the samples with a high proportion of eosinophilia [102–106]. Flow cytometric studies on granulocytes should be carried out within the first few hours following collection [102–106].

Analysis of monocytes: Although red cells and granulocytes are the most frequent cell types used for the diagnosis of PNH, several investigators have also analyzed peripheral blood monocytes for this purpose. Anti-CD14 and CD55 monoclonal antibodies are the favorite ones, though some laboratories have looked at the expression of CD48 and CD59 [102–106]. Analysis of the monocytes is accomplished by the use of FSC and SSC properties or CD45 expression and SSC properties.

Analysis of lymphocytes: Although lymphocytes, similar to other hematopoietic cells, show loss of GPI-linked proteins, they are infrequently used for the establishment of PNH diagnosis. The expression of CD55 and CD59 is variable on normal lymphocytes, and therefore these two markers may not provide reliable information in PNH flow cytometry studies. CD48 appears to be a more suitable marker, because it provides a clearer separation between normal and PNH lymphocytes [103–106]. Immunophenotypic studies of B-cells in PNH show that in the majority of patients with active PNH, B-cells are comprised primarily of naïve cells characterized with a $CD27^-$ IgG^- IGM^+ and IgD^+ phenotype [107].

Analysis of platelets: Expression of CD55 and CD59 has been studied on platelets in PNH patients [103–106, 108]. Platelets are usually gated based on their FSC and SSC properties and the use of non-GPI-linked CD molecules CD41 and/or CD61 as platelet markers. However, these studies are infrequent and their diagnostic significance has not been established. Frequent thromboembolic events in PNH strongly suggest an abnormal platelet function in these patients, probably due to the deficiency of CD55 and CD59. But, so far, no definitive platelet defect has been characterized and the underlying mechanisms remain unclear. However, a recent report indicates a profound platelet hyporeactivity in PNH patients based on clot formation, adhesion, and aggregation assays, concluding that the venous thromboembolism is probably induced by activation and dysregulation of plasma coagulation factors [91–96].

In general, peripheral blood is the sample of choice for immunophenotypic studies in PNH. Red cells and granulocytes are the recommended cell types, and CD55 and CD59 are the most commonly used markers for these studies. Samples with recent history of blood transfusions may provide ambiguous results. The following important points are provided by Richards and associates in their review article regarding flow cytometric studies in PNH [103–106]:

1. A small proportion of patients (about 5%) may display only a granulocyte PNH clone.
2. The chance of the detection of a red cell PNH clone is reduced after a severe episode of hemolysis.
3. In patients with severely hypoplastic marrow, there may be insufficient numbers of granulocytes for analysis.
4. In red cell analysis, deficiencies of at least two GPI-linked antigens are required to establish a PNH diagnosis, since congenital non-PNH single-antigen deficiencies have been observed rarely.

Molecular Genetics and Cytogenetics

As described earlier, the cornerstone of PNH diagnosis rests on the flow cytometry findings. Reported mutations in the *PIG-A* gene on chromosome Xp22 are numerous and heterogeneous, and further complicated by the presence of a pseudogene on chromosome 12. DNA sequencing is not generally available on a clinical basis. It must be kept in mind that the gene mutations found in PNH are acquired, not inherited, so they will only be found in the abnormal clone [14, 109]. Cytogenetic abnormalities in paroxysmal nocturnal hemoglobinuria usually occur in hematopoietic cells that are glycosylphosphatidylinositol-anchored protein (GPI-AP) positive [110]. Various chromosomal aberrations have been reported in up to 24% of patients with FA including trisomy 5, trisomy 6, trisomy 8, and monosomy 7 [111, 112].

Clinical Aspects

The disease may affect patients at any age, but the peak incidence is between 20 and 35 years [91, 92]. The severity of the clinical findings varies considerably from patient to patient and may include any of the following manifestations [91, 92, 95, 96]:

1. Acquired intravascular hemolysis demonstrated by hemoglobinemia, hemoglobinuria, hemosideriuria, and negative direct antiglobulin (Coombs') test. Plasma haptoglobulin level is low and plasma LDH level is often elevated.
2. Thrombosis of the relatively large veins in odd places, such as hepatic (Budd–Chiari syndrome), mesenteric, portal, or cerebral veins. The rate of venous thrombosis in western countries approaches 40% and is the major cause of death in PNH patients. Arterial thrombosis is rare.
3. Bone marrow hypoplasia leading to pancytopenia. As briefly mentioned earlier, in a significant proportion of PNH patients (up to 30%), there is a previous history of AA [101].

The possible association between PNH, MDS, and acute myeloid leukemia has been the subject of several case reports, although the incidence is relatively low at around 5% [92,

TABLE 7.4 Differential diagnoses in bone marrow aplasia.

Disorder	Bone marrow morphology	Immunophenotype	Cytogenetics and molecular genetics
Constitutional aplasias	Normo- to hypercellular at early stages, hypocellular marrow at later stages	Non-contributory	Frequent chromosomal breakage, sometimes −7, mutations in causative genes
Acquired AA	Hypocellular marrow	Often increased cytotoxic T-cells, strong association with HLA-DR2	Sometimes 5q−, −7, +6, +8, viral PCR
PNH	Hypocellular marrow	Loss of GPI-linked proteins, such as CD55 and CD59	Mutations in *PIG-A gene*
Hypocellular MDS	Hypocellular marrow with significant dysplastic changes, and sometimes increased blasts	Abnormal phenotypic patterns sometimes increased CD34+ and/or CD117+ cells	−7, +8, 5q−, 20q−, and other chromosomal aberrations
Hypoplastic AML	Hypocellular marrow with ≥20% blasts	Increased CD45^{dim}+ cells expressing myeloid markers, often CD34 and/or CD117	Frequent chromosomal aberrations involving 11q, 16q, or t(15;17), t(8;11), t(9;22), and others
Hypocellular hairy cell leukemia	Hypocellular marrow with the presence of hairy cells and often evidence of fibrosis	TRAP+, CD103+, CD25+, CD22+, 11c+	Not known

113–115]. Therapeutic approaches include iron and folic acid supplementation, red blood cell transfusion, and treatment with prednisone and androgen derivatives [92, 94]. Anticoagulation therapy is used for episodes of thrombosis. HSC transplantation has been used in selected cases [92, 116].

DIFFERENTIAL DIAGNOSIS

Morphologic features of bone marrow in advanced stages of constitutional marrow aplasias, acquired AA, and PNH are indistinguishable. Also, other bone marrow lesions, such as hypocellular MDS, hypoplastic AML, and hypocellular hairy cell leukemia, may morphologically mimic AA (Table 7.4). Clinical history and information regarding other clinicopathologic parameters are imperative for accurate diagnosis. It is important to remember that a proportion of patients with constitutional or acquired AA may eventually develop MDS or AML.

References

1. Hsi ED. (2007). *Hematopathology*. Churchill Livingstone, Philadelphia.
2. Foucar K. (2001). *Bone Marrow Pathology*, 2nd ed. ASCP Press, Chicago.
3. Naeim F. (1997). *Pathology of Bone Marrow*, 2nd ed. Williams & Wilkins, Baltimore.
4. Federman N, Sakamoto KM. (2005). The genetic basis of bone marrow failure syndromes in children. *Mol Genet Metab* **86**, 100–9.
5. Omine M, Kojima S, Nakao S, Mizoguchi H, Takaku F. (2005). International Consensus Conference on Bone Marrow Failure Syndromes: Review and recommendations. *Int J Hematol* **82**, 406–7.
6. Alter BP. (2002). Bone marrow failure syndromes in children. *Pediatr Clin North Am* **49**, 973–88.
7. Lieberman L, Dror Y. (2006). Advances in understanding the genetic basis for bone-marrow failure. *Curr Opin Pediatr* **18**, 15–21.
8. Marsh JC. (2005). Bone marrow failure syndromes. *Clin Med* **5**, 332–6.
9. Bagby GC, Lipton JM, Sloand EM, Schiffer CA. (2004). Marrow failure. *Hematology (Am Soc Hematol Educ Program)*, 318–36.
10. Tischkowitz MD, Hodgson SV. (2003). Fanconi anaemia. *J Med Genet* **40**, 1–10.
11. Kennedy RD, D'Andrea AD. (2005). The Fanconi Anemia/BRCA pathway: New faces in the crowd. *Genes Dev* **19**, 2925–40.
12. Shimamura A. (2007). Inherited aplastic anemia in children. *UpToDate.*
13. Kutler DI, Auerbach AD. (2004). Fanconi anemia in Ashkenazi Jews. *Fam Cancer* **3**, 241–8.
14. Akkari Y, Olson S. (2004). Fanconi Anemia: A Decade of Discoveries. *J Assoc Genet Technol* **30**, 48–53.
15. Yamashita T, Nakahata T. (2001). Current knowledge on the pathophysiology of Fanconi anemia: From genes to phenotypes. *Int J Hematol* **74**, 33–41.
16. Collins N, Kupfer GM. (2005). Molecular pathogenesis of Fanconi anemia. *Int J Hematol* **82**, 176–83.
17. D'Andrea AD. (2003). The Fanconi Anemia/BRCA signaling pathway: Disruption in cisplatin-sensitive ovarian cancers. *Cell Cycle* **2**, 290–2.
18. D'Andrea AD, Grompe M. (2003). The Fanconi anaemia/BRCA pathway. *Nat Rev Cancer* **3**, 23–34.
19. Polychronopoulou S, Koutroumba P. (2004). Telomere length variation and telomerase activity expression in patients with congenital and acquired aplastic anemia. *Acta Haematol* **111**, 125–31.
20. Taylor AM. (2001). Chromosome instability syndromes. *Best Pract Res Clin Haematol* **14**, 631–44. Gen.
21. Auerbach AD. (1997). Fanconi anemia: Genetic testing in Ashkenazi Jews. *Genet Test* **1**, 27–33.
22. Clark RD, Fletcher J, Petersen G. (1989). Conceiving a fetus for bone marrow donation: An ethical problem in prenatal diagnosis. *Prenat Diagn* **9**, 329–34.
23. Verlinsky Y, Rechitsky S, Schoolcraft W, Strom C, Kuliev A. (2001). Preimplantation diagnosis for Fanconi anemia combined with HLA matching. *JAMA* **285**, 3130–3.
24. Dokal I. (2000). The genetics of Fanconi's anaemia. *Baillieres Best Pract Res Clin Haematol* **13**, 407–25.
25. Kook H. (2005). Fanconi anemia: Current management. *Hematology* **10**(Suppl 1), 108–10.

26. D'Andrea AD. (2003). The Fanconi road to cancer. *Genes Dev* **17**, 1933–6.
27. Tischkowitz M, Dokal I. (2004). Fanconi anaemia and leukaemia – clinical and molecular aspects. *Br J Haematol* **126**, 176–91.
28. Alter BP. (2005). Fanconi's anemia, transplantation, and cancer. *Pediatr Transplant* (Suppl 7), 81–6.
29. Rosenberg PS, Socie G, Alter BP, Gluckman E. (2005). Risk of head and neck squamous cell cancer and death in patients with Fanconi anemia who did and did not receive transplants. *Blood* **105**, 67–73.
30. Mason PJ, Wilson DB, Bessler M. (2005). Dyskeratosis congenita – a disease of dysfunctional telomere maintenance. *Curr Mol Med* **5**, 159–70.
31. Walne AJ, Marrone A, Dokal I. (2005). Dyskeratosis congenita: A disorder of defective telomere maintenance? *Int J Hematol* **82**, 184–9.
32. Bessler M, Wilson DB, Mason PJ. (2004). Dyskeratosis congenita and telomerase. *Curr Opin Pediatr* **16**, 23–8.
33. Dokal I. (2000). Dyskeratosis congenita in all its forms. *Br J Haematol* **110**, 768–79.
34. Vulliamy TJ, Marrone A, Knight SW, Walne A, Mason PJ, Dokal I. (2006). Mutations in dyskeratosis congenita: Their impact on telomere length and the diversity of clinical presentation. *Blood* **107**, 2680–5.
35. Dokal I. (1999). Dyskeratosis congenita. *Br J Haematol* **105**(Suppl 1), 11–15.
36. Knight S, Vulliamy T, Copplestone A, Gluckman E, Mason P, Dokal I. (1998). Dyskeratosis Congenita (DC) Registry: Identification of new features of DC. *Br J Haematol* **103**, 990–6.
37. Shwachman H, Diamond LK, Oski FA, Khaw KT. (1964). The syndrome of pancreatic insufficiency and bone marrow dysfunction. *J Pediatr* **65**, 645–63.
38. Rothbaum R, Perrault J, Vlachos A, Cipolli M, Alter BP, Burroughs S, Durie P, Elghetany MT, Grand R, Hubbard V, Rommens J, Rossi T. (2002). Shwachman–Diamond syndrome: Report from an international conference. *J Pediatr* **141**, 266–70.
39. Shwachman H, Diamond LK, Oski FA, Khaw KT. (2004). Mutations of the SBDS gene are present in most patients with Shwachman–Diamond syndrome. *Blood* **104**, 3588–90.
40. Boocock GRB, Morrison JA, Popovic M. (2003). Mutations in SBDS are associated with Shwachman–Diamond syndrome. *Nat Genet* **33**, 97–101.
41. Elghetany MT, Alter BP. (2002). P53 protein overexpression in bone marrow biopsies of patients with Shwachman–Diamond syndrome has a prevalence similar to that of patients with refractory anemia. *Arch Pathol Lab Med* **126**, 452–5.
42. Kuijpers TW, Alders M, Tool AT, Mellink C, Roos D, Hennekam RC. (2005). Hematologic abnormalities in Shwachman–Diamond syndrome: Lack of genotype–phenotype relationship. *Blood* **106**, 356–61.
43. Da Costa L, Willig TN, Fixler J, Mohandas N, Tchernia G. (2001). Diamond–Blackfan anemia. *Curr Opin Pediatr* **13**, 10–15.
44. Vlachos A, Klein GW, Lipton JM. (2001). The Diamond Blackfan Anemia Registry: Tool for investigating the epidemiology and biology of Diamond–Blackfan anemia. *J Pediatr Hematol Oncol* **23**, 377–82.
45. Chen S, Warszawski J, Bader-Meunier B, Tchernia G, Da Costa L, Marie I. (2005). Dommergues JP; Societe Francaise d'Hematologie et d'Immunologie Pediatrique. Diamond–Blackfan anemia and growth status: The French Registry. *J Pediatr* **147**, 669–73.
46. Liu JM, Ellis SR. (2006). Diamond–Blackfan anemia: A paradigm for a ribosome-based disease. *Med Hypotheses* **66**, 643–8.
47. Hedberg VA, Lipton JM. (1988). Thrombocytopenia with absent radii. A review of 100 cases. *Am J Pediatr Hematol Oncol* **10**, 51–64.
48. Ihara K, Ishii E, Eguchi M, Takada H, Suminoe A, Good RA, Hara T. (1999). Identification of mutations in the c-mpl gene in congenital amegakaryocytic thrombocytopenia. *Proc Natl Acad Sci USA* **96**, 3132–6.
49. Ballmaier M, Germeshausen M, Schulze H, Cherkaoui K, Lang S, Gaudig A, Krukemeier S, Eilers M, Strauss G, Welte K. (2001). C-mpl mutations are the cause of congenital amegakaryocytic thrombocytopenia. *Blood* **97**, 139–46.
50. Thompson AA, Woodruff K, Feig SA, Nguyen LT, Schanen NC. (2001). Congenital thrombocytopenia and radio-ulnar synostosis: A new familial syndrome. *Br J Haematol* **113**, 866–70.
51. Keohane EM. (2004). Acquired aplastic anemia. *Clin Lab Sci* **17**, 165–71.
52. Brodsky RA, Jones RJ. (2005). Aplastic anaemia. *Lancet* **365**, 1647–56.
53. Schrier SL. (2007). Aplastic anemia: Pathogenesis; clinical manifestations; and diagnosis. *UpToDate.*
54. Nakao S, Feng X, Sugimori C. (2005). Immune pathophysiology of aplastic anemia. *Int J Hematol* **82**, 196–200.
55. Maciejewski JP, Risitano A, Kook H, Zeng W, Chen G, Young NS. (2002). Immune pathophysiology of aplastic anemia. *Int J Hematol* **76**(Suppl 1), 207–14.
56. Doney K, Pepe M, Storb R, Bryant E, Anasetti C, Appelbaum FR, Buckner CD, Sanders J, Singer J, Sullivan K. (1992). Immunosuppressive therapy of aplastic anemia: Results of a prospective, randomized trial of antithymocyte globulin (ATG), methylprednisolone, and oxymetholone to ATG, very high-dose methylprednisolone, and oxymetholone. *Blood* **79**, 2566–71.
57. Storb R, Etzioni R, Anasetti C, Appelbaum FR, Buckner CD, Bensinger W, Bryant E, Clift R, Deeg HJ, Doney K. (1994). Cyclophosphamide combined with antithymocyte globulin in preparation for allogeneic marrow transplants in patients with aplastic anemia. *Blood* **84**, 941–9.
58. Paquette RL, Tebyani N, Frane M, Ireland P, Ho WG, Champlin RE, Nimer SD. (1995). Long-term outcome of aplastic anemia in adults treated with antithymocyte globulin: Comparison with bone marrow transplantation. *Blood* **85**, 283–90.
59. Kim SW, Rice L, Champlin R, Udden MM. (1997). Aplastic anemia in eosinophilic fasciitis: Responses to immunosuppression and marrow transplantation. *Haematologia (Budap)* **28**, 131–7.
60. Kumar M, Goldman J. (2002). Severe aplastic anaemia and Grave's disease in a paediatric patient. *Br J Haematol* **118**, 327–9.
61. Risitano AM, Kook H, Zeng W, Chen G, Young NS, Maciejewski JP. (2002). Oligoclonal and polyclonal CD4 and CD8 lymphocytes in aplastic anemia and paroxysmal nocturnal hemoglobinuria measured by V beta CDR3 spectratyping and flow cytometry. *Blood* **100**, 178–83.
62. Risitano AM, Maciejewski JP, Green S, Plasilova M, Zeng W, Young NS. (2004). *In-vivo* dominant immune responses in aplastic anaemia: Molecular tracking of putatively pathogenetic T-cell clones by TCR beta-CDR3 sequencing. *Lancet* **364**, 355–64.
63. Zeng W, Maciejewski JP, Chen G, Risitano AM, Kirby M, Kajigaya S, Young NS. (2002). Selective reduction of natural killer T cells in the bone marrow of aplastic anaemia. *Br J Haematol* **119**, 803–9.
64. Maciejewski JP, Selleri C, Sato T, Anderson S, Young NS. (1995). Increased expression of Fas antigen on bone marrow CD34+ cells of patients with aplastic anaemia. *Br J Haematol* **91**, 245–52.
65. Dufour C, Capasso M, Svahn J, Marrone A, Haupt R, Bacigalupo A, Giordani L, Longoni D, Pillon M, Pistorio A, Di Michele P, Iori AP, Pongiglione C, Lanciotti M, Iolascon A. (2004). Associazione Italiana di Emato-Oncologia Pediatrica (AIEOP); Department of Hematology, Ospedale S. Martino, Genoa, Italy. Homozygosis for (12) CA repeats in the first intron of the human IFN-gamma gene is significantly associated with the risk of aplastic anaemia in Caucasian population. *Br J Haematol* **126**, 682–5.
66. Issaragrisil S. (2003). Aplastic anemia: low drug associations. *Curr Hematol Rep* **2**, 1–2.
67. Williame LM, Joos R, Proot F, Immesoete C. (1987). Gold-induced aplastic anemia. *Clin Rheumatol* **6**, 600–5.
68. Blain H, Hamdan KA, Blain A, Jeandel C. (2002). Aplastic anemia induced by phenytoin: A geriatric case with severe folic acid deficiency. *J Am Geriatr Soc* **50**, 396–7.
69. Wong IC, Lhatoo SD. (2000). Adverse reactions to new anticonvulsant drugs. *Drug Saf* **23**, 35–56.
70. Pellock JM. (2001). Felbamate in epilepsy therapy: Evaluating the risks. *Drug Saf* **21**, 225–39.

71. Wallerstein RO, Condit PK, Kasper CK, Brown JW, Morrison FR. (1969). Statewide study of chloramphenicol therapy and fatal aplastic anemia. *JAMA* **208**, 2045–50.
72. Muir KR, Chilvers CE, Harriss C, Coulson L, Grainge M, Darbyshire P, Geary C, Hows J, Marsh J, Rutherford T, Taylor M, Gordon-Smith EC. (2003). The role of occupational and environmental exposures in the aetiology of acquired severe aplastic anaemia: A case control investigation. *Br J Haematol* **123**, 906–14.
73. Powars D. (1965). Aplastic anemia secondary to glue sniffing. *N Engl J Med* **273**, 700–2.
74. Pyatt DW, Yang Y, Mehos B, Le A, Stillman W, Irons RD. (2000). Hematotoxicity of the chinese herbal medicine Tripterygium wilfordii hook f in CD34-positive human bone marrow cells. *Mol Pharmacol* **57**, 512–18.
75. Loge JP. (1965). Aplastic anemia following exposure to benzene hexachloride (lindane). *JAMA* **193**, 110–14.
76. Kurtzman G, Young N. (1989). Viruses and bone marrow failure. *Baillieres Clin Haematol* **2**, 51–67.
77. Brown KE, Tisdale J, Barrett AJ, Dunbar CE, Young NS. (1997). Hepatitis-associated aplastic anemia. *N Engl J Med* **336**, 1059–64.
78. Brown KE, Young NS. (1996). Parvoviruses and bone marrow failure. *Stem Cells* **14**, 151–63.
79. Fuhrer M, Durner J, Brunnler G, Gotte H, Deppner C, Bender-Gotze C, Albert E. (2006). HLA association is different in children and adults with severe acquired aplastic anemia. *Pediatr Blood Cancer* **48**, 186–91.
80. Usman M, Adil SN, Moatter T, Bilwani F, Arian S, Khurshid M. (2004). Increased expression of HLA DR2 in acquired aplastic anemia and its impact on response to immunosuppressive therapy. *J Pak Med Assoc* **54**, 251–4.
81. Rozman C, Marin P, Nomdedeu B, Montserrat E. (1987). Criteria for severe aplastic anaemia. *Lancet* **2**, 955–7.
82. Maciejewski JP, Selleri C. (2004). Evolution of clonal cytogenetic abnormalities in aplastic anemia. *Leuk Lymphoma* **45**, 433–40.
83. La Starza R, Matteucci C, Crescenzi B, Criel A, Selleslag D, Martelli MF, Van den Berghe H, Mecucci C. (1998). Trisomy 6 is the hallmark of a dysplastic clone in bone marrow aplasia. *Cancer Genet Cytogenet* **105**, 55–9.
84. Kaufman DW, Kelly JP, Issaragrisil S, Laporte JR, Anderson T, Levy M, Shapiro S, Young NS. (2006). Relative incidence of agranulocytosis and aplastic anemia. *Am J Hematol* **81**, 65–7.
85. Issaragrisil S, Kaufman DW, Anderson T, Chansung K, Leaverton PE, Shapiro S, Young NS. (2006). The epidemiology of aplastic anemia in Thailand. *Blood* **107**, 1299–307.
86. Young NS, Calado RT, Scheinberg P. (2006). Current concepts in the pathophysiology and treatment of aplastic anemia. *Blood* **108**, 2509–19.
87. Tichelli A, Socie G, Henry-Amar M, Marsh J, Passweg J, Schrezenmeier H, McCann S, Hows J, Ljungman P, Marin P, Raghavachar A, Locasciulli A, Gratwohl A, Bacigalupo A. (1999). Effectiveness of immunosuppressive therapy in older patients with aplastic anemia. European Group for Blood and Marrow Transplantation Severe Aplastic Anaemia Working Party. *Ann Intern Med* **130**, 193–201.
88. Marsh JC, Ball SE, Darbyshire P, Gordon-Smith EC, Keidan AJ, Martin A, McCann SR, Mercieca J, Oscier D, Roques AW, Yin JA. (2003). British Committee for Standards in Haematology. Guidelines for the diagnosis and management of acquired aplastic anaemia. *Br J Haematol* **123**, 782–801.
89. Mao P, Wang S, Wang S, Zhu Z, Liv Q, Xuv Y, Mo W, Ying Y. (2004). Umbilical cord blood transplant for adult patients with severe aplastic anemia using anti-lymphocyte globulin and cyclophosphamide as conditioning therapy. *Bone Marrow Transplant* **33**, 33–8.
90. Bacigalupo A, Bruno B, Saracco P, Di Bona E, Locasciulli A, Locatelli F, Gabbas A, Dufour C, Arcese W, Testi G, Broccia G, Carotenuto M, Coser P, Barbui T, Leoni P, Ferster A. (2000). Antilymphocyte globulin, cyclosporine, prednisolone, and granulocyte colony-stimulating factor for severe aplastic anemia: An update of the GITMO/EBMT study on 100 patients. European Group for Blood and Marrow Transplantation (EBMT) Working Party on Severe Aplastic Anemia and the Gruppo Italiano Trapianti di Midolio Osseo (GITMO). *Blood* **95**, 1931–4.
91. Omine M, Kinoshita T, Nakakuma H, Maciejewski JP, Parker CJ, Socie G. (2005). Paroxysmal nocturnal hemoglobinuria. *Int J Hematol* **82**, 417–21.
92. Ross WF. (2007). Diagnosis and treatment of paroxysmal nocturnal hemoglobinuria. *UpToDate.*
93. Ross WF. (2007). Pathogenesis of paroxysmal nocturnal hemoglobinuria. *UpToDate.*
94. Parker CJ. (2007). The pathophysiology of paroxysmal nocturnal hemoglobinuria. *Exp Hematol* **35**, 523–33.
95. Brodsky RA. (2006). New insights into paroxysmal nocturnal hemoglobinuria. *Hematology Am Soc Hematol Educ Program*, 24–8.
96. Luzzatto L, Gianfaldoni G. (2006). Recent advances in biological and clinical aspects of paroxysmal nocturnal hemoglobinuria. *Int J Hematol* **84**, 104–12.
97. Hernandez-Campo PM, Martin-Ayuso M, Almeida J, Lopez A, Orfao A. (2002). Comparative analysis of different flow cytometry-based immunophenotypic methods for the analysis of CD59 and CD55 expression on major peripheral blood cell subsets. *Cytometry* **50**, 191–201.
98. Tiu R, Maciejewski J. (2006). Immune pathogenesis of paroxysmal nocturnal hemoglobinuria. *Int J Hematol* **84**, 113–17.
99. Luzzatto L. (2006). Paroxysmal nocturnal hemoglobinuria: An acquired X-linked genetic disease with somatic-cell mosaicism. *Curr Opin Genet Dev* **16**, 317–22. Epub 2006 May 2. Gen.
100. Maciejewski JP, Risitano A, Sloand EM, Nunez O, Young NS. (2002). Distinct clinical outcomes for cytogenetic abnormalities evolving from aplastic anemia. *Blood* **99**, 3129–35.
101. Young NS, Maciejewski JP, Sloand E, Chen G, Zeng W, Risitano A, Miyazato A. (2002). The relationship of aplastic anemia and PNH. *Int J Hematol* **76**(Suppl 2), 168–72.
102. Nakao S, Sugimori C, Yamazaki H. (2006). Clinical significance of a small population of paroxysmal nocturnal hemoglobinuria-type cells in the management of bone marrow failure. *Int J Hematol* **84**, 118–22.
103. Richards SJ, Barnett D. (2007). The role of flow cytometry in the diagnosis of paroxysmal nocturnal hemoglobinuria in the clinical laboratory. *Clin Lab Med* **27**, 577–90.
104. Richards SJ, Rawstron AC, Hillmen P. (2000). Application of flow cytometry to the diagnosis of paroxysmal nocturnal hemoglobinuria. *Cytometry* **42**, 223–33.
105. Dworacki G, Sikora J, Mizera-Nyczak E, Trybus M, Mozer-Lisewska I, Czyz A, Zeromski J. (2005). Flow cytometric analysis of CD55 and CD59 expression on blood cells in paroxysmal nocturnal haemoglobinuria. *Folia Histochem Cytobiol* **43**, 117–20.
106. Hernandez-Campo PM, Almeida J, Sanchez ML, Malvezzi M, Orfao A. (2006). Normal patterns of expression of glycosylphosphatidylinositol-anchored proteins on different subsets of peripheral blood cells: A frame of reference for the diagnosis of paroxysmal nocturnal hemoglobinuria. *Cytometry B Clin Cytom* **70**, 71–81.
107. Richards SJ, Morgan GJ, Hillmen P. (2000). Immunophenotypic analysis of B cells in PNH: Insights into the generation of circulating naive and memory B cells. *Blood* **96**, 3522–8.
108. Maciejewski JP, Young NS, Yu M, Anderson SM, Sloand EM. (1996). Analysis of the expression of glycosylphosphatidylinositol anchored proteins on platelets from patients with paroxysmal nocturnal hemoglobinuria. *Thromb Res* **83**, 433–47.
109. Nishimura J, Murakami Y, Kiinoshita T. (1999). Paroxysmal nocturnal hemoglobinuria: An acquired genetic disease. *Am J Hematol* **62**, 175–82.
110. Sloand EM, Fuhrer M, Keyvanfar K, Mainwaring L, Maciejewski J, Wang Y, Johnson S, Barrett AJ, Young NS. (2003). Cytogenetic abnormalities in paroxysmal nocturnal haemoglobinuria usually occur in haematopoietic cells that are glycosylphosphatidylinositol-anchored protein (GPI-AP) positive. *Br J Haematol* **123**, 173–6.

111. Auewarakul CU, Tocharoentanaphol C, Wanachiwanawin W, Issaragrisil S. (2004). Monosomy 7 in patients with aplastic anemia and paroxysmal nocturnal hemoglobinuria with evolution into acute myeloid leukemia. *J Med Assoc Thai* **87**, 717–21. Gen.
112. Araten DJ, Swirsky D, Karadimitris A, Notaro R, Nafa K, Bessler M, Thaler HT, Castro-Malaspina H, Childs BH, Boulad F, Weiss M, Anagnostopoulos N, Kutlar A, Savage DG, Maziarz RT, Jhanwar S, Luzzatto L. (2001). Cytogenetic and morphological abnormalities in paroxysmal nocturnal haemoglobinuria. *Br J Haematol* **115**, 360–8.
113. Smith LJ. (2004). Paroxysmal nocturnal hemoglobinuria. *Clin Lab Sci* **17**, 172–7.
114. Wang H, Chuhjo T, Yasue S, Omine M, Nakao S. (2002). Clinical significance of a minor population of paroxysmal nocturnal hemoglobinuria-type cells in bone marrow failure syndrome. *Blood* **100**, 3897–902.
115. Meletis J, Terpos E. (2003). Recent insights into the pathophysiology of paroxysmal nocturnal hemoglobinuria. *Med Sci Monit* **9**, RA161–172.
116. Raiola AM, Van Lint MT, Lamparelli T, Gualandi F, Benvenuto F, Figari O, Mordini N, Berisso G, Bregante S, Frassoni F, Bacigalupo A. (2000). Bone marrow transplantation for paroxysmal nocturnal hemoglobinuria. *Haematologica* **85**, 59–62.

Myelodysplastic Syndromes

8

CHAPTER

Faramarz Naeim, P. Nagesh Rao and Wayne W. Grody

Myelodysplastic syndromes (MDS) are a group of hematologic disorders distinguished by clonal expansion of defective hematopoietic stem cells leading to abnormal maturation and peripheral blood cytopenia. These disorders are also known as refractory anemias (RA), dysmyelopoietic syndromes, and were originally labeled by some investigators as preleukemias. The peripheral cytopenias may be demonstrated as anemia, thrombocytopenia, granulocytopenia, or pancytopenia. Some categories of MDS show increased bone marrow blast cells and have a higher chance to eventually transform to acute myeloid leukemias (AMLs). The overall transformation rate to acute leukemia depends on the subtype of MDS, the presence or absence of chromosomal aberrations, and types of these abnormalities.

Classification of MDS by the World Health Organization (WHO) includes the following categories [1–5a] (Table 8.1):

- Refractory anemia (RA)
- Refractory anemia with ringed sideroblasts (RARS)
- Refractory cytopenia with multilineage dysplasia (RCMD)
- Refractory cytopenia with multilineage dysplasia and ringed sideroblasts (RCMD-RS)
- Refractory anemia with excess blasts (RAEB)
- MDS associated with isolated del(5q)
- MDS unclassifiable.

In a recently revised draft of the WHO classification, *myelodysplastic syndromes in children* was added to the list (updated WHO classification in press [5b]).

The WHO classification represents the clonal forms of MDS. However, non-clonal myelodysplastic changes have been observed in a variety of conditions, such as severe inflammatory states, viral infections, autoimmune disorders, megaloblastic anemia, exposure to arsenic, status post-chemotherapy, and endocrine dysfunctions.

ETIOLOGY AND PATHOGENESIS

There are two major categories of MDS: primary (with no known cause) and secondary (usually post-treatment chemotherapy or irradiation). The etiology and pathogenesis of the primary MDS are not clearly understood. Some familial clustering has been reported, but no causative germline mutations have been identified. Clonality of the underlying marrow failure has been supported by various molecular and cytogenetic techniques, such as karyotyping, X-chromosome inactivation studies, and fluorescence *in situ* hybridization (FISH) analysis.

Development of MDS is probably a multistep process with an initial genetic insult to the multipotent stem cells leading to the development of an abnormal clone [6–9]. The abnormal clone is the precursor of morphologically dysplastic and dysfunctional hematopoietic cells with a tendency to die prematurely. Excessive apoptosis (programmed cell death) of the hematopoietic precursors, particularly at the early stages, has been proposed as the primary mechanism for the bone marrow hypercellularity and peripheral cytopenia in patients with MDS [10–12]. According to some investigators, however, progression of the disease is accompanied by a decline in apoptosis

TABLE 8.1 WHO classification of MDS.*

Type	Blood findings	Bone marrow findings
Refractory anemia (RA)	Anemia, abnormal erythrocytes No or rare blasts	Dysplastic erythropoiesis <5% blasts <15% ringed sideroblasts
Refractory anemia with ringed sideroblasts (RARS)	Anemia, abnormal erythrocytes No blasts	Dysplastic erythropoiesis ≥15% ringed sideroblasts <5% blasts
Refractory cytopenia with multilineage dysplasia (RCMD)	Bicytopenia or pancytopenia No or rare blasts No Auer rods No absolute monocytosis	Multilineage dysplasia <5% blasts No Auer rods <15% ringed sideroblasts
Refractory cytopenia with multilineage dysplasia and ringed sideroblasts (RCMD-RS)	Bicytopenia or pancytopenia No or rare blasts No Auer rods No absolute monocytosis	Multilineage dysplasia <5% blasts No Auer rods ≥15% ringed sideroblasts
Refractory anemia with excess blasts-1 (RAEB-1)	Cytopenia(s) <5% blasts No Auer rods No absolute monocytosis	Unilineage or multilineage Dysplasia 5–9% blasts No Auer rods
Refractory anemia with excess blasts-2 (RAEB-2)	Cytopenia(s) 5–19% blasts Auer rods ± No absolute monocytosis	Unilineage or multilineage Dysplasia 10–19% blasts Auer rods ±
MDS associated with isolated del(5q)	Anemia Normal or increased platelets <5% blasts No Auer rods	Micromegakaryocytes Isolated del(5q) <5% blasts No Auer rods
Myelodysplastic syndrome, unclassifiable (MDS-U)	Cytopenia No or rare blasts No Auer rods	Unilineage myeloid Dysplasia <5% blasts No Auer rods

* Adapted from Ref. [1].

and increased levels of anti-apoptotic bcl-2 protein in the bone marrow progenitor cells in advanced MDS, such as RAEB. There are also reports of reversing the ratio of pro-apoptotic and anti-apoptotic proteins as the disease progresses. The possible contributing factors in the acceleration of apoptosis in the early developmental stages of MDS could be genetic damage and/or an altered marrow microenvironment.

Unlike bcl-2 which is an anti-apoptotic factor, the *C-MYC* gene is a pro-apoptotic regulator and seems to play a role in the pathogenesis of MDS at the early stages, such as RA and RARS [13–15].

There are several other genes that appear to be dysregulated in MDS (Figure 8.1). These include cell cycle regulatory genes, such as *EVI-1*, growth factors and angiogenesis genes, such as *TGF*-α and -β and *VEGF*, receptor tyrosine kinase genes, such as *FLT3*, immunoregulatory cytokine genes, such as *IFN*-γ and *TNF*-α, and genes regulating DNA methylation [16–18]. Abnormal DNA methylation of calcitonin and *p15* has been reported in MDS [19, 20].

The report of clinical autoimmune disorders in about 10% of MDS patients raises the possibility that bone marrow failure in MDS is immune mediated [21]. The autoimmune model for pathophysiology of MDS suggests an autoimmune T-cell attack to the bone marrow target cells causing an overproduction of pro-apoptotic cytokines, such as IFN-γ [22].

The ineffective hematopoiesis in MDS may involve one or several hematopoietic lines, resulting in anemia, thrombocytopenia, granulocytopenia, or pancytopenia. The lymphoid lineage may occasionally be involved. In such cases, evolution of acute leukemia at the later stages may be of a lymphoblastic type or biphenotypic [23, 24]. Lymphoid blast transformation is usually of B-cell type. The involvement of the lymphoid lineage in the myelodysplastic process may also cause immune dysfunction, such as cell-mediated

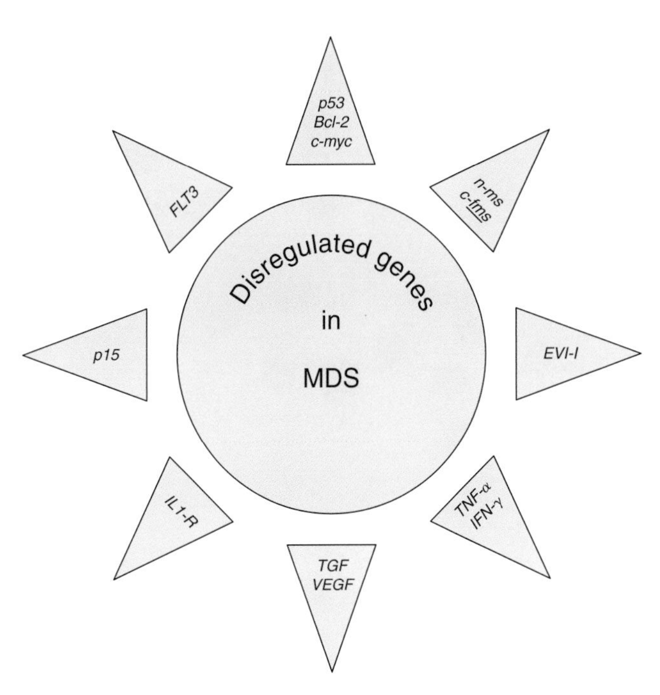

FIGURE 8.1 Numerous gene products appear to play important roles in pathogenesis of MDS.

suppression of bone marrow stem cells or deficient NK (natural killer) activities.

GENERAL MORPHOLOGIC FEATURES

The ineffective hematopoiesis in MDS is demonstrated by mono- or pancytopenia and abnormal morphology in one or more hematopoietic lines in bone marrow and peripheral blood [1, 25–27]. The bone marrow in patients with primary MDS is usually hyper- or normocellular, whereas patients with secondary MDS may show a variable marrow cellularity ranging from 5% to almost 100%. MDS bone marrow may show increased reticulin fibers with a higher frequency in the secondary MDS.

Bone marrow biopsy sections of MDS patients usually show some degree of topographical alterations, such as the presence of erythroid clusters next to the bone trabeculae, loss of sinusoidal orientation of megakaryocytes and their placement next to bone, and centrally located aggregates of myeloid precursors (Figure 8.2) [27, 28]. The morphologic appearance of aggregates of immature myeloid cells is referred to as "abnormal localization of immature precursors" (ALIP) (Figure 8.3). ALIP is defined as clusters of five or more myeloblasts and/or early immature myeloid cells, located in the marrow tissue away from bone trabeculae. More than three ALIP clusters per biopsy section are required to be diagnostically significant [1, 28]. Presence of ALIP, however, is not exclusive to MDS and has been observed in other hematologic conditions, such as myeloproliferative disorders and status post-bone marrow transplantation or chemotherapy [1, 28].

Signs of an inflammatory response, such as lymphoid aggregates, areas of edema, extravasation of erythrocytes, increased mast cells, plasma cells, and macrophages, disrupted sinusoids, and patchy or sometimes diffuse fibrosis are frequent findings. Occasionally, there is evidence of hemophagocytosis and/or presence of sea blue histiocytes [25–27].

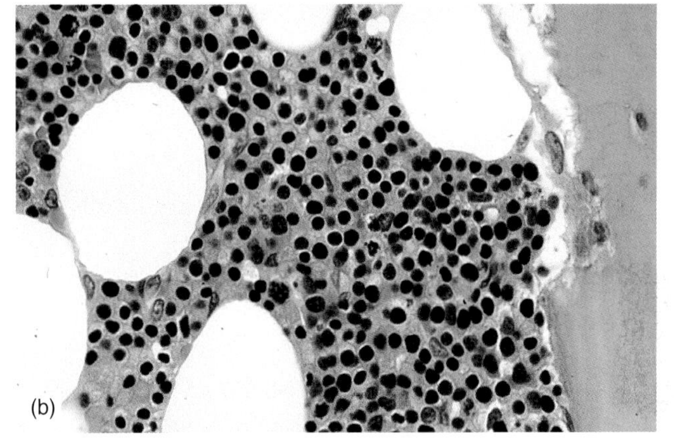

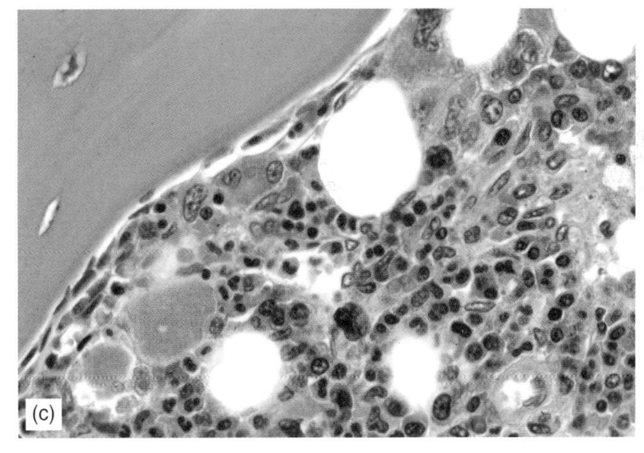

FIGURE 8.2 (a) Bone marrow section from a patient with MDS demonstrating hypercellularity with paratrabecular localization of the erythroid precursors and presence of a lymphoid aggregate. Higher power views demonstrate paratrabecular localization of erythroid precursors (b) and megakaryocytes (c).

The accurate assessment of dysplastic cytologic features in blood and bone marrow smears depends on the quality of the smear preparations. Slides should be made from a fresh specimen and properly stained with preferably <2 h of exposure to anticoagulants. The recommended threshold for significant dysplasia is at least 10% for each hematopoietic lineage [1]. The morphologic features of these dysplastic changes are discussed later.

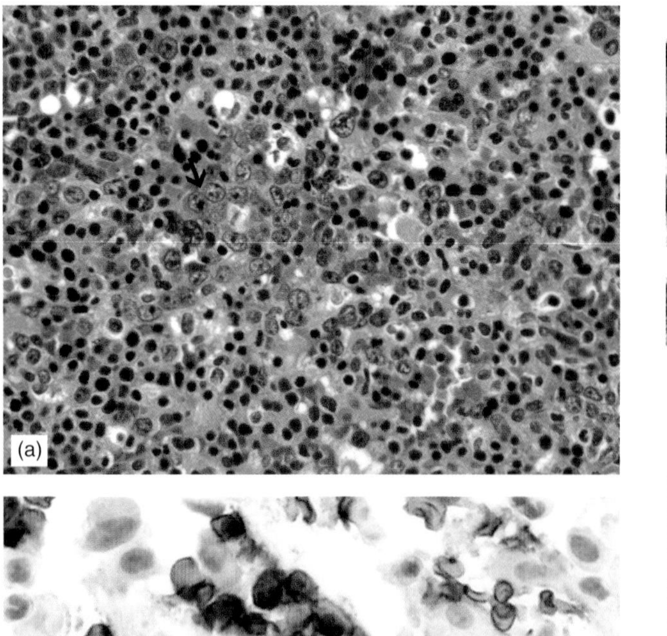

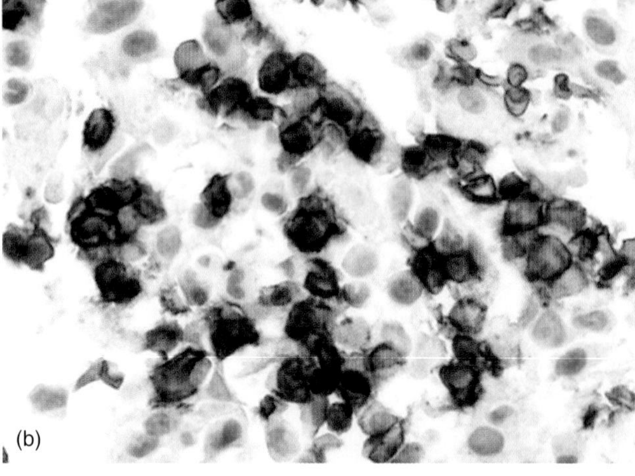

FIGURE 8.3 A hypercellular bone marrow biopsy section (a) showing several clusters of immature cells, referred to as "abnormal localization of immature precursors" (ALIP). The immunohistochemical stain (b) shows clusters of myeloperoxidase-positive immature cells.

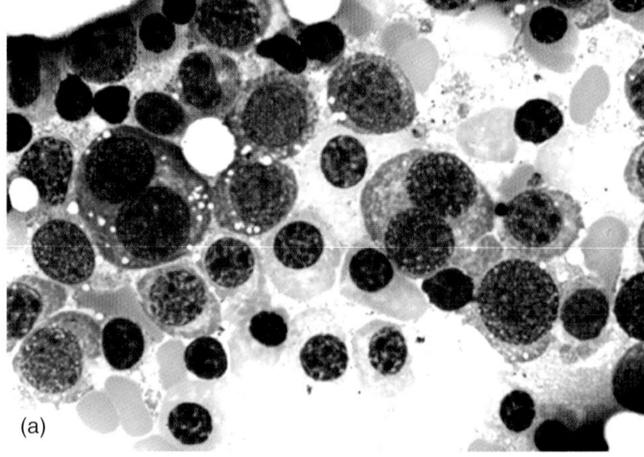

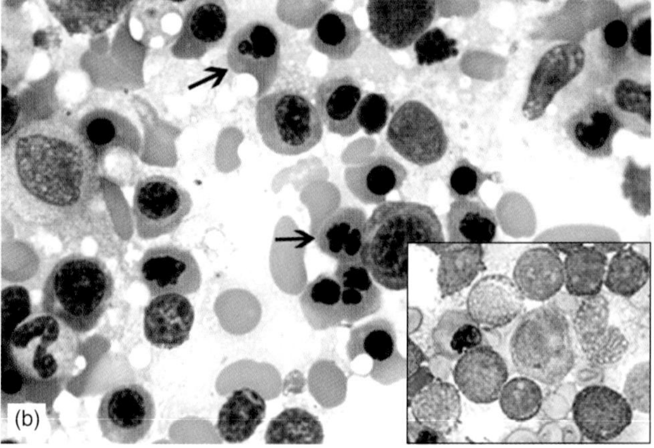

FIGURE 8.4 A bone marrow smear from a patient with RA showing dysplastic binucleated early erythroid precursors with vacuolated cytoplasm (a). PAS stain shows coarse PAS-positive granules in these cells (inset). Dysplastic late erythroid precursors with irregular or multilobated nuclei (arrows) are demonstrated in (b).

Dyserythropoiesis

Dysplastic features of the erythroid precursors in bone marrow smears include megaloblastic changes, irregular nuclear shape, nuclear fragmentation and budding, multinucleation, nuclear bridging, cytoplasmic vacuolization, poor hemoglobinization, and presence of ringed sideroblasts and periodic acid-Schiff (PAS)-positive cytoplasmic globules (Figures 8.4 and 8.5). In biopsy sections, erythroid colonies may be seen next to the bone trabeculae.

Blood smears show a wide variety of abnormal erythrocyte morphology, such as macro-ovalocytosis, microcytosis, schistocytosis, basophilic stippling, and the presence of teardrop-shaped red blood cells and Howell-Jolly bodies (Figure 8.6). Occasionally, nucleated red blood cells are present.

Dysgranulopoiesis

The granulocytic precursors may show abnormal variations in size, cytoplasmic granularity, and nuclear configuration (Figures 8.7–8.10) [1, 27, 29]. Abnormal staining of primary granules, hypergranularity or hypogranularity, is commonly observed in promyelocytes and myelocytes. Irregular distribution of the cytoplasmic basophilia may be present and the cytoplasm in the perinuclear area may stain lighter than that in the periphery. The more mature granulocytic cells may depict a marked variation in size and decreased or absent secondary granules. There may be coarse basophilic (pseudo-Chediak–Higashi) granules. Nuclear hyposegmentation (pseudo-Pelger-Huet anomaly) or hypersegmentation, and other forms of abnormal nuclear morphology, such as ringed (doughnut-shaped) nuclei, may be present. Eosinophils may be increased or show dysplastic changes, such as abnormal nuclear segmentation or abnormal granulation [30, 31]. Studies of bone marrow basophils on patients with MDS are very limited. However, basophilia has been observed in some MDS patients, and one report demonstrates lack of abnormal basophilic function in a group of patients with MDS [32, 33].

Abnormal Megakaryocytes and Platelets

Megakaryocytes may show multiple separated nuclei, hypo- or hyperlobated nuclei, vacuolated cytoplasm, and

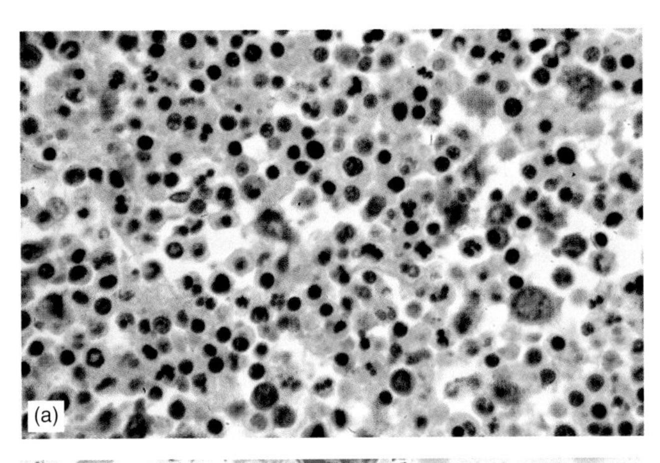

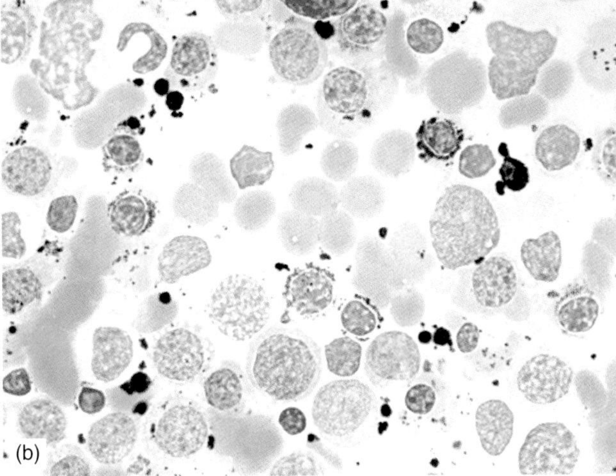

FIGURE 8.5 A hypercellular bone marrow biopsy section from a patient with RA showing erythroid preponderance with dysplastic changes (a). Numerous ringed sideroblasts are present in the bone marrow stained with Prussian blue (b).

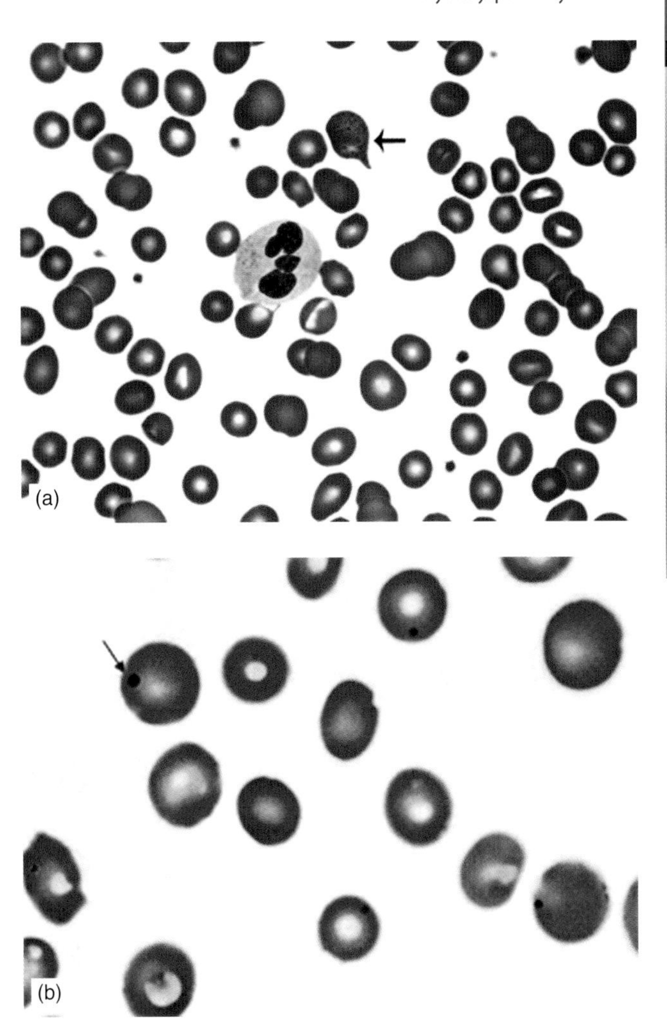

FIGURE 8.6 Patients with MDS often show abnormal erythrocyte morphology in peripheral blood smears, such as anisopoikilocytosis, teardrop (arrow), basophilic stippling (a), or Howell-Jolly bodies (b, arrow).

giant abnormal cytoplasmic granules. Mono- and binuclear megakaryocytes are frequently seen (Figures 8.11 and 8.12) [1, 27, 29]. Sometimes, it is difficult to distinguish the mononuclear micromegakaryocytes from stromal cells or macrophages. In biopsy sections, megakaryocytes may appear in clusters or localized close to bone trabeculae. Immunohistochemical stains are helpful in distinguishing the dysplastic megakaryocytes from other bone marrow elements. Megakaryocytes express CD36, CD41, CD61, and factor VIII.

Blood smears show pleomorphic platelets with the presence of giant forms. They may show hypogranulation or abnormal granules [1, 29]. Megakaryocytic fragments, bare megakaryocytic nuclei, and sometimes micromegakaryocytes may be present [1, 29].

IMMUNOPHENOTYPIC STUDIES

Flow Cytometry

The dysplastic hematopoietic cells in MDS may show altered expression of the CD molecules. These changes may be detected by flow cytometry using a pattern-recognition approach and the measurement of the intensity, lack or aberrant expression of certain CD molecules. Although flow cytometry has not been utilized routinely for establishment of the diagnosis of MDS, numerous reports support its implementation as an accessory tool in evaluation of patients with a clinical history suggestive of MDS. These recommendations are based on the following difficulties in establishing the diagnosis [34, 35]:

1. A significant proportion of cases with MDS do not show increased myeloblasts or ringed sideroblasts.
2. Approximately half of the MDS patients may not show cytogenetic abnormalities.
3. Dysplastic changes are not pathognomonic of MDS and have been observed in a wide variety of non-clonal conditions, such as viral infections, post-chemotherapy, heavy metal toxicity, and folate or vitamin B12 deficiencies.

In addition, it has been shown that some of the phenotypic alterations may correlate with the cytogenetic

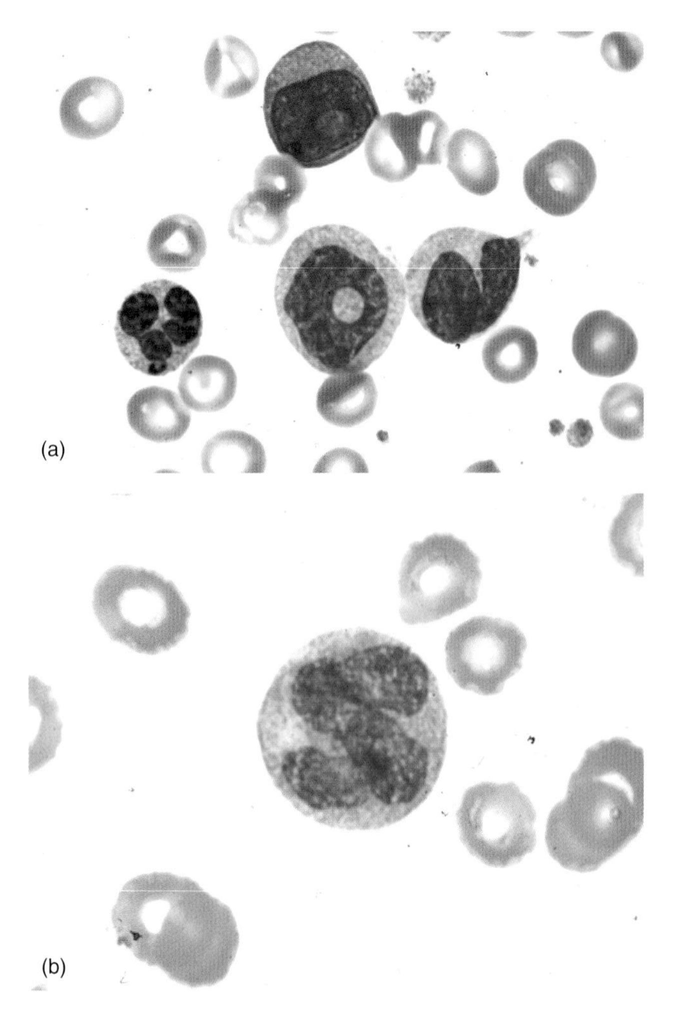

FIGURE 8.7 Blood smears showing a blast and an abnormal cell with doughnut-shaped nucleolus (a) and a dysplastic monocyte (b).

findings or clinical behavior of the disease. Therefore, flow cytometry is recommended as an accessory diagnostic workup in patients suspected for MDS.

The frequency of the hypogranularity of the myeloid cells in MDS often leads to a lower side scatter (SSC) in the flow cytometric dot plot analysis (Figure 8.13). There are also reports of increased expression of CD11a and CD66, reduced expression of CD10 and CD116, and aberrant expression of CD56 on the granulocytes in MDS [35, 36]. There may be an increased proportion of CD34-positive blast cells with frequent co-expression of CD4, CD11b, CD15, and/or CD56 (Figure 8.14) and over-expression of CD13, CD33, CD117, or CD133 [35, 36–38]. Also, a high HLA-DR and low CD11b on myeloblasts in MDS patients may be indicative of early conversion to acute leukemia [39]. There is a report of higher frequency of expression of CD7 and terminal deoxynucleotidyl transferase (TdT) (>60%) by the bone marrow blasts of MDS patients [40].

Erythroid precursors may show higher expression of CD105 and lower expression of CD71, as well as loss of A, B, and H antigens [36, 41]. Platelets may show decreased expression of CD41 and CD61. Several reports indicate that the number of hematogones (normal precursor B-cells), particularly type 1 (TdT^+, $CD34^+$, $CD10^+$, $CD19^+$), is reduced in the bone marrow of MDS patient [37, 42]. A flow cytometric scoring system has been proposed for the diagnosis of MDS by Cherian and associates [43].

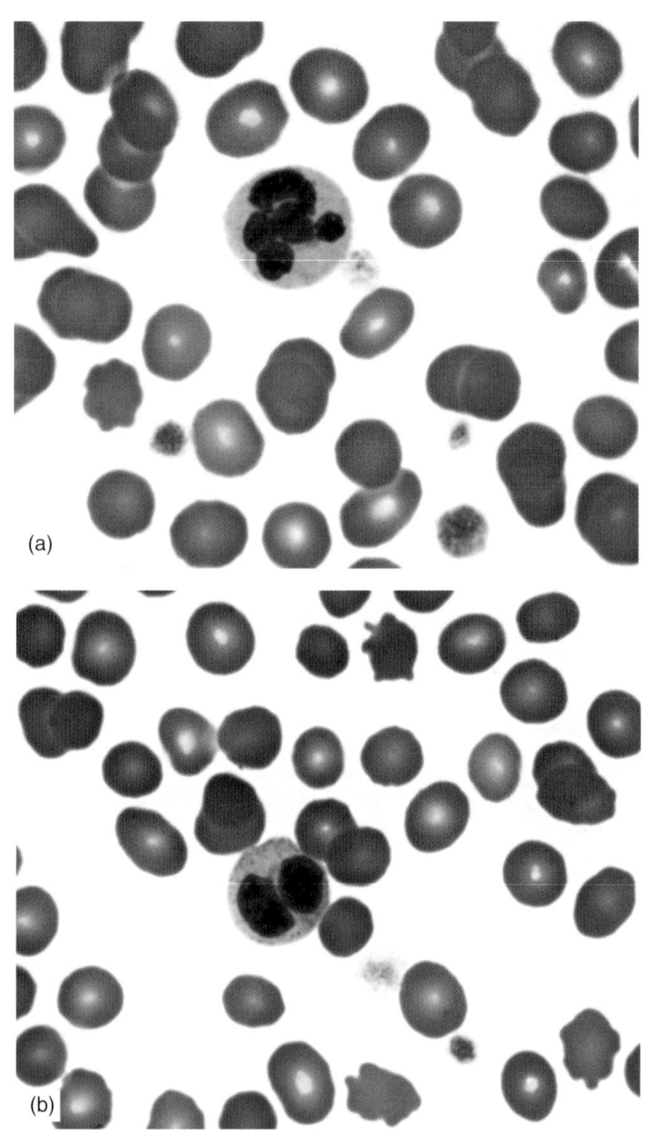

FIGURE 8.8 Blood smears showing a hypogranular and hypersegmented neutrophil (a) and a hyposegmented neutrophil (b).

Immunohistochemical Studies

Immunohistochemical stains on bone marrow biopsies of patients suspected for MDS may provide helpful information regarding the following matters:

1. Evaluation of the topographical alterations and estimation of the M:E ratio by using monoclonal antibodies against hemoglobin and/or glycophorin A molecules for erythroid precursors and myeloperoxidase for myeloid precursors.
2. Detection of clusters of immature cells (ALIP) and estimation of blast cell numbers by using the blast-associated markers such as CD34 and CD117.

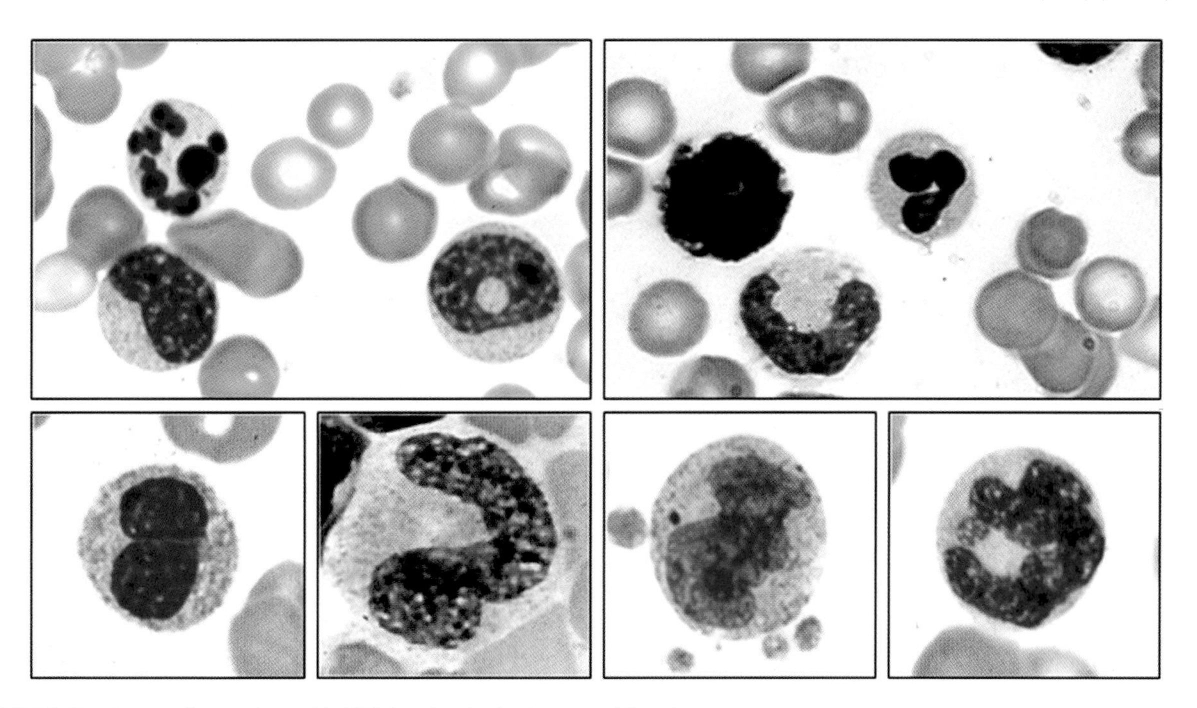

FIGURE 8.9 Blood smears from patients with MDS showing dysplastic neutrophils and monocytes.

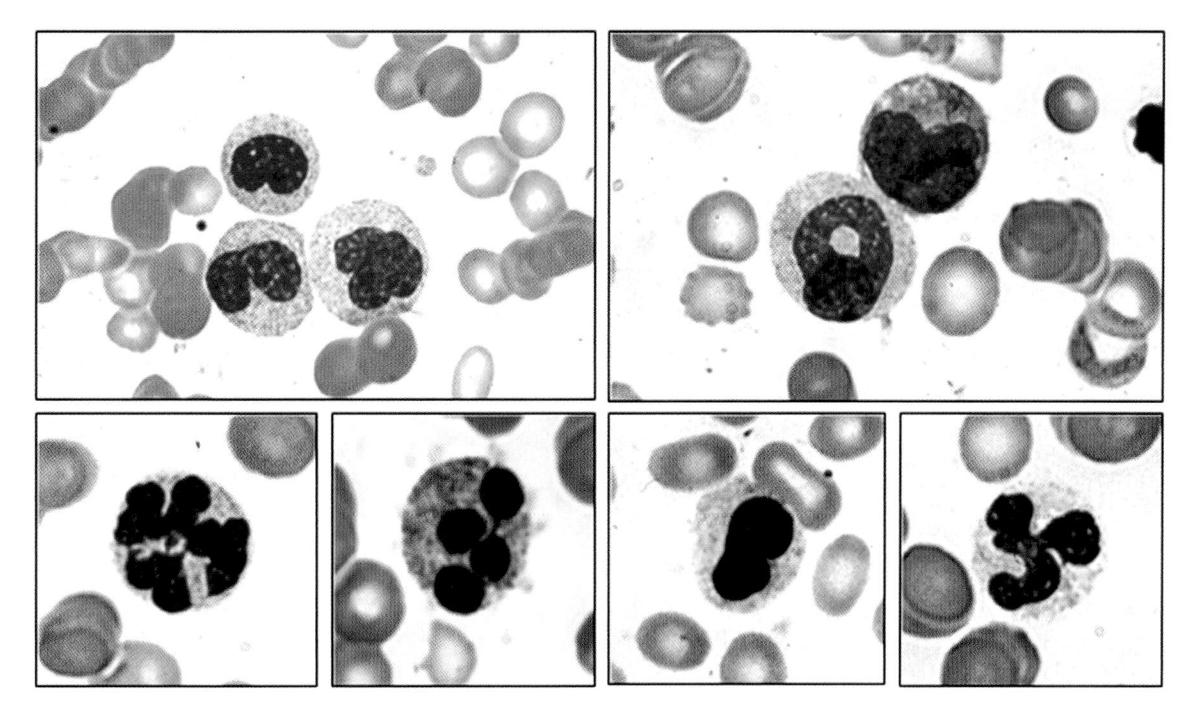

FIGURE 8.10 Blood smears from patients with MDS showing dysplastic neutrophils and monocytes.

3. Screening for the presence of micromegakaryocytes and topographical alterations of megakaryocytes by utilizing monoclonal antibodies against CD31, CD61, and factor VIII.
4. Evaluation of the monocytic component of the bone marrow by evaluating the results of CD68 and lysozyme stains.

Immunohistochemical stains are also occasionally used to evaluate the nature of the lymphoid aggregates, which are frequently observed in bone marrow biopsy sections of MDS patients. Sometimes these aggregates are morphologically atypical or are located next to the bone trabeculae, raising the possibility of a lymphoproliferative disorder.

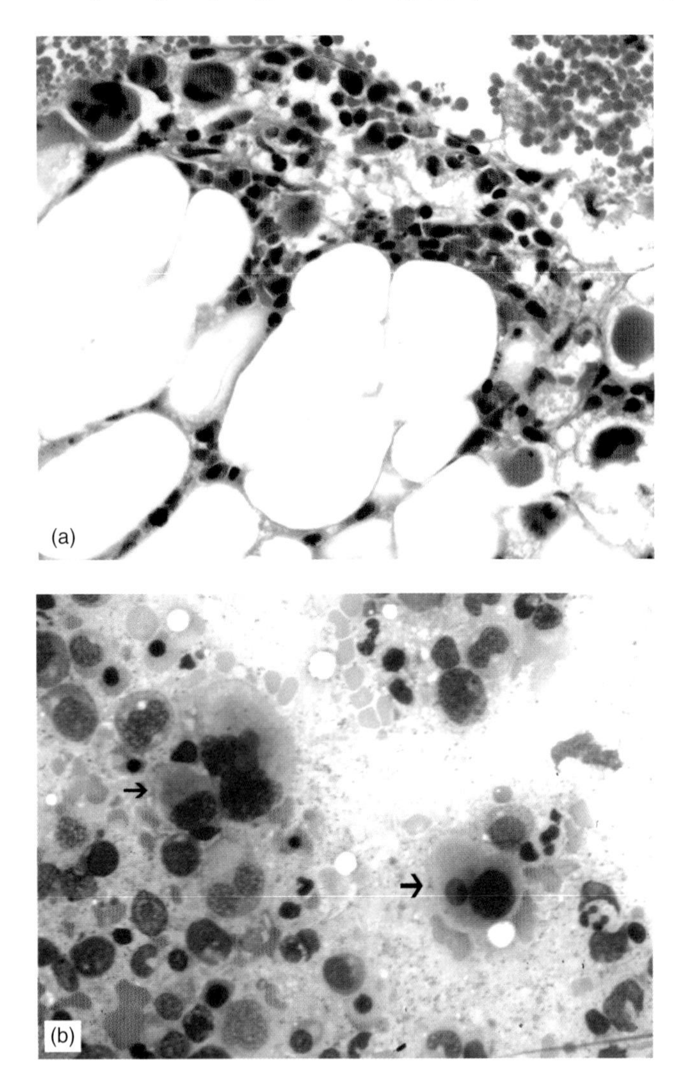

FIGURE 8.11 Clot section (a) and bone marrow smear (b) from a patient with 5q− syndrome demonstrating numerous micromegakaryocytes.

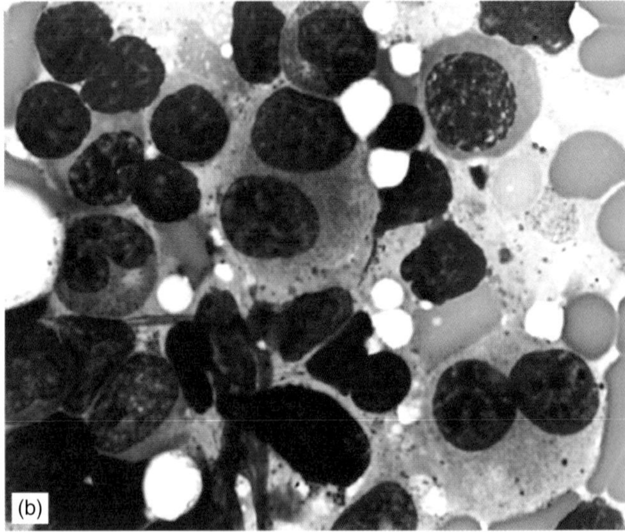

FIGURE 8.12 Micromegakaryocytes are frequent bone marrow findings in patients with MDS: (a and b) bone marrow smears.

MOLECULAR STUDIES

The diagnosis of MDS is typically based on clinical history and cell morphology in blood and bone marrow. Although clearly owing to one or more genetic defects in precursor stem cells, these are not yet sufficiently characterized to become targets for molecular diagnostic study. As such, the predominant molecular techniques used in MDS diagnosis are of the cytogenetic variety (FISH) to supplement the important standard karyotypic findings (see later). As already discussed, molecular techniques can be used to determine clonality of the underlying process, such as by characterization of polymorphic markers on the active and inactive X-chromosomes (applicable only in females). However, this is primarily a research tool and not directly relevant to the clinical diagnosis or management of a particular patient.

However, a number of individual point mutations, deletions, or epigenetic alterations have been observed in certain oncogenes, tumor suppressor genes, and signaling factors. The most frequent are point mutations in N-*ras* (found in 15–20% of MDS cases), tandem duplication mutation in the *FLT3* gene (5%), promoter methylation of *p15* (30–50%), inactivating and deletion mutations of *p53* (5–10%), and missense mutations in *AML1* [44].

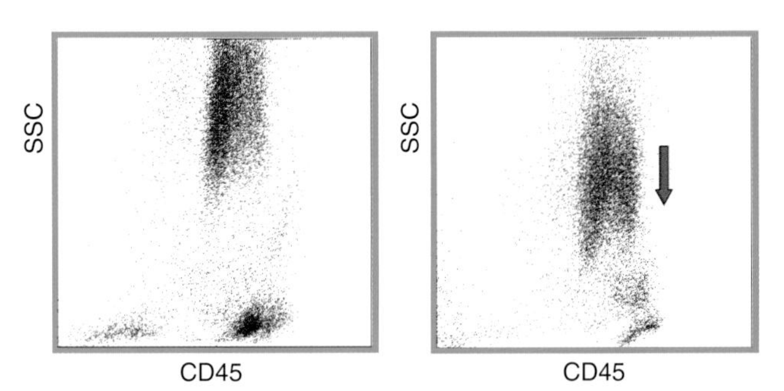

FIGURE 8.13 Dot plot analysis of side scatter (SSC) and CD45 of peripheral blood samples of a normal control (a) and a patient with MDS (b). There is a drop (arrow) in the SSC of the granulocytic population in the patient due to the hypogranularity of the granulocytes.

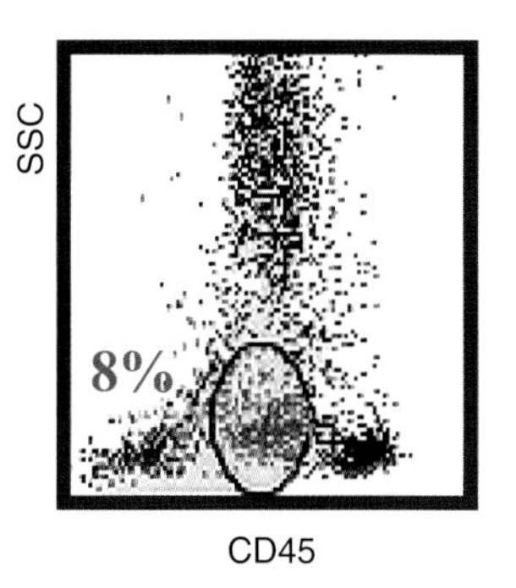

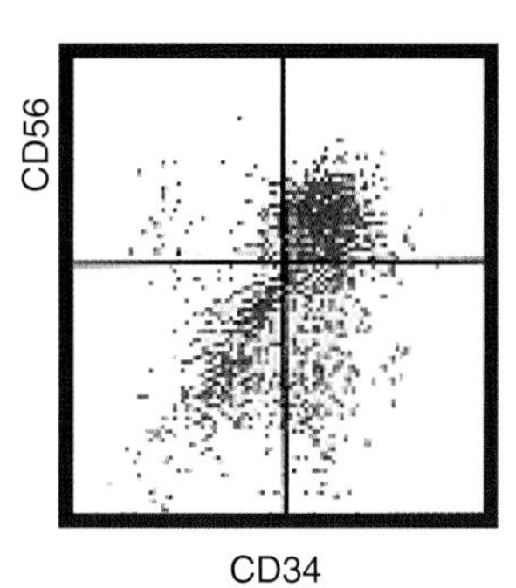

FIGURE 8.14 Flow cytometry of bone marrow from a patient with RAEB showing a population of blast cells co-expressing CD34 and CD56.

Inactivation of *p53* is of course a common occurrence in many types of cancer, but in MDS it appears to be an indicator of late stage or imminent progression to AML [45]. Again, although these findings may have important implications for our understanding of MDS pathogenesis and progression, they are not necessary or routinely used as targets for diagnosis.

CYTOGENETICS

Chromosomal changes occur in about 30–50% within the diverse subtypes of MDS and are the strongest independent prognostic indicators. These changes range from balanced translocations to unbalanced karyotypes with numerical or structural gains and losses as well as to complex aberrant karyotypes. The complex karyotypes are characterized by three or more chromosomal abnormalities and show an extremely unfavorable prognosis (Table 8.2) [46].

Conventional cytogenetics, FISH panel testing, and multiplex FISH (M-FISH) have analytical limitations. A combination of all three techniques should delineate the overwhelming majority of cytogenetic abnormalities in a bone marrow sample suspected for MDS.

Several recurrent and well-established cytogenetic changes have been described in MDS, and the detection of these changes can greatly facilitate diagnosis, prognosis, followup, and treatment of patients [47]. Although chromosomal abnormalities occur in almost half of *de novo* cases, aberrations are observed in up to 95% of secondary MDS. Most chromosomal defects in MDS are nonspecific, and with the exception of 5q−, none are specifically associated with any particular MDS subtypes [48].

Observed in nearly 50% of patients, chromosomal deletions are the most common defects in both *de novo* and secondary MDS. Deletions are generally interstitial, rather than terminal, and frequently occur in 5q, 7q, 20q, 11q, 13q, 12p, and 17p (Figures 8.15–8.19). Although a deletion observed as a sole abnormality is associated with low-risk MDS, deletions observed along with other abnormalities are associated with more advanced cases [48, 49].

Monosomies, trisomies, and unbalanced translocations are the next most common aberrations occurring

TABLE 8.2 Chromosomal abnormalities in MDS.

Type of abnormality	Chromosome	Prognosis
Numerical	−5	Good
	−7	Poor
	+8	Intermediate
	+13	
	+14	
	+15	
	+21	
Structural	der(1;7)(q10;p10)	
	t(3;21)(q26;q22)	
	ins(3;3)(q26;q21q26)	
	inv(3)(q21q26)	
	del(5)(q12-q31 or q31-35)	Good
	t(6;9)(p23;q34)	
	del(7)(q22)	Poor
	del(11q)	
	del(12)((p11p13)	
	del(13)(q12q14)	
	iso(17q)	Poor
	del(20)(q11q13) or	Poor
	del(20q)(q11.2)	
	idic(X)(q13)	
Normal karyotype		Good
Complex karyotype	3 or more abnormalities	Poor

in 15% of patients (Figures 8.20–8.23). The most common monosomies in MDS involve chromosomes 5, 7, and Y. Deletions and monosomies cause loss of one allele of a tumor suppressor gene with the subsequent submicroscopic deletion of the second allele on the homologous chromosome [43]. This recessive mechanism inactivates the cell's ability to control the cell cycle, DNA repair, and apoptosis [47, 48]. Although balanced translocations are relatively common aberrations in AMLs, they are very rare in MDS.

The most common of the chromosomal aberrations is represented by the chromosome 5q interstitial deletion. Deletion 5q occurs either as an isolated abnormality or accompanied by additional abnormalities, and accounts for up to 28% of all cytogenetic abnormalities in MDS with an overall frequency of approximately 15%. Prognosis of patients with primary MDS and isolated del(5q) is more favorable than that of those carrying an additional chromosomal abnormality. Patients with complex karyotypes have a poor prognosis with estimated survival of <6 months [50, 51].

The region of deletion in 5q chromosome is highly variable, but the most critically deleted region is about 1.5 Mb in size between bands 5q31 and 5q33. Only 5q33 deletions correspond to the 5q− syndrome and lead to a mild type of MDS, whereas 5q31 deletions are reported in other *de novo* and secondary subtypes and exhibit a more aggressive course.

The long arm of chromosome 5 has genes coding for many hematopoietic growth factors and growth factor

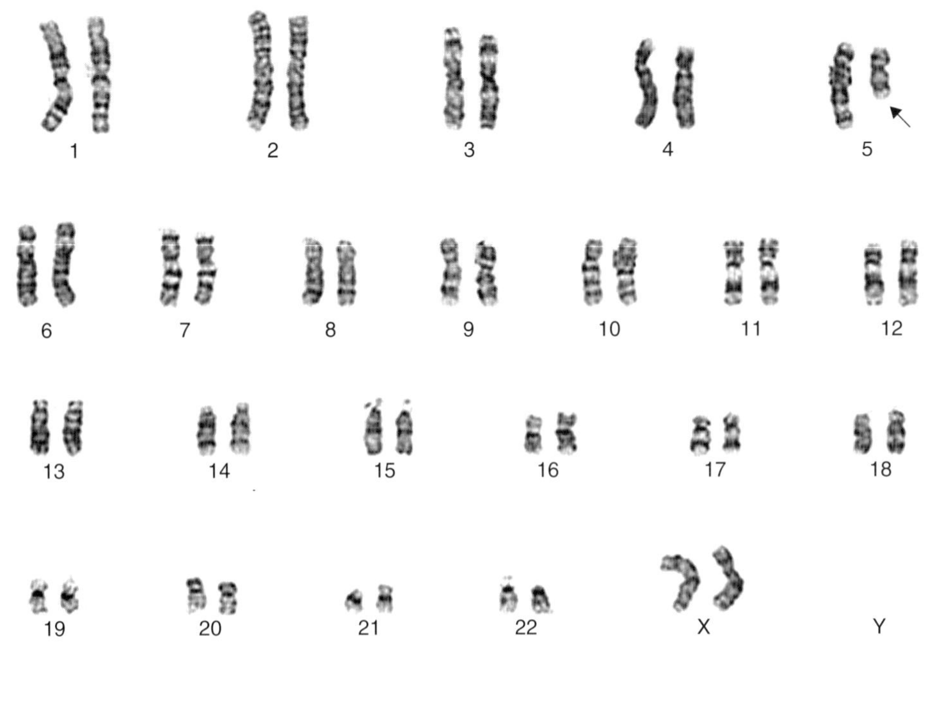

FIGURE 8.15 A G-banded female karyotype with a deletion of 5q.

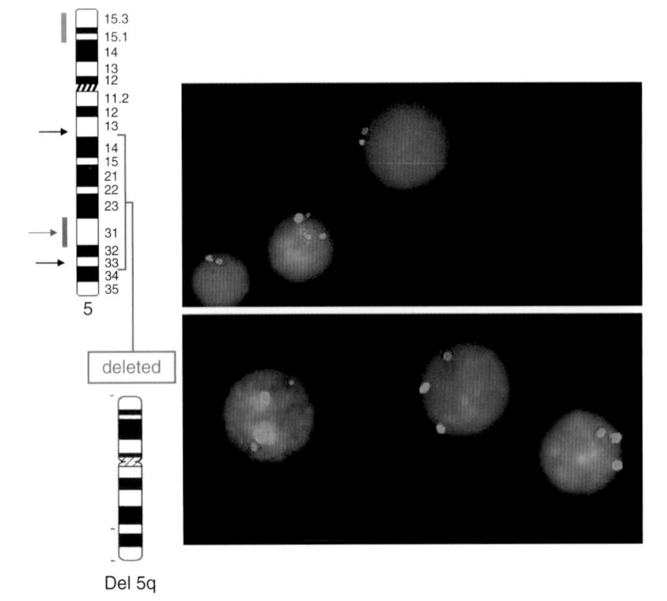

FIGURE 8.16 Ideogram of chromosome 5 showing the commonly deleted region in the long arm (red bracket). FISH studies on interphase cells show monosomy 5 (top right) and deleted 5q (bottom right).

receptors including IL-3, IL-4, IL-5, IL-9, and GMCSF. Genome mapping studies of the commonly deleted region have identified about 40 genes with 33 of them being expressed by hematopoietic progenitor stem cells [52]. Some of these genes are also described to be tumor suppressor genes, transcriptional regulators, cytokines, and growth factors [53–55]. Although a tumor suppressor gene is postulated to reside in this region, a critical gene(s) related to leukemogenesis has not yet been definitively identified. Patients with the 5q deletion have distinct clinical and pathological features that include anemia, the presence of dysplastic megakaryocytes in the bone marrow, and an indolent clinical course (see the following section).

Among patients who have MDS with a 5q deletion, the presence of one or more additional chromosomal abnormalities is associated with a more aggressive clinical course and considerably poorer overall survival as compared with patients who only demonstrate the isolated 5q deletion [49, 51].

7q−/−7 has been observed in all MDS subtypes, though it is much more common in advanced forms. 7q−/−7 occurs as a sole chromosomal abnormality in 1% of cases. 7q−/−7 is more common in secondary MDS, seen in up to 60% of the patients, and is therefore considered a secondary event in pathogenesis of the disease. Monosomy 7 is the most common chromosomal defect in bone marrow of patients with constitutional syndromes (e.g. Fanconi's anemia, type I neurofibromatosis, and severe congenital neutropenia) that predispose them to myeloid disorders [56]. Also, a recently described pediatric monosomy 7 syndrome presented with hepatosplenomegaly, leukocytosis, thrombocytopenia, male predominance, and an unfavorable outcome. Patients harboring deletions in the 7q31 to 7q36 regions have an inferior response to chemotherapy and shorter survival than those with deletions in the 7q22 region [57, 58].

Deletion of the long arm of chromosome 20 occurs in 5% of *de novo* and 7% of secondary MDS. This incidence might be an underestimation, since monosomy 20 and unbalanced translocations involving chromosome 20 occur as frequently as deletions. Although the critical region seems to be 20q11.2 to 20q12, deletions are rather large and involve most of the long arm of chromosome 20. Patients with del(20q) as a sole abnormality are in the low-risk MDS categories (RARS and RA), whereas those presenting with this deletion as a part of a complex karyotype (3 or more abnormalities) have a poor prognosis [59].

Deletion of the short arm of chromosome 17 encompasses not only simple deletions, but also unbalanced translocations, iso chromosome 17q, and (rarer) monosomies. Del(17q)

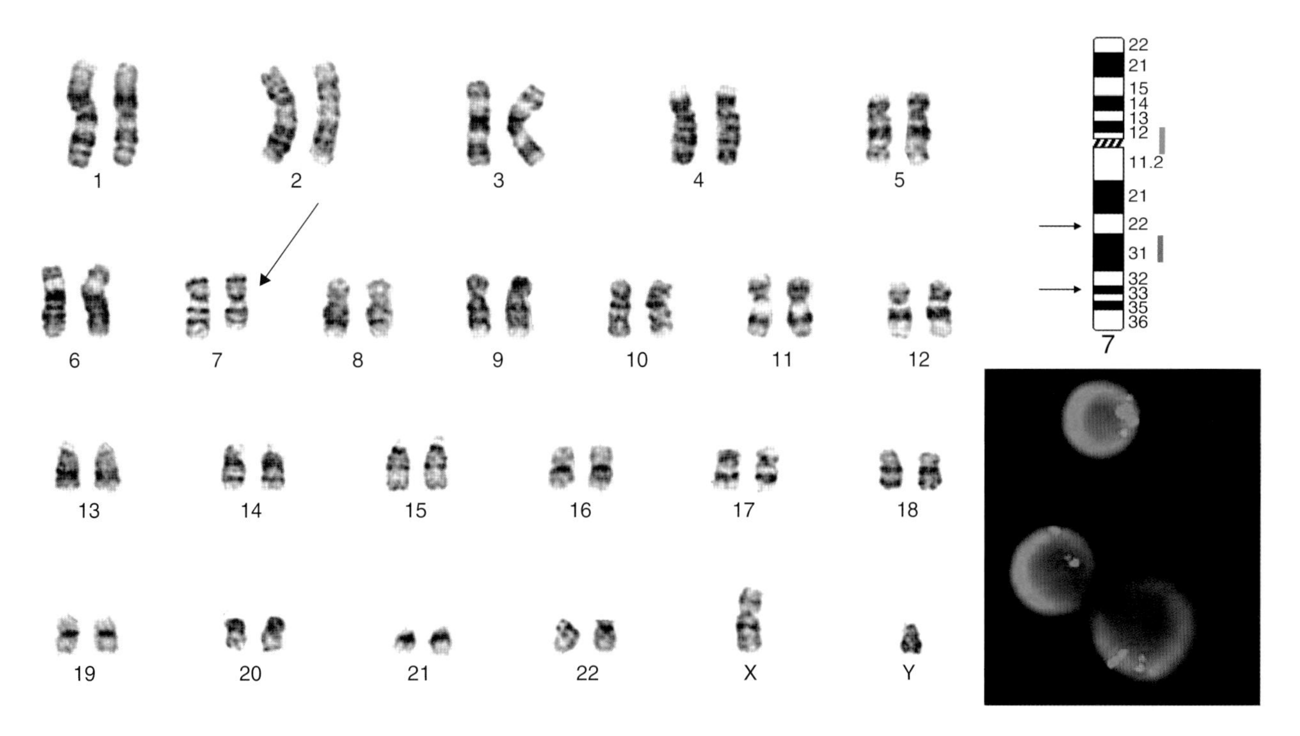

FIGURE 8.17 A G-banded karyotype showing a deletion of 7q, an ideogram of 7q with the two critical regions of deletion (black arrows), and a panel of interphase cells (bottom right) exhibiting 7q deletion by FISH.

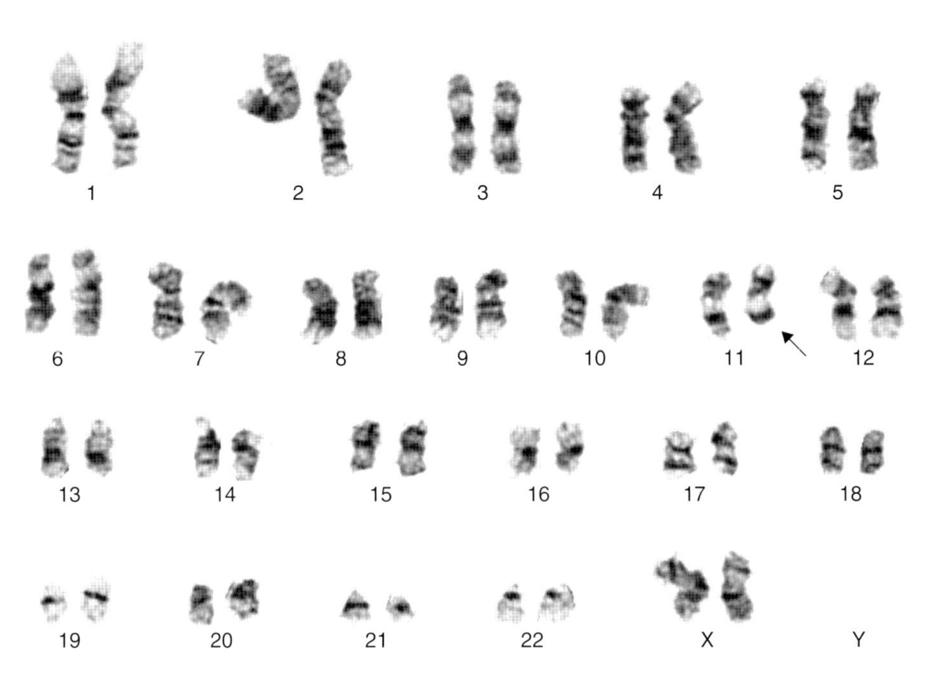

FIGURE 8.18 A G-banded karyotype with a deletion of 11q.

is rather rare in *de novo* MDS (~7%) but occurs more frequently in secondary MDS [60, 61]. Despite its heterogeneity, all the above-mentioned aberrations of the short arm of chromosome 17 lead to the loss of one p53 allele. Mutation or submicroscopic deletion of the other p53 allele occurs in 70% of the patients and cause inactivation of the gene.

Loss of the Y chromosome is observed in about 10% of MDS patients. It also occurs in about 7% of the elderly men without any hematological disorder. Therefore, MDS diagnosis cannot be based on the loss of chromosome-Y alone. When biological and clinical parameters point to an MDS diagnosis, loss of the Y chromosome identifies patients with a favorable clinical outcome.

Interstitial deletions or balanced translocations involving band 12p13 are found in about 5% of patients with RAEB. These patients usually belong to an intermediate-risk cytogenetic category for MDS. However, recent studies suggest that 12p13 aberrations signify a clinical outcome similar to that of patients included within the low-risk category [59].

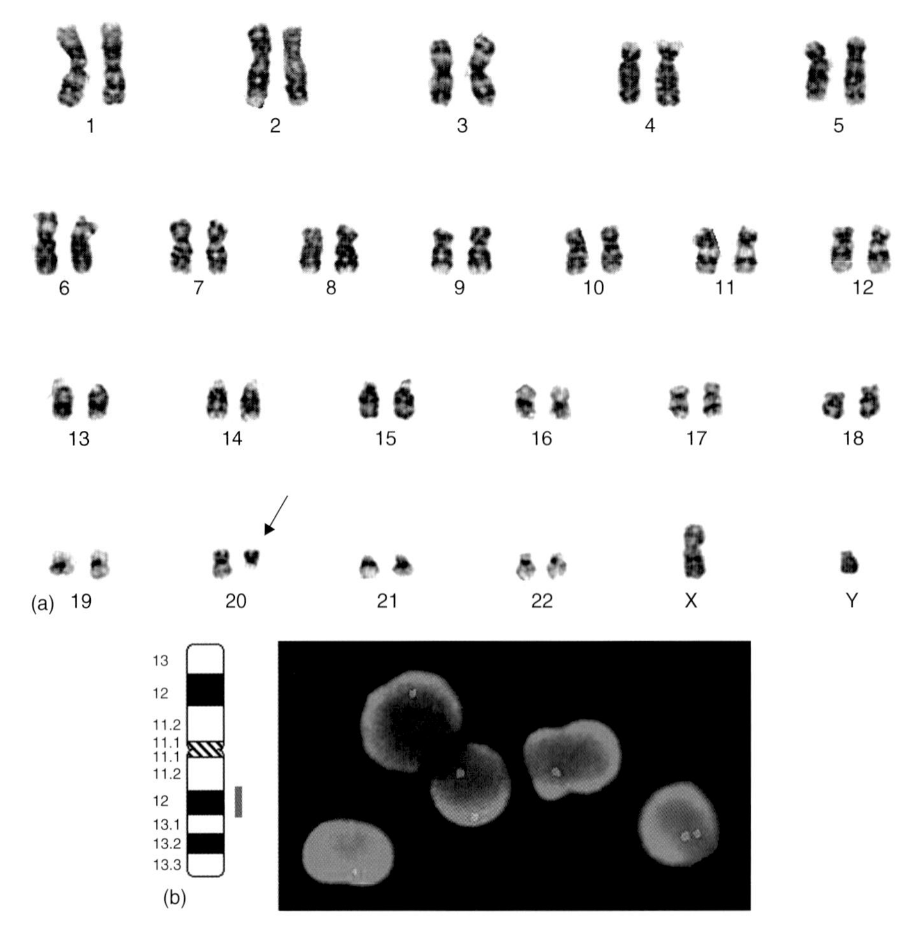

FIGURE 8.19 A G-banded karyotype showing a deletion of 20q (a) and the corresponding FISH studies (b).

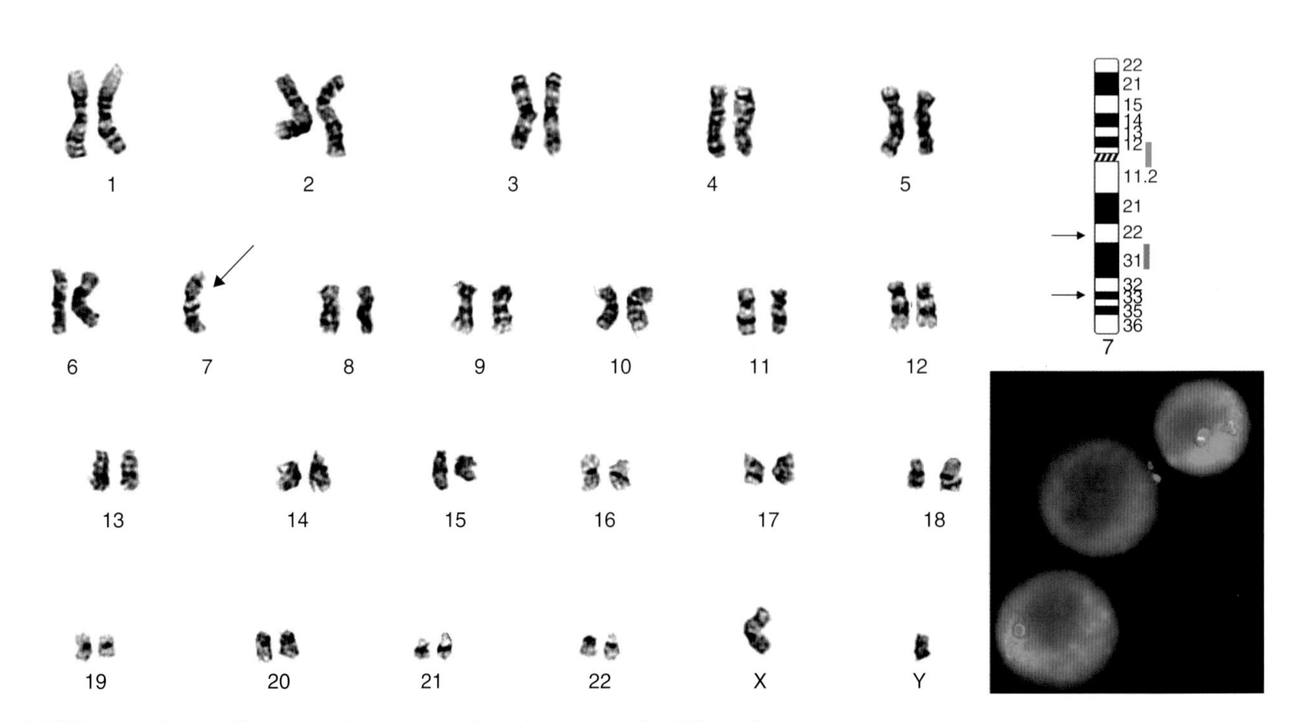

FIGURE 8.20 A G-banded karyotype with monosomy 7 and the corresponding FISH studies.

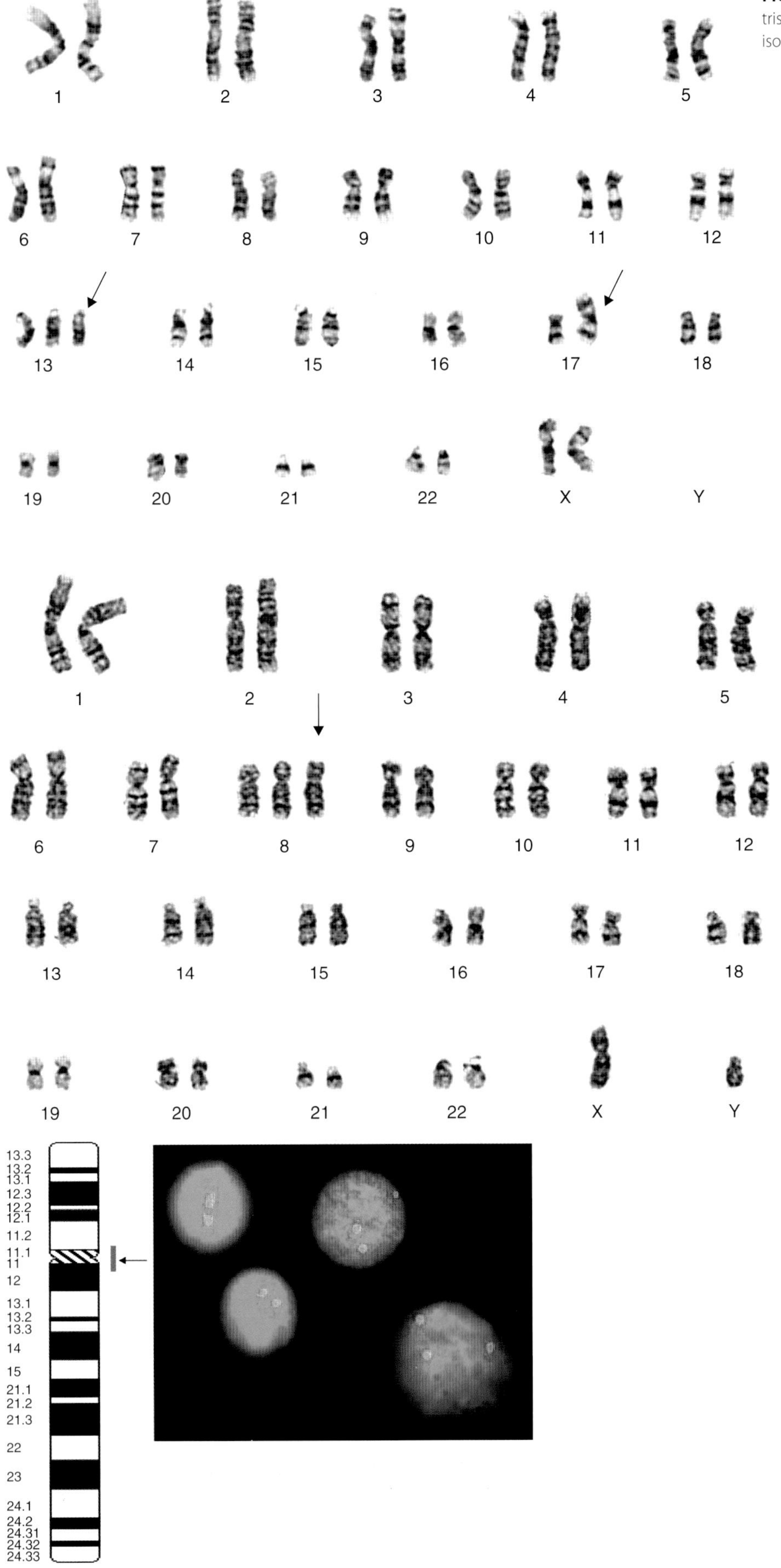

FIGURE 8.21 A G-banded karyotype showing trisomy 13 and deletion of 17p resulting from an isochromosome of 17q.

FIGURE 8.22 A G-banded karyotype with trisomy 8 and the corresponding FISH studies.

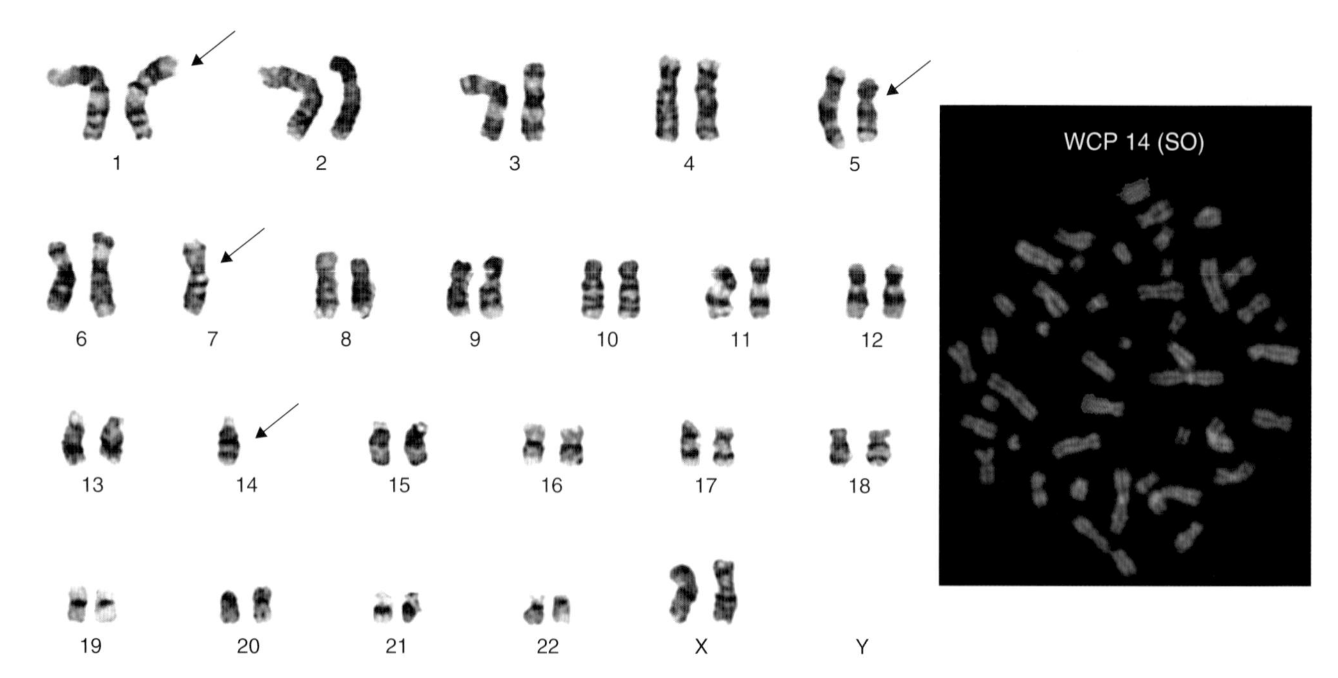

FIGURE 8.23 A G-banded karyotype showing a deletion of 1p, monosomy 7, and an unbalanced 5;14 translocation (left). The 5;14 translocation confirmed by whole chromosome paint of chromosome 14 (red signal) showing an unbalanced 5;14 translocation resulting in 5q deletion (right).

Several other chromosomal aberrations are observed in MDS, but are not specific to the disease. Trisomy 8 occurs in 10% of all MDS cases, but can be found in other clonal hematological disorders also. Trisomy 8 is more often associated with RARS and RAEB. Chromosome 3 rearrangements, typically translocations or inversions, occur in 2–5% of patients with MDS (also in AML). Chromosome 3 changes are frequently associated with −7/7q and 5q−, and are associated with short survival and a poor response to chemotherapy. Aberrations within 11q23 (the *MLL* gene locus) are found in 5% of MDS patients. Further karyotypic defects (e.g. rearrangements of the long arm of chromosome 3) occur in 5–10% of *de novo* MDS cases.

Based on cytogenetic abnormalities, MDS patients have been divided into three prognostic categories. Patients in the first, which is low-risk category, exhibit a normal karyotype, deletion of long arm of chromosome 5 as a sole abnormality, or harbor an isolated deletion of the long arm of chromosome 20. Patients with either a deletion of the short arm of chromosome 12 or trisomy 8 are categorized as an intermediate-risk group. Finally, the presence of complex karyotypes, monosomy 7, deletion of the short arm of chromosome 17, rearrangements involving chromosome 3, and aberrations of the long arm of chromosome 11 (*MLL*) indicates a high-risk group of MDS patients [49, 59, 62].

About 12% of patients with MDS with normal cytogenetics progress to AML, whereas 50% of those with chromosomal changes will progress to AML. Generally, therapy-related MDS (t-MDS) are much more clinically aggressive than primary MDS and this characteristic is reflected in the karyotypes. At least 80% of patients with t-MDS have chromosomally abnormal clones in the marrow and in a vast majority of cases these clones contain multiple abnormalities.

FISH and Other Technologies

It must be emphasized that no aberration is MDS-specific. However, if the aberration is one of those typically found in MDS, the diagnosis is strengthened. Lack of cytogenetic abnormality, however, is of no help in the diagnostic workup because one-third of patients with MDS have no chromosome aberrations. Also, there can be diagnostic problems due to difficulties in obtaining adequate bone marrow aspirate smears for evaluation of cytology. However, cytogenetic data can be obtained with FISH from nondividing or terminally differentiated cells, or from poor samples that contain too few cells for routine karyotyping studies. Detection of chromosomal abnormalities within interphase nuclei can be achieved by hybridizing with an appropriate selection of probes to the nuclei (Figures 8.16, 8.17, 8.19, 8.20, and 8.22). This technique permits the direct correlation of cytogenetic and cytologic features (e.g. trisomy 8 in a hypogranular neutrophil), which enables cytologists to differentiate malignant from benign conditions in equivocal cases [63].

One of the other advantages of FISH is that one can combine this method with cytology and/or immunohistochemistry to examine the cytogenetic pattern of specific cell populations to monitor the effects of therapy and to detect minimal residual disease. Most of the studies using FISH for the identification of lineage involvement in MDS indicate that, in most cases, the pluripotent stem cell is not affected because the lymphoid cells usually do not contain the chromosomal abnormality [64, 65]. Molecular cytogenetic techniques, such as M-FISH and spectral karyotyping (SKY), may allow for the comprehensive evaluation of the complex karyotypes. Although these techniques are used mostly in research, it is possible to analyze the origin of marker chromosomes, reveal cryptic rearrangements, and determine recurrent breakpoints and the structure of derivative chromosomes.

CLASSIFICATION

According to the WHO classification, there are eight categories of MDS (Table 8.1) [1, 66–68]. These categories, based on the clinical course and survival rate, are divided into two major risk groups. The low-risk groups include RA with or without ringed sideroblasts and 5q− syndrome. RCMD and RAEB are considered high risk. A brief pathologic description of each MDS subclass is presented later, and the most frequent chromosomal aberrations are demonstrated in Table 8.3.

Refractory Anemia

Refractory anemia (RA) is a low-risk MDS with mono-lineage dysplasia characterized by anemia, dyserythropoiesis, and low percentage of blasts in bone marrow and peripheral blood (Figure 8.24). The degree of dysplasia in the erythroid precursors varies and may include megaloblastic changes, multinucleation, nuclear bridging, nuclear fragmentation or budding, cytoplasmic vacuolization, and abnormal hemoglobinization. Bone marrow is often hypercellular and frequently shows erythroid preponderance [1, 25, 29]. Myeloblasts account for <5% in the bone marrow and <1% in the peripheral blood. Ringed sideroblasts, if present, account for <15% of the erythroid precursors. The granulocytic and megakaryocytic lines are either normal or show minimal dysplastic changes. In the peripheral blood, red blood cells often show some degree of anisopoikilocytosis with reduced polychromasia (reticulocytes). In rare cases, instead of erythroid lineage, dysplasia is demonstrated in the megakaryocytic or granulocytic lineages.

According to a WHO recommendation, in order to establish the diagnosis of RA, all other etiological possibilities for erythroid abnormalities should be excluded [1]. These possibilities include drug and chemical exposure, immunologic disorders, viral infections, congenital abnormalities, and vitamin deficiencies. Also, there should be an observation period of at least 6 months if cytogenetic and molecular studies show no evidence of clonal disorder. Development of multilineage dysplasia and/or an increase in the percent of blast cells are suggestive of progression of the disease into a more aggressive type of MDS [1, 66, 68].

Refractory anemia represents about 5–10% of the MDS cases; it usually affects elderly people and has no known etiology so far.

TABLE 8.3 Incidence of chromosomal aberrations in MDS subclasses.

Subgroup	Common alterations	Frequency (%)
RA	del(5q)	50
	−7	10–15
	−8	20
RARS	+8	30
	del(5q)	25
	del(11q)	10
	del(20q)	10–15
RAEB	−5/del(5q)	35–40
	−7/del(7q)	30–35
	+8	20

Refractory Anemia with Ringed Sideroblasts

Refractory anemia with ringed sideroblasts (RARS) is a low-risk MDS characterized by anemia, dyserythropoiesis, and presence of 15% or more ringed sideroblasts in the erythroid precursors [1, 66, 68]. Myeloblasts are not present in peripheral blood and account for <5% of the total cells in the bone marrow. Similar to RA, the degree of erythroid dysplasia in RARS varies and may include megaloblastic changes, multinucleation, nuclear bridging, fragmentation or budding, cytoplasmic vacuolization, and abnormal hemoglobinization [1, 25, 29]. The granulocytic and megakaryocytic lines are either normal or show minimal dysplastic changes. Bone marrow is hypercellular and shows erythroid preponderance with the presence of 15% or more ringed sideroblasts in erythroid precursors (Figure 8.25). Ringed sideroblasts represent nucleated red cells with precipitated iron particles in their mitochondria. Since mitochondria are often located around the nucleus, iron stain shows positively stained (siderotic) granules surrounding the nucleus like a ring. Ringed sideroblast is defined as an iron-containing erythroid precursor (sideroblast) with 10 or more siderotic granules encircling one-third or more of the nuclear parameter [1]. The bone marrow iron stain often shows increased iron stores with numerous hemosiderin-laden macrophages.

In the peripheral blood, red blood cells often show some degree of anisopoikilocytosis with reduced polychromasia (reticulocytes). A dimorphic morphologic pattern is common with a mixture of normochromic and hypochromic red cells. There is no absolute monocytosis (monocyte count $<1 \times 10^9$/L) [1, 25, 29].

Approximately 10% of the MDS cases are of RARS type. RARS is primarily a disease of old age, more frequently occurring in men than in women. Etiology of RARS is not known.

Refractory Cytopenia with Multilineage Dysplasia

Refractory cytopenia with multilineage dysplasia (RCMD) is characterized by the presence of dysplasia in two or more hematopoietic lineages and corresponding cytopenias [1]. The spectrum of dysplastic changes is discussed earlier in the section "General Morphologic Features." At least 10% of the precursor cells in two or more hematopoietic lineages show dysplastic changes (Figure 8.26) [1]. Bone marrow is usually hypercellular and contains <5% myeloblasts. Occasional blasts

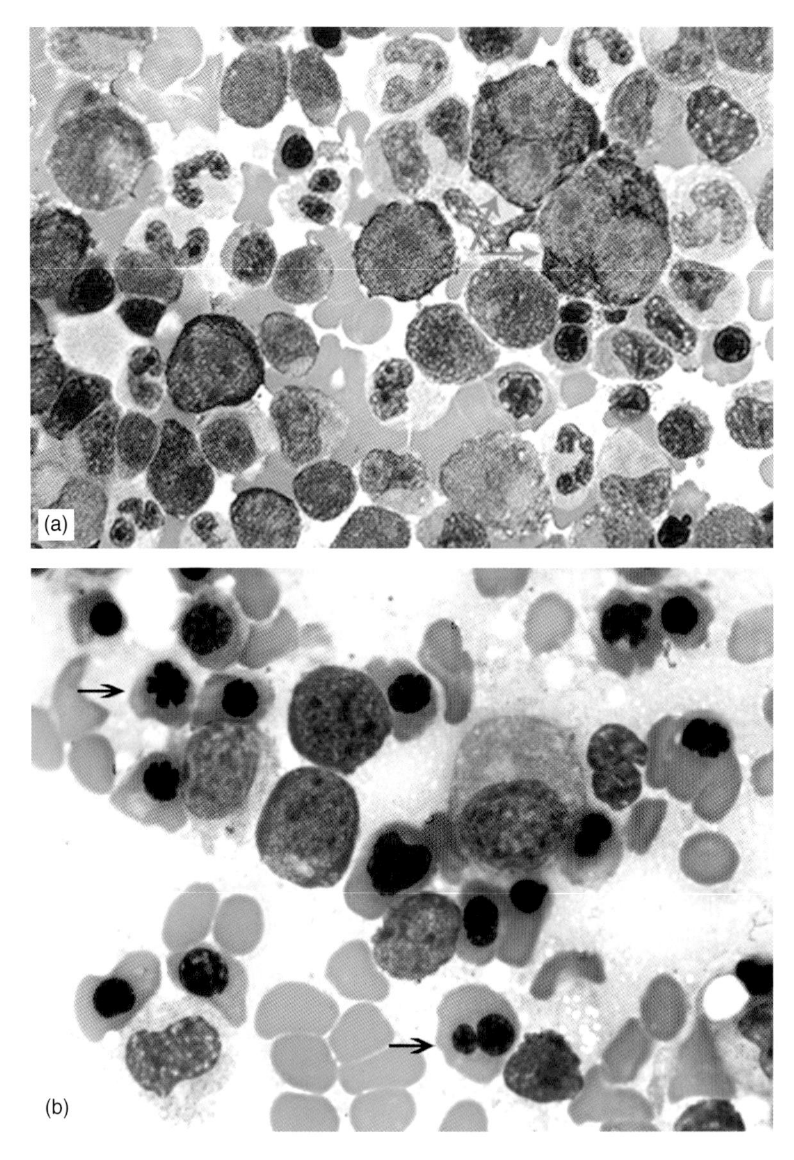

FIGURE 8.24 Refractory anemia: bone marrow smears demonstrating early dysplastic binucleated erythroid forms (a, arrows) and dysplastic late erythroid precursors with irregular or lobated nuclei (b, arrows).

may be present in the peripheral blood. There is no absolute monocytosis (monocyte count $<1 \times 10^9$/L).

Iron stores are often increased and ringed sideroblasts may be present, but they account for <15% of the erythroid precursors. The condition is referred to as RCMD-RS when ringed sideroblasts are 15% or more [1].

RCMD and RCMD-RS account for about 15–24% and 10–15% of MDS cases, respectively. They usually affect elderly people and have no known etiology so far.

Refractory Anemia with Excess Blasts

Refractory anemia with excess blasts (RAEB) is characterized by multilineage dysplasia and increased myeloblasts (5–19%) in the bone marrow and/or peripheral blood (1–19%) [1, 66, 69]. The spectrum of dysplastic changes is discussed earlier in the section "General Morphologic Features." At least 10% of the precursor cells in two or more hematopoietic lineages show dysplastic changes. Bone marrow is usually hypercellular and shows myeloid preponderance and left shift with increased blasts accounting for 5–19% of the bone marrow cells (Figures 8.27 and 8.28) [1, 25, 29]. The centrally located small aggregates of blasts, referred to as ALIP, are often present in the biopsy sections [1, 29]. The bone marrow biopsy in <15% of the cases is hypocellular. Peripheral blood smears show multilineage dysplasia, myeloid left shift, and the presence of 1–19% myeloblasts. There is no absolute monocytosis (monocyte count $<1 \times 10^9$/L).

Refractory anemia with excess blasts is divided into two groups: RAEB-1 and RAEB-2 [1].

RAEB-1 refers to the group with 5–9% blasts in the bone marrow and/or <5% blasts in the peripheral blood. The disorder is called RAEB-2 if the bone marrow blasts are 10–19% of the total cells and/or there are 5–19% blasts in the peripheral blood.

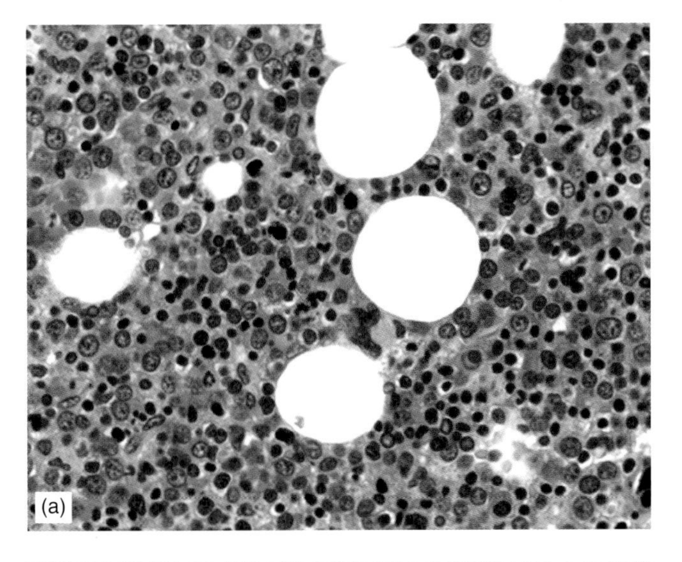

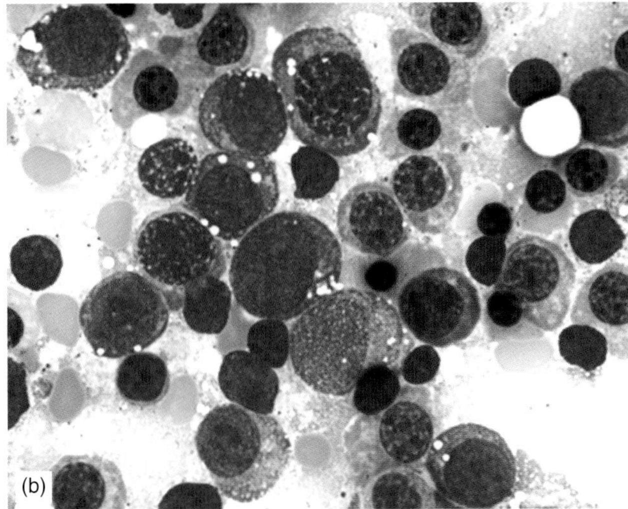

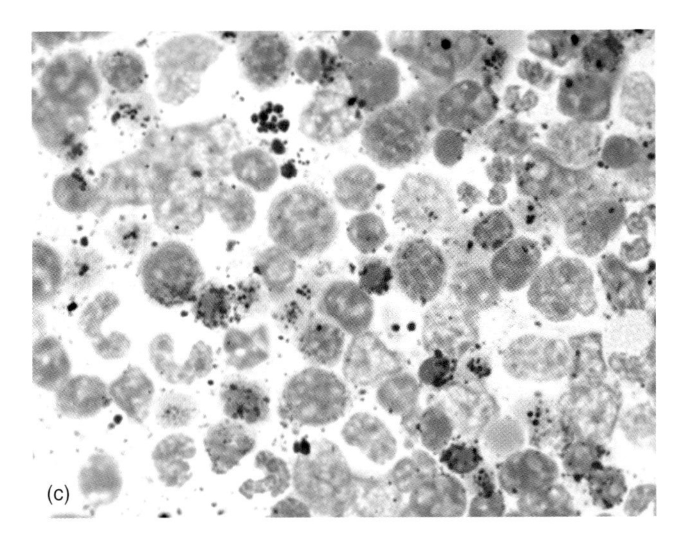

FIGURE 8.25 Refractory anemia with ringed sideroblasts: bone marrow biopsy reveals hypercellularity and erythroid preponderance (a); bone marrow smear shows erythroid preponderance and dysplastic vacuolated early erythroid precursors (b); and numerous ringed sideroblasts are present by iron stain (c).

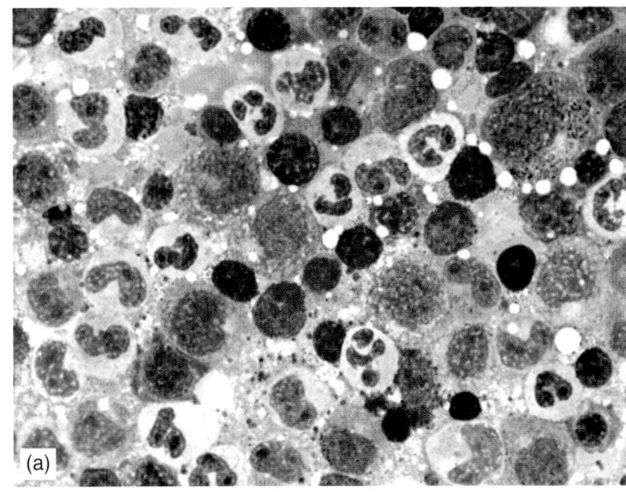

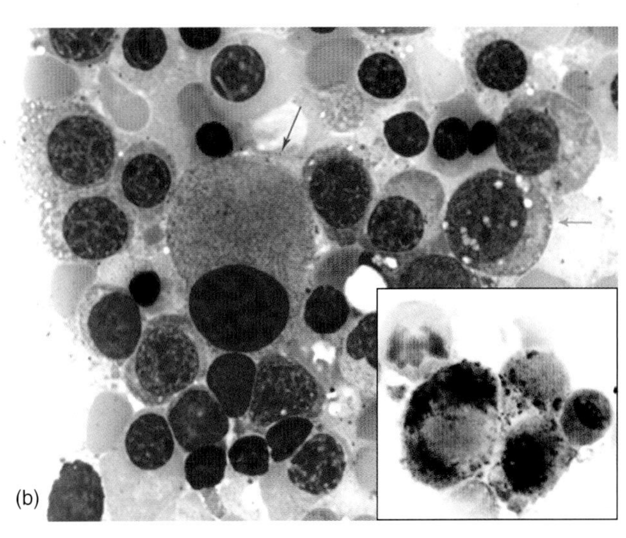

FIGURE 8.26 Bone marrow smears from a patient with refractory cytopenia with multilineage dysplasia demonstrating erythroid dysplasia (green arrow) and hypogranular neutrophils (a) and a micromegakaryocyte (b, blue arrow). The inset demonstrates coarse PAS-positive granules in a micromegakaryocyte and an erythroid precursor.

Refractory anemia with excess blasts accounts for about 30–40% of the MDS cases, and it usually affects patients older than 50 years. Etiology of RAEB is not known.

The 5q− Syndrome

5q− syndrome [MDS associated with isolated del(5q31-33) chromosome abnormality] is characterized by RA, which is usually macrocytic, normal to elevated platelet counts, modest leukopenia, and the presence of numerous micromegakaryocytes in the bone marrow (Figures 8.15 and 8.29) [1, 70, 71]. Bone marrow biopsies are usually hypercellular and show myeloid preponderance. Bone marrow smears show small megakaryocytes with mono- or hypolobated nuclei and <5% blasts [1, 25, 29]. Erythroid precursors often show dysplastic changes. Blood smears show macrocytosis, mild leukopenia, and, sometimes, occasional blasts.

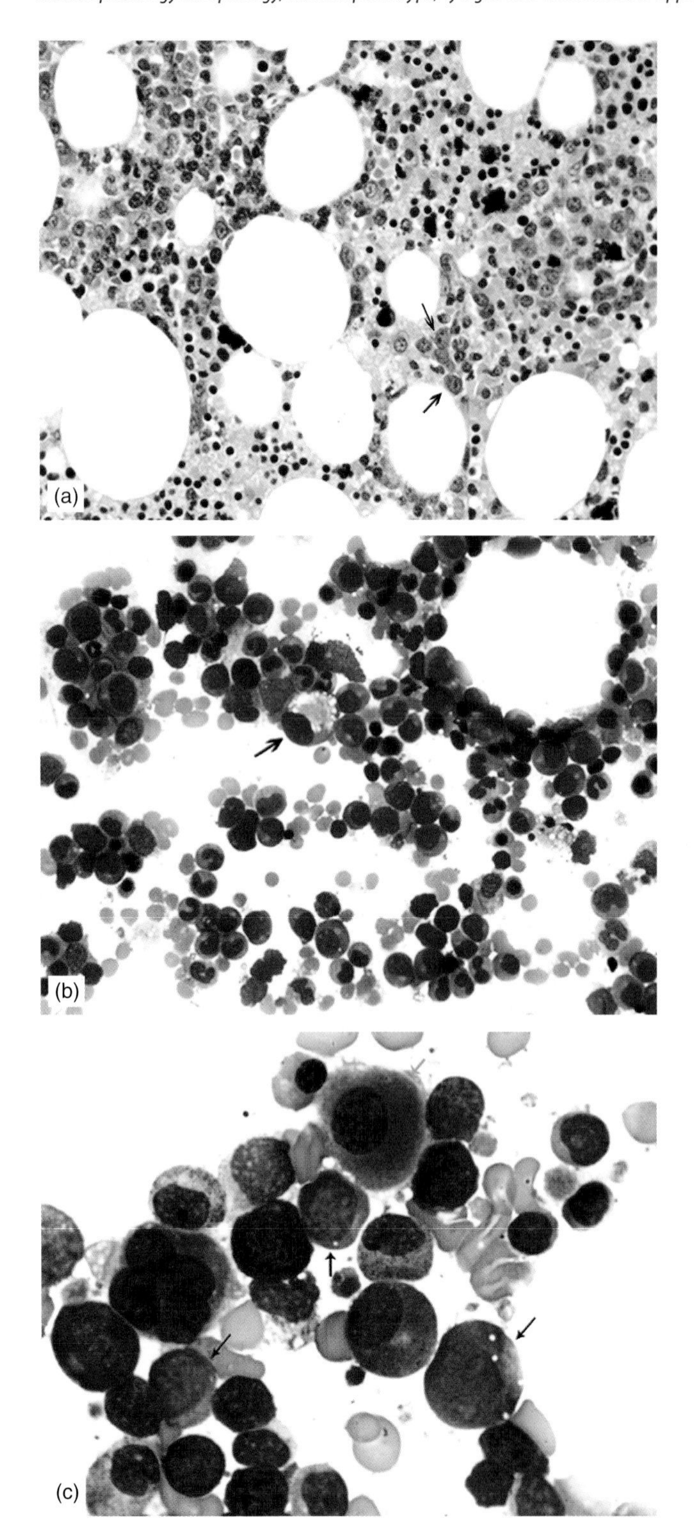

FIGURE 8.27 Bone marrow biopsy section is left shifted with increased number of immature cells (a, arrows). Bone marrow smears showing increased blasts (black arrows) and micromegakaryocytes (green arrows): (b) low power and (c) high power.

The etiology of this syndrome is not known, but the deleted region on the long arm of chromosome 5 (1.5 Mb at 5q31-q32) is the home of a number of important genes, such as IL-9 and EGR-1 [70, 71]. However, so far, a tumor suppressor gene responsible for MDS has not been identified on this chromosomal region.

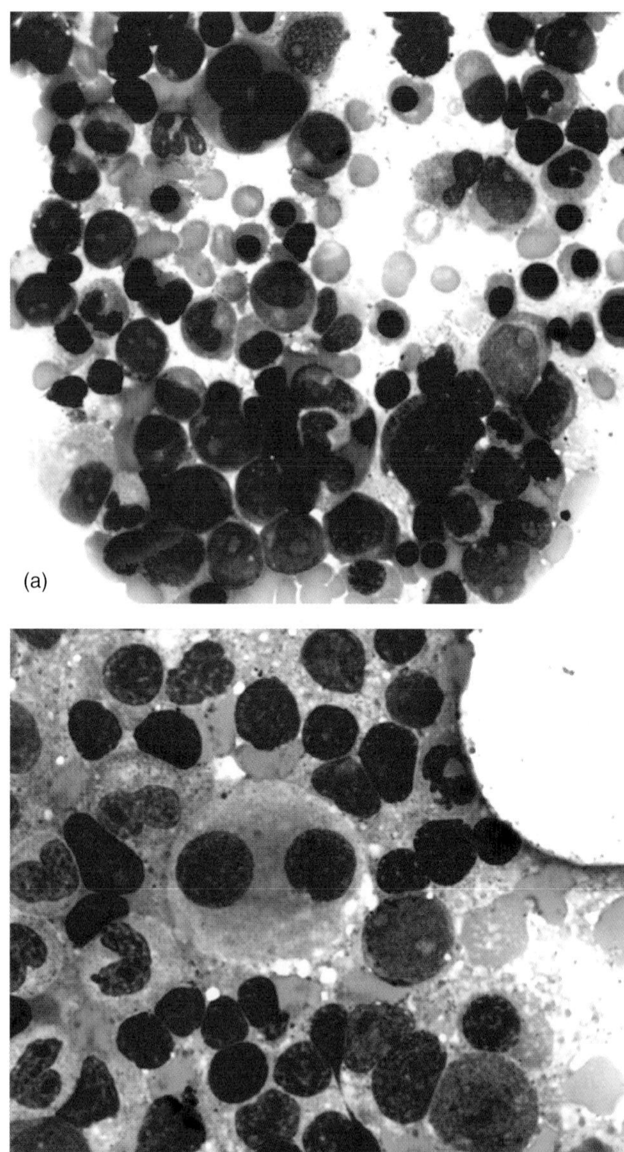

FIGURE 8.28 Refractory anemia with excess blasts: Bone marrow smears show increased proportion of myeloblasts and a trinucleated micromegakaryocyte (a) and a binucleated micromegakaryocyte (b).

5q− syndrome is considered a subtype of low-risk MDS, which predominantly affects elderly women. It accounts for about 10% of the MDS cases [70].

MDS, Unclassifiable

This term is recommended by WHO for those MDS cases that lack proper features to fall into one of the well-defined, above-mentioned categories, such as RA, RARS, RCMD, and RAEB [1]. Patients in this category usually show dysplastic changes in the granulocytic and/or megakaryocytic

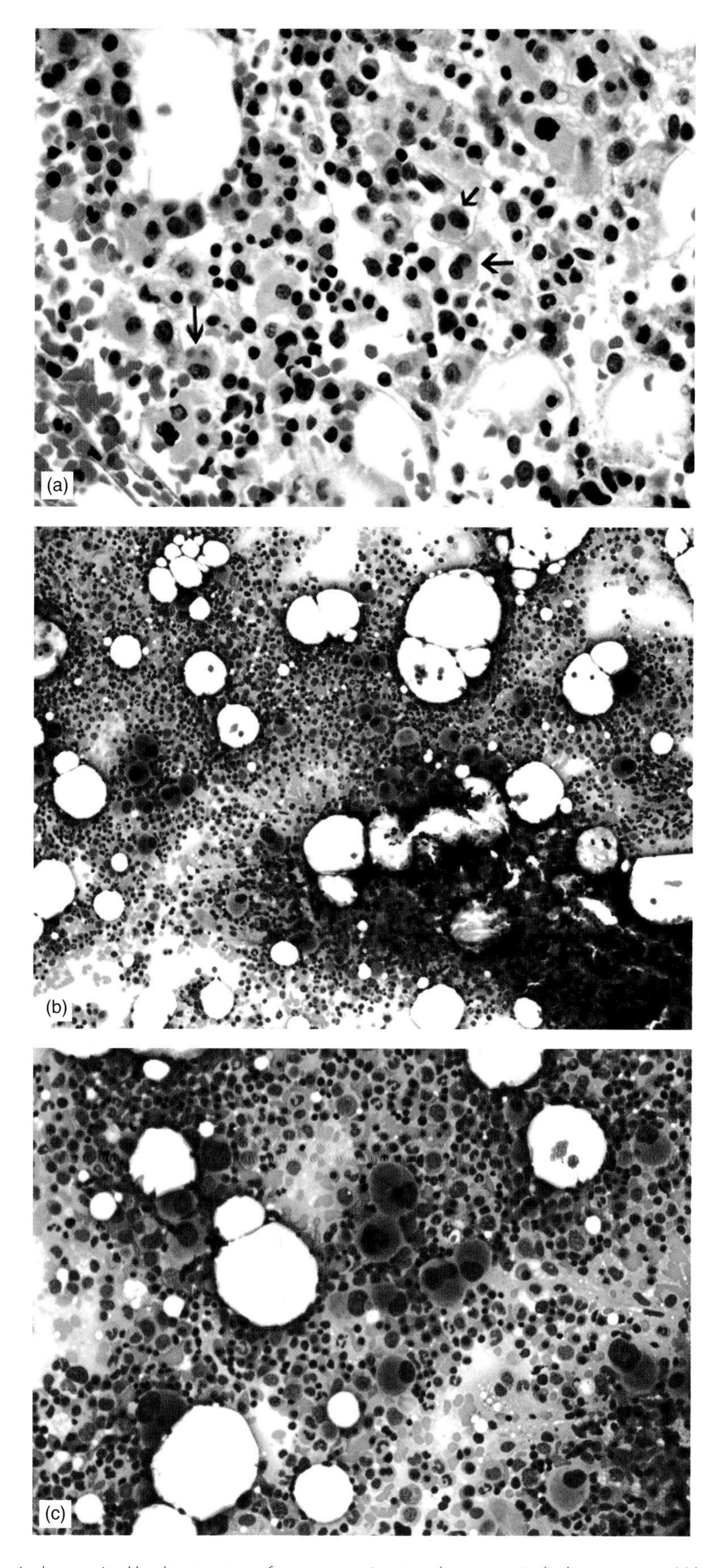

FIGURE 8.29 5q− syndrome is characterized by the presence of numerous micromegakaryocytes in the bone marrow: (a) biopsy section and (b) bone marrow smear.

lineages along with granulocytopenia and/or thrombocytopenia. Bone marrow is frequently hypercellular, but it could be normo- or hypocellular. There is no evidence of increased blasts.

OTHER TYPES OF MDS

Therapy-Related (Secondary) MDS

The development of MDS following cytotoxic chemotherapy and/or radiotherapy has been extensively investigated and reported in the literature [72–75]. The emergence of t-MDS is usually associated with a long latency period and is particularly seen following the use of alkalating agents. The pathologic manifestations of t-MDS are, in general, similar to those of primary MDS except for higher frequency of the following features [1, 72–75]:

1. High-risk variants
2. Bone marrow fibrosis
3. Bone marrow hypocellularity
4. Unclassifiable forms
5. Chromosomal aberrations (Table 8.4)
6. Transformation to acute leukemia.

t-MDS represents one spectrum of a broader syndrome now designated as *therapy-related AML and myelodysplasia* (see Chapter 12). It constitutes about 10–15% of the total MDS cases. The interval between initiation of therapy and the onset of MDS varies and depends on the type, duration, and dose of the therapeutic agent(s). This period in most studies ranges from 1 to 8 years, with a mean of 5 years [72–75]. The topoisomerase II inhibitors may occasionally cause MDS but more often are associated with *de novo* acute myelogenous leukemia without going through dysmyelopoiesis. Prolonged environmental or occupational exposure to benzene and benzene-derivative compounds may also lead to MDS.

TABLE 8.4 Comparison of chromosomal aberrations between primary and secondary MDS.

Abnormality	Primary MDS (%)	Secondary MDS (%)
del(5q)	10–20	20
Monosomy 7	10–15	30–50
Trisomy 8	15	10
Loss of 17p	3	10

Pediatric MDS

Myelodysplastic syndromes are rare in children, accounting for approximately 3–5% of all pediatric clonal hematologic disorders [76, 77]. They may also present themselves differently from the adult forms, particularly the MDS categories with <5% blasts. The following observations have been reported in pediatric patients with MDS [76–81]:

1. Hematopoietic dysplasia is frequently observed in a variety of conditions, such as infection, metabolic disorders, and nutritional deficiencies.
2. Neutropenia and thrombocytopenia are more frequently observed than anemia.
3. The rate of cytogenetic aberrations in low-risk MDS is higher in children (about 65%) than in adults (20–30%). Monosomy 7 is one of the most frequent findings.
4. The majority of the cases fall into the MDS/ myeloproliferative categories (see Chapter 10), particularly in patients younger than 5 years.

Hypocellular MDS

Hypocellular MDS accounts for about 5–10% of the MDS cases [82, 83]. This variant is usually therapy related and is often associated with more severe pancytopenia. Most investigators consider the diagnosis of hypocellular MDS when bone marrow cellularity is ≤25% of the age-matched normal range (Figure 8.30). Aplastic bone marrow conditions (such as aplastic anemia; Fanconi anemia, FA; and paroxysmal nocturnal hemoglobinuria, PNH) and hypocellular variants of hairy cell and acute leukemias are in the list of differential diagnosis [84]. Dysplastic hematopoiesis distinguishes hypocellular MDS from the bone marrow aplasia group, and blast counts of <20% separate this entity from acute leukemias. Dysplastic erythroid and myeloid cells are present in the blood smears and/or the bone marrow samples, and abnormal megakaryocytes are often identified. Estimation of blast number is facilitated by immunophenotypic studies by using blast-associated markers, such as CD34 and CD117.

Non-clonal Myelodysplasia

Non-clonal myelodysplastic changes have been observed in a variety of conditions, such as autoimmunity, infections, nutritional deficiencies, heavy metal intoxication, and post-chemotherapy and/or radiotherapy. Dysplastic changes in these conditions are often reversible upon elimination of the causative factors, and are not associated with chromosomal aberrations. Representative examples of non-clonal MDS are briefly discussed in the following sections.

Autoimmune Myelodysplasia

Myelodysplasia has been observed in a small proportion of patients with autoimmune disorders [21]. The dysplastic changes in these patients are not associated with chromosomal aberrations, and patients respond positively to

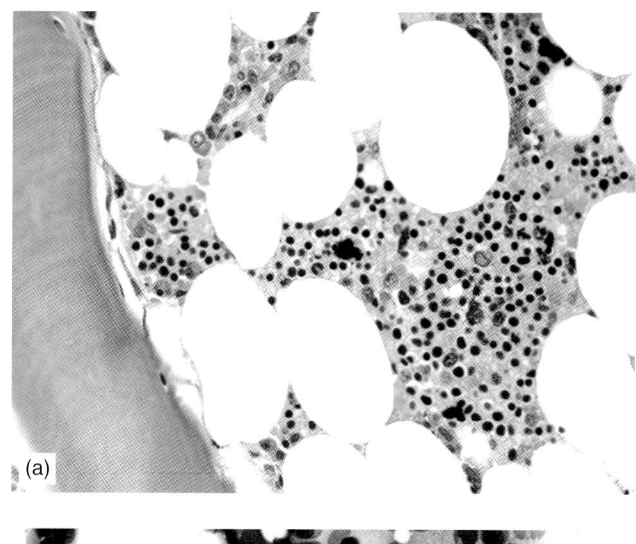

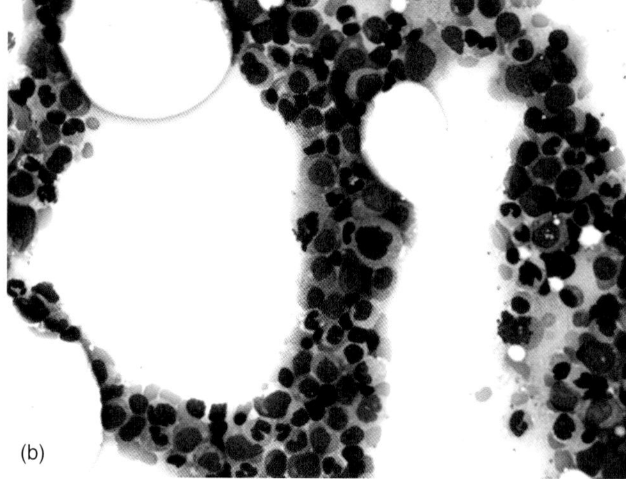

FIGURE 8.30 A bone marrow biopsy section (a) and a bone marrow smear (b) from a patient with hypocellular MDS.

immunosuppressive therapy. These patients are usually pancytopenic, show macrocytic anemia, with a variable marrow cellularity and dysplastic changes similar to those of classic MDS. Bone marrow blasts are under 5%.

However, as we mentioned earlier, about 10% of all MDS patients show clinical evidence of autoimmunity [22]. Because of this association, an autoimmune model has been suggested for pathogenesis of classic MDS, suggesting an autoimmune-induced apoptosis in this process.

HIV-Associated Myelodysplasia

Myelodysplastic changes observed in HIV-infected patients may be related to HIV, secondary infections or medications [85, 86]. Compared to classical MDS, patients with HIV-associated myelodysplasia more often show bone marrow hypocellularity, plasmacytosis, and eosinophilia [27, 87]. In these patients the degree of anemia and erythroid dysplasia is less severe, micromegakaryocytes are less frequent, blasts are <5%, no ringed sideroblasts are present, and cytogenetic studies are normal.

Paraneoplastic Myelodysplasia

Dysplastic changes similar to those of MDS have been reported in rare patients with solid tumors, such as carcinoma of colon, lung, kidney, prostate, and stomach [88–90]. The cause of dysplastic changes is not clear but does not appear to be drug-related. Production and release of growth factor-like proteins by the neoplastic cells is among the possibilities. Dysplastic changes are observed in both bone marrow and peripheral blood samples.

Myelodysplasia Associated with Heavy Metal Intoxication or Deficiency

There are occasional reports of myelodysplastic changes induced by arsenic and uranium intoxication, or copper deficiency [91, 92]. The dysplastic changes mimic RA or RCMD, and are often associated with anemia or pancytopenia. It is also interesting to know that arsenic trioxide, which acts through pro-apoptotic and anti-angiogenesis mechanisms, has been used to treat a variety of hematologic malignancies, including RAEB.

Myelodysplasia Associated with Chemotherapy or Irradiation

Myelodysplastic changes in bone marrow are common features of post-chemotherapy and radiotherapy. Bone marrow is hypocellular and shows multilineage dysplasia and myeloid left shift, mimicking hypocellular MDS [25–27].

CLINICAL ASPECTS

Clinical features of MDS represent bone marrow failure and cytopenia. Anemia, thrombocytopenia, and/or neutropenia may lead to symptoms such as fatigue, pallor, infection, bruising, and/or bleeding [1, 66–68]. But some patients may be asymptomatic at diagnosis. Establishment of the diagnosis is based on a multidisciplinary approach including morphologic evaluation and utilization of the accessory laboratory tests, such as immunophenotyping, cytogenetic analysis, molecular genetic studies, and *in vitro* colony growth assays [93].

Primary or *de novo* MDS is usually a disease of the elderly and is uncommon under the age of 50 years. The median age of onset is between 60 and 70 years with an estimated annual incidence of about 3.5–10 per 100,000 in the general population [1, 66–68]. t-MDS may arise at any age, usually 4–5 years after the initiation of chemotherapy or radiation therapy [72]. The percent of cytogenetic aberrations and risk of transformation to acute leukemia are significantly higher in t-MDS than in the primary MDS [59, 93]. A small proportion of MDS patients, roughly 4–5%, may develop blast transformation in extramedullary sites (granulocytic sarcoma), particularly skin. Evolution of MDS to granulocytic sarcoma is associated with poor prognosis.

According to the International Prognostic Scoring System (IPSS) for MDS patients (Tables 8.5 and 8.6), four distinctive risk groups are defined by low risk, intermediate

TABLE 8.5 The International Prognostic Scoring System (IPSS) for MDS.*

	Score			
	0	0.5	1.0	1.5
Prognostic variable				
% Blasts	<5	5–10	–	11–20
Karyotype**	Good	Intermediate	Poor	
Cytopenia(s)***	0–1	2–3		

*Adapted from Ref. [46].

**Good: Normal, Y, del(5q); poor: complex (≥3 abnormalities), and chromosome 7 abnormalities; intermediate: other abnormalities.

***Hemoglobin < 10 g/dL; Neutropenia < 1500/μL; Platelets < 100,000/μL.

TABLE 8.6 Survival and rate of transformation of MDS to acute leukemia according to the WHO subtypes.*

MDS subtype**	Median survival (years)	Evolution to AML (%)	IPSS score (%)***
RA	5.7	7.5	Low (57) Intermediate 1 (33) Intermediate 2 (10) High (0)
RARS	5.7	1.4	Low (96) Intermediate 1 (4) Intermediate 2 (0) High (0)
5q− syndrome	9.7	8	Low (61) Intermediate 1 (30) Intermediate 2 (9) High (0)
RCMD	2.7	10	Low (55) Intermediate 1 (40) Intermediate 2 (5) High (0)
RCMD-RS	2.6	13	Low (56) Intermediate 1 (36) Intermediate 2 (8) High (0)
RAEB-1	1.5	21	Low (0) Intermediate 1 (33) Intermediate 2 (55) High (12)
RAEB-2	0.8	34.5	Low (0) Intermediate 1 (0) Intermediate 2 (25) High (73)

*Adapted from Ref. [68].
**RA: refractory anemia; RARS: refractory anemia with ringed sideroblasts; RCMD: refractory cytopenia with multilineage dysplasia; RCMD-RS: refractory cytopenia with multilineage dysplasia and ringed sideroblasts; RAEB: refractory anemia with excess blasts.
*** Percent in each category.

1 risk, intermediate 2 risk, and high risk [94, 95]. The major parameters measured in the IPSS are percent blasts, nature of the chromosomal aberrations, and the extent of cytopenia (Table 8.5). These four groups show different survival rates and carry various risk levels for transformation to acute leukemia. For example, the overall survival time for the low-risk group is about 5.7 years; intermediate 1 risk group approximately 3.5 years; intermediate 2 risk group about 1.2 years; and high-risk group approximately 0.4 years [94].

At the present time, the only effective therapy available for MDS is hematopoietic stem cell transplantation. Other promising, newly developed therapeutic approaches include the utilization of DNA methyltransferase inhibitors, vascular endothelial growth inhibitors, and the use of thalidomide, arsenic trioxide, and anti-TNFα [95–97].

DIFFERENTIAL DIAGNOSIS

Diagnosis of MDS is based on a multidisciplinary clinicopathologic approach. It is accomplished by obtaining adequate, pertinent clinical and environmental histories, careful pathologic review of peripheral blood and bone marrow, immunophenotyping, cytogenetic analysis, and molecular genetic studies. It should be noted that a broad spectrum of hematologic disorders may mimic MDS, and should therefore be considered in the differential diagnosis (Table 8.7).

Disorders with Dysplastic Erythropoiesis

Congenital dyserythropoietic anemias are hereditary disorders with bone marrow erythroid hyperplasia and marked dyserythropoiesis, such as megaloblastic changes, and the presence of erythroid precursors with bi- and multilobular nuclei (see Chapter 23). Bone marrow morphologic features of congenital dyserythropoietic anemias may mimic those of RA and RARS [98]. In congenital dyserythropoietic anemias, ringed sideroblasts are usually absent, myeloid and megakaryocytic lineages are unremarkable, and there is no abnormal karyotype.

Megaloblastic anemia may share morphologic features, such as erythroid dysplastic and megaloblastic changes, macrocytosis and neutrophilic hypersegmentation, with MDS. Megaloblastic anemia is characterized by low levels of serum folate or vitamin B12 and lack of ringed sideroblasts. A mild myeloid left shift may be seen in the bone marrow of some cases of megaloblastic anemia, but in such cases the blast cells are usually <5%. Cytogenetic and molecular studies are normal in megaloblastic anemia (see Chapter 23).

Disorders with ringed sideroblasts are seen in rare cases of hereditary sideroblastic anemia, and patients with pyridoxine deficiency, zinc or alcohol toxicity, or as post-medication effect in some patients treated with chloramphenicol, cycloserine, or anti-tuberculosis drugs.

TABLE 8.7 Differential diagnosis of MDS.

Disorders with dysplastic erythropoiesis	
Congenital dyserythropoietic anemias	Inherited disorders with marked erythroid dysplasia and unremarkable myeloid series and megakaryocytes. Normal karyotypes.
Megaloblastic anemia	Reduced serum levels of folate or vitamin B12. Normal karyotype.
Disorders with ringed sideroblasts	Observed in hereditary sideroblastic anemia (rare) and patients with pyridoxine deficiency, zinc or alcohol toxicity, or as post-medication effect in some patients treated with chloramphenicol, cycloserine, or anti-tuberculosis drugs.
Acute erythroleukemia	Erythroid lineage accounts for >50% of the bone marrow cells and myeloblasts make up ≥20% of the non-erythroid component.
Hyperplastic bone marrows with myeloid preponderance	
Chronic myeloproliferative disorders	Peripheral blood cytosis, splenomegaly, no significant dyserythropoiesis, t(9;22) in CML, and *JAK2* mutations in others.
Transient myeloproliferative disorder	A transient neonatal condition in Down syndrome; usually disappears in 4–6 weeks.
Acute myeloid leukemias	Blasts ≥20%; frequent balanced chromosomal aberrations.
Hypoplastic bone marrows	
Aplastic anemias	No increased blasts; no significant dysplastic changes; mutated genes in FA; mutated *PIG-A* gene in PNH with loss GPI-linked proteins (CD55, CD59).
Hypoplastic AML	Blasts ≥20%; frequent balanced chromosomal aberrations.
Non-clonal myelodysplasia	Observed in viral infections, autoimmune disorders, paraneoplastic syndromes, heavy metal intoxication, and post-chemotherapy and radiation therapy. Normal karyotype.

Erythroleukemia shares many morphologic features with RAEB and at times a distinction between these two entities is difficult. However, in erythroleukemia, the erythroid lineage accounts for >50% of the bone marrow cells and myeloblasts make up ≥20% of the non-erythroid component (see Chapter 11).

Hyperplastic Bone Marrows with Myeloid Preponderance

Chronic myeloproliferative disorders (CMPD) may share overlapping bone marrow findings with some variants of MDS, such as 5q− syndrome, RCMD, and RAEB-1, either because of the presence of micromegakaryocytes or myeloid left shift. The accelerated phase of chronic myeloid leukemia (CML) may mimic RAEB-2. However, in CMPD, there is peripheral blood cytosis (thrombocytosis, granulocytosis, erythrocytosis, or all together), often with evidence of leukoerythroblastosis (presence of immature myeloid and erythroid cells in the blood) and no significant dysplastic changes. Splenomegaly is a frequent clinical presentation in myeloproliferative disorders but rare in MDS. As is discussed in Chapter 10, the bone marrow disorders under the WHO classification of myeloproliferative/myelodysplastic diseases show features of both CMPD and MDS.

Transient myeloproliferative disorder in Down syndrome is a neonatal condition which may mimic RAEB or AML with increased blasts in bone marrow and the presence of blasts in peripheral blood. This condition is transient and usually disappears within 4–6 weeks (see Chapter 22).

Acute myelogenous leukemias with relatively low blast counts (20–25%) are at times difficult to distinguish from RAEB-2, particularly when the blasts are dysplastic and do not fall into the classical morphologic criteria defined for normal blasts. Another challenge in estimating percent blasts is inadequate bone marrow aspirate smears (dry tap) due to fibrosis. In such cases, immunohistochemical stains for CD34 and CD117 may help to estimate the proportion of myeloblasts in the biopsy sections.

For the determination of percent blasts, the WHO recommends 500 and 200 differential counts for the bone marrow and blood smears, respectively [1]. Also, according to the WHO recommendation, disorders with evidence of known AML-associated cytogenetic abnormalities, such as t(8;21) and inv(16) should be considered as AML even if the blast count is <20%. Both MDS and AML show frequent cytogenetic abnormalities. While in MDS most aberrations represent unbalanced chromosomal aberrations (deletions, monosomies, and trisomies), AMLs often show balanced (reciprocal) chromosomal changes, such as translocations or inversions.

Hypoplastic Bone Marrows

Aplastic anemias, both acquired and constitutional forms, have overlapping features with hypocellular MDS. Aplastic anemia usually lacks significant dysplasia or evidence of increased blasts, but may occasionally show chromosomal aberrations. Most of the acquired aplastic anemias with cytogenetic abnormalities probably represent hypocellular MDS. Fanconi anemia is associated with FA complementary gene groups, and PNH shows evidence of mutations of the *PIGA* gene and loss of the expression of GPI-linked proteins, such as CD55 and CD59 on the hematopoietic cells (see Chapter 7).

Hypocellular AML shares many morphologic features with hypocellular MDS except for the higher percentage

(≥20%) of blast cells. Multilineage dysplasia is commonly found in hypocellular MDS, but it may be present or absent in hypocellular AML. Both lesions show frequent cytogenetic abnormalities.

Non-clonal Myelodysplasia

Non-clonal myelodysplastic changes are associated with conditions such as viral infections, autoimmune disorders, paraneoplastic syndromes, heavy metal intoxication, and post-chemotherapy and radiation therapy. These changes may be very similar to those of MDS. The clinical history and lack of chromosomal aberrations help to distinguish this broad entity from MDS.

References

1. Jaffe ES, Harris NL, Stein H, Vardiman JW. (2001). *Pathology and Genetics. Tumors of Haematopoietic and Lymphoid Tissues*. IARC Press, Lyon.
2. Catenacci DV, Schiller GJ. (2005). Myelodysplastic syndromes: A comprehensive review. *Blood Rev* **19**, 301–19.
3. Hofmann WK, Koeffler HP. (2005). Myelodysplastic syndrome. *Annu Rev Med* **56**, 1–16.
4. Cilloni D, Messa F, Carturan S, Arruga F, Defilippi I, Messa E, Gottardi E, Saglio G. (2004). Myelodysplastic syndromes. *Ann NY Acad Sci* **1028**, 400–8.

5a. Bennett JM. (2005). A comparative review of classification systems in myelodysplastic syndromes (MDS). *Semin Oncol* **32**(4 Suppl 5), S3–10.

5b. Swerdlow SH, Campo E, Harris NL, Jaffe ES, Pileri SA, Stein H, Thiele J. *WHO Classification of Tumours of Haematopoietic and Lymphoid Tissues*, 4th ed., in press.

6. Cherian S, Bagg A. (2006). The genetics of the myelodysplastic syndromes: Classical cytogenetics and recent molecular insights. *Hematology* **11**, 1–13.
7. Nishino HT, Chang CC. (2005). Myelodysplastic syndromes: Clinicopathologic features, pathobiology, and molecular pathogenesis. *Arch Pathol Lab Med* **129**, 1299–310.
8. Strom SS, Gu Y, Gruschkus SK, Pierce SA, Estey EH. (2005). Risk factors of myelodysplastic syndromes: A case–control study. *Leukemia* **19**, 1912–18.
9. Pellagatti A, Fidler C, Wainscoat JS, Boultwood J. (2005). Gene expression profiling in the myelodysplastic syndromes. *Hematology* **10**, 281–7.
10. Mhyre AJ, Deeg HJ. (2007). Control of hematopoiesis and apoptosis in MDS: More than FLIPing the coin. *Leuk Res* **31**, 747–9.
11. Westwood NB, Mufti GJ. (2003). Apoptosis in the myelodysplastic syndromes. *Curr Hematol Rep* **2**, 186–92.
12. Greenberg PL. (1998). Apoptosis and its role in the myelodysplastic syndromes: Implications for disease natural history and treatment. *Leuk Res* **22**, 1123–36.
13. Papaggeli PC, Kortsaris AC, Matsouka PT. (2003). Aberrant methylation of c-myc and c-fos protooncogenes and p53 tumor suppressor gene in myelodysplastic syndromes and acute non-lymphocytic leukemia. *J BUON* **8**, 341–50.
14. Hoffman B, Liebermann DA, Selvakumaran M, Nguyen HQ. (1996). Role of c-myc in myeloid differentiation, growth arrest and apoptosis. *Curr Top Microbiol Immunol* **211**, 17–27.
15. Mathew S, Lorsbach RB, Shearer P, Sandlund JT, Raimondi SC. (2000). Double minute chromosomes and c-MYC amplification in a child with secondary myelodysplastic syndrome after treatment for acute lymphoblastic leukemia. *Leukemia* **14**, 1314–15.
16. Keith T, Araki Y, Ohyagi M, Hasegawa M, Yamamoto K, Kurata M, Nakagawa Y, Suzuki K, Kitagawa M. (2007). Regulation of angiogenesis in the bone marrow of myelodysplastic syndromes transforming to overt leukaemia. *Br J Haematol* **137**, 206–15.
17. Platzbecker U, Meredyth-Stewart M, Ehninger G. (2007). The pathogenesis of myelodysplastic syndromes (MDS). *Cancer Treat Rev* **October**.
18. Silver RT, Bennett JM, Deininger M, Feldman E, Rafri S, Silverstein RL, Solberg LA, Spivak JL. (2004). The second international congress on myeloproliferative and myelodysplastic syndromes. *Leuk Res* **28**, 979–85.
19. Aggerholm A, Holm MS, Guldberg P, Olesen LH, Hokland P. (2006). Promoter hypermethylation of p15INK4B, HIC1, CDH1, and ER is frequent in myelodysplastic syndrome and predicts poor prognosis in early-stage patients. *Eur J Haematol* **76**, 23–32.
20. Fenaux P. (2005). Inhibitors of DNA methylation: Beyond myelodysplastic syndromes. *Nat Clin Pract Oncol* **2**(Suppl 1), S36–44.
21. Voulgarelis M, Giannouli S, Ritis K, Tzioufas AG. (2004). Myelodysplasia-associated autoimmunity: Clinical and pathophysiologic concepts. *Eur J Clin Invest* **34**, 690–700.
22. Giannouli S, Tzoanopoulos D, Ritis K, Kartalis G, Moutsopoulos HM, Voulgarelis M. (2004). Autoimmune manifestations in human myelodysplasia: A positive correlation with interferon regulatory factor-1 (IRF-1) expression. *Ann Rheum Dis* **63**, 578–82.
23. Ogata K, Yoshida Y. (2005). Clinical implications of blast immunophenotypes in myelodysplastic syndromes. *Leuk Lymphoma* **46**, 1269–74.
24. Disperati P, Ichim CV, Tkachuk D, Chun K, Schuh AC, Wells RA. (2006). Progression of myelodysplasia to acute lymphoblastic leukaemia: Implications for disease biology. *Leuk Res* **30**, 233–9.
25. His ED. (2007). *Hematopathology*. Churchill Livingstone, Philadelphia.
26. Foucar K. (2001). *Bone Marrow Pathology*, 2nd ed. ASCP Press, Chicago.
27. Naeim F. (1997). *Pathology of Bone Marrow*, 2nd ed. Williams & Wilkins, Baltimore.
28. Reddy VV. (2001). Topics in bone marrow biopsy pathology: Role of marrow topography in myelodysplastic syndromes and evaluation of post-treatment and post-bone marrow transplant biopsies. *Ann Diagn Pathol* **5**, 110–20.
29. Naeim F. (2001). *Atlas of Bone Marrow and Blood Pathology*. W.B. Saunders, Philadelphia.
30. Kuroda J, Kimura S, Akaogi T, Hayashi H, Yamano T, Sasai Y, Horiike S, Taniwaki M, Abe T, Kobayashi Y, Kondo M. (2000). Myelodysplastic syndrome with clonal eosinophilia accompanied by eosinophilic pulmonary interstitial infiltration. *Acta Haematol* **104**, 119–23.
31. Ando J, Tamayose K, Sugimoto K, Oshimi K. (2002). Late appearance of t(1;19)(q11;q11) in myelodysplastic syndrome associated with dysplastic eosinophilia and pulmonary alveolar proteinosis. *Cancer Genet Cytogenet* **139**, 14–17.
32. Matsushima T, Handa H, Yokohama A, Nagasaki J, Koiso H, Kin Y, Tanaka Y, Sakura T, Tsukamoto N, Karasawa M, Itoh K, Hirabayashi H, Sawamura M, Shinonome S, Shimano S, Miyawaki S, Nojima Y, Murakami H. (2003). Prevalence and clinical characteristics of myelodysplastic syndrome with bone marrow eosinophilia or basophilia. *Blood* **101**, 3386–90.
33. Ma SK, Chan JC, Wan TS, Chan AY, Chan LC. (1998). Myelodysplastic syndrome with myelofibrosis and basophilia: Detection of trisomy 8 in basophils by fluorescence *in-situ* hybridization. *Leuk Lymphoma* **31**, 429–32.
34. Benesch M, Deeg HJ, Wells D, Loken M. (2004). Flow cytometry for diagnosis and assessment of prognosis in patients with myelodysplastic syndromes. *Hematology* **9**, 171–7.
35. Miller DT, Stelzer GT. (2001). Contributions of flow cytometry to the analysis of the myelodysplastic syndrome. *Clin Lab Med* **21**, 811–28.
36. Malcovati L, Delia Porta MG, Lunghi M, Pascutto C, Vanelli L, Travaglino E, Maffioli M, Bernasconi P, Lazzarino M, Invernizzi R, Cazzola M. (2005). Flow cytometry evaluation of erythroid and

myeloid dysplasia in patients with myelodysplastic syndrome. *Leukemia* **19**, 776–83.
37. Ogata K, Kishikawa Y, Satoh C, Tamura H, Dan K, Hayashi A. (2006). Diagnostic application of flow cytometric characteristics of CD34+ cells in low-grade myelodysplastic syndromes. *Blood* **108**, 1037–44.
38. Monreal MB, Pardo ML, Pavlovsky MA, Fernandez I, Corrado CS, Giere I, Sapia S, Pavlovsky S. (2006). Increased immature hematopoietic progenitor cells CD34+ /CD38dim in myelodysplasia. *Cytometry B Clin Cytom* **70**, 63–70.
39. Mittelman M, Karcher DS, Kammerman LA, Lessin LS. (1993). High Ia (HLA-DR) and low CD11b (Mo1) expression may predict early conversion to leukemia in myelodysplastic syndromes. *Am J Hematol* **43**, 165–71.
40. Font P, Subira D, Mtnez-Chamorro C, Castanon S, Arranz E, Ramiro S, Gil-Fernandez JJ, Lopez-Pascual J, Alonso A, Perez-Saenz MA, Alaez C, Renedo M, Bias C, Escudero A, Fdez-Ranada JM. (2006). Evaluation of CD7 and terminal deoxynucleotidyl transferase (TdT) expression in CD34+ myeloblasts from patients with myelodysplastic syndrome. *Leuk Res* **30**, 957–63.
41. Porta MG, Malcovati L, Invernizzi R, Travaglino E, Pascutto C, Maffioli M, Galli A, Boggi S, Pietra D, Vanelli L, Marseglia C, Levi S, Arosio P, Lazzarino M, Cazzola M. (2006). Flow cytometry evaluation of erythroid dysplasia in patients with myelodysplastic syndrome. *Leukemia* **20**, 549–55.
42. Sternberg A, Killick S, Littlewood T, Hatton C, Peniket A, Seidl T, Soneji S, Leach J, Bowen D, Chapman C, Standen G, Massey E, Robinson L, Vadher B, Kaczmarski R, Janmohammed R, Clipsham K, Carr A, Vyas P. (2005). Evidence for reduced B-cell progenitors in early (low-risk) myelodysplastic syndrome. *Blood* **106**, 2982–91.
43. Cherian S, Moore J, Bantly A, Vergilio JA, Klein P, Luger S, Bagg A. (2005). Peripheral blood MDS score: A new flow cytometric tool for the diagnosis of myelodysplastic syndromes. *Cytometry B Clin Cytom* **64**, 9–17.
44. Hirai H. (2003). Molecular mechanisms of myelodysplastic syndrome. *Jpn J Clin Oncol* **33**, 153–60.
45. Sugimoto K, Hirano N, Toyoshima H, Chiba S, Mano H, Takaku F, Yazaki Y, Hirai H. (1993). Mutations of the p53 gene in myelodysplastic syndrome (MDS) and MDS-derived leukemia. *Blood* **81**, 3022–6.
46. Greenberg P, Cox C, LeBeau MM, Fenaux P, Morel P, Sanz G, Sanz M, Vallespi T, Hamblin T, Oscier D, Ohyashiki K, Toyama K, Aul C, Mufti G, Bennett J. (1997). International scoring system for evaluating prognosis in myelodysplastic syndromes. *Blood* **89**, 2079–88.
47. Shali W, Helias C, Fohrer C, Struski S, Gervais C, Falkenrodt A, Leymarie V, Lioure B, Raby P, Herbrecht R, Lessard M. (2006). Cytogenetic studies of a series of 43 consecutive secondary myelodysplastic syndromes/acute myeloid leukemias: Conventional cytogenetics, FISH, and multiplex FISH. *Cancer Genet Cytogenet* **168**, 133–45.
48. Panani AD, Pappa V. (2005). Hidden chromosome 8 abnormalities detected by FISH in adult primary myelodysplastic syndromes. *In Vivo* **19**, 979–81.
49. Bernasconi P, Boni M, Cavigliano PM, Calatroni S, Giardini I, Rocca B, Zappatore R, Dambruoso I, Caresana M. (2006). Clinical relevance of cytogenetics in myelodysplastic syndromes. *Ann NY Acad Sci* **1089**, 395–410.
50. Nishino H, Chang C. (2005). Myelodysplastic syndromes: Clinicopathologic features, pathobiology, and molecular pathogenesis. *Arch Pathol Lab Med* **129**, 1299–310.
51. Giagounidis AA, Germing U, Strupp C, Hildebrandt B, Heinsch M, Aul C. (2005). Prognosis of patients with del(5q)MDS and complex karyotype and the possible role of lenalidomide in this patient subgroup. *Ann Hematol* **84**, 569–71.
52. Boultwood J, Fidler C, Strickson AJ, Watkins F, Gama S, Kearney L, Tosi S, Kasprzyk A, Cheng JF, Jaju RJ, Wainscoat JS. (2002). Narrowing and genomic annotation of the commonly deleted region of the 5q− syndrome. *Blood* **99**, 4638–41.
53. Van den Berghe H, Cassiman JJ, David G, Fryns JP, Michaux JL, Sokal G. (1974). Distinct haematological disorder with deletion of the long arm of no. 5 chromosome. *Nature* **251**, 427–38.
54. van Leeuwen BH, Martinson ME, Webb GC, Young IG. (1989). Molecular organization of the cytokine gene cluster, involving the human IL-3, IL-4, IL-5, and GM-CSF genes, on human chromosome 5. *Blood* **73**, 1142–8.
55. Bartlett J, Dredge K, Dalgleish A. (2004). The evolution of thalidomide and its IMiD derivatives as anticancer agents. *Nat Rev Cancer* **4**, 314–22.
56. Occhipinti E, Correa H, Yu L, Craver R. (2005). Comparison of two new classifications for pediatric myelodysplastic and myeloproliferative disorders. *Pediatr Blood Cancer* **44**, 240–4.
57. Brezinova J, Zemanova Z, Ransdorfova S, Pavlistova L, Babicka L, Houskova L, Melichercikova J, Siskova M, Cermak J, Michalova K. (2007). Structural aberrations of chromosome 7 revealed by a combination of molecular cytogenetic techniques in myeloid malignancies. *Cancer Genet Cytogenet* **173**, 10–16.
58. Gonzalez MB, Gutierrez NC, Garcia JL, Schoenmakers EF, Sole F, Calasanz MJ, San Miguel JF, Hernandez JM. (2004). Heterogeneity of structural abnormalities in the 7q31.3 approximately q34 region in myeloid malignancies. *Cancer Genet Cytogenet* **150**, 136–43.
59. Hasle H, Baumann I, Bergstrasser E, Fenu S, Fischer A, Kardos G, Kerndrup G, Locatelli F, Rogge T, Schultz KR, Stary J, Trebo M, van den Heuvel-Eibrink MM, Harbott J, Nollke P, Niemeyer MC. (2004). The International Prognostic Scoring System (IPSS) for childhood myelodysplastic syndrome (MDS) and juvenile myelomonocytic leukemia (JMML). *Leukemia* **18**, 2008–14.
60. Kikukawa M, Aoki N, Mori M. (1998). A case of myelodysplastic syndrome with an intronic point mutation of the p53 tumour suppressor gene at the splice donor site. *Br J Haematol* **100**, 564–6.
61. Castro PD, Liang JC, Nagarajan L. (2000). Deletions of chromosome 5ql3.3 and 17p loci cooperate in myeloid neoplasms. *Blood* **95**, 2138–43.
62. Hackanson B, Robbel C, Lubbert P. (2005). *In vivo* effects of decitabine in myelodysplasia and acute myeloid leukemia: Review of cytogenetic and molecular studies. *Ann Hematol* **84**(Suppl 13), 32–8.
63. van Lom K, Hagemeijer A, Smit EM, Lowenberg B. (1993). *In situ* hybridization on May-Grunwald Giemsa-stained bone marrow and blood smears of patients with hematologic disorders allows detection of cell-lineage-specific cytogenetic abnormalities. *Blood* **82**, 884–8.
64. Gerristen WR, Donohue J, Bauman J, Jhanwar SC, Kernan NA, Castro-Malaspina H, O'Reilly RJ, Bourhis JH. (1992). Clonal analysis of myelodysplastic syndrome: Monosomy 7 is expressed in the myeloid lineage, but not in the lymphoid lineage as detected by fluorescent *in situ* hybridization. *Blood* **80**, 217–24.
65. Fenaux P, Morel P, Lai JL. (1996). Cytogenetics of myelodysplastic syndromes. *Semin Hematol* **33**, 127–38.
66. Cheson BD, Bennett JM, Kantarjian H, Pinto A, Schiffer CA, Nimer SD, Lowenberg B, Beran M, de Witte TM, Stone RM, Mittelman M, Sanz GF, Wijermans PW, Gore S, Greenberg PL. (2000). World Health Organization (WHO) international working group. Report of an international working group to standardize response criteria for myelodysplastic syndromes. *Blood* **96**, 3671–4.
67. Greenberg P, Anderson J, de Witte T, Estey E, Fenaux P, Gupta P, Hamblin T, Hellstrom-Lindberg E, List A, Mufti G, Neuwirtova R, Ohyashiki K, Oscier D, Sanz G, Sanz M, Willman C. (2000). Problematic WHO reclassification of myelodysplastic syndromes. Members of the International MDS Study Group. *J Clin Oncol* **18**, 3447–52.
68. Germing U, Gattermann N, Strupp C, Aivado M, Aul C. (2000). Validation of the WHO proposals for a new classification of primary myelodysplastic syndromes: A retrospective analysis of 1600 patients. *Leuk Res* **24**, 983–92.
69. Germing U, Strupp C, Kuendgen A, Aivado M, Giagounidis A, Hildebrandt B, Aul C, Haas R, Gattermann N. (2006). Refractory anaemia with excess of blasts (RAEB): Analysis of reclassification according to the WHO proposals. *Br J Haematol* **132**, 162–7.

70. Boultwood J, Lewis S, Wainscoat JS. (1994). The 5q− syndrome. *Blood* **84**, 3253–60.
71. Giagounidis AA, Germing U, Aul C. (2006). Biological and prognostic significance of chromosome 5q deletions in myeloid malignancies. *Clin Cancer Res* **12**, 5–10.
72. Rund D, Krichevsky S, Bar-Cohen S, Goldschmidt N, Kedmi M, Malik E, Gural A, Shafran-Tikva S, Ben-Neriah S, Ben-Yehuda D. (2005). Therapy-related leukemia: Clinical characteristics and analysis of new molecular risk factors in 96 adult patients. *Leukemia* **19**, 1919–28.
73. Rund D, Ben-Yehuda D. (2004). Therapy-related leukemia and myelodysplasia: Evolving concepts of pathogenesis and treatment. *Hematology* **9**, 179–87.
74. Pedersen-Bjergaard J, Andersen MK, Christiansen DH, Nerlov C. (2002). Genetic pathways in therapy-related myelodysplasia and acute myeloid leukemia. *Blood* **99**, 1909–12.
75. Pedersen-Bjergaard J, Andersen MK, Christiansen DH. (2000). Therapy-related acute myeloid leukemia and myelodysplasia after high-dose chemotherapy and autologous stem cell transplantation. *Blood* **95**, 3273–9.
76. Locatelli F, Zecca M, Pession A, Maserati E, De Stefano P, Severi F. (1995). Myelodysplastic syndromes: The pediatric point of view. *Haematologica* **80**, 268–79.
77. Niemeyer CM, Kratz CP, Hasle H. (2005). Pediatric myelodysplastic syndromes. *Curr Treat Options Oncol* **6**, 209–14.
78. Hasle H, Niemeyer CM, Chessells JM, Baumann I, Bennett JM, Kerndrup G, Head DR. (2003). A pediatric approach to the WHO classification of myelodysplastic and myeloproliferative diseases. *Leukemia* **17**, 277–82.
79. Rytting ME. (2004). Pediatric myelodysplastic syndromes. *Curr Hematol Rep* **3**, 173–7.
80. Gadner H, Haas OA. (1992). Experience in pediatric myelodysplastic syndromes. *Hematol Oncol Clin North Am* **6**, 655–72.
81. McKenna RW. (2004). Myelodysplasia and myeloproliferative disorders in children. *Am J Clin Pathol* **122**(Suppl), S58–69.
82. Marisavljevic D, Cemerikic V, Rolovic Z, Boskovic D, Colovic M. (2005). Hypocellular myelodysplastic syndromes: Clinical and biological significance. *Med Oncol* **22**, 169–75.
83. Goyal R, Qawi H, Ali I, Dar S, Mundle S, Shetty V, Mativi Y, Allampallam K, Lisak L, Loew J, Venugopal P, Gezer S, Robin E, Rifkin S, Raza A. (1999). Biologic characteristics of patients with hypocellular myelodysplastic syndromes. *Leuk Res* **23**, 357–64.
84. Barrett J, Saunthararajah Y, Molldrem J. (2000). Myelodysplastic syndrome and aplastic anemia: Distinct entities or diseases linked by a common pathophysiology? *Semin Hematol* **37**, 15–29.
85. Katsarou O, Terpos E, Patsouris E, Peristeris P, Viniou N, Kapsimali V, Karafoulidou A. (2001). Myelodysplastic features in patients with long-term HIV infection and haemophilia. *Haemophilia* **7**, 47–52.
86. Thiele J, Zirbes TK, Bertsch HP, Titius BR, Lorenzen J, Fischer R. (1996). AIDS-related bone marrow lesions – myelodysplastic features or predominant inflammatory-reactive changes (HIV-myelopathy)? A comparative morphometric study by immunohistochemistry with special emphasis on apoptosis and PCNA-labeling. *Anal Cell Pathol* **11**, 141–57.
87. Kaloutsi V, Kohlmeyer U, Maschek H, Nafe R, Choritz H, Amor A, Georgii A. (1994). Comparison of bone marrow and hematologic findings in patients with human immunodeficiency virus infection and those with myelodysplastic syndromes and infectious diseases. *Am J Clin Pathol* **101**, 123–9.
88. Sans-Sabrafen J, Buxo-Costa J, Woessner S, Florensa L, Besses C, Malats N, Porta M. (1992). Myelodysplastic syndromes and malignant solid tumors: Analysis of 21 cases. *Am J Hematol* **41**, 1–4.
89. Castello A, Coci A, Magrini U. (1992). Paraneoplastic marrow alterations in patients with cancer. *Haematologica* **77**, 392–7.
90. Raz I, Shinar E, Polliack A. (1984). Pancytopenia with hypercellular bone marrow – a possible paraneoplastic syndrome in carcinoma of the lung: A report of three cases. *Am J Hematol* **16**, 403–8.
91. Kumar N, Elliott MA, Hoyer JD, Harper Jr. CM, Ahlskog JE, Phyliky RL. (2005). "Myelodysplasia," myeloneuropathy, and copper deficiency. *Mayo Clin Proc* **80**, 943–6.
92. Rezuke WN, Anderson C, Pastuszak WT, Conway SR, Firshein SI. (1991). Arsenic intoxication presenting as a myelodysplastic syndrome: A case report. *Am J Hematol* **36**, 291–3.
93. Malcovati L, Porta MG, Pascutto C, Invernizzi R, Boni M, Travaglino E, Passamonti F, Arcaini L, Maffioli M, Bernasconi P, Lazzarino M, Cazzola M. (2005). Prognostic factors and life expectancy in myelodysplastic syndromes classified according to WHO criteria: A basis for clinical decision making. *J Clin Oncol* **23**, 7594–603.
94. Greenberg P, Cox C, LeBeau MM, Fenaux P, Morel P, Sanz G, Sanz M, Vallespi T, Hamblin T, Oscier D, Ohyashiki K, Toyama K, Aul C, Mufti G, Bennett J. (1997). International scoring system for evaluating prognosis in myelodysplastic syndromes. *Blood* **89**(6), 2079–88.
95. Greenberg PL, Baer MR, Bennett JM, Bloomfield CD, De Castro CM, Deeg HJ, Devetten MP, Emanuel PD, Erba HP, Estey E, Foran J, Gore SD, Millenson M, Navarro WH, Nimer SD, O'Donnell MR, Saba HI, Spiers K, Stone RM, Tallman MS. (2006). Clinical application and proposal for modification of the International Working Group (IWG) response criteria in myelodysplasia. *Blood* **108**, 419–25.
96. Steensma DP, Bennett JM. (2006). The myelodysplastic syndromes: Diagnosis and treatment. *Mayo Clin Proc* **81**, 104–30.
97. Greenberg PL, Baer MR, Bennett JM, Bloomfield CD, De Castro CM, Deeg HJ, Devetten MP, Emanuel PD, Erba HP, Estey E, Foran J, Gore SD, Millenson M, Navarro WH, Nimer SD, O'Donnell MR, Saba HI, Spiers K, Stone RM, Tallman MS. (2006). Myelodysplastic syndromes clinical practice guidelines in oncology. *J Natl Compr Canc Netw* **4**, 58–77.
98. Heimpel H, Anselstetter V, Chrobak L, Denecke J, Einsiedler B, Gallmeier K, Griesshammer A, Marquardt T, Janka-Schaub G, Kron M, Kohne E. (2003). Congenital dyserythropoietic anemia type II: Epidemiology, clinical appearance, and prognosis based on long-term observation. *Blood* **102**, 4576–81.

Chronic Myeloproliferative Diseases

9

CHAPTER

Faramarz Naeim, P. Nagesh Rao and Wayne W. Grody

Chronic myeloproliferative diseases (CMPD), similar to myelodysplastic syndrome (MDS), are a group of hematologic disorders distinguished by clonal expansion of abnormal hematopoietic stem cells at different levels leading to a hypercellular marrow with excessive terminal proliferation of the hematopoietic cells and peripheral blood granulocytosis, erythrocytosis, and/or thrombocytosis. This hyperproliferative process, in certain conditions, is associated with bone marrow fibrosis and extramedullary hematopoiesis. The extramedullary hematopoiesis along with excess sequestration of the hematopoietic cells in the spleen often leads to massive splenomegaly, one of the clinical hallmarks of CMPD. Hepatomegaly, though less frequent, may be also present [1–6a].

Morphologic features shared by various types of CMPD include (Figure 9.1):

Bone Marrow

1. Hypercellular bone marrow with mono- or multilineage hyperplasia, predominance of mature cells, and no significant dyserythropoiesis or dysgranulopoiesis.
2. Megakaryocytosis, often present in clusters with abnormal morphology.
3. Dilated sinuses containing clusters of hematopoietic cells.
4. Frequent focal or diffuse marrow fibrosis.
5. Frequent osteosclerosis.
6. Frequent basophilia and/or eosinophilia.

Blood

1. Granulocytosis, erythrocytosis, and/or thrombocytosis, often with a leukoerythroblastic picture.
2. Tear-drop-shaped erythrocytes.
3. Giant platelets.
4. Frequent basophilia and/or eosinophilia.
5. Lack of significant dysgranulopoiesis.
6. Lack of toxic granulation in neutrophils.

The classification of CMPD by the World Health Organization (WHO) includes the following categories [1]:

1. Chronic myelogenous leukemia (CML).
2. Chronic neutrophilic leukemia (CNL).
3. Chronic eosinophilic leukemia (CEL) and hypereosinophilic syndrome (HES).
4. Polycythemia vera (PV).
5. Chronic idiopathic myelofibrosis (with extramedullary hematopoiesis) (CIMF).
6. Essential thrombocythemia (ET).
7. Chronic myeloproliferative disease, unclassifiable (CMPD-U).

The category of *mast cell diseases* has been added into the above classification in a recently revised draft of the WHO classification (updated WHO classification in press [6b]). In this book, mast cell disorders are discussed separately in Chapter 20.

The exact mechanisms of the lymphoproliferative process in these disorders are not well understood. The hypersensitivity of the affected stem cells to certain growth factors and/or defective negative regulatory feedback mechanisms may play a role. However, recent investigations suggest abnormalities in tyrosine kinase genes as the central core to the pathogenesis of CMPD. The classical example of tyrosine kinase involvement in CMPD is the fusion of *ABL1* and *BCR* genes t(9;22)(q34;q11.2); the Philadelphia chromosome] in chronic

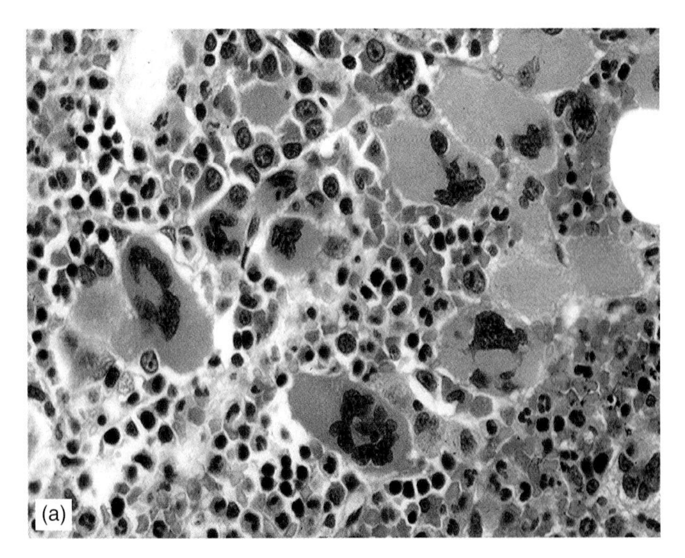

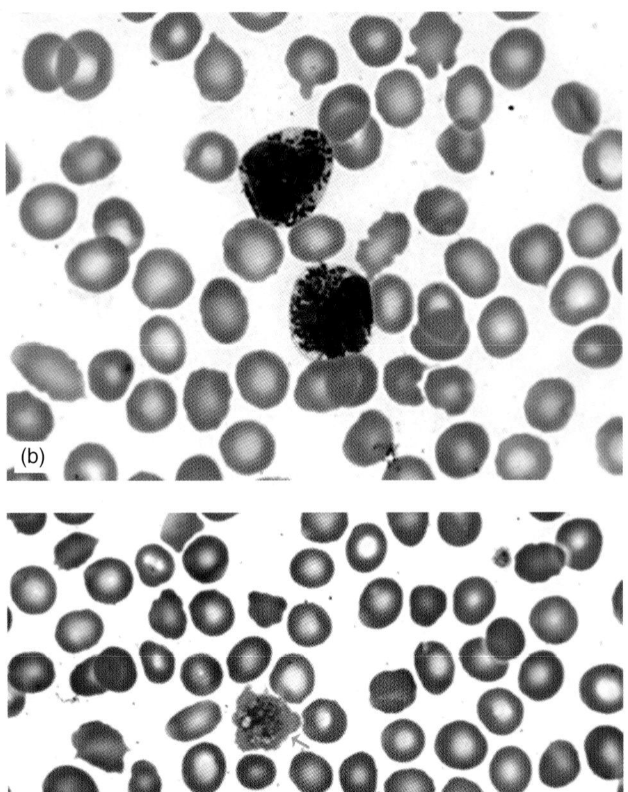

FIGURE 9.1 Chronic myeloproliferative disorders share some morphologic features such as (a) bone marrow hypercellularity and increased megakaryocytes with atypical features, (b) basophilia, and (c) the presence of giant platelets (arrow).

myelogenous leukemia. Several tyrosine kinase genes, other than *ABL1*, have been identified, such as *ABL2*, *PDGFRA*, *PDGFRB*, *FGFR1*, and *JAK2*. The fusion of *FIP1L1A-PDGFR1* genes and the activating V617F mutation in the *JAK2* gene have been recently reported in a significant proportion of Philadelphia-negative CMPD cases [7–9].

TABLE 9.1 Chromosomal aberrations in chronic myeloproliferative disorders.

Chromosome abnormality	Frequency (%)	Prognosis
Trisomy 1q	8	
4q12 deletion (CHIC2)		Response to Imanitib
5q33 aberrations (PDGFB)		Response to Imanitib
Monosomy 7	5	
Trisomy 8	16	Good prognosis
8p11 translocations		Poor prognosis
Trisomy 9 or 9p	10	Unclear
12p aberrations	3	
13q deletion	7	Unclear to good
20q deletion	9	Good

TABLE 9.2 Incidence of chromosomal abnormalities in subtypes of chronic myeloproliferative disorders.

Subtype*	Frequency (%)
CML	95
PV	34
CIMF	40
HES	7–12
ET	<3

*CML: chronic myelogenous leukemia, PV: polycythemia vera, CIMF: chronic idiopathic myelofibrosis, HES: hypereosinophilic syndrome, and ET: essential thrombocythemia.

The spectrum of cytogenetic aberrations in CMPD is heterogeneous, ranging from numerical gains and losses to structural changes including unbalanced translocations (Table 9.1) [10, 11]. Chromosomal gains and losses rather than balanced translocations appear to be common in CMPD. Standard karyotyping along with FISH is important in establishing the diagnosis and may provide very useful information for disease outcome.

Cytogenetic abnormalities in myeloproliferative disorder (MPD) subtypes other than CML occur at different frequencies ranging from 3% to 40%, depending on the subtype (Table 9.2). Compared to CML, the other MPD subtypes are more clinically and cytogenetically heterogeneous. In fact, at least 27 different chromosomal anomalies have been associated with MPD. Unlike the "Philadelphia chromosome" in CML, there is no pathognomonic chromosomal abnormality associated with the MPDs.

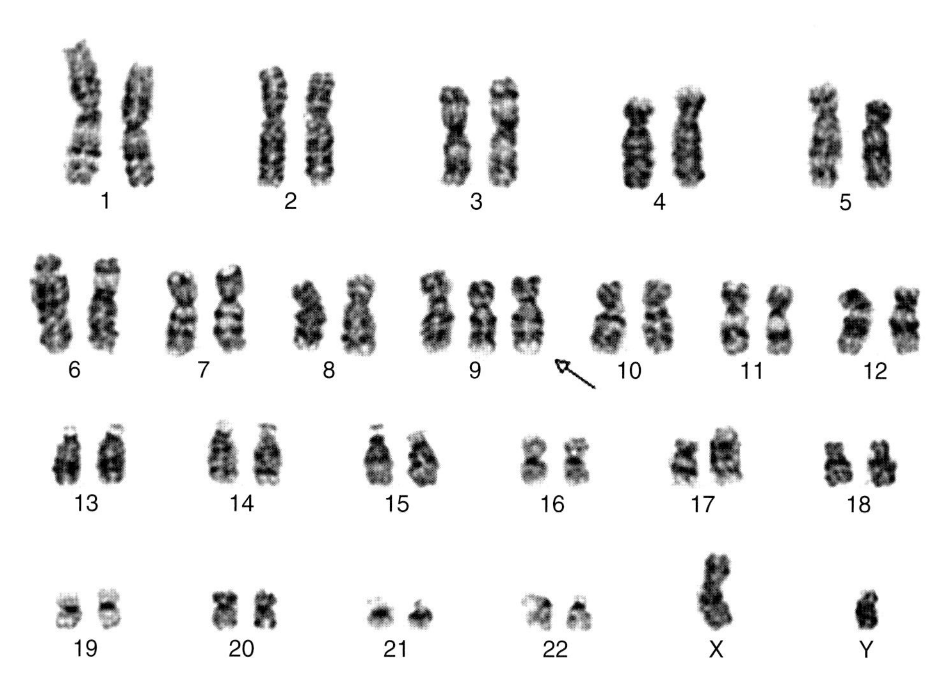

FIGURE 9.2 A G-banded karyotype with trisomy 9.

Chromosomal abnormalities are seen in 30–40% of patients with PV and IMF and seem to indicate a poor prognosis. On the other hand, chromosomal abnormalities are rare in patients with ET (about 5–6%). In cases suspicious for ET, cytogenetic studies are used for exclusion of other hematological malignancies associated with increased megakaryopoiesis such as 5q-syndrome or AML with inversion chromosome 3q.

Consistent acquired changes seen at diagnosis include deletion of the long arm of chromosome 20, del(13q), trisomy 8 and 9, and duplication of parts of 1q (Figures 9.2 and 9.3). Furthermore, del(20q), trisomy 8, and dupl(lq) all arise in multipotent progenitor cells. The molecular mapping of 20q deletions and, to some extent, 13q deletions has identified a number of candidate target genes, although no mutations have yet been found. Finally, translocations associated with the rare 8p11 myeloproliferative syndrome (Figure 9.4) and other atypical MPDs have permitted the identification of a number of novel fusion proteins involving fibroblast growth factor receptor-1 (*FGFR1*). Chromosomal anomalies are found most frequently in chronic idiopathic myelofibrosis, CIMF (up to 50%), followed by PV, whereas anomalies in ET and CEL are so infrequent that cytogenetics can be omitted when the diagnosis is clear [11–13]. The most common structural chromosomal anomalies of MPD in order of frequency are t(9;22)(q34;q11.2), del(20)(q11q13), del(13)(q12q14), del(5)(q13q33), and del(12)(p12). The most common numeric anomalies are loss of Y, +8, +9, and 7. Only the t(9;22) (or variant 9;22) is diagnostic of any specific type of MPD (CML). Relatively strong associations are observed for the del(13) in CIMF, the t(5;12)(q33;p13) in CEL, and the del(20), +8, and +9 in PV [11, 13]. Balanced translocations are rare [10, 11]. Molecular cytogenetic techniques in CMPD have suggested that some abnormalities may be more common than originally thought, whereas molecular studies are likely to detect the possible role of candidate genes implicated in the neoplastic process [11, 13–22].

The observation of a subclone or stem line with multiple chromosomal anomalies is often an indication of disease progression or clonal evolution [23]. This evolution is seen in at least 12% of patients with MPD. Multiple clones have been observed in 2.1% of patients with hematologic malignancies and in 1.8% of patients with MPD. The observation of multiple clones is most common among patients with a clone harboring t(9;22), del(20q), or +8 [13].

CHRONIC MYELOGENOUS LEUKEMIA

Also referred to as chronic myeloid, myelocytic, or granulocytic leukemia, CML was the first malignant disorder reported in association with a chromosomal aberration, the Philadelphia chromosome (Ph^1). CML has been the front runner in the understanding of molecular mechanisms in hematopoietic malignancies and target treatment approaches [24–26]. It is characterized by clonal expansion of bone marrow stem cells leading to selective granulocytic hyperplasia with or without thrombocytosis and the presence of its cytogenetic hallmark Ph^1, which is the result of balanced reciprocal t(9;22)(q34;q11.2) chromosomal translocation, resulting in a short 22q (Ph^1). CML demonstrates an evolutionary process with different clinicopathological stages of *chronic*, *accelerated*, and *acute* (blast transformation) phases.

Etiology and Pathogenesis

Ionizing radiation has been reported as a possible inducing agent in the development of CML based on the observation of increased incidence of CML 5 to 10 years after radiation

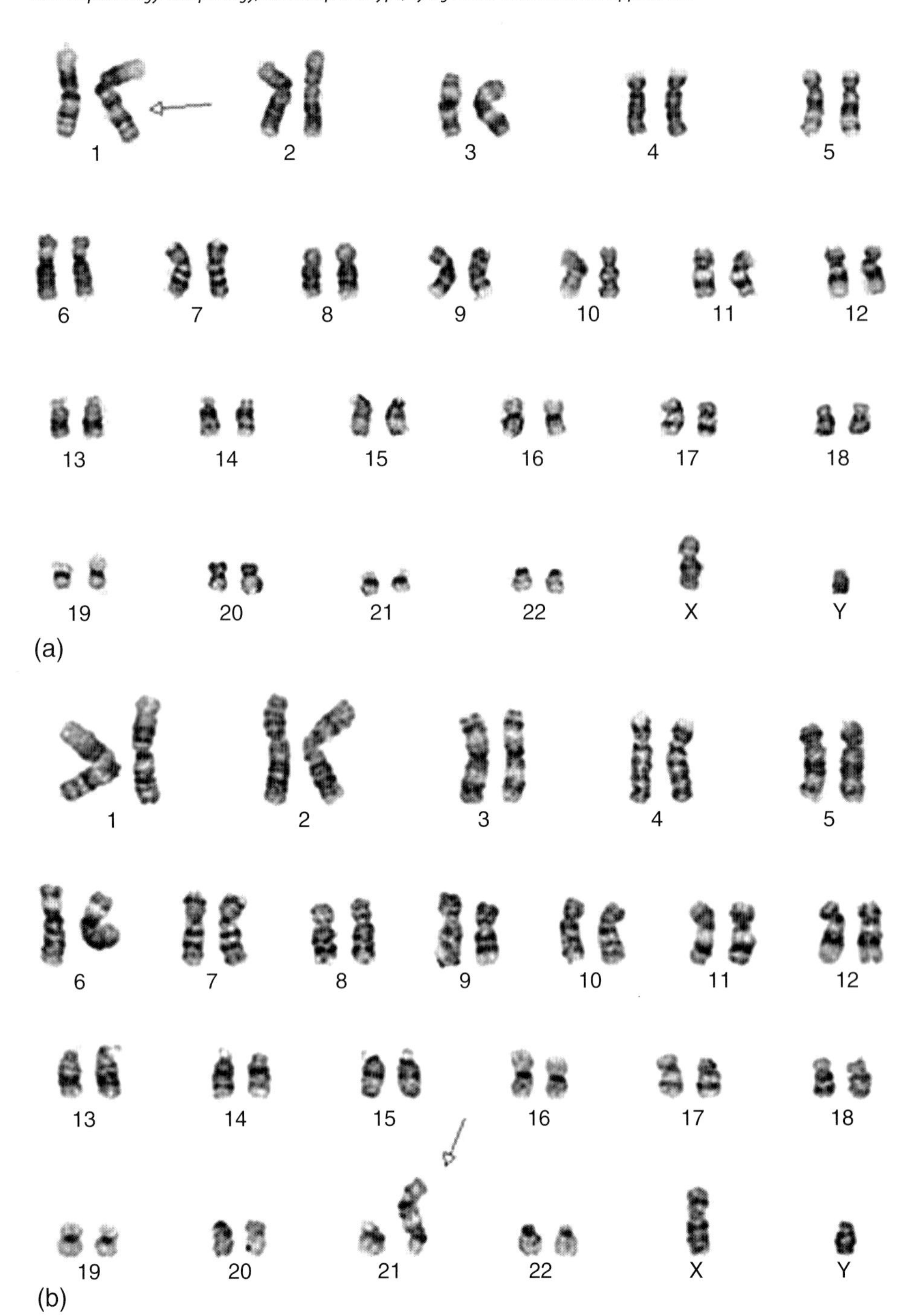

FIGURE 9.3 Duplication of 1q: (a) interstitial duplication of 1q and (b) whole arm duplication of 1q.

exposure. Also, possibility of a susceptibility gene has been raised based on the increased frequency of HLA-Cw3 and -Cw4 in CML patients, though hereditary disposition does not seem to play a significant role [27, 28]. Only very rare familial childhood forms have been reported [29, 30].

The pathognomonic hallmark for CML, as mentioned, is t(9;22)(q34;q11.2) (*Ph*[1]), which is observed in hematopoietic cells but not in bone marrow stromal cells. This chromosomal translocation creates a *BCR-ABL1* fusion gene with three different principal protein products, based on the site of the breakpoint on chromosome 22. All three *BCR-ABL1* protein products (p190, p210, and p230) demonstrate increased tyrosine kinase activity and are not detected in normal hematopoietic cells. A wide variety of pathogenic effects have been contributed to the *BCR-ABL1* gene fusion products including [26]:

1. Insensitivity of *Ph*[1] positive hematopoietic progenitor cells to growth-inhibiting regulatory cytokines.
2. Mitogenic activity of the *BCR-ABL1* fusion proteins on hematopoietic cells.
3. Resistance of *Ph*[1] positive hematopoietic cultured cell lines to apoptosis. The *in vitro* reports regarding this matter remain controversial.

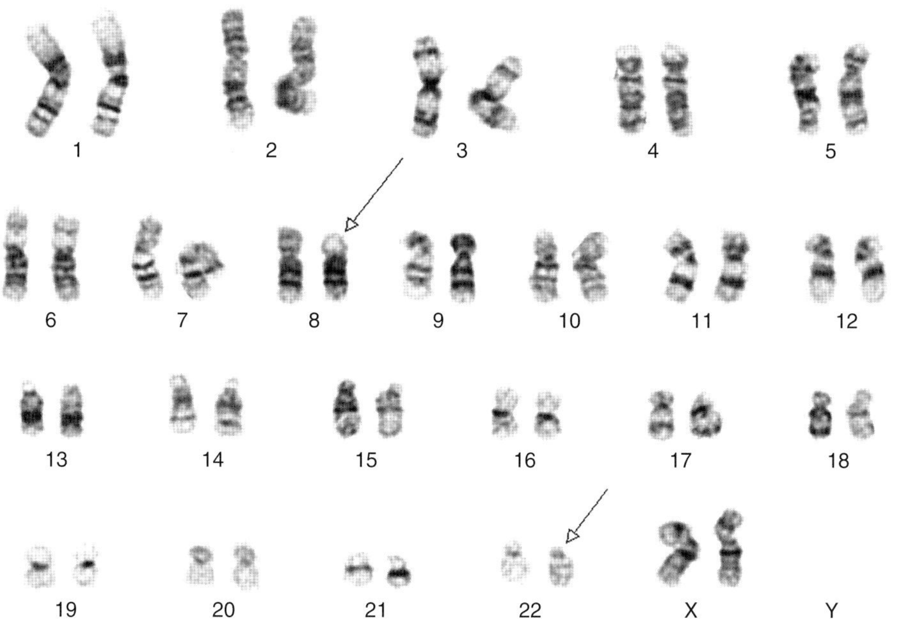

FIGURE 9.4 A G-banded karyotype with an 8;22 translocation involving the 8p11 band and 22q11.2 (red arrows).

4. Decreased adherence of *Ph*[1] positive hematopoietic progenitors to bone marrow stroma and fibronectin, leading to increased circulation of myeloid cells.
5. Genetic instability of *Ph*[1] positive hematopoietic progenitors leading to progression from a chronic phase to blast crisis.
6. Ability of the *BCR-ABL1* fusion proteins to promote leukemogenesis in mice.

Pathology

Morphology and Laboratory Findings

Bone marrow sections are hypercellular with marked myeloid preponderance and mild to moderate myeloid left shift (Figure 9.5). The paratrabecular myeloid regions are expanded with less mature forms next to the bone trabeculae and more mature forms closer to the center. Eosinophilia is a frequent finding. Megakaryocytes are usually increased and often appear in clusters. There may be patchy or diffuse fibrosis along with osteosclerosis. The extent of fibrosis to some degree correlates with the number of megakaryocytes. Scattered or clusters of pseudo-Gaucher cells (histiocytes) with abundant wrinkled cytoplasm are often present (Figure 9.5c) [1–4, 31, 32].

Bone marrow smears are highly cellular with an elevated M:E ratio of usually >10:1. Eosinophils are increased, and basophilia is a frequent feature but is usually <20% (Figure 9.6). Dysgranulopoiesis may be present but is not prominent. There may be mild to moderate myeloid left shift, but myeloblasts are usually below 5%. Megakaryocytes often show dysplastic changes, including the presence of numerous micromegakaryocytes ("dwarf" forms) as well as many large, bizarre, multilobulated forms, individually or in clusters. Pseudo-Gaucher cells are usually found attached to or in the vicinity of the stromal tissue fragments. Sometimes, their cytoplasm appears light blue by Wright's stain; hence referred to as "sea-blue histiocytes" (Figure 9.6b). Sea-blue histiocytes and pseudo-Gaucher cells are loaded with phagocytic particles and cell membrane debris due to the increased turnover of the bone marrow cells [1–4, 31, 32].

Blood smears show marked leukocytosis with a white blood cell count of often >100,000/μL. The morphologic findings of the myeloid cells mimic bone marrow with the presence of myeloid left shift and a wide spectrum of myeloid precursors, including myeloblasts and promyelocytes (Figure 9.7). Similar to in the bone marrow smears, myeloblasts in blood smears are usually below 5%. Usually, myelocytes are more numerous than metamyelocytes (myelocyte bulge). Absolute basophilia and eosinophilia is common, and there may be absolute monocytosis, but the monocytes in differential count usually do not exceed 3%. Scattered nucleated red blood cells may be present. Platelet count is normal or elevated but occasionally reduced during the chronic phase. Neutrophils and bands show reduced alkaline phosphatase activity by cytochemical stains [known as "leukocyte alkaline phosphatase (LAP) score"]. The elevated levels of serum lactic dehydrogenase, uric acid, and vitamin B_{12} are also frequently observed [1–4, 31, 32].

Splenomegaly is common. The red pulp is diffusely expanded with infiltration of the cords and sinuses filled with mature and immature myeloid cells (Figure 9.8). The malpighian corpuscles (white pulp) are reduced in size and number or are completely absent [33].

The recent implementation of effective targeted therapy in CML, such as treatment with imatinib, has created a necessity for follow-up bone marrow and molecular studies. The evaluation of post-therapy bone marrow samples shows

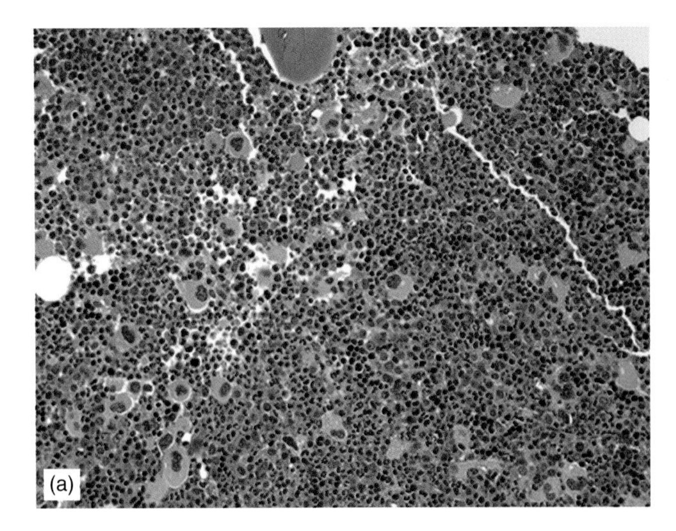

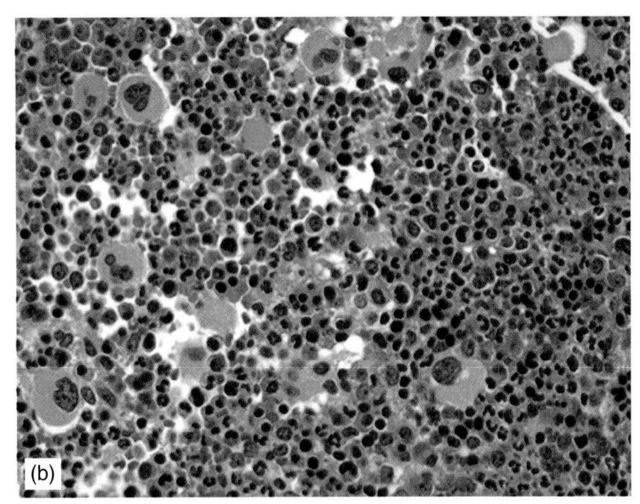

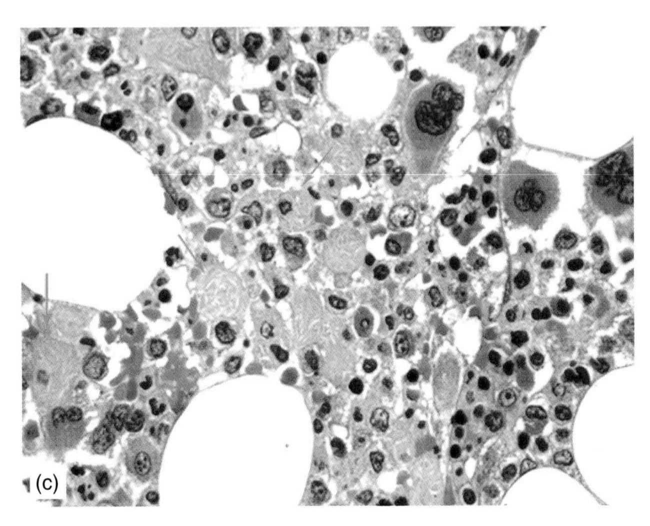

FIGURE 9.5 Bone marrow biopsy sections in patients with CML are hypercellular, show marked myeloid preponderance and increased megakaryocytes with small forms: (a) low power and (b) high power. A small aggregate of histiocytes (pseudo-Gaucher cells) is demonstrated in (c) (arrows).

progressive changes toward normal morphology. However, the post-therapy bone marrow samples may show certain morphologic features such as:

1. Frequent presence of non-diagnostic lymphoid aggregates, sometimes paratrabecular, consisting of a mixture of B and T lymphocytes.
2. Frequent presence of histiocytic aggregates (pseudo-Gaucher cells).
3. Bone marrow hypocellularity, particularly in cases with long history of treatment. The degree of hypocellularity in some instances is so severe that the bone marrow biopsy sections resemble aplastic anemia.

Accelerated phase of CML is often associated with a decline in the patient's clinical condition along with certain laboratory findings. The diagnosis of CML in accelerated phase (CML-AP), according to the WHO recommendation, is based on the presence of one or more of the following (Table 9.3) (Figure 9.9) [1]:

1. The presence of 10–19% blasts in blood or bone marrow samples.
2. Basophilia of ≥20%.
3. Persistent thrombocytopenia of ≤100,000/μL or thrombocytosis of ≥1,000,000/μL.
4. Progressive splenomegaly and/or increasing leukocyte count.
5. Cytogenetic or molecular evidence of clonal evolution.

Increased marrow fibrosis, marked megakaryocytosis with the presence of large clusters or sheets of megakaryocytes, and severe dysgranulopoiesis are all suggestive of CML-AP. CML-AP is a transient phase between the chronic phase and blast transformation [34].

Blast transformation (blast crisis) refers to the evolution of CML into acute leukemia. The exact mechanisms involved in this evolutionary process are not well understood, but recent studies suggest that the unrestricted activity of the *BCR-ABL1* fusion gene may play an important role [34–36]. Enhanced proliferation and differentiation arrest, the characteristic features of blast transformation in CML, seem to be dependent upon the cooperation of *BCR-ABL1* with the *p53* (17p13) and *RB1* (13q14) genes that appear to be directly or indirectly dysregulated in this process.

According to the recommendation of WHO, the diagnosis of CML in blast crisis (CML-BC) is made when (Table 9.3) (Figures 9.10 and 9.11) [1]:

1. Blasts are ≥20% of bone marrow nucleated cells or peripheral blood differential count.
2. Large foci or clusters of blasts are present in the bone marrow biopsy sections.
3. There is evidence of extramedullary tissue infiltration by blast cells.

CML-BC represents all morphologic features required for the diagnosis of acute leukemia. Blasts are ≥20% of the bone marrow smear or blood smear differential counts or appear in sheets or large clusters in the bone

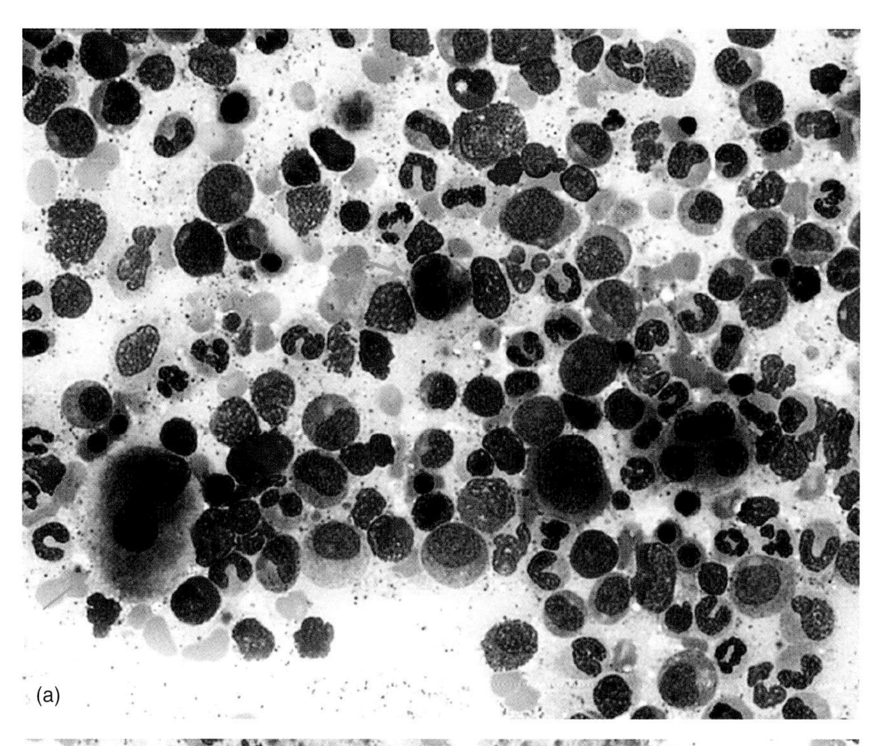

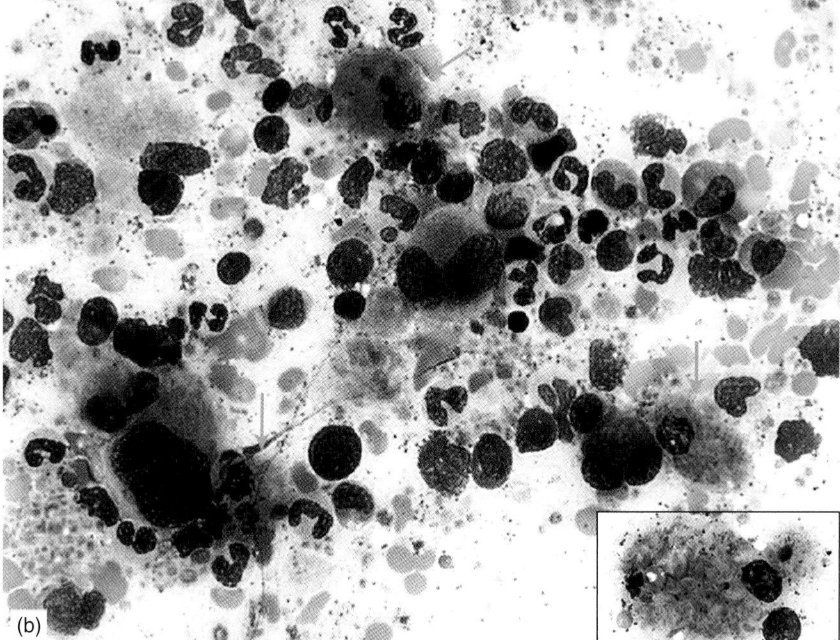

FIGURE 9.6 Bone marrow smears in a patient with CML showing myeloid preponderance, eosinophilia, and the presence of small megakaryocytes (a, arrows). Several sea-blue histiocytes are demonstrated (b, arrows and inset).

marrow biopsy and/or clot sections. There may be some degree of dysmyelopoiesis or bone marrow fibrosis.

Extramedullary CML-BC may involve any tissue, but is frequently observed in spleen, lymph node, skin, and central nervous system. Morphologic features are similar to other acute leukemic infiltrations. There is often the presence of immature eosinophils (eosinophilic myelocytes) which may provide a hint that blasts are of myeloid lineage.

The blast cells in approximately 70% of CML-BC cases are of myeloid (non-lymphoid) origin and express granulocytic, monocytic-, erythroid-, and/or megakaryocytic-associated CD molecules in immunophenotypic studies [35]. In roughly 30% of the CML cases, blast transformation is of lymphoid lineage. In our experience, a significant proportion of CML-BC cases consist of myeloblasts with the aberrant expression of lymphoid-associated CD molecules or blasts with biphenotypic features.

Immunophenotypic Studies

Flow cytometry is primarily used in CML-AP and CML-BC for the estimation of blast counts and their lineage assignments. However, there is some evidence of abnormal expression of CD molecules on mature and immature myeloid cells in patients with CML in chronic phase (CML-CP). For example, there are reports of reduced density of CD16

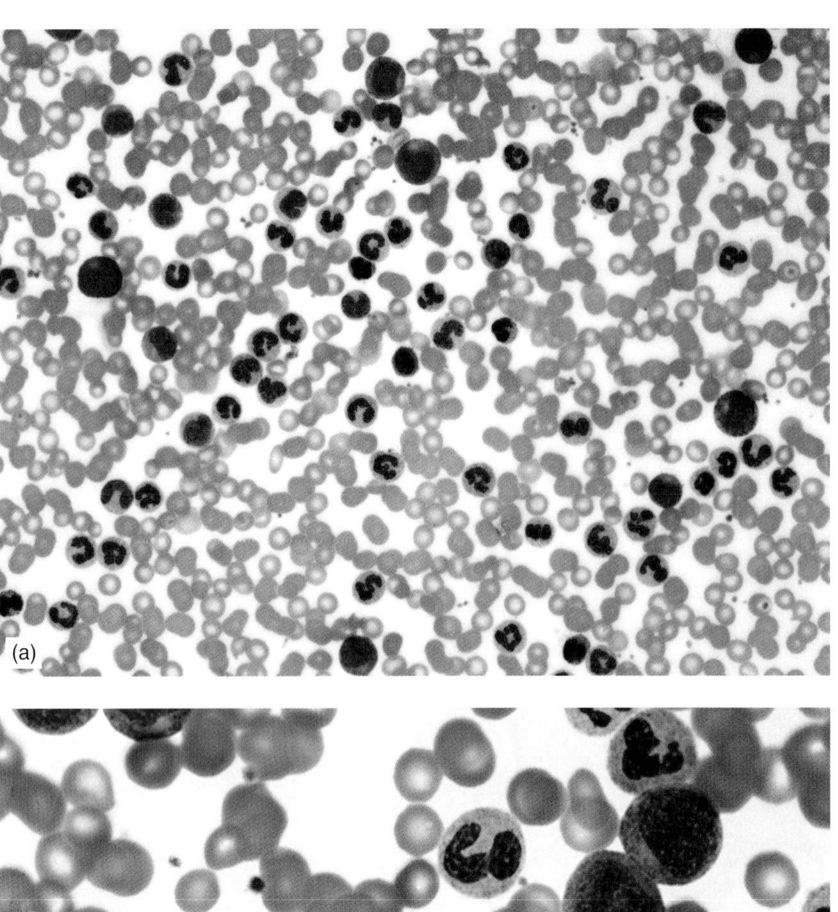

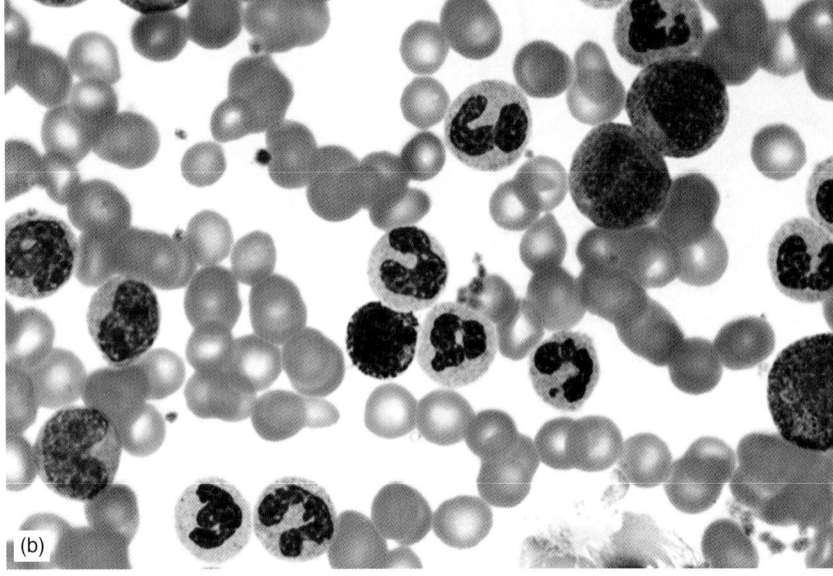

FIGURE 9.7 Peripheral blood smears of patients with CML show marked leukocytosis with myeloid left shift and high proportion of myelocytes and metamyelocytes (a) and (b); basophils are often present (b).

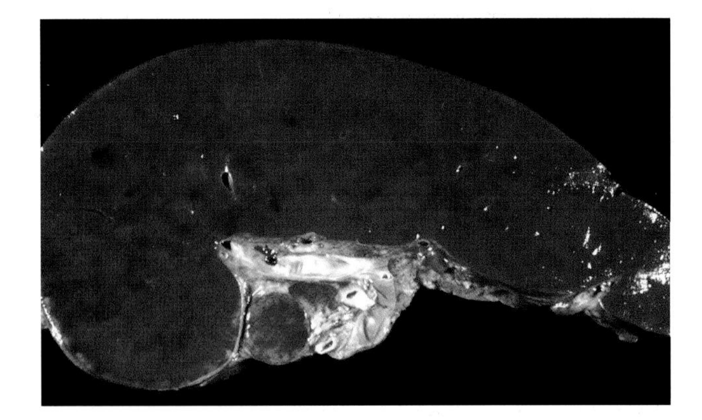

FIGURE 9.8 Splenomegaly is a frequent clinical finding in CML. Splenic involvement is usually diffuse.

expression on the neutrophils and an increased proportion of CD7+ myeloid stem cells (CD34+, CD33+) in patients with CML-CP [37].

A few flow cytometric studies have shown a significant reduction in the average telomere length of leukocytes in CML-AP and CML-BC patients. Also, there is a report of higher S-phase values in the leukocytes of patients with CML-AP (9±3) than patients with CML-CP (5±2) or normal controls (<1%) by flow cytometry [38]. In this study, the CML-CP cases with S-phase values of >7% evolved into accelerated phase (AP) within 18 months. It is possible to estimate the basophil counts in the bone marrow or the peripheral blood by looking at the $CD45^{dim}$, cells that in addition to CD13 and CD33 express CD22.

Blasts in AP and blast crisis of CML are often of myeloid origin (>70% of the cases) and in order of frequency

TABLE 9.3 Evolution of chronic myelogenous leukemia (CML).*

Stage	Characteristics
Chronic phase	Leukocytosis (often >100,000/μL) Hypercellular marrow with marked myeloid preponderance Blasts <10% Basophilia <20% Eosinophilia Megakaryocytosis with the presence of micromegakaryocytes Splenomegaly Low LAP score t(9;22)(q34;q11.2) *BCR-ABL1* fusion by FISH and/or RT-PCR
Accelerated phase	Hypercellular marrow with myeloid left shift Blasts 10–19% Basophilia ≥20% Thrombocytopenia <100,000/μL, or thrombocytosis >1,000,000/μL Increasing WBC count Increasing spleen size Additional cytogenetic abnormalities
Blast crisis (phase)	Blasts ≥20% Sheets or large clusters of blasts in the bone marrow biopsy Extramedullary tissue infiltration by blast cells

*Adapted from Ref. [1].

express CD33, CD13, CD11c, CD36, CD34, CD117, and CD15. Lymphoid blast transformation is usually of precursor B with the expression of TdT, CD19, CD10, and less frequently, CD20. Precursor T blast transformation is a rare event. As mentioned earlier, not infrequently, CML blasts show aberration expression of CD molecules or evidence of biphenotypic features (see Chapter 13) [39, 40]. In our experience, the most frequent lymphoid-associated markers aberrantly expressed on myeloblasts are CD7, CD56, and CD19, and the most frequent myeloid-associated markers aberrantly expressed on lymphoblasts are CD15, CD13, and CD33.

Immunohistochemical stains are helpful for the estimation of blast counts and identification of their lineage, particularly when there is no access to flow cytometry or there is inadequate marrow aspirate (dry tap). The following markers are frequently used: CD34 (blasts); CD117 (myeloblasts); TdT (lymphoblasts); CD31, CD61, and factor VIII for megakaryoblasts; glycophorin A and hemoglobin A for erythroid precursors; myeloperoxidase; lysozyme and CD68 for granulocytic and monocytic lineages; CD10, CD20, and CD79a for B lymphocytes; and CD2, CD3, CD5, and CD7 for T lymphocytes.

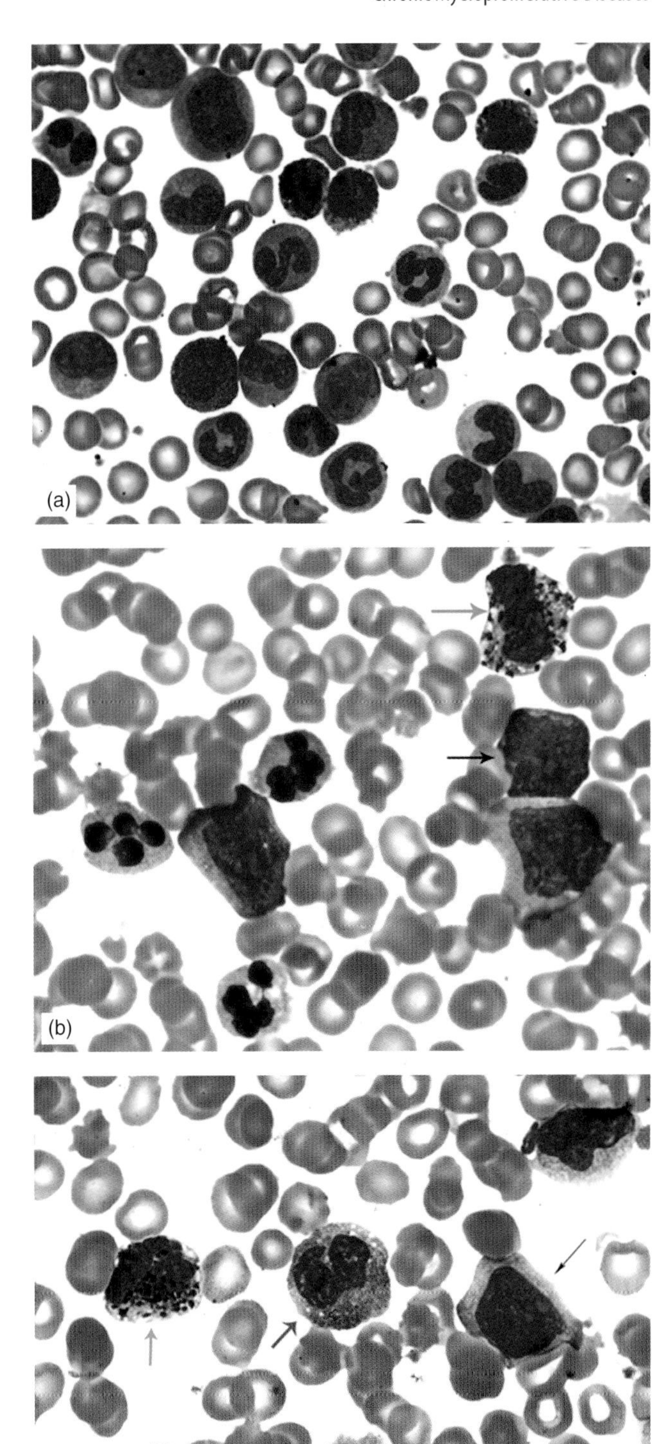

FIGURE 9.9 Peripheral blood smear of a patient with CML-AP showing myeloid left shift (a, intermediate power), basophils (green arrows) and blasts (black arrows) (b and c, high power). An eosinophil (red arrow) is demonstrated in (c).

Molecular Studies

As noted earlier, detection of the *BCR-ABL1* fusion gene is the hallmark of CML in diagnosis, monitoring, and targeted therapy. While traditionally this has been done by cytogenetic

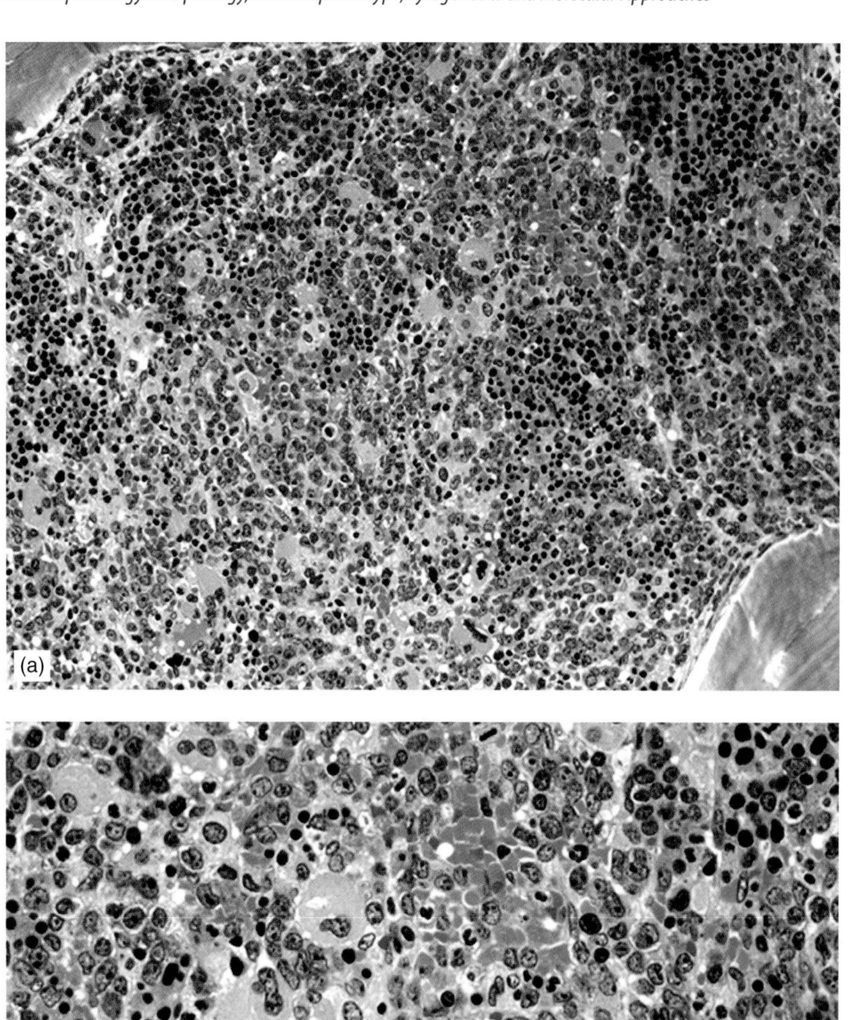

FIGURE 9.10 Bone marrow biopsy section of a patient with a history of CML showing increased blasts consistent with blast transformation: (a) low power and (b) high power.

analysis, and more recently molecular cytogenetic (FISH) testing, the most sensitive and quantitative approach is by molecular methods. The sensitivity of classical cytogenetics is limited by the number of cells cultured to the number of metaphases counted. Moreover, about 5% of CML cases have t(9;22) translocations that may not be visible under the light microscope and will thus be missed by this approach. Molecular testing, by any one of a number of available methods, should be able to detect such "cryptic" translocations and will therefore approach 100% sensitivity for initial diagnosis [41].

The translocation of the *BCR* gene on chromosome 22 to the *ABL1* oncogene on chromosome 9 produces a fusion protein with constitutive tyrosine kinase activity which is much higher than that found in normal myeloid cells. *BCR* stands for "breakpoint cluster region," reflecting the fact that the point of breakage at that site can occur over a fairly broad region. While many subtle variants are possible, the important rearrangements for clinical diagnosis are the major breakpoint (*M-BCR*) which produces the p210 gene product and is found primarily in CML and the minor breakpoint (*m-BCR*) which produces the p190 gene product and is

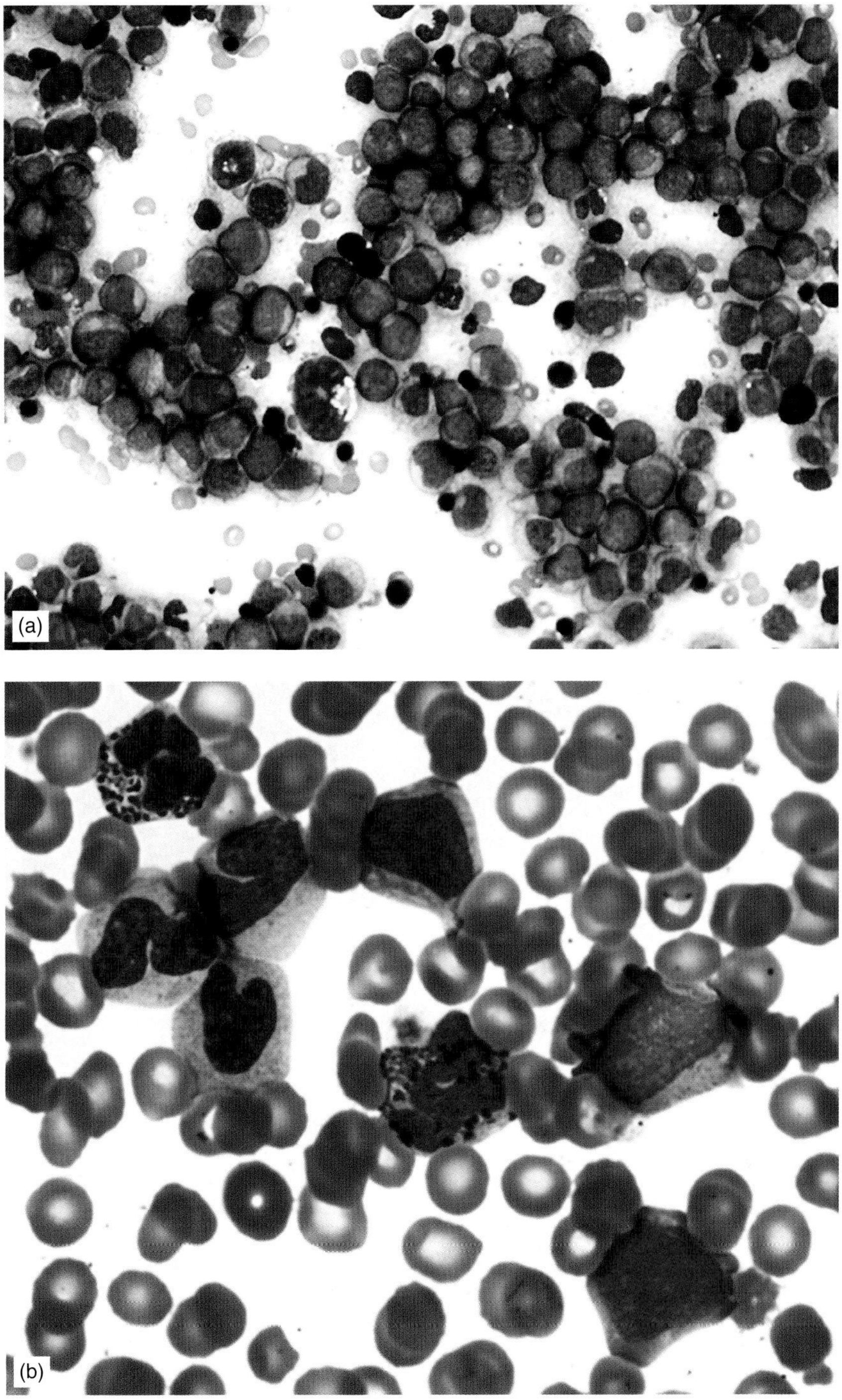

FIGURE 9.11 Bone marrow (a) and blood (b) smears of a patient with CML in blast transformation.

typical of acute lymphoblastic leukemia, more commonly seen in adult patients with that disease. However, some overlap is seen in a minority of cases, owing to alternative splicing of RNA transcripts [42]. One clear advantage of molecular testing is that it can readily distinguish between these isoforms.

The first molecular test for the *BCR-ABL1* translocation to enter wide use was the Southern blot, in which genomic DNA extracted from the patient's blood or bone marrow cells is digested with restriction endonucleases, subjected to agarose gel electrophoresis, blotted onto a nylon filter, and hybridized with a radioactively labeled DNA probe complementary to the *BCR* gene. The principle behind the assay is that a translocation of *BCR* on chromosome 22q11.2 to chromosome 9q34 (*ABL1*) will place it in a different milieu of restriction enzyme cleavage sites, producing a band shift on the resulting autoradiogram. The anomalous band(s) may be of higher or lower molecular weight than the unrearranged (germline) bands; in most cases the latter will also be seen because of the presence of non-malignant cells in a mixed specimen and/or the retention of one of the

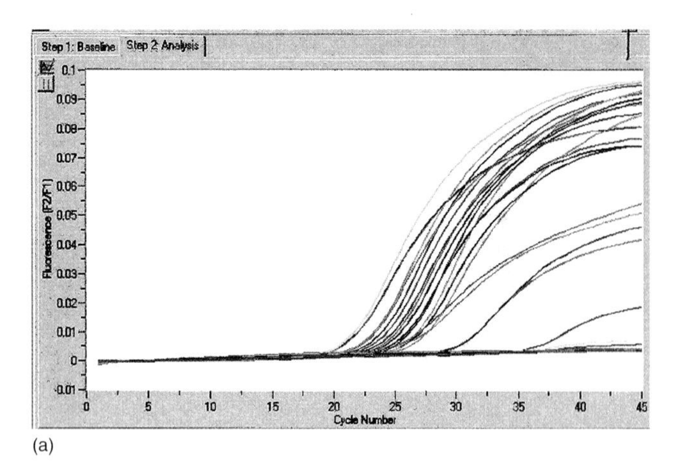

(a)

LightCycler Quantification Report

User: Administrator LightCycler ID: 2623

Filename: C:\LightCycler3\Users\Administrator\Data\081606.BCR-ABL.FL.ABT

LC Run Version: 5.10 LCDA Version: 3.5.28

This experiment was run on Aug 16, 2006 by Administrator.

Sample Information

Rotor Position	Sample Name	Rep. of...	Sample Type*	Known Conc.	Calculated Conc. (Units)	Crossing Point	Sample Comments
1	Standard I		S	2,070	1,985	21.11	X̄ = 1977 Error = −4.5%
2	Repli. of Standard I	1	S	2,070	1,969	21.13	
3	Standard II		S	70.00	75.45	25.28	X̄ = 76.48 Error = +9.2%
4	Repli. of Standard II	3	S	70.00	77.51	25.25	
5	Standard III		S	1.770	1.878	29.99	X̄ = 1.705 Error = −3.7%
6	Repli. of Standard III	5	S	1.770	1.533	30.25	
7	Positive control-G6PDH		P		29.65	26.47	
8	Positive control-BCR-ABL		P		31.28	26.40	31.28/29.65 = 1.0550
9	Negative control-G6PDH		N				
10	Negative control-BCR-AB		N				
11	17631-G6PDH		U		565.7	22.71	0.0225/565.7 = 3.97×10⁻⁵
12	17631-BCR-ABL		U		0.0225	35.63	3.97×10⁻⁵/1.0550 = 3.76×10⁻⁵
13	17712-G6PDH		U		174.5	24.21	
14	17712-BCR-ABL		U				0
15	17711-G6PDH		U		18.92	27.04	
16	17711-BCR-ABL		U				0
17	17729-G6PDH		U		96.18	24.97	0.0410/96.18 = 4.26×10⁻⁴
18	17729-BCR-ABL		U		0.0410	34.86	4.26×10⁻⁴/1.0550 = 4.04×10⁻⁴

(b)

FIGURE 9.12 Data obtained from real-time PCR analysis of a series of patients with suspected, diagnosed, or treated CML using the LightCycler instrument (Roche Molecular Diagnostics, Indianapolis, IN). (a) Amplification curves produced from extracted cellular mRNA that was first reverse transcribed to create cDNA templates; the specimens producing curves that enter log phase at lower PCR cycle number had higher amounts of starting *BCR-ABL* fusion mRNA. (b) Quantitative readout of data from the LightCycler; specimens that are *BCR-ABL*-negative show no amplification product from this target (but do show output from the control *GAPDH* gene target), whereas those specimens that are *BCR-ABL*-positive show amplification products from both genes. Quantitation of the *BCR-ABL* fusion gene is done mathematically by comparing its amplification to that of the *GAPDH* control gene.

BCR alleles in the germline configuration even in the CML cells. By convention, two different restriction enzymes are used, and extra bands must be observed with both digests in order to be diagnostic; an extra band seen with only one of the enzymes could represent a benign restriction fragment length polymorphism. While useful and more sensitive than classical cytogenetic testing, the Southern blot has a number of disadvantages in this context: (1) it is laborious, time-consuming, and expensive; (2) it utilizes large amounts of hazardous radioisotopes for probe labeling (at least until the more recent advent of chemiluminescent probe labeling systems); (3) it is not very useful for detecting minimal residual disease because the bands produced by *BCR-ABL1*-positive cells present at <5% of the total cell population may be too faint to be visible; and (4) it is generally not capable of distinguishing between the major and minor breakpoint forms, unless highly specific DNA probes are used.

Because of these disadvantages, most laboratories eventually moved on to methods based on polymerase chain reaction (PCR). If one of the PCR primers is chosen to hybridize at the *BCR* site and the other at the *ABL1* site, an amplified product will be seen only if the two genes are fused, bringing the two primers into proximity. Otherwise, a completely blank result will be produced, because single PCR primers hybridized to widely distant (in this case, on different chromosomes) regions of the genome are incapable of supporting the exponential amplification needed to visualize a product. One catch of this approach is that the breakpoint region on chromosome 22 spans such a large area that it cannot be efficiently amplified from a genomic DNA target. Instead, messenger RNA must be isolated to serve as the template; since it has had the large intronic regions spliced out, it is of an amplifiable size. Most laboratories now use a real-time PCR system that is both highly sensitive and highly quantitative. It is also capable, depending upon how the primer pairs are chosen, of differentiating between the *M-BCR* and *m-BCR* forms, which can be of great importance in the differential diagnosis (in a newly ascertained patient) of acute leukemia versus the blast crisis phase of CML [43]. Examples of the data produced by such systems are shown in Figure 9.12. Some test systems generate absolute quantitation (e.g. in nanograms) of *BCR-ABL1* fusion gene, whereas others, such as that shown here, produce a relative quantitation by comparing *BCR-ABL1* to an internal control gene (in this case, *GAPDH*).

In our hands, the real-time PCR method is sensitive down to a level of 1 CML cell in 1 million normal cells. It is also highly quantitative, so that trends (upward or downward) in treated patients obtained from periodic monitoring can be used to assess minimal residual disease, relapse, or development of resistance to the newer pharmacogenetic therapies (see later) [44]. However, one must be cautious in striving for ever-increasing levels of sensitivity, since the most powerful PCR approaches are capable of detecting trace levels of *BCR-ABL1* transcripts even in healthy people who have never had CML [45].

The availability of such highly sensitive and quantitative methods has become even more important as we have entered the era of molecular targeted therapy for CML using specific inhibitors of the tyrosine kinase fusion protein (imatinib and its successors). While the vast majority of patients initially respond to these drugs, many later develop resistance, as evidenced by the creeping up of *BCR-ABL1* RT-PCR levels. (Very few treated patients, even if in apparent clinical and cytogenetic remission, actually go down to undetectable *BCR-ABL1* levels with these sensitive assays [46].) Most of these relapses are due to either amplification of the *BCR-ABL1* fusion gene or, more commonly, the development of point mutations affecting amino acid residues at the site of binding of the drug or at more distal sites causing allosteric effects. A total of about 20 such mutations have been described [47]. These can be detected by DNA sequencing or, as we have used, DNA microarray hybridization. Since certain mutations may render either similar

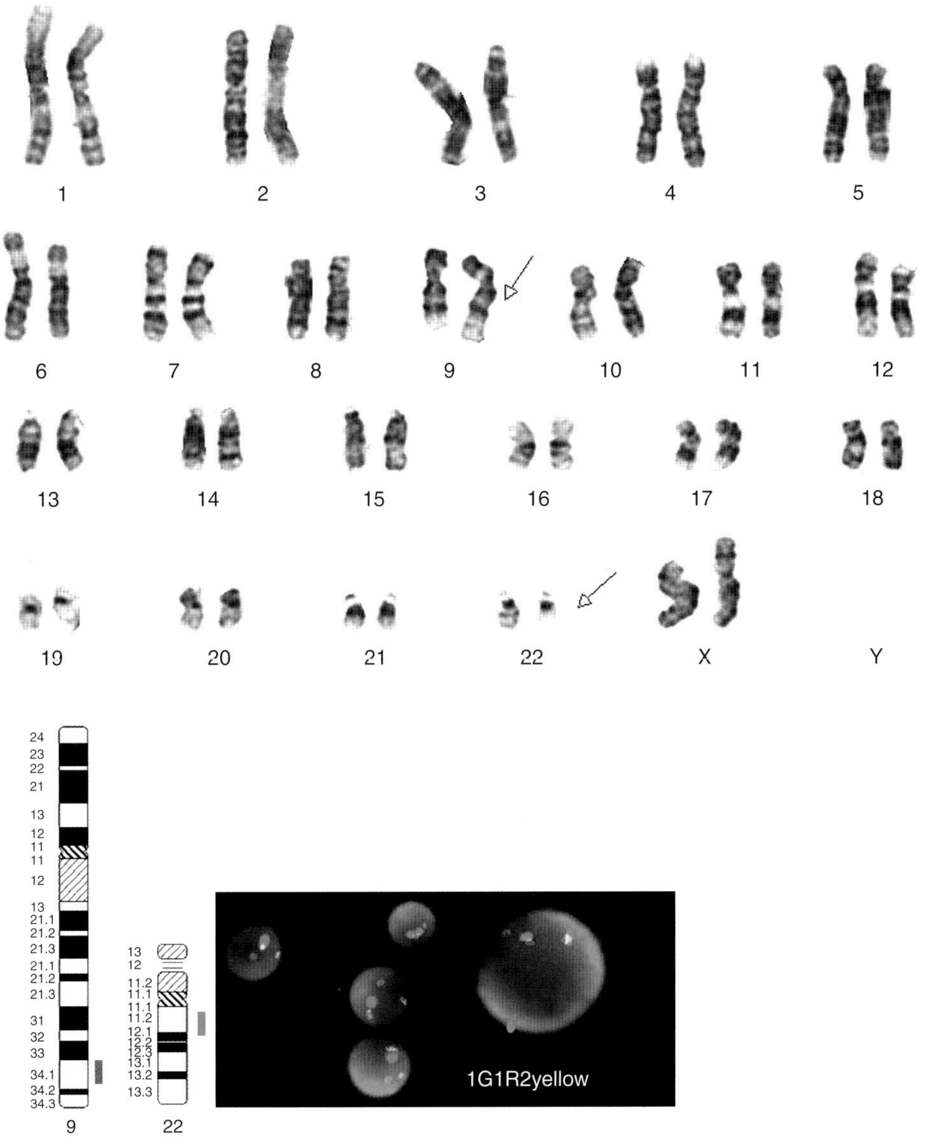

FIGURE 9.13 A G-banded karyotype with a classic t(9;22) showing the "Philadelphia" chromosome. An ideogram of chromosomes 9 and 22 showing the FISH probes for the *ABL* and *BCR* loci. The FISH panel shows cells with red *ABL*; green *BCR*, and yellow (Fusion) signals.

resistance or sensitivity to newer-generation drugs targeting the same protein, it will be important to accurately genotype these patients to guide management.

Cytogenetics

As described earlier, the *Ph* chromosome was the first consistent cytogenetic rearrangement found in a hematologic disease [48]. Banding techniques developed during the 1970s allowed for the identification of the *Ph* chromosome as being derived from a translocation between chromosomes 9 and 22, t(9;22)(q34;q11.2) (Figure 9.13) [49]. The translocation was subsequently described as resulting in the fusion of the *ABL1* protooncogene (a homolog of the Abelson murine leukemia virus oncogene) on chromosome 9q34 with a gene called *BCR* on chromosome 22q11.2 [50]. The *ABL1* gene encodes a tyrosine kinase that phosphorylates several proteins involved in signaling for cell proliferation, and the *BCR* gene encodes a 160-kDa phosphoprotein with kinase activity. The expression of the *BCR-ABL1* chimeric protein has an aberrant tyrosine kinase activity [51] and is leukemogenic [50–54].

The unambiguous presence of the *BCR-ABL1* fusion gene is required for a clinical diagnosis of CML and in typical cases remains the sole abnormality observed through most of the chronic phase. Approximately 90–95% of patients present with the t(9;22), whereas the remaining 5–10% of patients have cryptic or complex rearrangements but eventually fuse *BCR* and *ABL1*. Variant translocations with deletions at the involved breakpoints signify a poorer prognosis than the more common t(9;22) (Figure 9.14) [28, 29, 53]. FISH or molecular techniques can be used to establish diagnosis in cases where the t(9;22) cannot be identified by standard karyotyping.

CML-BC is often predicted by cytogenetic findings prior to pathologic changes; 75–80% of patients develop additional chromosome aberrations as the disease progresses.

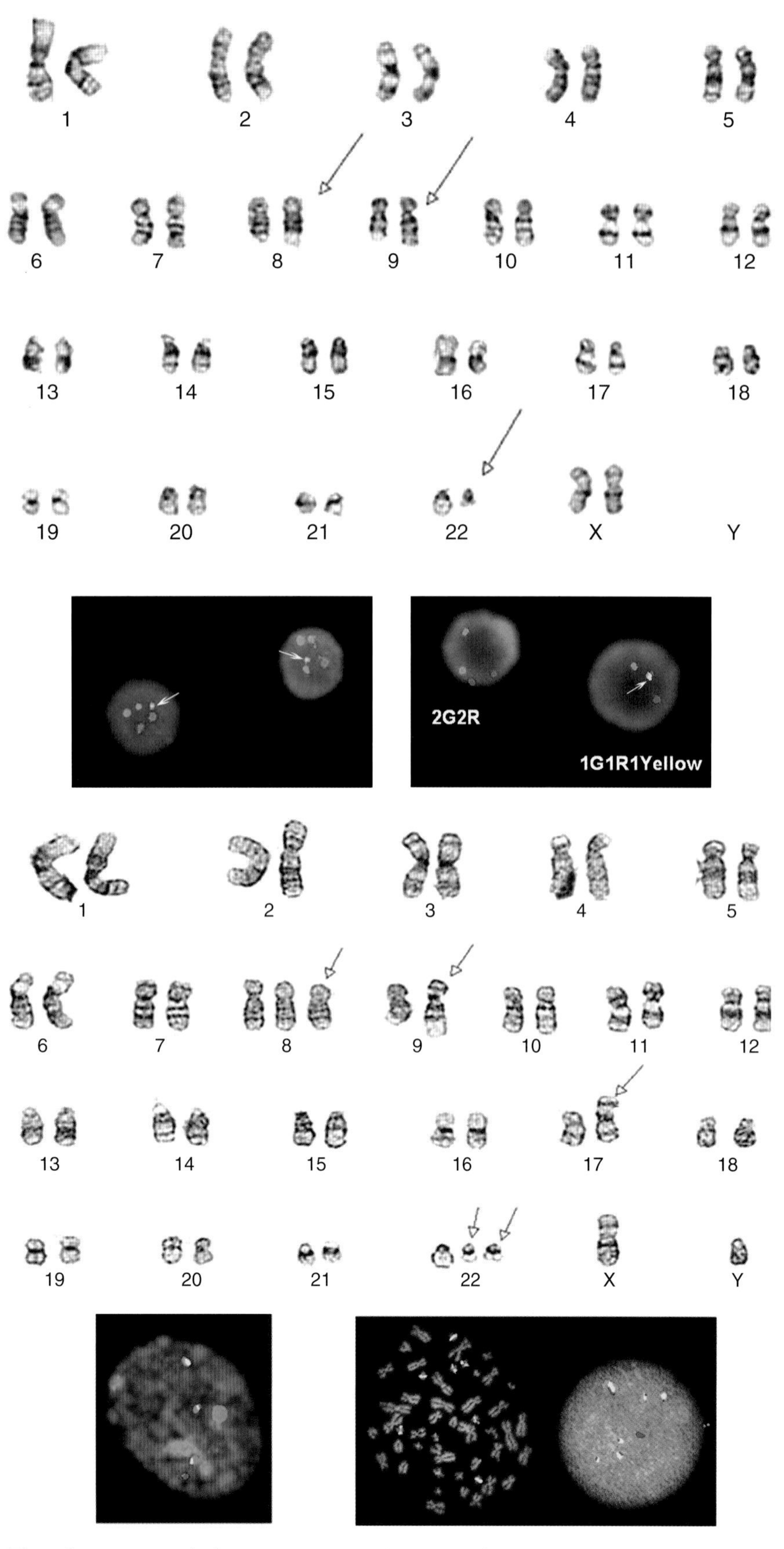

FIGURE 9.14 A G-banded karyotype showing a complex t(9;22;8) translocation with the "Ph" chromosome. Interphase FISH analyses on these cells (lower left) show one fusion signal (yellow) whereas the second fusion signal got rearranged on 9q and 8q subsequent to the initial 9;22 translocation. The lower right panel shows cells with a normal signal pattern (2 Red/2 Green) and a cell with a deletion of the reciprocal 22:9 fusion signal (1 Red/1 Green/1 Yellow).

FIGURE 9.15 A G-banded karyotype of CML-BC, with an extra *Ph* chromosome, Trisomy 8, and isochromosome 17q. The FISH panel on the left shows three fusion signals (yellow) representing the additional *Ph* chromosome. The left panel shows the gains of additional "Ph" chromosome during development of imanitib resistance.

The aberrations of chromosomes 8, 17, 19, and 22 are most often involved in disease evolution (major route), accounting for approximately 70% of patients with evolving disease. Trisomy 8, isochromosome 17, trisomy 19, or an extra *Ph* chromosome (derivative chromosome 22) is the most frequently observed secondary changes in blast crisis (Figure 9.15) [23, 55]. The remaining 30% of patients with evolving disease develop various secondary aberrations that may include trisomy 21, loss of the Y, monosomy 7 or 17, trisomy 17, or others [55, 56]. Genes known to have roles in transformation include *TP53*, *RB1*, *CDKN2A*, *INK4α*, *MINK*, *AML1*, and *EVL1*, although their role in transformation is currently unknown [23, 57].

Clinical Aspects

CML represents between 15% and 20% of adult leukemias with a median age at diagnosis of about 50 years. The only known risk factor is exposure to ionizing radiation [24, 25]. The disease is asymptomatic in over 30% of the cases and is only suspected by the elevated WBC counts on routine blood examinations. Frequent clinical symptoms include fatigue, pallor, night sweats, and weight loss. Splenomegaly is observed in 50–75% of the patients. Occasionally, the disease presents at the blast phase without prior clinical manifestation of the chronic phase. The diagnosis is established by the demonstration of t(9;22)(q34;q11) (*Ph*[1]) and/or *BCR-ABL1* fusion [49, 58, 59]. Transformation into accelerated or blast phase is usually associated with marked splenomegaly, severe anemia, and/or marked thrombocytopenia.

Molecular targeted therapy is now the recommended approach in the treatment of CML. The most exciting breakthrough in the treatment of CML has been the development of imatinib mesylate (IM, or Gleevec®) as an oral therapeutic agent. IM binds to a cleft between the N-terminal adenosine triphosphate binding domain and the C-terminal activation loop that forms the catalytic site of the Abl tyrosine kinase, locking the protein into the inactive conformation [54, 60]. Although IM appears to be extremely effective in CML, it has markedly reduced effectiveness in the acute leukemias. Patients with deletions at the *BCR-ABL1* breakpoint may not respond to therapy, and drug resistance can occur. IM resistance can occur via four main mechanisms. (1) The expression of the multidrug resistance P-glycoprotein increases drug efflux and decreases intracellular drug levels, thus decreasing drug effectiveness [61]. (2) The genomic amplification of the *BCR-ABL1* gene by gain of a second *Ph* chromosome or cellular aneuploidy is associated with resistance [62]. (3) The clonal evolution and development of chromosomal aberrations in addition to the t(9;22) may allow the clone to develop non-Bcr-Abl1-dependent growth mechanisms [63]. (4) Finally, *ABL1* gene mutations within the tyrosine kinase domain appear to prevent binding of IM to the protein [62].

Overall, the inhibitors of tyrosine kinase activity of Bcr-Abl have been more effective than the conventional drugs, such as interferon-alpha (IFN-α) combined with cytosine arabinoside (Ara-C). In the cases of no response or short-lived response to tyrosine kinase inhibitors, the possibility of bone marrow transplantation should be explored [64, 65].

CHRONIC NEUTROPHILIC LEUKEMIA

Chronic neutrophilic leukemia (CNL) is characterized by persistent peripheral blood neutrophilia, bone marrow hypercellularity, and hepatosplenomegaly [1, 66]. It is a rare condition which shares many morphological features with leukemoid reactions, but unlike leukemoid reactions, CNL is not associated with fever, infection, inflammatory process, or malignancy. In order to establish a diagnosis of CNL, all other myeloproliferative disorders and all causes of secondary (reactive) neutrophilia should be excluded (Table 9.4) [67, 68].

TABLE 9.4 WHO criteria for the diagnosis of chronic neutrophilic leukemia.*

1. *Requirements*
   a. Peripheral blood
      i. Persistent leukocytosis ≥25,000/μL.
      ii. Segmented neutrophils and bands >80% of the differential counts.
      iii. Blasts <1% of the blood cells.
   b. Bone marrow
      i. Hypercellular with marked granulocytic preponderance.
      ii. Myeloblasts <5% of the differential counts.
      iii. No significant dysgranulopoiesis.
   c. Hepatosplenomegaly
2. *Exclusions*
   a. All causes of physiologic and reactive neutrophilia, such as infections, inflammations, tissue damage (infarctions, burns), or other malignancies.
   b. All other chronic myeloproliferative diseases.
   c. Myelodysplastic syndromes.
   d. Myelodsyplastic/myeloproliferative disorders.

*Adapted from Ref. [1].

Etiology and Pathogenesis

The etiology and pathogenesis of CNL are not known. The frequent association of CNL with plasma cell myeloma in several reports may suggest release of cytokines by neoplastic plasma cells as the primary cause of neutrophilia [69]. However, clonality of CNL has been suggested by methylation studies of the X-linked hypoxanthine phosphoribosyl transferase gene and other probes [70]. Also, reports of the evolution of PV into CNL and the transformation of CNL to AML in certain cases support the clonal nature of this disorder [44]. Presence of the *JAK2* V617F tyrosine kinase mutation has been recently reported in CNL [72].

Pathology

Morphology

Bone marrow is hypercellular with marked granulocytic hyperplasia and an elevated M:E ratio approaching 10:1 or higher (Figure 9.16a). There is no evidence of increased blasts or promyelocytes. The erythroid line is unremarkable and megakaryocytes are either adequate or increased. No significant dysplastic changes are present. Bone marrow fibrosis is infrequent [1–4].

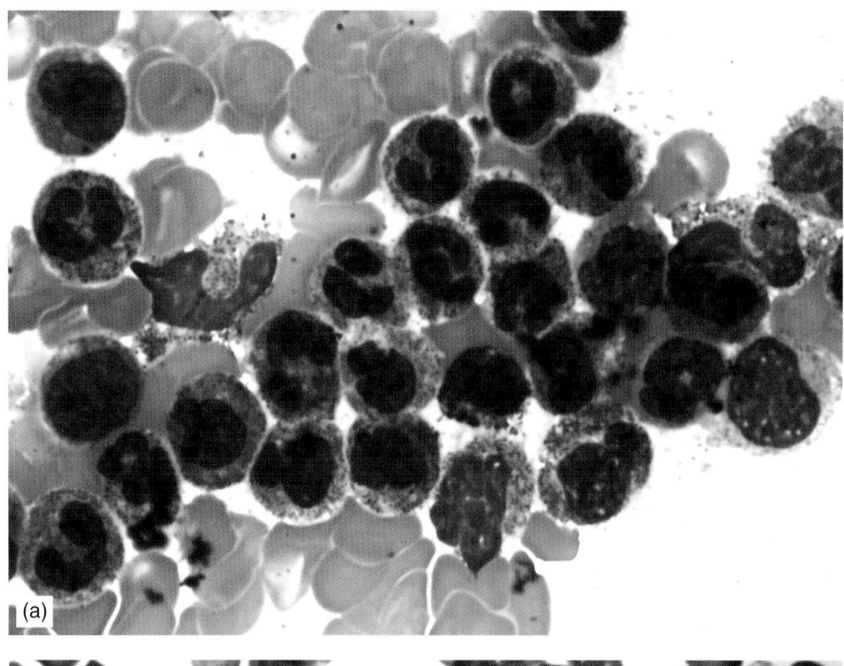

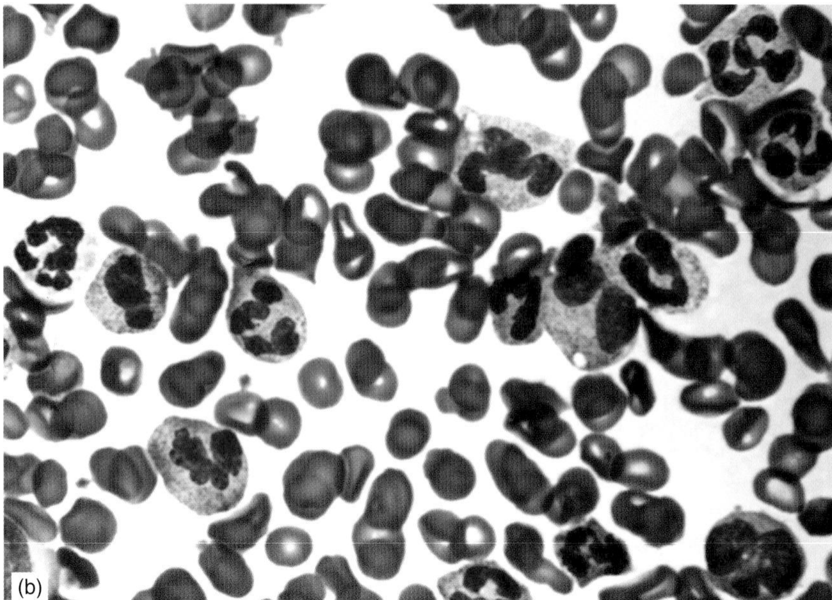

FIGURE 9.16 In chronic neutrophilic leukemia, bone marrow is hypercellular with marked preponderance of neutrophilic bands and segmented cells (a), and peripheral blood shows marked neutrophilia (b).

Blood smears show marked neutrophilia, usually ≥25,000/μL, with a modest myeloid left shift and presence of scattered (5–10%) myelocytes and metamyelocytes (Figure 9.16b). Promyelocytes are rare and myeloblasts are commonly absent. Neutrophils may show toxic granulation. The LAP score is often elevated. Mild anemia and/or thrombocytopenia may be present [1–4].

Splenomegaly and hepatomegaly are due to neutrophilic infiltration in the splenic red pulp and hepatic sinusoids and/or portal areas [1].

Immunophenotypic Studies

The immunophenotypic characteristics of the bands and neutrophils in CNL are similar to those of normal neutrophils and bands. So far, no aberrant expression or significant alteration of CD molecules have been reported.

Molecular Studies

No specific molecular markers exist for diagnosing CNL. Some cases show a *BCR-ABL1* fusion gene of the p230 isoform, detected as described earlier for CML (probably representing a variant of CML). Other cases may demonstrate the *JAK2* V617F mutation [71], described in detail later in this chapter.

Cytogenetic Results

No consistent chromosomal anomaly has been associated with CNL, and the primary genetic event is likely cryptic

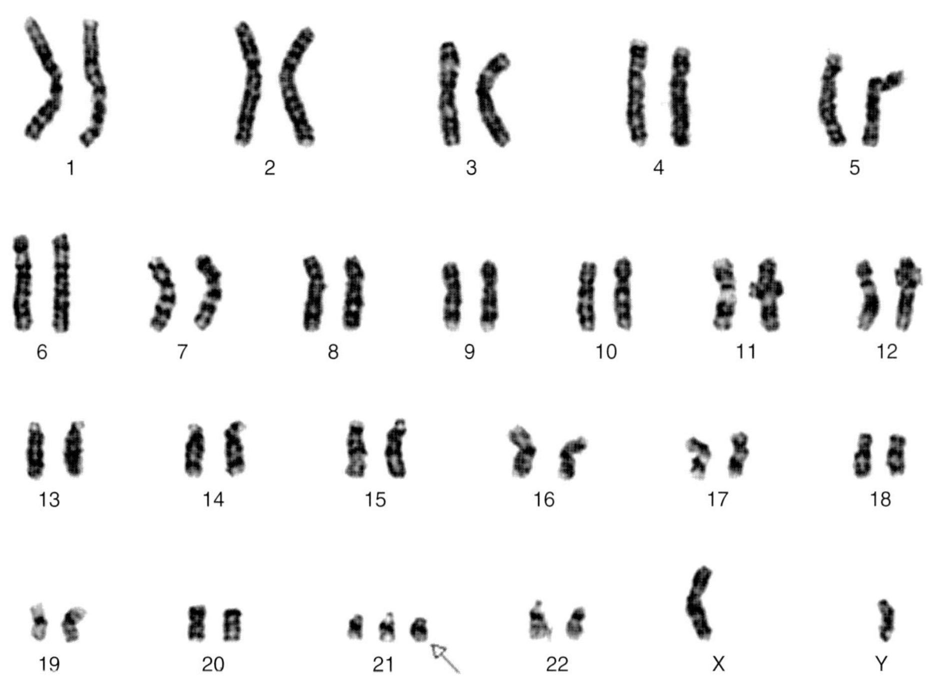

FIGURE 9.17 A G-banded karyotype with Trisomy 21.

(submicroscopic). Chromosomal anomalies reported to date may reflect secondary anomalies associated with chromosomal evolution in CNL. Sporadic reports of patients with +8, +9, del(20)(q11q13), del(11)(q14), +21 (Figure 9.17), and complex karyotypes are described in the literature [73, 74].

Clinical Aspects

CNL is a rare MPD which generally affects women and men over 60 years of age. In most cases, the disease behaves aggressively with a mean survival of $<$2 years [66, 73, 75]. The cause of death is often cerebral hemorrhage or infection. The transformation of CNL to AML has been reported [75]. There is also a report of CNL evolving from PV [76].

Due to the rarity of the disease, no standard therapeutic protocols are currently available. At the present time, allogeneic bone marrow transplantation appears to be the only potential cure. Certain drugs, such as hydroxyurea and IFN-α, may help to control granulocytosis and splenomegaly.

IDIOPATHIC HYPEREOSINOPHILIC SYNDROME AND CHRONIC EOSINOPHILIC LEUKEMIA

Idiopathic hypereosinophilic syndrome (HES) and chronic eosinophilic leukemia (CEL) represent overlapping persistent eosinophilic disorders with no known etiology, with a wide spectrum of clinical presentations ranging from indolent to aggressive clinical courses [1, 77–81]. HES is the preferred term when there is no evidence of clonality or increased blasts (Table 9.3). Neither HES nor CEL shows the *Ph*[1] chromosome or *BCR-ABL1* fusion gene.

Etiology and Pathogenesis

The etiology of HES and CEL is not known. The detection of *FIP1L1-PDGFRA* fusion gene in approximately 50% of the HES/CEL cases suggests a pathogenic role for tyrosine kinase activity of the fusion gene product in these disorders [82, 83]. The fusion of *FIP1L1* to *PDGFRA* is the result of a small interstitial deletion (only 800 kb in size) of the long arm of chromosome 4, del(4)(q12q12) [84]. Also, an acquired t(8;9)(p21–23;p23–24) has been reported in some cases of CEL [77, 85, 86]. This translocation fuses the *PMC1* gene to the *JAK2* gene (a tyrosine kinase), further supporting the role of tyrosine kinase activity in the pathogenesis of chronic MPDs, including HES/CEL.

Pathology

Morphology

The bone marrow is hypercellular and shows eosinophilic hyperplasia. Eosinophils counts may range from 10% to 70% of the bone marrow nucleated cells, with a mean of about 30% [1, 87, 88]. The maturation of eosinophils and myeloid cells in many instances (HES) is progressive and orderly without significant left shift or increased blasts (Figure 9.18). But blasts are increased ($>$5% and $<$20%) in the smaller proportion of the cases (CEL) (Table 9.5). Charcot–Leyden crystals are frequent findings. Charcot–Leyden crystals are colorless, long hexagonal, double-pointed, or needle-like lysophospholipase containing structures formed from the breakdown of eosinophils. Eosinophils may show dysplastic changes such as nuclear hypersegmentation or hyposegmentation, cytoplasmic vacuolization or hypogranularity, and/or abnormal eosinophilic granules [1, 87, 88]. However, both abnormal morphologic changes and

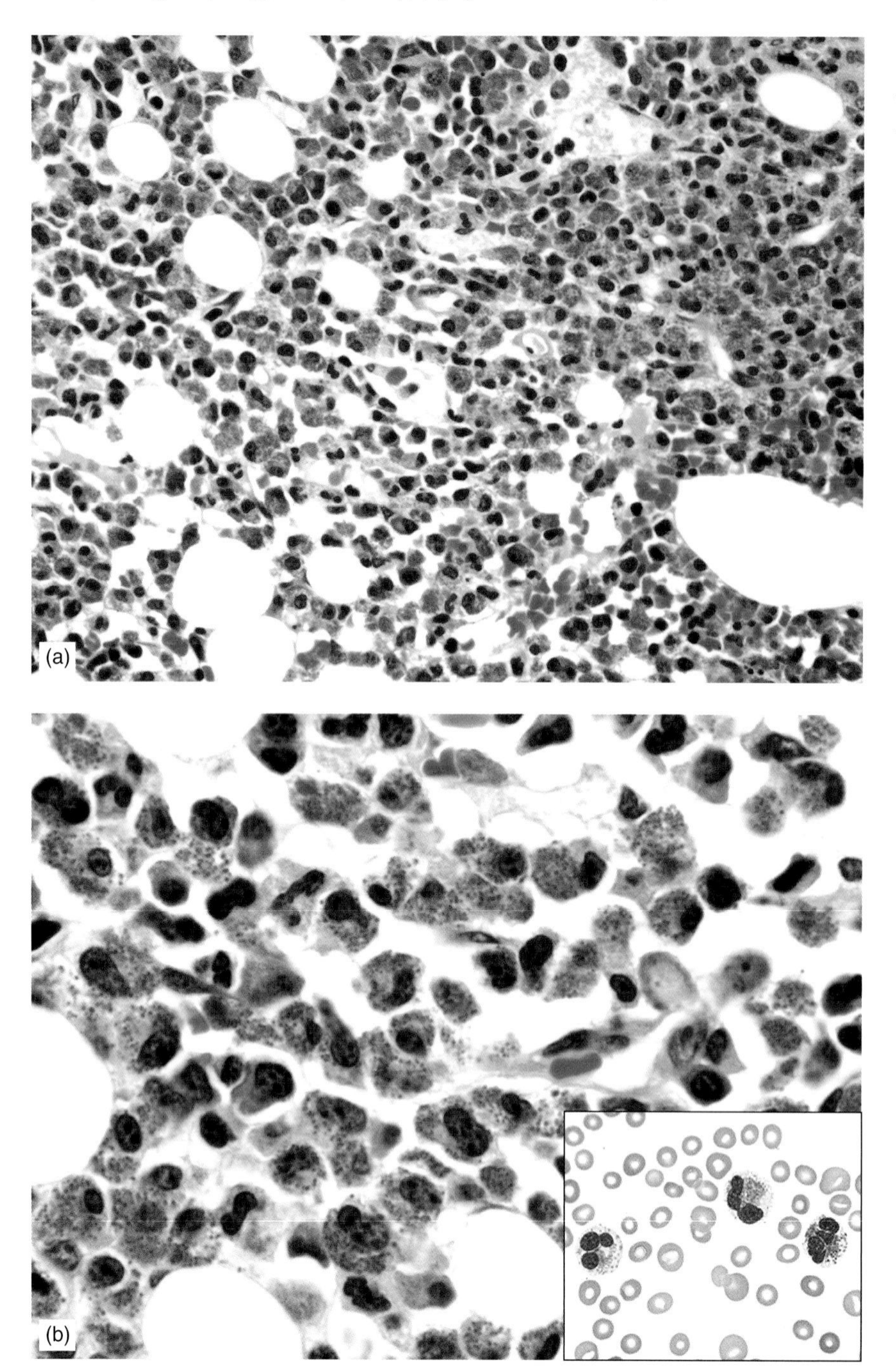

FIGURE 9.18 Bone marrow biopsy section (a and b) and blood smear (inset) of a patient with HES. Adapted from Naeim, F. (2001). *Atlas of Bone Marrow and Blood Pathology*, Saunders, by permission.

Charcot–Leyden crystals have been observed in cases of reactive eosinophilia. Myelofibrosis may be present but is not common.

The peripheral blood shows absolute eosinophilia (>1,500/μL) with or without neutrophilia, basophilia, myeloid left shift, or abnormal morphology. The leukocyte count is often moderately elevated (between 20,000 and 30,000/μL), and eosinophils in most instances account for 30–70% of the differential counts [1, 87, 88].

Eosinophilic infiltration may also be present in the extramedullary sites. The site of infiltration usually shows some degree of fibrosis, often with the presence of Charcot–Leyden crystals.

Immunophenotypic Studies

The eosinophils show different characteristic features than the neutrophils by flow cytometry. They appear as distinct clusters in FSC/SSC and CD45/SSC dot plot analyses, and their intensity of expression of myeloid-associated markers is different from neutrophilic granulocytes. A number of CD molecules are expressed on eosinophils, such as CD9 (leukocyte antigen MIC3), CD32 (FcgRII), CDw125 (IL-5 receptor alpha chain), and CD193 (chemokine receptor 3), but these molecules are not eosinophilic-specific and are also expressed by other leukocytes [89, 90].

Anti-CD34 and -CD117 monoclonal antibodies can be used to estimate the number of blasts by flow cytometry or immunohistochemical stains in blood samples, bone marrow aspirates, or biopsy sections.

TABLE 9.5 WHO criteria for the diagnosis of chronic eosinophilic leukemia and hypereosinophilic syndrome.*

1. *Requirements***
a. Persistent peripheral blood eosinophilia ≥1.5,000/μL.
b. Bone marrow eosinophilia.
c. Blasts <20% in blood or marrow.
2. *Exclusions*
a. All causes of reactive eosinophilia secondary to allergic, parasitic, infections, pulmonary, and collagen vascular diseases.
b. All neoplastic disorders with secondary, reactive eosinophilia, such as T-cell lymphoid malignancies, Hodgkin lymphoma, acute lymphoblastic leukemia/lymphoma, and mastocytosis.
c. Neoplastic disorders that eosinophils are a part of the neoplastic clone, such as chronic myeloproliferative diseases, myelodysplastic syndromes, and acute myelogenous leukemia.
d. Conditions associated aberrant expression or abnormal cytokine production of T lymphocytes.

*Adapted from Ref. [1].

**Diagnosis of chronic eosinophilic leukemia is made when items a–d are all excluded and if myeloid cells show clonal evolution by cytogenetic and/or molecular studies, or if the blast cells in the bone marrow are >5% and <20%, or >2% blasts are present in the peripheral blood.

Molecular Studies

The only specific molecular finding in HES/CEL is the *FIP1L1-PDGFRA* fusion gene, though it is only present in about half the cases [83, 84]. Since it is a tyrosine kinase, some patients have responded to imatinib therapy, so the finding has both therapeutic and diagnostic significance. And also like CML, resistance to the drug can arise from acquired mutations in the fusion gene, some of which may be sensitive to next-generation drugs [91].

Cytogenetics

Although no specific cytogenetic abnormalities have been associated with CEL, the presence of another clonal anomaly that is associated with MPD can help in the differential diagnosis between CEL and a reactive disease that involves the eosinophils. Cytogenetic anomalies in CEL often have been associated with a poor prognosis [92]. One important chromosomal aberration that has been linked with CEL is t(5;12) (Figure 9.19). The t(5;12) results in the fusion of the *PDGFRβ* tyrosine kinase gene on chromosome 5q33 and the *TEL* gene on chromosome 12p12. Various investigators have assigned the breakpoints within chromosome 5 as 5q33 or 5q31, and within chromosome 12 as p12 or p13; however, the investigators are most likely describing the same translocation [13]. Variant translocations that involve the *PDGFR* gene include the t(5;7)(q33;q11.2) and t(5;10)(q33;21.2). Fusion of the *FIP1L1* gene to the platelet-derived growth factor receptor alpha (*PDGFRα*) gene has recently been described in patients with HES [93]. These two genes lie very close to one another within chromosome band 4q12, so the fusion cannot be detected by karyotype. However,

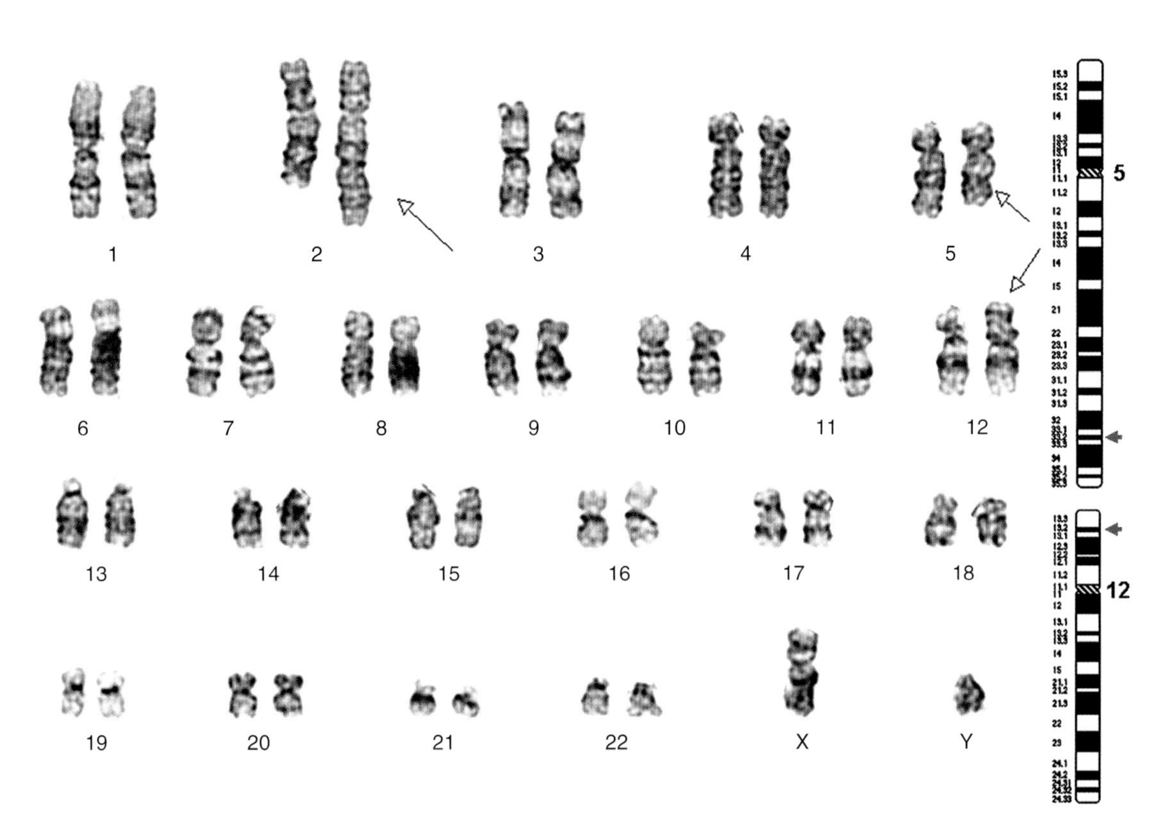

FIGURE 9.19 A G-banded karyotype with a balanced 5;12 translocation. (The 2q abnormality is a secondary event.)

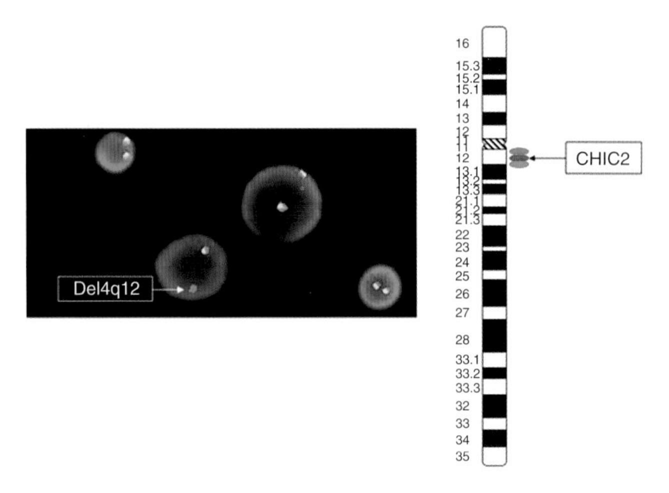

FIGURE 9.20 Deletion of the 4q12 region as identified by 3-color FISH.

the fusion results in a deletion of the intervening DNA sequences (CHIC2 deletion) (Figure 9.20), which can be detected *only* by FISH and is a target for IM treatment [93].

Clinical Aspects

HES and CEL involve men much more frequently than women (M:F ratio about 9:1). They are usually detected between the ages of 20 and 50 years and are rare in children. The most common clinical symptoms include fatigue, cough, dyspnea, myalgia, angioderma, rash, fever, and rhinitis [77–81]. The release of eosinophilic granules may damage the endocardium and the endothelial cells and lead to thrombus formation and emboli. Endocardial thrombosis and fibrosis may cause insufficiencies of the mitral or tricuspid valves [94].

The presence of blast cells in the peripheral blood, increased blasts in the bone marrow, multilineage dysplasia, marked splenomegaly, and cytogenetic aberrations are considered signs of adverse clinical outcome.

POLYCYTHEMIA VERA

Polycythemia vera (PV) is characterized by erythrocytosis or increased red blood cell mass with hyperviscosity, increased risk of thrombosis, and varying degrees of thrombocytosis, leukocytosis, and splenomegaly [1, 95]. PV, similar to the other myeloproliferative diseases, is the result of clonal expansion of a pleuripotent stem cell with the involvement of the myeloid lineages and a variable proportion of B lymphocytes. However, the majority of the T and NK cells do not seem to be affected.

The diagnosis of PV is established by complex clinical, laboratory, and morphologic features. The recommended criteria for the diagnosis of PV by WHO consist of two major categories [1].

Category A

1. Elevated RBC mass >25% above normal range, of Hb >18.5 g/dL in men and >16.5 g/dL in women.
2. No evidence of familial erythrocytosis or elevated erythropoietin (Epo) due to:
   a. Hypoxia (arterial pO_2 $\leq$ 92%)
   b. High-oxygen-affinity hemoglobin
   c. Truncated Epo receptor
   d. Inappropriate Epo production by tumors.
3. Splenomegaly.
4. Clonal genetic abnormalities other than *Ph*[1] or *BCR-ABL1* fusion gene.
5. Endogenous (erythropoietic-independent) erythroid colony formation *in vitro*.

Category B

1. Thrombocytosis >400,000/μL.
2. WBC >12,000/μL.
3. Bone marrow demonstrating panmyelosis with erythroid preponderance and megakaryocytosis.
4. Low serum erythropoietic levels.

The diagnosis of PV is made when A1 and A2 plus any other category A are present or when A1 and A2 plus any two of category B are present.

Etiology and Pathogenesis

The etiology of PV is not known. A genetic predisposition has been suggested based on the reports of PV in identical twins [96]. Also, a higher incidence of PV has been observed in Hiroshima atomic bomb survivors, US military personnel involved in the nuclear weapons tests, and persons with occupational exposure to chemical toxins.

The erythroid precursors seem to be erythropoietin (Epo) independent in tissue culture settings. The presence of erythroid colonies in the absence of exogenous Epo is considered an *in vitro* PV hallmark. The PV erythroid progenitors are also hypersensitive to Epo as well as to other hematopoietic growth factors and differentiate faster than their normal counterparts. In addition, it has been shown that the BFU-E cells from PV patients have increased sensitivity to insulin-like growth factor-1 (IGF-1). IGF-1 has an Epo-like activity and stimulates erythropoiesis [14].

One of the interesting current hypotheses in the pathogenesis of PV is the presence of a defect in transcription regulation that affects cytokine receptor signaling. It has been shown that the *JAK2* gene plays an important role in the EOP–EPO receptor signaling in erythropoiesis. *JAK2* mutation in mice during embryogenesis is lethal due to the lack of sufficient erythropoiesis. A unique clonal mutation in the *JAK2* gene has been reported, resulting in a valine to phenylalanine substitution at position 617, in 65–97% of PV patients [14].

Pathology

Morphology

Two distinct clinocopathologic phases have been described in PV: the polycythemic phase and the "spent" phase [1–4,

95, 97]. Polycythemic phase describes the earlier active phase of the disease when the bone marrow shows panmyelosis and blood displays erythrocytosis, sometimes in association with thrombocytosis and/or leukocytosis. Spent phase refers to the later stage of the disease when bone marrow is fibrotic or hypocellular, and there is evidence of extramedullary hematopoiesis, progressive splenomegaly, and anemia. In addition, a *transitional phase* has been described, referring to a process between these two phases characterized by erythrocytosis and bone marrow fibrosis. The morphologic features of the polycythemic and spent phases are described below.

Polycythemic phase is characterized by bone marrow hypercellularity and panmyelosis. Usually, there is marked erythroid preponderance and predominance of megakaryocytes (Figure 9.21). Erythropoiesis is normoblastic, and megakaryocytes are pleomorphic and have a tendency to appear in clusters and/or next to bone trabeculae. No significant dysplastic changes are noted in the granulocytic series and there is no evidence of increased myeloblasts. Basophilia is a common feature, and eosinophilia is not infrequent. In the majority of PV cases at this phase, bone marrow biopsies show no increase in reticulin fibers, though variable degrees of fibrosis are present in about 30% of the cases. Stainable iron is reduced or absent in the vast majority of cases [1–4, 95, 97].

The peripheral blood shows increased red cell mass and elevated hemoglobin levels. Erythrocytes are usually normochromic and normocytic, but sometimes are hypochromic and microcytic. The presence of deeply basophilic reticulocytes has been described. The activity of red cell glycolytic enzymes and the proportion of fetal hemoglobin are increased. There is often leukocytosis and thrombocytosis with basophilia and presence of giant, hypogranular platelets. There may be mild myeloid left shift but blasts are not usually found. The LAP score is often elevated.

"Spent" phase represents the late stage of PV and often is characterized by [1–4, 95, 97]:

1. Normalization of erythrocytosis and then progression to anemia.
2. Progressive splenomegaly.
3. Myelofibrosis and extramedullary hematopoiesis with leukoerythroblastic blood picture.
4. Myelodysplastic changes, sometimes with increased blasts.

There is some debate regarding the mechanism(s) involved in the evolution of PV to the spent phase. The original idea of the development of anemia in PV as the natural history of the disease due to bone marrow exhaustion has been challenged by the causative effects of chemotherapy, hemorrhage, and deficiencies of iron, vitamin B_{12}, or folic acid. Similarly, the development of myelodysplasia and/or marrow fibrosis may be more secondary to therapy than a naturally occurring phenomenon. For example, the incidence of myelofibrosis is higher in PV patients exposed to chemotherapy or radiation than those treated by phlebotomy.

Immunophenotypic Studies

No pathognomonic immunophenotypic features have been described in the blood or bone marrow of PV patients.

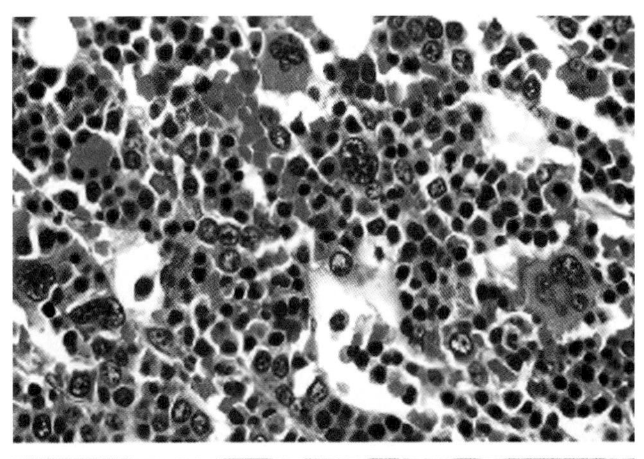

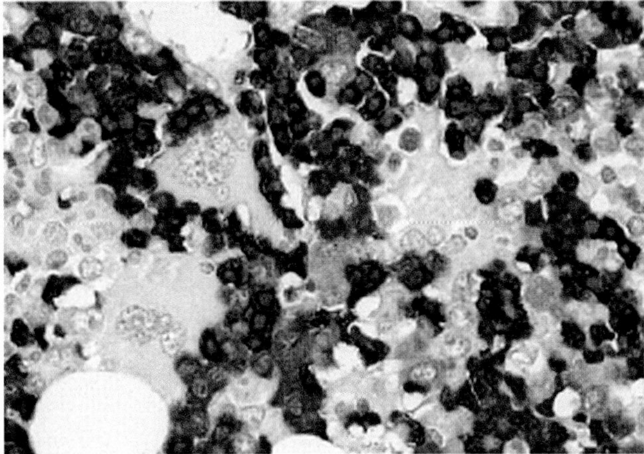

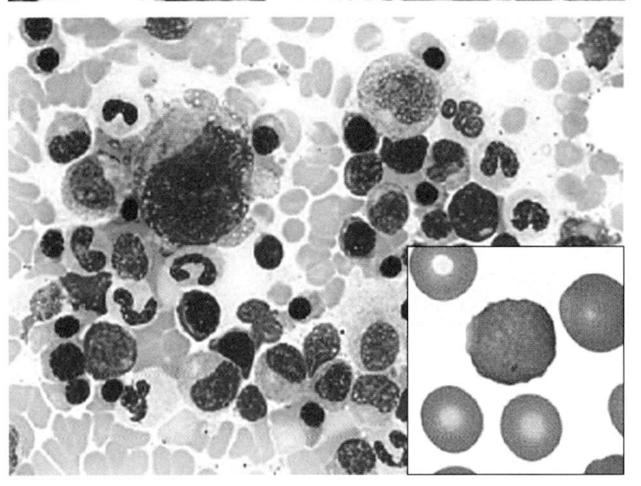

FIGURE 9.21 Bone marrow biopsy section (top) and smear (bottom) from a patient with PV demonstrating marked erythroid preponderance. The figure in the middle depicts hemoglobin A by immunohistochemical stains. The inset represents a deeply stained polychromatophilic erythrocyte.

Molecular Studies

A specific molecular marker for a number of MPD is a point mutation in the *JAK2* gene, V617F. Within this group, it is most frequently seen in PV, where some series find it in >90% of affected patients [98–100]. As such, it can be of help in the diagnosis of PV, and in its differential diagnosis from other disorders such as CML, chronic myelomonocytic leukemia, acute leukemia, and MDS. Since

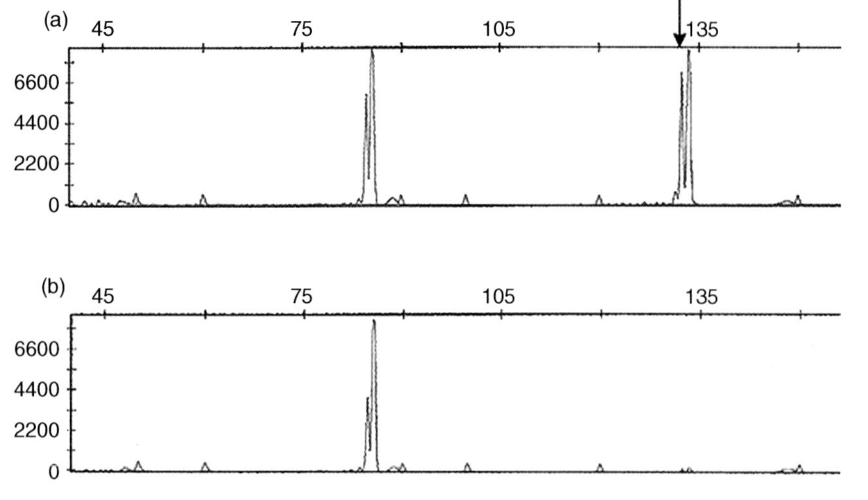

FIGURE 9.22 Detection of the *JAK2* mutation by allele-specific PCR followed by capillary electrophoresis of the amplification products. A positive sample (a) yields both the mutant PCR product (arrow) and the internal control normal gene product, whereas a negative sample (b) yields only the normal PCR product peak.

it is a single nucleotide change, any of a variety of straightforward assays can be used. Allele-specific PCR primers can be used to amplify both the mutant region of the gene (if present) and a nearby invariate portion of the gene (as an internal amplification control). The method also allows for semi-quantitative assessment of the proportion of *JAK2* mutant cells present, by comparing the heights of the two PCR peaks on capillary electrophoresis, which may have prognostic significance (Figure 9.22). Other methods in use include real-time PCR incorporating melting-curve analysis [99] and DNA sequencing or pyrosequencing [100]. Most of these methods are not very good at distinguishing homozygous from heterozygous *JAK2* mutations, which may also have some influence on clinical behavior. Homozygosity results from mitotic recombination in the *JAK2*-positive cells and/or loss-of-heterozygosity at the gene locus on chromosome 9p [98].

Another potential marker is the *MPL* gene for the thrombopoietin (TPO) receptor which shows reduced expression in platelets and megakaryocytes in PV [101].

Cytogenetics

In PV, cytogenetic results do not predict evolution of the disease, but they can provide clues to hematologic phenotype, duration of the disease, and consequences of myelosuppressive therapy [15]. A greater proportion of patients with advanced disease (and poorer prognosis) have chromosomally abnormal clones than patients with early stage PV [11, 15]. In addition, abnormal clones are more frequent among patients who have PV with myeloid metaplasia (78%) than among patients who have PV alone (19%) or PV with myelofibrosis (40%) [15]. The most common chromosomal anomalies at PV diagnosis are del(20)(q11q13), +8, and +9, with +8 and +9 often occurring together in the same clone. Additional abnormalities observed include del(1)(p11), del(3)(p11p14), t(1;6)(q11;p21), and t(1;7)(q10;p10) (Figure 9.23). In some patients with PV, a *de novo* leukemia or MDS develops; in these patients, chromosomal anomalies are more similar to the secondary disease than those associated with untreated PV. Still other patients with PV develop a chromosomally abnormal clone as a consequence of therapy. The most common chromosomal anomaly associated with therapy-related leukemia involves anomalies of chromosome 5 or 7 or both and unbalanced translocations derived from t(1;7)(q10;p10) [16].

Several studies have shown that an abnormal karyotype at diagnosis of PV is associated with a poor prognosis, while the proportion of patients with an abnormal karyotype increases during the course of the disease [15, 102, 103]. PV may progress to a terminal phase, which can involve transformation to myelofibrosis or acute leukemia. Almost all the PV patients who develop acute leukemia in late disease stages have chromosomal abnormalities. Trisomy 8 or 9 may persist in PV without further clonal evolution or leukemia development for up to 15 years, whereas other chromosomal abnormalities, such as −7 or 5q− or complex changes, may signal the terminal phase of the disease. The utilization of FISH methods in PV did not detect a substantially increased incidence rate of the chromosomal abnormalities, or did it reveal submicroscopic deletions in patients with a normal karyotype. However, both FISH and comparative genomic hybridization studies have revealed frequent abnormalities of chromosome 9, including gains in 9p.

Clinical Aspects

PV is more frequent in men than in women (about 2:1) with a peak incidence around 70–80 years of age. Most common complaints are non-specific and include headache, weakness, and dizziness. Approximately 5–20% of the patients may complain of arthritis. Itching, particularly after a warm bath, and erythromelalgia (erythroderma with burning pain of extremities) are common features. Although the reports of survival time for untreated patients range from 6 to 18 months, treated patients live much longer, usually exceeding 10 years. The major causes of death are thrombosis, transformation to MDS or AML, non-hematologic malignancies, and hemorrhage [104–107].

Thrombosis is either venous or arterial and appears to be primarily related to the patient's blood hyperviscosity. Transformation to MDS or AML in most instances is related

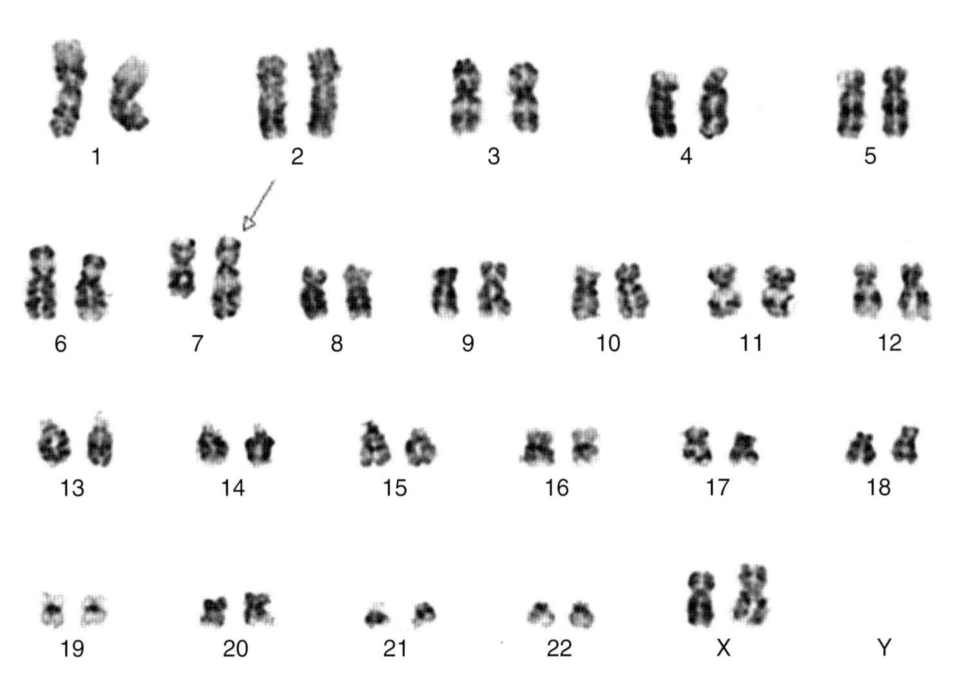

FIGURE 9.23 A G-banded unbalanced 1;7 translocation, resulting in three copies of 1q and loss of one copy of 7q.

to chemotherapy. In one report the incidence of AML in PV patients was 1.5%, 10%, and 13% for phlebotomy, ^{32}P, and chlorambucil therapy, respectively [105–107]. Approximately 3–12% of the patients may develop a secondary malignancy (carcinoma, non-Hodgkin lymphoma). The risk of secondary malignancies is much higher in patients who receive myelosuppressive therapy. Because of the high risk of development of malignancies, chlorambucil therapy has been discontinued. Other therapeutic modalities include IFN-α, anagrelide (a quinazolone derivative), and allopurinol [105–107].

CHRONIC IDIOPATHIC MYELOFIBROSIS

Chronic idiopathic myelofibrosis (CIMF) or agnogenic myeloid metaplasia or myelofibrosis with myeloid metaplasia is a clonal stem cell disorder and a subtype of CMPD characterized by myeloproliferation, atypical megakaryocytic hyperplasia, bone marrow fibrosis, extramedullary hematopoiesis, and marked splenomegaly [1, 108–110]. These changes often lead to anemia, leukoerythroblastosis (the presence of immature myeloid and erythroid cells in blood), and tear-drop-shaped red cells.

Etiology and Pathogenesis

The etiology of CIMF is not known. However, in a small proportion of the cases, development of CIMF has been linked to ionizing radiation, thorium dioxide, and petroleum derivatives, such as toluene and benzene [109]. Also, a mutation of the *GATA-1* transcription factor gene in mice may induce pictures similar to those of myelofibrosis.

Studies based on X-chromosome genes, karyotyping, or identification of *RAS* and *JAK2* gene mutations, are consistent with a clonal stem cell involvement [111]. Cytogenetic aberrations with chromosomal deletions, such as 13q−, 20q−, and 12p−, have been frequently reported in CIMF, as well as trisomies 8 and 9 [112]. Pathologic features consist of two fundamental components: myelofibrosis and extramedullary hematopoiesis.

Myelofibrosis appears to be a reactive process secondary to the activation of bone marrow stromal cells, particularly fibroblasts. Bone marrow fibroblasts in CIMF are polyclonal and structurally normal. Pathogenesis of myelofibrosis, increased number of stromal cells, and excess deposition of extracellular matrix proteins is probably mediated by the release of regulatory cytokines secondary to the clonal proliferation of the hematopoietic cells [111–113]. For example, transforming growth factor-beta (TGF-β) is a glycoprotein that is produced and released by endothelial cells, monocytes, and megakaryocytes. TGF-β enhances the production and release of extracellular matrix protein, such as collagen from fibroblasts [108–111]. A CIMF-like condition has been induced in mice by administration of high dose TPO. TPO induces megakaryocytosis and, therefore, increased release of TGF-β, resulting in bone marrow fibrosis [113–115].

Another possible mechanism for the induction of myelofibrosis in CIMF is increased rate of emperipolesis (entry of hematopoietic cells into megakaryocytic cytoplasm) observed by some investigators in experimental animal models [116]. Emperipolesis, which is probably induced by abnormal localization of P-selectin on the megakaryocytes, entraps the hematopoietic cells, causes cell damage, and releases growth factors leading to the activation of fibroblasts and fibrosis.

Extramedullary hematopoiesis is probably due to the abnormal release of hematopoietic precursors from bone marrow sinusoids into the circulation and their homing in other tissues, particularly spleen and liver. Bone marrow

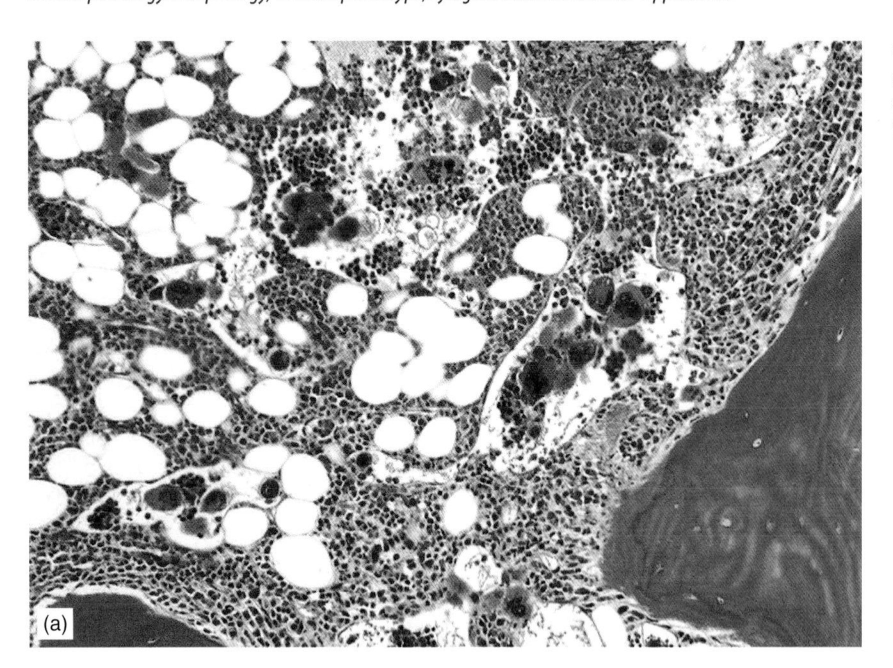

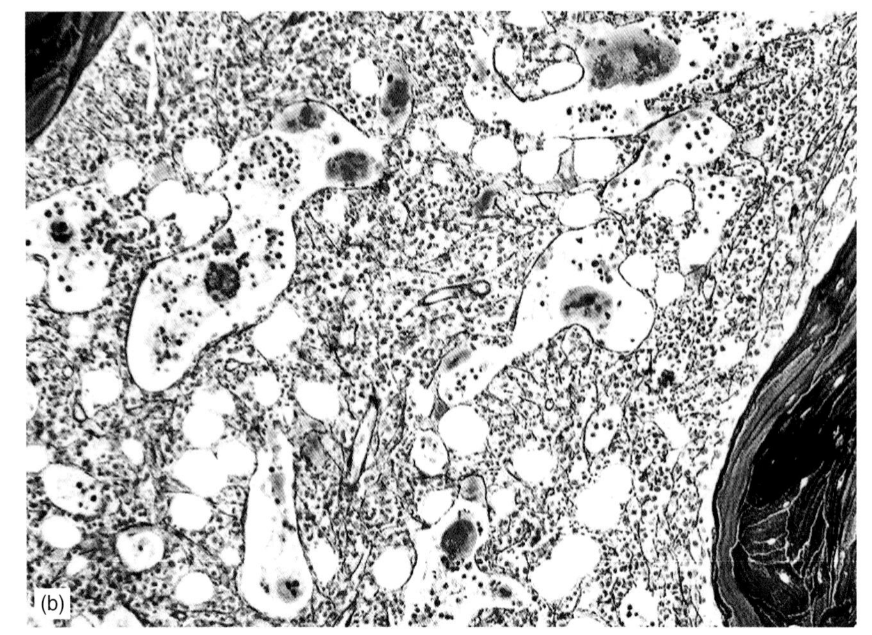

FIGURE 9.24 Biopsy sections in cellular phase of CIMF often show dilated hematopoietic-containing sinuses: (a) minimal amount of increased reticulin fibers and (b) reticulin stain.

fibrosis distorts the normal structure of the sinusoids, causing collapse of some and dilatation of others. These changes may disrupt the gatekeeping role of the sinusoids and cause the release of immature cells. In addition, the dilated sinusoids often show aggregates of hematopoietic precursors that could be released into the circulation.

Pathology

Morphology

The morphologic features of CIMF, such as bone marrow cellularity, extent of fibrosis, peripheral blood findings, and extramedullary hematopoiesis, vary considerably in different stages of the disease. The evolutionary process of CIMF in bone marrow and blood can be divided into two major phases: cellular and fibrotic [1–4, 117].

Prefibrotic stage or **cellular phase** represents the early stage of the disease, when bone marrow fibrosis is lacking or there is only a minimal amount of reticulin fibrosis (Figure 9.24). At this stage, the bone marrow, similar to the other chronic MPDs, is hypercellular and displays panmyelosis. Megakaryocytes are atypical and often appear in clusters around the sinusoids and/or bone trabeculae (Figure 9.25). Abnormal nuclear lobulation, naked nuclei, and large bizarre forms are frequent findings. Micromegakaryocytes are often present. There may be a myeloid left shift with a higher proportion of the intermediate cells and <10% myeloblasts. Lymphoid aggregates are present in up to 25% of the cases.

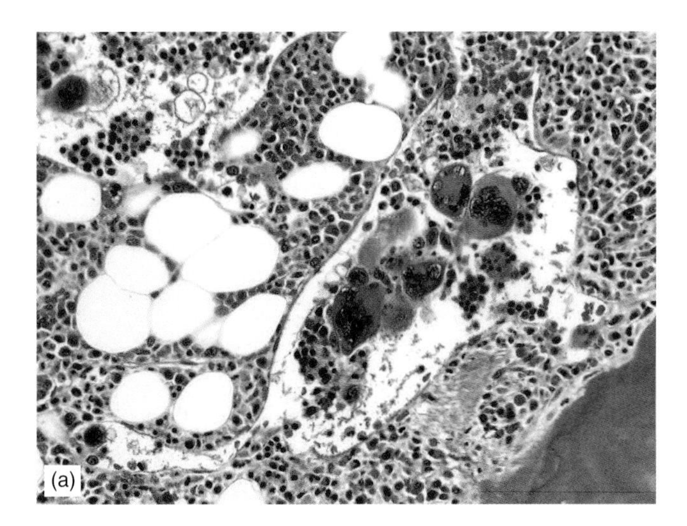

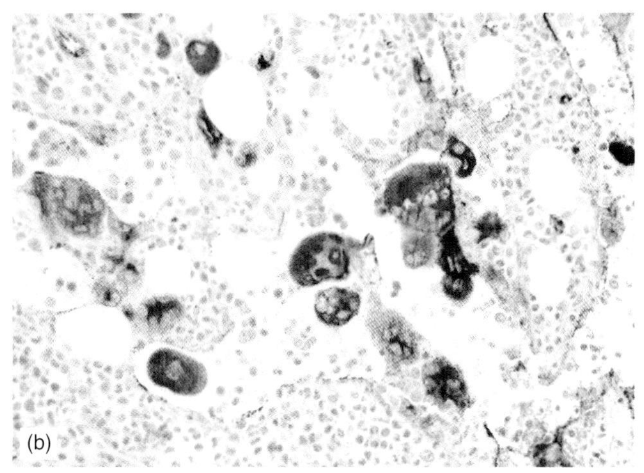

FIGURE 9.25 Biopsy section of cellular phase of CIMF demonstrating a dilated sinus containing numerous megakaryocytes and erythroid and myeloid precursors (a). Numerous megakaryocytes are demonstrated by immunohistochemical stain for factor VIII (b).

Leukocytosis, thrombocytosis, and mild to moderate anemia are the frequent peripheral blood findings. Anisopoikilocytosis is mild and occasional tear-drop-shaped red blood cells may be present. Leukoerythroblastosis is either mild or lacking. Giant platelets may be present.

Establishment of the diagnosis of CIMF at this stage is often difficult because of significant overlapping features with other chronic MPD, particularly idiopathic thrombocythemia. Their differential diagnoses are discussed at the end of this chapter.

Fibrotic stage represents the more advanced phase of the disease characterized by various degrees of fibrosis and reduced bone marrow cellularity (Figures 9.26 and 9.27). The extracellular matrix consists of excessive amounts of type I, III, IV, and V collagen, reticulin (glycoprotein coating of stromal cell strands), fibronectin, and laminin. Reticulin and type III collagen are the predominant components. Reticulin precedes the excessive collagen deposits and stains black by silver impregnation. Collagen appears bluish-green by trichrome stain and is usually detected in advanced fibrosis.

Osteosclerosis and dilatation of bone marrow sinusoids are common features. The dilated sinusoids often contain aggregates of hematopoietic precursors, including dysplastic megakaryocytes. The bone marrow in advanced stages of the disease is virtually replaced by a dense fibrous tissue with markedly reduced cellularity and scattered trapped dysplastic megakaryocytes and small islands of erythroid and myeloid precursors. The bone marrow aspiration is usually unsuccessful and results in a "dry" tap.

Blood examination reveals leukoerythroblastic morphology with the presence of immature myeloid and erythroid cells (Figure 9.28). Blasts may be present but are usually <5%. Tear-drop-shaped red cells (dacrocytes) are commonly present. Anemia is a frequent finding, and there is often mild to moderate leukocytosis. Hypersegmented neutrophils may be present. The platelet count may be increased or reduced with abnormal forms present. Bare megakaryocytic nuclei and micromegakaryocytes are often detected. Significant dysplastic changes in myeloid series and blasts >10% are suggestive of an AP. The LAP score is often increased, and serum levels of lactate dehydrogenase and uric acid may be elevated.

Extramedullary hematopoiesis is often observed in the spleen and liver, but is also seen in other sites, such as lymph nodes, lung, serosal surfaces, urogenital system, skin, and retroperitoneal and paraspinal spaces (Figure 9.29). In the spleen, the red pulp is involved with the presence of erythroid, myeloid, and megakaryocytic cells in the sinuses. The extent of red pulp involvement and the proportion of each hematopoietic lineage vary from case to case, but megakaryocytes are commonly prominent. Similarly, sinuses are the main sites of extramedullary hematopoiesis in the liver and the lymph nodes. The splenic cords and the hepatic parenchyma may show various degrees of fibrosis.

Immunophenotypic Studies

No pathognomonic immunophenotypic features have been described in the blood, bone marrow, or other tissues in patients with CIMF.

Molecular and Cytogenetic Studies

There are numerous reports regarding the association between *JAK2* mutation (*JAK2*^{V617F}) and *BCR-ABL*-negative CMPD [118]. In a large study of 157 patients with myelofibrosis with myeloid metaplasia, the rate of *JAK2* mutation was about 45% [118].

Although the proportion of cases of CIMF with abnormal karyotypes ranges from 30% to 75%, distinct recurrent chromosomal aberrations have been reported in 40–50% of patients. This discrepancy is mostly due to difficulty in sampling adequate numbers of quality metaphases from the few cells aspirated from fibrotic marrow [20, 119, 128]. Although no "specific" chromosome anomalies are observed in patients with CIMF, +1q, del(13q), del(20q), and +8 appear in approximately two-thirds of patients with pathologic karyotypes [20], and rarer anomalies include +9 and del(12p). The most common anomalies (del(13q) and translocations involving chromosome 13q14) likely interrupt the *RB1* gene, an important tumor-suppressor gene in retinoblastoma, osteosarcoma, and other solid tumors [15]. Although balanced

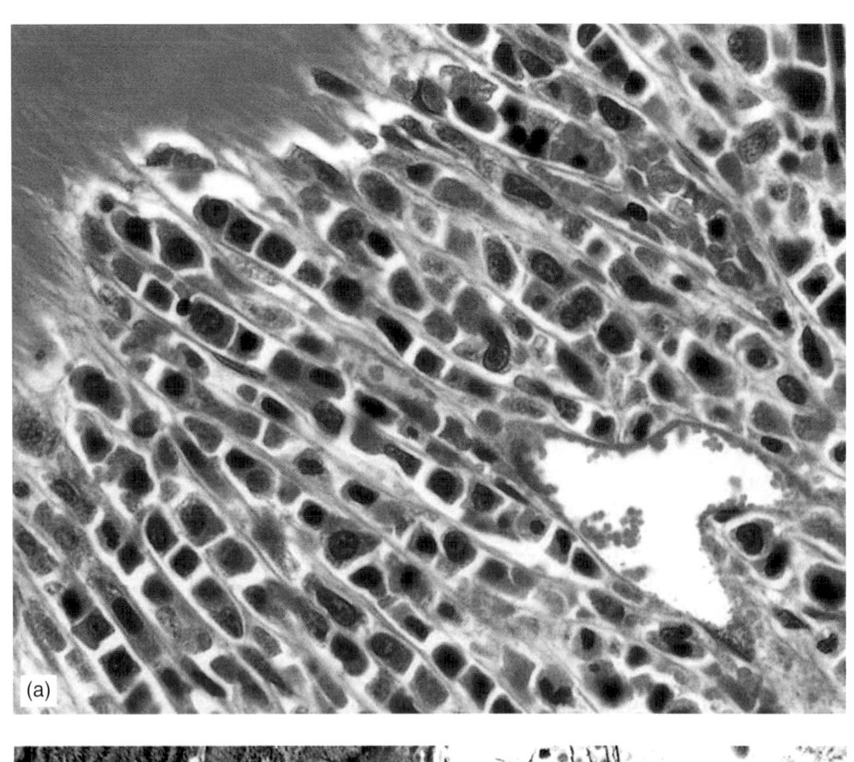

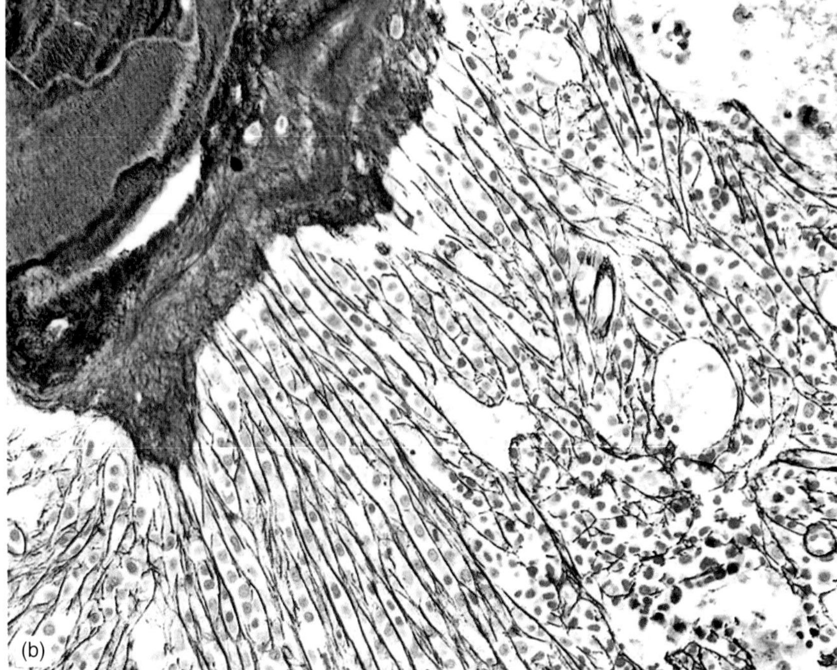

FIGURE 9.26 Biopsy section of a case of CIMF demonstrating separation of hematopoietic cells by delicate fibers (a) which are positive with reticulin stain (b).

translocations are uncommon, some reports document isolated cases with balanced translocations mostly involving chromosomes 1 and 12 with different partners [15, 119]. Specific cytogenetic abnormalities in CIMF are associated with significantly different survival outcomes [120, 121]. Prognostically favorable aberrations include 13q− and 20q−, whereas prognostically unfavorable clones may contain 12p− and +8 [122].

Clinical Aspects

The incidence of CIMF is approximately 1 per 100,000 with a median age of about 65 years. Men and women are equally affected. Marked splenomegaly is the hallmark, particularly in advanced stages. Non-specific symptoms such as fatigue, weight loss, night sweats, and fever are often present. Patients

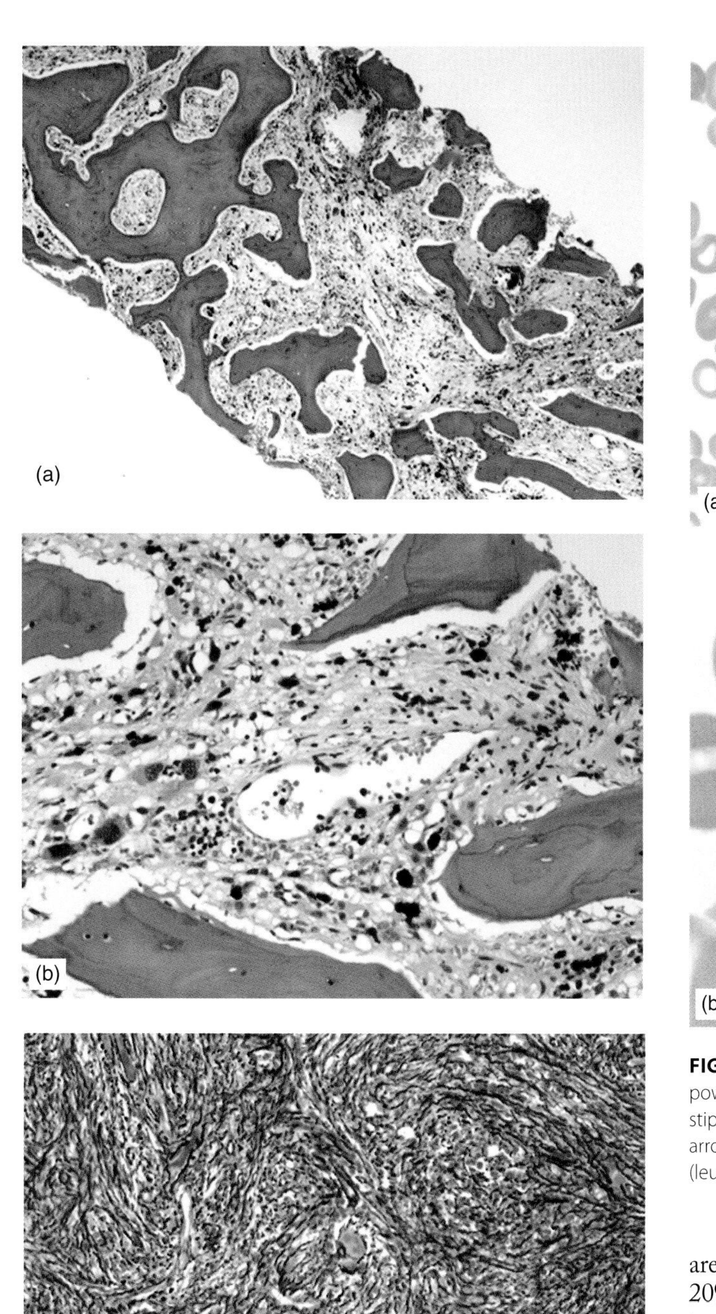

FIGURE 9.27 The advanced (fibrotic) stage of CIMF is characterized by extensive fibrosis and reduced number of hematopoietic cells: (a) low power, (b) high power, and (c) reticulin stain.

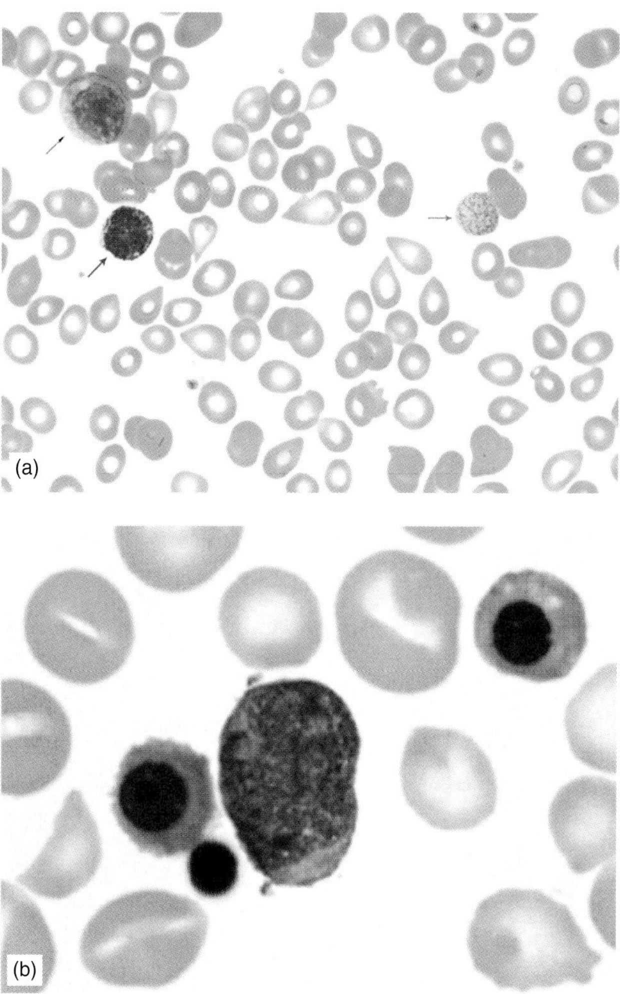

FIGURE 9.28 Peripheral blood smears from a patient with CIMF. (a) Low power demonstrating numerous tear-drop-shaped erythrocytes, basophilic stippling (blue arrow), a myelocyte (black arrow), and a basophil (red arrow). (b) High power showing two nucleated red cells and a myeloblast (leukoerythroblastosis).

frequently show anemia with abnormal (low or high) white cell and/or platelet counts. Serum lactate dehydrogenase levels are elevated [108–110].

Marked splenomegaly may lead to splenic infarction, portal hypertension, and thrombosis of the small portal veins. Advanced age, anemia, and chromosomal abnormalities are considered poor prognostic indicators. Anemia in about 20% of the patients is severe (Hb $<8\,g/dL$) and may be due to several factors, such as reduced bone marrow erythropoietic sites, ineffective erythropoiesis, splenic sequestration and destruction of red cells, autoimmune hemolysis, and/or bleeding [114].

Therapeutic approaches include treatment with hydroxyurea, splenectomy, splenic irradiation, allogeneic or autologous stem cell transplantation, and the use of antiangiogenic drugs [110].

ESSENTIAL THROMBOCYTHEMIA

Essential thrombocythemia (ET), or primary thrombocytosis, is a clonal stem cell disorder and a subtype of CMPD characterized by protracted thrombocytosis in the peripheral blood and increased number of megakaryocytes with atypical features in the bone marrow [1–4, 123–125].

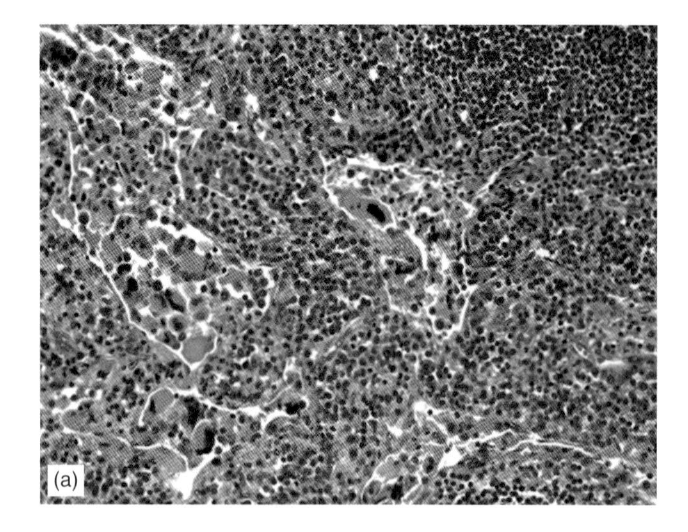

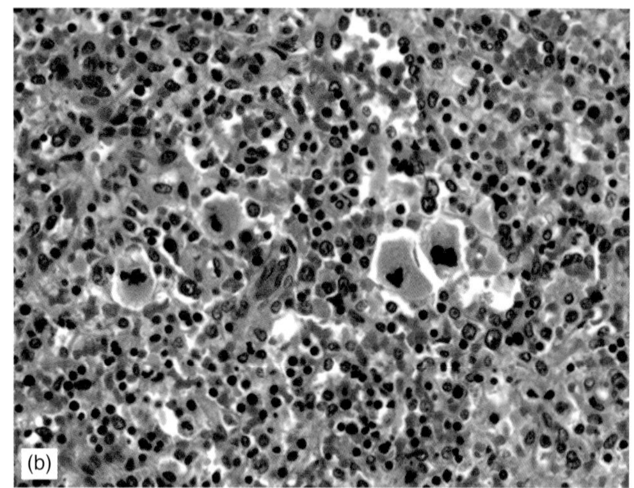

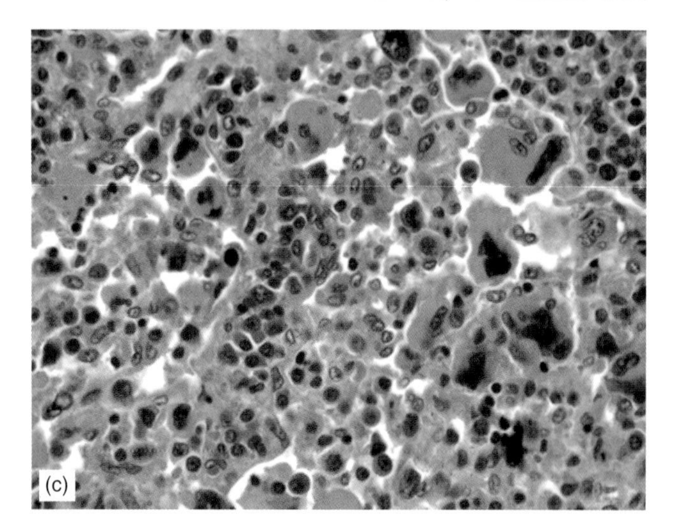

FIGURE 9.29 Lymph node: (a) low power, (b) high power, and spleen (c) sections demonstrating extramedullary hematopoiesis with numerous megakaryocytes and erythroid precursors.

Splenomegaly, thrombosis, and/or hemorrhagic events are frequent clinical manifestations. ET is diagnosed by the exclusion of reactive thrombocytosis (Table 9.6), familial thrombocytosis, and other subtypes of CMPD. The WHO criteria for the diagnosis of ET are presented in Table 9.7.

TABLE 9.6 Major differences between essential thrombocytosis (ET) and reactive thrombocytosis (RT).*

Features	ET	RT
Underlying disorder (inflammation, infection, malignancy, ischemia, tissue damage)	No	Yes
Digital or cerebrovascular ischemia	Yes	No
Arterial or venous thrombosis	Increased risk	No
Bleeding complications	Increased risk	No
Splenomegaly	May be present (40%)	No
Iron deficiency	No	May be present
Platelet function	May be abnormal	Normal
Large atypical megakaryocytes	Yes	No
Cytogenetic abnormalities	May be present	No
Plasma IL-6	Low	High
Plasma C-reactive protein	Low or normal	High
Spontaneous colony formation	Yes	No

*Adapted from Naeim, F. (1988). *Pathology of Bone Marrow*, 2nd ed. Williams & Wilkins, Baltimore and from Ref. [133].

TABLE 9.7 WHO criteria for the diagnosis of essential thrombocythemia.*

Positive criteria

1. Sustained platelet count of >600,000/μL in the peripheral blood.
2. Megakaryocytosis with the presence of enlarged forms in the bone marrow.

Criteria of exclusion

1. No evidence of polycythemia vera.
2. No evidence of chronic idiopathic myelofibrosis (lack of collagen fibrosis, minimal or lack of reticulin fibrosis).
3. No evidence of CML (no *Ph*[1] or *BCR/ABL* fusion gene).
4. No evidence of myelodysplastic syndrome [no del(5q), t(3;3) or inv(3q); no dysgranulopoiesis; few or lack of micromegakaryocytes].
5. No evidence of reactive thrombocytosis (no underlying inflammation, infection, cancer or tissue damage; no iron deficiency and no history of splenectomy).

*Adapted from Ref. [1].

Etiology and Pathogenesis

The etiology and pathogenesis of ET are not known. A number of genes such as polycythemia rubra vera-1 (*PRV-1*),

TPO and its receptor *c-MPL* (myeloproliferative leukemia virus oncogene) have been implicated in the pathogenesis of ET, but so far have not been conclusive [126]. Unlike ET, mutations of *TPO or c-MPL* have been associated with the familial ET, which is an autosomal dominant disorder. The presence of *JAK2* mutation in ET patients has been correlated to a higher frequency of transformation to PV in some studies [127].

Several approaches based on the X-linked DNA and transcript analysis have demonstrated clonal involvement development of ET [128]. However, more recent studies suggest heterogeneity in this process with the involvement of stem cells at different levels and evidence of polyclonal hematopoiesis in up to 50% of the cases [124, 129].

Pathology

Morphology

The bone marrow morphologic findings are highly variable and often demonstrate overlapping features with other MPD [1–4]. The morphologic hallmarks are lack or minimal reticulin fibrosis and increased number of megakaryocytes, including large or giant forms (Figures 9.30–9.33). These megakaryocytes may show hyperlobulated nuclei and/or appear in clusters or diffusely dispersed. The megakaryocytic clusters may be found around the sinusoids or close to the bone trabeculae. The bone marrow is usually normocellular for age or moderately hypercellular, but occasionally hypocellular.

The bone marrow smears show increased number of large megakaryocytes and the presence of platelet aggregates. Emperipolesis (internalization of hematopoietic cells) is a frequent finding in megakaryocytes (Figure 9.31). There is no evidence of increased myeloblasts or significant dysplastic changes in the granulocytic or erythroid lineages. Some cases may show marked erythroid preponderance mimicking PV.

The blood smears show marked thrombocytosis with marked variation in platelet size and the presence of giant forms (Figure 9.32). Megakaryocytic fragments may be present. The white blood cell count is normal or slightly elevated with normal differential counts or mild granulocytosis. Rarely, early granulocytic forms may be present. Red blood cells are normochromic and normocytic. Dacrocytes (tear-drop-shaped red cells) and leukoerythroblastosis are not characteristic features of ET.

In summary, morphologic findings in ET are not specific and may be seen in reactive thrombocytosis as well as other myeloproliferative disorders, particularly PV and cellular phase of CIMF. In a large recent study by the Polycythemia Vera Study Group (PVSG), the conclusion was that histologic criteria described in the WHO classification are difficult to apply to distinguish ET from prefibrotic myelofibrosis [130].

Immunophenotypic Studies

No pathognomonic immunophenotypic features have been described in the blood or bone marrow of patients with ET.

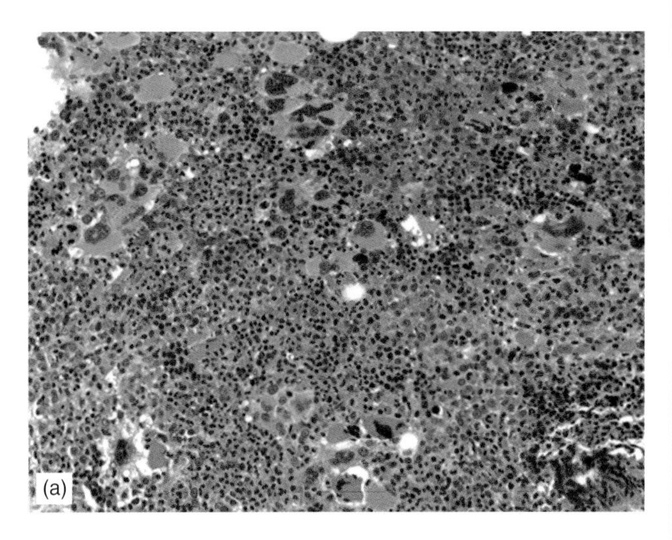

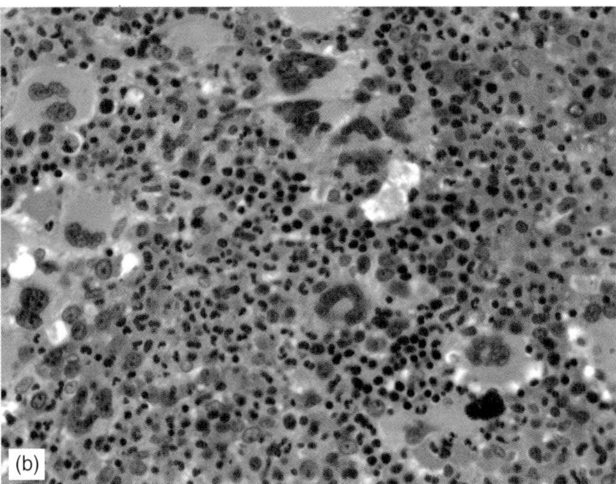

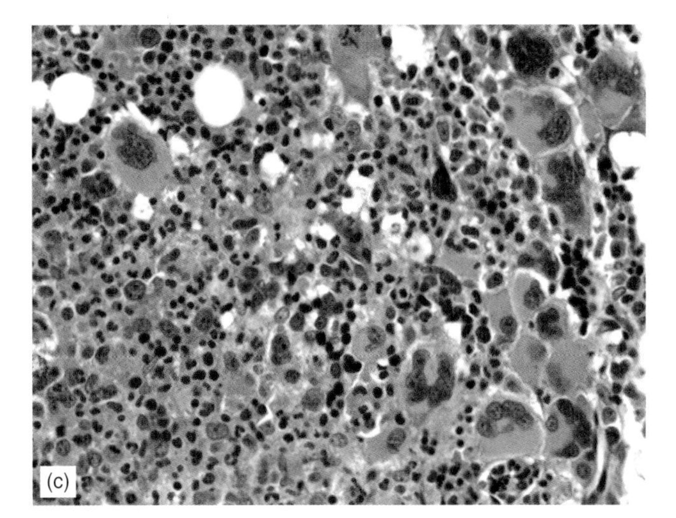

FIGURE 9.30 Bone marrow biopsy sections of a patient with ET demonstrating marked hypercellularity with clusters of megakaryocytes, including large and bizarre forms: (a) low power, (b), and (c) high power.

Molecular Studies

There are no specific molecular markers for ET, but the V617F mutation in *JAK2* has been found as in other MPD,

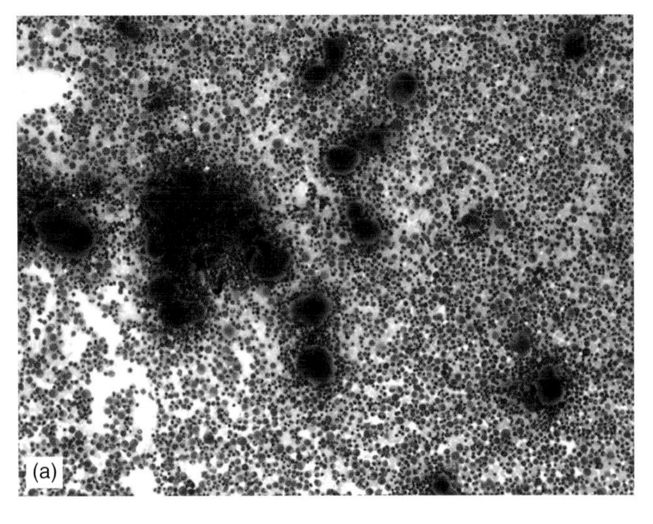

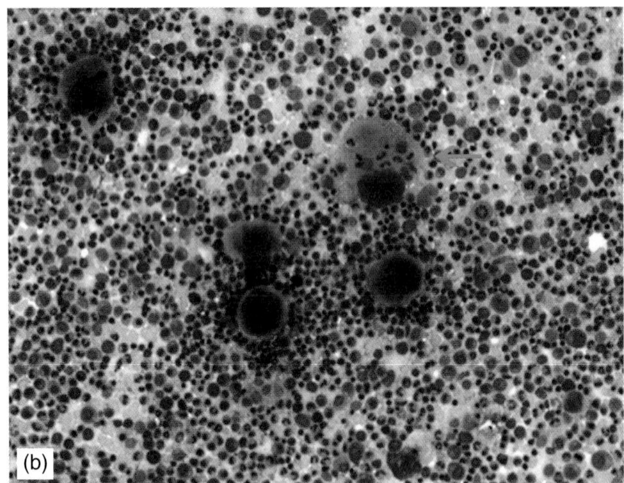

FIGURE 9.31 Bone marrow smears of a patient with ET demonstrating aggregates of megakaryocytes (a) low power and megakaryocytes with emperipolesis (b, arrow).

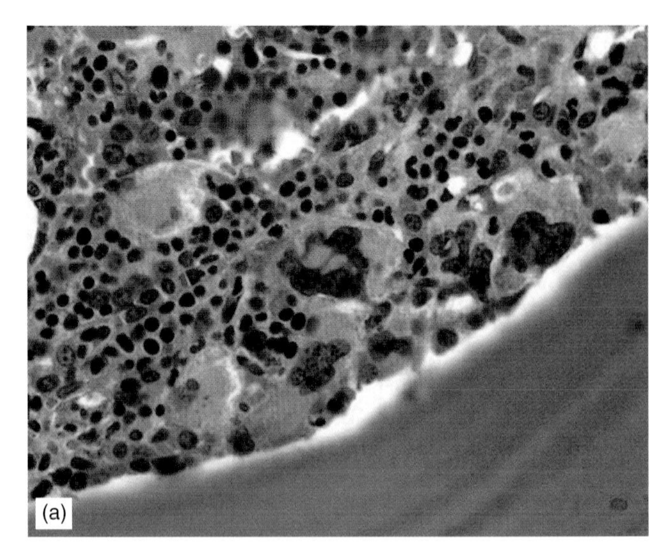

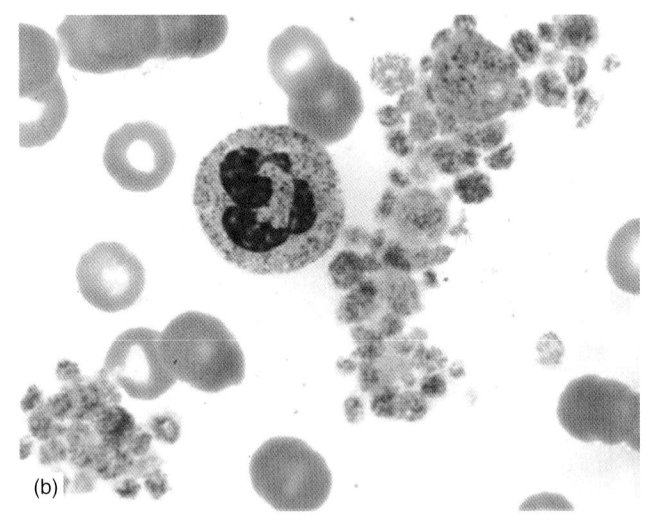

FIGURE 9.32 Bone marrow biopsy section of a patient with ET demonstrating a paratrabecular megakaryocytic aggregate (a) and platelet aggregates in blood smear (b).

though not as frequently as in PV, typically ranging from 30% to 60% [130, 131]. In the rare familial cases, sequencing of the TPO or TPO-receptor genes to search for inherited mutations may be useful [132].

Cytogenetic Results

No consistent chromosomal anomaly was associated with ET, but chromosomal abnormalities have been observed in about 5–7% of patients. Anomalies in ET may have developed as a consequence of therapy or as a *de novo* leukemic clone [15, 17]. However, one group detected a low percentage (<10%) of trisomy 8 and/or 9 in about 55% of their patients by FISH [18].

Clinical Aspects

The incidence of ET is approximately 2.5 per 100,000 with a median age of about 60 years. Women are affected more than men with a female to male ratio of about 2. Up to 50% of the patients are asymptomatic. The remaining patients may show splenomegaly (25–40%) and/or other clinical symptoms such as headache, lightheadedness, syncope, erythromelalgia, transient visual disturbances, and thrombohemorrhagic incidents (15–25%). A history of prior thrombosis has an adverse prognostic value. Some ET patients with markedly elevated platelet counts may show acquired von Willebrand deficiency with an increased tendency of bleeding. ET in approximately 2–4% of the patients may transform to PV, CIMF, or AML after a median follow-up of about 10 years [124, 125, 133–135].

Hydroxyurea, anagrelide, interferon, and pipobroman are the agents most frequently used for the treatment of ET to reduce the platelet count and the risk of thrombosis [124, 125].

DIFFERENTIAL DIAGNOSIS

At the beginning of this chapter we displayed a long list of morphologic findings shared by various types of CMPD

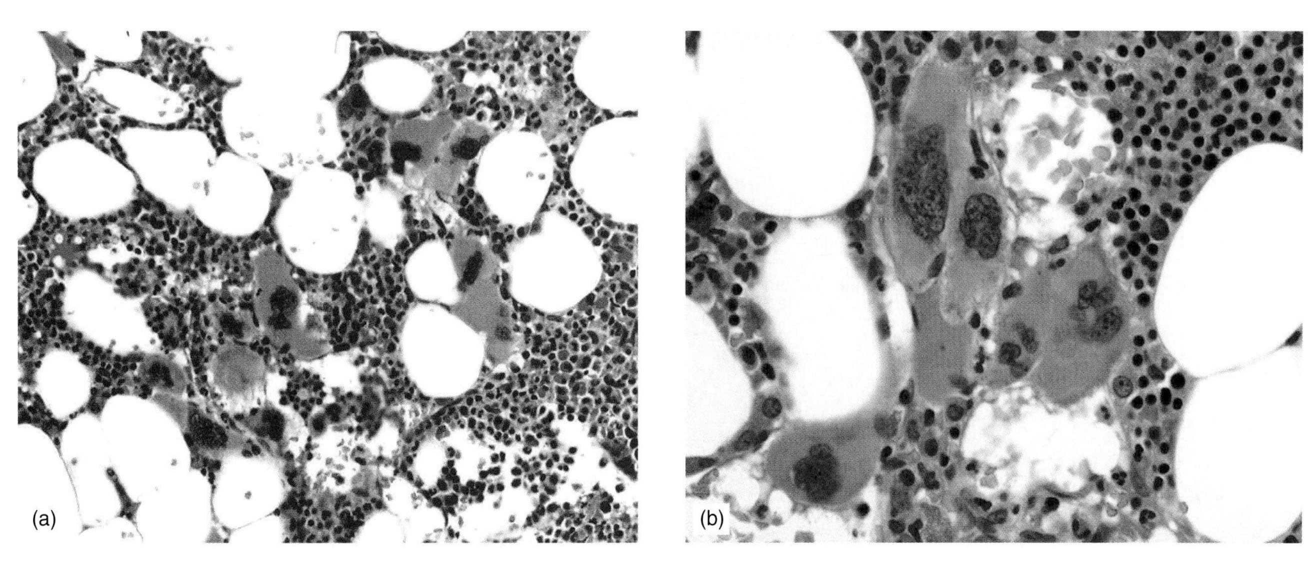

FIGURE 9.33 Bone marrow biopsy sections of a patient with ET demonstrating increased megakaryocytes, including large and bizarre forms: (a) low power and (b) high power.

TABLE 9.8 The major clinicopathologic features in chronic myeloproliferative disorders.*

Findings	PV	ET	CIMF	CML	CNL	CEL
Bone marrow						
Hypercellularity	+	±	Variable	+++	+	+
Increased myelopoiesis	+	−	+	+++	+++	+++
Increased erythropoiesis	+	−	±	−	−	−
Increased megakaryocytes	+	+	+	+	+	−
Fibrosis	Variable	Minimal	+ to +++	+	−	−
Osteosclerosis	−	−	+	+	−	−
Peripheral blood						
Increased WBC	+	+	++	+++	+++	++
Basophilia	±	±	+	++	−	−
Eosinophila	±	±	±	+	−	+++
RBC abnormalities	±	±	+++	±	−	−
Leukoerythroblastosis	−	−	+++	+	−	−
Thrombocytosis	+	+++	±	+	±	±
Abnormal platelets	±	+++	++	++	±	±
LAP (NAP) score	Elevated	Normal	Elevated	Low	Elevated	?
Splenomegaly	+	+	++++	+++	+	±
Cytogenetics	+8, +9, del(20q)	+8, +9	+8, +9,del(20q)	t(9;22)	+8, +9	t(5;12)
Molecular studies	*JAK2*	*JAK2*	*JAK2*	*BCR-ABL*	*JAK2*	*PDGFRA*

*Adapted from Ref. [3].

[1, 136, 137]. Diagnosis of several CMPD subtypes, such as CNL, CEL, and ET, are based on the exclusion of a wide variety of reactive conditions as well as other subtypes of CMPD (see Tables 9.4, 9.5, and 9.7). A summary of the distinctive features of CMPD subtypes is presented in Table 9.8). Following are pathologic features that may help to distinguish the four major different types of CMPD: PV, CML, CIMF, and ET.

Bone Marrow Findings

Fibrosis is the hallmark of the advanced stage of CIMF, while it is absent or minimal in ET. Other types of CMPD may show various degrees of fibrosis. Development of marrow fibrosis in CML may indicate an AP or blast transformation.

Megakaryocytes are commonly increased and show abnormal morphology in various subtypes of CMPD.

Dwarf forms or micromegakaryocytes are frequent in CML, giant megakaryocytes with frequent emperipolesis are common in ET, and dysplastic clusters of megakaryocytes are frequently found in the dilated sinusoids in early stages of CIMF.

Eosinophilia and **basophilia** are most common in CML and least frequent in ET and PV. Progressive increase in basophils in CML patients is suggestive of an AP.

Peripheral Blood Findings

Leukoerythroblastosis is considered the morphologic hallmark of CIMF, but it may also be seen in CML in chronic, accelerated, or blastic phases. Leukoerythroblastosis is not a feature of PV or ET.

Thrombocytosis is a frequent finding in all types of CMPD, but its persistent elevation ≥600,000/μL is more characteristic of ET. Giant platelets and megakaryocytic fragments have been observed more frequently in ET than in the others.

Granulocytosis is marked (>50,000/μL) and left-shifted with a significant number of myelocytes and metamyelocytes in CML. It is mild to moderate, but often left-shifted in CIMF, and less frequently in PV or ET.

Dacrocytes (tear-drop-shaped red blood cells) are the morphologic indicators of bone marrow fibrosis and, therefore, the most frequent finding in CIMF. Dacrocytosis is often associated with other red blood cell morphologic abnormalities, such as anisopoikilocytosis and the presence of schistocytes (fragmented red blood cells).

Molecular and Cytogenetics

At the molecular level, the finding of a *BCR-ABL1* fusion gene can help to separate CML from the other conditions. Conversely, detection of the *JAK2* mutation can distinguish PV, ET, and other non-CML MPDs from CML. Molecular methods that allow specific subtyping of the three *BCR-ABL1* isoforms (p190, p210, and p230) can assist in distinguishing ALL from the blast crisis of CML, as well as rare conditions such as CNL.

References

1. Jaffe ES, Harris NL, Stein H, Vardiman JW. (2001). *Pathology and Genetics. Tumors of Haematopoietic and Lymphoid Tissues*. IARC Press, Lyon.
2. His ED. (2007). *Hematopathology*. Churchill Livingstone, Philadelphia.
3. Foucar K. (2001). *Bone Marrow Pathology*, 2nd ed. ASCP Press, Chicago.
4. Naeim F. (1997). *Pathology of Bone Marrow*, 2nd ed. Williams & Wilkins, Baltimore.
5. Skoda R, Prchal JT. (2005). Chronic myeloproliferative disorders – Introduction. *Semin Hematol* **42**, 181–3.

6a. Murray J. (2006). Myeloproliferative disorders. *Clin Med* **5**, 328–32.

6b. Swerdlow SH, Campo E, Harris NL, Jaffe ES, Pileri SA, Stein H, Thiele J. *WHO Classification of Tumours of Haematopoietic and Lymphoid Tissues*, 4th ed., in press.

7. De Keersmaecker K, Cools J. (2006). Chronic myeloproliferative disorders: A tyrosine kinase tale. *Leukemia* **20**, 200–5.
8. Levine RL, Gilliland DG. (2007). JAK-2 mutations and their relevance to myeloproliferative disease. *Curr Opin Hematol* **14**, 43–7.
9. Lasho TL, Pardanani A, McClure RF, Mesa RA, Levine RL, Gilliland DG, Tefferi A. (2006). Concurrent MPL515 and JAK2V617F mutations in myelofibrosis: Chronology of clonal emergence and changes in mutant allele burden over time. *Br J Haematol* **135**, 683–7.
10. Haferlach T, Kern W, Schnittger S, Schoch C. (2004). Modern diagnostics in chronic myeloproliferative diseases (CMPDs). *Ann Hematol* **83**(Suppl 1), S59–61.
11. Adeyinka A, Dewald GW. (2003). Cytogenetics of chronic myeloproliferative disorders and related myelodysplastic syndromes. *Hematol Oncol Clin North Am* **17**, 1129–49.
12. Bacher U, Haferlach T, Kern W, Hiddemann W, Schnittger S, Schoch C. (2005). Conventional cytogenetics of myeloproliferative diseases other than CML contribute valid information. *Ann Hematol* **84**, 250–7.
13. Bain BJ. (1996). Eosinophilic leukaemias and the idiopathic hypereosinophilic syndrome. *Br J Haematol* **95**, 2–9.
14. Bench AJ, Pahl HL. (2005). Chromosomal abnormalities and molecular markers in myeloproliferative disorders. *Semin Hematol* **42**, 196–205.
15. Diez-Martin JL, Graham DL, Petitt RM, Dewald GW. (1991). Chromosome studies in 104 patients with polycythemia vera. *Mayo Clin Proc* **66**, 287–99.
16. Morrison-DeLap SJ, Kuffel DG, Dewald GW, Letendre L. (1986). Unbalanced 1;7 translocation and therapy-induced hematologic disorders: A possible relationship. *Am J Hematol* **21**, 39–47.
17. Hsiao HH, Ito Y, Sashida G, Ohyashiki JH, Ohyashiki K. (2005). *De novo* appearance of der(1;7)(q10;p10) is associated with leukemic transformation and unfavorable prognosis in essential thrombocythemia. *Leuk Res* **29**, 1247–52.
18. Elis A, Amiel A, Manor Y, Tangi I, Fejgin M, Lishner M. (1996). The detection of trisomies 8 and 9 in patients with essential thrombocytosis by fluorescence *in situ* hybridization. *Cancer Genet Cytogenet* **92**, 14–17.
19. Strasser-Weippl K, Steurer M, Kees M, Augustin F, Tzankov A, Dirnhofer S, Fiegl M, Simonitsch-Klupp I, Gisslinger H, Zojer N, Ludwig H. (2006). Prognostic relevance of cytogenetics determined by fluorescent *in situ* hybridization in patients having myelofibrosis with myeloid metaplasia. *Cancer* **107**, 2801–6.
20. Kozubek S, Lukasova E, Mareckova A, Skalnikova M, Kozubek M, Bartova E, Kroha V, Krahulcova E, Slotova J. (1999). The topological organization of chromosomes 9 and 22 in cell nuclei has a determinative role in the induction of t(9,22) translocations and in the pathogenesis of t(9,22) leukemias. *Chromosoma* **108**, 426–35.
21. Wu B, Zhou S, Song L, Liu X. (2002). Clinical significance of dual color-dual fusion translocation fluorescence *in situ* hybridization in the detection of bcr/abl fusion gene. *Zhonghua Zhong Liu Za Zhi* **24**, 364–6.
22. Lugo TG, Witte ON. (1989). The *BCR-ABL* oncogene transforms Rat-1 cells and cooperates with v-myc. *Mol Cell Biol* **9**, 1263–70.
23. Mitelman F. (1993). The cytogenetic scenario of chronic myeloid leukemia. *Leukocyte Lymphoma* **11**, 11–15.
24. Melo JV, Hughes TP, Apperley JF. (2003). Chronic myeloid leukemia. *Hematology Am Soc Hematol Educ Program*, 132–52.
25. O'Brien S, Tefferi A, Valent P. (2004). Chronic myelogenous leukemia and myeloproliferative disease. *Hematology Am Soc Hematol Educ Program*, 146–62.
26. Van Etten RA. (2007). Cellular and molecular biology of chronic myelogenous leukemia. *UpToDate.*
27. Rosas-Cabral A, Irigoyen L, Alvarado L, Vela-Ojeda J, Ayala-Sanchez M, Tripp-Villanueva F, Sanchez E, Gonzalez-Llaven J, Gariglio P. (2003). HLA CW3 and HLA CW4 have a protective effect on acquisition of chronic myeloid leukemia on Mexican patients. *Rev Invest Clin* **55**, 423–8.
28. Kuruvilla J, Gupta V, Gill KS, Lipton JH. (2003). Association of familial leukemia with HLA Cw3: Is it real? *Leuk Lymphoma* **44**, 309–11.
29. Kajtar B, Deak L, Kalasz V, Pajor L, Molnar L, Mehes G. (2005). Multiple constitutional chromosome translocations of familial nature

in Philadelphia chromosome-positive chronic myeloid leukemia: A report on a unique case. *Int J Hematol* **82**, 347–50.
30. Skoda R, Prchal JT. (2005). Lessons from familial myeloproliferative disorders. *Semin Hematol* **42**, 266–73.
31. Khonglah Y, Basu D, Dutta TK. (2002). Bone marrow trephine biopsy findings in chronic myeloid leukemia. *Malays J Pathol* **24**, 37–43.
32. Thiele J, Kvasnicka HM, Orazi A. (2005). Bone marrow histopathology in myeloproliferative disorders – current diagnostic approach. *Semin Hematol* **42**, 184–95.
33. Enriquez P, Neiman R. (1976). *The Pathology of Spleen. A Fundamental Approach*. ASCP Press, Chicago.
34. Giles FJ, Cortes JE, Kantarjian HM, O'Brien SM. (2004). Accelerated and blastic phases of chronic myelogenous leukemia. *Hematol Oncol Clin North Am* **18**, 753–74.
35. Ilaria Jr. RL. (2005). Pathobiology of lymphoid and myeloid blast crisis and management issues. *Hematology Am Soc Hematol Educ Program*, 188–94.
36. Calabretta B, Perrotti D. (2004). The biology of CML blast crisis. *Blood* **103**, 4010–22.
37. Normann AP, Egeland T, Madshus IH, Heim S, Tjønnfjord GE. (2003). CD7 expression by CD34+ cells in CML patients, of prognostic significance? *Eur J Haematol* **71**, 266–75.
38. Tripathi AK, Chaturvedi R, Ahmad R, Asim M, Sawlani KK, Singh MK, Tripathi P, Tekwani BL. (2003). Flow cytometric analysis of aneuploidy and S-phase fraction in chronic myeloid leukemia patients: Role in early detection of accelerated phase. *Leuk Res* **27**, 899–902.
39. Yen CC, Liu JH, Wang WS, Fan FS, Chiou TJ, Tai CJ, Yang MH, Chao TC, Hsiao LT, Chen PM. (2000). Immunophenotypic and genotypic characteristics of chronic myelogenous leukemia in blast crisis. *Zhonghua Yi Xue Za Zhi* Taipei **63**, 785–91.
40. Murase T, Suzuki R, Tashiro K, Morishima Y, Nakamura S. (1999). Blast crisis of chronic myelogenous leukemia exhibiting immunophenotypic features of a myeloid/natural killer cell precursor. *Int J Hematol* **69**, 89–91.
41. Van Etten RA. (2006). Molecular genetics of chronic myelogenous leukemia. *UpToDate*.
42. Van Rhee F, Hochhaus A, Lin F, Melo JV, Goldman JM, Cross NC. (1996). P190 *BCR-ABL* mRNA is expressed at low levels in p210-positive chronic myeloid and acute lymphoblastic leukemias. *Blood* **87**, 5213–17.
43. Nashed AL, Rao KW, Gulley ML. (2003). Clinical applications of *BCR-ABL* molecular testing in acute leukemia. *J Mol Diagn* **5**, 63–72.
44. Branford S, Hughes TP, Rudzki Z. (1999). Monitoring chronic myeloid leukemia therapy by real-time quantitative PCR in blood is a reliable alternative to marrow cytogenetics. *Br J Haematol* **107**, 587–99.
45. Biernaux C, Loos M, Sels A, Huez G, Strychmans P. (1995). Detection of major bcr-abl gene expression at a very low level in blood cells of some healthy individuals. *Blood* **86**, 3118–22.
46. Barbany G, Hoglund M, Simonsson B. (2002). Complete molecular remission in chronic myelogenous leukemia after imatinib mesylate therapy. *N Engl J Med* **347**, 539–40.
47. Shah NP, Nicoll JM, Nagar B, Gorre ME, Paquette RL, Kuriyan J, Sawyers CL. (2002). Multiple *BCR-ABL* kinase domain mutations confer polyclonal resistance to the tyrosine kinase inhibitor imatinib mesylate (STI571) in chronic phase and blast crisis chronic myeloid leukemia. *Cancer Cell* **2**, 117–25.
48. Nowell PC. (2007). Discovery of the Philadelphia chromosome: A personal perspective. *J Clin Invest* **117**, 2033–5.
49. Rowley JD. (1973). A new consistent chromosomal abnormality in chronic myelogenous leukaemia identified by quinacrine fluorescence and Giemsa staining. *Nature* **243**, 290–3.
50. Shtivelman E, Lifshitz B, Gale RP, Canaani E. (1985). Fused transcript of abl and bcr genes in chronic myelogenous leukaemia. *Nature* **315**, 550–4.
51. Lugo TG, Pendergast AM, Muller AJ, Witte ON. (1990). Tyrosine kinase activity and transformation potency of bcr-abl oncogene products. *Science* **247**, 1079–82.
52. Daley GQ, Van Etten RA, Baltimore D. (1990). Induction of chronic myelogenous leukemia in mice by the P210bcr/abl gene of the Philadelphia chromosome. *Science* **247**, 824–30.
53. Lee DS, Lee YS, Yun YS, Kim YR, Jeong SS, Lee YK, She CJ, Yoon SS, Shin HR, Kim Y, Cho HI. (2003). A study on the incidence of ABL gene deletion on derivative chromosome 9 in chronic myelogenous leukemia by interphase fluorescence *in situ* hybridization and its association with disease progression. *Genes Chromosomes Cancer* **37**, 291–9.
54. Rumpel M, Friedrich T, Deininger MW. (2003). Imatinib normalizes bone marrow vascularity in patients with chronic myeloid leukemia in first chronic phase. *Blood* **101**, 4641–3.
55. Cortes J, O'Dwyer ME. (2004). Clonal evolution in chronic myelogenous leukemia. *Hematol Oncol Clin North Am* **18**, 671–84.
56. Dorkeld F, Bernheim A, Dessen P, Huret JL. (1999). A database on cytogenetics in haematology and oncology. *Nucleic Acids Res* **27**, 353–4.
57. Gordon MY, Goldman JM. (1996). Cellular and molecular mechanisms in chronic myeloid leukaemia: Biology and treatment. *Br J Haematol* **95**, 10–20.
58. Faderl S, Talpaz M, Estrov Z, O'Brien S, Kurzrock R, Kantarjian HM. (1999). The biology of chronic myeloid leukemia. *N Engl J Med* **341**, 164–72.
59. Nakamura Y, Hirosawa S, Aoki N. (1993). Consistent involvement of the 3′ half part of the first BCR intron in adult Philadelphia-positive leukaemia without M-bcr rearrangement. *Br J Haematol* **83**, 53–7.
60. Nagar B, Bornmann WG, Pellicena P, Schindler T, Veach DR, Miller WT, Clarkson B, Kuriyan J. (2002). Crystal structures of the kinase domain of c-Abl in complex with the small molecule inhibitors PD173955 and imatinib (STI-571). *Cancer Res* **62**, 4236–43.
61. Illmer T, Schaich M, Platzbecker U, Freiberg-Richter J, Oelschlagel U, von Bonin M, Pursche S, Bergemann T, Ehninger G, Schleyer E. (2004). P-glycoprotein-mediated drug efflux is a resistance mechanism of chronic myelogenous leukemia cells to treatment with imatinib mesylate. *Leukemia* **18**, 401–8.
62. Gorre ME, Mohammed M, Ellwood K, Hsu N, Paquette R, Rao PN, Sawyers CL. (2001). Clinical resistance to STI-571 cancer therapy caused by *BCR-ABL* gene mutation or amplification. *Science* **293**, 876–80.
63. Hochhaus A, Kreil S, Corbin AS, La Rosee P, Muller MC, Lahaye T, Hanfstein B, Schoch C, Cross NC, Berger U, Gschaidmeier H, Druker BJ, Hehlmann R. (2002). Molecular and chromosomal mechanisms of resistance to imatinib (STI571) therapy. *Leukemia* **16**, 2190–6.
64. Kumar L. (2006). Chronic myelogenous leukaemia (CML): An update. *Natl Med J India* **19**, 255–63.
65. Maness LJ, McSweeney PA. (2004). Treatment options for newly diagnosed patients with chronic myeloid leukemia. *Curr Hematol Rep* **3**, 54–61.
66. Elliott MA, Hanson CA, Dewald GW, Smoley SA, Lasho TL, Tefferi A. (2005). WHO-defined chronic neutrophilic leukemia: A long-term analysis of 12 cases and a critical review of the literature. *Leukemia* **19**, 313–17.
67. Reilly JT. (2002). Chronic neutrophilic leukaemia: A distinct clinical entity? *Br J Haematol* **116**, 10–18.
68. Bohm J, Schaefer HE. (2002). Chronic neutrophilic leukaemia: 14 new cases of an uncommon myeloproliferative disease. *J Clin Pathol* **55**, 862–4.
69. Dincol G, Nalcaci M, Dogan O, Aktan M, Kucukkaya R, Agan M, Dincol K. (2002). Coexistence of chronic neutrophilic leukemia with multiple myeloma. *Leuk Lymphoma* **43**, 649–51.
70. Bohm J, Kock S, Schaefer HE, Fisch P. (2003). Evidence of clonality in chronic neutrophilic leukaemia. *J Clin Pathol* **56**, 292–5.
71. Katsuki K, Shinohara K, Takeda K, Ariyoshi K, Yamada T, Kameda N, Takahashi T, Nawata R, Shibata S, Asano Y, Okamura S. (2000). Chronic neutrophilic leukemia with acute myeloblastic transformation. *Jpn J Clin Oncol* **30**, 362–5.
72. Mc Lornan DP, Percy MJ, Jones AV, Cross NC, Mc Mullin MF. (2005). Chronic neutrophilic leukemia with an associated V617F JAK2 tyrosine kinase mutation. *Haematologica* **90**, 1696–7.

73. Elliott MA, Dewald GW, Tefferi A, Hanson CA. (2001). Chronic neutrophilic leukemia (CNL): A clinical, pathologic and cytogenetic study. *Leukemia* **15**, 35–40.
74. Choi IK, Kim BS, Lee KA, Ryu S, Seo HY, Sul H, Choi JG, Sung HJ, Park KH, Yoon SY, Oh SC, Seo JH, Choi CW, Shin SW, Yoon SY, Cho Y, Kim YK, Kim YH, Kim JS. (2004). Efficacy of imatinib mesylate (STI571) in chronic neutrophilic leukemia with t(15;19): Case report. *Am J Hematol* **77**, 366–9.
75. Elliott MA. (2004). Chronic neutrophilic leukemia: A contemporary review. *Curr Hematol Rep* **3**, 210–17.
76. Billio A, Venturi R, Morello E, Rosanelli C, Pescosta N, Coser P. (2001). Chronic neutrophilic leukemia evolving from polycythemia vera with multiple chromosome rearrangements: A case report. *Haematologica* **86**, 1225–6.
77. Gotlib J. (2005). Molecular classification and pathogenesis of eosinophilic disorders: 2005 update. *Acta Haematol* **114**, 7–25.
78. Roufosse F, Goldman M, Cogan E. (2006). Hypereosinophilic syndrome: Lymphoproliferative and myeloproliferative variants. *Semin Respir Crit Care Med* **27**, 158–70.
79. Abramson N. (2006). Chronic eosinophilic leukemia/hypereosinophilic syndrome and acute leukemia. *J Clin Oncol* **24**, 1647.
80. Gleich GJ, Leiferman KM. (2005). The hypereosinophilic syndromes: Still more heterogeneity. *Curr Opin Immunol* **17**, 679–84.
81. Wilkins HJ, Crane MM, Copeland K, Williams WV. (2005). Hypereosinophilic syndrome: An update. *Am J Hematol* **80**, 148–57.
82. Tanaka Y, Kurata M, Togami K, Fujita H, Watanabe N, Matsushita A, Maeda A, Nagai K, Sada A, Matsui T, Takahashi T. (2006). Chronic eosinophilic leukemia with the FIP1L1-PDGFRalpha fusion gene in a patient with a history of combination chemotherapy. *Int J Hematol* **83**, 152–5.
83. Cools J, Stover EH, Gilliland DG. (2005). Detection of the FIP1L1-PDGFRA fusion in idiopathic hypereosinophilic syndrome and chronic eosinophilic leukemia. *Meth Mol Med* **125**, 177–87.
84. La Starza R, Specchia G, Cuneo A, Beacci D, Nozzoli C, Luciano L, Aventin A, Sambani C, Testoni N, Foppoli M, Invernizzi R, Marynen P, Martelli MF, Mecucci C. (2005). The hypereosinophilic syndrome: Fluorescence *in situ* hybridization detects the del(4)(q12)-FIP1L1/ PDGFRA but not genomic rearrangements of other tyrosine kinases. *Haematologica* **90**, 596–601.
85. Roche-Lestienne C, Lepers S, Soenen-Cornu V, Kahn JE, Lai JL, Hachulla E, Drupt F, Demarty AL, Roumier AS, Gardembas M, Dib M, Philippe N, Cambier N, Barete S, Libersa C, Bletry O, Hatron PY, Quesnel B, Rose C, Maloum K, Blanchet O, Fenaux P, Prin L, Preudhomme C. (2005). Molecular characterization of the idiopathic hypereosinophilic syndrome (HES) in 35 French patients with normal conventional cytogenetics. *Leukemia* **19**, 792–8.
86. Reiter A, Walz C, Watmore A, Schoch C, Blau I, Schlegelberger B, Berger U, Telford N, Aruliah S, Yin JA, Vanstraelen D, Barker HF, Taylor PC, O'Driscoll A, Benedetti F, Rudolph C, Kolb HJ, Hochhaus A, Hehlmann R, Chase A, Cross NC. (2005). The t(8;9)(p22;p24) is a recurrent abnormality in chronic and acute leukemia that fuses PCM1 to JAK2. *Cancer Res* **65**, 2662–7.
87. Flaum MA, Schooley RT, Fauci AS, Gralnick HR. (1981). Clinicopathologic correlation of the idiopathic hypereosinophilic syndrome. I. Hematologic manifestations. *Blood* **58**, 1012–20.
88. Schooley RT, Flaum MA, Gralnick HR, Fauci AS. (1981). A clinicopathologic correlation of the idiopathic hypereosinophilic syndrome. II. Clinical manifestations. *Blood* **58**, 1021–6.
89. Azuma M, Nakamura Y, Sano T, Okano Y, Sone S. (2001). Adhesion molecule expression on eosinophils in idiopathic eosinophilic pneumonia. *Eur Respir J* **9**, 2494–500.
90. Berki T, David M, Bone B, Losonczy H, Vass J, Nemeth P. (2001). New diagnostic tool for differentiation of idiopathic hypereosinophilic syndrome (HES) and secondary eosinophilic states. *Pathol Oncol Res* **7**, 292–7.
91. Cools J, Stover EH, Boulton CL, Gotlib J, Legare RD, Amaral SM, Curley DP, Duclos N, Rowan R, Kutok JL, Lee BH, Williams IR, Coutre SE, Stone RM, DeAngelo DJ, Marynen P, Manley PW, Meyer T, Fabbro D, Neuberg D, Weisberg E, Griffin JD, Gilliland DG. (2003). PKC412 overcomes resistance to imatinib in a murine model of FIP1L1-PDGFR-alpha-induced myeloproliferative disease. *Cancer Cell* **3**, 459–69.
92. Oliver JW, Deol I, Morgan DL, Tonk VS. (1998). Chronic eosinophilic leukemia and hypereosinophilic syndromes. Proposal for classification, literature review, and report of a case with a unique chromosomal abnormality. *Cancer Genet Cytogenet* **107**, 111–17.
93. Pardanani A, Ketterling RP, Brockman SR, Flynn HC, Paternoster SF, Shearer BM, Reeder TL, Li CY, Cross NC, Cools J, Gilliland DG, Dewald GW, Tefferi A. (2003). CHIC2 deletion, a surrogate for FIP1L1-PDGFRA fusion, occurs in systemic mastocytosis associated with eosinophilia and predicts response to imatinib mesylate therapy. *Blood* **102**, 3093–6.
94. Gotlib J, Cools J, Malone III JM, Schrier SL, Gilliland DG, Coutre SE. (2004). The FIP1L1-PDGFRalpha fusion tyrosine kinase in hypereosinophilic syndrome and chronic eosinophilic leukemia: Implications for diagnosis, classification, and management. *Blood* **103**, 2879–91.
95. Thiele JM, Kvasnicka HM. (2005). Diagnosis of polycythemia vera based on bone marrow pathology. *Curr Hematol Rep* **4**, 218–23.
96. Gregg XT, Prchal JT. (2005). Recent advances in the molecular biology of congenital polycythemias and polycythemia vera. *Curr Hematol Rep* **4**, 238–42.
97. Thiele J, Kvasnicka HM. (2005). Diagnostic impact of bone marrow histopathology in polycythemia vera (PV). *Histol Histopathol* **20**, 317–28.
98. James C, Ugo V, Casadevall N, Constantinescu SN, Vainchenker W. (2005). A JAK2 mutation in myeloproliferative disorders: Pathogenesis and therapeutic and scientific prospects. *Trends Mol Med* **11**, 546–54.
99. Murugesan G, Aboudola S, Szpurka H, Verbic MA, Maciejewski JP, Tubbs RR, His ED. (2006). Identification of the JAK2 V617F mutation in chronic myeloproliferative disorders using FRET probes and melting curve analysis. *Am J Clin Pathol* **125**, 625–33.
100. Grainer TC. (2006). Diagnostic assays for the JAK2 V617F mutation in chronic myeloproliferative disorders. *Am J Clin Pathol* **125**, 651–3.
101. Moliterno AR, Hankins WD, Spivak JL. (1998). Impaired expression of the thrombopoietin receptor by platelets from patients with polycythemia vera. *N Engl J Med* **338**, 572.
102. Tefferi A, Meyer RG, Wyatt WA, Dewald GW. (2001). Comparison of peripheral blood interphase cytogenetics with bone marrow karyotype analysis in myelofibrosis with myeloid metaplasia. *Br J Haematol* **115**, 316–19.
103. Dingli D, Schwager SM, Mesa RA, Li CY, Dewald GW, Tefferi A. (2006). Presence of unfavorable cytogenetic abnormalities is the strongest predictor of poor survival in secondary myelofibrosis. *Cancer* **106**, 1985–9.
104. Streiff MB, Smith B, Spivak JL. (2002). The diagnosis and management of polycythemia vera in the era since the Polycythemia Vera Study Group: A survey of American Society of Hematology members' practice patterns. *Blood* **99**, 1144–9.
105. Tefferi A, Spivak JL. (2005). Polycythemia vera: Scientific advances and current practice. *Semin Hematol* **42**, 206–20.
106. Stuart BJ, Viera AJ. (2004). Polycythemia vera. *Am Fam Physician* **69**, 2139–44.
107. Tefferi A. (2007). Diagnostic approach to the patient with polycythemia. *UpToDate*
108. Barosi G, Hoffman R. (2005). Idiopathic myelofibrosis. *Semin Hematol* **42**, 248–58.
109. Tefferi A. (2006). New insights into the pathogenesis and drug treatment of myelofibrosis. *Curr Opin Hematol* **13**, 87–92.
110. Tefferi A. (2000). Myelofibrosis with myeloid metaplasia. *N Engl J Med* **342**, 1255–65.
111. Jacobson RJ, Salo A, Fialkow PJ. (1978). Agnogenic myeloid metaplasia: A clonal proliferation of hematopoietic stem cells with secondary myelofibrosis. *Blood* **51**, 189–94.
112. Reilly JT, Snowden JA, Spearing RL, Fitzgerald PM, Jones N, Watmore A, Potter A. (1997). Cytogenetic abnormalities and their

prognostic significance in idiopathic myelofibrosis: A study of 106 cases. *Br J Haematol* **98**, 96–102.
113. Buschle M, Janssen JW, Drexler H, Lyons J, Anger B, Bartram CR. (1988). Evidence for pluripotent stem cell origin of idiopathic myelofibrosis: Clonal analysis of a case characterized by a N-ras gene mutation. *Leukemia* **2**, 658–60.
114. Tefferi A. (2007). Clinical manifestations and diagnosis of agnogenic myeloid metaplasia (chronic idiopathic myelofibrosis). *UpToDate.*
115. Ahmed A, Chang CC. (2006). Chronic idiopathic myelofibrosis: Clinicopathologic features, pathogenesis, and prognosis. *Arch Pathol Lab Med* **130**, 1133–43.
116. Schmitt A, Jouault H, Guichard J, Wendling F, Drouin A, Cramer EM. (2000). Pathologic interaction between megakaryocytes and polymorphonuclear leukocytes in myelofibrosis. *Blood* **96**, 1342–7.
117. Tiele J, Kvasnicka HM. (2005). Hematopathologic findings in chronic idiopathic myelofibrosis. *Semin Oncol* **32**, 380–94.
118. Tefferi A, Lasho TL, Schwager SM, Steensma DP, Mesa RA, Li CY, Wadleigh M, Gary Gilliland D. (2005). The JAK2(V617F) tyrosine kinase mutation in myelofibrosis with myeloid metaplasia: Lineage specificity and clinical correlates. *Br J Haematol* **131**, 320–8.
119. Djordjevic V, Stankovic M, Nikolic A, Antonijevic N, Rakicevic LJ, Divac A, Radojkovic M. (2006). PCR amplification on whole blood samples treated with different commonly used anticoagulants. *Pediatr Hematol Oncol* **23**, 517–21.
120. Demory JL, Dupriez B, Fenaux P, Lai JL, Beuscart R, Jouet JP, Deminatti M, Bauters F. (1988). Cytogenetic studies and their prognostic significance in agnogenic myeloid metaplasia: A report on 47 cases. *Blood* **72**, 855–9.
121. Cervantes F, Barosi G. (2005). Myelofibrosis with myeloid metaplasia: Diagnosis, prognostic factors, and staging. *Semin Oncol* **32**, 395–402.
122. Odenike O, Tefferi A. (2005). Conventional and new treatment options for myelofibrosis with myeloid metaplasia. *Semin Oncol* **32**, 422–31.
123. Tefferi A. (2006). Essential thrombocythemia: Scientific advances and current practice. *Curr Opin Hematol* **13**, 93–8.
124. Harrison CN. (2002). Current trends in essential thrombocythaemia. *Br J Haematol* **117**, 796–808.
125. Tefferi A. (2007). Diagnosis and clinical manifestations of essential theombocythemia. *UpToDate.*
126. Martini M, Teofili L, Larocca LM. (2006). Overexpression of PRV-1 gene in polycythemia rubra vera and essential thrombocythemia. *Meth Mol Med* **125**, 265–73.
127. Wolanskyj AP, Lasho TL, Schwager SM, McClure RF, Wadleigh M, Lee SJ, Gilliland DG, Tefferi A. (2005). JAK2 mutation in essential thrombocythaemia: Clinical associations and long-term prognostic relevance. *Br J Haematol* **131**, 208–13.
128. Fialkow PJ, Faguet GB, Jacobson RJ, Vaidya K, Murphy S. (1981). Evidence that essential thrombocythemia is a clonal disorder with origin in a multipotent stem cell. *Blood* **58**, 916–19.
129. Harrison CN, Gale RE, Machin SJ, Linch DC. (1999). A large proportion of patients with a diagnosis of essential thrombocythemia do not have a clonal disorder and may be at lower risk of thrombotic complications. *Blood* **93**, 417–24.
130. Wilkins BS, Erber WN, Bareford D, Buck G, Wheatley K, East CL, Paul B, Harrison CN, Green AR, Campbell PJ. (2008). Bone marrow pathology in essential thrombocythemia: Interobserver reliability and utility for identifying disease subtypes. *Blood* **111**, 60–70.
131. Wolanskyj AP, Lasho TL, Schwager SM, McClure RF, Wadleigh M, Lee SJ, Gilliland DG, Tefferi A. (2005). JAK2 V617F mutation in essential thrombocythemia: Clinical association and long-term prognostic relevance. *Br J Haematol* **131**, 320–8.
132. Randi ML, Fabris F, Vio C, Fiorlami A. (1987). Familial thrombocythemia and/or thrombocytosis: Apparently a rare disorder. *Acta Haematol* **78**, 63.
133. Schafer AI. (2004). Thrombocytosis. *N Engl J Med* **350**, 1211–19.
134. Finazzi G, Harrison C. (2005). Essential thrombocythemia. *Semin Hematol* **42**, 230–8.
135. Tsimberidou AM, Giles FJ. (2002). Essential thrombocythemia (ET): Moving from palliation to cure. *Hematology* **7**, 315–23.
136. Tefferi A, Elliott MA, Pardanani A. (2006). Atypical myeloproliferative disorders: Diagnosis and management. *Mayo Clin Proc* **81**, 553–63.
137. Michiels JJ. (2004). Bone marrow histopathology and biological markers as specific clues to the differential diagnosis of essential thrombocythemia, polycythemia vera and prefibrotic or fibrotic agnogenic myeloid metaplasia. *Hematol J* **5**, 93–102.

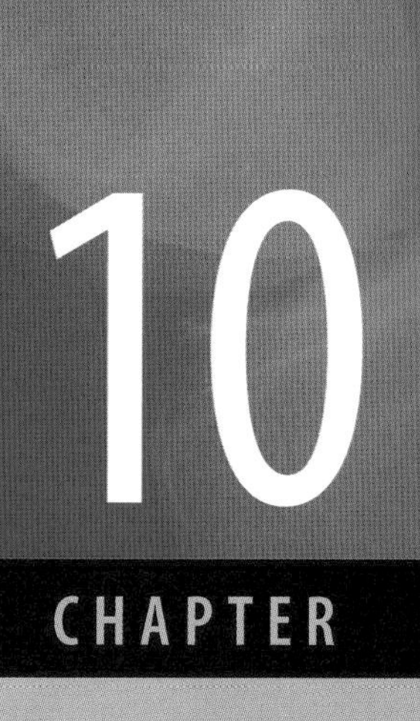

10 CHAPTER

Myelodysplastic/ Myeloproliferative Diseases

Faramarz Naeim and P. Nagesh Rao

The myelodysplastic/myeloproliferative disorders (MDS/MPD) are a group of hematologic disorders distinguished by clonal expansion of abnormal hematopoietic stem cells that share clinicopathological features of both MDS and chronic myeloproliferative disorders (CMPD) [1–4]. They are characterized by hypercellular bone marrows with excessive terminal proliferation of one or more hematopoietic lineages as well as dysplastic changes. This combination of proliferation and dysplasia may lead to increased production of one or more lineages (cytosis) and decreased production of other lineages simultaneously. Myeloid preponderance and left shift are common bone marrow findings, but blast cells are <20% in the bone marrow and/or the peripheral blood samples.

According to the WHO classification, these disorders are divided into the following major groups [1]:

- Chronic myelomonocytic leukemia
- Atypical chronic myeloid leukemia
- Juvenile myelomonocytic leukemia
- MDS/MPD, unclassifiable.

CHRONIC MYELOMONOCYTIC LEUKEMIA

Chronic myelomonocytic leukemia (CMML) is a clonal hematopoietic disorder characterized by both dysplastic and proliferative features including persistent monocytosis (>1,000/μL) of at least 3 months, bone marrow hypercellularity with myeloid preponderance, myelomonocytic dysplasia and left shift with <20% blasts (including promonocytes) in the peripheral blood or bone marrow. Cytogenetic and molecular studies are negative for *Ph*[1] and/or *BCR/ABL1* fusion gene (Table 10.1) [1–5].

Etiology and Pathogenesis

The etiology of CMML is not known. Most CMML cases lack cytogenetic or molecular changes that could explain the pathophysiology of this disorder. A small proportion of patients with CMML (about 2–3%), particularly

TABLE 10.1 The WHO proposed criteria for the diagnosis of chronic myelomonocytic leukemia.*

1.	Peripheral blood monocytosis >1,000/μL.
2.	Presence of myeloid blasts and promonocytes in the peripheral blood and/or the bone marrow; <20% of the differential counts.
3.	No *Ph*[1] or *BCR/ABL* fusion gene.
4.	Dysplastic myelopoiesis. If myelodysplasia is minimal or absent: (a) presence of an acquired clonal cytogenetic abnormality, or (b) persistent monocytosis for at least 3 months, and (c) exclusion of all other causes of monocytosis.

*From Ref. [1].

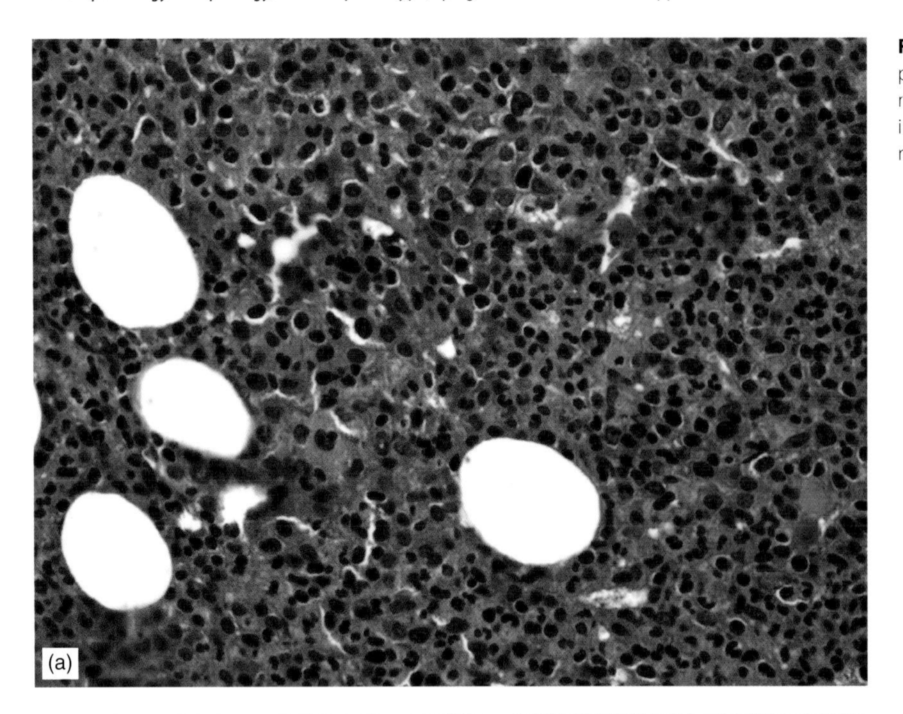

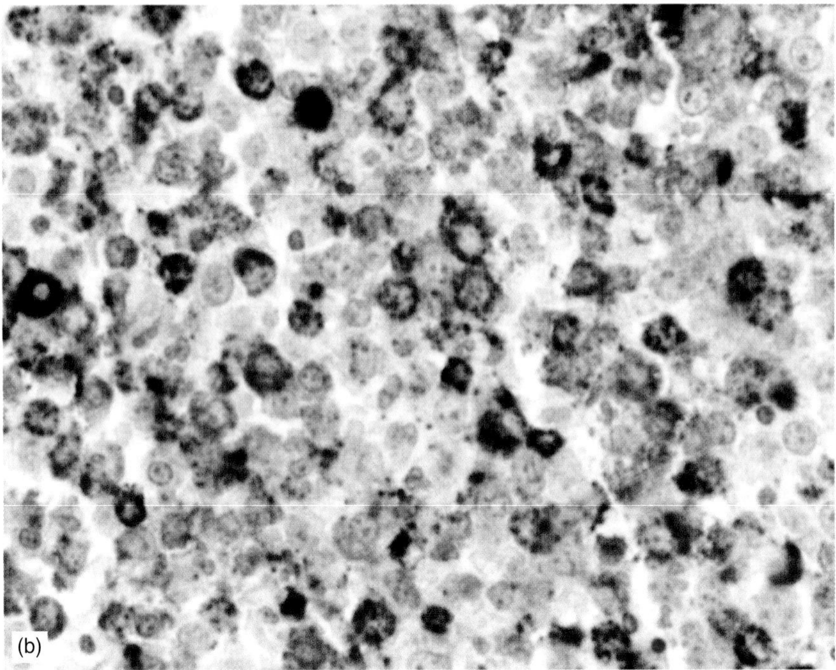

FIGURE 10.1 Bone marrow biopsy section from a patient with CMML demonstrating a hypercellular marrow with myeloid preponderance (a). The immunohistochemical stain for CD68 shows numerous positive monocytic/histiocytic cells (b).

in cases with eosinophilia, demonstrate t(5;12)(q33;p13) [6]. This translocation leads to the fusion of the *TEL* and platelet-derived growth factor receptor (*PDGFβR*) genes. *K-RAS* and *N-RAS* mutations are reported in approximately 50% of CMML cases [7, 8]. Elevated levels of survivin (inhibitor of apoptosis) have been detected in patients with CMML [9]. A spontaneous *in vitro* GM colony formation has been frequently observed, and there is evidence of supra-normal granulocyte/macrophage colony forming units (CFU-GM) in CMML patients [5].

Pathology

Morphology

The bone marrow biopsy sections are mostly (>75%) hypercellular and display myeloid preponderance and left shift with increased number of immature myelomonocytic precursors (Figure 10.1) [1–3, 10–12]. There are reports of relatively frequent monocytic nodules in the bone marrow biopsy sections in patients with CMML (Figure 10.2) [13, 14]. Bone marrow fibrosis is reported in about one-third of the patients.

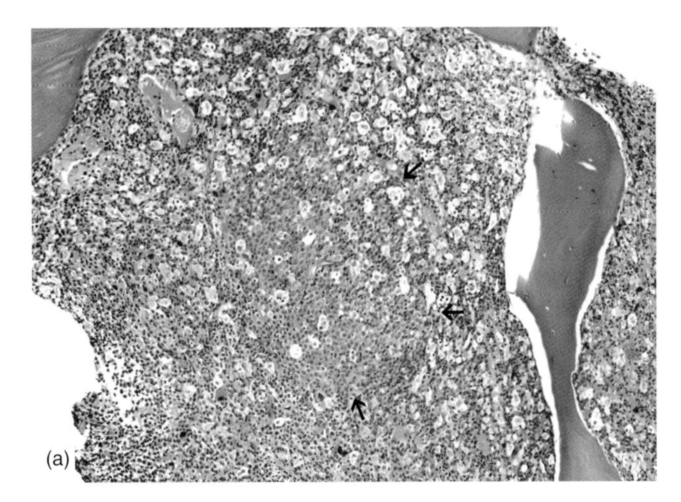

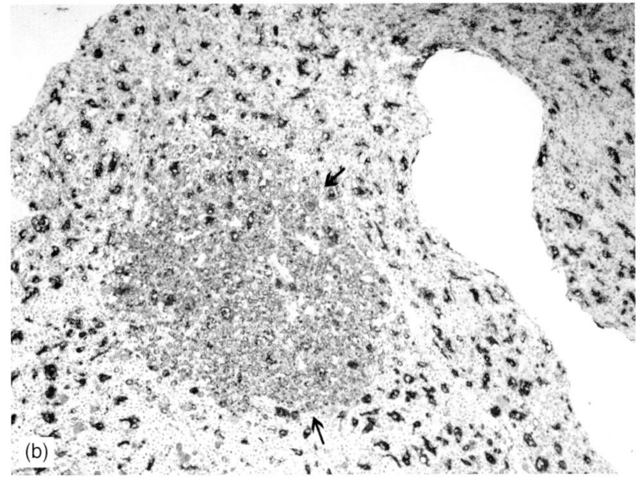

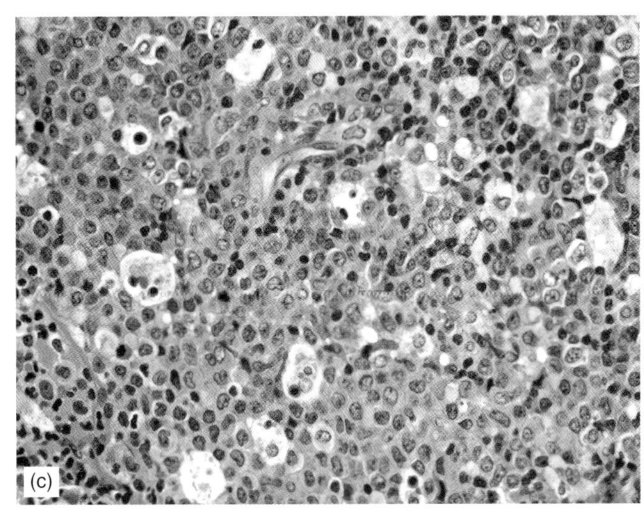

FIGURE 10.2 Monocytic nodules have been observed in bone marrow biopsy sections of patients with CMML. (a and b) represent low and high power fields (H&E stains). Arrow is pointed toward the monocytic nodule (b). The monocytic nodule is highlighted by immunohistochemical stain for CD68 (c, arrow). Courtesy of Sophie Song, M.D., Ph.D., Department of Pathology, UCLA.

The morphologic identification of monocytic precursors is much easier in bone marrow aspirate smears than in biopsy sections. The bone marrow smears often show dysplastic changes in the monocytic and the granulocytic series, such as bizarre morphology, nuclear hyper- or hyposegmentation, and cytoplasmic hypo- or hypergranularity (Figure 10.3). The total number of myeloblasts, monoblasts, and promonocytes is <20%. Auer rods may be detected in some cases. Erythroid dysplasia such as megaloblastic changes, irregular nuclei, nuclear fragments, and ringed sideroblasts are frequently observed. Micromegakaryocytes are often present. Eosinophilia is observed in a small proportion of CMML cases, particularly in association with t(5;12)(q33;p13) [1, 6].

The peripheral blood reveals monocytosis of >1,000/μL. In most instances the monocyte count is between 1,000 and 5,000/μL, but occasionally it may exceed 50,000/μL (Figure 10.3). CMML was divided into two myeloproliferative and myelodysplastic subtypes based on the WBC count [15]. There is often granulocytosis, with various degrees of anemia and/or thrombocytopenia. Monocytes and granulocytes are left-shifted and dysplastic with the presence of metamyelocytes, myelocytes, promyelocytes, and promonocytes (Figure 10.4a). The serum lysozyme levels are elevated. Myeloid blast cells and promonocytes are often <5%, but always <20% of the leukocyte differential count. Mild eosinophilia and/or basophilia are present. Usually, there is mild to moderate thrombocytopenia and anemia. Cases with over 13,000 leukocytes/μL are considered of myeloproliferative subtype (Figure 10.5). However, significant clinical or biological differences between these two groups are debatable [1, 16–18].

Extramedullary involvement is observed in the spleen and sometimes in the liver and the lymph nodes. There seems to be a correlation between elevated WBC and splenomegaly. The myelomonocytic cells infiltrate into the splenic red pulp [19] and hepatic and lymph node sinuses.

The following subcategories are distinguished by the WHO [1]:

CMML-1: Blasts <5% in the peripheral blood and <10% in the bone marrow.

CMML-2: Blasts between 5% and 19% in the peripheral blood and/or between 10% and 19% in the bone marrow (Figure 10.6).

CMML with eosinophilia: When the eosinophil count is >1,500/μL. This category should be further divided into CMML-1 or CMML-2 according to the blast counts.

Immunophenotype and Cytochemical Stains

Flow cytometric studies of the peripheral blood or bone marrow often provide valuable information regarding (1) myelomonocytic precursors, (2) the presence of aberrant expression of CD molecules, and (3) the estimation of blast counts [20, 21].

Monocytes are positive for CD4, CD13, CD14, CD15, CD64, CD11 (b and c), HLA-DR, and may show aberrant expression of CD molecules, such as CD56 (Figure 10.7). Granulocytic cells are distinguished by coexpression of CD10, CD13, CD33, CD15, CD16, CD11c, and myeloperoxidase (MPO). They may lose some of these expressions, such as CD10 or CD33, and may show an abnormal dot plot clustering pattern due to hypogranularity. The blast population in the blood or bone marrow is detected as a dimly CD45-positive cluster which may express

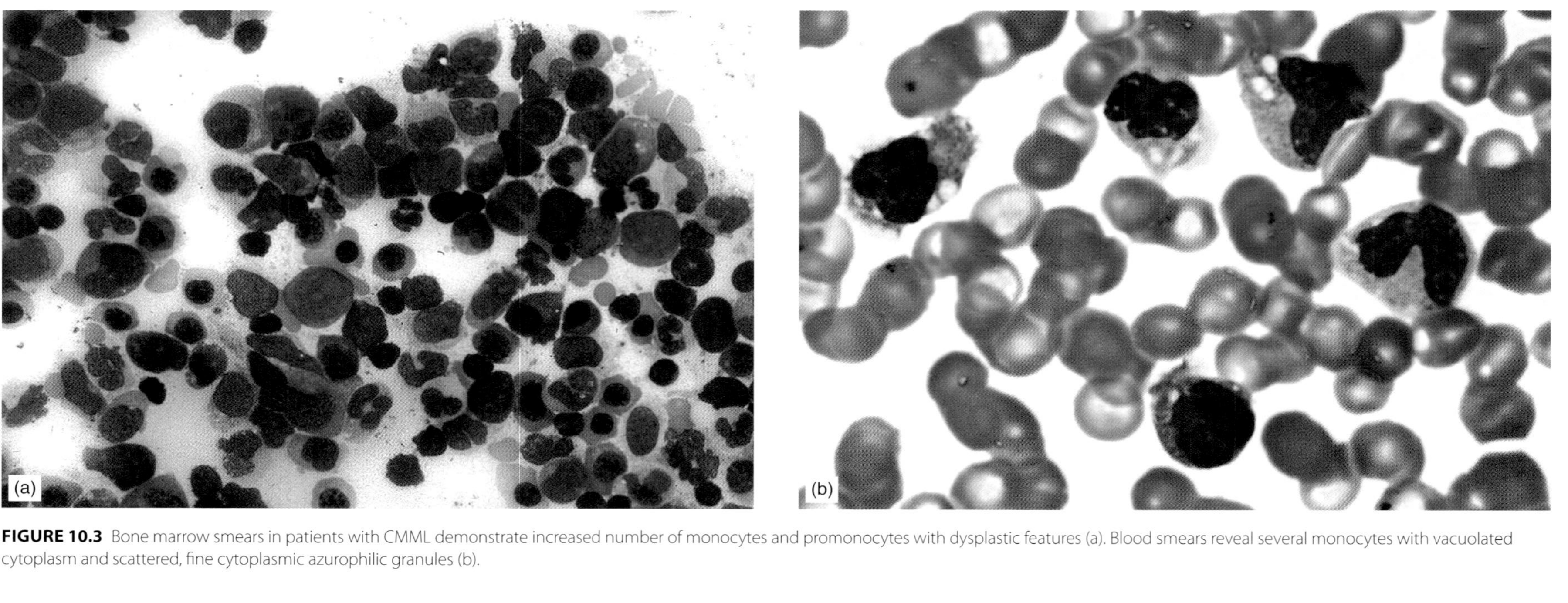

FIGURE 10.3 Bone marrow smears in patients with CMML demonstrate increased number of monocytes and promonocytes with dysplastic features (a). Blood smears reveal several monocytes with vacuolated cytoplasm and scattered, fine cytoplasmic azurophilic granules (b).

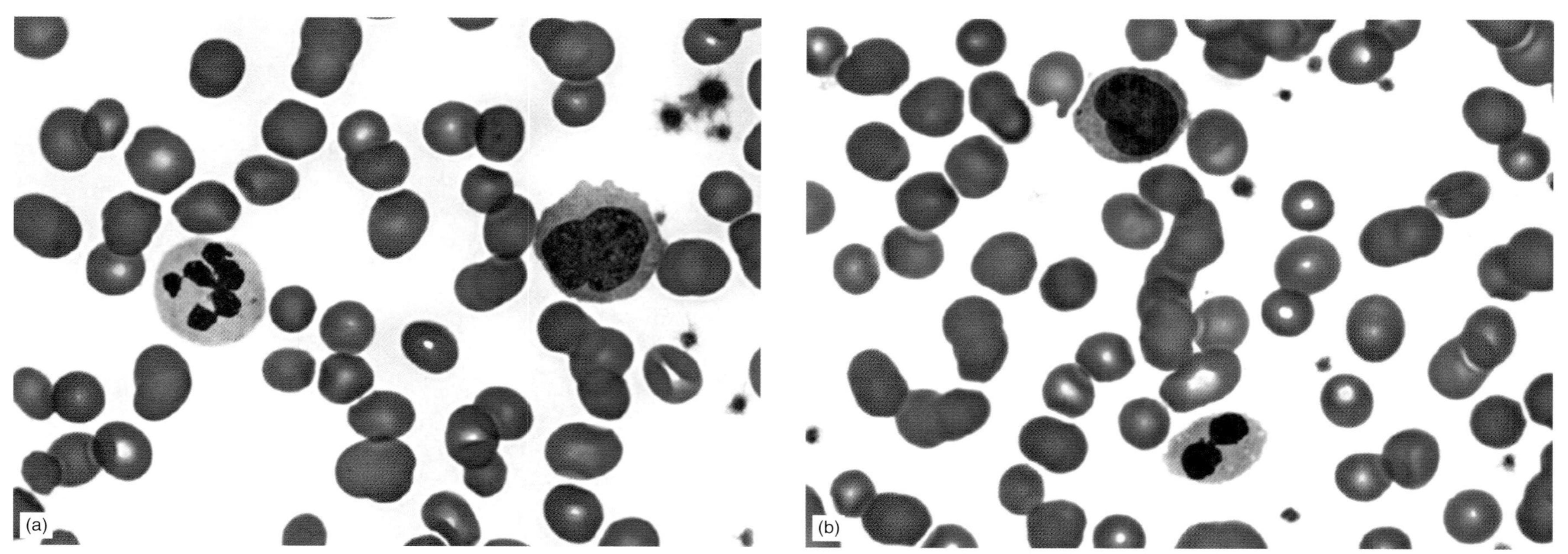

FIGURE 10.4 Dysplastic granulocytes are often present in the blood smears of patients with CMML. A hypersegmented and hypogranular neutrophil is demonstrated in (a) and a hyposegmented hypogranular neutrophil is depicted in (b).

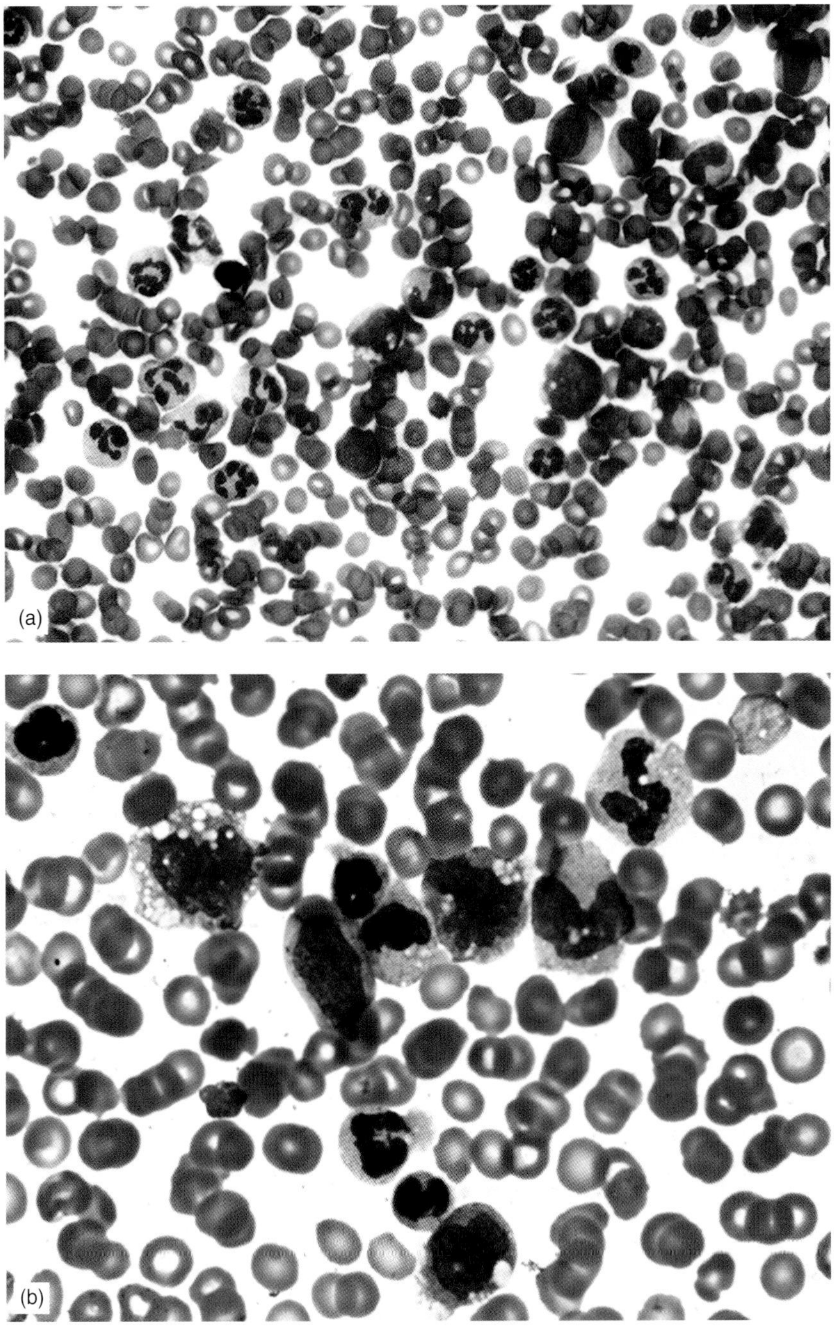

FIGURE 10.5 (a) Low and (b) high power field of a blood smear demonstrating the hyperproliferative variant of CMML.

CD34 and/or CD117 as well as other granulocytic- and/or monocytic-associated molecules.

Immunohistochemical stains, such as stains for glycophorin A, hemoglobin A, MPO, CD68, lysozyme, CD34, and CD117, provide information regarding the M:E ratio, monocytic component, and increased blasts (Figure 10.8).

Cytochemical stains, such as MPO and non-specific esterase, are sometimes useful for estimation of the granulocytic and monocytic components and the lineage confirmation of the blast cells.

Molecular and Cytogenetic Studies

At the molecular level, *K-RAS* and *N-RAS* mutations are reported in approximately 50% of the CMML cases [7, 8]. There is also a report of the elevated levels of *survivin* in patients with CMML [9]. Fusion of the *TEL* and *PDGFβR* genes [t(5;12)(q33;p13)] has been demonstrated in the CMML subtype with eosinophilia [6]. The gene mutation profiles in various clonal myelogenous disorders are presented in Figure 10.9. Results suggest that the molecular pathology of CMML is closer to MDS than CMPD.

Approximately 20–40% of the patients with CMML show cytogenetic abnormalities including +8 (Figure 10.10), −7, del(7q), i(7q), and structural abnormalities of 12p [1, 22]. Pentasomy of chromosome 8, trisomy of chromosome 19, monosomy of chromosome 15, isochromosome 14q, and t(1;3)(p36;q21) have been reported in occasional cases [23–27]. As mentioned earlier, t(5;12)(q33;p13) is

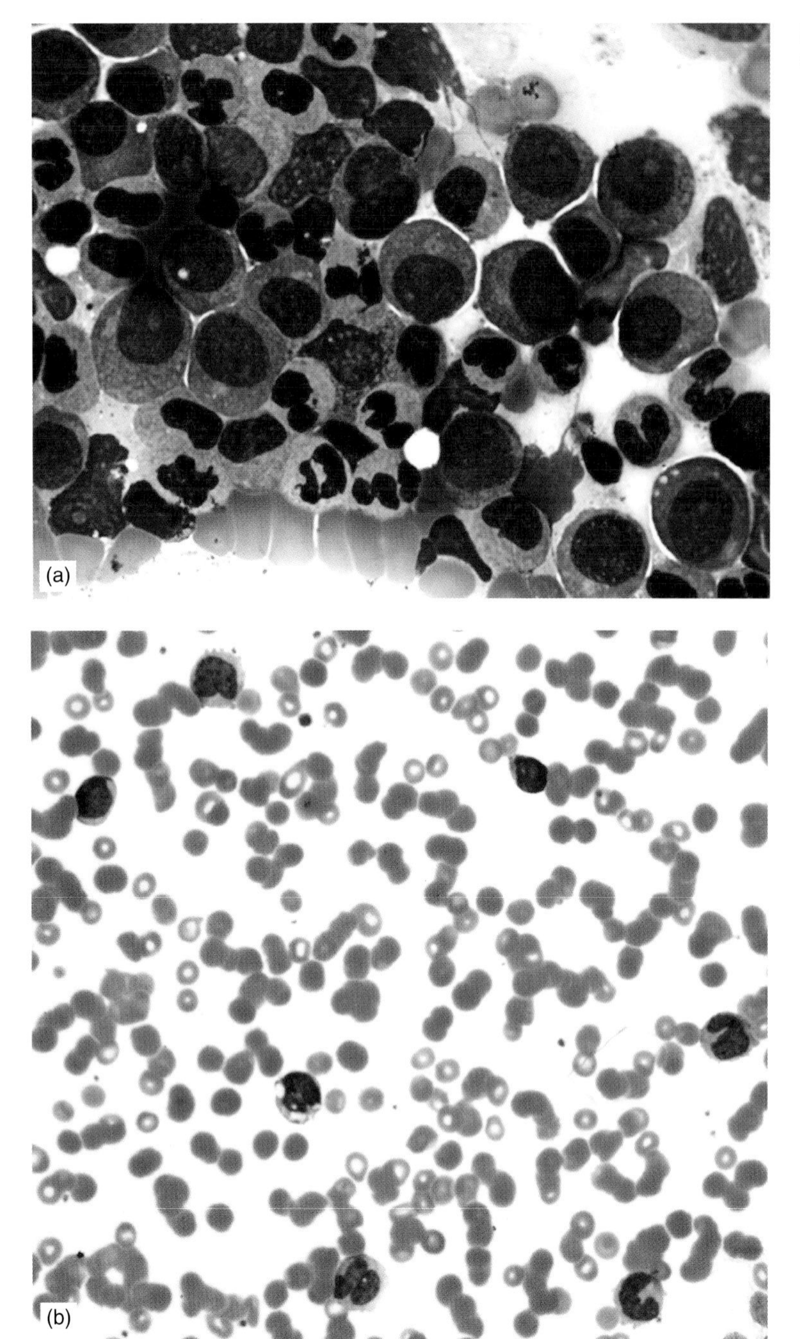

FIGURE 10.6 (a) Bone marrow and (b) blood smears representing CMML type 2.

frequently associated with CMML with eosinophilia [28, 29] (Figure 10.11). 11q23 abnormalities are infrequent.

Clinical Aspects

Chronic myelomonocytic leukemia is a disease of the elderly, with a median age of about 70 years. It affects men more than women, with a male:female ratio of 2–3:1 and an estimated incidence of about 4 per 100,000 per year [5, 30, 31]. Childhood CMML is relatively infrequent [32]. Clinical manifestations are related to anemia, thrombocytopenia, and splenomegaly. Some patients may demonstrate autoimmune disorders, such as vasculitis, pyoderma, and idiopathic thrombocytopenia [33]. Others may develop skin infiltration and serous effusions [34]. The reported unfavorable prognostic factors include low hemoglobin levels, low platelet counts, high percentage of marrow blasts (>10%), lymphocytosis (>2,500/μL), elevated serum lactate dehydrogenase (LDH) and β2-microglobulin levels, and abnormal cytogenetics. Approximately 15–30% of CMML patients progress to acute leukemia [16, 35]. Evidence of

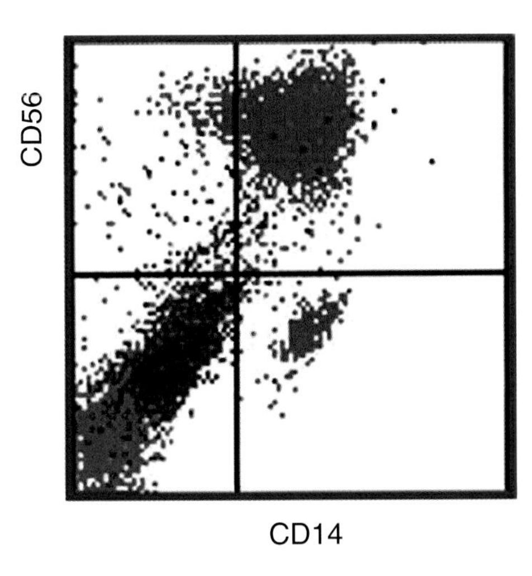

FIGURE 10.7 Flow cytometric analysis of monocytes in patients with CMML may show aberrant expression of CD56 or other markers. In this figure monocytes (red cluster, upper right quadrant) show coexpression of CD14 and CD56.

erythrophagocytosis has been suggested as an indicator of evolving blast transformation [36]. The median survival is about 2 years. Therapeutic approaches include conventional chemotherapy, such as hydroxycarbamide (hydroxyurea) and allogeneic bone marrow transplantation (BMT). Patients with t(5;12) appear to respond to Gleevec. BMT is currently the only curative option with a 5-year survival rate of about 20%. The therapeutic trial of farnesyl transferase inhibitors (inactivating RAS protein) has opened an avenue in the area of targeted biological therapy [5].

ATYPICAL CHRONIC MYELOID LEUKEMIA

Atypical chronic myeloid leukemia (aCML) is a clonal hematopoietic disorder characterized by both dysplastic and proliferative features including persistent granulocytosis with left shift, bone marrow hypercellularity with dysplastic hematopoiesis, myeloid preponderance, and left shift [1, 37–39]. Cytogenetic and molecular studies are negative for *Ph¹* and *BCR/ABL1* fusion gene (Table 10.2). This disorder shows considerable overlapping of clinicopathological features with both CML and CMML. It is distinguished from CML by older median age, lower levels of granulocytosis, multilineage dysplasia, lack of basophilia, and absence of *Ph¹* and *BCR/ABL1* fusion gene. Unlike CMML, aCML shows no absolute monocytosis and a milder myeloid left shift.

Etiology and Pathogenesis

The etiology and pathogenesis of aCML are not known. Some investigators suggest that aCML and CMML are closely related and may represent different spectra of the same disorder [37, 38, 40]. Reports of t(5;12)(q33;p13) and t(5;10)(q33;q22) in occasional aCML cases raise the possibility of a pathogenic role for the *PDGFβR* gene [37, 41].

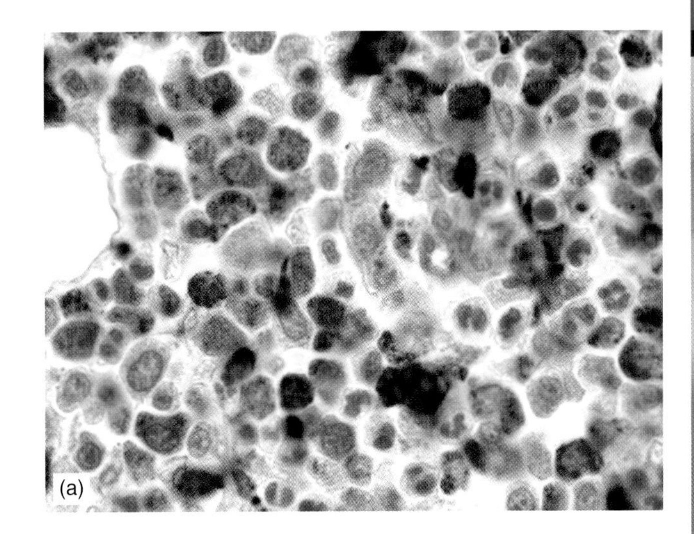

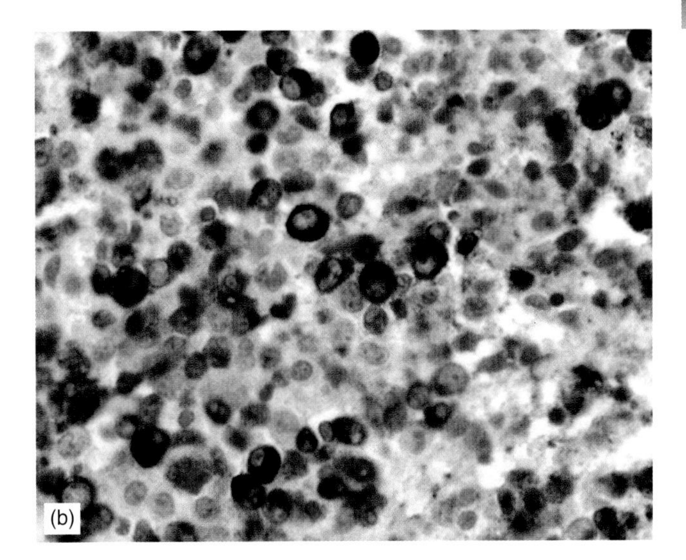

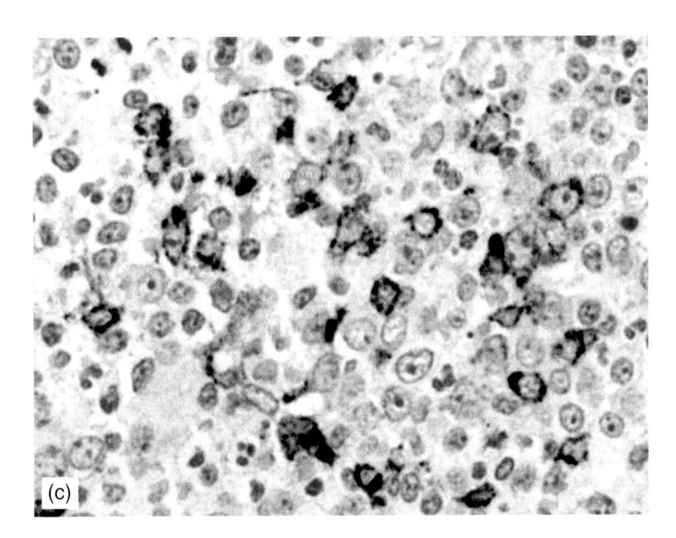

FIGURE 10.8 Immunohistochemical stains of a bone marrow biopsy involved with CMML demonstrating myelomonocytic precursors. Monocytes and macrophages express CD68 (a); myeloid precursors are strongly MPO-positive (b). Scattered CD34-positive cells represent blast cells (c).

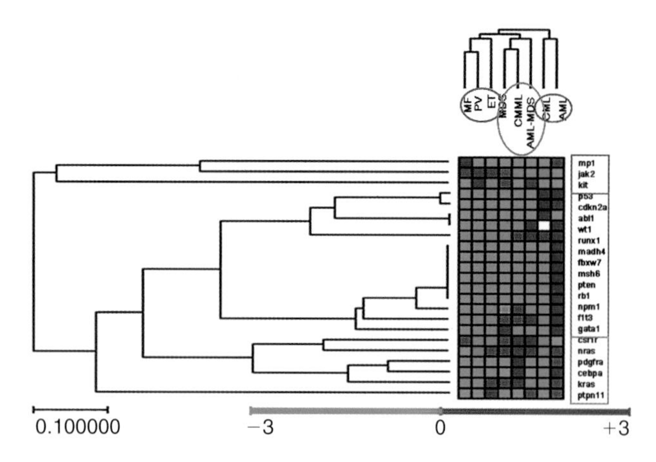

FIGURE 10.9 Schematic correlation of gene mutation profiles in various clonal myelogenous disorders. Data used from http://www.sanger.ac.uk/genetics/CGP/cosmic.www.sanger.ac.uk/genetics/CGP/cosmic. MF: myelofibrosis, PV: polycythemia vera, ET: essential thrombocythemia, MDS: myelodysplastic syndrome, CMML: chronic myelomonocytic leukemia, AML: acute myeloid leukemia, and CML: chronic monocytic leukemia. Courtesy of Dejun Shen, M.D., Ph.D.

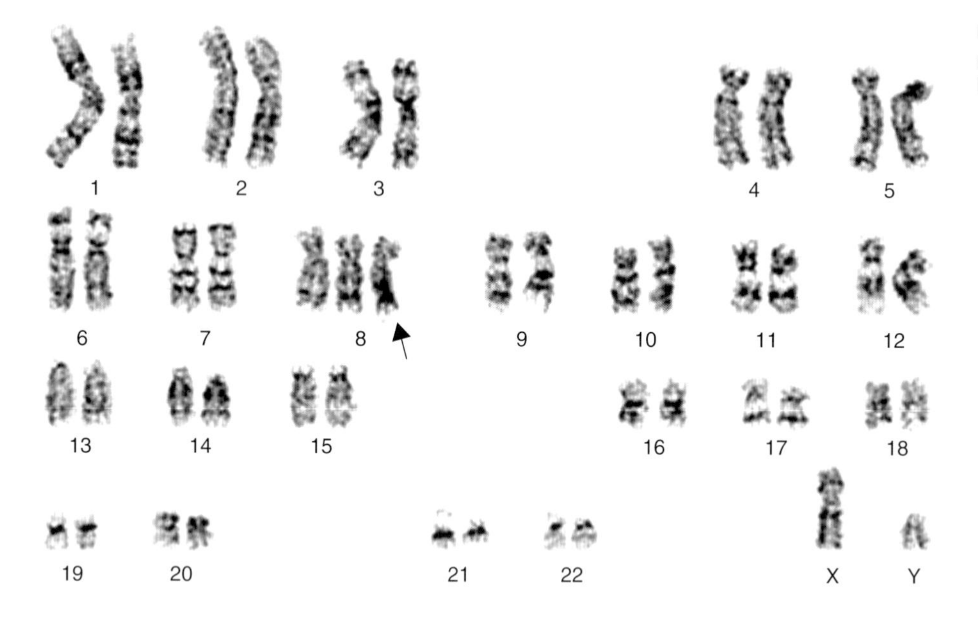

FIGURE 10.10 Trisomy 8 (arrow) in a patient with chronic myelomonocytic leukemia.

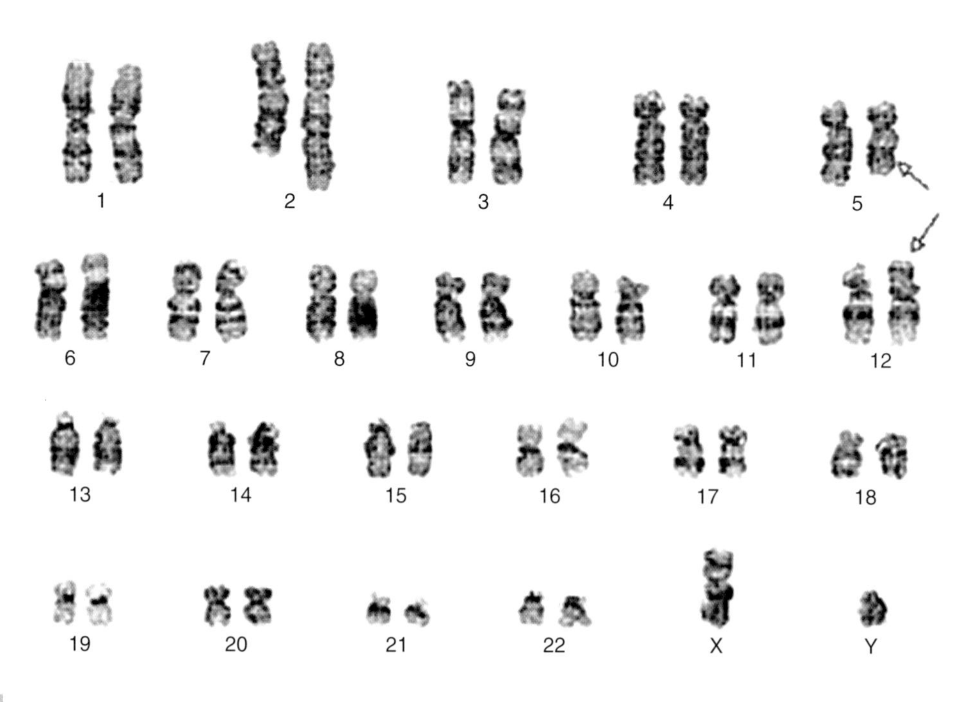

FIGURE 10.11 Translocation 5;12 (arrows) in a patient with chronic myelomonocytic leukemia.

TABLE 10.2 The WHO proposed criteria for the diagnosis of atypical chronic myeloid leukemia*

1.	Peripheral blood granulocytosis with increased number of mature and immature forms.
2.	Prominent dysgranulopoiesis.
3.	No *Ph*[1] or *BCR/ABL* fusion gene.
4.	No or minimal absolute monocytosis.
5.	No or minimal absolute basophilia.
6.	Bone marrow hypercellularity with myeloid preponderance and left shift with dysgranulopoiesis, with or without other hematopoietic dysplasias.
7.	Fewer than 20% blasts in the blood or bone marrow.

*From Ref. [1].

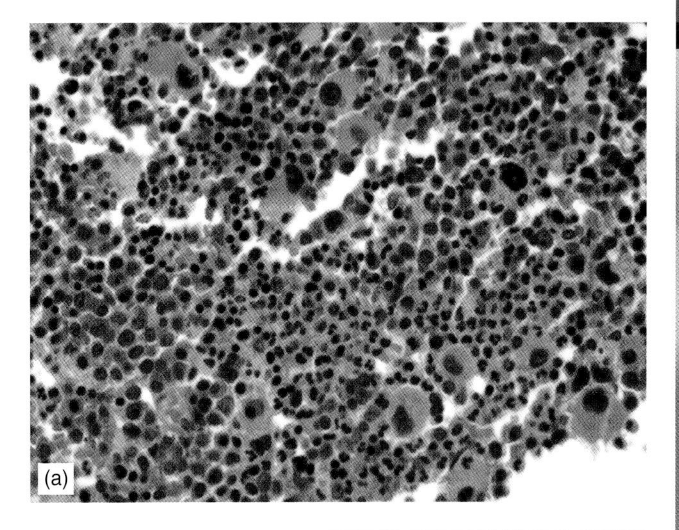

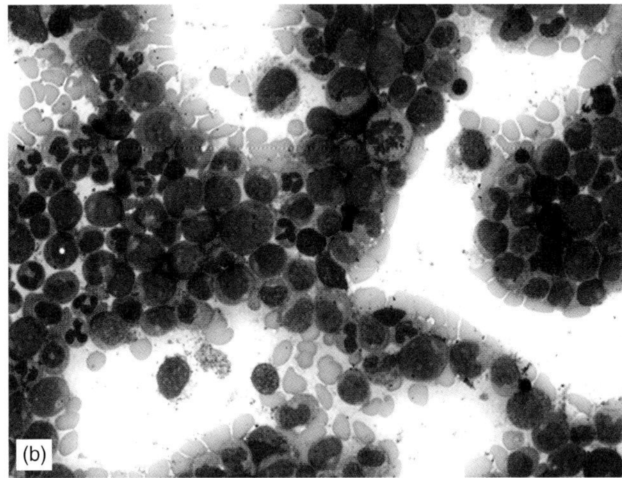

FIGURE 10.12 Bone marrow biopsy section (a) and bone marrow smear (b) of patients with aCML show myeloid preponderance and left shift.

Pathology

Morphology

The bone marrow biopsy and clot sections are hypercellular and show myeloid preponderance and left shift (Figure 10.12a). The bone marrow smears show an elevated M:E ratio, often >10:1 with dysgranulopoiesis and left shift. Hyposegmentation (pseudo Pelger-Huet) or hypersegmentation of the neutrophils, bizarre nuclear morphology of granulocytic precursors, and cytoplasmic hypo- or hypergranulation are frequently noted (Figure 10.12b) [1, 10–12]. Myeloblasts range from 1% to 10% but occasionally may reach up to 19%. Dyserythropoiesis with or without the presence of abnormal megakaryocytes are frequently observed. There is no evidence of monocytosis. The overall bone marrow morphologic features mimic those of CML, except for more significant dysplasia and lack of basophilia. Some cases may show increased reticulin fibers.

The peripheral blood shows elevated WBC, usually ranging from 30,000 to 90,000/μL, but in occasional cases exceeding 100,000/μL. The leukocytosis is primarily due to the increased number of neutrophilic granulocytes which are also left-shifted (Figure 10.12c). Granulocytic precursors account for about 10–20% or more of the leukocytes, but myeloblasts are always <10% and often range from 0% to 10%. Dysplastic granulopoiesis, as mentioned earlier, is always present. There is no or minimal absolute monocytosis and basophilia [1]. There is a variable degree of anemia which may be associated with abnormal morphology, such as anisopoikilocytosis and/or macrocytosis. Thrombocytopenia is a frequent feature.

Immunophenotype and Cytochemical Stains

No specific immunophenotypic features have been described for aCML. The leukocyte alkaline phosphatase (LAP) score is variable and ranges from low to high depending on the cases.

Molecular and Cytogenetic Studies

Atypical CML is negative for Philadelphia (*Ph*[1]) chromosome and shows no evidence of *BCR–ABL1* rearrangement. A high frequency of *RAS* mutations is reported in bcr/abl-negative CML [42]. Occasional cases of aCML show t(5;12)(q33; p13) or t(5;10)(q33;q22) involving the *PDGFβR* gene [37, 41].

Clinical Aspects

Atypical CML is a disorder of older adults with apparently no sex predominance. The incidence of aCML is not yet established. Clinical manifestations, similar to CMML, are related to anemia, thrombocytopenia, and splenomegaly [1, 43, 44]. The median survival is <2 years with 20–40% chance of evolving to acute myeloid leukemia. Therapeutic approaches include conventional chemotherapy, such as hydroxycarbamide. Allogeneic BMT is potentially curative for eligible patients.

JUVENILE MYELOMONOCYTIC LEUKEMIA

Juvenile myelomonocytic leukemia (JMML) is a clonal hematopoietic disorder of early childhood characterized by hepatosplenomegaly, granulocytosis, and monocytosis with left shift and dysplastic changes, elevated hemoglobin F levels, and frequent skin involvement. JMML shares considerable pathologic features with CMML [1, 45–48]. It is relatively rare, usually affects children under the age of 4, and accounts for <2% of all hematologic malignancies in children [49].

Etiology and Pathogenesis

The etiology of JMML is not known. The deregulation of GM-CSF signal transduction through the *RAS* (retrovirus-associated sequence) pathway appears to play an important role in the pathogenesis of this disorder [50–52]. Mutations of *RAS* and the *NF1* genes have been reported in about 30% of patients with JMML [50–54]. The *NF1* gene encodes neurofibromin, a guanosine triphosphatase protein, which is a RAS inhibitor. Mutations in *PTPN11*, which encodes the protein tyrosine phosphatase Shp-2, are common in JMML. It has been suggested that these mutations may induce hypersensitivity of hematopoietic progenitors to GM-CSF [55].

Pathology

Morphology and Laboratory Findings

The WHO criteria for the diagnosis of JMML are presented in Table 10.3 [1]. The bone marrow samples are cellular and display myeloid preponderance and left shift with increased number of immature myelomonocytic precursors (Figure 10.13) [1, 10–12]. However, the total number of myeloblasts, monoblasts, and promonocytes is <20%. Dysplastic changes

TABLE 10.3 The WHO proposed criteria for the diagnosis of juvenile myelomonocytic leukemia*.

1.	Peripheral blood monocytosis >1,000/μL.
2.	Presence of myeloid blasts and promonocytes in the peripheral blood and/or the bone marrow; <20% of the differential counts.
3.	No *Ph¹* or *BCR/ABL* fusion gene.
4.	Plus two or more of the following: (a) Elevated levels of hemoglobin F. (b) Presence of immature granulocytes in the peripheral blood. (c) WBC >10,000/μL. (d) Clonal chromosomal aberrations. (e) Hypersensitivity of myeloid precursors to GM-CSF *in vitro*.

*From Ref. [1].

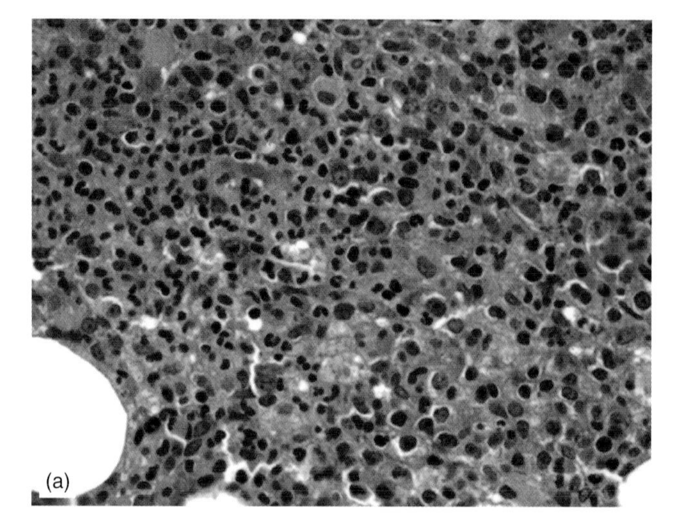

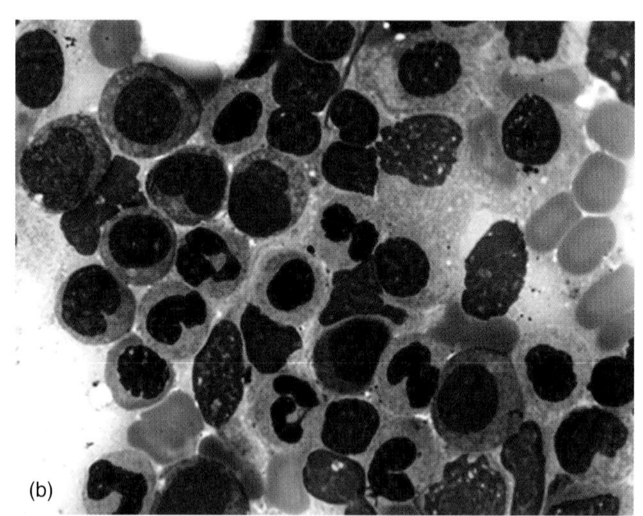

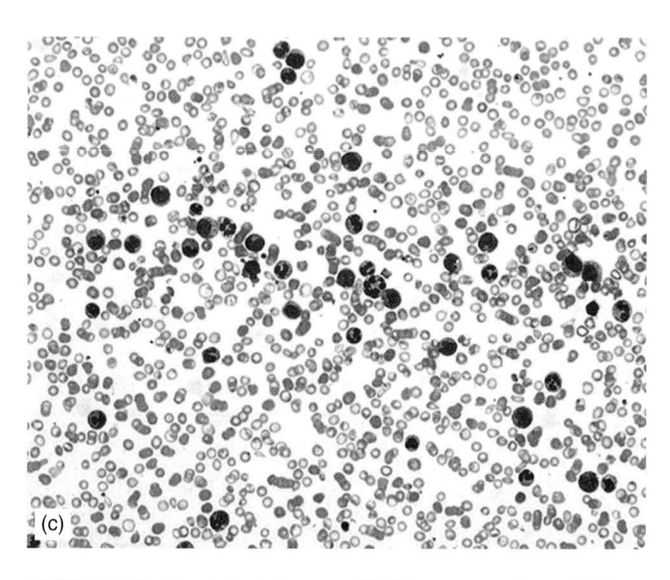

FIGURE 10.13 Morphologic features of JMML are very similar to CMML, characterized by the preponderance of myelomonocytic cells in the bone marrow (a: biopsy section; b: bone marrow smear) and absolute monocytosis in the peripheral blood (c).

are frequently observed in the monocytic and the granulocytic series, such as bizarre morphology, nuclear hyper- or hyposegmentation, and cytoplasmic hypo- or hypergranularity, but Auer rods are not present. Erythroid dysplasia is often minimal or lacking. Megakaryocytes may be reduced or show some degree of dysplastic changes including presence of micromegakaryocytes. Eosinophilia and basophilia are rare.

The peripheral blood reveals monocytosis and granulocytosis, often with various degrees of anemia and/or thrombocytopenia. Monocytes and granulocytes are left-shifted and dysplastic with the presence of metamyelocytes, myelocytes, promyelocyes, and promonocytes (Figure 10.12c) [1, 10–12]. The serum lysozyme levels are elevated. Myeloid blast cells and promonocytes are often <5% and never >19% of the leukocyte differential count. The average leukocyte count is about 30,000/μL. There is often some degree of anisopoikilocytosis. Macrocytosis is a frequent feature and nucleated red blood cells are often present. Eosinophilia and basophilia are rare. The hemoglobin F levels are elevated, the glucose-6-phosphatase activity is increased, and there is a high incidence of antinuclear (50%) and anti-IgG (40%) antibodies. There may be evidence of polyclonal hypergammaglobulinemia [1].

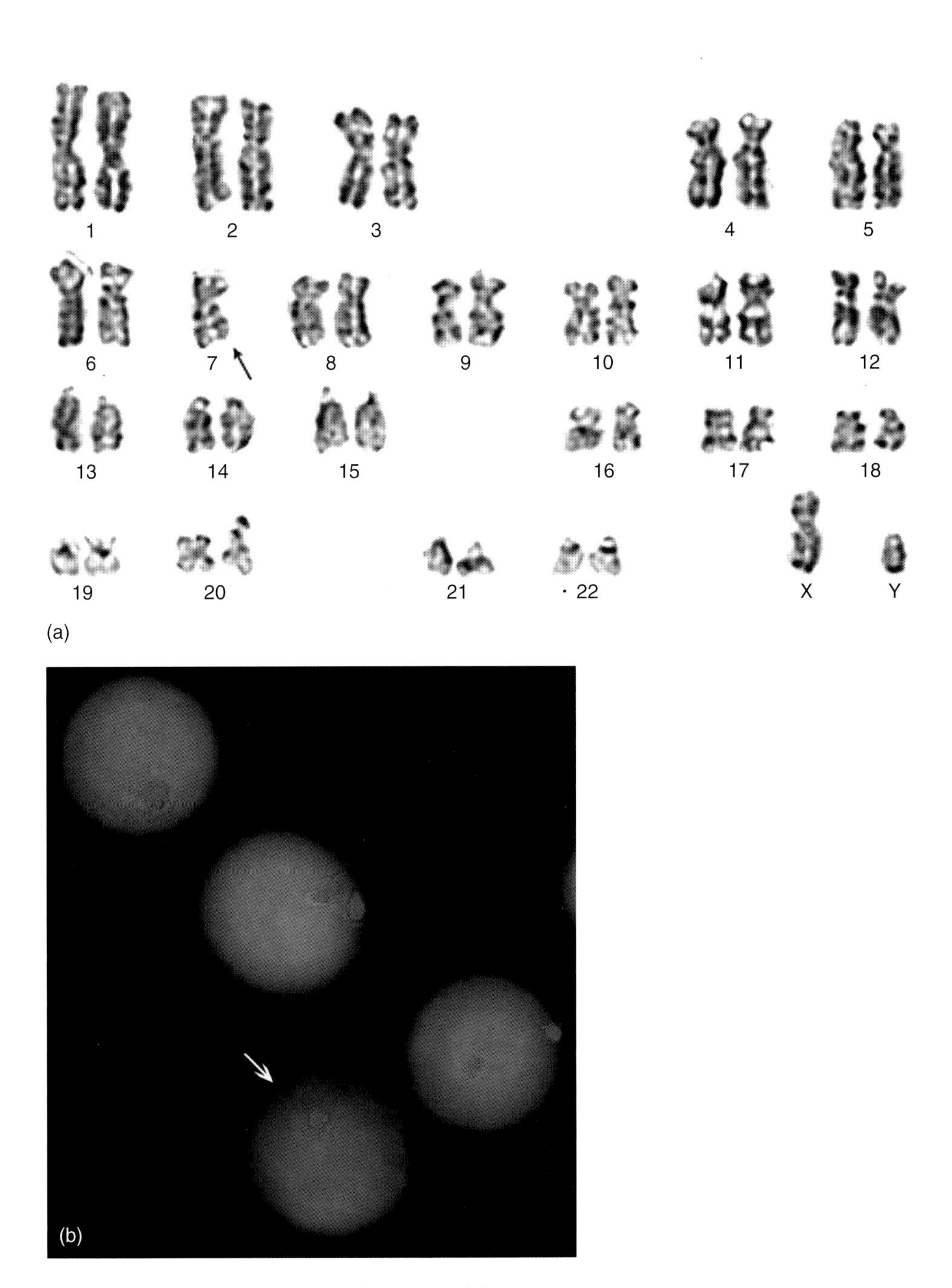

FIGURE 10.14 Monosomy 7 (arrow) in a child with JMML: (a) karyotype and (b) FISH.

TABLE 10.4 Clinicopathologic features of chronic myeloid leukemia (CML), atypical chronic myeloid leukemia (aCML), chronic myelomonocytic leukemia (CMML), and juvenile myelomonocytic leukemia (JMML).

Features	CML	aCML	CMML	JMML
Average age (year)	46	57	72	<4
Male:female	>1	>1	>1	>1
Splenomegaly	+++	++	+	+
Blood				
Average leukocyte count	>100,000/μL	60,000/μL	35,000/μL	30,000/μL
Absolute monocytosis	Often no	Often no	Yes	Yes
Basophilia	Often yes	No	No	No
Myeloid precursors	+++	++	++	++
LAP	Reduced	Variable	Variable	Variable
Anemia	Present	Present	Present	Present
Elevated hemoglobin F	No	No	No	Yes
Platelet count	Variable	Reduced	Reduced	Reduced
Bone Marrow				
Cellularity	Increased	Increased	Increased	Increased
Myeloid preponderance	Yes	Yes	Yes	Yes
Myeloid left shift	Yes	Yes	Yes	Yes
Monocytosis	No	No	Yes	Yes
Significant dysplasia	No	Yes	Yes	Yes

*Adapted from Refs [1, 37].

Extramedullary involvement is a frequent feature with the infiltration of the myelomonocytic cells in the dermis, the lung parenchyma, the hepatic sinusoids, and the splenic red pulp [1].

Immunophenotype and Cytochemical Stains

Similar to MDS and CMML, dysplastic myelomonocytic cells in JMML may show abnormal expression of CD molecules, such as aberrant expression of CD56 by monocytes or reduced expression of CD10 by neutrophils. The elevated number of monocytic cells in bone marrow and peripheral blood can be established by flow cytometry, immunohistochemistry, and/or cytochemical stains. Monocytes express CD4, CD14, CD16, HLA-DR, CD64, and CD68, and show strong positive reactions with lysozyme and non-specific esterase stains. LAP scores may be reduced.

Molecular and Cytogenetic Studies

As mentioned earlier, mutations of *RAS*, *NF1*, and *PTNP11* genes are frequently detected in patients with JMML [50–54, 56]. Recently, quantitative measurements of *RAS* and *PTPN11* were made by an allele-specific polymerase chain reaction (PCR) assay called TaqMan, and increased levels were correlated with relapse of JMML in transplanted patients [54]. Methylation of *p15*, which is a frequent finding in patients with MDS (78%), is a rare event (17%) in JMML patients [57].

Cytogenetic aberrations are non-specific. Monosomy of chromosome 7 is the most frequent cytogenetic abnormality (Figure 10.14) [58, 59]. Rare cases with t(3;12)(q21-22; p13.3) or der(15)t(3;15)(q13.1; q26) have been reported [60].

Clinical Aspects

Juvenile myelomonocytic leukemia is a rare early childhood hematologic disorder with roughly 0.6 new cases per year per million children at risk, accounting for <2% of hematologic malignancies in children [45–49]. The majority of patients are under the age of 4 years with a male:female ratio of about 2:5 [61–63]. Splenomegaly, hepatomegaly, lymphadenopathy, and skin rashes are noted in >90%, 80%, 70%, and 35% of the patients, respectively [61]. The cutaneous manifestations include neoplastic infiltration, eczema, xanthoma, and café-au-lait spots. JMML shows a high frequency (7–14%) of association with neurofibromatosis type 1 [45–49]. The prognosis is poor, but affected infants younger than 1 year of age appear to do better than older children [64]. Elevated hemoglobin F levels (>15%) and low platelet counts (<33,000/μL) are amongst the unfavorable prognostic indicators [64, 65].

Allogeneic BMT is the only available cure with approximately 50% 5-year event-free survival rate [66, 67].

Recent reports on the effects of zoledronic acid (ZOL), a blocker of RAS activity, are promising [52].

The entity *infantile (childhood) monosomy 7 syndrome* shares most of the clinicopathologic features of CMML [68–70]. Both disorders affect children at early ages (often <1 year old), show male predominance, show an association with neurofibromatosis type 1, and similar frequency of *RAS* gene mutation. Also, monosomy 7 is the most frequent chromosomal abnormality in JMML. However, children with JMML who lack monosomy 7 often display elevated levels of hemoglobin F. It seems that infantile monosomy 7 represents a cytogenetic subtype of JMML.

DIFFERENTIAL DIAGNOSIS

The chronic myeloproliferative/myelodysplastic disorders show significant overlapping of morphologic features among themselves and with CML. CML patients are usually younger and show much more severe leukocytosis than patients with CMML or aCML. Basophilia is a common feature in CML but not present in CMML or aCML [43]. Dysplastic myelopoiesis is a characteristic feature of aCML, CMML, and JMML, whereas it is insignificant in CML. Monocytosis is the hallmark of CMML and JMML and is lacking in CML and aCML. *Ph*[1] and/or *BCR/ABL* fusion gene are present in CML but negative in aCML, CMML, and JMML. JMML is a disease of early childhood (usually under the age of 4) and is commonly associated with skin rashes and elevated hemoglobin F levels [49, 70, 71]. There is a high frequency (7–14%) of neurofibromatosis type 1 in JMML patients. The major clinicopathologic features of CMML, aCML, and JMML are compared with one another and with CML in Table 10.4.

References

1. Jaffe ES, Harris NL, Stein H, Vardiman JW. (2001). *Pathology and Genetics of Tumors of Haematopoietic and Lymphoid Tissues*. IARC Press, Lyon.
2. Vardiman JW. (2003). Myelodysplastic syndromes, chronic myeloproliferative diseases, and myelodysplastic/myeloproliferative diseases. *Semin Diagn Pathol* **20**, 154–79.
3. Vardiman JW. (2004). Myelodysplastic/myeloproliferative diseases. *Cancer Treat Res* **121**, 13–43.
4. Bennett JM. (2003). The myelodysplastic/myeloproliferative disorders: The interface. *Hematol Oncol Clin North Am* **17**, 1095–100.
5. Bowen DT. (2005). Chronic myelomonocytic leukemia, lost in classification. *Hematol Oncol* **23**, 26–33.
6. Liu S, Li C, Bo L, Dai Y, Xiao Z, Wang J. (2004). AML1/RUNX1 fusion gene and t(5;21)(q13;q22) in a case of chronic myelomonocytic leukemia with progressive thrombocytopenia and monocytosis. *Cancer Genet Cytogenet* **152**, 172–4.
7. Parikh C, Subrahmanyam R, Ren R. (2006). Oncogenic NRAS rapidly and efficiently induces CMML- and AML-like diseases in mice. *Blood* **108**, 2349–57.
8. Janssen JW, Steenvoorden AC, Lyons J, Anger B, Böhlke JU, Bos JL, Seliger H, Bartram CR. (1987). RAS gene mutations in acute and chronic myelocytic leukemias, chronic myeloproliferative disorders, and myelodysplastic syndromes. *Proc Natl Acad Sci USA* **84**, 9228–32.
9. Invernizzi R, Travaglino E, Benatti C, Malcovati L, Della Porta M, Cazzola M, Ascari E. (2006). Survivin expression, apoptosis and proliferation in chronic myelomonocytic leukemia. *Eur J Haematol* **76**, 494–501.
10. His ED. (2007). *Hematopathology*. Churchill Livingstone, Philadelphia.
11. Foucar K. (2001). *Bone Marrow Pathology*, 2nd ed. ASCP Press, Chicago.
12. Naeim F. (1997). *Pathology of Bone Marrow*, 2nd ed. Williams & Wilkins, Baltimore.
13. Chen YC, Chou JM, Letendre L, Li CY. (2005). Clinical importance of bone marrow monocytic nodules in patients with myelodysplasia, retrospective analysis of 21 cases. *Am J Hematol* **79**, 329–31.
14. Chen YC, Chou JM, Ketterling RP, Letendre L, Li CY. (2003). Histologic and immunohistochemical study of bone marrow monocytic nodules in 21 cases with myelodysplasia. *Am J Clin Pathol* **120**, 874–81.
15. Bennett JM, Catovsky D, Daniel MT, Flandrin G, Galton DA, Gralnick H, Sultan C, Cox C. (1994). The chronic myeloid leukaemias, guidelines for distinguishing chronic granulocytic, atypical chronic myeloid, and chronic myelomonocytic leukaemia. Proposals by the French–American–British Cooperative Leukaemia Group. *Br J Haematol* **87**, 746–54.
16. Onida F, Kantarjian HM, Smith TL, Ball G, Keating MJ, Estey EH, Glassman AB, Albitar M, Kwari MI, Beran M. (2002). Prognostic factors and scoring systems in chronic myelomonocytic leukemia, a retrospective analysis of 213 patients. *Blood* **99**, 840–9.
17. Onida F, Beran M. (2004). Chronic myelomonocytic leukemia: Myeloproliferative variant. *Curr Hematol Rep* **3**, 218–26.
18. Germing U, Gattermann N, Minning H, Heyll A, Aul C. (1998). Problems in the classification of CMML – Dysplastic versus proliferative type. *Leuk Res* **22**, 871–8.
19. Steensma DP, Tefferi A, Li CY. (2003). Splenic histopathological patterns in chronic myelomonocytic leukemia with clinical correlations, reinforcement of the heterogeneity of the syndrome. *Leuk Res* **27**, 775–82.
20. Xu Y, McKenna RW, Karandikar NJ, Pildain AJ, Kroft SH. (2005). Flow cytometric analysis of monocytes as a tool for distinguishing chronic myelomonocytic leukemia from reactive monocytosis. *Am J Clin Pathol* **124**, 799–806.
21. Gorczyca W. (2004). Flow cytometry immunophenotypic characteristics of monocytic population in acute monocytic leukemia (AML-M5), acute myelomonocytic leukemia (AML-M4), and chronic myelomonocytic leukemia (CMML). *Methods Cell Biol* **75**, 665–77.
22. Toyama K, Ohyashiki K, Yoshida Y, Abe T, Asano S, Hirai H, Hirashima K, Hotta T, Kuramoto A, Kuriya S. (1993). Clinical implications of chromosomal abnormalities in 401 patients with myelodysplastic syndromes: A multicentric study in Japan. *Leukemia* **7**, 499–508.
23. Hamey Y, Dean N, Catalano JV, Campbell LJ. (1998). Pentasomy of chromosome 8 in chronic myelomonocytic leukemia. *Cancer Genet Cytogenet* **103**, 164–6.
24. Daskalakis M, Mauritzson N, Johansson B, Bouabdallah K, Onida F, Kunzmann R, Muller-Berndorff H, Schmitt-Graff A, Lubbert M. (2006). Trisomy 19 as the sole chromosomal abnormality in proliferative chronic myelomonocytic leukemia. *Leuk Res* **30**, 1043–7.
25. Zamora L, Espinet B, Salido M, Florensa L, Woessner S, Pedro C, Serrtano S, Sole F. (2002). Monosomy 15 in chronic myelomonocytic leukemia. Description of a case and review of the literature. *Cancer Genet Cytogenet* **134**, 165–7.
26. Flaherty L, Jarvis A, Harris M, Tyrrell V, Smith A. (1998). A case of chronic myelomonocytic leukemia with isochromosome 14q. *Cancer Genet Cytogenet* **101**, 134–7.
27. Fujisawa S, Harano H, Yamazaki E, Motomura S, Mohri H, Ishigatsubo Y. (2000). Chronic myelomonocytic leukemia with t(1;3)(p36;q21) and a synchronous gastric cancer. *Am J Med Sci* **319**, 258–60.

28. Golub TR, Barker GF, Lovett M, Gilliland DG. (1994). Fusion of PDGF receptor beta to a novel ets-like gene, tel, in chronic myelomonocytic leukemia with t(5;12) chromosomal translocation. *Cell* **77**, 307–16.
29. Baranger L, Szapiro N, Gardais J, Hillion J, Derre J, Francois S, Blanchet O, Boasson M, Berger R. (1994). Translocation t(5;12)(q31-q33;pl2-pl3), a non-random translocation associated with a myeloid disorder with eosinophilia. *Br J Haematol* **88**, 343–7.
30. Cortes J. (2003). CMML: A biologically distinct myeloproliferative disease. *Curr Hematol Rep* **2**, 202–8.
31. Bennett JM. (2002). Chronic myelomonocytic leukemia. *Curr Treat Options Oncol* **3**, 221–3.
32. Niemeyer CM, Arico M, Basso G, Biondi A, Cantu Rajnoldi A, Creutzig U, Haas O, Harbott J, Hasle H, Kerndrup G, Locatelli F, Mann G, Stollmann-Gibbels B, van't Veer-Korthof ET, van Wering E, Zimmermann M. (1997). Chronic myelomonocytic leukemia in childhood, a retrospective analysis of 110 cases. European Working Group on Myelodysplastic Syndromes in Childhood (EWOG-MDS). *Blood* **89**, 3534–43.
33. Hamidou MA, Boumalassa A, Larroche C, El Kouri D, Bletry O, Grolleau JY. (2001). Systemic medium-sized vessel vasculitis associated with chronic myelomonocytic leukemia. *Semin Arthritis Rheum* **31**, 119–26.
34. Strupp C, Germing U, Trommer I, Gattermann N, Aul C. (2000). Pericardial effusion in chronic myelomonocytic leukemia (CMML), a case report and review of the literature. *Leuk Res* **24**, 1059–62.
35. Germing U, Strupp C, Aivado M, Gattermann N. (2002). New prognostic parameters for chronic myelomonocytic leukemia. *Blood* **100**, 731–2.
36. Etzell J, Lu CM, Browne LW, Wang E. (2005). Erythrophagocytosis by dysplastic neutrophils in chronic myelomonocytic leukemia and subsequent transformation to acute myeloid leukemia. *Am J Hematol* **79**, 340–2.
37. Martiat P, Michaux JL, Rodhain J. (1991). Philadelphia-negative (Ph−) chronic myeloid leukemia (CML), comparison with Ph+ CML and chronic myelomonocytic leukemia. The Groupe Francais de Cytogenetique Hematologique. *Blood* **78**, 205–11.
38. Kurzrock R, Kantarjian HM, Shtalrid M, Gutterman JU, Talpaz M. (1990). Philadelphia chromosome-negative chronic myelogenous leukemia without breakpoint cluster region rearrangement, a chronic myeloid leukemia with a distinct clinical course. *Blood* **75**, 445–52.
39. Oscier D. (1997). Atypical chronic myeloid leukemias. *Pathol Biol (Paris)* **45**, 587–93.
40. Cogswell PC, Morgan R, Dunn M, Neubauer A, Nelson P, Poland-Johnston NK, Sandberg AA, Liu E. (1989). Mutations of the ras protooncogenes in chronic myelogenous leukemia, a high frequency of ras mutations in bcr/abl rearrangement-negative chronic myelogenous leukemia. *Blood* **74**, 2629–33.
41. Schwaller J, Anastasiadou E, Cain D, Kutok J, Wojiski S, Williams IR, LaStarza R, Crescenzi B, Sternberg DW, Andreasson P, Schiavo R, Siena S, Mecucci C, Gilliland DG. (2001). H4(D10S170), a gene frequently rearranged in papillary thyroid carcinoma, is fused to the platelet-derived growth factor receptor beta gene in atypical chronic myeloid leukemia with t(5;10)(q33;q22). *Blood* **97**, 3910–18.
42. van der Plas DC, Grosveld G, Hagemeijer A. (1991). Review of clinical, cytogenetic, and molecular aspects of Ph-negative CML. *Cancer Genet Cytogenet* **52**, 143–56.
43. Galton DA. (1992). Haematological differences between chronic granulocytic leukaemia, atypical chronic myeloid leukaemia, and chronic myelomonocytic leukaemia. *Leuk Lymphoma* **7**, 343–50.
44. Costello R, Sainty D, Lafage-Pochitaloff M, Gabert J. (1997). Clinical and biological aspects of Philadelphia-negative/BCR-negative chronic myeloid leukemia. *Leuk Lymphoma* **25**, 225–32.
45. Niemeyer CM, Kratz C. (2003). Juvenile myelomonocytic leukemia. *Curr Oncol Rep* **5**, 510–15.
46. Kratz CP, Niemeyer CM. (2005). Juvenile myelomonocytic leukemia. *Hematology* **10**(Suppl 1), 100–3.
47. Emanuel PD. (2004). Juvenile myelomonocytic leukemia. *Curr Hematol Rep* **3**, 203–9.
48. Niemeyer CM, Kratz C. (2003). Juvenile myelomonocytic leukemia. *Curr Treat Options Oncol* **4**, 203–10.
49. Chang YH, Jou ST, Lin DT, Lu MY, Lin KH. (2004). Differentiating juvenile myelomonocytic leukemia from chronic myeloid leukemia in childhood. *J Pediatr Hematol Oncol* **26**, 236–42.
50. Ohtsuka Y, Manabe A, Kawasaki H, Hasegawa D, Zaike Y, Watanabe S, Tanizawa T, Nakahata T, Tsuji K. (2005). RAS-blocking bisphosphonate zoledronic acid inhibits the abnormal proliferation and differentiation of juvenile myelomonocytic leukemia cells *in vitro*. *Blood* **106**(9), 3134–41.
51. Reimann C, Arola M, Bierings M, Karow A, van den Heuvel-Eibrink MM, Hasle H, Niemeyer CM, Kratz CP. (2006). A novel somatic K-Ras mutation in juvenile myelomonocytic leukemia. *Leukemia* **20**, 1637–8.
52. Flotho C, Kratz C, Niemeyer CM. (2007). Targeting RAS signaling pathways in juvenile myelomonocytic leukemia. *Curr Drug Targets* **8**, 715–25.
53. Flotho C, Steinemann D, Mullighan CG, Neale G, Mayer K, Kratz CP, Schlegelberger B, Downing JR, Niemeyer CM. (2007). Genome-wide single-nucleotide polymorphism analysis in juvenile myelomonocytic leukemia identifies uniparental disomy surrounding the NF1 locus in cases associated with neurofibromatosis but not in cases with mutant RAS or PTPN11. *Oncogene* **26**, 5816–21.
54. Archambeault S, Flores NJ, Yoshimi A, Kratz CP, Reising M, Fischer A, Noellke P, Locatelli F, Sedlacek P, Flotho C, Zecca M, Emanuel PD, Castleberry RP, Niemeyer CM, Bader P, Loh ML. (2007). Development of an allele-specific minimal residual disease assay for patients with juvenile myelomonocytic leukemia. *Blood*. **November 13**.
55. Chan RJ, Leedy MB, Munugalavadla V, Voorhorst CS, Li Y, Yu M, Kapur R. (2005). Human somatic PTPN11 mutations induce hematopoietic-cell hypersensitivity to granulocyte-macrophage colony-stimulating factor. *Blood* **105**, 3737–42.
56. Metzner A, Horstmann MA, Fehse B, Ortmeyer G, Niemeyer CM, Stocking C, Mayr GW, Jücker M. (2007). Gene transfer of SHIP-1 inhibits proliferation of juvenile myelomonocytic leukemia cells carrying KRAS2 or PTPN11 mutations. *Gene Ther* **14**, 699–703.
57. Hasegawa D, Manabe A, Kubota T, Kawasaki H, Hirose I, Ohtsuka Y, Tsuruta T, Ebihara Y, Goto Y, Zhao XY, Sakashita K, Koike K, Isomura M, Kojima S, Hoshika A, Tsuji K, Nakahata T. (2005). Methylation status of the p15 and p16 genes in paediatric myelodysplastic syndrome and juvenile myelomonocytic leukaemia. *Br J Haematol* **128**, 805–12.
58. Hall GW. (2002). Cytogenetic and molecular genetic aspects of childhood myeloproliferative/myelodysplastic disorders. *Acta Haematol* **108**, 171–9.
59. Hasle H, Aricò M, Basso G, Biondi A, Cantù Rajnoldi A, Creutzig U, Fenu S, Fonatsch C, Haas OA, Harbott J, Kardos G, Kerndrup G, Mann G, Niemeyer CM, Ptoszkova H, Ritter J, Slater R, Starý J, Stollmann-Gibbels B, Testi AM, van Wering ER, Zimmermann M. (1999). Myelodysplastic syndrome, juvenile myelomonocytic leukemia, and acute myeloid leukemia associated with complete or partial monosomy 7, European Working Group on MDS in Childhood (EWOG-MDS). *Leukemia* **13**, 376–85.
60. Tosi S, Mosna G, Cazzaniga G, Giudici G, Kearney L, Biondi A, Privitera E. (1997). Unbalanced t(3;12) in a case of juvenile myelomonocytic leukemia (JMML) results in partial trisomy of 3q as defined by FISH. *Leukemia* **11**, 1465–8.
61. Arico M, Biondi A, Pui CH. (1997). Juvenile myelomonocytic leukemia. *Blood* **90**, 479–88.
62. Smith FO, Sanders JE. (1999). Juvenile myelomonocytic leukemia: What we don't know. *J Pediatr Hematol Oncol* **21**, 461–3.
63. Hasle H, Niemeyer CM, Chessells JM, Baumann I, Bennett JM, Kerndrup G, Head DR. (2003). A pediatric approach to the WHO classification of myelodysplastic and myeloproliferative diseases. *Leukemia* **17**, 277–82.

64. Passmore SJ, Chessells JM, Kempski H, Hann IM, Brownbill PA, Stiller CA. (2003). Paediatric myelodysplastic syndromes and juvenile myelomonocytic leukaemia in the UK, a population-based study of incidence and survival. *Br J Haematol* **121**, 758–67.
65. Hasle H, Baumann I, Bergsträsser E, Fenu S, Fischer A, Kardos G, Kerndrup G, Locatelli F, Rogge T, Schultz KR, Starý J, Trebo M, van den Heuvel-Eibrink MM, Harbott J, Nöllke P, Niemeyer CM European Working Group on Childhood MDS (2004). The International Prognostic Scoring System (IPSS) for childhood myelodysplastic syndrome (MDS) and juvenile myelomonocytic leukemia (JMML). *Leukemia* **18**, 2008–14.
66. Zang DY, Deeg HJ, Gooley T, Anderson JE, Anasetti C, Sanders J, Myerson D, Storb R, Appelbaum F. (2000). Treatment of chronic myelomonocytic leukaemia by allogeneic marrow transplantation. *Br J Haematol* **110**, 217–22.
67. Yusuf U, Frangoul HA, Gooley TA, Woolfrey AE, Carpenter PA, Andrews RG, Deeg HJ, Appelbaum FR, Anasetti C, Storb R, Sanders JE. (2004). Allogeneic bone marrow transplantation in children with myelodysplastic syndrome or juvenile myelomonocytic leukemia, the Seattle experience. *Bone Marrow Transplant* **33**, 805–14.
68. Sieff CA, Chessells JM, Harvey BA, Pickthall VJ, Lawler SD. (1981). Monosomy 7 in childhood, a myeloproliferative disorder. *Br J Haematol* **49**, 235–49.
69. Hutter JJ, Hecht F, Kaiser-McCaw B, Hays T, Baranko P, Cohen J, Durie B. (1984). Bone marrow monosomy 7, hematologic and clinical manifestations in childhood and adolescence. *Hematol Oncol* **2**, 5–12.
70. Gadner H, Haas OA. (1992). Experience in pediatric myelodysplastic syndromes. *Hematol Oncol Clin North Am* **6**, 655–72.
71. Gassas A, Doyle JJ, Weitzman S, Freedman MH, Hitzler JK, Sharathkumar A, Dror Y. (2005). A basic classification and a comprehensive examination of pediatric myeloproliferative syndromes. *J Pediatr Hematol Oncol* **27**, 192–6.

Acute Myeloid Leukemia

11

CHAPTER

Faramarz Naeim and
P. Nagesh Rao

GENERAL CONSIDERATIONS

Acute myeloid leukemia (AML) represents a group of hematopoietic neoplasms derived from the bone marrow precursors of myeloid lineage. The neoplastic process is the result of clonal proliferation of an aberrant, committed stem cell at the level of CFU-S or later stages of differentiation leading to the accumulation of immature forms without, or with limited, maturation. Other terms used to denote AML include acute non-lymphoid leukemia (ANLL), acute myelogenous leukemia, and acute myeloblastic leukemia. The current WHO classification of AML is presented in Table 11.1 [1]. According to this classification, AML is divided into four major categories as:

1. AML with recurrent genetic abnormalities
2. AML with multilineage dysplasia
3. AML and myelodysplastic syndromes (MDS), therapy related
4. AML not otherwise categorized.

Etiology and Pathogenesis

The etiology of AML is not clearly understood. It has been demonstrated that environmental factors and family background play important roles in the development of AML. Three major environmental insults have been implicated in the increased incidence of AML: (1) ionizing radiation, (2) chemotherapeutic agents, and (3) occupational exposure to chemicals [1–6].

Ionizing radiation induces DNA damage leading to chromosomal breaks which may cause mutations, deletions, and translocations. The extent of this damage depends on the type of radiation, the amount and rate of absorption, distribution of the absorbed energy in the tissue, and the intervals between the radiation exposures [7–10]. The incidence of AML in atomic bomb survivors has been estimated to be as high as 24-fold than in the control population. This increase in leukemia incidence in the atomic bomb survivors started to show up 3 years after the radiation exposure, reached its peak at 6–8 years, and leveled off after 20 years [11–13]. The cumulative mortality studies of patients irradiated for the treatment of ankylosing spondylitis have shown a 10-fold increase in the incidence of acute leukemia. Also, radiation therapy in malignancies such as Hodgkin lymphoma and thyroid cancer has been associated with a higher incidence of AML, particularly when radiation is administered in combination with chemotherapy [7, 8].

Alkylating agents and topoisomerase type II inhibitors are amongst the most potent chemical factors in the development of acute leukemia and make up the bulk of the subcategory of therapy-related AMLs (t-AMLs) in the WHO classification (discussed later). The cumulative risk of drug-induced AML has been reported to range from 10% to 17% within 4–9 years from the beginning of chemotherapy or the combination of chemotherapy and radiation in patients with plasma cell myeloma, ovarian cancer, or Hodgkin lymphoma [14–22]. The latency period of alkalating-induced AML is usually 4–6 years. The topoisomerase type II inhibitor-associated AML has an overall shorter latency period, usually <3 years (discussed later). Also, immunosuppressive therapy in transplant patients and in patients with immune-associated disorders may increase the risk of AML [23, 24].

Occupational exposure to petroleum products (such as benzene), insecticides, and other

TABLE 11.1 Classification of AML according to the WHO.*

1. AML with recurrent genetic abnormalities (a) AML with t(8;21) (b) AML with abnormal eosinophils and inv(16) or t(16;16) (c) Acute promyelocytic leukemia with t(15;17) or variants (d) AML with 11q23 (MLL) abnormalities
2. AML with multilineage dysplasia (a) Following myelodysplastic syndrome or myelodysplastic/ myeloproliferative disorder (b) Without antecedent myelodysplastic syndrome
3. AML and myelodysplastic syndrome, therapy related (a) Alkylating agent related (b) Topoisomerase type II inhibitor related (c) Other types
4. AML not otherwise categorized (a) AML minimally differentiated (b) AML without maturation (c) AML with maturation (d) Acute myelomonocytic leukemia (e) Acute monoblastic and monocytic leukemia (f) Acute erythroid leukemia (g) Acute megakaryoblastic leukemia (h) Acute basophilic leukemia (i) Acute panmyelosis (j) Myeloid sarcoma

*Adapted from Ref. [1].

organic solvents increases the risk of AML [25–28]. The incidence of clonal chromosomal abnormalities is significantly higher in patients exposed to chemical solvents than the unexposed population. The most frequent aberrations include −5/del(5q), −7/del(7q), +8, and +21. Cigarette smoking, particularly in individuals over the age of 60, has shown a two-fold increase in the risk of AML [29–31].

Certain familial disorders are associated with a higher risk of AML. There is a 10- to 20-fold increased chance of leukemia, particularly AML, in patients with Down syndrome (trisomy 21) [32]. Many of these patients show an acquired mutation of the GATA-1 transcription factor which plays a role in megakaryocytic development. A significant proportion of AMLs in Down syndrome patients is of megakaryoblastic subtype. The incidence of AML is also high in inherited disorders with defective DNA repair, such as Bloom's syndrome, Fanconi's anemia, Wiscott–Aldrich syndrome, neurofibromatosis, Kostmann's syndrome (infantile agranulocytosis), and Diamond–Blackfan anemia [33, 34]. A rare constitutional trisomy 8 syndrome (with dysmorphic facial features and abnormal skeletal muscles) has also been associated with AML [35, 36].

Although retroviruses have been demonstrated to play a role in leukemogenesis of AML in experimental animals, no clear association has been found between AML and retroviruses in humans. The two major mechanisms of leukemogenesis by retroviruses in animal models are [28, 37–42]:

1. Encoding an oncogene that leads to leukemic transformation
2. Inappropriate activation of expression of a gene adjacent to its integration site.

Leukemogenesis, similar to most other cancer developments, appears to be a multistep process involving structural and functional changes in a cascade of genes leading to the clonal expansion of defective stem cells. These genetic alterations often include mutations of oncogenes and/or loss of tumor suppressor genes. The specific genetic events in the process of leukemogenesis are not currently well understood, though it has been suggested that at least two mutations are required: one leading to a proliferative advantage and the other causing impairment of the maturation process (the "two-hit" hypothesis). The following examples represent the multistep concept of leukemogenesis in AML.

In the chronic phase of chronic myeloid leukemia (CML), leukemic cells show t(9;22) resulting in the *BCR/ABL1* fusion gene. As CML progresses to the accelerated phase and then blast transformation, additional genetic abnormalities, such as mutation of *p53* (a tumor suppressor gene), evolve.

The high frequency of AML in patients with MDS strongly supports the two-hit hypothesis for leukemogenesis. MDS represents the first step, or the first hit, with frequent detectable chromosomal aberrations, including −5/del(5q), −7/del(7q), and +8. Evolution to AML, often with additional molecular and/or cytogenetic changes, depicts the final stage, or the second hit. There are studies indicating that the remission bone marrow samples from AML patients with t(8;21)(q22;q22) and the *RUNX1/RUNXT1* fusion transcript carry the abnormal fusion for several years after the completion of chemotherapy. These observations suggest that the *RUNX1/RUNXT1* may not be sufficient for the development of AML by itself, and additional mutation(s) are necessary [37, 40, 43]. Some of the patients with severe congenital neutropenia and a documented nonsense mutation in the G-CSF receptor develop AML, raising the possibility that this kind of mutation causes resistance to apoptosis, allowing more time for additional mutation(s) to occur.

A significant proportion of AMLs are associated with specific recurrent cytogenetic abnormalities such as t(8;21)(q22;q22), inv(16)(p13q22), t(15;17)(q11;q12), and 11q23 abnormalities. The products of the fusion genes resulting from chromosomal rearrangements play an important role in the evolution of AML (discussed later).

Pathology

Morphology

Acute myeloid leukemia refers to neoplasm of non-lymphoid hematopoietic progenitor cells. Therefore, it consists of subtypes representing various myeloid differentiations such as myeloblasts, promyelocytes, monoblasts, promonocytes, erythroblasts, and megakaryoblasts [1–4, 44]. In general, myeloblasts are the most predominant precursor cells in AML categories. The WHO requirement for the diagnosis of AML is the presence of 20% or more blast cells in the bone marrow or blood differential counts [1]. In addition to myeloblasts, "blast" counts in certain categories of AML may include monoblasts, megakaryoblasts, promonocytes,

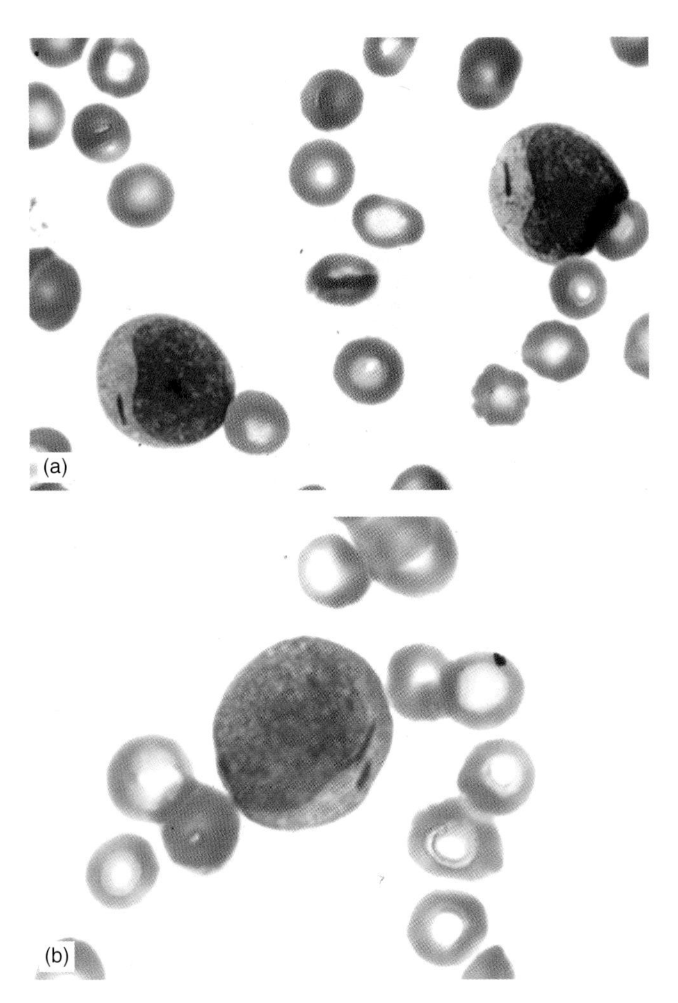

FIGURE 11.1 Peripheral blood smears (a and b) showing myeloblasts with Auer rods.

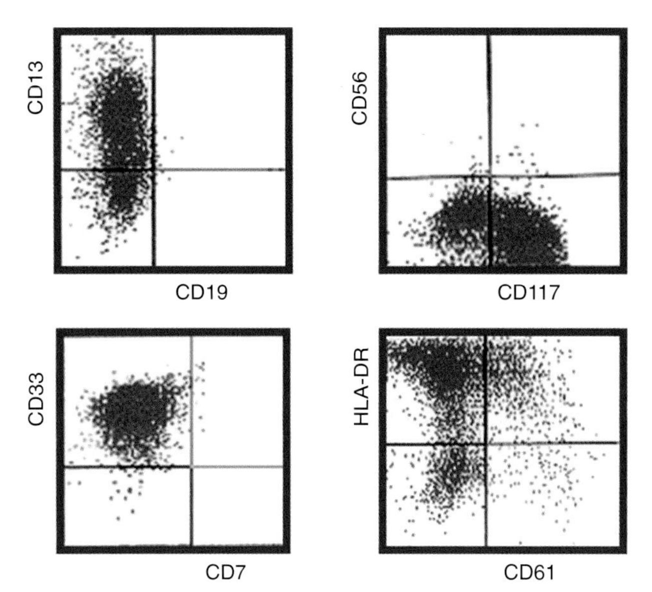

FIGURE 11.2 An example of dot plot analysis of an AML by flow cytometry. The blasts express myeloid-associated markers CD13, CD33, and CD117 and are positive for HLA-DR.

or promyelocytes. Erythroblasts are excluded from the blast count. In certain conditions, such as in the category of AML with recurrent genetic abnormalities, the requirement for ≥20% blasts may be sidestepped. Three morphologic types of myeloblasts have been described: Types I, II, and III. Type I myeloblasts contain no cytoplasmic granules, type II myeloblasts contain ≥20% cytoplasmic azurophilic granules, and type III myeloblasts contain 20 cytoplasmic azurophilic granules. Auer rods may be present (Figure 11.1). Myeloblasts are positive for CD13, CD33, and HLA-DR, and may express CD117, CD34, and myeloperoxidase (MPO). Promyelocytes are overall larger and carry larger quantities of azurophilic granules than myeloblasts. They depict a well-developed Golgi system and a round or an oval nucleus, which is often eccentric. Type III myeloblasts and promyelocytes share overlapping morphologic features, and therefore their distinction at times is difficult. Myeloblasts are HLA-DR-positive and may express CD34, whereas promyelocytes are negative for HLA-DR and CD34, but positive for CD13, CD33, and MPO and may express CD117. The morphologic features of monoblasts, promonocytes, erythroblasts, and megakaryoblasts are described later in the appropriate sections.

In general, the bone marrow biopsy/clot sections are hypercellular and show diffuse infiltration of the bone marrow by immature myeloid cells. Occasionally, the bone marrow is hypocellular. Blood examination may reveal anemia, leukopenia, and/or thrombocytopenia. Myeloid left shift is a common feature, and often a variable number of blasts are present. However, in some cases, at the time of bone marrow diagnosis, the peripheral blood smears may show no evidence of blast cells (aleukemic leukemia).

Immunophenotype and Cytochemical Stains

Immunophenotyping is an important component of bone marrow and blood evaluations in the current diagnostic work-up of acute leukemias for the following reasons [45, 46]:

1. Lineage assignment to distinguish AML from acute lymphoblastic leukemia (ALL) and to assign the leukemia to the proper subcategories, such as myeloblastic, monoblastic, erythroblastic, or megakaryoblastic (Figure 11.2).
2. Blast enumeration to confirm the presence of ≥20% blast cells in the bone marrow or blood and to evaluate post-treated samples for residual disease.
3. Search for aberrant expressions of CD molecules to use for the detection of residual disease and possible prognostic values.
4. Detection of ambiguous (bilineal, biphenotypic) leukemias.

To fulfill the above-listed missions, the flow cytometry and immunohistochemical laboratories utilize a panel of monoclonal antibodies. This panel, which may vary considerably from one laboratory to another, should contain antibodies against CD molecules that help identify (1) blast hematopoietic cells, (2) cells of myeloid lineage, and (3) cells of lymphoid lineage. For example, the following panel is currently used for flow cytometric studies of

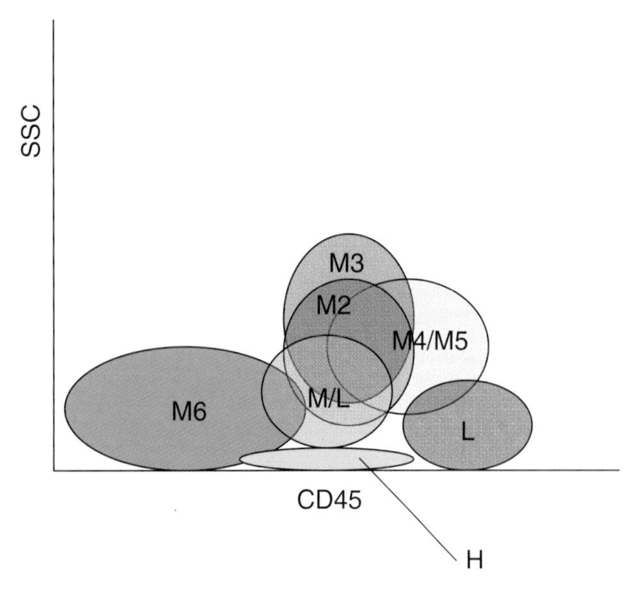

FIGURE 11.3 A diagram of SSC/CD45 features of hematopoietic malignancies by flow cytometry. H: hematogones, M/L: acute myeloid leukemias (minimally differentiated, without maturation) and acute lymphoid leukemias, L: mature lymphoid malignancies, M2: acute myeloid leukemia with maturation, M3: acute promyelocytic leukemia, M4/M5: acute myelomonocytic and acute monocytic leukemia, M6: erythroleukemia and multiple myeloma.

acute leukemia at the VA Greater Los Angeles Healthcare System:

Hematopoietic blasts	CD34, CD45, HLA-DR, and TdT
B-lineage	CD10, CD19, CD20, CD22, cytoplasmic CD22, and cytoplasmic CD79a
T-lineage	CD2, CD3, CD5, CD7, and cytoplasmic CD3
Myelomonocytic precursors	CD13, CD14, CD33, CD36, CD64, cytoplasmic CD13, and cytoplasmic MPO
Megakaryoblasts	CD41 and CD61
Erythroblasts	Glycophorin A (GPA) and CD71
Others	CD56, CD38

Four-color flow cytometry is the accepted standard of practice in most hematopathology laboratories. The side scatter (SSC)/CD45 characteristics of different types of hematopoietic malignancies are presented in Figure 11.3.

Most special cytochemical stains have been replaced by immunophenotyping in most laboratories. However, certain stains are still being used in the differential diagnosis and the classification of acute leukemia [47, 48].

Myeloperoxidase Stain: Myeloperoxidase is a lysosomal enzyme present in granulocytic and monocytic cells (Figure 11.4). MPO is expressed in neutrophilic and eosinophilic lineages in all stages of maturation, but in basophils it is more often detected in the immature forms. The mature basophils are usually negative for MPO. The intensity of MPO staining is less in monocytes than in granulocytes. Erythroid precursors and lymphocytes are MPO-negative. A peroxidase isoenzyme has been detected by electron microscopy in the dense tubular system of platelets and megakaryocytes, but by conventional techniques, these cells are MPO-negative [49].

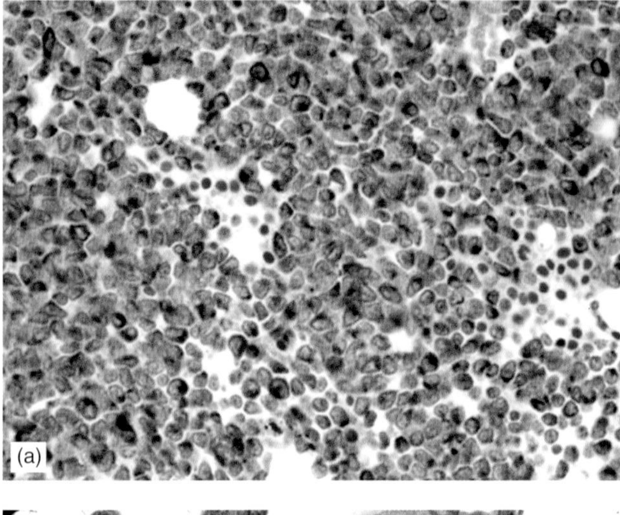

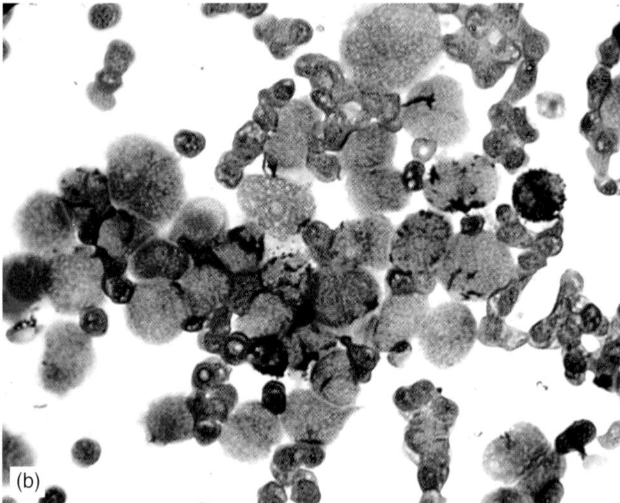

FIGURE 11.4 Immunohistochemical (a) and cytochemical (b) stains for MPO demonstrating numerous MPO-positive cells.

Myeloperoxidase activity declines rather rapidly. Air-dried unstained smears should be stored at cool temperatures, in the dark, and be used within 1–2 weeks.

Sudan Black B Stain: Sudan Black B is a lipophilic dye that stains the granulocytic series [47]. The pattern of reactivity of Sudan Black B in the granulocytic series is similar to that of MPO (Figure 11.5a). Monocytes are either negative or weakly positive with this stain. Lymphocytes, erythroid cells, megakaryocytes, and platelets are usually Sudan Black B-negative. Unlike MPO, Sudan Black B is stable, and therefore archival cytologic materials could be used for staining. Sudan Black B stain does not work in paraffin sections.

Periodic Acid-Schiff Reaction: Periodic acid-Schiff reaction in hematopoietic cells is primarily due to the presence of cytoplasmic glycogen. The granulocytic lineage and plasma cells show diffuse, fine PAS-positive granules, whereas dysplastic erythroid precursors (Figure 11.5b) and sometimes blasts in acute lymphoid leukemia, monocytic

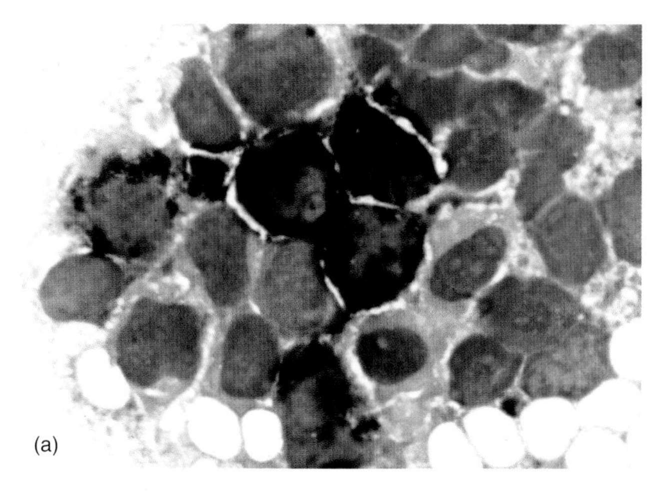

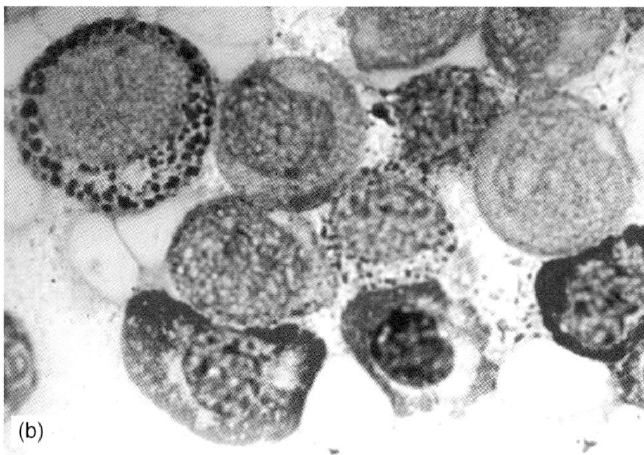

FIGURE 11.5 Bone marrow smears. (a) Cytochemical stains for Sudan Black B demonstrates dense, darkly stained cytoplasmic granules in myeloid cells. (b) Dysplastic erythroid precursors show coarse PAS-positive cytoplasmic granules.

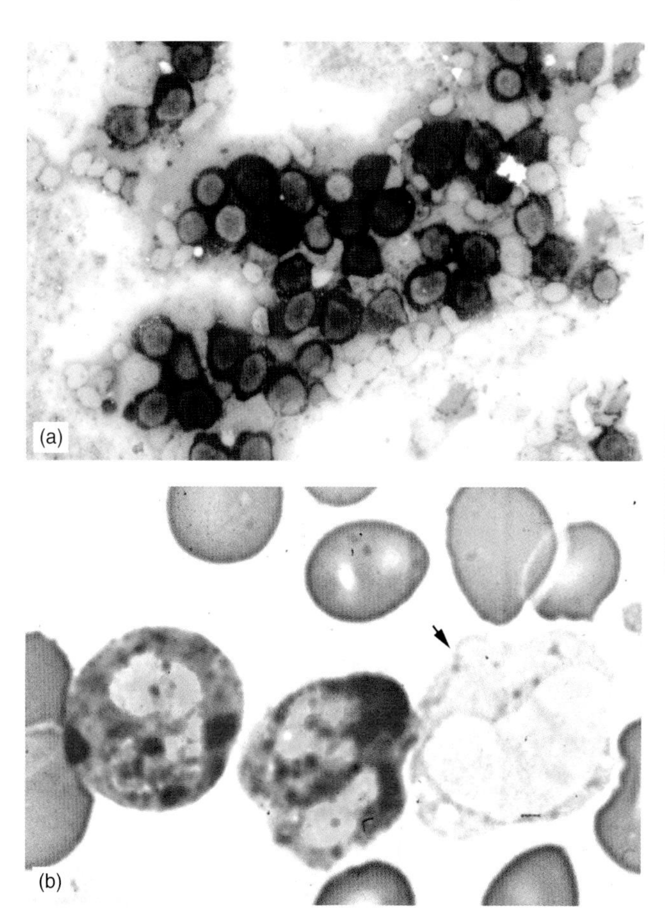

FIGURE 11.6 (a) Bone marrow smear showing numerous cells positive for an NSE stain. (b) Granulocytes stain for naphthol AS-D chloroacetate. A monocyte with a few cytoplasmic granules is present (arrow).

leukemia, and megakaryocytic leukemia show coarse PAS-positive cytoplasmic granules.

Alpha-Naphthyl Butyrate Esterase: Alpha-naphthyl butyrate esterase, also known as *non-specific esterase* (NSE), is a monocytic marker (Figure 11.6a) [47]. This stain is helpful in distinguishing acute leukemias with monocytic differentiation, as well as histiocytic lesions. However, lymphoblasts, erythroblasts, and megakaryoblasts may also show a few cytoplasmic-positive granules. Granulocytic series are negative for an NSE stain. An NSE activity is fluoride sensitive.

Naphthol AS-D Acetate Esterase: Naphthol AS-D acetate esterase is demonstrated in all stages of maturation in the granulocytic and monocytic series [47]. Lymphoblasts, erythroblasts, and megakaryoblasts may also show a few punctuate cytoplasmic-positive granules. The enzyme activity is inhibited by sodium fluoride in monocytes but not in granulocytes.

Naphthol AS-D Chloroacetate: Naphthol AS-D chloroacetate is primarily expressed in the granulocytic series (Figure 11.6b) and mast cells [47]. Other hematopoietic elements are essentially negative, though some monocytes, megakaryocytes, lymphoid and erythroid cells and their leukemic counterparts may show a weak reaction. Naphthol AS-D chloroacetate is very stable and is demonstrated in archival cytologic materials and paraffin-embedded tissue sections.

Molecular and Cytogenetic Studies

A garden variety of molecular genetic and cytogenetic abnormalities have been reported in AMLs. Some of these abnormalities are recurrent and more specific, such as [t(8;21)(q22;q22);(*RUNX1/RUNXT1*)], [with t(15;17)(q11;q12); (*PML/RAR*α)], [inv(16)(p13q22) or t(16;16)(p13;q22); (*CBF*β*/MYH11*)], and 11q23 (*MLL*) abnormalities, and others are non-specific [50–52]. The non-specific chromosomal abnormalities include translocations, trisomies, monosomies, deletions, and other structural changes which are described later with each AML subtype.

Clinical Aspects

The overall incidence of AML is about 3 per 100,000 population per year. The median age for AML onset is 60 years, with a male:female ratio of about 1 [53, 54]. The clinical symptoms are related to cytopenias and include weakness, fatigue, recurrent infections, and hemorrhagic episodes, such as gum bleeding or ecchymoses. Bone pain is infrequent.

Extramedullary infiltration (chloroma, granulocytic sarcoma) is occasionally seen, particularly in AMLs with monocytic differentiation. The most frequent extramedullary sites are skin, gum, and liver, but lymph nodes, intestinal tract, female reproductive systems, mediastinum, or other sites may also be involved. On rare occasions, the extramedullary involvement may be the very first presenting symptom.

Cytogenetic results are the most informative indicators of prognosis. The favorable karyotypes include t(8;21), t(15;17), and structural changes in 16q. Karyotypes with adverse clinical outcomes are monosomy 5 or 7, del(5q), and abnormal structural changes in 3q26. Resistant disease after first course of chemotherapy (>15% blasts in the bone marrow) also indicates poor prognosis. Five-year survival for the favorable prognostic category has been reported to be 70% with a 33% chance of relapse, whereas the figures for the poor prognostic category are 15% and 78%, respectively [55–57].

AML WITH RECURRENT GENETIC ABNORMALITIES

Several specific chromosomal translocations have been identified in AML patients, some of which are strongly associated with certain morphologic features. The most prominent translocations are t(8;21), t(15;17), t(16;16), or inv(16), and translocations involving the long arm of chromosome 11 (11q23).

A recently revised draft of the WHO classification has added several cytogenetic aberrations into this group, such as t(9;11)(p22;q23), t(1;22)(p13;q13), t(9;22)(q34;q11.2), and AML with mutation of *CEBPA* (new WHO classification in press [58b]). These entities are briefly discussed at the end of this chapter under Other Recurrent Genetic Abnormalities.

AML with t(8;21)(q22;q22);(*RUNX1/RUNXT1*)

Acute myeloid leukemia with t(8;21) is a relatively common leukemia accounting for about 18% of all cases of AML with cytogenetic abnormalities and 40% of AMLs with maturation (discussed later). In this balanced translocation, *RUNX1* (runt-related transcription factor 1), *AML1* gene, on the long arm of chromosome 21 (q22), fuses with the *RUNXT1* (acute myelogenous leukemia 1 translocation 1; also named MTG8, ETO or CBFA2T1) gene on the long arm of chromosome 8 (q22). This fusion results in an *RUNX1/RUNXT1* chimeric product.

Etiology and Pathogenesis

The etiology of AML with t(8;21) is not clear. Molecular studies suggest that altered transcriptional regulation and reduced apoptosis play important roles in the pathogenesis of this leukemia [58, 59]. The AML1 protein in conjunction with core-binding factor beta (CBFβ) forms a transcription factor. The AML/CBFβ transcription factor plays a regulatory role in a number of genes that are involved in myelogenesis and differentiation. Knockout of *AML1* or *CBFβ* gene in mice leads to defective hematopoiesis and embryonic death. Also, the *RUNX1/RUNXT1* fusion product reduces apoptosis by activating the expression of the anti-apoptosis gene *BCL-2* [60–62].

Pathology

Morphology

The morphologic features in a significant proportion of AML with t(8;21) are similar to those described in the category of AML with maturation (discussed later). Approximately 40% of AML with maturation show t(8;21) [1, 61, 63]. The myeloblasts are large, often with indented nuclei and basophilic cytoplasm. Type II and III myeloblasts are prominent and some blasts may contain large granules mimicking cytoplasmic granules seen in the Chediak–Higashi syndrome. Auer rods are frequent and also may be detected in the more mature myeloid forms. Promyelocytes, myelocytes, metamyelocytes, bands, and segmented neutrophils are present and often show dysplastic changes (Figure 11.7). Eosinophilia is common, and some cases may show increased bone marrow basophils and/or mast cells.

Approximately 7% of the AML cases with t(8;21) show morphologic features of acute myelomonocytic leukemia [64, 65] (discussed later). These patients depict peripheral blood monocytosis with the presence of immature forms and increased myeloblasts, monoblasts, and promonocytes in their bone marrows.

Rare cases of AML with t(8;21) may show blast counts of <20% in their blood or bone marrow.

Immunophenotype and Cytochemical Stains

Blasts and immature cells express myeloid-associated markers such as CD13, CD33, and CD117. CD34 is often positive and there may be aberrant expression of CD19 and/or CD56. Blast cells may show dim expression of TdT in a small proportion of cases. Because of the presence of type II and III myeloblasts and predominance of promyelocytes and myelocytes, the immature myeloid population on the CD45/SSC flow cytometry dot plot is moved up, demonstrating the granularity of the immature myeloid population. Blast cells are usually MPO- and/or NSE-positive.

Molecular and Cytogenetic Studies

The molecular detection of *RUNX1/RUNXT1* fusion is usually by fluorescence *in situ* hybridization (FISH) or reverse transcriptase polymerase chain reaction (RT-PCR) (Figure 11.8). It is important to know that some patients who have remained in continuous remission for a long period may still show *RUNX1/RUNXT1* mRNA in their leukocytes by RT-PCR techniques [43, 60, 67]. The clinical significance of the persistence of *RUNX1/RUNXT1* is not clear. However, the detection of *RUNX1/RUNXT1* fusion by itself may not indicate relapse or active disease.

The hallmark cytogenetic finding is the t(8;21)(q22;q22) (*RUNX1/RUNXT1*) (Figure 11.8). Complex translocations, such as t(8;21;14) or t(8;12;21), have been reported in occasional cases [68, 69]. Some patients may only show −Y or del(9q) [70]. In some cases, t(8;21) is cryptic and

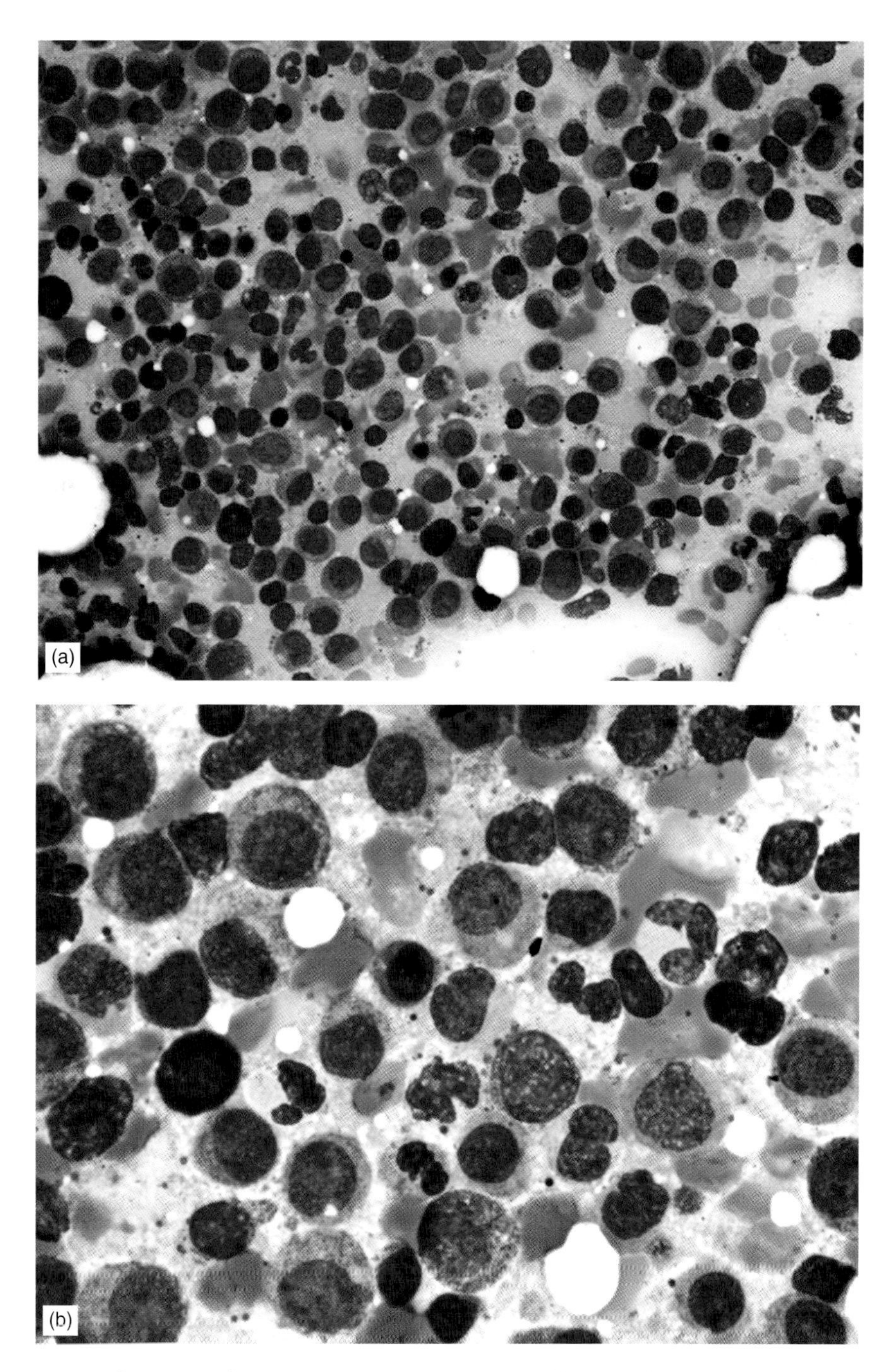

FIGURE 11.7 Bone marrow smear of a patient with t(8;21) AML demonstrating increased blasts with the presence of progressive maturation (acute myeloid leukemia with maturation, AML-M2): (a) low power and (b) high power.

undetectable by standard karyotyping. In such cases, FISH (Figure 11.8b) or RT-PCR technique is required to establish the *RUNX1/RUNXT1* rearrangement.

Clinical Aspects

Acute myeloid leukemia with t(8;21)(q22;q22) accounts for about 5–20% of AMLs in adults and is the most frequent AML in children [1, 60, 61]. The average age for adults is about 30 years, which is significantly lower than the average age for other types of AML. This leukemia has a favorable prognosis in adults [59, 71, 72]. The clinical outcome in children is poor.

Acute Promyelocytic Leukemia

Acute promyelocytic leukemia (APL) is one of the variants of AML associated with t(15;17)(q11;q12);(*PML/RAR*α) or other forms of chromosomal translocation involving the retinoic acid receptor-α (*RAR*α) gene. APL accounts for 5–10% of all AMLs [1, 73, 74].

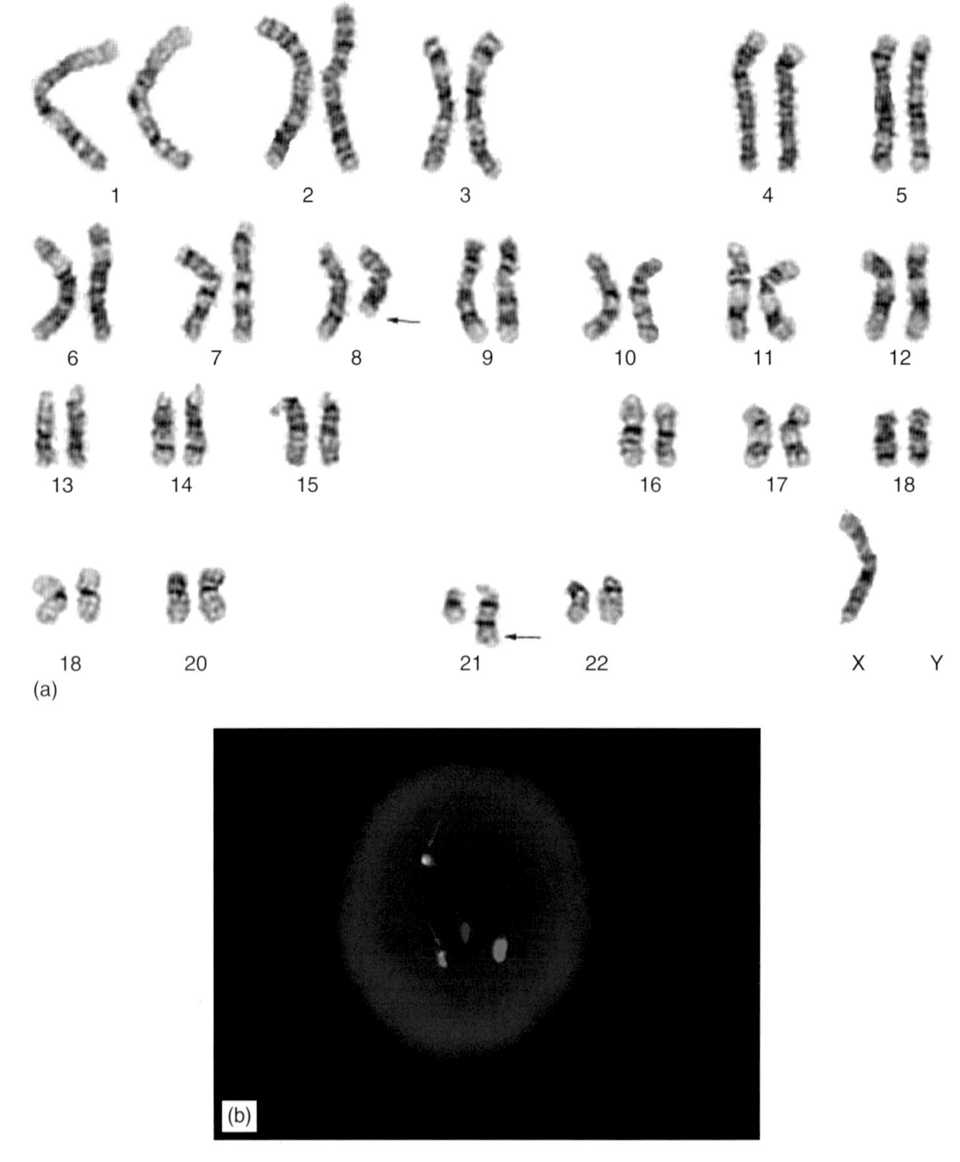

FIGURE 11.8 (a) G-banded karyotype showing t(8;21) (arrows). (b) The *RUNX1/RUNXT1* fusion is demonstrated by FISH (arrows).

Etiology and Pathogenesis

The etiology of APL is not known. Alteration in the *RARα* gene appears to be the key pathobiologic event in the development of APL. *RARα* is primarily expressed in hematopoietic cells. It binds to retinoic acid to regulate the transcription of genes that are important in the differentiation pathway in hematopoiesis. The t(15;17) leads to the production of *PML–RARα* fusion protein which is less sensitive to retinoic acid. This reduced sensitivity to retinoic acid may lead to persistent transcriptional repression, and therefore, prevention of further differentiation of promyelocytes [74–76]. The *PML–RARα* fusion protein is also capable of blocking normal *RARα*-mediated functions. It has been shown that transgenic mice expressing *PML–RARα* fusion protein in the myeloid progenitors in their bone marrow eventually develop a promyelocytic-type leukemia [60, 74]. *In vitro* studies of human stem cells transfected by a *PML–RARα* cDNA containing a retroviral vector have shown (1) a rapid induction of stem cell differentiation to promyelocytes, (2) maturation arrest at the promyelocyte stage, (3) preferential stem cell commitment to granulocytic differentiation, and (4) protection of apoptosis induced by the removal of hematopoietic growth factors [74, 77, 78]. The *PML* (promyelocytic leukemia) gene is suggested to encode a tumor suppressor protein essential for several signals in apoptosis and functions as a transcriptional co-activator with the *p53* tumor suppressor gene [74, 79, 80].

Pathology

Morphology

Two morphologic variants of APL have been described: APL with hypergranular promyelocytes and APL with microgranular (hypogranular) promyelocytes [1–4].

Hypergranular promyelocytes have a cytoplasm heavily loaded with azurophilic granules, which are often coarser

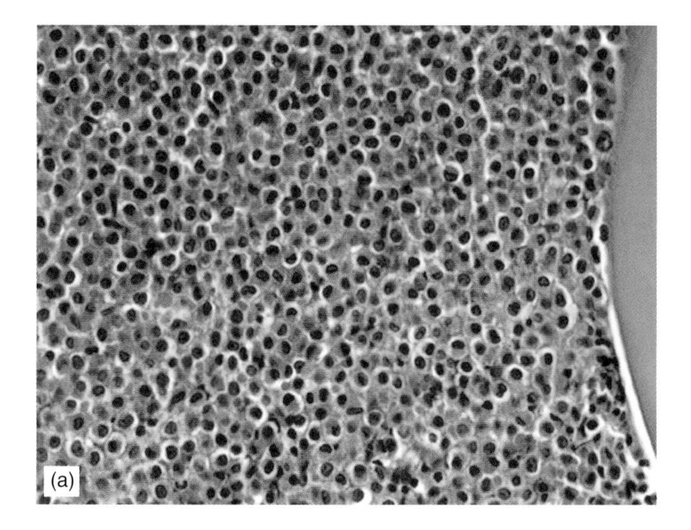

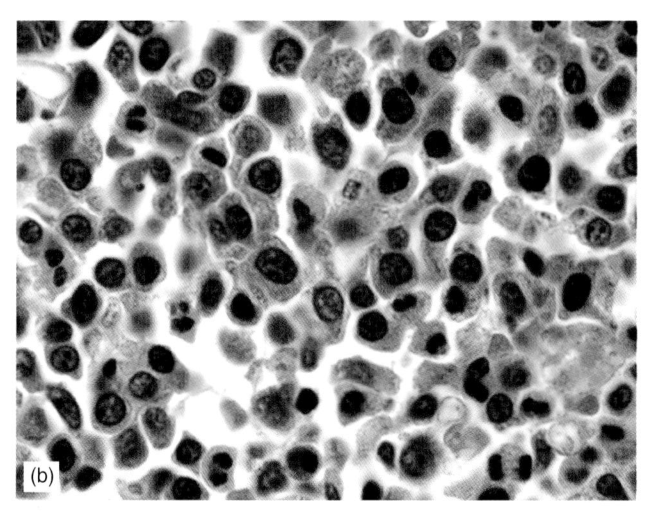

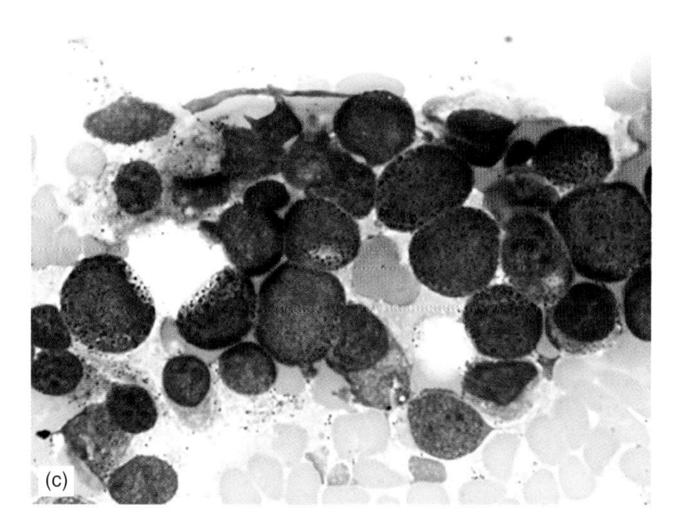

FIGURE 11.9 Acute promyelocytic leukemia. The bone marrow biopsy section shows sheets of immature cells with nuclear spacing and cytoplasmic granules (a and b). Numerous promyelocytes with numerous basophilic granules are demonstrated in the bone marrow smear (c).

and more numerous than the ones seen in normal promyelocytes (Figures 11.9 and 11.10). Auer rods are often present and in some cells appear in bundles (faggot cells) (Figures 11.10 and 11.11). Nuclei may appear round, irregular, folded, or dumbbell-shaped, but densely packed granules may obscure the visibility of the nuclei. The hypergranular promyelocytes are the predominant cells in the marrow, but smaller promyelocytes with basophilic cytoplasm and fewer azurophilic granules and microgranular promyelocytes are also present. Myeloblasts are less than promyelocytes and average around 10% of the bone marrow cells. The hypergranular variant, according to the literature, accounts for about 75–80% of the APLs. However, in our experience at the UCLA Medical Center and the VA Greater Los Angeles Healthcare System, we have seen more of microgranular variant than the hypergranular type.

In the remaining 20–25% of the cases of APL, the promyelocytes show abundant cytoplasm with lack of or sparse azurophilic granules. The azurophilic granules appear finer than the granules seen in the hypergranular variant (Figure 11.12). Auer rods and faggot cells may be present, but not so frequent as in the hypergranular subtype. The nuclei are predominantly bilobed, but folded and convoluted forms are often present, mimicking monocytic features. A small proportion of bone marrow cells may consist of myeloblasts, hypergranular promyelocytes, and small hyperbasophilic promyelocytes.

The bone marrow biopsy sections are hypercellular and show clusters and/or sheets of immature myeloid cells with abundant granular cytoplasm and nuclear spacing. Nuclei are commonly irregular or folded, and the nuclear chromatin is fine, often with prominent nucleoli.

The peripheral blood smears often show leukocytosis with the presence of atypical promyelocytes. A marked elevation of leukocyte count is seen more frequently in the microgranular variant.

Immunophenotype and Cytochemical Stains

The promyelocytes in APL show a homogenous expression of CD33 and often partial or dim expression of CD13 and CD15 (Figures 11.9–11.13). They are either negative for CD34 and CD117 or show partial expression. APL cells may also display aberrant expression of CD2, CD9, and/or CD56. The t(11;17);(*PLZF/RARα*) variant is often CD13+ and CD56+, whereas the t(5;17);(*NPM/RARα*) subtype is usually negative for CD13 and CD56 [81, 82]. HLA-DR and CD14 are commonly negative in APL cells. Promyelocytes are strongly MPO- and Sudan Black B-positive and -negative or weakly positive for NSE.

Molecular and Cytogenetic Studies

The genetic hallmark of APL is a translocation involving the *RARα* gene. Four major translocations with the involvement of the *RARα* gene have been associated with APL [74–76, 82–84]. These include:

t(15;17)(q23;q12);(*PML;RARα*)

t(11;17)(q23;q12);(*PLZF;RARα*)

t(11;17)(q23;q12);(*NuMA;RARα*)

t(5;17)(q23;q12);(*NPM;RARα*).

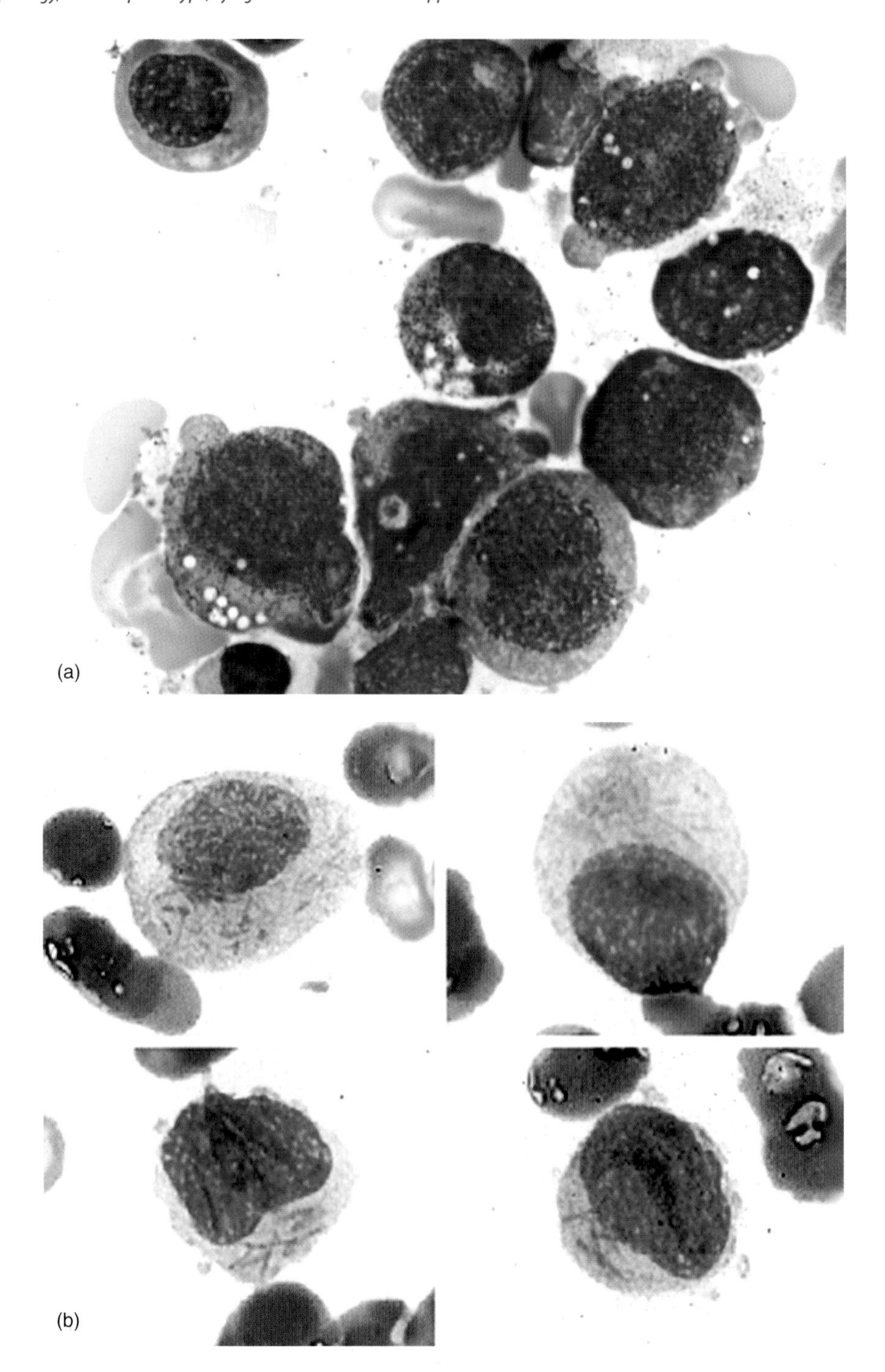

FIGURE 11.10 Acute promyelocytic leukemia. (a) Bone marrow smear demonstrating several dysplastic promyelocytes. (b) Promyelocytes with several Auer rods (faggot cells) are demonstrated (blood smears).

The most common translocation is t(15;17) which is associated with the expression of PML–RARα fusion protein (Figure 11.14a). The t(11;17)(q23;q11.12) variant represents fusion of the *RARα* gene with the *PLZF* (promyelocytic leukemia zinc finger) gene (Figure 11.15). PLZF protein is expressed in myeloid lineages and its expression is downregulated during differentiation. A rare variant, t(11;17) (q13; q11.12), involves the *NuMA* (nuclear matrix-mitotic aparatus protein) gene [84, 85]. Translocation of (5;17)(q35;q11.12) is another rare variant which involves the nucleophosmin (*NPM*) gene [74, 86]. This gene plays a role in the regulation of ribosomal nuclear processing and transport. Other infrequently reported genetic abnormalities include *STAT5b/RARα* fusion, der(7)(7;8)(q34;q21), del (6p23), partial long arm deletion of chromosome 17, and complex four-way variant t(15;17) [87–91].

FISH and RT-PCR studies (Figures 11.14b and 11.16) are routinely performed for the detection of *PML/RARα*

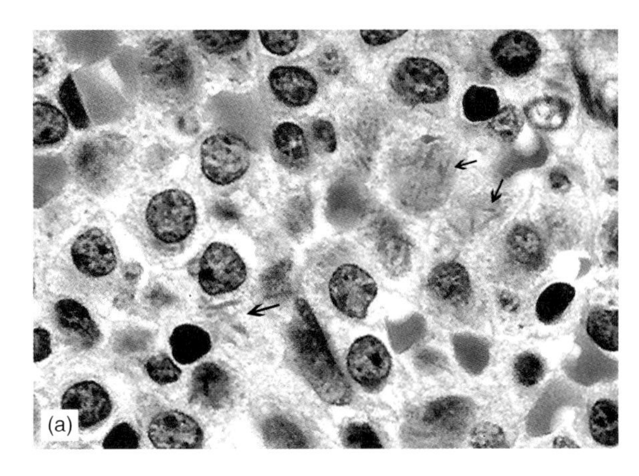

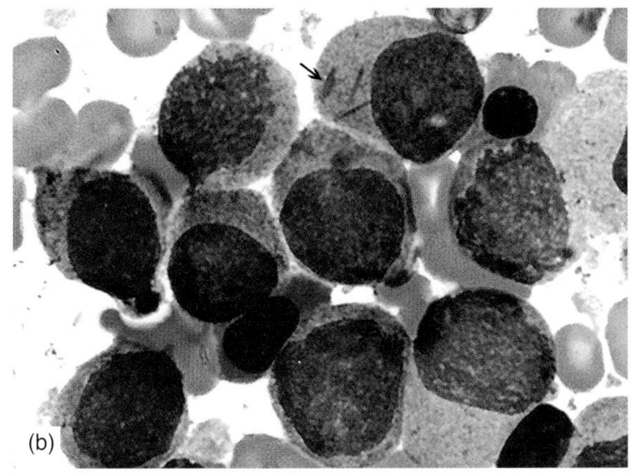

FIGURE 11.11 Bone marrow biopsy section (a) and smear (b) of a patient with acute promyelocytic leukemia showing abnormal promyelocytes with several Auer rods (faggot cells) (arrows).

fusion either to establish the diagnosis of promyelocytic leukemia or to rule out residual disease [86, 92, 93].

Clinical Aspects

Acute promyelocytic leukemia is primarily seen in young adults and middle-aged patients, but it may occur at any age. Clinical symptoms are related to complications of cytopenia and disseminated intravascular coagulopathy (DIC) [84, 94]. Weakness/fatigue, infection, and hemorrhagic episodes are often complications of anemia, granulocytopenia, and thrombocytopenia, respectively. DIC is either present at diagnosis or detected soon after chemotherapy. DIC is a serious complication which may lead to cerebrovascular or pulmonary hemorrhage in up to 40% of patients. The risk is reported to be higher in the microgranular variant of APL. Three major factors may contribute to the mechanism of DIC: (1) release of tissue factor which is involved in the activation of factor X through factor VII, (2) release of cancer procoagulants which activate factor X independent of factor VII, and (3) increased expression of annexin II receptor on leukemic promyelocytes [94, 95]. Annexin II receptor binds plasminogen and increases plasmin formation [95].

Acute promyelocytic leukemia is one of the favorable types of AML. Favorable prognostic factors include age under 30 years, initial leukocyte count <10,000/μL, and platelet count >40,000/μL [94, 96]. There are studies suggesting that the expression of CD56 on the leukemic promyelocytes, methylation of *p15* kinase inhibitor gene, and t(11;17);(*PLZF/RARα*) are associated with less favorable prognosis [94, 96, 97].

All-*trans* retinoic acid (ATRA) is a highly effective therapeutic agent [98–100]. It accelerates the terminal differentiation of leukemic promyelocytes and induces clinical remission. For complete molecular remission and long-term survival, a combination of ATRA and cytotoxic chemotherapy is necessary. APL patients with t(11;17);(*PLZF/RARα*) do not respond to ATRA.

AML with inv(16)(p13q22) or t(16;16)(p13;q22); (*CBFβ/MYH11*)

Acute myeloid leukemia with inv(16)(p13q22) or t(16;16) (p13;q22) is one of the variants of AML characterized by fusion of the *CBFβ/MYH11* genes. This leukemia depicts myelomonocytic differentiation with the presence of abnormal eosinophils. It accounts for about 10% of all AMLs [1].

Etiology and Pathogenesis

The etiology of AML with structural abnormalities of chromosome 16 and *CBFβ/MYH11* fusion is not known. It has been suggested that the CBFβ/MYH11 (core-binding factor, beta subunit/smooth muscle myosin heavy chain 11) fusion protein inhibits the function of the *AML1/CBFβ* transcription factor leading to the repression of transcription [101–103]. The *CBFβ/MYH11* fusion gene disrupts the normal transcription factor activity of CBF functions as a class II mutation. In addition, most of these patients are known to possess mutually exclusive mutations of the receptor tyrosine kinases (RTKs), *c-KIT*, and *FLT3*, as well as *RAS* genes. These sets of mutations provide a paradigm for the "two-hit" hypothesis of leukemogenesis [103].

Pathology

Morphology

The bone marrow samples show myeloid left shift, increased number of immature myelomonocytic cells, and the presence of atypical eosinophilic precursors (Figures 11.17 and 11.18). Myeloblasts (including types II and III), monoblasts, and promonocytes usually account for ≥20% of the total marrow cells, but occasionally may be less. Eosinophils usually constitute >5% of the marrow differential counts and appear to be a part of the leukemic clone. Some of the eosinophilic promyelocytes, myelocytes, and metamyelocytes contain large purple-violet granules in addition to eosinophilic granules [1–4]. These atypical granules are rarely found in more mature eosinophils. In rare cases of AML with inv(16)(p13q22) or t(16;16)(p13;q22), the atypical eosinophils may not be present, or instead of both myeloid and monocytic differentiation, the acute leukemia may represent only myeloid or only monocytic features.

The bone marrow biopsy and clot sections are usually hypercellular with increased immature myeloid forms and increased blasts. There is often evidence of eosinophilia.

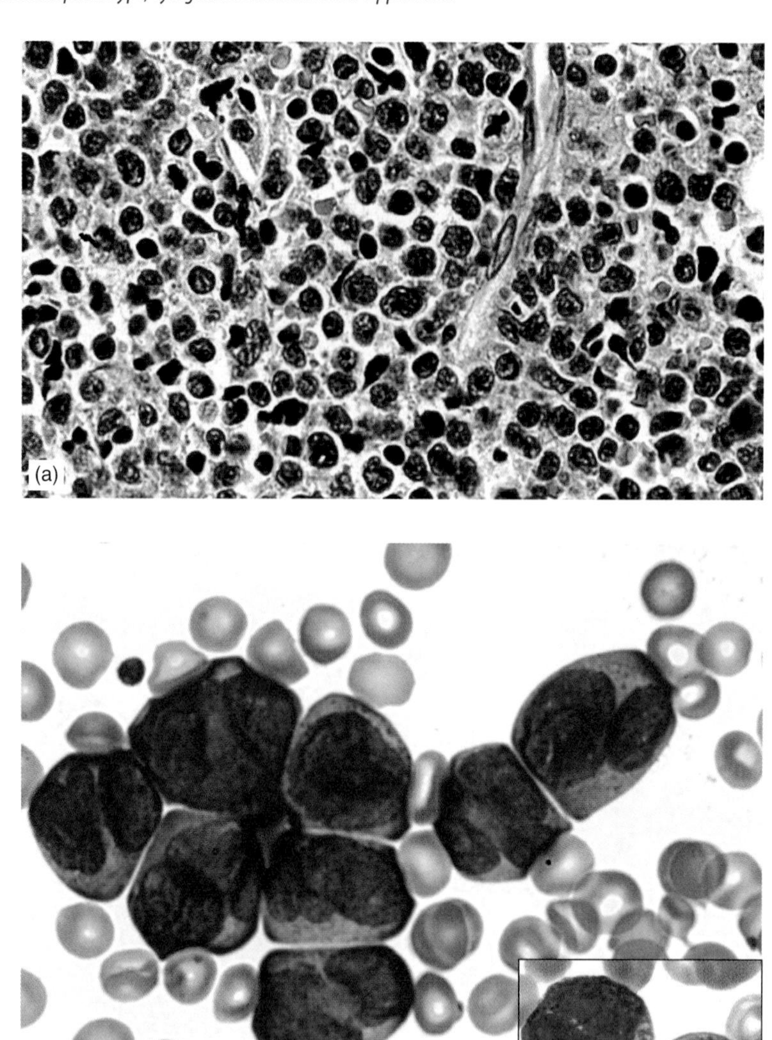

FIGURE 11.12 Acute promyelocytic leukemia, hypogranular variant. Bone marrow biopsy section (a), bone marrow smear (b), and peripheral blood smear (inset) showing hypogranular promyelocytes with convoluted nuclei.

The blood smears may show eosinophilia with the presence of blasts and promonocytes. Absolute monocytosis is a frequent finding.

Immunophenotype and Cytochemical Stains

The immature myelomonocytic population expresses CD13 and CD33 with a partial expression of CD11c, CD14, CD15, CD34, CD36, CD64, CD117, HLA-DR, and MPO by flow cytometry. Immunohistochemical stains such as MPO, lysozyme, and CD68 are used for the evaluation of the bone marrow myelomonocytic component, and CD34 and CD117 stains are often helpful for the estimation of blast cells. An aberrant expression of CD2 has been frequently observed.

The cytochemical stains show strong MPO positivity for the granulocytic lineage and various degrees of diffuse cytoplasmic NSE staining for the monocytic population. The abnormal eosinophils may be weakly positive for naphthol AS-D chloroacetate esterase.

Molecular and Cytogenetic Studies

CBFβ/MYH11 fusion with inv(16)(p13q22) or t(16;16)(p13;q22) is the characteristic genetic feature of this leukemia (Figures 11.19 and 11.20) [1, 103, 104]. The fusion transcript is detected by RT-PCR. Karyotyping and FISH studies reveal chromosome 16 paracentric inversion or translocation between the two chromosome 16s. Other associated cytogenetic abnormalities include trisomies 8, 21, and 22 as well as the loss of Y-chromosome deletion of the long arm of chromosome Y [104–106].

Clinical Aspects

Acute myeloid leukemia with structural abnormalities of chromosome 16 and *CBFβ/MYH11* fusion mostly occur in middle-aged patients, but it may occur at any age. The leukemia is associated with a favorable prognosis with a complete remission rate of >90%. In a large retrospective

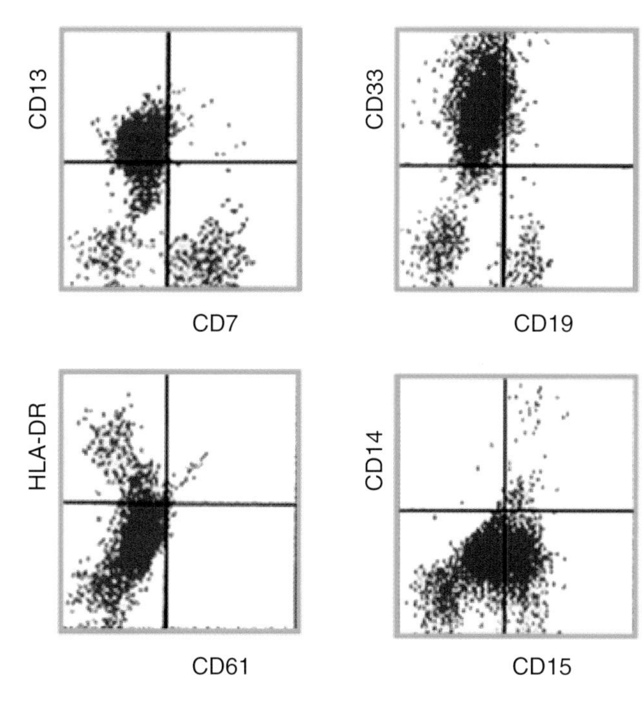

FIGURE 11.13 Acute promyelocytic leukemia. Flow cytometric studies reveal expression of CD13, CD15, and CD33 in a large population of cells. Only a small number of cells are positive for HLA-DR or CD14.

clinical survey of 110 patients conducted by the French AML Intergroup, the median age at diagnosis was 34 years with a female:male ratio of slightly >1 [104]. The estimated overall survival at 3 years was 58%. The adverse prognostic factors include age >35, elevated WBC of ≥120,000/μL and thrombocytopenia of ≤30,000/μL. A combination of cytotoxic drugs, such as daunorubicin and cytarabine with or without mitoxantrone, has been used for induction therapy.

AML with 11q23 (*MLL*) Abnormalities

The 11q23 (*MLL*) abnormalities are frequently observed in AMLs with monocytic differentiation, such as acute myelomonocytic and acute monocytic leukemias.

Over 73 different recurring translocations involving more than 50 partner genes have been reported involving 11q23 in acute leukemias [60, 94, 107].

Etiology and Pathogenesis

The etiology of AML with 11q23 abnormalities is not known. These abnormalities involve the *MLL* (mixed-lineage leukemia) gene with translocation breakpoints in between exons 5 and 11. The MLL protein has a potential DNA-binding site (AT-hook) which is able to bind to DNA and regulate the expression of genes that are important in hematopoiesis, including the development of myelomonocytic lineages [60, 94]. The translocation of the *MLL* gene results in a chimeric gene product that may play a role in leukemogenesis. Chimeric mice with t(9;11);(*AML/AF9*) have been shown to develop AML [108].

Pathology

Morphology

Monocytic differentiation is one of the hallmarks of AMLs with 11q23 abnormalities. Most of the cases fall into the category of acute myelomonocytic or acute monocytic leukemia with increased numbers of monoblasts and promonocytes in the bone marrow or peripheral blood (Figure 11.21) (discussed later). Monoblasts have variable amounts of dark-blue cytoplasm with no or few azurophilic granules or vacuoles. The nuclei are round or slightly indented or folded. The nuclear chromatin is finely dispersed and one or more prominent nucleoli are present. Promonocytes have more abundant cytoplasm which is less basophilic. They may contain few cytoplasmic azurophilic granules or vacuoles. The nuclei are irregular, folded, or convoluted with fine nuclear chromatin and often inconspicuous nucleoli.

Immunophenotype and Cytochemical Stains

The monoblasts and promonocytes express CD4 (dim), CD11c, and CD14. CD34 and CD117 are usually negative. Myeloblasts often express CD13, CD33, and CD117 and may express CD34. CD36, CD64, and HLA-DR are usually expressed on both myeloblasts and monocytic precursors.

Immunohistochemical stains such as MPO, lysozyme, and CD68 are used for the evaluation of the bone marrow myelomonocytic component, and CD34 and CD117 stains are often helpful in the estimation of myeloblasts.

The cytochemical stains show strong MPO positivity for the granulocytic lineage and various degrees of diffuse cytoplasmic NSE staining for the monocytic population.

Molecular and Cytogenetic Studies

The gene involved in the translocation of 11q23 is the *MLL* (also known as *ALL1* or *HRX*) gene [60, 109, 110]. Over 73 different translocations involving the *MLL* gene have been reported in both acute myeloid and lymphoid leukemias [60, 107 111, 112–116]. The rearrangement of 11q23 is common in patients with acute myelomonocytic and acute monocytic leukemias, particularly in children [60, 117–119]. Approximately 75% of acute leukemias in infants under 1 year show 11q23 abnormalities [110, 120, 121]. These leukemias are of acute lymphoblastic type or AML with monocytic differentiation. The translocation of (4;11)(q21;q23); (*AF4;MLL*) is the common translocation in children with pre-B ALL [122], and t(9;11)(p22;q23); (*MLLT3;MLL*) or t(10;11)(p12;q23);(*ABI1;MLL*) have been reported in association with adult myeloid leukemias [113, 114].

Clinical Aspects

Myeloid leukemias with abnormalities of 11q23 account for about 5% of total AMLs. In infants younger than

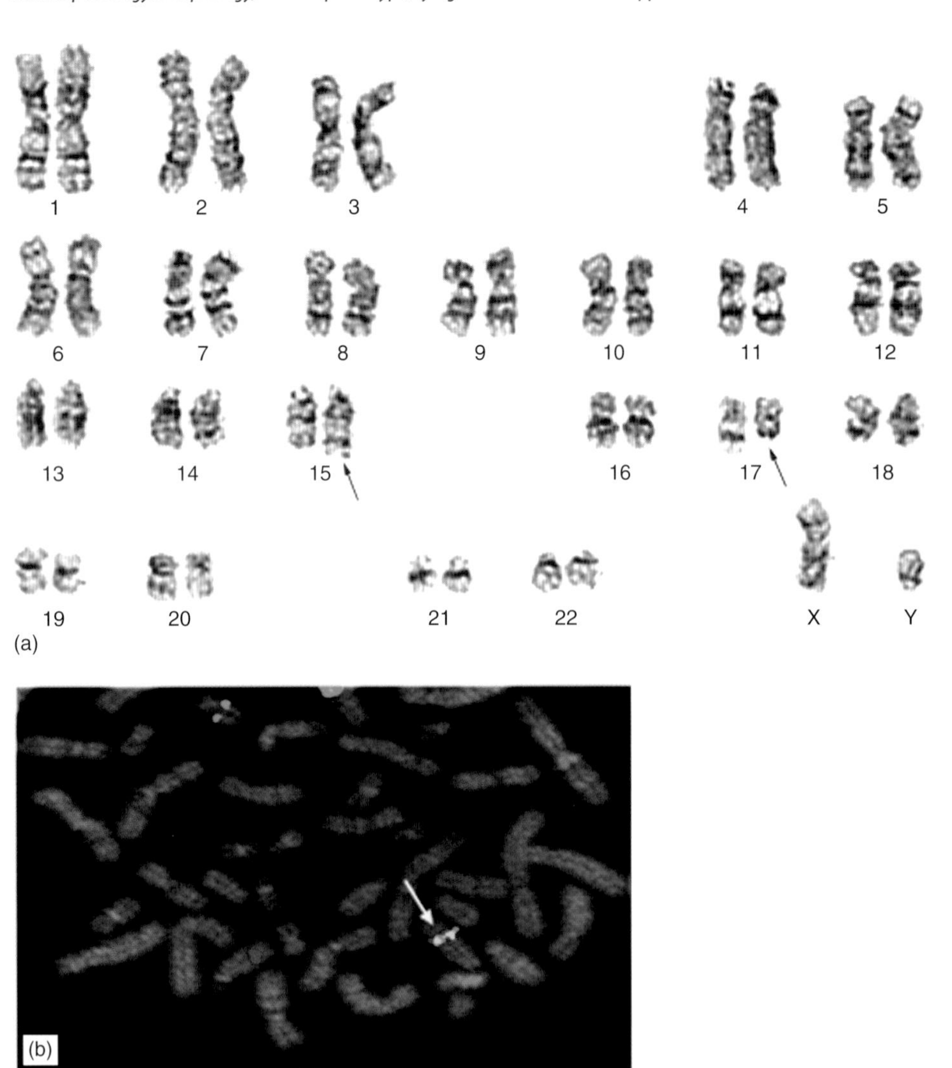

FIGURE 11.14 (a) G-banded karyotype showing t(15;17) (arrows). (b) The *PML–RARα* fusion is demonstrated by a single fusion PML (red) and RARα (green) FISH probe (arrow).

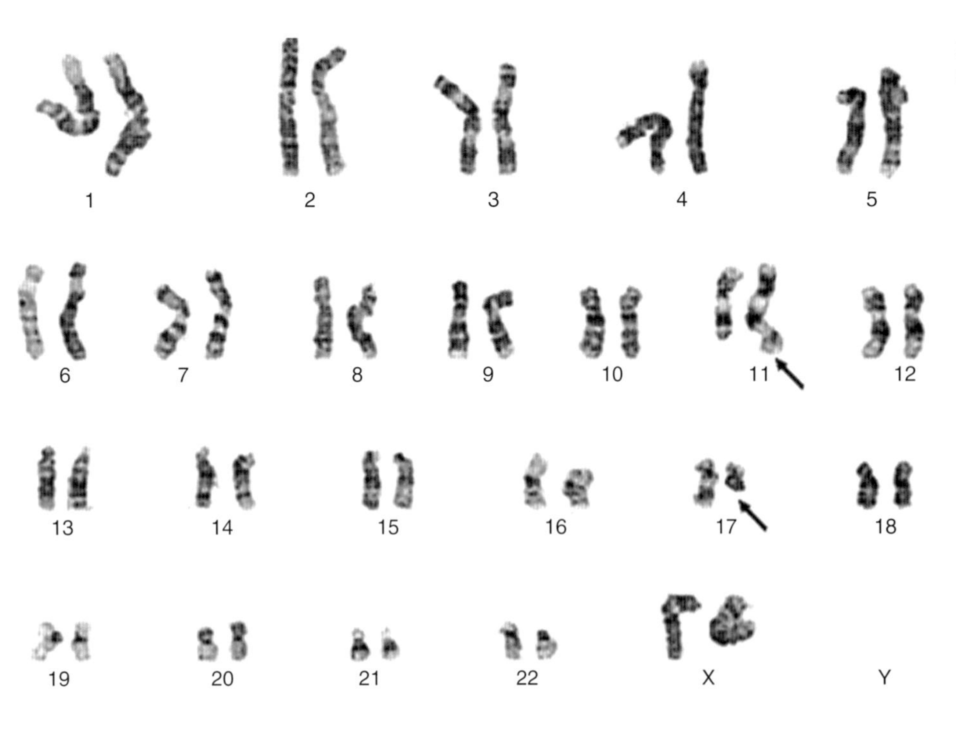

FIGURE 11.15 G-banded karyotype showing t(11;17) (arrows).

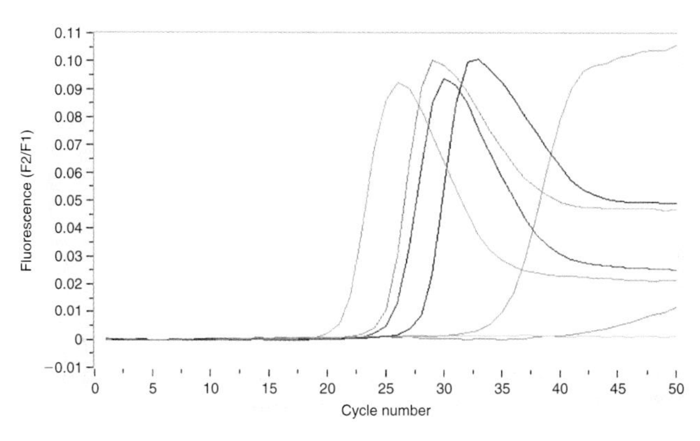

FIGURE 11.16 RT-PCR technique demonstrating a positive signal (the blue curve) in a patient with *PML/RARα* fusion (acute promyelocytic leukemia). Other curves are positive controls. The flat green and blue lines are negative controls.

1 year, up to 75% of all acute leukemias involve translocation of 11q23. These leukemias are primarily of acute myelomonocytic, acute monocytic, or acute lymphoid types. The translocation of 11q23 in adults is often associated with therapy-related leukemias, particularly after treatment with topoisomerase II inhibitors (discussed later). The leukemias with 11q23 abnormalities appear to fall into the category of leukemias with intermediate survival rate.

Differential Diagnosis

Cytogenetic aberrations are the main pathognomonic markers in AMLs with recurrent genetic abnormalities. However, certain morphologic and immunophenotypic features are helpful in distinguishing these leukemias. For example, a significant proportion of AMLs with t(8;21) are similar to those described in the category of AML with maturation, with the presence of type II and III myeloblasts and predominance of promyelocytes and myelocytes (discussed later).

Acute promyelocytic leukemia, particularly the microgranular variant, may mimic acute leukemias with monocytic differentiation. The leukemic promyelocytes, unlike monocytic cells, often show several Auer rods, are strongly MPO and Sudan Black B-positive, and do not express CD4, CD14, or HLA-DR (Table 11.2).

Acute myeloid leukemia with inv(16)(p13q22) or t(16;16)(p13;q22) depicts myelomonocytic differentiation with the presence of abnormal eosinophils. Most of the acute leukemias with 11q23 (*MLL*) abnormalities are morphologically of myelomonocytic or monocytic types.

AML WITH MULTILINEAGE DYSPLASIA

Acute myeloid leukemia with multilineage dysplasia is defined in the WHO classification as an acute leukemia with the presence of ≥20% myeloid blasts in bone marrow or blood and the evidence of multilineage dysplasia [1]. Dysplastic changes occur in more than one lineage with ≥50 cells in each lineage affected. Megakaryocytic dysplasia is a

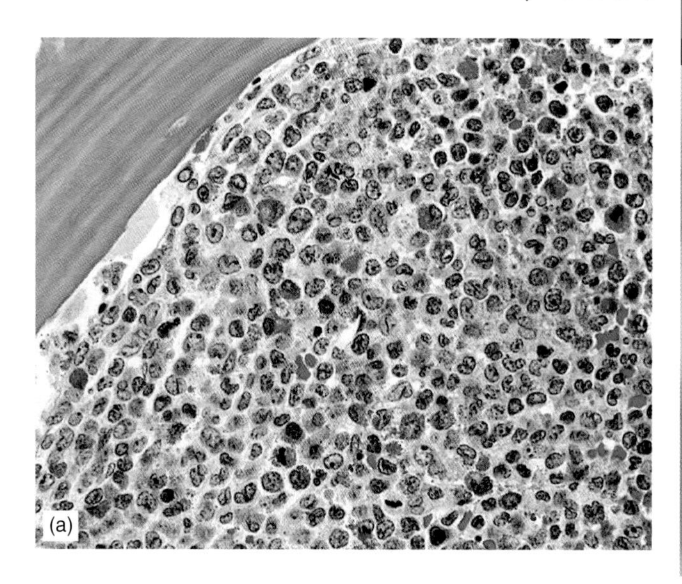

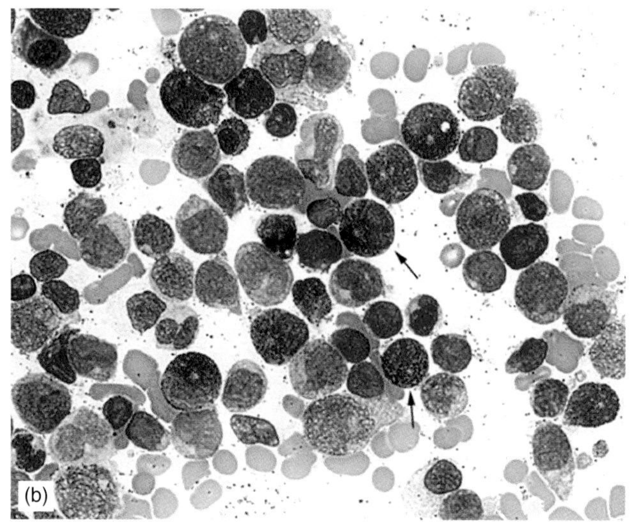

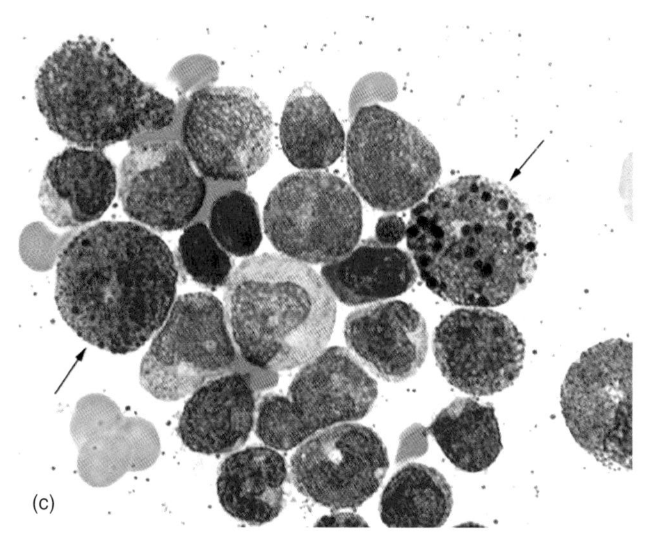

FIGURE 11.17 Bone marrow biopsy section (a) and bone marrow smear (b and c) from a patient with acute myelomonocytic leukemia with atypical eosinophilia and inv(16)(p13q22). Atypical eosinophils contain a mixture of eosinophilic and basophilic granules (arrows).

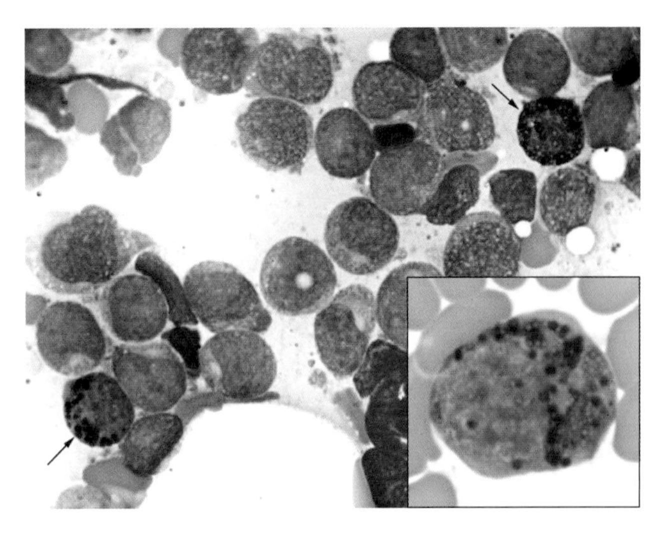

FIGURE 11.18 Bone marrow smear from a patient with acute myelomonocytic leukemia with atypical eosinophilia and inv(16)(p13q22). Atypical eosinophils contain a mixture of eosinophilic and basophilic granules (arrows and inset).

dominant feature. Dysplastic changes may precede the development of acute leukemia but remain as a part of the picture.

Etiology and Pathogenesis

A significant proportion of patients with AML and multilineage dysplasia follow a history of MDS, and therefore share the etiology and pathogenesis of MDS (see Chapter 8).

Pathology

Morphology

The presence of ≥20% blasts and significant multilineage dysplasia are the diagnostic hallmarks for this category. Dysplastic changes may involve all non-lymphoid hematopoietic cells, but often affect megakaryocytic and myeloid lineages (Figure 11.22). At least 50% of the cells in each lineage must show dysplastic morphology [1, 123].

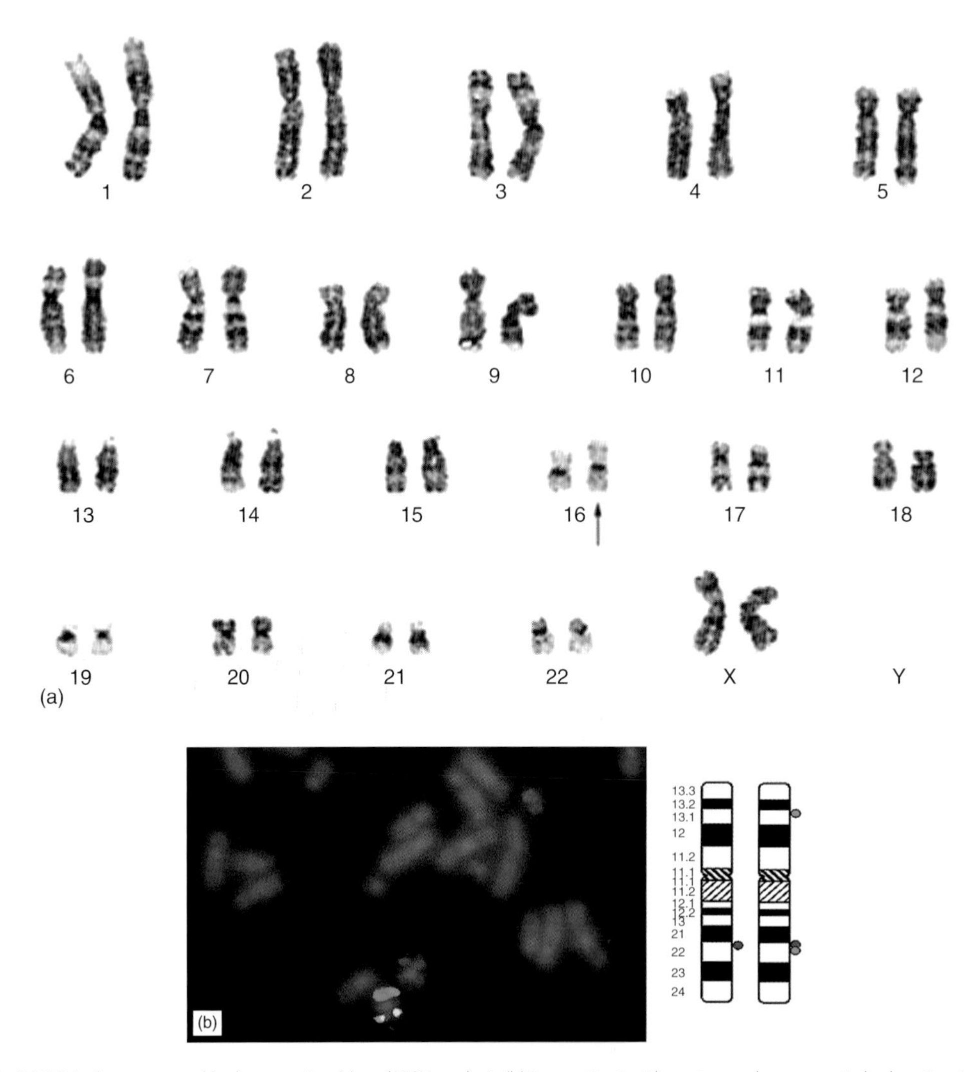

FIGURE 11.19 A t(16;16) is demonstrated by karyotyping (a) and FISH analysis (b) in a patient with acute myelomonocytic leukemia with atypical eosinophilia.

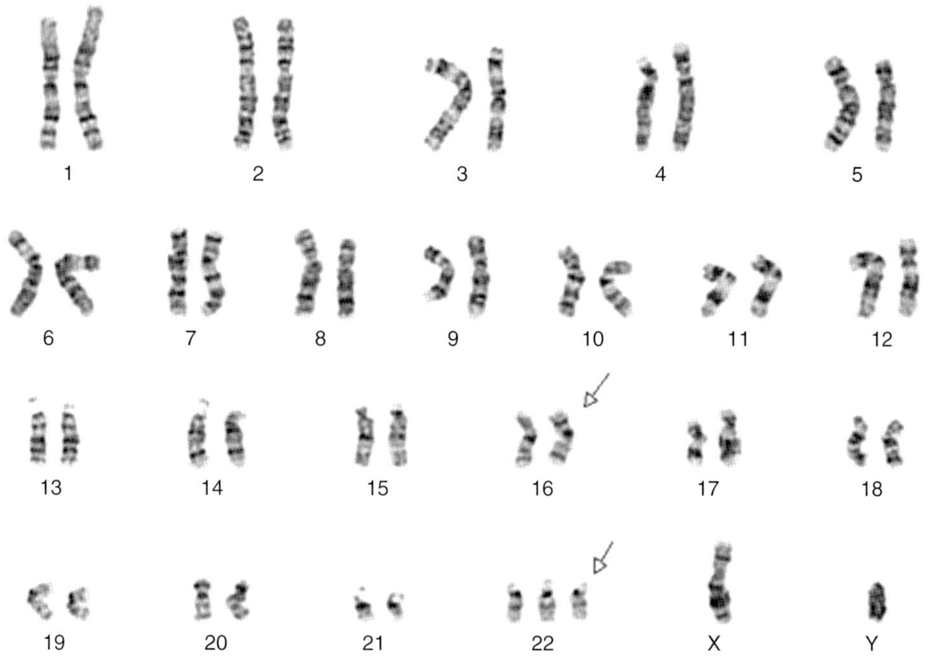

FIGURE 11.20 Karyotyping of bone marrow cells in a patient with acute myelomonocytic leukemia demonstrating inv(16)(p13q22) and trisomy 22.

Micromegakaryocytes with hypogranular cytoplasm, hypolobated nuclei, and/or mono- or binucleated forms are prominent. Large bizarre megakaryocytes may also be present. Neutrophilic series are often hypogranular and hyposegmented or may show abnormal segmentation. Erythropoiesis may appear megaloblastic or show nuclear budding or fragmentation. Ringed sideroblasts may be present.

Pancytopenia is a frequent finding in blood examinations. Blasts are often present in various numbers and dysplastic changes are observed, and these may be more obvious in the peripheral blood smears than the bone marrow smears. Biopsy and clot sections are usually hypercellular and show myeloid left shift with increased blasts.

Immunophenotype

Flow cytometry shows a population of blast cells expressing myeloid-associated markers. CD34 and CD117 are frequently expressed. In some cases, blasts may show aberrant expression of CD4, CD7, and/or CD56. The hypogranularity of the myeloid cells may lead to a lower SSC in flow cytometric dot plot preparations. The myeloid precursors may show increased expression of CD11a and CD66 and reduced expression of CD10 and CD116.

Immunohistochemistry may help to estimate the blast component and myeloid proportion by using CD34, CD117, and MPO stains.

Molecular and Cytogenetic Studies

No recurrent molecular or cytogenetic changes have been reported in this group. A garden variety of chromosomal deletions, monosomies, and trisomies have been observed, such as −5/del(5q), −7/del(7q), del(11q) (Figure 11.23), del(12p), del(20) (Figure 11.25), and trisomy 8, 9, 11, 18, 19, and 21. Most of these changes are also frequently associated with MDS and chronic myeloproliferative disorders (see Chapters 8 and 9). Non-specific translocations, such as t(1;7)(q10;p10), t(3;21)(q26;q22), and t(6;9)(p23;q34), have also been occasionally observed [123–127]. The possibility of distinguishing AML with multilineage dysplasia from other subtypes based on gene-expression profiling has been suggested [128].

Clinical Aspects

Acute myeloid leukemia with multilineage dysplasia is a disease of the elderly with a median age of 60 years [53, 54]. Cytogenetic risks, age, and multilineage dysplasia have been reported to correlate inversely with the overall survival in the AML patients [129–131].

Differential Diagnosis

Acute myeloid leukemia with multilineage dysplasia is distinguished from refractory anemia with excess blasts (RAEB) by the presence of ≥20% blasts in bone marrow and/or peripheral blood. It has morphologic overlapping features with acute erythroid leukemia (erythroid/myeloid), but unlike erythroid leukemia, in this leukemia the bone marrow erythroid component is not ≥50%. The requirement of significant multilineage dysplasia of >50% of the affected cells distinguishes this leukemia from most other types of AML.

AML AND MDS, THERAPY RELATED

Therapy-related AML (t-AML) and therapy-related MDS (t-MDS) represent spectrums of a progressive clonal hematopoietic disorder which is evolved following cytotoxic chemotherapy and/or irradiation [1, 132–134]. The

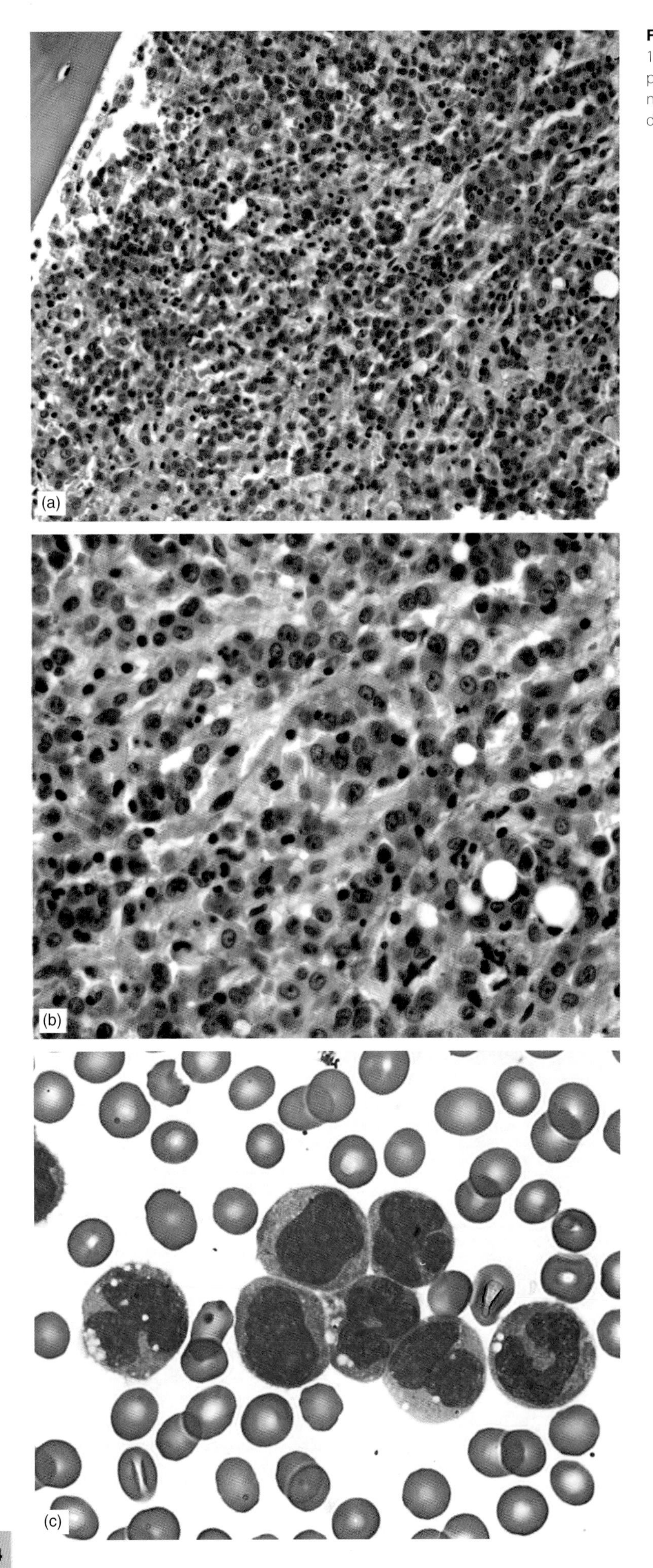

FIGURE 11.21 Acute monocytic leukemia in a patient with 11q23 abnormality. Biopsy section (a: low power and b: high power) reveals numerous immature cells with convoluted nuclei and fine nuclear chromatin. Peripheral blood smear (c) demonstrates abnormal promonocytes and monocytes.

TABLE 11.2 Clinicopathologic features of therapy-related AML.*

Features	Alkylating agents	Topoisomerase II inhibitors
Latency period	4–5 years	<3 years
Preceded MDS	Often present	Often absent
AML subtype	Variable	Mostly monocytic; sometimes promyelocytic or other types
Cytogenetics	Deletions: Often del(5) and del(7)	Translocations: t(9;11); t(6;11); t(15;17); t(8;21); t(3;21); t(6;9)
Median survival time	<8 months	>8 months

* Adapted from Ref. [4].

reason for chemotherapy or irradiation is usually a primary malignancy. In a report by the University of Chicago of 306 patients with t-MDS and t-AML, 25% had Hodgkin lymphoma, 23% had non-Hodgkin lymphoma, and 38% had solid tumors as the primary malignancies [132]. Breast cancer was the most common solid tumor accounting for 10% of the total cases. Approximately 6% of the patients had no prior malignancy and underwent cytotoxic chemotherapy for autoimmune disorders.

The latency period between the initiation of chemotherapy and/or the irradiation and development of t-MDS or t-AML ranges from several months to several years. Overall, the latency period is shorter in patients treated with

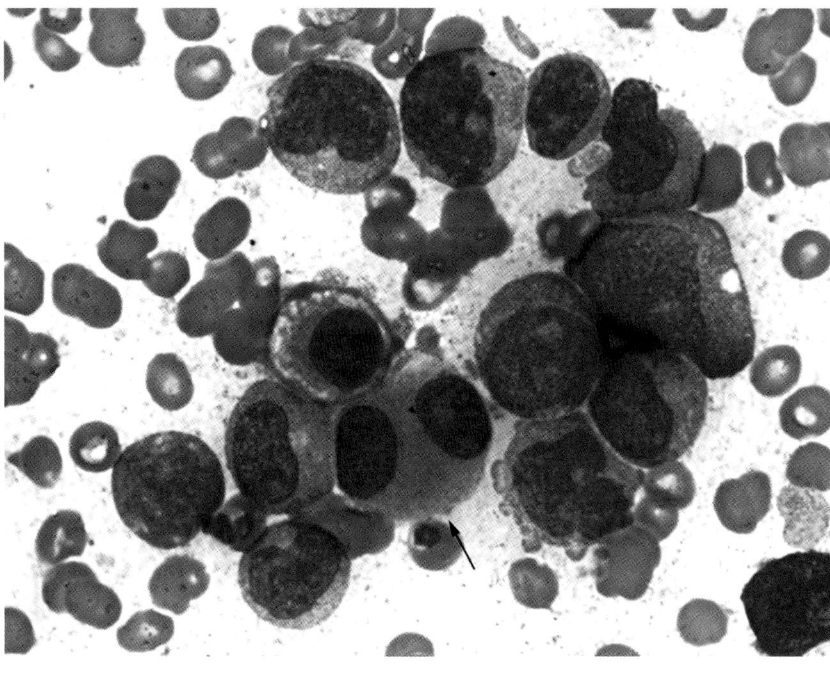

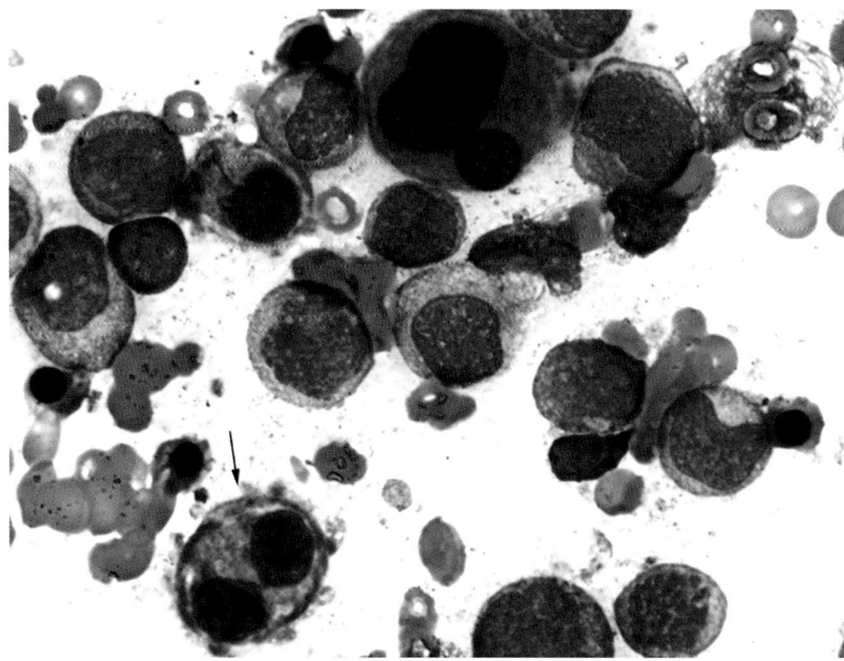

FIGURE 11.22 Acute myeloid leukemia with multilineage dysplasia. Bone marrow smears demonstrating dysplastic immature myelomonocytic cells and binucleated micromegakaryocytes (arrows).

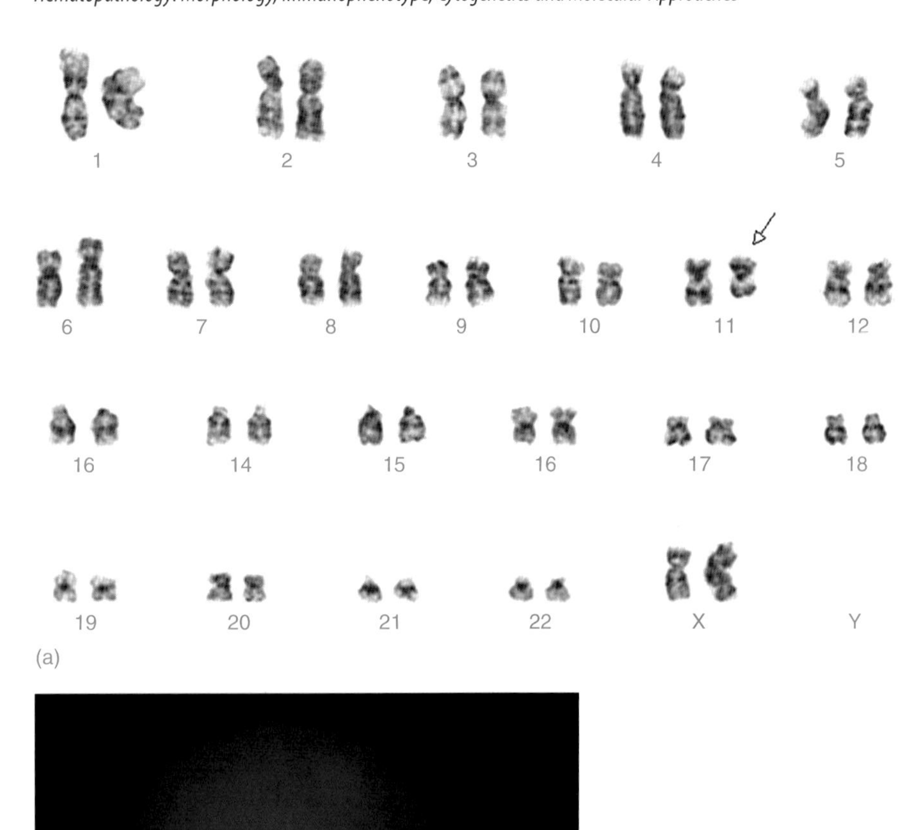

FIGURE 11.23 G-banded karyotype (a, arrow) and FISH analysis (b) showing del(11q) in a patient with acute myeloid leukemia with multilineage dysplasia.

topoisomerase II inhibitors than those treated with alkylating agents or radiation, and longer in younger patients and patients with a non-malignant primary diagnosis [1, 132, 133]. The presence of more than one prior malignancy before diagnosis of t-MDS/t-AML in patients raises the possibility of a constitutional defect predisposing to t-MDS/t-AML.

The primary feature separating t-MDS from t-AML is the percentage of blast counts which is <20% in t-MDS and ≥20% in t-AML. Some of the patients, particularly those treated with topoisomerase II inhibitors, may bypass the MDS phase (Table 11.3).

TABLE 11.3 Features distinguishing acute promyelocytic leukemia from acute monocytic leukemia.

Features	AML-M3	AML-M5
Azurophilic granules	More frequent	Less frequent
Auer rods	Frequent	Rare
MPO stain	Strong	Weak
NSE	Negative/weak	Strong
CD4	Negative	Often positive
CD14	Negative	Positive
HLA-DR	Negative	Positive

Alkylating Agent/Radiation-Related AML

Alkylating agent/radiation-related AML has a latency period of about 5–6 years and is usually (>70%) preceded by MDS [1, 133, 135]. The average time for progression from MDS to AML is about 5 months [132]. The occurrence rate appears to be dependent on the age of the patient and the total accumulative dose of the chemotherapeutic agents and/or radiation [1].

Pathology

Morphology

The characteristic morphologic features are dysplastic hematopoiesis and increased blasts [1, 136]. Dysplastic changes are usually multilineage and involve myeloid, erythroid, and megakaryocytic series. Hypogranulation and abnormal segmentation of the granulocytic cells, megaloblastic changes in the erythroid series with ringed sideroblasts, and the presence of micromegakaryocytes are frequent findings. Bone marrow basophilia is sometimes present. Blasts (including promonocytes) are increased (≥20%) and often depict dysplastic changes. Morphologically, most t-AML cases correspond to AML with maturation, but a minority of the cases fit into acute myelomonocytic, acute monocytic, acute erythroleukemia, or acute megakaryocytic leukemia (discussed later).

Bone marrow is often hypercellular, but in about 25% of the cases is hypocellular. Bone marrow fibrosis may be present in one-fourth of the cases.

The peripheral blood may show anemia or pancytopenia with aniso-poikilocytosis and leukoerythroblastic features and presence of blast cells.

Immunophenotype

Flow cytometry shows a population of blast cells expressing myeloid-associated markers. CD34 and CD117 are frequently expressed. In some cases, blasts may show aberrant expression of CD4, CD7, and/or CD56. The hypogranularity of the myeloid cells may lead to a lower SSC in the flow cytometric dot plot preparations. Similar to the MDS cases, there may be increased expression of CD11a and CD66 and reduced expression of CD10 and CD116 in the granulocytic cells.

Immunohistochemistry may help to estimate the blast component and the myeloid proportion of the bone marrow cells by using CD34, CD117, and MPO stains.

Molecular and Cytogenetic Studies

Over 90% of the alkylating agent-/radiation-related AMLs show clonal chromosomal aberrations, most frequently involving loss of all or part of chromosome 5, chromosome 7, or both (see Chapter 8). Balanced chromosomal translocations are rare and mostly involve 11q23 or 21q22 (Table 11.2). Some reports show an association between radiation t-AML and t(15;17) or inv(16) [137].

Clinical Aspects

The latency period between the diagnosis of the primary disease and the occurrence of t-AML appears to be longer in the younger patients and patients who have been treated by alkylating agents for non-malignant conditions, such as autoimmune disorders. In the report of the University of Chicago series [132], the median latency period was 82 months and 130 months for the patients ≤50 years of age and patients with a non-malignant primary diagnosis, respectively. The overall median latency period for the entire t-AML patient population was 65 months with a median survival of 6.9 months. Patients with chromosomal deletion of 5 and/or 7 had a shorter median survival time than those with chromosomal translocations.

Topoisomerase II Inhibitor–Related AML

Topoisomerase II inhibitor–related AML generally has a shorter latency period than the alkylating agents–related AML, ranging from 1 to 3 years [1, 132, 133, 135]. Antracyclines, doxorubicin, etoposide, epipodophyllotoxins, and teniposide are among the major drugs targeting DNA-topoisomerase II.

Pathology

Morphology

The morphologic features most commonly represent acute myelomonocytic or acute monocytic leukemias (discussed later). But some cases may present morphologic and cytogenetic findings consistent with APL [134, 137]. An antecedent myelodysplastic phase is usually lacking.

Immunophenotype

See immunophenotypic features of APL, acute myelomonocytic leukemia, and acute monocytic leukemia in this chapter.

Molecular and Cytogenetic Studies

Topoisomerase II inhibitor–related AML is commonly associated with chromosomal translocations, particularly involving 11q23 and the *MLL* gene [138, 139]. The 11q23-associated cytogenetic changes include del(11q23), (6;11)(q27;q23), t(9;11)(p22;q23), (10;11)(p12;q23), and t(11;19)(q23;p13.1) [138]. Other cytogenetic abnormalities such as t(15;17)(q11;q12), (3;21)(q26;q22), t(8;21)(q22;q22), t(6;9)(p23;q34), and t(8;16)(p11.2;p13) have also been reported in topoisomerase II inhibitor–related AMLs [140–145].

Clinical Aspects

The overall latency period is shorter and the median survival time is longer for the topoisomerase II inhibitor–related AML than for the alkylating agent–related AML.

AMLs NOT OTHERWISE CATEGORIZED

This category represents acute leukemias that are excluded from all previously described subclasses. The primary distinguishing features of leukemias in this category include lineage differentiation and the extent of maturation based on morphological, immunophenotypic, and cytochemical characteristics. Therefore, it includes most of the AML subtypes defined by the French–American–British (FAB) classification including M0, M1, M2, M4, M5, M6, and M7 variants plus acute basophilic leukemia (ABL) and acute panmyelosis with myelofibrosis (APMF).

The criteria for diagnosis of acute leukemia in this category are similar to the previously described subclasses and is based on the WHO requirement of the presence of 20% or more blast cells in the bone marrow or blood differential counts. In addition to myeloblasts, "blast" counts in certain categories of AML may include monoblasts, megakaryoblasts, promonocytes, or promyelocytes. WHO recommends to count 500 nucleated cells on the bone marrow smears and/or 200 on the blood smears in order to establish a diagnosis of AML.

AML, Minimally Differentiated

Minimally differentiated AML (AML-M0) is defined as an AML with no morphologic or cytochemical evidence of myeloid differentiation based on conventional light microscopic examinations [1]. The myeloid lineage in this category is established by immunophenotypic characteristics and/or ultrastructural studies.

Etiology and Pathogenesis

The etiology and pathogenesis of minimally differentiated AML are not known. No recurrent cytogenetic or molecular abnormalities have been reported in this leukemia, although >50% of the cases may show a variety of chromosomal aberrations (discussed later).

Pathology

Morphology

The leukemic blasts lack features of morphologic differentiation [1–4]. They are often medium sized with scant nongranular basophilic cytoplasm, round or slightly irregular nuclei, fine chromatin, and one or more prominent nucleoli (Figure 11.24). Type II and III myeloblasts are absent or extremely rare (<3%), and Auer rods are not present. Special cytochemical stains, such as MPO, Sudan Black B, and NSE, are negative (≤3% blasts show positive staining). Bone marrow is usually hypercellular and packed with leukemic blast cells, but remnants of normal hematopoietic cells may be noted. The presence of residual normal-maturing myeloid precursors may create a morphologic pattern mimicking AML with maturation (AML-M2). The distinguishing features between AML with minimal differentiation and AML with maturation are lack of type I and II myeloblasts, absence of Auer rods, and negative cytochemical staining in the former (discussed later).

Immunophenotype

Flow cytometric studies reveal expression of myeloid-associated markers, such as CD13, CD33, and/or CD117. CD34 and HLA-DR are often positive, whereas CD38 is negative in a significant proportion of the cases. The CD34+/CD38– phenotype reflects the very early stage of differentiation in the leukemic blast cells. Most reports show cytoplasmic MPO expression by flow cytometry in leukemia cells in the majority of the cases [146–149]. Monocytic-associated markers, such as CD11c and CD14, are usually negative. Approximately 50% of the cases may express TdT and/or CD7 and about 20% are positive for CD56 [150–152]. The blast cells in occasional cases may express CD2, CD10, or CD19, but cytoplasmic CD3, CD22, and CD79a are negative.

Immunohistochemical stains are negative for CD68, CD3, and CD20, but may show positive reaction for MPO in the blast population.

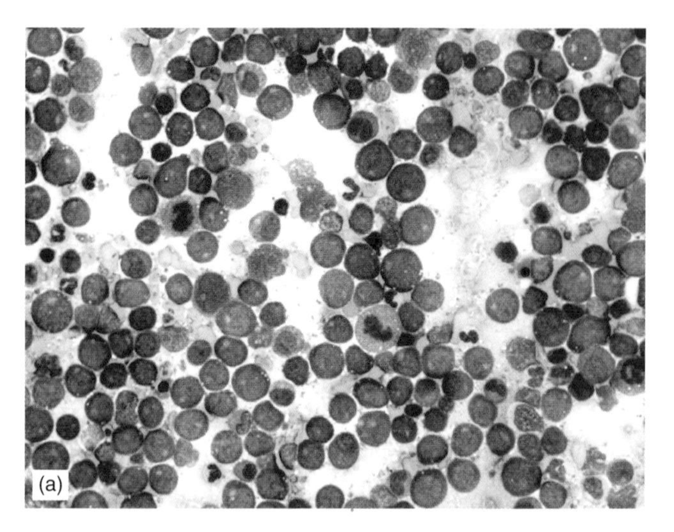

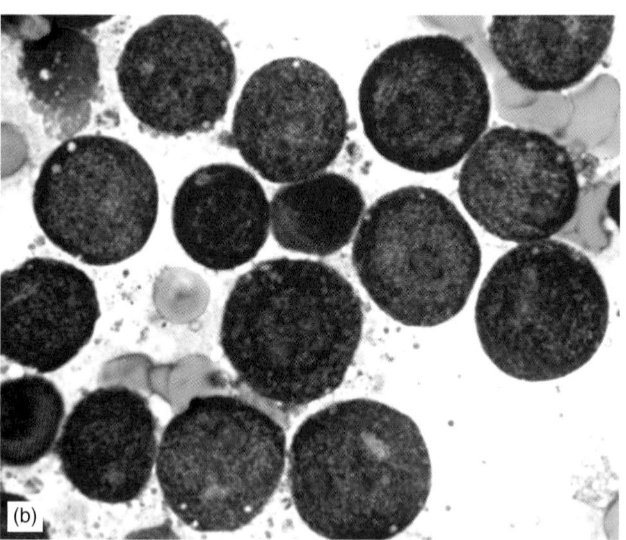

FIGURE 11.24 Acute myeloid leukemia, minimally differentiated. Bone marrow smears (a, low power; b, high power) showing numerous blasts with round nuclei, fine nuclear chromatin, and dark-blue cytoplasm. No cytoplasmic granules or Auer rods are present. These blasts were negative for MPO, Sudan Black B, and NSE, but expressed CD33 and CD117 by flow cytometry.

Molecular and Cytogenetic Studies

A high frequency of mutation in the *RUNX1(AML1)* gene has been reported in AML-M0 patients [153–156]. Occasional cases may show BCR/ABL transcript, but RUNX1/RUNXT1 or CBFβ/MYH11 transcripts are not found. There is no evidence of T-cell receptor (TCR) or immunoglobulin gene rearrangements.

Cytogenetic aberrations are variable and have been reported in ~50% of the cases. Trisomy 4, 8, 11, 13, and 14 and monosomy 7 are among the most common reported abnormalities [157]. Cases with (9;22)(q34;q11.2), t(11;12)(q23;q24), inv(3)(q21q26), and del(20)q11 (Figure 11.25) are reported less frequently [158–163].

Clinical Aspects

This category of leukemia accounts for <5% of all AMLs in most reported studies. The affected patients are usually older than 60 years, and the male:female ratio is about 2 [164]. The prognosis is poor with a median survival of <6 months [165]. In one report, the coexpression of CD7 and CD56 was associated with poorer prognosis in patients younger than 46 years [150].

Differential Diagnosis

ALL, AML without maturation, acute monoblastic leukemia, acute megakaryoblastic leukemia (AMKL), and occasionally large cell lymphoma are amongst the list of differential diagnosis. The blast cells in minimally differentiated AML express myeloid-associated markers (such as CD13, CD33, CD117) and show <3% positivity for MPO and Sudan Black B in routine cytochemical stains.

AML without Maturation

Acute myeloid leukemia without maturation (AML-M1) is defined as an acute leukemia with no significant myeloid maturation and ≥90% blast cells in the non-erythroid population [1]. The myeloid nature of blast cells is confirmed by positive (≥3%) staining for MPO and/or Sudan Black B by cytochemical techniques as well as expression of myeloid-associated markers by immunophenotypic studies.

Etiology and Pathogenesis

The etiology and pathogenesis of AML without maturation are not known. No recurrent cytogenetic or molecular abnormalities have been observed for this leukemia, although trisomy 13 has been reported in some cases [166].

Pathology

Morphology

The morphologic features of the blast cells overlap with those described in AML with minimal differentiation except for (1) the presence of blast cells with some azurophilic cytoplasmic granules and (2) positive MPO and/or Sudan Black B cytochemical staining in ≥3% of the blast cells (Figure 11.26) [1–4]. Auer rods are not present.

Similar to most other AMLs, bone marrow is hypercellular and packed with blasts. Variable degrees of marrow fibrosis may be present in a minority of the cases.

Immunophenotype

Flow cytometric studies reveal the expression of myeloid-associated markers, such as CD13, CD33, and/or CD117. CD34 and HLA-DR and cytoplasmic MPO are often positive. Monocytic-associated markers, such as CD11c and CD14, are usually negative. Also, cytoplasmic CD3, CD22, and CD79a are negative.

Immunohistochemical stains are negative for CD68, CD3, and CD20 but may show positive reaction for MPO in the blast population.

Molecular and Cytogenetic Studies

There are reports suggesting a reciprocal exchange between *D12S158* at 12p13.3 and the *MYH11* gene at 16p13 in AML-M1 leukemia [167].

Cytogenetic aberrations are variable and include both numerical (aneuploidy) abnormalities and translocations. Trisomy 11, trisomy 13, and trisomy 14 as well as t(9;12)(q34;p13), t(11;19) (q23;p13), t(14;17) (q32; q11.2), der(12)t(12;17)(p13;q11.2), and der(16)t(16;20)(p13;p11.2) have been reported in this leukemic subtype [157, 166, 168, 169].

Clinical Aspects

AML without maturation accounts for 10–15% of all AMLs and is rare in children. The prognosis is poor, particularly in those with marked leukocytosis and increased circulating blasts.

Differential Diagnosis

The major differential diagnosis of AML without maturation includes ALL, minimally differentiated AML, acute monoblastic leukemia, and AMKL. The blast cells in AML without maturation express myeloid-associated markers (such as CD13, CD33, and CD117) and show ≥3% positivity for MPO and Sudan Black B cytochemical stains.

AML with Maturation

Acute myeloid leukemia with maturation (AML-M2) is defined as an acute leukemia with ≥20% blast cells in the bone marrow and/or peripheral blood and evidence of granulocytic maturation [1]. The maturing non-blast granulocytic cells account for ≥10% and monocytic cells are ≤20% of the bone marrow cells. The myeloid nature of blast cells is confirmed by positive (≥3%) staining for MPO and/or Sudan Black B by cytochemical technique, as well as expression of myeloid-associated markers by immunophenotypic studies.

Etiology and Pathogenesis

The etiology and pathogenesis of AML with maturation are not known. About 40% of the AML-M2 type shows association with t(8;21)(q22;q22) involving fusion of the *RUNX1* (*AML1*) and *RUNXT1*(*ETO*) genes. As mentioned earlier, loss of the *AML1* or *CBFβ* gene in mice leads to defective hematopoiesis and embryonic death. Also, *RUNX1/RUNXT1* fusion product reduces apoptosis by activating the expression of the anti-apoptosis gene *BCL-2* [170, 171].

Pathology

Morphology

The myeloblasts are large, often with indented nuclei and basophilic cytoplasm. Type II and III myeloblasts are prominent and some blasts may contain large granules mimicking cytoplasmic granules seen in the Chediak–Higashi syndrome [1–4, 172]. Auer rods are frequent and may also be detected in the more mature myeloid forms. Promyelocytes, myelocytes, metamyelocytes, bands, and segmented neutrophils are present and often show dysplastic changes (Figure 11.7). Eosinophilia is common, and some cases may show increased bone marrow basophils and/or mast cells.

The bone marrow biopsy sections are often hypercellular and packed with blasts. Occasionally, bone marrow

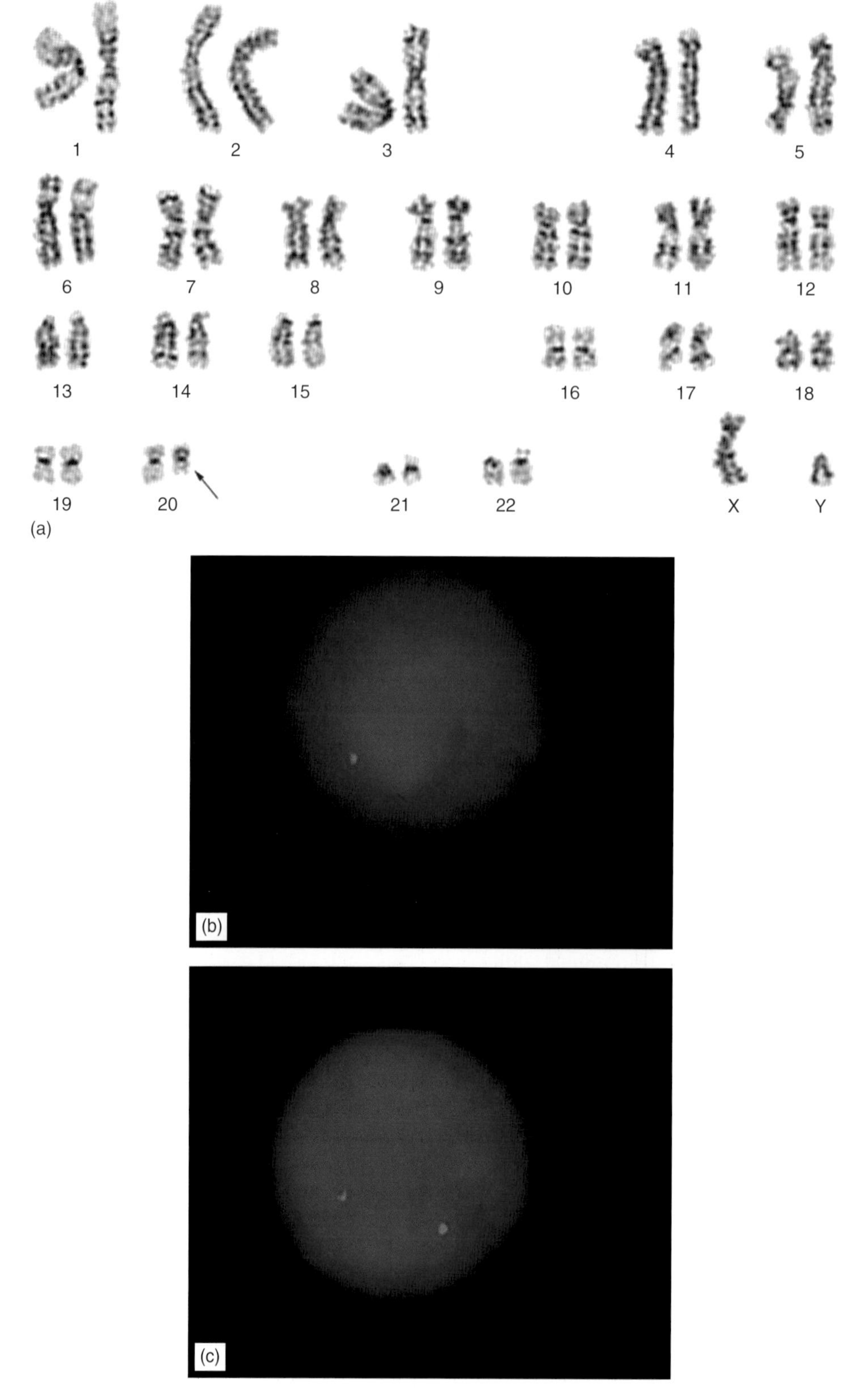

FIGURE 11.25 G-banded karyotype (a, arrow) and FISH analysis (b) showing del(20q) in a patient with AML-M0. (c) A normal cell with two red (20q-specific) signals .

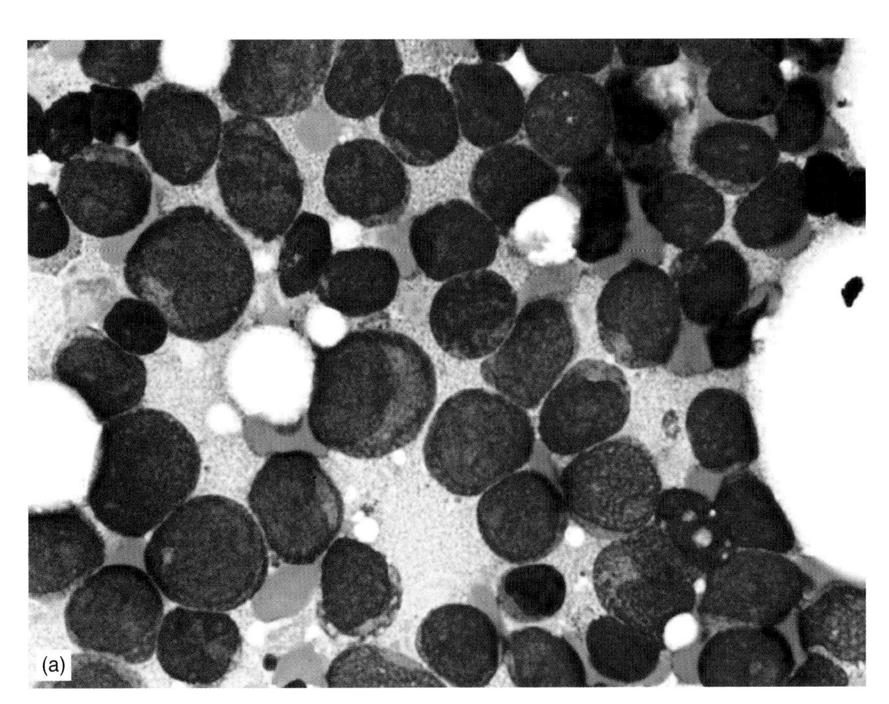

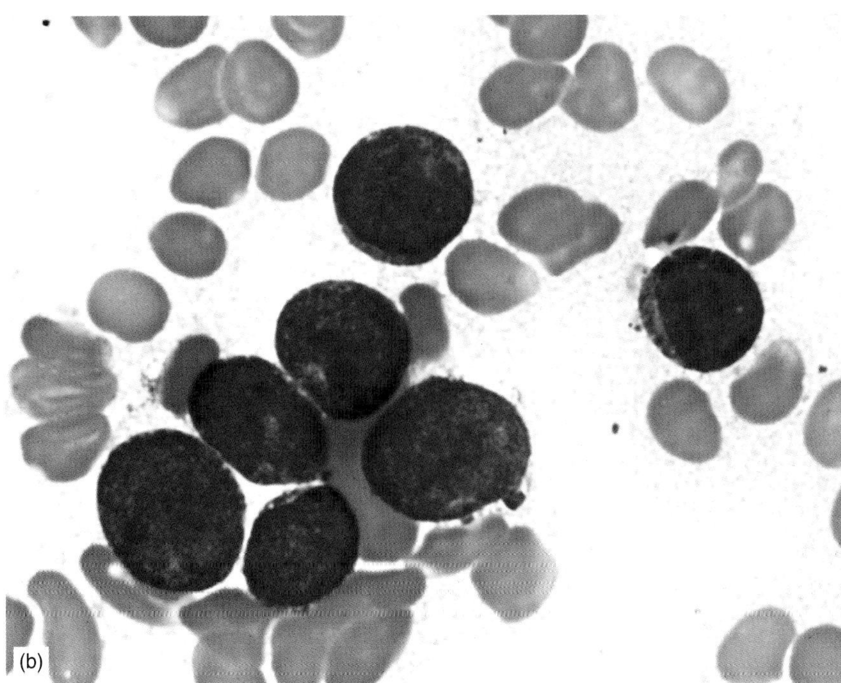

FIGURE 11.26 Acute myeloid leukemia without maturation. Bone marrow smears (a, low power; b, high power) showing numerous blasts with round nuclei, fine nuclear chromatin, and dark blue cytoplasm. There is no evidence of maturation. Auer rods are not present. More than 3% of the blast cells were positive for MPO and Sudan Black B.

may appear normocellular or hypocellular. Variable degrees of marrow fibrosis may be present in a minority of cases.

Immunophenotype

Flow cytometric studies reveal the expression of myeloid-associated markers, such as CD13, CD33, and/or CD117. CD34, HLA-DR, and cytoplasmic MPO are often positive. Monocytic-associated markers, such as CD11c and CD14, are usually negative or present in ≤20% of the bone marrow cells. The lymphoid-associated markers, such as CD3, CD20, CD22, CD79a, are usually negative, but there may be aberrant expression of CD19 and/or CD56 [173, 174]. Blast cells may show dim expression of TdT in a small proportion of cases.

Immunohistochemical stains are negative for CD3 and CD20 but may show positive reaction for MPO in the blast population. Scattered immature cells (<20%) may express CD68 in occasional cases.

Molecular and Cytogenetic Studies

As mentioned earlier in this chapter, t(8;21)(q22;q22) accounts for ~40% of the chromosomal aberrations in the cases of AML with maturation (Figure 11.8). For this reason, cytogenetic studies for t(8;21) and molecular monitoring by

RT-PCR are frequently used for diagnosis and detection of minimal residual disease in AML-M2 [175]. Other reported cytogenetic abnormalities include t(6;9)(p23;q34), t(2;9)(q14;p12), t(5;11)(q35;q13) (Figure 11.27), t(10;11) (p13;q14), t(8;19)(q22;q13), t(8;16)(p11;p13), del(12)(p11→p13), and various complex translocations [69, 140, 175–181].

Clinical Aspects

This category of leukemia accounts for 30–40% of all AMLs and occurs in both children and adults. It is the most frequent AML in children. The ones with t(8;21) have a more favorable prognosis in adults with an expected disease-free survival of about 2 years. The clinical outcome for children and cases with other types of chromosomal aberrations is poor.

Differential Diagnosis

The major differential diagnosis includes RAEB, APL, and acute myelomonocytic leukemia. Immunophenotypic and cytogenetic studies are helpful to reach to a definitive diagnosis. Blast cells in AML with maturation often express CD13, CD33, and/or CD117 and lack CD14 expression. They often show t(8;21)(q22;q22) by cytogenetic and molecular analyses.

Acute Myelomonocytic Leukemia

Acute myelomonocytic leukemia (AML-M4) is defined as an acute leukemia with increased immature granulocytic and monocytic cells. Myeloblasts, monoblasts, and promonocytes account for ≥20% of the total bone marrow nucleated cells and/or peripheral blood differential counts [1].

Etiology and Pathogenesis

The etiology and pathogenesis of acute myelomonocytic leukemia are not known. The CBFβ/MYH11 fusion protein, associated with the chromosome 16 aberrations observed in a majority of these patients, appears to induce granulocytic dysplasia in experimental animals [182, 183]. Also, the translocation of the *MLL* gene, associated with 11q23 aberrations, results in a chimeric gene product which may play a role in leukemogenesis [184, 185].

Pathology

Morphology

The bone marrow smears show myeloid left shift and increased number of immature myelomonocytic cells [1–4]. Myeloblasts show scant-to-moderate amounts of dark-blue cytoplasm, some of which containing various numbers of azurophilic granules (types II and III myeloblasts). Auer rods may be present. The nuclei are usually round or oval, but they may be irregular. The nuclear chromatin is fine, and multiple prominent nucleoli are often present.

Monoblasts are usually larger than myeloblasts (~40 μm) with abundant dark- to light-blue cytoplasm and scattered fine azurophilic granules. The nucleoli are round, oval, or folded, and the nuclear chromatin is fine. There is often a single large nucleolus, but multiple prominent nucleoli may be present. Promonocytes are larger than monocytes (~30–35 μm) and have abundant light blue to gray cytoplasm. Scattered cytoplasmic azurophilic granules and/or cytoplasmic vacuoles may be present (Figure 11.28). The nuclei are delicately folded or convoluted, often with a cerebriform pattern. The nuclear chromatin is fine, and nucleoli are present but not prominent. Bone marrow smears show an increased number (≥20%) of promonocytes, monoblasts, and myeloblasts (Figures 11.29 and 11.30).

The bone marrow biopsy sections are often hypercellular and packed with immature myelomonocytic cells (Figure 11.29). Variable degrees of marrow fibrosis may be present in a minority of the cases. Monocytic nodules may be present.

The peripheral blood smears show absolute monocytosis (often ≥5,000/μL) with the presence of promonocytes, left-shifted granulocytic series, and various numbers of circulating blasts (Figure 11.29).

In a significant proportion of patients (5–35%), there is evidence of extramedullary leukemic infiltration, such as involvement of skin, mucosal membranes, lymph nodes, liver, and/or spleen [1, 186].

Immunophenotype and Cytochemical Stains

The immature myelomonocytic population expresses CD13 and CD33 with partial expression of CD11c, CD14, CD15, CD34, CD36, CD64, CD117, HLA-DR, and MPO by flow cytometry (Figure 11.31) [187]. Immunohistochemical stains, such as MPO, lysozyme, and CD68, are used for the evaluation of the bone marrow myelomonocytic component, and CD34 and CD117 stains are often helpful for the estimation of blast cells. Monocytic precursors strongly express CD68 and lysozyme and are negative or weakly positive for MPO. Granulocytic precursors strongly express MPO but are negative for CD68 and show weak or moderately positive reactions for lysozyme. Aberrant expression of CD2, CD7, and/or CD56 has been observed frequently [188–190].

The cytochemical stains show strong MPO positivity for the granulocytic lineage and various degrees of diffuse cytoplasmic NSE staining for the monocytic population.

Molecular and Cytogenetic Studies

In addition to the chromosomal aberrations involving 11q23 and inv(16)(p13q22) [184, 191, 192], a number of nonspecific cytogenetic abnormalities have been reported in acute myelomonocytic leukemias. These include dup(1)(p31.2p36.2), t(1;3)(p36;q21), t(8;12)(q13;p13), t(9;21)(q13;q22), and t(6;7)(q23;q35) [192–197].

Clinical Aspects

Acute myelomonocytic leukemia accounts for about 15–25% of all AMLs. The median age is around 50 years, but it may occur at any age. The incidence is slightly more in males than in females. Similar to other acute leukemias, clinical symptoms are the result of bone marrow involvement and extramedullary infiltration by the leukemic cells. Fatigue,

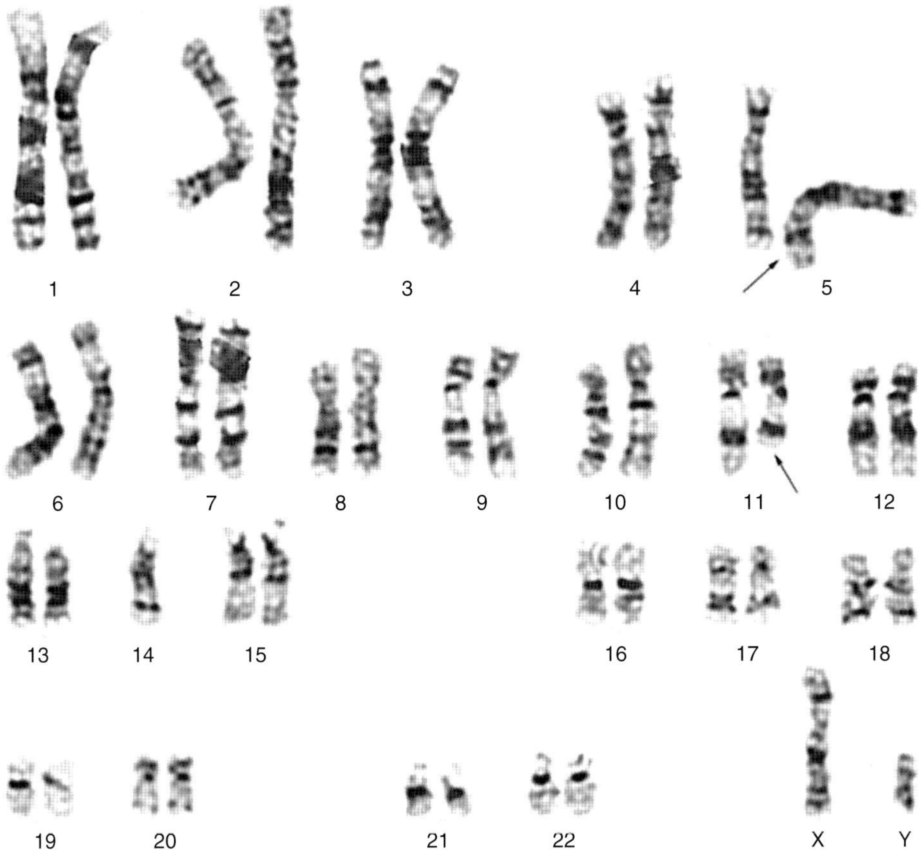

1 2 3 4 5 6 7 8 9 10 11 12 13 14 15 16 17 18 19 20 21 22 X Y

FIGURE 11.27 Translocation of 5;11 in a patient with acute myeloid leukemia with maturation.

fever, bleeding disorders, gingival hyperplasia, lymphadenopathy, hepatosplenomegaly, and skin involvement are among frequent clinical findings. As mentioned earlier in this chapter, patients with inv(16) have a favorable prognosis and those with translocation of 11q23 fall into the category of leukemias with intermediate survival rate [104]. Some studies show a correlation between the expression of CD56 by the leukemic cells and severe fatal hyperleukocytosis in patients with acute myelomonocytic leukemia [189]. Successful effect of treatment with NUP98–HOXD11 fusion transcripts and monitoring of minimal residual disease in patients with AML-M4 has been reported [198].

Differential Diagnosis

The differential diagnosis of acute myelomonocytic leukemia includes chronic myelomonocytic leukemia (CMML), AML with maturation, APL, and acute monocytic leukemia. The diagnosis of acute myelomonocytic leukemia is established by the demonstration of the sum of ≥20% myeloblasts and monoblastic/promonocytes in the bone marrow or peripheral blood. It is distinguished from the microgranular variant of APL by the expression of NSE, CD4, CD14, and HLA-DR and by the absence of t(15;17) (Table 11.3).

Acute Monoblastic and Acute Monocytic Leukemias

Acute monoblastic and acute monocytic leukemias are acute leukemias in which ≥80% of the leukemic cells are of monocytic lineage consisting of monoblasts, promonocytes, and monocytes. When monoblasts are the major cellular component (≥80% of the leukemic cells) the term "acute monoblastic leukemia" (AML-M5a) is used, and when promonocytes and monocytes account for most of the leukemic cells (≥80%), the condition is referred to as "acute monocytic leukemia" (AML-M5b) [1].

Etiology and Pathogenesis

The etiology and pathogenesis of acute leukemias of monocytic lineage are not known. As mentioned earlier, translocation of the *MLL* gene, associated with 11q23 aberrations, results in a chimeric gene product which may play a role in the development of acute leukemias of monocytic lineage. Up to 10% of t-AMLs, particularly the topoisomerase II inhibitor–related type, are associated with 11q23 aberrations, suggesting a causative role for chemotherapy/radiation in the development of this disorder.

Pathology

Morphology

Monoblasts usually show abundant dark- to light-blue cytoplasm with no or a few scattered fine azurophilic granules (Figures 11.32 and 11.33). The nucleoli are round, oval, or folded, and the nuclear chromatin is fine. There is often a single large nucleolus, but multiple prominent nucleoli may be present. Promonocytes have abundant light-blue to gray cytoplasm (Figure 11.34). Scattered cytoplasmic

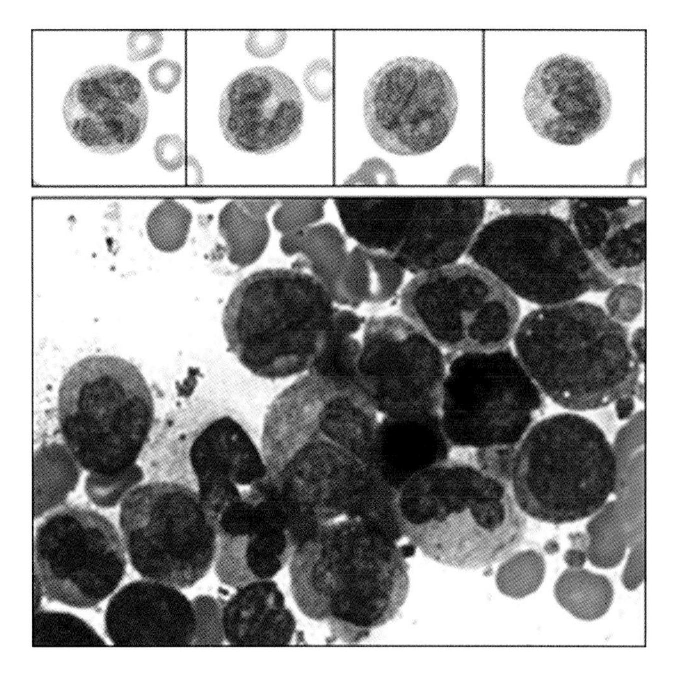

FIGURE 11.28 Promonocytes are counted as blast cells in myeloid leukemias and are prominent in acute myelomonocytic and monocytic leukemias.

azurophilic granules and/or cytoplasmic vacuoles may be present. The nuclei are delicately folded or convoluted, often with a cerebriform pattern. The nuclear chromatin is fine and nucleoli are present but not prominent. Auer rods are rare. Granulocytic precursors account for ≤20% of the bone marrow non-erythroid nucleated cells.

The bone marrow biopsy sections are often hypercellular and packed with blasts and immature myelomonocytic cells. Variable degrees of marrow fibrosis may be present in a minority of cases.

The peripheral blood smears show absolute monocytosis (often ≥5,000/μL) with the presence of monoblasts, promonocytes, and monocytes (Figure 11.33b). The proportion of monocytes and promonocytes in blood smears may sometimes be much greater than the blast cells as compared to the bone marrow smears.

Extramedullary leukemic infiltration by leukemia cells is relatively common, such as involvement of gum, skin (Figure 11.35), central nervous system, lymph nodes, liver, and/or spleen [199].

Immunophenotype and Cytochemical Stains

The major difference between acute myelomonocytic leukemia and acute monocytic leukemia is the proportion of monocytic versus granulocytic precursors [1]. Morphologic features alone may not be sufficiently clear to make such a distinction, and in most instances, there is a need for immuonophenotyping in order to separate these two entities from each other.

The immature monocytic population expresses CD4, CD11c, CD15, CD14, CD36, CD64, and HLA-DR by flow cytometry [187, 200]. Aberrant expression of CD56 may be present (Figure 11.36). A small proportion of leukemic cells may also express CD13, CD33, CD117, and/or weak MPO. CD34 is usually negative. Immunohistochemical stains are usually positive for lysozyme and CD68 and may show a weak or focal reactivity for MPO.

The cytochemical stains show strong diffuse cytoplasmic NSE staining. Monoblasts are usually MPO-negative, but promonocytes may show a weak positive reaction.

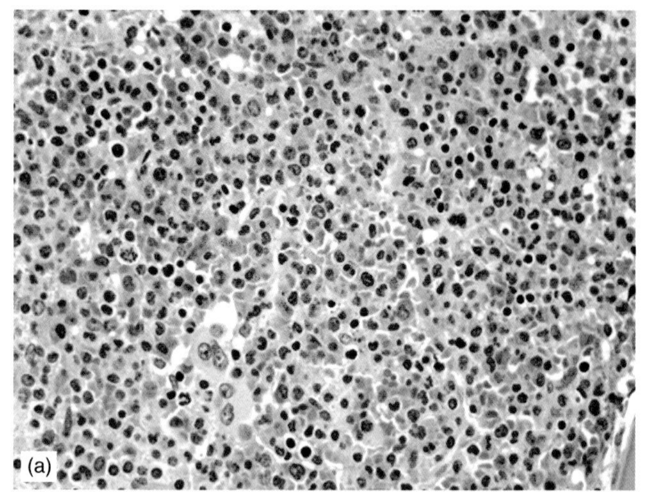

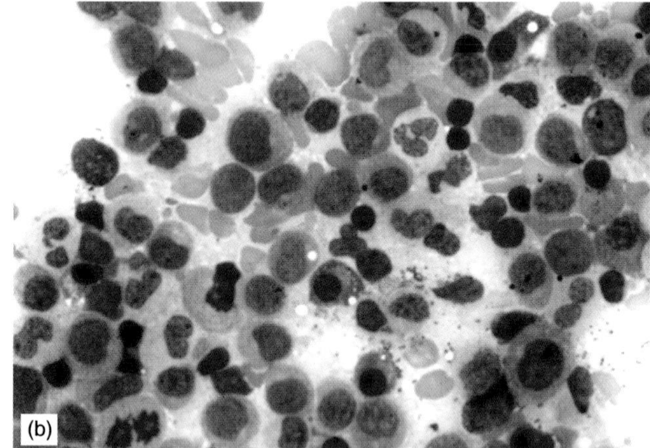

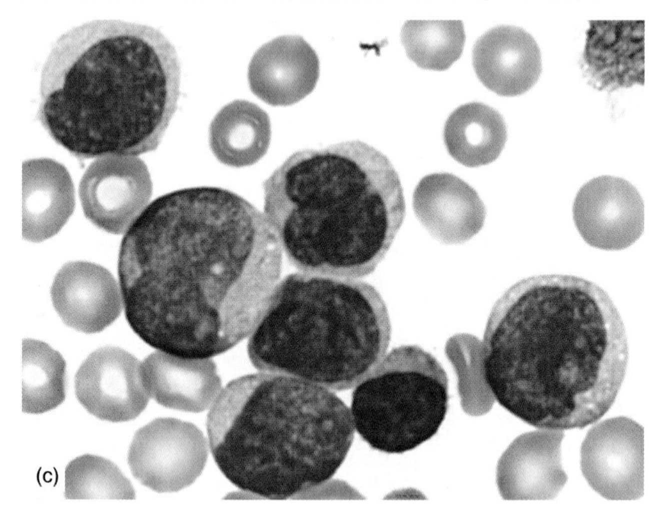

FIGURE 11.29 Acute myelomonocytic leukemia. Biopsy section (a) demonstrating a hypercellular marrow with numerous immature cells with round or irregular nuclei and fine chromatin. Some immature cells show one or more prominent nucleoli. Bone marrow smear depicts myeloid left shift with increased immature myelomonocytic cells and blasts (b). Blood smear shows several blasts and immature monocytic cells (c).

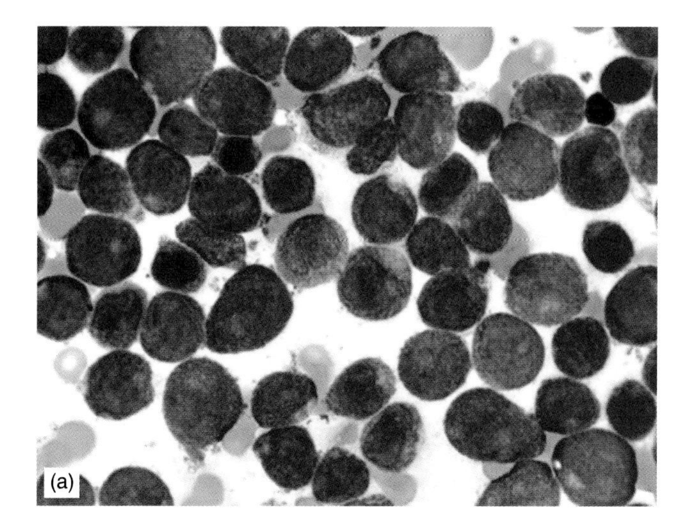

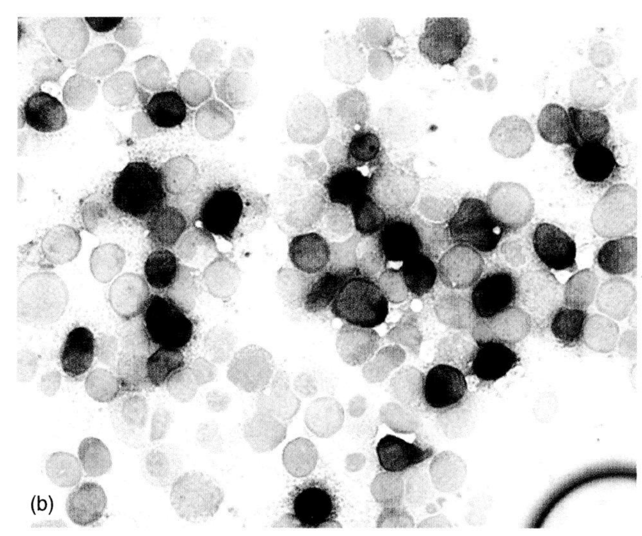

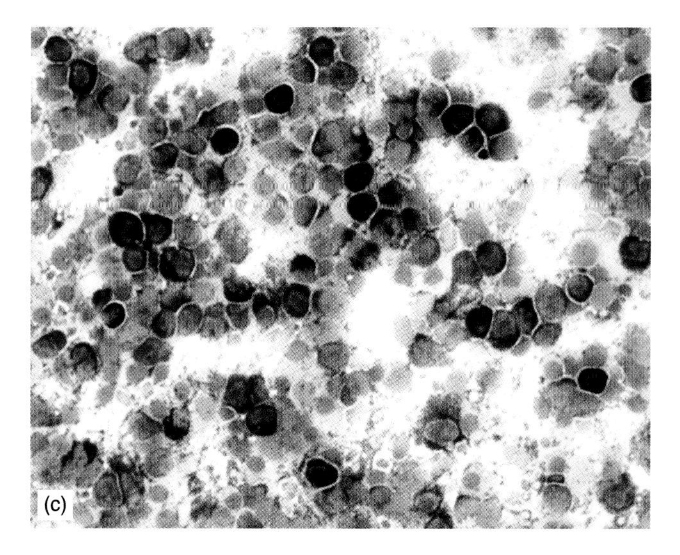

FIGURE 11.30 Acute myelomonocytic leukemia. Bone marrow smear showing increased blasts and immature myeloid cells with round or convoluted nuclei (a). Monoblasts/promonocytes are positive for an NSE stain (b) and myeloblasts are positive for MPO (c).

Molecular and Cytogenetic Studies

Overall, the frequency of chromosomal aberrations, such as translocation of 11q23 and trisomy 8, is significantly higher in monoblastic (M5a) type than the monocytic (M5b) type. In one large study, the incidence of 11q23 (*MLL*) aberrations in M5a and M5b was reported as 33.3% and 15.9%, respectively [185]. Reported cytogenetic abnormalities and affected genes include [83, 201–207]:

t(11;17)(q23;q21); (*MLL-AF17*)

t(10;11)(p11.2;q23);(*ABI-1;MLL*)

t(11;20)(p15;q11.2)(*NUP98-TOP1*)

ins(X;11)(q24;q23q13);(*Septing6-MLL*)

t(8;16)(q11;p13);(*MOZ-CBP*) (Figure 11.37)

(5;11)(q31;q23q23);(*GRAF-MLL*).

Acute monocytic leukemia with t(8:16)(p11;p13) is rare and has been mostly reported in infants and children, often with a bleeding tendency and disseminated intravascular coagulopathy (DIC). The leukemia cells may demonstrate hemophagocytosis and may morphologically mimic the microgranular variant of acute promyelocytic leukemia [204].

Clinical Aspects

Acute leukemias of monocytic lineage account for about 3–6% of all AMLs. The median age is around 50 years, but it may occur at any age. The incidence is higher in men than in women (male:female ratio is about 1.8). Clinical symptoms are the result of bone marrow involvement and extramedullary infiltration by the leukemic cells. Fatigue, fever, bleeding disorders, gingival hyperplasia, lymphadenopathy, CNS involvement, and hepatosplenomegaly are among the frequent symptoms. In one major study, the monocytic type (M5b) had a better clinical outcome than the monoblastic type (M5a) with a reported 3-year disease-free survival of 28% and 18%, respectively [208]. In another study, the complete remission rate and disease-free survival did not differ significantly between patients with M5a and M5b [209].

Differential Diagnosis

The differential diagnosis of acute monoblastic leukemia (AML-M5a) includes ALL, minimally differentiated AML, AML without maturation, AMKL, and large cell lymphoma.

Acute monocytic leukemia (AML-M5b) should be distinguished from chronic myelomonocytic leukemia (CMML), acute myelomonocytic leukemia, and microgranular variant of APL.

Leukemia cells of monocytic/monoblastic origin may express NSE, lysozyme, CD4, CD14, CD64, and/or CD68.

Acute Erythroid Leukemias

Acute erythroid leukemia (AML-M6) is defined as a subtype of AML with predominance of erythroid precursors (≥50% of bone marrow nucleated cells should be of

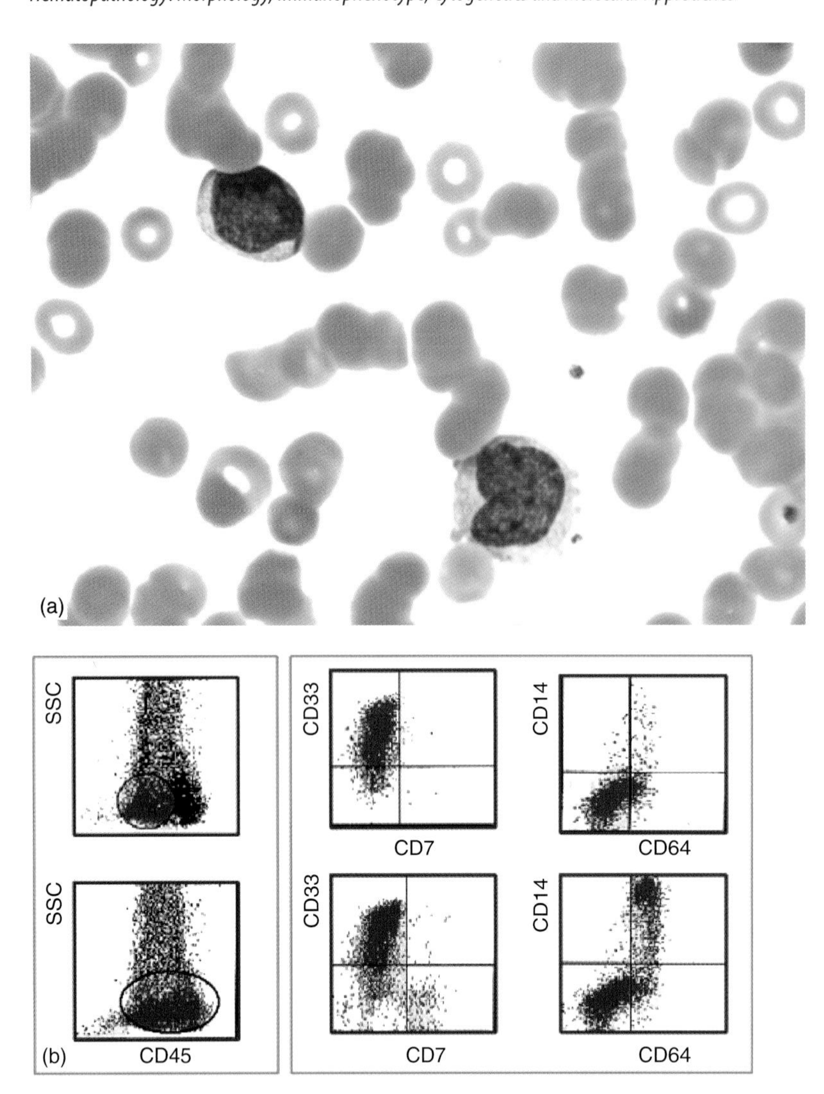

FIGURE 11.31 Acute myelomonocytic leukemia. Peripheral blood smear showing a blast and a promonocyte (a). Flow cytometry demonstrates two gates (b). The upper gate, CD45^{dim} population, expresses CD13 and CD33 and is negative for CD14. The lower gate, CD45^{dim} and CD45^{strong}, demonstrates an additional population of monocytes expressing CD14.

erythroid origin) [1]. Acute erythroid leukemia is divided into two morphologic categories: (1) erythroleukemia, consisting of myeloblasts and erythroid precursors, and (2) pure erythroid leukemia [1, 210, 211].

Etiology and Pathogenesis

The etiology and pathogenesis of erythroid leukemia are not known. Recent studies suggest that the loss of splicing function of the hematopoietic transcription factor Spi-1/PU.1 may play a role in the pathogenesis of erythroid leukemia [212]. A significant number of erythroid leukemias are therapy related or are the result of blast transformation in MDS, and therefore in part share the pathogenesis of t-AML and MDS [213, 214]. It has also been suggested that an inadequate supply of the mutagenic nucleotide of cytosine, possibly through impaired synthesis, could cause both the megaloblastic and the leukemic changes in erythroleukemia [215].

Pathology

Morphology

Multilineage dysplasias, particularly dyserythropoiesis, are common bone marrow features. These include megaloblastic changes, nuclear budding and fragmentation, multinuclearity, and basophilic stippling in the erythroid precursors. Hypogranulation, abnormal nuclear segmentation, and giant forms may be present in the granulocytic series. Micromegakaryocytes and megakaryocytes with separated nuclei are not infrequent [1, 214, 216–218]. Erythroid leukemia is divided into two subcategories according to the WHO classification.

Erythroleukemia (erythroid/myeloid, AML-M6a): This subtype is defined by at least 50% of the bone marrow nucleated cells being of erythroid origin and ≥20% myeloblasts in the non-erythroid component (Figures 11.38 and 11.39) [1, 216, 217]. The erythroid series are dysplastic and left shifted, but are usually found in all stages of maturation. Myeloblasts may be of type I, II, or III or a mixture of all three types. Occasionally, Auer rods are present. Bone marrow iron stores are often increased and ringed sideroblasts may be present. This morphologic category accounts for the majority of acute erythroid leukemias.

Bone marrow biopsy sections are usually hypercellular with clusters or sheets of immature cells and marked reduction in the normal hematopoietic components.

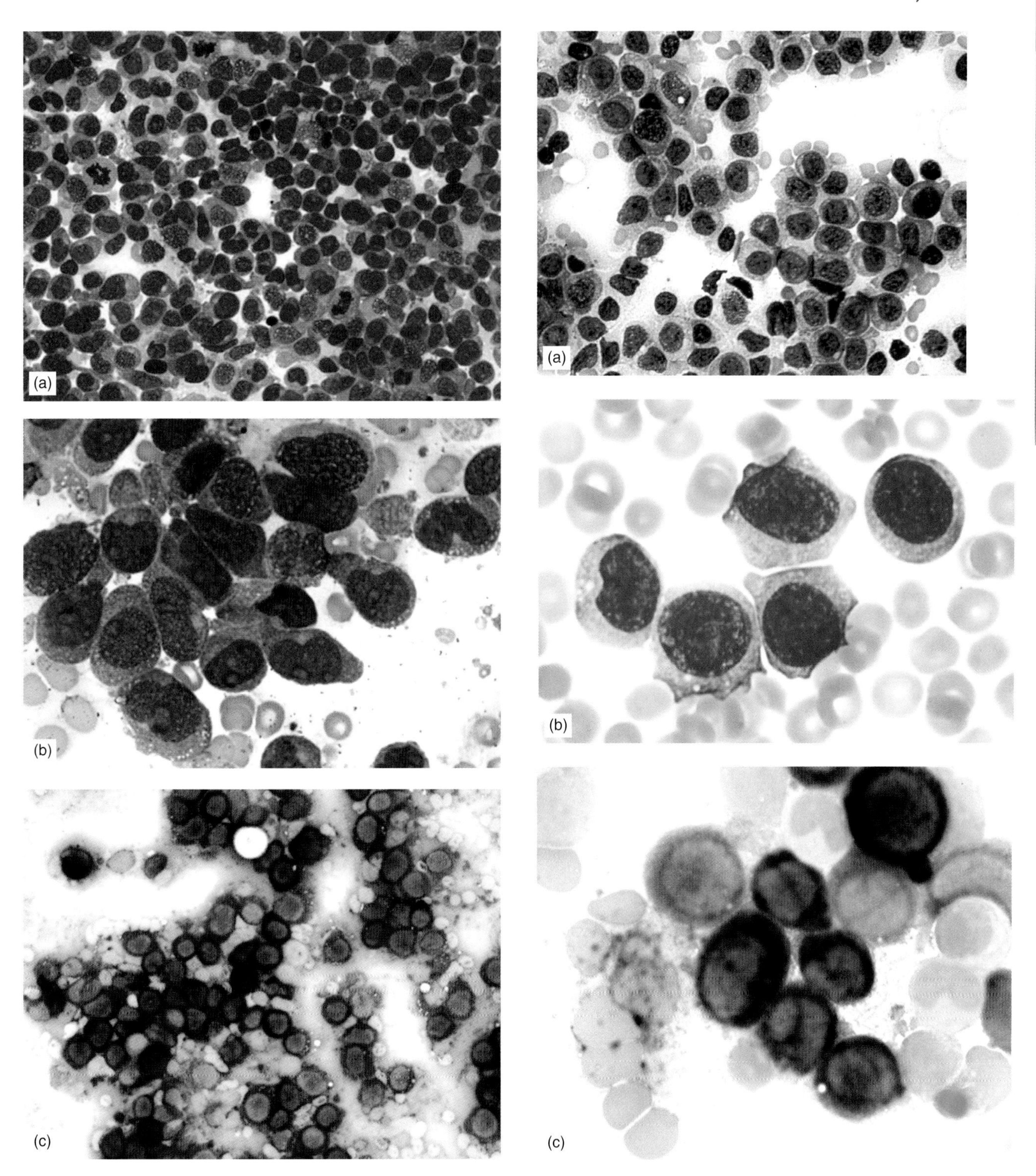

FIGURE 11.32 Acute monoblastic leukemia. Bone marrow smear demonstrates sheets of blast cells with variable amounts of finely vacuolated, blue cytoplasm, round or irregular nuclei, fine chromatin, and one or more prominent nucleoli: (a) low power and (b) high power. The majority of blast cells stain for NSE (c).

FIGURE 11.33 Acute monoblastic leukemia. Bone marrow smear demonstrates numerous blast cells with abundant blue cytoplasm, round nuclei, fine chromatin, and one or more prominent nucleoli (a). Similar blast cells are present in the peripheral blood smear (b). Blasts are positive with NSE stain (c).

Blood smears show aniso-poikilocytosis with the presence of schistocytes, tear-drops, and macrocytes. Basophilic stippling is present. Granulocytic series may show hypogranulation and hyposegmentation. Giant and/or hypogranular platelets are often present. Various numbers of nucleated red blood cells and blasts are often present.

The WHO requirement for the diagnosis of erythroleukemia leaves significant overlapping features with RA

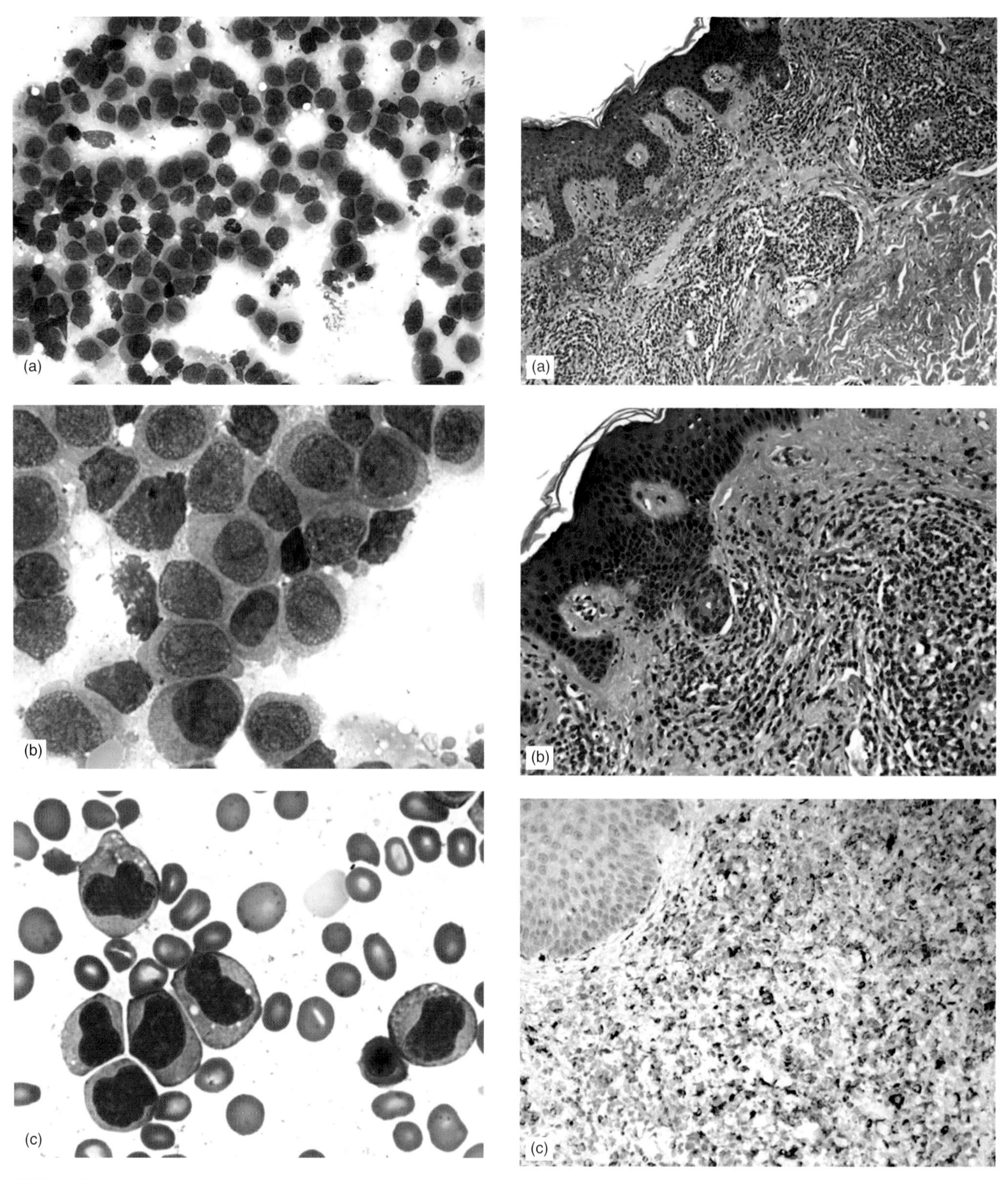

FIGURE 11.34 Acute monocytic leukemia. Bone marrow smear showing increased number of promonocytes with scattered blasts: (a) low power and (b) high power. Peripheral blood smears demonstrate several atypical monocytes (c).

FIGURE 11.35 Dermal infiltration of leukemic cells in a patient with acute monocytic leukemia: (a) low power and (b) high power. Many of the leukemic cells are positive for CD68 by immunohistochemistry (c).

and RAEB. If we follow the WHO requirements, a simple calculation tells us that any bone marrow sample with dysplastic changes, ≥50% and ≤80% erythroid precursors, and 4% or more myeloblasts in the total bone marrow cells is qualified for the diagnosis of acute erythroleukemia. For example, a bone marrow sample with 51% erythroid precursors and 10% myeloblasts in the total bone marrow cells should be diagnosed as erythroleukemia, whereas a bone

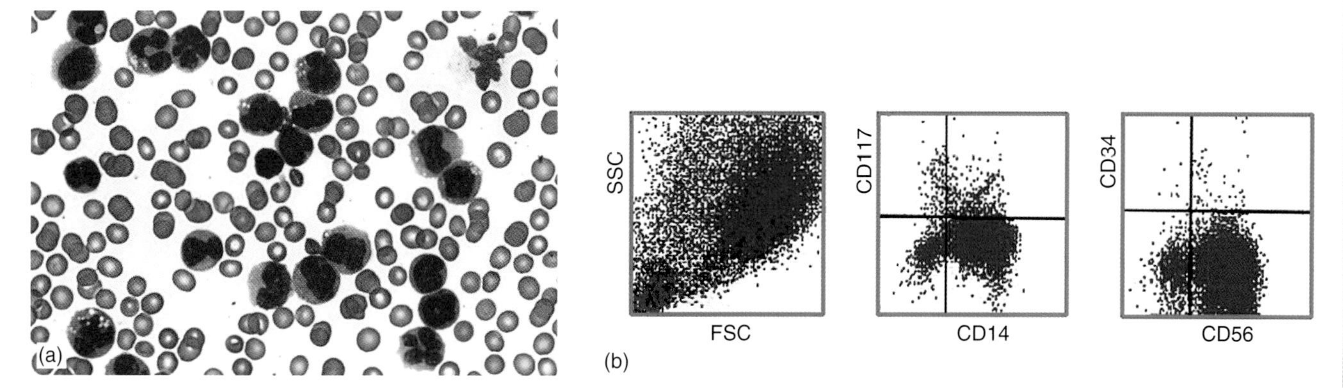

FIGURE 11.36 (a) Peripheral blood smear from a patient with acute monocytic leukemia demonstrating numerous atypical monocytes and promonocytes (bone marrow smears showed >20% blasts and promonocytes). (b) Flow cytometric studies reveal a large population of monocytic cells which are positive for CD14 and negative for CD34 and CD117. They also aberrantly express CD56.

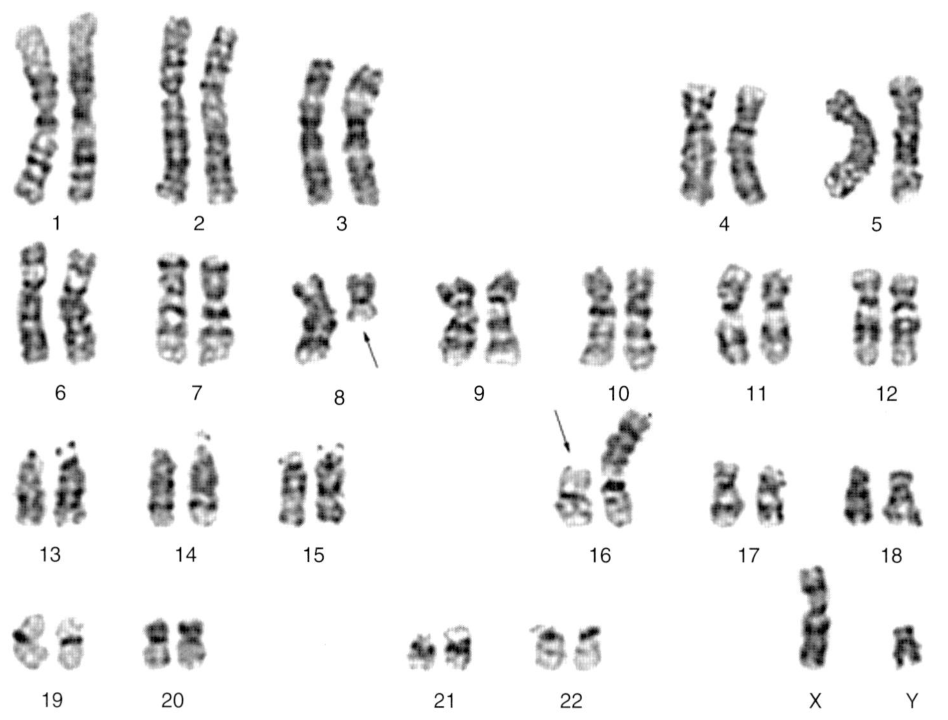

FIGURE 11.37 G-banded karyotype reveals t(8;16) in a patient with acute monocytic leukemia.

marrow sample with 49% erythroid precursors and 15% myeloblasts is called RAEB II and not an acute leukemia! This problem with the diagnostic criteria for acute erythroleukemia has been raised in the literature. Further clinical investigations are needed for clarification of the status of erythroleukemias with low myeloblast counts [211].

Pure Erythroid Leukemia (AML-M6b): This category is defined as a disorder with >80% of bone marrow nucleated cells consisting of erythroid precursors (Figures 11.40 and 11.41). These cells are predominantly erythroblasts (pronormoblasts) with dark-blue cytoplasm, round nuclei, fine nuclear chromatin, and one or more prominent nucleoli. The cytoplasm often contains poorly demarcated vacuoles [1, 218]. Evidence of dyserythropoiesis and/or megaloblastic changes is often present. Pure erythroid leukemia is far less frequent than the erythroleukemia (erythroid/myeloid) type.

Immunophenotype and Cytochemical Stains

The erythroid precursors are often positive for GPA, hemoglobin A, and CD71 (transferring receptor) (Figure 11.42). The very early cells may lack expression of one or more of these markers. HLA-DR and CD34 are often negative. The erythroid precursors do not express myeloid-, monocytic-, or megakaryocytic-associated markers. They are negative for cytochemical MPO, Sudan Black B, and NSE stains, but often show globular or coarsely granular cytoplasmic PAS positivity.

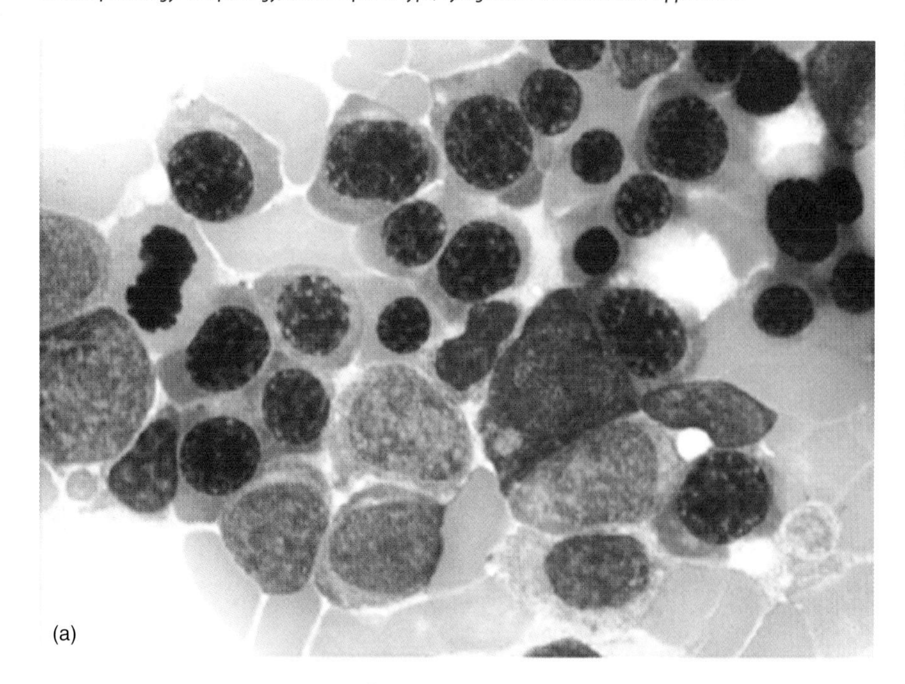

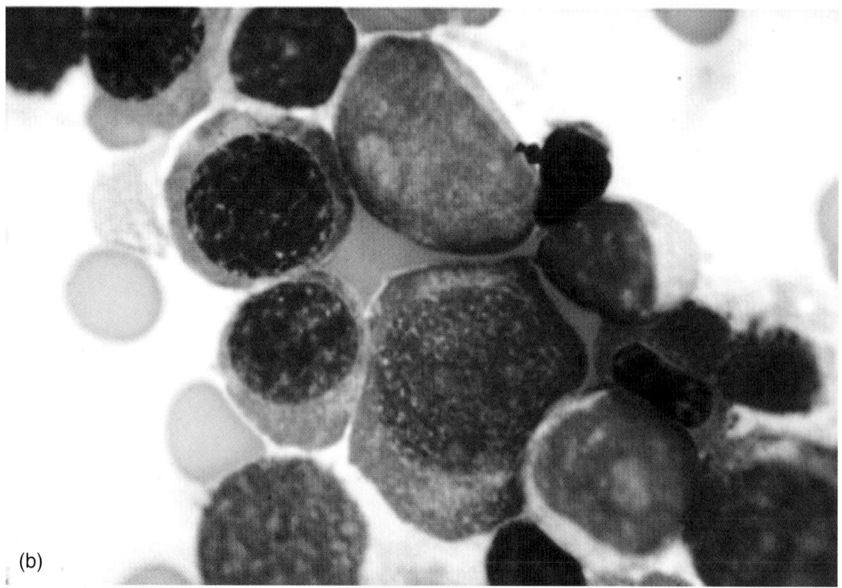

FIGURE 11.38 Erythroid/myeloid leukemia (erythroleukemia (AML-M6a)). Bone marrow smears show erythroid preponderance and left shift with increased myeloblasts: (a) low power and (b) high power.

The myeloblasts in the AML-M6a variant, similar to the myeloblasts in other AMLs, express myeloid-associated markers, such as CD13, CD33, CD117, and MPO, and are often positive for CD34 and HLA-DR.

Molecular and Cytogenetic Studies

Acute erythroid leukemia shares many cytogenetic features with MDS (see Chapter 8). Partial loss or monosomy of chromosomes 5 and 7 is the most frequent chromosomal aberration reported in acute erythroid leukemias, followed by abnormalities of chromosomes 8, 16, and 21 [219–221]. Some of the cases of pure erythroid leukemia are found to be associated with a *BCR-ABL1* fusion. Complex chromosomal abnormalities are frequent findings.

Clinical Aspects

Acute erythroid leukemias account for about 5% of all AMLs. The vast majority (>90%) are of erythroid/myeloid (AML-M6a) subtype. Pure erythroid leukemia is rare. The median age is around 57 years, ranging from 20 to 80 years [222]. The incidence is higher in men than in women (male: female ratio is about 2:1). A significant proportion (up to 50%) of acute erythroid leukemia represents either therapy-related or evolution in patients with a history of MDS [213, 222]. Severe anemia, usually with granulocytopenia and/or thrombocytopenia, is a common feature.

Acute erythroid leukemia is an aggressive disease, but the erythroid/myeloid type (AML-M6a) does significantly better than the pure erythroid leukemia type (AML-M6b).

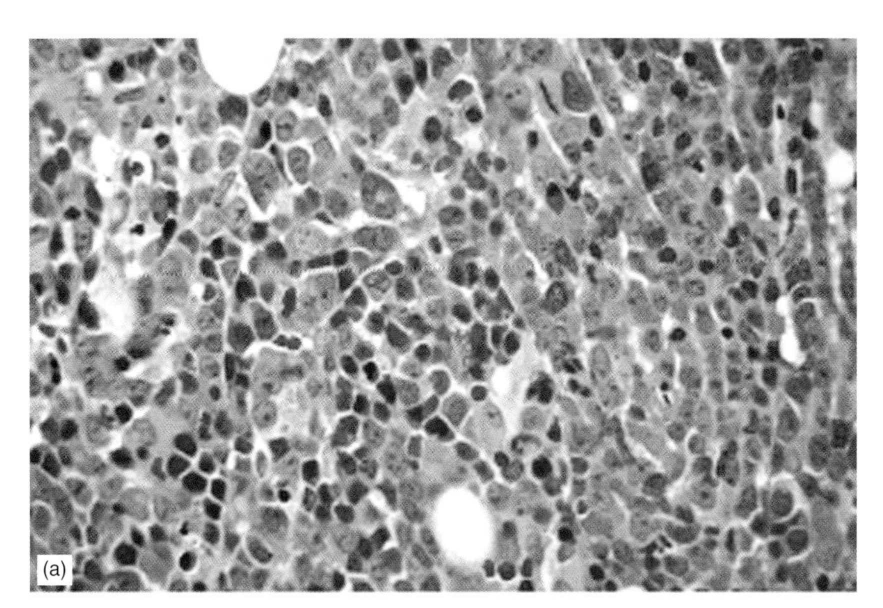

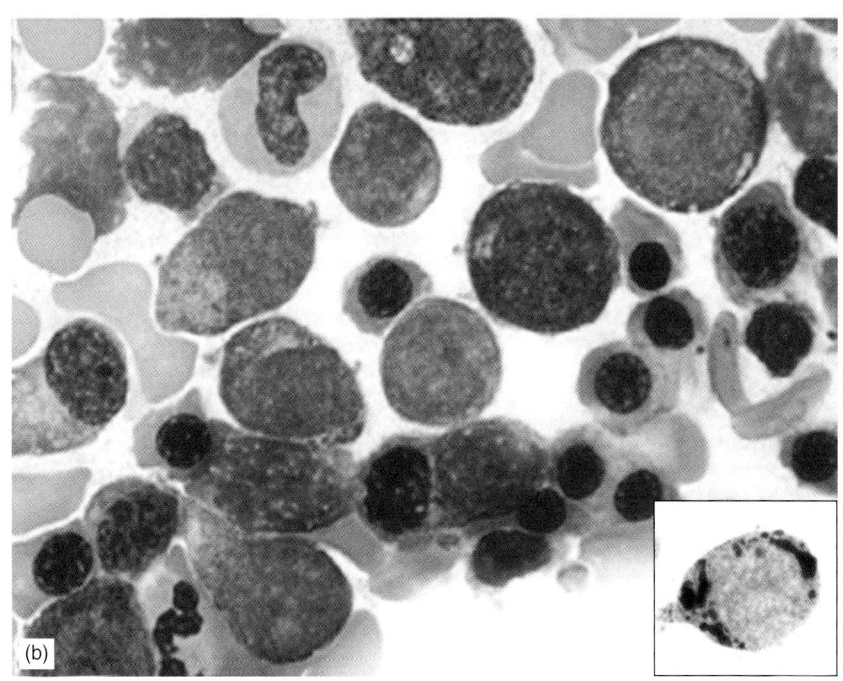

FIGURE 11.39 Erythroid/myeloid leukemia (erythroleukemia, AML-M6a). (a) Bone marrow biopsy section showing hypercellular marrow with increased immature cells. (b) Bone marrow smears show erythroid preponderance and left shift with increased myeloblasts. Inset shows an erythroid precursor with coarse PAS positive cytoplasmic granules.

In one report the average survival time for the AML-M6a was 30 months compared to 3 months for the AML-M6b [222].

Differential Diagnosis

The distinguishing feature of acute erythroleukemia (AML-M6a) from RAEB is the proportion of erythroid component in the bone marrow. In AML-M6a, 50% or more bone marrow nucleated cells are of erythroid lineage and ≥20% of the non-erythroid population consists of myeloblasts. AML with multilineage dysplasia should also be included in the differential diagnosis. According to the WHO recommendation, if ≥50% of the myeloid or megakaryocytic lineages show dysplasia, the case should be classified as AML with multilineage dysplasia.

Pure erythroid leukemia (AML-M6b) should be distinguished from megaloblastic anemia. The erythroid left shift and dysplastic changes are not so severe in megaloblastic anemia as in pure erythroid leukemia. Besides, in megaloblastic anemia, there is often evidence of vitamin B12 or folate deficiency, whereas serum levels of vitamin B12 and folate are normal or elevated in pure erythroid leukemia. The differential diagnosis of pure erythroid leukemia also includes ALL, minimally differentiated AML, AML without maturation, and AMKL.

Acute Megakaryoblastic Leukemia

Acute megakaryoblastic leukemia (AML-M7) is defined as an AML in which megakaryoblasts account for ≥50 of the total blast cells [1]. Similar to the other types of AML, total

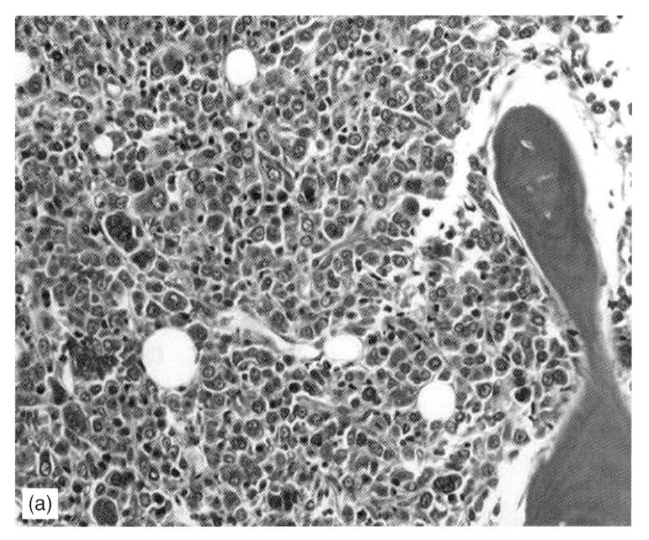

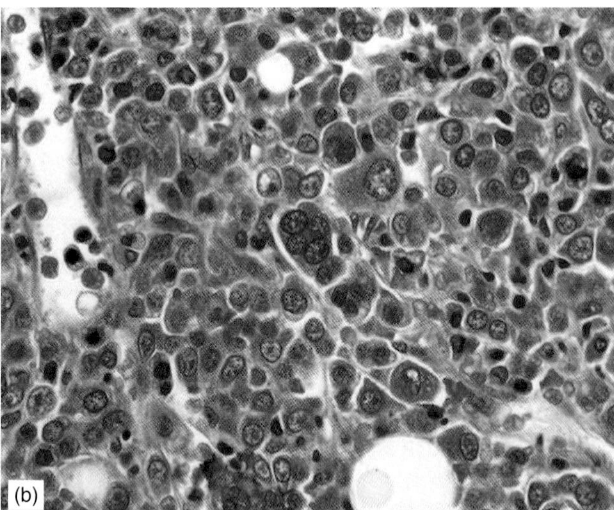

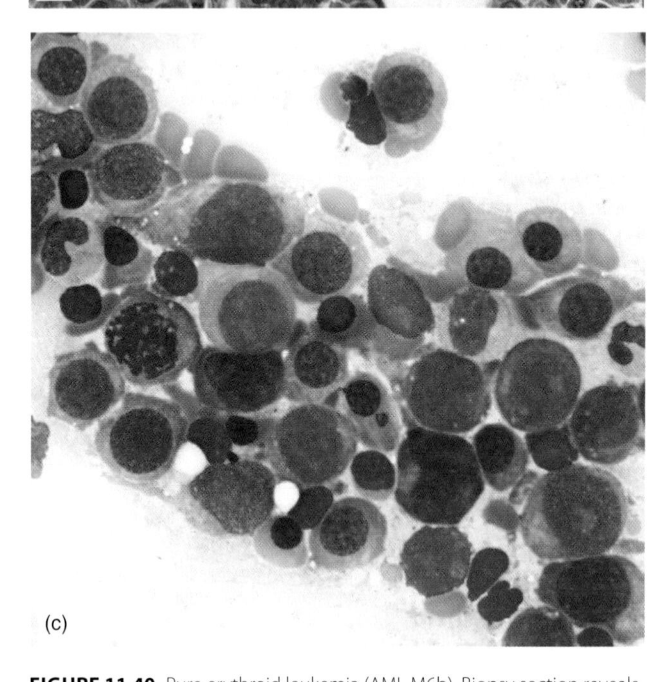

FIGURE 11.40 Pure erythroid leukemia (AML-M6b). Biopsy section reveals a markedly hypercellular marrow with sheets of blasts and immature erythroid cells: (a) low power and (b) high power. Bone marrow smear shows erythroid preponderance with increased erythroblasts (c).

blasts comprise ≥20% of the bone marrow nucleated cells or peripheral blood differential counts.

Etiology and Pathogenesis

The etiology and pathogenesis of AMKL are not known. The strong association of Down syndrome with AMKL has raised the possibility that overexpression or mutation of certain hematopoietic regulatory genes located on chromosome 21 may be involved in the development of this leukemia. The strongest candidate gene for leukemogenesis is *RUNX1* (*AML1*) on chromosome 21q22.3. RUNX1 is a transcription factor required for the production of all hematopoietic precursors. The translocation of the *RUNX1* gene is observed as t(8;21) in AML-M2, t(12;21) in childhood ALL, and t(3;21) in some cases of CML in blast crisis [223]. In addition, acquired mutations of the *GATA1* gene have been detected in the vast majority of patients with AML-M7 and Down syndrome. *GATA1* is located on the X chromosome (Xp11.23) and is a member of the transcription factor family that regulates proliferation and differentiation of hematopoietic cells [223].

Pathology

Morphology

Megakaryoblasts are the predominant component of the blast population in the bone marrow and/or peripheral blood (Figures 11.43 and 11.44) [1–4]. They are markedly pleomorphic, ranging from small, round cells with scanty cytoplasm and inconspicuous nucleoli, resembling hematogones, to large cells with abundant cytoplasm and prominent nucleoli. They often display cytoplasmic blebs or psuedopods and may appear in clusters mimicking metastatic tumors. Since megakaryoblasts may resemble hematogones, lymphoblasts, type 1 myeloblasts, or metastatic tumors and do not react with a specific cytochemical stain, often it is necessary to perform immunophenotypic and/or ultrastructural studies to be able to assign their megakaryocytic lineage.

The peripheral blood smears show circulating micromegakaryocytes, megakaryocytic fragments, atypical giant platelets, and blasts. Red blood cells may show anisopoikilocytosis and scattered tear-drop shapes. Granulocytes may show dysplastic changes, such as hypogranulation and abnormal segmentation of their nuclei. Leukoerythroblastosis is uncommon.

Bone marrow fibrosis, and as a consequence, dry tap (failure of bone marrow aspiration) is a common feature. Bone marrow cellularity varies depending on the extent of fibrosis, but there is evidence of increased blast cells, either in clusters or as diffuse interstitial infiltration. There is also evidence of increased megakaryocytes with dysplastic morphology.

Immunophenotype and Cytochemical Stains

Megakaryoblasts carry one or more platelet glycoproteins (GP). These include CD41 (GPIIb/IIIa), CD42 (GPIb), and CD62 (GPIIIa), which are expressed both on the surface membrane and in the cytoplasm (Figure 11.44b) [224]. Therefore, both flow cytometry and immunohistochemical stains are helpful in their identification. Megakaryoblasts may also express CD36 and factor VIII. HLA-DR, CD45,

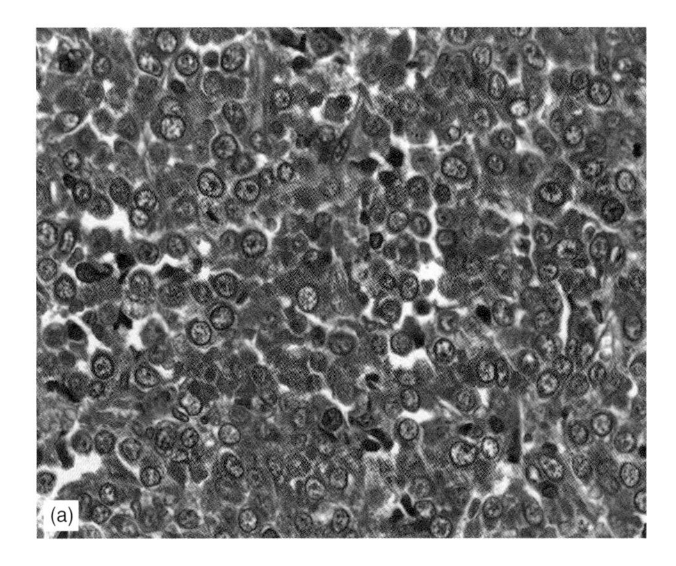

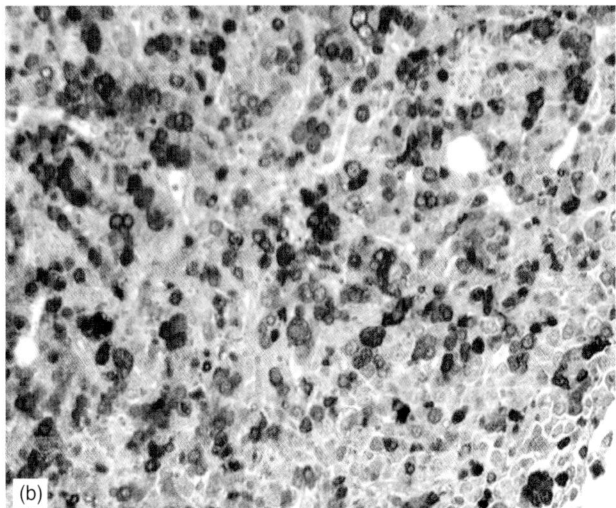

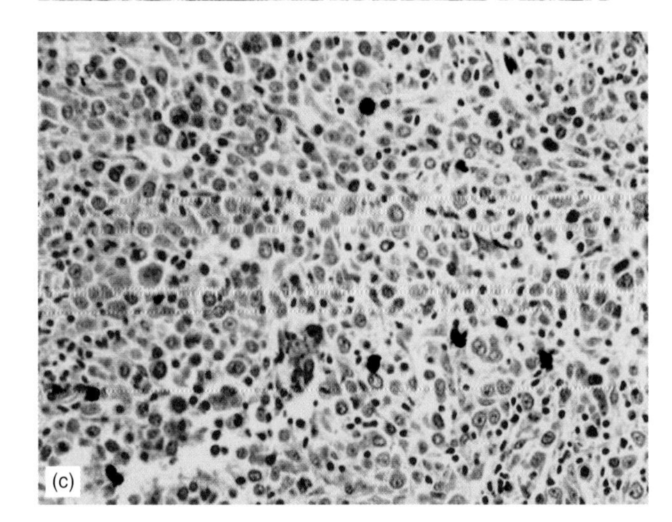

FIGURE 11.41 Pure erythroid leukemia (AML-M6b). Biopsy section reveals a markedly hypercellular marrow with sheets of blasts and immature cells (a). Immunohistochemical stain for hemoglobin A shows numerous positive cells (b). MPO stain is negative (c).

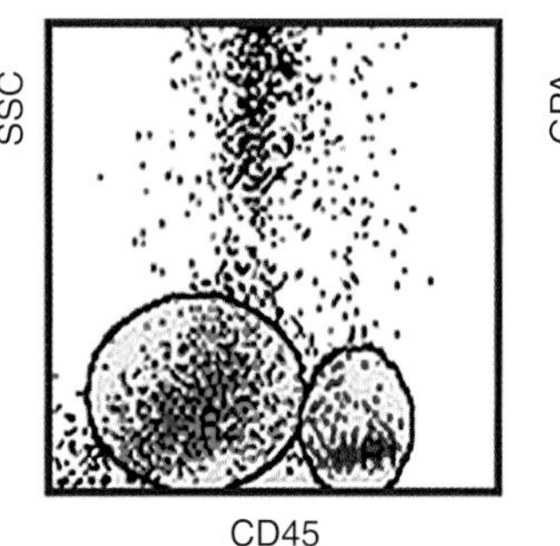

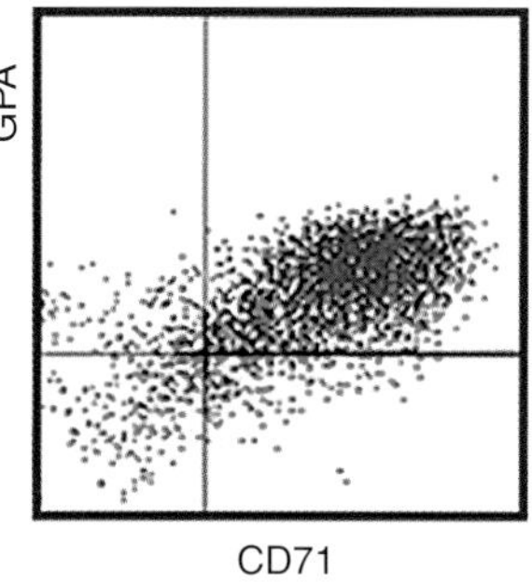

FIGURE 11.42 Pure erythroid leukemia (AML-M6b). Flow cytometry of the bone marrow cells shows a large population of CD45-negative cells (green gate) expressing CD71 and glycophorin A (GPA).

TdT, and lymphoid-associated markers are often negative, except for aberrant expression of CD7 which may be present in some cases [224, 225].

No conventional cytochemical stain is specific for megakaryoblasts. However, megakaryoblasts may show diffuse and/or globular PAS reaction and punctuate NSE staining. MPO and Sudan Black B are negative, but ultrastructural cytochemical staining or immunostaining reveals expression of platelet peroxidase. Megakaryoblasts also express acid phosphatase.

Molecular and Cytogenetic Studies

In addition to trisomy 21 in Down syndrome, t(1;22)(p13;q13);(*RBM15-MKL1*) has also been reported in *de novo* AMKL (Figure 11.45) [226–228]. In addition to OTT-MAL transcript, there are reports of activated *JAK2* and *GATA1* mutations in AMKL [223, 229]. Chromosomal aberrations, such as −7/7q−, −5/5q−, and trisomy 8 have been reported particularly in the therapy-related type. Other chromosomal aberrations include t(9;22)(q34;q11.2) and abnormalities of 3q21→q26, 17q22, 11q14→21, 21q21→22, and 16q22→23 [230–233]. Occasional cases of highly complex chromosomal abnormalities have been reported [234].

Clinical Aspects

AMKL represents 3–5% of the AMLs. It occurs in all ages with two distribution peaks: children between 1 and 3 years old and adults [235–238]. A significant proportion of affected children have Down syndrome [223, 238, 239]. In children with Down syndrome, the incidence of AML is 46-fold greater than that in the normal age group. AMKL accounts for at least 50% of the AML cases in patients with Down syndrome [234]. AMKL has also been associated with mediastinal germ cell tumors in young adult males. Hepatosplenomegaly is rare in adults but frequently observed in children, particularly in association with t(1;22). This translocation has distinctive clinicopathologic features including onset in infancy, extensive bone marrow fibrosis with clustering of leukemic blasts mimicking metastatic tumors, and aggressive clinical course.

Differential Diagnosis

The differential diagnosis of AMKLs includes chronic idiopathic myelofibrosis, minimally differentiated AML,

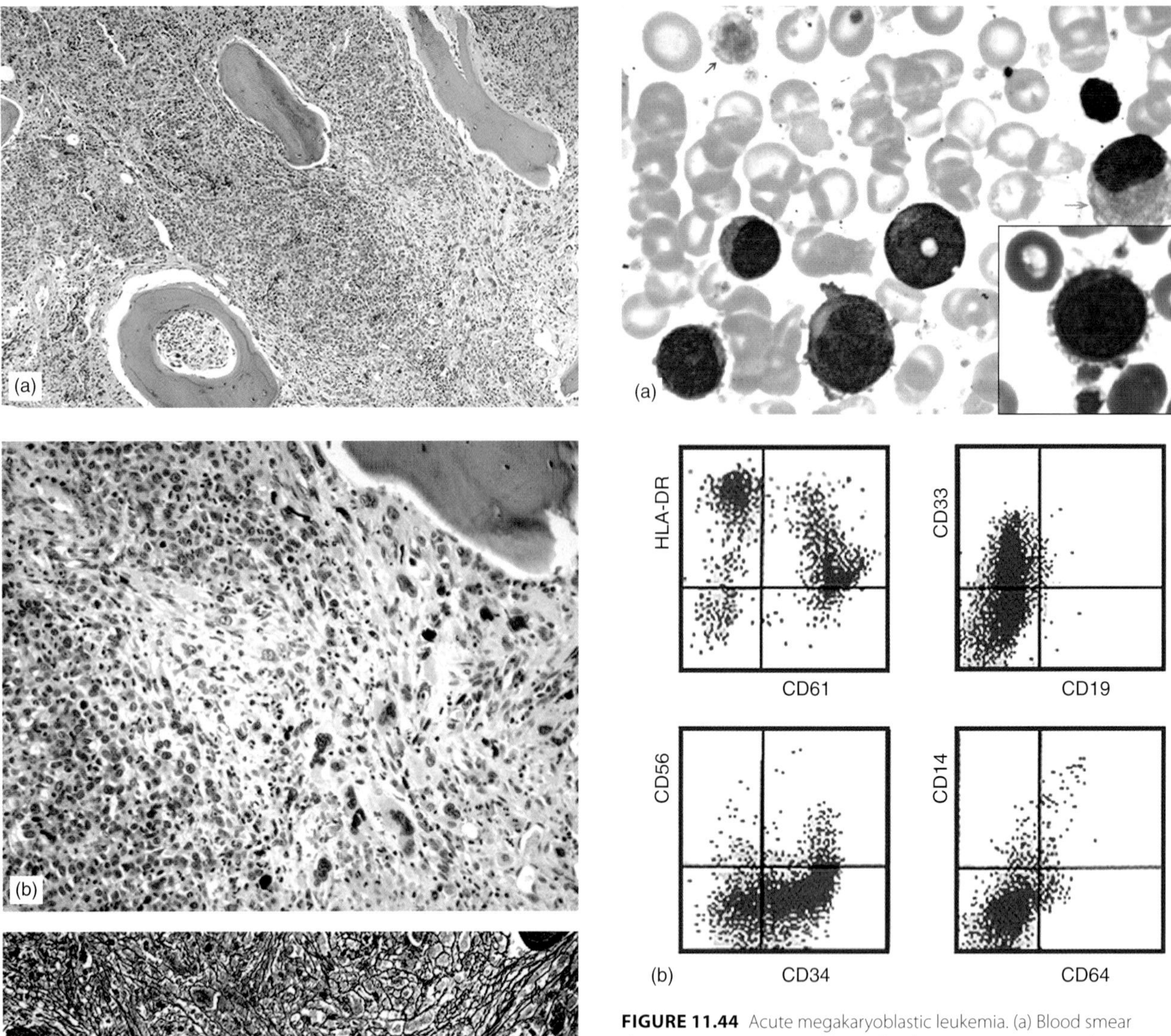

FIGURE 11.43 Acute megakaryoblastic leukemia. Bone marrow biopsy sections show large clusters of immature cells with areas of fibrosis and increased megakaryocytes: (a) low power and (b) high power. Reticulin stain reveals marked reticulin fibrosis (c).

FIGURE 11.44 Acute megakaryoblastic leukemia. (a) Blood smear demonstrating numerous blast cells of various sizes, some with cytoplasmic blebs. A micromegakaryocyte (green arrow) and a giant platelet (red arrow) are present. (b) Flow cytometry of peripheral blood demonstrates a population of cells expressing CD34 and CD61 consistent with megakaryoblasts.

AML without maturation, ALL, APMF, and metastatic tumors.

The presence of megakaryoblastic clusters embedded in the fibrotic bone marrow may mimic metastatic carcinoma, non-Hodgkin lymphoma, neuroblastoma, or rhabdomyosarcoma.

The diagnosis of AMKL is suggested by certain clinicopathological features, such as Down syndrome, myelofibrosis with no or minimal leukoerythroblastosis, presence of megakaryoblasts in the peripheral blood, and/or evidence of t(1;22). The diagnosis is confirmed when at least 50% of the total blast population is of megakaryocytic origin evidenced by the presence of platelet myeloperoxidase (PPO) by electron microscopy and/or expression of CD41, CD42, CD61, and factor VIII.

Acute Basophilic Leukemia

Acute basophilic leukemia is a rare type of AML with basophilic differentiation [1, 240–242]. ABL is either *de novo* or the result of basophilic blast transformation in CML.

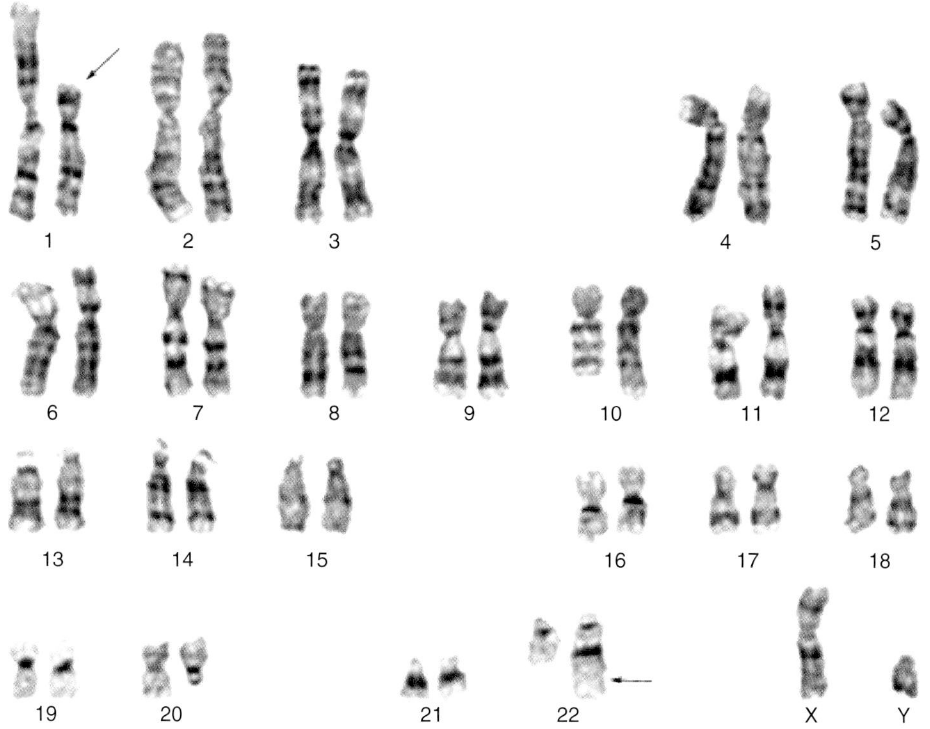

FIGURE 11.45 Acute megakaryoblastic leukemia. G-banded karyotype demonstrating t(1;22).

Etiology and Pathogenesis

The etiology and pathogenesis of ABL are not known.

Pathology

Morphology

The blasts show a high nuclear:cytoplasmic ratio with variable amounts of cytoplasm containing coarse basophilic granules. The nuclei are round, oval, or bilobed with a fine chromatin and one or more prominent nucleoli. Scattered mature basophils are often present. Electron microscopy reveals characteristic features of basophilic granules [1].

Bone marrow biopsy sections appear hypercellular with large clusters or sheets of blast cells and marked reduction in the normal hematopoietic components. There may be dysplastic erythropoiesis in the bone marrow. A variable degree of reticulin fibrosis is often present, particularly in advanced stages of the disease.

Blood smears usually show the presence of blasts with basophilic granules and mature basophils. Anisopoikilocytosis and thrombocytopenia may be present.

Immunophenotype and Cytochemistry

The blast cells express some of the myeloid-associated markers, such as CD13, CD15, and CD33. They are often positive for CD34 and HLA-DR and may express CD9, CD25, CD117, or CD203c [240, 243]. The cytoplasmic basophilic granules may stain with toluidine blue but are usually negative for MPO and NSE.

Molecular and Cytogenetic Studies

No specific chromosomal aberration is known for ABL. There are sporadic reports of abnormalities of deletion of 12p, t(6;9)(p23;q34), t(8;21)(q22;q22), and t(X;6)(p11;q23) [244–246]. Cases of basophilic transformation in CML show t(9;22)(q34;q11.2).

Clinical Aspects

Acute basophilic leukemia is a rare disease probably representing 1% of all AMLs. It occurs at any age including infancy. Clinical findings may include anemia, skin rashes, hepatosplenomegaly, and gastric ulcers.

Differential Diagnosis

Differential diagnosis includes blast transformation of CML, certain subtypes of AML with basophilia, particularly the ones associated with t(6;9) or abnormalities of 12p, and occasional cases of ALL with coarse azurophilic granules.

Acute Panmyelosis with Myelofibrosis

Previously referred to as *acute myelofibrosis*, *acute myelosclerosis*, or *acute myelodysplasia with myelofibrosis*, APMF is an acute leukemia with panmyeloid proliferation, increased blasts, and bone marrow fibrosis [1, 247, 248]. This entity shares significant overlapping features with AMKL, such as bone marrow fibrosis, unaspirable marrow, dysplastic megakaryocytes, and the presence of megakaryoblasts. However, in AMKL, megakaryoblasts comprise the predominant proportion of the blast population, whereas in APMF majority of the blasts are of non-megakaryocytic origin [1, 249].

Etiology and Pathogenesis

The etiology and pathogenesis of APMF are not known.

Pathology

Morphology

Bone marrow fibrosis is one of the morphologic hallmarks. Marrow fibrosis often leads to unaspirable bone marrow (dry tap) and inadequate marrow smears. The bone marrow cellularity varies depending on the extent of fibrosis. There is evidence of increased blast cells, either in clusters or as diffuse interstitial infiltration [1, 248, 249]. There is also evidence of increased megakaryocytes with dysplastic morphology, including micromegakaryocytes and nonlobulated and/or hypolobulated forms.

The peripheral blood smears may show absent to mild anisopoikilocytosis with macrocytes and occasional teardrop shapes. Granulocytes may show dysplastic changes, such as hypogranulation and abnormal segmentation of their nuclei. Abnormal platelets may be present. Circulating blasts are variable, ranging from a few to a frank leukemic picture [1, 248].

Immunophenotype and Cytochemical Stains

The majority of the blast cells express CD34, often with one or more myeloid-associated markers, such as CD13, CD33, and/or CD117 and cytoplasmic MPO [250]. Only a minority of blast cells express platelet-associated molecules, such as CD41, CD42, or CD61. Cytochemical stains may show presence of MPO-positive blast cells.

Molecular and Cytogenetic Studies

Monosomy 7 and deletion of 5q or 7q are the most frequent chromosomal aberrations in this condition. Also, interstitial deletion of the long arm of chromosome 11 has been reported [250, 251].

Clinical Aspects

APMF is a rare type of AML often presenting with marked cytopenia. Splenomegaly is minimal or absent. It has an unfavorable prognosis with a median survival of <1 year in some reports [248, 251].

Differential Diagnosis

As mentioned earlier, this entity shares many overlapping features with AMKL, such as bone marrow fibrosis, unaspirable marrow, dysplastic megakaryocytes, and the presence of megakaryoblasts. However, unlike AMKL, in APMF the majority of the blasts are of non-megakaryocytic origin. The differential diagnosis also includes chronic idiopathic myelofibrosis, minimally differentiated AML, AML without maturation, ALL, and metastatic tumors.

Granulocytic Sarcoma

Granulocytic sarcoma (chloroma) refers to extramedullary tumors of myeloid precursors. The term "chloroma" was originally used in response to the green color of the tumor due to the presence of MPO in the tumor cells. Granulocytic sarcoma is associated with CML, CML in blast crisis, and *de novo* AML. It either develops during the active phase of the disease or represents relapse without evidence of recurrent disease in the blood or the bone marrow. The most frequent sites of involvement include skin, lymph nodes, respiratory system, gastrointestinal tract, CNS, and subperiosteal structures of the skull, ribs, vertebrae, and pelvis.

Granulocytic sarcomas are composed of various proportions of immature and mature myeloid cells (Figure 11.46). Some tumors resemble CML and consist predominantly of mature granulocytic cells, and others similar to AML show the predominance of myeloblasts and immature myeloid cells. The ones with blasts and immature cells may resemble lymphomas or non-hematopoietic malignancies. The presence of immature eosinophils is a distinctive feature of granulocytic sarcomas. Tumor cells in these lesions express myeloid-associated molecules in the biopsy sections, such as MPO, NES, and/or lysozyme.

OTHER TYPES OF AML

Acute Myeloid Leukemia With Chromosome 3 Aberrations and Thrombocytosis

Approximately 3% of AML cases show cytogenetic abnormalities of 3q along with thrombocytosis. The cytogenetic aberrations include the inv(3)(q21q26), t(3:3)(q21;q26), and t(5;3)(q14;q21q26). In some cases thrombocytosis may exceed 1,000,000/μL. Bone marrow samples consistently show increased numbers of megakaryocytes, including micromegakaryocytes.

Other Recurrent Genetic Abnormalities

Other recurrent genetic abnormalities associated with AML include t(1;22)(p13;q13)(*RBM15;MKL1*), t(9;22)(q34;q11.2) (*BCR;ABL1*), AML with C/EBPα mutation, and AML with mutated nucleoplasmin (*NPM*) gene. The t(9;22) is seen in CML in blast transformation, but less frequently may be present *de novo* without a history of CML.

The t(1;22), as mentioned earlier, has distinctive clinicopathologic features including onset in infancy, extensive bone marrow fibrosis with clustering of leukemic blasts (usually megakaryoblasts) mimicking metastatic tumors, and aggressive clinical course [252–254].

C/EBPα (CCAAT/enhancer-binding protein alpha) is a critical regulator for early myeloid differentiation. Mutations in C/EBPα occur in 10% of patients with AML, leading to the expression of a 30 kDa dominant-negative isoform (C/EBPα 30) [255–257].

The mutations of the *NPM* gene result in aberrant cytoplasmic localization of the NPM protein (NPMc+). These mutations occur in 25–35% of adult AML that show normal karyotype [258–260]. Patients with NPM mutations show high remission induction rates and improved survival [259].

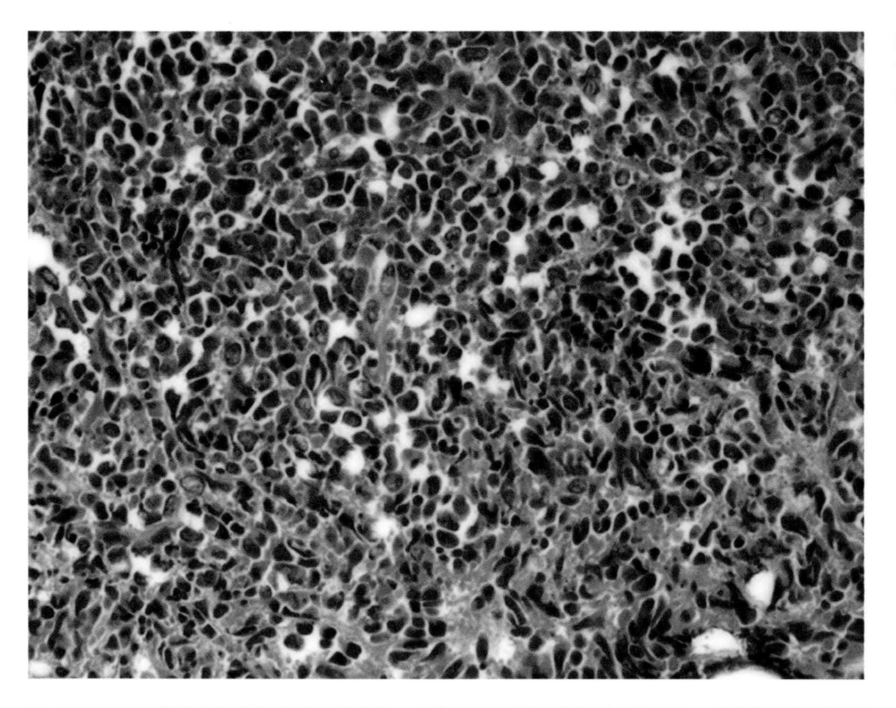

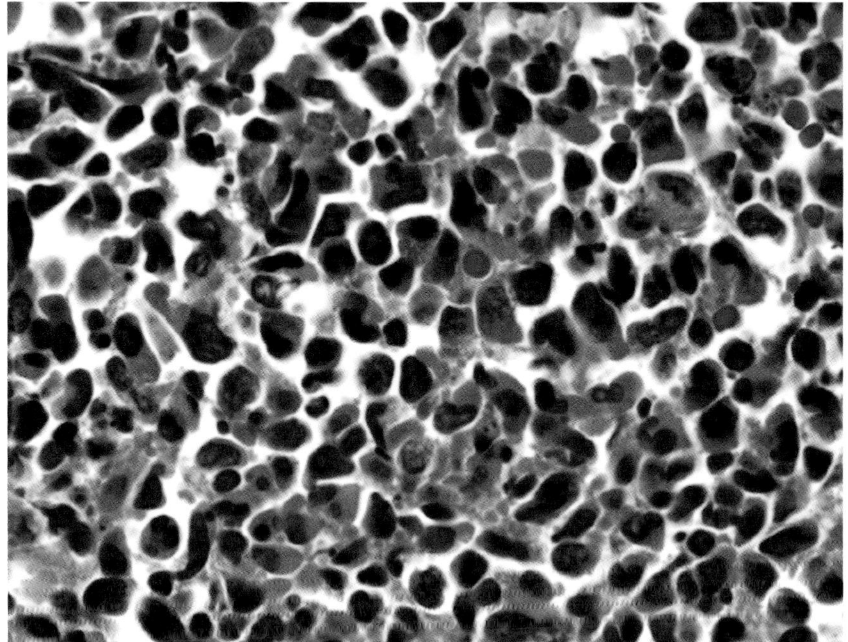

FIGURE 11.46 Soft tissue infiltration of immature myeloid cells representing granulocytic sarcoma (chloroma).

Hypoplastic Acute Leukemia

Hypoplastic acute leukemia, or hypocellular acute leukemia, is a leukemic condition characterized by marked bone marrow hypocellularity. The etiology and pathogenesis of this leukemia are not known, but bone marrow toxicity may play a contributing role. In a study by Gladson and Naeim [261], a history of alcohol abuse was demonstrated in 30% of the patients, potential exposure to toxic chemicals in 20%, and history of chemotherapy or radiation therapy due to a second malignancy in 20%. This type of leukemia tends to involve elderly people, primarily men.

The bone marrow is hypocellular, usually <30% of the average cellularity in the normal matching age group. Blast cells are prominent and account for ≥20% of the bone marrow cells (Figure 11.47). They display scant-to-moderate amounts of cytoplasm, round or oval nuclei with fine nuclear chromatin, and one or more prominent nucleoli. In most studies, blast cells appear to be of myeloid origin by the presence of cytoplasmic azurophilic granules, Auer rods, positive reactions for MPO, Sudan Black B, or expression of myeloid-associated molecules, such as CD13, CD33, and CD117.

The differential diagnoses include aplastic anemia and hypocellular MDS. Diagnosis is made based on the presence of ≥20% blasts in a hypocellular marrow.

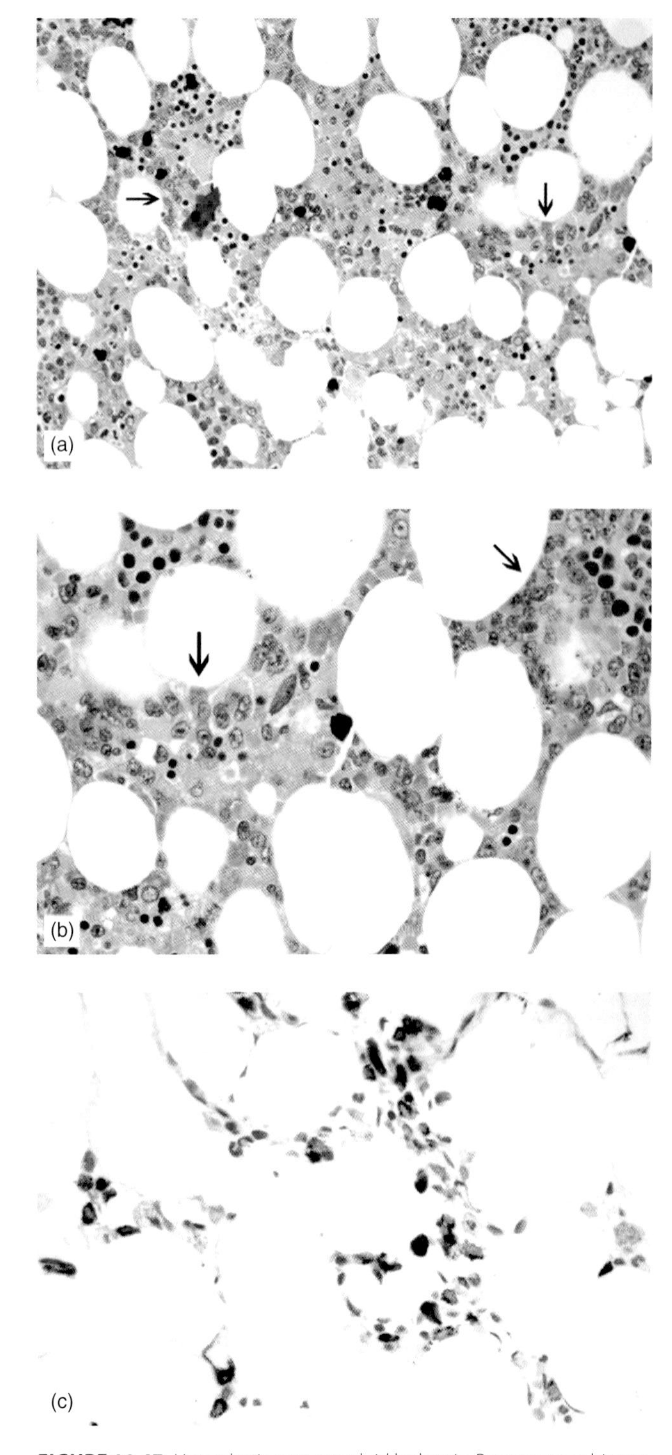

FIGURE 11.47 Hypoplastic acute myeloid leukemia. Bone marrow biopsy section showing hypocellularity and increased blasts (a, low power; b, high power). Immunohistochemical stain for CD34 reveals numerous positive cells (c).

References

1. Jaffe ES, Harris NL, Stein H, Vardiman JW. (2001). *Pathology and Genetics. Tumors of Haematopoietic and Lymphoid Tissues*. IARC Press, Lyon.
2. His ED. (2007). *Hematopathology*. Churchill Livingstone, Philadelphia.
3. Foucar K. (2001). *Bone Marrow Pathology*, 2nd ed. ASCP Press, Chicago.
4. Naeim F. (1997). *Pathology of Bone Marrow*, 2nd ed. Williams & Wilkins, Baltimore.
5. Bowen DT. (2006). Etiology of acute myeloid leukemia in the elderly. *Semin Hematol* **43**, 82–8.
6. Warner JK, Wang JC, Hope KJ, Jin L, Dick JE. (2004). Concepts of human leukemic development. *Oncogene* **23**, 7164.
7. Roldán Schilling V, Fernández Abellán P, Domínguez Escribano JR, Rivas González C, Mut Barberá E, Calatayud Cendra R. (1998). Acute leukemias after treatment with radioiodine for thyroid cancer. *Haematologica* **83**, 767–8.
8. Pedersen-Bjergaard J. (1992). Radiotherapy- and chemotherapy-induced myelodysplasia and acute myeloid leukemia. A review. *Leuk Res* **16**, 61–5.
9. Haran-Ghera N. (1989). Radiation induced deletion of chromosome 2 in myeloid leukemogenesis. *Curr Top Microbiol Immunol* **149**, 35–41.
10. Linos A, Gray JE, Orvis AL, Kyle RA, O'Fallon WM, Kurland LT. (1980). Low-dose radiation and leukemia. *N Engl J Med* **302**, 1101–5.
11. Kamada N. (1969). The effects of radiation on chromosomes of bone marrow cells. II. Studies on bone marrow chromosomes of atomic bomb survivors in Hiroshima. *Nippon Ketsueki Gakkai Zasshi* **32**, 236–48.
12. Hamilton HB, Brody JA. (1975). Review of thirty years study of Hiroshima and Nagasaki atomic bomb survivors. III. Future research and health surveillance. A. Health surveillance studies. *J Radiat Res (Tokyo)* **16**(Suppl), 138–48.
13. Auxier JA. (1975). A physical dose estimates for A-bomb survivors. Studies at Oak Ridge, U.S.A. *J Radiat Res (Tokyo)* **16**(Suppl), 1–11.
14. Felix CA. (1998). Secondary leukemias induced by topoisomerase-targeted drugs. *Biochim Biophys Acta* **1400**, 233–55.
15. Smith SM, Le Beau MM, Huo D, Karrison T, Sobecks RM, Anastasi J, Vardiman JW, Rowley JD, Larson RA. (2003). Clinical-cytogenetic associations in 306 patients with therapy-related myelodysplasia and myeloid leukemia: The University of Chicago series. *Blood* **102**, 43–52.
16. Pedersen-Bjergaard J, Andersen MK, Christiansen DH. (2000). Therapy-related acute myeloid leukemia and myelodysplasia after high-dose chemotherapy and autologous stem cell transplantation. *Blood* **95**, 3273–9.
17. Giles FJ, Koeffler HP. (1994). Secondary myelodysplastic syndromes and leukemias. *Curr Opin Hematol* **1**, 256–60.
18. Rosner F, Grünwald HW, Zarrabi MH. (1979). Acute leukemia as a complication of cytotoxic chemotherapy. *Int J Radiat Oncol Biol Phys* **5**, 1705–7.
19. Bergsagel DE, Bailey AJ, Langley GR, MacDonald RN, White DF, Miller AB. (1979). The chemotherapy on plasma-cell myeloma and the incidence of acute leukemia. *N Engl J Med* **301**, 743–8.
20. Greene MH, Boice JD Jr, Greer BE, Blessing JA, Dembo AJ, Knoche E, McLeod HL, Graubert TA. (2006). Pharmacogenetics of alkylator-associated acute myeloid leukemia. *Pharmacogenomics* **7**, 719–29.
21. Bennett JM, Kaminski MS, Leonard JP, Vose JM, Zelenetz AD, Knox SJ, Horning S, Press OW, Radford JA, Kroll SM, Capizzi RL. (2005). Assessment of treatment-related myelodysplastic syndromes and acute myeloid leukemia in patients with non-Hodgkin lymphoma treated with tositumomab and iodine I131 tositumomab. *Blood* **105**, 4576–82.
22. Roulston D, Anastasi J, Rudinsky R, Nucifora G, Zeleznik-Le N, Rowley JD, McGavran L, Tsuchida M, Hayashi Y. (1995). Therapy-related acute leukemia associated with t(11q23) after primary acute myeloid leukemia with t(8;21): A report of two cases. *Blood* **86**, 3613–14.
23. Grünwald HW, Rosner F. (1979). Acute leukemia and immunosuppressive drug use: A review of patients undergoing immunosuppressive therapy for non-neoplastic diseases. *Arch Intern Med* **139**, 461–6.
24. Traweek ST, Slovak ML, Nademanee AP, Brynes RK, Niland JC, Forman SJ. (1996). Myelodysplasia and acute myeloid leukemia occurring after autologous bone marrow transplantation for lymphoma. *Leuk Lymphoma* **20**, 365–72.

25. Snyder R, Kalf GF. (1994). A perspective on benzene leukemogenesis. *Crit Rev Toxicol* **24**, 177–209.
26. Natelson EA. (2007). Benzene-induced acute myeloid leukemia: A clinician's perspective. *Am J Hematol* **82**, 826–30.
27. Pyatt D. (2004). Benzene and hematopoietic malignancies. *Clin Occup Environ Med* **4**, 529–55.
28. Van Maele-Fabry G, Duhayon S, Lison D. (2007). A systematic review of myeloid leukemias and occupational pesticide exposure. *Cancer Causes Control* **18**, 457–78.
29. Lichtman MA. (2007). Cigarette smoking, cytogenetic abnormalities, and acute myelogenous leukemia. *Leukemia* **21**, 1137–40.
30. Thomas X, Chelghoum Y. (2004). Cigarette smoking and acute leukemia. *Leuk Lymphoma* **45**, 1103–9.
31. Sandler DP, Shore DL, Anderson JR, Davey FR, Arthur D, Mayer RJ, Silver RT, Weiss RB, Moore JO, Schiffer CA. (1993). Cigarette smoking and risk of acute leukemia: Associations with morphology and cytogenetic abnormalities in bone marrow. *J Natl Cancer Inst* **85**, 1994–2003.
32. Drabkin HA, Erickson P. (1995). Down syndrome and leukemia, an update. *Prog Clin Biol Res* **393**, 169–76.
33. Segel GB, Lichtman MA. (2004). Familial (inherited) leukemia, lymphoma, and myeloma: An overview. *Blood Cells Mol Dis* **32**, 246–61.
34. Osato M, Yanagida M, Shigesada K, Ito Y. (2001). Point mutations of the RUNx1/AML1 gene in sporadic and familial myeloid leukemias. *Int J Hematol* **74**, 245–51.
35. Seghezzi L, Maserati E, Minelli A, Dellavecchia C, Addis P, Locatelli F, Angioni A, Balloni P, Miano C, Cavalli P, Danesino C, Pasquali F. (1996). Constitutional trisomy 8 as first mutation in multistep carcinogenesis: Clinical, cytogenetic, and molecular data on three cases. *Genes Chromosomes Cancer* **17**, 94–101.
36. Maserati E, Aprili F, Vinante F, Locatelli F, Amendola G, Zatterale A, Milone G, Minelli A, Bernardi F, Lo Curto F, Pasquali F. (2002). Trisomy 8 in myelodysplasia and acute leukemia is constitutional in 15–20% of cases. *Genes Chromosomes Cancer* **33**, 93–7.
37. Stock W, Thirman MJ. (2006). Molecular biology of acute promyelocytic leukemia. *UpTo Date.*
38. Moreau-Gachelin F. (2006). Lessons from models of murine erythroleukemia to acute myeloid leukemia (AML): Proof-of-principle of cooperativity in AML. *Haematologica* **91**, 1644–52.
39. Fröhling S, Scholl C, Bansal D, Huntly BJ. (2007). HOX gene regulation in acute myeloid leukemia: CDX marks the spot? *Cell Cycle* **6**, 2241–5.
40. Yamagata T, Maki K, Mitani K. (2005). Runx1/AML1 in normal and abnormal hematopoiesis. *Int J Hematol* **82**, 1–8.
41. Troke PJ, Kindle KB, Collins HM, Heery DM. (2006). MOZ fusion proteins in acute myeloid leukaemia. *Biochem Soc Symp* **73**, 23–39.
42. Steffen B, Muller-Tidow C, Schwable J, Berdel WE, Serve H. (2005). The molecular pathogenesis of acute myeloid leukemia. *Crit Rev Oncol Hematol* **56**, 195–221.
43. Nucifora G, Larson RA, Rowley JD. (1993). Persistence of the 8;21 translocation in patients with acute myeloid leukemia type M2 in long-term remission. *Blood* **82**, 712–15.
44. Naeim F. (2001). *Atlas of Bone Marrow and Blood Pathology*. W.B Saunders, Philadelphia.
45. Weir EG, Borowitz MJ. (2001). Flow cytometry in the diagnosis of acute leukemia. *Semin Hematol* **38**, 124–38.
46. Raanani P, Ben-Bassat I. (2004). Detection of minimal residual disease in acute myelogenous leukemia. *Acta Haematol* **112**, 40–54.
47. Hayhoe FGJ, Quaglino D. (1988). *Haematological Cytochemistry*, 2nd ed. Churchill Livingstone, Edinburgh.
48. Dunphy CH. (1999). Comprehensive review of adult acute myelogenous leukemia: Cytomorphological, enzyme cytochemical, flow cytometric immunophenotypic, and cytogenetic findings. *J Clin Lab Anal* **13**, 19–26.
49. Breton-Gorius J, Vanhaeke D, Pryzwansky KB, Guichard J, Tabilio A, Vainchenker W, Carmel R. (1984). Simultaneous detection of membrane markers with monoclonal antibodies and peroxidatic activities in leukaemia: Ultrastructural analysis using a new method of fixation preserving the platelet peroxidase. *Br J Haematol* **58**, 447–58.
50. Glassman AB. (2000). Chromosomal abnormalities in acute leukemias. *Clin Lab Med* **20**, 39–48.
51. McKenzie SB. (2005). Advances in understanding the biology and genetics of acute myelocytic leukemia. *Clin Lab Sci* **18**, 28–37.
52. Rucker FG, Bullinger L, Schwaenen C, Lipka DB, Wessendorf S, Frohling S, Bentz M, Miller S, Scholl C, Schlenk RF, Radlwimmer B, Kestler HA, Pollack JR, Lichter P, Dohner K, Dohner H. (2006). Disclosure of candidate genes in acute myeloid leukemia with complex karyotypes using microarray-based molecular characterization. *J Clin Oncol* **24**, 3887–94.
53. Yanada M, Suzuki M, Kawashima K, Kiyoi H, Kinoshita T, Emi N, Saito H, Naoe T. (2005). Long-term outcomes for unselected patients with acute myeloid leukemia categorized according to the World Health Organization classification: A single-center experience. *Eur J Haematol* **74**, 418–23.
54. Bao L, Wang X, Ryder J, Ji M, Chen Y, Chen H, Sun H, Yang Y, Du X, Kerzic P, Gross SA, Yao L, Lv L, Fu H, Lin G, Irons RD. (2006). Prospective study of 174 *de novo* acute myelogenous leukemias according to the WHO classification: Subtypes, cytogenetic features and FLT3 mutations. *Eur J Haematol* **77**, 35–45.
55. Wheatley K, Burnett AK, Goldstone AH, Gray RG, Hann IM, Harrison CJ, Rees JK, Stevens RF, Walker H. (1999). A simple, robust, validated and highly predictive index for the determination of risk-directed therapy in acute myeloid leukaemia derived from the MRC AML 10 trial. United Kingdom Medical Research Council's Adult and Childhood Leukaemia Working Parties. *Br J Haematol* **107**, 69–79.
56. Slovak ML, Kopecky KJ, Cassileth PA, Harrington DH, Theil KS, Mohamed A, Paietta E, Willman CL, Head DR, Rowe JM, Forman SJ, Appelbaum FR. (2000). Karyotypic analysis predicts outcome of preremission and postremission therapy in adult acute myeloid leukemia: A Southwest Oncology Group/Eastern Cooperative Oncology Group Study. *Blood* **96**, 4075–83.
57. Byrd JC, Mrózek K, Dodge RK, Carroll AJ, Edwards CG, Arthur DC, Pettenati MJ, Patil SR, Rao KW, Watson MS, Koduru PR, Moore JO, Stone RM, Mayer RJ, Feldman EJ, Davey FR, Schiffer CA, Larson RA, Bloomfield CD Cancer and Leukemia Group B (CALGB 8461) (2002). Pretreatment cytogenetic abnormalities are predictive of induction success, cumulative incidence of relapse, and overall survival in adult patients with *de novo* acute myeloid leukemia: Results from Cancer and Leukemia Group B (CALGB 8461). *Blood* **100**, 4325–36.

58a. Davis JN, McGhee L, Meyers S. (2003). The ETO (MTG8) gene family. *Gene* **303**, 1–10.

58b. Swerdlow SH, Campo E, Harris NL, Jaffe ES, Pileri SA, Stein H, Thiele J. (In press). 4th ed. *WHO Classification of Tumours of Haematopoietic and Lymphoid Tissues.*

59. Peterson LF, Zhang DE. (2004). The 8;21 translocation in leukemogenesis. *Oncogene* **23**, 4255–62.
60. Le Beau MM, Larson RA. (2006). Cytogenetics in acute myeloid leukemia. *UpTo Date .*
61. Nucifora G, Rowley JD. (1994). The AML1 and ETO genes in acute myeloid leukemia with a t(8;21). *Leuk Lymphoma* **14**, 353–62.
62. Boyapati A, Yan M, Peterson LF, Biggs JR, Le Beau MM, Zhang DE. (2007). A leukemia fusion protein attenuates the spindle checkpoint and promotes aneuploidy. *Blood* **109**, 3963–71.
63. Kita K, Shirakawa S, Kamada N. (1994). Cellular characteristics of acute myeloblastic leukemia associated with t(8;21)(q22;q22). The Japanese Co-operative Group of Leukemia/Lymphoma. *Leuk Lymphoma* **13**, 229–34.
64. Auger MJ, Ross FM, Mackie MJ. (1991). 8;21 translocation with duplication of the der(21) in a patient with myelomonocytic leukemia. *Cancer Genet Cytogenet* **51**, 139–41.
65. Kwong YL, Ching LM, Liu HW, Lee CP, Pollock A, Chan LC. (1993). 8;21 translocation and multilineage involvement. *Am J Hematol* **43**, 212–16.

66. Jurlander J, Caligiuri MA, Ruutu T, Baer MR, Strout MP, Oberkircher AR, Hoffmann L, Ball ED, Frei-Lahr DA, Christiansen NP, Block AM, Knuutila S, Herzig GP, Bloomfield CD. (1996). Persistence of the RUNX1/RUNXT1 fusion transcript in patients treated with allogeneic bone marrow transplantation for t(8;21) leukemia. *Blood* **88**, 2183–91.
67. Kozu T, Miyoshi H, Shimizu K, Maseki N, Kaneko Y, Asou H, Kamada N, Ohki M. (1993). Junctions of the AML1/MTG8(ETO) fusion are constant in t(8;21) acute myeloid leukemia detected by reverse transcription polymerase chain reaction. *Blood* **82**, 1270–6.
68. Ishida F, Ueno M, Tanaka H, Makishima H, Suzawa K, Hosaka S, Hidaka E, Ishikawa M, Yamauchi K, Kitano K, Kiyosawa K. (2002). t(8;21;14)(q22;q22;q24) is a novel variant of t(8;21) with chimeric transcripts of AML1-ETO in acute myelogenous leukemia. *Cancer Genet Cytogenet* **132**, 133–5.
69. Farra C, Awwad J, Valent A, Lozach F, Bernheim A. (2004). Complex translocation (8;12;21): A new variant of t(8;21) in acute myeloid leukemia. *Cancer Genet Cytogenet* **155**, 138–42.
70. Peniket AJ. (2005). Del(9q) acute myeloid leukaemia: Clinical and cytological characteristics and prognostic implications. *Br J Haematol* **130**, 969.
71. Lai YY, Qiu JY, Jiang B, Lu XJ, Huang XJ, Zhang Y, Liu YR, Shi HL, Lu DP. (2005). Characteristics and prognostic factors of acute myeloid leukemia with t (8; 21) (q22; q22). *Zhongguo Shi Yan Xue Ye Xue Za Zhi* **13**, 733–40.
72. Cho EK, Bang SM, Ahn JY, Yoo SM, Park PW, Seo YH, Shin DB, Lee JH. (2003). Prognostic value of AML1/ETO fusion transcripts in patients with acute myelogenous leukemia. *Korean J Intern Med* **18**, 13–20.
73. Lowenberg B, Griffin JD, Tallman MS. (2003). Acute myeloid leukemia and acute promyelocytic leukemia. *Hematology Am Soc Hematol Educ Program*, 82–101.
74. Stock W, Thirman MJ. (2006). Molecular biology of acute promyelocytic leukemia. *UpTo Date.*
75. He LZ, Guidez F, Tribioli C, Peruzzi D, Ruthardt M, Zelent A, Pandolfi PP. (1998). Distinct interactions of PML-RARalpha and PLZF-RARalpha with co-repressors determine differential responses to RA in APL. *Nat Genet* **18**, 126–35.
76. Mueller BU, Pabst T, Fos J, Petkovic V, Fey MF, Asou N, Buergi U, Tenen DG. (2006). ATRA resolves the differentiation block in t(15;17) acute myeloid leukemia by restoring PU.1 expression. *Blood* **107**, 3330–8.
77. Breitman TR, Collins SJ, Keene BR. (1981). Terminal differentiation of human promyelocytic leukemic cells in primary culture in response to retinoic acid. *Blood* **57**, 1000–4.
78. Robertson KA, Emami B, Mueller L, Collins SJ. (1992). Multiple members of the retinoic acid receptor family are capable of mediating the granulocytic differentiation of HL-60 cells. *Mol Cell Biol* **12**, 3743–9.
79. Guo A, Salomoni P, Luo J, Shih A, Zhong S, Gu W, Pandolfi PP. (2000). The function of PML in p53-dependent apoptosis. *Nat Cell Biol* **2**, 730–6.
80. Zhong S, Salomoni P, Ronchetti S, Guo A, Ruggero D, Pandolfi PP. (2000). Promyelocytic leukemia protein (PML) and Daxx participate in a novel nuclear pathway for apoptosis. *J Exp Med* **191**, 631–40.
81. Paietta E. (2003). Expression of cell-surface antigens in acute promyelocytic leukaemia. *Best Pract Res Clin Haematol* **16**, 369–85.
82. Sainty D, Liso V, Cantù-Rajnoldi A, Head D, Mozziconacci MJ, Arnoulet C, Benattar L, Fenu S, Mancini M, Duchayne E, Mahon FX, Gutierrez N, Birg F, Biondi A, Grimwade D, Lafage-Pochitaloff M, Hagemeijer A, Flandrin G. (2000). A new morphologic classification system for acute promyelocytic leukemia distinguishes cases with underlying PLZF/RARA gene rearrangements. Group Français de Cytogénétique Hématologique, UK Cancer Cytogenetics Group and BIOMED 1 European Community-Concerted Action "Molecular Cytogenetic Diagnosis in Haematological Malignancies". *Blood* **96**, 1287–96.
83. Dubé S, Fetni R, Hazourli S, Champagne M, Lemieux N. (2003). Rearrangement of the MLL gene and a region proximal to the RARalpha gene in a case of acute myelocytic leukemia M5 with a t(11;17)(q23;q21). *Cancer Genet Cytogenet* **145**, 54–9.
84. Lo-Coco F, Ammatuna E. (2006). The biology of acute promyelocytic leukemia and its impact on diagnosis and treatment. *Hematology Am Soc Hematol Educ Program*, 156–61.
85. Grimwade D, Biondi A, Mozziconacci MJ, Hagemeijer A, Berger R, Neat M, Howe K, Dastugue N, Jansen J, Radford-Weiss I, Lo Coco F, Lessard M, Hernandez JM, Delabesse E, Head D, Liso V, Sainty D, Flandrin G, Solomon E, Birg F, Lafage-Pochitaloff M. (2000). Characterization of acute promyelocytic leukemia cases lacking the classic t(15;17): Results of the European Working Party. Groupe Français de Cytogénétique Hématologique, Groupe de Français d'Hematologie Cellulaire, UK Cancer Cytogenetics Group and BIOMED 1 European Community-Concerted Action "Molecular Cytogenetic Diagnosis in Haematological Malignancies". *Blood* **96**, 1297–308.
86. Falini B, Martelli MP, Bolli N, Bonasso R, Ghia E, Pallotta MT, Diverio D, Nicoletti I, Pacini R, Tabarrini A, Galletti BV, Mannucci R, Roti G, Rosati R, Specchia G, Liso A, Tiacci E, Alcalay M, Luzi L, Volorio S, Bernard L, Guarini A, Amadori S, Mandelli F, Pane F, Lo-Coco F, Saglio G, Pelicci PG, Martelli MF, Mecucci C. (1999). Immunohistochemistry predicts nucleophosmin (NPM) mutations in acute myeloid leukemia. *Blood* **108**, 1999–2005.
87. Dong S, Tweardy DJ. (2002). Interactions of STAT5b-RARalpha, a novel acute promyelocytic leukemia fusion protein, with retinoic acid receptor and STAT3 signaling pathways. *Blood* **99**, 2637–46.
88. Vial JP, Mahon FX, Pigneux A, Notz A, Lacombe F, Reiffers J, Bilhou-Nabera C. (2003). Derivative (7)t(7;8)(q34;q21) a new additional cytogenetic abnormality in acute promyelocytic leukemia. *Cancer Genet Cytogenet* **140**, 78–81.
89. Nakase K, Wakita Y, Minamikawa K, Yamaguchi T, Shiku H. (2000). Acute promyelocytic leukemia with del(6)(p23). *Leuk Res* **24**, 79–81.
90. Engel E. (1977). Partial long arm deletion of a chromosome 17 in acute promyelocytic leukemia. *Arch Intern Med* **137**, 697.
91. Yoo SJ, Seo EJ, Lee JH, Seo YH, Park PW, Ahn JY. (2006). A complex, four-way variant t(15;17) in acute promyelocytic leukemia. *Cancer Genet Cytogenet* **167**, 168–71.
92. O'Connor SJ, Evans PA, Morgan GJ. (1999). Diagnostic approaches to acute promyelocytic leukaemia. *Leuk Lymphoma* **33**, 53–63.
93. Reiter A, Lengfelder E, Grimwade D. (2004). Pathogenesis, diagnosis and monitoring of residual disease in acute promyelocytic leukaemia. *Acta Haematol* **112**, 55–67.
94. Larson Ra. (2006). Clinical features and treatment of acute promyelocytic leukemia. *UpTo Date* .
95. Menell JS, Cesarman GM, Jacovina AT, McLaughlin MA, Lev EA, Hajjar KA. (1999). Annexin II and bleeding in acute promyelocytic leukemia. *N Engl J Med* **340**, 994–1004.
96. Asou N, Adachi K, Tamura J, Kanamaru A, Kageyama S, Hiraoka A, Omoto E, Akiyama H, Tsubaki K, Saito K, Kuriyama K, Oh H, Kitano K, Miyawaki S, Takeyama K, Yamada O, Nishikawa K, Takahashi M, Matsuda S, Ohtake S, Suzushima H, Emi N, Ohno R. (1998). Analysis of prognostic factors in newly diagnosed acute promyelocytic leukemia treated with all-trans retinoic acid and chemotherapy. Japan Adult Leukemia Study Group. *J Clin Oncol* **16**, 78–85.
97. Murray CK, Estey E, Paietta E, Howard RS, Edenfield WJ, Pierce S, Mann KP, Bolan C, Byrd JC. (1999). CD56 expression in acute promyelocytic leukemia: A possible indicator of poor treatment outcome? *J Clin Oncol* **17**, 293–7.
98. Lengfelder E, Saussele S, Weisser A, Buchner T, Hehlmann R. (2005). Treatment concepts of acute promyelocytic leukemia. *Crit Rev Oncol Hematol* **56**, 261–74.
99. Zhou GB, Chen SJ, Chen Z. (2005). Acute promyelocytic leukemia: A model of molecular target based therapy. *Hematology* **10**(Suppl 1), 270–80.
100. Sirulnik LA, Stone RM. (2005). Acute promyelocytic leukemia: Current strategies for the treatment of newly diagnosed disease. *Clin Adv Hematol Oncol* **3**, 391–7.

101. Shigesada K, van de Sluis B, Liu PP. (2004). Mechanism of leukemogenesis by the inv(16) chimeric gene CBFB/PEBP2B-MHY11. *Oncogene* **23**, 4297–307.
102. Lukasik SM, Zhang L, Corpora T, Tomanicek S, Li Y, Kundu M, Hartman K, Liu PP, Laue TM, Biltonen RL, Speck NA, Bushweller JH. (2002). Altered affinity of CBF beta SMMHC for Runx1 explains its role in leukemogenesis. *Nat Struct Biol* **9**, 674–9.
103. Reilly JT. (2005). Pathogenesis of acute myeloid leukaemia and inv(16)(p13;q22): A paradigm for understanding leukaemogenesis? *Br J Haematol* **128**, 18–34.
104. Delaunay J, Vey N, Leblanc T, Fenaux P, Rigal-Huguet F, Witz F, Lamy T, Auvrignon A, Blaise D, Pigneux A, Mugneret F, Bastard C, Dastugue N, Van den Akker J, Fiere D, Reiffers J, Castaigne S, Leverger G, Harousseau JL, Dombret HFrench Acute Myeloid Leukemia Intergroup; Groupe Ouest-Est des Leucemies Aigues Myeoblastiques; Leucemies Aigues Myeoblastiques de l'Enfant; Acute Leukemia French Association; Bordeaux-Grenoble-Marseille-Toulouse cooperative groups (2003). Prognosis of inv(16)/t(16;16) acute myeloid leukemia (AML): A survey of 110 cases from the French AML Intergroup. *Blood* **102**, 462–9. Epub 2003 Mar 20.
105. Marlton P, Keating M, Kantarjian H, Pierce S, O'Brien S, Freireich EJ, Estey E. (1995). Cytogenetic and clinical correlates in AML patients with abnormalities of chromosome 16. *Leukemia* **9**, 965–71.
106. Wong KF, Kwong YL. (1999). Trisomy 22 in acute myeloid leukemia: A marker for myeloid leukemia with monocytic features and cytogenetically cryptic inversion 16. *Cancer Genet Cytogenet* **109**, 131–3.
107. Poppe B, Vandesompele J, Schoch C, Lindvall C, Mrozek K, Bloomfield CD, Beverloo HB, Michaux L, Dastugue N, Herens C, Yigit N, De Paepe A, Hagemeijer A, Speleman F. (2004). Expression analyses identify MLL as a prominent target of 11q23 amplification and support an etiologic role for MLL gain of function in myeloid malignancies. *Blood* **103**, 229–35.
108. LinksCorral J, Lavenir I, Impey H, Warren AJ, Forster A, Larson TA, Bell S, McKenzie AN, King G, Rabbitts TH. (1996). An Mll-AF9 fusion gene made by homologous recombination causes acute leukemia in chimeric mice: A method to create fusion oncogenes. *Cell* **85**, 853–61.
109. Rowley JD. (1993). Rearrangements involving chromosome band 11q23 in acute leukaemia. *Semin Cancer Biol* **4**, 377–85.
110. Bernard OA, Berger R. (1995). Molecular basis of 11q23 rearrangements in hematopoietic malignant proliferations. *Genes Chromosomes Cancer* **13**, 75–85.
111. Zatkova A, Schoch C, Speleman F, Poppe B, Mannhalter C, Fonatsch C, Wimmer K. (2006). GAB2 is a novel target of 11q amplification in AML/MDS. *Genes Chromosomes Cancer* **45**, 798–807.
112. Zatkova A, Ullmann R, Rouillard JM, Lamb BJ, Kuick R, Hanash SM, Schnittger S, Schoch C, Fonatsch C, Wimmer K. (2004). Distinct sequences on 11q13.5 and 11q23-24 are frequently coamplified with MLL in complexly organized 11q amplicons in AML/MDS patients. *Genes Chromosomes Cancer* **39**, 263–76.
113. Casillas JN, Woods WG, Hunger SP, McGavran L, Alonzo TA, Feig SA. Children's Cancer Group (2003). Prognostic implications of t(10;11) translocations in childhood acute myelogenous leukemia: A report from the Children's Cancer Group. *J Pediatr Hematol Oncol* **25**, 594–600.
114. Klaus M, Schnittger S, Haferlach T, Dreyling M, Hiddemann W, Schoch C. (2003). Cytogenetics, fluorescence *in situ* hybridization, and reverse transcriptase polymerase chain reaction are necessary to clarify the various mechanisms leading to an MLL-AF10 fusion in acute myelocytic leukemia with 10;11 rearrangement. *Cancer Genet Cytogenet* **144**, 36–43.
115. Wechsler DS, Engstrom LD, Alexander BM, Motto DG, Roulston D. (2003). A novel chromosomal inversion at 11q23 in infant acute myeloid leukemia fuses MLL to CALM, a gene that encodes a clathrin assembly protein. *Genes Chromosomes Cancer* **36**, 26–36.
116. Van Limbergen H, Poppe B, Janssens A, De Bock R, De Paepe A, Noens L, Speleman F. (2002). Molecular cytogenetic analysis of 10;11 rearrangements in acute myeloid leukemia. *Leukemia* **16**, 344–51.
117. Berger R, Bernheim A, Sigaux F, Daniel MT, Valensi F, Flandrin G. (1982). Acute monocytic leukemia chromosome studies. *Leuk Res* **6**, 17–26.
118. Rowley JD. (1983). Consistent chromosome abnormalities in human leukemia and lymphoma. *Cancer Invest* **1**, 267–80.
119. Kaneko Y, Maseki N, Takasaki N, Sakurai M, Hayashi Y, Nakazawa S, Mori T, Sakurai M, Takeda T, Shikano T, *et al.* (1986). Clinical and hematologic characteristics in acute leukemia with 11q23 translocations. *Blood* **67**, 484–91.
120. Sorensen PH, Chen CS, Smith FO, Arthur DC, Domer PH, Bernstein ID, Korsmeyer SJ, Hammond GD, Kersey JH. (1994). Molecular rearrangements of the MLL gene are present in most cases of infant acute myeloid leukemia and are strongly correlated with monocytic or myelomonocytic phenotypes. *J Clin Invest* **93**, 429–37.
121. Cimino G, Lo Coco F, Biondi A, Elia L, Luciano A, Croce CM, Masera G, Mandelli F, Canaani E. (1993). ALL-1 gene at chromosome 11q23 is consistently altered in acute leukemia of early infancy. *Blood* **82**, 544–6.
122. Uckun FM, Downing JR, Chelstrom LM, Gunther R, Ryan M, Simon J, Carroll AJ, Tuel-Ahlgren L, Crist WM. (1994). Human t(4;11)(q21;q23) acute lymphoblastic leukemia in mice with severe combined immunodeficiency. *Blood* **84**, 859–65.
123. Gahn B, Haase D, Unterhalt M, Drescher M, Schoch C, Fonatsch C, Terstappen LW, Hiddemann W, Büchner T, Bennett JM, Wörmann B. (1996). *De novo* AML with dysplastic hematopoiesis: Cytogenetic and prognostic significance. *Leukemia* **10**, 946–51.
124. Haferlach T, Schoch C, Löffler H, Gassmann W, Kern W, Schnittger S, Fonatsch C, Ludwig WD, Wuchter C, Schlegelberger B, Staib P, Reichle A, Kubica U, Eimermacher H, Balleisen L, Grüneisen A, Haase D, Aul C, Karow J, Lengfelder E, Wörmann B, Heinecke A, Sauerland MC, Büchner T, Hiddemann W. (2003). Morphologic dysplasia in *de novo* acute myeloid leukemia (AML) is related to unfavorable cytogenetics but has no independent prognostic relevance under the conditions of intensive induction therapy: Results of a multiparameter analysis from the German AML Cooperative Group studies. *J Clin Oncol* **21**, 256–65.
125. Oyarzo MP, Lin P, Glassman A, Bueso-Ramos CE, Luthra R, Medeiros LJ. (2004). Acute myeloid leukemia with t(6;9)(p23;q34) is associated with dysplasia and a high frequency of flt3 gene mutations. *Am J Clin Pathol* **122**, 348–58.
126. de Souza Fernandez T, Ornellas MH, Otero de Carvalho L, Tabak D, Abdelhay E. (2000). Chromosomal alterations associated with evolution from myelodysplastic syndrome to acute myeloid leukemia. *Leuk Res* **24**, 839–48.
127. Shen Y, Xue Y, Li J, Pan J, Wu Y. (2003). Clinical, cytogenetic and dual-color FISH studies on five cases of myelodysplastic syndrome or acute myeloid leukemia patients with 1;7 translocation. *Chin Med J (Engl)* **116**, 231–4.
128. Tsutsumi C, Ueda M, Miyazaki Y, Yamashita Y, Choi YL, Ota J, Kaneda R, Koinuma K, Fujiwara S, Kisanuki H, Ishikawa M, Ozawa K, Tomonaga M, Mano H. (2004). DNA microarray analysis of dysplastic morphology associated with acute myeloid leukemia. *Exp Hematol* **32**, 828–35.
129. Arber DA, Stein AS, Carter NH, Ikle D, Forman SJ, Slovak ML. (2003). Prognostic impact of acute myeloid leukemia classification. Importance of detection of recurring cytogenetic abnormalities and multilineage dysplasia on survival. *Am J Clin Pathol* **119**, 672–80.
130. Meckenstock G, Aul C, Hildebrandt B, Heyll A, Germing U, Wehmeier A, Giagounidis A, Suedhoff T, Burk M, Soehngen D, Schneider W. (1998). Dyshematopoiesis in *de novo* acute myeloid leukemia: Cell biological features and prognostic significance. *Leuk Lymphoma* **29**, 523–31.
131. Ferrara F, Palmieri S, Pocali B, Pollio F, Viola A, Annunziata S, Sebastio L, Schiavone EM, Mele G, Gianfaldoni G, Leoni F.

(2002). *De novo* acute myeloid leukemia with multilineage dysplasia: Treatment results and prognostic evaluation from a series of 44 patients treated with fludarabine, cytarabine and G-CSF (FLAG). *Eur J Haematol* **68**, 203–9.

132. Smith SM, Le Beau MM, Huo D, Karrison T, Sobecks RM, Anastasi J, Vardiman JW, Rowley JD, Larson RA. (2003). Clinical-cytogenetic associations in 306 patients with therapy-related myelodysplasia and myeloid leukemia: The University of Chicago series. *Blood* **102**, 43–52.
133. Rund D, Ben-Yehuda D. (2004). Therapy-related leukemia and myelodysplasia: Evolving concepts of pathogenesis and treatment. *Hematology* **9**, 179–87.
134. Beaumont M, Sanz M, Carli PM, Maloisel F, Thomas X, Detourmignies L, Guerci A, Gratecos N, Rayon C, San Miguel J, Odriozola J, Cahn JY, Huguet F, Vekhof A, Stamatoulas A, Dombret H, Capote F, Esteve J, Stoppa AM, Fenaux P. (2003). Therapy-related acute promyelocytic leukemia. *J Clin Oncol* **21**, 2123–37.
135. Ellis M, Ravid M, Lishner M. (1993). A comparative analysis of alkylating agent and epipodophyllotoxin-related leukemias. *Leuk Lymphoma* **11**, 9–13.
136. Michels SD, McKenna RW, Arthur DC, Brunning RD. (1985). Therapy-related acute myeloid leukemia and myelodysplastic syndrome: A clinical and morphologic study of 65 cases. *Blood* **65**, 1364–72.
137. Andersen MK, Larson RA, Mauritzson N, Schnittger S, Jhanwar SC, Pedersen-Bjergaard J. (2002). Balanced chromosome abnormalities inv(16) and t(15;17) in therapy-related myelodysplastic syndromes and acute leukemia: Report from an international workshop. *Genes Chromosomes Cancer* **33**, 395–400.
138. Secker-Walker LM, Moorman AV, Bain BJ, Mehta AB. (1998). Secondary acute leukemia and myelodysplastic syndrome with 11q23 abnormalities. EU Concerted Action 11q23 Workshop. *Leukemia* **12**, 840–4.
139. Atlas M, Head D, Behm F, Schmidt E, Zeleznik-Le NH, Roe BA, Burian D, Domer PH. (1998). Cloning and sequence analysis of four t(9;11) therapy-related leukemia breakpoints. *Leukemia* **12**, 1895–902.
140. Alsabeh R, Brynes RK, Slovak ML, Arber DA. (1997). Acute myeloid leukemia with t(6;9) (p23;q34): Association with myelodysplasia, basophilia, and initial CD34 negative immunophenotype. *Am J Clin Pathol* **107**, 430–7.
141. Pedersen-Bjergaard J, Johansson B, Philip P. (1994). Translocation (3;21)(q26;q22) in therapy-related myelodysplasia following drugs targeting DNA-topoisomerase II combined with alkylating agents, and in myeloproliferative disorders undergoing spontaneous leukemic transformation. *Cancer Genet Cytogenet* **76**, 50–5.
142. Andersen MK, Christiansen DH, Jensen BA, Ernst P, Hauge G, Pedersen-Bjergaard J. (2001). Therapy-related acute lymphoblastic leukaemia with MLL rearrangements following DNA topoisomerase II inhibitors, an increasing problem: Report on two new cases and review of the literature since 1992. *Br J Haematol* **114**, 539–43.
143. Pedersen-Bjergaard J, Philip P, Larsen SO, Andersson M, Daugaard G, Ersbøll J, Hansen SW, Hou-Jensen K, Nielsen D, Sigsgaard TC. (1993). Therapy-related myelodysplasia and acute myeloid leukemia. Cytogenetic characteristics of 115 consecutive cases and risk in seven cohorts of patients treated intensively for malignant diseases in the Copenhagen series. *Leukemia* **7**, 1975–86.
144. Pui CH, Relling MV, Rivera GK, Hancock ML, Raimondi SC, Heslop HE, Santana VM, Ribeiro RC, Sandlund JT, Mahmoud HH. (1995). Epipodophyllotoxin-related acute myeloid leukemia: A study of 35 cases. *Leukemia* **9**, 1990–6.
145. Quesnel B, Kantarjian H, Bjergaard JP, Brault P, Estey E, Lai JL, Tilly H, Stoppa AM, Archimbaud E, Harousseau JL. (1993). Therapy-related acute myeloid leukemia with t(8;21), inv(16), and t(8;16): A report on 25 cases and review of the literature. *J Clin Oncol* **11**, 2370–9.
146. Béné MC, Bernier M, Casasnovas RO, Castoldi G, Doekharan D, van der Holt B, Knapp W, Lemez P, Ludwig WD, Matutes E, Orfao A, Schoch C, Sperling C, van't Veer MB. (2001). Acute myeloid leukaemia M0: Haematological, immunophenotypic and cytogenetic characteristics and their prognostic significance: An analysis in 241 patients. *Br J Haematol* **113**, 737–45.
147. Kaleem Z, Crawford E, Pathan MH, Jasper L, Covinsky MA, Johnson LR, White G. (2003). Flow cytometric analysis of acute leukemias. Diagnostic utility and critical analysis of data. *Arch Pathol Lab Med* **127**, 42–8.
148. Kaleem Z, White G. (2001). Diagnostic criteria for minimally differentiated acute myeloid leukemia (AML-M0). Evaluation and a proposal. *Am J Clin Pathol* **115**, 876–84.
149. Costello R, Mallet F, Chambost H, Sainty D, Arnoulet C, Gastaut JA, Olive D. (1999). The immunophenotype of minimally differentiated acute myeloid leukemia (AML-M0): Reduced immunogenicity and high frequency of CD34+/CD38− leukemic progenitors. *Leukemia* **13**, 1513–18.
150. Suzuki R, Murata M, Kami M, Ohtake S, Asou N, Kodera Y, Tomonaga M, Masaki Y, Kusumoto S, Takeuchi J, Matsuda S, Hirai H, Yorimitsu S, Hamajima N, Seto M, Shimoyama M, Ohno R, Morishima Y, Nakamura S. (2003). Prognostic significance of CD7+ CD56+ phenotype and chromosome 5 abnormalities for acute myeloid leukemia M0. *Int J Hematol* **77**, 482–9.
151. Kotylo PK, Seo IS, Smith FO, Heerema NA, Fineberg NS, Miller K, Greene ME, Chou P, Orazi A. (2000). Flow cytometric immunophenotypic characterization of pediatric and adult minimally differentiated acute myeloid leukemia (AML-M0). *Am J Clin Pathol* **113**, 193–200.
152. Venditti A, Del Poeta G, Buccisano F, Tamburini A, Aronica G, Bruno A, Cox-Froncillo MC, Maffei L, Simone MD, Papa G, Amadori S. (1997). Biological pattern of AML-M0 versus AML-Ml: Response. *Blood* **89**, 345–6.
153. Roumier C, Eclache V, Imbert M, Davi F, MacIntyre E, Garand R, Talmant P, Lepelley P, Lai JL, Casasnovas O, Maynadie M, Mugneret F, Bilhou-Naberra C, Valensi F, Radford I, Mozziconacci MJ, Arnoulet C, Duchayne E, Dastugue N, Cornillet P, Daliphard S, Garnache F, Boudjerra N, Jouault H, Fenneteau O, Pedron B, Berger R, Flandrin G, Fenaux P, Preudhomme C. Groupe Francais de Cytogenetique Hematologique (GFCH). Groupe Français d'Hématologie Cellulaire (GFHC) (2003). M0 AML, clinical and biologic features of the disease, including AML1 gene mutations: A report of 59 cases by the Groupe Français d'Hématologie Cellulaire (GFHC) and the Groupe Français de Cytogénétique Hématologique (GFCH). *Blood* **101**, 1277–83.
154. Taketani T, Taki T, Takita J, Tsuchida M, Hanada R, Hongo T, Kaneko T, Manabe A, Ida K, Hayashi Y. (2003). AML1/RUNX1 mutations are infrequent, but related to AML-M0, acquired trisomy 21, and leukemic transformation in pediatric hematologic malignancies. *Genes Chromosomes Cancer* **38**, 1–7.
155. Harada H, Harada Y, Niimi H, Kyo T, Kimura A, Inaba T. (2004). High incidence of somatic mutations in the AML1/RUNX1 gene in myelodysplastic syndrome and low blast percentage myeloid leukemia with myelodysplasia. *Blood* **103**, 2316–24.
156. Cammenga J, Niebuhr B, Horn S, Bergholz U, Putz G, Buchholz F, Löhler J, Stocking C. (2007). RUNX1 DNA-binding mutants, associated with minimally differentiated acute myelogenous leukemia, disrupt myeloid differentiation. *Cancer Res* **67**, 537–45.
157. Klaus M, Haferlach T, Schnittger S, Kern W, Hiddemann W, Schoch C. (2004). Cytogenetic profile in *de novo* acute myeloid leukemia with FAB subtypes M0, Ml, and M2: A study based on 652 cases analyzed with morphology, cytogenetics, and fluorescence *in situ* hybridization. *Cancer Genet Cytogenet* **155**, 47–56.
158. González García JR, Garcés Ruíz OM, Delgado Lamas JL, Ramífrez-Dueñas ML. (1997). Two different Philadelphia chromosomes in a cell line from an AML-M0 patient. *Cancer Genet Cytogenet* **98**, 111–14.
159. Chen S, Xue Y, Zhu X, Wu Y, Pan J. (2007). Minimally differentiated acute myeloid leukemia (9;22) translocation showing primary

multi-drug resistance and expressing multiple multidrug-resistant proteins. *Acta Haematol* **118**, 38–41.
160. Wieser R, Schreiner U, Wollenberg B, Neubauer A, Fonatsch C, Rieder H. (2001). Masked inv(3)(q21q26) in a patient with minimally differentiated acute myeloid leukemia. *Haematologica* **86**, 214–15.
161. Fujisawa S, Tanabe J, Harano H, Kanamori H, Motomura S, Mohri H, Ishigatsubo Y. (1999). Acute minimally differentiated myeloid leukemia (M0) with inv(3)(q21q26). *Leuk Lymphoma* **35**, 627–30.
162. Cox MC, Scanzani A, Del Poeta G, Venditti A, Panetta P, Derme V, Sgro R, Masi M, Amadori S. (2000). A novel t(11;12)(q23-24;q24) in a case of minimally-differentiated acute myeloid leukemia (AML-M0). *Cancer Genet Cytogenet* **118**, 76–9.
163. Tecimer C, Loy BA, Martin AW. (1999). Acute myeloblastic leukemia (M0) with an unusual chromosomal abnormality: Translocation (1;14)(p13;q32). *Cancer Genet Cytogenet* **111**, 175–7.
164. Cascavilla N, Melillo L, D'Arena G, Greco MM, Carella AM, Sajeva MR, Perla G, Matera R, Minervini MM, Carotenuto M. (2000). Minimally differentiated acute myeloid leukemia (AML M0): Clinico-biological findings of 29 cases. *Leuk Lymphoma* **37**, 105–13.
165. Segeren CM, de Jong-Gerrits GC, van't Veer MB. (1995). AML-M0: Clinical entity or waste basket for immature blastic leukemias? A description of 14 patients. Dutch Slide Review Committee of Leukemias in Adults. *Ann Hematol* **70**, 297–300.
166. McGrattan P, Alexander HD, Humphreys MW, Kettle PJ. (2002). Tetrasomy 13 as the sole cytogenetic abnormality in acute myeloid leukemia M1 without maturation. *Cancer Genet Cytogenet* **135**, 192–5.
167. La Starza R, Wlodarska I, Matteucci C, Falzetti D, Baens M, Martelli MF, Van den Berghe H, Marynen P, Mecucci C. (1998). Rearrangement between the MYH11 gene at 16p13 and D12S158 at 12p13 in a case of acute myeloid leukemia M1 (AML-M1). *Genes Chromosomes Cancer* **23**, 10–15.
168. Ahmad F, Dalvi R, Mandava S, Das BR. (2007). Acute myelogeneous leukemia (M0/M1) with novel chromosomal abnormality of t(14;17)(q32; q11.2). *Am J Hematol* **82**, 676–8.
169. Ma SK, Wan TS, Chan LC, Chiu EK. (2000). Hand-mirror blasts, AML-M1, and der(1)t(1;19)-(p13;p13.1). *Leuk Res* **24**, 95–6.
170. Klampfer L, Zhang J, Zelenetz AO, Uchida H, Nimer SD. (1996). The RUNX1/RUNXT1 fusion protein activates transcription of BCL-2. *Proc Natl Acad Sci USA* **93**, 14059–64.
171. Frank RC, Sun X, Berguido FJ, Jakubowiak A, Nimer SD. (1999). The t(8;21) fusion protein, RUNX1/RUNXT1, transforms NIH3T3 cells and activates AP-1. *Oncogene* **18**, 1701–10.
172. Powari M, Varma N, Varma S, Komal HS. (2000). Pseudo-Chediak Higashi anomaly in an Indian patient with acute myeloid leukemia (AML-M2). *Am J Hematol* **65**, 324–5.
173. Kozlov I, Beason K, Yu C, Hughson M. (2005). CD79a expression in acute myeloid leukemia t(8;21) and the importance of cytogenetics in the diagnosis of leukemias with immunophenotypic ambiguity. *Cancer Genet Cytogenet* **163**, 62–7.
174. Di Bona E, Sartori R, Zambello R, Guercini N, Madeo D, Rodeghiero F. (2002). Prognostic significance of CD56 antigen expression in acute myeloid leukemia. *Haematologica* **87**, 250–6.
175. Tobal K, Liu Yin JA. (1998). Molecular monitoring of minimal residual disease in acute myeloblastic leukemia with t(8;21) by RT-PCR. *Leuk Lymphoma* **31**, 115–20.
176. de Oliveira FM, Tone LG, Simões BP, Falcão RP, Brassesco MS, Sakamoto-Hojo ET, dos Santos GA, Marinato AF, Jácomo RH, Rego EM. (2007). Acute myeloid leukemia (AML-M2) with t(5;11)(q35;q13) and normal expression of cyclin D1. *Cancer Genet Cytogene* **172**, 154–7.
177. Specchia G, Mestice A, Clelia Storlazzi T, Anelli L, Pannunzio A, Grazia Roberti M, Rocchi M, Liso V. (2001). A novel translocation t(2;9)(q14;p12) in AML-M2 with an uncommon phenotype: Myeloperoxidase-positive and myeloid antigen-negative. *Leuk Res* **25**, 501–7.
178. Xue Y, Niu C, Chen S, Wang Y, Guo Y, Xie X, Lu D, Li P. (2000). Two cases of AML (M2) with a t(8;19)(q22;q13): A new cytogenetic variant. *Cancer Genet Cytogenet* **118**, 154–8.
179. Vieira L, Oliveira V, Ambrosio AP, Marques B, Pereira AM, Hagemeijer A, Boavida MG. (2001). Translocation (8;17;15;21)(q22;q23;ql5;q22) in acute myeloid leukemia (M2). A four-way variant of t(8;21). *Cancer Genet Cytogenet* **128**, 104–7.
180. Lahortiga I, Belloni E, Vazquez I, Agirre X, Larrayoz MJ, Vizmanos JL, Valganon M, Zudaire I, Saez B, Mateos MC, Di Fiore PP, Calasanz MJ, Odero MD. (2005). NUP98 is fused to HOXA9 in a variant complex t(7;11;13;17) in a patient with AML-M2. *Cancer Genet Cytogenet* **157**, 151–6.
181. Xue Y, Xu L, Chen S, Fu J, Guo Y, Li J, Wu Y, Pan J, Lu D. (2001). t(8;21;8)(p23;q22;q22): A new variant form of t(8;21) translocation in acute myeloblastic leukemia with maturation. *Leuk Lymphoma* **42**, 533–7.
182. Poirel H, Radford-Weiss I, Rack K, Troussard X, Veil A, Valensi F, Picard F, Guesnu M, Leboeuf D, Melle J. (1995). Detection of the chromosome 16 CBF beta-MYH11 fusion transcript in myelomonocytic leukemias. *Blood* **85**, 1313–22.
183. O'Reilly J, Chipper L, Springall F, Herrmann R. (2000). A unique structural abnormality of chromosome 16 resulting in a CBF beta-MYH11 fusion transcript in a patient with acute myeloid leukemia, FAB M4. *Cancer Genet Cytogenet* **121**, 52–5.
184. Lafay-Cousin L, Soenen V, Mazingue F, Preudhomme C, Laï JL, Andrieux J. (2004). Chromosomal insertion involving MLL in childhood acute myeloblastic leukemia (M4). *Cancer Genet Cytogenet* **150**, 153–5.
185. Schoch C, Schnittger S, Klaus M, Kern W, Hiddemann W, Haferlach T. (2003). AML with 11q23/MLL abnormalities as defined by the WHO classification: Incidence, partner chromosomes, FAB subtype, age distribution, and prognostic impact in an unselected series of 1897 cytogenetically analyzed AML cases. *Blood* **102**, 2395–402.
186. Cooper CL, Loewen R, Shore T. (2000). Gingival hyperplasia complicating acute myelomonocytic leukemia. *J Can Dent Assoc* **66**, 78–9.
187. Gorczyca W. (2004). Flow cytometry immunophenotypic characteristics of monocytic population in acute monocytic leukemia (AML-M5), acute myelomonocytic leukemia (AML-M4), and chronic myelomonocytic leukemia (CMML). *Meth Cell Biol* **75**, 665–77.
188. Khanlari B, Buser A, Lugli A, Tichelli A, Dirnhofer S. (2003). The expression pattern of CD56 (N-CAM) in human bone marrow biopsies infiltrated by acute leukemia. *Leuk Lymphoma* **44**, 2055–9.
189. Novotny JR, Nuckel H, Duhrsen U. (2006). Correlation between expression of CD56/NCAM and severe leukostasis in hyperleukocytic acute myelomonocytic leukaemia. *Eur J Haematol* **76**, 299–308.
190. Cruse JM, Lewis RE, Pierce S, Lam J, Tadros Y. (2005). Aberrant expression of CD7, CD56, and CD79a antigens in acute myeloid leukemias. *Exp Mol Pathol* **79**, 39–41.
191. Zhang L, Alsabeh R, Mecucci C, La Starza R, Gorello P, Lee S, Lill M, Schreck R. (2007). Rare t(1;11)(q23;p15) in therapy-related myelodysplastic syndrome evolving into acute myelomonocytic leukemia: A case report and review of the literature. *Cancer Genet Cytogenet* **178**, 42–8.
192. Fu JF, Liang DC, Yang CP, Hsu JJ, Shih LY. (2003). Molecular analysis of t(X;11)(q24;q23) in an infant with AML-M4. *Genes Chromosomes* **38**, 253–9.
193. Wong KF, Wong ML, Tu SP. (2006). Dup(1)(p31.2p36.2) in acute myelomonocytic leukemia. *Cancer Genet Cytogenet* **165**, 83–4.
194. Xinh PT, Tri NK, Nagao H, Nakazato H, Taketazu F, Fujisawa S, Yagasaki F, Chen YZ, Hayashi Y, Toyoda A, Hattori M, Sakaki Y, Tokunaga K, Sato Y. (2003). Breakpoints at 1p36.3 in three MDS/AML(M4) patients with t(1;3)(p36;q21) occur in the first intron and in the 5′region of MEL1. *Genes Chromosomes Cancer* **36**, 313–6.
195. Yamamoto K, Nagata K, Tsurukubo Y, Inagaki K, Ono R, Taki T, Hayashi Y, Hamaguchi H. (2002). Translocation (8;12)(q13;p13)

during disease progression in acute myelomonocytic leukemia with t(11;19)(q23;p13.1). *Cancer Genet Cytogenet* **137**, 64–7.

196. Paulien S, Maarek O, Daniel MT, Berger R. (2002). A novel translocation, t(9;21)(q13;q22) rearranging the RUNX1 gene in acute myelomonocytic leukemia. *Ann Genet* **45**, 67–9.
197. Nagel S, Kaufmann M, Scherr M, Drexler HG, MacLeod RA. (2005). Activation of HLXB9 by juxtaposition with MYB via formation of t(6;7)(q23;q36) in an AML-M4 cell line (GDM-1). *Genes Chromosomes Cancer* **42**, 170–8.
198. Terui K, Kitazawa J, Takahashi Y, Tohno C, Hayashi Y, Taketani T, Taki T, Ito E. (2003). Successful treatment of acute myelomonocytic leukaemia with NUP98-HOXD11 fusion transcripts and monitoring of minimal residual disease. *Br J Haematol* **120**, 274–6.
199. Chen L, Rodgers TR, Chaffins ML, Maeda K. (2005). Acute monocytic leukemia with cutaneous manifestation. *Arch Pathol Lab Med* **129**, 425–6.
200. Yang DT, Greenwood JH, Hartung L, Hill S, Perkins SL, Bahler DW. (2005). Flow cytometric analysis of different CD14 epitopes can help identify immature monocytic populations. *Am J Clin Pathol* **124**, 930–6.
201. Mo J, Lampkin B, Perentesis J, Poole L, Bao L. (2006). Translocation (8;18;16)(p11;q21;p13). A new variant of t(8;16)(p11;p13) in acute monoblastic leukemia: Case report and review of the literature. *Cancer Genet Cytogenet* **165**, 75–8.
202. Suzukawa K, Shimizu S, Nemoto N, Takei N, Taki T, Nagasawa T. (2005). Identification of a chromosomal breakpoint and detection of a novel form of an MLL-AF17 fusion transcript in acute monocytic leukemia with t(11;17)(q23;q21). *Int J Hematol* **82**, 38–41.
203. Panagopoulos I, Kitagawa A, Isaksson M, Morse H, Mitelman F, Johansson B. (2004). MLL/GRAF fusion in an infant acute monocytic leukemia (AML M5b) with a cytogenetically cryptic ins(5;11)(q31;q23q23). *Genes Chromosomes Cancer* **41**, 400–4.
204. Tasaka T, Matsuhashi Y, Uehara E, Tamura T, Kakazu N, Abe T, Nagai M. (2004). Secondary acute monocytic leukemia with a translocation t(8;16)(p11;p13): Case report and review of the literature. *Leuk Lymphoma* **45**, 621–5.
205. Kim HJ, Ki CS, Park Q, Koo HH, Yoo KH, Kim EJ, Kim SH. (2003). MLL/SEPTIN6 chimeric transcript from inv ins(X;11)(q24;q23q13) in acute monocytic leukemia: Report of a case and review of the literature. *Genes Chromosomes Cancer* **38**, 8–12.
206. Chen S, Xue Y, Chen Z, Guo Y, Wu Y, Pan J. (2003). Generation of the NUP98-TOP1 fusion transcript by the t(11;20) (p15;q11) in a case of acute monocytic leukemia. *Cancer Genet Cytogenet* **140**, 153.
207. Morerio C, Rosanda C, Rapella A, Micalizzi C, Panarello C. (2002). Is t(10;11)(p11.2;q23) involving MLL and ABI-1 genes associated with congenital acute monocytic leukemia? *Cancer Genet Cytogenet* **139**, 57–9.
208. Tallman MS, Kim HT, Paietta E, Bennett JM, Dewald G, Cassileth PA, Wiernik PH, Rowe JM. Eastern Cooperative Oncology Group (2004). Acute monocytic leukemia (French-American-British classification M5) does not have a worse prognosis than other subtypes of acute myeloid leukemia: A report from the Eastern Cooperative Oncology Group. *J Clin Oncol* **22**, 1276–86.
209. Liu LB, Li L, Zou P. (2006). Comparison of cytogenetics and clinical manifestations between M(5a) and M(5b) of acute monocytic leukemia. *Zhongguo Shi Yan Xue Ye Xue Za Zhi* **14**, 654–7.
210. Park S, Picard F, Dreyfus F. (2002). Erythroleukemia: A need for a new definition. *Leukemia* **16**, 1399–401.
211. Selby DM, Valdez R, Schnitzer B, Ross CW, Finn WG. (2003). Diagnostic criteria for acute erythroleukemia. *Blood* **101**, 2895–6.
212. Delva L, Gallais I, Guillouf C, Denis N, Orvain C, Moreau-Gachelin F. (2004). Multiple functional domains of the oncoproteins Spi-1/PU.l and TLS are involved in their opposite splicing effects in erythroleukemic cells. *Oncogene* **23**, 4389–99.
213. Domingo-Claros A, Larriba I, Rozman M, Irriguible D, Vallespi T, Aventin A, Ayats R, Milk F, Sole F, Florensa L, Gallart M, Tuset E, Lopez C, Woessner S. (2002). Acute erythroid neoplastic proliferations. A biological study based on 62 patients. *Haematologica* **87**, 148–53.
214. Park S, Picard F, Guesnu M, Maloum K, Leblond V, Dreyfus F. (2004). Erythroleukaemia and RAEB-t: A same disease? *Leukemia* **18**, 888–90.
215. Parry TE. (2005). On the pathogenesis of erythroleukaemia (H0493). *Leuk Res* **29**, 119–21.
216. Hasserjian RP, Howard J, Wood A, Henry K, Bain B. (2001). Acute erythremic myelosis (true erythroleukaemia): A variant of AML FAB-M6. *J Clin Pathol* **54**, 205–9.
217. McCloskey SM, McMullin MF, Morris TC, Markey GM. (2004). Bone marrow architecture in acute myeloid/erythroid leukaemia. *Br J Haematol* **126**, 1.
218. Huang Q. (2004). Pure erythroid leukemia. *Arch Pathol Lab Med* **128**, 241–2.
219. Lessard M, Struski S, Leymarie V, Flandrin G, Lafage-Pochitaloff M, Mozziconacci MJ, Talmant P, Bastard C, Charrin C, Baranger L, Helias C, Cornillet-Lefebvre P, Mugneret F, Cabrol C, Pages MP, Fert-Ferret D, Nguyen-Khac F, Quilichini B, Barin C, Berger R. On behalf of the Groupe Francophone de Cytogenetique Hematologique (GFCH). The Groupe Francais d'Hematologie Cellulaire (GFHC) (2005). Cytogenetic study of 75 erythroleukemias. *Cancer Genet Cytogenet* **163**, 113–22.
220. Athanasiadou A, Saloum R, Gaitatzi M, Anagnostopoulos A, Fassas A. (2001). Isolated pentasomy of chromosome 8 in erythroleukemia. *Leuk Lymphoma* **42**, 1409–12.
221. Cigudosa JC, Odero MD, Calasanz MJ, Sole F, Salido M, Arranz E, Martinez-Ramirez A, Urioste M, Alvarez S, Cervera JV, MacGrogan D, Sanz MA, Nimer SD, Benitez J. (2003). *De novo* erythroleukemia chromosome features include multiple rearrangements, with special involvement of chromosomes 11 and 19. *Genes Chromosomes Cancer* **36**, 406–12.
222. Mazzella FM, Kowal-Vern A, Shrit MA, Wibowo AL, Rector JT, Cotelingam JD, Collier J, Mikhael A, Cualing H, Schumacher HR. (1998). Acute erythroleukemia: Evaluation of 48 cases with reference to classification, cell proliferation, cytogenetics, and prognosis. *Am J Clin Pathol* **110**, 590–8.
223. Gurbuxani S, Vyas P, Crispino JD. (2004). Recent insights into the mechanisms of myeloid leukemogenesis in Down syndrome. *Blood* **103**, 399–406. Epub 2003 Sep 25.
224. Helleberg C, Knudsen H, Hansen PB, Nikolajsen K, Kjaersgaard E, Ralfkiaer E, Johnsen HE. (1997). CD34+ megakaryoblastic leukaemic cells are CD38−, but CD61+ and glycophorin A+: Improved criteria for diagnosis of AML-M7? *Leukemia* **11**, 830–4.
225. Gurbuxani S. (2006). CD36−a marker for drug sensitive acute megakaryoblastic leukemia. *Leuk Lymphoma* **47**, 2004–5.
226. Chan WC, Carroll A, Alvarado CS, Phillips S, Gonzalez-Crussi F, Kurczynski E, Pappo A, Emami A, Bowman P, Head DR. (1992). Acute megakaryoblastic leukemia in infants with t(1;22)(p13;q13) abnormality. *Am J Clin Pathol* **98**, 214–21.
227. Lewis M, Kaicker S, Strauchen J, Morotti R. (2007200716). Hepatic involvement in congenital acute megakaryoblastic leukemia: A case report with emphasis on the liver pathology findings. *Pediatr Dev Pathol* **1**.
228. Duchayne E, Fenneteau O, Pages MP, Sainty D, Arnoulet C, Dastugue N, Garand R, Flandrin G. Groupe Français d'Hématologie Cellulaire. Groupe Français de Cytogénétique Hématologique (2003). Acute megakaryoblastic leukaemia: A national clinical and biological study of 53 adult and childhood cases by the Groupe Français d'Hématologie Cellulaire (GFHC). *Leuk Lymphoma* **44**, 49–58.
229. Walters DK, Mercher T, Gu TL, O'Hare T, Tyner JW, Loriaux M, Goss VL, Lee KA, Eide CA, Wong MJ, Stoffregen EP, McGreevey L, Nardone J, Moore SA, Crispino J, Boggon TJ,

Heinrich MC, Deininger MW, Polakiewicz RD, Gilliland DG, Druker BJ. (2006). Activating alleles of JAK3 in acute megakaryoblastic leukemia. *Cancer Cell* **10**, 65–75.

230. Roche-Lestienne C, Dastugue N, Richebourg S, Roquefeuil B, Dalle JH, Lai JL, Andrieux J. (2006). Acute megakaryoblastic leukemia with der(7)t(5;7)(q11;p11 approximately p12) associated with Down syndrome: A fourth case report. *Cancer Genet Cytogenet* **169**, 184–6.
231. Kobayashi K, Usami I, Kubota M, Nishio T, Kakazu N. (2005). Chromosome 7 abnormalities in acute megakaryoblastic leukemia associated with Down syndrome. *Cancer Genet Cytogenet* **158**, 184–7.
232. Dastugue N, Lafage-Pochitaloff M, Pagès MP, Radford I, Bastard C, Talmant P, Mozziconacci MJ, Léonard C, Bilhou-Nabéra C, Cabrol C, Capodano AM, Cornillet-Lefebvre P, Lessard M, Mugneret F, Pérot C, Taviaux S, Fenneteaux O, Duchayne E, Berger R. Groupe Français d'Hematologie Cellulaire (2002). Cytogenetic profile of childhood and adult megakaryoblastic leukemia (M7): A study of the Groupe Français de Cytogénétique Hématologique (GFCH). *Blood* **100**, 618–26.
233. Aktas D, Tuncbilek E, Cetin M, Hicsonmez G. (2001). Tetrasomy 8 as a primary chromosomal abnormality in a child with acute megakaryoblastic leukemia. A case report and review of the literature. *Cancer Genet Cytogenet* **126**, 166–8.
234. Toretsky JA, Everly EM, Padilla-Nash HM, Chen A, Abruzzo LV, Eskenazi AE, Frantz C, Ried T, Stamberg J. (2003). Novel translocation in acute megakaryoblastic leukemia (AML-M7). *J Pediatr Hematol Oncol* **25**, 396–402.
235. Paredes-Aguilera R, Romero-Guzman L, Lopez-Santiago N, Trejo RA. (2003). Biology, clinical, and hematologic features of acute megakaryoblastic leukemia in children. *Am J Hematol* **73**, 71–80.
236. Tallman MS, Neuberg D, Bennett JM, Francois CJ, Paietta E, Wiernik PH, Dewald G, Cassileth PA, Oken MM, Rowe JM. (2000). Acute megakaryocytic leukemia: The Eastern Cooperative Oncology Group experience. *Blood* **96**, 2405–11.
237. Cripe LD, Hromas R. (1998). Malignant disorders of megakaryocytes. *Semin Hematol* **35**, 200–9.
238. Gassmann W, Loffler H. (1995). Acute megakaryoblastic leukemia. *Leuk Lymphoma* **18**(Suppl 1), 69–73.
239. Lorsbach RB. (2004). Megakaryoblastic disorders in children. *Am J Clin Pathol* **122**(Suppl), S33–46.
240. Duchayne E, Demur C, Rubie H, Robert A, Dastugue N. (1999). Diagnosis of acute basophilic leukemia. *Leuk Lymphoma* **32**, 269–78.
241. Gupta R, Jain P, Anand M. (2004). Acute basophilic leukemia: Case report. *Am J Hematol* **76**, 134–8.
242. Bernini JC, Timmons CF, Sandler ES. (1995). Acute basophilic leukemia in a child. Anaphylactoid reaction and coagulopathy secondary to vincristine-mediated degranulation. *Cancer* **75**, 110–14.
243. Staal-Viliare A, Latger-Cannard V, Didion J, Grégoire MJ, Lecompte T, Jonveaux P, Rio Y. (2007). CD203c /CD117-, an useful phenotype profile for acute basophilic leukaemia diagnosis in cases of undifferentiated blasts. *Leuk Lymphoma* **48**, 439–41.
244. Shin SY, Koo SH, Kwon KC, Park JW, Ko CS, Jo DY. (2007). Monosomy 7 as the sole abnormality of an acute basophilic leukemia. *Cancer Genet Cytogenet* **172**, 168–71.
245. Seth T, Vora A, Bhutani M, Ganessan K, Jain P, Kochupillai V. (2004). Acute basophilic leukemia with t(8;21). *Leuk Lymphoma* **45**, 605–8.
246. Dastugue N, Duchayne E, Kuhlein E, Rubie H, Demur C, Aurich J, Robert A, Sie P. (1997). Acute basophilic leukaemia and translocation t(X;6)(p11;q23). *Br J Haematol* **98**, 170–6.
247. Sultan C, Sigaux F, Imbert M, Reyes F. (1981). Acute myelodysplasia with myelofibrosis: A report of eight cases. *Br J Haematol* **49**, 11–16.
248. Thiele J, Kvasnicka HM, Schmitt-Graeff A. (2004). Acute panmyelosis with myelofibrosis. *Leuk Lymphoma* **45**, 681–7.
249. Orazi A, O'Malley DP, Jiang J, Vance GH, Thomas J, Czader M, Fang W, An C, Banks PM. (2005). Acute panmyelosis with myelofibrosis: An entity distinct from acute megakaryoblastic leukemia. *Mod Pathol* **18**, 603–14.
250. Suvajdzic N, Marisavljevic D, Kraguljac N, Pantic M, Djordjevic V, Jankovic G, Cemerikic-Martinovic V, Colovic M. (2004). Acute panmyelosis with myelofibrosis: Clinical, immunophenotypic and cytogenetic study of twelve cases. *Leuk Lymphoma* **45**, 1873–9.
251. Thiele J, Kvasnicka HM, Zerhusen G, Vardiman J, Diehl V, Luebbert M, Schmitt-Graeff A. (2004). Acute panmyelosis with myelofibrosis: A clinicopathological study on 46 patients – including histochemistry of bone marrow biopsies and follow-up. *Ann Hematol* **83**, 513–21.
252. Ma Z, Morris SW, Valentine V, Li M, Herbrick JA, Cui X, Bouman D, Li Y, Mehta PK, Nizetic D, Kaneko Y, Chan GC, Chan LC, Squire J, Scherer SW, Hitzler JK. (2001). Fusion of two novel genes, RBM15 and MKL1, in the t(1;22)(p13;q13) of acute megakaryoblastic leukemia. *Nat Genet* **28**, 220–1.
253. Hsiao HH, Yang MY, Liu YC, Hsiao HP, Tseng SB, Chao MC, Liu TC, Lin SF. (2005). RBM15-MKL1 (OTT-MAL) fusion transcript in an adult acute myeloid leukemia patient. *Am J Hematol* **79**, 43–5.
254. Kawaguchi H, Hitzler JK, Ma Z, Morris SW. (2005). RBM15 and MKL1 mutational screening in megakaryoblastic leukemia cell lines and clinical samples. *Leukemia* **19**, 1492–4.
255. Geletu M, Balkhi MY, Peer Zada AA, Christopeit M, Pulikkan JA, Trivedi AK, Tenen DG, Behre G. (2007). Target proteins of C/EBPalphap30 in AML: C/EBPalphap30 enhances sumoylation of C/EBPalphap42 via up-regulation of Ubc9. *Blood* **110**, 3301–9.
256. Pabst T, Mueller BU. (2007). Transcriptional dysregulation during myeloid transformation in AML. *Oncogene* **26**, 6829–37.
257. Baldus CD, Mrózek K, Marcucci G, Bloomfield CD. (2007). Clinical outcome of *de novo* acute myeloid leukaemia patients with normal cytogenetics is affected by molecular genetic alterations: A concise review. *Br J Haematol* **137**, 387–400.
258. Bolli N, Nicoletti I, De Marco MF, Bigerna B, Pucciarini A, Mannucci R, Martelli MP, Liso A, Mecucci C, Fabbiano F, Martelli MF, Henderson BR, Falini B. (2007). Born to be exported: COOH-terminal nuclear export signals of different strength ensure cytoplasmic accumulation of nucleophosmin leukemic mutants. *Cancer Res* **67**, 6230–7.
259. Brown P, McIntyre E, Rau R, Meshinchi S, Lacayo N, Dahl G, Alonzo TA, Chang M, Arceci RJ, Small D. (2007). The incidence and clinical significance of nucleophosmin mutations in childhood AML. *Blood* **110**, 979–85.
260. Chen W, Rassidakis GZ, Medeiros LJ. (2006). Nucleophosmin gene mutations in acute myeloid leukemia. *Arch Pathol Lab Med* **130**, 1687–92.
261. Gladson CL, Naeim F. (1986). Hypocellular bone marrow with increased blasts. *Am J Hematol* **21**, 15–22.

The Neoplasms of Precursor Lymphoblasts

12
CHAPTER

Faramarz Naeim, P. Nagesh Rao and Wayne W. Grody

Lymphoblastic neoplasms represent leukemias or lymphomas of precursor lymphoid cells – blast cells committed to lymphoid differentiation. Clinically, they are divided into two major categories: lymphoblastic lymphoma (LBL) and acute lymphoblastic leukemia (ALL). Typically, LBL represents a neoplastic process involving extramedullary lymphoid tissues with ≤25% bone marrow involvement, and ALL involves >25% of bone marrow with or without extramedullary lesions [1, 2]. Since LBL and ALL share considerable clinicopathological features and present similar biological properties, they are considered the same disease. In the WHO classification, LBL and ALL are lumped together as precursor lymphoid neoplasms and are divided into two major categories: (1) precursor B-lymphoblastic leukemia/lymphoma and (2) precursor T-lymphoblastic leukemia/lymphoma [1]. In addition to these two categories, there are rare types of precursor acute lymphoblastic leukemia/lymphoma, such as NK-cell type, granular ALL, hypoplastic ALL, and ALL with eosinophilia, which are briefly discussed at the end of this chapter.

PRECURSOR B-LYMPHOBLASTIC LEUKEMIA/LYMPHOMA

Precursor B-lymphoblastic leukemia/lymphoma (B-ALL/B-LBL) may initially present itself as ALL with the involvement of bone marrow and/or blood, or LBL with the involvement of the lymphoid and/or other extramedullary tissues. In a significant proportion of the cases, however, both bone marrow and extramedullary tissues are involved. ALL and LBL are regarded as different clinical presentations of the same disease.

Etiology and Pathogenesis

The etiology and pathogenesis of B-ALL and B-LBL are not clearly understood. Epidemiological studies suggest that exposure to ionizing radiation, certain chemicals, viruses or bacteria, or other environmental factors may play a role in the development of some subcategories of B-ALL/B-LBL [3–6]. For example, studies of children with precursor B-ALL show that the affected children had limited social contacts during their infancy, or had not received certain vaccinations, particularly for *Haemophilus influenzae* [7]. Chromosomal instability syndromes, such as Fanconi anemia and Ataxia-telangiectasia, are considered risk factors [8–11].

The common chromosome translocations in precursor B-ALL often arise prenatally and appear to play a role in the pathogenesis of the disease [12–15]. For example, t(12;21)(p13;q22) fuses the *ETV6* and *AML1* genes together. The breakpoints in the *ETV6* and *AML1* genes occur randomly, meaning that each patient's leukemic cells have a unique breakpoint in the DNA sequence [12, 13]. Analyses of pairs of identical twins with concordant acute lymphoblastic leukemia and t(12;22) have shown that leukemic cells from both twins share identical breakpoints in the *ETV6* and *AML1* genes. The assumption is that in these twins a preleukemic clone with t(12;22) has been developed, which after birth has transformed into a full-blown leukemia [7]. For this transformation to occur, there may be a need for some additional postnatal events supporting the "two-hit" model of leukemogenesis for the development of B-ALL [14, 15].

Chromosomal translocations result in the development of chimeric genes with fusion protein products. These products may play an important role in the development of leukemia. For example, *BCR/ABL1* fusion gene products of t(9;22) in precursor B-ALL patients affect the RAS/MAPK pathway of signal transduction and promote leukemogenesis (see later).

Pathology

Morphology

The bone marrow biopsy and clot sections are usually hypercellular and are diffusely infiltrated by sheets of uniformly appearing blast cells [1, 2, 16, 17]. These cells have scanty basophilic cytoplasm with round, oval, or indented nuclei; finely dispersed nuclear chromatin; and prominent or indistinct nucleoli. In some cases, the leukemic blasts may appear pleomorphic with variable amounts of cytoplasm or may show convoluted nuclei. Mitotic figures are variable, but often easily detectable. Bone marrow fibrosis and osteoporosis are sometimes present. Fibrosis may be mild, extensive, focal, or diffuse, and it may lead to unsuccessful bone marrow aspiration (dry tap). It is more frequent in the B-ALL than in the T-ALL. Bone marrow necrosis may be present in some cases. Necrosis is usually of the coagulative type with the preservation of the basic outline of the necrotic cells.

Bone marrow smears and touch preparations show numerous blast cells, which are often small with scanty non-granular blue cytoplasm, fine chromatin, and indistinct nuclei (Figure 12.1). But less frequently, the blast cells are larger and more pleomorphic and show variable amounts of cytoplasm which may display vacuolization or azurophilic granules (5–10% of the cases), and one or more prominent nucleoli. Some cases may show cytoplasmic tails (pseudopods) referred to

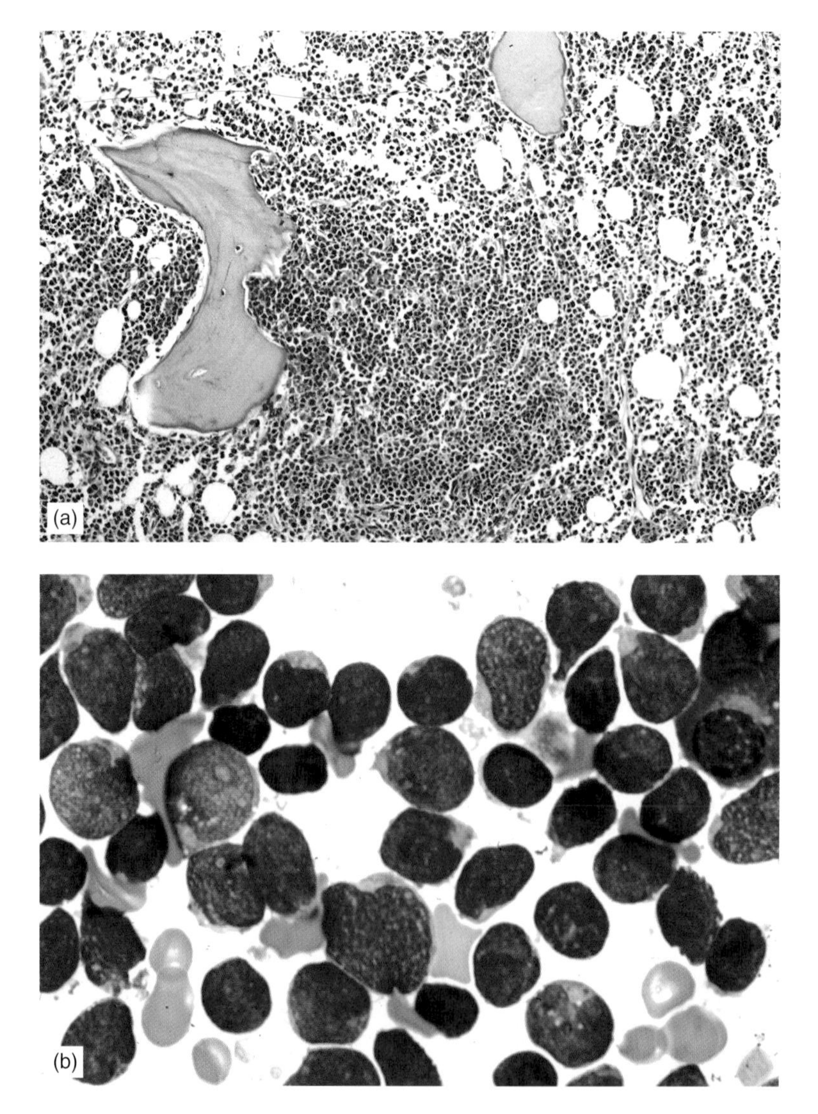

FIGURE 12.1 Bone marrow biopsy section (a) and bone marrow smear (b) of a patient with acute lymphoblastic leukemia.

as *hand-mirror* cells (Figure 12.2). In the case of bone marrow necrosis, the aspirated necrotic cells appear as a mixture of smudge cells, bare degenerated nuclei, and cell debris with an increased amorphous, basophilic background material.

Lymphoblasts may also be present in the peripheral blood smears in variable numbers (Figure 12.1b). They account for the majority of leukocytes in patients with WBC > 10,000/μL. Approximately 20% of patients at the time of diagnosis present with a leukocyte count exceeding 50,000/μL. Anemia, granulocytopenia, and/or thrombocytopenia are common features.

The involvement of lymph nodes and other tissues is usually diffuse with total or partial effacement of the normal architecture and morphologic features similar to those described earlier in the bone marrow biopsy sections.

Immunophenotype and Cytochemical Stains

The precursor B-lymphoblasts characteristically express nuclear terminal deoxynucleotidyl transferase (TdT) and CD79a, are weakly positive for CD45, and lack the expression of surface membrane immunoglobulin (SIg) [1, 2, 18]. These cells are divided into three phenotypic subcategories [2] (Table 12.1):

- Early precursor B
- Intermediate precursor B (representing common ALL)
- Late precursor B (representing pre-B-ALL).

The early precursor B-cells express cytoplasmic CD22, HLA-DR, and usually CD19. The leukemia cells of intermediate precursor B-cells (common ALL) are positive

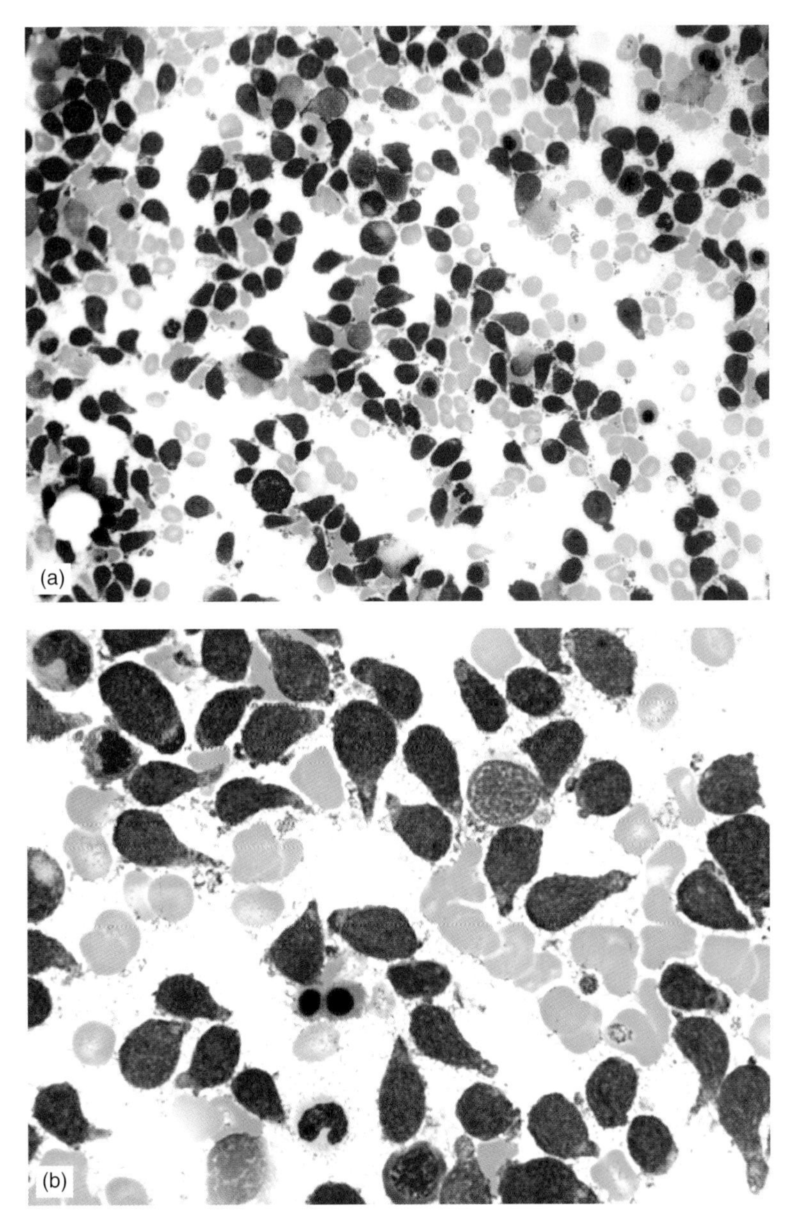

FIGURE 12.2 Bone marrow smear demonstrating the hand-mirror variant of acute lymphoblastic leukemia: (a) low power and (b) high power.

for CD10 (Figure 12.3), and cells representing the late precursor B-ALL are CD20-positive and express cytoplasmic μ heavy chain (Table 12.1).

Expression of CD34 has been observed in about 40% of the cases of B-ALL, particularly in the early and intermediate precursor categories. Aberrant expression of myeloid markers, such as CD13, CD15, CD33, and CD68, has been observed in some cases of B-ALL, particularly in association with certain chromosome translocations. For example, blast cells in patients with t(12;21) may show coexpression of CD13 and/or CD33, and leukemic cells in cases with t(4;11) and 11q23 abnormalities may show coexpression of CD15 and/or CD68.

Precursor B-cells are negative for MPO but may show coarse PAS-positive cytoplasmic granules. In some cases blast cells may show punctuate or focal positive reaction for NSE, or light gray staining for Sudan Black B stain.

TABLE 12.1 Immunophenotypic characteristics of precursor B-lymphoblastic leukemia/lymphoma.

Stage	Immunophenotype
Early precursor	HLA-DR, TdT, cCD22, CD79a, CD19
Intermediate precursor (common)	HLA-DR, TdT, cCD22, CD79a, CD19, CD10, CD20 (variable)
Late precursor	HLA-DR, TdT (variable), cCD22, CD79a, CD19, CD10, CD20, cytoplasmic μ

Cytogenetic and Molecular Studies

Cytogenetic abnormalities in B-ALL/B-LBL are considered among the most useful prognostic indicators. They are often associated with distinct immunophenotypic features. Reciprocal chromosomal translocations, hyperdiploidy, and hypodiploidy are the most common cytogenetic abnormalities in ALL/LBL patients [19–25] (Table 12.2). The major recurrent chromosomal translocations include t(9;22) (Philadelphia chromosome-positive ALL), t(4;11), t(1;19), and t(12;21).

Philadelphia Chromosome, t(9;22)(q34;q11.2)

This abnormality is observed in about 5% of children and 20% of adults with B-ALL/B-LBL [22, 25–27]. It is the most frequent rearrangement in adult ALL. Molecular studies, such as reverse transcriptase polymerase chain reaction (RT-PCR), for the detection of *BCR/ABL1* rearrangement should be used when there is non-diagnostic cytogenetic analysis, and for monitoring patients under therapy in a more quantitative manner. Over 50% of ALL patients with Philadelphia chromosome (*Ph*[1]) have additional chromosomal abnormalities such as monosomy 7. *Ph*[1] positivity is associated with poor prognosis. For example, the likelihood of remaining in remission for 3 years has been reported to be 17% for adult patients with *Ph*[1] compared to 48% in patients with no *Ph*[1] [22, 28].

Molecular studies of *BCR/ABL1* fusion in t(9;22) reveal two distinct subgroups giving rise to two types of fusion proteins weighing 185–190 and 210 kDa [22, 29–31].

TABLE 12.2 Cytogenetic abnormalities and prognosis in precursor B-lymphoblastic neoplasms.

Cytogenetics	Genes	Frequency	Prognosis
t(9;22)(q34;q11.2)	*BCR/ABL1*	25%, adults 3–5%, children	Poor
t(4;11)(q21;q23)	*AF4/MLL**	5%, children	Poor
t(1;19)(q23;p13.3)	*PBX1**/E2A*	6%, children	Poor
t(12;21)(p13;q22)	*ETV6(TEL)/AML1*	25%, children 3–4%, adults	Favorable
Hyperdiploidy		20–25%	Favorable
Hypodiploidy		5%	Poor

Alternative designations:
**ALL* or *HRX*.
***PRL*.

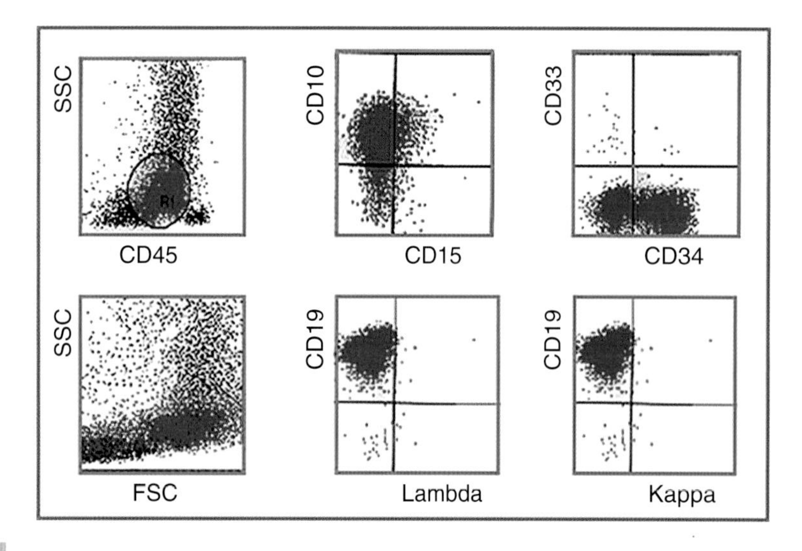

FIGURE 12.3 Bone marrow flow cytometric study of a patient with precursor B-acute lymphoblastic leukemia. The blast cells are dimly CD45-positive and express CD10, CD19, and CD34.

Approximately 30–50% of adult ALL patients show *BCR-ABL1* rearrangement similar to that observed in chronic myelogenous leukemia (CML). The translocation breaks occur within the major-breakpoint cluster region (referred to as M-bcr) of the *BCR* gene and the *ABL1* gene, leading to a chimeric gene which encodes for a 210-kDa fusion protein (p210) [22]. In children and approximately 50–70% of adults with t(9;22), the breakpoint occurs further downstream in the *BCR* gene, referred to as the minor bcr (m-bcr). This fusion gene encodes smaller fusion proteins ranging from 185 to 190 kDa (p185) [22].

Both fusion proteins, p185 and p210, exhibit tyrosine kinase activity similar to that of the native *ABL*-encoded enzyme, but at a higher level. In addition, the native protein is found in both the nucleus and the cytoplasm, whereas the fusion protein is exclusively cytoplasmic. As in CML, these proteins are appealing candidates for molecular-targeted therapies (e.g. imatinib).

At the molecular level, the *BCR-ABL1* fusion event can be detected in a qualitative manner by Southern blot or quantitatively and extremely sensitively by real-time RT-PCR. The standard Southern blot procedure employs hybridization with a radiolabeled or chemiluminescent *BCR* probe. If the translocation has occurred, novel bands or junction fragments will be observed in addition to the unrearranged germline bands when informative restriction endonucleases are used (Figure 12.4). It is customary to run two lanes, with two different restriction enzymes, since apparent rearrangement with just one could be due to a benign polymorphism (RFLP) rather than a true translocation.

More recently, the cumbersome Southern blot procedure has been largely replaced by PCR methods because of their much greater efficiency, quantitative accuracy, and sensitivity. However, the Southern blot can easily target large stretches of DNA, whereas the expanse of the breakpoint cluster region in ALL (as well as in CML) is too large to be covered reliably by a DNA-targeted primer set. Instead, the target chosen is the *BCR-ABL1* fusion transcript from which long introns have been spliced out to yield a target of more

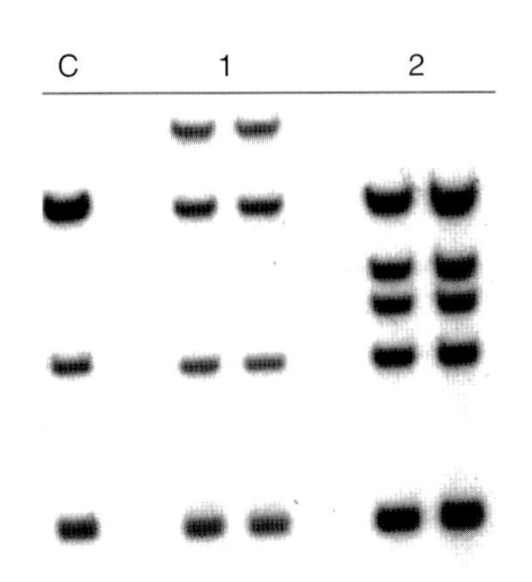

FIGURE 12.4 Southern blot analysis of *BCR-ABL1* translocation in two patients (lanes 1and 2) with Ph[1]-positive adult ALL. Lane C is a negative control sample showing the position of the germline (unrearranged) DNA pattern using this particular restriction endonuclease. Both patients show extra, non-germline hybridizing bands, indicative of a rearrangement caused by the chromosome 9;22 translocation.

manageable size [32]. Naturally, this RNA-based test requires a reverse transcriptase (RT) step in order to generate a DNA target which can then be amplified by PCR. This introduces a potential confounding variable in the assay, owing to the lability of RNA. Blood or bone marrow specimens collected for this test must reach the laboratory in an expeditious manner (in our laboratory, we reject specimens that are >48 h old) and should begin RNA extraction and processing immediately.

Moreover, the *BCR-ABL1* quantity detected is typically given in relative terms, compared to the mRNA of a standard "housekeeping" gene such as *G6PDH*. Sensitivities of BCR-ABL1 mRNA per 10,000 or 100,000 control gene mRNAs allow a rough extrapolation of the number of leukemic cells relative to the normal cells in the specimen (based on the assumption – which may not always be true – that the stability of the two transcripts is roughly equal). Sensitivities at this level, assuming that they are accurate and reproducible, are suitable for the detection of minimal residual disease in treated patients, and monitoring of tumor loads over long-term therapy. However, caution must be exercised in the setup of this assay so that it is not *too* sensitive, since *BCR-ABL1* fusion events have been detected in blood and tissues of healthy individuals, including children [33]. In our laboratory, we are skeptical of any apparently positive result that does not appear or reach its logarithmic "crossing point" until after PCR cycle number 40 or 45 (Figure 12.5).

t(4;11)(q21;q23)

This translocation, which results in a chimeric fusion gene, *AF4/MLL* is observed in about 5% of patients with ALL (Figure 12.6) [34, 35]. The characteristic clinicopathological features associated with t(4;11) are [22, 36–39]:

- High leukocyte count, often ≥200,000/μL
- Early precursor B-cell type with lack of expression of CD10 and CD20 (Table 12.1)
- Frequent coexpression of myeloid-associated markers CD15 and CD65 (Figure 12.7)
- Poor prognosis.

In t(4;11)(q21;q23), the *AF4* gene on chromosome 4 fuses to the *MLL* (or *ALL1*) gene on chromosome 11 (Figures 12.8 and 12.9) [34, 35]. The *AF4/MLL* fusion gene is transcribed into a hybrid mRNA. This transcript can be detected by RT-PCR techniques for establishing the diagnosis or monitoring the residual disease.

The majority of infants with ALL have t(4;11)(q21; q23) or other abnormalities of 11q23.

t(1;19)(q23;p13.3)

This translocation is primarily seen in childhood pre-B-ALL (late precursor B) with a rate ranging from 6% to 30% in various reports [22, 40–43]. The t(1;19) is less frequent in adult pre-B-ALL [22]. Two forms of t(1;19) have been reported: a reciprocal translocation and an unbalanced form as der(19)t(1;19)(q23;p13) (Figure 12.10) [44]. The biological behavior and clinical course of the disease appear to be similar for both forms. The leukemic cells represent late precursor

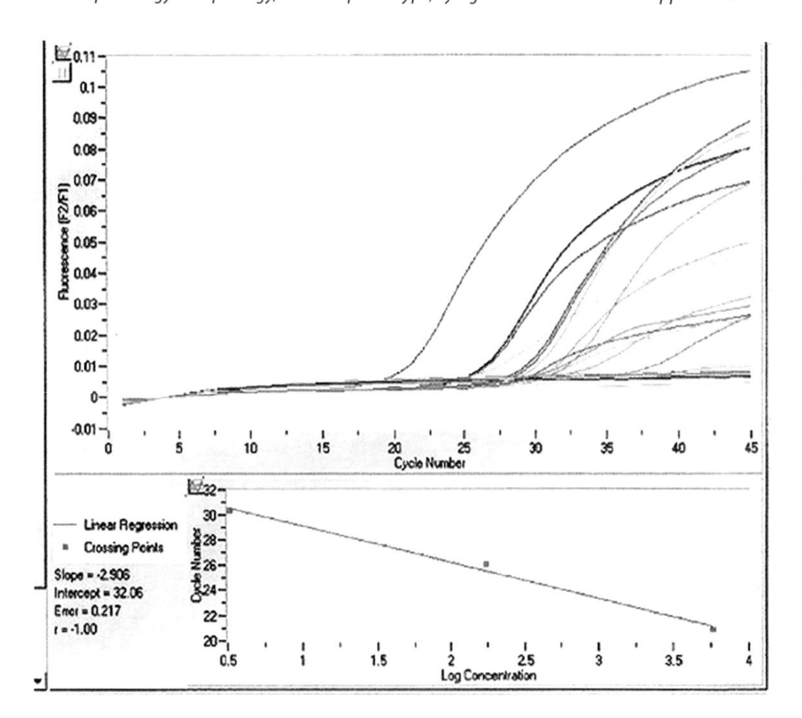

FIGURE 12.5 RT-PCR analysis of a series of patients showing the presence of the *bcr-abl* fusion mRNA target, using the Roche LightCycler instrument. In general, the lower the PCR cycle number (*x*-axis) at which the amplification reaches its logarithmic phase, the higher the amount of starting *bcr-abl* target sequence in the specimen. Very late-rising and/or low-rising signals should be interpreted with caution.

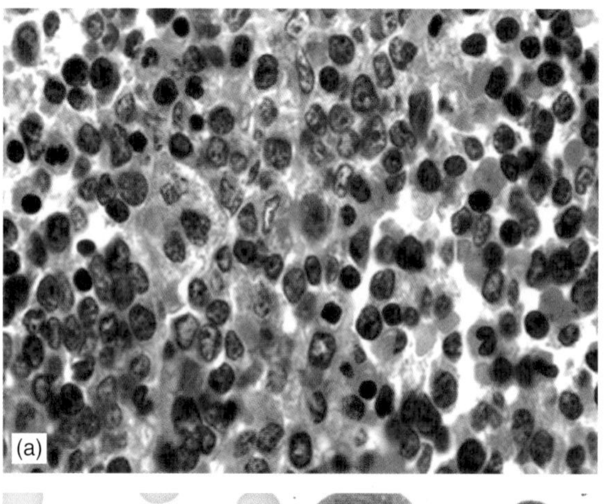

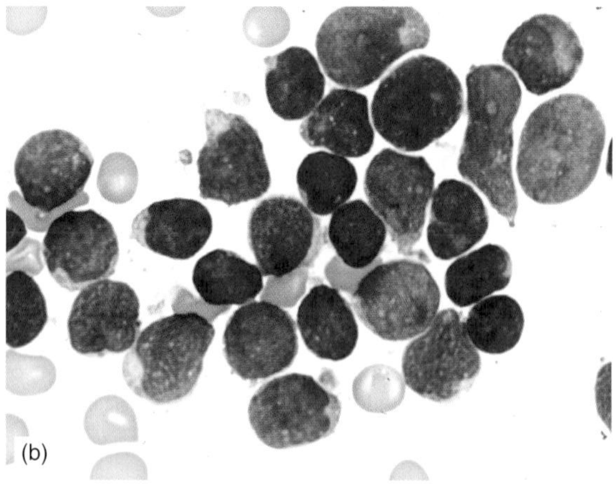

FIGURE 12.6 Bone marrow clot section (a) and smear (b) from a case of acute lymphoblastic leukemia with t(4;11). The blasts are of various sizes and show scanty cytoplasm.

B-cells and are often negative for CD34 and express CD10, CD19, CD20, and cytoplasmic μ [44, 45].

The t(1;19)(q23;p13) leads to a chimeric gene as the result of the fusion of the *PBX1* and *E2A* genes from chromosomes 1 and 19, respectively [43]. The leukemogenic effects of the *PBX1/E2A* fusion protein are not well understood.

Most clinical studies suggest that ALL patients with t(1;19) have a poor prognosis [22, 41, 42, 46].

t(12;21)(p13;q22)

This translocation is one of the most frequent genetic abnormalities in childhood ALL, accounting for about 25% of the cases [22, 47, 48]. It is less frequent in adults, occurring in 3–4% of the patients with B-ALL/B-LBL. The t(12;21) leads to the fusion of the *TEL* (*ETV6*) gene on chromosome 12 with the *AML1* (*RUNX1*) gene on chromosome 21 [22, 47, 48]. It is interesting to know that the persistence of the *TEL-AML1* (*ETV6-RUNX1*) transcript has been reported in some patients with t(12;21) ALL in clinically long-term remission [49].

Because of the similarity of the size and banding patterns of 12p and 21q, the t(12;21) in routine karyotyping is cryptic and its detection may be difficult. Therefore, RT-PCR and FISH are recommended techniques for the detection of this translocation (Figure 12.11) [50]. The dual-color dual-fusion FISH probes in addition to finding the expected 12;21 translocation can also detect extra *RUNX1* signals without the *ETV6-RUNX1* fusion, indicative of the existence of cells with a hyperdiploid karyotype or gene amplification. It is important to recognize the distinction between polysomy 21 (often seen in high hyperdiploidy with good prognosis) (Figure 12.12) and *RUNX1* amplification (associated with poor prognosis) (Figure 12.13). In the

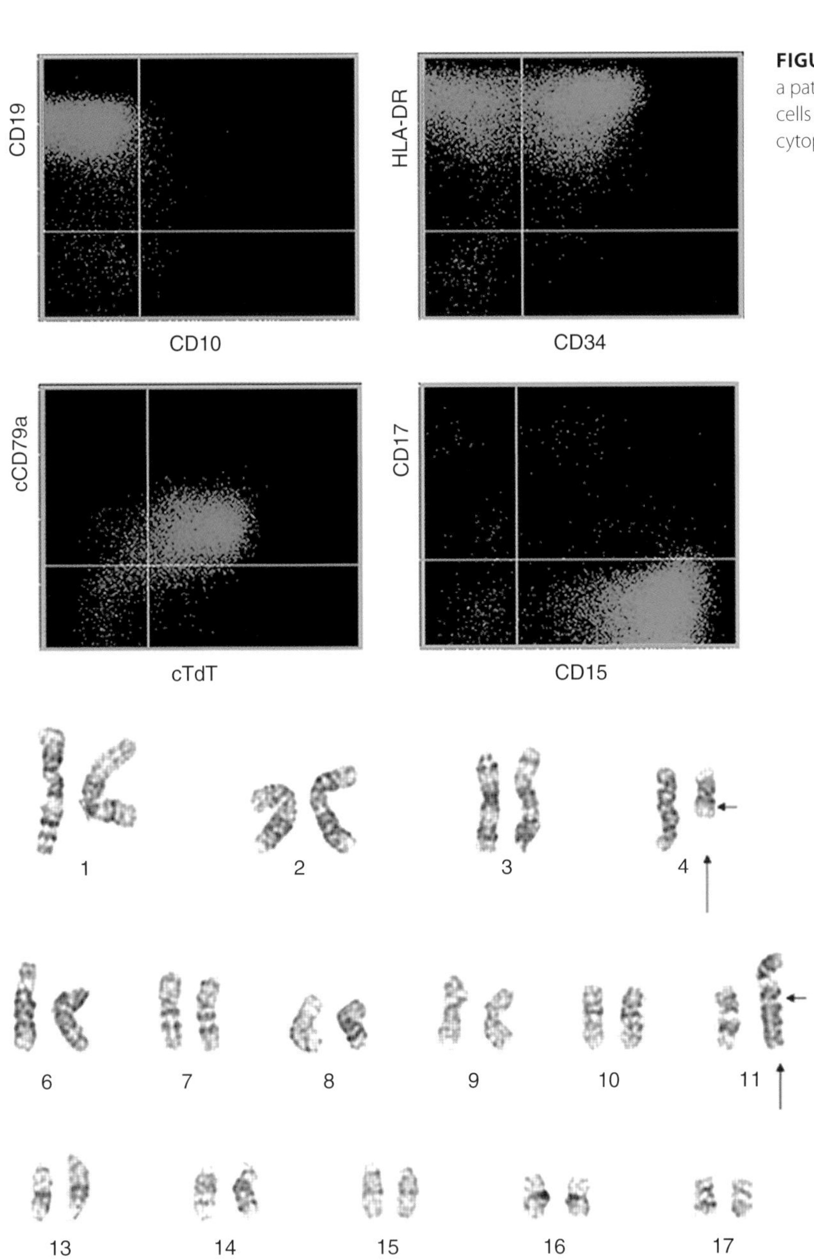

FIGURE 12.7 Flow cytometric analysis of bone marrow from a patient with t(4;11) acute lymphoblastic leukemia. The blast cells are CD10-negative but express CD15, CD19, CD34, HLA-DR, cytoplasmic CD79a and TdT.

FIGURE 12.8 G-banded karyotype of bone marrow of a patient with acute lymphoblastic leukemia demonstrating t(4;11)(q21;q23) (arrows).

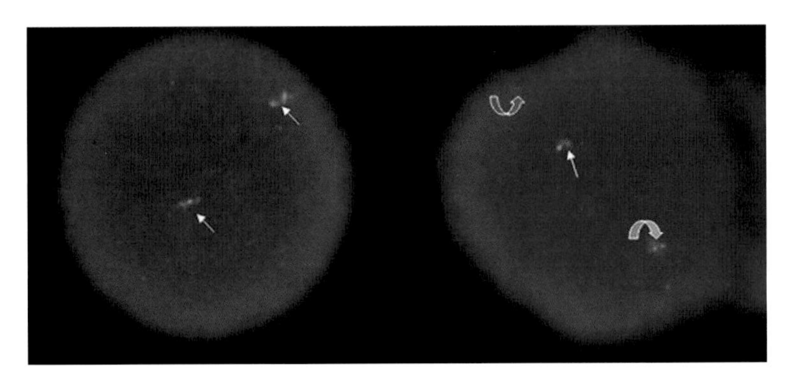

FIGURE 12.9 FISH analysis of bone marrow of a patient with acute lymphoblastic leukemia. The arrows on the left image show normal control and the curved arrows on the right image demonstrate t(4;11).

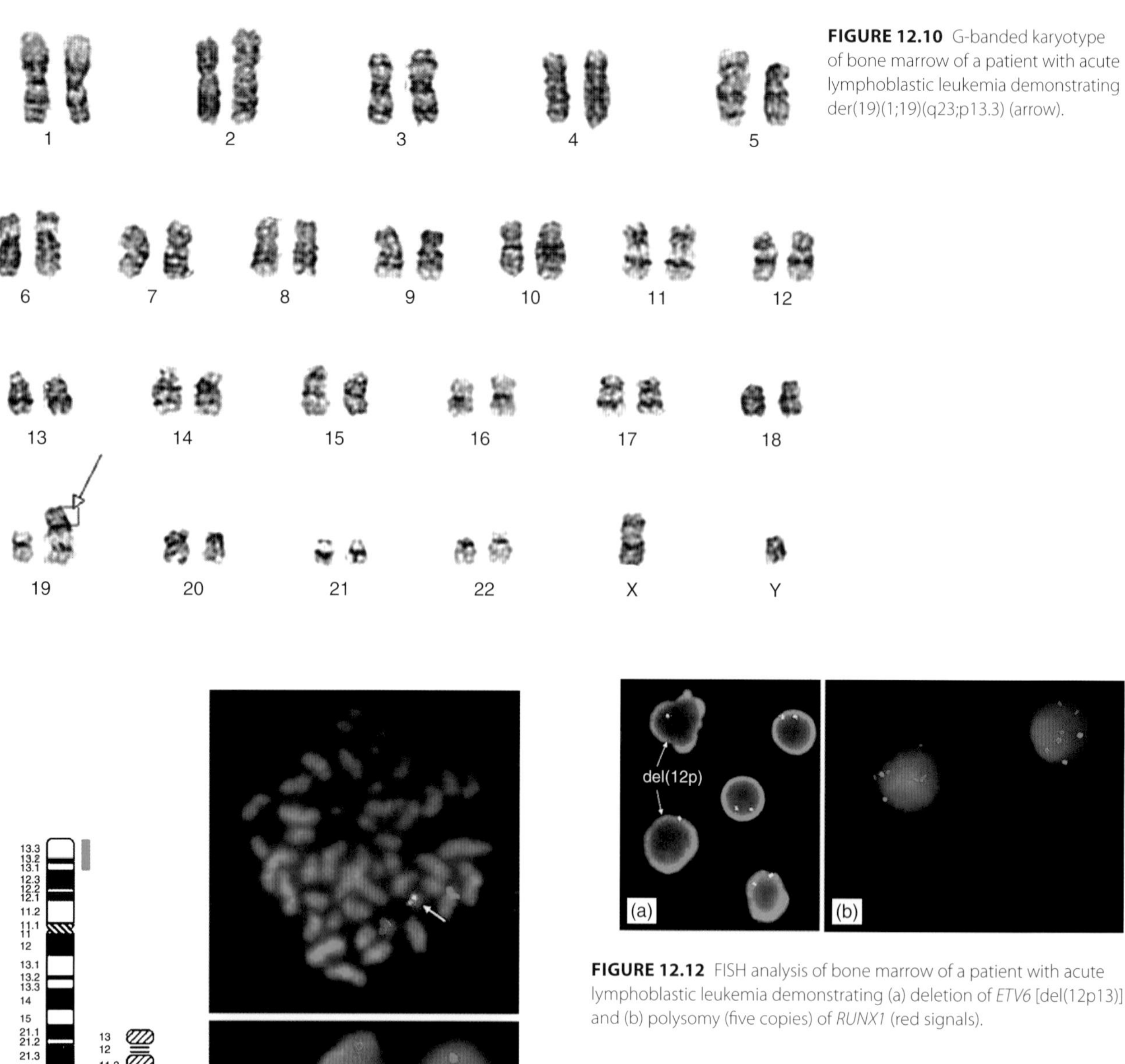

FIGURE 12.10 G-banded karyotype of bone marrow of a patient with acute lymphoblastic leukemia demonstrating der(19)(1;19)(q23;p13.3) (arrow).

FIGURE 12.11 FISH analysis of bone marrow of a patient with acute lymphoblastic leukemia demonstrating t(12;21)(p13;q22) (arrows).

FIGURE 12.12 FISH analysis of bone marrow of a patient with acute lymphoblastic leukemia demonstrating (a) deletion of *ETV6* [del(12p13)] and (b) polysomy (five copies) of *RUNX1* (red signals).

latter the gene signals are clustered, numbering greater than five or more copies.

The leukemic patients with t(12;21) have a favorable prognosis. They have a much higher 5-year event-free survival than those without this translocation (in one report 91% versus 65%, respectively) [22, 47, 48].

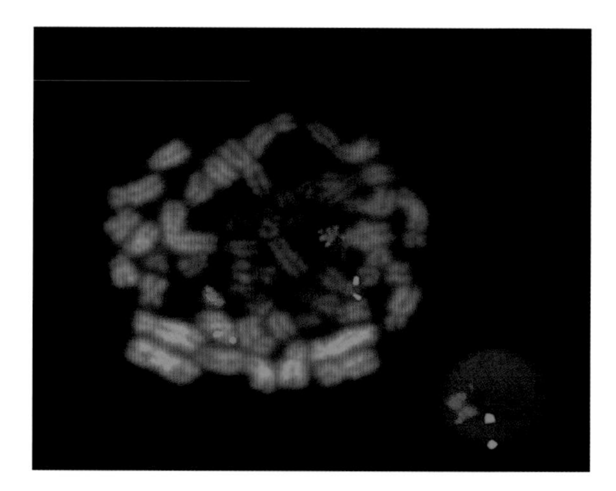

FIGURE 12.13 FISH analysis of bone marrow of a patient with acute lymphoblastic leukemia demonstrating *ETV6* (green signals) and amplification of the *RUNXT1* (red signals).

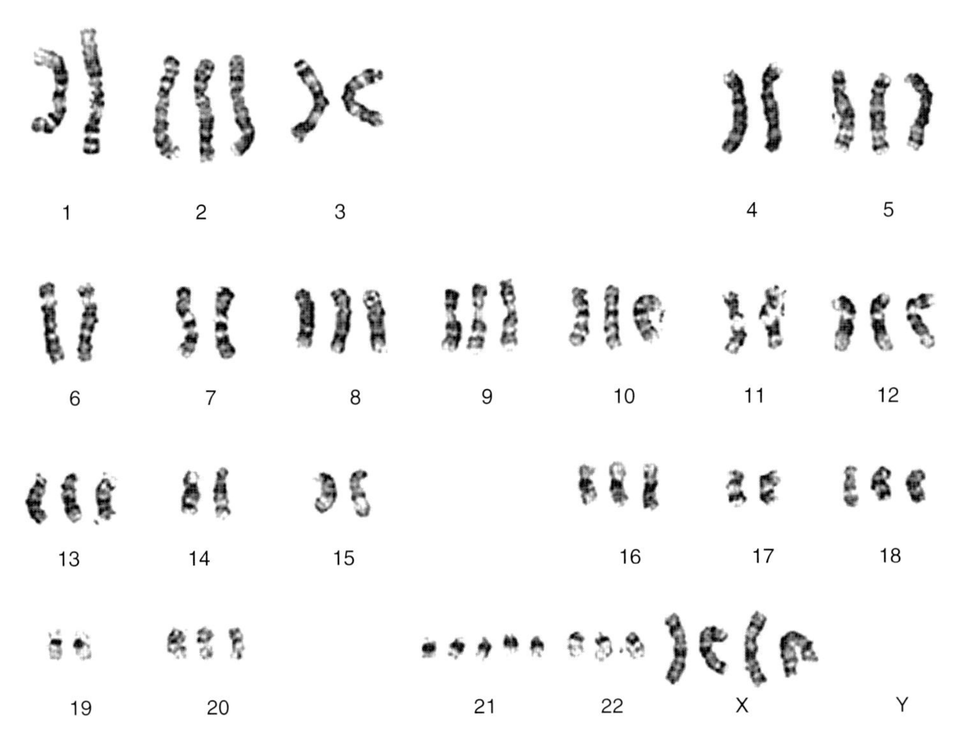

FIGURE 12.14 Hyperdiploid karyotype 62, XX, +X, +X, +2, +5, +8, +9, +10, +12, +13, +16, +18, +20, +21 ×3, +22.

Hyperdiploidy

Hyperdiploidy is divided into two subcategories: (1) hyperdiploidy with ≤50 chromosomes (between 46 and 50 chromosomes) and (2) hyperdiploidy with >50 (usually between 51 and 65) chromosomes (Figure 12.14). Hyperdiploidy is reported in up to 50% of children with precursor B-ALL [2], but it is far less frequent in adult patients. The most common chromosomal additions include chromosomes 21 (often multiple copies), 4, 6, 10, 14, 17, 18, 20, X, and duplication of 1q and isochromosome 17q [51, 52]. Hyperdiploidy indicates favorable prognosis, particularly in association with trisomy of chromosomes 4, 6, and 10 [2, 51, 52].

Hypodiploidy

This condition refers to having chromosome numbers of <45 and DNA index of <1 (Figure 12.15). Several reports indicate poor prognosis in association with hypodiploidy in both children and adult patients with ALL [52–56]. In a large study [56] children with hypodiploid ALL were divided into three major groups: near-haploid (23–29 chromosomes), low hypodiploid (33–39 chromosomes), and high hypodiploid (42–45 chromosomes). Survival analysis showed a poor outcome for the near-haploid and low hypodiploid groups.

Immunoglobulin Gene Rearrangement Clonality

Though not often needed to confirm the diagnosis of these malignancies, precursor B-ALL and LBL should in most cases demonstrate clonal rearrangements of their immunoglobulin genes. Because they are precursor B-cell lesions, they often will not have completed the full ontological sequence of rearrangements, from heavy to light chain. In other words, their transformation to malignancy occurred before their normal maturation could be completed. In such cases one may see rearrangement of the heavy chain genes (IGH) but not of the light chains (IGK or IGL). Therefore, a greater proportion of these clonal cases will be detected if one uses probes or primers specific for the IGH region. Moreover, some of these cells will not yet have produced antibodies detectable at the protein level, yet the genetic analysis can detect the precursor molecular signature of clonality.

It is important to keep in mind that, unlike the translocations and aneuploidies listed above, immunoglobulin gene rearrangement is a normal process in all B-lymphocytes. It is the fundamental mechanism by which a finite (though diverse) number of genes can be made to encode an almost infinite number of antibody species. By rearranging DNA at these loci (e.g. the heavy chain (IGH) locus on chromosome 14), a specific variable (V) gene (out of the 45 or so available) is brought into contiguity with a specific diversity (D) gene and a joining (J) gene; the VDJ complex is then brought together with a specific constant (C) gene by an RNA splicing event. The structural order of these gene families from 5′ to 3′ on chromosome 14 is V–D–J–C.

As is the case for *BCR-ABL1* detection, the older Southern blot methods, using DNA probes typically directed at the J-region genes of the IGH region (Figure 12.16), have largely been replaced by PCR approaches. However, as this region is also quite large, it cannot be covered comprehensively by most series of PCR primer pairs, and so some rearrangements in other areas will go undetected (false negatives). Also, antibodies can further diversify themselves through somatic hypermutation of the variable genes, and if any of these changes occur at a primer hybridization site, they can further reduce the efficiency and sensitivity of the assay [57]. For these reasons, many laboratories, including our own, perform an initial screen by PCR, but if that is negative, reflex to the Southern blot procedure. Most PCR approaches detect 70–80% of B-cell neoplasms,

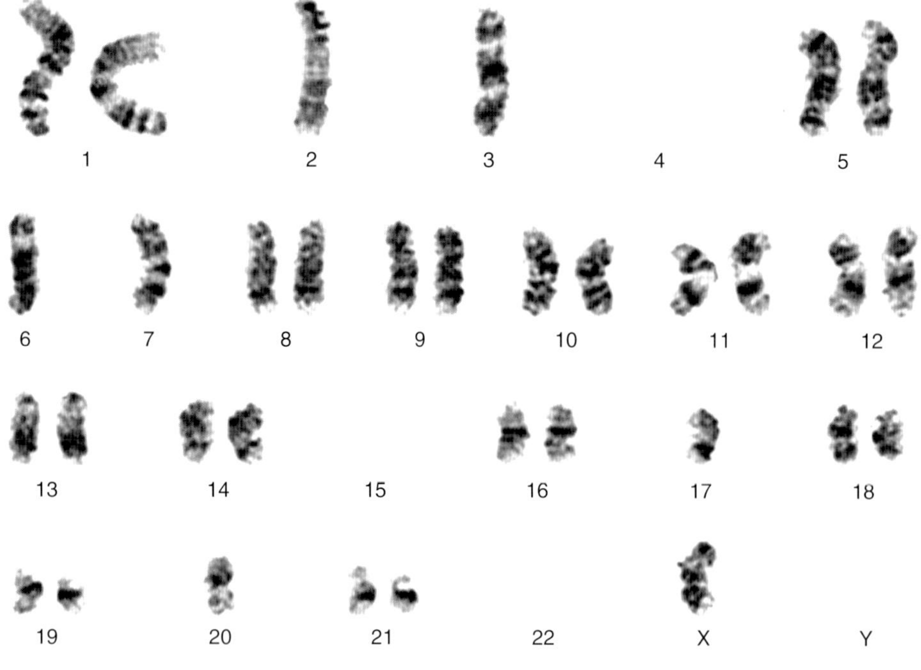

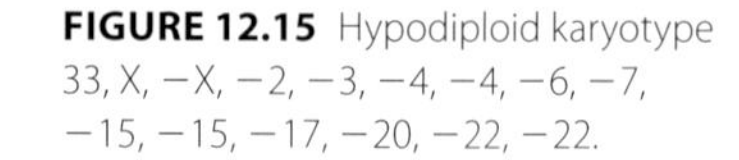

FIGURE 12.15 Hypodiploid karyotype 33, X, −X, −2, −3, −4, −4, −6, −7, −15, −15, −17, −20, −22, −22.

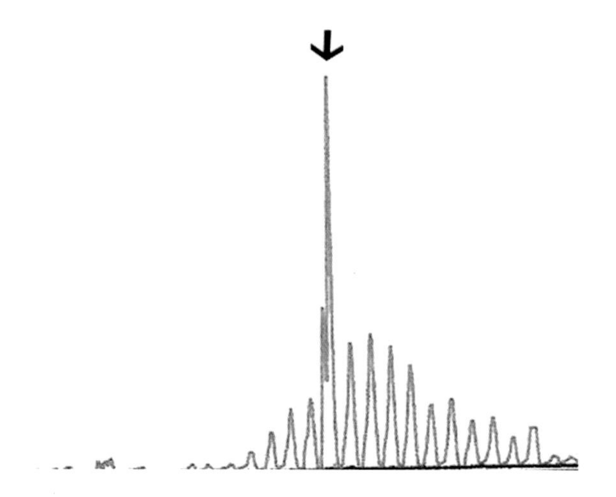

FIGURE 12.16 PCR analysis for immunoglobulin heavy chain clonality. Results are shown for the framework 1 primer set only, illustrating a clonal peak (arrow) superimposed on a polyclonal background population, a pattern often seen in leukemia specimens.

though the sensitivity can be increased somewhat by adding more primer sets for fuller coverage of the region [58]. The most commonly used one is a set of three primer pairs, designated frameworks 1, 2, and 3. Whether done by PCR or Southern blot, the basic principle is the same: a clonal population will generally produce a predominant (or at least visible) DNA pattern (peak or band) above the background of polyclonal population, which should show only germline bands (on Southern blot) or a continuous smear of rearranged bands (on PCR). Germline bands will usually still be seen even in a B-cell malignancy because of the phenomenon of allelic exclusion: the suppression of rearrangement of *IGH* genes on the opposite allele once it has occurred. In addition, the laboratory will rarely receive a "pure" neoplastic lymphoid specimen; there will virtually always be accompanying benign or non-lymphoid cells in the sample, which will contribute the germline bands. It should also be noted that up to 15% of T-ALLs may rearrange their *IGH* genes, so detection of clonality by this method is not an absolute proof of the lymphocyte subclass of origin.

Clinical Aspects

The incidence of precursor B-ALL in the United States approaches 3 per 100,000 population. The peak incidence is between 2 and 5 years of age, affecting boys more than girls. Acute lymphoblastic leukemia/lymphoma is the most common form of cancer in children, comprising about 30% of all childhood malignancies. Up to 85% of childhood ALLs and 40% of childhood lymphomas are of precursor B-cell type [1, 2, 7, 58]. Lymphoblastic lymphoma is defined by the presence of <25% blasts in the bone marrow and evidence of a mediastinal mass or lymphadenopathy.

Precursor B-LBL is uncommon in adults and accounts for ≤1% of lymphomas. It occurs in younger individuals, usually under 35 years of age. Lymphadenopathy and cutaneous involvement are the most frequent presentations [59]. Skin lesions may be multifocal.

The presenting clinical symptoms are often non-specific and secondary to bone marrow/lymphoid tissue infiltration and pancytopenia, such as fever, bleeding, bone pain, and lymphadenopathy.

Favorable prognostic factors include [1–3, 7]:

- Age younger than 1 or older than 10 years
- White blood cell count in normal range or <50,000/μL

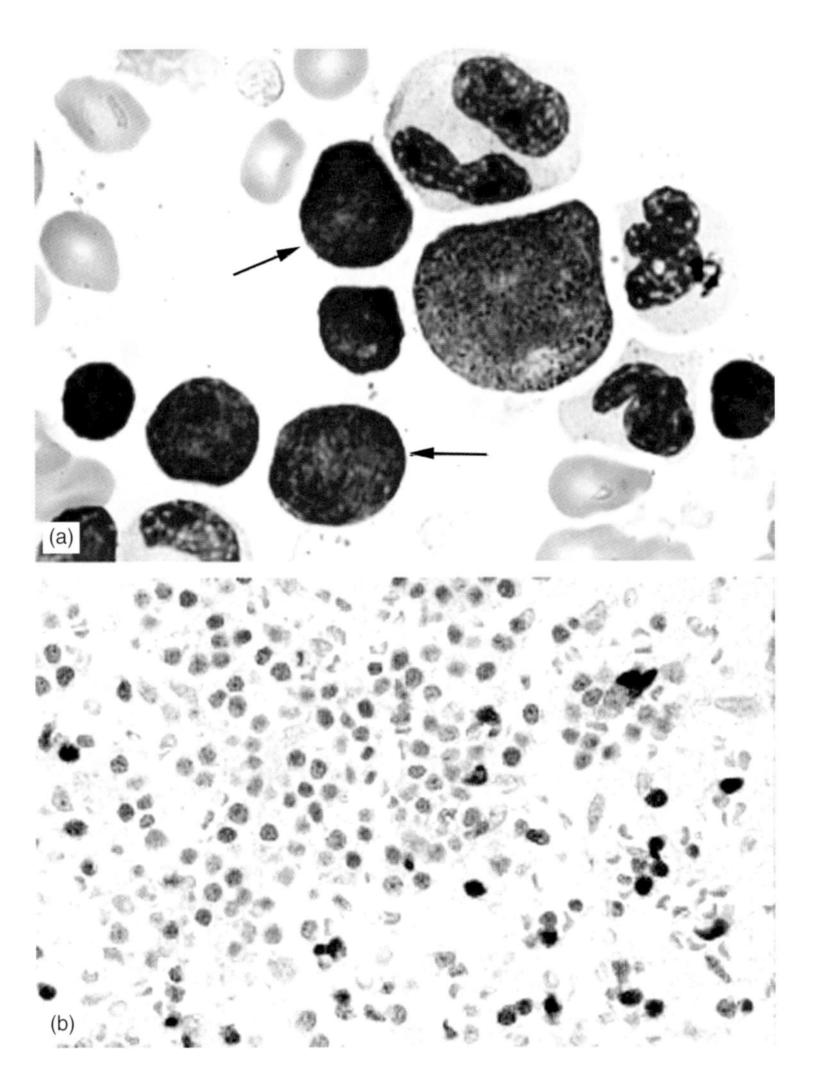

FIGURE 12.17 (a) Bone marrow smear showing scattered hematogones (arrows). (b) Immunohistochemical stain for TdT on a bone marrow biopsy section demonstrating scattered positive cells representing hematogones.

- Hyperdiploidy
- t(12;21)(p13;q22).

The overall response to therapy is significantly better in children than in adults. The current 5-year survival rate for precursor B ALL in children is approaching 85%.

Differential Diagnosis

The differential diagnosis of precursor B-ALL includes hematogone hyperplasia; precursor T-ALL; various types of acute myeloid leukemia such as minimally differentiated AML, AML without maturation, and megakaryoblastic leukemia; and metastatic small (round) cell tumors such as neuroblastoma.

Hematogones represent the normal bone marrow precursor B-cells. These cells consist of a heterogenous population. The earlier forms often express TdT, CD34, CD10, and CD19. The more mature forms lose CD34 and TdT and gain CD20 expression. These cells may morphologically resemble lymphoblasts, but usually show somewhat denser nuclear chromatin and absent or inconspicuous nucleoli (Figure 12.17). Hematogones display a distinctive SSC versus CD45 pattern in flow cytometric studies (Figure 12.18). They are CD45^{dim} and appear as the tail of the mature lymphocytes (CD45^{strong}), together creating a triangular shape. They are found below the blast cells on the SSC/CD45 dot plot. Hematogones account for about 5–10% of the bone marrow cells in children and <5% of the bone marrow cells in adults, but they may be increased in various conditions such as iron deficiency anemia, immune-associated thrombocytopenia, and the following cytotoxic chemotherapy. One of the major difficulties is the distinction between postchemotherapy hematogone hyperplasia and residual disease in children with precursor B-ALL. Both hematogones and residual ALL cells may express CD34, CD10, CD19, and TdT, but they usually have different flow cytometric dot plot patterns. Also, hematogones are scattered through the bone marrow, usually do not form clusters (Figure 12.17b) and are polyclonal, and do not show cytogenetic aberrations; whereas residual precursor B-ALL cells often appear in clusters, are monoclonal, and may show cytogenetic aberrations.

The differential diagnosis of precursor B-LBL includes precursor T-LBL, Burkitt lymphoma, blastic variant

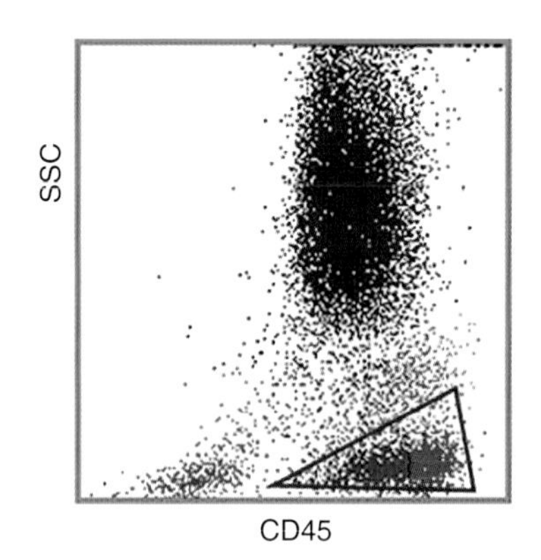

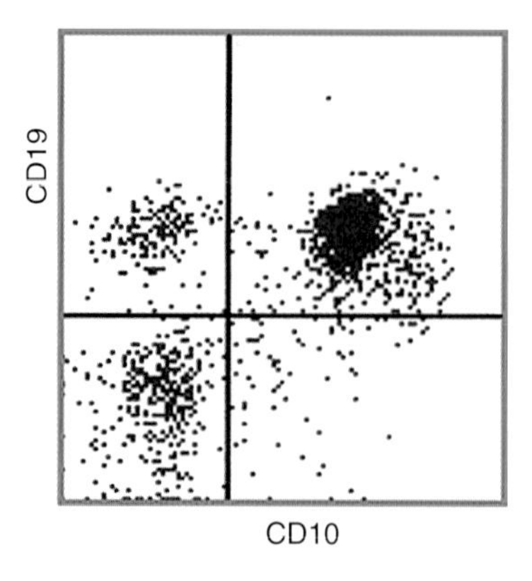

FIGURE 12.18 Flow cytometry of bone marrow hematogones. Hematogones are CD45^{dim} (green population) and usually located on the left side of the lymphocytes (red population) on SSC/CD45 dot plot analysis, as a tail of a triangle. They are B-cells and express CD10.

of mantle cell lymphoma, granulocytic sarcoma, and metastatic small (round cell) tumors. Burkitt and blastic mantle cell lymphomas are TdT negative and have characteristic cytogenetic abnormalities (see Chapter 15).

Precursor B-ALL is distinguished from precursor T-ALL based on their immunophenotypic and cytogenetic characteristics. AML blasts express myeloid-associated markers, such as CD13, CD33, and CD117, and may show positive cytochemical staining for MPO and Sudan Black B.

Metastatic round cell tumors are negative for lymphoid- and myeloid-associated CD molecules and positive for markers that are expressed by the primary tumor.

PRECURSOR T-LYMPHOBLASTIC LEUKEMIA/ LYMPHOBLASTIC LYMPHOMA

Precursor T-lymphoblastic leukemia/lymphoma (T-ALL/ T-LBL) may initially present itself as ALL with the involvement of bone marrow and/or blood, or LBL with the involvement of the lymphoid and/or other extramedullary tissues [1, 2]. In a significant proportion of cases, however, both bone marrow and extramedullary tissues are involved. T-ALL and T-LBL are regarded as different clinical presentations of the same disease.

Etiology and Pathogenesis

The etiology and pathogenesis of precursor T-ALL and T-LBL are not known. Epidemiological studies, as briefly discussed earlier, point to chromosomal instability and exposure to ionizing radiation, certain chemicals, viruses, or bacteria as risk factors for the development of T-ALL.

The most frequent chromosomal translocations in precursor T-ALL result in the activation of proto-oncogenes such as *TAL1-STIL, LYL1, LMO1, LMO2, TAN*, and *MYC*, usually by positioning these genes next to the regulatory elements of the T-cell receptor (TCR) gene [24, 60, 61]. These genes are considered to be transcription regulators and their inappropriate expression may play a role in leukemogenesis [62].

Pathology

Morphology

The morphologic features of precursor T-ALL and T-LBL are similar to those of precursor B-ALL and B-LBL. The bone marrow biopsy and clot sections usually are hypercellular and diffusely infiltrated by sheets of uniformly appearing blast cells (Figure 12.19). These cells have scanty cytoplasm with round, oval, or indented nuclei; finely dispersed nuclear chromatin; and prominent or indistinct nucleoli. In some cases, the leukemic blast cells may appear pleomorphic with variable amounts of cytoplasm or may show convoluted nuclei. Mitotic figures are more frequently observed in T-ALL than in B-ALL [1, 2]. Bone marrow fibrosis and osteoporosis are sometimes present. Fibrosis may be mild, extensive, focal, or diffuse, and it may lead to unsuccessful bone marrow aspiration (dry tap). It is more frequent in B-ALL than in T-ALL. Large areas of bone marrow necrosis are infrequent. Necrosis is usually of the coagulative type with the preservation of the basic outline of the necrotic cells.

Similar to the precursor B-ALL, bone marrow smears and touch preparations show numerous blast cells that are often small with scanty non-granular blue cytoplasm, fine chromatin pattern, and indistinct nuclei. In a minority of the cases, the blast cells are pleomorphic and may show azurophilic granules.

Lymphoblasts may also be present in the peripheral blood smears in variable numbers. They account for the majority of leukocytes in patients with WBC >10,000/μL. Anemia, granulocytopenia, and/or thrombocytopenia are common features.

The involvement of lymph nodes and other tissues is usually diffuse with total or partial effacement of the normal architecture and morphologic features similar to those described earlier in the bone marrow biopsy sections. In some cases, the high rate of tumor cell turnover and necrosis may stimulate the macrophages. These macrophages with abundant pale, vacuolated cytoplasm and phagocytic cell debris are dispersed throughout the lymphomatous lesion, creating a "starry sky" pattern.

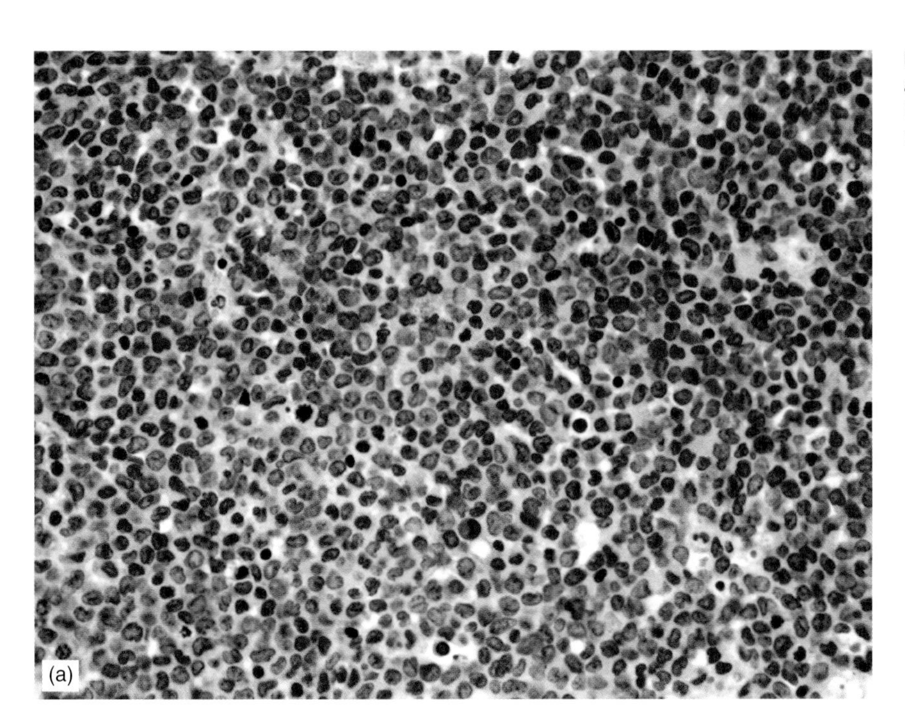

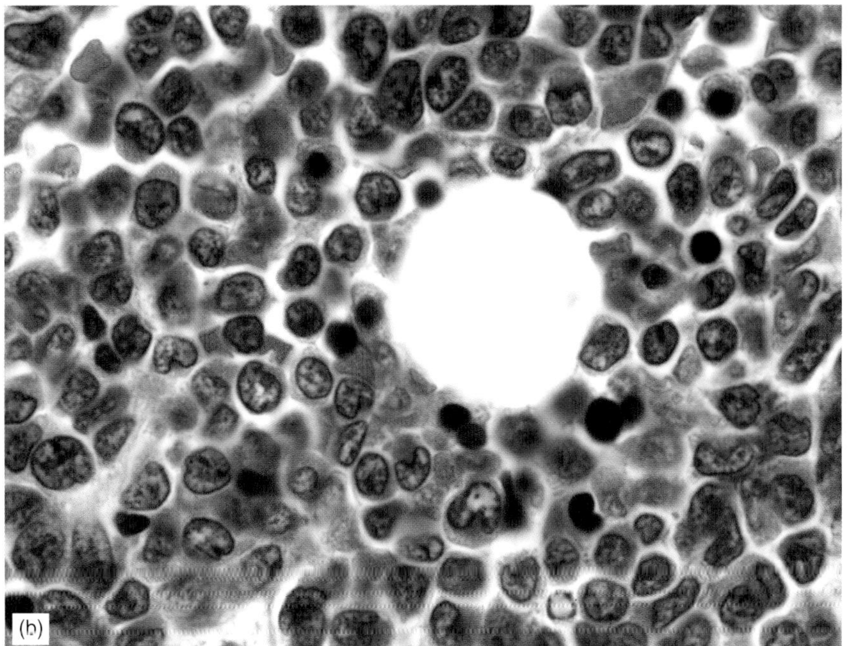

FIGURE 12.19 Bone marrow biopsy section from a patient with precursor T-acute lymphoblastic leukemia showing sheets of blast cells with irregular nuclei (a) low power and (b) high power.

Immunophenotype and Cytochemical Stains

The vast majority of precursor T-ALL/T-LBL cases characteristically express nuclear TdT, cytoplasmic CD3, and CD7 (Figures 12.19b and 12.20). The neoplastic T-blast cells are frequently CD34-positive and express weak CD45 [2, 63–67]. The precursor T-cells are divided into three phenotypic subcategories (Table 12.3):

- Early T-cell
- Intermediate thymocyte
- Late thymocyte.

The early precursor T-blast cells, in addition to TdT, CD7, and cytoplasmic CD3, express CD2 and CD38. The intermediate precursor cells characteristically express CD1a, surface CD3, CD4, and CD8, and the blasts of the late stage thymocyte category express CD4 or CD8 (Table 12.3). The intermediate precursor T-ALL/T-LBL type is the most frequent one (common type), accounting for over 50% of the cases. In general, cases of T-LBL may represent more of mature T-cell phenotype than those of T-ALL.

Approximately half of the T-LBL cases also express CD44 (homing receptor/cell adhesion molecule) [2]. Aberrant expression of myeloid-associated markers, such

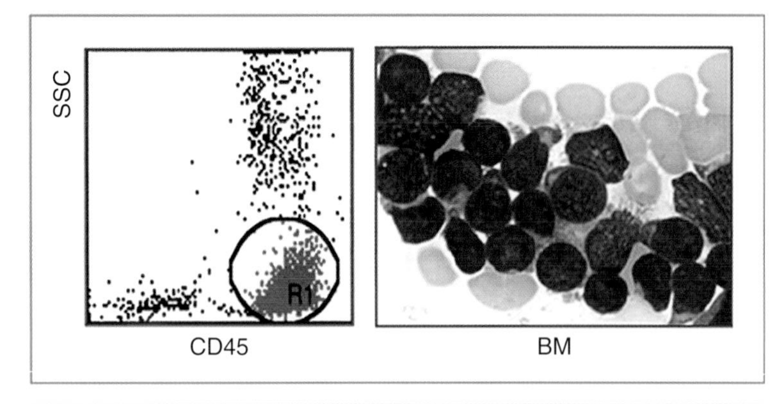

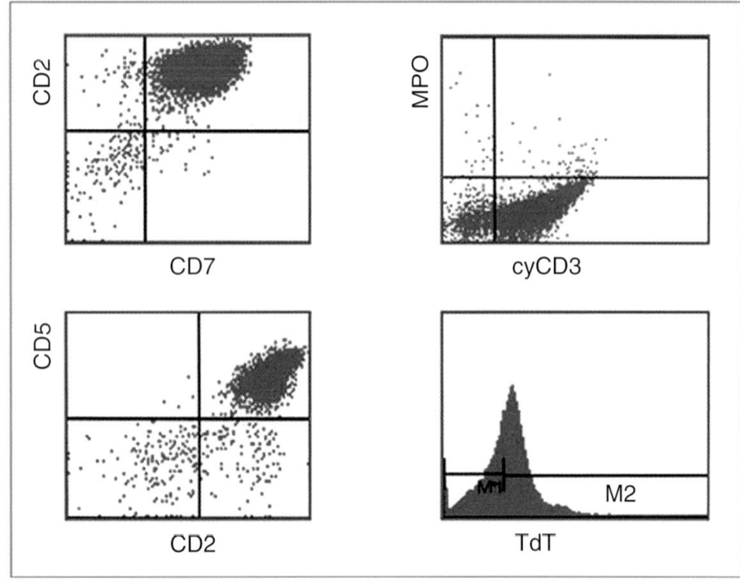

FIGURE 12.20 Flow cytometric analysis of bone marrow from a patient with precursor T-lymphoblastic leukemia demonstrating a population of CD45+ blast cells also expressing CD2, CD5, CD7, and cytoplasmic CD3 (cyCD3), and Id I.

TABLE 12.3 Immunophenotypic characteristics of T-cells in various stages of maturation.

Stage	Immunophenotype
Early precursor	CD7, TdT, cCD3
Intermediate precursor (common thymocyte)	CD7, TdT, cCD3, CD1a, CD2, CD5, CD38, variable CD4, variable CD8
Late stage (late thymocyte)	CD7, CD5, CD3, CD2, CD45,CD4 or CD8

as CD13 and CD33, is not infrequent [67, 68]. Occasional cases may also express CD117 or CD79a. A significant proportion of cases of T-ALL and T-LBL express CD10, whereas CD16 or CD56 expression is infrequent [58].

No lineage-specific cytochemical stains are available for precursor T-blast cells. These cells may show focal acid phosphatase or NSE reactions.

Cytogenetic and Molecular Studies

Various genes serve as partners in these translocations, such as *C-MYC, TAL1/SCL, LMO1(RBTN1)*, and *HOX11,* located on chromosomes 8q24, 1p32, 11p15, and 10q24, respectively [2, 24, 69]. Precursor T-cell neoplasms may show evidence of either or both *TCR* and *IGH* gene rearrangements [60, 61].

Quantitative chromosomal abnormalities include del(6q) (Figure 12.21), del(9p) (Figure 12.22), and trisomy 8 (Table 12.4). Approximately 30% of cases of T-ALL/T-LBL show translocations involving 14q11.2 (*TCRα/δ*) or 7q34 (*TCRβ/γ*) (Table 12.4), such as t(11;14)(p15;q11.2), (14;21)(q11.2;q22), and t(7;14))(q34;q11.2) (Figure 12.23) [22, 62, 70]. Inversion of chromosome 6 [inv(6)(p21.2q27)] has been reported in association with precursor T-ALL [71–73].

Generation of TCR molecules in T-cells occurs by much the same mechanism as the generation of immunoglobulin molecules in B-cells. Therefore, clonality of T-cell malignancies can be demonstrated by examining the rearrangement patterns of the TCR genes, in much the same way as is done with the immunoglobulin genes in the B-cell lesions. *TCR* rearrangements are somewhat more complicated, however, and there are advantages and disadvantages to the various target loci available. For example, the *TCR-β* genes produce a wider range of rearrangements and thus are more informative, but the region is so large that a high number of PCR primer sets are required to span it. The *TCR-γ* region, in contrast, is smaller and easily encompassed by a small number of primer sets, but the limited number of potential rearrangements increases the chance of the artifact known as pseudoclonality, in which amplification of a specimen in which T-cells are scanty will produce

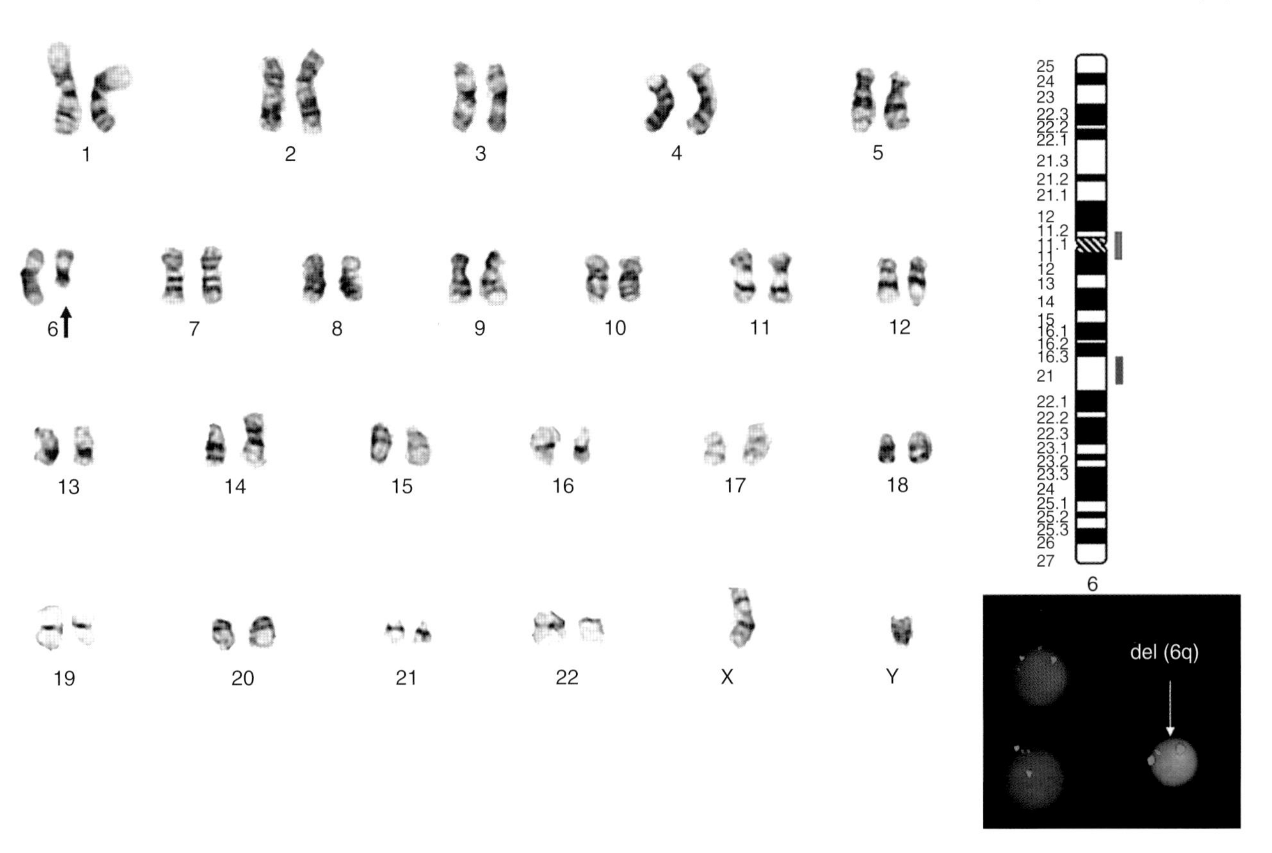

FIGURE 12.21 G-banded karyotype and FISH analysis demonstrating 46,XY,del(6)(q15).

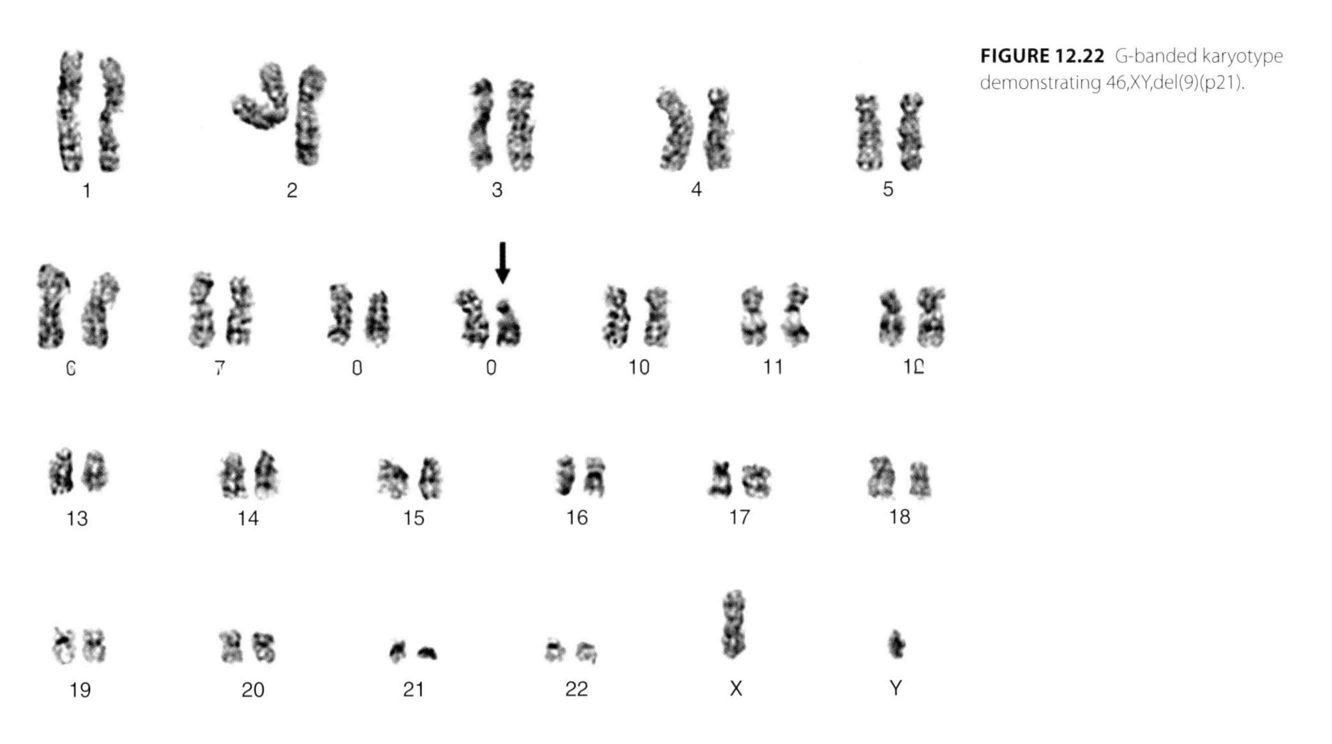

FIGURE 12.22 G-banded karyotype demonstrating 46,XY,del(9)(p21).

an apparent clonal rearrangement pattern even though the cells are neither malignant nor clonal. Still, TCR-γ PCR is most often used by virtue of its practicality [74] and because these genes customarily rearrange before the *TCR-β* genes do, but a negative result will often require reflex to Southern blot analysis in these disorders, at least for those laboratories that choose not to do the extensive TCR-β PCR coverage. On the plus side, *TCR* genes do not undergo somatic hypermutation as do the immunoglobulin genes, and so is not a source of false negatives. Molecular analysis is more often required in the precursor T-cell lesions because of the absence of surface molecules detectable by immunologic methods and flow cytometry. However, as is the case for the B-cell lesions, detection of *TCR* gene rearrangements does not absolutely prove T-cell origin, since a minority of B-ALLs will also rearrange these genes.

TABLE 12.4 Recurrent cytogenetic abnormalities in precursor T-lymphoblastic neoplasms.

Chromosomal aberrations	Affected genes
t(1;7)(p32;q35)	*TAL1/TCRβ*
t(1;14)(p32;q11.2)	*TAL1/TCRαδ*
T(7;10)(q35;q24)	*TCRβ/TAL3**
t(7;19)(q35;p13)	*TCRβ/TAL2***
t(8;14)(q24;q11.2)	*MYC/TCRαδ*
t(11;14)(p15;q11.2)	*TTG1/TCRαδ*
t(11;14)(p13;q11.2)	*TTG2/TCRαδ*
inv(14)(q11.2;q32)	*TRCαδ/TCL1*
del(1p32)	*TAL1*
del(6q)	
del(9p)	
+8	

Alternative designations:
**HOX11, SCL* or *TCL5.*
***TAN.*

It should also be noted that the gene rearrangement clonality studies as described here are less suitable for detecting minimal residual disease than the PCR-based methods that target translocations such as *BCR-ABL1* and the others listed in this chapter. The reason is that the translocations are not found in normal cells and non-pathologic states, so the PCR primers theoretically have no competition from normal targets that would lower the overall sensitivity of the assay. In contrast, *IGH* and *TCR* gene rearrangements are normal phenomena in all lymphocytes, and the same primer hybridization sites are present, with the same efficiency in both benign and neoplastic lymphocytes. This lowers the sensitivity of the detection of the malignant clone to a level of only 5–10%, compared to 1 in 100,000 or less for the PCR-based translocation assays.

A more recent technique, gene expression or mutation profiling by microarray hybridization, may provide valuable information regarding the biology of these disorders and their response to therapy (Figure 12.24) [75].

Clinical Aspects

Precursor T-cell neoplasms represent about 15% of ALLs in children and 25% in adults. Approximately 2% of adult non-Hodgkin lymphomas are precursor T-cell type [2, 7, 76, 77]. Most of the patients are adolescent or young adults with male preponderance, and the vast majority of the patients are at stage III or IV at the time of diagnosis. Anterior mediastinal mass and/or peripheral lymphadenopathy is detected in between 50% and 75% of the cases. Cervical, supraclavicular, and axillary lymph nodes are frequent targets. Extranodal tissues such as skin, testicle, or bone are involved less frequently. A high frequency of CNS involvement has been noted in patients with T-ALL. The overall prognosis of the precursor T-cell neoplasms is worse

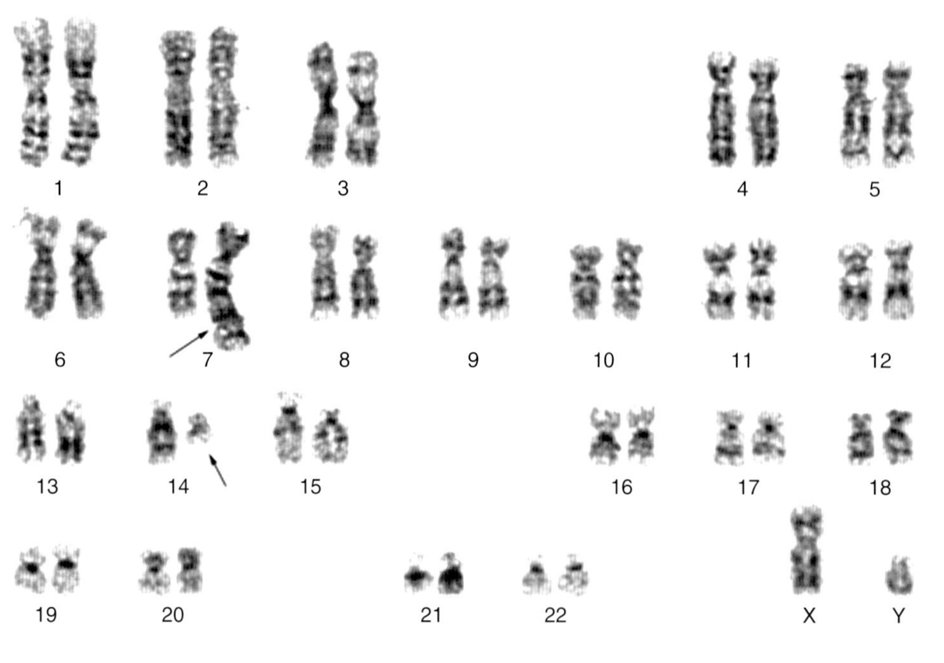

FIGURE 12.23 G-banded karyotype of bone marrow of a patient with acute precursor T-lymphoblastic leukemia demonstrating t(7;14) (q35;q11.2) (arrows).

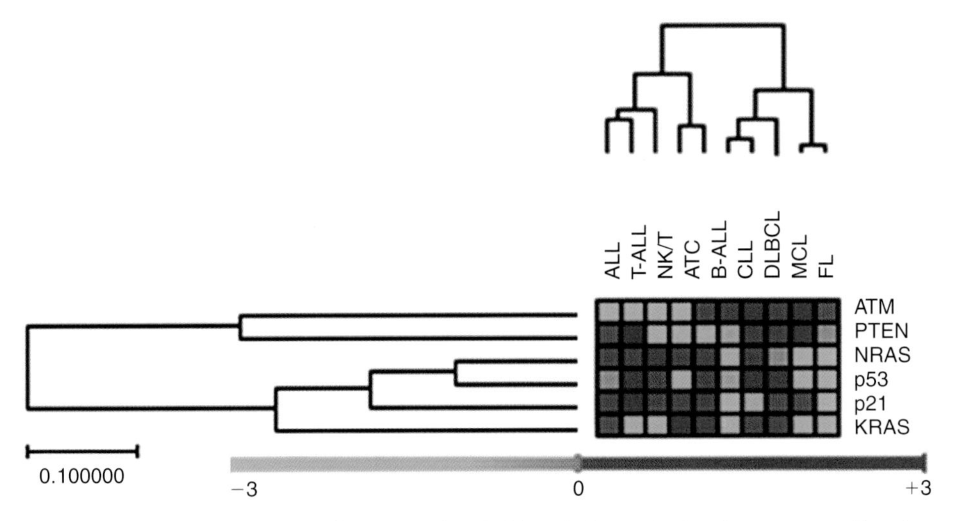

FIGURE 12.24 Schematic correlation of gene mutation profiles in acute lymphoblastic leukemias (ALL) and certain types of lymphomas. Data used from www.sanger.ac.uk/genetics/CGP/cosmic. Courtesy of Dr. Dejun Shen. T-ALL: precursor T-lymphoblastic leukemia/lymphoma, NK/T: NK/T-cell type lymphoma, ATC: adult T-cell leukemia/lymphoma, B-ALL: precursor B-acute lymphoblastic leukemia/lymphoma, DLBCL: diffuse large B-cell lymphoma, MCL: mantle cell lymphoma, and FL: follicular lymphoma.

than that of their B-cell counterparts. So far, no clear-cut correlation has been found between the prognosis and the immunophenotypic or cytogenetic results. Consolidation with high-dose therapy and autologous or allogeneic stem cell transplantation are considered in young patients [76, 77].

Differential Diagnosis

The differential diagnosis of precursor T-cell neoplasms includes hematogone hyperplasia; precursor B-ALL/B-LBL; various acute myeloid leukemias such as minimally differentiated AML, AML without maturation, and megakaryoblastic leukemia; and metastatic small (round) cell tumors such as neuroblastoma.

Hematogones represent the normal bone marrow precursor B-cells. The earlier hematogones may express TdT, but they coexpress TdT and lack the expression of T-cell markers. As was mentioned earlier, hematogones display a distinctive SSC versus CD45 pattern in flow cytometric studies (see Figure 12.15). Hematogones account for about 5–10% of the bone marrow cells in children and <5% of the bone marrow cells in adults. Hematogones are evenly distributed in the bone marrow interstitium, are polyclonal, and do not show cytogenetic aberrations.

The differential diagnosis of precursor T-LBL includes precursor B-LBL, Burkitt lymphoma, blastic variant of mantle cell lymphoma, granulocytic sarcoma, and metastatic small (round cell) tumors. Burkitt and blastic mantle cell lymphomas are TdT negative, express B-cell-associated CD molecules, and have characteristic cytogenetic abnormalities (see Chapter 15).

Precursor T-ALL is distinguished from precursor B-ALL based on their immunophenotypic and cytogenetic characteristics. AML blasts express myeloid-associated markers, such as CD13, CD33, and CD117; lack cytoplasmic CD3 expression; and may show positive cytochemical staining for MPO and Sudan Black B.

Metastatic round cell tumors are negative for lymphoid- and myeloid-associated CD molecules and positive for markers that are expressed by the primary tumor.

OTHER LYMPHOBLASTIC LEUKEMIA/ LYMPHOMA VARIANTS

NK-Cell Lymphoblastic Leukemia/ Lymphoma

The blast cells in a small proportion of ALL/LBL express CD56 and lack B-, T-, and myeloid-associated markers, do not show Ig and *TCR* gene rearrangements and are negative for EBV [79–81]. They are CD45^{dim} and may express CD2, CD7, TdT, and/or CD16, and often lack cytoplasmic azurophilic granules. These tumors are clinically aggressive and have a poor prognosis.

ALL with Cytoplasmic Granules

In a small proportion of patients with ALL (4–7%), particularly in children, lymphoblasts contain coarse azurophilic granules [82–86]. These granules are usually larger than the granules in myeloblasts and are MPO-negative by conventional light microscopic examination. However, they may display MPO positivity by electron microscopy. ALL with cytoplasmic granules is usually of precursor B-cell type.

ALL with Eosinophilia

Eosinophilia has been reported in rare cases of ALL [87–89]. These patients may show clinical symptoms of hypereosinophilic syndrome. ALL with eosinophilia is often

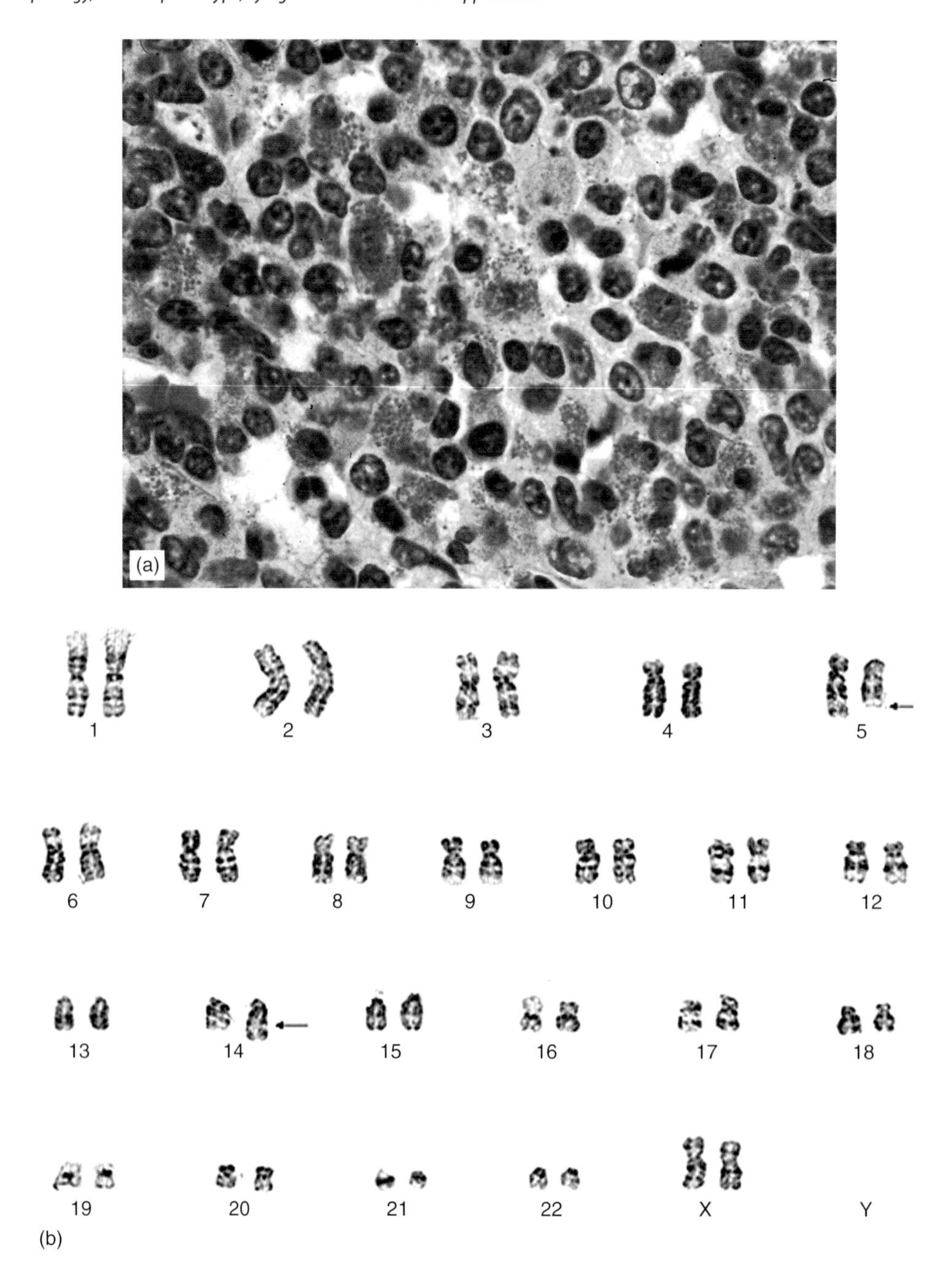

FIGURE 12.25 (a) Bone marrow biopsy section of a patient with precursor B-acute lymphoblastic leukemia and eosinophilia. (b) G-banded karyotyping showing t(5;14)(q31;q32).

of the precursor B-cell type and has been reported frequently in association with t(5;14)(q31;q32), involving *IL-3* (5q31), and *IGH* (14q32) genes (Figure 12.25). Increased expression of IL-3 is thought to be responsible for the marked eosinophilia. ALL with eosinophilia has an aggressive clinical course.

ALL Preceded by or Associated with Hypoplastic Marrow

Occasionally ALL patients may initially present with a hypoplastic marrow and pancytopenia mimicking aplastic anemia [90–92]. The hypoplastic marrow contains a variable number of lymphoblasts (Figure 12.26) and may eventually become packed with lymphoblasts with an obvious acute leukemia picture. This condition, unlike hypoplastic AML, is not usually associated with myelodysplastic changes and is often of precursor B-cell type.

ALL, Burkitt Type

Acute leukemia of Burkitt type in the WHO classification is considered a clinical spectrum (leukemic phase) of Burkitt lymphoma. The neoplastic cells are monomorphic medium-sized lymphoid cells with scanty deep blue and vacuolated cytoplasm and round and multiple nuclei. These

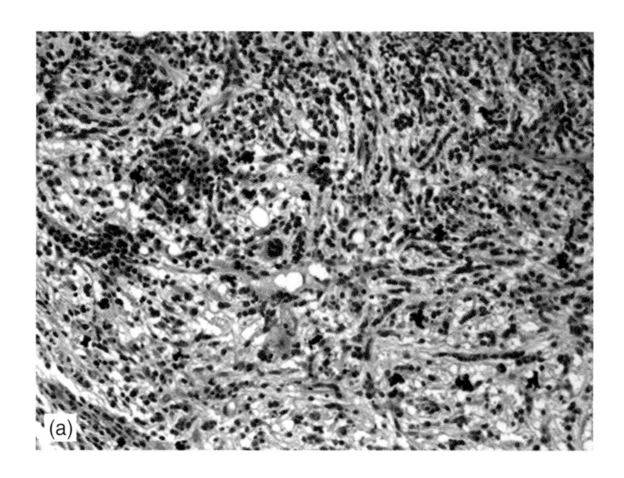

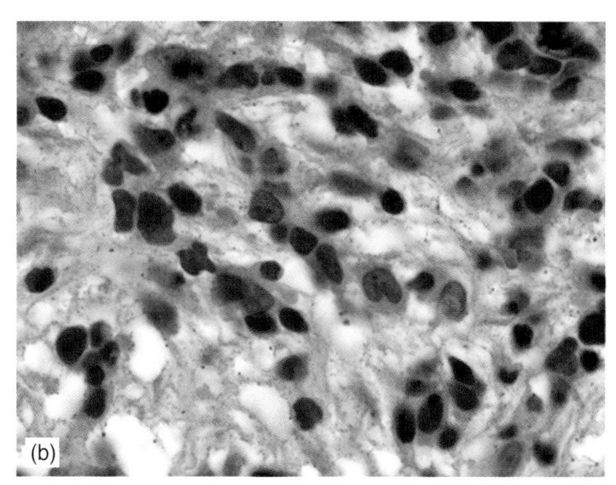

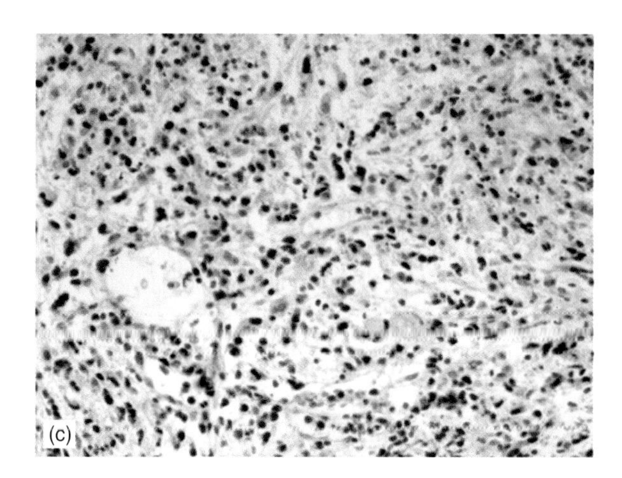

FIGURE 12.26 Bone marrow biopsy section from a patient with hypoplastic acute lymphoblastic leukemia (a) low power and (b) high power. Immunohistochemical stain showing numerous TdT positive cells (c).

cells express surface Ig and lack TdT and CD34 expression. Burkitt leukemia/lymphoma is discussed in Chapter 15 as a subcategory of mature B-cell neoplasms.

References

1. Jaffe ES, Harris NL, Stein H, Vardiman JW. (2001). *Pathology and Genetics. Tumors of Haematopoietic and Lymphoid Tissues.* IARC Press, Lyon.
2. Freedman AS, Harris NL. (2006). Clinical and pathologic features of precursor T and precursor B-lymphoblastic leukemia/lymphoma. *UpToDate.*
3. Greer JP, Foerster J, Rodgers GM, Paraskevas F, Glader B. (2004). *Wintrobe's Clinical Hematology*, 11th ed. Lippington Williams & Wilkins, Philadelphia.
4. Infante-Rivard C, Siemiatycki J, Lakhani R, Nadon L. (2005). Maternal exposure to occupational solvents and childhood leukemia. *Environ Health Persp* **113**, 787–92.
5. McNally RJ, Eden TO. (2004). An infectious aetiology for childhood acute leukaemia: A review of the evidence. *Br J Haematol* **127**, 243–63.
6. Gorini G, Stagnaro E, Fontana V, Miligi L, Ramazzotti V, Nanni O, Rodella S, Tumino R, Crosignani P, Vindigni C, Fontana A, Vineis P, Costantini AS. (2007). Alcohol consumption and risk of leukemia: A multicenter case-control study. *Leukemia Res* **31**, 379–86.
7. Greaves M. (2002). Childhood leukaemia. *BMJ* **324**, 283–7.
8. Tischkowitz M, Dokal I. (2004). Fanconi anaemia and leukaemia–clinical and molecular aspects. *Br J Haematol* **126**, 176–91.
9. Whitlock JA. (2006). Down syndrome and acute lymphoblastic leukaemia. *Br J Haematol* **135**, 595–602.
10. Haidar MA, Kantarjian H, Manshouri T, Chang CY, O'Brien S, Freireich E, Keating M, Albitar M. (2000). ATM gene deletion in patients with adult acute lymphoblastic leukemia. *Cancer* **88**, 1057–62.
11. Rao SR, Iyer RS, Gladstone B, Advani SH. (1993). Ataxia telangiectasia with acute lymphoblastic leukemia. *Indian Pediatr* **30**, 257–61.
12. Sabaawy HE, Azuma M, Embree LJ, Tsai HJ, Starost MF, Hickstein DD. (2006). TEL-AML1 transgenic zebrafish model of precursor B cell acute lymphoblastic leukemia. *P Natl Acad Sci USA* **103**, 15166–71.
13. Zelent A, Greaves M, Enver T. (2004). Role of the TEL-AML1 fusion gene in the molecular pathogenesis of childhood acute lymphoblastic leukaemia. *Oncogene* **23**, 4275–83.
14. Maia AT, van der Velden VH, Harrison CJ, Szczepanski T, Williams MD, Griffiths MJ, van Dongen JJ, Greaves MF. (2003). Prenatal origin of hyperdiploid acute lymphoblastic leukemia in identical twins. *Leukemia* **17**, 2202–6.
15. Teuffel O, Betts DR, Dettling M, Schaub R, Schäfer BW, Niggli FK. (2004). Prenatal origin of separate evolution of leukemia in identical twins. *Leukemia* **18**, 1624–9.
16. Naeim F. (2001). *Atlas of Bone Marrow and Blood Pathology.* W.B. Saunders, Philadelphia.
17. His ED. (2007). *Hematopathology.* Churchill Livingstone, Philadelphia, PA.
18. Krampera M, Perbellini O, Vincenzi C, Zampieri F, Pasini A, Scupoli MT, Guarini A, De Propris MS, Coustan-Smith E, Campana D, Foa R, Pizzolo G. (2006). Methodological approach to minimal residual disease detection by flow cytometry in adult B-lineage acute lymphoblastic leukemia. *Haematologica* **91**, 1109–12.
19. Mullighan CG, Flotho C, Downing JR. (2005). Genomic assessment of pediatric acute leukemia. *Cancer J* **11**, 268–82.
20. Armstrong SA, Look AT. (2005). Molecular genetics of acute lymphoblastic leukemia. *J Clin Oncol* **23**, 6306–15.
21. Berger R. (1991). Molecular cytogenetics of acute lymphoblastic leukemia. *Nouv Rev Fr Hematol* **33**, 86–91.
22. Le Beau MM, Larson RA. (2007). Cytogenetics in acute lymphoblastic leukemia. *UpToDate.*
23. Hayne CC, Winer E, Williams T, Chaves F, Khorsand J, Mark HF. (2006). Acute lymphoblastic leukemia with 4;11 translocation analyzed by a multi-modal strategy of conventional cytogenetics, FISH, morphology, flow cytometry and molecular genetics, and review of the literature. *Exp Mol Pathol* **81**, 62–71.
24. Carroll WL, Bhojwani D, Min DJ, Raetz E, Relling M, Davies S, Downing JR, Willman CL, Reed JC. (2003). Pediatric acute lymphoblastic leukemia. *Hematol Am Soc Hematol Educ Prog*, 102–31.
25. Jones LK, Saha V. (2005). Philadelphia positive acute lymphoblastic leukaemia of childhood. *Br J Haematol* **130**, 489–500.

26. Ribeiro RC, Abromowitch M, Raimondi SC, Murphy SB, Behm F, Williams DL. (1987). Clinical and biologic hallmarks of the Philadelphia chromosome in childhood acute lymphoblastic leukemia. *Blood* **70**(4), 948–53.
27. Westbrook CA, Hooberman AL, Spino C, Dodge RK, Larson RA, Davey F, Wurster-Hill DH, Sobol RE, Schiffer C, Bloomfield CD. (1992). Clinical significance of the BCR-ABL fusion gene in adult acute lymphoblastic leukemia: A Cancer and Leukemia Group B study (8762). *Blood* **80**, 2983–90.
28. Wetzler M, Dodge RK, Mrózek K, Carroll AJ, Tantravahi R, Block AW, Pettenati MJ, Le Beau MM, Frankel SR, Stewart CC, Szatrowski TP, Schiffer CA, Larson RA, Bloomfield CD. (1999). Prospective karyotype analysis in adult acute lymphoblastic leukemia: The Cancer and Leukemia Group B experience. *Blood* **93**, 3983–93.
29. Anastasi J, Feng J, Dickstein JI, Le Beau MM, Rubin CM, Larson RA, Rowley JD, Vardiman JW. (1996). Lineage involvement by BCR/ABL in Ph+ lymphoblastic leukemias: chronic myelogenous leukemia presenting in lymphoid blast vs Ph+ acute lymphoblastic leukemia. *Leukemia* **10**, 795–802.
30. Fainstein E, Marcelle C, Rosner A, Canaani E, Gale RP, Dreazen O, Smith SD, Croce CM. (1987). A new fused transcript in Philadelphia chromosome positive acute lymphocytic leukaemia. *Nature* **330**, 386–8.
31. Clark SS, McLaughlin J, Timmons M, Pendergast AM, Ben-Neriah Y, Dow LW, Crist W, Rovera G, Smith SD, Witte ON. (1988). Expression of a distinctive BCR-ABL oncogene in Ph1-positive acute lymphocytic leukemia (ALL). *Science* **239**, 775–7.
32. Van Dongen JJ, Macintyre EA, Gabert JA, *et al.* (2003). Standardized RT-PCR analysis of fusion gene transcripts from chromosome aberrations in acute leukemia for detection of minimal residual disease: report of the BIOMED-1 concerted action: Investigation of minimal residual disease in acute leukemia. *Leukemia* **13**, 1901–28.
33. Bose S, Deininger M, Gora-Tybor J, *et al.* (1998). The presence of typical and atypical BCR-ABL fusion genes in leukocytes of normal individuals: Biologic significance and implications for the assessment of minimal residual disease. *Blood* **92**, 3362–7.
34. Puccetti E, Beissert T, Güller S, Li JE, Hoelzer D, Ottmann OG, Ruthardt M. (2003). Leukemia-associated translocation products able to activate RAS modify PML and render cells sensitive to arsenic-induced apoptosis. *Oncogene* **22**, 6900–8.
35. Wu LX, Xu JH, Wu GH, Chen YZ. (2003). Inhibitory effect of curcumin on proliferation of K562 cells involves down-regulation of p210(bcr/abl) initiated Ras signal transduction pathway. *Acta Pharmacol Sinic* **24**, 1155–60.
36. Frestedt JL, Hilden JM, Kersey JH. (1996). AF4/FEL, a gene involved in infant leukemia: sequence variations, gene structure, and possible homology with a genomic sequence on 5q31. *DNA Cell Biol* **15**, 669–78.
37. Hilden JM, Frestedt JL, Moore RO, Heerema NA, Arthur DC, Reaman GH, Kersey JH. (1995). Molecular analysis of infant acute lymphoblastic leukemia: MLL gene rearrangement and reverse transcriptase-polymerase chain reaction for t(4;11)(q21;q23). *Blood* **86**, 3876–82.
38. Parkin JL, Arthur DC, Abramson CS, McKenna RW, Kersey JH, Heideman RL, Brunning RD. (1982). Acute leukemia associated with the t(4;11) chromosome rearrangement: Ultrastructural and immunologic characteristics. *Blood* **60**, 1321–31.
39. Pui CH, Frankel LS, Carroll AJ, Raimondi SC, Shuster JJ, Head DR, Crist WM, Land VJ, Pullen DJ, Steuber CP. (1991). Clinical characteristics and treatment outcome of childhood acute lymphoblastic leukemia with the t(4;11)(q21;q23): A collaborative study of 40 cases. *Blood* **77**, 440–7.
40. Chen CS, Sorensen PH, Domer PH, Reaman GH, Korsmeyer SJ, Heerema NA, Hammond GD, Kersey JH. (1993). Molecular rearrangements on chromosome 11q23 predominate in infant acute lymphoblastic leukemia and are associated with specific biologic variables and poor outcome. *Blood* **81**, 2386–93.
41. Abdelhaleem M. (2007). Frequent but nonrandom expression of myeloid markers on de novo childhood acute lymphoblastic leukemia. *Exp Mol Pathol* **83**, 138–41.
42. Raimondi SC, Behm FG, Roberson PK, Williams DL, Pui CH, Crist WM, Look AT, Rivera GK. (1990). Cytogenetics of pre-B-cell acute lymphoblastic leukemia with emphasis on prognostic implications of the t(1;19). *J Clin Oncol* **8**, 1380–8.
43. Piccaluga PP, Malagola M, Rondoni M, Ottaviani E, Testoni N, Laterza C, Visani G, Pileri SA, Martinelli G, Baccarani M. (2006). Poor outcome of adult acute lymphoblastic leukemia patients carrying the (1;19)(q23;p13) translocation. *Leukemia Lymphoma* **47**, 469–72.
44. Uckun FM, Sensel MG, Sather HN, Gaynon PS, Arthur DC, Lange BJ, Steinherz PG, Kraft P, Hutchinson R, Nachman JB, Reaman GH, Heerema NA. (1998). Clinical significance of translocation t(1;19) in childhood acute lymphoblastic leukemia in the context of contemporary therapies: A report from the Children's Cancer Group. *J Clin Oncol* **16**, 527–35.
45. Berendes P, Hoogeveen A, van Dijk M, van Denderen J, van Ewijk W. (1995). Specific immunologic recognition of the tumor-specific E2A-PBX1 fusion-point antigen in t(1;19)-positive pre-B cells. *Leukemia* **9**, 1321–7.
46. Pui CH, Raimondi SC, Hancock ML, Rivera GK, Ribeiro RC, Mahmoud HH, Sandlund JT, Crist WM, Behm FG. (1994). Immunologic, cytogenetic, and clinical characterization of childhood acute lymphoblastic leukemia with the t(1;19) (q23; p13) or its derivative. *J Clin Oncol* **12**, 2601–6.
47. Michael PM, Levin MD, Garson OM. (1984). Translocation 1;19 – a new cytogenetic abnormality in acute lymphocytic leukemia. *Cancer Genet Cytogen* **12**, 333–41.
48. Levin MD, Michael PM, Garson OM, Tiedemann K, Firkin FC. (1984). Clinicopathological characteristics of acute lymphoblastic leukemia with the 4;11 chromosome translocation. *Pathology* **16**, 63–6.
49. Douet-Guilbert N, Morel F, Le Bris MJ, Herry A, Le Calvez G, Marion V, Berthou C, De Braekeleer M. (2003). Translocation (12;21) followed by insertion of chromosome 3 material in the derivative chromosome 12 in a case of childhood acute lymphoblastic leukemia. *Cancer Genet Cytogen* **142**, 120–3.
50. Rubnitz JE, Pui CH, Downing JR. (1999). The role of TEL fusion genes in pediatric leukemias. *Leukemia* **13**, 6–13.
51. Endo C, Oda M, Nishiuchi R, Seino Y. (2003). Persistence of TEL-AML1 transcript in acute lymphoblastic leukemia in long-term remission. *Pediatr Int* **45**, 275–80.
52. Eguchi-Ishimae M, Eguchi M, Tanaka K, Hamamoto K, Ohki M, Ueda K, Kamada N. (1998). Fluorescence in situ hybridization analysis of 12;21 translocation in Japanese childhood acute lymphoblastic leukemia. *Jpn J Cancer Res* **89**, 783–8.
53. Harrison CJ. (2001). The detection and significance of chromosomal abnormalities in childhood acute lymphoblastic leukaemia. *Blood Rev* **15**, 49–59.
54. Oláh E, Balogh E, Kajtár P, Pajor L, Jakab Z, Kiss C. (1997). Diagnostic and prognostic significance of chromosome abnormalities in childhood acute lymphoblastic leukemia. *Ann NY Acad Sci* **824**, 8–27.
55. Nachman JB, Heerema NA, Sather H, Camitta B, Forestier E, Harrison CJ, Dastugue N, Schrappe M, Pui CH, Basso G, Silverman LB, Janka-Schaub GE. (2007). Outcome of treatment in children with hypodiploid acute lymphoblastic leukemia. *Blood* **110**, 1112–15.
56. Das PK, Sharma P, Koutts J, Smith A. (2003). Hypodiploidy of 37 chromosomes in an adult patient with acute lymphoblastic leukemia. *Cancer Genet Cytogen* **145**, 176–8.
57. Bagg A, Braziel RM, Arber DA, *et al.* (2002). Immunoglobulin heavy chain gene analysis in lymphomas: A multi-center study demonstrating the heterogeneity of performance of polymerase chain reaction assays. *J Mol Diagn* **4**, 81–9.
58. Bagg A. (2006). Immunoglobulin and T-cell receptor gene rearrangements: Minding your B's and T's in assessing lineage and clonality in neoplastic lymphoproliferative disorders. *J Mol Diagn* **8**, 426–9.
59. Harrison CJ, Moorman AV, Broadfield ZJ, Cheung KL, Harris RL, Reza Jalali G, Robinson HM, Barber KE, Richards SM, Mitchell CD, Eden TO, Hann IM, Hill FG, Kinsey SE, Gibson BE, Lilleyman J, Vora A, Goldstone AH, Franklin IM, Durrant J, Martineau M. Childhood and Adult Leukaemia Working Parties (2004). Three distinct

subgroups of hypodiploidy in acute lymphoblastic leukaemia. *Br J Haematol* **125**, 552–9.
60. Millot F, Robert A, Bertrand Y, Mechinaud F, Laureys G, Ferster A, Brock P, Rohrlich P, Mazingue F, Plantaz D, Plouvier E, Pacquement H, Behar C, Rialland X, Chantraine JM, Guilhot F, Otten J. (1997). Cutaneous involvement in children with acute lymphoblastic leukemia or lymphoblastic lymphoma. The Children's Leukemia Cooperative Group of the European Organization of Research and Treatment of Cancer (EORTC). *Pediatrics* **100**, 60–4.
61. Graux C, Cools J, Michaux L, Vandenberghe P, Hagemeijer A. (1989). Cytogenetics and molecular genetics of T-cell acute lymphoblastic leukemia: From thymocyte to lymphoblast. *Leukemia* **20**, 1496–510.
62. Hara J, Benedict SH, Yumura K, Ha-Kawa K, Gelfand EW. (1989). Rearrangement of variable region T cell receptor gamma genes in acute lymphoblastic leukemia. V gamma gene usage differs in mature and immature T cells. *J Clin Invest* **83**, 1277–83.
63. Hara J, Benedict SH, Champagne E, Takihara Y, Mak TW, Minden M, Gelfand EW. (1988). T cell receptor delta gene rearrangements in acute lymphoblastic leukemia. *J Clin Invest* **82**, 1974–82.
64. Wang J, Jani-Sait SN, Escalon EA, Carroll AJ, de Jong PJ, Kirsch IR, Aplan PD. (2000). The t(14;21)(q11, 2;q22) chromosomal translocation associated with T-cell acute lymphoblastic leukemia activates the BHLHB1 gene. *Proc Natl Acad Sci USA* **97**, 3497–502.
65. Nadler LM, Reinherz EL, Weinstein HJ, D'Orsi CJ, Schlossman SF. (1980). Heterogeneity of T-cell lymphoblastic malignancies. *Blood* **55**, 806–10.
66. Reinherz EL, Nadler LM, Sallan SE, Schlossman SF. (1979). Subset derivation of T-cell acute lymphoblastic leukemia in man. *J Clin Invest* **64**, 392–7.
67. Cossman J, Chused TM, Fisher RI, Magrath I, Bollum F, Jaffe ES. (1983). Diversity of immunological phenotypes of lymphoblastic lymphoma. *Cancer Res* **43**, 4486–90.
68. Sheibani K, Nathwani BN, Winberg CD, Burke JS, Swartz WG, Blayney D, van de Velde S, Hill LR, Rappaport H. (1987). Antigenically defined subgroups of lymphoblastic lymphoma. Relationship to clinical presentation and biologic behavior. *Cancer* **60**, 183–90.
69. Lewis RE, Cruse JM, Sanders CM, Webb RN, Tillman BF, Beason KL, Lam J, Koehler J. (2006). The immunophenotype of pre-TALL/LBL revisited. *Exp Mol Pathol* **81**(2), 162–5.
70. Khalidi HS, Chang KL, Medeiros LJ, Brynes RK, Slovak ML, Murata-Collins JL, Arber DA. (1999). Acute lymphoblastic leukemia. Survey of immunophenotype, French-American-British classification, frequency of myeloid antigen expression, and karyotypic abnormalities in 210 pediatric and adult cases. *Am J Clin Pathol* **111**, 467–76.
71. Thandla S, Aplan PD. (1997). Molecular biology of acute lymphocytic leukemia. *Semin Oncol* **24**, 45–56.
72. Heerema NA, Sather HN, Sensel MG, Kraft P, Nachman JB, Steinherz PG, Lange BJ, Hutchinson RS, Reaman GH, Trigg ME, Arthur DC, Gaynon PS, Uckun FM. (1998). Frequency and clinical significance of cytogenetic abnormalities in pediatric T-lineage acute lymphoblastic leukemia: A report from the Children's Cancer Group. *J Clin Oncol* **16**, 1270–8.
73. Wong KF, Leung JN. (2005). Precursor T-lymphoblastic leukemia with an inv(6)(p21.2q27). *Cancer Genet Cytogen* **158**, 192–3.
74. Greiner TC, Rubocki RJ. (2002). Effectiveness of capillary electrophoresis using fluorescent-labeled primers in detecting T-cell receptor gamma gene rearrangements. *J Mol Diagn* **4**, 137–43.
75. Chiaretti S, Li X, Gentleman R, Vitale A, Vignetti M, Mandelli F, Ritz J, Foa R. (2004). Gene expression profile of adult T-cell acute lymphocytic leukemia identifies distinct subsets of patients with different response to therapy and survival. *Blood* **103**, 2771–8.
76. Matutes E. (2007). Adult T-cell leukaemia/lymphoma. *J Clin Pathol* **60**, 1373–7.
77. Bassan R, Gatta G, Tondini C, Willemze R. (2004). Adult acute lymphoblastic leukaemia. *Crit Rev Oncol Hematol* **50**, 223–61.
78. Oshimi K, Kawa K, Nakamura S, Suzuki R, Suzumiya J, Yamaguchi M, Kameoka J, Tagawa S, Imamura N, Ohshima K, Kojya S, Iwatsuki K, Tokura Y, Sato E, Sugimori HNK-cell Tumor Study Group (2003). NK-cell neoplasms in Japan. *Hematology* **10**, 237–45.
79. Oshimi K. (2003). Leukemia and lymphoma of natural killer lineage cells. *Int J Hematol* **78**, 18–23.
80. Gloeckner-Hofmann K, Ottesen K, Schmidt S, Nizze H, Feller AC, Merz H. (2000). T-cell/natural killer cell lymphoblastic lymphoma with an unusual coexpression of B-cell antigens. *Ann Hematol* **79**, 635–9.
81. Pombo-de-Oliveira MS, Campos MM, Bossa YE, Alencar DM, Curvello C, Agudelo DP, Mendonca N, Pereira E, Macedo-Silva ML. (2004). Acute leukemia with natural killer cells antigens in Brazilian children. *Leukemia Lymphoma* **45**, 739–43.
82. Pitman SD, Huang Q. (2007). Granular acute lymphoblastic leukemia: A case report and literature review. *Am J Hematol* **82**, 834–7.
83. Cap J, Babusikova O, Kaiserova E, Panzer-Grunmayer R. (2000). Granular acute lymphoblastic leukemia in a 15-year-old boy. *Med Oncol* **17**, 144–6.
84. Jain P, Kumar R, Gujral S, Kumar A, Singh A, Jain Y, Dubey S, Anand M, Arya LS. (2000). Granular acute lymphoblastic leukemia with hypereosinophilic syndrome. *Ann Hematol* **79**, 272–4.
85. Fulcher JW, Allred TJ, Kulharya A, Satya-Prakash KL, Seigler M, Neibarger D, Mazzella FM. (2006). Granular acute lymphoblastic leukemia in adults: Report of a case and review of the literature. *South Med J* **99**, 894–7.
86. Lima M. (2004). Laboratory diagnosis of large granular lymphocytic leukemia. *Lab Hematol* **10**, 148–9.
87. Wilson F, Tefferi A. (2005). Acute lymphocytic leukemia with eosinophilia: Two case reports and a literature review. *Leukemia Lymphoma* **46**, 1045–50.
88. Bae SY, Yoon SY, Huh JH, Sung HJ, Choi IK. (2007). Hypereosinophilia in biphenotypic (B-cell/T-cell) acute lymphoblastic leukemia. *Leukemia Lymphoma* **48**(7), 1417–19.
89. Rezk S, Wheelock L, Fletcher JA, Oliveira AM, Keuker CP, Newburger PE, Xu B, Woda BA, Miron PM. (2006). Acute lymphocytic leukemia with eosinophilia and unusual karyotype. *Leukemia Lymphoma* **47**, 1176–9.
90. Krober SM, Horny HP, Steinke B, Kaiserling E. (2003). Adult hypocellular acute leukaemia with lymphoid differentiation. *Leukemia Lymphoma* **44**, 1797–801.
91. Ishikawa K, Seriu T, Watanabe A, Hayasaka K, Takeda O, Sato T, Takahashi I, Suzuki T, Nishinomiya F, Sato W. (1995). Detection of neoplastic clone in the hypoplastic and recovery phases preceding acute lymphoblastic leukemia by in vitro amplification of rearranged T-cell receptor delta chain gene. *J Pediatr Hematol Oncol* **17**, 270–5.
92. Liang R, Cheng G, Wat MS, Ha SY, Chan LC. (1993). Childhood acute lymphoblastic leukaemia presenting with relapsing hypoplastic anaemia: Progression of the same abnormal clone. *Br J Haematol* **83**, 340–2.

Acute Leukemias of Ambiguous Lineage

13

CHAPTER

Faramarz Naeim

The addition of cytochemical and immunophenotypic techniques to the standard morphologic evaluation of leukemic blast cells has led to the growing recognition of acute leukemias with ambiguous lineage assignment [1, 2]. According to the WHO classification, these leukemias fall into two major categories [1]:

1. Acute leukemias which lack sufficient evidence of morphologic, cytochemical, and immunophenotypic features of lineage differentiation (acute undifferentiated leukemia).
2. Acute leukemias which have morphologic and/or immunophenotypic characteristics of both myeloid and lymphoid lineages, or both T- and B-lymphoid cells (acute bilineal and acute biphenotypic leukemias) (Figure 13.1).

ETIOLOGY AND PATHOGENESIS

The etiology and pathogenesis of this heterogeneous group of leukemias are not known. However, recent reports suggest that the mixed-lineage leukemia (*MLL*) gene may play an important role in the development of acute leukemias of ambiguous lineage. For example, studies by Ono and associates suggest that *MLL* fusion products are essential to immortalize hematopoietic progenitors [3]. These investigators demonstrated that MLL-SEPT6 fusion protein with activated FMS-like tyrosine kinase 3 (*FLT3*) together could induce acute biphenotypic leukemia in mice. The *MLL* gene (also called *ALL1* or *HRX*) has been mapped on chromosome 11 (11q23). More than 30 partner genes have been identified in association with 11q23 translocations such as t(4;11), t(9;11), and t(11;19) [3, 4]. Fusion of

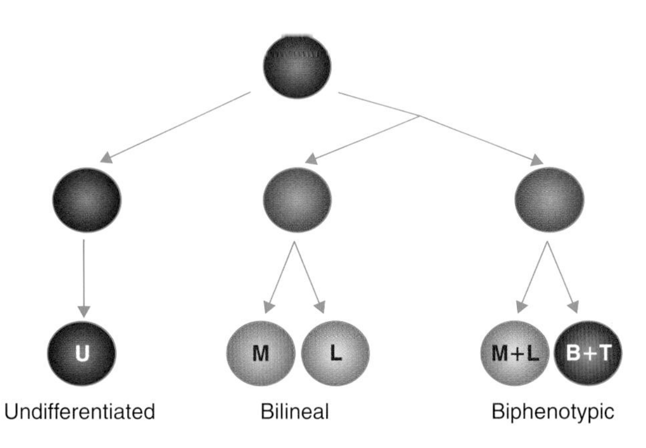

FIGURE 13.1 Scheme of clonal development of acute leukemias of undifferentiated (U), bilineal and biphenotypic types.

MLL with a translocation partner leads to the development of leukemias in experimental animals [3].

PATHOLOGY

Morphology

Morphologic features of acute leukemias of ambiguous lineage are variable. The ones that fall into the category of acute undifferentiated leukemia show no morphologic features of lymphoid or myeloid differentiation. Blasts consist of primitive undifferentiated cells with no cytoplasmic granules or lineage-associated cytochemical and immunophenotypic features. The bilineal leukemias consist of two distinct populations of blast cells representing both lymphoid and myeloid lineages (Figure 13.2). Lymphoblasts are usually smaller with scanty cytoplasm, no cytoplasmic granules, and less prominent nucleoli; and blasts of myeloid origin (myeloblasts and monoblasts) are larger with more abundant cytoplasm, variable amounts of cytoplasmic granules, and prominent nucleoli.

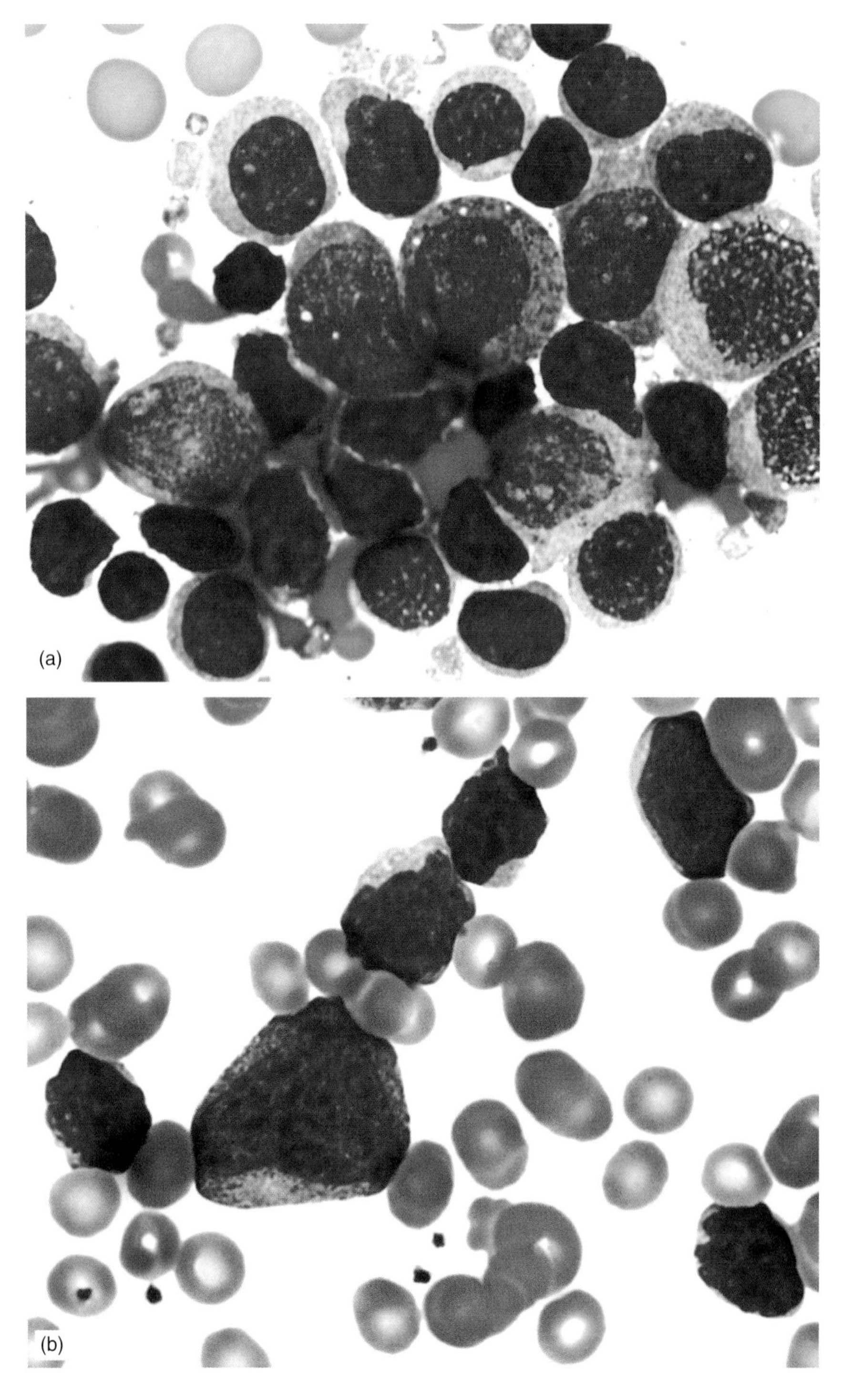

FIGURE 13.2 Bone marrow (a) and peripheral blood (b) smears of a patient with bilineal leukemia. Two distinct populations of leukemic cells (larger and smaller) are present.

These two populations demonstrate distinct cytochemical and immunophenotypic properties (see later).

Immunophenotype and Cytochemical Stains

The primary basis for the characterization of acute leukemias with ambiguous lineage is their immunophenotypic features. Although the term *biphenotypic* leukemia has been used widely in the literature, the general consensus is that this term should be used more restrictively and only when the leukemic blast cells co-express several markers of both myeloid and lymphoid or B- and T-cells. A scoring system has been proposed by the European Group for the Immunologic Classification of Leukemia (EGIL) for the lineage assignments of leukemias [5] (Table 13.1). The most lineage-specific markers are considered to be cytoplasmic CD22, cytoplasmic CD79a, and cytoplasmic μ for B-cells; CD3 and TCR for T-cells; and MPO for myeloid cells; each receiving a score of 2. The least lineage-specific markers are considered to be TdT and CD24 for B-cells; TdT, CD7, and CD1a for T-cells; and CD14, CD15, and CD64 for myeloid cells; each receiving a score of 0.5. In this scoring system, leukemia is considered biphenotypic when the total scores of the CD markers co-expressed by the leukemic cells are >2 in more than one lineage-associated category. For example, if the same blast cell population in a patient with acute leukemia co-expresses CD79a and CD19 (B-cell markers with total score of 3) and myeloperoxidase and CD15 (myeloid markers with total score of 2.5), the leukemia is considered biphenotypic.

In spite of the recommendation of the WHO for the scoring system proposed by EGIL, there are still some ambiguities regarding the interpretation of phenotypic results for acute leukemias with ambiguous lineage. For example, the required minimal percentage of positive cells co-expressing myeloid and lymphoid markers has not been defined, nor has the required intensity of the expression of myeloid- and lymphoid-associated markers been clarified.

The vast majority of the acute leukemias with aberrant expression of lymphoid or myeloid markers are not qualified to be called biphenotypic or bilineal. They are mostly considered to be AMLs with aberrant expression of lymphoid markers or ALLs with aberrant expression of myeloid markers (Table 13.2).

TABLE 13.1 Scoring system for biphenotypic acute leukemias.*

Score	B-lymphoid	T-lymphoid	Myeloid
2	cyCD79a	CD3 (s/cy)	MPO
	cyIgM	Anti-TCR	
	cyCD22		
1	CD19	CD2	CD117
	CD20	CD5	CD13
	CD10	CD8	CD33
		CD10	CD65
0.5	TdT	TdT	CD14
	CD24	CD7	CD15
		CD1a	CD64

*Adapted from Ref. [5] and European Group for the Immunologic Classification of Leukemia (EGIL).
cy: cytoplasmic; s: surface.

Acute Leukemias with Expression of Aberrant Markers

This category consists of:

A. Acute myeloid leukemias with aberrant expression of lymphoid markers (AML + L) [6–10]. In this type of leukemia the total immunophenotypic score for the myeloid markers is >2 whereas that for lymphoid-associated markers is ≤2 (Figure 13.3). In our experience the most frequent aberrant lymphoid markers expressed by AML cells are CD7, CD19, and CD56. Other frequently reported lymphoid markers in AMLs include CD2, CD22, and CD79a.

B. Acute lymphoid leukemias with aberrant expression of myeloid markers (ALL + M) [11–13]. In this type of leukemia the total score for the lymphoid markers is >2 whereas that for myeloid-associated markers is <2 (Figure 13.4). In our experience the most frequent aberrant myeloid markers expressed by ALL cells are CD13, CD15, and CD33. Other frequently reported myeloid markers in ALLs include CD14, CD65, and CD66c.

C. Acute B-cell lymphoid leukemias with aberrant expression of T-cell markers (B-ALL + T) or acute T-cell lymphoid leukemias with aberrant expression of B-cell markers (T-ALL + B) [14].

TABLE 13.2 Acute leukemias with aberrant markers, biphenotypic features, or bilineal populations.

1 Acute leukemias with aberrant expression: a. AMLs expressing lymphoid-associated markers (AML + L) such as CD7, CD19, and CD56 b. ALLs expressing myeloid-associated markers (ALL + M) such as CD13, CD15, and CD33
2 Acute biphenotypic leukemias: One population of leukemic cells co-expressing markers of both myeloid and lymphoid lineages or both B- and T-cells with scores of >2 for each lineage
3 Acute bilineal (mixed lineage) leukemias: Two separate populations of leukemic cells representing two distinct lineages

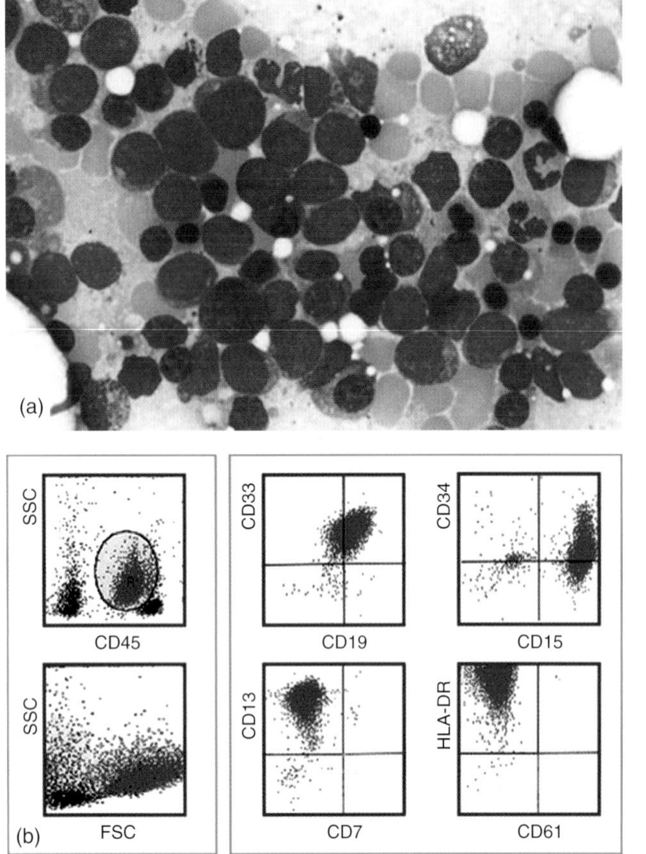

FIGURE 13.3 Acute myeloid leukemia with aberrant expression of CD19. (a) Bone marrow smear showing numerous blast cells and (b) flow cytometry revealing a population of $CD13^+$, $CD15^+$, $CD33^+$, and $CD34^+$ cells with aberrant expression of CD19.

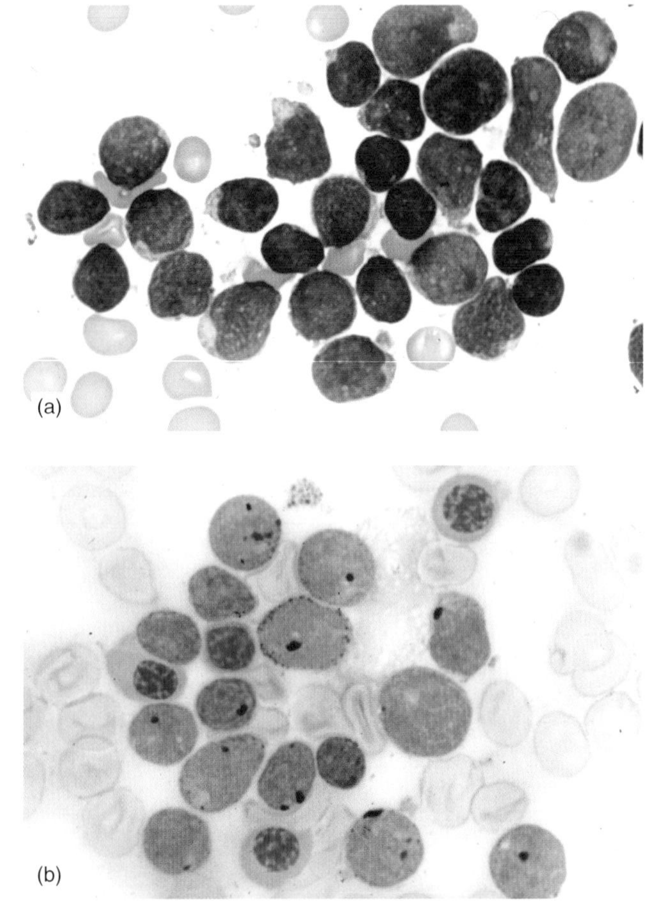

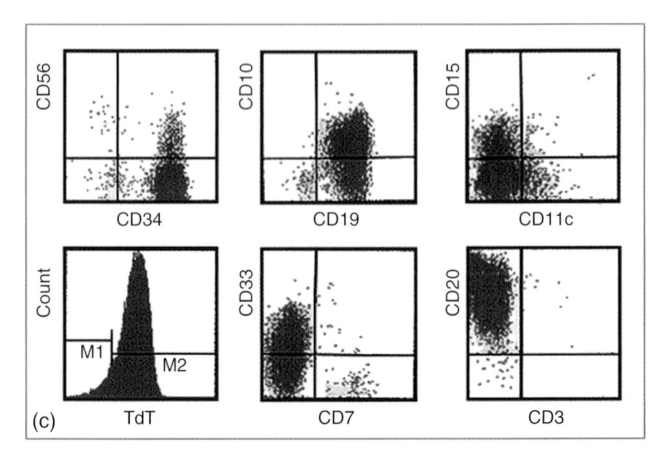

FIGURE 13.4 Acute lymphoid leukemia with aberrant expression of CD33. (a) Bone marrow smear showing numerous blast cells, (b) blasts containing coarse PAS^+ cytoplasmic granules, and (c) flow cytometry revealing a population of $CD10^+$, $CD19^+$, $CD20^+$, $CD34^+$, and TdT^+ cells with aberrant expression of CD15 (partial) and CD33.

Biphenotypic Acute Leukemia

In biphenotypic acute leukemia the population of leukemic cells co-expresses both myeloid and lymphoid or both B- and T-cell markers (Figures 13.5 and 13.6). The immunophenotypic score for each lineage is >2 [7, 15–19] (Table 13.1). Occasionally, the leukemic population may demonstrate a trilineage immunophenotypic picture, co-expressing a combination of myeloid, B-, and T-cell markers (triphenotypic) [20].

Bilineal Acute Leukemia

In bilineal acute leukemia two distinct populations of blast cells are present, each population representing a distinct lineage association. Most of the bilineal acute leukemias consist of two populations of myeloblasts and lymphoblasts, but rarely they may display a mixture of B- and T-lymphoblasts (Figures 13.7 and 13.8). The presence of two populations of blast cells in bilineal acute leukemia is often synchronous; both populations of blast cells are present at the same time. On rare occasions, there may be two simultaneous neoplastic processes in two separate sites. For example, the bone marrow may be involved with ALL and the CNS is infiltrated by AML [21].

Asynchronous bilineal acute leukemias are more frequently reported, and in those cases, one lineage switches to another during the disease process [22–29]. The predominant pattern in most studies is ALL switching to AML in relapse. The frequency of lineage switch in relapse is about 7%. Lineage switch may represent relapse of the original

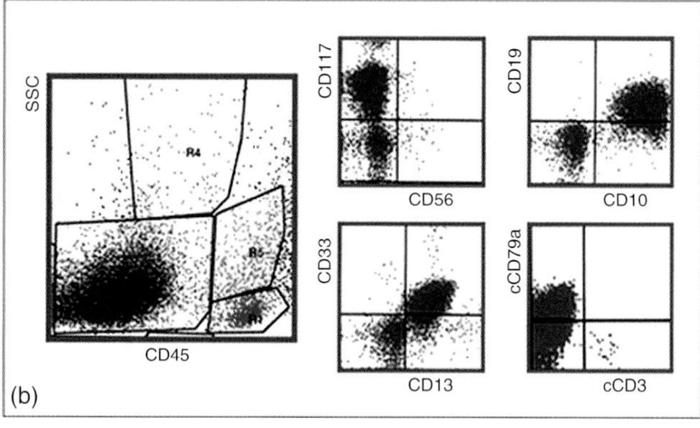

FIGURE 13.5 Biphenotypic acute leukemia. (a) Bone marrow smear demonstrating numerous pleomorphic blast cells with round or irregular nuclei and fine chromatin and (b) flow cytometry showing a population of CD45-negative cells co-expressing myeloid and B-cell markers.

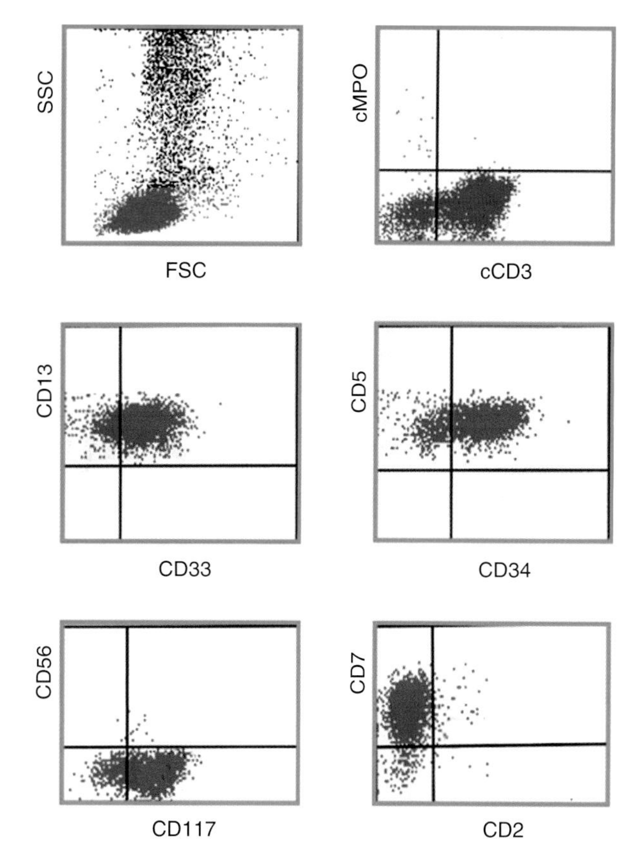

FIGURE 13.6 Flow cytometric analysis of a case of acute biphenotypic leukemia demonstrating CD34 positivity and co-expression of myeloid and T-cell markers.

clone or development of a second new clone. An increased risk of the development of AML has been reported in patients with ALL who receive intensive chemotherapy, particularly in patients with 11q23 abnormalities or the Philadelphia chromosome.

There are also occasional reports of switching from bilineal to biphenotypic acute leukemias and *vice versa* [30].

Cytochemical stains such as MPO, Sudan Black B, and NSE are helpful in distinguishing blasts of myeloid origin. Myeloblasts are often MPO and Sudan Black B positive and monoblasts/promonocytes usually express NSE.

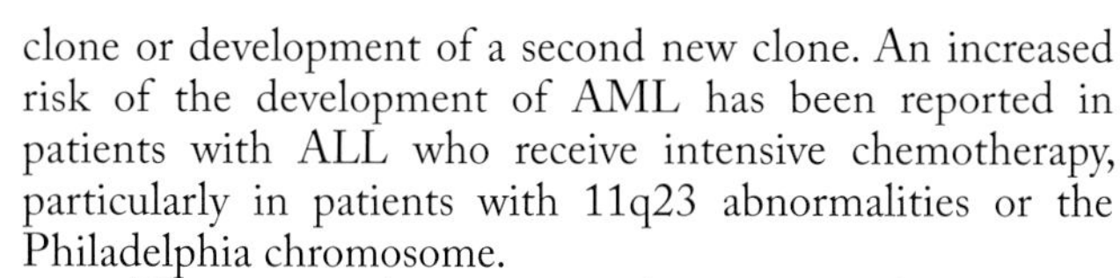

Molecular and Cytogenetic Studies

IgH and *TCRγ* gene rearrangements and cyclin A1 and *HOXA9* gene expression have been reported in biphenotypic acute leukemias [31]. Translocation of 11q23 and the Philadelphia chromosome are the most frequent cytogenetic abnormalities observed in biphenotypic and bilineal acute leukemias of B-precursor/myeloid type. More than 30 partner genes have been identified in association with 11q23 translocations such as t(4;11), t(9;11), and t(11;19) [3, 28, 32, 33]. The Philadelphia chromosome may be a part of complex cytogenetic abnormalities, such as combination of t(9;22) and del(7) or t(2;9;22) (Figure 13.9) [34]. The T-precursor/myeloid biphenotypic or bilineal acute leukemias may be associated with t(5;18)(q31;q23) and t(3;12)(p25;q24.3) [35, 36]. Also, trisomy 10 has been reported in association with acute biphenotypic leukemia [37].

CLINICAL ASPECTS

Biphenotypic acute leukemias probably represent 4–8% of all acute leukemias [20]. The bilineal acute leukemias are less frequent. They occur at any age, but are more frequent in adults. Most of these leukemias are of precursor B-cell/myeloid type. They usually have an aggressive clinical course, probably due to the frequency of unfavorable karyotypes such as 11q23 abnormalities or the Philadelphia chromosome [38, 39]. These leukemias often show low complete remission rate, high recurrence rate, and short disease-free and overall survival time (Table 13.3).

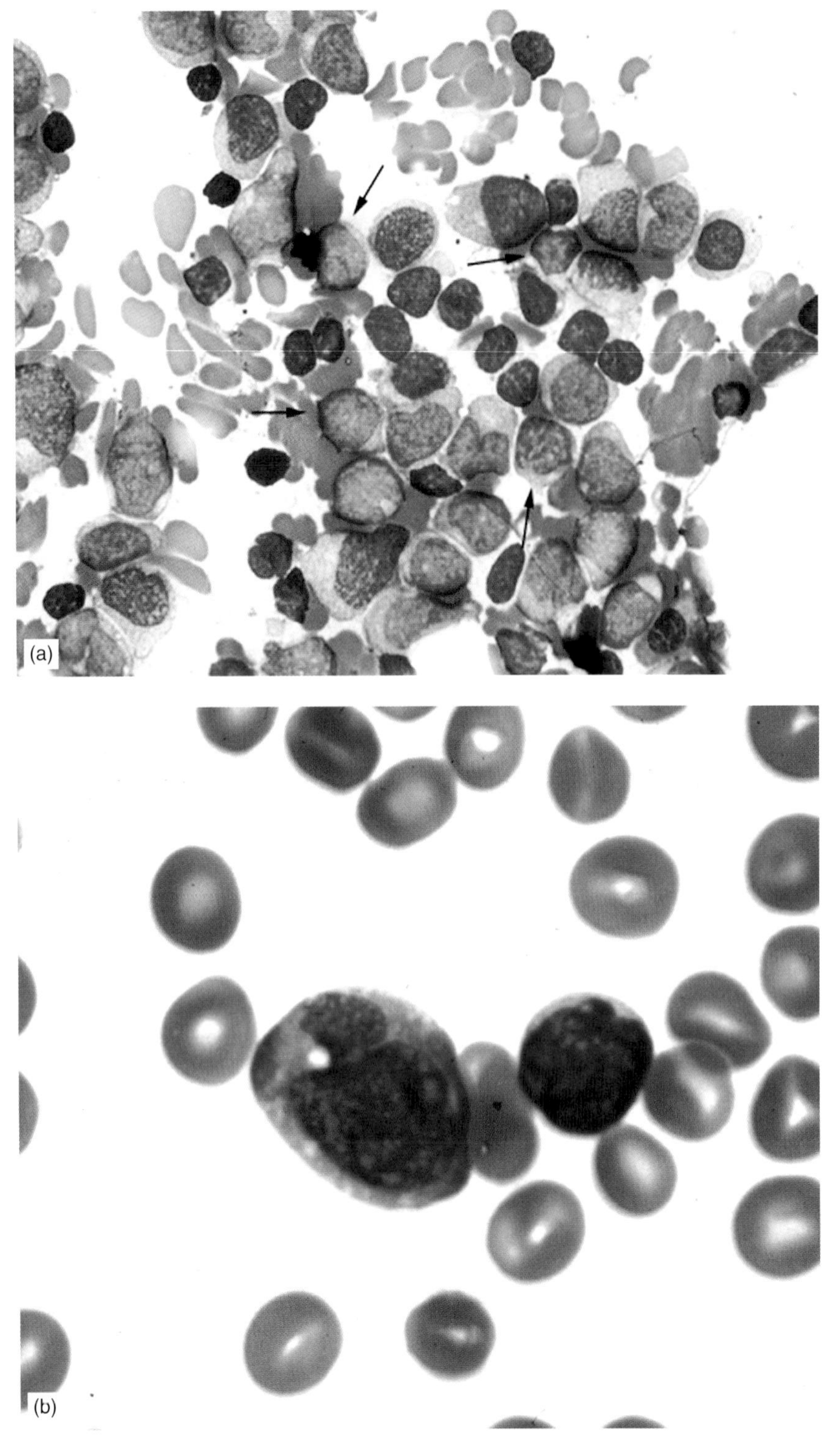

FIGURE 13.7 Acute bilineal leukemia. Bone marrow smear showing scattered small blasts with round nuclei and scanty cytoplasm (arrows) and numerous larger blasts with round or irregular nuclei and abundant cytoplasm (a). Blood smears revealing a small blast with scanty cytoplasm and inconspicuous nucleoli and a larger blast with abundant cytoplasm, irregular nucleus, and prominent nucleoli (b).

DIFFERENTIAL DIAGNOSIS

The diagnosis of biphenotypic acute leukemia is established by immunophenotypic studies. These leukemias should be distinguished from AMLs with aberrant expression of lymphoid-associated markers (AML + L) and ALLs with aberrant expression of myeloid-associated markers (ALL + L).

Bilineal acute leukemias may show morphologic evidence of two separate leukemia populations such as larger and smaller blasts. But diagnosis is confirmed by the evidence of two populations of blast cells distinctly expressing different lineage-associated markers by immunophenotypic studies.

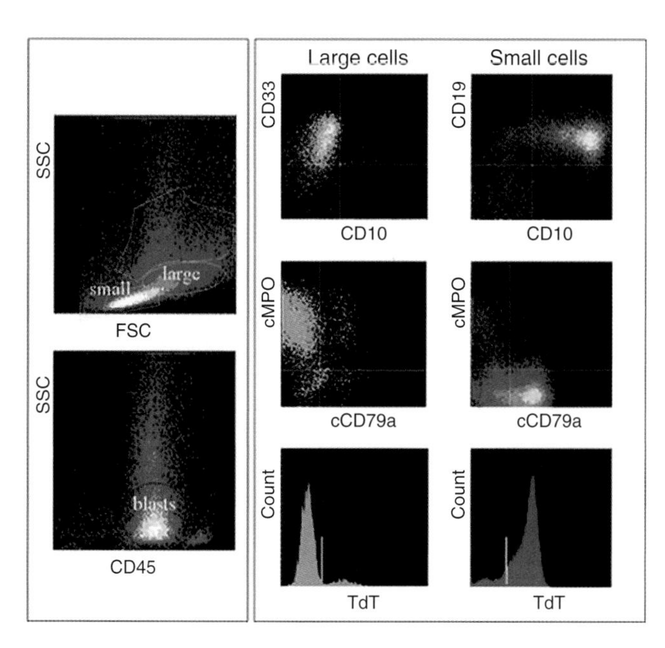

FIGURE 13.8 Acute bilineal leukemia. Flow cytometric studies revealing two distinct populations of larger and smaller blast cells with immunophenotypic features consistent with acute myeloid and acute lymphoid leukemias, respectively.

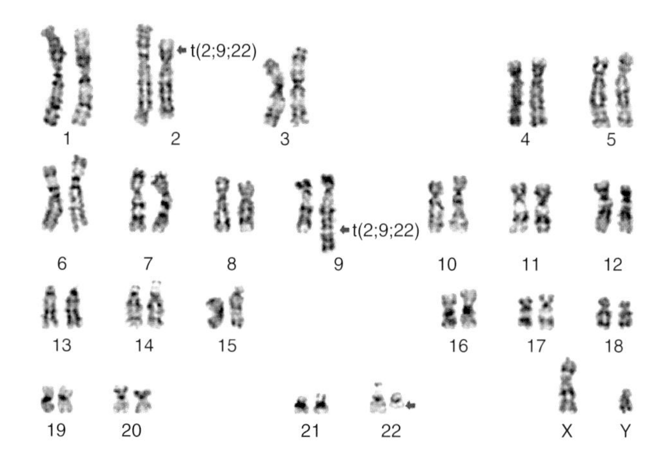

FIGURE 13.9 G-banded karyotype in a patient with CML in blast transformation. There is a three-way translocation involving chromosomes 2, 9, and 22 (arrows).

TABLE 13.3 Clinical and cytogenetic features of biphenotypic acute leukemias.

Clinical features
About 4–8% of acute leukemias
Poor prognosis
Low rate of complete remission
High recurrence rate
Short disease-free and overall survival time
Cytogenetic abnormalities
Precursor B-cell/myeloid
t(9;22), 11q23 (MLL) abnormalities
Precursor T-cell/myeloid
t(8;13), t(3;12)
Other less frequent abnormalities
14q32, t(3;7), t(5;18), t(6;14), t(12;17)

References

1. Jaffe ES, Harris NL, Stein H, Vardiman JW. (2001). *Pathology and Genetics. Tumors of Haematopoietic and Lymphoid Tissues*. IARC Press, Lyon.
2. Naeim F. (1997). *Pathology of Bone Marrow*, 2nd ed. Williams & Wilkins, Baltimore.
3. Ono R, Nakajima H, Ozaki K, Kumagai H, Kawashima T, Taki T, Kitamura T, Hayashi Y, Nosaka T. (2005). Dimerization of MLL fusion proteins and FLT3 activation synergize to induce multiple-lineage leukemogenesis. *J Clin Invest* **115**, 919–29.
4. Gozzetti A, Calabrese S, Raspadori D, Crupi R, Tassi M, Bocchia M, Fabbri A, Lauria F. (2007). Concomitant t(4;11) and t(1;19) in a patient with biphenotypic acute leukemia. *Cancer Genet Cytogenet* **177**, 81–2.
5. Matutes E, Morilla R, Farahat N, Carbonell F, Swansbury J, Dyer M, Catovsky D. (1997). Definition of acute biphenotypic leukemia. *Haematologica* **82**, 64–6.
6. Cruse JM, Lewis RE, Pierce S, Lam J, Tadros Y. (2005). Aberrant expression of CD7, CD56, and CD79a antigens in acute myeloid leukemias. *Exp Mol Pathol* **79**, 39–41.
7. Frater JL, Yaseen NR, Peterson LC, Tallman MS, Goolsby CL. (2003). Biphenotypic acute leukemia with coexpression of CD79a and markers of myeloid lineage. *Arch Pathol Lab Med* **127**, 356–9.
8. Kozlov I, Beason K, Yu C, Hughson M. (2005). CD79a expression in acute myeloid leukemia t(8;21) and the importance of cytogenetics in the diagnosis of leukemias with immunophenotypic ambiguity. *Cancer Genet Cytogenet* **163**, 62–7.
9. He G, Wu D, Sun A, Xue Y, Jin Z, Qiu H, Miao M, Tang X, Fu Z, Chen Z. (2007). CytCD79a expression in acute leukemia with t(8;21): Biphenotypic or myeloid leukemia? *Cancer Genet Cytogenet* **17**, 76–7.
10. Al-Qurashi FH, Owaidah T, Iqbal MA, Aljurf M. (2004). Trisomy 4 as the sole karyotypic abnormality in a case of acute biphenotypic leukemia with T-lineage markers in minimally differentiated acute myelocytic leukemia. *Cancer Genet Cytogenet* **150**, 66–9.
11. Vitale A, Guarini A, Ariola C, Meloni G, Perbellini O, Pizzuti M, De Gregoris C, Mettivier V, Pastorini A, Pizzolo G, Vignetti M, Mandelli F, Foà R. (2007). Absence of prognostic impact of CD13 and/or CD33 antigen expression in adult acute lymphoblastic leukemia. Results of the GIMEMA ALL 0496 trial. *Haematologica* **92**, 342–8.
12. Lewis RE, Cruse JM, Sanders CM, Webb RN, Tillman BF, Beason KL, Lam J, Koehler J. (2006). The immunophenotype of pre-TALL/LBL revisited. *Exp Mol Pathol* **81**, 162–5.
13. Kalina T, Vaskova M, Mejstrikova E, Madzo J, Trka J, Stary J, Hrusak O. (2005). Myeloid antigens in childhood lymphoblastic leukemia: Clinical data point to regulation of CD66c distinct from other myeloid antigens. *BMC Cancer* **12**(5(1)), 38.
14. Lau LG, Tan LK, Koay ES, Ee MH, Tan SH, Liu TC. (2004). Acute lymphoblastic leukemia with the phenotype of a putative B-cell/T-cell bipotential precursor. *Am J Hematol* **77**, 156–60.
15. Aribi A, Bueso-Ramos C, Estey E, Estrov Z, O'Brien S, Giles F, Faderl S, Thomas D, Kebriaei P, Garcia-Manero G, Pierce S, Cortes J, Kantarjian H, Ravandi F. (2007). Biphenotypic acute leukaemia: A case series. *Br J Haematol* **138**, 213–16.
16. Nakagawa Y, Hasegawa M, Kurata M, Yamamoto K, Abe S, Inoue M, Takemura T, Hirokawa K, Suzuki K, Kitagawa M. (2005). Expression

of IAP-family proteins in adult acute mixed lineage leukemia (AMLL). *Am J Hematol* **78**, 173–80.

17. Butcher BW, Wilson KS, Kroft SH, Collins RH Jr, Bhushan V. (2005). Acute leukemia with B-lymphoid and myeloid differentiation associated with an inv(5)(q13q33) in an adult patient. *Cancer Genet Cytogenet* **157**, 62–6.
18. Rubio MT, Dhedin N, Boucheix C, Bourhis JH, Reman O, Boiron JM, Gallo JH, Lheritier V, Thomas X, Fiere D, Vernant JP. (2003). Adult T-biphenotypic acute leukaemia: Clinical and biological features and outcome. *Br J Haematol* **123**, 842–9.
19. Weir EG, Ali Ansari-Lari M, Batista DA, Griffin CA, Fuller S, Smith BD, Borowitz MJ. (2007). Acute bilineal leukemia: A rare disease with poor outcome. *Leukemia* **21**, 2264–70.
20. Lee PS, Lin CN, Liu C, Huang CT, Hwang WS. (2003). Acute leukemia with myeloid, B-, and natural killer cell differentiation. *Arch Pathol Lab Med* **127**, E93–E95.
21. Ikarashi Y, Kakihara T, Imai C, Tanaka A, Watanabe A, Uchiyama M. (2004). Double leukemias simultaneously showing lymphoblastic leukemia of the bone marrow and monocytic leukemia of the central nervous system. *Am J Hematol* **75**, 164–7.
22. Pane F, Frigeri F, Camera A, Sindona M, Brighel F, Martinelli V, Luciano L, Selleri C, Del Vecchio L, Rotoli B, Salvatore F. (1996). Complete phenotypic and genotypic lineage switch in a Philadelphia chromosome-positive acute lymphoblastic leukemia. *Leukemia* **10**, 741–5.
23. Shende AC, Zaslav AL, Redner A, Bonagura VR, Hatam L, Paley C, Lanzkowsky P. (1995). A *de novo* lineage switch from B-cell acute lymphoblastic leukemia to acute myelocytic leukemia: A case report. *Am J Hematol* **50**, 75–7.
24. Cuneo A, Balboni M, Piva N, Carli MG, Tomasi P, Previati R, Negrini M, Scapoli G, Spanedda R, Castoldi G. (1994). Lineage switch and multilineage involvement in two cases of pH chromosome-positive acute leukemia: Evidence for a stem cell disease. *Haematologica* **79**, 76–82.
25. Krawczuk-Rybak M, Zak J, Jaworowska B. (2003). A lineage switch from AML to ALL with persistent translocation t(4;11) in congenital leukemia. *Med Pediatr Oncol* **41**, 95–6.
26. Lounici A, Cony-Makhoul P, Dubus P, Lacombe F, Merlio JP, Reiffers J. (2000). Lineage switch from acute myeloid leukemia to acute lymphoblastic leukemia: Report of an adult case and review of the literature. *Am J Hematol* **65**, 319–21.
27. Stasik C, Ganguly S, Cunningham MT, Hagemeister S, Persons DL. (2006). Infant acute lymphoblastic leukemia with t(11;16)(q23;p13.3) and lineage switch into acute monoblastic leukemia. *Cancer Genet Cytogenet* **168**, 146–9.
28. Stass S, Mirro J, Melvin S, Pui CH, Murphy SB, Williams D. (1984). Lineage switch in acute leukemia. *Blood* **64**, 701–6.
29. Mantadakis E, Danilatou V, Stiakaki E, Paterakis G, Papadhimitriou S, Kalmanti M. (2005). T-cell acute lymphoblastic leukemia relapsing as acute myelogenous leukemia. *Blood Cancer* **27**, 551–3.
30. Ciolli S, Leoni F, Caporale R, Carbone A, Francia di Celle P, Foa R, Ferrini PR. (1993). Mixed acute leukemia with genotypic lineage switch: A case report. *Leukemia* **7**, 1061–5.
31. Golemovic M, Sucic M, Zadro R, Mrsic S, Mikulic M, Labar B, Rajic LJ, Batinic D. (2006). IgH and TCRgamma gene rearrangements, cyclin A1 and HOXA9 gene expression in biphenotypic acute leukemias. *Leuk Res* **30**, 211–21.
32. Yamaguchi M, Yamamoto K, Miura O. (2003). Aberrant expression of the LHX4 LIM-homeobox gene caused by t(1;14)(q25;q32) in chronic myelogenous leukemia in biphenotypic blast crisis. *Genes Chromosomes Cancer* **38**, 269–73.
33. Jiang JG, Roman E, Nandula SV, Murty VV, Bhagat G, Alobeid B. (2005). Congenital MLL-positive B-cell acute lymphoblastic leukemia (B-ALL) switched lineage at relapse to acute myelocytic leukemia (AML) with persistent t(4;11) and t(1;6) translocations and JH gene rearrangement. *Leuk Lymphoma* **46**, 1223–7.
34. Monma F, Nishii K, Ezuki S, Miyazaki T, Yamamori S, Usui E, Sugimoto Y, Lorenzo VF, Katayama N, Shiku H. (2006). Molecular and phenotypic analysis of Philadelphia chromosome-positive bilineage leukemia: Possibility of a lineage switch from T-lymphoid leukemic progenitor to myeloid cells. *Cancer Genet Cytogenet* **164**, 118–21.
35. Salamanchuk Z, Jakobczyk M, Mensah P, Skotnicki AB. (2001). Novel translocation (5;18)(q31;q23) in biphenotypic acute leukemia. *Cancer Genet Cytogenet* **131**, 92–3.
36. Dunphy CH, Batanian JR. (1999). Biphenotypic hematological malignancy with T-lymphoid and myeloid differentiation: Association with t(3;12)(p25;q24.3). Case report and review of the literature. *Cancer Genet Cytogenet* **114**, 51–7.
37. Inoue T, Fujiyama Y, Kitamura S, Andoh A, Hodohara K, Bamba T. (2000). Trisomy 10 as the sole abnormality in biphenotypic leukemia. *Leuk Lymphoma* **39**, 405–9.
38. Killick S, Matutes E, Powles RL, Hamblin M, Swansbury J, Treleaven JG, Zomas A, Atra A, Catovsky D. (1999). Outcome of biphenotypic acute leukemia. *Haematologica* **84**, 699–706.
39. Altman AJ. (1990). Clinical features and biological implications of acute mixed lineage (hybrid) leukemias. *Am J Pediatr Hematol Oncol* **12**, 123–33.

14

CHAPTER

Lymphoid Malignancies of Non-precursor Cells: General Considerations

Faramarz Naeim,
P. Nagesh Rao
and
Wayne W. Grody

In the following chapters, all categories of lymphoid malignancies except those of precursor B- and T-cell lymphoblastic leukemias/lymphomas are discussed. These tumors, according to the WHO classification, are divided into the following major categories [1]:

1. Mature B-cell neoplasms
2. Mature T- and NK-cell neoplasms
3. Hodgkin lymphoma.

Mature B-cell neoplasms comprise a wide spectrum of lymphoid malignancies representing clonal proliferation of B-lymphocytes at various stages of differentiation, from the early naïve B-cells to the end-stage mature plasma cells. These disorders may primarily involve bone marrow and peripheral blood (leukemia), lymphoid or extramedullary tissues (lymphoma), or both. They comprise >85% of all lymphoid neoplasms and are divided into numerous categories [1, 2]. The most frequent types are diffuse large B-cell lymphoma and follicular lymphoma accounting for about 30% and 20%, respectively, of all non-Hodgkin lymphoid malignancies.

Mature T-cell neoplasms represent the post-thymic stages of maturation. These tumors, along with NK-cell neoplasms, account for >15% of all lymphoid malignancies [1, 2]. Similar to the mature B-cell neoplasms, T- and NK-cell malignancies may primarily involve bone marrow and peripheral blood (leukemia), lymphoid or extramedullary tissues (lymphoma), or more commonly both.

Hodgkin lymphoma represents a heterogenous but distinct clinicopathologic entity characterized by the presence of neoplastic Reed–Sternberg cells and their variants in a background of an admixture of reactive inflammatory cells, preferential involvement of lymph nodes, and frequent occurrence in young adults. In most instances, the neoplastic cells appear to be of B-cell lineage. Hodgkin lymphoma comprises about 30% of all lymphomas and is divided into two major categories: (1) nodular-lymphocyte-predominant and (2) classical Hodgkin lymphomas [3].

This chapter presents an overview of non-Hodgkin and Hodgkin lymphoid malignancies.

ETIOLOGY AND PATHOGENESIS

The etiology of lymphomas is not fully understood. A number of environmental factors are associated with increased incidence of lymphomas. An elevated risk of the development of non-Hodgkin lymphoma has been associated with exposure to pesticides, fertilizers, organic solvents, epoxy glues, and wood dust [4–6]. Ultraviolet light and hair dyes have also been implicated in some studies [4].

A number of viruses, such as Epstein–Barr virus (EBV), human T-cell lymphotrophic virus (HTLV-1), and human herpes virus-8 (HHV-8), play a crucial role in the development of certain categories of lymphomas [7–14]. Epstein–Barr virus has been detected in endemic African and in a fraction of sporadic forms of Burkitt lymphomas. Also, association of EBV infection with B-cell lymphomas has been observed in immunosuppressive conditions, such as AIDS, and in immunosuppressive therapy after organ transplantation [8–12]. Infection of EBV in lymphoid tissue is frequently documented in a significant proportion of cases with classical Hodgkin lymphoma. The EBV genome in the infected lymphoid cells usually exists as an extrachromosomal circular episome in the nucleus. The EBV infection promotes lymphocytic proliferation with clonal expansion of the EBV genome. The lymphocytic proliferation, in certain conditions, may increase the chance of chromosomal aberrations. HTLV-1 has been detected in 100% of the cases of adult T-cell leukemia/lymphoma. The HTLV-1 transregulatory product, HTLV-1-Tax, activates certain host genes such as IL-2, PDGF, GM-CSF, and cyclin-dependent kinase inhibitor p21/waf1 [13, 14]. Human herpes virus-8 has been detected in the primary effusion lymphomas in virtually 100% of the cases [15]. The mechanism of viral carcinogenesis in the primary effusion lymphomas is not fully understood. There is also a strong association between gastric mucosa-associated lymphoid tissues (MALT) type lymphoma and *Helicobacter pylori* [16–18].

Lymphomagenesis appears to be a multistep process associated with a number of genetic changes affecting proto-oncogenes and tumor suppressor genes [19–22]. The cytogenetic alterations in lymphoma cells are of two types: balanced and unbalanced chromosomal changes [23].

Reciprocal chromosomal translocations are the characteristic features of lymphoid malignancies. Translocations in non-Hodgkin lymphoma of B-cell types involve chromosomal breakpoints within immunoglobulin (Ig) loci while T-cell receptor (TCR) loci are commonly implicated in T-cell lymphomas. Breakpoints within the Ig loci are often located within the joining (J) and switch (S) sequences. Chromosomal translocations in non-Hodgkin lymphoma, in addition to Ig or TCR loci, involve proto-oncogenes in proximity to the chromosomal recombination sites. Usually, in these translocations, the structure of the proto-oncogene is not affected. The juxtaposition of the regulatory DNA sequences derived from the partner chromosome changes the pattern of expression of the oncogene. Less commonly, in certain types of lymphomas, such as anaplastic large cell and MALT, the chromosomal translocation results in the juxtaposition of two genes, creating a chimeric gene coding for a new protein [24, 25].

Unbalanced chromosomal abnormalities are usually preceded by balanced translocations and appear to represent the second genetic hit in the clonal evolution of the lymphoid malignancies. For example, in addition to t(14;18), loss of 6q has been described in follicular lymphomas and appears to predict poor prognosis [26].

Inactivation of tumor suppressor genes has been observed in some lymphomas and is usually the result of point mutation of one allele and/or chromosomal deletion of the second allele. Deletions and mutations of the *p53* tumor suppressor gene have been associated with chronic lymphocytic leukemia or small lymphocytic, follicular, Burkitt, and mantle cell lymphomas [27–31].

PATHOLOGY

Morphology

Lymphoid malignancies have extremely diverse morphologic features. This diversity, to some degree, correlates with the stage of their differentiation. The precursor B-cell goes through several maturation and evolutionary steps, such as naïve lymphocyte, mantle cell, follicular B-blast, centroblast, centrocyte, marginal zone B-cell, B-immunoblast, and plasma cell, and each step depicts some characteristic morphologic features (Figure 14.1). The non-Hodgkin lymphoid malignancies, particularly of B-cell types, show two major patterns of lymph node involvement: follicular and diffuse (Figure 14.2).

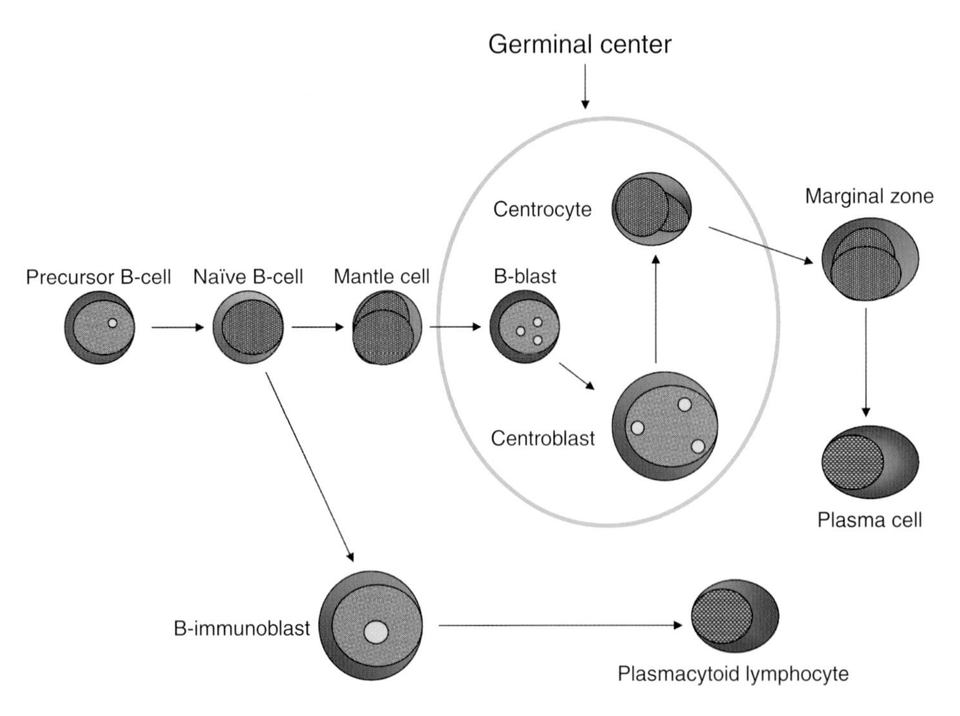

FIGURE 14.1 Scheme of B-cell differentiation. Adapted from Harris NL, Ref. [1].

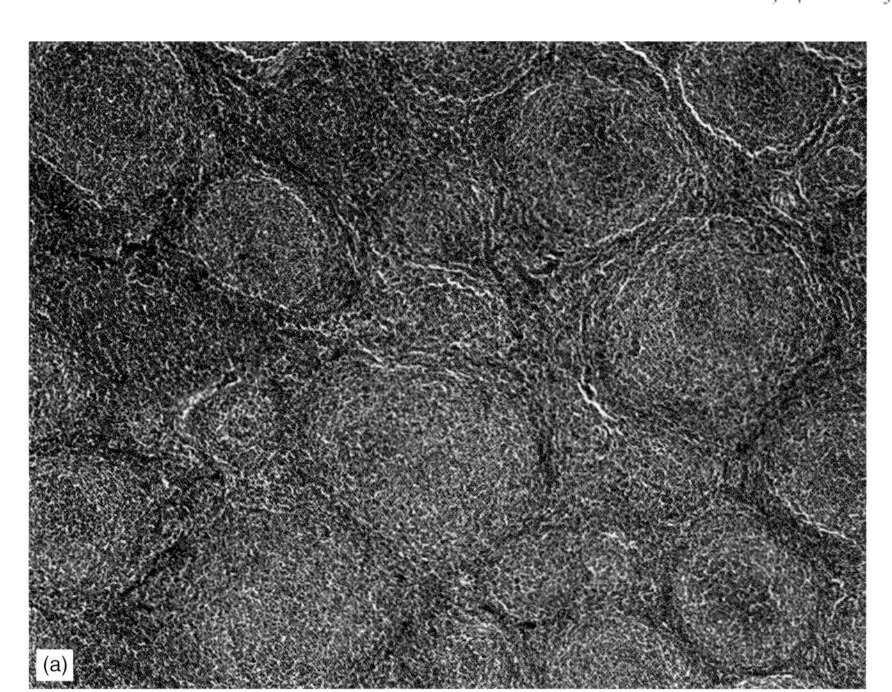

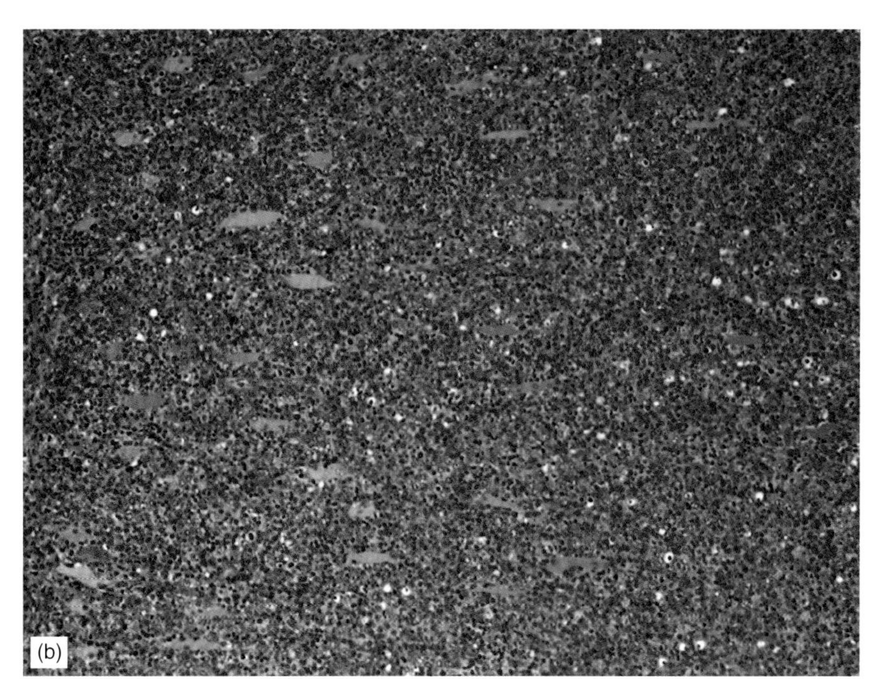

FIGURE 14.2 Lymph node sections demonstrating follicular (a) and diffuse (b) patterns of lymphoma.

The T-cell lymphoid malignancies are virtually all diffuse. They also comprise a diverse morphology ranging from small lymphocytes to large, bizarre anaplastic tumor cells. The NK-cell neoplasms usually show cytoplasmic granules and represent a subgroup of large granular lymphocytes (LGLs). Large granular lymphocytes are divided into two major categories: T- and NK-cell types. These two categories are morphologically indistinguishable, but one (T-cell type) demonstrates *TCR* gene rearrangement and the other (NK-cell type) does not (Figure 14.3).

Bone marrow involvement in the mature B- and T-cell neoplasms may appear in different patterns such as diffuse, paratrabecular, nodular, interstitial, or mixed. Diffuse

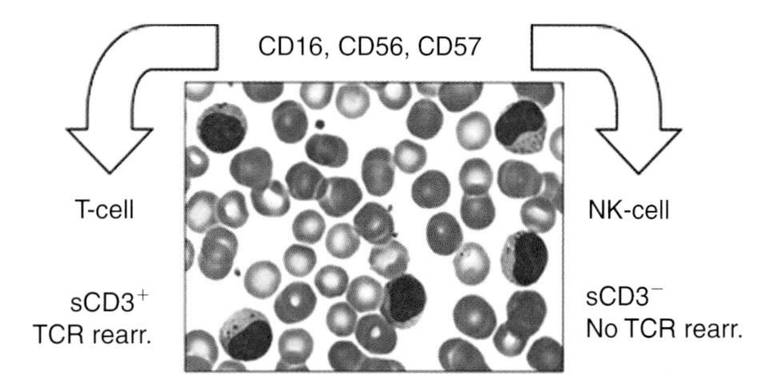

FIGURE 14.3 Large granular lymphocytes (LGLs) are of two types: T- and NK-cells. T-cell LGLs express surface CD3 (sCD3) and demonstrate T-cell receptor (TCR) gene rearrangement. NK-cell LGLs are negative for sCD3 and do not show *TCR* gene rearrangement.

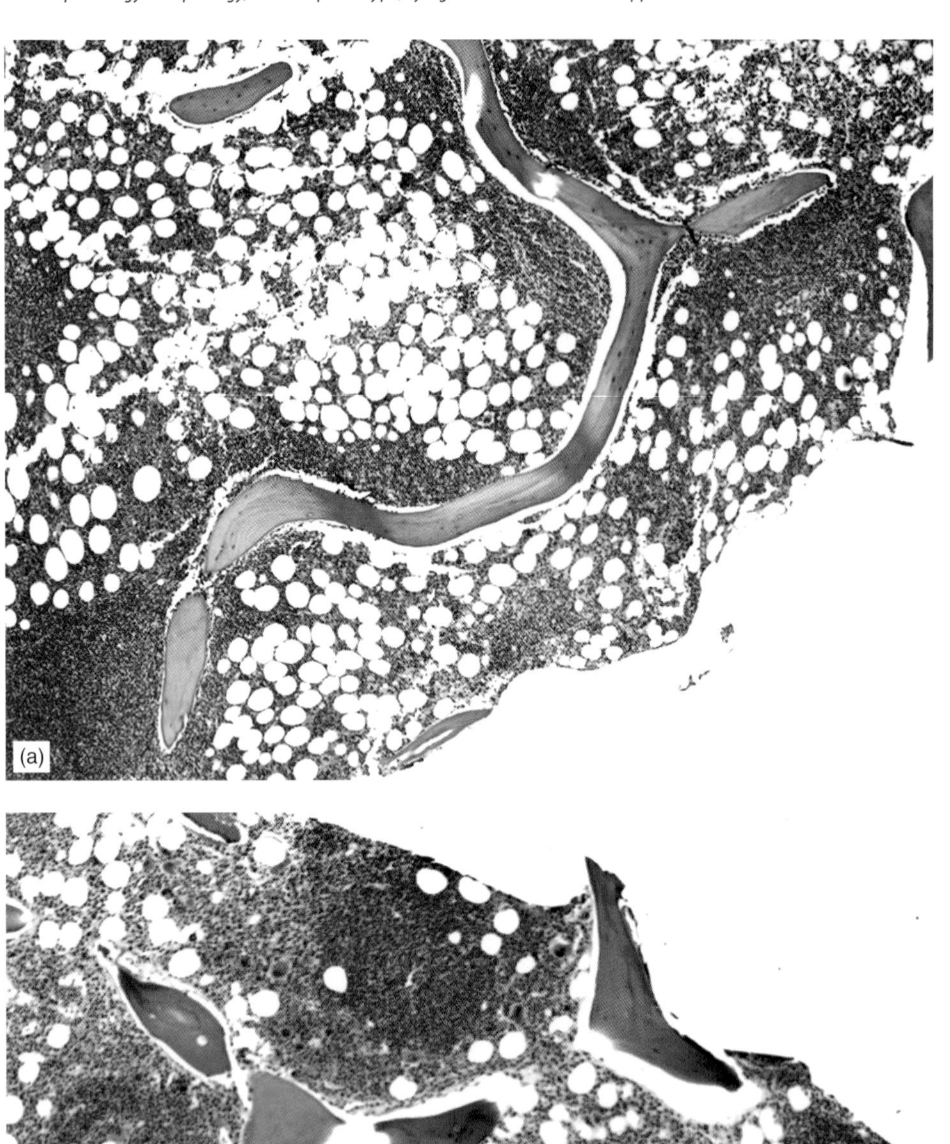

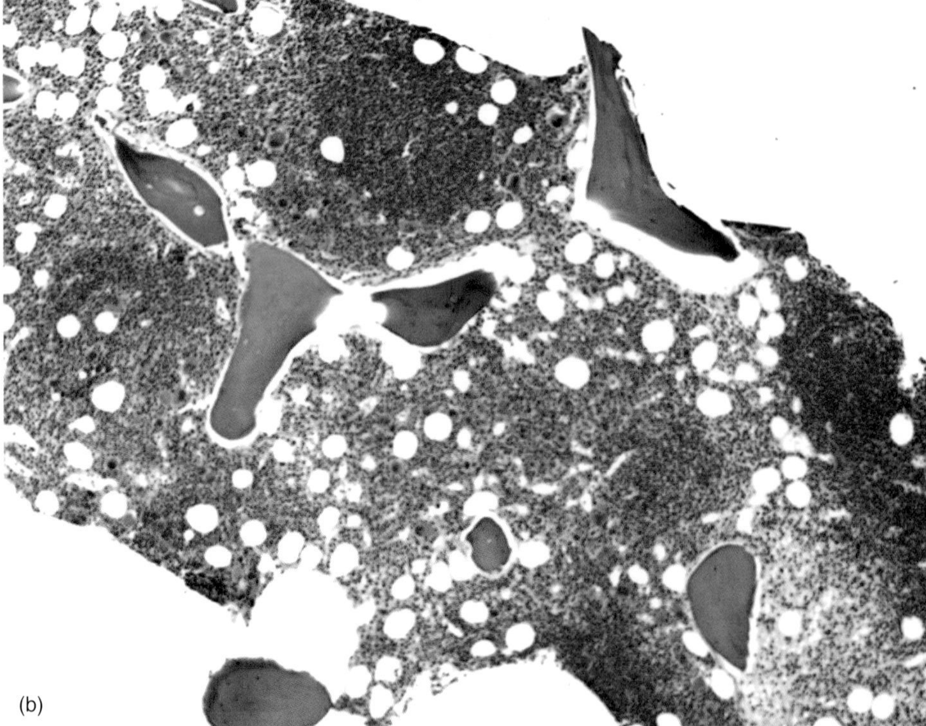

FIGURE 14.4 Bone marrow biopsy sections showing paratrabecular (a) and nodular (b) lymphomatous involvement.

involvement is defined as sheets of space-occupying lymphoid cells without the formation of well-defined nodular aggregates. Paratrabecular disease refers to aggregates of lymphoma cells next to the bony trabeculae (Figure 14.4a). Nodular involvement consists of well-defined non-paratrabecular aggregates (nodules) of lymphoma cells (Figure 14.4b) [32]. Interstitial involvement is demonstrated by infiltration of lymphoma cells into the fatty tissue without obvious obliteration of the bone marrow architecture on low-power examination (Figure 14.5).

The morphologic diversity in Hodgkin lymphomas is the result of both the number and the type of Reed–Sternberg and Hodgkin cells as well as the proportion of the lymphocytic component in the background of mixed

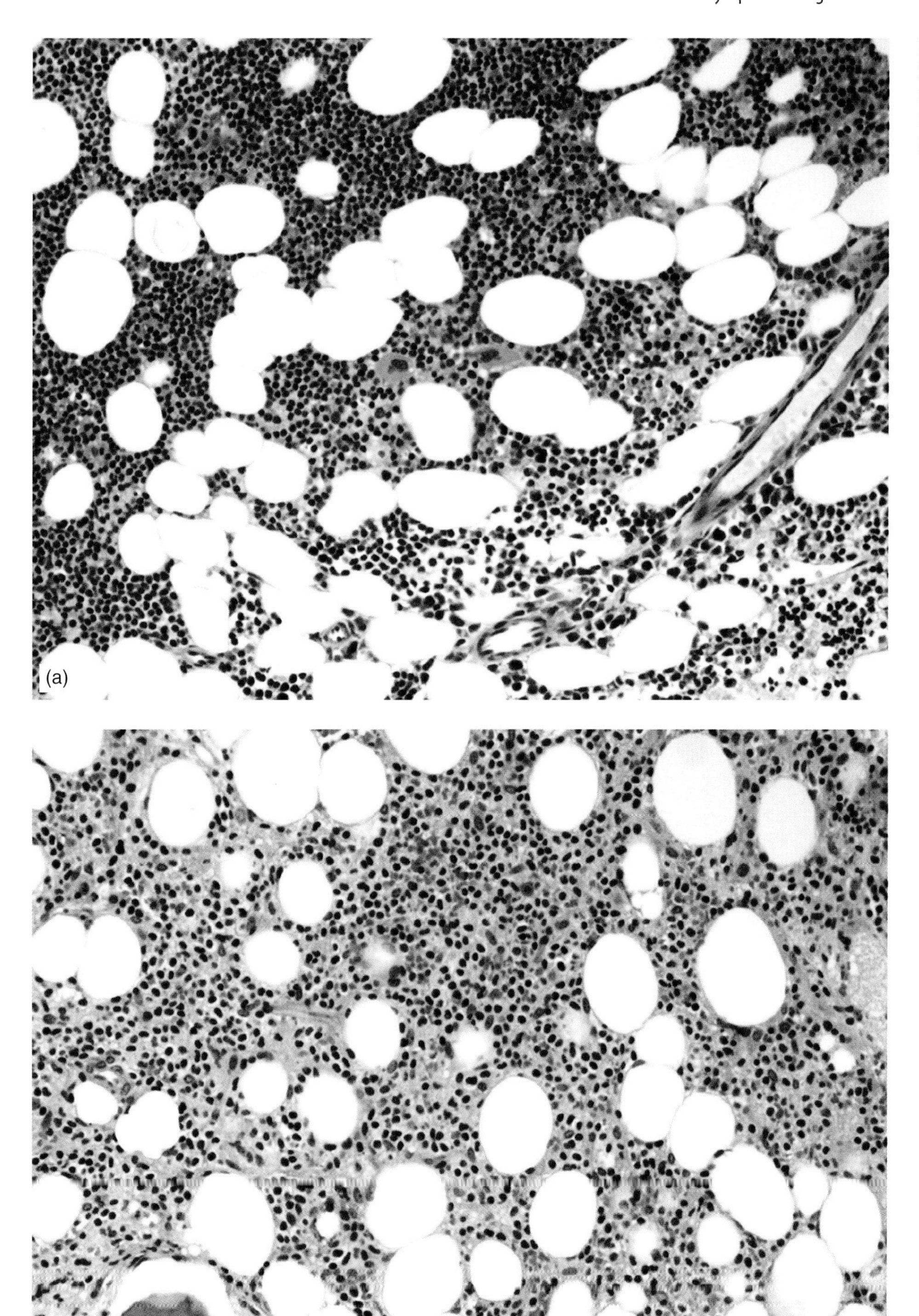

FIGURE 14.5 Bone marrow biopsy sections showing interstitial involvement in patients with chronic lymphocytic (a) and hairy cell leukemias (b).

inflammatory cells. Involvement of bone marrow in Hodgkin lymphoma is significantly less frequent than that in non-Hodgkin lymphoma. This involvement is often nodular and is associated with variable degrees of fibrosis.

The WHO classification of non-Hodgkin lymphoid malignancies is primarily based on morphologic and immunophenotypic features of the neoplastic cells (Table 14.1) [1], whereas that of Hodgkin lymphoma is based on morphologic features (Table 14.1).

Mature B-cell and T-cell lymphoid malignancies, plasma cell neoplasms, and Hodgkin lymphomas are discussed separately in Chapters 15, 16, and 17, respectively.

TABLE 14.1 The WHO classification of mature lymphoid malignancies and Hodgkin lymphoma.*

Mature B-cell
Chronic lymphocytic leukemia/small lymphocytic lymphoma
B-cell prolymphocytic leukemia
Lymphoplasmacytic lymphoma
Splenic marginal zone lymphoma
Hairy cell leukemia
Plasma cell neoplasms
Extranodal marginal zone B-cell lymphoma of mucosa-associated lymphoid tissue
Nodal marginal zone B-cell lymphoma
Follicular lymphoma
Mantle cell lymphoma
Diffuse large B-cell lymphoma
Mediastinal (thymic) large B-cell lymphoma
Intravascular large B-cell lymphoma
Primary effusion lymphoma
Burkitt lymphoma/leukemia
Mature T- and NK-cells
Leukemic/disseminated
T-cell prolymphocytic leukemia
T-cell large granular lymphocytic leukemia
Aggressive NK-cell leukemia
Adult T-cell leukemia/lymphoma
Cutaneous
Mycosis fungoides
Sezary syndrome
Primary cutaneous anaplastic large cell lymphoma
Lymphomatoid populosis
Other extranodal
Extranodal NK/T-lymphoma, nasal type
Entropathy-type T-cell lymphoma
Hepatosplenic T-cell lymphoma
Subcutaneous panniculitis-like T-cell lymphoma
Nodal
Angioimmunoblastic T-cell lymphoma
Peripheral T-cell lymphoma, unspecified
Anaplastic large cell lymphoma
Hodgkin lymphoma
Nodular-lymphocyte-predominant Hodgkin lymphoma
Classical Hodgkin lymphoma
Nodular sclerosis classical Hodgkin lymphoma
Mixed cellularity classical Hodgkin lymphoma
Lymphocyte-rich classical Hodgkin lymphoma
Lymphocyte-depleted classical Hodgkin lymphoma

*Adapted from Ref. [1].

TABLE 14.2 Major characteristic immunophenotypic features in small mature B-cell lymphoid malignancies.

Type of lymphoid malignancy	Immunophenotype	
	Positive	Negative
CLL/SLL	CD5, CD19, CD23	CD10, FMC7, CD103, bcl-1
Mantle cell	CD5, CD19, CD43, FMC7, bcl-1	CD10, CD23, bcl-6
Marginal zone	CD19, CD29, CD79a	CD5, CD10, CD23, bcl-1
Follicular	CD10, CD19, CD20, bcl-2, bcl-6	CD5, CD43, bcl-1
Lymphoplasmacytic	CD19, CD20, CD22, CD79a	CD5, CD43, bcl-1
Hairy cell	CD19, CD11c, FMC7, CD22, CD103	CD5, CD10, CD23, bcl-1
Prolymphocytic	CD19, CD20, FMC7, CD22, CD79a	CD23, CD10, and usually CD5
Burkitt	CD10, CD19, CD20, CD22, bcl-6, KI-67	CD5, CD23, CD34, TdT, bcl-2

Immunophenotype

The mature B-cell neoplasms are negative for CD34 and TdT and express various B-cell-associated antigens according to their stage of maturation (Table 14.2). For example, neoplastic cells in chronic lymphocytic leukemia or small lymphocytic lymphoma correspond to naïve B-cells and co-express CD5, CD19, CD20, and CD23; whereas lymphoid cells in mantle cell lymphoma are positive for CD5, CD19, and CD20, but negative for CD23. Neoplasms of follicular-center-cell origin express CD10, but are negative for CD5; marginal zone B-cell lymphomas do not express CD5, CD10, and CD23.

The mature T- and NK-cell neoplasms are also negative for CD34 and TdT. The mature T-cell neoplasms express the entire TCR–CD3 complex and show both cytoplasmic and surface positivity for CD3. These tumors are mostly of the helper T-cell type and therefore express CD4. NK-cell neoplasms express only the ε chain of CD3 in the cytoplasm; they are negative for CD4, but may express CD8. Both T- and NK-cells express CD2 and CD7. The large granular cells of T- or NK-cell origin commonly express CD16, CD56, and/or CD57 and show positive reactions for cytotoxic proteins such as perforin, TIA-1, and granzyme B (see Figure 14.3).

The immunophenotypic features of Reed–Sternberg cells and their variants are different in nodular-lymphocyte-predominant and classical Hodgkin lymphomas. These cells in nodular-lymphocyte-predominant Hodgkin lymphoma are positive for CD45 and often express B-cell-associated markers such as CD20 and CD79a, but are negative for CD15 and CD30. In classical Hodgkin lymphoma,

TABLE 14.3 Major characteristic cytogenetic features of small mature B-cell lymphoid malignancies.

Type of lymphoid malignancy	Chromosomal abnormalities	Genes
CLL/SLL	del(17)(p13) del(11)(q22) Trisomy 12 del(13)(q14)	*p53* *ATM*
Mantle cell	t(11;14)(q13;q32)	*BCL-1/IGH*
Marginal zone Splenic Extranodal	del(7q) Trisomy 3 t(11;18)(q21;q21)	*BIRC3/MALT1*
Follicular	t(14;18)(q32;q21)	*IGH/BCL-2*
Lymphoplasmacytic	del(6)(q21→q23) t(9;14) (p13;q32)	*PAX5/IGH*
Hairy cell	Non-specific	
B-prolymphocytic	14q32 abnormalities	*IGH*
Burkitt	t(8;14)(q24;q32) t(2;8)(p12;q24) t(8;22)(q24;q11.2)	*cMYC/IGH* *IGK/cMYC* *cMYC/IGL*

Reed–Sternberg cells and their variants are negative for CD45 and express CD15 and CD30.

Cytogenetic and Molecular Studies

The hallmark of mature lymphoid malignancies is reciprocal chromosomal translocation. In most translocations, an intact proto-oncogene is located next to the IGH or TCR genes. Quantitative chromosomal changes such as trisomies and deletions also occur, which in some cases involve tumor suppressor genes. As an example, Table 14.3 demonstrates the major characteristic cytogenetic features observed in mature small B-cell lymphoid neoplasms [23].

The most common proto-oncogenes involved in B-cell neoplasms are *BCL1 (CCND1)*, *BCL-2*, *BCL-6*, *MYC*, and *PAX 5* [33–35]. Of these, the one that is most often tested at the molecular level is *BCL-2*. The others are either too uncommon or too heterogeneous in their molecular breakpoints to be efficiently detected by PCR or Southern blot and are better addressed by karyotype or fluorescence *in situ* hybridization (FISH) to detect the translocation, or by immunohistochemistry to detect increased expression of the protein product. Similar to the *BCR-ABL* translocation discussed in Chapter 11, the t(14;18) translocation of *BCL-2* may be detected by either Southern blot or PCR. For the latter, direct DNA-based PCR is adequate, unlike BCR-ABL testing which requires RT-PCR. The *ALK* and *NFKB-2* genes are involved in certain types of T-cell malignancies such as anaplastic large cell and cutaneous T-cell lymphomas, respectively [19, 34, 35]. The most common tumor suppressor gene involved in lymphoid malignancies is *p53*. A large proportion of molecular changes in lymphoid malignancies are readily and inexpensively detected at the protein level (immunohistochemistry and/or flow cytometry) than molecular techniques.

The mature B-cell and plasma-cell disorders naturally demonstrate clonal immunoglobulin gene rearrangements, but since they are usually readily diagnosed by routine means, molecular analysis is not often necessary. As the T-cell malignancies may have fewer protein markers to choose from, clonality studies may be helpful to both diagnose and differentiate them from B-cell lesions. *ALK* is an important alternative target in some cases of T-cell malignancy, but again it is usually easier to detect overexpression of the protein by immunochemical means, or to detect the common t(2;5) translocation or the variants by an appropriate FISH probe, which is actually more sensitive than PCR [36]. As for Hodgkin disease, although gene rearrangement studies have been invaluable in research to prove the B-cell origin of this tumor, the malignant cells (Reed–Sternberg) are usually too scanty in the specimen to produce a robust clonal signal in routine analysis [37].

The more detailed cytogenetic and molecular characteristics of specific subtypes of the lymphoid neoplasms with illustrations are presented in Chapters 15 through 18.

CLINICAL ASPECTS

The lymphoid neoplasms of non-precursor cells represent the fifth most common malignancy in the United States accounting for about 4% of all cancers. They are more common in men than in women and are more frequent in adults than in children with a steady increase in incidence with age. In general, the lymphoid malignancies in children are more commonly extranodal than nodal and clinically more aggressive. The overwhelming majority of lymphoid malignancies are of B-cell lineage accounting for >85% of the cases. All types of T/NK-cell neoplasms make up about 15% of the total lymphoid malignancies (Table 14.4).

The Ann Arbor staging system originally developed for the staging of Hodgkin lymphomas has been extended to non-Hodgkin lymphoid malignancies. This staging system is based on the location(s), number of involved sites, and presence or absence of systemic symptoms (Table 14.5). Since in the majority of cases of non-Hodgkin lymphoma the disease is disseminated at the time of diagnosis, the staging system is less useful in non-Hodgkin lymphoma than it is in Hodgkin lymphoma.

An international prognostic index (IPI) has been proposed for patients with non-Hodgkin lymphoma. This scoring system is based on the following factors, which were found to show a reverse correlation with relapse-free survival [38]:

Age >60 years

Elevated serum lactate dehydrogenase (LDH)

TABLE 14.4 Frequency of non-Hodgkin lymphoid malignancies.[a]

Type of lymphoma	Approximate frequency (%)
Diffuse large B-cell	31
Follicular	22
MALT	8
Mature T-cell (except ALCL[b])	8
CLL/SLL[c]	7
Mantle cell	6
Mediastinal large B-cell	2.5
ALCL	2.5
Burkitt	2.5
Nodal marginal zone	2
Lymphoplasmacytic	1
Others	7.5

[a]Adapted from Refs. [1, 38].

[b]Anaplastic large cell lymphoma.

[c]Chronic lymphocytic leukemia/small lymphocytic lymphoma.

TABLE 14.5 The Ann Arbor staging for lymphomas.

Stage I	Involvement of a single node region or a single extralymphatic organ or site (Stage 1E).
Stage II	Two or more involved lymph node regions on the same side of the diaphragm, or with localized involvement of an extralymphatic organ or site (IIE).
Stage III	Lymph node involvement on both sides of the diaphragm, or with localized involvement of an extralymphatic organ or site (IIE), or spleen (IIIS), or both (IIIES).
Stage IV	Presence of diffuse or disseminated involvement of one or more extralymphatic organs, with or without associated lymph node involvement.
Systemic symptoms	
A	Asymptomatic.
B	Presence of fever, sweats, or weight loss >10% of body weight.

Eastern Cooperative Oncology Group (ECOG) performance status >2

Ann Arbor clinical stage III or IV

Number of involved extranodal disease sites >1.

Each factor receives a score of 1. Non-Hodgkin lymphomas with a total IPI score of 0–1 are considered low risk, and those with IPI scores of 2, 3, and >3 are in the categories of low intermediate risk, high intermediate risk, and high risk, respectively.

TABLE 14.6 The prognostic categories for lymphoid malignancies.*

B-cell lineage	T-cell lineage
Indolent	
CLL/SLL	LGL, T-cell type
Hairy cell leukemia	Mycosis fungoides and Sezary syndrome
Splenic marginal zone	Primary cutaneous anaplastic large cell
MALT type leukemia/lymphoma	Smoldering and chronic adult T-cell
Nodal marginal zone	
Follicular lymphoma, grades I and II	
Lymphoplasmacytic	
Aggressive	
Plasma cell myeloma	Prolymphocytic
Mantle cell	Peripheral T-cell, unspecified
Follicular lymphoma, grade III	Angioimmunoblastic
Diffuse large cell	T/NK-cell of nasal type
Primary mediastinal large cell	Anaplastic large cell lymphoma
	Hepatosplenic T-cell
	Subcutaneous panniculitis-like
Highly aggressive	
Precursor B-lymphoblastic	Precursor T-lymphoblastic
Burkitt	Adult T-cell leukemia/lymphoma

*Adapted from Ref. [39].

A clinicopathologic prognostic scheme was proposed by Hiddemann *et al.* [39], in which the non-Hodgkin lymphoid malignancies were divided into three major groups of indolent, aggressive, and highly aggressive tumors (Table 14.6).

DIFFERENTIAL DIAGNOSIS

The presence of a wide variety of mature lymphoid malignancies and numerous subcategories makes their accurate diagnosis and classification very challenging. In spite of significant clarification in diagnostic criteria and classification, advances in technology, and availability of accessory tools in numerous diagnostic centers, there is still evidence of significant discordance in the diagnosis and classification of lymphoid malignancies among pathologists [40–42]. For example, in one study, the agreement rate in diagnosis/classification of lymphomas among the pathologists ranged from <50% in cases of Burkitt-like lymphoma and lymphoplasmacytic lymphoma to >85% in cases of small mature B-cell lymphoid malignancies, diffuse large B-cell lymphoma, and other lymphoid malignancies [40]. Some general practical points helpful in differential diagnosis of mature lymphoid neoplasms are presented as follows:

1. Morphologic features in many instances are not sufficient to separate the B-cell from the T-cell disorders. Some subcategories of small mature B-cell malignancies, such as follicular lymphoma, marginal zone B-cell lymphoma, and mantle cell lymphoma, may morphologically mimic one another. Large granular lymphocytes of NK- and T-cell origins are morphologically indistinguishable. Therefore, it is highly recommended that immunophenotypic, cytogenetic (more specifically FISH), and sometimes molecular studies be part of the routine diagnostic work-up of lymphoid malignancies.
2. Node-based lymphomas should be distinguished from garden varieties of reactive lymphadenopathies such as follicular lymphoma from follicular hyperplasia, Hodgkin lymphoma from viral infections, and drug-induced lymphadenitis from T-cell lymphoma.
3. Cutaneous lymphoid malignancies may mimic various inflammatory dermal lesions.
4. Burkitt lymphoma/leukemia should be distinguished from precursor B- and T-cell lymphoblastic leukemias/lymphomas.
5. Anaplastic large cell and non-Hodgkin lymphomas share overlapping morphologic features.

References

1. Jaffe ES, Harris NL, Stein H, Vardiman JW. (2001). *Pathology and Genetics. Tumors of Haematopoietic and Lymphoid Tissues*. IARC Press, Lyon.
2. Freedman AS, Friedberg JW. (2006). Classification of lymphomas. *UpToDate*.
3. Nakatsuka S, Aozasa K. (2006). Epidemiology and pathologic features of Hodgkin lymphoma. *Int J Hematol* **83**, 391–7.
4. Dreiher J, Kordysh E. (2006). Non-Hodgkin lymphoma and pesticide exposure: 25 years of research. *Acta Haematol* **116**, 153–64.
5. Greer JP, Foerster J, Rodgers GM, Paraskevas F, Glader B. (2004). *Wintrobe's Clinical Hematology*, 11th ed. Lippington Williams & Wilkins, Philadelphia.
6. Ekstrom-Smedby K. (2006). Epidemiology and etiology of non-Hodgkin lymphoma – a review. *Acta Oncol* **45**, 258–71.
7. Grulich AE, Vajdic CM. (2005). The epidemiology of non-Hodgkin lymphoma. *Pathology* **37**, 409–19.
8. Andersson J. (2006). Epstein–Barr virus and Hodgkin's lymphoma. *Herpes* **13**, 12–16.
9. Gandhi MK. (2006). Epstein–Barr virus-associated lymphomas. *Expert Rev Anti Infect Ther* **4**, 77–89.
10. Cesarman E, Mesri EA. (2006). Pathogenesis of viral lymphomas. *Cancer Treat Res* **131**, 49–88.
11. Gandhi MK, Khanna R. (2005). Viruses and lymphoma. *Pathology* **37**, 420–33.
12. Parekh S, Ratech H, Sparano JA. (2003). Human immunodeficiency virus-associated lymphoma. *Clin Adv Hematol Oncol* **1**, 295–301.
13. Shuh M, Beilke M. (2005). The human T-cell leukemia virus type 1 (HTLV-1): New insights into the clinical aspects and molecular pathogenesis of adult T-cell leukemia/lymphoma (ATLL) and tropical spastic paraparesis/HTLV-associated myelopathy (TSP/HAM). *Microsc Res Tech* **68**, 176–96.
14. Taylor G. (2007). Molecular aspects of HTLV-I infection and adult T-cell leukaemia/lymphoma. *J Clin Pathol* **60**, 1392–6.
15. Ahmed N, Heslop HE. (2006). Viral lymphomagenesis. *Curr Opin Hematol* **13**, 254–9.
16. Algood HM, Cover TL. (2006). *Helicobacter pylori* persistence: An overview of interactions between *H. pylori* and host immune defenses. *Clin Microbiol Rev* **19**, 597–613.
17. McGee DJ, Mobley HL. (2000). Pathogenesis of *Helicobacter pylori* infection. *Curr Opin Gastroenterol* **16**, 24–31.
18. Fischbach W, Chan AO, Wong BC. (2005). *Helicobacter pylori* and gastric malignancy. *Helicobacter* **10**(Suppl 1), 34–9.
19. Karenko L, Hahtola S, Ranki A. (2007). Molecular cytogenetics in the study of cutaneous T-cell lymphomas (CTCL). *Cytogenet Genome Res* **118**, 353–61.
20. Landgren O, Caporaso NE. (2007). New aspects in descriptive, etiologic, and molecular epidemiology of Hodgkin's lymphoma. *Hematol Oncol Clin North Am* **21**, 825–40.
21. Bench AJ, Erber WN, Follows GA, Scott MA. (2007). Molecular genetic analysis of haematological malignancies II: Mature lymphoid neoplasms. *Int J Lab Hematol* **29**, 229–60.
22. Felberbaum RS. (2005). The molecular mechanisms of classic Hodgkin's lymphoma. *Yale J Biol Med* **78**, 203–10.
23. Campbell LJ. (2005). Cytogenetics of lymphomas. *Pathology* **37**, 493–507.
24. Nakagawa M, Seto M, Hosokawa Y. (2006). Molecular pathogenesis of MALT lymphoma: Two signaling pathways underlying the antiapoptotic effect of API2-MALT1 fusion protein. *Leukemia* **20**(6), 929–36.
25. Staber PB, Vesely P, Haq N, Ott RG, Funato K, Bambach I, Fuchs C, Schauer S, Linkesch W, Hrzenjak A, Dirks WG, Sexl V, Bergler H, Kadin ME, Sternberg DW, Kenner L, Hoefler G. (2007). The oncoprotein NPM-ALK of anaplastic large-cell lymphoma induces JUNB transcription via ERK1/2 and JunB translation via mTOR signaling. *Blood* **110**, 3374–83.
26. Levine EG, Bloomfield CD. (1990). Cytogenetics of non-Hodgkin's lymphoma. *J Natl Cancer Inst Monogr* **10**, 7–12.
27. Kerbauy FR, Colleoni GW, Saad ST, Regis Silva MR, Correa Alves A, Aguiar KC, Albuquerque DM, Kobarg J, Seixas MT, Kerbauy J. (2004). Detection and possible prognostic relevance of p53 gene mutations in diffuse large B-cell lymphoma. An analysis of 51 cases and review of the literature. *Leuk Lymphoma* **45**, 2071–8.
28. Weinkauf M, Christopeit M, Hiddemann W, Dreyling M. (2007). Proteome- and microarray-based expression analysis of lymphoma cell lines identifies a p53-centered cluster of differentially expressed proteins in mantle cell and follicular lymphoma. *Electrophoresis* **28**, 4416–26.
29. Bai M, Tsanou E, Skyrlas A, Sainis I, Agnantis N, Kanavaros P. (2007). Alterations of the p53, Rb and p27 tumor suppressor pathways in diffuse large B-cell lymphomas. *Anticancer Res* **27**, 2345–52.
30. Yu D, Carroll M, Thomas-Tikhonenko A. (2007). p53 status dictates responses of B lymphomas to monotherapy with proteasome inhibitors. *Blood* **109**, 4936–43.
31. Van Dyke T. (2007). p53 and tumor suppression. *N Engl J Med* **356**, 79–81.
32. Arber DA, George TI. (2005). Bone marrow biopsy involvement by non-Hodgkin's lymphoma: Frequency of lymphoma types, patterns, blood involvement, and discordance with other sites in 450 specimens. *Am J Surg Pathol* **29**, 1549–57.
33. Bagg A. (2005). Molecular diagnosis in lymphoma. *Curr Hematol Rep* **4**, 313–23.
34. Hegde U, Wilson WH. (2001). Gene expression profiling of lymphomas. *Curr Oncol Rep* **3**, 243–9.
35. Catalano A, Iland H. (2005). Molecular biology of lymphoma in the microarray era. *Pathology* **37**, 508–22.
36. Cataldo KA, Jalal SM, Law ME, *et al.* (1999). Detection of t(2;5) in anaplastic large cell lymphoma: Comparison of immunohistochemical studies, FISH, and RT-PCR in paraffin-embedded tissue. *Am J Surg Pathol* **23**, 1386–92.
37. Re D, Kuppers R, Diehl V. (2005). Molecular pathogenesis of Hodgkin's lymphoma. *J Clin Oncol* **23**, 6379–86.

38. The Non-Hodgkin's Lymphoma Classification Project (1997). A clinical evaluation of the International Lymphoma Study Group classification of non-Hodgkin's lymphoma. *Blood* **89**, 3909–18.
39. Hiddemann W, Longo DL, Coiffier B, Fisher RI, Cabanillas F, Cavalli F, Nadler LM, De Vita VT, Lister TA, Armitage JO. (1996). Lymphoma classification – the gap between biology and clinical management is closing. *Blood* **88**, 4085–9.
40. No authors listed. (1969). Histopathological definition of Burkitt's tumour. *Bull World Health Organ* **40**, 601–7.
41. El-Zimaity HM, Wotherspoon A, de Jong D. Houston MALT lymphoma Workshop (2005). Interobserver variation in the histopathological assessment of MALT/MALT lymphoma: Towards a consensus. *Blood Cells Mol Dis* **34**, 6–16.
42. Bernhards J, Fischer R, Hübner K, Schwarze EW, Georgii A. (1992). Histopathological classification of Hodgkin's lymphomas. Results from the reference pathology of the German Hodgkin Trial. *Ann Oncol* **3**(Suppl 4), 31–3.

15 CHAPTER

Mature B-Cell Neoplasms

Faramarz Naeim, P. Nagesh Rao and Wayne W. Grody

Mature B-cell neoplasms represent a wide spectrum of lymphoid malignancies developed from clonal proliferation of B-lymphocytes at various stages of differentiation from the early naïve B-cells to the end-stage mature plasma cells (Figure 15.1). These lymphoproliferative disorders may primarily involve bone marrow and peripheral blood (leukemia), lymphoid or extramedullary tissues (lymphoma), or both. They are more frequent than T-cell malignancies and comprise >85% of all lymphoid tumors [1, 2].

Mature B-cell neoplasms demonstrate clonal immunoglobulin gene rearrangements by either PCR or Southern blot. Unlike the precursor B-cell lesions, they will usually show rearrangements in their light chain genes as well as in the heavy chain genes, but since the IgH locus has become the standard screen for all B-cell clonality studies, it is easier to begin with this one from a laboratory management perspective. Any equivocal results can be confirmed by examining the light chain genes, most commonly J_K. As noted in Chapter 14, most of these cases are readily diagnosed by histologic and immunochemical methods, so gene rearrangement studies are usually not needed. However, they will be introduced in some detail here, since the basic principles of detection are the same for all the B-cell-derived lymphoid neoplasms to be covered subsequently.

The hallmark of B-cell leukemia/lymphoma diagnosis by gene rearrangement studies is the demonstration of clonal patterns of rearrangement in the DNA of the immunoglobulin loci. These rearrangements are a normal part of the maturation of all B-lymphocytes, as a means to expand the repertoire of produced antibody specificities, so that a finite number of genes can ultimately encode an almost limitless number of antibodies capable of responding to the wide variety of epitopic insults to which the body is prone during life. Once a particular rearrangement occurs in an individual lymphocyte (typically on only one allele, the remaining is fixed by the principle of "allelic exclusion"), all the daughter cells of that lymphocyte will contain the same rearrangement pattern forming a subclone. Because of the tremendous number of possible rearrangements, the subclones in a normal or reactive (nonmalignant) specimen (blood, bone marrow, lymph node, etc.) will span the entire range of DNA patterns with no single subclone represented in a high enough percentage to be detectable by the standard methods of analysis. In contrast, most malignant lesions are assumed to be clonal, having descended from a transforming event in a single progenitor cell, and the specimen should therefore exhibit a single predominant rearrangement pattern, one present at sufficiently high proportion to stand out from the background polyclonal "smear." It should be noted that very few specimens sent to the clinical laboratory are "pure" tumor; most will contain a background of incidental benign cells or, in the case of tissue biopsies (e.g. skin), a complement of nonlymphoid cells (e.g. fibroblasts) that do not undergo immunoglobulin gene rearrangements in any circumstances. These background cells, along with the typically unrearranged allele seen even in malignant lymphocytes (representing half the total target DNA), are responsible for the background polyclonal "smear" seen in both Southern blot and PCR assays and for the persistence of the "germline bands" seen on Southern blot (see examples in figures).

Although all of the immunoglobulin gene loci undergo rearrangement, convention in the field has coalesced around the detection of heavy chain rearrangements (located on chromosome

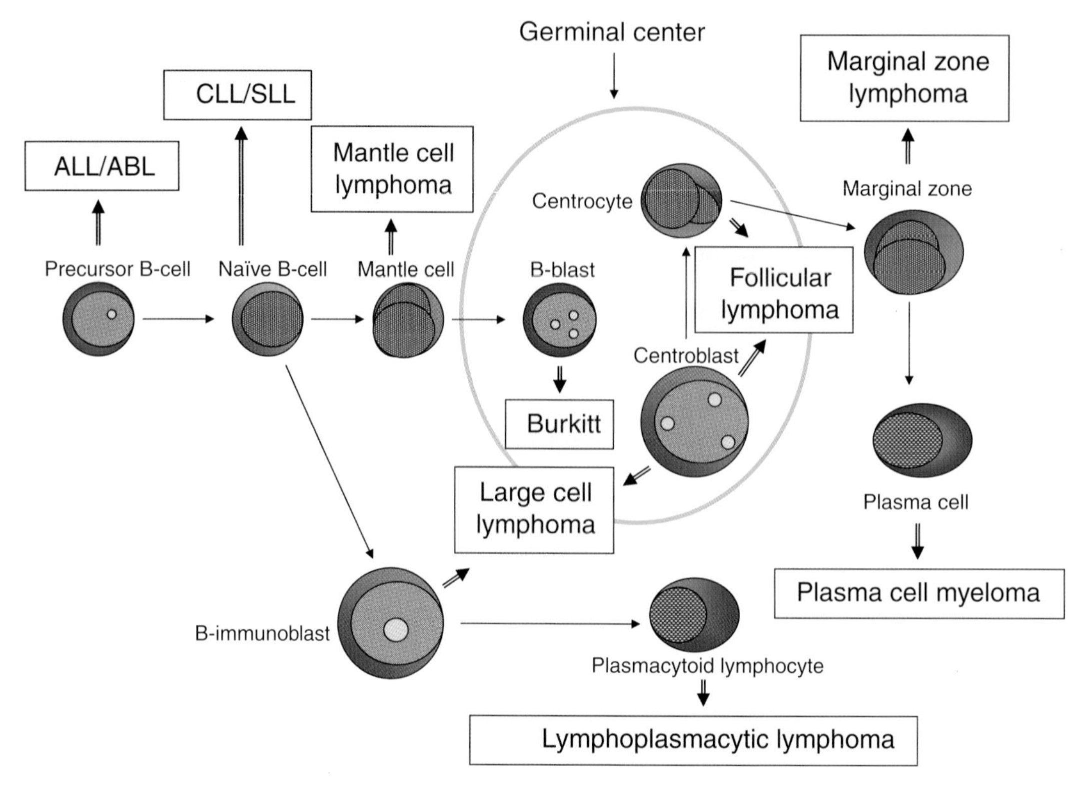

FIGURE 15.1 Scheme of B-cell differentiation and associated B-cell lymphomas. Adapted from Harris NL, Ref. [1].

14q32) as the first-line test, as this locus typically rearranges first in lymphocyte ontogeny and will thus be demonstrable in the highest proportion of neoplasms [3]. As noted earlier, the kappa light chain locus can then be examined subsequently for confirmation or further subclassification. Southern blot analysis was the first method devised for detection of immunoglobulin gene rearrangements, and in contrast to most other areas of the molecular pathology laboratory, it has yet to be entirely replaced by more modern PCR-based methods. The reason is that PCR cannot cover such a large region of gene comprehensively even with the use of multiplex primer sets typically directed at the V and J consensus sequences. Most laboratories now use a set of three "framework" primer pairs, designated FR1, FR2, and FR3 [4]. Depending on the subtype of B-cell tumor, the combined set will detect between 50% and 90% of clonal neoplasms. False-negative results can be due to the rearrangement breakpoints being out of coverage range of the primer sets used or because of interference with primer-target hybridization as a result of somatic hypermutation in the immunoglobulin genes themselves, another normal process for expanding antibody diversity. The latter phenomenon is especially frequent in follicular lymphomas (FLs), so that standard PCR methods have the lowest sensitivity (as low as 50%) for detecting clonality in these lesions [5]. In contrast, Southern blot methods using probes directed at the J_H region should detect >95% of clonal B-cell lesions assuming the specimen sent to the laboratory contains a sufficient proportion (at least 5–10%) of the malignant cells in question.

Gene rearrangement studies in B-cell lesions are performed for several purposes:

1. To distinguish malignant (clonal) lesions from reactive (nonclonal, polyclonal) lesions.
2. To determine likely cell of origin (B-cell versus T-cell) of the lesion.
3. To determine if a second (synchronous or metachronous) lesion represents the same or a different (new) clone.
4. To detect or monitor minimal residual disease (MRD) after initial therapy.

The first of these aims is accomplished by observing clonally rearranged band(s) separate from the persisting germline bands on Southern blot of extracted DNA hybridized with a probe directed at the J_H region. The rearranged band(s) can be anywhere in the lane, of either higher or lower molecular weight than the germline band(s); the location of the latter is determined by running a control sample (e.g. human placental DNA) that does not undergo gene rearrangement in the adjacent lane for each restriction endonuclease digest (Figure 15.2). Because of the chance that a single nongermline band could also result from a benign restriction fragment-length polymorphism (RFLP) present in this gene region, convention dictates that rearranged band(s) must be seen with at least two out of the three restriction enzymes used to call the specimen clonal [6]. One must also be careful to distinguish extra-germline artifacts

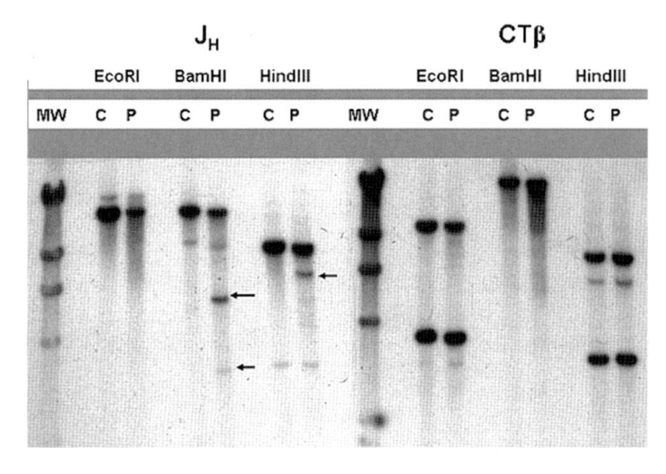

FIGURE 15.2 Southern blot analysis for clonality of immunoglobulin gene rearrangements in a B-cell lymphoma. Control (human placental) DNA (C) and patient (P) DNA were digested with three different restriction endonucleases (EcoRI, BamHI, and HindIII) and hybridized with either the J_H immunoglobulin heavy chain probe (left blot) or the CTβ T-cell receptor probe (right blot) and subjected to autoradiography. Nongermline rearranged bands are seen in the patient's DNA with two of the restriction digests (arrows) hybridized with the J_H probe only; no rearranged bands are seen on the CTβ blot, consistent with clonal B-cell origin.

due to partial restriction endonuclease digestion from true rearrangements. Alternatively, one can use a PCR assay comprised of the three framework primer sets. In this case, reactive lesions will demonstrate only a rounded and symmetrical polyclonal "smear" of amplicons representing the full range of benign rearrangements in the specimen, while a clonal lesion will demonstrate one or two sharp spikes representing an amplicon of discrete size derived from a clone of cells containing the same rearrangement. The clonal spikes may be freestanding or superimposed on the background polyclonal smear depending upon the percentage of clonal cells in the submitted specimen (Figure 15.3). Because the PCR method may miss a significant proportion of clonal B-cell lesions, for the reasons already discussed, a negative result does not exclude the possibility of a clonal B-cell population in the specimen. For this reason, in our laboratory we use the PCR method as an initial screen only: if it is "positive", the case is signed out as such, but if it is "negative," we reflex to Southern blot analysis for the definitive interpretation (except for paraffin-embedded tissue samples, which are not amenable to Southern blot and must be signed out based on the PCR results alone along with a prominent disclaimer about the potentially low sensitivity).

The second purpose of molecular and cytogenetic studies is to determine the cell of origin in a neoplastic process. While on the surface appearing relatively straightforward, here too, one must be mindful of exceptions to the rule. Though one may generally assume that lymphocytes with rearranged immunoglobulin heavy chain genes are B-cells, a small percentage of T-cell malignancies may rearrange their immunoglobulin genes, and vice versa [7]. In cases where doubt persists as to cell of origin, the analysis of J_K rearrangements may be informative, since these are more specific (though less sensitive) for B-cell malignancies.

The third purpose of these studies is the source classification of recurrent or multifocal lesions. As the gene rearrangement patterns observed on Southern blot or PCR are generally unique for a particular clone (out of the myriad possible rearrangements), they essentially establish a sort of DNA "fingerprint" for that clone. If analysis of a lymphoid lesion at another body site or a recurrence subsequent to treatment shows the identical pattern, it can be assumed to represent a metastasis or recurrence of the primary lesion [8]. Identity is assessed by looking for a clonal amplicon peak at the same molecular weight (length in base pairs) in PCR analysis (Figure 15.4) or a banding pattern (size and position) on Southern blot. This finding may have impact on the choice of therapy directed at the secondary lesion. Moreover, it is a conclusion not readily achieved by more traditional methods, since two lesions that appear to have similar histology or immunophenotype may in fact represent separate clones, whereas lesions with different histological and immunologic features could still represent evolution of the same clone.

The fourth purpose of gene rearrangement analysis, the detection of MRD, is actually more reliable in theory than in practice. Although one would like to be able to detect MRD as early and sensitively as possible after treatment in order to more proactively decide on subsequent management [9], the routine techniques described here for initial clonality determination at the time of diagnosis are not so ideal for sensitive detection of MRD after treatment. As noted, Southern blot analysis will detect a clone only if it represents 5–10% of the total cells in the specimen. And while one may always think of PCR as orders of magnitude more sensitive than Southern blot, that is not the case here. PCR is incredibly powerful at picking up a unique DNA sequence present in trace amount in a sample containing mostly nonhomologous sequences, as is the case when detecting the *BCR-ABL1* fusion gene in chronic myelogenous leukemia (see Chapter 9). The reason is that the primers are directed at the unique fusion sequence which should not be present in a normal cell population. But we have already stated that immunoglobulin gene rearrangement is a normal process that occurs in all lymphocytes, both benign and malignant, and the primers we use to assay clonality are based on consensus sequences which (barring somatic hypermutation) should not be significantly different between individuals, whether they have leukemia/lymphoma or not. Thus, the framework primers are competing for hybridization with both the clonal rearrangement and all the other benign rearrangements present in the sample. Since the thermodynamics of the hybridizations are essentially equivalent, there is no reason for the assay to preferentially pick out the clonal rearrangement, unless it is present in a predominant proportion of cells in the specimen, which will not be the case in MRD situations. In order to preferentially target the rearrangement in the malignant cells, primers must be designed that are patient specific and are based on DNA sequencing of the clone that was detected at the time of diagnosis. Then, typically using highly sensitive and quantitative real-time PCR approaches, the same clone can be identified after treatment even when the same is present in only trace amounts because the primers are (theoretically) not competing for all the other benign rearrangements in the

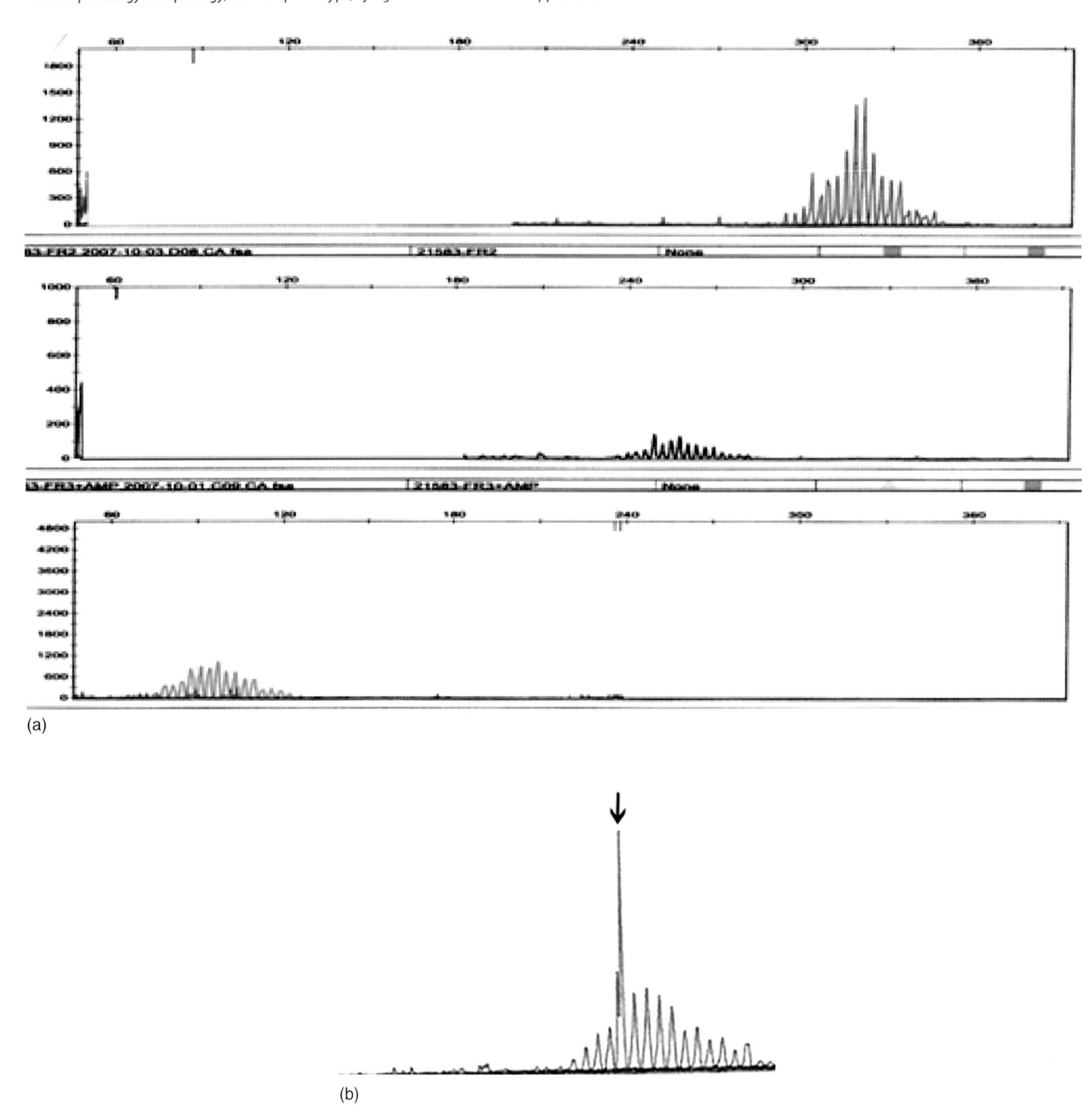

FIGURE 15.3 Examples of immunoglobulin gene clonality analysis by PCR. (a) Negative study showing only a polyclonal amplicon pattern with framework primers 1 (blue), 2 (black), and 3 (green). (b) Clonal peak (arrow) superimposed on a background polyclonal "smear" produced with the framework 1 primer set.

specimen [10]. Unfortunately, these methods are not generally available in routine clinical laboratories, so this application is generally relegated to special research situations.

The newest avenue of molecular testing of B-cell malignancies is gene profiling using gene expression–based microarrays. By detecting up- or downregulation of hundreds or thousands of genes in parallel, these lesions can be subclassified as to diagnosis and behavior in ways not possible by the more traditional methods [11, 12]. But, like the MRD approaches discussed earlier, these methods are at present restricted to the research setting. One possible compromise for the clinical laboratory is to choose one or a few of the more dramatically correlated genes in the microarray, such as *ZAP-70*, and simply test for those to achieve an informative approach of the entire array [11, 13].

Mature B-cell neoplasms are divided into the following categories [1]:

Chronic lymphocytic leukemia/small lymphocytic lymphoma (CLL/SLL)

B-cell prolymphocytic leukemia (B-PLL)

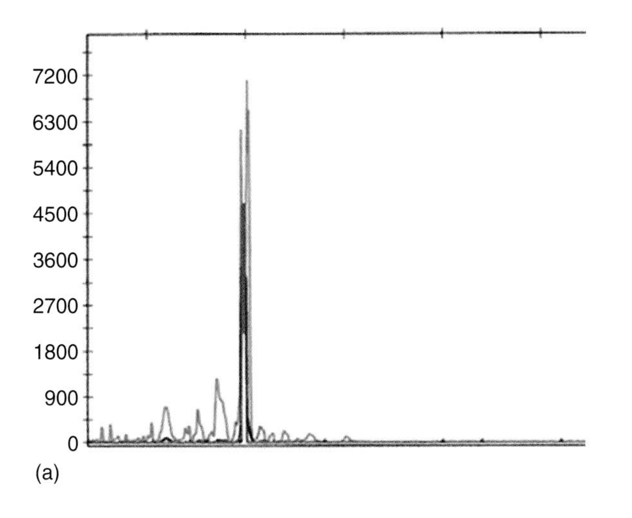

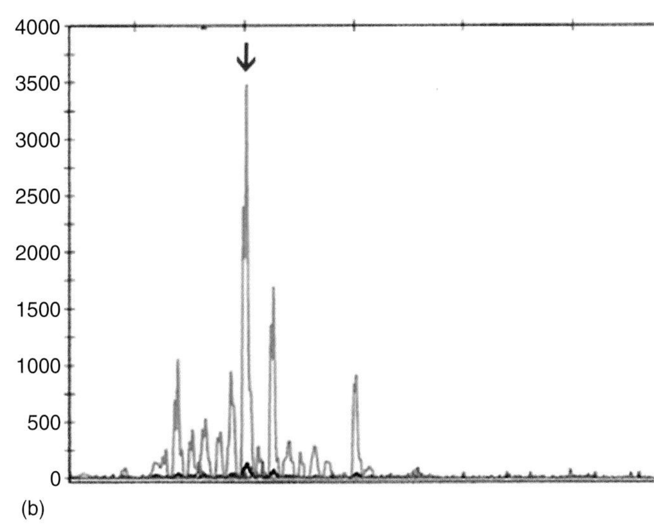

FIGURE 15.4 Immunoglobulin heavy chain clonality study in a patient with diffuse large B-cell lymphoma using PCR (framework 3 primer set results shown). (a) Clonal peak at 100 bp at the time of initial diagnosis. (b) The results at start of relapse. An amplicon peak of the same size is seen, suggesting that this is a recurrence of the initial clone, though this time there is a more prominent background polyclonal population.

Lymphoplasmacytic lymphoma (LPL)

Splenic marginal zone lymphoma (SMZL)

Extranodal marginal zone B-cell lymphoma of mucosa-associated lymphoid tissue

Nodal marginal zone B-cell lymphoma (nodal MZL)

Hairy cell leukemia (HCL)

Plasma cell disorders

Follicular lymphoma (FL)

Mantle cell lymphoma (MCL)

Diffuse large B-cell lymphoma (DLBCL)

Mediastinal (thymic) large B-cell lymphoma (MLBCL)

Intravascular large B-cell lymphoma (IVLBCL)

Primary effusion lymphoma (PEL)

Other variants of large B-cell lymphoma

Burkitt lymphoma/leukemia (BL)

All the above categories except for the plasma cell disorders are discussed in this chapter. Chapter 16 deals with the plasma cell disorders.

CHRONIC LYMPHOCYTIC LEUKEMIA/SMALL LYMPHOCYTIC LYMPHOMA

Chronic lymphocytic leukemia/small lymphocytic lymphoma is a lymphoproliferative disorder of small, mature B-lymphocytes primarily involving peripheral blood, bone marrow, and lymph nodes. The neoplastic cells are presumably derived from the naïve B-cells, characteristically coexpressing CD5 and CD23 (Figure 15.1). CLL/SLL is divided into two overlapping categories [1, 2, 14]:

1. CLL, primarily involving the bone marrow and the peripheral blood with or without lymph node involvement.
2. SLL, primarily involving lymph nodes with <30% bone marrow involvement and no peripheral blood lymphocytosis (non-leukemic).

Etiology and Pathogenesis

The etiology and pathogenesis of CLL/SLL are not known. A genetic background appears to play a role in the development of this disorder. Although CLL is considered the most common leukemia in the Western hemisphere, the incidence is very low in China and Japan. The incidence has remained low in the Japanese who have settled in Hawaii and the United States, suggesting a genetic role rather than environmental factors [15]. Certain genetic polymorphisms may predispose patients to CLL [16]. Familial risk of the development of CLL has been repeatedly discussed in the literature, including a series of 32 CLL families reported by the National Cancer Institute [17–21]. The familial susceptibility to CLL is observed in 5–10% of the patients [21]. No definitive environmental and occupational risk factors have been identified for CLL. Studies of the atomic bomb survivors during the period of 1950–1987 did not show a significant change in the incidence of CLL [22].

Pathology

Morphology

Peripheral blood and bone marrow lymphocytosis are morphologic hallmarks for the diagnosis of CLL. The majority of the lymphocytes are small and mature, appearing with a round nucleus, clumped chromatin, inconspicuous nucleoli, and scanty basophilic cytoplasm. Usually, a smaller proportion of lymphoid cells (<10%) consist of prolymphocytes. Prolymphocytes are larger than lymphocytes with more abundant cytoplasm and a prominent nucleolus. About 30% of the patients with CLL show a white blood cell count of >100,000/μL.

The proposed criteria by the International Workshop on CLL (IWCLL) are as follows [23]:

1. A sustained peripheral blood lymphocyte count of >10,000/μL with most of the cells being mature-appearing lymphocytes.
2. A bone marrow aspirate showing >30% lymphocytes.
3. Peripheral blood lymphocytes that have a B-cell phenotype consistent with CLL (i.e. weak expression of surface Ig, CD5+, and rosette formation with mouse erythrocytes).

According to the IWCLL, the diagnosis of CLL is confirmed if criteria 1 and 2 or 1 and 3 are present. If the peripheral blood lymphocyte count is <10,000/μL, then both criteria 2 and 3 must be present in order to make a diagnosis of CLL.

The National Cancer Institute–sponsored Working Group (NCI-WG) has recommended the following criteria for the diagnosis of CLL [24, 25]:

1. Peripheral blood lymphocyte count of >5,000/μL with <55% of the cells being atypical (prolymphocytes). The cells should:
   a. Express B-cell-specific differentiation antigens (CD19, CD20, CD23) and be positive for CD5, without other pan-T-cell markers.
   b. Express restricted kappa or lambda light chain surface Ig, confirming monoclonality.
   c. Express low density surface Ig.
2. Bone marrow aspirate showing >30% lymphocytes.

Patients with a blood count of ≤5,000/μL may subsequently develop CLL. Follow-up white blood cell counts and flow cytometric studies are recommended for the adult patients with persistent lymphocytosis of 3,000–5,000/μL when immunophenotypic features are consistent with CLL. There is no evidence that early diagnosis of CLL in asymptomatic patients with mild lymphocytosis grants clinical benefits [26].

The blood smears show evidence of absolute lymphocytosis with >90% of lymphocytes consisting of small cells with round nucleus, coarse chromatin, indistinct or absent nucleoli, and scanty non-granular cytoplasm (Figure 15.5). Smudge and basket cells are frequently present, particularly in cases with high lymphocyte count (Figure 15.5b and c). These cells represent degenerated and destructed lymphocytes. Approximately 15% of the CLL patients show *atypical* morphologic features, such as increased proportion of prolymphocytes (PL; >10% but <55%), or presence of >15% lymphocytes with cleaved nuclei, or lymphoplasmacytic morphology (Figure 15.6). Cases with >10% and <55% prolymphocytes are referred to as CLL/PLL.

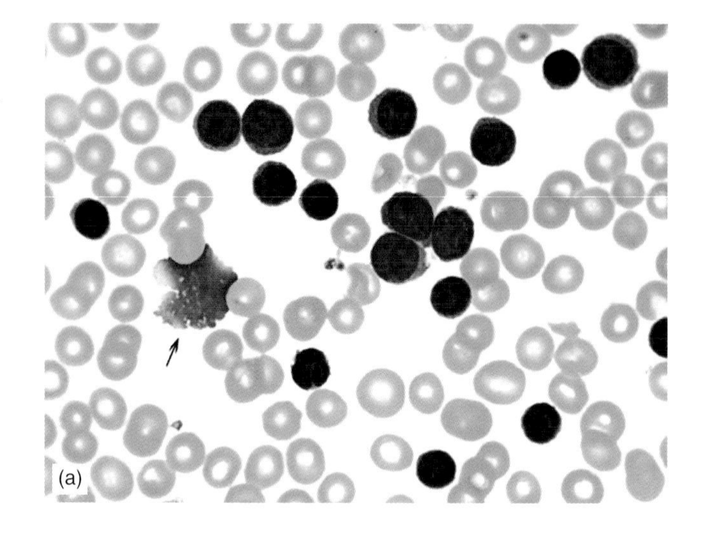

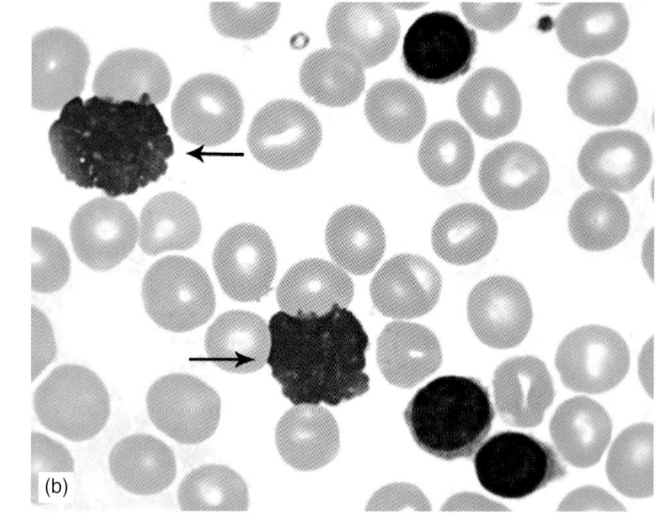

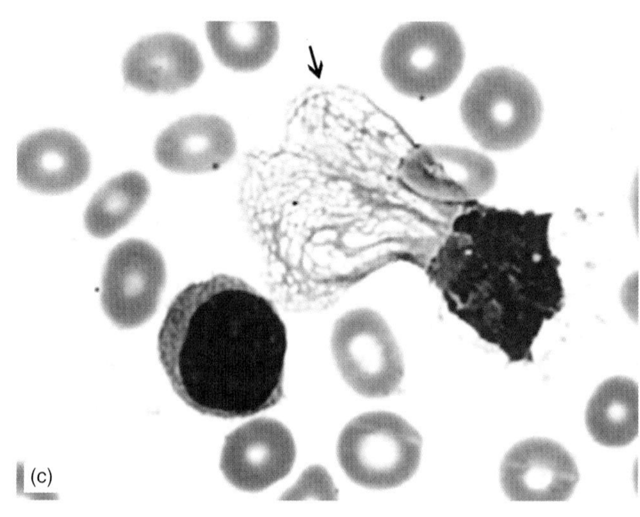

FIGURE 15.5 Blood smear of a patient with chronic lymphocytic leukemia demonstrating lymphocytosis with the presence of smudge cells (a and b, arrows) and a basket cell (c, arrow).

The pattern of involvement in the bone marrow biopsy sections is interstitial, nodular, diffuse, or a combination of these (Figures 15.7 and 15.8). The diffuse pattern is usually seen in the advanced stages of the disease. The lymphoid

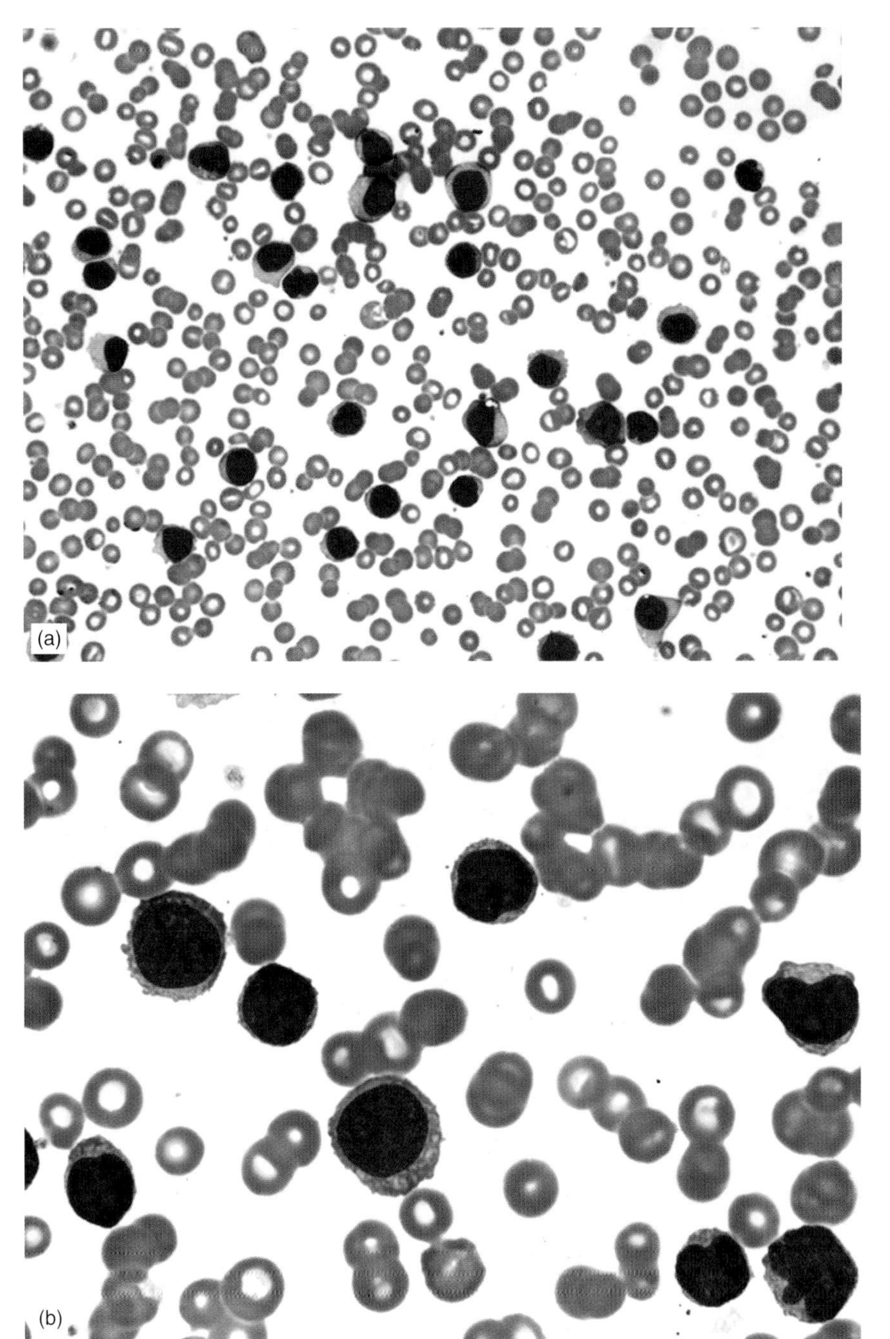

FIGURE 15.6 Blood smear of a patient demonstrating a mixture of lymphocytes and prolymphocytes: (a) low power and (b) high power views.

infiltrates consist of small, round lymphocytes with scattered prolymphocytes and larger cells called paraimmunoblasts. The lymphocyte count in the bone marrow smears in CLL is >30%, whereas in SLL it is ≤30%.

The affected lymph nodes show architectural effacement with a diffuse infiltration by small lymphocytes. Characteristically, there are ill-defined paler areas with the predominance of prolymphocytes and paraimmunoblasts (Figure 15.9). These areas are called "proliferation centers" or "pseudofollicles". Pseudofollicules are less frequently observed in the bone marrow and spleen. In some cases, there is a predominance of atypical lymphoid cells, such as lymphocytes with irregular nuclei or lymphoplasmacytic cells. These cases may mimic mantle cell or LPLs. In the spleen, white pulp is the primary site of involvement, but the red pulp is also frequently involved.

Immunophenotype

The CLL cells express CD5, CD19, CD23, CD43, and HLA-DR and are weakly positive for surface Ig, CD20, and CD11c (Figures 15.10 and 15.11). The expression of CD22, CD79a, and FMC7 is weak or absent. The CLL cells are negative for CD10 and BCL-1, and in some cases may express CD25, CD38, ZAP-70, and CD79b [27–31].

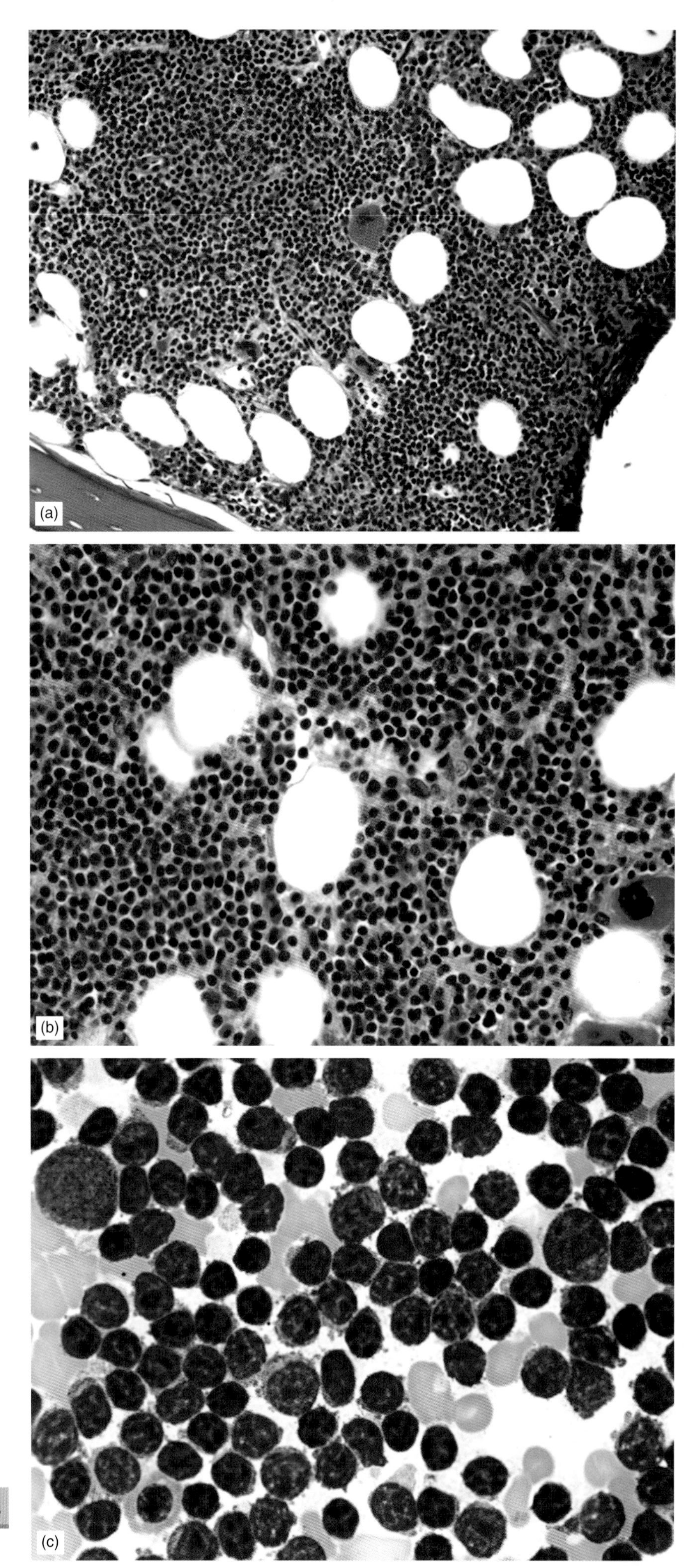

FIGURE 15.7 Bone marrow biopsy section (a and b) and bone marrow smear (c) showing involvement with chronic lymphocytic leukemia.

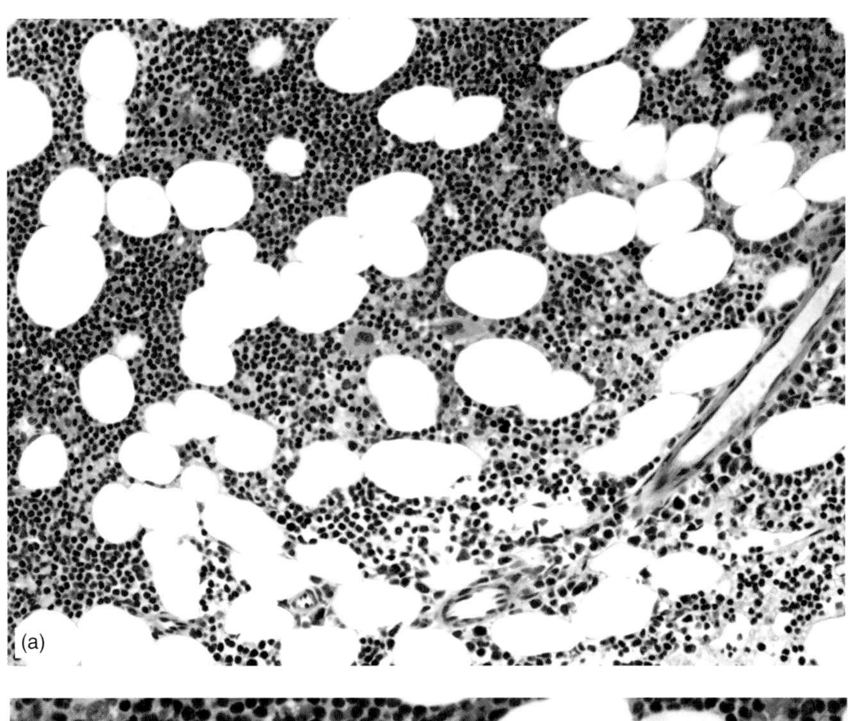

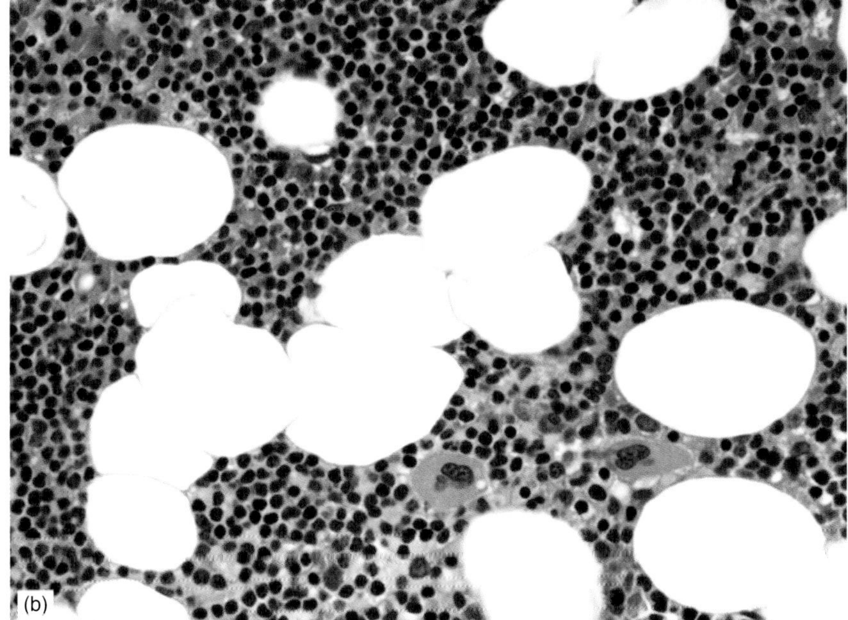

FIGURE 15.8 Bone marrow biopsy section demonstrating interstitial lymphoid infiltrate in a patient with chronic lymphocytic leukemia: (a) low power and (b) high power views.

A small proportion of CLL cases may totally lack the expression of CD5 or CD23 or partially express one or both of these markers. Also, some CLL cases may express strong FMC7 or CD20. These cases represent CLL with *atypical* immunophenotypic features.

The expression of CD38 and/or ZAP-70 is associated with an aggressive clinical course (see the following section) [32, 33]. CD38 is an ectoenzyme 45 kD transmembrane glycoprotein which appears to contribute to proliferative potential of B-CLL cells, enhancing clinical aggressiveness of the disease [34]. ZAP-70 (zeta-chain associated protein of 70 kD) is an intracellular tyrosine kinase which is involved in TCR signaling and is highly expressed in B-CLL cells which have unmutated IgV_H genes [35]. The expression of ZAP-70 is strong in normal T- and NK-cells and weak or negative in normal B-cells. Therefore, most flow cytometry laboratories use the patient's T- or NK-cells as internal positive controls, and the patient's normal B-cells (CD19+, CD5−) as internal negative control. ZAP-70 is considered positive when it is expressed by at least 20% of the CLL cells (Figure 15.12).

An immunophenotypic scoring system has been proposed for the diagnosis of CLL (Table 15.1). According to this scoring system, each of the following results is scored as for 1: weak expression of surface Ig, CD5 positivity, CD23 positivity, weak expression of CD22 or CD79a, and lack of

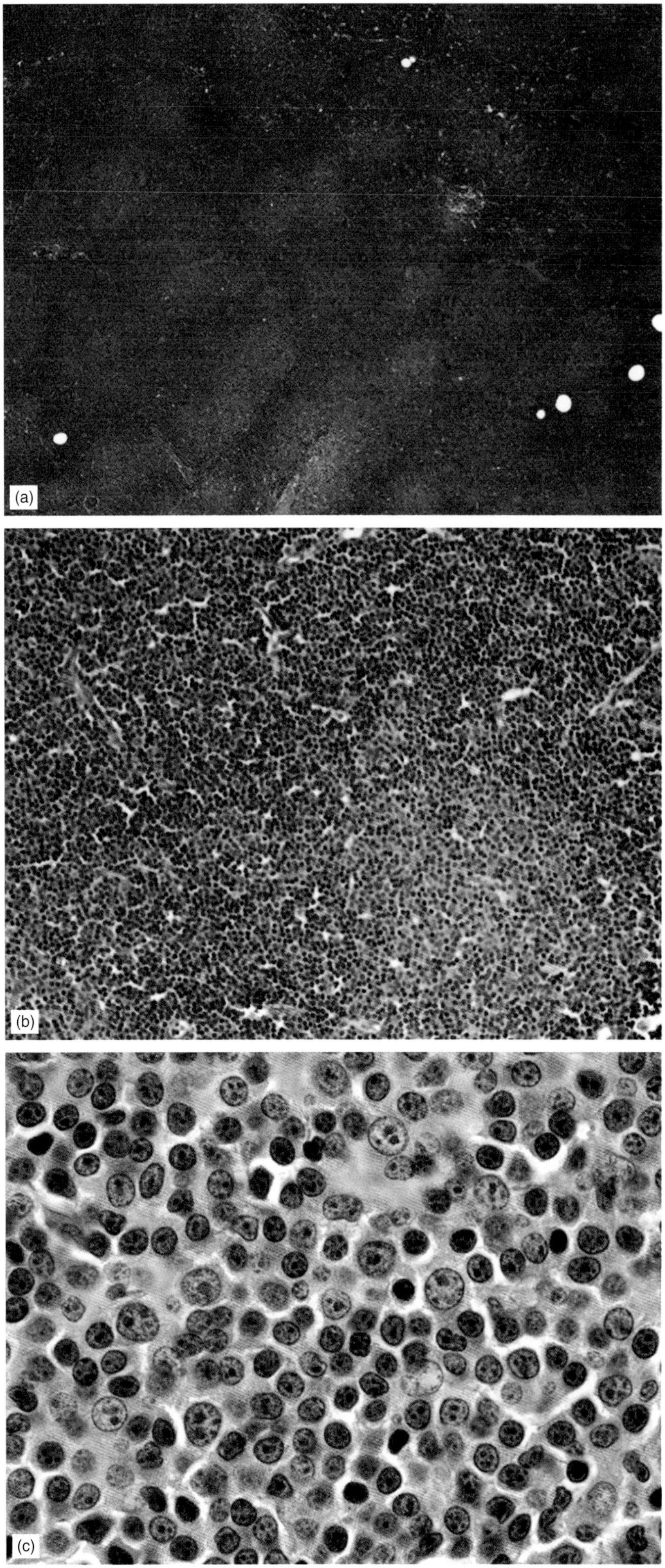

(a)

(b)

(c)

FIGURE 15.9 Lymph node section of a patient with chronic lymphocytic leukemia demonstrating proliferating centers or pseudofollicles (a and b, pale areas) consisting of a mixture of lymphocytes, prolymphocytes, and paraimmunoblasts (c).

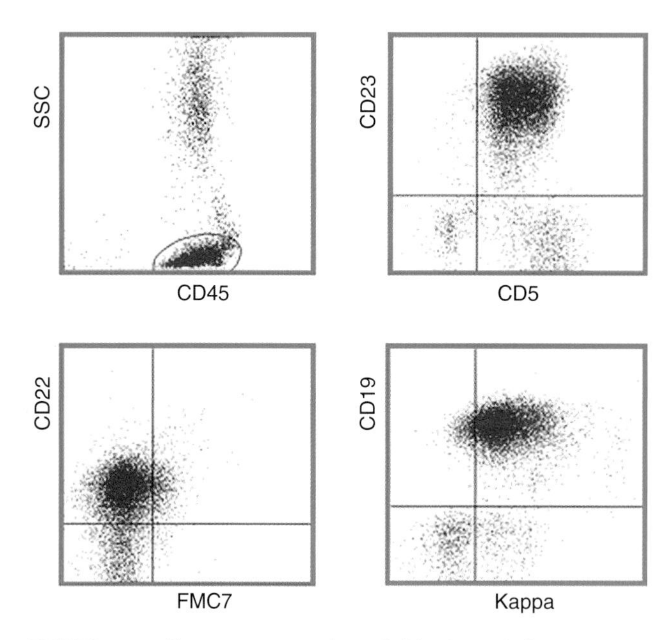

FIGURE 15.10 Flow cytometric analysis of a blood sample from a patient with CLL. The tumor cells are B-cells and characteristically coexpress CD5 and CD23.

expression of FMC7. Total scores of 4 or 5 are consistent with the diagnosis of CLL [36].

Aberrant expression of CD2, CD7, CD13, CD10, and CD34 has been reported in 10–30% of patients with CLL/SLL [37].

Cytogenetic and Molecular Studies

In CLL patients, karyotypic abnormalities tend to increase in frequency and number during the course of the disease. Chromosomal translocations, thought to occur mainly during the gene rearrangement process and common in other lymphoid malignancies, are rare in CLL. When translocations are found, they tend to result in a genetic loss rather than in the formation of a fusion gene or overexpression of an oncogene. These facts raise the pathogenetic possibility of missing tumor suppressor genes. Conventional cytogenetics detects structural chromosome abnormalities in about 40–50% of CLL patients [38–43].

Among patients with abnormal karyotypes, as many as 65% have one chromosome abnormality, 25% have two abnormalities, and the remainder have more complex abnormalities (6%) [4]. A 13q14 deletion is the most common finding (36–50% of the patients) [38, 39]; this deletion is believed to be a primary event in B-CLL, as it is present in a majority of the tumor cells and is frequently the sole abnormality (Figures 15.13 and 15.15). The second most common abnormality, and the most common abnormality to be detected by conventional cytogenetics, is trisomy 12 (11–21% of the patients) (Figures 15.14 and 15.15). Trisomy 12 usually displays an excess of large lymphocytes identifying the CLL mixed-cell-type variant of the FAB classification [42]. Trisomy 12 may be a secondary event in the course of CLL because it is typically identified in a minority of the tumor cells. Trisomy 12 is predominantly associated with unmutated V_H genes and seems to be associated with advanced or atypical cases of CLL. Less frequent primary aberrations in CLL include 14q32 rearrangements (up to 21%), 11q22.3 deletion (9–15%), and a 17p13 deletion (7–12%) (Figures 15.16 and 15.17) [41]. Other less

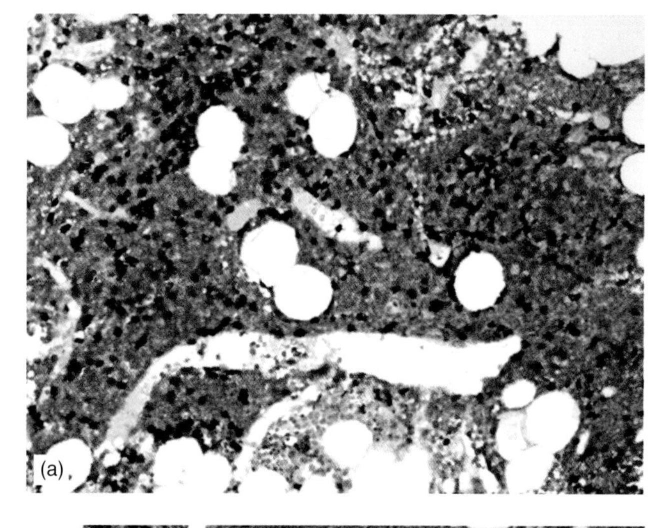

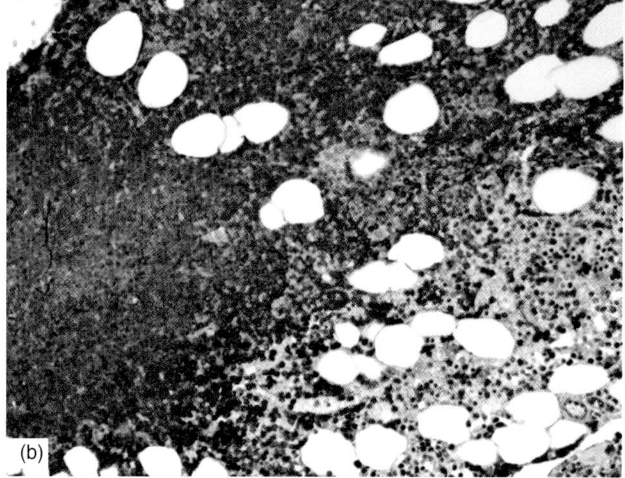

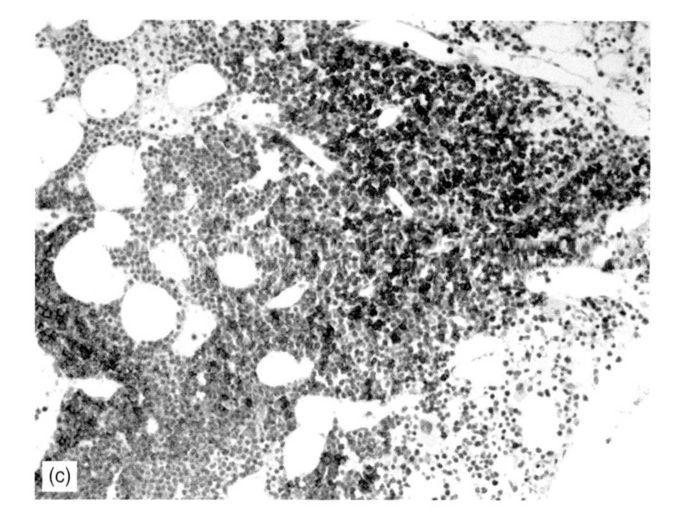

FIGURE 15.11 Immunohistochemical stains of a bone marrow biopsy section from a patient with CLL. (a) Dual staining demonstrating sheets of CD20+ cells (red) and scattered CD3+ cells (brown). The tumor cells also express CD5 (b) and CD23 (c).

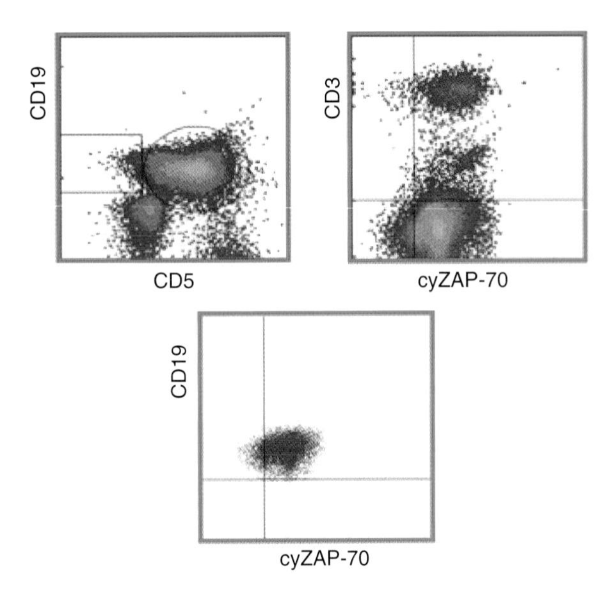

FIGURE 15.12 Flow cytometric analysis of ZAP-70. A large population of B-cells are CD5+, CD3–, and coexpress CD19 and ZAP-70.

TABLE 15.1 The immunophenotypic scoring system for the diagnosis of chronic lymphocytic leukemia/small lymphocytic lymphoma.*

Marker	Expression	Score**	Expression	Score
Surface Ig	Weak	1	Strong	0
CD5	Positive	1	Negative	0
CD23	Positive	1	Negative	0
CD22/CD79a	Weak or negative	1	Strong	0
FMC7	Negative	1	Positive	0

*Adapted from Refs [1, 26].

**Total scores in CLL/SLL are usually >3 and are often <3 in other lymphomas.

frequent chromosome abnormalities also occur (e.g. complex karyotypes).

A chromosome 6q deletion occurs in 7% of all CLL patients (as a primary event in 4%) and represents a cytogenetic and clinicobiological entity that exhibits a distinct phenotypic and hematologic profile. Patients with del(6q) usually present with a relatively high WBC count, classical immunophenotype, and CD38 positivity, which are associated with acceleration to the more aggressive prolymphocytic leukemia (PLL) [40]. Therefore, del(6q) patients require immediate therapy to achieve remission [42].

Chromosomal aberrations are not always detected in CLL patients' B-cells because it has been well noted that B-CLL cells are unresponsive to most lymphocyte mitogens and are extremely difficult to maintain in culture. For this reason, molecular cytogenetic techniques are more sensitive for the detection of clinically significant chromosome abnormalities than standard chromosome analysis. For example, fluorescence *in situ* hybridization (FISH) serves to unravel cryptic chromosomal aberrations that may not otherwise be detected due to the low mitotic index achieved in cultures obtained from the samples of most CLL patients even in the presence of B-cell mitogens [39, 41, 42, 44]. Additionally, when metaphases can be obtained, they are often so poor in quality that many aberrations escape detection. Therefore, FISH performed in conjunction with conventional cytogenetics is the methodology of choice for these disorders. All molecular cytogenetic techniques (i.e. FISH, CGH, and array CGH) have increased the detection rate of CLL to 80% [43, 44]. By using these techniques, as many as 65% of patients have one chromosome abnormality, 25% show two chromosome abnormalities, and the remaining 10% have more complex abnormalities. The recommended FISH panel for CLL detection consists of 11q22.3 (*ATM* gene), 13q14 (D13S319), *IGH* locus-specific probe (14q32), the centromere of chromosome 12 (D12Z3), and

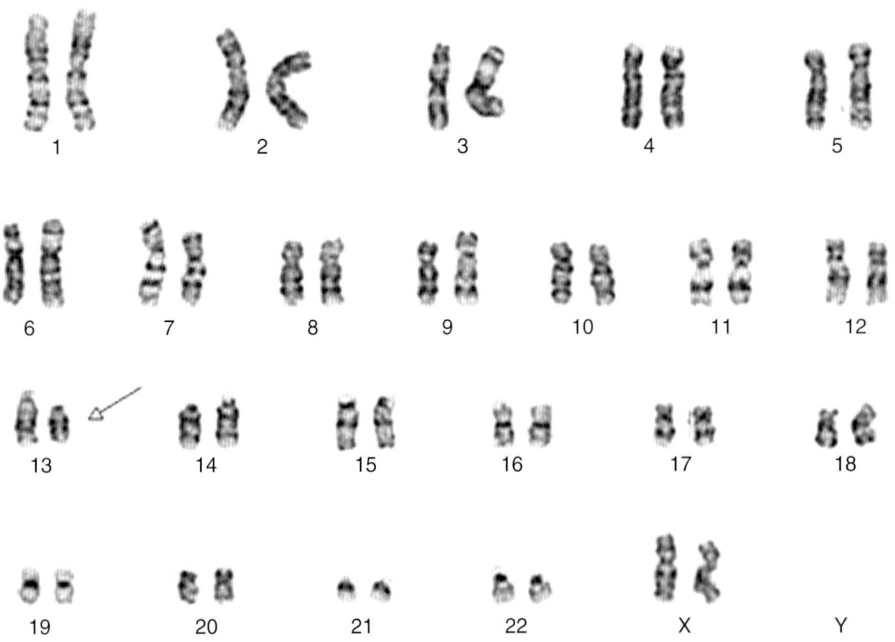

1 2 3 4 5
6 7 8 9 10 11 12
13 14 15 16 17 18
19 20 21 22 X Y

FIGURE 15.13 Bone marrow karyotype of a patient with CLL demonstrating 46,XX,del(13)(q12q14).

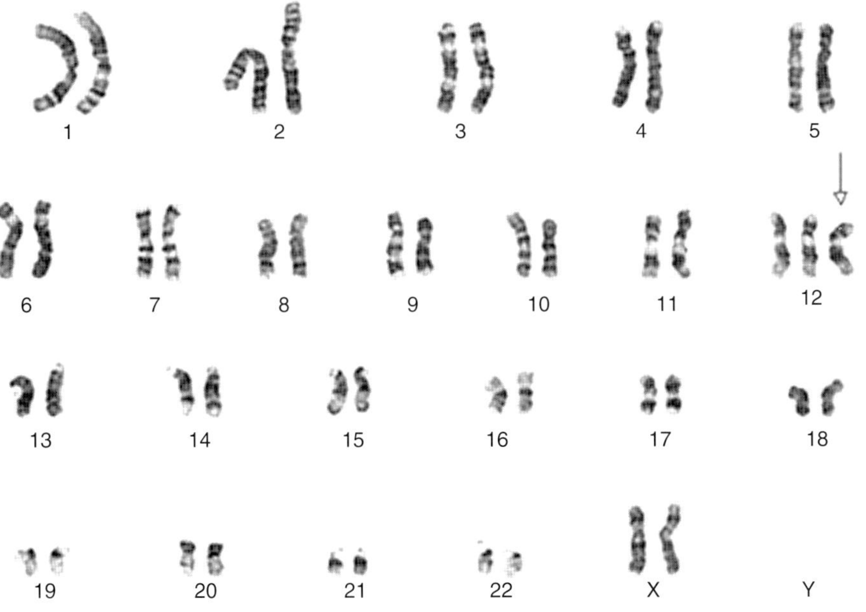

FIGURE 15.14 Bone marrow karyotype of a patient with CLL demonstrating 47, XX,+12

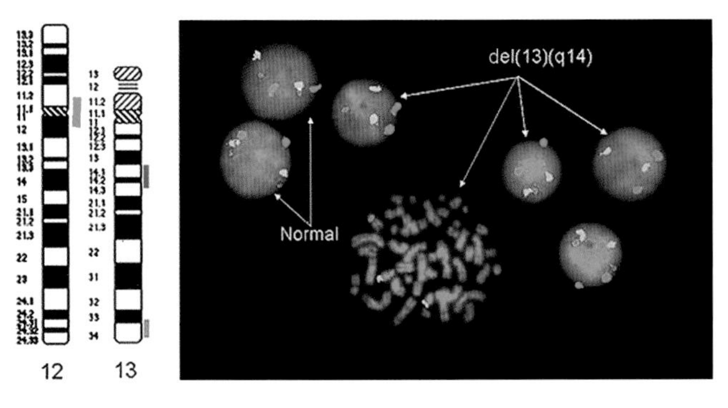

FIGURE 15.15 FISH analysis of cells from a patient with CLL showing two normal aqua (13q34), three green (Trisomy chromosome 12), and one red (deleted 13q14) signals.

17p13.1 (*p53* gene). In addition, there are recommendations to add 6q21 probe to the panel [41, 42].

The mutational status of the *IGVH* genes divides CLL into two major subtypes: mutated and unmutated [13, 45]. Approximately 15% of the CLL patients show no evidence of V_H gene mutation (unmutated). In general, these patients have an aggressive clinical course and advanced disease stage. The unmutated group shows a strong association with overexpression of ZAP-70 protein which could be detected by flow cytometry (see the following section).

Clinical Aspects

The most common type of leukemia in the Western countries is CLL/SLL accounting for about 40% of all leukemias in patients above 65 years of age. This disorder is extremely rare under the age of 30 years, but 20–30% of the patients are diagnosed under the age of 55 years [26, 46, 47]. The male:female ratio is about 2:1. Although the presence of familial aggregates of CLL has been well documented, the mode of inheritance is not known [20]. There is a sevenfold increase in the risk of CLL in first-degree relatives.

Approximately 25% of the patients are free of symptoms, and the CLL is an incidental finding during a routine blood examination. About 5–10% of the patients show systematic symptoms, such as weight loss, fever, night sweats, and/or extreme fatigue. Physical examination may reveal lymphadenopathy, splenomegaly, and hepatomegaly in approximately 85%, 50%, and 14% of the patients, respectively [47]. Autoimmune complications, primarily hemolytic anemia and thrombocytopenia, occur in up to 25% of the CLL/SLL patients [48].

The natural history of CLL/SLL is extremely variable with survival times ranging from 2 to 20 years [47]. Overall, the response rate to therapy and survival is better in women than in men [49]. Also, patients with atypical morphologic and/or immunophenotypic features tend to have a more aggressive clinical course. In general, presence and extent of lymphadenopathy, splenomegaly, hepatomegaly, anemia, and thrombocytopenia are the major clinical parameters that correlate with prognosis. Two major clinical staging systems have been developed by Rai *et al.* and Binet *et al.* [50, 51]. The original staging system proposed by Rai consisted of six stages from 0 to 5. This staging system was later (1987) modified and simplified to three major groups of low risk, intermediate risk, and high risk (Table 15.2) [26, 47]. In the Binet staging system, there are five designated sites of involvement demonstrated by cervical, axillary, and inguinal lymphadenopathies (unilateral or bilateral), splenomegaly, and hepatomegaly. Anemia and/or thrombocytopenia represent an advanced stage (Table 15.3).

Several biomarkers are indicative of aggressive clinical course in CLL/SLL. These include expression of CD38 and ZAP-70, unmutated *IGVH* (V_H) genes, del(11q22.3), and del(17p13.1). Several studies indicate a significantly less favorable median survival for patients with unmutated V_H and CD38-positve CLL cells than for patients with mutated V_H and CD38-negative tumor

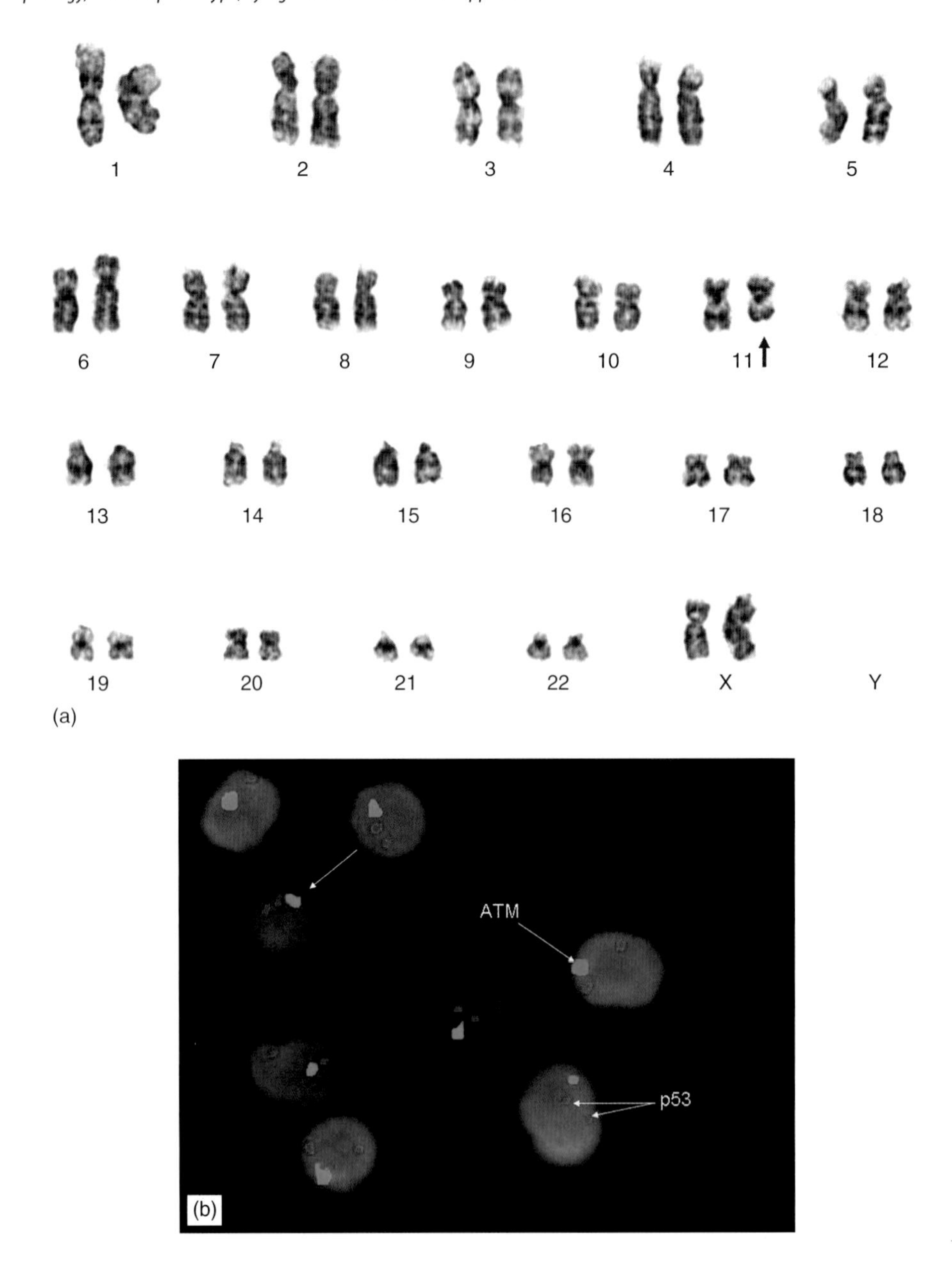

FIGURE 15.16 (a) Bone marrow karyotype of a patient with CLL with a 46,XX,del(11)(q22.1) and (b) FISH analysis with the ATM probe (green signals) and TP53 (red signals) on the same sample reveals a deletion of the 11q22 locus.

cells [52–55]. There is a strong correlation between unmutated V_H and overexpression of ZAP-70 in CLL cells. Also, there is an association between ZAP-70 expression and del(11q22.3) [56].

Cytogenetics is also helpful in predicting the course of CLL [42–44]. In fact, chromosomal abnormalities are independent predictors of disease regression and survival. Patients with diploid karyotypes or a 13q deletion as a sole abnormality have the best prognosis and a benign clinical course (median survival 79–133 months) [39, 40, 44]. The presence of the del (6q) or trisomy 12 usually has an adverse effect on patient's survival and results in intermediate prognosis (median survival 33–114 months) [39]. Patients with 11q22-23 (median survival 13–79 months), 17p13 deletion (median survival 9–32 months), and complex karyotypes have the worst prognosis [13, 40–51, 57].

Although data related to chromosomal abnormalities is important in determining a diagnosis and prognosis for CLL patients, it is also useful for additional applications, such as detecting MRD, and possibly indicating potential target sites for therapeutic interventions. Jahrsdorfer *et al.* showed that cytogenetic status correlates with the biological behavior of B-CLL *in vitro* [38]. Poor prognostic cytogenetics was associated with more rapid spontaneous apoptosis *in vitro*, lower immunogenicity, and higher lactose dehydrogenase (LDH). Good prognostic cytogenetics was associated

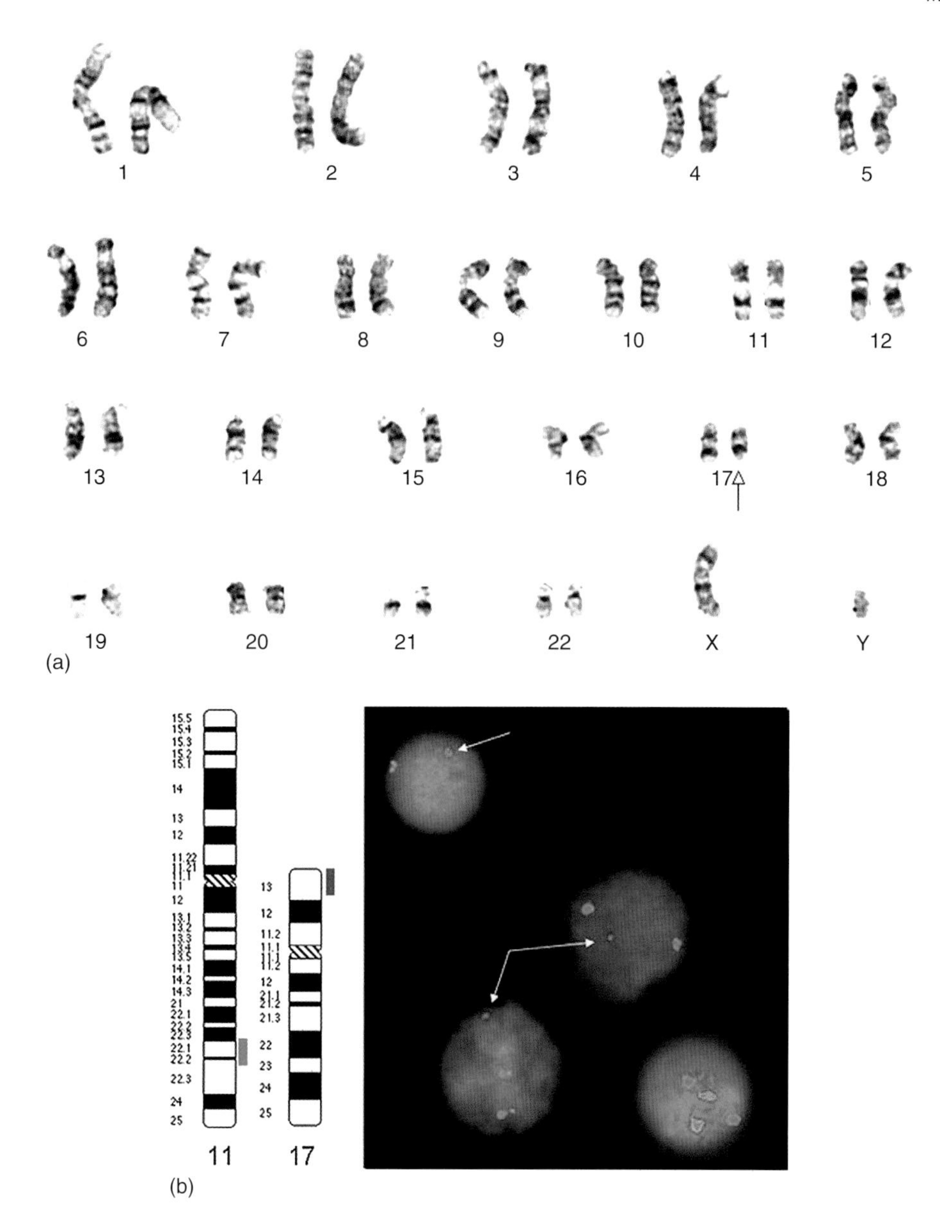

FIGURE 15.17 Bone marrow karyotype (a) and FISH analysis (b) demonstrating del(17)(p11.2).

with less spontaneous apoptosis, higher *BCL-2* levels, stronger immunogenicity, and lower levels of LDH [38].

A number of serologic parameters such as elevated levels of β2-macroglobulin soluble CD23 are reported in association with an aggressive clinical course in CLL patients. The prognostic factors in CLL/SLL are summarized in Table 15.3.

Several treatment modalities are available for patients with CLL/SLL [26, 58]. The therapeutic approaches are mainly based on the patient age and physical status, stage of the disease and cytogenetic findings. A "watch and wait" approach may be chosen for patients in early stage of disease or low-risk category, whereas patients with advanced or high-risk disease usually receive treatment. Alkylating agents, purine analogs, and monoclonal antibodies (such as rituximab) are frequently utilized in the therapeutic protocols.

Transformation of CLL to a More Aggressive Disease (Richter Syndrome)

Development of a high-grade non-Hodgkin lymphoma in patients with CLL was first described by Richter in 1928 (Figure 15.18). The term "Richter syndrome" was later referred to the transformation of CLL to a wide variety of more aggressive lymphoid malignancies, such as large cell lymphoma, prolymphocytic leukemia, lymphoblastic lymphoma, Hodgkin lymphoma, and plasma cell myeloma [59–62]. The incidence of Richter syndrome in CLL is about 5–10% with prolymphocytic leukemia being the most frequent type of transformation. The transformed cells may arise from the original CLL clone or may represent a new neoplastic clone, and the sequence-specific quantitative PCR methods discussed earlier can be used to distinguish between these two possibilities. The exact mechanism(s)

TABLE 15.2 The Rai and Binet staging systems in chronic lymphocytic leukemia/small lymphocytic lymphoma.*

Staging system	Features	Frequency (%)
Modified Rai (1987)		
Low risk (stage 0)	Lymphocytosis only	30
Intermediate risk (stages I and II)	Lymphadenopathy and/or hepatosplenomegaly	60
High risk (stages III and IV)	Hemoglobin <11 g/dL and/or platelet count <100,000/μL	10
Binet (1981)		
Stage A	<3 lymphoid areas**	60
Stage B	>3 lymphoid areas	30
Stage C	Hemoglobin <10 g/dL and/or platelet counts <100,000/μL	10

*Adapted from Refs [26, 47].

**Lymphoid areas are designated: Unilateral or bilateral cervical lymphadenopathy, and inguinal lymph; unilateral or bilateral axillary lymphadenopathy; unilateral or bilateral inguinal lymphadenopathy; splenomegaly; hepatomegaly.

TABLE 15.3 Prognostic factors in chronic lymphocytic leukemia/small lymphocytic lymphoma.*

Factor	Low risk	High risk
Gender	Female	Male
Clinical stage		
Binet	A	C
Rai	0	III and IV
Lymphocyte morphology	Typical	Atypical
Bone marrow involvement	Non-diffuse	Diffuse
Elevated levels of serum β2-macroglobulin and CD23	Not present	Present
CD38 expression	Negative	Positive
ZAP-70	Negative	Positive
IgVH gene status	Mutated	Unmutated
Cytogenetics	Normal or del(13q14)	del(17p13) or del(11)(q22)

*Adapted from Ref. [26].

of Richter transformation is not well understood. Multiple genetic abnormalities such as p53 mutation, deletion of retinoblastoma gene (13q14), increased copy number of *C-MYC* and decreased expression of *c-MYB* gene have been described [59]. Trisomy 12 and 11q aberrations are more frequent in Richter syndrome than in the overall CLL population [59, 60].

In most instances, Richter transformation is associated with the development of systemic symptoms, such as fever, weight loss, and night sweats, and/or a rapid organomegaly, such as increased lymphadenopathy, splenomegaly, and/or hepatomegaly. The site of transformation is usually lymph node or bone marrow, and occasionally extranodal/extramedullary sites, such as the skin, gastrointestinal tract, and central nervous system. Richter transformation is associated with a rapid clinical deterioration and a low response rate to therapeutic strategies. The median survival duration has been reported between 5 and 8 months [59].

Differential Diagnosis

The differential diagnosis comprises conditions associated with absolute peripheral blood lymphocytosis, increased proportion of CD5+ B-cells, and the presence of a monoclonal population of B-cells.

Chronic polyclonal B-cell lymphocytosis is a rare reactive lymphoproliferative disorder often observed in middle-aged women with a history of heavy smoking. The absolute blood lymphocyte count ranges from 4,000 to 20,000/μL with the presence of activated and binucleated lymphocytes (see Chapter 19). The majority of the lymphocytes are B-cells. These cells, unlike CLL cells, are polyclonal and lack CD5 expression.

CD5+ B-lymphocytes comprise a subset of the B-cells in normal individuals. In most studies, they account for up to 25% of the B-cells, though there are reports claiming that up to 47% of the normal B-cells may coexpress CD5 [63, 64]. A more recent study by Gupta and associates reported a mean percentage of about 12% CD5+ B-cells in the peripheral blood and bone marrow of the normal individuals. The expression of CD5 is dim on the normal B-cells and brighter on the CLL cells [65]. The recommended cutoff point for the detection of minimal residual CLL in treated patients is ≥25% CD5+ B-cells in the peripheral blood or in the bone marrow samples.

Monoclonal B-cell expansion in the elderly is a condition reported in about 3.5% of healthy individuals above 65 years of age with no evidence of absolute lymphocytosis or history of lymphoid malignancy. The immunophenotypic features of these monoclonal B-cells can be divided into two major groups: CLL-like and non-CLL-like.

The CLL-like phenotype is characterized by CD19+, CD23+, CD5+, FMC7−, and CD10−. The CD20 expression may be dim (typical) or strong (atypical). This phenotype has been observed in up to 13.5% of healthy relatives of patients with CLL [65, 66].

The non-CLL-like phenotype is characterized by CD19+, CD20+, CD23−, CD5−, FMC7+, and CD10−. This phenotype has been referred to as monoclonal B-lymphocytosis of undetermined significance (MLUS) by some investigators [67].

Mantle cell lymphoma and LPL share overlapping morphologic features with CLL. The neoplastic cells in MCL express BCL-1 protein and are usually negative for CD23. The cytogenetic hallmark for MCL is t(11;14). The immunophenotype of LPL is characterized by the lack of expression of CD5, CD10, and CD23 and the expression of surface and cytoplasmic IgM. A significant proportion

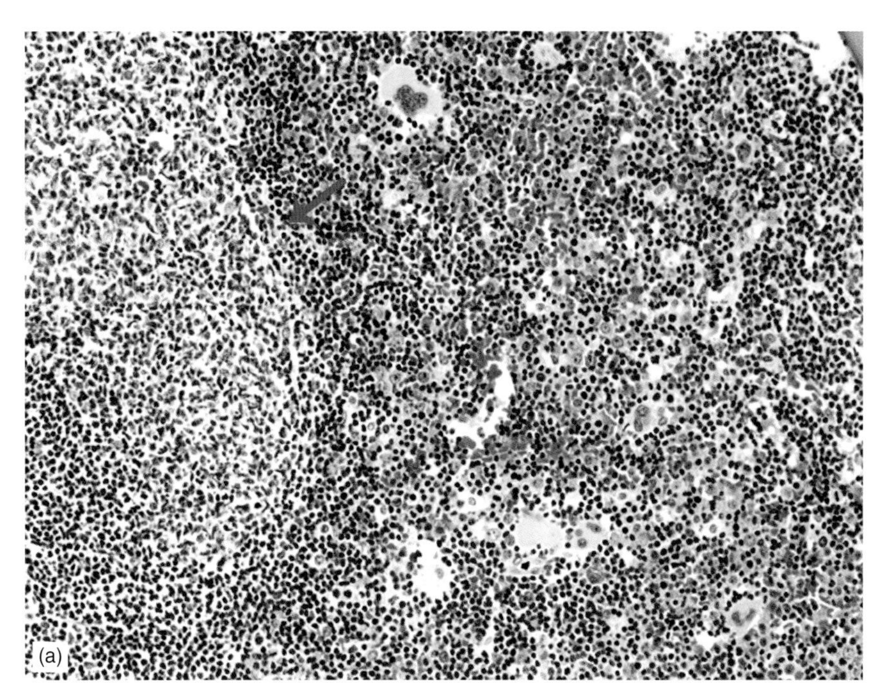

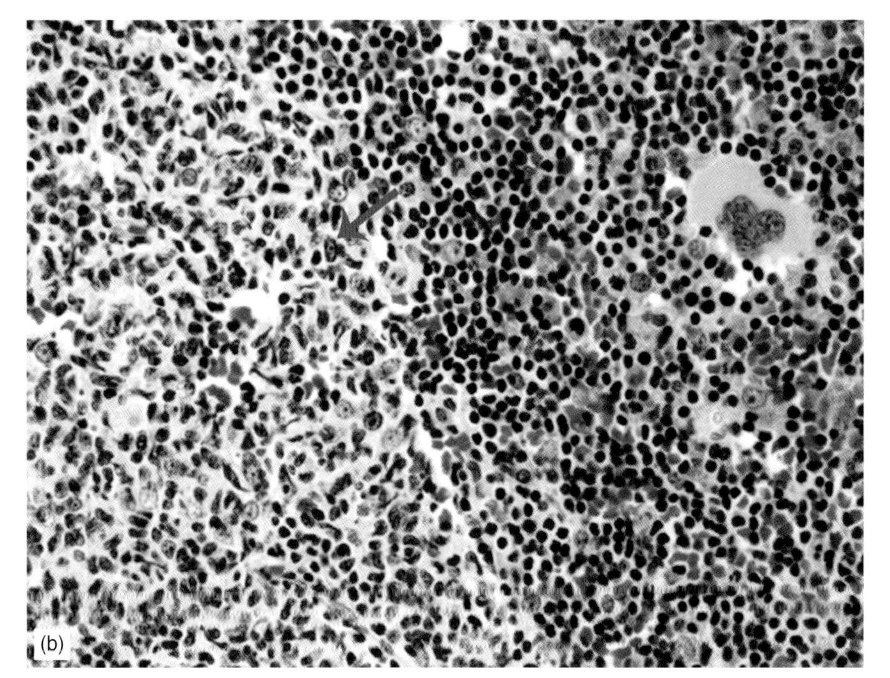

FIGURE 15.18 Richter syndrome. Bone marrow biopsy section from a patient with CLL demonstrating a focal area (arrow) of transformation to large cell lymphoma: (a) low power and (b) high power views.

of patients with LPL show del(6)(q21→q23) or t(9;14) (discussed later). A comparison of immunophenotypic and cytogenetic features in CLL/SLL, LPL, and MCL is presented in Table 15.4.

B-CELL PROLYMPHOCYTIC LEUKEMIA

B-cell prolymphocytic leukemia (B-PLL) is a rare lymphoproliferative disorder characterized by the clonal proliferation of prolymphocytes primarily involving blood, bone marrow, and spleen [1, 68–70]. Prolymphocytes are medium-sized cells with variable amounts of light basophilic cytoplasm, usually round nucleus, moderately condensed chromatin, and a prominent nucleolus (Figure 15.19). In prolymphocytic leukemia, prolymphocytes account for >55% of the lymphoid cells. B-cell leukemias with prolymphocyte-like features are divided into three groups:

1. *De novo* B-PLL.
2. Prolymphocytic leukemia evolved from the transformation of CLL [71].

TABLE 15.4 Comparison of immunophenotypic and cytogenetic features in chronic lymphocytic leukemia/small lymphocytic lymphoma (CLL/SLL), lymphoplasmacytic lymphoma (LPL), and mantle cell lymphoma (MCL).

Immunophenotype	CLL/SLL	LPL	MCL
CD5	+	−	+
CD10	−	−	−
CD19	+	+	+
CD20	Dim	+	+
CD22	Dim	+	+
CD23	+	+/− (dim)	−
CD79a	Dim	+	+
FMC7	−	+	+
BCL-1	−	−	+
Cytogenetics	del(17)(p13) del(11)(q22) Trisomy12 del(13)(q14)	t(9;14)	t(11;14)

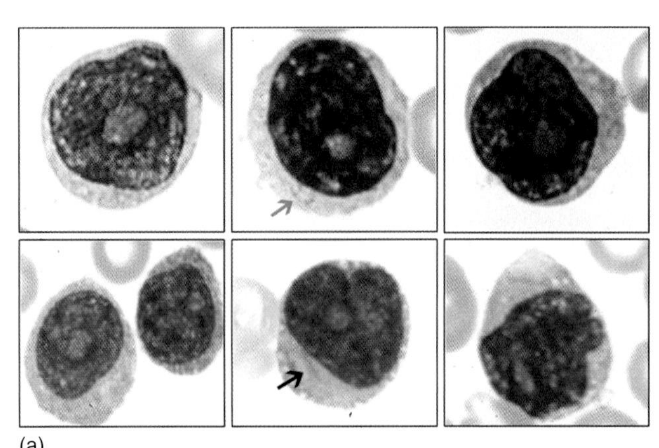

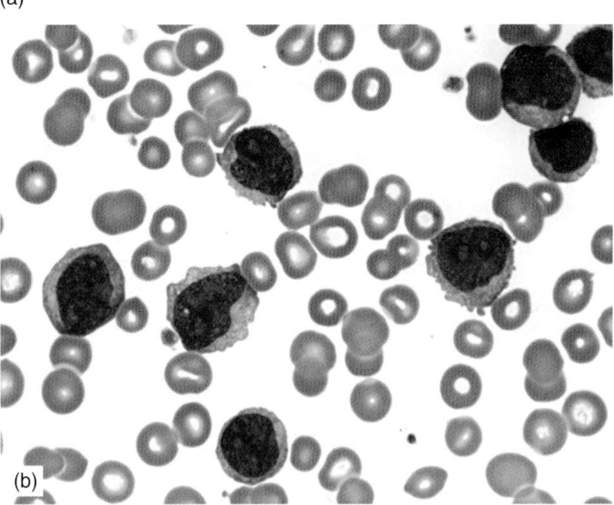

FIGURE 15.19 Prolymphocytes in the blood smear of patients with prolymphocytic leukemia (a and b). Some prolymphocytes may show cytoplasmic granules (a, green arrow) or inclusions (a, black arrow).

3. Leukemic phase of MCL with prolymphocytic morphology and evidence of t(11;14) [70, 71].

In this section the *de novo* B-PLL is discussed.

Etiology and Pathogenesis

The etiology and pathogenesis of B-PLL are not known.

Pathology

Morphology

Peripheral blood, bone marrow, and spleen are the major sites of involvement. There is a marked peripheral blood lymphocytosis (usually >100,000/mL) with the presence of >55% prolymphocytes. Prolymphocytes comprise >55% of the lymphoid cells in the peripheral blood. Prolymphocytes are larger and contain more cytoplasm than CLL cells (Figure 15.19). In most cases, they display a round nucleus with moderately condensed chromatin and prominent nucleolus (Figure 15.20). In rare cases, however, the nucleus is irregular or indented, and there are more than one prominent nucleoli. Occasionally, a small proportion of prolymphocytes may show cytoplasmic granules or inclusions [74]. The bone marrow biopsy sections often show a diffuse infiltration by the neoplastic prolymphocytes.

Splenic involvement is a frequent finding. Both white and red pulps are extensively infiltrated by prolymphocytes. The white pulp is markedly expanded and the red pulp is diffusely or patchily involved, often creating a mixture of diffuse and nodular patterns. Some of the extended white pulp nodule may show smaller lymphoid cells in the center surrounded by larger cells in the periphery [1].

Immunophenotype

B-cell prolymphocytic leukemia cells express surface Ig (IgM with light chain restriction) and B-cell-associated CD molecules, such as CD19, CD20, CD22, CD79a, and FMC7. They are negative for CD10 and CD23. CD5 is positive in approximately 30% of the cases [1], and over 50% of the cases express CD38 and/or ZAP-70 [68].

Cytogenetic and Molecular Studies

The chromosomal aberration of t(11;14)(q13;q32) frequently reported in leukemias with prolymphocytic morphology now, in most instances, is considered to represent the leukemic phase of a subtype of MCL [68, 72, 73]. Trisomy 12, del(11)(q22), and del(13)(q14) that are typically reported in CLL/SLL have also been reported in a number of B-PLL cases. However, some of these cases may represent the PLL transformation of CLL/SLL [71]. Several studies report karyotypic abnormalities involving 17p13 and 8q24, the sites of the *p53* and *c-MYC* genes, respectively (Figure 15.21). Loss of heterozygosity at 17p13 has been reported in 53% of the B-PLL patients [75, 76]. In one study, a complex karyotype was reported by demonstrating der(14)t(14;17) and t(2;8), involving both *p53* and *c-MYC* genes [75, 76].

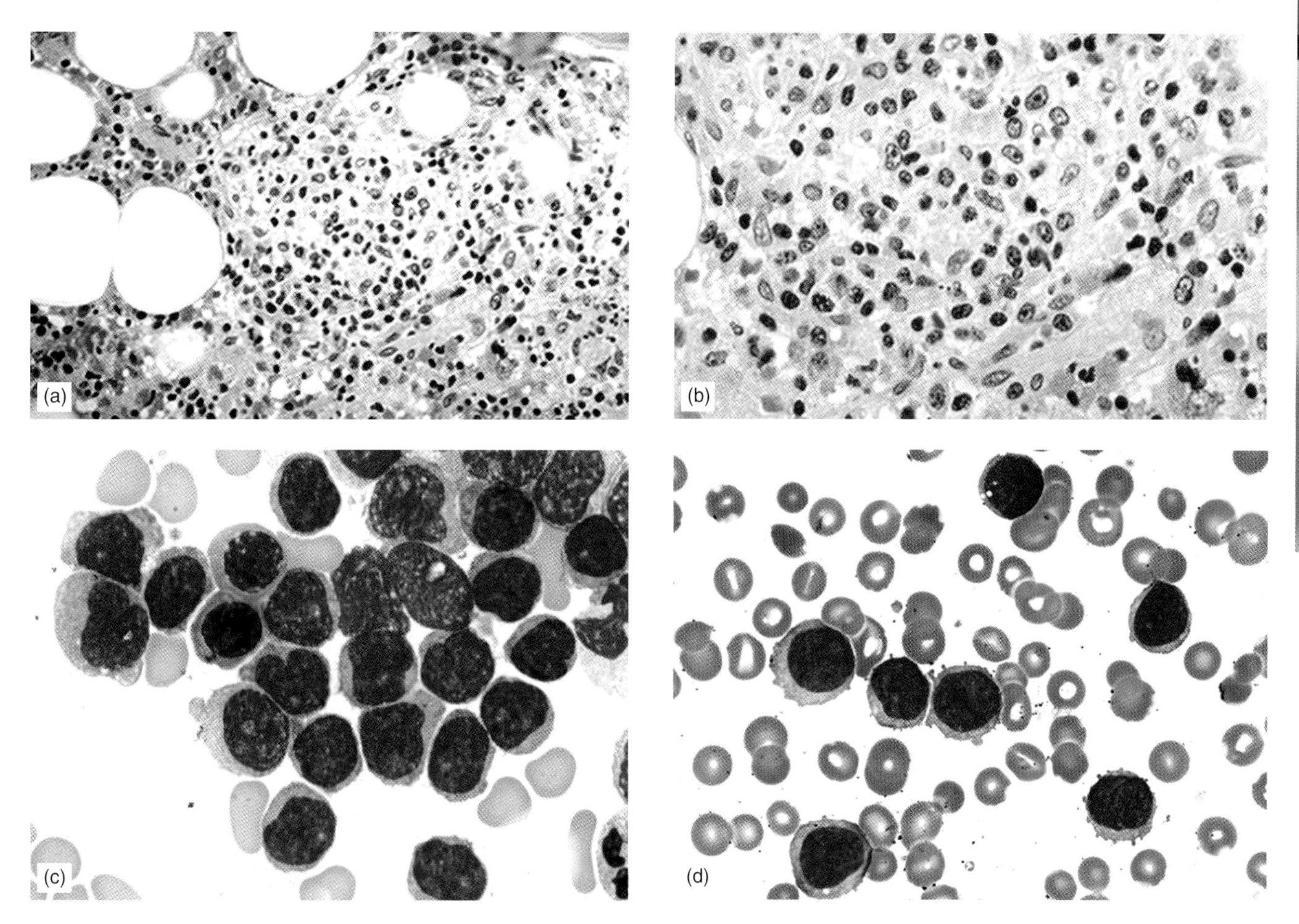

FIGURE 15.20 Prolymphocytic leukemia. (a and b) Bone marrow biopsy section, (c) bone marrow smear, and (d) blood smear. From Ref. [91] by permission.

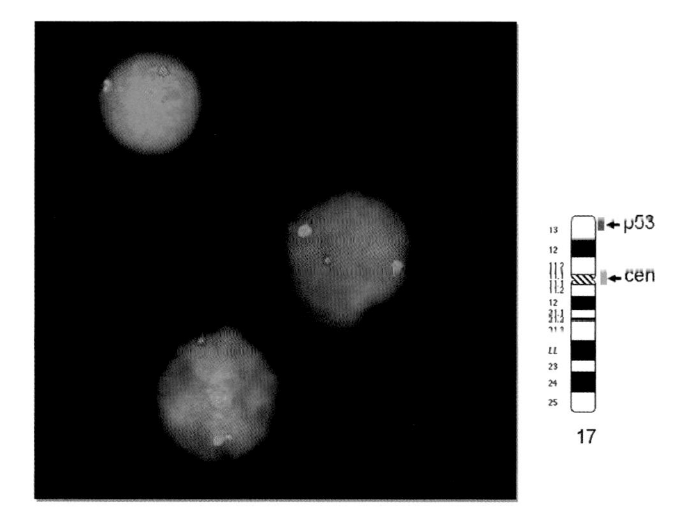

FIGURE 15.21 Deletion of *p53* (red signal) in B-cell prolymphocytic leukemia.

Other reported cytogenetic abnormalities include del(6q), t(6;12), and t(8;14) [68, 77]. B-PLL will usually show clonal immunoglobulin gene rearrangements just like the other B-cell malignancies.

Clinical Aspects

B-PLL is extremely rare and accounts for about 1% of all chronic lymphoid leukemias. It tends to affect elderly patients, usually >60 years old. The male:female ratio is >1. The characteristic features of B-PLL include a markedly elevated lymphocyte count, often >100,000/μL, massive splenomegaly, and minimal or no peripheral lymphadenopathy [68–70]. Anemia and thrombocytopenia are frequent findings. Serous effusions and central nervous system involvements are infrequent [68, 78]. B-PLL has an aggressive clinical course with a median survival time of about 3 years [79]. A complete response to alemtuzumab (anti-CD52 antibody) has been recently reported in a patient with B-PLL [80]. Transformation of B-PLL to diffuse large cell lymphoma (Richter syndrome) has been observed [81].

Differential Diagnosis

The differential diagnosis of B-PLL is with atypical CLL (CLL/PLL), HCL variant, MCL, and splenic marginal zone B-cell lymphoma.

The characteristic feature of B-PLL is >55% prolymphocytes in the peripheral blood lymphocyte count, whereas in CLL/PLL the prolymphocytes account for 10–55% of the lymphoid cells (Figure 15.22). The immunophenotype of CLL/PLL in most instances is similar to that of CLL, show coexpression of CD5 and CD23 and lack of FMC7, whereas the leukemic cells in *de novo* B-PLL lack CD5 and CD23 expression, but express FMC7.

Splenomegaly, similar to B-PLL, is one of the clinical hallmarks of HCL. But the lymphocyte count in HCL is low, normal, or modestly elevated. In a variant of HCL, leukemic cells morphologically mimic prolymphocytes. But the hairy cells are positive for tartrate-resistant acid phosphatase (TRAP) and express CD103, whereas B-PLL cells are negative for TRAP and CD103.

As briefly discussed earlier, the neoplastic cells in a group of patients with splenomegaly and markedly elevated peripheral blood lymphocyte count show morphologic features identical to prolymphocytes with cytogenetic evidence of t(11;14). These disorders were originally considered as B-PLL, but now most investigators consider these as a variant of splenic MCL.

Splenic marginal zone B-cell lymphomas similar to B-PLL usually show massive splenomegaly. But the peripheral blood lymphocyte count is usually much lower, the neoplastic cells are morphologically different from prolymphocytes (see the following section).

LYMPHOPLASMACYTIC LYMPHOMA/ WALDENSTROM MACROGLOBULINEMIA

Lymphoplasmacytic lymphoma/Waldenstrom macroglobulinemia (LPL/WM) is a mature B-cell lymphoproliferative disorder consisting of small lymphocytes, plasmacytoid lymphocytes, and plasma cells with the production of monoclonal IgM [1, 2, 82]. The primary sites of the involvement are bone marrow, lymph nodes, and spleen.

Etiology and Pathogenesis

The etiology and pathogenesis of the LPL/WM are not known. In several studies, the development of LPL/WM has been attributed to hepatitis C infection [83–85]. Occupational exposure to leather, rubber, dyes, and paints has been associated with LPL/WM in occasional cases [86]. Identification of family clusters and detection of the disease in identical twins suggest a genetic predisposition [87–89].

Recent studies demonstrate that in the majority of LPL/WM cases, class switch recombination (CSR) does not occur, suggesting that the neoplastic cells are either constitutively unable to carry out CSR or are prevented from doing so [90].

Pathology

Morphology and Laboratory Findings

The bone marrow biopsy sections show a nodular or diffuse infiltration of the marrow by lymphocytes, plasmacytoid lymphocytes, and plasma cells in various proportions. Scattered prolymphocytes and immunoblasts are usually present (Figures 15.22–15.24) [91–93]. The mixed lymphoplasmacytic population is more clearly demonstrated in the bone marrow smears, which may also show increased numbers of mast cells. Plasma cells may show Ig-containing nuclear inclusions (Dutcher bodies) or cytoplasmic inclusions (Russell bodies). Circulating neoplastic cells may be seen in the peripheral blood smears, some with plasmacytoid features, but lymphocyte count is not as high as observed in CLL (Figure 15.23c). The red blood cells show rouleaux formation or evidence of agglutination (cryoglobulinemia). Moderate to severe anemia is noted in up to 80% of the patients. Monoclonal IgM serum levels are usually >3 g/dL (75% kappa light chain restricted).

The lymph node biopsy sections show a diffuse involvement with sheets of lymphocytes admixed with plasmacytoid lymphocytes and plasma cells. Dutcher and/or Russell bodies may be present. The sinuses are often open with histiocytes engulfing immunoglobulin molecules. Prolymphocytes and immunoblasts are dispersed throughout the lesion, but proliferation centers, characteristics of CLL/SLL, are often lacking. Splenic involvement is often diffuse with the infiltration of both white and red pulps [1].

Immunophenotype

In the majority of LPL/WM cases, the neoplastic cells show characteristic immunophenotypic features of post-germinal B-cells and express CD19, CD20, CD22, CD79a, and FMC7 [1, 2, 81, 94]. These cells express plasma-cell-associated markers CD38 and CD138, are negative for CD5 and CD10, but may show dim CD23 expression [94]. They are IgM-positive but lack IgD expression [1, 81].

Cytogenetic and Molecular Studies

Deletion of 6q21-q23 is the most common chromosomal aberration in LPL/WM, reported in 40–70% of the patients (Figure 15.25) [84, 95]. Some patients may demonstrate t(9;14)(p13;q32). This translocation puts *PAX5* and *IGH* genes in juxtaposition. *PAX5* encodes B-cell-specific activator protein (BSAP), which is an important regulator of B-cell proliferation and differentiation. Other reported chromosomal abnormalities include trisomies or structural aberrations of chromosomes 10, 11, 12, 15, 20, and 21 [96, 97]. The LPL/WM cells show clonal rearrangement of immunoglobulin heavy and light chains with somatic mutation of the V-region genes [98]. As noted earlier, documentation of somatic V-region hypermutation, a property particularly of more mature B-cells that have passed through the germinal center, requires subcloning and sequencing studies, which are not routinely available in most clinical molecular pathology laboratories. Mutation of the *p53* tumor suppressor gene has also been described in some cases.

Clinical Aspects

LPL/WM is a rare disorder accounting for 1–1.5% of the lymphoid malignancies. The median age is about 65 years with <1% of the patients under 40 years of age.

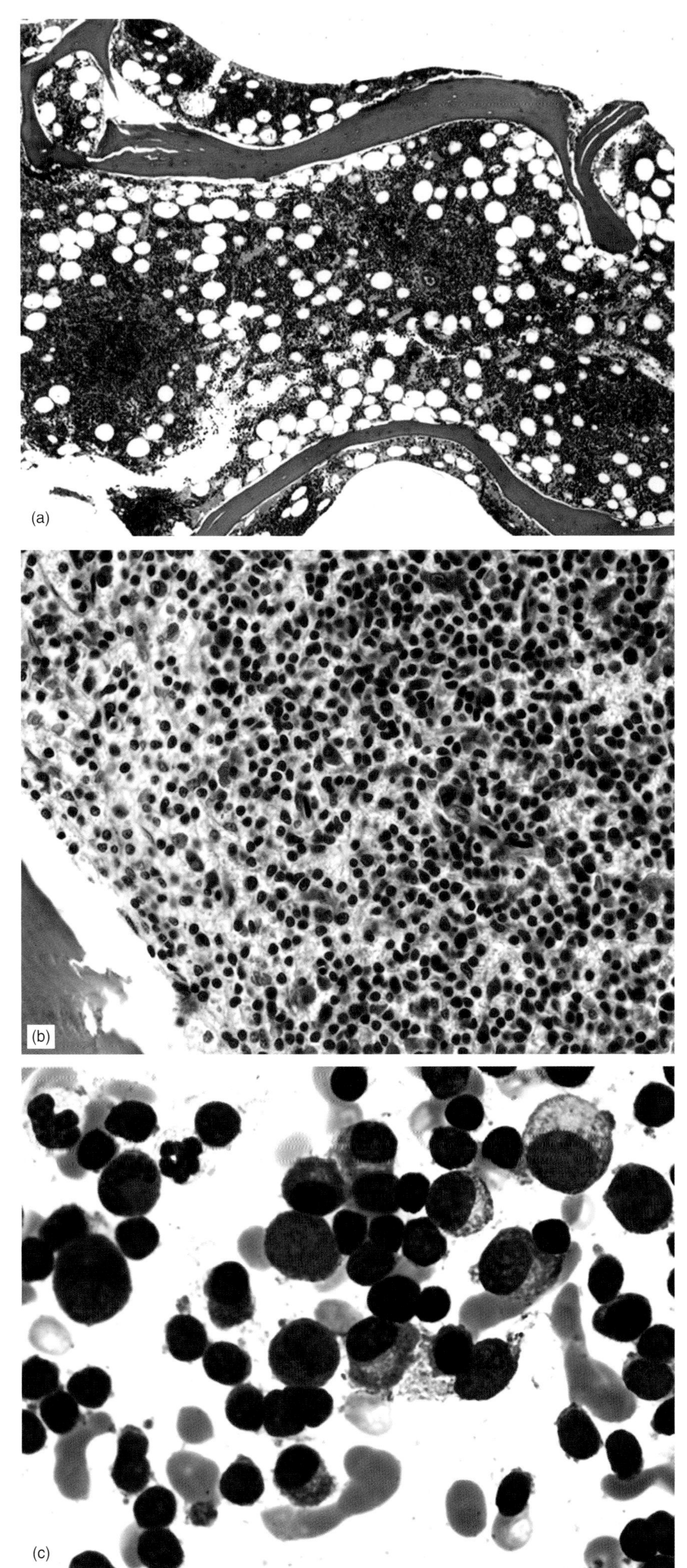

FIGURE 15.22 Lymphoplasmacytic lymphoma. The biopsy section demonstrates a nodular bone marrow involvement with a lymphoplasmacytic infiltrate: (a) low power and (b) high power views. The bone marrow smear shows a mixture of lymphocytes, plasmacytoid lymphocytes, and plasma cells (c).

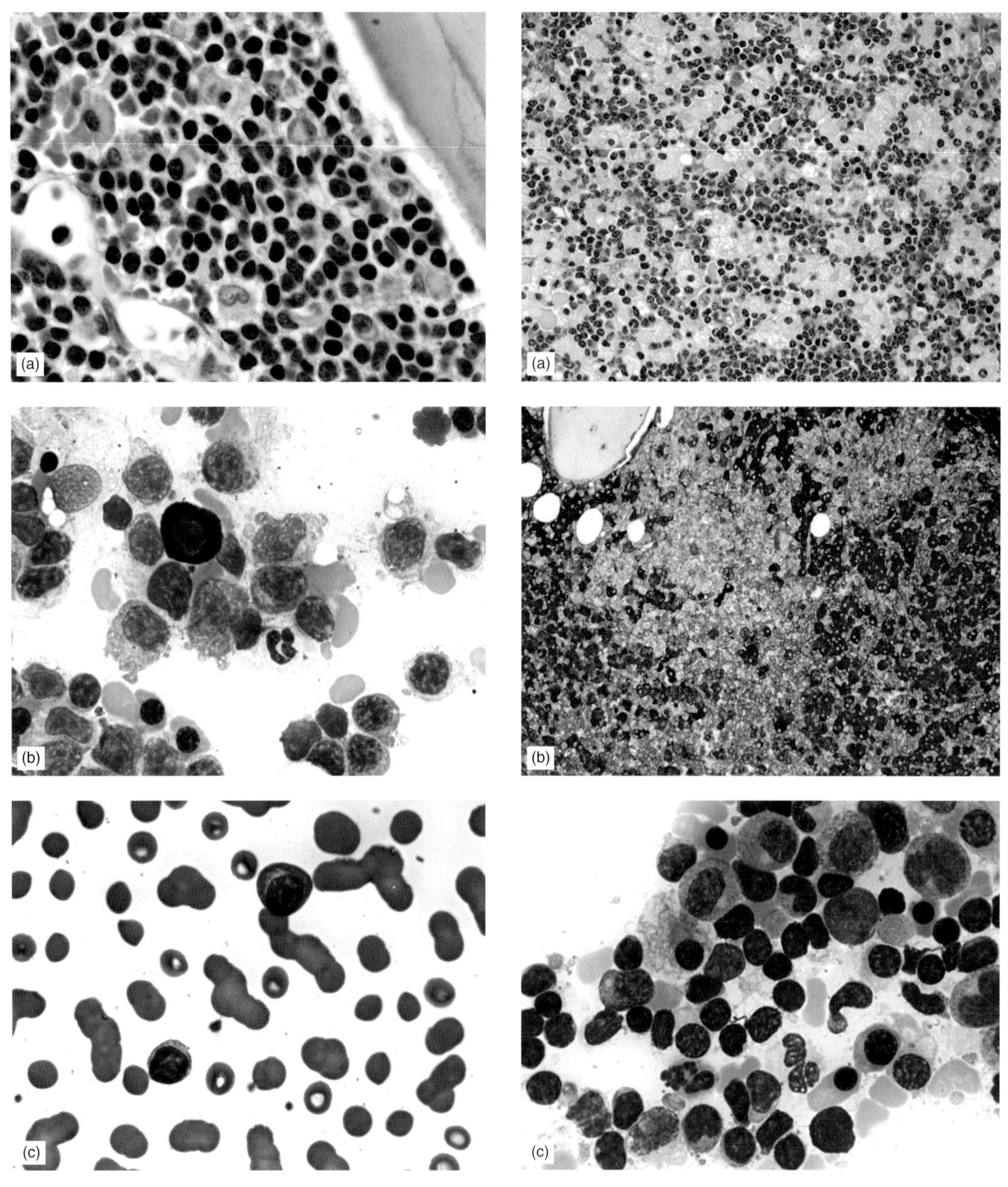

FIGURE 15.23 Lymphoplasmacytic lymphoma. The biopsy section and bone marrow smear demonstrate a lymphoplasmacytic infiltrate (a and b). The peripheral blood smear shows circulating plasma cells and red cell rouleaux formation (c).

FIGURE 15.24 Lymphoplasmacytic lymphoma. Bone marrow biopsy section demonstrates a lymphoplasmacytic infiltrate with aggregates of plasma cells with abundant pale vacuolated cytoplasm (a). These plasma cells express cytoplasmic Ig kappa light chain by immunoperoxidase stain (b). Bone marrow smear displays a lymphoplasmacytic infiltrate (c). From Ref. [91] by permission.

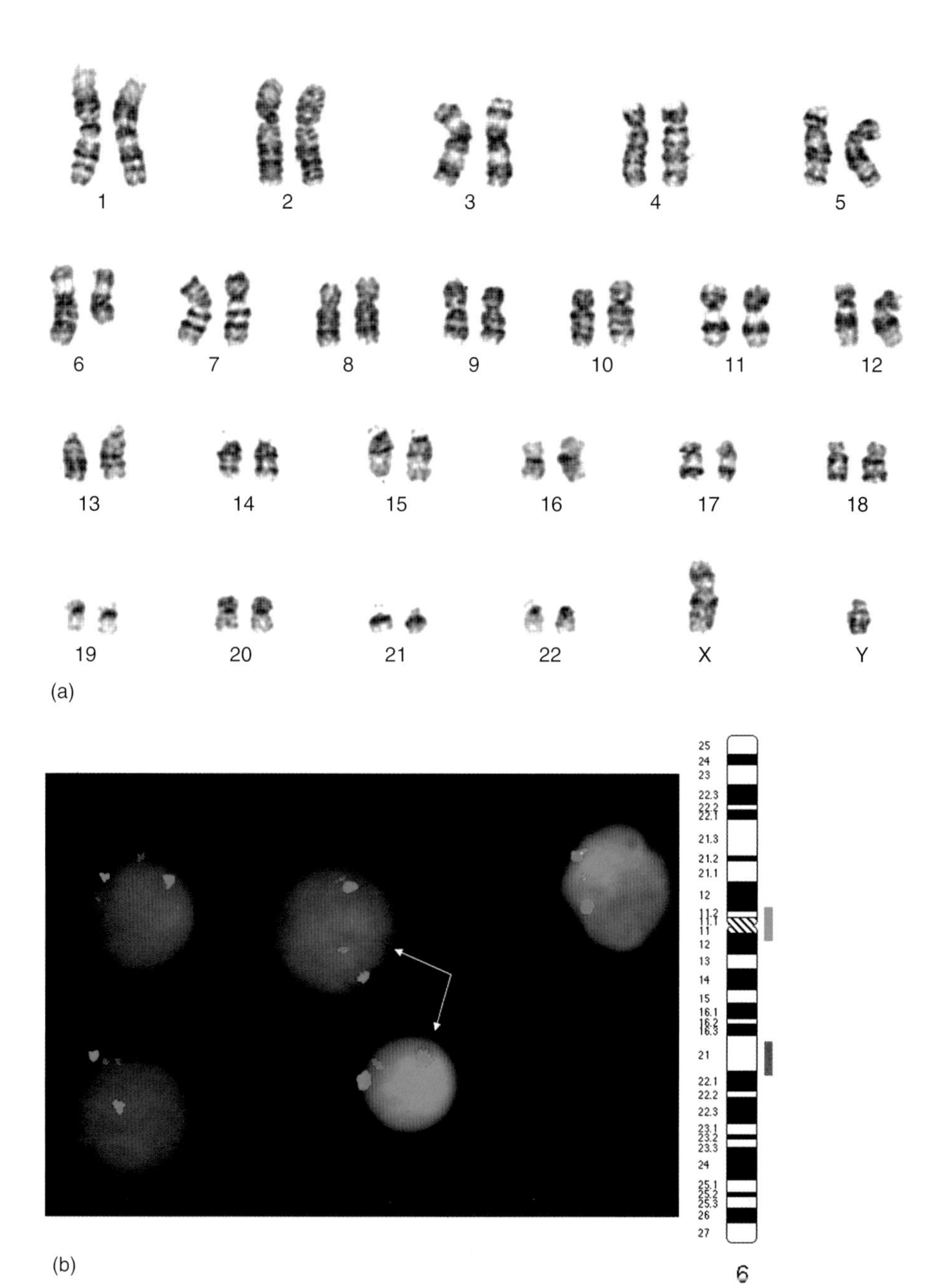

FIGURE 15.25 G-banded karyotype (a) with a deletion of 6q and confirmed on interphase cells by FISH (b) with a 6q-specific probe (red signal) (arrows).

Men account for approximately 60% of the patients [82, 99, 100].

Clinical symptoms include fatigue, bleeding, and hyperviscosity-related neuropathy, such as headache, vertigo, blurring or loss of vision, diplopia, or ataxia. Bone pain is rare and <5% of the patients have lytic bone lesions. Lymphadenopathy and splenomegaly are reported in 4–30% of the patients. Cryoglobulinemia, autoimmune hemolytic anemia and/or thrombocytopenia, and coagulopathies may occur [101]. Amyloidosis and erythematous urticarial skin vasculitis (Schnitzler syndrome) have been reported in some LPL/WM patients [102].

The Third International Workshop on Waldenstrom's Macroglobulinemia recommended treatment for patients with hemoglobin levels <10 g/dL, platelet count <100,000/mL, bulky adenopathy or organomegaly, symptomatic hyperviscosity, amyloidosis, and cryoglobulinemia [103]. A variety of therapeutic regimens are available including rituximab (anti-CD20) and combination chemotherapy with or without rituximab [103–105].

Differential Diagnosis

The differential diagnosis of LPL includes small B-cell lymphomas, such as CLL/SLL and MCL, plasma cell myeloma, and monoclonal gammopathy of undetermined significance (MGUS) [105]. The immunophenotypic and cytogenetic features of LPL/WM, CLL, and MCL are presented in Table 15.5. Unlike CLL cells, the neoplastic cells of LPL are negative for CD5 and CD23 and may express CD138. The major distinguishing features of MCL are CD5+, BCL-1+, CD23−, and t(11;14). LPL cells are CD5 negative and show no t(11;14), but frequently demonstrate del(6q21-23). Plasma cell

TABLE 15.5 Morphologic, immunophenotypic, and cytogenetic characteristics of lymphoplasmacytic lymphoma (LPL), splenic marginal zone lymphoma (SMZL), and hairy cell leukemia (HCL).

Features	LPL	SMZL	HCL
Spleen			
Primary involved area	White pulp	White pulp	Red pulp
Pattern	Nodular	Nodular	Diffuse
Bone marrow	Nodular, diffuse or interstitial	Mostly intrasinusoidal	Mostly interstitial
Cytology	Small lymphocytes, plasmacytoid lymphocytes	Medium-sized lymphocytes, polar villous projections	Medium-sized lymphocytes, hairy projections
Immunophenotype	IgM+, CD19+, CD20+, CD22+, CD79+, FMC7+, CD23±, CD5−, CD10−, CD25−, CD103−, TRAP−	IgM+, IgD+, CD19+, CD20+, CD22+, CD79a+, CD5−, CD10−, CD23−, CD25−, CD103−, TRAP−	IgM+, CD19+, CD20+, CD22+, CD79a+, FMC7+, CD11c+, CD25+, CD103+, TRAP+, CD5−, CD10−, CD23−
Cytogenetics	del(6)	del(7)(q21→q32)	None

disorders are comprised predominantly of plasma cells, rarely involve lymph nodes or spleen, and show a non-IgM monoclonal serum paraprotein (see Chapter 16).

SPLENIC MARGINAL ZONE LYMPHOMA

Splenic marginal zone lymphoma (SMZL) or splenic lymphoma with villous lymphocytes is a B-cell neoplasm of small to medium-sized lymphocytes presumably arising from the marginal zone of the splenic white pulp. Characteristic features include splenomegaly, bone marrow infiltration with intrasinusoidal pattern, moderate peripheral blood lymphocytosis with the presence of villous lymphocytes, and a relative indolent course [1, 2, 106–110].

There are two other types of marginal zone-related B-cell non-Hodgkin lymphomas: nodal marginal zone B-cell lymphoma and extranodal marginal zone B-cell lymphoma of mucosa-associated lymphoid tissue (MALT) type. These two distinct clinicopathologic entities are discussed separately in the following sections.

Etiology and Pathogenesis

The etiology and pathogenesis of SMZL are not known. The possibility of environmental factors initiating *in vivo* somatic mutation of *IGVH* gene has been entertained [111]. The increased incidence of SMZL in patients infected with hepatitis C virus suggests a role for a viral antigen epitope in the B-cell selection and the development of SMZL [1, 112].

Allelic loss of chromosome 7q (deletions of 7q22-32) is frequently observed in SMZL [113]. These alterations lead to dysregulation of cyclin-dependent kinase 6 (*CDK6*) gene, possibly playing a role in the pathogenesis of SMZL. Also, the 7q deletion may play a role in the inactivation of *p53* tumor suppressor gene.

Pathology

Morphology

The spleen is enlarged with a median weight of 1,750g. The cut surface usually shows multiple gray-tan nodules of various sizes. Diffuse involvement is rare [1, 2, 110]. In the early stages of involvement, the splenic sections show enlarged white pulps with the expansion of marginal zones merging into one another. In the center of the nodules, usually there is a remnant of germinal center surrounded by mantle cells and expanded marginal zones (Figure 15.26). Mantle cells are small with scanty cytoplasm and slightly irregular nuclei. The cells surrounding mantle cells in the marginal zone are larger with more dispersed chromatin and abundant pale cytoplasm. Admixed with these cells are scattered centroblasts and immunoblasts. There is infiltration of neoplastic cells into the red pulp (Figure 15.27). In later stages, eventually, the white pulp expansion and red pulp infiltration create sheets of neoplastic cells, making the separation of white and red pulps unclear [114].

The hilar splenic lymph nodes are commonly involved. The peripheral lymph nodes are affected less frequently. There is partial effacement of nodal architecture with infiltrating neoplastic nodules. Some of the nodules may contain a central reactive follicle. Sinuses are usually spared. Some cases may show complete effacement of the nodal architecture.

Bone marrow is commonly involved. The pattern of involvement is intrasinusoidal, interstitial, nodular, paratrabecular, or a combination of these (Figures 15.28 and 15.29). The intrasinusoidal pattern is highly characteristic of SMZL. The bone marrow smears show the presence of atypical small to medium-sized lymphocytes with abundant cytoplasm, round or irregular nuclei, and condensed chromatin. Some of the lymphocytes may show villous cytoplasmic projections, which are often polar (Figure 15.29c).

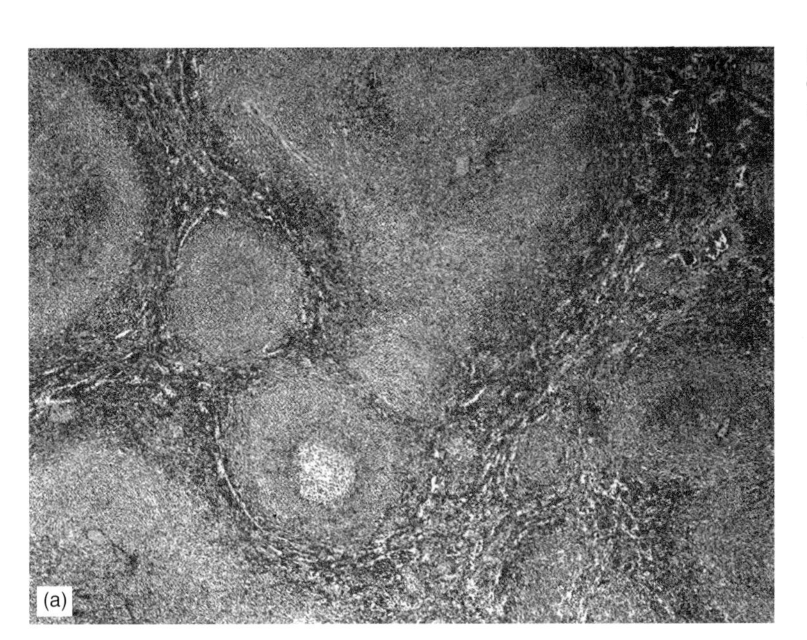

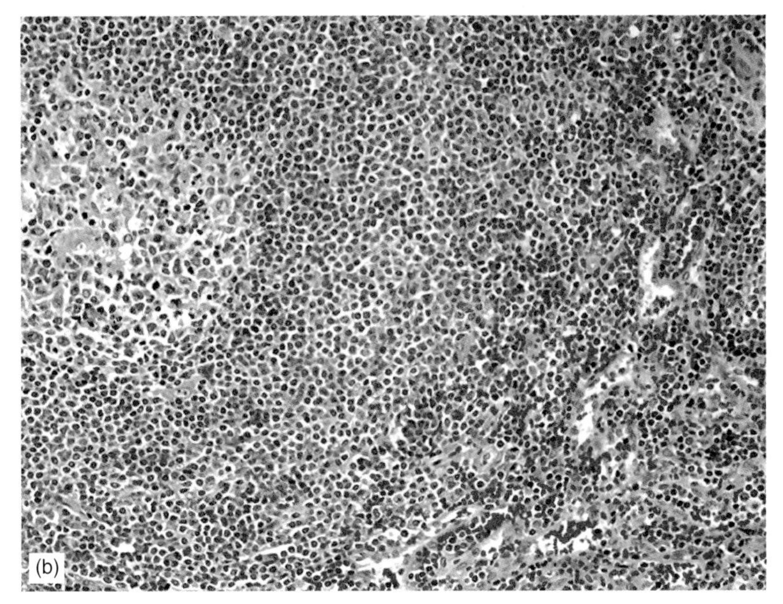

FIGURE 15.26 Splenic marginal zone B-cell lymphoma demonstrating the expansion of marginal zone of the white pulp in the spleen: (a) low power and (b) high power views.

There is usually a moderate peripheral blood lymphocytosis with various proportions of atypical lymphoid cells. These cells are morphologically similar to those described in the bone marrow smears and may or may not show polar villous projections. Those without villous projections may appear plasmacytoid. Scattered larger cells with prominent nucleoli may be present.

The liver is involved in the majority of the cases. The portal tracts are the predominant sites of infiltration. Skin, pleura, and soft tissue involvements have been rarely reported.

Immunophenotype

The neoplastic cells express surface IgM and IgD and show positivity for B-cell-associated markers, such as CD19, CD20, CD79a, FMC7, and *PAX5* [1, 2, 110]. They are negative for CD5, CD10, CD23, CD25, CD103, BCL-1, and BCL-6. They express BCL-2 and usually lack the expression of CD43 and TRAP. Results for DBA44 staining are variable.

Cytogenetic and Molecular Studies

As mentioned earlier, allelic loss of chromosome 7q (deletions of 7q22-32) and dysregulation of the *CDK6* gene have been reported in some cases of SMZL [113]. Abnormalities of chromosome 14q32, harboring the *IGH* gene, have also been reported in SMZL in the forms of t(6;14) and t(9;14) (Figure 15.30). The earlier reports of SMZL with t(11;14) and *BCL-1* gene rearrangement are now believed to represent MCL. Trisomy 3 is observed in some cases of SMZL, but it is more frequent in the nodal and extranodal types (Figure 15.31). Abnormalities of *p53* gene are reported in about 17% of the patients. Most SMZL cases have been associated with multiple somatic immunoglobulin variable region hypermutations, but a subset has been

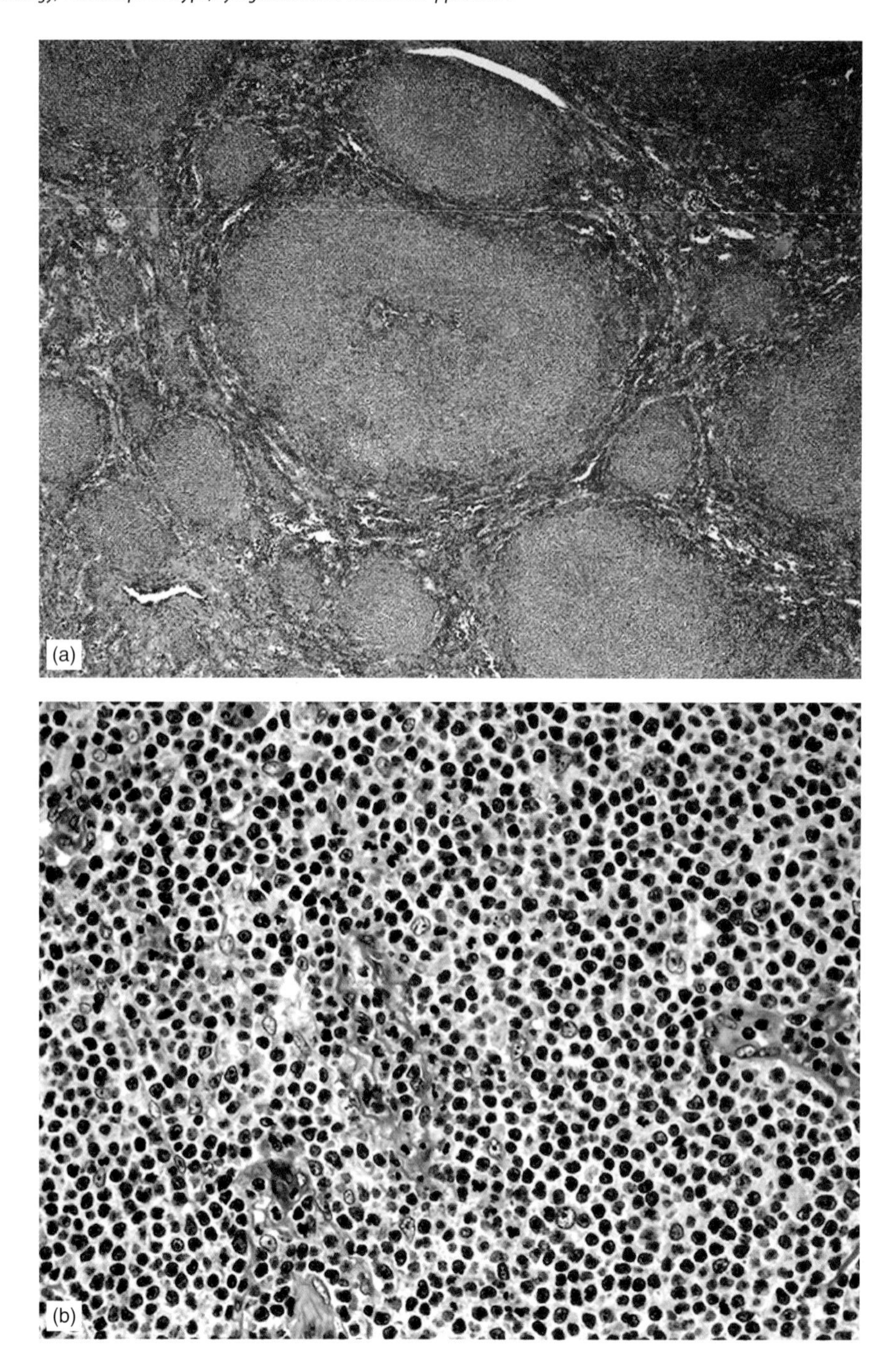

FIGURE 15.27 Splenic marginal zone B-cell lymphoma demonstrating the expansion of marginal zone into the splenic red pulp (a). The high power view shows monocytoid lymphocytes with variable amounts of cytoplasm (b).

identified without mutations that may have a different clinical course [113, 115].

Clinical Aspects

SMZL accounts for about 1–2% of non-Hodgkin lymphomas and constitutes 8–14% of lymphomas in the surgically removed spleens [106–110]. The median age is 65 years with a male:female ratio of about 1 or 1:2. Moderate to massive splenomegaly is a common feature. Hepatomegaly is observed in some patients, but peripheral lymphadenopathy is rare. Patients usually develop mild to moderate degrees of anemia, thrombocytopenia, and neutropenia which could be attributed to bone marrow infiltration and splenic sequestration. The majority of the patients show absolute lymphocytosis. Serum immunoglobulin studies reveal a small IgM or IgG spike, usually <3 g/dL in about 50% of the cases. Transformation to large cell lymphoma occasionally occurs.

Autoimmune conditions such as autoimmune hemolytic anemia, immune thrombocytopenia, lupus anticoagulants, rheumatoid arthritis, and biliary cirrhosis have been observed in association with SMZL [106–110].

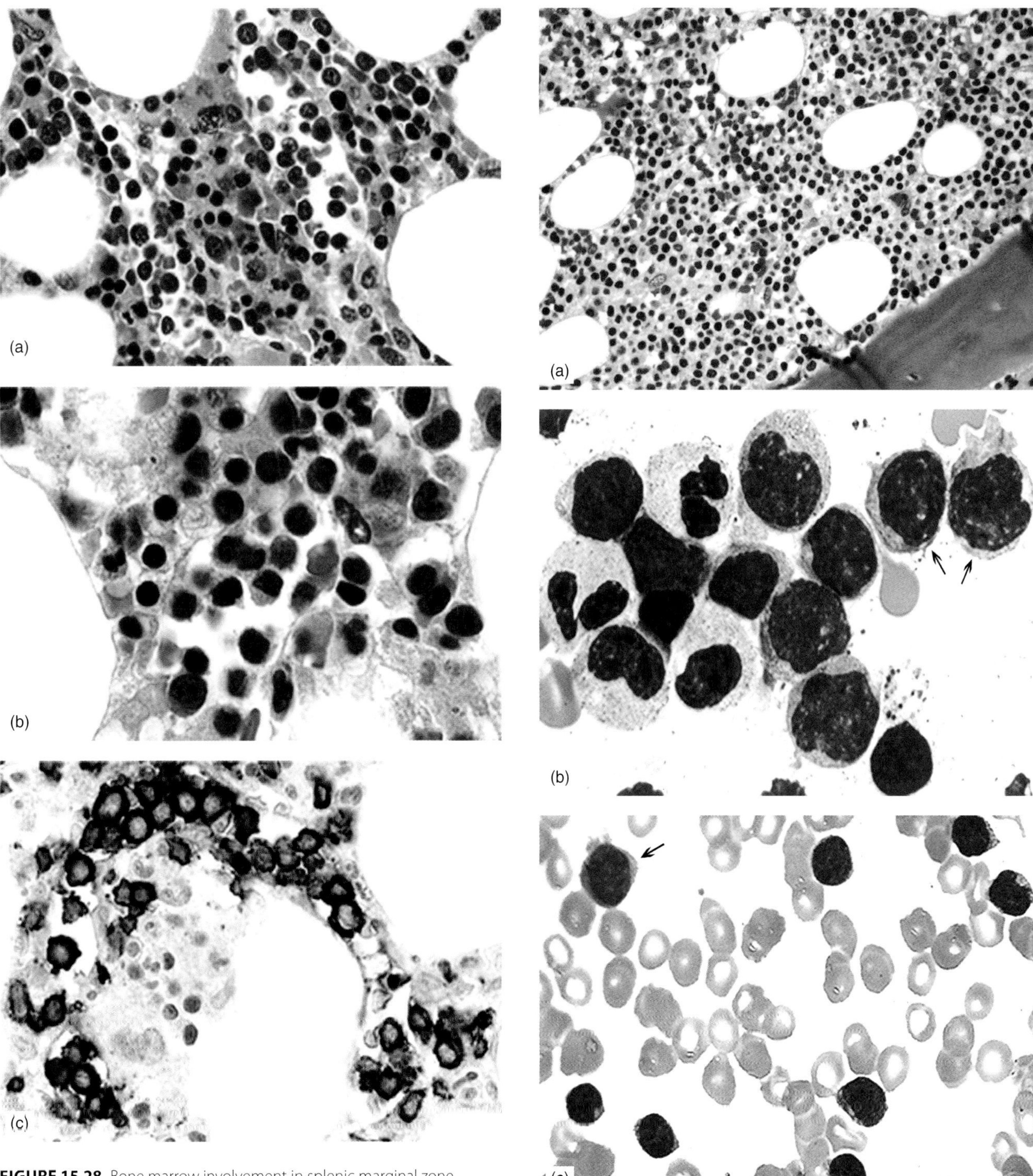

FIGURE 15.28 Bone marrow involvement in splenic marginal zone B-cell lymphoma is often sinusoidal: (a) low power and (b) high power. Immunohistochemical stains show clusters of CD20$^+$ cells within the sinusoids (c). From Ref. [91] by permission.

FIGURE 15.29 Bone marrow involvement in splenic marginal zone B-cell lymphoma: (a) biopsy section, (b) bone marrow smear, and (c) blood smear. Some lymphocytes show polar cytoplasmic projections.

The SMZL is considered a low-risk lymphoma with an indolent clinical course. The overall survival is >70% at 10 years with complete remission rate of 80%. The treatment strategies include (1) no therapy, (2) splenic irradiation, (3) splenectomy, (4) chemotherapy, and (5) treatment with anti-CD20 [110, 116–118].

Differential Diagnosis

The differential diagnosis of SMZL includes HCL, LPL, MCL, FL, and CLL/SLL. A summary of morphologic, immunophenotypic, and cytogenetic characteristics of SMZL, HCL, and LPL is presented in Table 15.5.

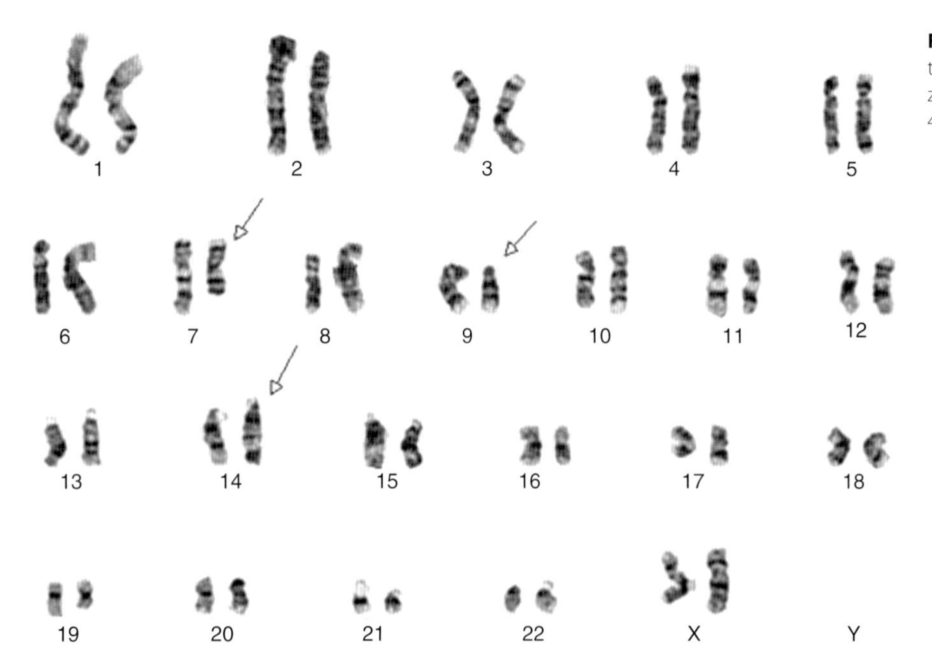

FIGURE 15.30 G-banded karyotype of the tumor cells in a patient with splenic marginal zone B-cell lymphoma demonstrating 46,XX,del(7)(q32), t(9;14)(p13;q32).

SMZL involves the splenic white pulp and often displays a nodular pattern, whereas HCL diffusely involves the red pulp with atrophy of the white pulp. Bone marrow fibrosis and interstitial infiltration are common features in HCL, whereas in SMZL marrow fibrosis is infrequent and intrasinusoidal infiltration is the characteristic feature. Unlike SMZL cells, HCL cells are TRAP-positive and express CD25 and CD103.

Overall SMZL cells are larger and have more abundant cytoplasm than the neoplastic cells in LPL. The IgM serum levels are usually <3g/dL in SMZL and >3g/dL in LPL. SMZL cells express IgM and IgD, whereas LPL cells are only IgM-positive. The primary cytogenetic abnormalities in SMZL are deletions of 7q22-32 and trisomy 3, whereas del(q21-q23) and t(9;14) are the cytogenetic characteristics of LPL.

The neoplastic cells in MCL are usually smaller and unlike SMZL cells, express CD5 and BCL-1. MCL has a much higher frequency of peripheral lymphadenopathy than SMZL. The cytogenetic hallmark of MCL is t(11;14).

The neoplastic cells in FL consist of a mixture of smaller centrocytes and larger centroblasts. These cells, unlike SMZL cells, express CD10 and BCL-6. The cytogenetic hallmark of FL is t(14;18).

CLL/SLL primarily consists of small mature lymphocytes with scanty cytoplasm. These cells, unlike SMZL cells, coexpress CD5 and CD23, are usually negative for FMC7, show dim expression of CD20, CD22, and CD79a, and have different cytogenetic profiles (see Table 15.5).

EXTRANODAL MARGINAL ZONE B-CELL LYMPHOMA OF MUCOSA-ASSOCIATED LYMPHOID TISSUE

Extranodal marginal zone B-cell lymphoma of mucosa-associated lymphoid tissue (MALT lymphoma) represents a neoplastic B-cell infiltrate which extends to the adjacent epithelium [1, 2, 119–122]. The cells are polymorphic, consisting of small lymphocytes, marginal zone cells, monocytoid cells, and often plasma cells. Scattered immunoblasts and centroblast-like cells are also present.

Etiology and Pathogenesis

A strong association has been found between MALT lymphoma and certain autoimmune disorders and infections, such as Sjogren syndrome, Hashimoto thyroiditis, hepatitis C virus infection, *Helicobacter gastritis*, and *Borrelia afzelii* infection of skin [123–126]. According to Isaacson, the reactive MALT lymphoma in autoimmune conditions or infections in stomach, salivary gland, thyroid gland, lung, and other tissues provide the substrate for the development of lymphoma. In most instances, treatment of *H. pylori* in gastric MALT lymphoma results in regression of early lesions [123]. Similarly, a proportion of cutaneous MALT-type lymphomas may regress by antibiotic therapy for Borrelia infection [125, 126].

The juxtaposition of *BIRC3* (formerly *API2*)(11q21) and *MALT1*(18q21) genes associated with t(11;18), the most common structural abnormality observed in MALT lymphoma (discussed later), may play a role in the pathogenesis of this disorder [122, 127]. The *API2-MALT* translocation increases NF-kappaB activation. NF-kappaB is a transcription factor that plays a role in preventing TNF-α-induced cell death [128, 129].

Pathology

Morphology

The most common site of MALT lymphoma is the gastrointestinal tract accounting for about 50% of the cases. Other sites of involvement in order of frequency include lung, salivary

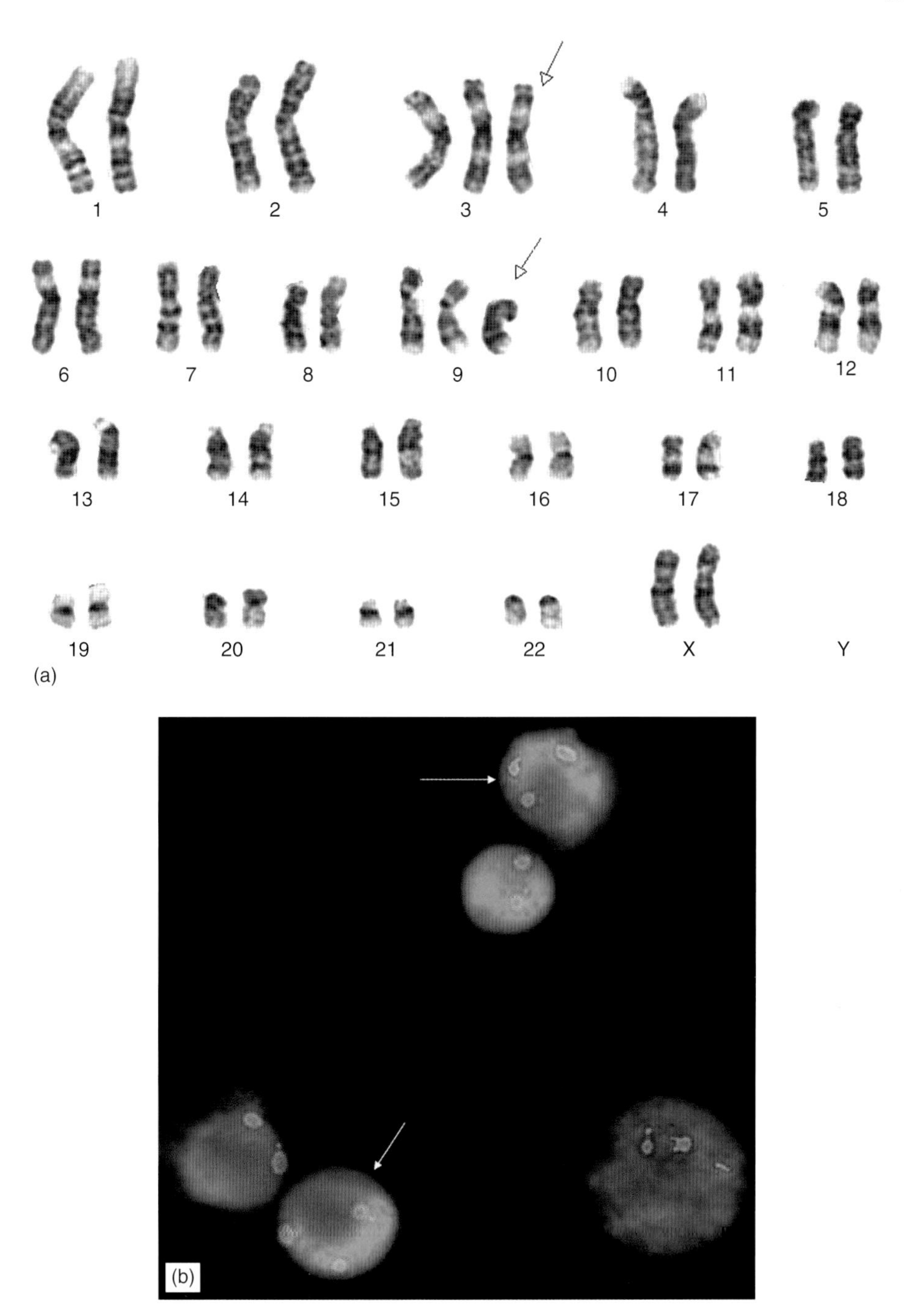

FIGURE 15.31 (a) G-banded karyotype of the tumor cells in a patient with splenic marginal zone B-cell lymphoma demonstrating 48,XX, +3, +9. Trisomy 3 is also demonstrated by FISH analysis (b).

glands, ocular adenexae, skin, thyroid, and breast [1, 2, 122, 130–132]. The involved tissues show a lymphomatous infiltrate around reactive follicles spreading into the surrounding areas. The MALT lymphoma cells are polymorphous consisting of various proportions of small lymphocytes, marginal zone (centrocyte-like) cells, monocytoid B-cells, and plasma cells. Scattered blast cells (centroblast- or immunoblast-like) are often noted, but if they are in sheets or large clusters, a diagnosis of large cell lymphoma should be made. Occasional follicles may show "colonization" by marginal zone or monocytoid B-cells. The epithelial tissue is characteristically infiltrated by the neoplastic cells, forming *lymphoepithelial lesions* (Figure 15.32). Lymphoepithelial lesion is defined as infiltrative aggregates of ≥3 marginal zone cells in the epithelium with distortion or destruction of the epithelial structure [1].

Lymph node or bone marrow involvement is reported in up to 25% of the cases in some studies. Peripheral blood is usually not involved in initial stages.

Immunophenotype

The immunophenotypic features of MALT lymphoma are similar to SMZL, except that MALT lymphoma cells express surface IgM (less often IgG or IgA) but lack IgD,

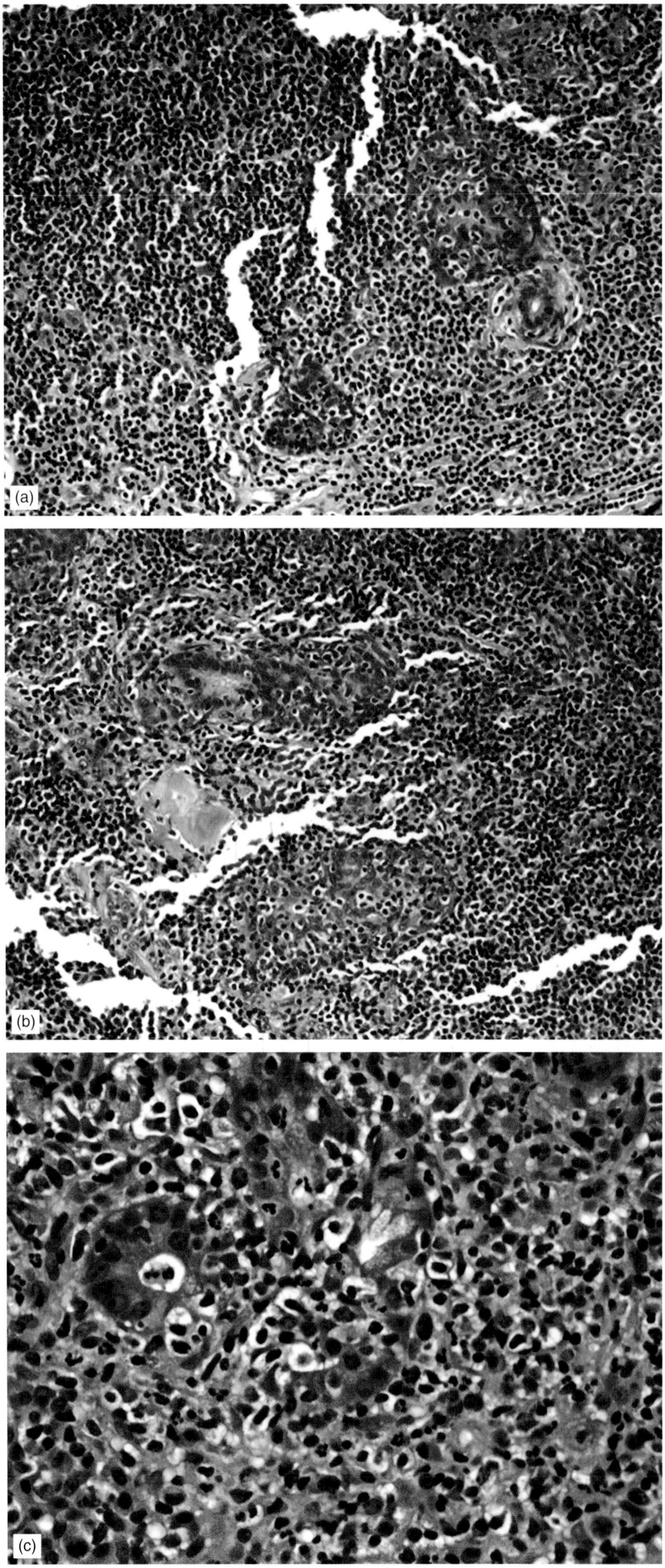

FIGURE 15.32 Lymphoepithelial lesions in the case of gastric MALT-type lymphoma: (a) low power, (b) intermediate power, and (c) high power views.

(a)

(b)

(c)

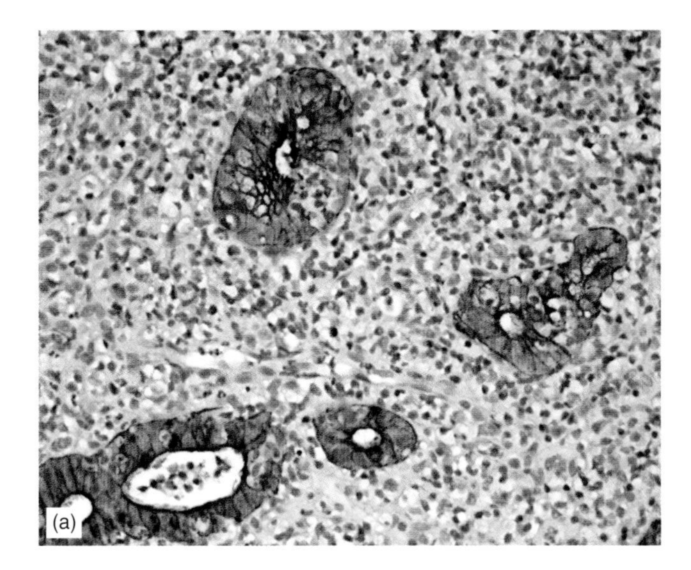

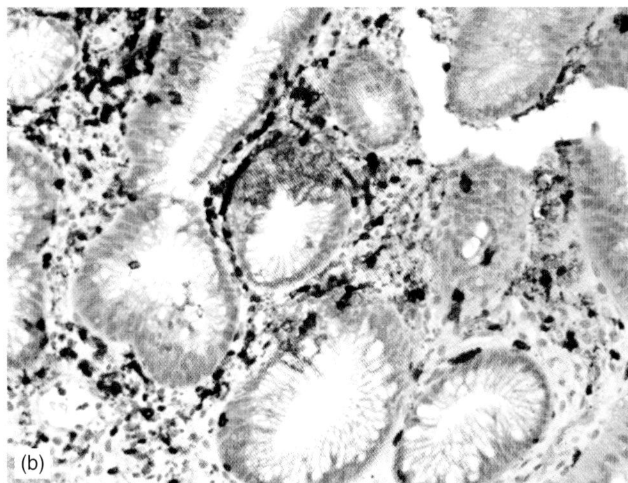

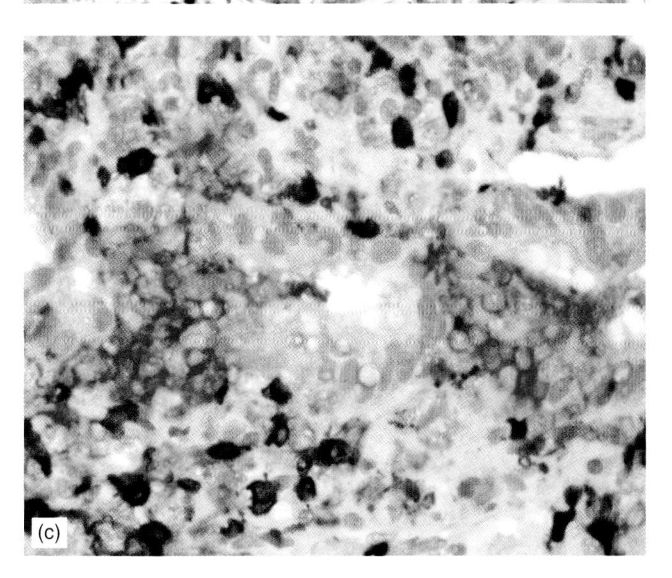

FIGURE 15.33 Immunohistochemical stains in MALT-type lymphoma. (a) A cytokeratin stain highlights glandular structures infiltrated by lymphocytes. (b) The infiltrating lymphocytes are predominantly CD20+ (red stain): (b) intermediate power and (c) high power views.

and about 40–60% of the cases show cytoplasmic Ig pointing to plasmacytic differentiation. The MALT lymphoma cells are positive for CD19, CD20, and CD79a, and negative for CD5, CD10, CD23, and BCL-1 (Figure 15.33) [1, 2, 121, 122].

Cytogenetic and Molecular Studies

The most common cytogenetic structural abnormalities reported in MALT lymphoma are t(11;18)(q21;q21) and trisomy 3 found in about 25–35% of the cases (Figure 15.34) [133, 134]. A less frequent nonrandom translocation is t(1;14)(p22;q32), which has been reported in association with gastric and lung MALT lymphomas [129]. These two translocations have not been observed in SMZL or nodal marginal zone B-cell lymphoma. Histological transformation of MALT lymphoma to large cell lymphoma has been associated with t(6;14)(p21;q32) pointing to the alteration of cyclin D3 expression (*CCND3*) in this process [135].

Important negative findings are lack of involvement of the *BCL-1* and *BCL-2* genes and absence of t(11;14)(q13;q32) and t(14;18)(q32;q21). Occasional cases may show *BCL-6* rearrangements involving chromosome 3q27 [136]. Naturally, these lesions should also show clonal immunoglobulin gene rearrangements, though this finding is often of less diagnostic significance than the other markers listed here.

Trisomy 3 is the most common numerical cytogenetic abnormality in MALT lymphoma reported in up to 60% of the cases.

Clinical Aspects

MALT lymphoma accounts for about 5% of all non-Hodgkin lymphomas. The most frequent site of involvement is stomach (50%) followed by lung (14%), salivary glands (14%), ocular adenexae (12%), skin (11%), thyroid glands (4%), and breast (4%) [1, 2, 130–132].

There is no age preference, but there is a slight female predominance. Symptoms are related to the site of involvement, such as abdominal pain (peptic ulcer disease) in the case of gastric lymphoma or Sjogren's syndrome in the case of salivary gland involvement. Systemic "B" symptoms are infrequent. Involvement of multiple mucosal sites at the time of initial diagnostic workup has been reported in up to one-third of the patients.

MALT lymphomas have a high rate of complete remission with 80% survival rate at ≥10 years. The therapeutic approaches include a variety of combinations of antibiotics, surgery, radiation, and chemotherapy, though universally accepted optimal therapeutic regimens based on the results of controlled studies have not yet been established [137].

Differential Diagnosis

The differential diagnosis of MALT lymphoma includes a variety of reactive lymphoproliferative disorders and small

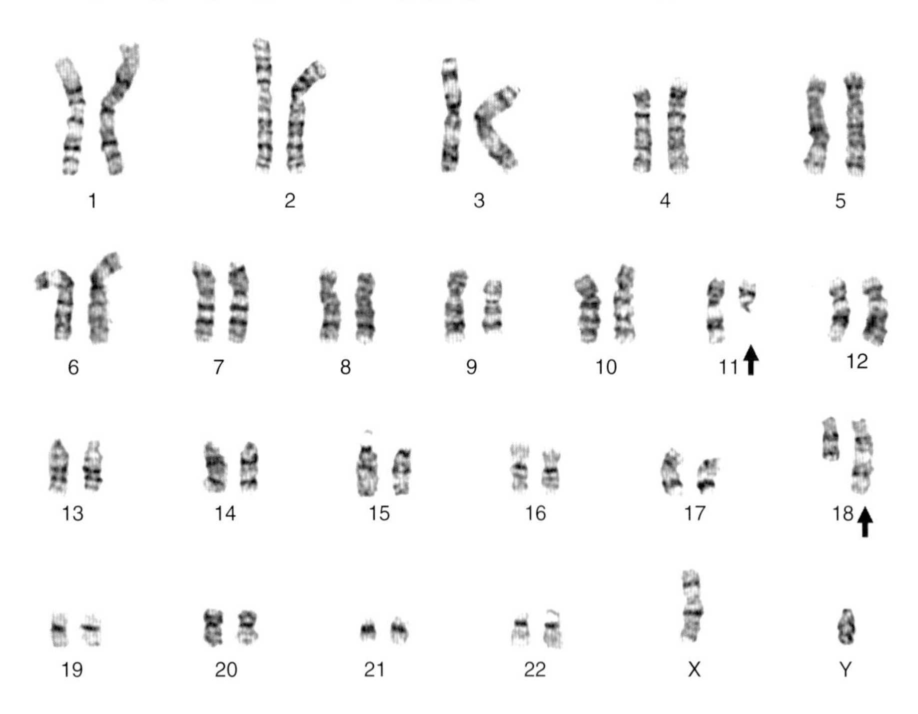

FIGURE 15.34 G-banded karyotype of tumor cells from a patient with MALT lymphoma demonstrating 46,XY,t(11;18)(q21;q21).

B-cell lymphomas. MALT lymphoma is distinguished from *H. pylori* gastritis, lymphoepithelial sialadenitis, and Hashimoto thyroiditis by the presence of destructive lymphoid infiltration of the epithelium and evidence of a monotypic B-cell population by immunophenotypic studies or molecular genetic analyses. Features distinguishing MALT lymphoma from other small B-cell lymphomas are similar to those previously discussed in SMZL.

NODAL MARGINAL ZONE B-CELL LYMPHOMA

Nodal marginal zone B-cell lymphoma (nodal MZL) is a rare, primary node–based lymphoma of marginal zone type with no evidence of extranodal disease. The diagnosis of nodal MZL should not be made in patients with Sjogren's syndrome, MALT lymphoma, or when another type of low-risk lymphoma is present in the same lymph node [1, 2, 138].

Etiology and Pathogenesis

The etiology and pathogenesis of nodal MZL are not known.

Pathology

Morphology

The neoplastic cells primarily consist of marginal zone and monocytoid B-cells infiltrating interfollicular areas of the involved lymph node. Various proportions of plasma cells and scattered blasts (centroblasts and immunoblasts) are usually present. Follicular colonization may be present. Two morphologic subtypes have been distinguished [2]:

1. Lymph nodes with aggregates of monocytoid B-cells with a parafollicular, perivascular, and perisinusoidal distribution. Germinal centers and mantle zones are preserved. The majority of these patients show evidence of MALT lymphoma, and therefore the lymph node involvement should be considered a secondary process.
2. Lymph nodes with marginal zone B-cells infiltrating and expanding around follicles with germinal centers and shrinkage or disappearance of the mantle zones.

Bone marrow is involved in 30% of the cases. Peripheral blood involvement is rare.

Immunophenotype

Most cases show immunophenotypic features similar to MALT lymphoma. Some cases, similar to SMZL, may show IgD expression.

Cytogenetic and Molecular Studies

Trisomy 3 is observed in some cases of nodal and extranodal marginal zone B-cell lymphomas (see Figure 15.31). Immunoglobulin gene rearrangements will be present but are nonspecific.

Clinical Aspects

Nodal MZL is rare, accounting for <2% of all non-Hodgkin lymphomas [139, 140]. The median age is around 50 years. Patients show localized or generalized lymphadenopathy with no systemic "B" symptoms. Median survival exceeds 12 years.

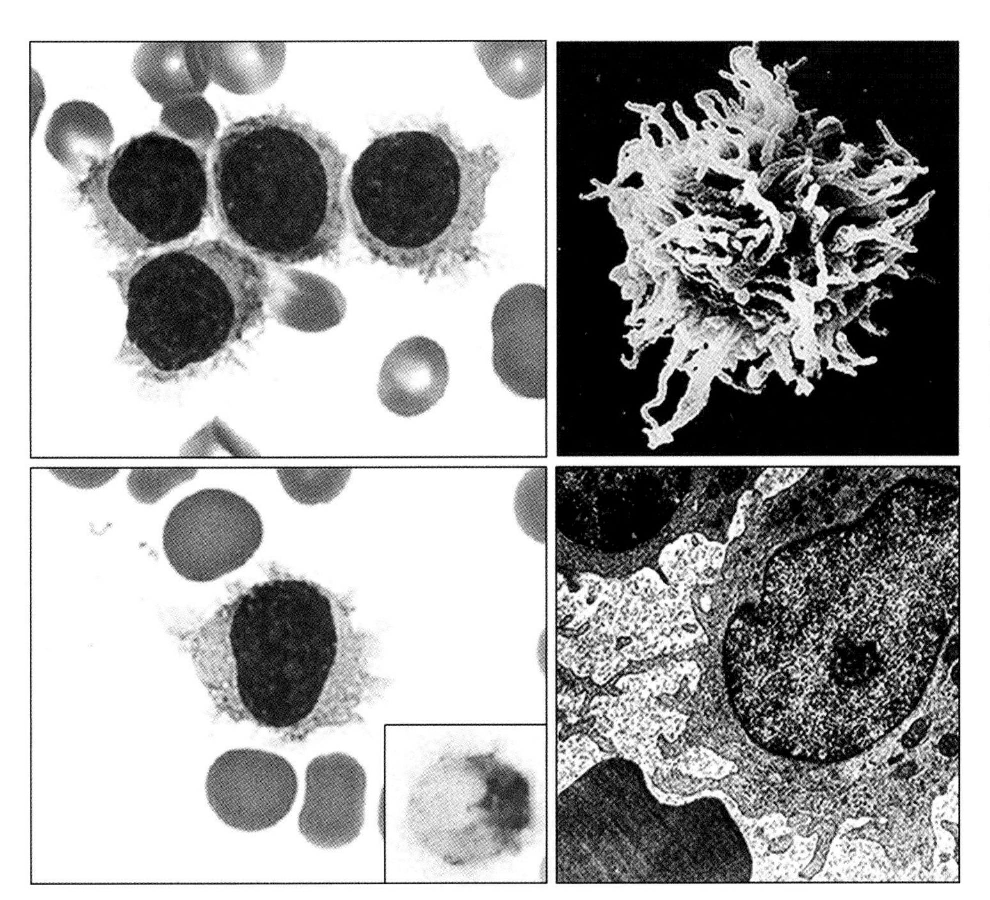

FIGURE 15.35 Hairy cell leukemia. Peripheral blood smear demonstrates neoplastic cells with cytoplasmic hairy projections (left). The inset shows a hair cell positive for tartrate-resistant acid phosphatase. Features of scanning (right top) and transmission (right bottom) electron microscopy are demonstrated. From Naeim F. (1997). *Atlas of Bone Marrow and Blood Pathology*, W.B. Saunders, Philadelphia; and Naeim F. (1980). Cytoskeletal of redistribution of surface membrane receptors in hairy cell leukemia. *Am J Clin Pathol* **74**, 660–3, by permission. From Ref. [91] by permission.

HAIRY CELL LEUKEMIA

Hairy cell leukemia, or leukemic reticuloendotheliosis, is an indolent mature B-cell lymphoid leukemia characterized by the proliferation of medium-sized lymphocytes with cytoplasmic "hairy" projections involving blood, bone marrow, spleen, and occasionally other tissues [1, 141, 142].

Etiology and Pathogenesis

The etiology and pathogenesis of HCL are not known. Environmental factors such as ionizing radiation and organic chemicals have been suggested as possible causes [143–145]. A report of siblings affected with HCL and sharing the same HLA haplotype has raised the possibility of genetic predisposition in the development of this disease in some familial cases [146]. Immunoglobulin gene rearrangement studies suggest an extrafollicular origin for the HCL cells, probably a marginal zone B-cell origin [147–149].

Pathology

Morphology

Hairy cells are larger than mature lymphocytes and show abundant pale blue cytoplasm, often with ill-defined border (Figure 15.35) [1, 141, 142]. Cells with characteristic elongated (hairy) cytoplasmic projections are frequently identified. The nuclei are round, oval, folded, indented, or dumbbell-shaped. The nuclear chromatin is condensed but finer than in CLL cells. Rare cells may show prominent nucleoli. The hairy cytoplasmic projections are easily detected by phase contrast microscopy or by scanning and transmission electron microscopy.

Peripheral blood examination in the majority of the patients reveals pancytopenia with the presence of various proportions of hairy cells. Only about 10% of the patients show leukocytosis, and in such cases, the hairy cells account for the majority of the leukocytes. Occasionally, leukocytosis exceeds 100,000/μL. Monocytopenia is one of the characteristic features of HCL.

The bone marrow involvement is usually interstitial or diffuse with patchy, densely cellular areas composed of neoplastic cells (Figure 15.36). Nodular involvement is rare. In most cases, the bone marrow is hypercellular with a cellularity of >50–90%. But occasionally, the bone marrow may appear markedly hypocellular (<25%) simulating aplastic anemia (Figure 15.37). The HCL cells are relatively uniform with abundant clear cytoplasm and round, oval, or irregular nuclei, often without prominent nucleoli or presence of mitotic figures. The presence of abundant cytoplasm creates a nuclear spacing or "fried egg" pattern. Extravasated red blood cells are frequently seen.

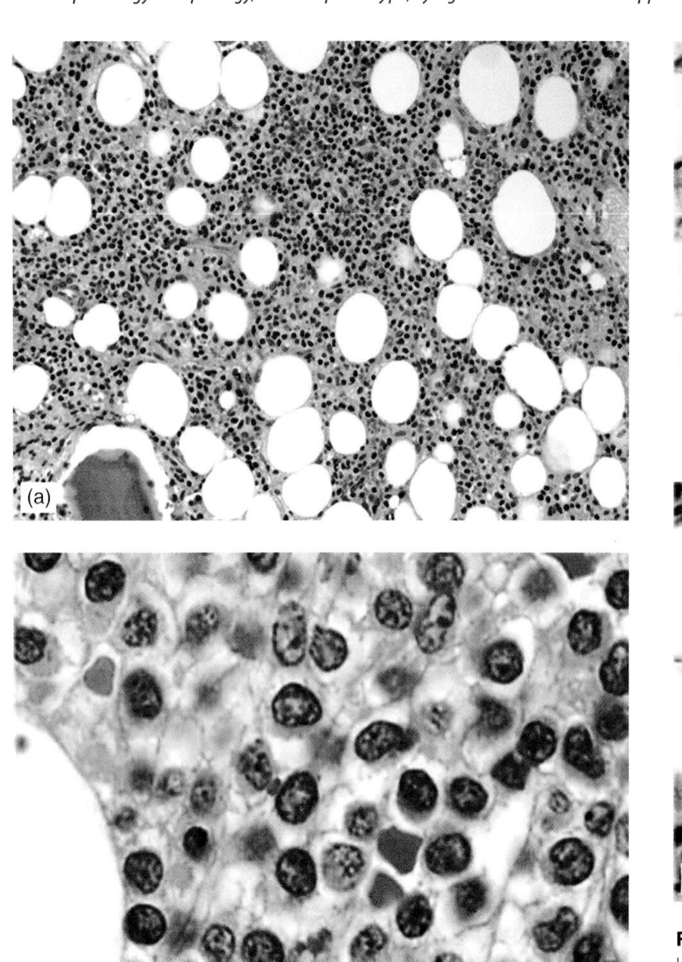

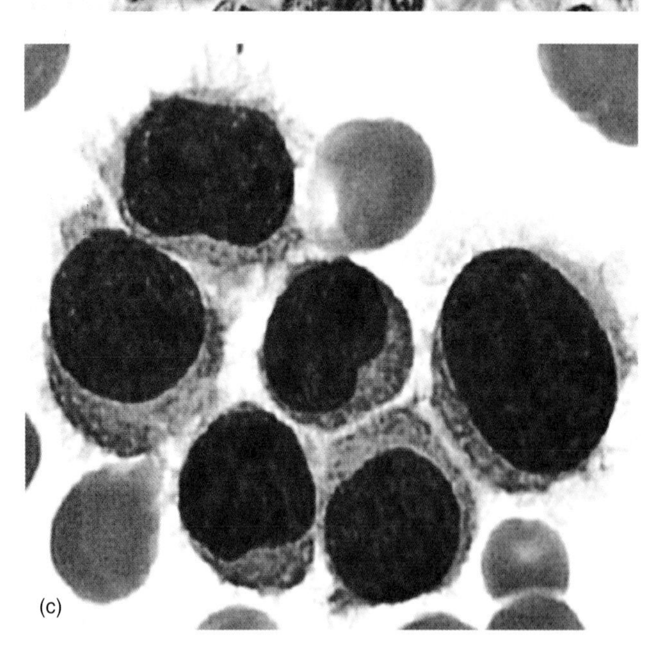

FIGURE 15.36 Hairy cell leukemia. Bone marrow biopsy section demonstrates an interstitial leukemic infiltrate (a). The tumor cells show nuclear spacing and a "fried egg" pattern (b). Bone marrow smear shows numerous tumor cells (c), some with cytoplasmic projections.

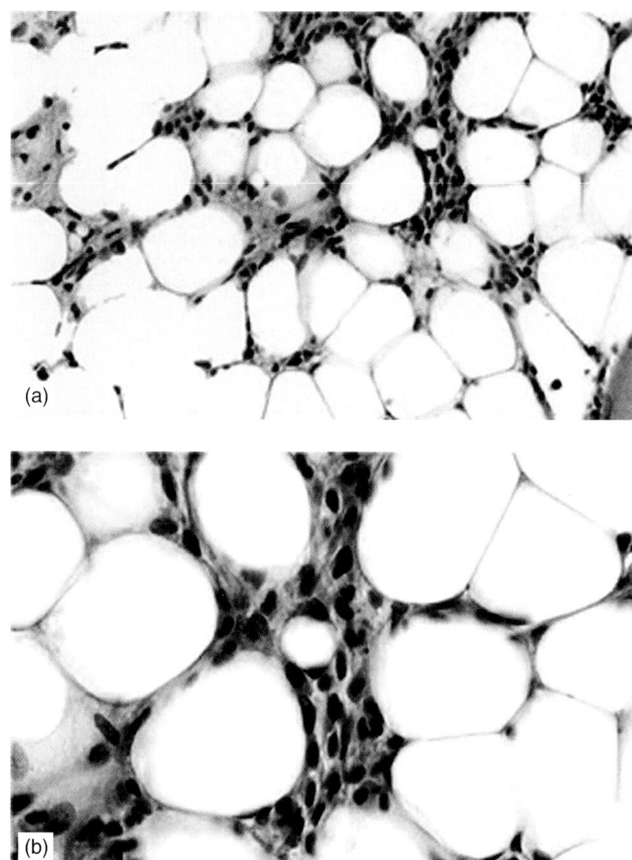

FIGURE 15.37 Hairy cell leukemia. The hypocellular variant of hairy cell leukemia may mimic aplastic anemia in bone marrow biopsy sections. Interstitial or focal hairy cell infiltration is sometimes overlooked (a and b). From Ref. [91] by permission.

The bone marrow biopsy sections often reveal moderate to marked increase in reticulin fibers leading to unsuccessful bone marrow aspiration (dry tap) (Figure 15.38). The reticulin fibers tend to surround the individual or small clusters of tumor cells. It has been shown that the hairy cells produce and assemble a fibronectin meshwork, which is in part responsible for the bone marrow fibrosis. In the vast majority of the cases, diagnosis of HCL is made by examination of bone marrow biopsy sections. In rare occasions, particularly when the bone marrow involvement is focal, the initial biopsy sections may not be diagnostic and deeper cuts or additional bone marrow biopsy is required.

Splenic involvement is one of the characteristic features of HCL, and splenomegaly is one of the most prominent clinical features. However, about 20% of the patients may lack significant splenomegaly. HCL infiltrates red pulp cords and sinusoids in a diffuse pattern (Figure 15.39). Scattered red blood cell lakes surrounded by hairy cells are often present. The white pulp is atrophic, and occasional small normal lymphoid aggregates may be present.

HCL involves the liver in up to 50% and the lymph nodes in about 15% of the cases. The hepatic infiltration is both portal and sinusoidal and the lymph node involvement is usually paracortical (Figure 15.40). Involvement of other tissues, such as skin and lung, is rare.

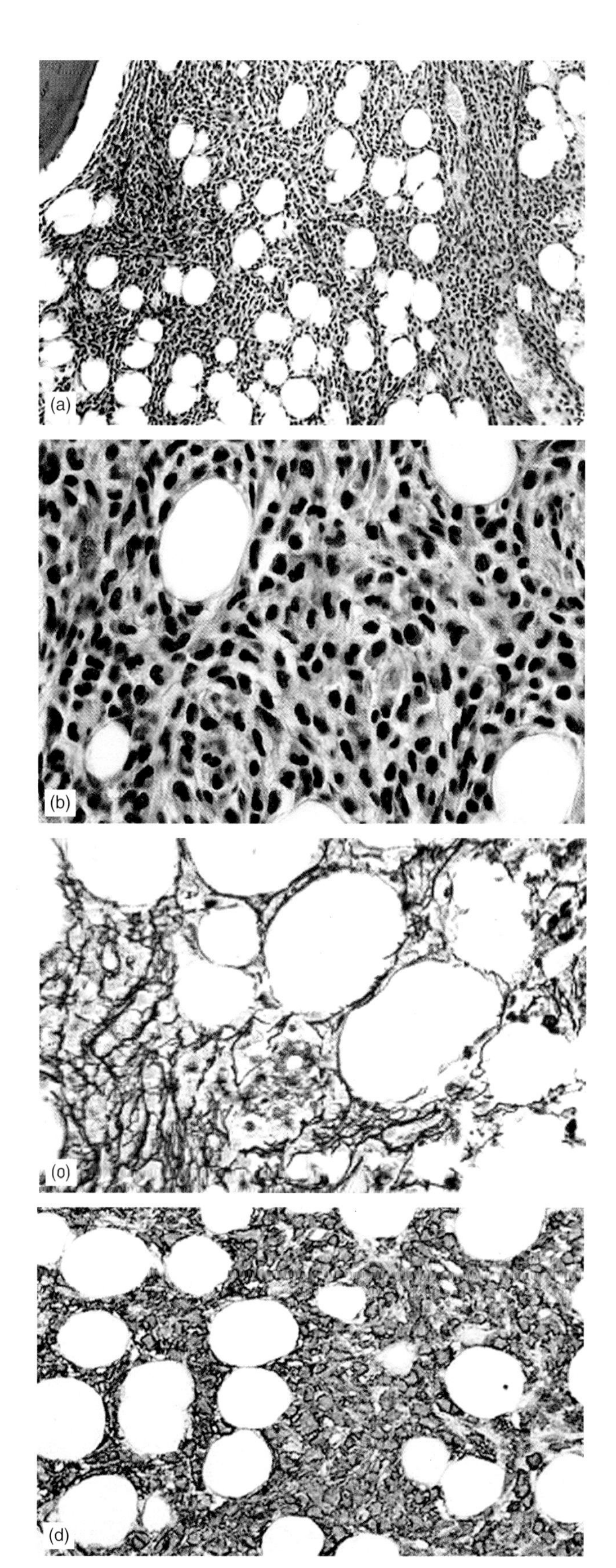

FIGURE 15.38 Hairy cell leukemia. Bone marrow biopsy section demonstrates an interstitial leukemic infiltrate: (a) low power and (b) high power. The reticulin stain shows increased reticulin fibers (c) and the tumor cells express DBA44 by immunohistochemical stain (d).

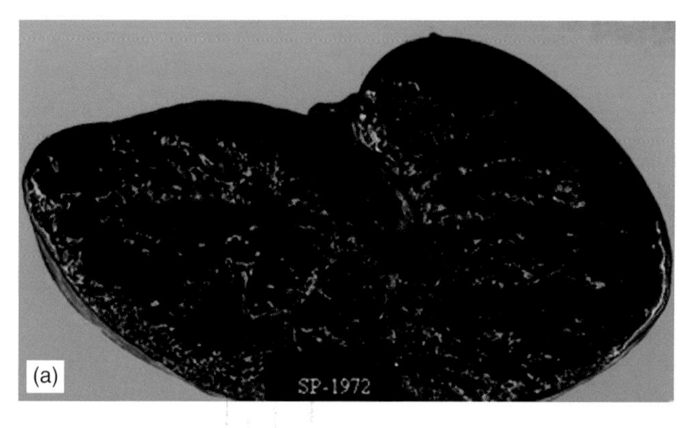

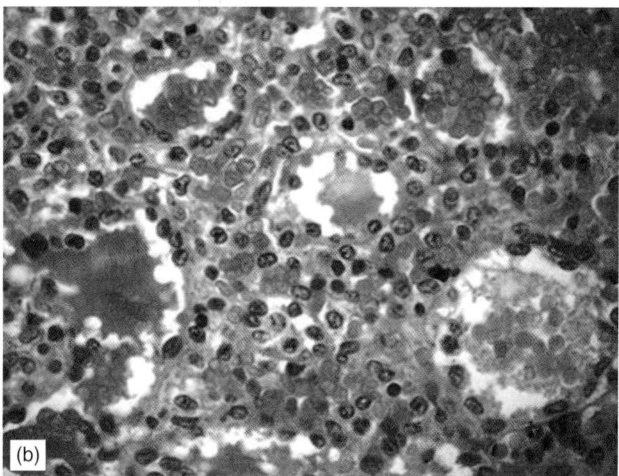

FIGURE 15.39 Splenic infiltration in hairy cell leukemia is diffuse (a) and involves the red pulp with the presence of red blood cell lakes (b).

Immunophenotype and Cytochemical Stains

Hairy cells are SIg^+ (kappa or lambda) and express B-cell-associated markers, such as CD19, CD20, CD22, and CD79a. They characteristically coexpress CD25 and CD103 and are strongly positive for CD11c and FMC7 [150, 151]. The coexpression of CD25 and CD103 is rather unique for the HCL among the B-cell neoplasms (Figure 15.41). Immunohistochemical staining for DBA44 is also helpful in distinguishing hairy cells, though this antigen has been expressed by other lymphoid neoplasms [152]. Hairy cells are typically negative for CD5, CD10, CD23, CD27, and CD38 [153].

Hairy cells express an isoenzyme of acid phosphatase resistant to tartaric acid [141, 152]. Although the presence of TRAP is considered characteristic for hairy cells, occasionally cells in other lymphoid malignancies, such as Sezary syndrome, certain lymphomas, T-cell prolymphocytic leukemia, and Hodgkin lymphoma, may show TRAP positivity [141]. Gaucher cells may also express TRAP. Cytochemical TRAP stains have been routinely used for the diagnosis of HCL for many years, and now anti-TRAP antibodies are available for immunohistochemical studies.

None of the immunophenotypic and cytochemical markers are specific for HCL, but the expression of CD25, CD103, CD11c, FMC7, TRAP in a B-cell lymphoproliferative disorder is consistent with HCL if clinicopathologic findings are supportive of such diagnosis.

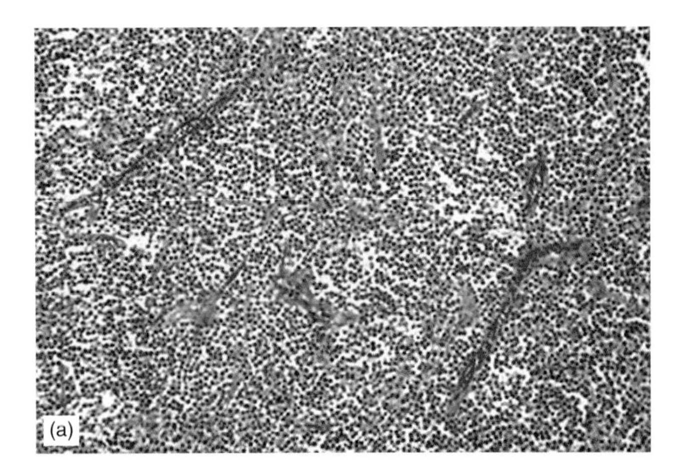

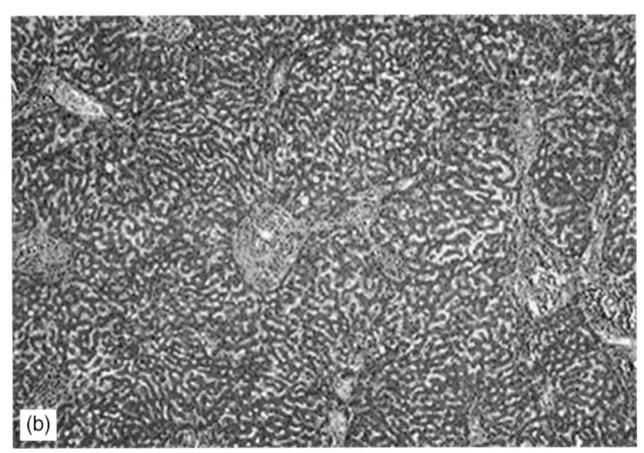

FIGURE 15.40 Hairy cell leukemia. (a) A lymph node biopsy section showing diffuse infiltration by hairy cells. (b) A liver biopsy showing infiltration of the portal areas and sinuses by the leukemic cells.

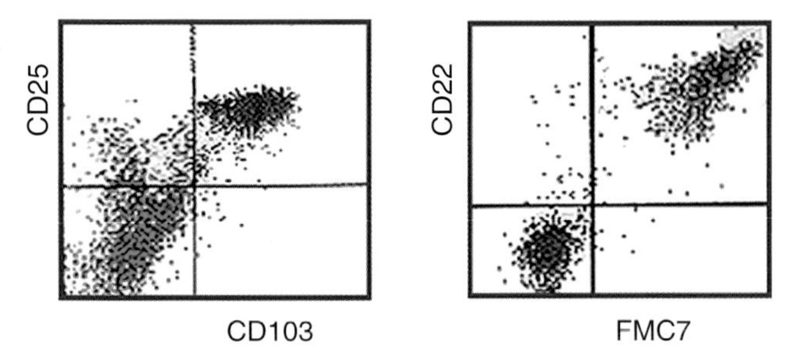

FIGURE 15.41 Hairy cell leukemia. Flow cytometry demonstrates expression of CD25, CD103, CD22, and FMC by the tumor cells. From Ref. [91] by permission.

TABLE 15.6 Clinical and laboratory findings in hairy cell leukemia.

Average age	55 years
Male:female ratio	3–5:1
Physical findings	
Splenomegaly	80%
Hepatomegaly	20–30%
Petechia and ecchymosis	30%
Lymphadenopathy	15–20%
Laboratory findings	
Anemia	85%
Thrombocytopenia	80%
Neutropenia	80%
Monocytopenia	80%
Hypergammaglobulinemia	20%

Cytogenetic and Molecular Studies

No specific cytogenetic or molecular aberrations have been found in HCL. Random chromosomal abnormalities such as 5q−, 6q−, 11q−, 17q− and trisomies 3, 4, 5, 12, and 18 have been reported [154–156]. Abnormalities of chromosome 5, most commonly trisomy 5 and interstitial deletions of band 5q13, have been observed in up to 40% of the cases [157].

As a disease of B-cells, HCL should show clonal immunoglobulin gene rearrangements [158]. The majority of HCL cases show mutation of the *VH* genes and express activation-induced cytidine deaminase, a molecule essential for somatic mutation and isotype switch [148]. Unlike most other B-cell neoplasms, HCL frequently expresses multiple Ig isotypes [159].

Clinical Aspects

HCL is relatively rare and accounts for about 2% of all lymphoid leukemias [1, 159]. The median age is about 55 years with a marked male predominance. Clinical symptoms are mostly related to splenomegaly and pancytopenia and include abdominal fullness or discomfort, fatigue, weight loss, fever, bruising, and bleeding [160–162]. Splenomegaly is present in about 80% of the HCL cases. Hepatomegaly is observed in 25–40% of the patients, but lymphadenopathy is not a major clinical feature and is found in 10–20% of the patients. Table 15.6 demonstrates a summary of the clinical and laboratory findings in HCL. Evidence of defective cell–mediated immunity has been demonstrated in HCL by several investigators [163–165].

Hairy cell leukemia is an indolent leukemia with an overall 12-year survival rate of >85% [166–168]. There are reports indicating that HCL patients without splenomegaly tend to remain free from significant neutropenia, have an excellent survival rate, and are usually older than the patients with splenomegaly. There is also some evidence that the morphology of the hairy cells may correlate with prognosis. For example, there are reports indicating that the HCL cells with oval nuclei are associated with better prognosis than the ones with convoluted or indented nuclei and that the nuclear convolution is more frequently associated with marrow fibrosis and severe pancytopenia [169, 170]. Therapeutic modalities include interferon alpha, purine analogs such as pentostatin (2′-DCF) and cladribine (2-CdA), splenectomy, and rituximab (anti-CD20) therapy [171–173].

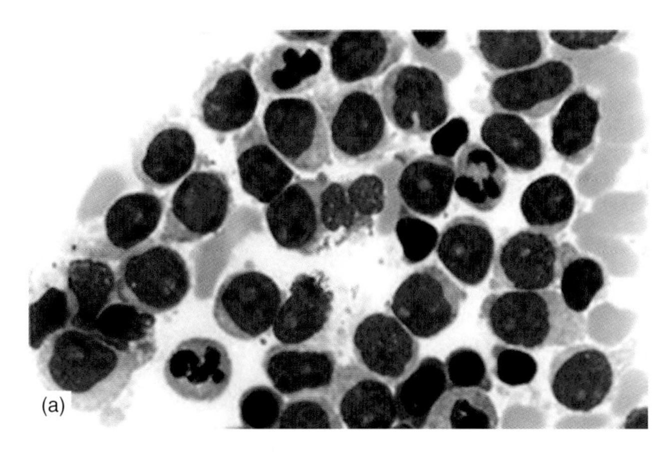

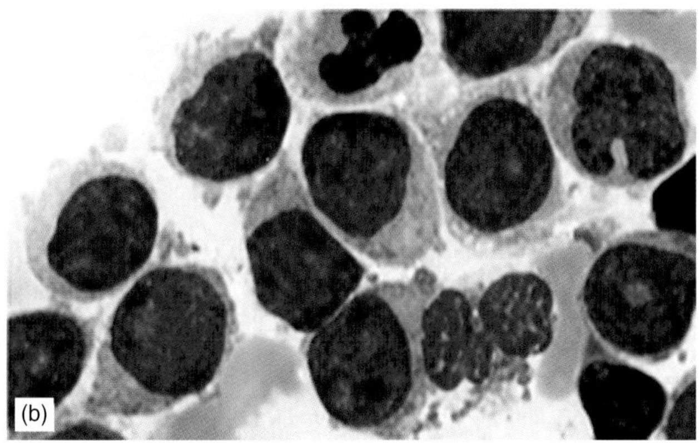

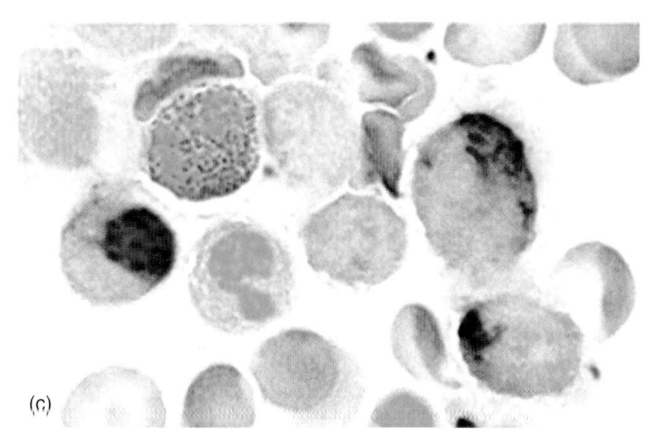

FIGURE 15.42 Hairy cell leukemia, morphologic variant. Bone marrow smear (a: low power; b: high power views) shows prolymphocyte-like hairy cells. These cells are TRAP-positive (c). From Ref. [91] by permission.

Hairy Cell Leukemia Variants

Rare cases of HCL show morphologic features intermediate between those of hairy cells and prolymphocytes (Figure 15.42). This variant is characterized by elevated white blood cell count of usually >50,000/mL and lack monocytopenia. The neoplastic cells have abundant basophilic cytoplasm with cytoplasmic projections, moderately dense heterochromatin, and a prominent nucleolus. These cells, in contrast to typical HCL cells, lack the expression of CD25 and may also be negative for CD103 and/or TRAP [174, 175].

An unusual, rare morphologic variant of HCL has been reported in which the tumor cells have multilobated nuclei [176]. A blastic variant of HCL has been described in which the blast cells are TRAP-positive and show fine cytoplasmic projections [177].

Differential Diagnosis

The differential diagnosis of HCL includes SMZL, PLL, and atypical CLL/SLL. HCL involves the splenic red pulp in a diffuse pattern with atrophy of the white pulp, whereas SMZL involves the white pulp and often has a nodular pattern. Bone marrow fibrosis and interstitial infiltration are common features of HCL, whereas in SMZL bone marrow fibrosis is infrequent and intrasinusoidal infiltration is the characteristic feature. Unlike SMZL cells, HCL cells are TRAP-positive and express CD25 and CD103 (see Table 15.6).

Hairy cell variant with prolymphocytic features is distinguished from PLL and atypical CLL (CLL/PLL) by the coexpression of CD25 and CD103 and TRAP positivity. CLL/PLL cells often express CD5 and CD23 and have different cytogenetic profiles.

FOLLICULAR LYMPHOMA

Follicular lymphoma is the second most common lymphoma and represents a neoplasm of follicle center B-cells consisting of a mixture of centrocytes and centroblasts [1, 2, 93]. The pattern of lymph node involvement is at least partially follicular. Other terminologies used for this lesion are "follicle center lymphoma" and "follicular center cell lymphoma."

Etiology and Pathogenesis

The etiology of FL is not known. Inhibition of apoptosis as the result of overexpression of BCL-2 appears to play a critical role in the pathogenesis of this disorder [178–180]. Approximately 75–90% of patients with FL demonstrate t(14;18)(q32;q21) resulting in the juxtaposition of the *BCL-2* gene on chromosome 18 into the *IGH* heavy chain locus on chromosome 14. This translocation leads to constitutive expression of BCL-2 giving the transformed cells an extended survival and growth advantage. The t(14;18) alone, however, does not seem to be sufficient for the development of lymphoma. It keeps the transformed cells alive and therefore at risk for subsequent cytogenetic alterations necessary for the development of a fully malignant phenotype to occur [178].

Published studies demonstrate methylation of androgen receptor, SHP1, and death-associated protein kinase genes in FL [181]. By contrast, methylation of the cyclin-dependent kinase inhibitors p15, p16, and p57 is uncommon in FL and may indicate an important step toward the transformation of FL to a more aggressive lymphoma.

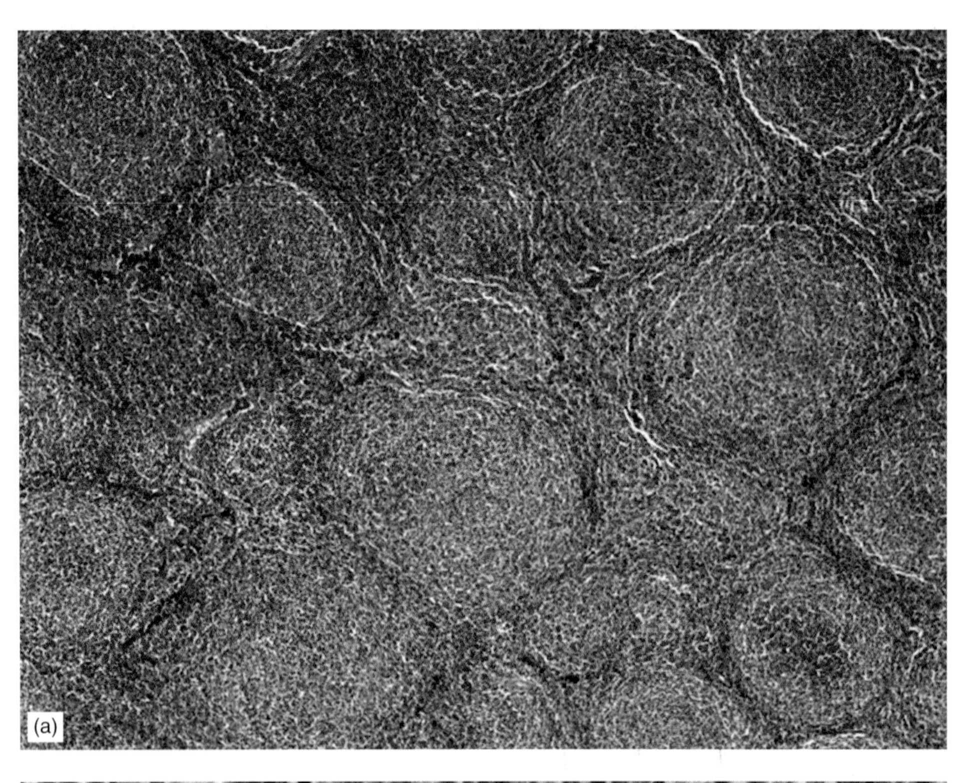

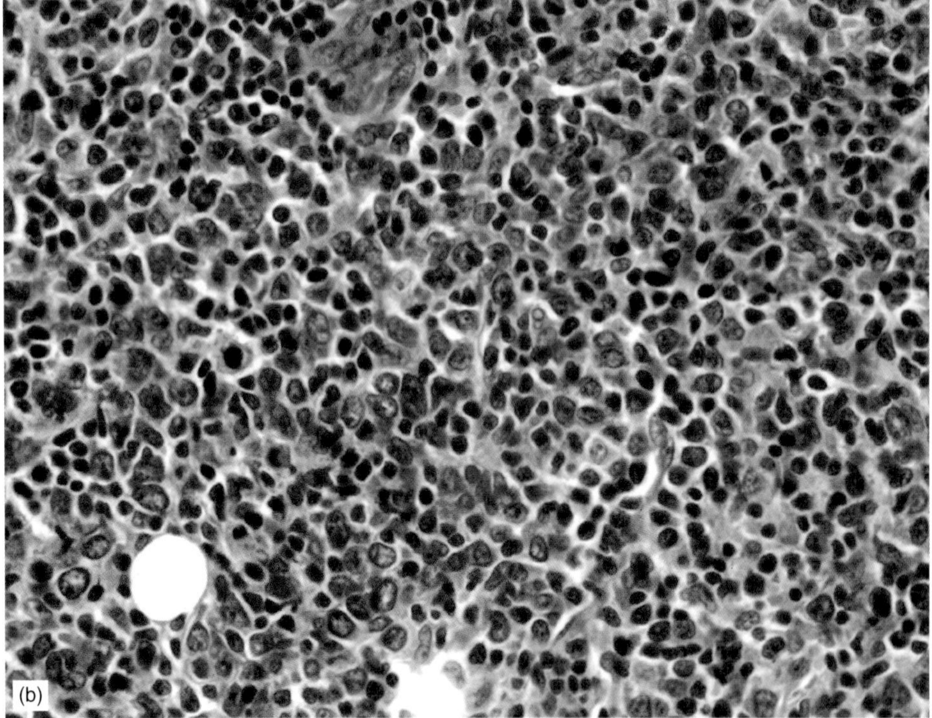

FIGURE 15.43 Follicular lymphoma; lymph node section. (a) Low power view demonstrating back-to-back follicles of various sizes and remnants of mantle zones. (b) High power view showing a mixed population of centrocytes and centroblasts.

Pathology

Morphology

The involved lymph nodes show effacement of the nodal architecture with a lymphoproliferative process which displays a predominantly follicular pattern (Figures 15.43 and 15.44) [1, 2, 93]. The neoplastic follicles are densely packed against one another, are often ill-defined, and show loss of or minimal mantle zone areas. These follicles show no polarity or tingible body macrophages and consist of various proportions of centrocytes and centroblasts interspersed with T-cells and follicular dendritic cells. Centrocytes are small to medium sized, show scant pale cytoplasm, irregular (angulated, twisted, convoluted) nucleus, and inconspicuous nucleoli [1]. Centrocytes are the predominant cells in the majority of the cases. Centroblasts are large transformed

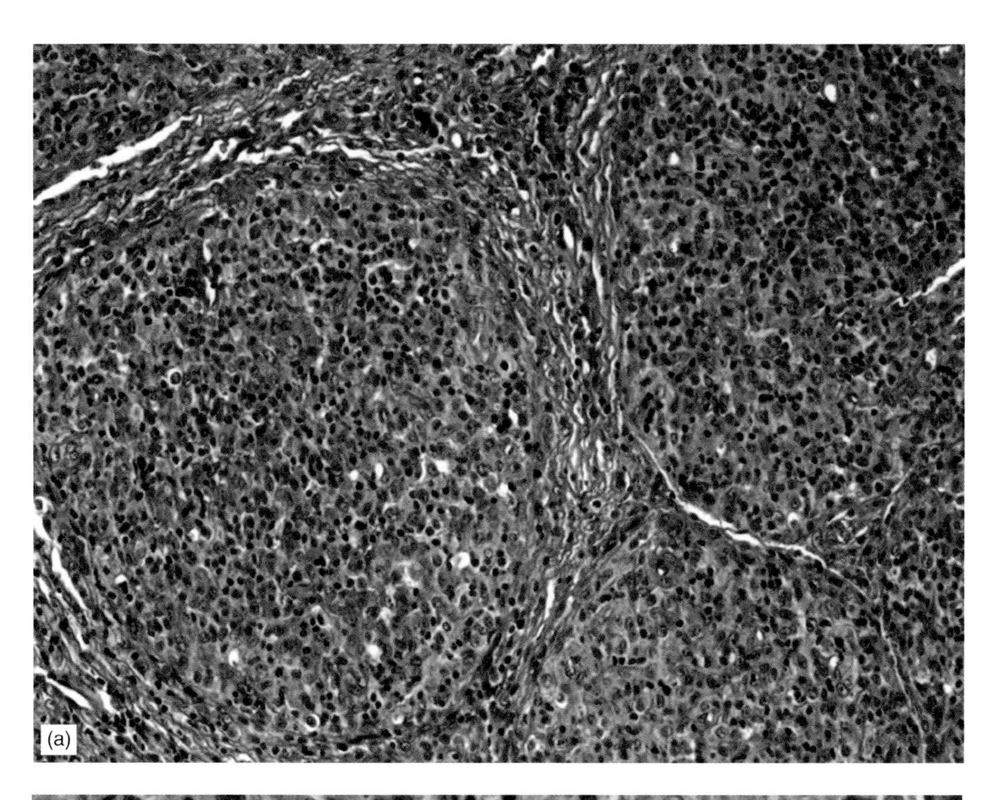

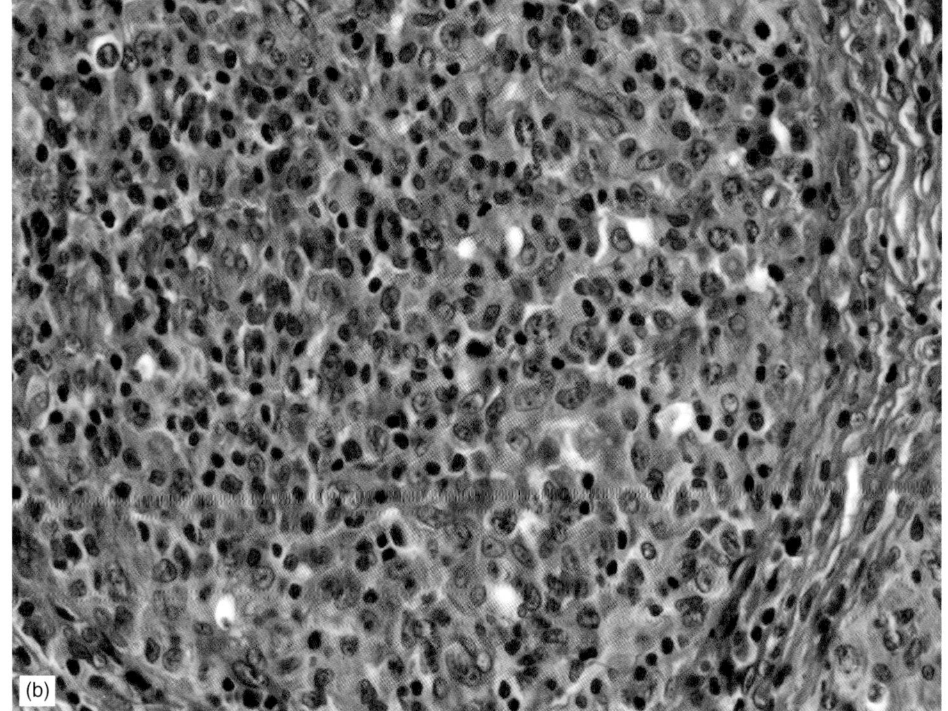

FIGURE 15.44 Follicular lymphoma; lymph node section. (a) Intermediate power view demonstrating back-to-back follicles with loss of mantle zones. (b) High power view showing a mixed population of centrocytes and centroblasts.

cells with small amounts of basophilic cytoplasm, round, oval, or slightly irregular nuclei, vesicular chromatin, and one to three nucleoli located close to the nuclear membrane. In some cases, centroblasts may show significant atypical features, such as hyperchromatic or markedly irregular nuclei. Follicular dendritic cells are also large and have a round nucleus and a vesicular chromatin, but unlike centroblasts, they often appear in doublets or show double nuclei, their nucleoli are centrally located, and they have ill-defined pale cytoplasm. These cells express CD21 and CD23.

Areas of diffuse involvement may be present, often associated with fibrosis. The neoplastic cells are usually present in the interfollicular areas and are easily identified by immunohistochemical stains (CD10+, BCL-6+). Discrete clusters of marginal zone monocytoid B-cells may be present in about 10% of the cases [1]. The pattern of involvement is divided into three categories:

1. Follicular: >75% of the involved tissue shows follicular pattern.
2. Follicular and diffuse: 25–75% of the involved tissue shows follicular pattern.
3. Minimally follicular: <25% of the involved tissue shows follicular pattern.

The following grading system has been recommended for FLs based on the proportion of centroblasts per 40× high-power microscopic field (hpf):

Grade 1	0–5 centroblasts/hpf
Grade 2	6–15 centroblasts/hpf
Grade 3	>15 centroblasts/hpf
3a	Centrocytes present
3b	Solid sheets of centroblasts

According to the WHO recommendations, variations in the pattern or grading observed in different areas of the involved tissue should be mentioned in the pathology report.

Bone marrow involvement is observed in 40–45% of the cases and is typically paratrabecular (Figures 15.45 and 15.46) [1, 2, 93, 182]. The lymphomatous aggregates usually do not show follicular configuration and consist of a mixture of centrocytes and centroblasts. There may be discordance between morphologic findings of the involved bone marrow and the lymph node in the same patient. In such cases, the grade of bone marrow involvement is often less than that of the involved lymph node. For example, in a patient with a grade 3 FL in the lymph node, the bone marrow involvement may appear as grade 1 or 2. Peripheral blood involvement is a frequent finding with the presence of atypical lymphoid cells with nuclear clefts or notches (Figure 15.46b).

Immunophenotype

The neoplastic cells of FL express B-cell-associated molecules, such as CD19, CD20, CD22, and CD79a. They are positive for CD10, cytoplasmic BCL-2, and nuclear BCL-6 in the majority of the cases and typically negative for CD5, CD11c, and CD43 (Figures 15.47 and 15.48) [1, 2, 182–184]. Expression of CD23 is variable and the Ki-67 fraction is low. A tight meshwork of $CD21^+$ and $CD23^+$ cells is present in the neoplastic follicles representing follicular dendritic cells.

The higher grade FLs may lack the expression of BCL-2 but may express CD43 [1, 2].

Cytogenetic and Molecular Studies

Cytogenetic abnormality is a common feature in FL. The most common chromosomal translocation in this disorder is t(14;18)(q32;q21), which has been observed in approximately 85% of the cases (Figure 15.49) [178]. This translocation places *BCL-2* gene on chromosome 18 next to the *IGH* heavy chain locus on chromosome 14 [178]. The resultant overexpression of BCL-2 confers a survival advantage on these B-cells, which now defies apoptosis. The continuing accumulation of these cells, rather than their rapid proliferation, is the hallmark of this low-grade lymphoma. In some patients, a subsequent genetic event involving a gene for cell

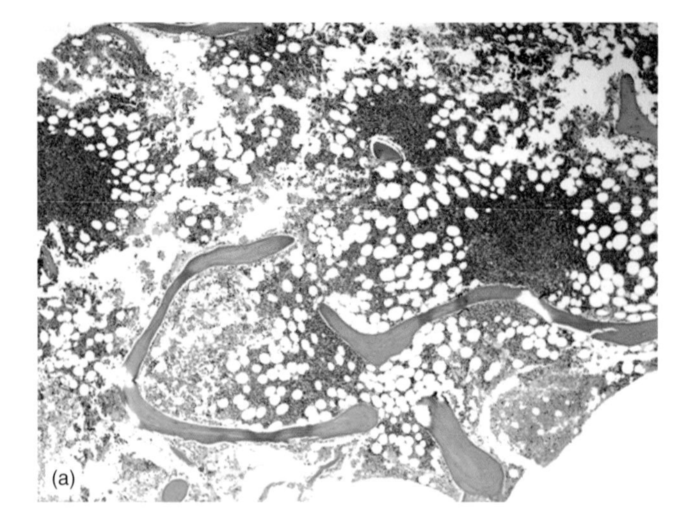

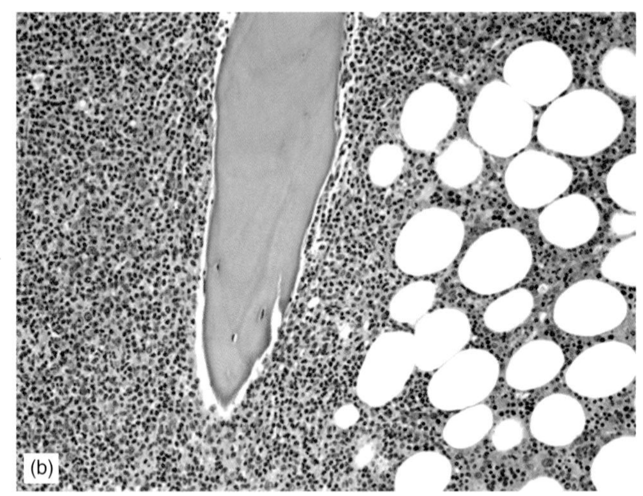

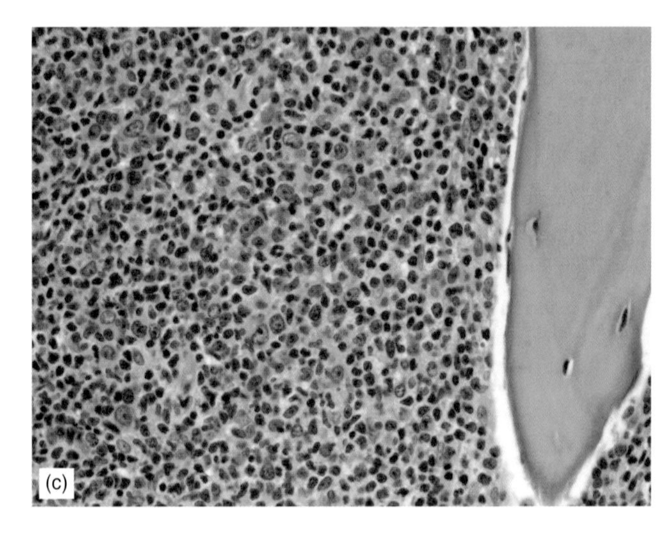

FIGURE 15.45 Follicular lymphoma. Bone marrow section showing paratrabecular lymphoid infiltrates: (a) low power, (b) intermediate power, and (c) high power views.

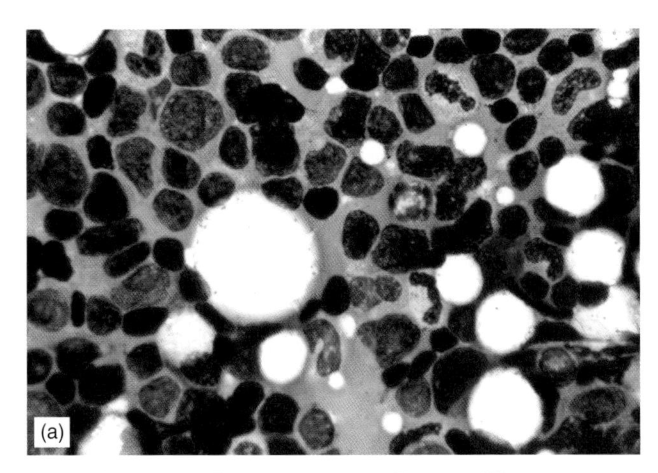

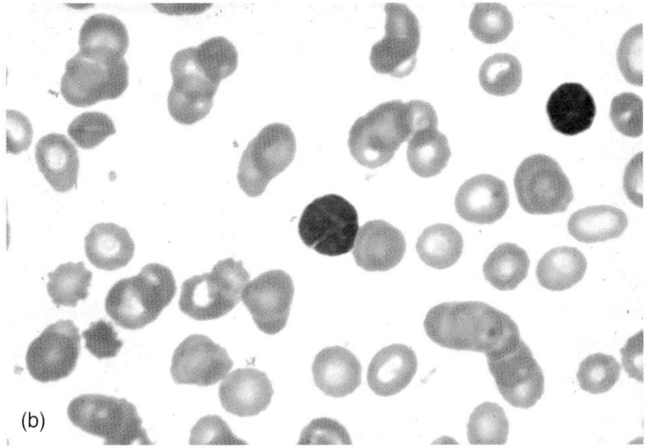

FIGURE 15.46 Bone marrow (a) and peripheral blood smears (b) of a patient with follicular lymphoma demonstrating atypical lymphocytes with irregular or cleaved nuclei.

proliferation (e.g. ras) may transform indolent disease into a more aggressive lymphoma. The BCL-2 protein is not normally expressed in germinal center cells. Variant translocations have been described, t(2;18)(p12;q21) and t(18;22)(q21;q11.2), involving the *IGK* or *IGL* genes, respectively, rather than *IGH*. There have also been complex translocations described involving a third chromosome breakpoint in addition to 14q32 and 18q21 [185].

In addition to the G-banding techniques, this translocation is detected by molecular methods using Southern blot, qualitative PCR testing, or quantitative (real-time) PCR (Figure 15.50). Recent studies have shown prognostic correlation with the level of *BCL-2* fusion genes using quantitative PCR methods [186, 187]. Most cases demonstrate a pattern of translocation involving the major breakpoint region (MBR) of the *BCL-2* gene with a smaller proportion involving the minor cluster region (MCR). Primers specific for MBR and MCR are required for detection of each. And about 25% of cases involve other breakpoints that will not be detected by either primer set and require additional custom primers [188]. That is one reason why the cytogenetic test for t(14;18) will actually pick up more cases than the molecular test [189].

The presence of t(14;18) is not specific for FL and has been found in 15–20% of diffuse large B-cell and MALT lymphomas. Most FL patients show, in addition to t(14;18), clonal evolution events leading to additional chromosomal aberrations such as +8, +7, +12, +18 or abnormalities of 3q, 6q, 13q, and 17q (Table15.7) [1, 190–193]. The 3q27 abnormality involves the *BCL-6* gene and is present in about 15% of the cases. A more aggressive variant of FL (grade 3b) is less frequently associated with t(14;18), but often carries the t(3;14) involving the *BCL-6* oncogene. The 17p13 abnormalities involve the *p53* gene is often associated with the transformation of FL [1]. In addition, gene expression profiling is beginning to be applied to FL (Figure 15.51) [194].

Clinical Aspects

FL is the second most common lymphoma comprising between 22% and 35% of the non-Hodgkin lymphomas in the Western Europe and United States, respectively. It is the most common indolent lymphoma in Western countries accounting for up to 70% of all low-grade lymphomas [2].

The median age at diagnosis is about 60 years with slight female predominance. Painless peripheral lymphadenopathies in the cervical, axillary, and inguinal regions are the major clinical presentations. Mediastinal and hilar lymph nodes, bone marrow, spleen, and liver are also frequently involved. Involvement of the central nervous system is rare. Systemic "B" symptoms are present in about 20% of the cases and patients may rarely present primary extranodal disease involving skin or other tissues. The clinical course is variable and primarily depends on the stage and grade of the disease. The median survival is 7–10 years for patients with grade 1 or 2 and stage III to IV diseases [2].

An international prognostic index (IPI) has been proposed for FL, which includes the following five adverse prognostic factors [195, 196]:

1. Age >60 years
2. Stage III or IV
3. Hemoglobin level <12 g/dL
4. Number of involved nodal areas >4
5. Elevated serum lactate dehydrogenase (LDH).

Patients with FL fall into three major groups: low risk with 0 to 1 adverse factors (36%), intermediate risk with 2 adverse factors (37%), and high risk with 3 or more adverse factors (27%) [195–197].

Therapeutic modalities range from involved field radiotherapy for stages I and II to combination chemotherapy for advanced stages [198]. Rituximab has also been added to the therapeutic regimens, particularly in relapsed or refractory lymphomas [199]. Autologous and allogeneic stem cell transplantations have been attempted for patients with recurrent FL, though limitations such as high recurrence rate and risk of secondary MDS still exist [200]. The anti-idiotype vaccination in FL is under evaluation [178].

Variants of Follicular Lymphoma

Two major variants of FL are recognized: primary cutaneous follicle center cell lymphoma and diffuse follicle center cell lymphoma.

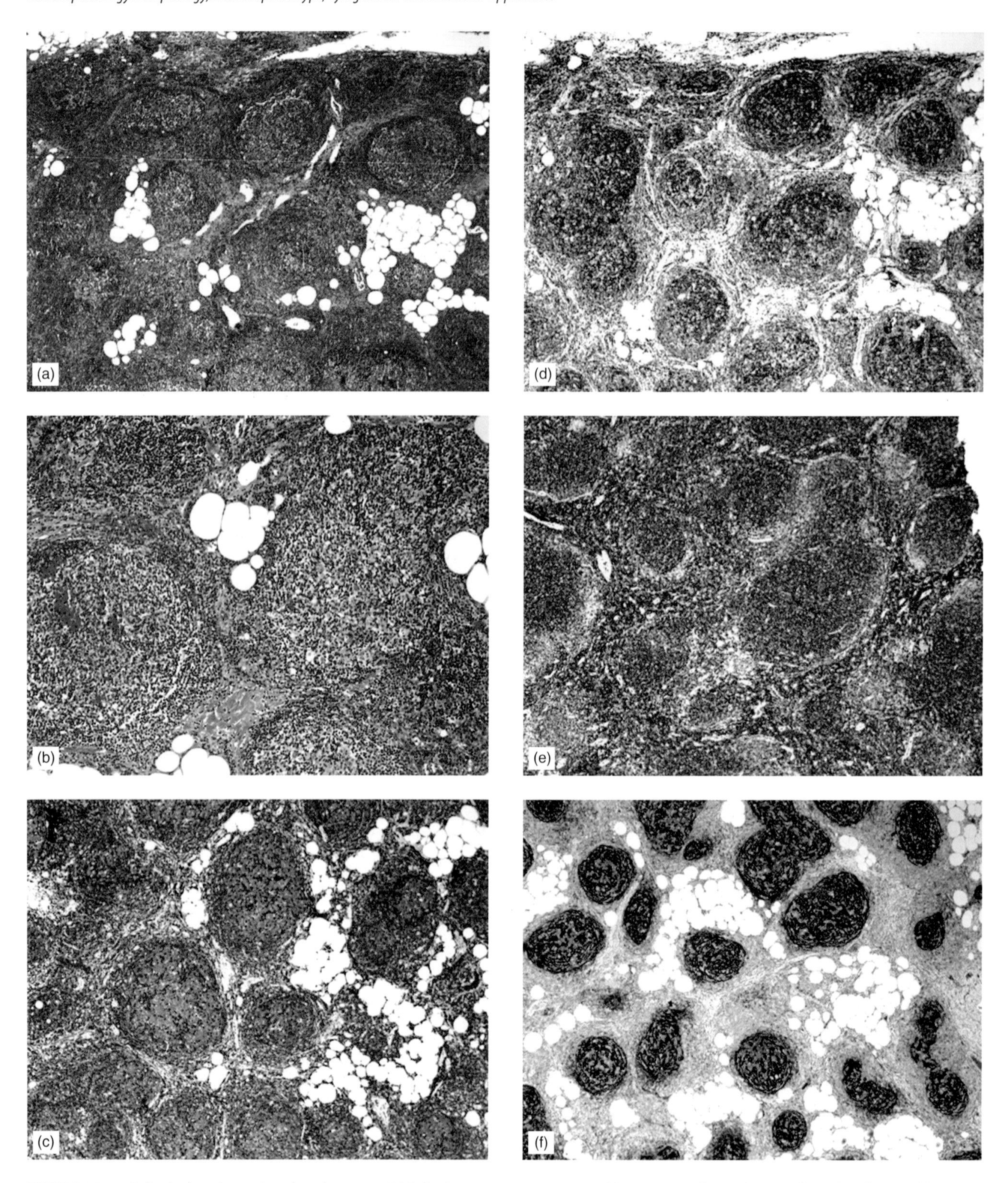

FIGURE 15.47 Follicular lymphoma; lymph node section. (a) Follicular structures are separated from one another with a rim of mantle cells resembling reactive follicles. (b) Higher power view demonstrates lack of tingible body macrophages and some degree of fibrosis. (c) Dual immunohistochemical stains for CD3 (brown) and CD20 (red) show predominance of $CD20^{+}$ cells within the follicles. These cells are positive for CD10 and BCL-2 (d and e, respectively). CD21 stain shows a meshwork of interfollicular dendritic cells (f).

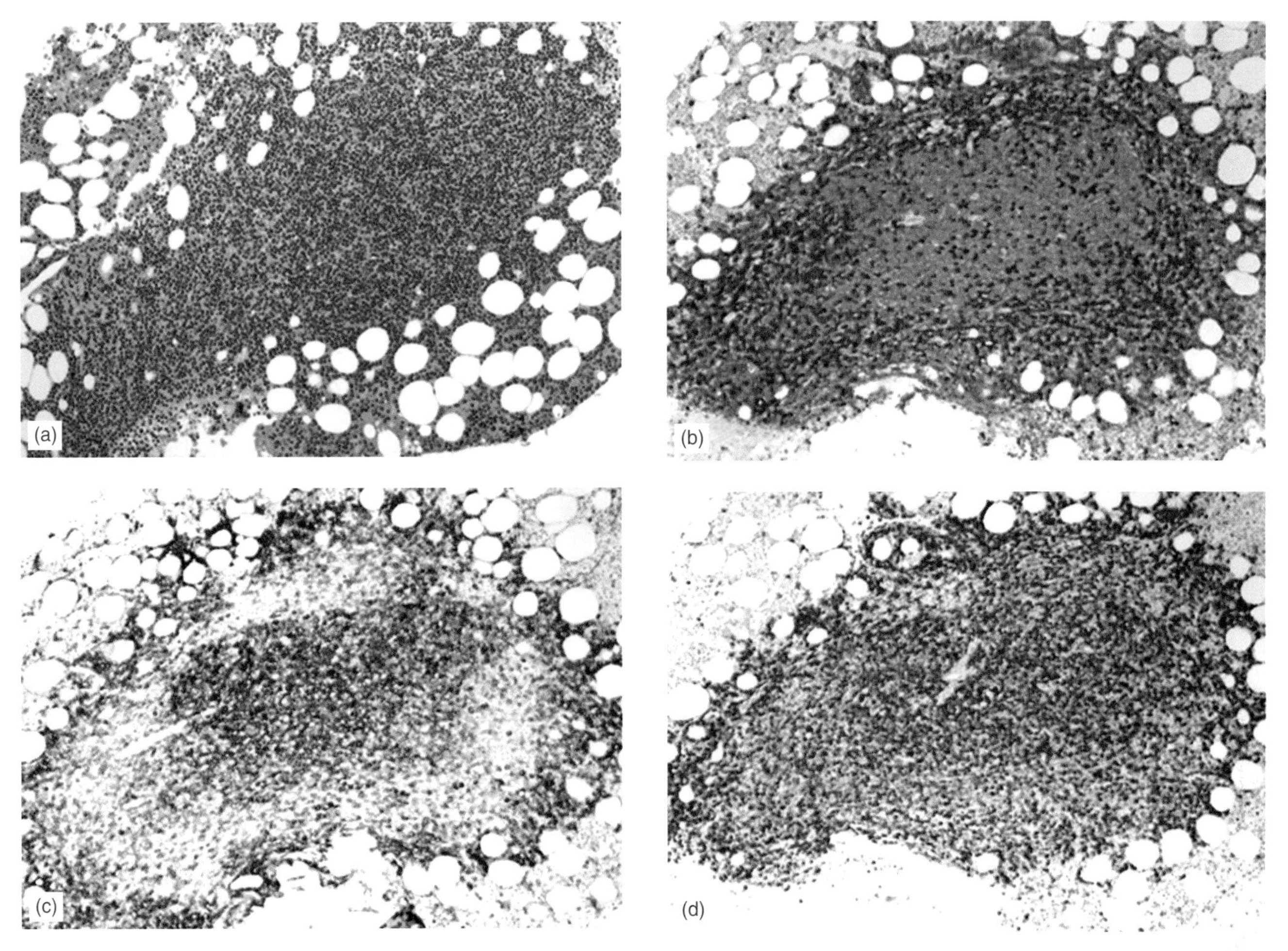

FIGURE 15.48 Bone marrow lymphomatous involvement. (a) A lymphoid aggregate with expansion into the surrounding fatty tissue. (b) Dual immunohistochemical stains for CD3 (brown) and CD20 (red) show CD20+ cells in the center surrounded with CD30+ cells. The CD20+ cells also express CD10 (c) and bcl-2 (d).

Primary cutaneous follicle center cell lymphoma is a B-cell cutaneous lymphoid malignancy usually occurring on the head, neck, and trunk. The lesions tend to be limited to skin without nodal involvement. The lymphomatous infiltrate consists of a mixture of centrocytes and centroblasts and appear partially follicular. The tumor cells are often positive for CD10 and BCL-6 and negative for BCL-2 [201]. Approximately 6% of the cases demonstrate t(14;18) and 6% show extracutaneous progression [202, 203].

Diffuse follicle center lymphoma is a rare lymphoma primarily consisting of centrocytes with a diffuse infiltrating pattern [1]. It is divided into two grades based on the percent of centroblasts per high-power microscopic field (hpf): grade 1 with 0–5 centroblasts/hpf and grade 2 with 6–15 centroblasts/hpf. The neoplastic cells show immunophenotypic features similar to the FL, including expression of CD10, bcl-2, and bcl-6.

Differential Diagnosis

The major differential diagnosis is between FL and reactive follicular hyperplasia (RFH). Reactive follicles are usually separated by interfollicular areas, are surrounded by a mantle zone, show polarity, contain tingible body macrophages, and lack BCL-2 expression. Neoplastic follicles are often back to back or merging, show minimal or no mantle zone, appear monomorphic with loss of polarity, lack tingible body macrophages, and express BCL-2 (Table 15.8). There is evidence of interfollicular infiltration by the presence of CD10 and BCL-6 positive cells in these areas.

Follicular colonization in MALT lymphomas may mimic FL. However, the clinical setting and immunophenotypic features of MALT lymphoma (CD10−, BCL-6−) are different from those of FL (CD10+, BCL-6+). Also, MCLs with nodular or follicular patterns may simulate FL (see the following section).

MANTLE CELL LYMPHOMA

Mantle cell lymphoma (MCL) is an aggressive B-cell neoplasm consisting of small, mature centrocyte-like cells with

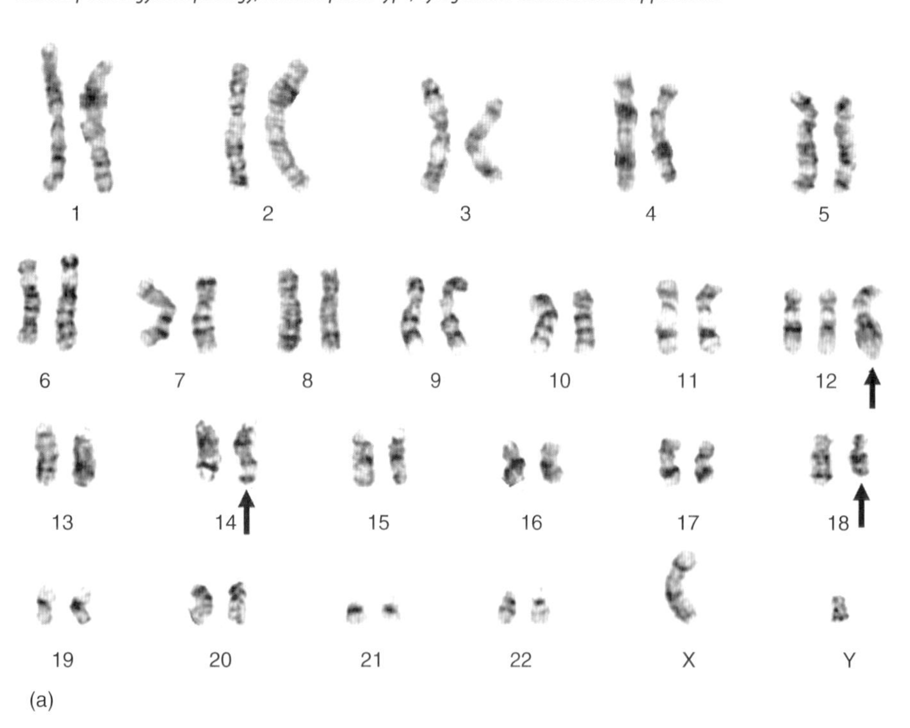

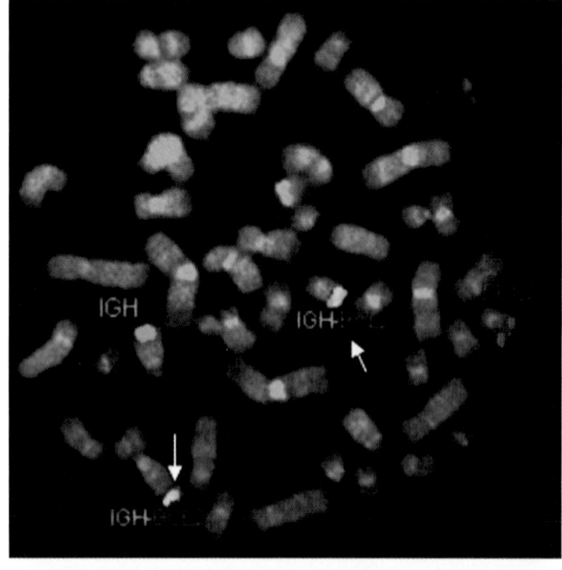

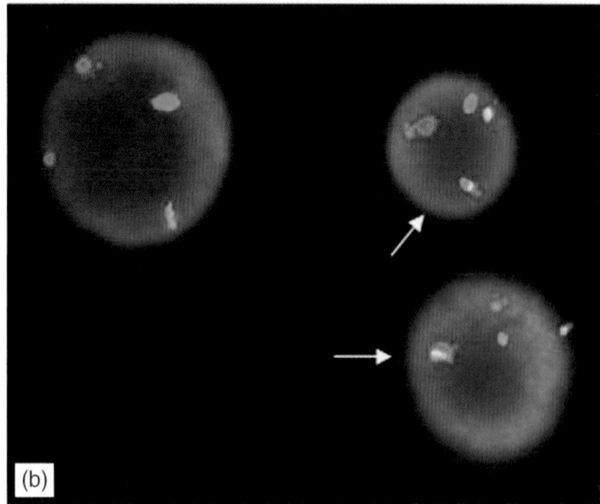

FIGURE 15.49 Karyotype (a) and FISH (b) analysis of lymphoma cells from a patient with follicular lymphoma demonstrating 47,XY, +12,t(14;18)(q32;q21) and *IGH-BCL-2* fusion.

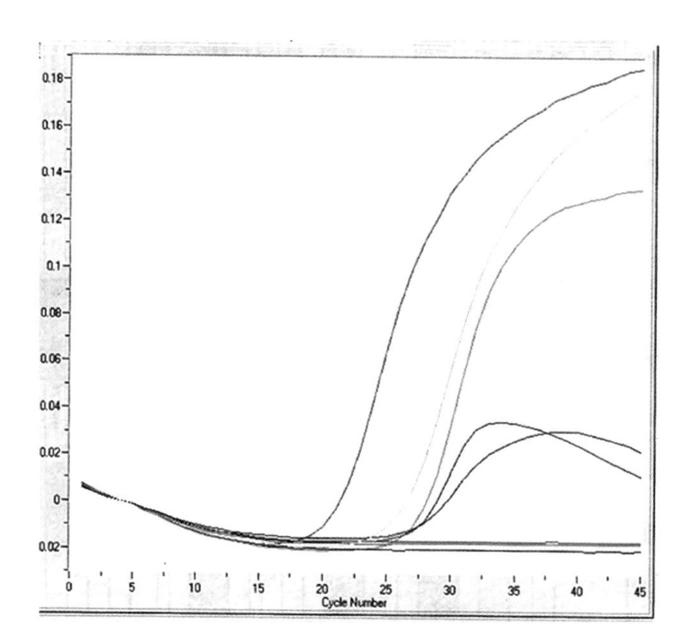

FIGURE 15.50 *BCL-2* fusion gene quantitation by real-time PCR. Patient 1 is positive for *BCL-2*, demonstrating amplification of the target with mid-log point at about cycle 30 (blue curve). The tan curve is the amplification of the control reference gene in the specimen, tPA. Patient 2 is negative for *BCL-2*, demonstrating no amplification of this target (red curve). The tan curve represents the tPA control target for Patient 2, confirming that there are no inhibitors of PCR in the specimen.

TABLE 15.7 Cytogenetic abnormalities in follicular lymphoma.*

Abnormalities	Frequency (%)
Structural	
t(14;18)(q32;q21)	78
3q27-28	16
17p**	15
del(6q)	13
1q12-21	13
1p21-22	10
10q22-24	10
Numerical	
+X	21
+7	20
+18	20
+12/dup(12q)	10

*Adapted from Tilly H, Rossi A, Stamatoullas A, Lenormand B, Bigorgne C, Kunlin A, Monconduit M, Bastard C. (1994). Prognostic value of chromosomal abnormalities in follicular lymphoma. *Blood* **84**, 1043–9.

**Associated with a worse prognosis.

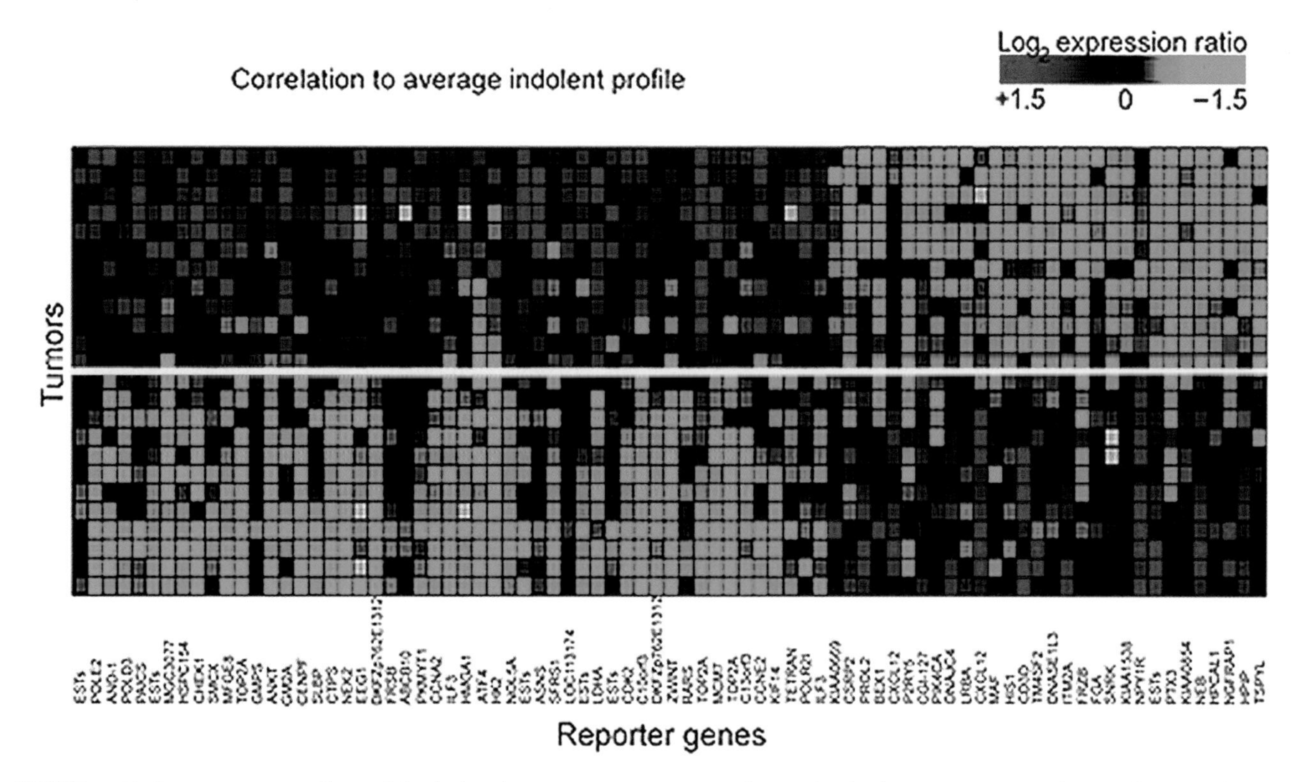

FIGURE 15.51 Gene expression profiling in follicular lymphomas using a microassay technique. Results demonstrate two populations of patients: those with aggressive disease are segregated above the solid yellow line and those with indolent clinical course are placed below it. From Ref. [194] by permission. This research was originally published in *Blood*.

TABLE 15.8 Morphologic features distinguishing follicular lymphoma (FL) from reactive follicular hyperplasia (RFH).

Features	FL	RFH
Follicles	Back to back or merging, often with loss of mantle zone	Separated, with preservation of mantle zone
	Lack of polarity	Presence of polarity
	Lack of tingible body macrophages	Presence of tingible body macrophages
	Monoclonal	Polyclonal
	Commonly BCL-2 positive	BCL-2 negative
	Low Ki-67 fraction	High Ki 67 fraction
Interfollicular areas	CD10 positive cells present	CD10 positive cells absent
	BCL-6 positive cells present	BCL-6 positive cells absent

scant cytoplasm and slightly irregular nuclei. MCL lacks centroblasts and immunoblasts [1, 2, 204, 205].

Etiology and Pathogenesis

The etiology of MCL is not known. The *BCL-1* (*cyclin D1*, abbreviated *CCND1*) gene located on the chromosome 11q13 apparently plays an important role in the pathogenesis of MCL [205, 206]. *BCL-1* is an important regulator of the G1 phase of the cell cycle. The chromosomal translocation t(11;14)(q13;q32) leads to the upregulation of *BCL-1* leading to the inhibition of the suppressive effect of Rb and p27, amplification of the cyclin-dependent kinase-4 (CDK-4), deletion of CDK inhibitor p16, and overexpression of p16 transcriptional repressor BMI-1, causing dysregulation of the cell cycle [205–207].

Pathology

Morphology

The affected lymph node shows effacement of the nodal architecture with an infiltrative process with a histologic pattern that may be diffuse, vaguely nodular, marginal zone pattern, or a combination of these (Figures 15.52–15.54) [1, 2, 208]. Most commonly, the involved lymph node shows transitional areas between diffuse and nodular patterns, but occasionally nodular pattern is predominant. In the marginal zone pattern, the neoplastic cells expand the mantle zone area surrounding a germinal center [205]. This pattern is more common in spleen. A prominent meshwork of follicular dendritic cells is usually present.

The cytologic features in typical MCL cases consist of monotonous small to medium-sized lymphocytes with scant cytoplasm, slightly irregular nucleus, condensed chromatin, and inconspicuous nucleoli (Figure 15.52). Hyalinized small vascular structures are frequently present. Centroblasts and immunoblasts are typically absent, though centroblasts may be present in remnants of germinal centers. In some cases, a proportion of the neoplastic cells may show more abundant cytoplasm and appear like monocytoid B-cells. The neoplastic cells in occasional cases may mimic CLL/SLL with small lymphocytes, round nucleus, and condensed chromatin. A prolymphocyte-like variant with marked leukocytosis mimicking PLL has also been described. These patients usually have splenomegaly and demonstrate clinicopathological features very similar to PLL, except for the demonstration of t(11;14).

The blastoid variant of MCL is usually referred to the cases in which the neoplastic cells resemble lymphoblasts, with dispersed chromatin, prominent nucleoli, and high mitotic figures (Figure 15.55) [208–210]. A pleomorphic type consisting of large cells with oval or irregular nuclei has also been reported [1, 211]. A summary of the cytologic variants of MCL is presented in Table 15.9.

Extranodal involvement is frequent in MCL and often involves bone marrow, peripheral blood, spleen, liver, and gastrointestinal tract [212–215]. Bone marrow infiltration is reported in 50–80% of the cases. The pattern of involvement is often a combination of nodular, interstitial, and paratrabecular. Isolated paratrabecular infiltrations are rare. The neoplastic cells may also be present in the peripheral blood in various numbers creating a condition mimicking CLL. The spleen is affected in 30–50% of the cases with the primary involvement of the white pulp. The white pulp nodules are expanded and often are merged, sometimes surrounding the residual germinal centers. The extent of splenic red pulp involvement is variable. The most common clinical presentation of gastrointestinal involvement in MCL is *lymphomatous polyposis* with the presence of multiple intestinal lymphoid polyps [216].

Immunophenotype

The MCL cells express B-cell-associated markers such as CD19, CD20, CD22, and CD79a with surface IgM^+ and IgD^+. The Ig lambda light chain restriction is observed more often than the kappa light chain. The neoplastic cells are also positive for CD5, FMC7, CD43, BCL-1, and BCL-2, and lack of the expression of CD10, BCL-6, and often CD23 [1, 207, 217–219]. In some cases, CD23 is weakly expressed. The combination of BCL-1 and CD5 expression is characteristic immunophenotypic feature of MCL (Figure 15.54). The Ki-67 labeling is typically low,

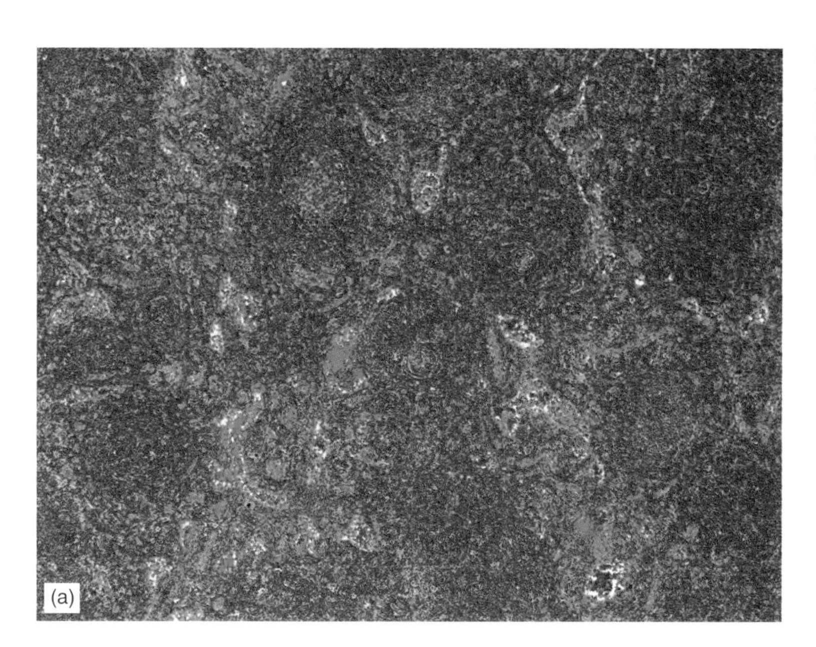

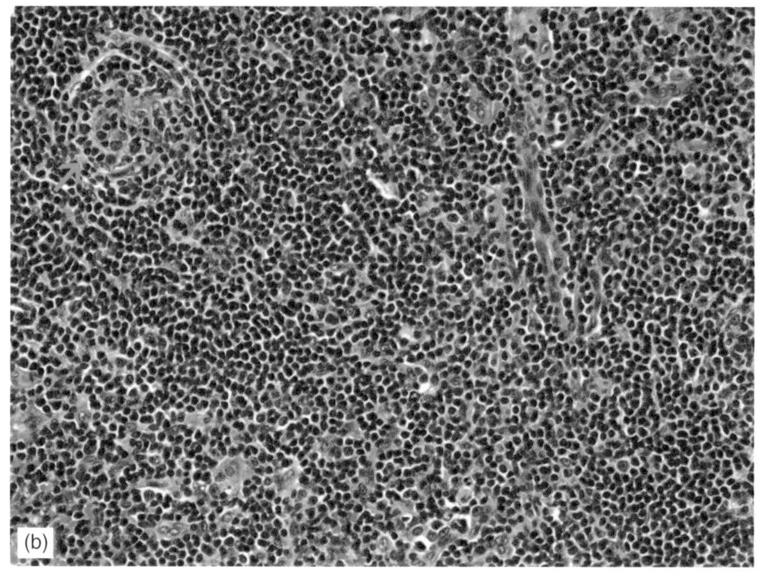

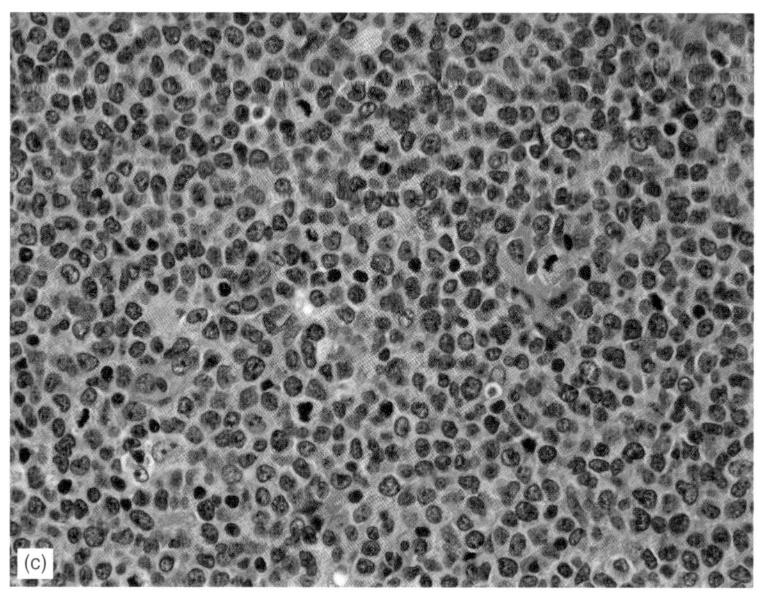

FIGURE 15.52 Mantle cell lymphoma. Lymph node section demonstrates expansion of mantle zones with remnants of follicular structures: (a) low power and (b) intermediate power views. Tumor cells are monomorphic and show some mitotic figures: (c) high power view.

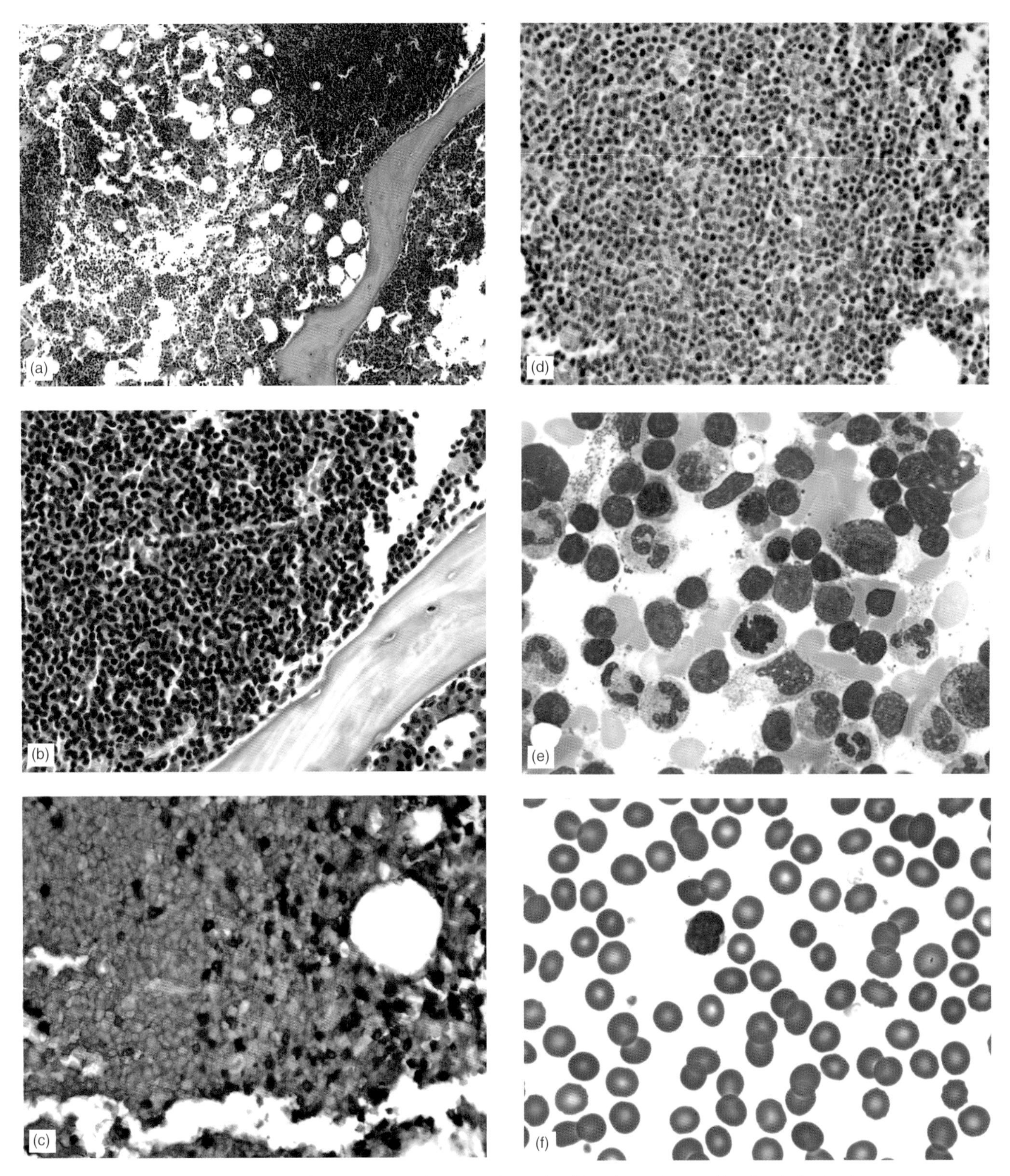

FIGURE 15.53 Bone marrow biopsy section demonstrating involvement with mantle cell lymphoma. Immunohistochemical stains show tumor cells (a: low power, and b: high power) expressing CD20 (c, red) and BCL-1 (d). Bone marrow smear shows numerous atypical lymphoid cells (e) and blood smear demonstrates a small lymphocyte with irregular nucleus (f).

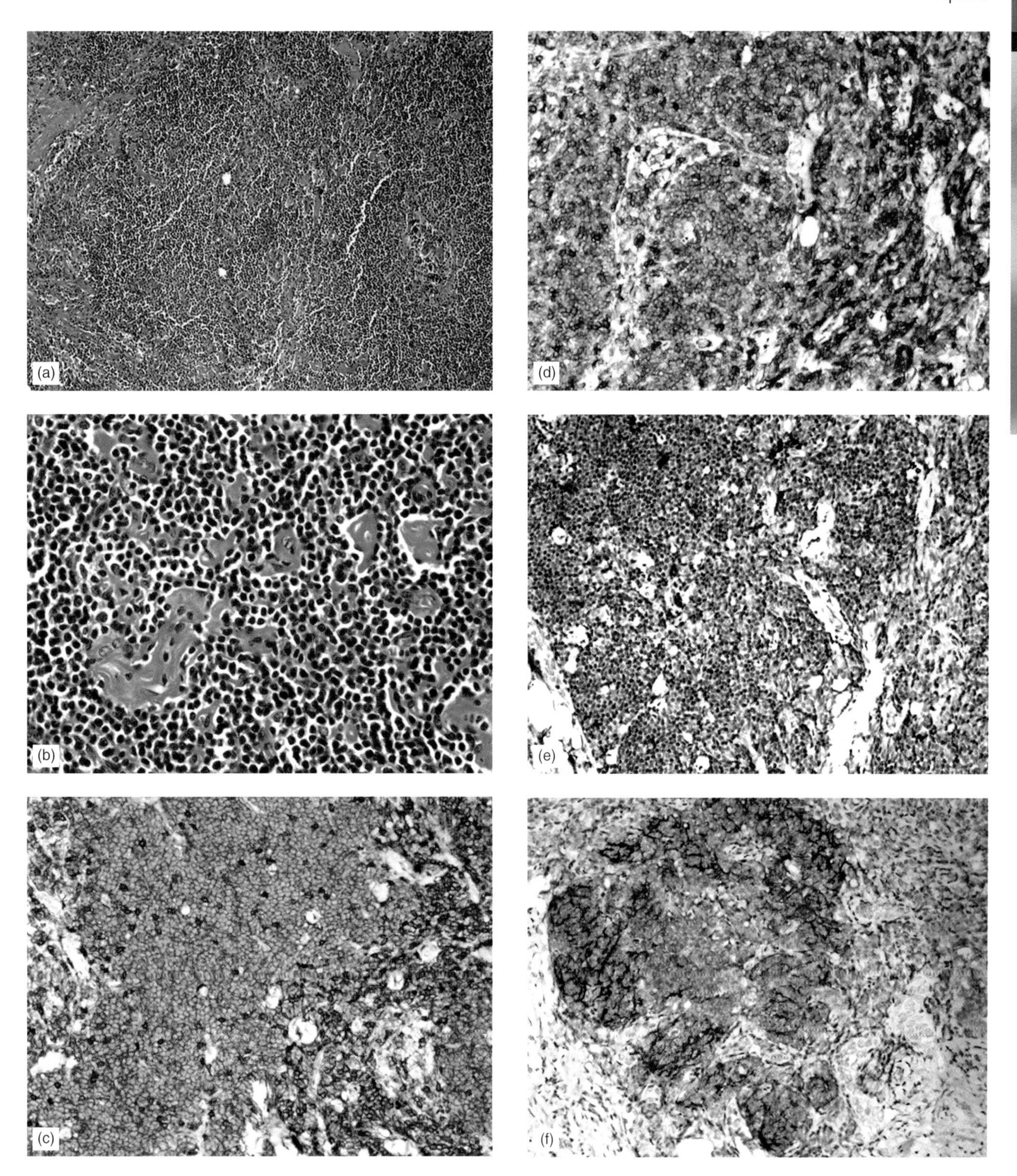

FIGURE 15.54 Mantle cell lymphoma. A lymph node section demonstrates sheets of small lymphocytes in a hyalinized fibrotic stroma: (a) low power and (b) high power views. Immunohistochemical stains show that these lymphocytes are predominantly B-cells (c: CD20+ = red, CD3+ = brown) and express CD5 (d) and BCL-1 (e). Aggregates of follicular dendritic cells are demonstrated expressing CD21 (f).

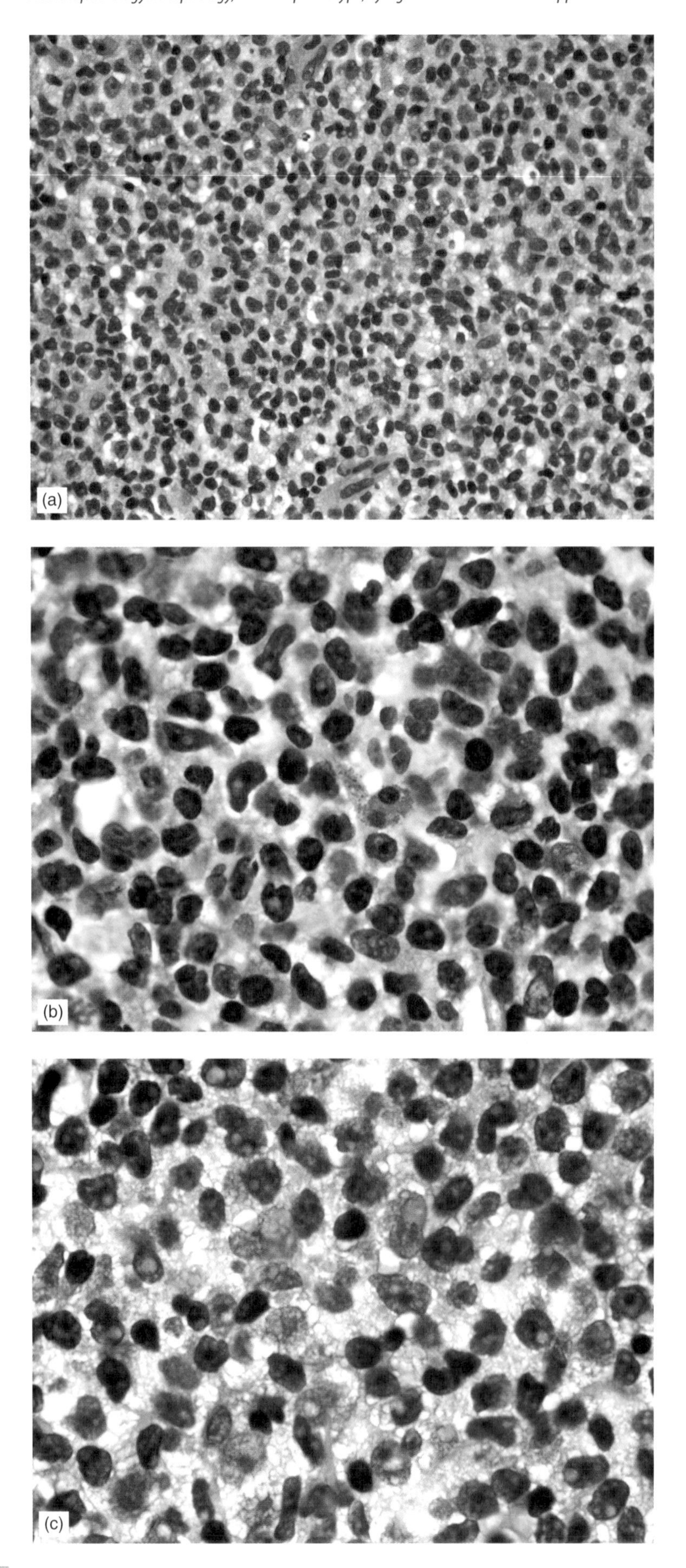

FIGURE 15.55 Blastoid variant of mantle cell lymphoma. Sheets of neoplastic cells demonstrating variable amounts of vacuolated cytoplasm, irregular nuclei with finely dispersed chromatin, and prominent nucleoli: (a) low power, (b) intermediate power, and (c) high power views.

TABLE 15.9 Cytologic variants of mantle cell lymphoma.*

Typical	Small to medium-sized lymphocytes with scant cytoplasm, slightly irregular nucleus, and condensed chromatin.
CLL-like	Small lymphocytes with scant cytoplasm, round nucleus, and condensed chromatin.
Monocytoid B-cell	Prominent foci of medium-sized cells with abundant pale cytoplasm.
Prolymphocyte-like	Medium-sized cells with variable amounts of cytoplasm, round or slightly irregular nucleus, relatively condensed chromatin and single prominent nucleus.
Blastoid variants	
Classic	Cells resembling lymphoblasts with high mitotic figure (>10/10 hpf).
Pleomorphic	Pleomorphic large cells with variable amounts of cytoplasm, cleaved or oval nucleus. Nucleoli may be prominent.

*Adapted from Ref. [10].

except for the blastoid variant of MCL. The presence of a prominent meshwork of follicular dendritic cells is demonstrated by the expression of CD21 and/or CD35.

Cytogenetic and Molecular Studies

The characteristic cytogenetic alteration in MCL is the t(11;14)(q13;q32) (Figure 15.56). Classical cytogenetic studies detect t(11;14) in up to 65% of MCLs [220, 221]. But recent studies using FISH techniques have shown that this translocation is present in nearly all MCLs [222]. The t(11;14) is extremely rare in other lymphomas. It has been detected in occasional atypical cases of chronic lymphocytic leukemia (CLL) associated with an aggressive clinical course, as well as some cases of prolymphocytic leukemia, but in fact, these cases may represent atypical forms of MCL. Approximately 5% of patients with multiple myelomas show t(11;14) (see Chapter 16).

The main feature of MCL is overexpression of cyclin D1, which is due to the translocation of *BCL-1* to the heavy chain locus [223]. Cyclin D1 is a positive regulator of the G1/S cell-cycle restriction point and the overexpression induces increased cycling of the cells. Even if the overexpression of cyclin D1 is the main feature of MCL, it is probably not the sole oncogenic feature, as overexpression of the gene in mice fails to induce lymphomagenesis [224].

The breakpoints on 11q13 occur along a wide region with only 35% of these within a restricted region (major translocation cluster, abbreviated MTC). Thus, a PCR-based assay for the t(11;14) (*BCL1-IGH*) can detect the translocation only in a small fraction of MCL cases at genomic level (the translocation cannot be amplified at RNA level because there is no fusion product, only a deregulated expression of an otherwise normal *CCND1*), and PCR has a limited diagnostic role because of its high false-negative rate [225]. FISH is now considered the technique of choice for searching for the t(11;14) during the diagnostic workup [226]. A new gene named *PRAD1* located approximately 120 kb downstream of the MTC breakpoint was first identified in studies on parathyroid adenomas with inversion in chromosome 11, and this gene was considered to be a putative oncogene deregulated by t(11;14). In further studies, the *PRAD1* sequence was recognized as having a high degree of homology with cyclins, and the new member in that gene family was renamed *CCND1* encoding for the cyclin D1 protein [227]. In t(11;14), the coding region of *CCND1* is structurally intact, but the chromosomal rearrangement positioning the *CCND1* gene adjacent to the enhancer region of the immunoglobulin heavy chain gene results in upregulation of *CCND1* and in increased expression of the cyclin D1 protein. In normal lymphoid cells, the RNA and protein levels of cyclin D1 are extremely low or absent [228]. Although the oncogenic mechanism of cyclin D1 is not completely understood, the constant expression of cyclin D1 has an important role in the pathogenesis of MCL because cyclin D1 promotes the progression of cells through the main commitment checkpoint in G1- to S-phase of the cell cycle [229].

Although C-MYC mRNA overexpression has been found in a subset of MCLs, no structural gene alterations of the *C-MYC* seem to be involved in the pathogenesis of MCL. Some studies have shown p53 mutations and p53 protein overexpression to occur also in the aggressive variants of MCL, where the effect of overexpressed cyclin D1 may be enhanced by loss of cyclin-CDK inhibition as a result of the p53 mutation [230]. Structural and numerical centrosome abnormalities have been described to take place at a much higher frequency in MCL in comparison with other lymphoma subtypes, and this might explain their often near-tetraploid karyotype, especially among blastoid variants [231].

Several studies using comparative genomic hybridization (CGH) and/or FISH to characterize alterations in the frequency of DNA copy number sequences in MCL have been published [232, 233]. These studies have confirmed a characteristic profile of chromosomal changes in MCL different from that in other lymphomas. For example, in addition to trisomy 12, deletions in 13q, 11q, 6q, and 17p are the most frequent aberrations in CLL, whereas studies by CGH showed gains of chromosomes X, 1q, 7, and 3, together with losses in chromosomes 6q, X, and 1p in DLBCL [234]. In contrast, in MCL the most frequent chromosomal imbalances are gains of chromosomes 3q (49–70%), 8q (22–30%), and 12q (20–30%) and the most frequent losses of chromosomes 1p (24–33%), 6q (27–37%), 9p (16–41%), 11q (22–31%), and 13q (41–69%). DNA amplifications and the total number of copy number changes have been reported to be higher in the blastoid variant than in the common variant of MCL. In line with findings on the gene level, loss of chromosome 17p has been found to correlate with the rate of *p53* inactivation, gain of 12q with the frequency of *CDK4* amplification, and high level amplification of chromosome 10p12-13 with BMI-1 amplification. Interestingly, further studies on the deletion 11q by FISH showed deletion of 11q22-23 in as many as 49% of MCL samples [235]. This region is known to harbor a number of tumor-associated genes, including the ataxia telangiectasia mutated (*ATM*) gene. These commonly affected

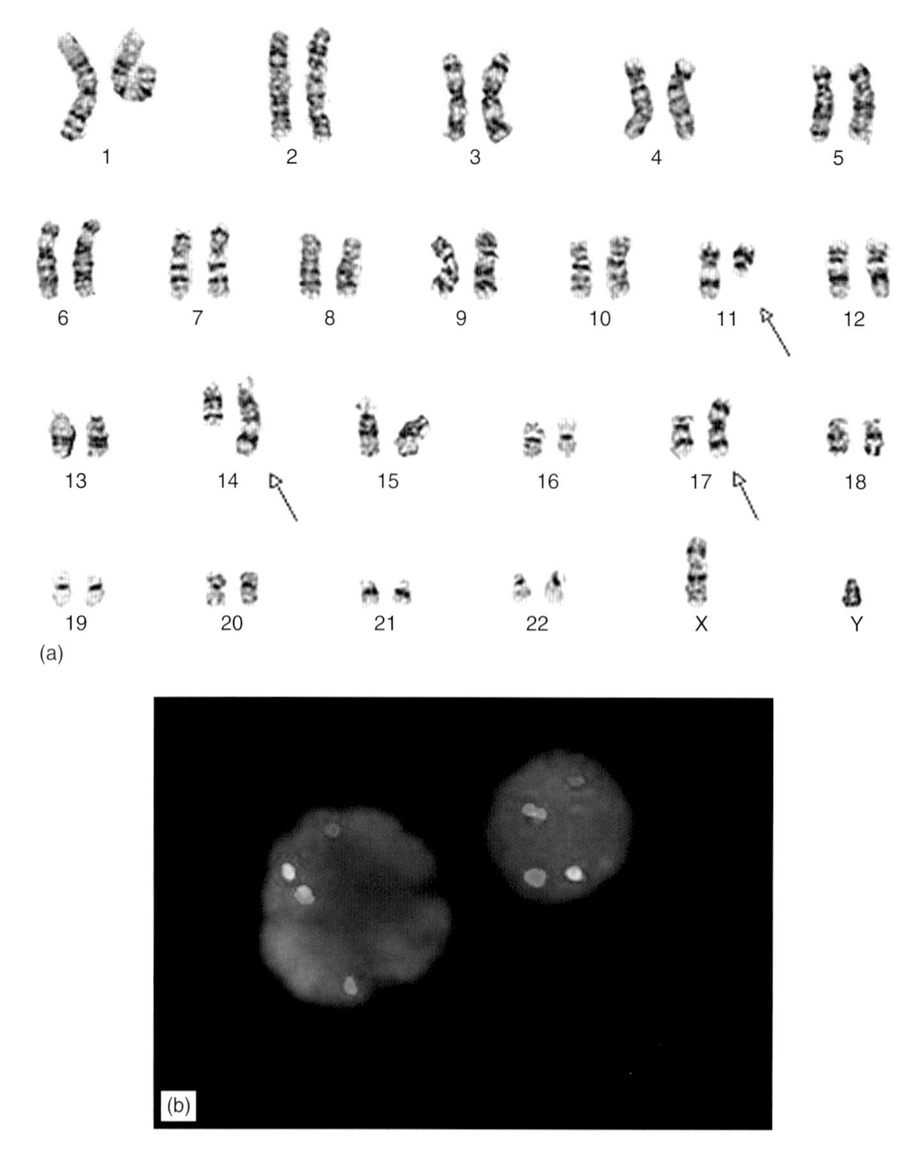

FIGURE 15.56 Karyotype of neoplastic cells from a patient with mantle cell lymphoma demonstrating 46, XY, t(11;14)(q13;q32), iso(17q) (a). FISH analysis shows cells with two fusion signals consistent with 11;14 translocation (b).

chromosomal regions may also harbor unknown oncogenes or tumor suppressor genes.

De Vos and colleagues found that several cell-cycle-related genes were deregulated when typical MCLs were compared with blastoid MCLs, such as *CDK4* gene, which was upregulated in the blastoid MCLs [236, 237]. *CDK4* cooperates with cyclin D1 in the progression through the G1/S checkpoint. Other cytogenetic abnormalities that were found when comparing typical and blastoid MCLs were deletions of *p53* and *p16*. Tumor cell proliferation has been shown to be associated with decreased survival time [238] and, as discussed earlier, is also predictive of blastoid transformation [239].

Clinical Aspects

MCL accounts for 5–10% of non-Hodgkin lymphomas in the United States and Europe. The median age is 63 years with a male:female ratio of about 3:1. Lymphadenopathy and splenomegaly (30–50%) are the major clinical findings and about 30% of the patients show systemic "B" symptoms [212–215]. MCL is considered an aggressive lymphoma with a median survival of 3–4 years. Patients with peripheral blood involvement, high mitotic figures, blastoid variants, and complex cytogenetic abnormalities show more aggressive clinical course [1, 240]. In one report, blastoid transformation occurred in 35% of patients with MCL with a median survival time of 4 months following transformation [241]. Therapeutic modalities such as combination chemotherapy and rituximab may improve the response rate, but without cure in most cases [242–245].

Differential Diagnosis

The differential diagnosis includes CLL/SLL, FL, PLL, and marginal zone-related B-cell lymphomas. There are overlapping cytologic features between CLL/SLL and MCL. Also, the vaguely nodular patterns in MCL and the pseudo-follicles in CLL/SLL may resemble one another. Unlike CLL, the neoplastic cells of MCL are monomorphic and lack prolymphocytes and paraimmunoblasts, usually lack the expression of CD23, express FMC7, CD79a, and BCL-1 nuclear protein, and demonstrate t(11;14).

The vague nodular pattern in MCL may mimic follicular pattern in FL. Unlike FL, the neoplastic cells of MCL lack centroblasts and immunoblasts, do not express CD10, are negative for t(14;18), express CD5, and demonstrate t(11;14).

MCL with monocytoid B-cell morphology may simulate marginal zone lymphomas of the spleen, lymph node, or extranodal sites. Marginal zone B-cells lack the expression of CD5, CD10, and CD23 and are negative for t(11;14).

The reported cases of PLL with t(11;14) are now considered *prolymphocytoid variant* of MCL [72, 246].

DIFFUSE LARGE B-CELL LYMPHOMA

Diffuse large B-cell lymphoma (DLBCL) is the most common non-Hodgkin lymphoma consisting of large transformed neoplastic B-lymphocytes with prominent nucleoli, diffusely infiltrating the involved tissues [1, 247, 248].

Etiology and Pathogenesis

The etiology of DLBCL is not known. Immunodeficiency is a significant risk factor often in association with EBV infection. DLBCL is either *de novo* or the result of transformation of a less aggressive lymphoid malignancy [1]. DLBCL has also been reported in association with certain autoimmune disorders and chronic inflammatory conditions, such as Sjogren syndrome, rheumatoid arthritis and hepatitis C, and with HHV-8-positive multicentric Castleman disease [249].

The pathogenesis of this disorder appears to be complex with the involvement of a variety of genes including *BCL-6*, *PIM1*, *MYC*, *RoH/TTF* (*ARHH*), and *PAX5*. *BCL-6* located at the 3q27 breakpoint may contribute to the lymphomagenesis by mediating the germinal center B-cell phenotype through transcriptional repression of the DNA-damage sensor ATR [250, 251]. The recent genetic profiling studies suggest two major groups in DLBCL: germinal center B-cell type and activated B-cell type. These two groups show different genetic profiles and chromosomal aberrations, and therefore probably different pathogenic pathways [252].

Pathology

Morphology

The neoplastic lymphoid cells are typically large cells (larger than a macrophage) and in most instances resemble centroblasts or immunoblasts or a mixture of the two [1, 2]. The involved lymph nodes show partial or total effacement of the nodal architecture with diffuse infiltration of large atypical neoplastic cells (Figure 15.57). Bone marrow involvement is interstitial, diffuse, or nodular (Figure 15.58). The neoplastic cells may mimic clusters of megaloblasts or metastatic carcinoma. Peripheral blood smears may show the presence of neoplastic large cells. The most common extranodal primary site is the gastrointestinal tract, but skin, CNS, bone, spleen, liver, and other organs may be a primary site. Several morphologic variants have been described including centroblastic, immunoblastic, anaplastic, and T-cell/histiocyte-rich types [1, 2, 253].

Centroblastic variant: This variant is the most frequent type and consists of medium- to large-sized cells with scant amphophilic/basophilic cytoplasm, round or oval nucleus, fine nuclear chromatin, and several nucleoli bound to nuclear membrane (Figure 15.57). Multilobated (>3 lobes) centroblasts may be present and sometimes may create a polymorphic appearance. Some cases may show a mixture of centroblasts and immunoblasts.

Immunoblastic variant: This variant comprises about 10% of DLBCL and represents diffuse lymphomas with >90% immunoblasts. Immunoblasts are defined as large cells with abundant basophilic cytoplasm, round or oval nucleus, fine chromatin, and a prominent central nucleolus (Figure 15.59) [1, 254, 255]. Plasmacytoid features may be present. This variant is commonly seen in immune-compromised patients.

Anaplastic variant: This variant consists of atypical, large cells with bizarre pleomorphic nuclei, some of which may resemble Reed–Sternberg cells (Figure 15.60) [1, 256, 257]. These cells may appear in large clusters, resembling metastatic carcinoma. Sinusoidal pattern has been observed. The anaplastic variant of DLBCL does not share the distinctive clinicopathologic features of the T/null-anaplastic large cell lymphoma.

T-cell/histiocyte-rich variant: This is referred to those cases with heavy background of T-cells and often histiocytes admixed with scattered neoplastic large lymphocytes (Figure 15.61) [1, 258–261]. The neoplastic B-cells account for <10% of the total cells and may resemble Reed–Sternberg cells or variants. Histiocytes may appear epithelioid.

Immunophenotype

The neoplastic cells of DLBCL are commonly positive for B-cell-associated molecules, such as CD19, CD20, CD22, and CD79a, and express surface Ig (mostly IgM) (Figure 15.62) [1, 261, 262]. The BCL-6 nuclear protein is detected in about 70% of the cases, and 25–50% of the cases express BCL-2 [263]. Expression of CD5 or CD10 has been reported in 10% and 25–50% of the cases, respectively. The DLBCL cells are negative for BCL-1 [264]. The germinal center B-cell type of DLBCL is often associated with the expression of CD10 and BCL-6, whereas the activated B-cell type is usually positive for MUM1 (multiple myeloma1) and BCL-2 [265]. The majority of the cases of anaplastic variant of DLBCL are CD30-positive. The Ki-67 expression is high (>40%). Some cases of immunoblastic variant of DLBCL, particularly the AIDS-related ones, express MUM1 [265]. Occasional cases may demonstrate expression of the plasma-cell-associated marker CD138.

Cytogenetic and Molecular Studies

Numerous chromosomal aberrations, point mutations, and deletions have been described in DLBCL suggesting a

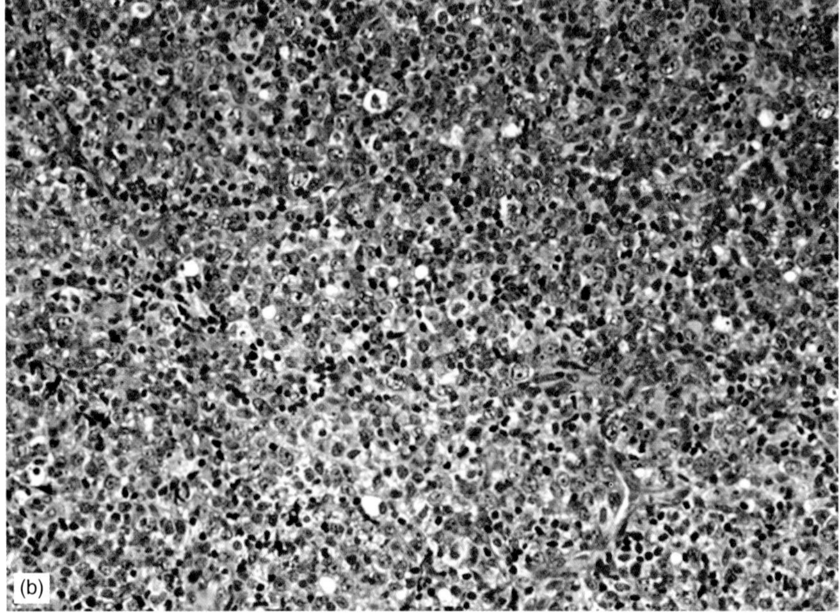

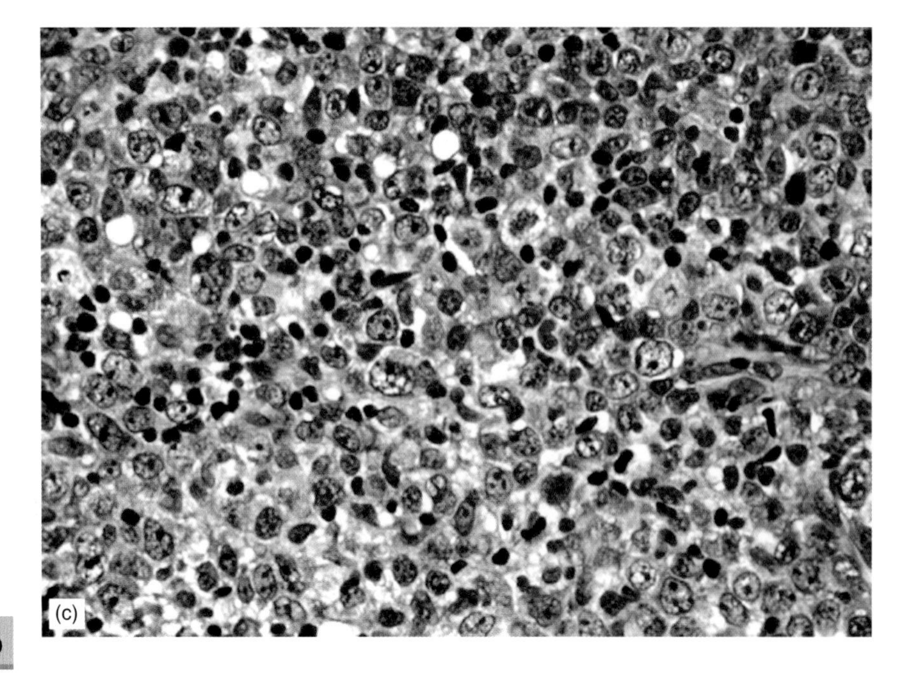

FIGURE 15.57 Lymph node section demonstrating diffuse large cell lymphoma: (a) low power, (b) intermediate power, and (c) high power views.

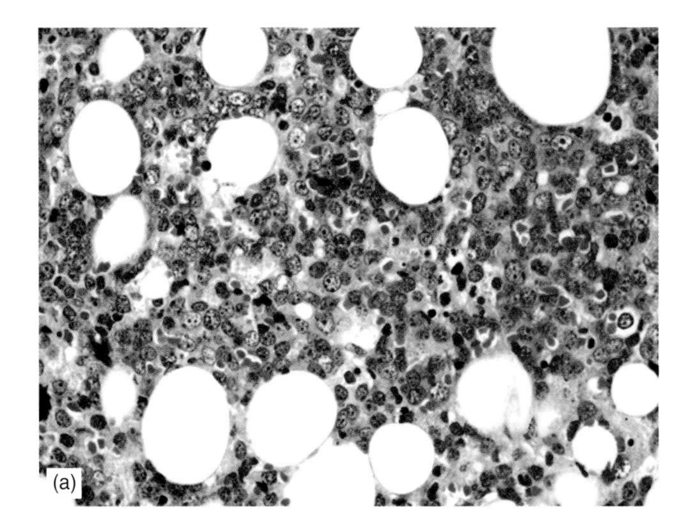

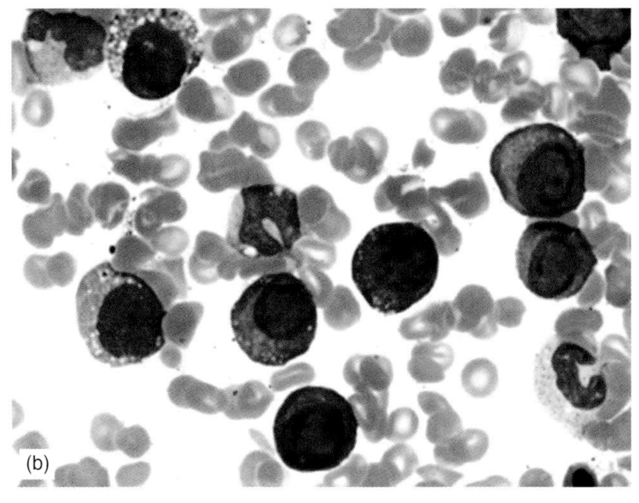

FIGURE 15.58 Bone marrow (a) and blood (b) involvement in a patient with diffuse large B-cell lymphoma.

heterogenous genetic background. The chromosomal translocation t(14;18)(q32;q21.3) involving *BCL-2* is cytogenetically found in 20–30% of the cases [266]. However, overexpression of the *BCL-2* protein is found in 25–50% of the cases, suggesting that other mechanisms for the BCL-2 overexpression are also involved [267]. Translocations involving the locus for *BCL-6* at 3q27 are found in almost a third of the cases [268]. Furthermore, mutations in the 5′ regulatory region of the *BCL-6* gene are also seen, implying that *BCL-6* may be involved in the majority of DLBCL [269]. Clonal karyotypic abnormalities have been reported in up to 87% of the cases of DLBCL. The most commonly involved breakpoints are 14q32, 3q27, and 18q21, which are sites for *IGH*, *BCL-6*, and *BCL-2*, respectively. Other frequent chromosomal abnormalities include 1q36, 8q24, 3p21, 1p22, and 6q21 chromosomal bands. Complex cytogenetic abnormalities are frequent. *c-MYC* rearrangement (8q24) is rare and usually observed in immunodeficient patients. Mutations of *p53* and homozygous deletions at 9p21 are reported in lymphoma cases with large cell transformation [270].

Translocations affecting the immunoglobulin gene sites at 14q32 (*IGH*), 22q11.2 (*IGL*), and 2p12 (*IGK*) have been identified in approximately half of DLBCL cases. These include t(14;18)(q32;q21), t(8;14)(q24;q32) or t(8;22)(q24;q11.2), t(3;14)(q27;q32), t(3;22)(q27;q11.2), and other rearrangements involving 14q32. By conventional cytogenetics, FISH and CGH, the most common abnormalities observed in DLBCL are gains of chromosomes 2p (*REL*), 8q24 (*c-MYC*), chromosome 12 and 18q, losses of chromosomes 6q and 17p, and rearrangements of 3q27 (*BCL-6*) and 14q32 (*IGH*).

Genomic profiling studies demonstrate two major subgroups: one group shows characteristic features of germinal center B-cells (germinal center B-cell-like, or GCB-like) and the other group displays a profile similar to the activated B-cells (activated B-cell-like, or ABC-like) [252]. Gene profiling studies by Bea and associates in three groups of patients, DLBCL of GCB-like, DLBCL of ABC-like, and mediastinal large B-cell lymphoma, demonstrated significant differences in the frequency of particular chromosomal aberrations. For example, the group of DLBCL–GCB displayed frequent gains of chromosomes 12q, whereas the group of DLBCL–ABC had frequent trisomy 3, gains of chromosomes 3q and 18q21-q22, and losses of chromosome 6q21-q22. Other molecular profiling studies of DLBCL have revealed distinct subgroups with independent predictors of prognosis [271].

Clinical Aspects

DLBCL is the most common histological type and accounts for 30–40% of all non-Hodgkin lymphomas [2, 247, 248, 272]. The median age is about 64 years. The male:female ratio is slightly >1. Most patients present with a rapidly enlarging mass, usually in the neck or abdomen. Extranodal involvement is observed in up to 40% of the patients, involving gastrointestinal tract, skin, CNS, bone, testis, liver, spleen, lung, and other organs. Approximately 20% of the patients are at stage 1 and about 40% show disseminated disease at the time of diagnosis. Bone marrow involvement is seen in 10–20% of the cases. The serum LDH levels are elevated in about 50% of the cases and approximately 30% of the patients show systemic "B" symptoms. A significant proportion of DLBCLs are the result of transformation of less aggressive lymphomas, such as CLL/SLL, LPL, SMZL, MALT lymphoma, and FL.

This lymphoma is considered an aggressive lymphoma. The IPI is highly predictive of the patients' clinical outcome. The clinical parameters used in the IPI include age >60, elevated serum LDH, ECOG performance status ≥2, stage III or IV, and number of involved extranodal site >1. The 5-year survival rate is 26% and 73% for the high- and low-risk IPI groups, respectively [2].

Other adverse prognostic factors include the expression of CD5 and survivin protein [264, 273]. Survivin is the product of *BIRC5* gene and inhibitor of apoptosis. A gain in different regions of chromosome 3q has been associated with shorter survival [252]. On the contrary, expression of BCL-6 nuclear protein has been reported to correlate with longer overall survival [274].

Therapeutic regimens include combination chemotherapy (e.g. CHOP) with or without involved field radiation for early stages and CHOP or other alternative

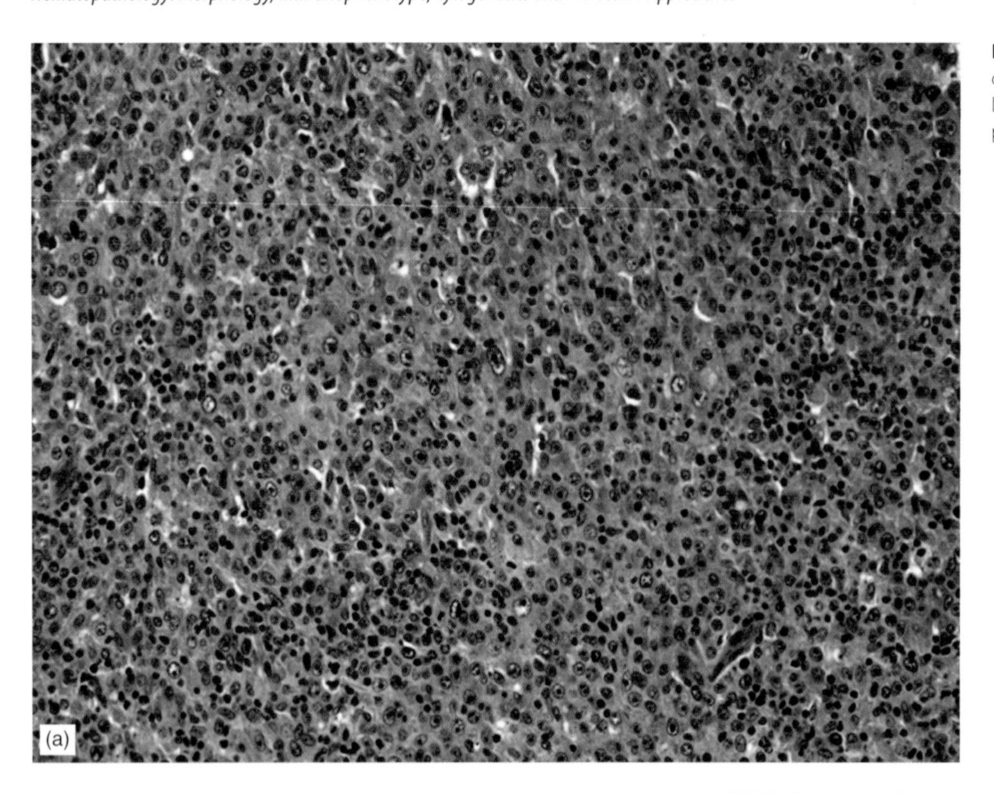

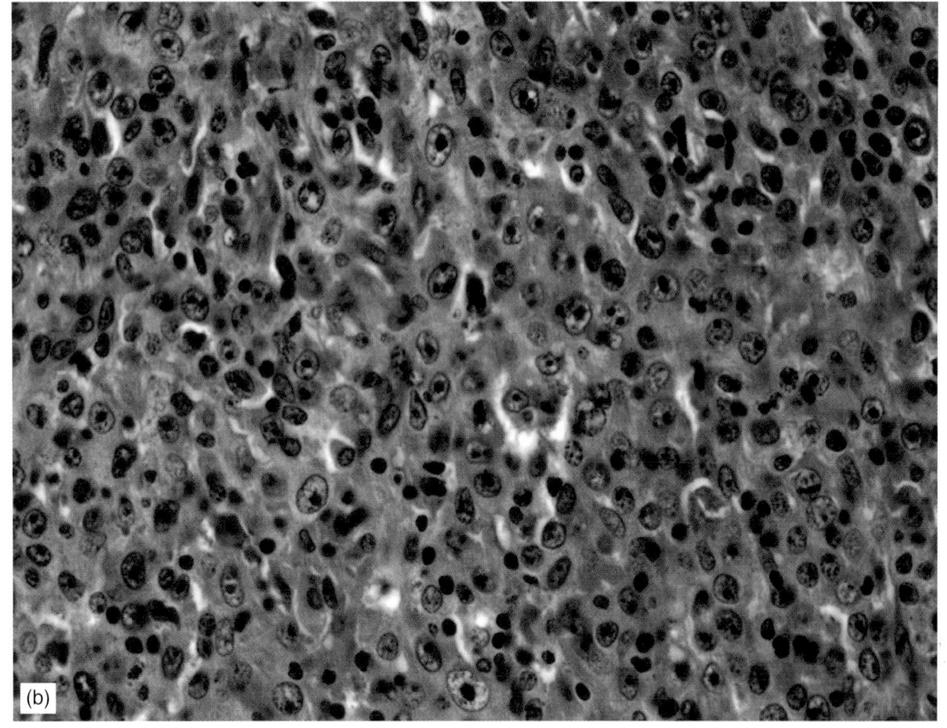

FIGURE 15.59 Lymph node section demonstrating diffuse large cell lymphoma, immunoblastic type: (a) low power and (b) high power views.

combination chemotherapy regimens with or without rituximab, or autologous transplantation in advanced or recurrent disease [2, 275, 276].

Differential Diagnosis

The centroblastic variant of DLBCL may show overlapping morphologic features with Burkitt-like lymphoma. In general, the expression of Ki-67, CD10, and p53 is higher, and BCL-2 and adhesion molecules are lower in the neoplastic cells of Burkitt-like lymphoma than the DLBCL cells (Table 15.10) [277].

The extranodal immunoblastic and plasmablastic variants of DLBCL may mimic plasma cell myeloma. The differential diagnosis of T-cell/histiocyte-rich DLBCL includes lymphocyte predominance and mixed

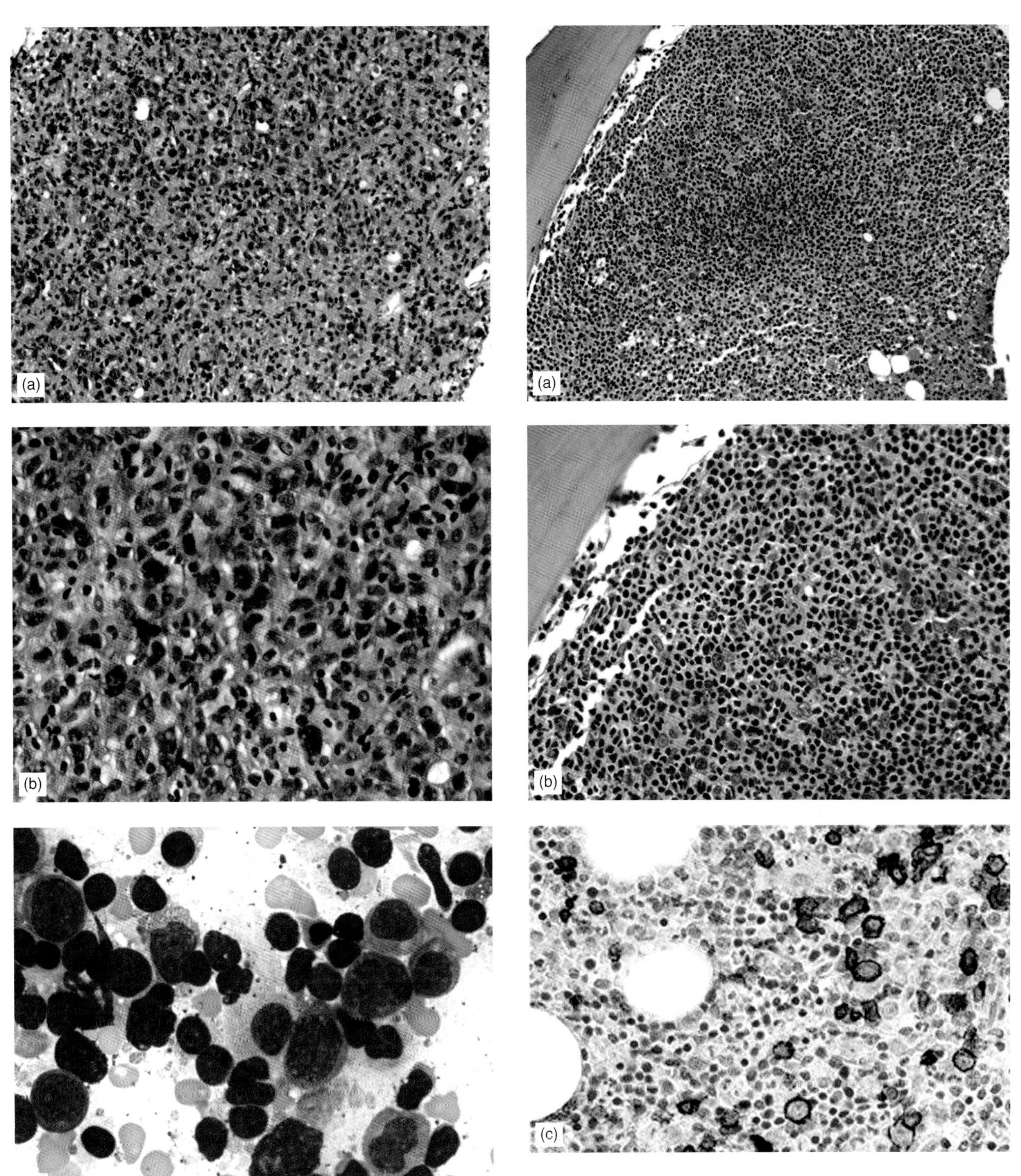

FIGURE 15.60 Bone marrow lymphomatous involvement in a patient with anaplastic large B-cell lymphoma: (a) low power and (b) high power views of biopsy section. Anaplastic cells are demonstrated in bone marrow smear (c).

FIGURE 15.61 T-cell rich large B-cell lymphoma. Bone marrow biopsy section demonstrating scattered large cells mixed with a large number of small lymphocytes: (a) low power and (b) high power views. Large cells are CD20-positive and small cells are CD20-negative (c).

cellularity Hodgkin lymphomas. The anaplastic variant of DLBCL shares morphologic features with T/null-anaplastic large cell lymphoma and classical Hodgkin lymphomas.

MEDIASTINAL LARGE B-CELL LYMPHOMA

Mediastinal (thymic) large B-cell lymphoma (MLBCL) is a distinct clinicopathologic subtype of DLBCL primarily involving the thymus [1, 278–280].

TABLE 15.10 Phenotypic characteristics of diffuse large B-cell lymphoma (DLBCL) and Burkitt-like lymphoma (BLL).*

Expression	DLBCL (%)	BLL (%)
CD19, CD20, and CD22	100	100
CD10	27	85
Mean Ki-67 fraction	53	88
p53	16	54
BCL-2	53	15
CD11a	87	38
CD18	85	10
CD44	86	8

*Adapted from Ref. [277].

Etiology and Pathogenesis

The etiology and pathogenesis of MLBCL are not known. Rare cases are associated with human herpesvirus 6 (HHV-6), raising the possibility that this virus may play a pathogenic role in a minority of the cases [278].

Pathology

Morphology

The neoplasm consists of large cells with variable amounts of cytoplasm, which is often pale or clear (Figure 15.63). The nuclear features may resemble centroblasts or large centrocytes, often with the presence of large cells with multilobated nuclei. Immunoblast-like cells may be dominant in minority of the cases. Reed–Sternberg-like cells may be present, as well as scattered eosinophils. Mitotic figures are frequent. Sclerosis is a common feature separating solid nests of tumor cells by thick hyalinized bands of connective tissue [2, 280–282].

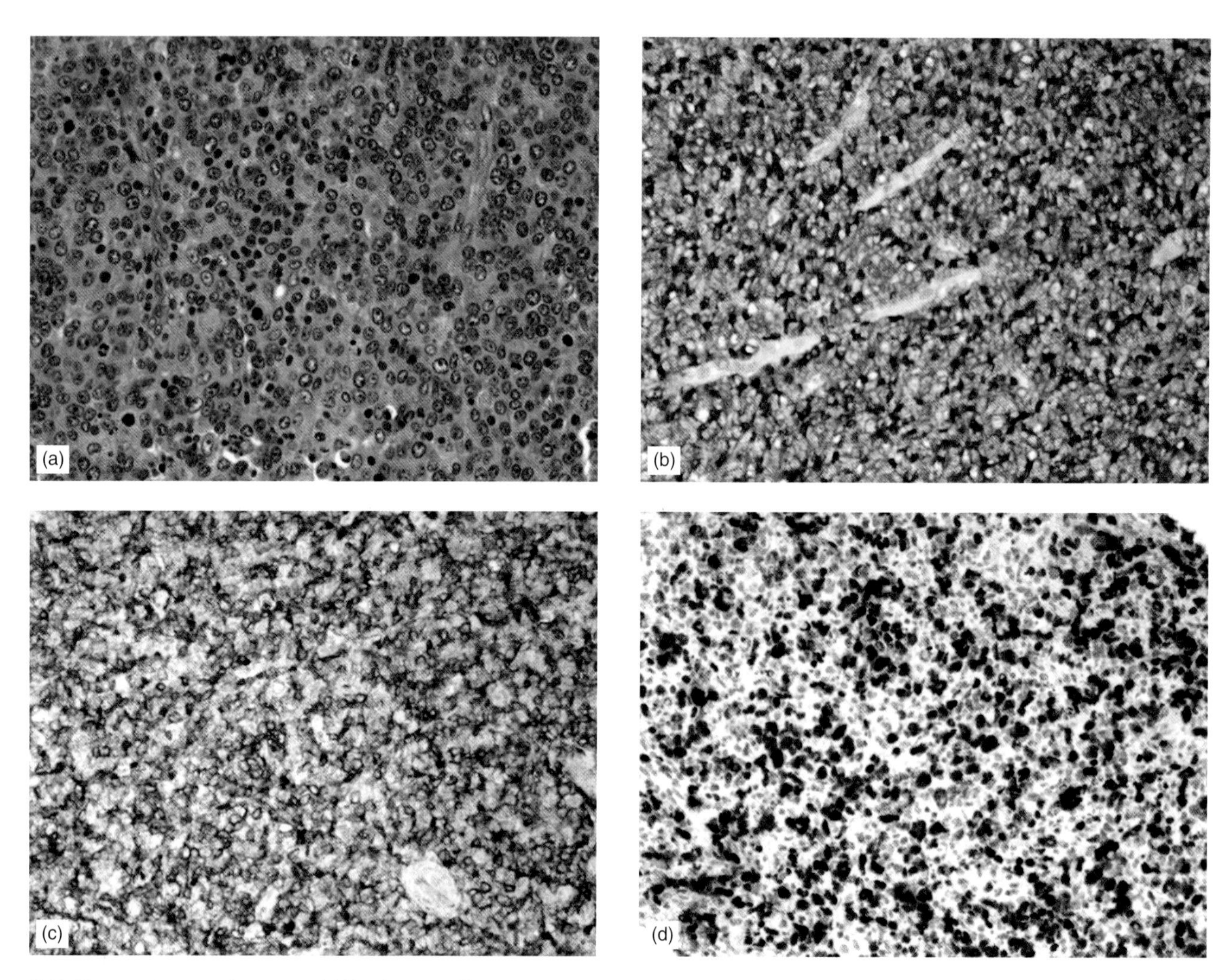

FIGURE 15.62 Immunophenotype of diffuse large B-cell lymphoma. The neoplastic cells show round nuclei with open chromatin and prominent nucleoli (a) and express CD20 (b, red), CD79a (c), and a high percentage of Ki-67 (d).

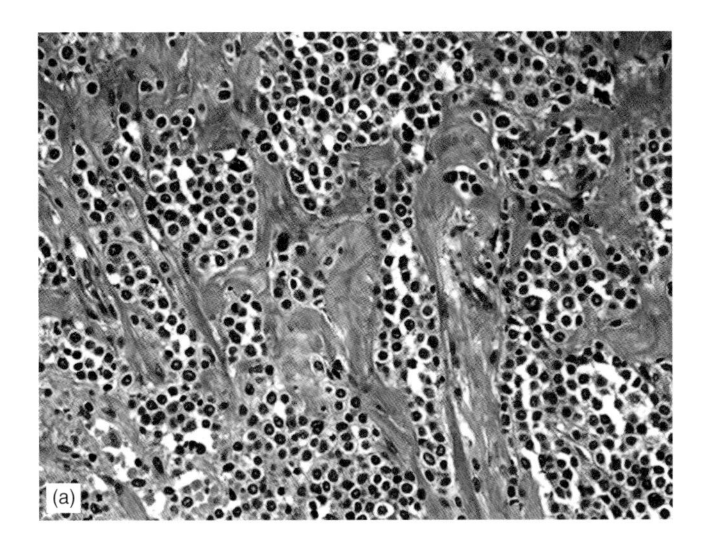

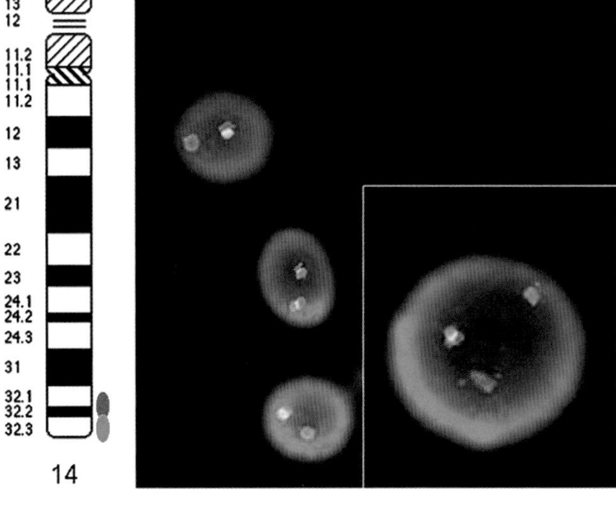

FIGURE 15.64 FISH analysis of neoplastic cells in a patient with primary mediastinal large B-cell lymphoma demonstrating *IGH* gene rearrangement (split red-green signals) (inset). Two of three other cells show 5'IGvH (green signal) deletion.

FIGURE 15.63 Primary mediastinal B-cell lymphoma. Aggregate of large neoplastic large cells are separated by thick bands of hyalinized connective tissue: (a) low power and (b) high power views. The neoplastic cells express CD20 by immunohistochemical technique (c).

Immunophenotype

The neoplastic cells express B-cell-associated markers, such as CD19, CD20, CD22, and CD79a, and they may also express CD23 and CD30 [1, 2, 283, 284]. CD5 may be positive in minority of the cases and CD10 is negative. The multilobulated and Reed–Sternberg-like cells, unlike classic Reed–Sternberg cells, are negative for CD15 but express CD45.

Cytogenetic and Molecular Studies

The most frequent chromosomal abnormalities in MLBCL are gains of chromosomes 9p21 and 2p14-p16 [252]. Immunoglobulin gene rearrangement studies are usually positive, but *BCL-2*, *BCL-6*, and *MYC* genes lack rearrangements (Figure 15.64). *MAL*, a gene that encodes a protein associated with lipid rafts in the T-cells and epithelial cells, is overexpressed in majority of the MLBCL cases [283]. The *REL* gene, located at chromosomal position 2p16, is also frequently amplified [285].

Clinical Aspects

Primary MLBCL comprises about 7% of DLBCLs and 2.4% of all non-Hodgkin lymphomas [2]. Women are more affected than men, and the median age is around 40 years. The affected patients usually show a bulky anterior mediastinal mass originating in thymus. Superior vena cava syndrome is reported in over 50% of the cases. Relapses are often extranodal involving liver, gastrointestinal tract, CNS, and other organs. An event-free 10-year survival is about 50% [286]. Therapeutic regimens include CHOP or other alternative combination chemotherapy regimens followed by either involved field or modified mantle field radiation therapy.

Differential Diagnosis

The differential diagnosis of MLBCL includes nodular sclerosis Hodgkin lymphoma, thymoma, and other mediastinal masses, such as seminoma and melanoma. The neoplastic cells of MLBCL express B-cell-associated markers and CD45, and lack the expression of CD15, cytokeratin, PLAP, and HMB-45.

OTHER VARIANTS OF LARGE B-CELL LYMPHOMA

Intravascular Large B-Cell Lymphoma

Intravascular large B-cell lymphoma (IVLBCL) is a rare and aggressive variant of extranodal large B-cell lymphoma characterized by the presence of aggregates of large neoplastic B-cells within the lumina of small to medium-sized blood vessels [287–289]. The lack of expression of adhesion molecules, such as CD29 (beta 1 integrin) and CD54 (ICAM-1), in some cases of IVLBCL suggests that the intravascular pattern is secondary to a defect in homing receptors in the tumor cells [1, 290, 291].

Morphology

The neoplastic cells are large with large vesicular nucleus and prominent nucleoli. They appear as intraluminal clusters in the small to medium-sized vessels of many organs (Figure 15.65) [287–289, 292]. The tumor clusters may be associated with fibrin thrombi.

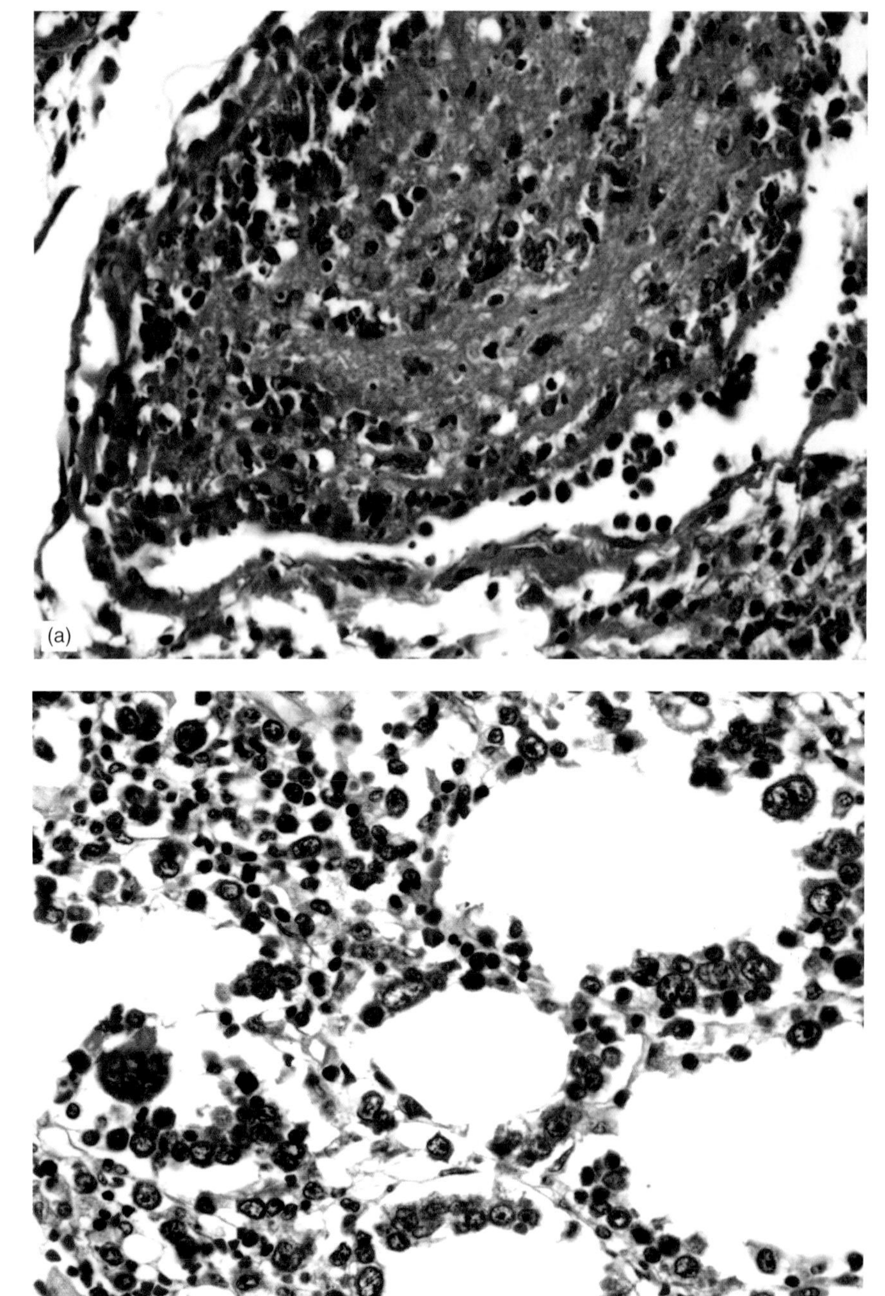

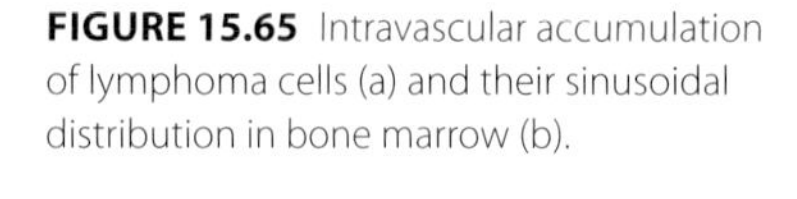

FIGURE 15.65 Intravascular accumulation of lymphoma cells (a) and their sinusoidal distribution in bone marrow (b).

TABLE 15.11 Immunophenotypic features of intravascular large B-cell lymphoma.*

Phenotype	% Positive
CD5	38
CD10	13
CD19	85
CD20	96
CD23	4
BCL-2	91
BCL-6	26
MUM1/IRF4	95
BCL-1	0
κ chain	71
λ chain	18
EBER (EBV)	0

*Adapted from Ref. [287].

Immunophenotype

The tumor cells are commonly positive for B-cell-associated molecules, such as Ig, CD19, CD20, CD22, and CD79a, and often express BCL-2 and MUM1 [287]. A smaller proportion of the cases may express CD5, CD10, and/or BCL-6. The BCL-1 expression is negative (Table 15.11).

Cytogenetic and Molecular Studies

No recurrent cytogenetic abnormalities have been reported so far. *BCL-6* rearrangement have been reported in sporadic cases [293]. Most cases show clonal *Ig* gene rearrangements.

Clinical Aspects

IVLBCL is an aggressive systemic disorder. Clinical symptoms are mostly secondary to the occlusion of the small to medium-sized vessels in various organs. Clinical manifestations may include skin lesions, neurological symptoms, nephritic syndrome, and disseminated intravascular coagulation (DIC). Hepatosplenomegaly, anemia, and thrombocytopenia are reported in over 70% of the patients [287]. In a recent report of 96 patients with IVLBCL, a correlation was found between CD5+, CD10– phenotype, and poor prognosis [287].

Primary Effusion Lymphoma

Primary effusion lymphoma is a rare variant of large cell lymphoma characterized by malignant serous effusions without detectable tumor masses [1, 294]. Rare cases may show solid organ involvements, such as skin and heart [295, 296]. It has been primarily observed in AIDS patients and is associated with HHV-8 [294, 297–299]. However, rare cases of HHV-8-negative PEL have been reported [300, 301]. The majority of cases are also co-infected with EBV [1, 294]. Pleural, pericardial, and peritoneal cavities are the most frequent sites of involvement.

Morphology

The cytocentrifuge slide preparations of the lymphomatous effusions show large neoplastic cells with immunoblastic, plasmablastic, and/or anaplastic features (Figure 15.66) [1, 294, 302]. These cells have variable amounts of basophilic cytoplasm. A perinuclear pale area may be present (plasmablastic). Nuclei are round or irregular and show prominent nucleoli. Binucleated or multilobated nuclei may be present and some tumor cells may resemble Reed–Sternberg cells.

Immunophenotype

The tumor cells express CD45 but are usually negative for surface and cytoplasmic Ig and B-cell-associated markers, such as CD19, CD20, and CD79a [1, 302–304]. They often express CD30, CD38, and CD138, and in some cases may show aberrant expression of cytoplasmic CD3. Immunohistochemical stains are positive for HHV-8-associated latent protein.

Cytogenetic and Molecular Studies

No recurrent cytogenetic abnormalities have been reported. Although the tumor cells often lack expression of membrane or cytoplasmic Ig, *Ig* genes are rearranged and mutated, so molecular studies are more necessary here than in many of the other B-cell lymphomas [1, 303]. HHV-8 viral genomes are detected in virtually all patients, and most cases show EBV infection demonstrated by EBER [249, 305]. Some cases may also show *TCR* gene rearrangement. A case of PEL of the pericardial cavity carrying t(1;22)(q21;q11.2) and t(14;17)(q32;q23) has been reported [306].

Clinical Aspects

The majority of patients are men and have contracted HIV. Clinical symptoms are secondary to the effusions [303]. Some patients may demonstrate Kaposi sarcoma or multicentric Castleman disease [249]. Most patients do not respond to conventional chemotherapy and have a short survival time, usually <6 months.

Plasmablastic Lymphoma

Plasmablastic lymphoma is a rare type of LBCL reported in the HIV-infected patients [307–309]. In the majority of the cases, the neoplastic cells are positive for EBV and HHV-8. Plasmablastic lymphoma is often preceded by multicentric or plasmablastic variant of Castleman disease.

Morphology

The tumor cells are large and resemble immunoblasts or atypical immature plasma cells with variable amounts of amphophilic or basophilic cytoplasm, round or irregular

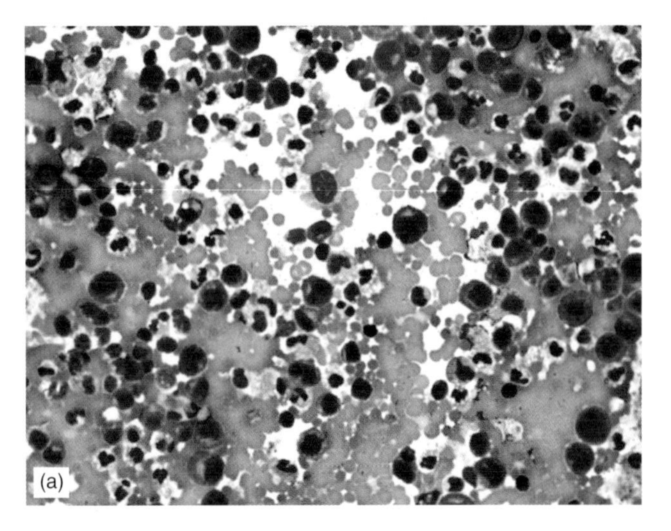

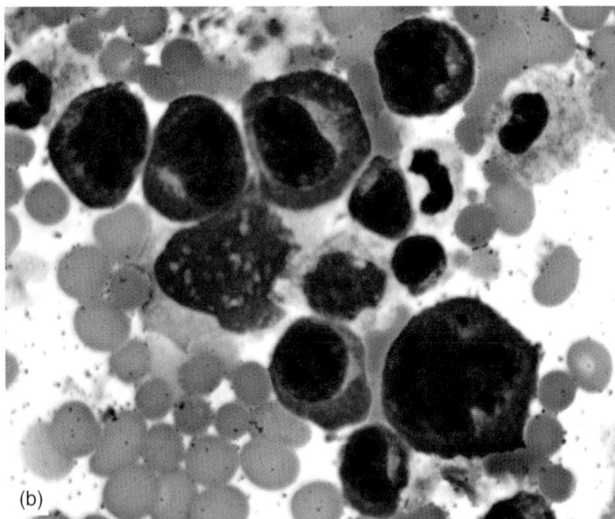

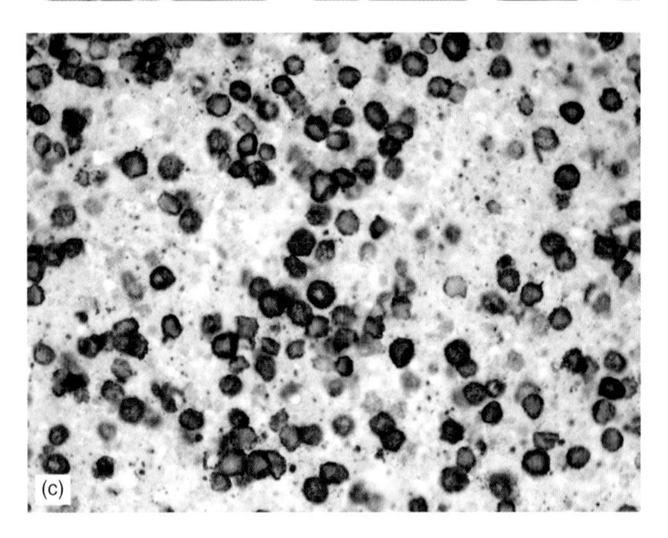

FIGURE 15.66 Primary effusion lymphoma. Pericardial effusion demonstrating large plasmablastic cells with dark-blue cytoplasm and irregular nuclei mixed with red cells and inflammatory cells: (a) low power and (b) high power views. These cells express CD138 by immunohistochemical stains (c).

vesicular nucleus and one or more prominent nucleoli (Figure 15.67) [1, 310].

Immunophenotype

These cells, similar to PEL cells, are usually negative for B-cell-associated markers, such as CD19, CD20, and CD79a, but positive for CD38 and CD138. They often lack CD45 expression but show a high Ki-67 fraction [1, 249]. Immunohistochemical stains are positive for HHV-8-associated latent protein (Figure 15.67).

Cytogenetic and Molecular Studies

No recurrent cytogenetic abnormalities have been reported. The *Ig* genes are clonally rearranged. HHV-8 viral genomes are detected in virtually all patients, and most cases show EBV infection demonstrated by EBER.

Clinical Aspects

This tumor typically involves the oral cavity of AIDS patients, though extra-oral cases have also been reported [307–309, 311]. The affected patients may demonstrate Kaposi sarcoma or multicentric Castleman disease [249].

Anaplastic Lymphoma Kinase Positive DLBCL

Anaplastic lymphoma kinase (ALK)-positive DLBCL is a rare, recently defined lymphoma comprised of large anaplastic cells with immunoblastic and/or plasmablastic morphologic features [256, 312, 313]. The tumor cells usually have an intrasinusoidal distribution and express cytoplasmic ALK, CD138, CD45, EMA, and monoclonal cytoplasmic light chain and weak IgA [256, 312, 313]. Other B-cell-associated markers are usually negative, as well as CD3, CD30, CD56, and TIA-1. Cytogenetic studies may reveal t(2;5)(p23;q35);(*ALK/NPM1*) or t(2;17)(p23;q23); (*ALK/CLTC*) [256, 312].

DLBCL Associated with Chronic Inflammation

Certain chronic inflammations such as autoimmune disorders, chronic pyothorax, and hepatitis C virus are sometimes associated with DLBCL [314–316]. The neoplastic cells in these cases express CD20 and TIA-1 and demonstrate clonal *IGH* and/or *TCR* gene rearrangements.

Primary Central Nervous System DLBCL

Primary central nervous system (CNS) lymphomas are aggressive tumors confined to the CNS [317–319]. These lymphomas are usually of large B-cell type and are distinguished from nodal DLBCL by high expression of regulators of the protein response (UPR), and interleukin-4 (IL-4) [319]. More recent studies show upregulation of the ECM (extracellular matrix)-related osteopontin gene, SPP1 in primary CNS large B-cell lymphomas [317].

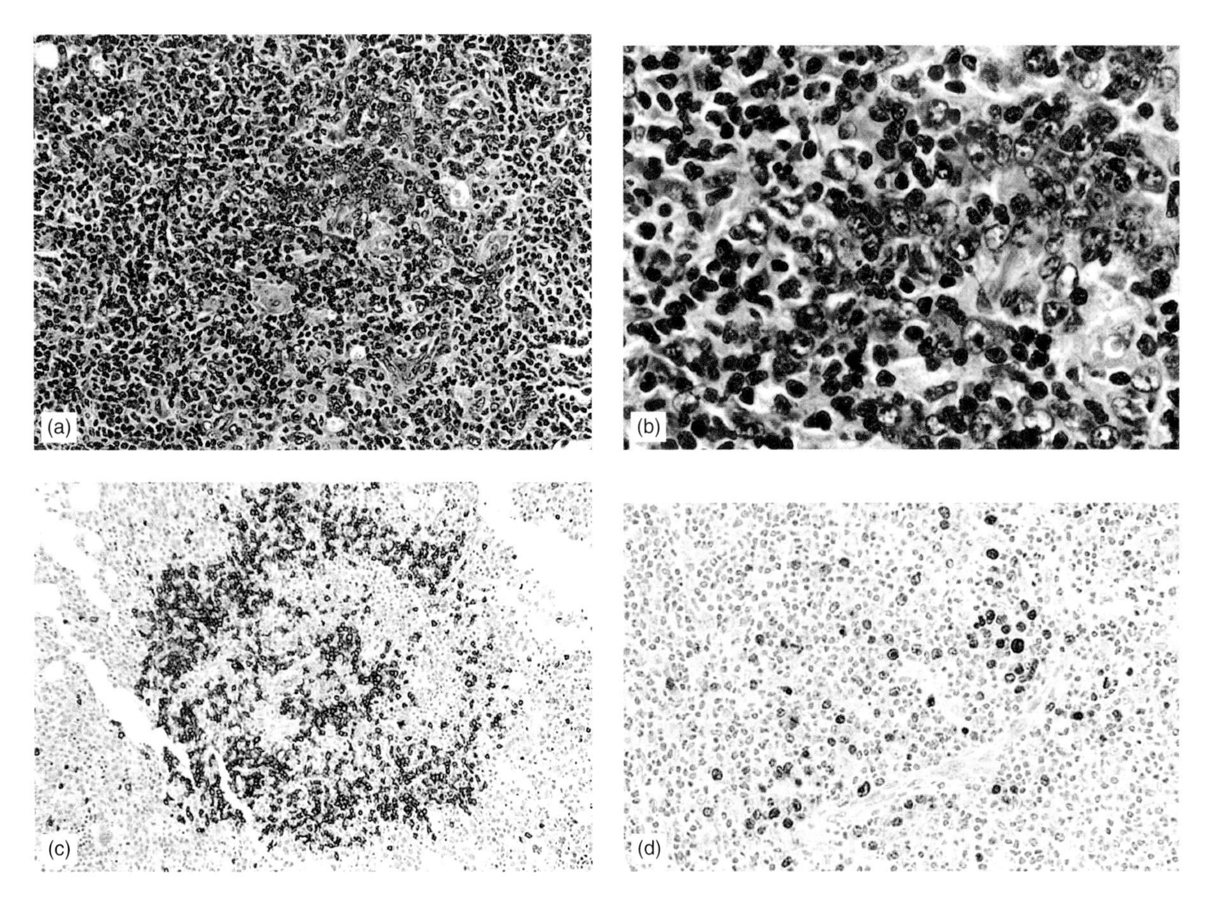

FIGURE 15.67 Plasmablastic lymphoma in a patient with AIDS. Aggregates of immature cells (plasmablasts) are present: (a) low power and (b) high power views. These cells express CD20 (c) and are positive for HHV-8 (d).

Primary Cutaneous DLBCL

Primary cutaneous DLBCL, leg type, is one of the categories in cutaneous lymphomas defined by the World Health Organization–European Organization for Research and Treatment of Cancer Classification [320–324]. These tumors in addition to the expression of B-cell-associated markers are often CD10+ and BCL-6+ and may demonstrate t(14;18) translocation by PCR [321].

BURKITT LYMPHOMA

Burkitt lymphoma/leukemia is a highly aggressive mature B-cell lymphoid malignancy consisting of endemic, sporadic, and immunodeficiency-associated variants [323–325]. All variants demonstrate chromosomal rearrangements involving *C-MYC* oncogene and share morphologic and immunophenotypic features, but differ in clinical and geographic presentations [326].

Etiology and Pathogenesis

The strong association of EBV with the endemic variant of BL suggests an important etiologic role for this virus. Virtually, all patients with endemic BL show the EBV genome in their neoplasm. The assumption is that recurrent or chronic viral (EBV, HIV), bacterial, or parasitic infections (malaria) lead to the development of lymphoma due to defective T-cell regulation of EBV-infected B-cells [1, 327] and increased chance of *C-MYC* translocation. In the sporadic and immunodeficiency-associated BL, EBV infection is detected in <30% and 25–40% of the tumors, respectively [1].

C-MYC overexpression plays an important role in the pathogenesis of BL. Numerous genes are induced by over expression of *C-MYC* including *cyclin D2*, *TRAP1* (apoptosis gene), *HLA-DRBI* histocompatibility gene, and the mitochondrial *HSPD1* (heat shock 60 kD protein). *C-MYC* overexpression also regulates a number of CDKs, such as CDK2 and CDK4, and other cell cycle-related products including *CCN D1*, *p27*, and *p53* [323, 328, 329]. In summary, the upregulation of *C-MYC* oncogene promotes cell cycle progression, inhibits the differentiation process, alters cellular metabolism, reduces cell adhesion, and increases telomerase activity (Figure 15.68).

Pathology

Morphology

The involved tissues are diffusely infiltrated by sheets of monotonous medium-sized neoplastic lymphoid cells. The

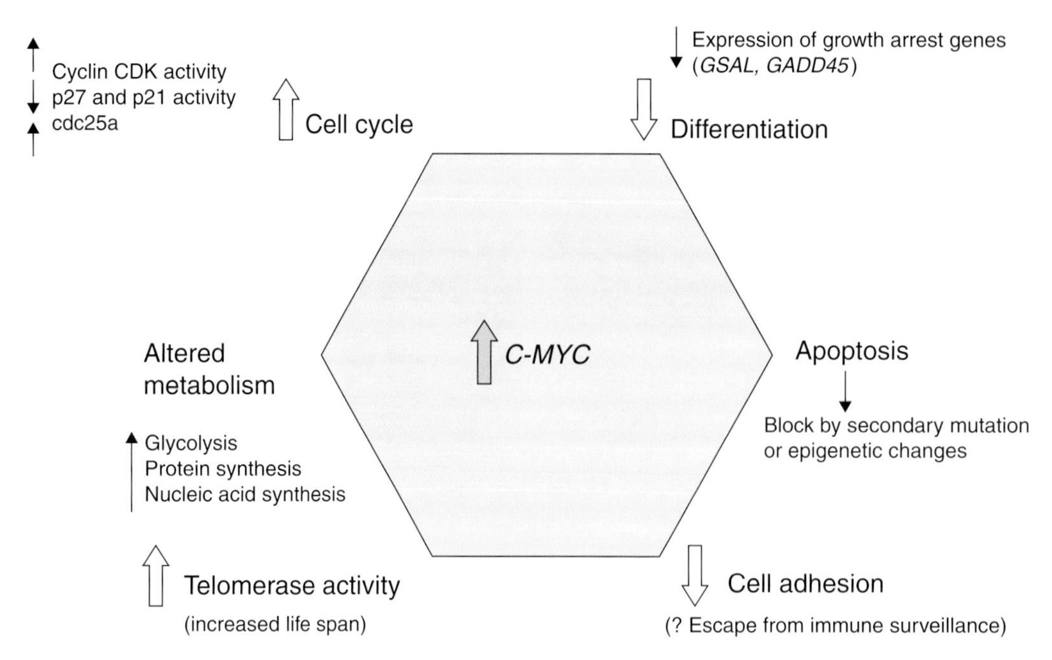

FIGURE 15.68 Effects of *C-MYC* overexpression in various aspects of cell function. Adapted from Hecht JL, Aster JC. (2000). Molecular biology of Burkitt's lymphoma. *J Clin Oncol* **18**, 3707–21.

typical (classical) cytologic features are uniformity, high nuclear:cytoplasmic ratio, basophilic and often vacuolated cytoplasm, round nucleus with clumped chromatin and relatively clear parachromatin, and multiple centrally located nucleoli [1, 2, 93, 324]. Mitotic figures are numerous as well as a high rate of apoptosis. Scattered macrophages with abundant, pale, cytoplasm-containing cell debris are present, creating a "starry sky" pattern (Figure 15.69). BL may infiltrate bone marrow and/or peripheral blood and present a leukemic picture (Figure 15.70).

The WHO has defined two morphologic variants [1]:

1. *Atypical Burkitt/Burkitt-like*: The neoplastic cells in atypical Burkitt or Burkitt-like variant are pleomorphic and show variation in the nuclear size and shape. The nucleoli are fewer in number, but more prominent [1, 277].
2. *BL with plasmacytoid differentiation*: Some neoplastic cells display plasmacytic differentiation with basophilic cytoplasm, eccentrically located nucleus, and a single central nucleolus. This variant is more common in patients with immunodeficiency.

Immunophenotype and Cytochemical Stains

Neoplastic cells of BL express membrane IgM, are membrane Ig light chain restricted, and are positive for B-cell-associated markers, such as CD19, CD20, CD22, and CD79a (Figures 15.71 and 15.72) [1, 2, 93, 330]. They express CD10, CD43, and BCL-6 and lack expression of CD5, CD23, and BCL-2. Virtually all the viable neoplastic cells express Ki-67. The EBV-positive cases (e.g. endemic type) express CD21. BL cells usually lack the expression of adhesion molecules, such as CD11a, CD11c, CD18, and CD44. CD34 and TdT are negative. The BL cells show positive staining for oil red O (Figure 15.73).

Cytogenetic and Molecular Studies

The genetic hallmark of BL is the translocation of *MYC* at chromosome 8q24 [1, 2, 93, 330]. This translocation observed in a majority of the cases is between *MYC* and the Ig heavy chain region on chromosome 14 [t(8;14)(q24;q32)] (Figure 15.74), but less frequently, it may involve Ig light chain regions on chromosomes 2 (kappa) and 22 (lambda), t(2;8)(p12;q24) and t(8;22)(q24;q11.1).

In the endemic cases, the breakpoint on chromosome 14 involves the joining region of Ig heavy chain, and the breakpoint on chromosome 8 lies outside the *c-MYC* gene, whereas in the sporadic BL, the breakpoints are in the heavy chain switch region and inside the *c-MYC* gene [2]. In fact, the heterogeneity of the molecular breakpoints makes this translocation difficult to detect by standard PCR approaches, and cytogenetics or FISH are preferred [331, 332]. Mutation of *BCL-6* gene has been reported in 25–50% of the BL cases. Most endemic cases and 25–40% of the cases associated with AIDS contain EBV genome [2].

Clinical Aspects

Three major clinical forms of BL have been described: endemic, sporadic, and immunodeficiency associated [2, 323–325]. The differences are primarily based on epidemiology and clinical presentations. The male:female ratio is about 2–4:1 in all three forms. The endemic and sporadic forms are most common in children, and affected children with the endemic form are younger than the patients with sporadic form.

Endemic BL occurs in equatorial Africa. It is the most common childhood malignancy in this region with a peak incidence at 4–7 years. There is some correlation between the incidence of endemic BL and geographical

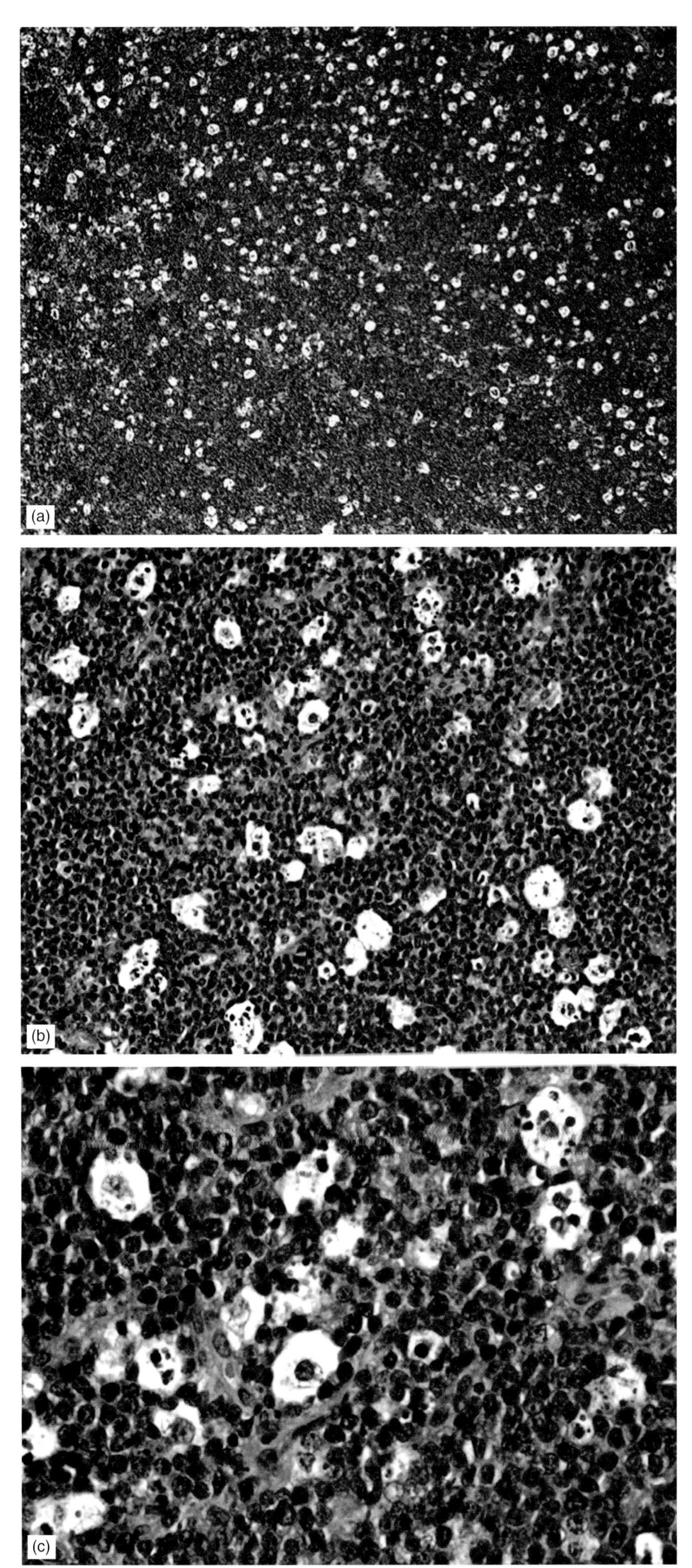

FIGURE 15.69 Burkitt lymphoma. A lymph node section demonstrating a diffuse infiltration by monomorphic lymphoid cells with high nuclear: cytoplasmic ratio, basophilic cytoplasm, round nucleus with clumped chromatin and relatively clear parachromatin, and multiple centrally located nucleoli. Scattered macrophages with abundant pale cytoplasm-containing cell debris are present, creating a "starry sky" pattern: (a) low power, (b) intermediate power, and (c) high power views.

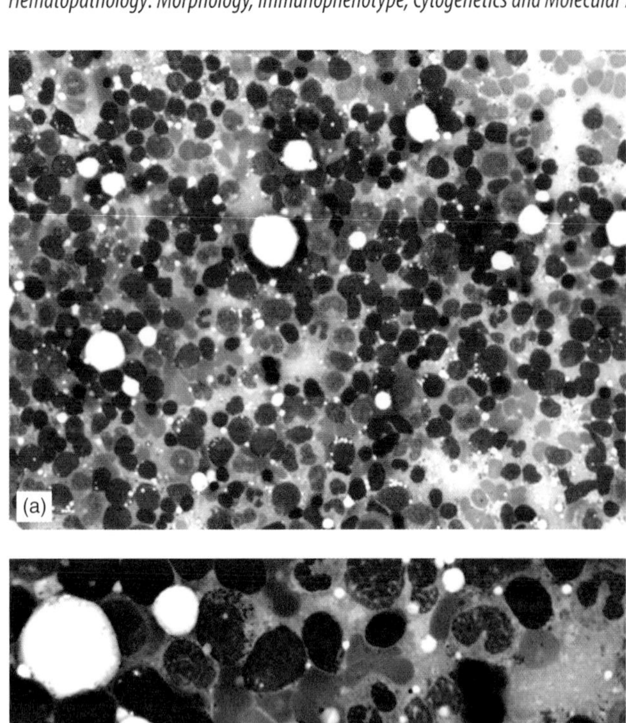

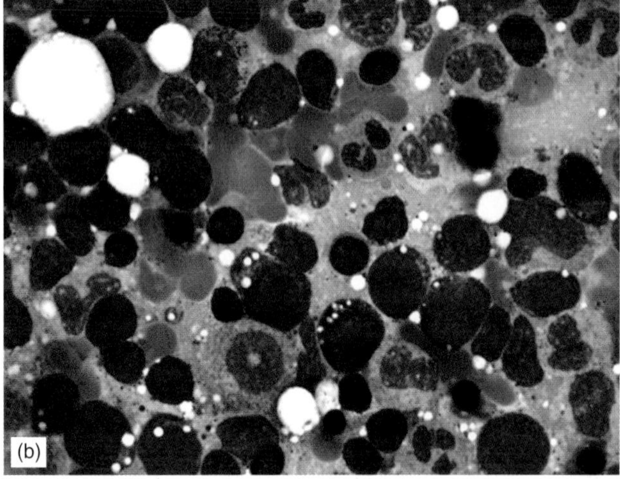

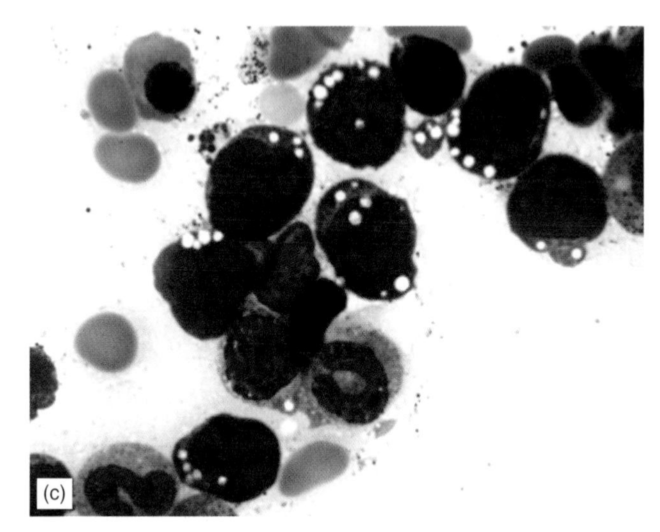

FIGURE 15.70 Burkitt lymphoma. Bone marrow smear showing monomorphic neoplastic lymphoid cells with high nuclear:cytoplasmic ratio, basophilic vacuolated cytoplasm, and round nuclei: (a) low power, (b) intermediate power, and (c) high power views.

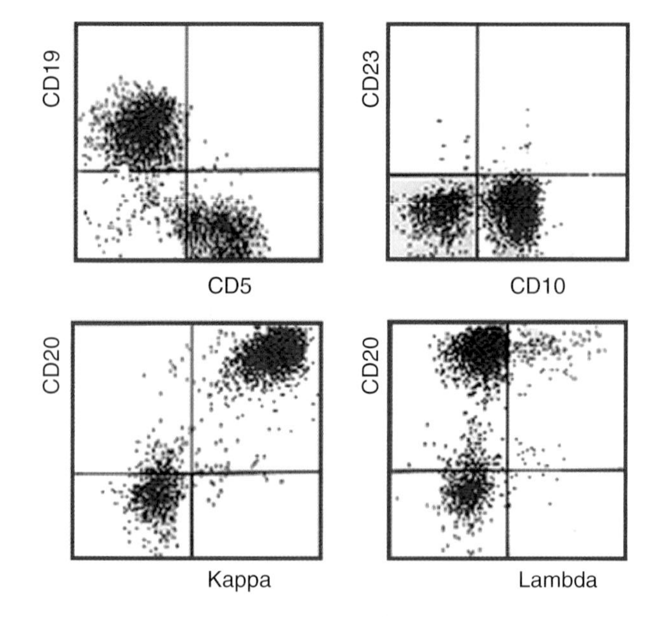

FIGURE 15.71 Flow cytometric analysis of neoplastic cells of Burkitt lymphoma demonstrates a population of B-cells expressing CD10, CD19, and CD20 with kappa light chain restriction.

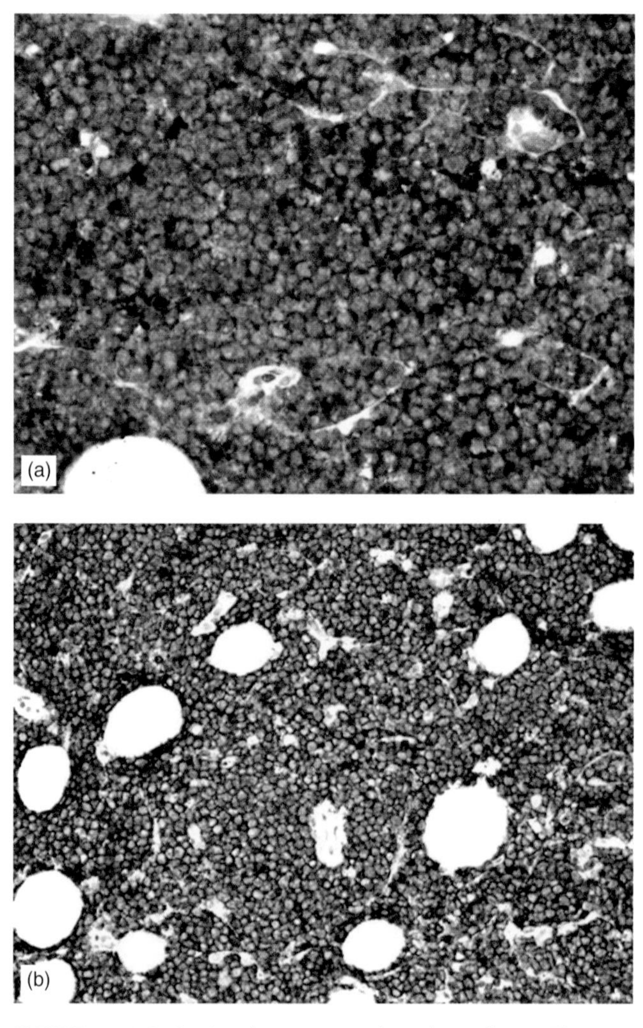

FIGURE 15.72 Burkitt lymphoma: immunohistochemical stains. The tumor cells express CD10 (a), CD20 (b), Ki-67 (c), and BCL-6 (d).

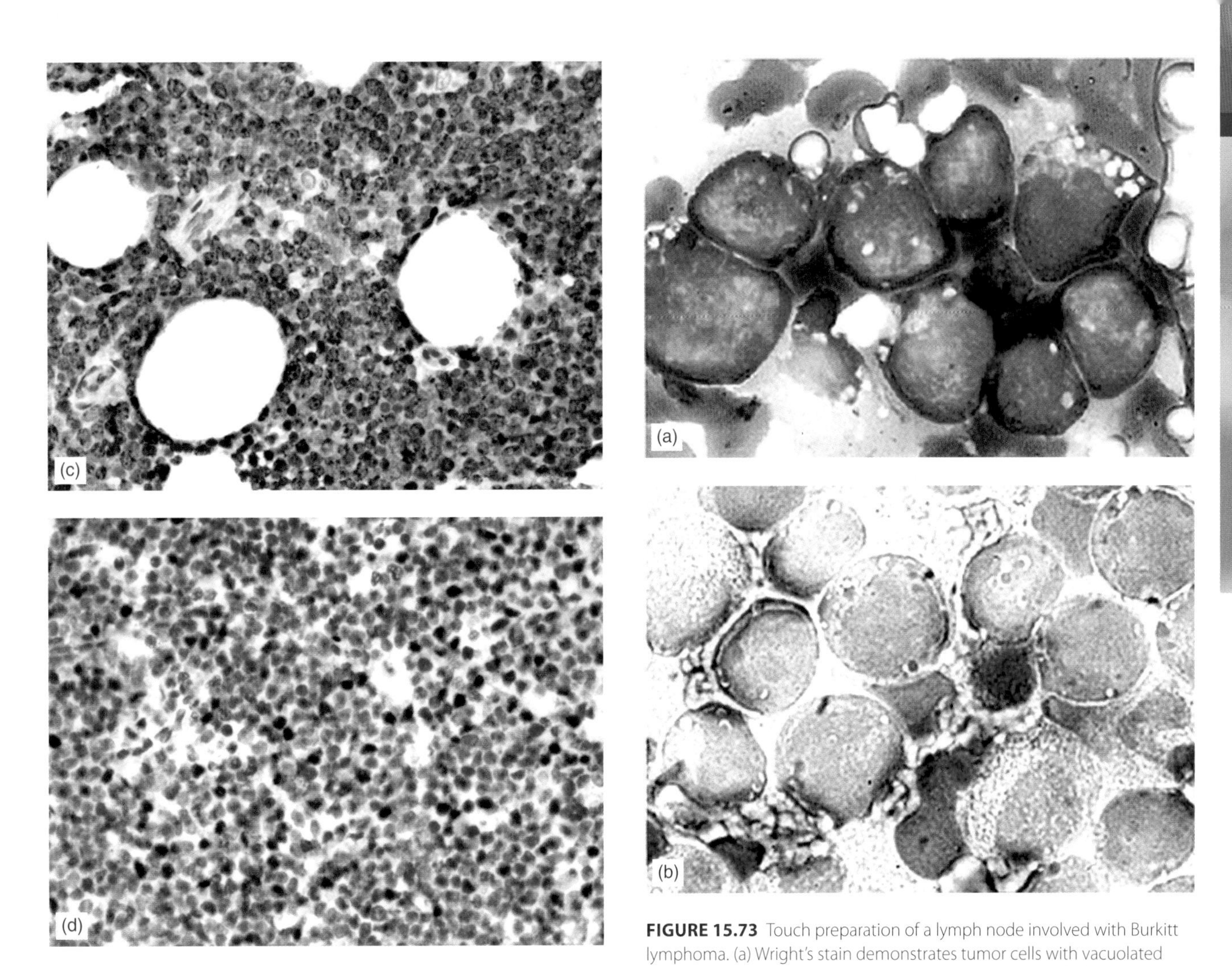

FIGURE 15.72 *(Continued)*

FIGURE 15.73 Touch preparation of a lymph node involved with Burkitt lymphoma. (a) Wright's stain demonstrates tumor cells with vacuolated cytoplasm. (b) Oil red O stain shows numerous orange-red dot-like deposits within the cytoplasmic vacuoles. From Ref. [91] by permission.

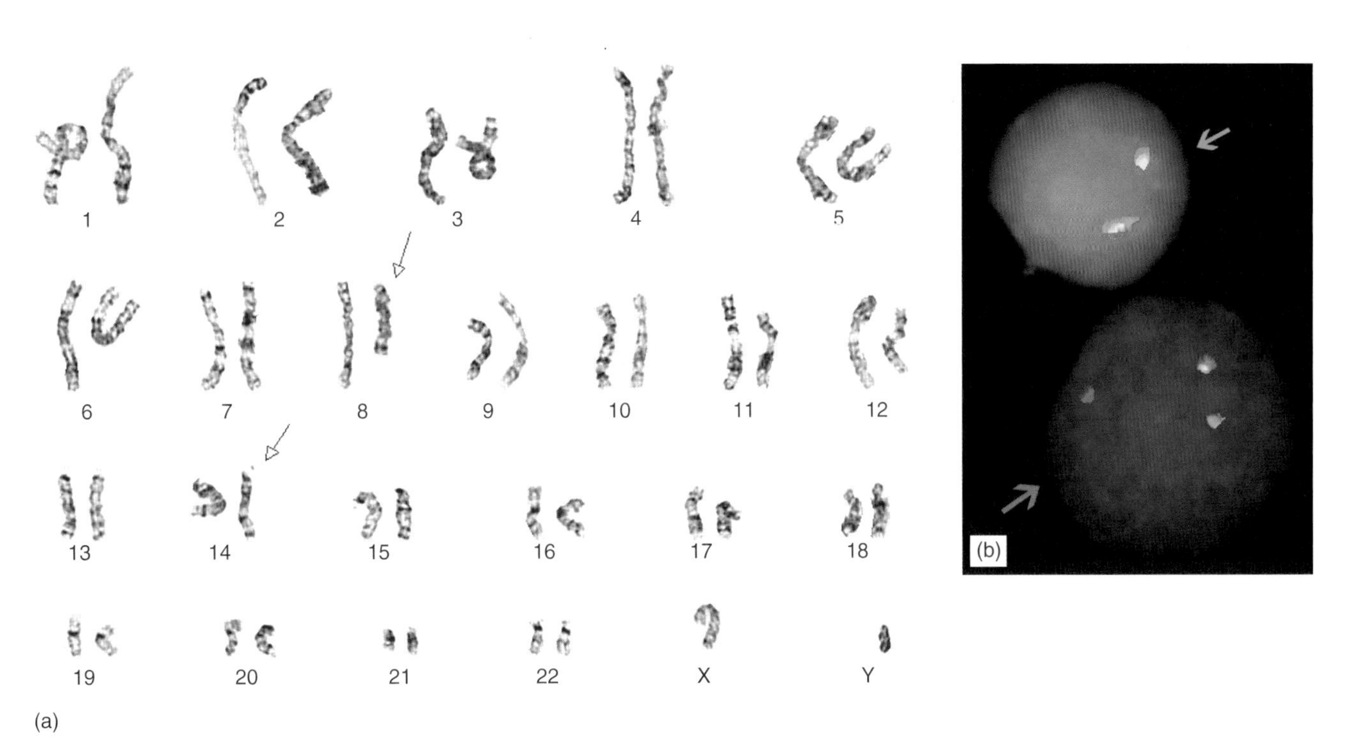

FIGURE 15.74 Burkitt lymphoma. (a) Karyotype demonstrating 46,XY,t(8;14)(q24;q32). (b) FISH analysis with split *C-MYC* signals.

distribution of endemic malaria. The disease usually presents as a mass in the jaw or facial bones and spreads to other extranodal sites, such as bone marrow, peripheral blood, meninges, testis, ovary, kidney, and breast.

Sporadic BL is seen all over the world accounting for about 1–2% of all lymphomas in the United States and Western Europe. It occurs mainly in children and young adults and represents 30–50% of all childhood lymphomas [1, 2]. The most common clinical presentation is abdominal mass with ascites involving stomach, distal ileum, cecum, and/or mesentery. Bone marrow, peripheral blood, kidney, testis, ovary, breast, and CNS may be involved.

Immunodeficiency-associated BL is primarily observed in patients with HIV infection. Post-transplant-associated BL is less frequent. This form more often involves lymph nodes but may involve bone marrow and peripheral blood and presents as acute leukemia.

Rapidly growing mass with elevated serum LDH levels are characteristic features of BL. Approximately 70% of patients are in advanced stages of the disease at diagnosis. BL is considered as a highly aggressive tumor. Involvement of the bone marrow and CNS, tumor size >10 cm in diameter, and elevated serum LDH are considered poor prognostic factors. Response to intensive combination chemotherapy with complete remission approaches 90% in early stages and 60–80% in advanced stages of the disease. Allogeneic bone marrow transplantation has been utilized in patients not responding to chemotherapy [2, 333, 334].

Differential Diagnosis

The differential diagnosis includes precursor B acute lymphoblastic leukemia/lymphoma and DLBCL. Precursor B-cell neoplasms are often TdT- and CD34-positive and lack the expression of membrane Ig heavy or light chains, whereas BL cells are negative for CD34 and TdT and express membrane Ig. Some cases of BL are borderline and share some of the morphologic features of DLBCL by demonstrating larger cells or an admixture of centroblast- or immunoblast-like cells. Also, a minority of DLBCL cases may demonstrate t(8;14) and C-MYC deregulation. The following recommendations are provided for the distinction of BL from DLBCL in borderline cases (Table 15.10) [277, 335]:

1. In tumors with borderline morphologic features, diagnosis of BL is made when the Ki-67-positive fraction is at least 99% in viable tumor cells.
2. Tumors with typical morphologic features of DLBCL with high Ki-67 fraction or t(8;14) and morphologically borderline cases with low Ki-67 fraction should be classified as DLBCL.

References

1. Jaffe ES, Harris NL, Stein H, Vardiman JW. (2001). *Pathology and Genetics. Tumors of Haematopoietic and Lymphoid Tissues*. ARC Press, Lyon.
2. Freedman AS, Friedberg JW. (2006). Classification of lymphomas. *UpToDate*.
3. Korsmeyer SJ, Hieter PA, Revetch JV, Poplack DG, Waldmann TA, Leder P. (1981). Developmental hierarchy of immunoglobulin gene rearrangements in human leukemic pre-B cells. *Proc Natl Acad Sci USA* **78**, 7096–100.
4. Abdel-Reheim FA, Edwards E, Arber DA. (1996). Utility of a rapid polymerase chain reaction panel for the detection of molecular changes in B-cell lymphoma. *Arch Pathol Lab Med* **120**, 357–63.
5. Bagg A, Braziel RM, Arber DA, Bijwaard KE, Chu AY. (2002). Immunoglobulin heavy chain gene analysis in lymphomas: A multi-center study demonstrating heterogeneity of performance of polymerase chain reaction assays. *J Mol Diagn* **4**, 81–9.
6. Cossman J, Zehnbauer B, Garrett CT, Smith LJ, Williams M, Jaffe ES, Hanson LO, Love J. (1991). Gene rearrangements in the diagnosis of lymphoma/leukemia: Guidelines for use based on a multi-institutional study. *Am J Clin Pathol* **95**, 347–54.
7. Arber DA. (2000). Molecular diagnostic approach to non-Hodgkin's lymphoma. *J Mol Diagn* **2**, 178–90.
8. Libra M, de Re V, Gloghini A, Navolanic PM, Carbone A, Boiocchi M. (2004). Second primary lymphoma or recurrence: A dilemma solved by VDJ rearrangement analysis. *Leuk Lymphoma* **45**, 1539–43.
9. Van Dongen JJ, Seriu T, Panzer-Grumayer ER, Biondi A, Pongers-Willemse MJ, Corral L. (1998). Prognostic value of minimal residual disease in acute lymphoblastic leukaemia in childhood. *Lancet* **352**, 1731–8.
10. Van der Velden VHJ, Hochhaus A, Cazzaniga G, Szczepanski T, Gabert J, van Dongen JJM. (2003). Detection of minimal residual disease in hematologic malignancies by real-time quantitative PCR: Principles, approaches, and laboratory aspects. *Leukemia* **17**, 1013–34.
11. Rosenwald A, Alizadeh AA, Widhopf G, Simon R, Davis RE, Yu X, Yang L, Pickeral OK, Rassenti LZ, Powell J, Botstein D, Byrd JC, Grever MR, Cheson BD, Chiorazzi N, Wilson WH, Kipps TJ, Brown PO, Staudt LM. (2001). Relation of gene expression phenotype to immunoglobulin mutation genotype in B cell CLL. *J Exp Med* **194**, 1639.
12. Jelinek DF, Tschumper RC, Stolovitzky GA, Iturria SJ, Tu Y, Lepre J, Shah N, Kay NE. (2003). Identification of global gene expression signature of B cell CLL. *Mol Cancer Res* **1**, 346.
13. Rassenti LZ, Huynh L, Toy TL, Chen L, Keating MJ, Gribben JG, Neuberg DS, Flinn IW, Rai KR, Byrd JC, Kay NE, Greaves A, Weiss A, Kipps TJ. (2004). ZAP-70 compared with immunoglobulin heavy-chain gene mutation status as a predictor of disease progression in chronic lymphocytic leukemia. *N Engl J Med* **351**, 893–901.
14. Byrd JC, Stilgenbauer S, Flinn IW. (2004). Chronic lymphocytic leukemia. *Hematology Am Soc Hematol Educ Program*, 163–83.
15. Rai KR, Keating MJ. (2006). Pathophysiology and cytogenetics of chronic lymphocytic leukemia. *UpToDate*.
16. Jonsson V, Houlston RS, Catovsky D, Yuille MR, Hilden J, Olsen JH, Fajber M, Brandt B, Sellick G, Allinson R, Wiik A. (2005). CLL family "Pedigree 14" revisited: 1947–2004. *Leukemia* **19**, 1025–8.
17. Ishibe N, Sgambati MT, Fontaine L, Goldin LR, Jain N, Weissman N, Marti GE, Caporaso NE. (2001). Clinical characteristics of familial B-CLL in the National Cancer Institute Familial Registry. *Leuk Lymphoma* **42**, 99–108.
18. Goldin LR, Slager SL. (2007). Familial CLL: Genes and environment. *Hematology Am Soc Hematol Educ Program*, 339–45.
19. Goldin LR, Caporaso NE. (2007). Family studies in chronic lymphocytic leukaemia and other lymphoproliferative tumours. *Br J Haematol* **139**, 774–9.
20. Sellick GS, Catovsky D, Houlston RS. (2006). Familial chronic lymphocytic leukemia. *Semin Oncol* **33**, 195–201.
21. Catovsky D. (2004). Definition and diagnosis of sporadic and familial chronic lymphocytic leukemia. *Hematol Oncol Clin North Am* **18**, 783–94.

22. Preston DL, Kusumi S, Tomonaga M, Izumi S, Ron E, Kuramoto A, Kamada N, Dohy H, Matsuo T, Matsuo T. (1994). Cancer incidence in atomic bomb survivors. Part III. Leukemia, lymphoma and multiple myeloma, 1950–1987. *Radiat Res* **137**(Suppl 2), S68–97.
23. International Workshop on Chronic Lymphocytic Leukemia. (1989). Chronic lymphocytic leukemia: Recommendations for diagnosis, staging, and response criteria. *Ann Intern Med* **110**, 236–8.
24. Cheson BD, Bennett JM, Grever M, Kay N, Keating MJ, O'Brien S, Rai KR. (1996). National Cancer Institute-sponsored Working Group guidelines for chronic lymphocytic leukemia: Revised guidelines for diagnosis and treatment. *Blood* **87**, 4990–7.
25. Hallek M, Cheson BD, Catovsky D, Caligaris-Cappio F, Dighiero G, Dohner H, Hillmen P, Keating MJ, Montserrat E, Rai KR, Kipps TJ. (2008). Guidelines for the diagnosis and treatment of chronic lymphocytic leukemia: A report from the International Workshop on Chronic Lymphocytic Leukemia (IWCLL) updating the National Cancer Institute-Working Group (NCI-WG) 1996 guidelines. *Blood* **16** [Epub ahead of print].
26. Oscier D, Fegan C, Hillmen P, Illidge T, Johnson S, Maguire P, Matutes E, Milligan D. (2004). Guidelines. Working Group of the UK CLL Forum. British Committee for Standards in Haematology. Guidelines on the diagnosis and management of chronic lymphocytic leukaemia. *Br J Haematol* **125**, 294–317.
27. Habib LK, Finn WG. (2006). Unsupervised immunophenotypic profiling of chronic lymphocytic leukemia. *Cytometry B Clin Cytom* **70**, 124–35.
28. Huang JC, Finn WG, Goolsby CL, Variakojis D, Peterson LC. (1999). CD5− small B-cell leukemias are rarely classifiable as chronic lymphocytic leukemia. *Am J Clin Pathol* **111**, 123–30.
29. Frater JL, McCarron KF, Hammel JP, Shapiro JL, Miller ML, Tubbs RR, Pettay J, His ED. (2001). Typical and atypical chronic lymphocytic leukemia differ clinically and immunophenotypically. *Am J Clin Pathol* **116**, 655–64.
30. McCarron KF, Hammel JP, His ED. (2000). Usefulness of CD79b expression in the diagnosis of B-cell chronic lymphoproliferative disorders. *Am J Clin Pathol* **113**, 805–13.
31. Schlette E, Medeiros LJ, Keating M, Lai R. (2003). CD79b expression in chronic lymphocytic leukemia. Association with trisomy 12 and atypical immunophenotype. *Arch Pathol Lab Med* **127**, 561–6.
32. Nola M, Pavletic SZ, Weisenburger DD, Smith LM, Bast MA, Vose JM, Armitage JO. (2004). Prognostic factors influencing survival in patients with B-cell small lymphocytic lymphoma. *Am J Hematol* **77**, 31–5.
33. Rai KR, Wasil T, Iqbal U, Driscoll N, Patel D, Janson D, Mehrotra B. (2004). Clinical staging and prognostic markers in chronic lymphocytic leukemia. *Hematol Oncol Clin North Am* **18**, 795–805.
34. Morabito F, Damle RN, Deaglio S, Keating M, Ferrarini M, Chiorazzi N. (2006). The CD38 ectoenzyme family: Advances in basic science and clinical practice. *Mol Med* **12**, 342–4.
35. Rassenti LZ, Kipps TJ. (2006). Clinical utility of assessing ZAP-70 and CD38 in chronic lymphocytic leukemia. *Cytometry B Clin Cytom* **70**, 209–13.
36. Moreau EJ, Matutes E, A'Hern RP, Morilla AM, Morilla RM, Owusu-Ankomah KA, Seon BK, Catovsky D. (1997). Improvement of the chronic lymphocytic leukemia scoring system with the monoclonal antibody SN8 (CD79b). *Am J Clin Pathol* **108**, 378–82.
37. Kampalath B, Barcos MP, Stewart C. (2003). Phenotypic heterogeneity of B cells in patients with chronic lymphocytic leukemia/small lymphocytic lymphoma. *Am J Clin Pathol* **119**, 824–32.
38. Jahrsdorfer B, Wooldridge JE, Blackwell SE, Taylor CM, Link BK, Weiner GJ. (2005). Good prognosis cytogenetics in B-cell chronic lymphocytic leukemia is associated *in vitro* with low susceptibility to apoptosis and enhanced immunogenicity. *Leukemia* **19**, 759–66.
39. Herholz H, Kern W, Schnittger S, Haferlach T, Dicker F, Haferlach C. (2007). Translocations as a mechanism for homozygous deletion of 13q14 and loss of the ATM gene in a patient with B-cell chronic lymphocytic leukemia. *Cancer Genet Cytogenet* **174**, 57–60.
40. Glassman AB, Hayes KJ. (2005). The value of fluorescence *in situ* hybridization in the diagnosis and prognosis of chronic lymphocytic leukemia. *Cancer Genet Cytogenet* **158**, 88–91.
41. Aoun P, Blair HE, Smith LM, Dave BJ, Lynch J, Weisenburger DD, Pavletic SZ, Sanger G. (2004). Fluorescence *in situ* hybridization detection of cytogenetic abnormalities in B-cell chronic lymphocytic leukemia/small lymphocytic lymphoma. *Leuk Lymphoma* **45**, 1595–603.
42. Cuneo A, Rigolin GM, Bigoni R, De Angeli C, Veronese A, Cavazzini F, Bardi A, Roberti MG, Tammiso E, Agostini P, Ciccone M, Della Porta M, Tieghi A, Cavazzini L, Negrini M, Castoldi G. (2004). Chronic lymphocytic leukemia with 6q− shows distinct hematological features and intermediate prognosis. *Leukemia* **18**, 476–83.
43. La Starza R, Barba G, Matteucci C, Crescenzi B, Romoli S, Pierini V, Beacci D, Cantaffa R, Martelli MF, Mecucci C. (2006). Chronic lymphocytic leukaemia. Is terminal del(14)(q24) a new marker for prognostic stratification? *Leuk Res* **30**, 1569–72.
44. Dicker F, Schnittger S, Haferlach T, Kern W, Schoch C. (2006). Immunostimulatory oligonucleotide-induced metaphase cytogenetics detect chromosomal aberrations in 80% of CLL patients: A study of 132 CLL cases with correlation to FISH, IgVH status, and CD38 expression. *Blood* **108**, 3152–60.
45. Wiestner A, Rosenwald A, Barry TS, Wright G, Davis RE, Henrickson SE, Zhao H, Ibbotson RE, Orchard JA, Davis Z, Stetler-Stevenson M, Raffeld M, Arthur DC, Marti GE, Wilson WH, Hamblin TJ, Oscier DG, Staudt LM. (2003). ZAP-70 expression identifies a chronic lymphocytic leukemia subtype with unmutated immunoglobulin genes, inferior clinical outcome, and distinct gene expression profile. *Blood* **101**, 4944–51.
46. Freedman AS, Harris AL. (2006). Clinical and pathologic features of small lymphocytic, prolymphocytic, and lymphoplasmacytic lymphomas. *UpToDate.*
47. Rai KR, Keating MJ. (2006). Clinical manifestation and diagnosis of chronic lymphocytic leukemia. *UpToDate.*
48. Hamblin TJ. (2006). Autoimmune complications of chronic lymphocytic leukemia. *Semin Oncol* **33**, 230–9.
49. Catovsky D, Fooks J, Richards S. (1989). Prognostic factors in chronic lymphocytic leukaemia: The importance of age, sex and response to treatment in survival. A report from the MRC CLL 1 trial. MRC Working Party on Leukaemia in Adults. *Br J Haematol* **72**, 141–9.
50. Rai KR, Sawitsky A, Cronkite EP, Chanana AD, Levy RN, Pasternack BS. (1975). Clinical staging of chronic lymphocytic leukemia. *Blood* **46**, 219–34.
51. Binet JL, Auquier A, Dighiero G, Chastang C, Piguet H, Goasguen J, Vaugier G, Potron G, Colona P, Oberling F, Thomas M, Tchernia G, Jacquillat C, Boivin P, Lesty C, Duault MT, Monconduit M, Belabbes S, Gremy F. (1981). A new prognostic classification of chronic lymphocytic leukemia derived from a multivariate survival analysis. *Cancer* **48**, 198–206.
52. Hamblin TJ. (2007). Prognostic markers in chronic lymphocytic leukaemia. *Best Pract Res Clin Haematol* **20**, 455–68.
53. Montserrat E. (2006). New prognostic markers in CLL. *Hematology Am Soc Hematol Educ Program*, 279–84
54. Rassenti LZ, Kipps TJ. (2006). Clinical utility of assessing ZAP-70 and CD38 in chronic lymphocytic leukemia. *Cytometry B Clin Cytom* **70**, 209–13.
55. Tobin G, Rosenquist R. (2005). Prognostic usage of V(H) gene mutation status and its surrogate markers and the role of antigen selection in chronic lymphocytic leukemia. *Med Oncol* **22**, 217–28.
56. Dickinson JD, Gilmore J, Iqbal J, Sanger W, Lynch JC, Chan J, Bierman PJ, Joshi SS. (2006). 11q22.3 deletion in B-chronic lymphocytic leukemia is specifically associated with bulky lymphadenopathy and ZAP-70 expression but not reduced expression of adhesion/cell surface receptor molecules. *Leuk Lymphoma* **47**, 231–44.
57. Grever MR, Lucas DM, Dewald GW, Neuberg DS, Reed JC, Kitada S, Flinn IW, Tallman MS, Appelbaum FR, Larson RA, Paietta E, Jelinek DF, Gribben JG, Byrd JC. (2007). Comprehensive assessment of genetic and molecular features predicting outcome in patients with

chronic lymphocytic leukemia: Results from the US Intergroup Phase III Trial E2997. *J Clin Oncol* **25**, 799–804.
58. Rai KR, Keating MJ. (2006). Treatment of chronic lymphocytic leukemia. *UpToDate*.
59. Tsimberidou AM, Keating MJ. (2005). Richter syndrome: Biology, incidence, and therapeutic strategies. *Cancer* **103**, 216–28.
60. Tsimberidou AM, Keating MJ. (2006). Richter's transformation in chronic lymphocytic leukemia. *Semin Oncol* **33**, 250–6.
61. Nakamura N, Abe M. (2003). Richter syndrome in B-cell chronic lymphocytic leukemia. *Pathol Int* **53**, 195–203.
62. Giles FJ, O'Brien SM, Keating MJ. (1988). Chronic lymphocytic leukemia in (Richter's) transformation. *Semin Oncol* **25**, 117–25.
63. Maloum K, Sutton L, Baudet S, Laurent C, Bonnemye P, Magnac C, Merle-Béral H. (2002). Novel flow-cytometric analysis based on BCD5+ subpopulations for the evaluation of minimal residual disease in chronic lymphocytic leukaemia. *Br J Haematol* **119**, 970–5.
64. Sanchez ML, Almeida J, Gonzalez D, Gonzalez M, Garcia-Marcos MA, Balanzategui A, Lopez-Berges MC, Nomdedeu J, Vallespi T, Barbon M, Martin A, de la Fuente P, Martin-Nunez G, Fernandez-Calvo J, Hernandez JM, San Miguel JF, Orfao A. (2003). Incidence and clinicobiologic characteristics of leukemic B-cell chronic lymphoproliferative disorders with more than one B-cell clone. *Blood* **102**, 2994–3002.
65. Gupta R, Jain P, Deo SV, Sharma A. (2004). Flow cytometric analysis of CD5+ B cells: A frame of reference for minimal residual disease analysis in chronic lymphocytic leukemia. *Am J Clin Pathol* **121**, 368–72.
66. Ghia P, Prato G, Scielzo C, Stella S, Geuna M, Guida G, Caligaris-Cappio F. (2004). Monoclonal $CD5^+$ and $CD5^-$ B-lymphocyte expansions are frequent in the peripheral blood of the elderly. *Blood* **103**, 2337–42.
67. Wang C, Amato D, Rabah R, Zheng J, Fernandes B. (2002). Differentiation of monoclonal B lymphocytosis of undetermined significance (MLUS) and chronic lymphocytic leukemia (CLL) with weak CD5 expression from CD5(−) CLL. *Leuk Res* **26**, 1125–9.
68. Krishnan B, Matutes E, Dearden C. (2006). Prolymphocytic leukemia. *Semin Oncol* **33**, 257–63.
69. Stone RM. (1990). Prolymphocytic leukemia. *Hematol Oncol Clin North Am* **4**, 457–71.
70. Absi A, His E, Kalaycio M, Absi A, His E, Kalaycio M. (2005). Prolymphocytic leukemia. *Curr Treat Options Oncol* **6**, 197–208.
71. Schlette E, Bueso-Ramos C, Giles F, Glassman A, Hayes K, Medeiros LJ. (2001). Mature B-cell leukemias with more than 55% prolymphocytes. A heterogeneous group that includes an unusual variant of mantle cell lymphoma. *Am J Clin Pathol* **115**, 571–81.
72. Veillon DM, Nordberg ML, Glass J, Sattar T, Cotelingam JD. (2001). Prolymphocytic leukemia or prolymphocytic transformation of mantle cell lymphoma. *Am J Clin Pathol* **116**, 781–2.
73. Ruchlemer R, Parry-Jones N, Brito-Babapulle V, Attolico I, Wotherspoon AC, Matutes E, Catovsky D. (2004). B-prolymphocytic leukaemia with t(11; 14) revisited: A splenomegalic form of mantle cell lymphoma evolving with leukaemia. *Br J Haematol* **125**, 330–6.
74. Robinson DS, Melo JV, Andrews C, Schey SA, Catovsky D. (1985). Intracytoplasmic inclusions in B prolymphocytic leukaemia: Ultrastructural, cytochemical, and immunological studies. *J Clin Pathol* **38**, 897–903.
75. Lens D, Coignet LJ, Brito-Babapulle V, Lima CS, Matutes E, Dyer MJ, Catovsky D. (1999). B cell prolymphocytic leukaemia (B-PLL) with complex karyotype and concurrent abnormalities of the p53 and c-MYC gene. *Leukemia* **13**, 873–6.
76. Lens D, De Schouwer PJ, Hamoudi RA, Abdul-Rauf M, Farahat N, Matutes E, Crook T, Dyer MJ, Catovsky D. (1997). p53 abnormalities in B-cell prolymphocytic leukemia. *Blood* **89**, 2015–23.
77. Kuriakose P, Perveen N, Maeda K, Wiktor A, Van Dyke DL. (2004). Translocation (8;14)(q24;q32) as the sole cytogenetic abnormality in B-cell prolymphocytic leukemia. *Cancer Genet Cytogenet* **150**, 156–8.
78. Andrieu V, Encaoua R, Carbon C, Couvelard A, Grange MJ. (1998). Leukemic pleural effusion in B-cell prolymphocytic leukemia. *Hematol Cell Ther* **40**, 275–8.
79. Shvidel L, Shtalrid M, Bassous L, Klepfish A, Vorst E, Berrebi A. (1999). B-cell prolymphocytic leukemia: A survey of 35 patients emphasizing heterogeneity, prognostic factors and evidence for a group with an indolent course. *Leuk Lymphoma* **33**, 169–79.
80. Chaar BT, Petruska PJ. (2007). Complete response to alemtuzumab in a patient with B prolymphocytic leukemia. *Am J Hematol* **82**, 417.
81. Grange MJ, Andrieu V, Chemlal K. (1999). Richter's syndrome in a patient with B prolymphocytic leukemia. *Haematologica* **84**, 461–641.
82. Dimopoulos MA, Galani E, Matsouka C. (1999). Waldenstrom's macroglobulinemia. *Hematol Oncol Clin North Am* **13**, 1351–66.
83. Veneri D, Aqel H, Franchini M, Meneghini V, Krampera M. (2004). Prevalence of hepatitis C virus infection in IgM-type monoclonal gammopathy of uncertain significance and Waldenstrom macroglobulinemia. *Am J Hematol* **77**, 421.
84. Pozzato G, Mazzaro C, Crovatto M, Modolo ML, Ceselli S, Mazzi G, Sulfaro S, Franzin F, Tulissi P, Moretti M. (1994). Low-grade malignant lymphoma, hepatitis C virus infection, and mixed cryoglobulinemia. *Blood* **84**, 3047–53.
85. Lin P, Medeiros LJ. (2005). Lymphoplasmacytic lymphoma/Waldenstrom macroglobulinemia: An evolving concept. *Adv Anat Pathol* **12**, 246–55.
86. Williamson LM, Greaves M, Worters JR, Harling CC. (1989). Waldenstrom's macroglobulinaemia: Three cases in shoe repairers. *Br Med J* **298**, 498–9.
87. Fine JM, Muller JY, Rochu D, Marneux M, Gorin NC, Fine A, Lambin P. (1986). Waldenstrom's macroglobulinemia in monozygotic twins. *Acta Med Scand* **220**, 369–73.
88. Renier G, Ifrah N, Chevailler A, Saint-Andre JP, Boasson M, Hurez D. (1989). Four brothers with Waldenstrom's macroglobulinemia. *Cancer* **64**, 1554–9.
89. Treon SP, Hunter ZR, Aggarwal A, Ewen EP, Masota S, Lee C, Santos DD, Hatjiharissi E, Xu L, Leleu X, Tournilhac O, Patterson CJ, Manning R, Branagan AR, Morton CC. (2006). Characterization of familial Waldenstrom's macroglobulinemia. *Ann Oncol* **17**, 488–94.
90. Kriangkum J, Taylor BJ, Strachan E, Mant MJ, Reiman T, Belch AR, Pilarski LM. (2006). Impaired class switch recombination (CSR) in Waldenstrom macroglobulinemia (WM) despite apparently normal CSR machinery. *Blood* **107**, 2920–7.
91. Naeim F. (1997). *Pathology of Bone Marrow*, 2nd ed. Williams & Wilkins, Baltimore.
92. His ED. (2007). *Hematopathology*. Churchill Livingstone, Philadelphia.
93. Knowels DM. (2001). *Neoplastic Hematopathology*, 2nd ed. Lippincott Williams & Wilkins, Philadelphia.
94. Konoplev S, Medeiros LJ, Bueso-Ramos CE, Jorgensen JL, Lin P. (2005). Immunophenotypic profile of lymphoplasmacytic lymphoma/Waldenstrom macroglobulinemia. *Am J Clin Pathol* **124**, 414–20.
95. Schop RF, Van Wier SA, Xu R, Ghobrial I, Ahmann GJ, Greipp PR, Kyle RA, Dispenzieri A, Lacy MQ, Rajkumar SV, Gertz MA, Fonseca R. (2006). 6q deletion discriminates Waldenstrom macroglobulinemia from IgM monoclonal gammopathy of undetermined significance. *Cancer Genet Cytogenet* **169**, 150–3.
96. Liu YC, Miyazawa K, Sashida G, Kodama A, Ohyashiki K. (2006). Deletion (20q) as the sole abnormality in Waldenstrom macroglobulinemia suggests distinct pathogenesis of 20q11 anomaly. *Cancer Genet Cytogenet* **169**, 69–72.
97. Terre C, Nguyen-Khac F, Barin C, Mozziconacci MJ, Eclache V, Leonard C, Chapiro E, Farhat H, Bouyon A, Rousselot P, Choquet S, Spentchian M, Dubreuil P, Leblond V, Castaigne S. (2006). Trisomy 4, a new chromosomal abnormality in Waldenstrom's macroglobulinemia: A study of 39 cases. *Leukemia* **20**, 1634–6.
98. Aoki H, Takishita M, Kosaka M, Saito S. (1995). Frequent somatic mutations in D and/or JH segments of Ig gene in Waldenstrom's macroglobulinemia and chronic lymphocytic leukemia (CLL) with Richter's syndrome but not in common CLL. *Blood* **85**, 1913–19.
99. Dimopoulos MA, Panayiotidis P, Moulopoulos LA, Sfikakis P, Dalakas M. (2000). Waldenstrom's macroglobulinemia: Clinical features, complications, and management. *J Clin Oncol* **18**, 214–26.

100. Kimby E, Treon SP, Anagnostopoulos A, Dimopoulos M, Garcia-Sanz R, Gertz MA, Johnson S, LeBlond V, Fermand JP, Maloney DG, Merlini G, Morel P, Morra E, Nichols G, Ocio EM, Owen R, Stone M, Blade J. (2006). Update on recommendations for assessing response from the Third International Workshop on Waldenstrom's macroglobulinemia. *Clin Lymphoma Myeloma* **6**, 380–3.
101. Menke MN, Feke GT, McMeel JW, Branagan A, Hunter Z, Treon SP. (2006). Hyperviscosity-related retinopathy in Waldenstrom macroglobulinemia. *Arch Ophthalmol* **124**, 1601–6.
102. Kyle RA, Rajkumar SV. (2007). Clinical manifestations and diagnosis of Waldenstrom's macroglobulinemia. *UpToDate.*
103. Treon SP, Gertz MA, Dimopoulos M, Anagnostopoulos A, Blade J, Branagan AR, Garcia-Sanz R, Johnson S, Kimby E, Leblond V, Fermand JP, Maloney DG, Merlini G, Morel P, Morra E, Nichols G, Ocio EM, Owen R, Stone MJ. (2006). Update on treatment recommendations from the Third International Workshop on Waldenstrom's macroglobulinemia. *Blood* **107**, 3442–6.
104. Johnson SA. (2006). Advances in the treatment of Waldenstrom's macroglobulinemia. *Expert Rev Anticancer Ther* **6**, 329–34.
105. Pangalis GA, Kyrtsonis MC, Kontopidou FN, Siakantaris MP, Dimopoulou MN, Vassilakopoulos TP, Tzenou T, Kokoris S, Dimitriadou E, Kalpadakis C, Tsalimalma K, Tsaftaridis P, Panayiotidis P, Angelopoulou MK. (2005). Differential diagnosis of Waldenstrom's macroglobulinemia and other B-cell disorders. *Clin Lymphoma* **5**, 235–40.
106. Shaye OS, Levine AM. (2006). Marginal zone lymphoma. *J Natl Compr Canc Netw* **4**, 311–18.
107. Oscier D, Owen R, Johnson S. (2005). Splenic marginal zone lymphoma. *Blood Rev* **19**, 39–51.
108. Dogan A, Isaacson PG. (2003). Splenic marginal zone lymphoma. *Semin Diagn Pathol* **20**, 121–7.
109. Thieblemont C, Felman P, Callet-Bauchu E, Traverse-Glehen A, Salles G, Berger F, Coiffier B. (2003). Splenic marginal-zone lymphoma: A distinct clinical and pathological entity. *Lancet Oncol* **4**, 95–103.
110. Franco V, Florena AM, Iannitto E. (2003). Splenic marginal zone lymphoma. *Blood* **101**, 2464–72.
111. Zhu D, Orchard J, Oscier DG, Wright DH, Stevenson FK. (2002). V(H) gene analysis of splenic marginal zone lymphomas reveals diversity in mutational status and initiation of somatic mutation *in vivo. Blood* **100**, 2659–61.
112. Marasca R, Vaccari P, Luppi M, Zucchini P, Castelli I, Barozzi P, Cuoghi A, Torelli G. (2001). Immunoglobulin gene mutations and frequent use of VH1-69 and VH4-34 segments in hepatitis C virus-positive and hepatitis C virus-negative nodal marginal zone B-cell lymphoma. *Am J Pathol* **159**, 253–61.
113. Algara P, Mateo MS, Sanchez-Beato M, *et al.* (2002). Analysis of the IgV(H) somatic mutations in splenic marginal zone lymphoma defines a group of unmutated cases with frequent 7q deletion and adverse clinical course. *Blood* **99**, 1299–304.
114. Mollejo M, Algara P, Mateo MS, Sánchez-Beato M, Lloret E, Medina MT, Piris MA. (2002). Splenic small B-cell lymphoma with predominant red pulp involvement: A diffuse variant of splenic marginal zone lymphoma? *Histopathology* **40**, 22–30.
115. Tierens A, Delabie J, Pittaluga S, *et al.* (1998). Mutation analysis of the rearranged immunoglobulin heavy chain genes of marginal zone cell lymphomas indicates an origin from different marginal zone B lymphocyte subsets. *Blood* **91**, 2381–6.
116. Matutes E, Oscier D, Montalban C, Berger F, Callet-Bauchu E, Dogan A, Felman P, Franco V, Iannitto E, Mollejo M, Papadaki T, Remstein ED, Salar A, Solé F, Stamatopoulos K, Thieblemont C, Traverse-Glehen A, Wotherspoon A, Coiffier B, Piris MA. (2008). Splenic marginal zone lymphoma proposals for a revision of diagnostic, staging and therapeutic criteria. *Leukemia* **22**, 487–95.
117. Thieblemont C. (2005). Clinical presentation and management of marginal zone lymphomas. *Hematology Am Soc Hematol Educ Program*, 307–13.
118. Bertoni F, Zucca E. (2005). State-of-the-art therapeutics: Marginal-zone lymphoma. *J Clin Oncol* **23**, 6415–20.
119. De Wolf-Peeters C. (2000). Marginal zone lymphomas, MALT and splenic types. *Adv Clin Path* **4**, 190–2.
120. Zucca E, Bertoni F, Roggero E, Cavalli F. (2000). The gastric marginal zone B-cell lymphoma of MALT type. *Blood* **96**, 410–19.
121. Bacon CM, Du MQ, Dogan A. (2006). MALT lymphoma: A practical guide for pathologists. *J Clin Pathol* **209**, 344–51.
122. Cavalli F, Isaacson PG, Gascoyne RD, Zucca E. (2001). MALT lymphomas. *Hematology Am Soc Hematol Educ Program* **20**, 241–58.
123. Wundisch T, Kim TD, Thiede C, Morgner A, Alpen B, Stolte M, Neubauer A. (2003). Etiology and therapy of *Helicobacter pylori* - associated gastric lymphomas. *Ann Hematol* **82**, 535–45.
124. Mariette X. (2001). Lymphomas complicating Sjogren's syndrome and hepatitis C virus infection may share a common pathogenesis: Chronic stimulation of rheumatoid factor B cells. *Ann Rheum Dis* **60**, 1007–10.
125. Tsuji K, Suzuki D, Naito Y, Sato Y, Yoshino T, Iwatsuki K. (2005). Primary cutaneous marginal zone B-cell lymphoma. *Eur J Dermatol* **15**, 480–3.
126. Bogle MA, Riddle CC, Triana EM, Jones D, Duvic M. (2005). Primary cutaneous B-cell lymphoma. *J Am Acad Dermatol* **53**, 479–84.
127. Inagaki H. (2007). Mucosa-associated lymphoid tissue lymphoma: Molecular pathogenesis and clinicopathological significance. *Pathol Int* **57**, 474–84.
128. Noels H, van Loo G, Hagens S, Broeckx V, Beyaert R, Marynen P, Baens M. (2007). A novel TRAF6 binding site in MALT1 defines distinct mechanisms of NF-kappaB activation by API2middle dot-MALT1 fusions. *J Biol Chem* **282**, 10180–9.
129. Sagaert X, Laurent M, Baens M, Wlodarska I, De Wolf-Peeters C. (2006). MALT1 and BCL10 aberrations in MALT lymphomas and their effect on the expression of BCL10 in the tumour cells. *Mod Pathol* **19**, 225–32.
130. Abbondanzo SL. (2001). Extranodal marginal-zone B-cell lymphoma of the salivary gland. *Ann Diagn Pathol* **5**, 246–54.
131. Cho-Vega JH, Vega F, Rassidakis G, Medeiros LJ. (2006). Primary cutaneous marginal zone B-cell lymphoma. *Am J Clin Pathol* **125**(Suppl), S38–49.
132. Wannesson L, Cavalli F, Zucca E. (2005). Primary pulmonary lymphoma: Current status. *Clin Lymphoma Myeloma* **6**, 220–7.
133. Tai YC, Tan JA, Peh SC. (2004). 18q21 Rearrangement and trisomy 3 in extranodal B-cell lymphomas: A study using a fluorescent *in situ* hybridisation technique. *Virchows Arch* **445**, 506–14.
134. Schreuder MI, Hoefnagel JJ, Jansen PM, van Krieken JH, Willemze R, Hebeda KM. (2005). FISH analysis of MALT lymphoma-specific translocations and aneuploidy in primary cutaneous marginal zone lymphoma. *J Pathol* **205**, 302–10.
135. Sonoki T, Harder L, Horsman DE, Karran L, Taniguchi I, Willis TG, Gesk S, Steinemann D, Zucca E, Schlegelberger B, Solé F, Mungall AJ, Gascoyne RD, Siebert R, Dyer MJ. (2001). Cyclin D3 is a target gene of t(6;14)(p21.1;q32.3) of mature B-cell malignancies. *Blood* **98**, 2837–44
136. Min DL, Zhou XY, Yang WT, Lu HF, Zhang TM, Zhen AH, Cao PZ, Shi DR. (2005). Point mutation of 5′ noncoding region of BCL-6 gene in primary gastric lymphomas. *World J Gastroenterol* **11**, 51–5.
137. Cavalli F, Isaacson PG, Gascoyne RD, Zucca E. (2001). MALT lymphomas. *Hematology Am Soc Hematol Educ Program*, 241–58.
138. Mollejo M, Camacho FI, Algara P, Ruiz-Ballesteros E, Garcia JF, Piris MA. (2005). Nodal and splenic marginal zone B cell lymphomas. *Hematol Oncol* **23**, 108–18.
139. Arcaini L, Paulli M, Boveri E, Magrini U, Lazzarino M. (2003). Marginal zone-related neoplasms of splenic and nodal origin. *Haematologica* **88**, 80–93.
140. Maes B, De Wolf-Peeters C. (2002). Marginal zone cell lymphoma – an update on recent advances. *Histopathology* **40**, 117–26.
141. Naeim F. (1988). Hairy cell leukemia: Characteristics of the neoplastic cells. *Hum Pathol* **19**, 375–88.

142. Sharpe RW, Bethel KJ. (2006). Hairy cell leukemia: Diagnostic pathology. *Hematol Oncol Clin North Am* **20**, 1023–49.
143. Hardell L, Eriksson M, Nordstrom M. (2002). Exposure to pesticides as risk factor for non-Hodgkin's lymphoma and hairy cell leukemia: Pooled analysis of two Swedish case–control studies. *Leuk Lymphoma* **43**, 1043–9.
144. Clavel J, Conso F, Limasset JC, Mandereau L, Roche P, Flandrin G, Hémon D. (1996). Hairy cell leukaemia and occupational exposure to benzene. *Occup Environ Med* **53**, 533–9.
145. Clavel J, Hémon D, Mandereau L, Delemotte B, Séverin F, Flandrin G. (1996). Farming, pesticide use and hairy-cell leukemia. *Scand J Work Environ Health* **22**, 285–93.
146. Villemagne B, Bay JO, Tournilhac O, Chaleteix C, Travade P. (2005). Two new cases of familial hairy cell leukemia associated with HLA haplotypes A2, B7, Bw4, Bw6. *Leuk Lymphoma* **46**, 243–5.
147. Vanhentenrijk V, Tierens A, Wlodarska I, Verhoef G, Wolf-Peeters CD. (2004). V(H) gene analysis of hairy cell leukemia reveals a homogeneous mutation status and suggests its marginal zone B-cell origin. *Leukemia* **18**, 1729–32.
148. Forconi F, Sahota SS, Raspadori D, Ippoliti M, Babbage G, Lauria F, Stevenson FK. (2004). Hairy cell leukemia: At the crossroad of somatic mutation and isotype switch. *Blood* **104**, 3312–17.
149. Tiacci E, Liso A, Piris M, Falini B. (2006). Evolving concepts in the pathogenesis of hairy-cell leukaemia. *Nat Rev Cancer* **6**, 437–48.
150. Matutes E. (2006). Immunophenotyping and differential diagnosis of hairy cell leukemia. *Hematol Oncol Clin North Am* **20**, 1051–63.
151. Babusíková O, Tomová A. (2003). Hairy cell leukemia: Early immunophenotypical detection and quantitative analysis by flow cytometry. *Neoplasma* **50**, 350–6.
152. Went PT, Zimpfer A, Pehrs AC, Sabattini E, Pileri SA, Maurer R, Terracciano L, Tzankov A, Sauter G, Dirnhofer S. (2005). High specificity of combined TRAP and DBA.44 expression for hairy cell leukemia. *Am J Surg Pathol* **29**, 474–8.
153. Forconi F, Raspadori D, Lenoci M, Lauria F. (2005). Absence of surface CD27 distinguishes hairy cell leukemia from other leukemic B-cell malignancies. *Haematologica* **90**, 266–8.
154. Han T, Sadamori N, Block AM, Xiao H, Henderson ES, Emrich L, Sandberg AA. (1988). Cytogenetic studies in chronic lymphocytic leukemia, prolymphocytic leukemia and hairy cell leukemia: A progress report. *Nouv Rev Fr Hematol* **30**, 393–5.
155. Sambani C, Trafalis DT, Mitsoulis-Mentzikoff C, Poulakidas E, Makropoulos V, Pantelias GE, Mecucci C. (2001). Clonal chromosome rearrangements in hairy cell leukemia: Personal experience and review of literature. *Cancer Genet Cytogenet* **129**, 138–44.
156. Sucak GT, Ogur G, Topal G, Ataoglu O, Cankus G, Haznedar R. (1998). del(17)(q25) in a patient with hairy cell leukemia: A new clonal chromosome abnormality. *Cancer Genet Cytogenet* **100**, 152–4.
157. Haglund U, Juliusson G, Stellan B, Gahrton G. (1994). Hairy cell leukemia is characterized by clonal chromosome abnormalities clustered to specific regions. *Blood* **83**, 2637–45.
158. Cleary ML, Wood GS, Warnke R, *et al.* (1984). Immunoglobulin gene rearrangements in hairy cell leukemia. *Blood* **64**, 99.
159. Goodman GR, Bethel KJ, Saven A. (2003). Hairy cell leukemia: An update. *Curr Opin Hematol* **10**, 258–66.
160. Hoffman MA. (2006). Clinical presentations and complications of hairy cell leukemia. *Hematol Oncol Clin North Am* **20**, 1065–73.
161. Cawley JC. (2006). The pathophysiology of the hairy cell. *Hematol Oncol Clin North Am* **20**, 1011–21.
162. Wanko SO, de Castro C. (2006). Hairy cell leukemia: An elusive but treatable disease. *Oncologist* **11**, 780–9.
163. Van De Corput L, Falkenburg JH, Kester MG, Willemze R, Kluin-Nelemans JC. (1999). Impaired expression of CD28 on T cells in hairy cell leukemia. *Clin Immunol* **93**, 256–62.
164. Van De Corput L, Falkenburg JH, Kluin-Nelemans JC. (1998). T-cell dysfunction in hairy cell leukemia: An updated review. *Leuk Lymphoma* **30**, 31–9.
165. Kluin-Nelemans JC, Kester MG, Oving I, Cluitmans FH, Willemze R, Falkenburg JH. (1994). Abnormally activated T lymphocytes in the spleen of patients with hairy-cell leukemia. *Leukemia* **8**, 2095–101.
166. Ravandi F, Jorgensen JL, O'Brien SM, Verstovsek S, Koller CA, Faderl S, Giles FJ, Ferrajoli A, Wierda WG, Odinga S, Huang X, Thomas DA, Freireich EJ, Jones D, Keating MJ, Kantarjian HM. (2006). Eradication of minimal residual disease in hairy cell leukemia. *Blood* **107**, 4658–62. [Epub 2006].
167. Gidron A, Tallman MS. (2006). Hairy cell leukemia: Towards a curative strategy. *Hematol Oncol Clin North Am* **20**, 1153–62.
168. Else M, Ruchlemer R, Osuji N, Del Giudice I, Matutes E, Woodman A, Wotherspoon A, Swansbury J, Dearden C, Catovsky D. (2005). Long remissions in hairy cell leukemia with purine analogs: A report of 219 patients with a median follow-up of 12.5 years. *Cancer* **104**, 2442–8.
169. Bartl R, Frisch B, Hill W, Burkhardt R, Sommerfeld W, Sund M. (1983). Bone marrow histology in hairy cell leukemia. Identification of subtypes and their prognostic significance. *Am J Clin Pathol* **79**, 531–45.
170. Naeim F. (1982). Clinicopathological subtypes in hairy-cell leukemia. *Am J Clin Pathol* **78**, 80–5.
171. Habermann TM. (2006). Splenectomy, interferon, and treatments of historical interest in hairy cell leukemia. *Hematol Oncol Clin North Am* **20**, 1075–86.
172. Thomas DA, Ravandi F, Kantarjian H. (2006). Monoclonal antibody therapy for hairy cell leukemia. *Hematol Oncol Clin North Am* **20**, 1125–36.
173. Robak T. (2006). Current treatment options in hairy cell leukemia and hairy cell leukemia variant. *Cancer Treat Rev* **32**, 365–76.
174. Cessna MH, Hartung L, Tripp S, Perkins SL, Bahler DW. (2005). Hairy cell leukemia variant: Fact or fiction. *Am J Clin Pathol* **123**, 132–8.
175. Chen YH, Tallman MS, Goolsby C, Peterson L. (2006). Immunophenotypic variations in hairy cell leukemia. *Am J Clin Pathol* **125**, 251–9.
176. Hanson CA, Ward PC, Schnitzer B. (1989). A multilobular variant of hairy cell leukemia with morphologic similarities to T-cell lymphoma. *Am J Surg Pathol* **13**, 671–9.
177. Diez Martin JL, Li CY, Banks PM. (1987). Blastic variant of hairy-cell leukemia. *Am J Clin Pathol* **87**, 576–83.
178. Winter JN, Gascoyne RD, Van Besien K. (2004). Low-grade lymphoma. *Hematology Am Soc Hematol Educ Program*, 203–20.
179. Bende RJ, Smit LA, van Noesel CJ. (2007). Molecular pathways in follicular lymphoma. *Leukemia* **21**, 18–29.
180. de Jong D. (2005). Molecular pathogenesis of follicular lymphoma: A cross talk of genetic and immunologic factors. *J Clin Oncol* **23**, 6358–63.
181. Hayslip J, Montero A. (2006). Tumor suppressor gene methylation in follicular lymphoma: A comprehensive review. *Mol Cancer* **5**, 44.
182. Schmidt B, Kremer M, Gotze K, John K, Peschel C, Hofier H, Fend F. (2006). Bone marrow involvement in follicular lymphoma: Comparison of histology and flow cytometry as staging procedures. *Leuk Lymphoma* **47**, 1857–62.
183. Iancu D, Hao S, Lin P, Anderson SK, Jorgensen JL, McLaughlin P, Medeiros LJ. (2007). Follicular lymphoma in staging bone marrow specimens: Correlation of histologic findings with the results of flow cytometry immunophenotypic analysis. *Arch Pathol Lab Med* **131**, 282–7.
184. Perea G, Altés A, Bellido M, Aventín A, Bordes R, Ayats R, Remacha AF, Espinosa I, Briones J, Sierra J, Nomdedéu JF. (2004). Clinical utility of bone marrow flow cytometry in B-cell non-Hodgkin lymphomas (B-NHL). *Histopathology* **45**, 268–74.
185. Bentley G, Palutke M, Mohamed AN. (2005). Variant t(14;18) in malignant lymphoma: A report of seven cases. *Cancer Genet Cytogenet* **157**, 12–17.
186. Arcaini L, Colombo N, Bernasconi P, Calatroni S, Passamonti F, Orlandi E, Bonfichi M, Burcheri S, Porta MD, Rumi E, Montanari

F, Algarotti A, Pascutto C, Lazzarino M. (2006). Role of the molecular staging and response in the management of follicular lymphoma patients. *Leuk Lymphoma* **47**, 1018–22.
187. Rambaldi A, Carlotti E, Oldani E, Delia Starza I, Baccarani M, Cortelazzo S, Lauria F, Arcaini L, Morra E, Pulsoni A, Rigacci L, Rupolo M, Zaja F, Zinzani PL, Barbui T, Foa R. (2005). Quantitative PCR of bone marrow BCL2/IgH+ cells at diagnosis predicts treatment response and long-term outcome in follicular non-Hodgkin lymphoma. *Blood* **105**, 3428–33.
188. Albinger-Hegyi A, Hochreutener B, Abdou MT, Hegyi I, Dours-Zimmermann MT, Kurrer MO, Heitz PU, Zimmermann DR. (2002). High frequency of t(14;18)-translocation breakpoints outside of major breakpoint and minor cluster regions in follicular lymphomas: Improved polymerase chain reaction protocols for their detection. *Am J Pathol* **160**, 823–32.
189. Einerson RR, Kurtin PJ, Dayharsh GA, Kimlinger TK, Remstein ED. (2005). FISH is superior to PCR in detecting t(14;18)(q32;q21)-IgH/bcl-2 in follicular lymphoma using paraffin-embedded tissue samples. *Am J Clin Pathol* **124**, 421–9.
190. Knutsen T. (1998). Cytogenetic changes in the progression of lymphoma. *Leuk Lymphoma* **31**, 1–19.
191. Yamamoto K, Ono K, Katayama Y, Shimoyama M, Matsui T. (2007). Derivative (3)t(3;18)(q27;q21)t(18;16)(q21;?) involving the BCL2 and BCL6 genes in follicular lymphoma with t(3;14;18)(q27;q32;q21). *Cancer Genet Cytogenet* **179**, 69–75.
192. Wong KF. (2007). Transformed follicular lymphoma with concurrent t(2;3), t(8;14) and t(14;18). *Cancer Genet Cytogenet* **173**, 68–70.
193. Wong KF, Chan JK. (2000). Follicular lymphoma with trisomy 18 and over-expression of BCL2 in the absence of t(14;18)(q32;q21). *Cancer Genet Cytogenet* **123**, 52–4.
194. Glas AM, Kersten MJ, Delahaye LJ, Witteveen AT, Kibbelaar RE, Velds A, Wessels LF, Joosten P, Kerkhoven RM, Bernards R, van Krieken JH, Kluin PM, van't Veer LJ, de Jong D. (2005). Gene expression profiling in follicular lymphoma to assess clinical aggressiveness and to guide the choice of treatment. *Blood* **105**, 301–7.
195. Solal-Celigny P. (2006). Follicular lymphoma international prognostic index. *Curr Treat Options Oncol* **7**, 270–5.
196. Solal-Celigny P. (2005). Prognosis of follicular lymphomas. *Clin Lymphoma* **6**, 21–5.
197. Luminari S, Federico M. (2006). Prognosis of follicular lymphomas. *Hematol Oncol* **24**, 64–72.
198. Aurora V, Winter JN. (2006). Follicular lymphoma: Today's treatments and tomorrow's targets. *Expert Opin Pharmacother* **7**, 1273–90.
199. Coiffier B. (2006). Monoclonal antibody as therapy for malignant lymphomas. *C R Biol* **329**, 241–54.
200. Brown JR, Feng Y, Gribben JG, Neuberg D, Fisher DC, Mauch P, Nadler LM, Freedman AS. (2007). Long-term survival after autologous bone marrow transplantation for follicular lymphoma in first remission. *Biol Blood Marrow Transplant* **13**, 1057–65.
201. Mirza I, Macpherson N, Paproski S, Gascoyne RD, Yang B, Finn WG, His ED. (2002). Primary cutaneous follicular lymphoma: An assessment of clinical, histopathologic, immunophenotypic, and molecular features. *J Clin Oncol* **20**, 647–55.
202. Bekkenk MW, Geelen FA, van Voorst Vader PC, Heule F, Geerts ML, van Vloten WA, Meijer CJ, Willemze R. (2000). Primary and secondary cutaneous CD30(+) lymphoproliferative disorders: A report from the Dutch Cutaneous Lymphoma Group on the long-term follow-up data of 219 patients and guidelines for diagnosis and treatment. *Blood* **95**, 3653–61.
203. Cerroni L, Kerl H. (2001). Primary cutaneous follicle center cell lymphoma. *Leuk Lymphoma* **42**, 891–900.
204. Bertoni F, Zucca E, Cavalli F. (2004). Mantle cell lymphoma. *Curr Opin Hematol* **11**, 411–18.
205. Campo E, Raffeld M, Jaffe ES. (1999). Mantle-cell lymphoma. *Semin Hematol* **36**, 115–27.
206. Fernandez V, Hartmann E, Ott G, Campo E, Rosenwald A. (2005). Pathogenesis of mantle-cell lymphoma: All oncogenic roads lead to dysregulation of cell cycle and DNA damage response pathways. *J Clin Oncol* **23**, 6364–9.
207. O'Connor OA. (2007). Mantle cell lymphoma: Identifying novel molecular targets in growth and survival pathways. *Hematology Am Soc Hematol Educ Program*, 270–6.
208. Fisher RI. (1996). Mantle-cell lymphoma: Classification and therapeutic implications. *Ann Oncol* **7**(Suppl 6), S35–39.
209. Oliveira FM, Tone LG, Simões BP, Rego EM, Araújo AG, Falcão RP. (2007). Blastoid mantle cell lymphoma with t(2;8) (p12;q24). *Leuk Lymphoma* **48**, 2079–82.
210. Pott C, Schrader C, Brüggemann M, Ritgen M, Harder L, Raff T, Tiemann M, Dreger P, Kneba M. (2005). Blastoid variant of mantle cell lymphoma: Late progression from classical mantle cell lymphoma and quantitation of minimal residual disease. *Eur J Haematol* **74**, 353–8.
211. Zoldan MC, Inghirami G, Masuda Y, Vandekerckhove F, Raphael B, Amorosi E, Hymes K, Frizzera G. (1996). Large-cell variants of mantle cell lymphoma: Cytologic characteristics and p53 anomalies may predict poor outcome. *Br J Haematol* **93**, 475–86.
212. Goy A. (2007). Mantle cell lymphoma: Evolving novel options. *Curr Oncol Rep* **9**, 391–8.
213. Velders GA, Kluin-Nelemans JC, De Boer CJ, Hermans J, Noordijk EM, Schuuring E, Kramer MH, Van Deijk WA, Rahder JB, Kluin PM, Van Krieken JH. (1996). Mantle-cell lymphoma: A population-based clinical study. *J Clin Oncol* **14**, 1269–74.
214. Bosch F, López-Guillermo A, Campo E, Ribera JM, Conde E, Piris MA, Vallespí T, Woessner S, Montserrat E. (1998). Mantle cell lymphoma: Presenting features, response to therapy, and prognostic factors. *Cancer* **82**, 567–75.
215. Norton AJ, Matthews J, Pappa V, Shamash J, Love S, Rohatiner AZ, Lister TA. (1995). Mantle cell lymphoma: Natural history defined in a serially biopsied population over a 20-year period. *Ann Oncol* **6**, 249–56.
216. Foss HD, Stein H. (2000). Pathology of intestinal lymphomas. *Recent Results Cancer Res* **156**, 33–41.
217. Gong JZ, Lagoo AS, Peters D, Horvatinovich J, Benz P, Buckley PJ. (2001). Value of CD23 determination by flow cytometry in differentiating mantle cell lymphoma from chronic lymphocytic leukemia/small lymphocytic lymphoma. *Am J Clin Pathol* **116**, 893–7.
218. Dono M, Cerruti G, Zupo S. (2004). The CD5+ B-cell. *Int J Biochem Cell Biol* **36**, 2105–11.
219. Dunphy CH, Wheaton SE, Perkins SL. (1997). CD23 expression in transformed small lymphocytic lymphomas/chronic lymphocytic leukemias and blastic transformations of mantle cell lymphoma. *Mod Pathol* **10**, 818–22.
220. Vandenberghe E, De Wolf-Peeters C, van den Oord J, Wlodarska I, Delabie J, Stul M, Thomas J, Michaux JL, Mecucci C, Cassiman JJ, Van den Berghe H. (1991). Translocation (11;14): A cytogenetic anomaly associated with B cell lymphomas of non-follicle centre cell lineage. *J Pathol* **163**, 13–18.
221. de Boer CJ, van Krieken JH, Schuuring E, Kluin PM. (1997). Bcl-1/cyclin D1 in malignant lymphoma. *Ann Oncol* **8**, 109–17.
222. Katz RL, Caraway NP, Gu J, Jiang F, Pasco-Miller LA, Glassman AB, Luthra R, Hayes KJ, Romaguera JE, Cabanillas FF, Medeiros LJ. (2000). Detection of chromosome 11q13 breakpoints by interphase fluorescence *in situ* hybridization. A cell lymphoma. *Am J Clin Pathol* **114**, 248–57.
223. Stacey DW. (2003). Cyclin D1 serves as a cell cycle regulatory switch in actively proliferating cells. *Curr Opin Cell Biol* **15**, 158–63.
224. Lovec, *et al.* (1994). Cyclin D1/bcl-1 cooperates with myc genes in the generation of B-cell lymphoma in transgenic mice. *EMBO J* **13**(15), 3487–95.
225. Hui P, Howe JG, Crouch J, Nimmakayalu M, Qumsiyeh MB, Tallini G, Flynn SD, Smith BR. (2003). Real-time quantitative RT-PCR of cyclin D1 in mantel cell lymphoma: Comparison with FISH and immunohistochemistry. *Leuk Lymphoma* **44**, 1385–94.
226. Pinyol M, Campo E, Nadal A, Terol MJ, Jares P, Nayach I, Fernandez PL, Piris MA, Montserrat E, Cardesa A. (1996). Detection of the

bcl-1 rearrangement at the major translocation cluster in frozen and paraffin-embedded tissues of mantle cell lymphomas by polymerase chain reaction. *Am J Clin Pathol* **105**, 532–7.

227. Motokura T, Bloom T, Kim HG, Jüppner H, Ruderman JV, Kronenberg HM. (1991). Novel cyclin encoded by a bcl1-linked candidate oncogene. *Nature* **350**, 512–15.
228. Zukerberg LR, Yang WI, Arnold A, Harris NL. (1995). Cyclin D1 expression in non-Hodgkin's lymphomas detection by immunohistochemistry. *Am J Clin Pathol* **103**, 756–60.
229. Rimokh R, Berger F, Delsol G, Charrin C, Bertheas MF, French M, Garoscio M, Felman P, Coiffier B, Bryon PA. (1993). Rearrangement and overexpression of the BCL-l/PRAD-1 gene intermediate lymphocytic lymphomas and in t(11q13)-bearing leukemias. *Blood* **81**, 3063–7.
230. Greiner TC, Moynihan MJ, Chan WC, Lytle DM, Pedersen A, Anderson JR, Weisenburger DD. (1996). p53 Mutations in mantle cell lymphoma are associated with variant cytology and predict a poor prognosis. *Blood* **87**, 4302–10.
231. Khoury JD, Sen F, Abruzzo LV, Hayes K, Glassman A, Medeiros LJ. (2003). Cytogenetic findings in blastoid mantle cell lymphoma. *Hum Pathol* **34**, 1022–9.
232. Bigoni R, Cuneo A, Milani R, Roberti MG, Bardi A, Rigolin GM, Cavazzini F, Agostini P, Castoldi G. (2001). Secondary chromosome changes in mantle cell lymphoma: Cytogenetic and fluorescence *in situ* hybridization studies. *Leuk Lymphoma* **40**, 581–90.
233. Allen JE, Hough RE, Goepel JR, Bottomley S, Wilson GA, Alcock HE, Baird M, Lorigan PC, Vandenberghe EA, Hancock BW, Hammond DW. (2002). Identification of novel regions of amplification and deletion within mantle cell lymphoma DNA by comparative genomic hybridization. *Br J Haematol* **116**, 291–8.
234. Monni O, Joensuu H, Franssila K, Knuutila S. (1996). DNA copy number changes in diffuse large B-cell lymphoma – comparative genomic hybridization study. *Blood* **87**, 5269–78.
235. Monni O, Zhu Y, Franssila K, Oinonen R, Hoglund P, Elonen E, Joensuu H, Knuutila S. (1999). Molecular characterization of deletion at 11q22.1-23.3 in mantle cell lymphoma. *Br J Haematol* **104**, 665–71.
236. Beà S, Ribas M, Hernández JM, Bosch F, Pinyol M, Hernández L, García JL, Flores T, González M, López-Guillermo A, Piris MA, Cardesa A, Montserrat E, Miró R, Campo E. (1999). Increased number of chromosomal imbalances and high-level DNA amplifications in mantle cell lymphoma are associated with blastoid variants. *Blood* **93**, 4365–74.
237. De Vos , De Vos S, Krug U, *et al.* (2003). Cell cycle alterations in the blastoid variant of mantle cell lymphoma (MCL-BV) as detected by gene expression profiling of mantle cell lymphoma (MCL) and MCL-BV. *Diagn Mol Pathol* **12**, 35–43.
238. Rosenwald A, Wright G, *et al.* (2003). The proliferation gene expression signature is a quantitative integrator of oncogenic events that predicts survival in mantle cell lymphoma. *Cancer Cell* **3**, 185–97.
239. Räty R, Franssila K, Joensuu H, Teerenhovi L, Elonen E. (2002). Ki-67 expression level, histological subtype, and the International Prognostic Index as outcome predictors in mantle cell lymphoma. *Eur J Haematol* **69**, 11–20.
240. Onciu M, Schlette E, Medeiros LJ, Abruzzo LV, Keating M, Lai R. (2001). Cytogenetic findings in mantle cell lymphoma cases with a high level of peripheral blood involvement have a distinct pattern of abnormalities. *Am J Clin Pathol* **116**, 886–92.
241. Räty R, Franssila K, Jansson SE, Joensuu H, Wartiovaara-Kautto U, Elonen E. (2003). Predictive factors for blastoid transformation in the common variant of mantle cell lymphoma. *Eur J Cancer* **39**, 321–9.
242. Brody J, Advani R. (2006). Treatment of mantle cell lymphoma: Current approach and future directions. *Crit Rev Oncol Hematol* **58**, 257–65.
243. Williams ME, Densmore JJ. (2007). Biology and therapy of mantle cell lymphoma. *Curr Opin Oncol* **17**, 425–31.
244. Weigert O, Unterhalt M, Hiddemann W, Dreyling M. (2007). Current management of mantle cell lymphoma. *Drugs* **67**, 1689–702.
245. Ruan J, Leonard JP. (2006). Mantle cell lymphoma: Current concept in biology and treatment. *Cancer Treat Res* **131**, 141–59.
246. Smith MD, Singleton TP, Balaraman S, Jaiyesimi I, O'Malley B, Al-Saadi A, Mattson JC. (2004). Case report: Mantle cell lymphoma, prolymphocytoid variant, with leukostasis syndrome. *Mod Pathol* **17**, 879–83.
247. Friedberg JW, Fisher RI. (2006). Diffuse large B-cell NHL. *Cancer Treat Res* **131**, 121–40.
248. Ghesquieres H, Berger F, Felman P, Callet-Bauchu E, Bryon PA, Traverse-Glehen A, Thieblemont C, Baseggio L, Michallet AS, Coiffier B, Salles G. (2006). Clinicopathologic characteristics and outcome of diffuse large B-cell lymphomas presenting with an associated low-grade component at diagnosis. *J Clin Oncol* **24**, 5234–41.
249. Dupin N, Diss TL, Kellam P, Tulliez M, Du MQ, Sicard D, Weiss RA, Isaacson PG, Boshoff C. (2000). HHV-8 is associated with a plasmablastic variant of Castleman disease that is linked to HHV-8-positive plasmablastic lymphoma. *Blood* **95**, 1406–12.
250. Ranuncolo SM, Polo JM, Dierov J, Singer M, Kuo T, Greally J, Green R, Carroll M, Melnick A. (2007). Bcl-6 mediates the germinal center B cell phenotype and lymphomagenesis through transcriptional repression of the DNA-damage sensor ATR. *Nat Immunol* **8**, 705–14 [Epub].
251. Lossos IS. (2005). Molecular pathogenesis of diffuse large B-cell lymphoma. *J Clin Oncol* **23**, 6351–7.
252. Bea S, Zettl A, Wright G, Salaverria I, Jehn P, Moreno V, Burek C, Ott G, Puig X, Yang L, Lopez-Guillermo A, Chan WC, Greiner TC, Weisenburger DD, Armitage JO, Gascoyne RD, Connors JM, Grogan TM, Braziel R, Fisher RI, Smeland EB, Kvaloy S, Holte H, Delabie J, Simon R, Powell J, Wilson WH, Jaffe ES, Montserra E, Muller-Hermelink HK, Staudt LM, Campo E, Rosenwald A. (2005). Lymphoma/leukemia molecular profiling project. Diffuse large B-cell lymphoma subgroups have distinct genetic profiles that influence tumor biology and improve gene-expression-based survival prediction. *Blood* **106**, 3183–90.
253. Diebold J, Anderson JR, Armitage JO, Connors JM, Maclennan KA, Müller-Hermelink HK, Nathwani BN, Ullrich F, Weisenburger DD. (2002). Diffuse large B-cell lymphoma: A clinicopathologic analysis of 444 cases classified according to the updated Kiel classification. *Leuk Lymphoma* **43**, 97–104.
254. Engels EA, Pittaluga S, Whitby D, Rabkin C, Aoki Y, Jaffe ES, Goedert JJ. (2003). Immunoblastic lymphoma in persons with AIDS-associated Kaposi's sarcoma: A role for Kaposi's sarcoma-associated herpesvirus. *Mod Pathol* **16**, 424–9.
255. Mentzer SJ, Longtine J, Fingeroth J, Reilly JJ, DeCamp MM, O'Donnell W, Swanson SJ, Faller DV, Sugarbaker DJ. (1996). Immunoblastic lymphoma of donor origin in the allograft after lung transplantation. *Transplantation* **61**, 1720–5.
256. Reichard KK, McKenna RW, Kroft SH. (2007). ALK-positive diffuse large B-cell lymphoma: Report of four cases and review of the literature. *Mod Pathol* **20**, 310–19 [Epub].
257. Haralambieva E, Pulford KA, Lamant L, Pileri S, Roncador G, Gatter KC, Delsol G, Mason DY. (2000). Anaplastic large-cell lymphomas of B-cell phenotype are anaplastic lymphoma kinase (ALK) negative and belong to the spectrum of diffuse large B-cell lymphomas. *Br J Haematol* **109**, 584–91.
258. Aki H, Tuzuner N, Ongoren S, Baslar Z, Soysal T, Ferhanoglu B, Sahinler I, Aydin Y, Ulku B, Aktuglu G. (2004). T-cell-rich B-cell lymphoma: A clinicopathologic study of 21 cases and comparison with 43 cases of diffuse large B-cell lymphoma. *Leuk Res* **28**, 229–36.
259. El Weshi A, Akhtar S, Mourad WA, Ajarim D, Abdelsalm M, Khafaga Y, Bazarbashi S, Maghfoor I. (2007). T-cell/histiocyte-rich B-cell lymphoma: Clinical presentation, management and prognostic factors: Report on 61 patients and review of literature. *Leuk Lymphoma* **48**, 1764–73.
260. Abramson JS. (2006). T-cell/histiocyte-rich B-cell lymphoma: Biology, diagnosis, and management. *Oncologist* **11**, 384–92.

261. Haarer CF, Roberts RA, Frutiger YM, Grogan TM, Rimsza LM. (2006). Immunohistochemical classification of *de novo*, transformed, and relapsed diffuse large B-cell lymphoma into germinal center B-cell and nongerminal center B-cell subtypes correlates with gene expression profile and patient survival. *Arch Pathol Lab Med* **130**, 1819–24.
262. Chang CC, McClintock S, Cleveland RP, Trzpuc T, Vesole DH, Logan B, Kajdacsy-Balla A, Perkins SL. (2004). Immunohistochemical expression patterns of germinal center and activation B-cell markers correlate with prognosis in diffuse large B-cell lymphoma. *Am J Surg Pathol* **28**, 464–70.
263. Lossos IS, Jones CD, Warnke R, Natkunam Y, Kaizer H, Zehnder JL, Tibshirani R, Levy R. (2001). Expression of a single gene, BCL-6, strongly predicts survival in patients with diffuse large B-cell lymphoma. *Blood* **98**, 945–51.
264. Yamaguchi M, Seto M, Okamoto M, Ichinohasama R, Nakamura N, Yoshino T, Suzumiya J, Murase T, Miura I, Akasaka T, Tamaru J, Suzuki R, Kagami Y, Hirano M, Morishima Y, Ueda R, Shiku H, Nakamura S. (2002). *De novo* CD5+ diffuse large B-cell lymphoma: A clinicopathologic study of 109 patients. *Blood* **99**, 815–21.
265. Carbone A, Gloghini A, Larocca LM, Capello D, Pierconti F, Canzonieri V, Tirelli U, Dalla-Favera R, Gaidano G. (2001). Expression profile of MUM1/IRF4, BCL-6, and CD138/syndecan-1 defines novel histogenetic subsets of human immunodeficiency virus-related lymphomas. *Blood* **97**, 744–51.
266. Gascoyne RD. (1977). Pathologic prognostic factors in diffuse aggressive non-Hodgkin's lymphoma. *Hematol Oncol Clin North Am* **11**, 847–62.
267. Rantanen S, Monni O, Joensuu H, Franssila K, Knuutila S. (2001). Causes and consequences of BCL2 over expression in diffuse large B-cell lymphoma. *Leuk Lymphoma* **42**, 1089–98.
268. Barrans SL, O'Connor SJM, Evans PAS. (2002). Rearrangement of the BCL6 locus at 3q27 is an independent poor prognostic factor in nodal diffuse large B-cell lymphoma. *Br J Hematol* **117**, 322–32.
269. Migliazza A, Martinotti S, Chen W. (1995). Frequent somatic hypermutation of the 5′ noncoding region of the BCL-6 gene in B-cell lymphoma. *Proc Natl Acad Sci USA* **92**, 12520.
270. Mitani S, Kamata H, Fujiwara M, Aoki N, Okada S, Watanabe M, Tango T, Mori S. (2007). Missense mutation with/without nonsense mutation of the p53 gene is associated with large cell morphology in human malignant lymphoma. *Pathol Int* **57**, 430–6.
271. Rosenwald A, Wright G, Chan WC, Connors JM, Campo E, Fisher RI, Gascoyne RD, Muller-Hermelink HK, Smeland EB, Giltnane JM, Hurt EM, Zhao H, Averett L, Yang L, Wilson WH, Jaffe ES, Simon R, Klausner RD, Powell J, Duffey PL, Longo DL, Greiner TC, Weisenburger DD, Sanger WG, Dave BJ, Lynch JC, Vose J, Armitage JO, Montserrat E, López-Guillermo A, Grogan TM, Miller TP, LeBlanc M, Ott G, Kvaloy S, Delabie J, Holte H, Krajci P, Stokke T, Staudt LM. (2002). Lymphoma/leukemia molecular profiling project. The use of molecular profiling to predict survival after chemotherapy for diffuse large-B-cell lymphoma. *N Engl J Med* **346**, 1937–47.
272. Habermann TM. (2005). Diffuse large B-cell lymphoma. *Clin Adv Hematol Oncol* **3**(Suppl), 4–5.
273. Adida C, Haioun C, Gaulard P, Lepage E, Morel P, Briere J, Dombret H, Reyes F, Diebold J, Gisselbrecht C, Salles G, Altieri DC, Molina TJ. (2000). Prognostic significance of survivin expression in diffuse large B-cell lymphomas. *Blood* **96**, 1921–5.
274. Lossos IS, Morgensztern D. (2006). Prognostic biomarkers in diffuse large B-cell lymphoma. *J. Clin Oncol* **24**, 995–1007.
275. O'Connor OA, Hamlin P. (2006). New drugs for the treatment of advanced-stage diffuse large cell lymphomas. *Semin Hematol* **43**, 251–61.
276. Coiffier B. (2006). Standard treatment of advanced-stage diffuse large B-cell lymphoma. *Semin Hematol* **43**, 213–20.
277. Braziel RM, Arber DA, Slovak ML, Gulley ML, Spier C, Kjeldsberg C, Unger J, Miller TP, Tubbs R, Leith C, Fisher RI, Grogan TM. (2001). The Burkitt-like lymphomas: A Southwest Oncology Group study delineating phenotypic, genotypic, and clinical features. *Blood* **97**, 3713–20.
278. Kolonic SO, Dzebro S, Kusec R, Planinc-Peraica A, Dominis M, Jaksic B. (2006). Primary mediastinal large B-cell lymphoma: A single-center study of clinicopathologic characteristics. *Int J Hematol* **83**, 331–6.
279. Sekiguchi N, Nishimoto J, Tanimoto K, Kusumoto S, Onishi Y, Watanabe T, Kobayashi Y, Asamura H, Kagami Y, Matsuno Y, Tobinai K. (2004). Primary mediastinal large B-cell lymphoma: A single-institution clinical study in Japan. *Int J Hematol* **79**, 465–71.
280. Savage KJ. (2006). Primary mediastinal large B-cell lymphoma. *Oncologist* **11**, 488–95.
281. Devouassoux-Shisheboran M, Travis WD. (2007). Pathology of mediastinal tumors. *UpToDate.*
282. Tsai HW, Yen YS, Chang KC. (2006). Mediastinal large B-cell lymphoma with rosette formation mimicking thymoma and thymic carcinoid. *Histopathology* **49**, 93–5.
283. Pileri SA, Zinzani PL, Gaidano G, Falini B, Gaulard P, Zucca E, Sabattini E, Ascani S, Rossi M, Cavalli F. (2003). International Extranodal Lymphoma Study Group. Pathobiology of primary mediastinal B-cell lymphoma. *Leuk Lymphoma* **44**(Suppl 3), S21–26.
284. Calaminici M, Piper K, Lee AM, Norton AJ. (2004). CD23 expression in mediastinal large B-cell lymphomas. *Histopathology* **45**, 619–24.
285. Gilmore TD, Starczynowski DT, Kalaitzidis D. (2004). Relevant gene amplification in B-cell lymphomas? *Blood* **103**, 3243–4.
286. Hamlin PA, Portlock CS, Straus DJ, Noy A, Singer A, Horwitz SM, O'connor OA, Yahalom J, Zelenetz AD, Moskowitz CH. (2005). Primary mediastinal large B-cell lymphoma: Optimal therapy and prognostic factor analysis in 141 consecutive patients treated at Memorial Sloan Kettering from 1980 to 1999. *Br J Haematol* **130**, 691–9.
287. Murase T, Yamaguchi M, Suzuki R, Okamoto M, Sato Y, Tamaru J, Kojima M, Miura I, Mori N, Yoshino T, Nakamura S. (2007). Intravascular large B-cell lymphoma (IVLBCL): A clinicopathologic study of 96 cases with special reference to the immunophenotypic heterogeneity of CD5. *Blood* **109**, 478–85.
288. Chaukiyal P, Singh S, Woodlock T, Dolan JG, Bruner K. (2006). Intravascular large B-cell lymphoma with multisystem involvement. *Leuk Lymphoma* **47**, 1688–90.
289. Bouzani M, Karmiris T, Rontogianni D, Delimpassi S, Apostolidis J, Mpakiri M, Nikiforakis E. (2006). Disseminated intravascular B-cell lymphoma: Clinicopathological features and outcome of three cases treated with anthracycline-based immunochemotherapy. *Oncologist* **11**, 923–8.
290. Ferry JA, Harris NL, Picker LJ, Weinberg DS, Rosales RK, Tapia J, Richardson Jr. EP. (1988). Intravascular lymphomatosis (malignant angioendotheliomatosis). A B-cell neoplasm expressing surface homing receptors. *Mod Pathol* **1**, 444–52.
291. Ponzoni M, Ferreri AJ. (2006). Intravascular lymphoma: A neoplasm of "homeless" lymphocytes? *Hematol Oncol* **24**, 105–12.
292. Thomson JJ, Walt JV, Ireland R. (2007). Bone marrow trephine biopsy appearances of the intravascular subtype of diffuse large B-cell lymphoma. *Br J Haematol* **136**, 683.
293. Rashid R, Johnson RJ, Morris S, Dickinson H, Czyz J, O'Connor SJ, Owen RG. (2006). Intravascular large B-cell lymphoma associated with a near-tetraploid karyotype, rearrangement of BCL6, and a t(11;14)(q13;q32). *Cancer Genet Cytogenet* **171**, 101–4.
294. Navarro WH, Kaplan LD. (2006). AIDS-related lymphoproliferative disease. *Blood* **107**, 13–20.
295. Inoue S, Miyamoto T, Yoshino T, Yamadori I, Hagari Y, Yamamoto O. (2006). Primary effusion lymphoma with skin involvement. *J Clin Pathol* **59**, 1221–2.
296. Piccaluga PP, Vigna E, Placci A, Agostinelli C, Laterza C, Papayannidis C, Leone O, Martinelli G, Zinzani PL, Baccarani M, Pileri SA. (2006). Primary cardiac non-Hodgkin lymphoma presenting with atrial flutter and pericardial effusion. *Br J Haematol* **134**, 356.

297. Desai S, Freeman NJ. (2004). Primary effusion lymphoma (PEL) revisited: A hematologist's perspective. *Clin Adv Hematol Oncol* **2**, 60–1.
298. Carbone A, Gloghini A. (2008). KSHV/HHV8-associated lymphomas. *Br J Haematol* **140**, 13–24.
299. Carbone A, Gloghini A. (2006). HHV-8-associated lymphoma: State-of-the-art review. *Acta Haematol* **117**, 129–31.
300. Jenkins C, Sorour Y, Blake E, Elliot R, Al-Sabah AI, Green J. (2005). Human-immunodeficiency-virus-negative, human-herpes-virus-8-negative abdominal cavity primary effusion lymphoma. *Clin Oncol (R Coll Radiol)* **17**, 636–8.
301. Matsumoto Y, Nomura K, Ueda K, Satoh K, Yasuda N, Taki T, Yokota S, Horiike S, Okanoue T, Taniwaki M. (2005). Human herpesvirus 8-negative malignant effusion lymphoma: A distinct clinical entity and successful treatment with rituximab. *Leuk Lymphoma* **46**, 415–19.
302. Carbone A, Gloghini A. (2005). AIDS-related lymphomas: From pathogenesis to pathology. *Br J Haematol* **130**, 662–70.
303. John P, Doweiko JP, Groopman JE. (2007). AIDS-related lymphomas: Primary effusion lymphoma. *UpToDate*.
304. Chen YB, Rahemtullah A, Hochberg E. (2007). Primary effusion lymphoma. *Oncologist* **12**, 569–76.
305. Wakely Jr. PE, Menezes G, Nuovo GJ. (2002). Primary effusion lymphoma: Cytopathologic diagnosis using *in situ* molecular genetic analysis for human herpesvirus 8. *Mod Pathol* **15**, 944–50.
306. Fujiwara T, Ichinohasama R, Miura I, Sugawara T, Harigae H, Yokoyama H, Takahashi S, Tomiya Y, Yamada M, Ishizawa K, Kameoka J, Sasaki T. (2005). Primary effusion lymphoma of the pericardial cavity carrying t(1;22)(q21;q11) and t(14;17)(q32;q23). *Cancer Genet Cytogenet* **156**, 49–53.
307. Dong HY, Scadden DT, de Leval L, Tang Z, Isaacson PG, Harris NL. (2005). Plasmablastic lymphoma in HIV-positive patients: An aggressive Epstein–Barr virus-associated extramedullary plasmacytic neoplasm. *Am J Surg Pathol* **29**, 1633–41.
308. Teruya-Feldstein J. (2005). Diffuse large B-cell lymphomas with plasmablastic differentiation. *Curr Oncol Rep* **7**, 357–63.
309. Folk GS, Abbondanzo SL, Childers EL, Foss RD. (2006). Plasmablastic lymphoma: A clinicopathologic correlation. *Ann Diagn Pathol* **10**, 8–12.
310. Lin O, Gerhard R, Zerbini MC, Teruya-Feldstein J. (2005). Cytologic features of plasmablastic lymphoma. *Cancer* **105**, 139–44.
311. Tavora F, Gonzalez-Cuyar LF, Sun CC, Burke A, Zhao XF. (2006). Extra-oral plasmablastic lymphoma: Report of a case and review of literature. *Hum Pathol* **37**, 1233–6.
312. Rudzki Z, Rucińska M, Jurczak W, Skotnicki AB, Maramorosz-Kurianowicz M, Mruk A, Piróg K, Utych G, Bodzioch P, Srebro-Stariczyk M, Włodarska I, Stachura J. (2005). ALK-positive diffuse large B-cell lymphoma: Two more cases and a brief literature review. *Pol J Pathol* **56**, 37–45.
313. Gascoyne RD, Lamant L, Martin-Subero JI, Lestou VS, Harris NL, Müller-Hermelink HK, Seymour JF, Campbell LJ, Horsman DE, Auvigne I, Espinos E, Siebert R, Delsol G. (2003). ALK-positive diffuse large B-cell lymphoma is associated with Clathrin-ALK rearrangements: Report of 6 cases. *Blood* **102**, 2568–73.
314. Smedby KE, Baecklund E, Askling J. (2006). Malignant lymphomas in autoimmunity and inflammation: A review of risks, risk factors, and lymphoma characteristics. *Cancer Epidemiol Biomarkers Prev* **15**, 2069–77.
315. Aozasa K, Takakuwa T, Nakatsuka S. (2005). Pyothorax-associated lymphoma: A lymphoma developing in chronic inflammation. *Adv Anat Pathol* **12**, 324–31.
316. Andres E, Herbrecht R, Campos F, Marcellin L, Oberling F. (1997). Primary hepatic lymphoma associated with chronic hepatitis C. *Ann Med Interne (Paris)* **148**, 280–3.
317. Tun HW, Personett D, Baskerville KA, Menke DM, Jaeckle KA, Kreinest P, Edenfield B, Zubair AC, O'Neill BP, Lai WR, Park PJ, McKinney M. (2008). Pathway analysis of primary central nervous system lymphoma. *Blood* **111**, 3200–10.
318. Shuangshoti S, Assanasen T, Lerdlum S, Srikijvilaikul T, Intragumtornchai T, Torner PS. (2008). Primary central nervous system plasmablastic lymphoma in AIDS. *Neuropathol Appl Neurobiol* **34**, 245–7.
319. Rubenstein JL, Fridlyand J, Shen A, Aldape K, Ginzinger D, Batchelor T, Treseler P, Berger M, McDermott M, Prados M, Karch J, Okada C, Hyun W, Parikh S, Haqq C, Shuman M. (2006). Gene expression and angiotropism in primary CNS lymphoma. *Blood* **107**, 3716–23.
320. Slater DN. (2005). The new World Health Organization-European Organization for Research and Treatment of Cancer classification for cutaneous lymphomas: A practical marriage of two giants. *Br J Dermatol* **153**, 874–80.
321. Goodlad JR, Krajewski AS, Batstone PJ, McKay P, White JM, Benton EC, Kavanagh GM, Lucraft HH. (2003). Scotland and Newcastle Lymphoma Group. Primary cutaneous diffuse large B-cell lymphoma: Prognostic significance of clinicopathological subtypes. *Am J Surg Pathol* **27**, 1538–45.
322. Prince HM, Yap LM, Blum R, McCormack C. (2003). Primary cutaneous B-cell lymphomas. *Clin Exp Dermatol* **28**, 8–12.
323. Blum KA, Lozanski G, Byrd JC. (2004). Adult Burkitt leukemia and lymphoma. *Blood* **104**, 3009–20 [Epub 2004].
324. Ferry JA. (2006). Burkitt's lymphoma: Clinicopathologic features and differential diagnosis. *Oncologist* **11**, 375–83.
325. Bociek RG. (2005). Adult Burkitt's lymphoma. *Clin Lymphoma* **6**, 11–20.
326. Mossafa H, Damotte D, Jenabian A, Delarue R, Vincenneau A, Amouroux I, Jeandel R, Khoury E, Martelli JM, Samson T, Tapia S, Flandrin G, Troussard X. (2006). Non-Hodgkin's lymphomas with Burkitt-like cells are associated with c-Myc amplification and poor prognosis. *Leuk Lymphoma* **47**, 1885–93.
327. Jarrett RF. (2006). Viruses and lymphoma/leukaemia. *J Pathol* **208**, 176–86.
328. Bellan C, Lazzi S, De Falco G, Nyongo A, Giordano A, Leoncini L. (2003). Burkitt's lymphoma: New insights into molecular pathogenesis. *J Clin Pathol* **56**, 188–92.
329. Bell A, Rickinson AB. (2003). Epstein–Barr virus, the TCL-1 oncogene and Burkitt's lymphoma. *Trends Microbiol* **11**, 495–7.
330. Haralambieva E, Boerma EJ, van Imhoff GW, Rosati S, Schuuring E, Muller-Hermelink HK, Kluin PM, Ott G. (2005). Clinical, immunophenotypic, and genetic analysis of adult lymphomas with morphologic features of Burkitt lymphoma. *Am J Surg Pathol* **29**, 1086–94.
331. Basso K, Frascella E, Zanesco L, Rosolen A. (1999). Improved long-distance polymerase chain reaction for the detection of t(8;14)(q24;q32) in Burkitt's lymphomas. *Am J Pathol* **155**, 1479–85.
332. Dave SS, Fu K, Wright GW, Lam LT, Kluin P, Boerma EJ, Greiner TC, Weisenburger DD, Rosenwald A, Ott G, Müller-Hermelink HK, Gascoyne RD, Delabie J, Rimsza LM, Braziel RM, Grogan TM, Campo E, Jaffe ES, Dave BJ, Sanger W, Bast M, Vose JM, Armitage JO, Connors JM, Smeland EB, Kvaloy S, Holte H, Fisher RI, Miller TP, Montserrat E, Wilson WH, Bahl M, Zhao H, Yang L, Powell J, Simon R, Chan WC, Staudt LM. (2006). Lymphoma/leukemia molecular profiling project. Molecular diagnosis of Burkitt's lymphoma. *N Engl J Med* **354**, 2431–42.
333. Gerecitano J, Straus DJ. (2006). Treatment of Burkitt lymphoma in adults. *Expert Rev Anticancer Ther* **6**, 373–81.
334. Kasamon YL, Swinnen LJ. (2004). Treatment advances in adult Burkitt lymphoma and leukemia. *Curr Opin Oncol* **16**, 429–35.
335. Harris NL, Jaffe ES, Diebold J, Flandrin G, Muller-Hermelink HK, Vardiman J, Lister TA, Bloomfield CD. (1999). World Health Organization classification of neoplastic diseases of the hematopoietic and lymphoid tissues: Report of the Clinical Advisory Committee meeting-Airlie House, Virginia, November 1997. *J Clin Oncol* **17**, 3835–49.

16

CHAPTER

Plasma Cell Myeloma and Related Disorders

Faramarz Naeim,
P. Nagesh Rao
and
Wayne W. Grody

GENERAL CONSIDERATIONS

The monoclonal proliferation of plasma cells, also known as "monoclonal gammopathies," "plasma cell dyscrasias," or "paraproteinemias," consists of several clinicopathologic entities. It is characterized by a single class of immunoglobulin (Ig) product, referred to as "M-component" (monoclonal component), present in serum and/or urine protein electrophoresis (Figure 16.1) [1–3]. In addition to protein electrophoresis, other more sensitive and specific techniques, such as immunoelectrophoresis and immunofixation, have been used to detect the specific class or subtype of monoclonal Ig in these disorders (Figures 16.2 and 16.3).

The following categories are discussed in this chapter. Waldenstrom macroglobulinemia (lymphoplasmacytic lymphoma) has already been discussed in Chapter 15.

- Monoclonal gammopathy of undetermined significance (MGUS)
- Plasma cell myeloma (PCM) and variants
- Plasmacytoma and variants
- Monoclonal Ig deposit diseases
- Heavy chain diseases.

Etiology and Pathogenesis

The etiology of these related disorders is not known. Several environmental and genetic factors have been associated with the increased risk of PCM. Increased incidence of PCM has been observed in the Japanese survivors of atomic bomb and in radiologists exposed to long-term low doses of radiation [4–6]. Occupations in agriculture and metal industries and exposure to benzene and hair dyes have been reported to be in association with increased risk of PCM [7–9]. There are also reports linking obesity and diet to the increased risk of PCM [10, 11].

There are several reports of familial clusters of PCM affecting two or more first-degree relatives suggesting that hereditary factors may be involved in the development of monoclonal gammopathies in certain conditions [12–14]. Viral infections, such as hepatitis C, HIV, and HHV-8 (Kaposi-sarcoma-associated herpes virus), have been associated with increased risk of PCM [15–17].

Dysregulation in immune function may also play a role in the pathogenesis of monoclonal plasma cell disorders [18–20].

Molecular genetic studies have shed some light on the pathogenesis of these disorders (Figure 16.4) [21, 22]. A broad spectrum of abnormalities have been identified in the signaling pathways, cell cycle, apoptotic processes, and bone marrow microenvironment. For example, the monoclonal plasma cells are capable of producing IL-6 and IL-6 receptor, leading to autocrine stimulation [23–25]. IL-6 activates two pathways, the *JAK2-STAT3* and the *RAS-MAP*. The first pathway upregulates anti-apoptotic proteins Mcl-1 and Bcl-X, and the second pathway leads to the upregulation of transcription factors, such as ELK-1 and AP-1 [25–28].

Gene expression profiling studies have identified three genes which may play important roles in pathogenesis and clinical outcome of PCM: *RAN*, *CHC1L*, and *ZHX-2* [29]. The *RAN* gene maps to 12q24, is a member of the Ras family, and has a role in regulating chromosome condensation, spindle formation, and cell-cycle progression. The *CHC1L* gene maps

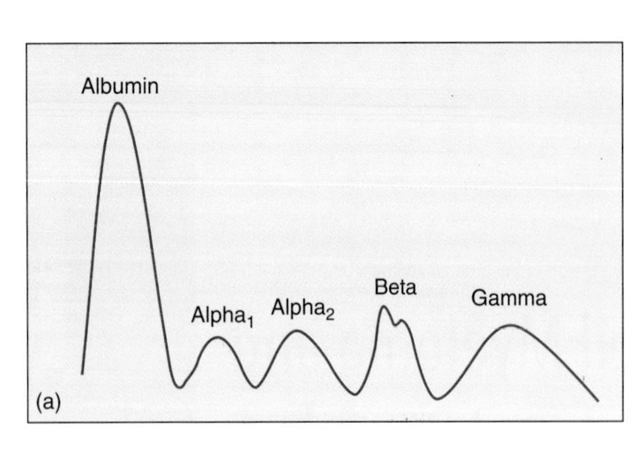

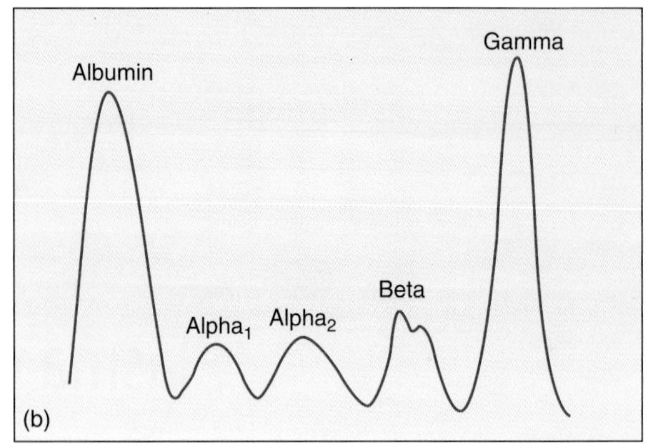

FIGURE 16.1 Schematics of serum protein electrophoresis: (a) demonstrates a normal profile and (b) shows a spike in the gamma region.

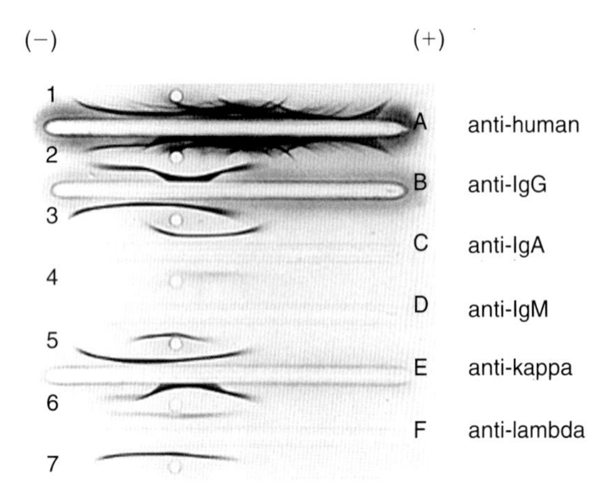

FIGURE 16.2 Serum immunoglobulin electrophoresis with a panel of IgG, IgA, IgM, kappa, and lambda antibodies demonstrating an IgG/kappa spike. Wells 1–7 contain control serum (odd number) and patient serum (even number). From Bossuyt X, Bogaerts A, Schiettekatte G, Blanckaert N. (1998). Serum protein electrophoresis and immunofixation by a semiautomated electrophoresis system. *Clin Chem* **44**, 944–9.

ELP G A M K L

FIGURE 16.3 Immunofixation analysis of a serum demonstrating IgA-kappa monoclonal gammopathy. Courtesy of Eugene Dinovo, Ph.D., Department of Veterans Affairs, Greater Los Angeles Healthcare System.

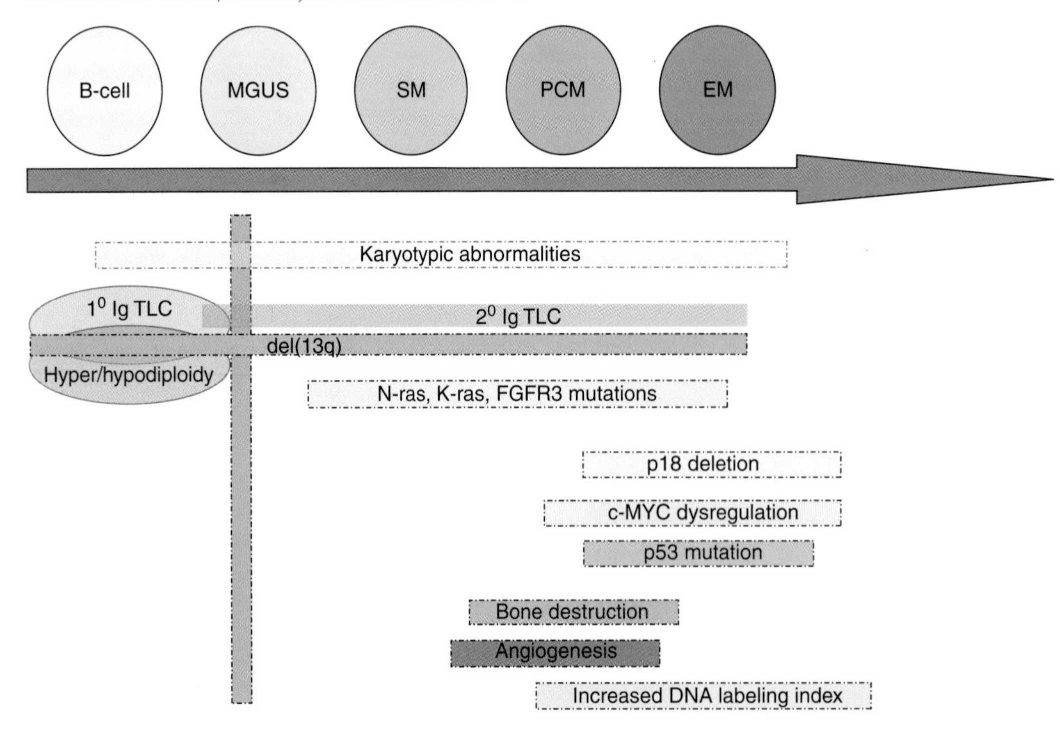

FIGURE 16.4 Schematic stages of molecular, cytogenetic, and clinical events in plasma cell disorders from monoclonal gammopathy with undetermined significance (MGUS) to smoldering myeloma (SM), plasma cell myeloma (PCM), and extramedullary myeloma (EM). Adapted from Ref. [21].

to 13q14.3 and appears to harbor a tumor suppressor gene for PCM. The *ZHX-2* gene maps to 8q24, near the *MYC* oncogene, and is a negative regulator of the NF-Y transcription factor. This factor is a master transcriptional regulator of numerous genes involved in cell-cycle control and proliferation.

The levels of cyclin D1, cyclin D2, or cyclin D3 mRNA in the tumor cells of almost all patients with MGUS and PCM are relatively high. Increased expression of one of the cyclin D proteins facilitates activation of CDK4, leading to inactivation of the *RB1* (retinoblastoma 1) gene, and consequently cell-cycle progression [21].

Pathology

Morphology

The morphologic hallmark of PCM and related disorders is an increase in bone marrow plasma cells. The neoplastic plasma cells have a tendency to appear in clusters or sheets in the biopsy sections and, unlike reactive plasmacytosis, are not located around the vascular structures (Figure 16.5). The bone marrow biopsy sections are usually hypercellular, but occasionally they may appear hypocellular (Figure 16.6). Prominent osteoclastic activity may be evident, adjacent to the bony trabeculae.

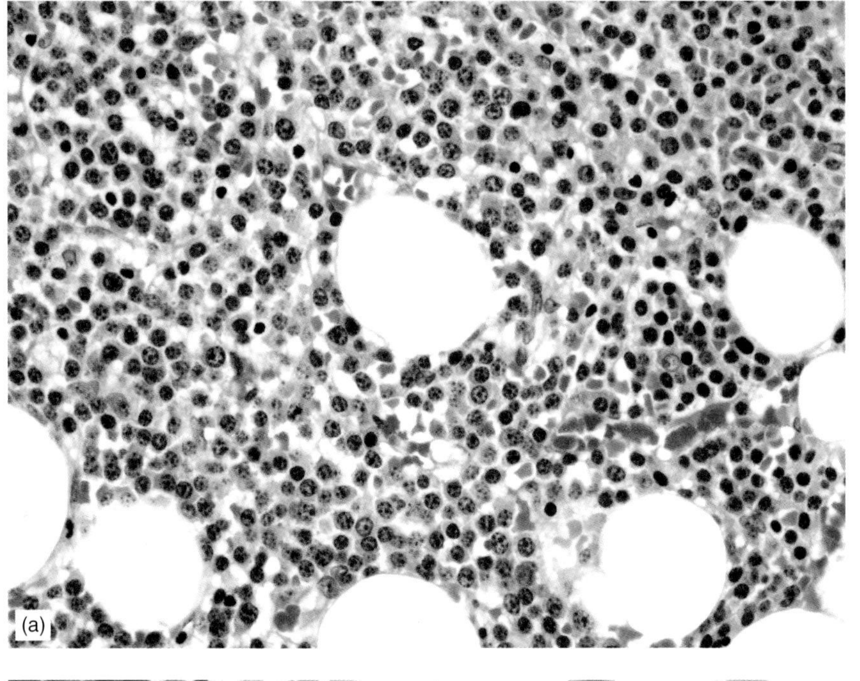

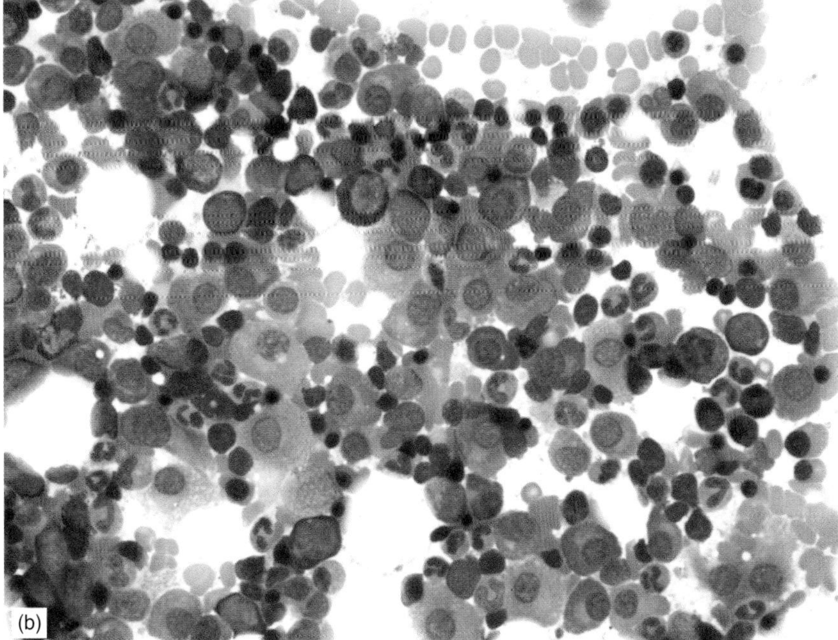

FIGURE 16.5 Plasma cell myeloma. Bone marrow biopsy section (a) and bone marrow smear (b) demonstrating numerous plasma cells.

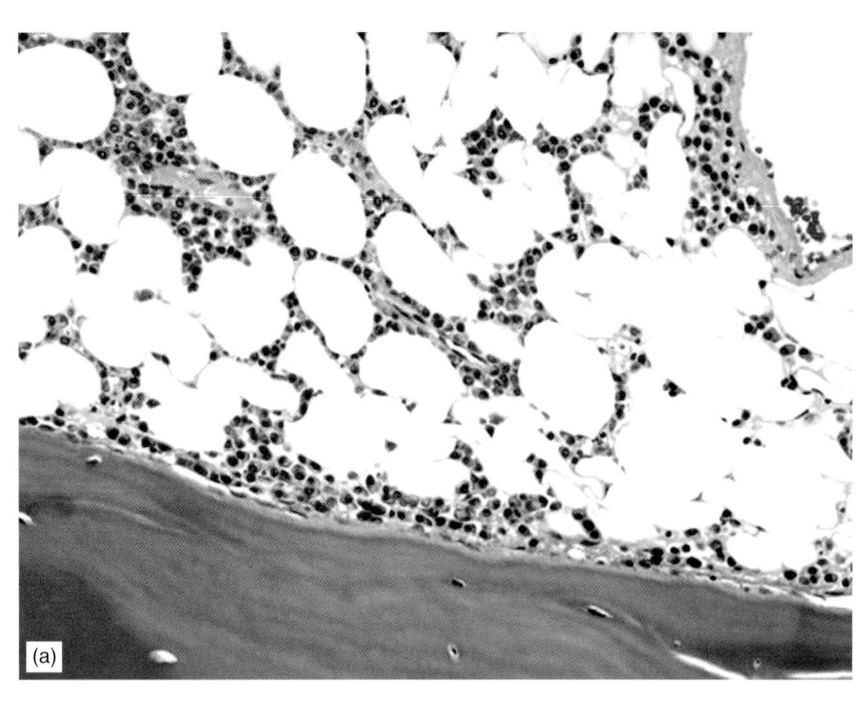

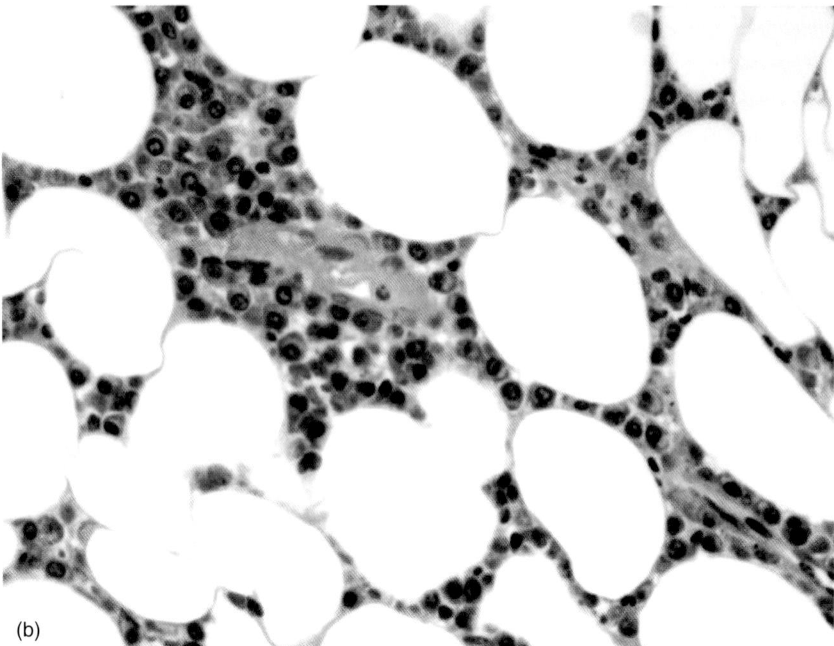

FIGURE 16.6 Hypocellular plasma cell myeloma with interstitial infiltration of plasma cells around fatty tissue: (a) low power and (b) high power views.

The cytologic features of plasma cells in the bone marrow smears may vary from normal-appearing mature plasma cells to immature and anaplastic forms (Figures 16.7–16.9). Plasmablasts show a high nuclear:cytoplasmic ratio, deep blue cytoplasm, with or without perinuclear hof, round or irregular nucleus, fine chromatin, and one or several prominent nucleoli (Figure 16.9). Multinucleated or multilobated plasma cells may be present. Cells with cherry-red, round cytoplasmic (Russell bodies) or nuclear (Dutcher bodies) inclusions, as well as cytoplasmic crystals may be present (Figures 16.10 and 16.11). Some plasma cells may appear like grapes and demonstrate numerous, round, Ig-containing cytoplasmic structures (Mott and Morula cells) (Figure 16.12). The blood smears may show rouleaux formation of the red blood cells or the presence of plasma cells in various proportions (Figure 16.9c). Solitary neoplasms of plasma cells (plasmacytoma) may involve bone or other extramedullary sites.

Immunophenotype

Plasma cells in myeloma and related disorders, similar to normal plasma cells, lack surface Ig and express cytoplasmic Ig, CD38, CD138, and CD79a. CD138 is more specific than CD38 in the detection of plasma cells (Figures 16.13 and 16.14) [30–34]. However, the CD138 molecules disintegrate and disappear fast and therefore may not be detected

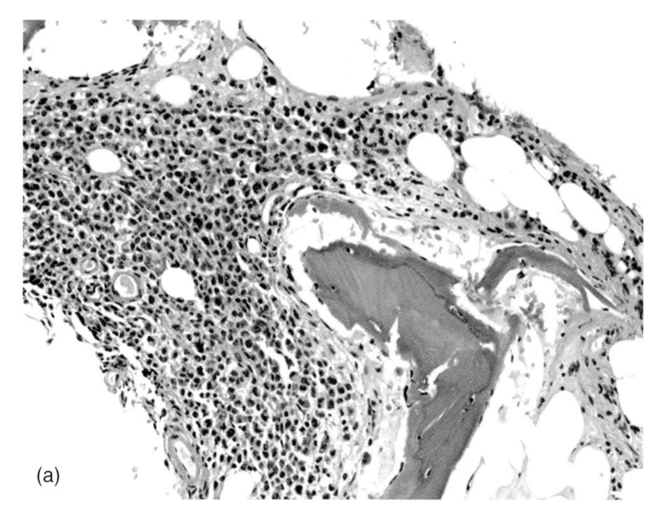

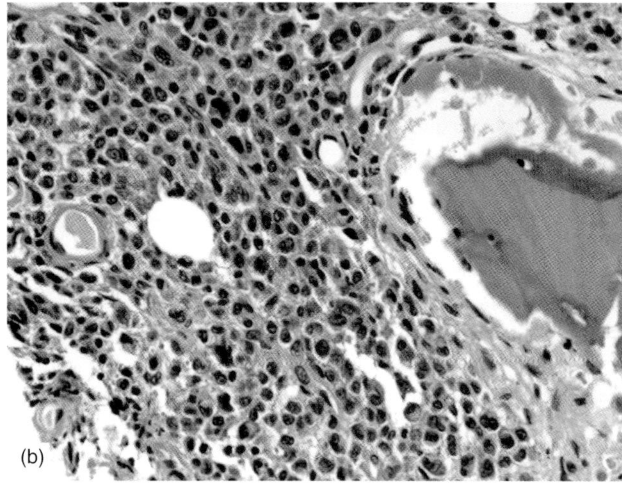

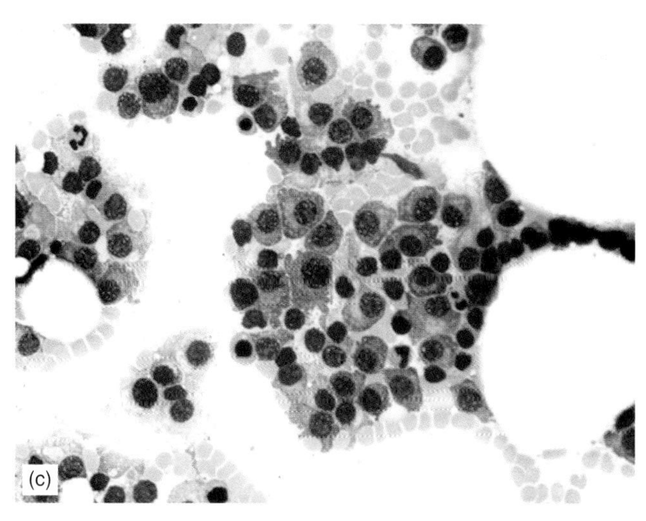

FIGURE 16.7 Plasma cell myeloma. Bone marrow biopsy section demonstrating a paratrabecular aggregate of plasma cells: (a) low power and (b) high power views. Bone marrow smear shows clusters of plasma cells (c).

by flow cytometry if the specimen is not prepared immediately. Also, most flow cytometry laboratories have experienced a significant loss in the number of plasma cells during the cell preparation for the flow cytometry procedures. Therefore, if the percentage of plasma cells in the bone marrow smears is not high, immunohistochemistry on the bone marrow biopsy sections is recommended.

Plasma cells in PCM and related disorders are monoclonal and express only one type of Ig light chain. The class of Ig is usually IgG, less frequently IgA, and occasionally IgD, IgE, or IgM. The neoplastic plasma cells often lack CD19 and CD20 expressions but may aberrantly express CD10, CD56, or, less frequently, myelomonocytic or T-cell-associated markers [30–34]. Both normal and neoplastic plasma cells may express CD117.

Cytogenetic and Molecular Studies

PCM is a genetically unstable malignancy of post-germinal center B-lineage cells. Chromosome analysis in PCM is hampered by the low proliferative fraction in most cases. Recent studies using cytokine-stimulated bone marrow cultures and FISH and microarray techniques have increased the proportion of informative cases.

Active myeloma is often preceded by an indolent phase of MGUS, where the plasma cells are already abnormal with an aneuploid DNA content. In fact, almost all myeloma tumors and most cases of MGUS are aneuploid as demonstrated by DNA content measurements using flow cytometry, conventional cytogenetics, or molecular cytogenetics (FISH) [35, 36].

By classical cytogenetics, only one-third of PCM patients have a complex abnormal karyotype. The remaining two-thirds have normal karyotypes [37, 38]. However, the observed normal karyotypes are often derived from the other non-neoplastic hematopoietic cells, rather than from abnormal plasma cells, because the plasma cells fail to grow. Plasma cells may fail to grow in culture for three major reasons. First, samples from PCM patients may fail to grow because of the low proliferative capacity of the myeloma cells [39]. Myeloma cells (especially in early myeloma) are stroma dependent, so removing the cells from their supportive microenvironment results in apoptosis and lack of growth. However, if myeloma cells have become stroma independent (i.e. in advanced stages of the disease), removal of the myeloma cells from their microenvironment can result in proliferation and an abnormal karyotype [39]. The second explanation for the laboratory's inability to obtain abnormal metaphases lies in the quality of the bone marrow aspirates received for cytogenetic studies. Aspirates frequently contain drastically fewer plasma cells than the corresponding smears used for morphological assessment since the number of tumor cells in a given specimen largely depends on the level of local bone marrow infiltration and the degree of sample dilution by blood [38]. For this reason, it is essential that the first few milliliters of the bone marrow drawn be sent for cytogenetic analysis. Also, the needle should be repositioned during aspiration, rather than simply continuing to withdraw marrow from the initial puncture site, to ensure that adequate numbers of abnormal cells are submitted to the laboratory. Finally, aspirates should be processed as soon as possible if FISH is requested. Several techniques have been created to selectively culture plasma cells [40], but these methods may not be helpful if the sample provided to the laboratory is of poor quality.

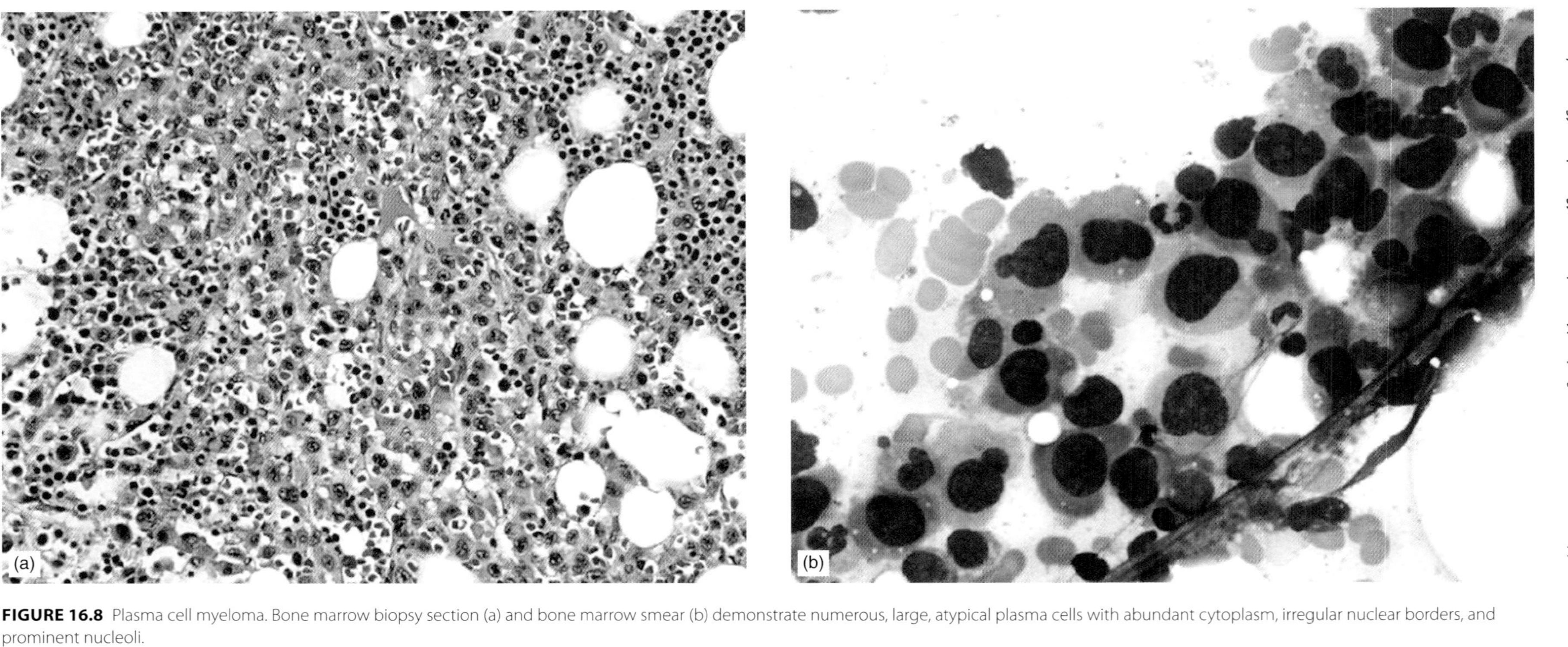

FIGURE 16.8 Plasma cell myeloma. Bone marrow biopsy section (a) and bone marrow smear (b) demonstrate numerous, large, atypical plasma cells with abundant cytoplasm, irregular nuclear borders, and prominent nucleoli.

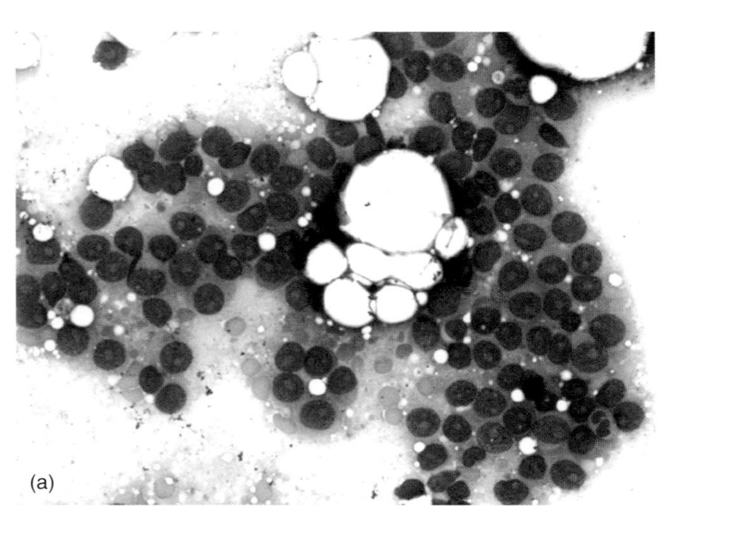

FIGURE 16.9 Plasmablastic myeloma. Bone marrow smear demonstrating cells with variable amounts of blue cytoplasm, round nuclei with finely dispersed chromatin, and a single prominent nucleolus: (a) low power and (b) high power views.

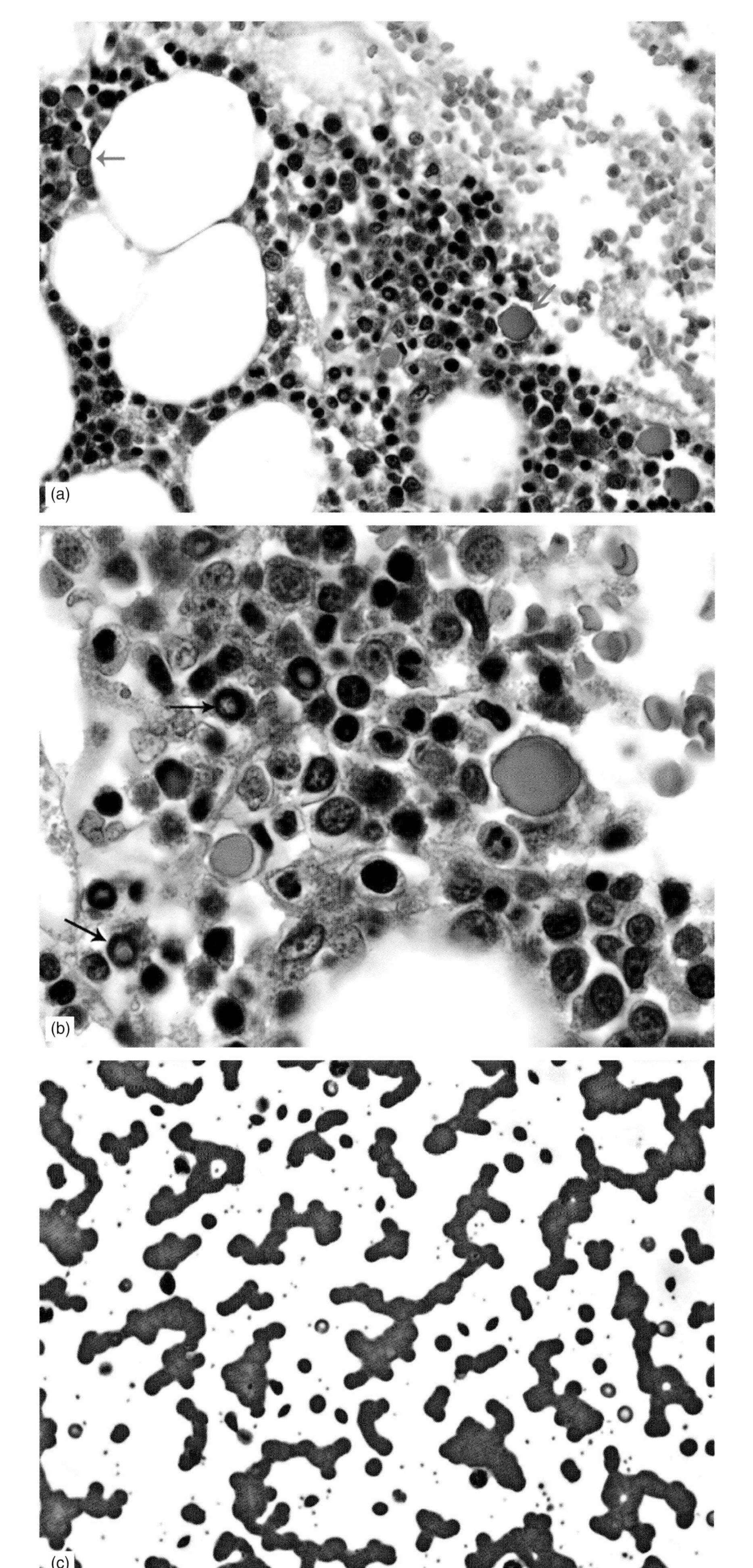

FIGURE 16.10 Plasma cell myeloma. Bone marrow biopsy section demonstrating plasma cells with nuclear inclusions (Dutcher bodies, black arrows) and cytoplasmic inclusions (Russell bodies, green arrows): (a) low power and (b) high power views. Blood smear showing formations (c).

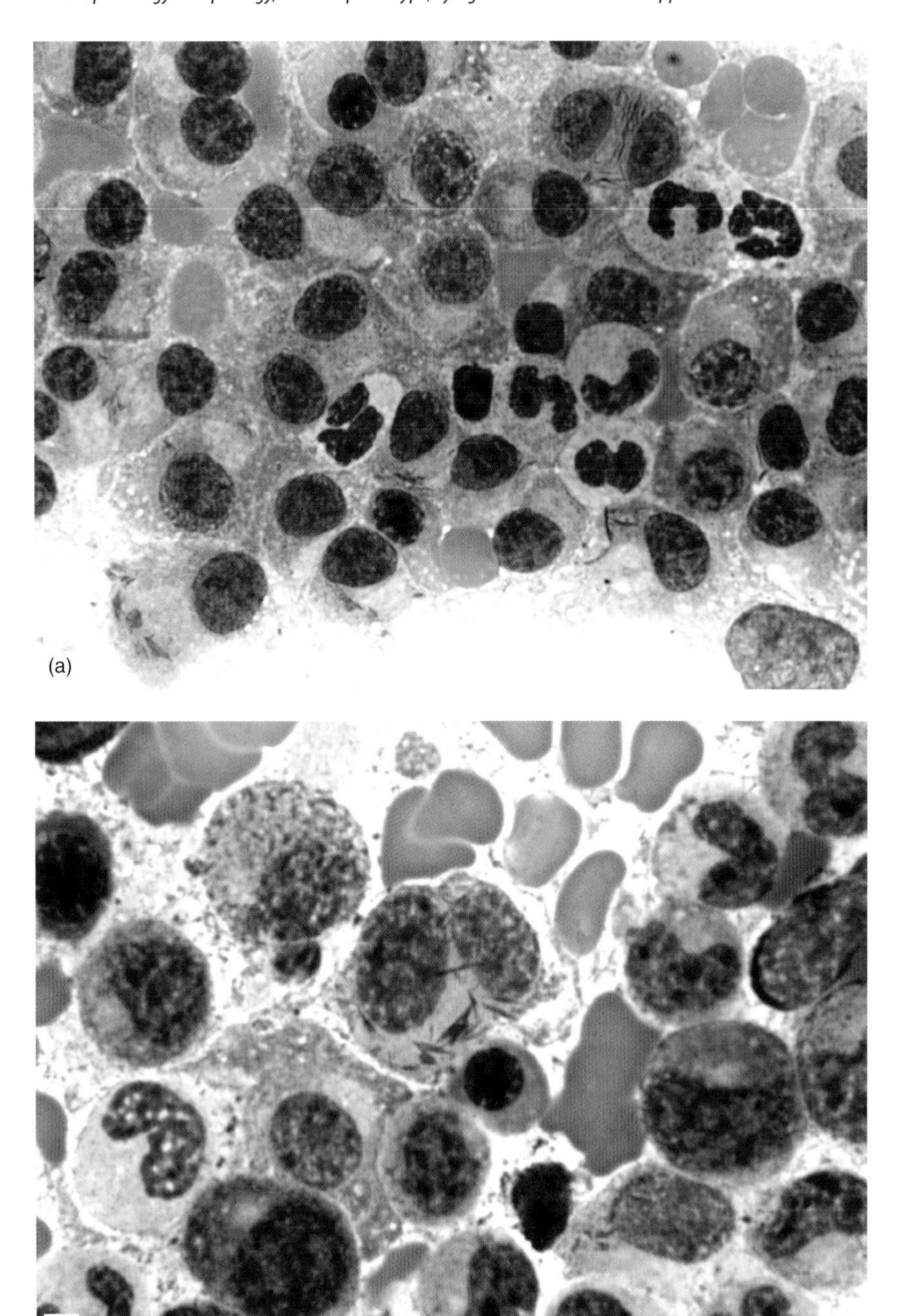

FIGURE 16.11 Plasma cell myeloma. Bone marrow smear demonstrating plasma cells with cytoplasmic needle-like immunoglobulin crystals: (a) low power and (b) high power views.

Several recurring aberrations are observed in karyotypically abnormal PCM. The majority of PCM cases demonstrate chromosomal aneuploidy. Four categories of aneuploidy can be defined by karyotyping:

1. Hypodiploidy (<46 chromosomes) (Figure 16.15).
2. Pseudodiploidy (46–47 chromosomes but with structural or numerical abnormalities).
3. Hyperdiploidy (>50 chromosomes, HRD) (Figure 16.16).
4. Near-tetraploidy (>75 chromosomes).

Multiple non-random trisomies are associated with hyperdiploid tumors [41], especially trisomies of odd-numbered chromosomes [40]. It was initially believed that trisomies were more common than monosomies in PCM, but the opposite is true. The most common trisomies are of chromosomes 3, 5, 7, 9, 11, 15, 19, and 21 [42]. The most common monosomies are of chromosomes 13, 14, 16, and 22. No specific numerical chromosomal abnormality is constant or predictive of disease progression. The prevalence of aneuploidy is independent of stages. Karyotypes are typically complex and exhibit >10 abnormalities in almost half of the patients and even >20 aberrations in about 10% of the cases. HRD, present in nearly 60% of PCM, is most often associated with trisomies of chromosomes 3, 5, 7, 9, 11, 15, 19, and 21 and represents one of the central genetic

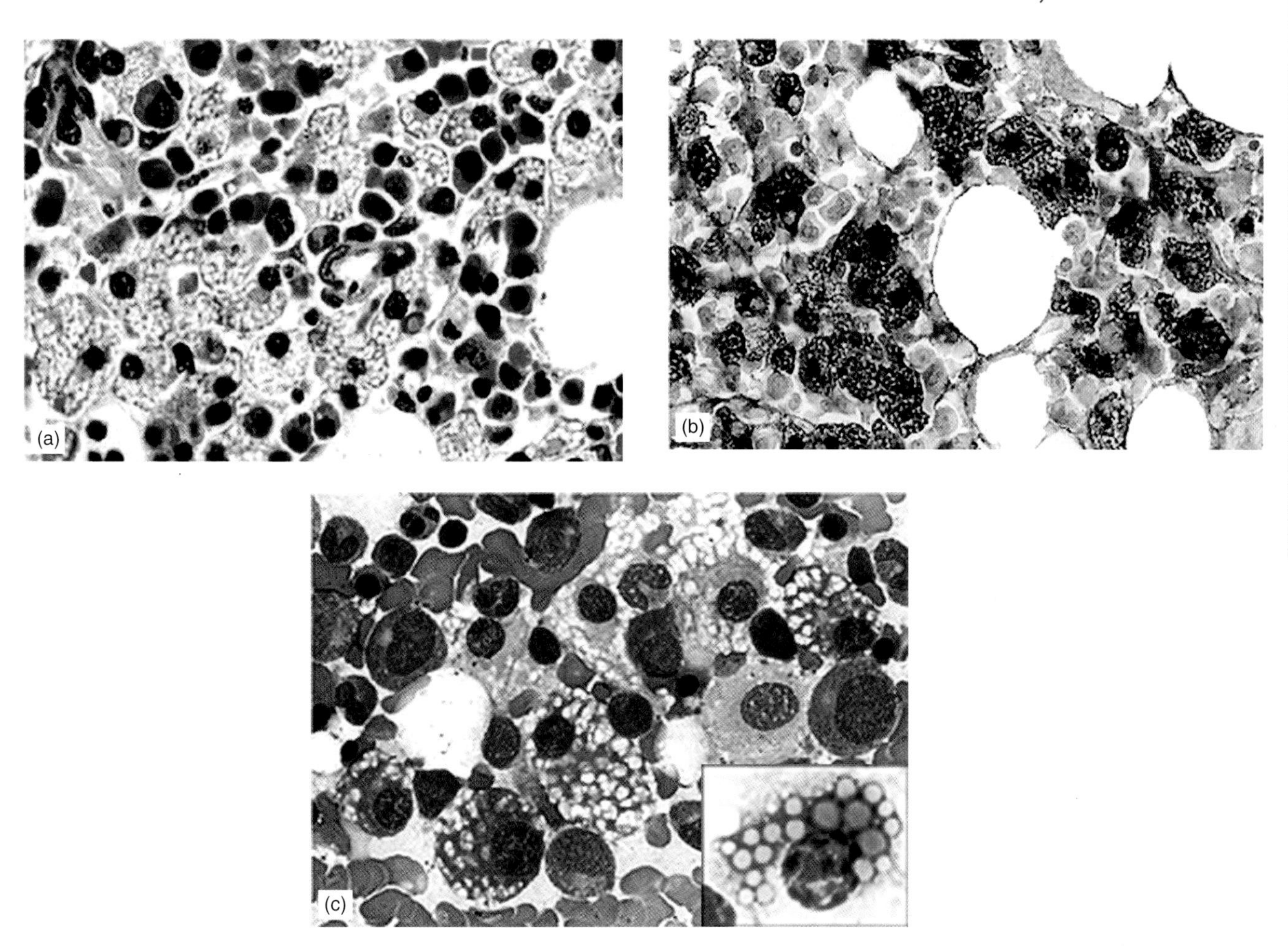

FIGURE 16.12 Plasma cell myeloma with cytoplasmic vacuoles. (a) Bone marrow biopsy section, (b) immunohistochemical stain for Ig kappa light chain, and (c) bone marrow smear. From Naeim F. (1997). *Pathology of Bone Marrow*, 2nd ed. Williams & Wilkins, Baltimore, by permission.

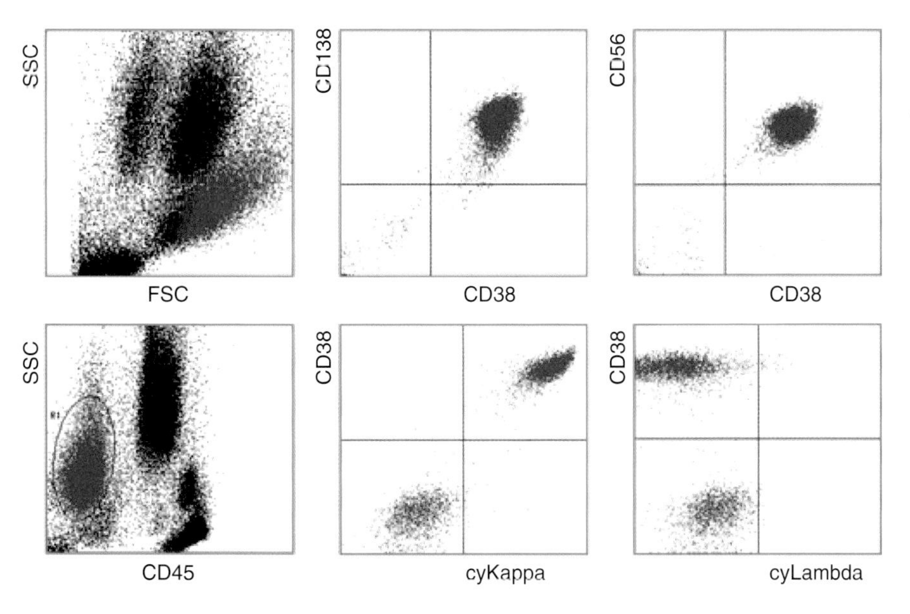

FIGURE 16.13 Flow cytometry of plasma cell myeloma demonstrating a population of cells (red) which are negative for CD45 and express CD38, CD138, CD56, and cytoplasmic kappa light chain.

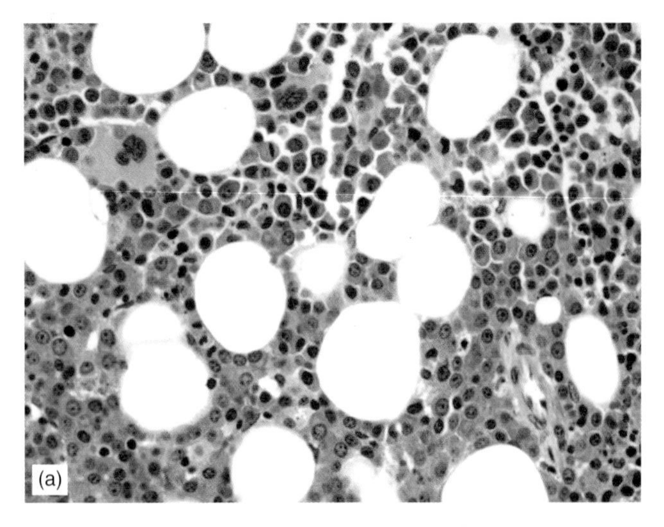

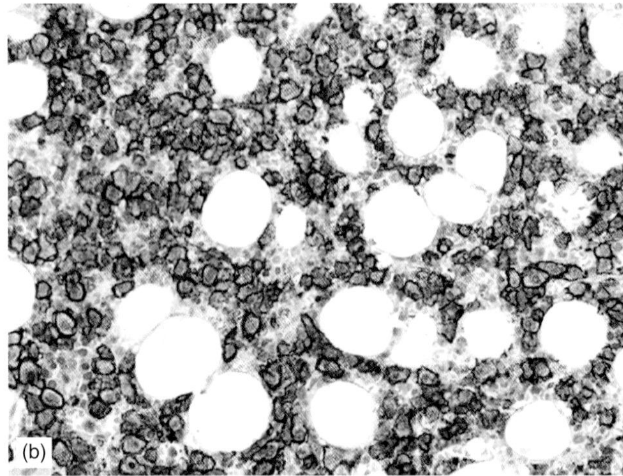

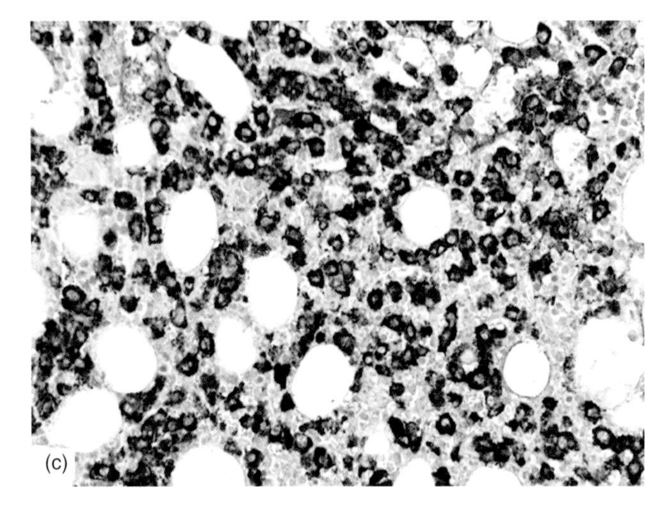

FIGURE 16.14 Plasma cell myeloma. Bone marrow biopsy section demonstrating a large population of plasma cells (a). Immunohistochemical stains show expression of CD138 (b) and kappa light chain (c) by the plasma cells.

pathways in the development of PCM, and this type of disease has been previously shown to have distinct gene expression signature [21]. Hyperdiploidy is seen in about 40% of PCM patients, with an incidence of 25% of the hypo- and pseudodiploid karyotypes. Triploidy and tetraploidy are very rare, usually not observed over 10%. Although the reasons are still unclear, the frequency of HRD has been shown to be more common in elderly patients [43].

The vast majority of PCM involves the translocation of the *IGH* gene on 14q32 to one of several non-random partners (~40 gene regions) and this is considered to be an initial event in the genesis of PCM (seen in about 40% of patients) [40]. Standard cytogenetic analyses identify abnormalities of 14q32 in up to 40% of cases (Figures 16.17–16.19), while this detection rate nearly doubles with interphase FISH studies. Illegitimate recombination of the *IGH* gene also occurs in MGUS but at a slightly lower frequency (~50%). Similar rearrangements of the *IGK* and *IGL* genes have been found to occur in a small subset of PCM and MGUS [44]. The rearrangements involving the *IGH* gene are commonly simple reciprocal translocations, but more complex recombination events, such as insertions and duplications, are also observed (Tables 16.1 and 16.2).

Cyclin D1 (most common), D2, and D3 genes (*CCND1*, *CCND2*, and *CCND3*) on chromosomes 11q13, 12p13, and 6p21, respectively, MAF family member genes (*c-MAF*, *MAFA*, and *MAFB*) on chromosomes 16q23, 8q24, and 20q11, respectively, and the fibroblast growth factor receptor 3 gene (*FGFR3*) on chromosome 4p16 are commonly observed as *IGH* translocation partners [1]. These translocations are markers for distinct subtypes of myeloma with important prognostic implications. *IGH* gene translocations are found more frequently in non-HRD tumors (70%) than in HRD tumors (20%). The t(11;14) (Figure 16.16) and t(4;14) are the most common *IGH* translocations followed by t(14;16) (Figure 16.19), and t(14;20) is the least common [38, 40]. The t(14;16)(q32;q23) and t(14;20)(q32;q11.2) result in the activation of *c-MAF* and *MAFB* proto-oncogenes, respectively, and are together seen in approximately 6% of cases. The reciprocal t(4;14)(p16;q32) translocation results in the hyperactivation of both the *FGFR3* and *MMSET* genes. Two cyclin D family members are activated by translocations in PCM: cyclin D1 by the t(11;14)(q13;q32) in 17% and *CCND3* by t(6;14)(p21;q32) (Figure 16.18) in 2%. The overall rate of 14q32 translocations, however, significantly increases with disease progression and reaches up to 90% in advanced tumors and human myeloma cell lines. *IGH* translocation and HRD act similarly through the upregulation of one of the cyclins (D1, D2, or D3) [39, 41]. Translocations between the Ig heavy chain locus and *CCND1*, *CCND3*, *c-MAF*, *MAFB*, *FGFR3*, and *MMSET* represent recurrent genetic lesions in approximately 40% of PCM [45, 46].

Although numerical and gross structural changes can be diagnosed by karyotyping without difficulty, small interstitial deletions, partial genomic gains, and cryptic translocations (e.g. *IGH* translocations) can be easily overlooked due to the karyotype's limited spatial resolution. Modern molecular-based techniques, such as CGH and FISH, allow the detection of genetic abnormalities independent of proliferating cells. With these methods, chromosomal aberrations are found in >90% of patients with PCM and most (if not all) patients with MGUS. Considering that analyses with interphase FISH and molecular genetic techniques reveal

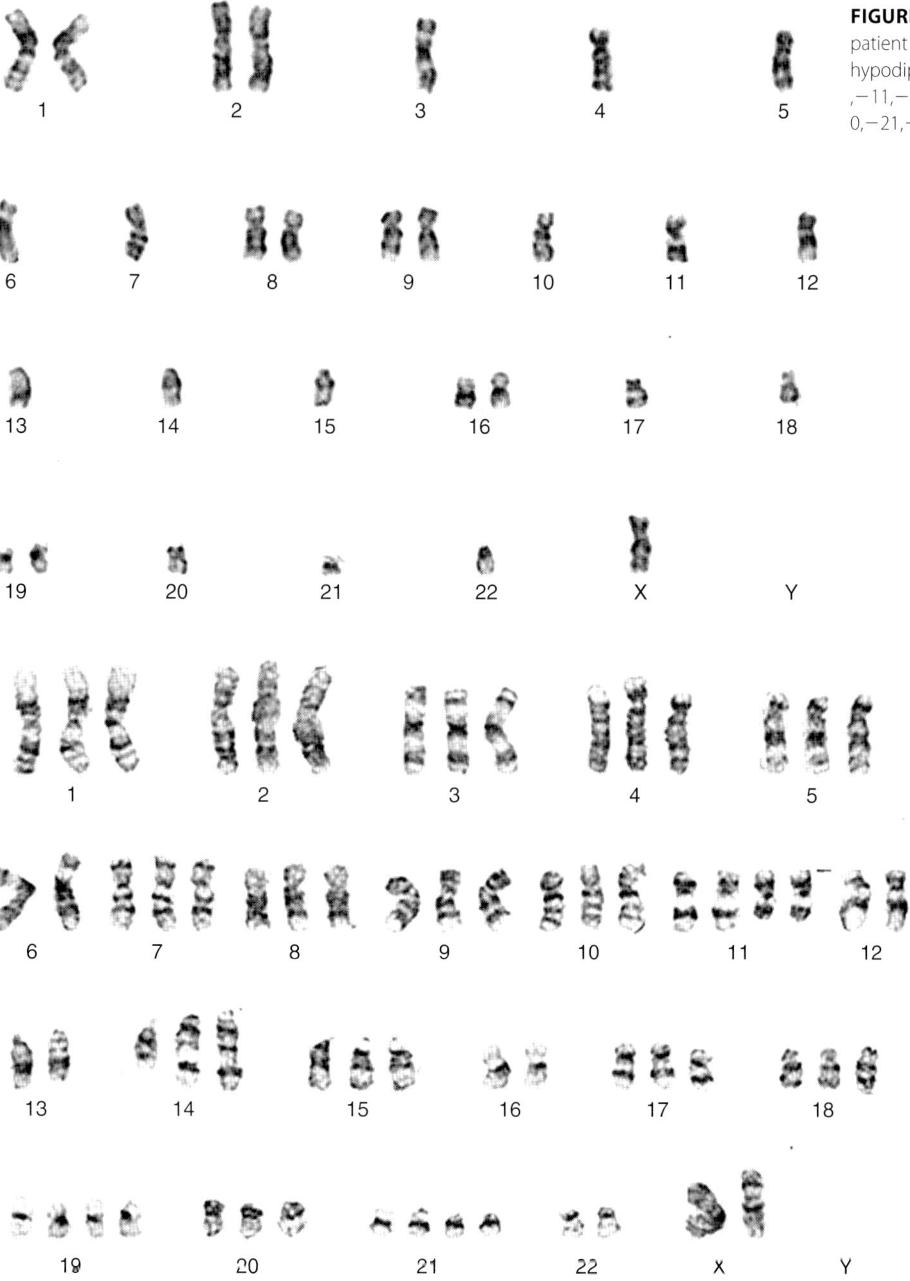

FIGURE 16.15 Bone marrow karyotype in a patient with plasma cell myeloma demonstrating hypodiploidy with 30,X,−X,−3,−4,−5,−6,−7,−10,−11,−12,−13,−14,−15,−17,del(17)(p13),−18,−20,−21,−22.

FIGURE 16.16 Bone marrow karyotype in a patient with plasma cell myeloma demonstrating hyperdiploidy with 66,XX,−X,−6,del(11)(q23), +del(11)(q23), −12,−13,der(14)t(11;14)(q13;q32)×2, −16, +19, +21,−22.

chromosomal abnormalities in close to 100% of PCM cases and in the majority of MGUS cases, the normal karyotypes are most likely not representative of the neoplastic clone.

FISH permits reliable identification of both translocations and small deletions or gains in PCM [4]. Most clinical laboratories currently test for 13q14 (*RB1*) and 17p13.1 (*TP53*) deletions as well as the primary translocations t(4;14)(p16.3;q32) and t(11;14)(q13;q32). Ploidy should be determined in all tumors; for example, disomy of chromosome band 13q14 in a near-tetraploid karyotype is functionally a deletion of the region. Polyploidy can be reliably excluded by the use of control probes mapped to genomic regions that rarely display aneuploidy (e.g. chromosomes 2, 10, and 12) [35, 37]. Finally, some laboratories test for t(6;14)(p21;q32), t(14;16)(q32;q23), and t(14;20), and detection of the most frequent chromosomal abnormalities (e.g. +1q, +9q, +11q) [36].

Cytogenetics is helpful in determining myeloma patients' clinical outcome [36]. Normal metaphases and normal FISH, HRD tumors, and *CCND1* gene activation are associated with a better prognosis [1]. However, patients with *c-MAF*, *MAFB*, or *FGFR3* activation, del(13q), del(17p), hypodiploidy, 1q abnormalities, or 9q trisomies are associated with a worse prognosis [35, 39].

A strong association between chromosome 1q abnormalities and the etiology of PCM disease has been suggested. Tandem duplications and jumping translocations of 1q21 occur frequently in this malignancy [46]. The gains of 1q is one of the most common abnormalities in PCM [42]. In addition, hyperdiploidy with 1q was found to have a less

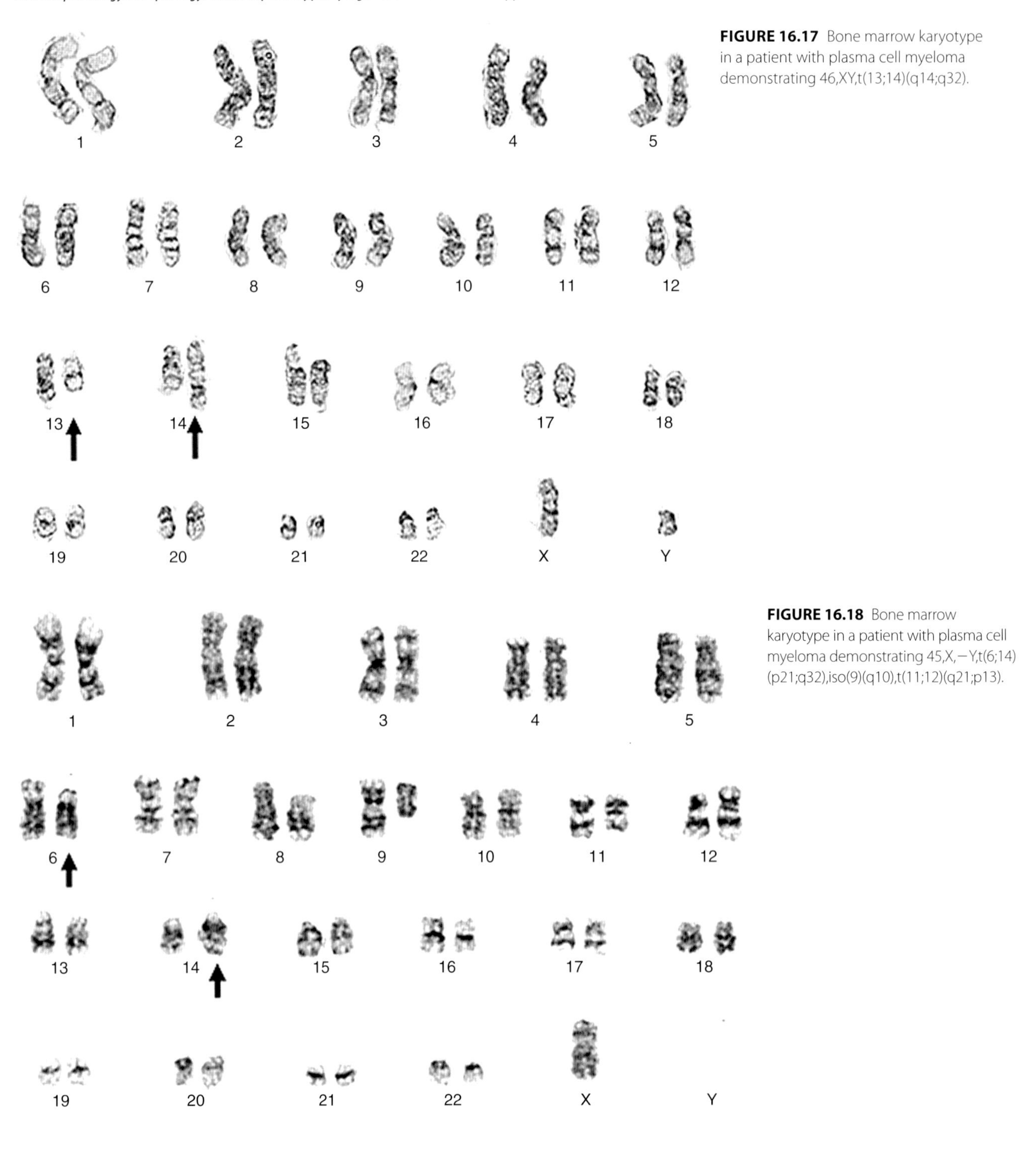

FIGURE 16.17 Bone marrow karyotype in a patient with plasma cell myeloma demonstrating 46,XY,t(13;14)(q14;q32).

FIGURE 16.18 Bone marrow karyotype in a patient with plasma cell myeloma demonstrating 45,X,−Y,t(6;14)(p21;q32),iso(9)(q10),t(11;12)(q21;p13).

favorable clinical outcome than hyperdiploidy lacking this feature [47].

Additional copies of 1q21 have been reported to accompany the progression of smoldering to overt PCM by molecular and FISH studies such as array-CGH (aCGH) and interphase FISH [48]. Also it has been noted that patients with gains or amplifications of 1q21 were linked to inferior survival, and thus these events may be linked to PCM pathogenesis and progression.

Among losses, monosomy or partial deletion of chromosome 13 (13q14) is the most common finding, occurring in 15–40% of the newly diagnosed cases (Figure 16.20). Patients with a 13q14 deletion have been reported to have significant reductions in the rate of response to conventional dose chemotherapy (41% versus 79%) and overall survival (24 versus >60 months) compared to patients without this deletion. When present, it was the most important independent variable associated with unfavorable outcome [49].

PCM and related disorders show clonal Ig gene rearrangements and high frequency of Ig VH gene somatic mutation. Approximately 5% of the cases of PCM may show more than one rearranged Ig bands best seen by Southern

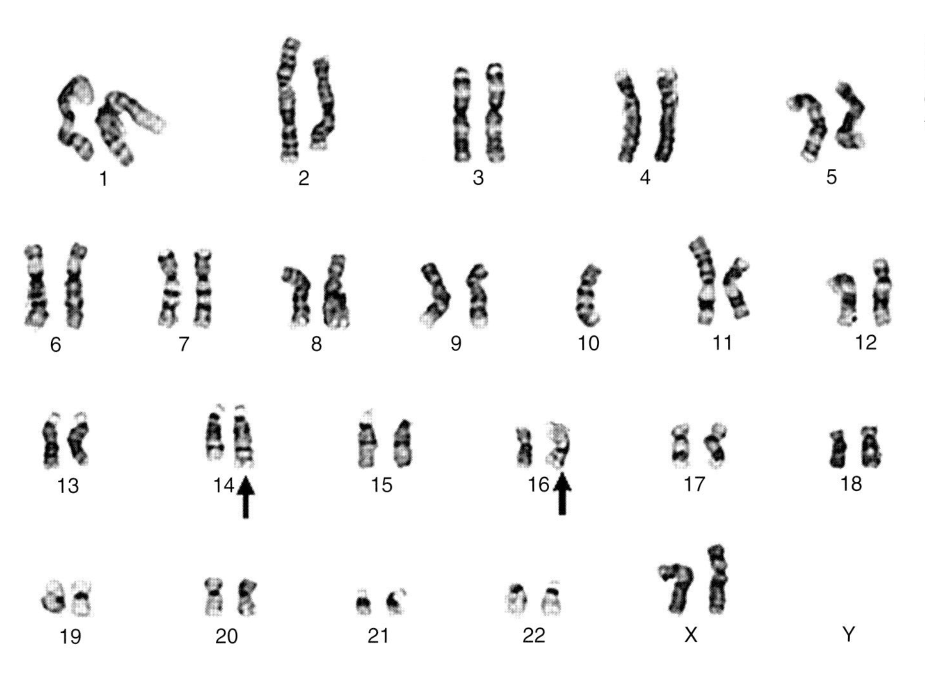

FIGURE 16.19 Bone marrow karyotype in a patient with plasma cell myeloma demonstrating 45,XX,t(2;11)(p11.2;p11.2),−10, t(14;16)(q32;q23).

TABLE 16.1 Chromosomal regions/genes as translocation partners with 14q32(IGH) in PCM/MGUS with known frequencies.

4p16	*FGFR3,WHSC1* (15%)
4p13	*ARHH*
6p23-25	*IRF4*
6p21	*CCND3* (15–20%)
8q24	*c-MYC*
11q13	*CCND1*
16q23	*MAF* (5–10%)
18q21	*BCL2*
20q11-q13	*MAFB*

TABLE 16.2 Frequency of *IGH* translocation and clinical outcome.

Abnormality	PCM and smoldering PCM (%)	PCM (%)	Prognosis
IGL translocations	<20	<20	Unknown
IGH translocations			
t(4;14)	35–50	50–70	Poor
t(11;14)	2–10	15	Poor
t(14;16)	15–30	15	Good
t(6;14)	2–5	5	Poor

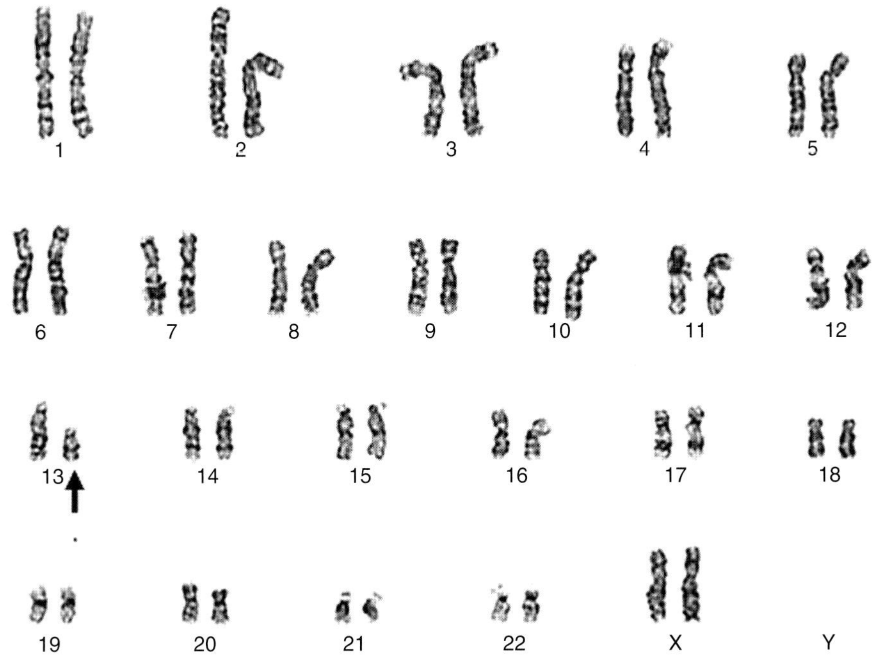

FIGURE 16.20 Bone marrow karyotype in a patient with plasma cell myeloma demonstrating 46,XX,del(13)(q13q22).

TABLE 16.3 Clinicopathological features of MGUS according to the International Myeloma Working Group.*

1. Serum M-protein level >3 g/dL
2. Bone marrow monoclonal plasma cells >10%
3. No evidence of other B-cell lymphoproliferative disorders
4. No evidence of organ or tissue impairment or bone lesions

*From Ref. [51].

blot [1]. Approximately 50% of the tumors carry the translocation of Ig heavy chain (*IGH*) locus (14q32) with various oncogenes and suppressor genes, such as *cyclin D1*, *cyclin D3*, *PGFR3*, *CHC1L*, *ZHX-2*, *MAF*, and *MAFB* at 11q13, 6p21, 4p16, 13q14.3, 8q24.3, 16q23, and 20q12, respectively [50]. Because the diagnosis of myeloma is straightforward, such gene rearrangement studies are usually not required. However, detection of clonality by PCR may be helpful for monitoring minimal residual disease.

The importance of cytogenetic and molecular features as determinants of outcome is being increasingly recognized. The present data support that PCM is characterized by marked inter- and intratumor cytogenetic heterogeneity that may account for the diverse clinical behavior of this neoplasm. Multivariate analysis of all cytogenetic and clinicopathologic features, including patient survival, is needed to identify the cytogenetic variables of independent prognostic significance and also to define the predictive role of the proposed cytogenetic classification for response to treatment and survival of patients with PCM.

MONOCLONAL GAMMOPATHY OF UNDETERMINED SIGNIFICANCE

The term "monoclonal gammopathy of undetermined significance" (MGUS) stands for a clinical condition defined by the presence of monoclonal Ig production without evidence of PCM, amyloidosis, Waldenstrom macroglobulinemia, or other related plasma cell or lymphoproliferative disorders (Table 16.3) [51]. Several other terms have been used for this condition, such as idiopathic, asymptomatic, non-myelomatous, cryptogenic, and benign monoclonal gammopathy. However, the term "benign monoclonal gammopathy" is inappropriate, because a significant proportion of patients with MGUS eventually develop PCM or other lymphoproliferative disorders. The cumulative probability of progression of MGUS to PCM or other lymphoproliferative disorders in one large study was 12% at 10 years, 25% at 20 years, and 30% at 25 years [52, 53]. Clearly, MGUS and PCM represent different time points along the same disease spectrum, and so far, no molecular or cytogenetic test can reliably distinguish them.

The incidence of MGUS increases by age. In a recent study, the prevalence of MGUS was 3.21% in individuals older than 50 and 5.3% in those older than 70 years [54]. It affects more African Americans than Caucasians. MGUS is usually an incidental finding detected by elevated total protein concentration on a routine blood test, followed by demonstration of a monoclonal spike by serum protein electrophoresis. The presence of light chain in the urine (Bence-Jones protein) is generally suggestive of PCM. The nature of serum M-component in MGUS is IgG in about 75%, IgM in 15%, and IgA in 10% of the cases. Plasma cells in bone marrow biopsy sections and smears appear mature and account for <10% of the total bone marrow nucleated cells (Figure 16.21). Immunophenotypic studies reveal a monoclonal population expressing Ig molecules corresponding to the patient's serum M-component and often an abnormal population of CD19− and CD56+ plasma cells. The elevated levels of monoclonal protein, presence of IgA or IgM class or an abnormal free light chain ratio, and a high percentage of plasma cells are predictors of MGUS progressing to a more aggressive B-cell lymphoproliferative disorder [55].

PLASMA CELL MYELOMA

Plasma cell myeloma (PCM) (multiple myeloma, myelomatosis, Kahler's disease) is a multifocal bone-marrow-based plasma cell neoplasm with the production of monoclonal Ig, often associated with bone destruction and osteolytic lesions, hypercalcemia, and anemia [1, 51]. PCM has been divided into several clinicopathologic entities:

- Asymptomatic myeloma (smoldering myeloma)
- Indolent myeloma
- Symptomatic myeloma (or symptomatic plasma cell myeloma)
- Non-secretory myeloma
- Plasma cell leukemia.

Asymptomatic (Smoldering) Myeloma

Smoldering myeloma represents the point of transition from MGUS to PCM without anemia, skeletal lesions, hypercalcemia, or renal insufficiency. The serum M-protein level is ≥3 g/dL and the bone marrow plasma cells are ≥10% but <30% (Table 16.4). These patients do not need treatment but should be followed up closely, because many of them eventually become symptomatic.

Indolent Myeloma

This category is described by the WHO but not included in the report of criteria for the classification of plasma cell disorders by the International Myeloma Working Group (IMWG) [51]. According to the WHO and Alexanian [1, 56], indolent myeloma is similar to smoldering myeloma in that there is no evidence of anemia, hypercalcemia, or renal insufficiency, but unlike smoldering myeloma there are up to three lytic bone lesions and the serum M-component is at intermediate levels (IgG >3 and <7 g/dL).

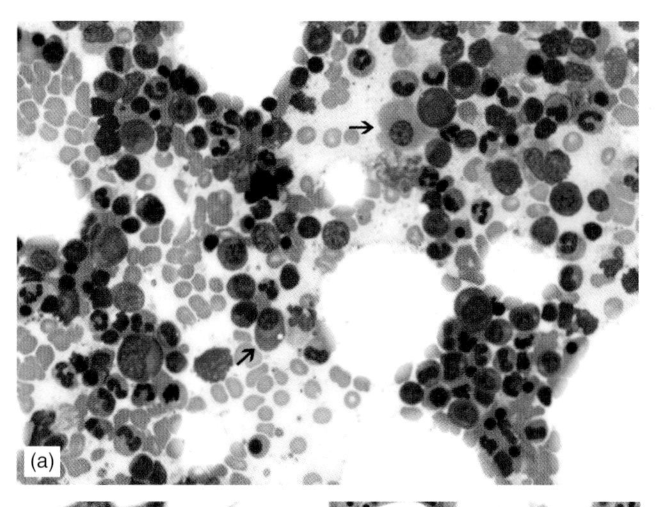

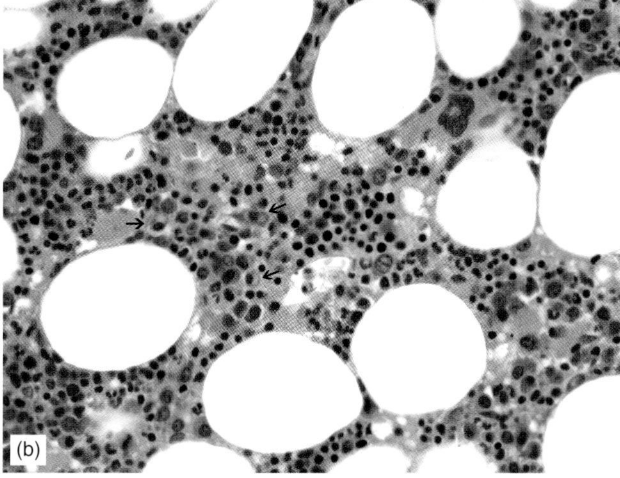

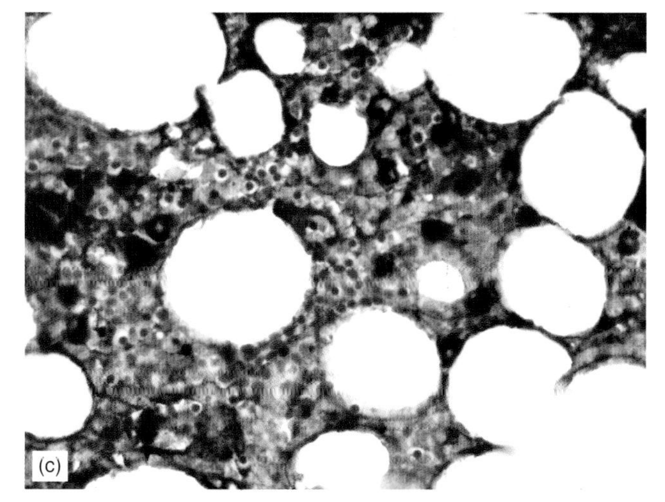

FIGURE 16.21 Monoclonal gammopathy of undetermined significance. Bone marrow smear (a) and biopsy section (b) demonstrate modest plasmacytosis. Plasma cells show kappa light chain restriction (brown) by dual immunohistochemical stains (c).

Symptomatic PCM

Symptomatic PCM, or myelomatosis, is characterized by monoclonal proliferation of plasma cells in the bone marrow and an M-protein production (see Figures 16.5–16.12).

TABLE 16.4 Criteria for asymptomatic myeloma (smoldering myeloma).*

Serum M-protein	≥3 g/dL
Bone marrow clonal plasma cells	≥10%
No related tissue or organ impairment or symptoms	

*From Ref. [51].

TABLE 16.5 Myeloma-related organ or tissue impairment.*

Elevated serum calcium levels	>0.25 mmol/L above the upper limit of normal or >2.75 mmol/L
Renal insufficiency	Creatinine >173 mmol/L
Anemia	Hemoglobin 2 g/dL below the lower limit of normal or <10 g/dL
Bone lesions	Lytic lesions or osteoporosis with compression fractures
Others	Symptomatic hyperviscosity, amyloidosis, recurrent bacterial infections

*From Ref. [51].

TABLE 16.6 Criteria for the diagnosis of plasma cell myeloma set by the International Myeloma Working Group.*

M-protein in serum and/or urine
Bone marrow clonal plasma cells or plasmacytoma**
Related organ or tissue impairment (end-organ damage, including bone lesions)***

*From Ref. [51].

**Immunophenotypic studies demonstrate monoclonal population of abnormal plasma cells.

***Some patients may have no symptoms but show evidence of organ or tissue impairment (see Table 15.3).

This neoplastic proliferation leads to bone destruction and pathological fractures, particularly in the spine and ribs [51]. Anemia, hypercalcemia, and renal insufficiency are other common features (Table 16.5). The criteria for the diagnosis of PCM proposed by the IMWG are shown in Table 16.6. According to the IMWG, the most critical criterion for symptomatic myeloma is the evidence of organ or tissue impairment manifested by anemia, hypercalcemia, lytic bone lesions (Figure 16.22), renal insufficiency, hyperviscosity, amyloidosis, or recurrent infections [51]. Unlike the WHO criteria for the diagnosis of PCM (Table 16.7), no level of serum or urine M-protein and no minimal percentage of clonal bone marrow plasma cells are

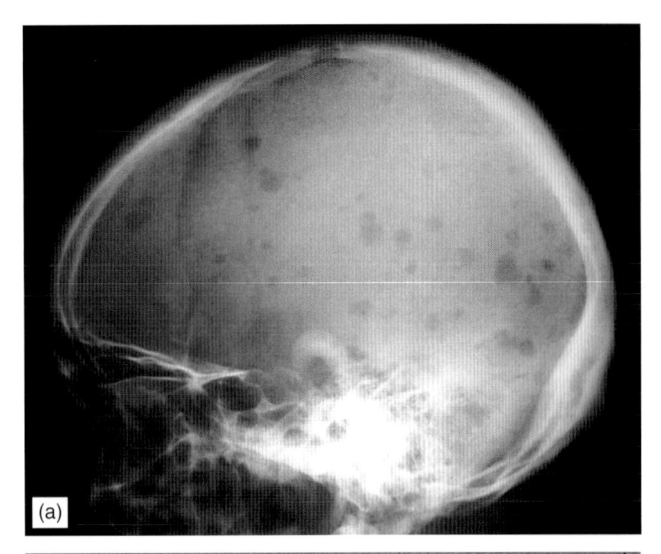

FIGURE 16.22 (a) Skull X-ray of a patient with plasma cell myeloma demonstrating punched out lytic lesions. (b) Gross specimen of a vertebrate showing several bone marrow lesions of plasma cell myeloma (arrows).

TABLE 16.7 WHO diagnostic criteria for plasma cell myeloma.*

Major criteria
A. Marrow plasmacytosis (>30%)
B. Plasmacytoma on biopsy
C. M-component:
i. Serum: IgG > 3 g/dL, IgA >2 g/dL
ii. Urine >1 g/24 h of Bence-Jones protein
Minor criteria
A. Marrow plasmacytosis (10–30%)
B. M-component present but less than above
C. Lytic bone lesions
D. Reduced normal Ig (<50% normal levels)
The diagnosis of plasma cell myeloma requires a minimum of one major and one minor criteria or three minor criteria which must include (A) and (B).

*From Ref. [1].

required in the proposed diagnostic criteria by the IMWG. Approximately 5% of patients with evidence of tissue or organ impairment may demonstrate <10% plasma cells in their bone marrow, and therefore based on the IMWG criteria, they should carry the diagnosis of PCM [51] and be treated appropriately.

Approximately 50% of the patients with PCM show IgG, 25% IgA M-proteins, and about 20% demonstrate only monoclonal light chain production [29, 30, 57]. IgD and IgM PCM are rare [58, 59]. Bence–Jones protein is detected in the urine of about 75% of patients. Over 95% of the patients with PCM show an M-protein in the serum or urine at the time of diagnosis [51]. Approximately 20% of patients show hypercalcemia. Conventional radiologic studies reveal lytic bone lesions, osteoporosis, or fractures in about 80% of the patients at diagnosis.

The bone marrow biopsies more often show focal lesions than diffuse, and therefore repeated biopsies may be required to establish the diagnosis. The morphologic features of the neoplastic plasma cells may vary from normal-appearing mature forms to blastic or anaplastic forms. As mentioned earlier, the neoplastic plasma cells, unlike normal plasma cells, lack the expression of CD19 and may show aberrant expression of CD56, CD10, and a number of other myeloid or T-cell-associated CD molecules [30–34].

An international staging system has been proposed based on the serum beta-2 microglobulin (β2M) and serum albumin levels [30, 60]:

Stage I: Serum β2M <3.5 mg/L and serum albumin ≥3.5 g/dL

Stage II: Neither stage I nor stage III

Stage III: Serum β2M ≥5.5 mg/L

The median survival times for patients with stages I, II, and III are 62, 44, and 29 months, respectively [30].

Hyperdiploidy and t(11;14)(q13;q32) are reported in association with favorable prognosis (median survival ≥50 months), and t(4;14)(p16;q32.3), t(14;16)(q32.3;q23), del(17p13) (locus for *p53*), and hypodiploidy are considered indicators of poor prognosis (median survival 25 months). The del(13q14) is associated with an intermediate prognosis with median survival of 42 months. In one flow cytometric study, patients with >10 circulating plasma cells (CD38+, CD45−) per 50,000 mononuclear cells had a significantly lower median survival than those with ≤10 circulating plasma cells [61]. Table 16.8 provides a list of the adverse prognostic factors. Conventional chemotherapy, autologous and allogeneic stem cell transplantation, and more recently targeted therapies have been used to increase the survival rate [29].

Non-secretory Myeloma

Non-secretory myeloma accounts for 1–5% of all myelomas and is characterized by the absence of detectable M-protein in the serum and urine [62, 63]. However, utilization of more sensitive techniques such as serum-free light chain assay may significantly reduce the number of these cases [64, 65]. The reports on non-secretory myelomas suggest a lower incidence of renal failure and hypogammaglobulinemia, lower median percentage of bone marrow plasma cells, higher incidence of neurological presentation, and longer survival than the secretory myelomas [66]. The therapeutic

TABLE 16.8 Adverse prognostic factors in plasma cell myeloma.*

Age	≥70 years
Serum albumin	<3 g/dL
Serum creatinine	≥2 mg/dL
Beta-2 microglobulin	>4 mg/L
Plasma cell labeling index	≥1%
Circulating plasma cells	>10 per 50,000 mononuclear cells
Serum calcium	≥11 mg/dL
Platelet count	<15,000/μL
Hemoglobin	<10 g/dL
Cytogenetics	t(4;14)(p16;q32.3) t(14;16)(q32.3;q23) del(17p13) (*p53* locus) hypodiploidy

* Adapted from Kyle RA, Gertz MA, Witzig TE, Lust JA, Lacy MQ, Dispenzieri A, Fonseca R, Rajkumar SV, Offord JR, Larson DR, Plevak ME, Therneau TM, Greipp PR. (2003). Review of 1027 patients with newly diagnosed multiple myeloma. *Mayo Clin Proc* **78**, 21–33, and Ref. [30].

approaches for non-secretory myeloma are similar to those for secretory PCM.

Plasma Cell Leukemia

Plasma cell leukemia is a rare event characterized by the presence of >2,000/μL of circulating plasma cells accounting for >20% of the white cell differential count (Figure 16.23) [1, 67–69]. Plasma cell leukemia is divided into two categories: primary and secondary. Primary plasma cell leukemia constitutes about 60% of the cases and is manifested *de novo* without evidence of previous history of PCM. Secondary plasma cell leukemia accounts for the remaining 40% and represents leukemic transformation in patients with a history of PCM. Patients with primary plasma cell leukemia are younger, have fewer lytic bone lesions and smaller amounts of serum M-protein, demonstrate higher incidence of hepatosplenomegaly, and show a longer survival than patients with secondary plasma cell leukemia [51]. The immunophenotype of primary plasma cell leukemia is frequently IgD, IgE, or light chain only [1], but clonal gene rearrangement studies targeting the IgM region will be sufficient to establish a monoclonal plasma cell disorder if needed.

PLASMACYTOMA

Plasmacytoma is a solitary neoplasm of plasma cells involving bone or extramedullary sites. Plasmacytoma demonstrates identical morphologic and immunophenotypic features of PCM [1].

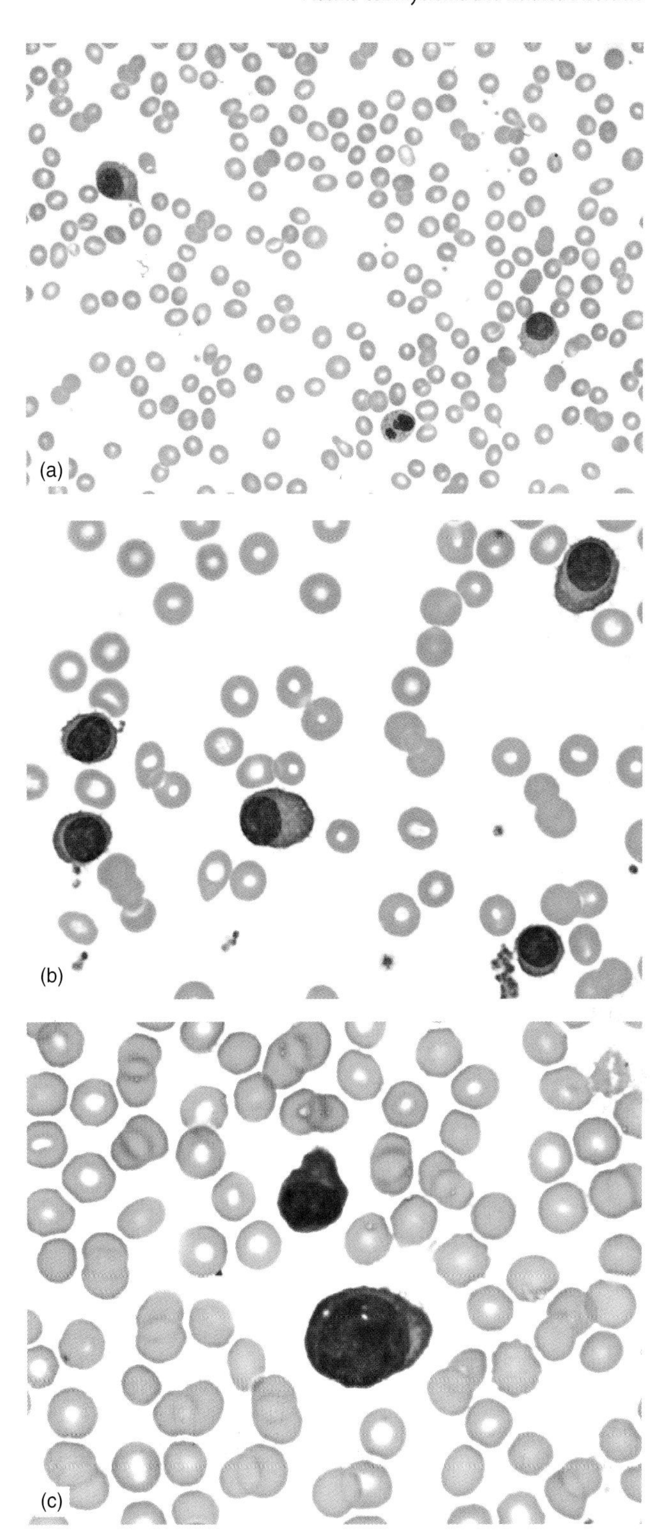

FIGURE 16.23 Blood smears of patients with plasma cell leukemia demonstrating circulating plasma cells with various morphologic features (a, b and c). © From Naeim F. (1997). *Pathology of Bone Marrow*, 2nd ed. Williams & Wilkins, Baltimore, by permission.

Solitary Plasmacytoma of Bone

Solitary plasmacytoma of bone is a rare condition and accounts for about 3–5% of the plasmacytic neoplasms. The

TABLE 16.9 Characteristics of plasmacytoma of bone and extramedullary sites.*

No M-protein in serum or urine**
Solitary bone or extramedullary tumor of monoclonal plasma cells
Bone marrow not consistent with plasma cell myeloma
Normal skeletal survey
No related organ or tissue impairment

*From Ref. [51].

**A small amount of M-component may be present in some cases.

median age is around 55 years and the male:female ratio is about 2:1. It often involves the axial skeleton, particularly thoracic vertebrae and ribs. Involvement of distal bones, particularly below the knees or elbows, is extremely rare [51, 70]. Bone pain at the site of the lesion is one of the most common presenting symptoms. Infiltration of the tumor cells into the surrounding soft tissue may result in a palpable mass. Radiologic studies show no evidence of additional lesions. Morphologic and immunophenotypic features are identical to those of PCM, but no serum or urine M-protein is detected, and there is no evidence of anemia, hypercalcemia, renal insufficiency, or other organ or tissue impairment (Table 16.9). Multiple solitary plasmacytomas without evidence of PCM occur in up to 5% of the cases [51]. Approximately 50% of patients with solitary plasmacytoma of bone eventually develop PCM [51]. Plasmacytomas of >5 cm in diameter have a greater chance of conversion to PCM. Radiotherapy is the treatment of choice.

Extramedullary Plasmacytoma

Extramedullary plasmacytoma is a monoclonal plasma cell neoplasm that arises outside the bone marrow. The most frequent site of involvement is the upper respiratory tract, including the nasal cavity and sinuses, nasopharynx, and larynx [51, 70–74], but any organ or tissue may be involved, such as gastrointestinal tract and urinary tracts, thyroid, male and female reproductive systems, parotid gland, lymph nodes, and central nervous system (Figure 16.24). The diagnosis is made based on the monoclonality of the plasma cell tumor and lack of evidence for PCM and serum or urine M-protein (Table 16.9). IgA is the most frequent immunophenotype, but again it is IgM that is typically assessed at the molecular level for monoclonality (Figure 16.25). Approximately 15% of the patients may eventually develop symptomatic PCM [74]. Surgery and/or radiation are the treatment of choice.

OSTEOSCLEROTIC MYELOMA

Osteosclerotic myeloma, or POEMS syndrome, is characterized by a combination of peripheral neuropathy (P), organomegaly (O), endocrinopathy (E), monoclonal plasma cell disorder (M), and skin changes (S) [75, 76]. Other frequent features include sclerotic bone lesions, Castleman's disease, papilledema, serous effusions, and thrombocytosis. The biopsy sections of sclerotic lesions show a monoclonal population of plasma cells. Plasmacytosis is usually modest (median 5%), but the bone marrow is often hypercellular with myeloid preponderance, increased megakaryocytes, and thick bone trabeculae. The M-protein is typically small and sometimes undetectable by routine serum protein electrophoresis [76]. The median age and survival reported in a Mayo Clinic study of 99 patients were 51 years and 165 months, respectively [76].

MONOCLONAL IMMUNOGLOBULIN DEPOSITION DISEASES

Monoclonal immunoglobulin deposition diseases are monoclonal gammopathies characterized by the deposition of Ig-derived proteins in the organs and tissues causing impairment of their function. These disorders are divided into two major groups: (1) disorders with the deposition of fibrillary proteins (primary amyloidosis) and (2) disorders with the deposition of an amorphous, non-fibrillary protein, known as monoclonal light and heavy chain deposition diseases.

Primary Amyloidosis

"Amyloidosis" is a general term referring to a heterogeneous group of disorders characterized by the extracellular deposition of fibrillar proteins with antiparallel beta-pleated sheet configuration on X-ray diffraction [77–85]. These fibrillar structures are identified on biopsy sections by an intense yellow-green fluorescence by thioflavine T and by binding to Congo red stain, leading to apple green birefringence under polarized light [77, 78]. So far, 23 different proteins have been identified that form fibrillar extracellular amyloid deposits in tissues that bind Congo red [79]. The major categories of amyloidosis include (1) primary amyloidosis, (2) secondary amyloidosis, (3) dialysis-related amyloidosis, (4) heritable amyloidosis, and (5) senile amyloidosis (Table 16.10).

Primary amyloidosis (AL) is referred to a specific type of amyloidosis in which the fibrillar protein is derived from monoclonal Ig light chains. Primary amyloidosis is considered a variant of monoclonal plasma cell proliferative disorder, and in about 10% of the cases it is associated with symptomatic PCM.

In approximately 75% of the cases, the fibrillary protein is derived from the variable region of lambda light chain. The kappa light chain is involved in the remaining 25%. There appears to be a correlation between the site of involvement and the involved variable region of the light chain. For example, the amyloid deposit in patients with dominant renal involvement is often derived from V lambda IV, whereas in patients with dominant cardiac involvement the amyloid deposit is derived from V lambda II or III [77–85]. A serum or urinary monoclonal protein could be

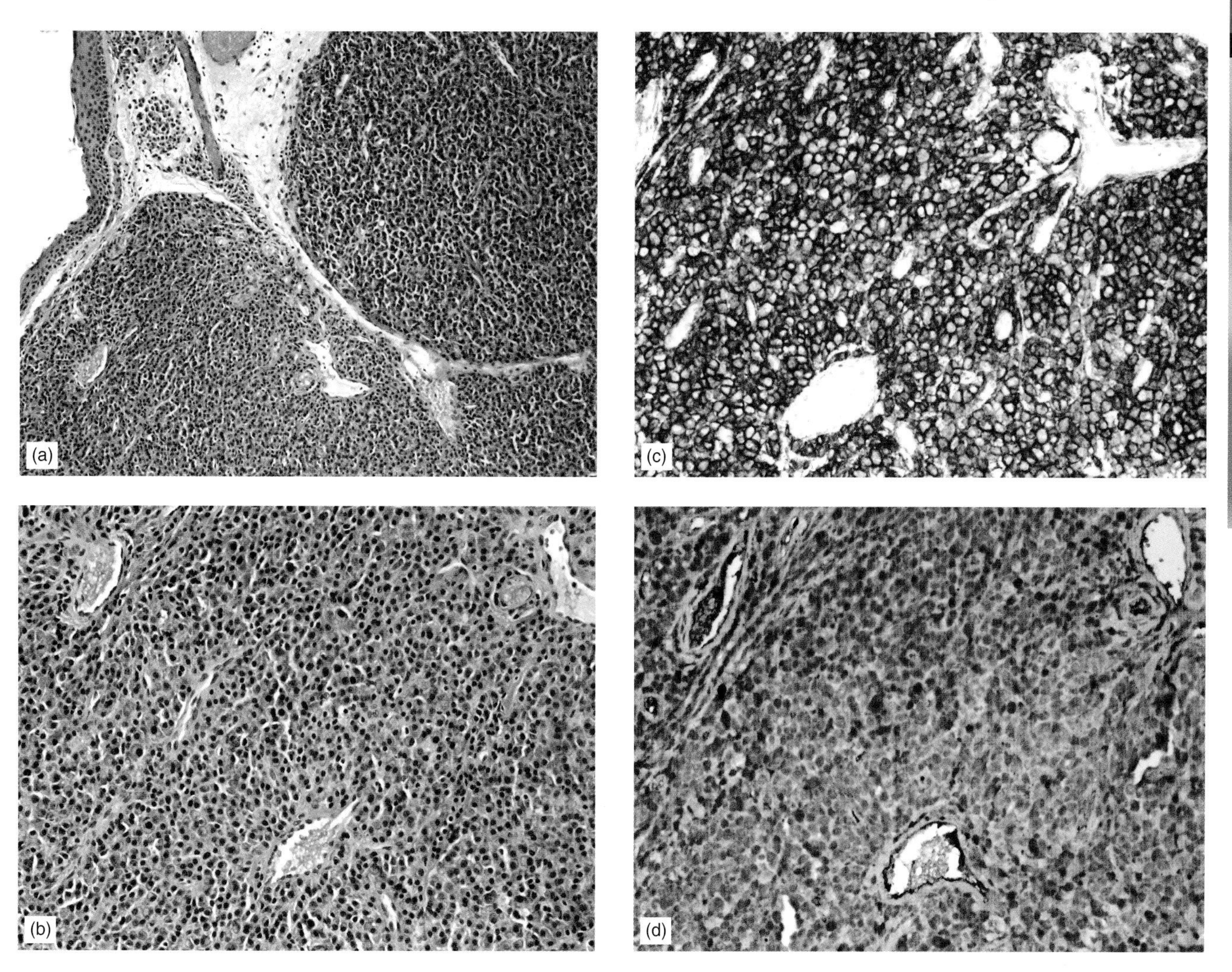

FIGURE 16.24 Solitary plasmacytoma of conjunctiva. Biopsy section demonstrating a large aggregate of plasma cells: (a) low power and (b) intermediate power. Immunohistochemical stains for CD138 (c) and lambda light chain (d, pink) demonstrate a monoclonal population of plasma cells. Courtesy of G. Pezeshkpour, M.D., Department of Pathology, VA Greater Los Angeles Healthcare System.

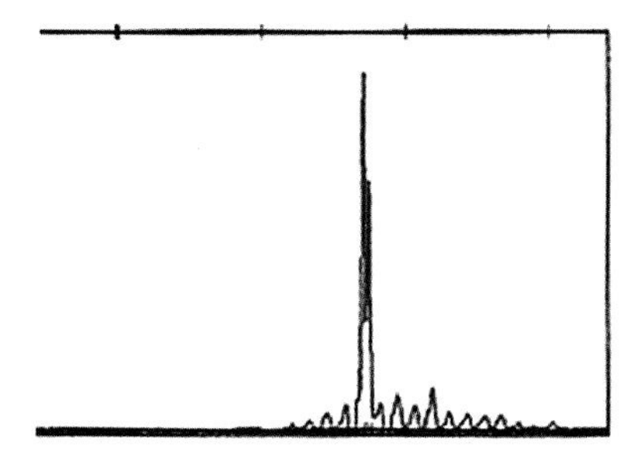

FIGURE 16.25 IgH PCR study of an extramedullary plasmacytoma showing a prominent clonal peak (framework 1 primer set) amid a weaker polyclonal background.

detected in over 85% of the patients using immunofixation techniques. Also, a protein known as serum amyloid P component (SAP) is detected by scintiography in patients with primary amyloidosis.

The diagnosis of amyloidosis is based on the biopsies obtained from the affected organs. Liver and kidney biopsies are positive in over 90% of the cases, followed by abdominal fat pad aspirate and biopsies of rectum, bone marrow (Figure 16.26), and skin [78].

Monosomy of chromosome 18 is the most common abnormality in primary amyloidosis followed by t(11;14)(q13;q32) and del(13q14) [85, 86].

The most frequent clinical symptoms in primary amyloidosis are (1) nephrotic syndrome with or without renal insufficiency, (2) cardiomyopathy, (3) peripheral neuropathy, (4) hepatomegaly, and (5) macroglossia [78]. Elevated serum β-2 microglobulin and bone marrow plasma cells $>10\%$, dominant cardiac involvement, and circulating plasma cells $>1\%$ correlate with poor prognosis [78–83]. The actuarial survival for 810 patients studied at the Mayo Clinic was 51% at 1 year, 16% at 5 years, and 4.7% at 10 years [85]. Progression to PCM is rare and in one large study it was reported in only 0.4% of the patients between 10 and 81 months [86]. Therapeutic approaches include chemotherapy, such as melphalan with or without prednisone or dexamethasone and stem cell transplantation.

The most common form of heritable amyloidosis is familial Mediterranean fever, an autosomal recessive autoinflammatory disorder characterized by periodic

TABLE 16.10 Major categories of amyloidosis.*

Amyloid protein	Precursor	Clinical status
AL	Ig light chain	Primary amyloidosis, local or systemic, associated with monoclonal plasma cell disorders
AA	Apolipoprotein AA	Secondary amyloidosis, systemic, associated with chronic infections
Aβ2M	Beta-2 microglobulin	Hemodialysis, systemic
AApoAI, AApoAII, AGel, ALys, ACys, and others	Apolipoprotein AI and AII, gelsolin lysozyme, crystatin C, and others	Familial, systemic
Ab, APro, ATTR, AMed	Ab protein precursor, prolactin, transthretin, lactadherin	Senile, local, or systemic

*Adapted from Ref. [79].

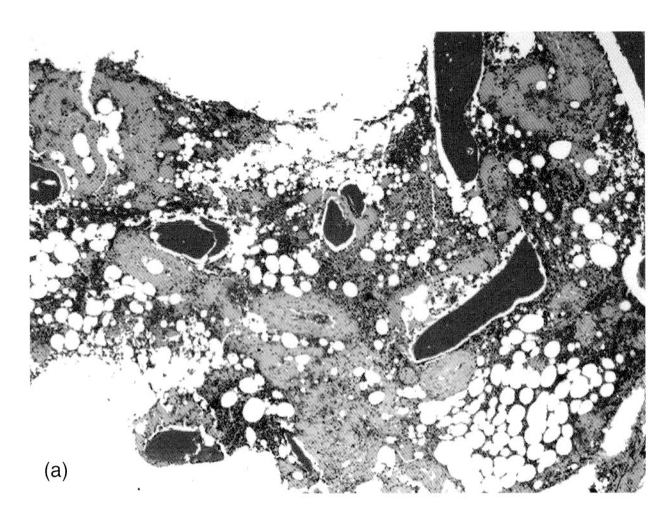

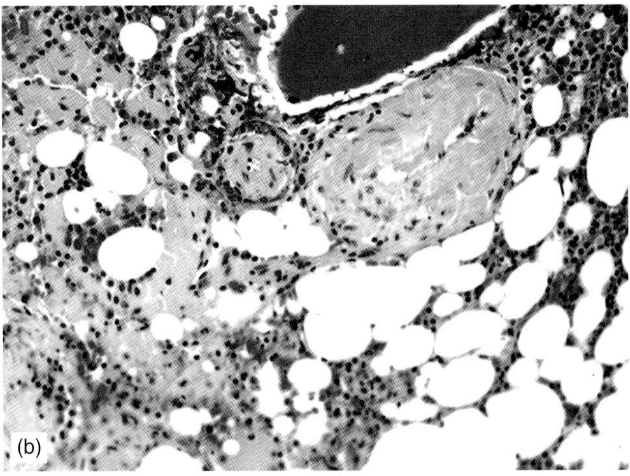

FIGURE 16.26 Bone marrow biopsy section demonstrating primary amyloidosis: (a) low power and (b) high power views.

fevers, abdominal pain (peritonitis), pleuritis, arthritis, pericarditis, and skin rash. It is most prevalent in individuals of Armenian descent, in whom the carrier frequency approaches 1 in 7 but is also found in Arabs, Jews, Persians, Italians, and Greeks [1, 87–90]. In addition to the periodic attacks of pain and fever, the life-threatening complication of the disease is amyloid deposition, particularly in the kidneys. However, this form of amyloidosis is to be distinguished from the others discussed here since it is neither related to plasma cell proliferation nor is it composed predominantly of Ig protein components. Differential diagnosis is made by the clinical history and genetic testing. The causative gene, *MEFV*, will often, but not always, demonstrate mutations in the homozygous or compound heterozygous state. Since over 60 different mutations have been reported, practical testing is usually limited to a subset of the more common ones found in Mediterranean populations [91]. Technical approaches include DNA sequencing (Figure 16.27a) and allele-specific DNA probe hybridization (Figure 16.27b).

Monoclonal Light and Heavy Chain Diseases

The light chain deposition disease (LCDD) and heavy chain deposition disease (HCDD) are clinical variants of monoclonal plasma cell disorders characterized by the deposition of abnormal light chain, heavy chain, or both in the tissues or organs [75, 92, 93]. The deposits, unlike primary amyloidosis, are not fibrillar, do not bind Congo red, and do not contain SAP. LCDD is more common than primary amyloidosis and often consists of kappa light chain (Table 16.11). The primary defect in LCDD appears to be mutations in the Ig light chain variable region, with predominant involvement of V_{kIV} of kappa light chain [1, 78, 29].

Deletion of the *CH1* constant domain and point mutation of variable regions of the heavy chain are the primary events in HCDD [1, 78, 93]. These events lead to premature secretion of heavy chain binding protein and increased tendency for tissue deposition. HCDD of IgG1 and IgG3 isotypes are associated with reduced complement activities [1, 94]. Many organs may be involved, including kidneys, liver, heart, nerves, and blood vessels. Approximately 85% of the cases show a detectable serum M-component.

Heavy Chain Diseases

The heavy chain disease is a monoclonal lymphoplasmacytic disorder characterized by the production of incomplete Ig molecules because of the lack of gamma chain binding sites for light chains. There are three major categories of heavy chain diseases: α, γ, and μ [1, 95, 96].

The α heavy chain disease (αHCD, Mediterranean lymphoma) is considered a variant of MALT-type lymphoma [1]. It occurs in older children and young adults and is associated with gastrointestinal symptoms, such as malabsorption, intestinal obstruction, and diarrhea [1, 97–99].

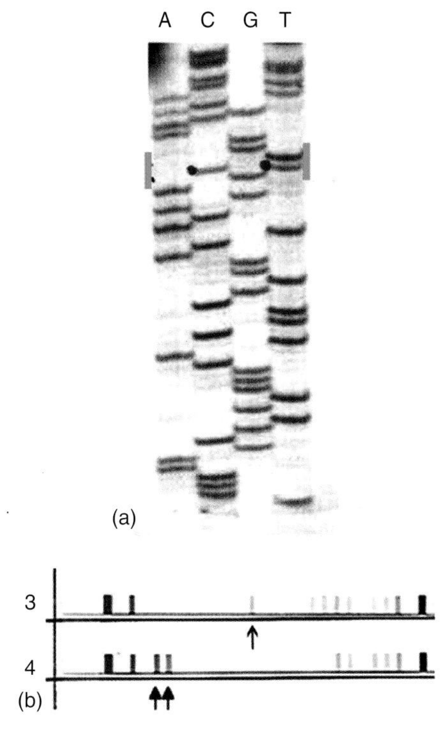

FIGURE 16.27 (a) DNA sequencing gel showing heterozygosity for the V726A mutation in the 2*MEFV* gene of a patient with familial Mediterranean fever (the second mutation was M694V). The red dots mark the nucleotide position showing both T (normal) and C (mutant). (b) Reverse hybridization strips using allele-specific oligonucleotide probes directed against the most prevalent mutations in the *MEFV* gene associated with familial Mediterranean fever. Patient #3 is homozygous for mutation K695R (single arrow), and patient #4 is compound heterozygous for mutations P369S and E148Q (double arrows).

TABLE 16.11 Comparison between primary amyloidosis and light chain deposition disease.

Type	Ig	Deposition	Congo red
Primary amyloidosis	Light chain, often λ	Fibrillary	Positive
Light chain deposition	Light chain, often κ	Non-fibrillary	Negative

αHCD is the most frequent heavy chain disease with the majority of the reported cases being from Middle East, North and South Africa, and the Far East. The pathologic features are more or less similar to those described in MALT lymphoma depicted by a mucosal infiltrate of centrocyte-like lymphocytes and plasma cells. An abnormal heavy chain protein is detected in the serum of 20–90% of the patients [1, 97]. Antibacterial therapy may completely resolve the disease in early stages. Some cases may eventually transform to large B-cell lymphoma. Despite the similar ethnic appellation, this disorder has nothing to do with familial Mediterranean fever, which is not a B-cell disorder.

The γ heavy chain disease (γHCD, Franklin disease) is a rare condition presenting with lymphadenopathy, splenomegaly, and hepatomegaly with lymphoplasmacytic infiltration similar to lymphoplasmacytic lymphoma [1, 100]. Immunofixation studies may show the presence of serum IgG without light chain. Some patients may demonstrate autoimmune disorders or chronic lymphocytic leukemia [100].

The μ heavy chain disease is a rare lymphoproliferative disorder with clonal IgM molecules with defective variable region [101–103]. Clinically, it resembles chronic lymphocytic leukemia and is often associated with hepatosplenomegaly [1]. The bone marrow is infiltrated by mature small lymphocytes admixed with vacuolated plasma cells [1]. Lymphadenopathy is unusual.

References

1. Jaffe ES, Harris NL, Stein H, Vardiman JW. (2001). *Pathology and Genetics. Tumors of Haematopoietic and Lymphoid Tissues.* IARC Press, Lyon.
2. Anderson KC, Shaughnessy Jr. JD, Barlogie B, Harousseau JL, Roodman GD. (2002). Multiple myeloma. *Hematology Am Soc Hematol Educ Program*, 214–40.
3. Veillon DM, Cotelingam JD. (2007). Pathologic studies useful for the diagnosis and monitoring of plasma cell dyscrasias. *Contrib Nephrol* **153**, 25–43.
4. Iwanaga M, Tagawa M, Tsukasaki K, Kamihira S, Tomonaga M. (2007). Prevalence of monoclonal gammopathy of undetermined significance: Study of 52,802 persons in Nagasaki City, Japan. *Mayo Clin Proc* **82**, 1474–9.
5. Neriishi K, Nakashima E, Suzuki G. (2003). Monoclonal gammopathy of undetermined significance in atomic bomb survivors: Incidence and transformation to multiple myeloma. *Br J Haematol* **121**, 405–10.
6. Dainiak N. (2002). Hematologic consequences of exposure to ionizing radiation. *Exp Hematol* **30**, 513–28.
7. Heineman EF, Olsen JH, Pottern LM, Gomez M, Raffn E, Blair A. (1992). Occupational risk factors for multiple myeloma among Danish men. *Cancer Causes Control* **3**, 555–68.
8. Fritschi L, Siemiatycki J. (2006). Occupation and malignant lymphoma: A population based case control study in Germany. *Occup Environ Med* **63**, 17–26.
9. Kyle RA, Rajkumar SV. (2007). Epidemiology of the plasma-cell disorders. *Best Pract Res Clin Haematol* **20**, 637–64.
10. Brown LM, Gridley G, Pottern LM, Baris D, Swanso CA, Silverman DT, Hayes RB, Greenberg RS, Swanson GM, Schoenberg JB, Schwartz AG, Fraumeni JF. (2001). Diet and nutrition as risk factors for multiple myeloma among blacks and whites in the United States. *Cancer Causes Control* **12**, 117–25.
11. Vlajinac HD, Pekmezović TD, Adanja BJ, Marinković JM, Kanazir MS, Suvajdzić ND, Colović MD. (2003). Case–control study of multiple myeloma with special reference to diet as risk factor. *Neoplasma* **50**, 79–83.
12. Altieri A, Chen B, Bermejo JL, Castro F, Hemminki K. (2006). Familial risks and temporal incidence trends of multiple myeloma. *Eur J Cancer* **42**, 1661–70.
13. Altieri A, Bermejo JL, Hemminki K. (2005). Familial risk for non-Hodgkin lymphoma and other lymphoproliferative malignancies by histopathologic subtype: The Swedish Family-Cancer Database. *Blood* **106**, 668–72.
14. Catovsky D. (2005). Familial multiple myeloma. *Haematologica* **90**, 3–4.
15. Amara S, Dezube BJ, Cooley TP, Pantanowitz L, Aboulafia DM. (2006). HIV-associated monoclonal gammopathy: A retrospective analysis of 25 patients. *Clin Infect Dis* **43**, 1198–205.

16. Lakatos PL, Fekete S, Horanyi M, Fischer S, Abonyi ME. (2006). Development of multiple myeloma in a patient with chronic hepatitis C: A case report and review of the literature. *World J Gastroenterol* **12**, 2297–300.
17. Malnati MS, Dagna L, Ponzoni M, Lusso P. (2003). Human herpesvirus 8 (HHV-8/KSHV) and hematologic malignancies. *Rev Clin Exp Hematol* **7**, 375–405.
18. Paglieroni T, Caggiano V, MacKenzie M. (1992). Abnormalities in immune regulation precede the development of multiple myeloma. *Am J Hematol* **40**, 51–5.
19. González D, van der Burg M, García-Sanz R, Fenton JA, Langerak AW, González M, van Dongen JJ, San Miguel JF, Morgan GJ. (2007). Immunoglobulin gene rearrangements and the pathogenesis of multiple myeloma. *Blood* **110**, 3112–21.
20. Prabhala RH, Neri P, Bae JE, Tassone P, Shammas MA, Allam CK, Daley JF, Chauhan D, Blanchard E, Thatte HS, Anderson KC, Munshi NC. (2006). Dysfunctional T regulatory cells in multiple myeloma. *Blood* **107**, 301–4.
21. Bergsagel PL, Kuehl WM. (2005). Molecular pathogenesis and a consequent classification of multiple myeloma. *J Clin Oncol* **23**, 6333–8.
22. Chng WJ, Glebov O, Bergsagel PL, Kuehl WM. (2007). Genetic events in the pathogenesis of multiple myeloma. *Best Pract Res Clin Haematol* **20**, 571–96.
23. Singhal S, Mehta J. (2006). Multiple myeloma. *Clin J Am Soc Nephrol* **1**, 1322–30.
24. Li QF, Wu CT, Duan HF, Sun HY, Wang H, Lu ZZ, Zhang QW, Liu HJ, Wang LS. (2007). Activation of sphingosine kinase mediates suppressive effect of interleukin-6 on human multiple myeloma cell apoptosis. *Br J Haematol* **138**, 632–9.
25. Löffler D, Brocke-Heidrich K, Pfeifer G, Stocsits C, Hackermüller J, Kretzschmar AK, Burger R, Gramatzki M, Blumert C, Bauer K, Cvijic H, Ullmann AK, Stadler PF, Horn F. (2007). Interleukin-6 dependent survival of multiple myeloma cells involves the Stat3-mediated induction of microRNA-21 through a highly conserved enhancer. *Blood* **110**, 1330–3.
26. Derenne S, Monia B, Dean NM, Taylor JK, Rapp MJ, Harousseau JL, Bataille R, Amiot M. (2002). Antisense strategy shows that Mcl-1 rather than Bcl-2 or Bcl-x(L) is an essential survival protein of human myeloma cells. *Blood* **100**, 194–9.
27. Cheung WC, Van Ness B. (2002). Distinct IL-6 signal transduction leads to growth arrest and death in B cells or growth promotion and cell survival in myeloma cells. *Leukemia* **16**, 1182–8.
28. Puthier D, Bataille R, Amiot M. (1999). IL-6 up-regulates mcl-1 in human myeloma cells through JAK/STAT rather than ras/MAP kinase pathway. *Eur J Immunol* **29**, 3945–50.
29. Harousseau JL, Shaughnessy Jr. J, Richardson P. (2004). Multiple myeloma. *Hematology Am Soc Hematol Educ Program*, 237–56.
30. Kyle RA, Rajkumar SV. (2007). Diagnosis and differential diagnosis of multiple myeloma. *UpToDate.*
31. Dingli D, Nowakowski GS, Dispenzieri A, Lacy MQ, Hayman SR, Rajkumar SV, Greipp PR, Litzow MR, Gastineau DA, Witzig TE, Gertz MA. (2006). Flow cytometric detection of circulating myeloma cells before transplantation in patients with multiple myeloma: A simple risk stratification system. *Blood* **107**, 3384–8.
32. Rawstron AC, Orfao A, Beksac M, Bezdickova L, Brooimans RA, Bumbea H, Dalva, K, Fuhler G, Gratama J, Hose D, Kovarova L, Lioznov M, Mateo G, Morilla R, Mylin AK, Omedé P, Pellat-Deceunynck C, Perez Andres M, Petrucci M, Ruggeri M, Rymkiewicz G, Schmitz A, Schreder M, Seynaeve C, Spacek M, de Tute RM, Van Valckenborgh E, Weston-Bell N, Owen RG, San Miguel JF, Sonneveld P, Johnsen HE; on behalf of the European Myeloma Network. (2008). Report of the European Myeloma Network on multiparametric flow cytometry in multiple myeloma and related disorders. *Haematologica* **93**, 431–8.
33. DiGiuseppe JA. (2007). Flow cytometric immunophenotyping of plasmacytic neoplasms. *Am J Clin Pathol* **127**, 172–4.
34. Kobayashi S, Hyo R, Amitani Y, Tanaka M, Hashimoto C, Sakai R, Tamura T, Motomura S, Maruta A. (2006). Four-color flow cytometric analysis of myeloma plasma cells. *Am J Clin Pathol* **126**, 908–15.
35. Stewart AK, Bergsagel PL, Greipp PR, Dispenzieri A, Gertz MA, Hayman SR, Kumar S, Lacy MQ, Lust JA, Russell SJ, Witzig TE, Zeldenrust SR, Dingli D, Reeder CB, Roy V, Kyle RA, Rajkumar SV, Fonseca R. (2007). A practical guide to defining high-risk myeloma for clinical trials, patient counseling and choice of therapy. *Leukemia* **21**, 529–34.
36. Zhan F, Sawyer J, Tricot G. (2006). The role of cytogenetics in myeloma. *Leukemia* **20**, 1484–6.
37. Dewald GW, Therneau T, Larson D, Lee YK, Fink S, Smoley S, Paternoster S, Adeyinka A, Ketterling R, Van Dyke DL, Fonseca R, Kyle R. (2005). Relationship of patient survival and chromosome anomalies detected in metaphase and/or interphase cells at diagnosis of myeloma. *Blood* **106**, 3553–8.
38. Liebisch P, Dohner H. (2006). Cytogenetics and molecular cytogenetics in multiple myeloma. *Eur J Cancer* **42**, 1520–9.
39. Chiecchio L, Protheroe RK, Ibrahim AH, Cheung KL, Rudduck C, Dagrada GP, Cabanas ED, Parker T, Nightingale M, Wechalekar A, Orchard KH, Harrison CJ, Cross NC, Morgan GJ, Ross FM. (2006). Deletion of chromosome 13 detected by conventional cytogenetics is a critical prognostic factor in myeloma. *Leukemia* **20**, 1610–17.
40. Zhan F, Huang Y, Colla S, Stewart JP, Hanamura I, Gupta S, Epstein J, Yaccoby S, Sawyer J, Burington B, Anaissie E, Hollmig K, Pineda-Roman M, Tricot G, van Rhee F, Walker R, Zangari M, Crowley J, Barlogie B, Shaughnessy Jr. JD. (2006). The molecular classification of multiple myeloma. *Blood* **108**, 2020–8.
41. Agnelli L, Fabris S, Bicciato S, Basso D, Baldini L, Morabito F, Verdelli D, Todoerti K, Lambertenghi-Deliliers G, Lombardi L, Neri A. (2007). Upregulation of translational machinery and distinct genetic subgroups characterise hyperdiploidy in multiple myeloma. *Br J Haematol* **36**, 565–73.
42. Liebisch P, Wendl C, Wellmann A, Kröber A, Schilling G, Goldschmidt H, Einsele H, Straka C, Bentz M, Stilgenbauer S, Döhner H. (2003). High incidence of trisomies 1q, 9q, and 11q in multiple myeloma: Results from a comprehensive molecular cytogenetic analysis. *Leukemia* **17**, 2535–7.
43. Nilsson T, Hoglund M, Lnehoff S, Rylander L, Turesson I, Westin J, Mitelman F, Jojansson B. (2003). A pooled analysis of karyotypic patterns, breakpoints and imbalances in 783 cytogenetically abnormal multiple myelomas reveals frequently involved chromosome segments as well as significant age- and sex-related differences. *Br J Haematol* **120**, 960–9.
44. Fonseca R, Bailey RJ, Ahman GJ, Rajkumar V, Hoyer JD, Lust JA, Kyle RA, Gertz MA, Greipp PR, Dewald GW. (2002). Genomic abnormalities in monoclonal gammopathy of undetermined significance. *Blood* **100**, 1417–24.
45. Fonseca R. Barlogie B. Bataille R, Bastard C, Bergsagel PL, Chesi M, Davies FE, Drach J, Greipp PR, Kirsch IR, Kuehl WM, Hernandez JM, Minvielle S, Pilarski LM, Shaughnessy Jr. JD, Stewart AK, Avet-Loiseau H. (2004). Genetics and cytogenetics of multiple myeloma: A workshop report. *Cancer Res* **64**, 1546–58.
46. Kuehl WM, Bergsagel PL. (2002). Multiple myeloma: Evolving genetic events and host interactions. *Nat Rev Cancer* **2**, 175–87.
47. Carrasco R, Tonon G, Huang Y, *et al.* (2006). High-resolution genomic profiles defines distinct clinicopathogenetic subgroups of multiple myeloma patients. *Cancer Cell* **4**, 313–25.
48. Hanamura I, Stewart JP, Huang Y. (2006). Frequent gain of chromosome band 1q21 in plasma cell dyscrasias detected by fluorescence *in situ* hybridization: Incidence increases from MGUS to relapsed myeloma and is related to prognosis and disease progression following tandem stem cell transplantation. *Blood* **108**, 1724–32.
49. Zojer N, Königsberg R, Ackermann J, Fritz E, Dallinger S, Krömer E, Kaufmann H, Riedl L, Gisslinger H, Schreiber S, Heinz R, Ludwig H, Huber H, Drach J. (2000). Deletion of 13q14 remains an independent

adverse prognostic variable in multiple myeloma despite its frequent detection by interphase fluorescence *in situ* hybridization. *Blood* **95**, 1925–30.

50. Fenk R, Haas R, Kronenwett R. (2004). Molecular monitoring of minimal residual disease in patients with multiple myeloma. *Hematology* **9**, 17–33.
51. The International Myeloma Working Group. (2003). Criteria for the classification of monoclonal gammopathies, multiple myeloma and related disorders: A report of the International Myeloma Working Group. *Br J Haematol* **121**, 749–57.
52. Kyle RA, Therneau TM, Rajkumar SV, Offord JR, Larson DR, Plevak MF, Melton III LJ. (2002). A long-term study of prognosis in monoclonal gammopathy of undetermined significance. *N Engl J Med* **346**, 564–9.
53. Kyle RA, Rajkumar SV. (2005). Monoclonal gammopathy of undetermined significance. *Clin Lymphoma Myeloma* **6**, 102–14.
54. Kyle RA, Therneau TM, Rajkumar SV, Larson DR, Plevak MF, Offord JR, Dispenzieri A, Katzmann JA, Melton III LJ. (2006). Prevalence of monoclonal gammopathy of undetermined significance. *N Engl J Med* **354**, 1362–9.
55. Blade J. (2006). Clinical practice, monoclonal gammopathy of undetermined significance. *N Engl J Med* **355**, 2765–70.
56. Alexanian R. (1980). Localized and indolent myeloma. *Blood* **56**, 521–5.
57. Kyle RA, Rajkumar SV. (2007). Clinical and laboratory manifestation of multiple myeloma. *UpToDate*.
58. Sinclair D. (2002). IgD myeloma: Clinical, biological and laboratory features. *Clin Lab* **48**, 617–22.
59. Annibali O, Petrucci MT, Del Bianco P, Gallucci C, Levi A, Foa R, Avvisati G. (2006). IgM multiple myeloma: Report of four cases and review of the literature. *Leuk Lymphoma* **47**, 1565–9.
60. Greipp PR, San Miguel J, Durie BG, Crowley JJ, Barlogie B, Bladé J, Boccadoro M, Child JA, Avet-Loiseau H, Kyle RA, Lahuerta JJ, Ludwig H, Morgan G, Powles R, Shimizu K, Shustik C, Sonneveld P, Tosi P, Turesson I, Westin J. (2005). International staging system for multiple myeloma. *J Clin Oncol* **23**, 3412–20.
61. Nowakowski GS, Witzig TE, Dingli D, Tracz MJ, Gertz MA, Lacy MQ, Lust JA, Dispenzieri A, Greipp PR, Kyle RA, Rajkumar SV. (2005). Circulating plasma cells detected by flow cytometry as a predictor of survival in 302 patients with newly diagnosed multiple myeloma. *Blood* **106**, 2276–9.
62. Blade J, Kyle RA. (1999). Nonsecretory myeloma, immunoglobulin D myeloma, and plasma cell leukemia. *Hematol Oncol Clin North Am* **13**, 1259–72.
63. Gafumbegete E, Richter S, Jonas L, Nizze H, Makovitzky J. (2004). Nonsecretory multiple myeloma with amyloidosis: A case report and review of the literature. *Virchows Arch* **445**, 531–6.
64. Shaw GR. (2006). Nonsecretory plasma cell myeloma – becoming even more rare with serum free light-chain assay: A brief review. *Arch Pathol Lab Med* **130**, 1212–15.
65. Drayson M, Tang LX, Drew R, Mead GP, Carr-Smith H, Bradwell AR. (2001). Serum free light-chain measurements for identifying and monitoring patients with nonsecretory multiple myeloma. *Blood* **97**, 2900–2.
66. Smith DB, Harris M, Gowland E, Chang J, Scarffe JH. (1986). Nonsecretory multiple myeloma: A report of 13 cases with a review of the literature. *Hematol Oncol* **4**, 307–13.
67. Saccaro S, Fonseca R, Veillon DM, Cotelingam J, Nordberg ML, Bredeson C, Glass J, Munker R. (2005). Primary plasma cell leukemia: Report of 17 new cases treated with autologous or allogeneic stem-cell transplantation and review of the literature. *Am J Hematol* **78**, 288–94.
68. Hayman SR, Fonseca R. (2001). Plasma cell leukemia. *Curr Treat Options Oncol* **2**, 205–16.
69. Costello R, Sainty D, Bouabdallah R, Fermand JP, Delmer A, Divine M, Marolleau JP, Gastaut JA, Olive D, Rousselot P, Chaibi P. (2001). Primary plasma cell leukaemia: A report of 18 cases. *Leuk Res* **25**, 103–7.
70. Dimopoulos MA, Hamilos G. (2002). Solitary bone plasmacytoma and extramedullary plasmacytoma. *Curr Treat Options Oncol* **3**, 255–9.
71. Saccaro S, Fonseca R, Veillon DM, Cotelingam J, Nordberg ML, Bredeson C, Glass J, Munker R, Mendenhall WM, Mendenhall CM, Mendenhall NP. (2003). Solitary plasmacytoma of bone and soft tissues. *Am J Otolaryngol* **24**, 395–9.
72. Strojan P, Soba E, Lamovec J, Munda A. (2002). Extramedullary plasmacytoma: Clinical and histopathologic study. *Int J Radiat Oncol Biol Phys* **53**, 692–701.
73. Galieni P, Cavo M, Pulsoni A, Avvisati G, Bigazzi C, Neri S, Caliceti U, Benni M, Ronconi S, Lauria F. (2000). Clinical outcome of extramedullary plasmacytoma. *Haematologica* **85**, 47–51.
74. Alexiou C, Kau RJ, Dietzfelbinger H, Kremer M, Spiess JC, Schratzenstaller B, Arnold W. (1999). Extramedullary plasmacytoma: Tumor occurrence and therapeutic concepts. *Cancer* **85**, 2305–14.
75. Dispenzieri A. (2005). POEMS syndrome. *Hematology Am Soc Hematol Educ Program* **5**, 360–7.
76. Dispenzieri A, Kyle RA, Lacy MQ, Rajkumar SV, Therneau TM, Larson DR, Greipp PR, Witzig TE, Basu R, Suarez GA, Fonseca R, Lust JA, Gertz MA. (2003). POEMS syndrome: Definitions and long-term outcome. *Blood* **101**, 2496–506.
77. Kyle RA. (2007). Diagnosis of primary (AL) amyloidosis. *UpToDate*.
78. Kyle RA, Rajkumar SV. (2007). Pathogenesis and clinical features of primary (AL) amyloidosis and light and heavy chain deposition diseases. *UpToDate*.
79. Buxbaum JN. (2004). The systemic amyloidoses. *Curr Opin Rheumatol* **16**, 67–75.
80. Musolin L, Reyna R, DeBord J. (2003). Primary amyloidosis (AL) presenting with nephrotic syndrome: A case report and discussion. *W V Med J* **99**, 28–30.
81. Gertz MA, Rajkumar SV. (2002). Primary systemic amyloidosis. *Curr Treat Options Oncol* **3**, 261–71.
82. Comenzo RL. (2000). Primary systemic amyloidosis. *Curr Treat Options Oncol* **1**, 83–9.
83. Perfetti V, Vignarelli MC, Casarini S, Ascari E, Merlini G. (2001). Biological features of the clone involved in primary amyloidosis (AL). *Leukemia* **15**, 195–202.
84. Grateau G. (2000). Amyloidosis physiopathology. *Joint Bone Spine* **67**, 164–70.
85. Hayman SR, Bailey RJ, Jalal SM, Ahmann GJ, Dispenzieri A, Gertz MA, Greipp PR, Kyle RA, Lacy MQ, Rajkumar SV, Witzig TE, Lust JA, Fonseca R. (2001). Translocations involving the immunoglobulin heavy-chain locus are possible early genetic events in patients with primary systemic amyloidosis. *Blood* **98**, 2266–8.
86. Harrison CJ, Mazzullo H, Ross FM, Cheung KL, Gerrard G, Harewood L, Mehta A, Lachmann HJ, Hawkins PN, Orchard KH. (2002). Translocations of 14q32 and deletions of 13q14 are common chromosomal abnormalities in systemic amyloidosis. *Br J Haematol* **117**, 427–35.
87. Lidar M, Livneh A. (2007). Familial Mediterranean fever: Clinical, molecular and management advancements. *Neth J Med* **65**, 318–24.
88. Bhat A, Naguwa SM, Gershwin ME. (2007). Genetics and new treatment modalities for familial Mediterranean fever. *Ann NY Acad Sci* **1110**, 201–8.
89. Yonem O, Bayraktar Y. (2007). Secondary amyloidosis due to FMF. *Hepatogastroenterology* **54**, 1061–5.
90. Onen F. (2006). Familial Mediterranean fever. *Rheumatol Int* **26**, 489–96.
91. Telatar M, Grody WW. (2000). Molecular genetic testing for familial Mediterranean fever. *Mol Genet Metab* **71**, 256–60.
92. Suzuki K. (2005). Light- and heavy-chain deposition disease (LHCDD): Difficulty in diagnosis and treatment. *Intern Med* **44**, 915–16.
93. Preud'homme JL, Aucouturier P, Touchard G, Striker L, Khamlichi AA, Rocca A, Denoroy L, Cogné M. (1994). Monoclonal immunoglobulin deposition disease (Randall type), relationship with structural abnormalities of immunoglobulin chains. *Kidney Int* **46**, 965–72.
94. Herzenberg AM, Lien J, Magil AB. (1996). Monoclonal heavy chain (immunoglobulin G3) deposition disease: Report of a case. *Am J Kidney Dis* **28**, 128–31.
95. Wahner-Roedler DL, Kyle RA. (2005). Heavy chain diseases. *Best Pract Res Clin Haematol* **18**, 729–46.

96. Witzig TE, Wahner-Roedler DL. (2002). Heavy chain disease. *Curr Treat Options Oncol* **3**, 247–54.
97. Salem PA, Estephan FF. (2005). Immunoproliferative small intestinal disease: Current concepts. *Cancer J* **11**, 374–82.
98. Al-Saleem T, Al-Mondhiry H. (2005). Immunoproliferative small intestinal disease (IPSID): A model for mature B-cell neoplasms. *Blood* **105**, 2274–80.
99. Fine KD, Stone MJ. (1999). Alpha-heavy chain disease, Mediterranean lymphoma, and immunoproliferative small intestinal disease: A review of clinicopathological features, pathogenesis, and differential diagnosis. *Am J Gastroenterol* **94**, 1139–52.
100. Wahner-Roedler DL, Witzig TE, Loehrer LL, Kyle RA. (2003). Gamma-heavy chain disease: Review of 23 cases. *Medicine (Baltimore)* **82**, 236–50.
101. Wahner-Roedler DL, Kyle RA. (1992). Mu-heavy chain disease: Presentation as a benign monoclonal gammopathy. *Am J Hematol* **40**, 56–60.
102. Franklin EC, Frangione B, Prelli F. (1976). The defect in mu heavy chain disease protein GLI. *J Immunol* **116**, 1194–5.
103. Frangione B, Franklin EC, Prelli F. (1976). Mu heavy-chain disease – a defect in immunoglobulin assembly, structural studies of the kappa chain. *Scand J Immunol* **5**, 623–7.

17 CHAPTER

Mature T-Cell and NK-Cell Neoplasms

Faramarz Naeim, P. Nagesh Rao, Sophie Song and Wayne W. Grody

Mature T- and NK-cell neoplasms represent a wide spectrum of lymphoid malignancies developed from the clonal proliferation of mature T- and NK-cells. These disorders may involve bone marrow and peripheral blood (leukemia), lymphoid or extramedullary tissues (lymphoma), or both. Mature T- and NK-cell neoplasms comprise <15% of all lymphoid tumors. According to the World Health Organization (WHO) classification, they are divided into three major categories: (1) leukemic or disseminated, (2) nodal, and (3) extranodal [1] (Table 17.1).

TABLE 17.1 WHO classification of mature T- and NK-cell neoplasms.*

Leukemic or disseminated
T-cell prolymphocytic leukemia
T-cell large granular lymphocytic leukemia
Aggressive NK-cell leukemia
Adult T-cell leukemia/lymphoma
Nodal
Peripheral T-cell lymphoma, unspecified
Angioimmunoblastic T-cell lymphoma
Anaplastic large cell lymphoma
Extranodal
Cutaneous
Blastic NK-cell lymphoma**
Mycosis fungoides/Sezary syndrome
Primary cutaneous anaplastic large cell lymphoma
Other extranodal
Extranodal NK/T-cell lymphoma, nasal type
Enteropathy-type T-cell lymphoma
Hepatosplenic T-cell lymphoma
Subcutaneous panniculitis-like T-cell lymphoma

*Adapted from Ref. [1].

**Known also as agranular hematodermic neoplasm it is apparently derived from plasmacytoid dendritic cells and not from NK-cells.

T-CELL PROLYMPHOCYTIC LEUKEMIA

T-cell prolymphocytic leukemia (T-PLL) is a sporadic and aggressive lymphoproliferative disorder of post-thymic T-cells characterized by a high peripheral blood lymphocyte count and infiltration of the bone marrow, spleen, liver, lymph nodes, and skin [1–3].

Etiology and Pathogenesis

The etiology and pathogenesis of T-PLL are not known. A high frequency of ataxia-telangiectasia (*ATM*) gene mutations suggests that *ATM* functions as a type of tumor suppressor gene [4–6]. Most T-PLL cases also show an aberrant T-cell receptor alpha (*TCRA*) gene rearrangement that activates *TCL1* or *MTCP1-B1* oncogenes [4].

Pathology

Morphology

T-PLL refers to a group of mature T-cell leukemias with a diverse morphology but similar clinical outcome. The observed morphologic variations include [1]:

1. Cells with typical prolymphocytic features: medium-sized lymphocytes with variable amount of non-granular basophilic cytoplasm; round, oval, or irregular nucleus; coarse chromatin; and a single prominent nucleolus (Figures 17.1 and 17.2). T-prolymphocytes often show cytoplasmic blebs. The prolymphocytic morphology accounts for about 70% of the cases.

2. Small lymphocytes often with irregular nuclei and indistinct nucleolus [7]. This morphologic subtype was previously referred to as T-CLL, but now it is included in T-PLL, because of similar biological behavior. It represents about 25% of the cases [1].

3. Approximately 5% of the cases may show lymphoid cells with cerebriform (Sezary-like) nuclei [1, 8, 9].

The peripheral blood and bone marrow are the primary sites of involvement. There is marked peripheral blood lymphocytosis, usually >100,000/μL, often with anemia and thrombocytopenia. The bone marrow is commonly infiltrated in a diffuse or nodular pattern. Splenic infiltration consists of the involvement of both white and red pulps. Skin is affected in about 20% of the cases with dense infiltration of the dermis without epidermal infiltration. The involved lymph nodes are diffusely infiltrated, primarily in the paracortical areas. The remnants of follicular structures may be present [1].

Immunophenotype

The neoplastic cells in the majority of T-PLL cases are of helper T phenotype, expressing CD4 and pan-T-cell markers CD2, CD3, CD5, and CD7. They lack the expression of CD1a and TdT. In approximately 15% of the cases neoplastic prolymphocytes are CD8+, and in about 25% of the cases they coexpress CD4 and CD8 (Figure 17.3) [1, 10].

Cytogenetic and Molecular Studies

Analogous to the immunoglobulin (IgH) receptor loci that are frequently affected by translocations in B-cell lymphomas, the T-cell receptor (TCR) gene loci are targeted by chromosomal breakpoints in approximately 30% of precursor T-cell lymphoblastic leukemias/lymphomas involving various translocation partners. Conventional cytogenetic studies have shown the presence of complex karyotypes and some recurrent chromosomal abnormalities: the most frequent being t(14;14)(q11.2;q32) (Figure 17.4), inv(14)(q11.2q32) (Figure 17.5), t(X;14)(q28;q11.2), i(8)(q10), and t(8;8)(p12;q11.2) [11].

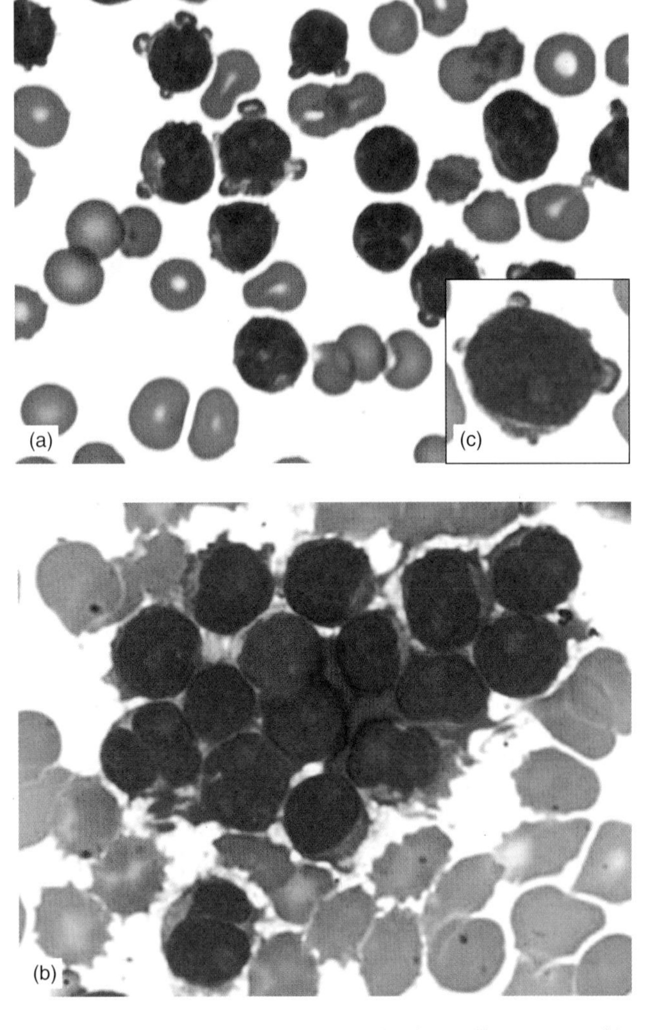

FIGURE 17.1 T-prolymphocytic leukemia. Blood (a) and bone marrow (b) smears demonstrating numerous prolymphocytes with cytoplasmic blebs (c) inset of (a).

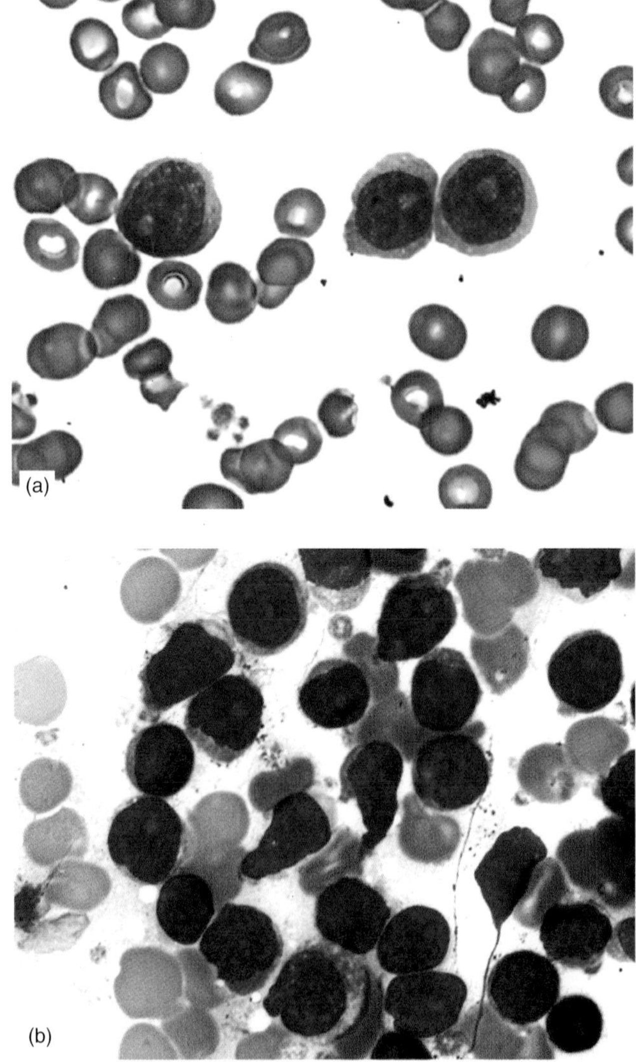

FIGURE 17.2 T-prolymphocytic leukemia. Blood (a) and bone marrow (b) smears demonstrating numerous prolymphocytes.

The inv(14)(q11.2q32) or t(14;14) (q11.2;q32) is reported in over 70% of the cases. These translocations juxtapose *TCRαδ* (14q11.2) and *TCL1* (14q32) genes. The abnormalities of chromosomes 8, 11, 14, and X in T-prolymphocytic leukemia

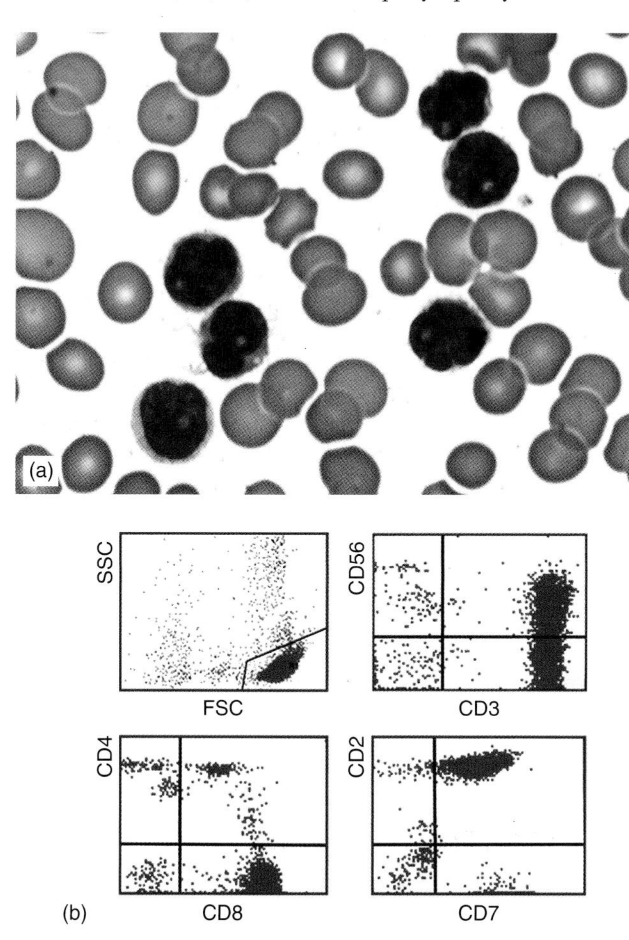

FIGURE 17.3 Flow cytometric analysis of a case of T-prolymphocytic leukemia demonstrating a less common cytotoxic phenotype. The neoplastic cells express CD2, CD3, CD5, CD8, and partial CD56.

have been identified by fluorescence *in situ* hybridization (FISH) studies. Other more consistent aberrations include the loss of 8p, 11q, 22q11, 13q, 6q, 9p, 12p, 11p11-p14, and 17p. In addition, four regions of gain at 8q, 14q32, 22q21-qter, and 6p are also observed. The Xq28 (*MTCP-1* gene locus) or the 14q32.1 regions are involved in translocations or inversions with the TCRαδ at 14q11.2. Translocations involving the TCR loci either TCRα/δ at 14q11.2 or TCRβ at 7q35 have been detected in 15–33% of the patients [12]. These translocations lead to a deregulated expression of the partner gene by juxtaposition with the regulatory region of one of the TCR loci. The TCR breakpoints in many cases resemble TCR recombination signals, implying that the genetic alteration occurred during TCR rearrangement. The inactivation of the *ATM* gene by deletion and mutation is consistently found in T-PLL. *ATM* mutations have been detected in over 50% of the T-PLL cases suggesting that *ATM* acts as a tumor suppressor gene [4].

Recent studies using FISH for loss of heterozygosity (LOH) and comparative genomic hybridization (CGH) analyses have identified losses of the 11q21-q23 region in most of the T-PLL cases. The *ATM* gene located in this region has been shown to be lost by these deletions (Figure 17.7). Other chromosomal aberrations include der(11)t(1;11)(q21;q23) (Figure 17.6); t(X;7)(q28;q35), t(X;14)(q28;q11), and t(3;22) (q21;q11.2) have also been reported [5, 11].

At the molecular level, the rearrangements of *TCR* and the involvement of *TCL1*, *MTCP1-B1*, and *ATM* genes are common findings. ATM mutations have been detected in over 50% of the T-PLL cases suggesting that ATM acts as a tumor suppressor gene [3–5]. Unfortunately, the *ATM* gene is extremely large, and detection of mutations, which requires extensive gene sequencing, is not routinely available for this purpose.

TCR gene rearrangements are detected in the same general manner as for immunoglobulin gene rearrangements in B-cell malignancies (see Chapter 15). However, they are often of more crucial importance to the case, since one does not have the advantage of surface immunoglobulin

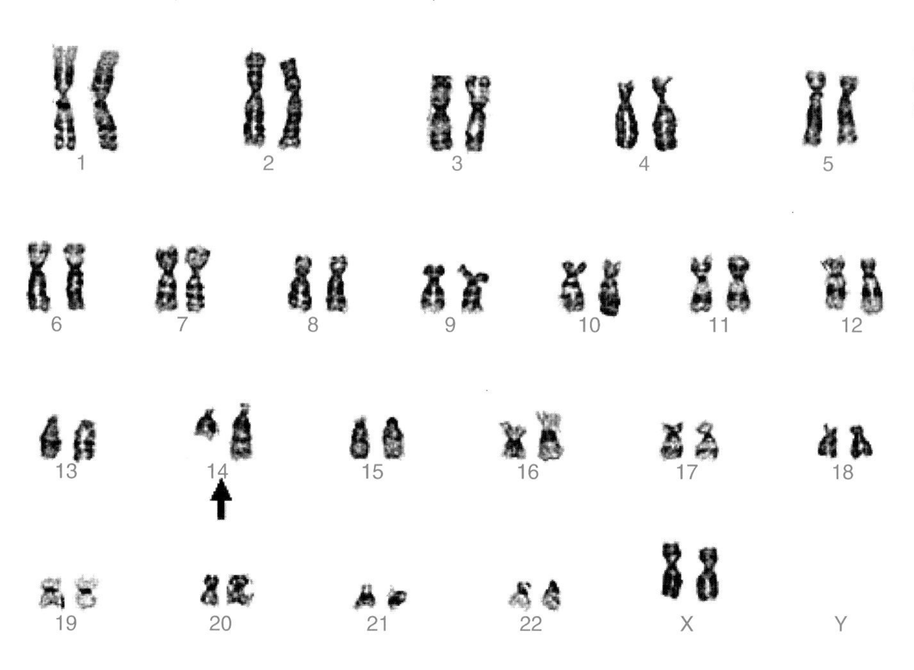

FIGURE 17.4 Karyotype of tumor cells in a patient with T-prolymphocytic leukemia demonstrating 46,XX,t(14;14)(q11.2;q32).

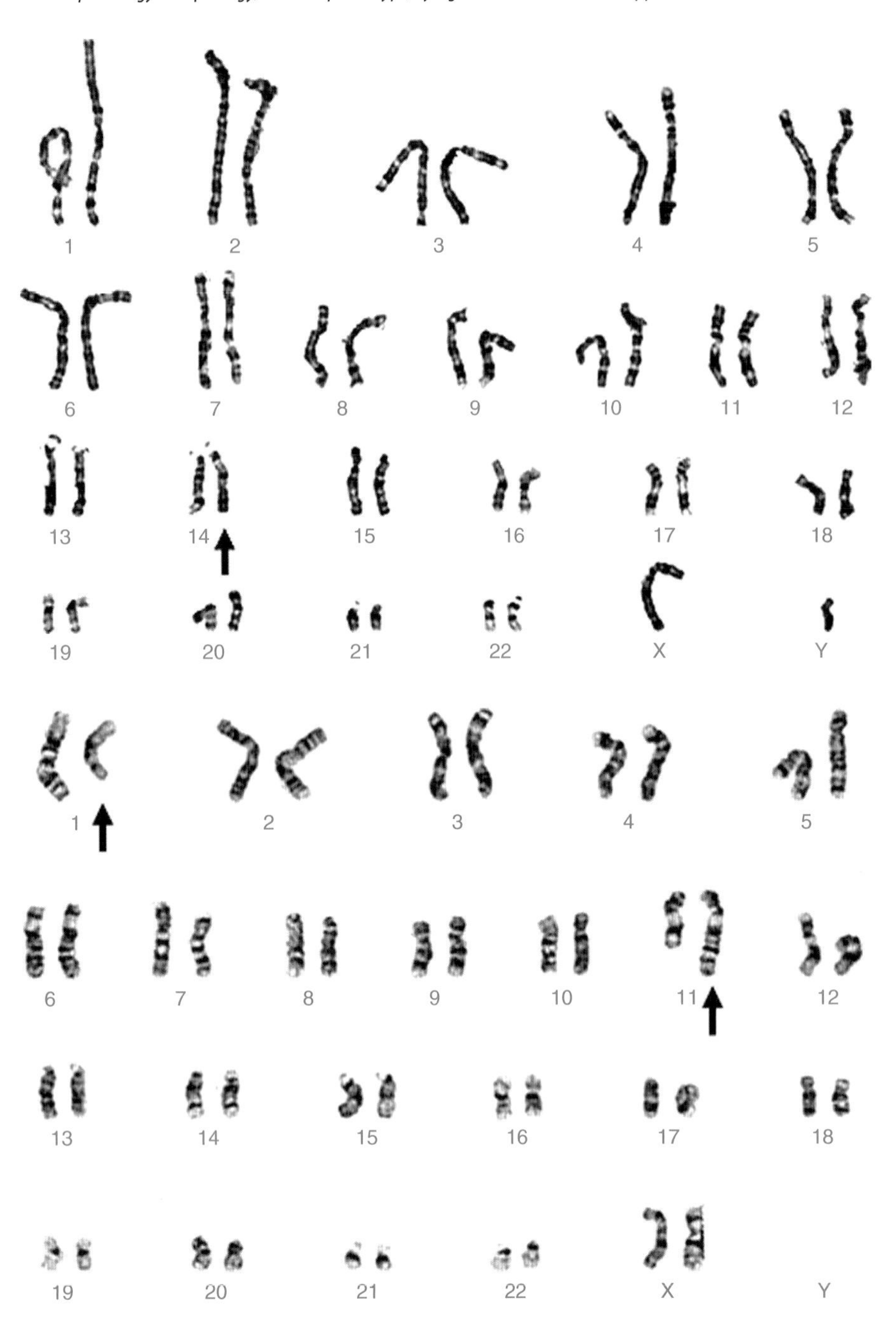

FIGURE 17.5 Karyotype of tumor cells in a patient with T-prolymphocytic leukemia demonstrating 46,XY,inv(14)(q11.2q32).

FIGURE 17.6 Karyotype of tumor cells in a patient with T-prolymphocytic leukemia demonstrating 46,XX,t(1;11)(q21;q23).

immunophenotyping (light-chain restriction) as ancillary evidence of clonality. Again, the Southern blot method will pick up a greater proportion of clonal rearrangements because it is capable of surveying a larger span of the target gene region. Most laboratories use a probe directed to the constant region of the beta-chain genes ($T_C\beta$) on chromosome 7q34. This is because the most circulating T-cells are of the $\alpha\beta$ type. But as laboratories have moved away from the cumbersome Southern blot procedure, various polymerase chain reaction (PCR) strategies have been developed. Unlike the J_H region targeted in B-cell lesions, however, the TCRβ region is very large and complex, and to cover it adequately (i.e. with pick-up rate approaching that of Southern blot) requires the use of a large number of primers [13] which can itself be quite cumbersome and difficult to interpret. However, the *TCR* genes do not undergo somatic hypermutation as do the immunoglobulin genes, eliminating an important cause of false-negative PCR results seen in the B-cell lesions. One compromise is to target the TCRγ locus, which has far fewer V and J genes, and thus a less complex array of rearrangements must be detected. Moreover, because the γ genes typically rearrange before the β genes, this approach should be equally sensitive as TCRβ PCR testing [14]. However, the relatively more limited number of TCRγ rearrangements can produce a type of false-positive result known as *pseudoclonality*; if the total number of T-cells in the submitted specimen is scant, preferential amplification of a small number of these cells that happen to have the same

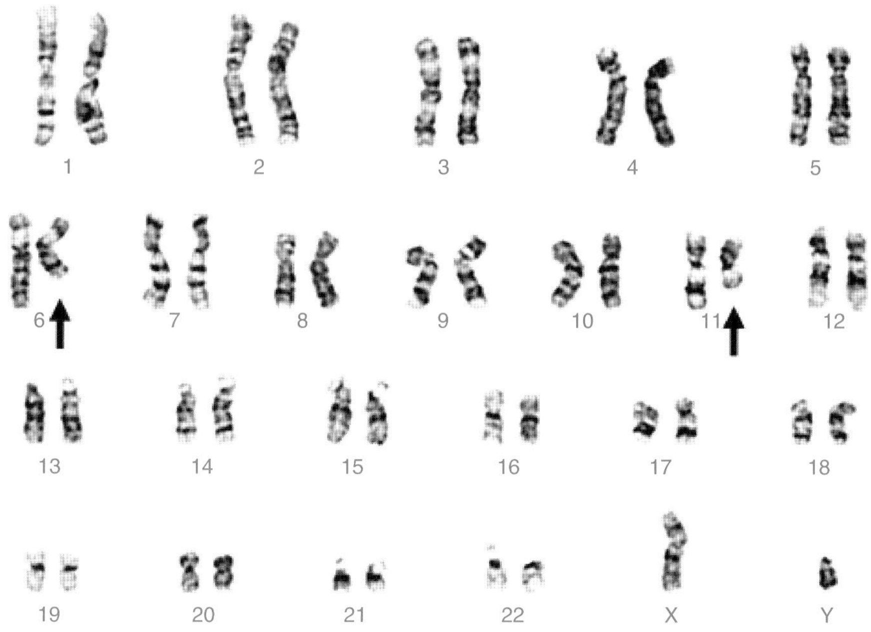

FIGURE 17.7 Karyotype of tumor cells in a patient with T-prolymphocytic leukemia demonstrating 46,XY,del(6)(q13q23),del(11)(q21).

rearrangement pattern can give the appearance of a clone when there is really none. This is unlikely to occur in a leukemia specimen, but may be seen in tissue samples such as skin biopsies.

Clinical Aspects

T-PLL is a rare sporadic T-cell lymphoproliferative disorder, often presented with hepatosplenomegaly and generalized lymphadenopathy [1, 3, 15]. Most of the patients are older than 50 years. The peripheral lymphocyte count is markedly elevated (usually >100,000/μL), commonly associated with anemia and thrombocytopenia. Skin infiltration and serous effusions may be observed. The disease typically has an aggressive course with a median survival of <1 year, though occasional cases with spontaneous remission have been reported [16]. An indolent form of T-PLL has also been reported with t(3;2)(q21;q11.2) and elevated serum β2-microglobulin [4]. Combination chemotherapy, and more recently treatment with monoclonal anti-CD52 antibodies (CAMPATH-1H), and allogeneic stem cell transplantation have been used with some responses [3, 17, 18].

Differential Diagnosis

The differential diagnosis includes all leukemic lymphoproliferative disorders that have prolymphocytic morphology or cerebriform nuclei. B-prolymphocytic leukemia and prolymphocytic variant of mantle cell lymphoma are of B-cell lineage with their own characteristic immunophenotypic features. The neoplastic cells of Sezary syndrome (SS) often lack CD7 expression, whereas T-PLL cells are typically CD7+. The tumor cells of adult T-cell leukemia/lymphoma (ATL) may mimic the cerebriform variant of T-PLL, but unlike the T-PLL cells, they are positive for human T-lymphotropic virus type I (HTLV-I).

LARGE GRANULAR LYMPHOCYTIC LEUKEMIAS AND RELATED DISORDERS

The large granular lymphocytes (LGLs) account for 8–15% of the peripheral blood lymphocytes (200–400/mL) characterized by abundant cytoplasm with azurophilic granules (Figure 17.8) [19]. The azurophilic granules contain cytolytic components such as perforin and granzymes. Perforin is a cytolytic protein that induces apoptosis by creating pores in the plasma membrane of the target cell. Granzymes are proteases that induce apoptosis in virus-infected cells. The LGLs are divided into two major categories: cytotoxic T- and NK-cells. The LGL T-cells typically express CD3, CD8, and CD57 and show *TCR* gene rearrangement; whereas the NK-cells express CD56, are negative for surface CD3, may express CD8, and do not show *TCR* gene rearrangement (Figure 17.9) [19–21].

Reactive (non-clonal) large granular lymphocytosis is a relatively frequent phenomenon and has been observed in various conditions, such as viral infections, collagen vascular disorders, myelodysplastic syndromes (MDS), non-Hodgkin lymphomas, hemophagocytic syndrome, and in patients with solid tumors [21–24].

The leukemic (clonal) LGL disorders are of either T- or NK-cell type and are characterized by persistent (≥6 months) large granular lymphocytosis and often evidence of infiltration of various organs such as bone marrow, spleen, and liver [1, 20, 21]. There is also an extranodal NK/T-cell lymphoma. According to the WHO, these disorders are classified as [1, 25]:

1. T-cell large granular lymphocytic (T-LGL) leukemia
2. Aggressive NK-cell leukemia
3. Extranodal NK/T-cell lymphoma, nasal type.

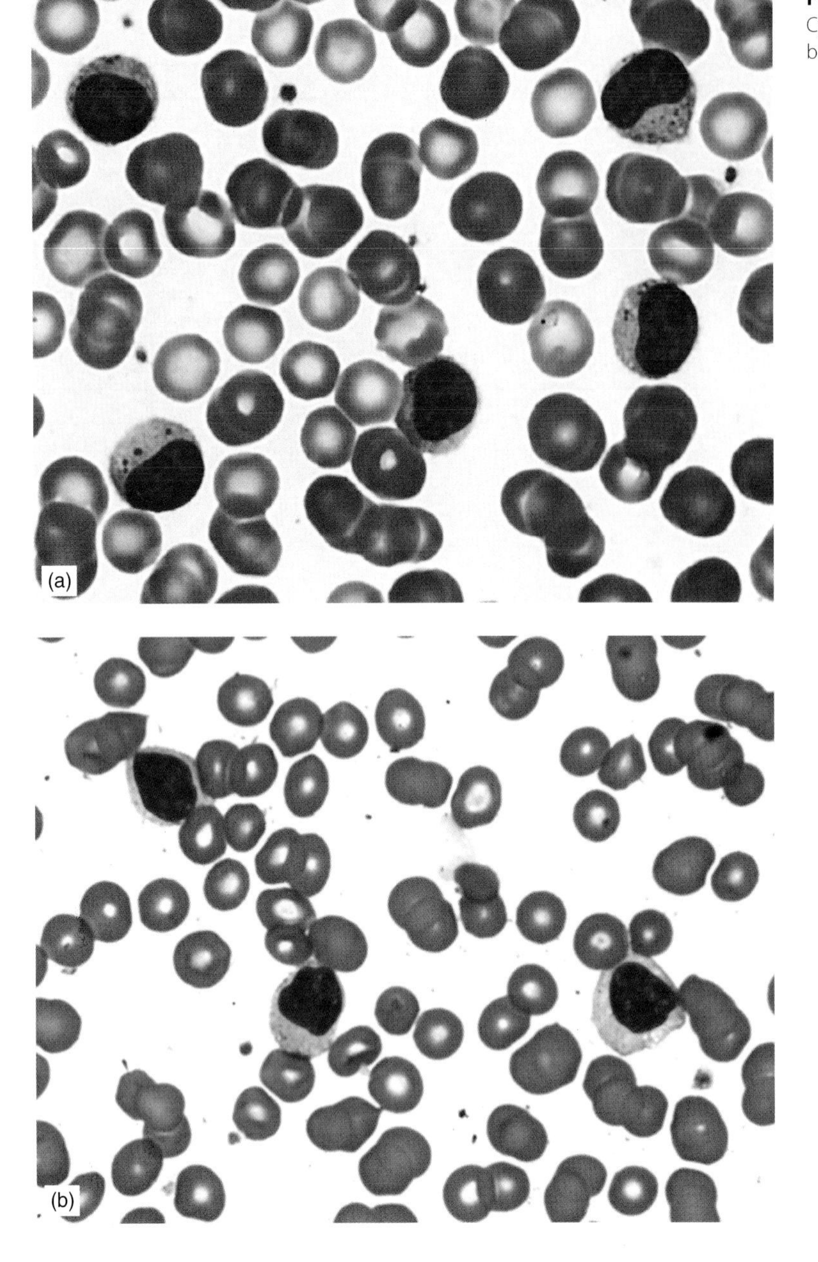

FIGURE 17.8 Large granular lymphocytic (LGL) leukemia. Cytoplasmic granules in LGL cells are usually clearly visible (a), but sometimes difficult to detect (b).

T-CELL LARGE GRANULAR LYMPHOCYTIC LEUKEMIA

The T-cell large granular lymphocytic (T-LGL) leukemia is a chronic lymphoproliferative disorder characterized by persistent T-cell large granular lymphocytosis (usually >2,000/μL), cytopenia, and strong association with rheumatoid arthritis [19–21, 25, 26].

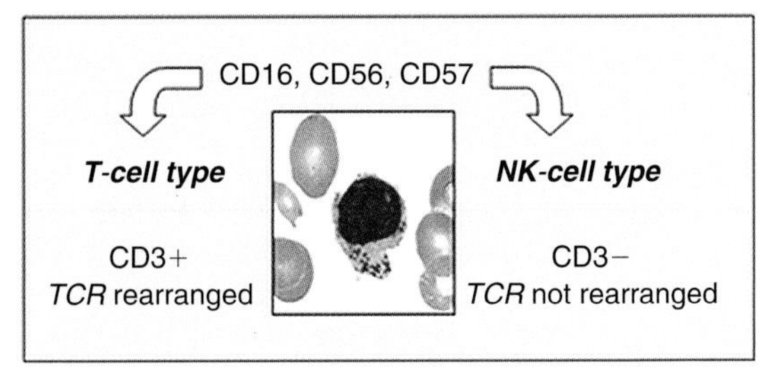

FIGURE 17.9 The major differences between T- and NK-cell types of LGLs.

Etiology and Pathogenesis

The etiology and pathogenesis of T-LGL leukemia are not known. The strong association of T-LGL leukemia with rheumatoid arthritis suggests that an antigen-driven mechanism may play a role in the initial steps of LGL expansion. Recent studies imply that certain cytokines, such as IL-12 and IL-15, induce proliferation and activation of LGL and, therefore, may play a role in LGL leukemogenesis

[27]. Also, the leukemic LGL cells are able to activate the PI3K–Akt pathway. This pathway regulates the balance between cell survival and apoptosis and has the capacity to block Fas-induced apoptosis [28].

Sera from approximately 50% of the patients with T-LGL leukemia react with HTLV-I/II envelope protein p21e (epitope BA21) in spite of the infrequent detection of the HTLV-II in these patients. These findings suggest the presence of a cellular or retroviral serum protein with homology to the BA21 epitope with a potential role in the LGL leukemogenesis [29, 30].

Pathology

Morphology

The T-LGL leukemia typically shows a persistent (>6 months) absolute large granular lymphocytosis of >2,000 µL, often associated with neutropenia or pancytopenia. The total lymphocyte count in most patients is modestly elevated (5,000–10,000 µL), but in about one-fourth of the cases it is within normal limits. In a minority of the patients (about 5%), the absolute LGL count is <1,000/µL, or the large lymphocytes lack cytoplasmic azurophilic granules, despite their CD3 and CD57 coexpression (see Figure 17.8) [1, 19–22, 25].

Granulocytopenia is observed in over 80% of the patients with approximately half of the patients demonstrating <500/µL absolute neutrophil counts [19, 31, 32]. Neutropenia is attributed to different possible mechanisms such as induction of apoptosis in neutrophils by Fas ligand secreted by the leukemic LGL cells, bone marrow infiltration, splenomegaly, or an autoimmune process [28, 32].

Anemia is observed in about 50% of the patients, which may be severe and transfusion dependent (about 20%). The possible mechanisms of anemia include an autoimmune process, splenomegaly, bone marrow infiltration, or pure red cell aplasia. T-LGL leukemia is reported as the most common underlying cause of the pure red cell aplasia. The inhibition of erythroid colony-forming units (CFU-E) and burst-forming units (BFU-E) has been observed by the LGL leukemic cells in patients with pure red cell aplasia [33].

Moderate thrombocytopenia is a frequent finding, due to an autoimmune process, bone marrow infiltration, or secondary to splenomegaly.

The bone marrow is involved in about 90% of the cases. The pattern of leukemic infiltration is usually interstitial and/or sinusoidal. The involvement of the spleen is a common feature with infiltration of the red pulp, often associated with white pulp hyperplasia secondary to the presence of an autoimmune condition (Felty's syndrome). The liver infiltration may involve portal and sinusoidal areas. The involvement of the lymph nodes and other organs is unusual.

Immunophenotype

The T-LGL leukemia cells demonstrate a mature post-thymic phenotype and typically express CD3 (surface and cytoplasmic), CD8, TCRαβ, CD16, CD57, CD122 (IL-2 receptor-β), and TIA-1 (Figure 17.10) [19–21]. There are minor variants which are characterized by the expression of CD4, CD26, CD56, or TCRγδ. The CD56+ subtype is often associated with an aggressive clinical course [26].

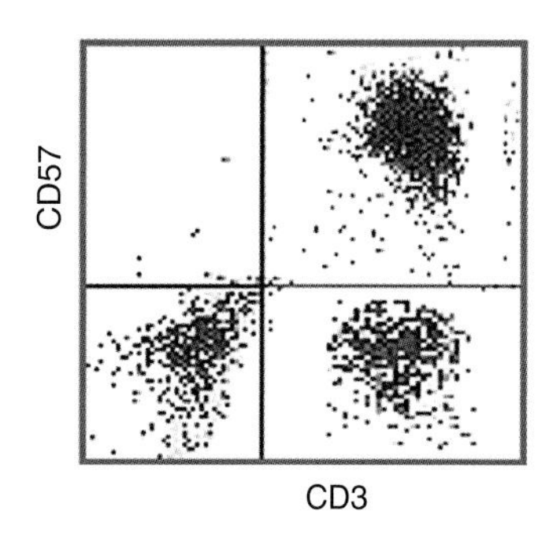

FIGURE 17.10 Flow cytometry of T-LGL leukemia showing coexpression of CD3 and CD57 by tumor cells.

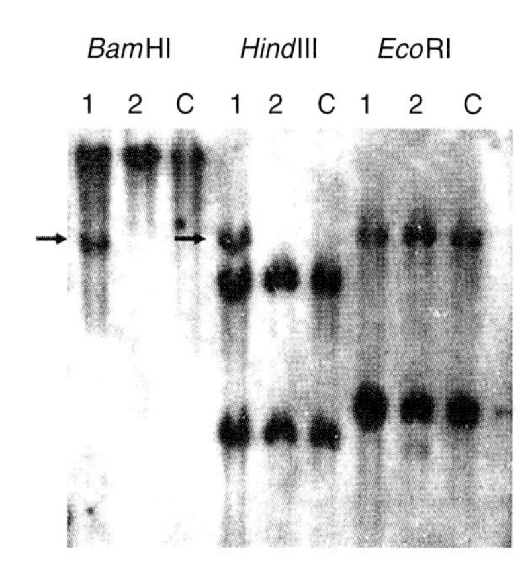

FIGURE 17.11 Clonal *TCRβ* gene rearrangement in a T-LGL leukemia detected by Southern blotting using the $T_C\beta$ probe. Non-germline, rearranged hybridization bands are seen for patient 1 with two of the three restriction enzymes used (*Bam*HI, *Hind*III, arrows), which is sufficient to diagnose T-cell clonality. Patient 2 shows only germline bands.

Several monoclonal antibodies are raised against the TCR variable domain and are available for the immunophenotypic analysis of TCR. Of these, monoclonal antibodies against the Vβ 13.1 region have been reported to be highly associated with T-LGL leukemia [34].

Molecular and Cytogenetic Studies

The T-LGL leukemia cells show *TCRβ* and/or *TCRγ* gene rearrangement by Southern blot or PCR. As noted earlier, TCRβ probes are used for Southern blot analysis (Figure 17.11), whereas the TCRγ region is most often used as the primary PCR target (Figure 17.12) because it is much smaller and thus easier to amplify with just a few primer sets than is TCRβ. In one study, DNA microarray analysis of T-LGL leukemias showed evidence of *IL-1β* gene activation, which was associated with the elevated serum levels of IL-1β in the majority of the patients [35, 36].

Various cytogenetic abnormalities have been reported in a minority of the T-LGL leukemia cases. These include inv(7)(p15q35) (Figure 17.13), inv(14)(q11.2q32), del(14)(q22-q24) (Figure 17.14), t(11;15)(q13;q22-24), inv(4)(p14q12), trisomy 8, and trisomy 14 [37, 38].

Clinical Aspects

The T-LGL leukemia accounts for about 85% of all large granular leukemias. The remaining 15% represent the NK type. The median age is around 60 years with only 10% of patients younger than 40 years [19, 39–41]. The clinical symptoms are primarily related to the patient's granulocytopenia and anemia, such as recurrent infections, fever, night sweats, and fatigue. Approximately 30% of the patients are asymptomatic at the time of diagnosis [42]. T-LGL leukemia has been frequently observed in association with connective tissue diseases, primarily rheumatoid arthritis [43–45].

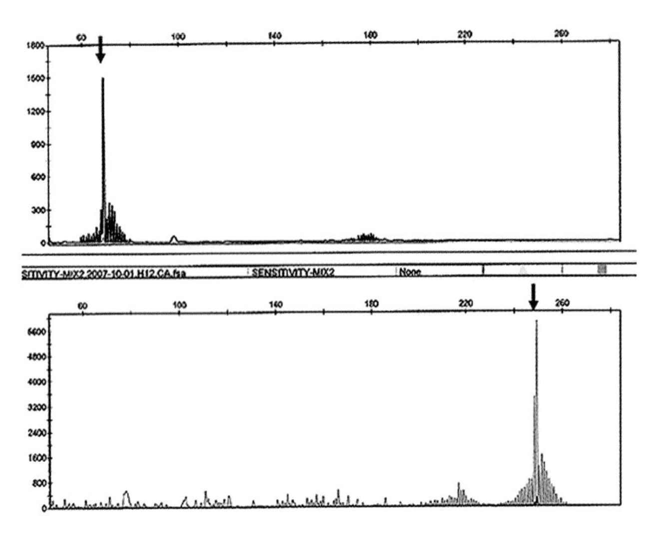

FIGURE 17.12 Clonal *TCRγ* gene rearrangement in a T-LGL leukemia. Discrete clonal PCR peaks are demonstrated above the polyclonal background smear in two of the four *TCRγ* gene regions targeted (Group 1 and AltVγ, arrows).

Rheumatoid factor, antinuclear antibodies, and circulating immune complexes are detected in 40–60% of the patients [22]. Rheumatoid arthritis has been reported in about 25% of the patients with T-LGL leukemia. Many of these patients present the triad combinations of neutropenia, rheumatoid arthritis, and splenomegaly (Felty's syndrome) [46]. Other associated disorders include B-cell lymphoid malignancies, non-Hodgkin lymphoma, thymoma, monoclonal gammopathy, and MDS [19].

The T-LGL leukemia is considered a chronic indolent disorder with reported median survival of >10 years [45]. However, patients with severe neutropenia, "B" symptoms, CD3+, and CD56+ have less favorable prognosis and require treatment [26]. Therapeutic approaches include low-dose chemotherapy with methotrexate, cyclophosphamide, or cytosporin as single agents or in combination with prednisone [19].

AGGRESSIVE NK-CELL LEUKEMIA

Aggressive NK-cell leukemia is characterized by systemic proliferation of LGLs of NK type (NK-LGL), strong Epstein–Barr virus (EBV) association, and an aggressive clinical course [1, 19, 47–49]. A rare indolent EBV-negative condition known as *chronic NK-cell lymphoproliferative disorder* has been described [26]. The neoplastic nature of this indolent condition is uncertain.

Etiology and Pathogenesis

The etiology of the aggressive NK-cell leukemia is not known. The strong association of this disorder with EBV

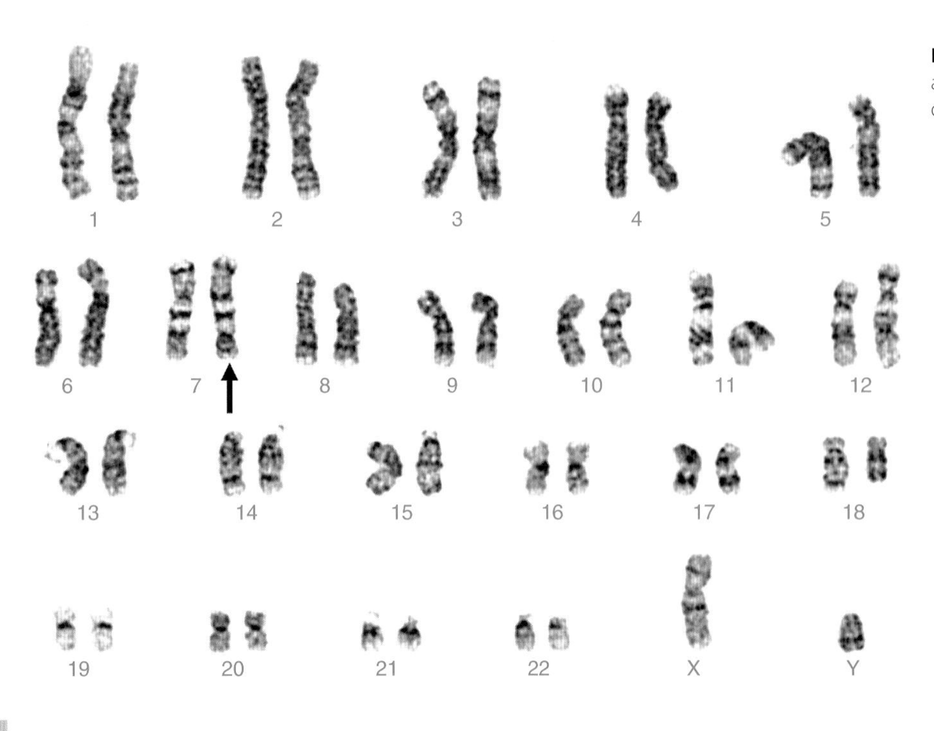

FIGURE 17.13 Karyotype of tumor cells in a patient with T-prolymphocytic leukemia demonstrating 46,XY,inv(7)(p15q35).

suggests that this virus may play a role in the pathogenesis of this disease. EBV infection has been reported in over 50% of the cases of NK-LGL leukemias reported in Japan [50]. EBV nuclear antigen 1 (EBNA-1) and EBV-encoded RNA 1 (EBER-1) have been detected in the leukemic cells by *in situ* hybridization techniques [19].

Pathology

Morphology

The peripheral blood shows an absolute lymphocytosis (usually >10,000/μL) with increased proportion of LGLs. These cells have abundant light blue cytoplasm containing azurophilic granules (see Figure 17.8). The nuclei are round, oval, or irregular and may appear pleomorphic or hyperchromatic. The nuclear chromatin is condensed and nucleoli are indistinct. The amount and the size of the azurophilic granules are variable. According to some observers, these cells are slightly larger than the normal LGLs seen in the peripheral blood [1]. Anemia is common and usually severe, and thrombocytopenia is frequent. In contrast to T-LGL leukemia, severe neutropenia is less common.

Bone marrow is almost always infiltrated by the neoplastic cells. The involvement is diffuse, focal, interstitial, or sinusoidal (Figure 17.15). The tumor cells may be mixed with normal hematopoietic cells and are sometimes difficult to detect. Scattered reactive histiocytes may be present, some showing hemophagocytosis [1, 51].

The extramedullary infiltrations may mimic *extranodal NK/T-cell lymphoma of the nasal type* and often show vascular involvement (angiocentric pattern) and areas of necrosis (see later). Most patients show splenic and hepatic involvement, often with massive hepatosplenomegaly. The leukemic infiltration in the spleen involves the red pulp and in the liver involves the portal areas and/or sinusoids [52]. Also, the gastrointestinal tract, lymph nodes, and cerebrospinal and peritoneal fluids may be involved.

Immunophenotype

The NK-LGL leukemia cells are typically CD2+, surface CD3−, cytoplasmic CD3+, CD4−, CD7+, CD8+, CD16+, CD56+, CD57±, and TCR− (Figure 17.16) [19–21]. They also express TIA-1 and granzyme B (see Figure 17.15b and c).

Molecular and Cytogenetic Studies

The NK-cells, in contrast to T-LGLs, do not demonstrate *TCRαβ* or *TCRγδ* gene rearrangements (Figure 17.17) [53]. The evidence of EBV infection in clonal episomal form has been observed in the majority of the cases, and EBNA-1 and EBER-1 can be detected by *in situ* hybridization techniques.

Various non-random cytogenetic abnormalities have been reported in patients with aggressive NK-cell leukemia including gains of 1p, 6p, 8, 11q, 12q, 17q, 19p, 20q, and Xp, and losses of 6q (Figure 17.18), 11q, 13q, and 17p [37, 54].

Clinical Aspects

In comparison with T-LGL leukemia, NK-LGL leukemia has a more aggressive clinical course and affects younger individuals with a median age of about 40 years. The disease presents with fever, night sweats, weight loss, anemia or pancytopenia, and often massive hepatosplenomegaly [47, 55, 56]. Multiorgan failure is the major cause of death with a survival of <1 year in most instances [22]. Chemotherapy has not been effective [19]. The rare cases of EBV-negative NK-cell lymphoproliferative disorder usually have an indolent clinical course [26].

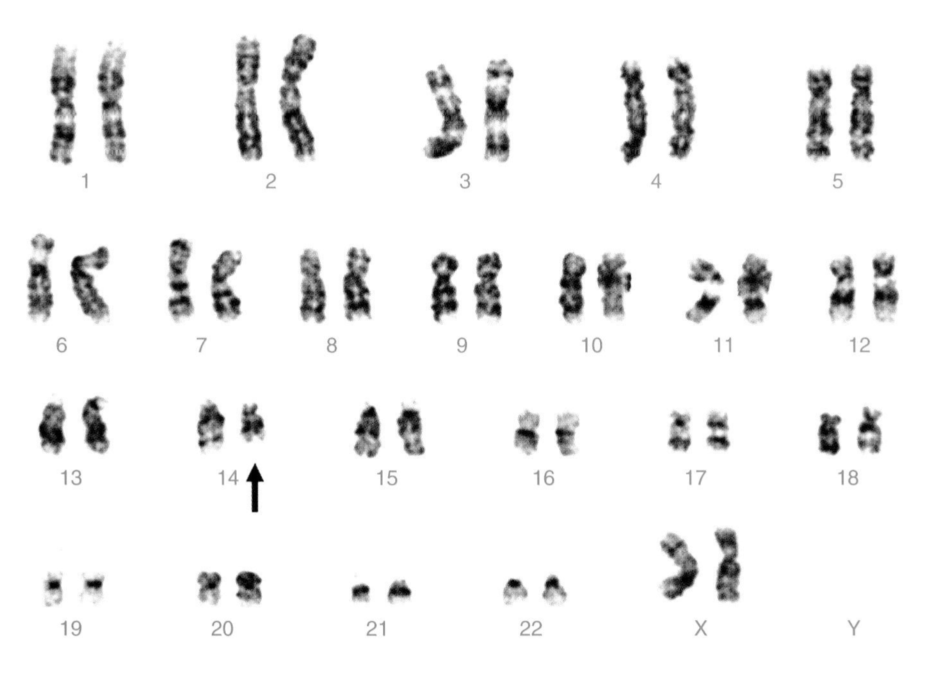

FIGURE 17.14 Karyotype of tumor cells in a patient with T-prolymphocytic leukemia demonstrating 46,XX,del(14)(q22→q24).

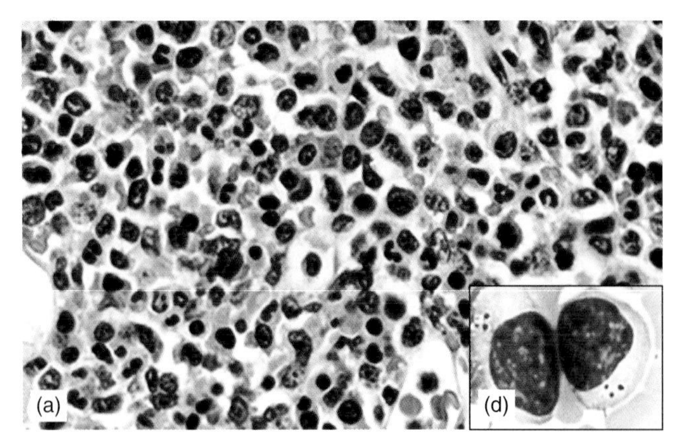

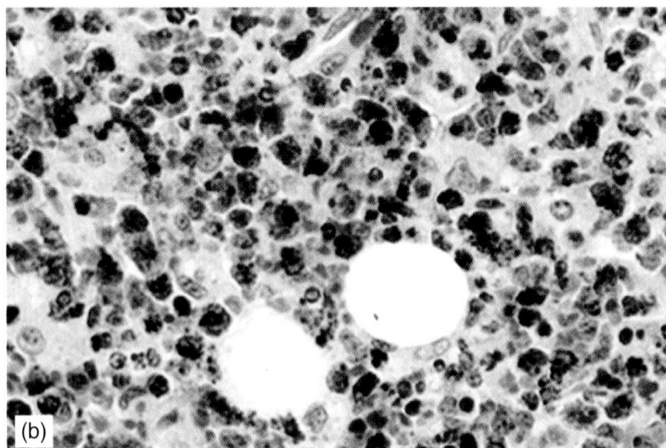

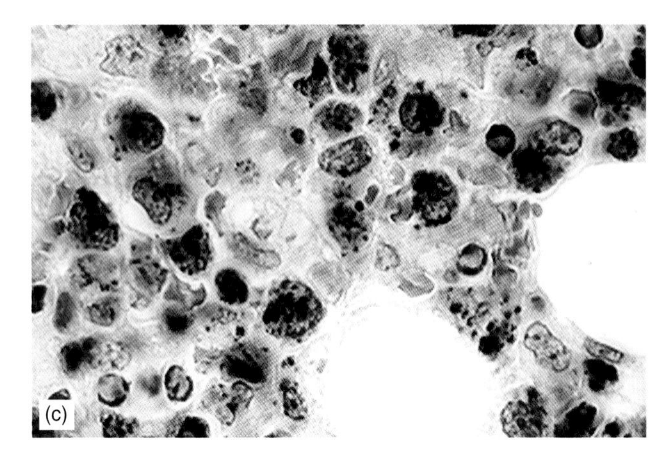

FIGURE 17.15 Bone marrow involvement in a patient with aggressive NK-cell leukemia. An interstitial atypical lymphoid infiltrate is noted (a). The tumor cells are highlighted by a TIA-1 histochemical stain (b and c). (d) Inset of (a) is a demonstration of two LGL cells by touch preparation. Adapted from Naeim F. (1997). *Pathology of Bone Marrow*, 2nd ed. Williams & Wilkins, Baltimore, by permission.

EXTRANODAL NK/T-CELL LYMPHOMA, NASAL TYPE

Extranodal NK/T-cell lymphoma, nasal type, is characterized by the strong EBV association and angiocentric infiltration, leading to vascular destruction and necrosis. The nasopharynx is the most common site of involvement,

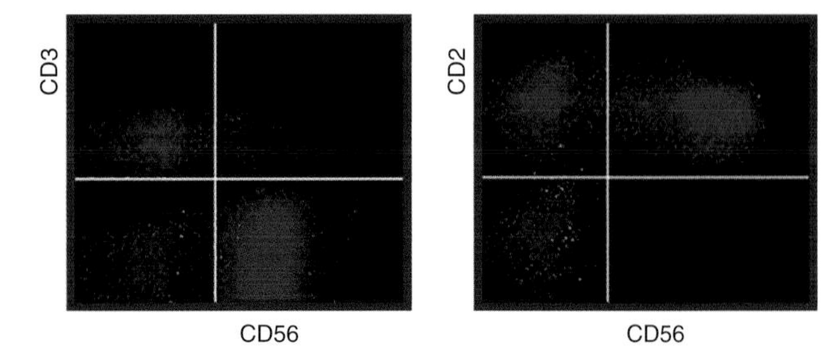

FIGURE 17.16 Flow cytometry of aggressive NK-cell leukemia. Tumor cells are positive for CD56 but lack the expression of surface CD3.

but other extranodal organs such as skin, gastrointestinal tract, testis, and soft tissues may be affected [1]. The NK-cell phenotype (CD56+, CD3−, CD8±, and EBV+) is the predominant cell type, but rare cases may demonstrate a cytotoxic T-cell phenotype (CD56−, CD3+, CD8+, and EBV+) [1, 57, 58]. This disorder was also referred to as lethal midline granuloma, polymorphic reticulosis, and malignant midline reticulosis.

Etiology and Pathogenesis

The etiology of extranodal NK/T-cell lymphoma, nasal type, is not known. The strong association of this disorder (90%) with EBV and evidence of a clonal episomal form strongly suggest a pathogenetic role for this virus [57].

Pathology

Morphology

The lymphomatous infiltrate is polymorphic with a diffuse or patchy involvement and an angiocentric and angiodestructive growth in over 85% of the cases (Figures 17.19 and 17.20) [58]. The ulceration of the overlying epithelium is common and may be associated with atypical hyperplasia of the adjacent epithelium, mimicking squamous cell carcinoma. The neoplasm consists of a mixture of small to large atypical lymphoid cells with variable amount of cytoplasm and irregular nuclei. The predominant infiltrating lymphoid cells may be large or small. The larger cells may show prominent nucleoli [49–51, 58]. Focal or confluent coagulative necrosis is common, often with the presence of apoptotic bodies [1]. Some of the tumor cells, particularly in cytologic preparations (such as touch preparation), may show cytoplasmic azurophilic granules. The lymphomatous infiltrate is often heavily admixed with inflammatory cells such as lymphocytes, plasma cells, eosinophils, and histiocytes, leading to a polymorphic pattern.

Immunophenotype

The NK-cell phenotype (surface CD3−, cytoplasmic CD3ε+, and CD56+) is the most common variant, with the vast majority of the cases being positive for CD2, TIA-1, granzyme B, and EBV [1, 60]. The NK-cell type is usually negative for other T- and NK-cell-associated markers

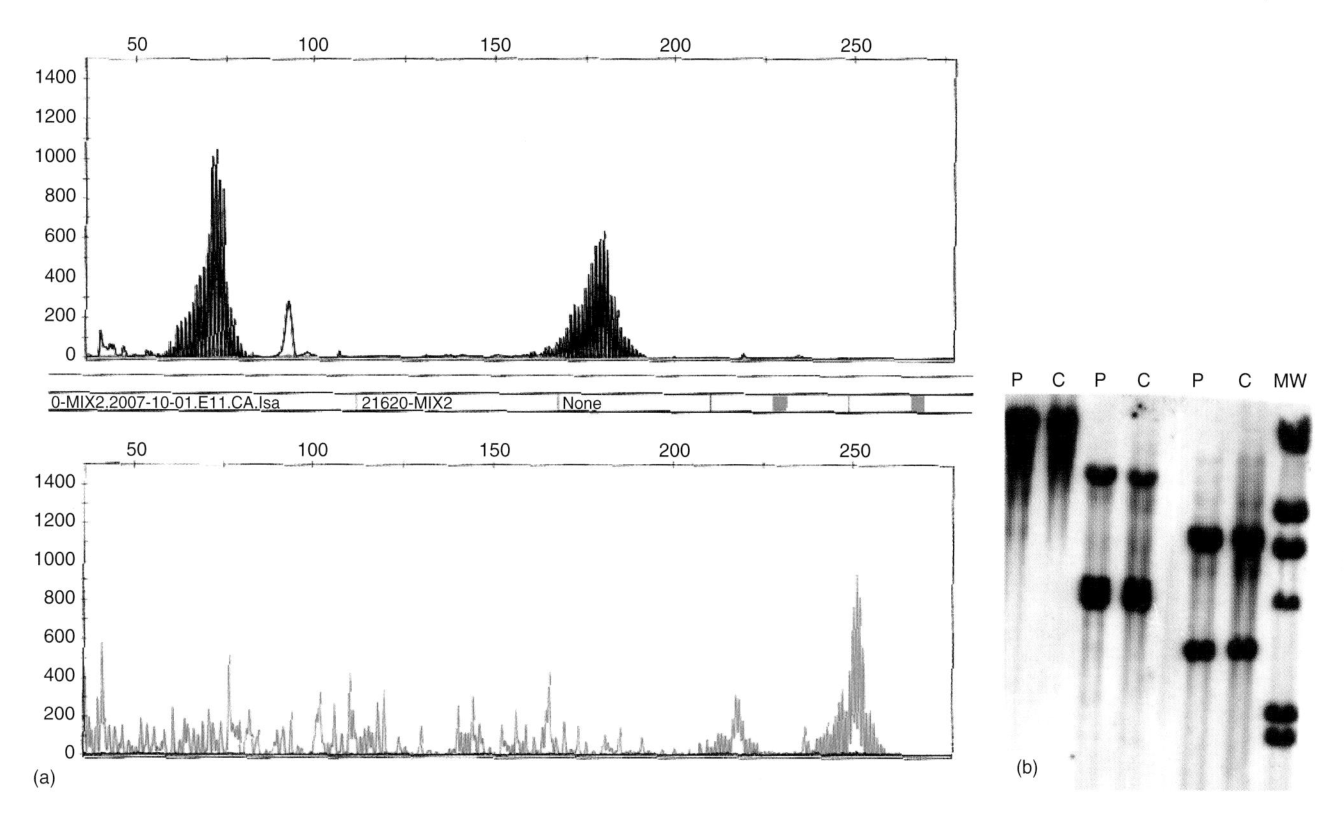

FIGURE 17.17 PCR (a) and Southern blot analysis (b) for clonal *TCR* gene rearrangements in an aggressive NK-cell leukemia, both of which are negative, showing only polyclonal or germline signals for (left-to-right, top-to-bottom) Group 1, Group 2, Vγ, and AltVγ by PCR, and hybridization signals for the patient (P) identical to the negative control (C) in the Southern blot (MW, molecular weight marker).

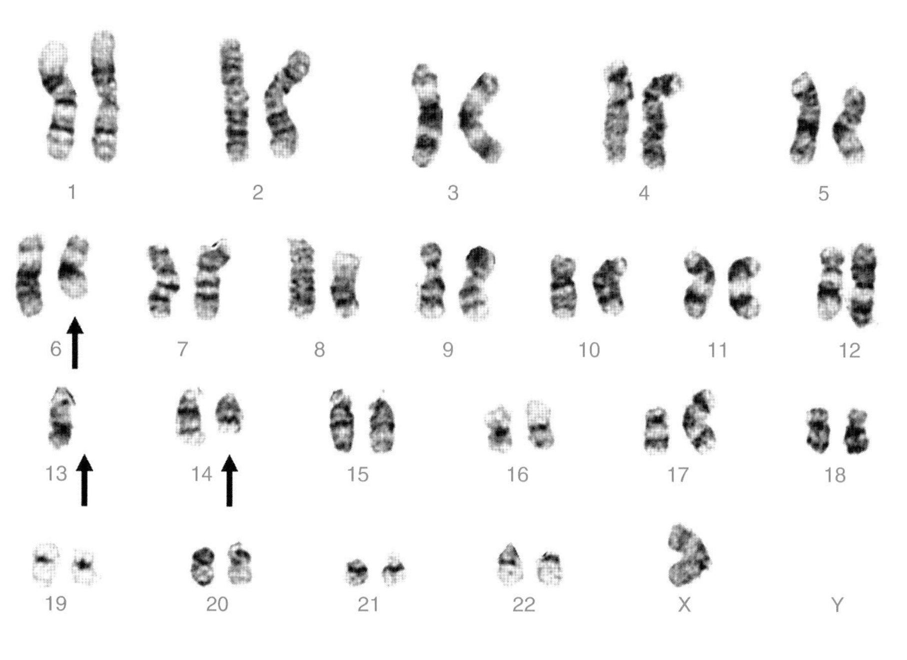

FIGURE 17.18 Karyotype of tumor cells in a patient with aggressive NK-cell leukemia demonstrating 44,X,−X,del(6)(q21),−13,del(14)(q22q24).

such as CD4, CD5, CD8, TCR (αβ or γδ), CD16, and CD57, but occasionally may express CD7 or CD57 [1, 58, 59]. Occasional cases are CD56−, but are positive for CD2, cytoplasmic CD3ε, EBV, TIA-1, and granzyme B.

Molecular and Cytogenetic Studies

The neoplastic cells do not demonstrate *TCRαβ* or *TCRγδ* gene rearrangements. The evidence of EBV infection in clonal episomal form has been observed in the vast majority

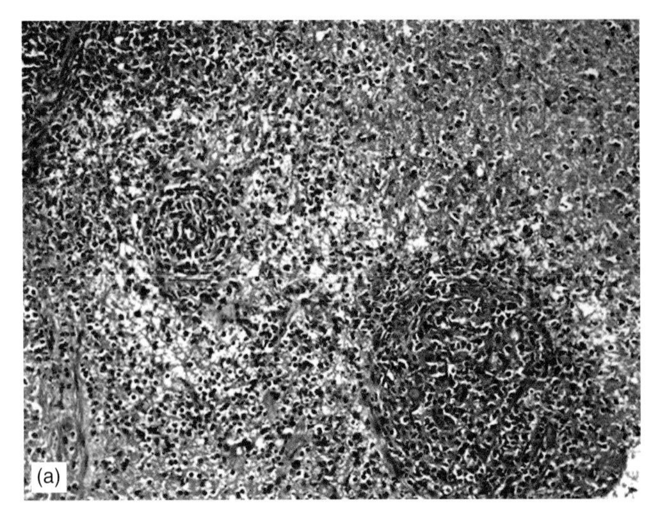

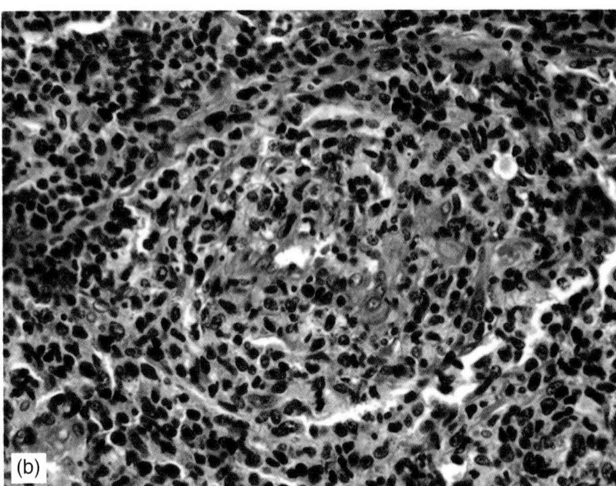

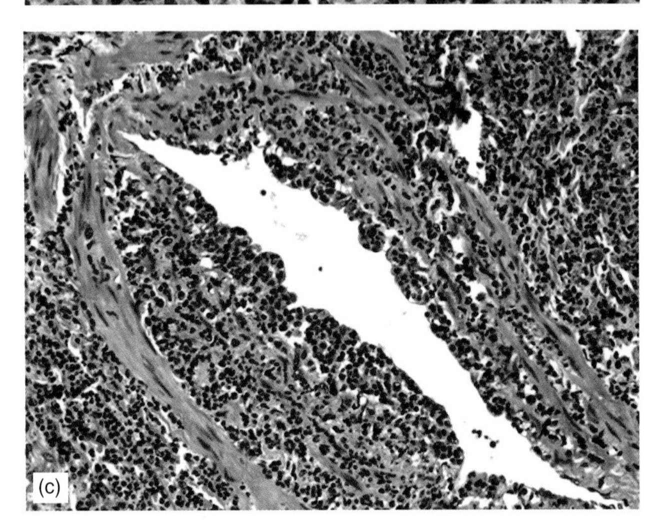

FIGURE 17.19 Extranodal NK/T-cell lymphoma, nasal type, is characterized by angiocentric and angiodestructive infiltration: (a) low power and (b) high power views. (c) The vascular infiltration may involve all the layers of the vascular wall.

of the cases, and EBNA-1 and EBER-1 can be detected by *in situ* hybridization techniques [61]. Although the *in situ* hybridization technique will demonstrate the presence of EBV RNA in essentially all of the tumor cells, proof of clonality, in the absence of clonal *TCR* gene rearrangements, requires examination of Southern blot patterns (fingerprints) of the EBV genome [14]. PCR approaches should be used with caution, given the high frequency of latent EBV infection in the general population.

The most frequent cytogenetic finding is del(6q21-23) (Figure 17.21). Other reported non-random aberrations are i(1q), i(6p), i(7q), del(13q) +8, +X, and 11q23 rearrangements [1, 37].

Clinical Aspects

Extranodal NK/T-cell lymphomas are more prevalent in Asia and Central and South America, and predominantly affect males in the fifth decade of their age. They are clinically divided into two categories: nasal and non-nasal.

The nasal NK/T-cell lymphomas occur in the nose and the upper respiratory and oral cavities including nasopharynx, paranasal sinuses, tonsils, and larynx [49]. Symptoms are local and may include nasal obstruction, bleeding, or destruction and perforation of the hard palate [62]. These tumors are locally invasive but infrequently show distant metastasis. Less than 10% of the patients demonstrate bone marrow involvement [49, 57].

The non-nasal NK/T-cell lymphomas are often multifocal, and dissemination occurs in the early stage of the disease. The primary sites of involvement include the skin, digestive system, spleen, and testis. Nodal involvement is rare. In general, the non-nasal tumors are more aggressive than the nasal types and respond poorly to therapy. The therapeutic approaches include radiation and combination chemotherapy. The reported complete remission rate for early stages of the disease ranges from 60% to 80% for radiotherapy and from 40% to 60% for chemotherapy, but about 50% of the patients may show relapse within the first year [47, 63–65]. The expression of multidrug resistance gene may play an important role in the overall poor response to chemotherapy in a significant proportion of the patients [66].

DIFFERENTIAL DIAGNOSIS

The clinicopathologic features of T-LGL leukemia, aggressive NK-cell leukemia, and extranodal NK/T-cell lymphoma are summarized in Table 17.2. T-LGL leukemia and aggressive NK-cell leukemia should be distinguished from secondary, reactive large granular lymphocytosis. The reactive T-LGL expansions have been reported in autoimmune connective tissue disorders, inflammatory skin disorders, lymphomas, hemophagocytic syndrome, and various viral infections such as EBV, HIV, and CMV [1, 22]. Increased number of circulating NK-cells has been observed in patients with viral infection, solid tumors, MDS, and atomic bomb survivors [67]. Reactive large granular lymphocytosis is polyclonal and is not associated with chromosomal aberrations.

The differential diagnosis for extranodal NK/T-cell lymphomas of nasal type includes CD56+ T-cell lymphomas such as hepatosplenic T-cell lymphoma and other

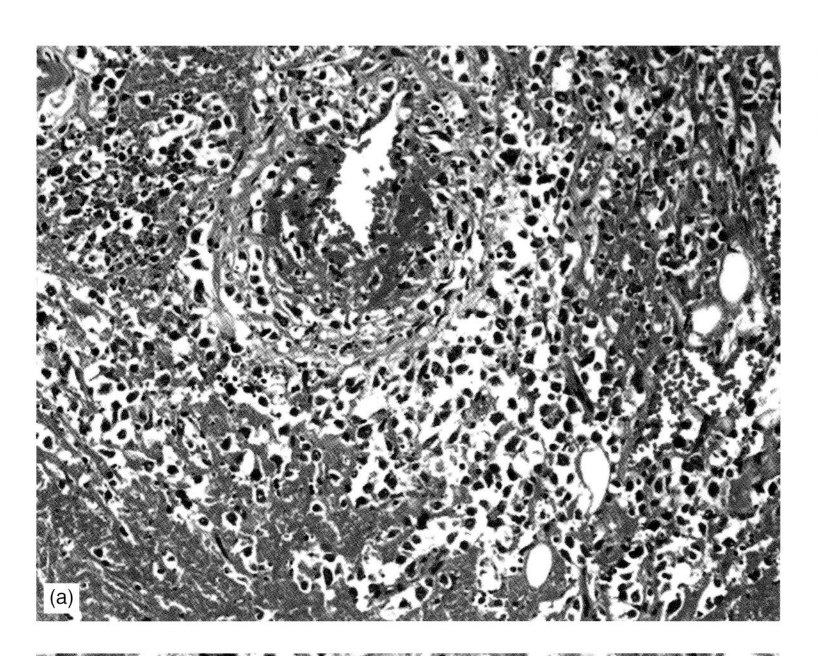

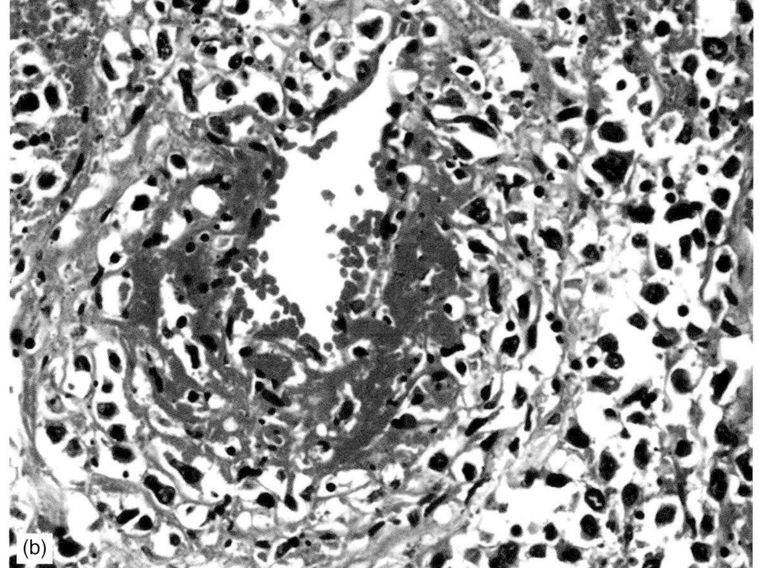

FIGURE 17.20 Extranodal NK/T-cell lymphoma, nasal type, is characterized by an angiodestructive process leading to obstruction and necrosis: (a) low power and (b) high power views.

peripheral T-cell lymphomas. The CD56+ T-cell lymphomas are negative for EBV, express surface CD3 and TCR, and show evidence of *TCR* gene rearrangement.

BLASTIC NK-CELL LYMPHOMA

This neoplasm is now considered to be of dendritic cell origin and, therefore, is discussed in Chapter 21.

FULMINANT EBV-POSITIVE T-CELL LYMPHOPROLIFERATIVE DISORDER

Fulminant EBV-positive T-cell lymphoproliferative disorder is a rare aggressive disorder presenting with fever, hepatosplenomegaly, and pancytopenia. The patients have a clinical history of recent viral-like upper respiratory illness. The age of the patients in one report consisting of five patients ranged from 2 to 37 years and all patients died within 7 days to 8 months [68, 69]. There was an infiltration of lymphocytes in the splenic sinusoids and hepatic portal cells. These cells were positive for CD3, βF1, EBER, and TIA-1, and negative for CD56. The infiltrating T-cells in some cases were CD4+, and in other cases were CD8+, or coexpressed CD4 and CD8. The PCR studies revealed clonal *TCRγ* gene rearrangement and a deletion in the EBV latent membrane protein-1 (*LMP-1*) gene.

ADULT T-CELL LEUKEMIA/LYMPHOMA

Adult T-cell leukemia/lymphoma (ATL) is an HTLV-I-associated peripheral T-cell lymphoid malignancy often

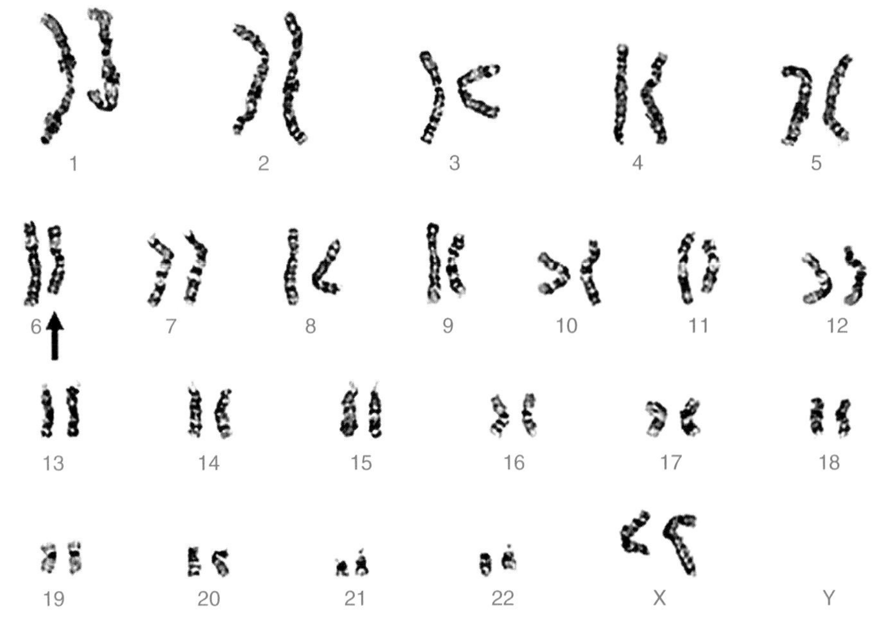

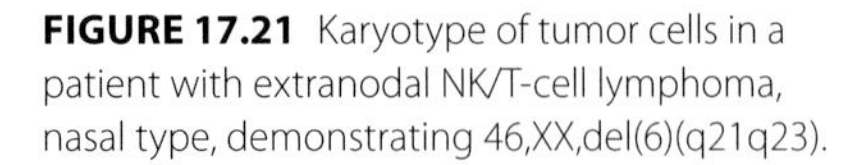

FIGURE 17.21 Karyotype of tumor cells in a patient with extranodal NK/T-cell lymphoma, nasal type, demonstrating 46,XX,del(6)(q21q23).

TABLE 17.2 Clinicopathological features of T-LGL leukemia, aggressive NK-cell leukemia, and extranodal NK/T-cell lymphoma, nasal type.

Features	T-LGL	NK-cell	Extranodal NK/T-cell
Median age (years)	60	40	50
Male:female	1	1	>1
Association	Rheumatoid arthritis	EBV	EBV
Immunophenotype	Surface CD3, CD8, TCR, CD57	CD3ε, CD56	CD3ε, CD56
Prognosis	Indolent	Aggressive	Aggressive

presenting as an acute leukemic onset and aggressive clinical course [70–72].

Etiology and Pathogenesis

HTLV-I has been implicated in the development of ATL [73–77]. A subpopulation of patients infected by HTLV-I, approximately 6% of males and 2% of females, eventually develop ATL after a long latent period [75]. The transmission of HTLV-I from infected to non-infected cells is via cell–cell interaction, which is apparently facilitated by ICAM-1 (CD56) [77]. The infected cells enter human body through three major routes: (1) sexual transmission, (2) breast feeding, and (3) parenteral transmission.

The HTLV-I *TAX* gene plays an important role in the leukemogenesis of the infected cells. The Tax protein (p40) induces proliferation and inhibits apoptosis of the HTLV-I-infected cells [78]. However, ATL cells do not always need *TAX* expression; *TAX* transcription has been detected in only 34% of the ATL cases by RT-PCR [79, 80]. Therefore, multistep genetic and epigenetic changes are implicated in the ATL leukemogenesis. For example, mutation of *p53*, deletion of *p16*, and upregulation of *TSLC1* genes are reported in ATL [81]. The aberrant methylation of certain genes such as *MELIS* and *EGR3* is an example of epigenetic changes in ATL [82, 83]. Recent reports suggest that the HTLV-I *HBZ* gene may play an important role in the regulation of viral replication and proliferation of infected T-cells [84].

Pathology

Morphology

The neoplastic cells are pleomorphic, medium- to large-sized with variable amount of amphophilic or basophilic non-granular cytoplasm, and hyperlobated (clover leaf) or convoluted nuclei with condensed chromatin and often distinct nucleoli (Figures 17.22 and 17.23). A small proportion of blast-like cells with dispersed chromatin and prominent nucleoli are usually present. Cells resembling Reed–Sternberg cells and multinucleated anaplastic giant cells with convoluted or cerebriform nuclei may be seen in tissue infiltrations [1, 85, 86].

Bone marrow involvement is often patchy and is often associated with osteoclastic activities, leading to hypercalcemia [76, 87]. The involved lymph nodes show diffuse infiltration with effacement of nodal architect and proliferation of endothelial venules. The skin infiltration usually involves the upper dermis, with frequent epidermal involvement and formation of tumor cell aggregates resembling Pautrier microabscesses [1, 86].

Immunophenotype

The neoplastic cells are typically of T-helper phenotype expressing CD4 and T-cell-associated markers CD2, CD3, and CD5 (Figure 17.24) [85, 86]. The majority of the cases

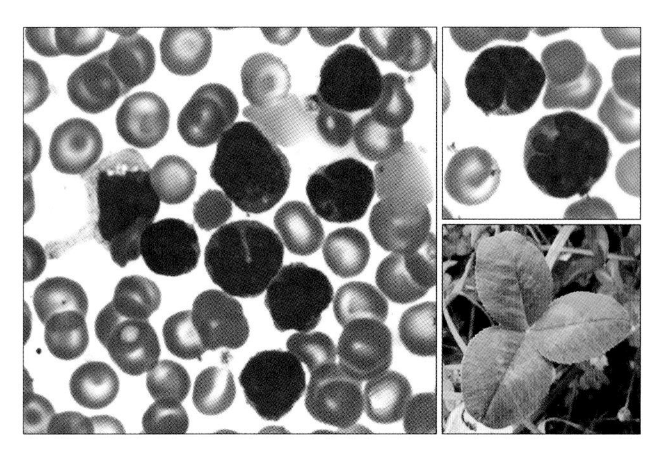

FIGURE 17.22 Adult T-cell leukemia/lymphoma. Blood smears showing atypical medium- to large-sized cells with amphophilic or basophilic non-granular cytoplasm and hyperlobated (clover leaf) nuclei.

lack CD7 expression but are positive for CD25 and CD52 [88, 89]. Occasional cases may be CD4−/CD8+ or CD4+/CD8+. TIA-1 and granzyme B are negative. The anaplastic large cells may express CD30, but they are negative for anaplastic lymphoma kinase (ALK) [1].

Cytogenetic and Molecular Studies

There is no distinct karyotypic or molecular abnormality in ATL. Cytogenetic analysis often shows a complex karyotype, particularly in the leukemic forms. Recurrent abnormalities include +3, +7, +21, monosomy X, loss of chromosome Y, and abnormalities of chromosomes 6 and 14q (at 14q11.2 and 14q32 breakpoints) (Figure 17.25) [85, 90, 91].

The neoplastic cells demonstrate *TCR* gene rearrangements with clonally integrated HTLV-I [92]. The detection and clonal pattern analysis of HTLV-I are quite specialized and are available only in selected reference laboratories [93]. The deletion of *p16* (multiple tumor suppressor 1) gene and the mutation of *p53* gene have been reported [94].

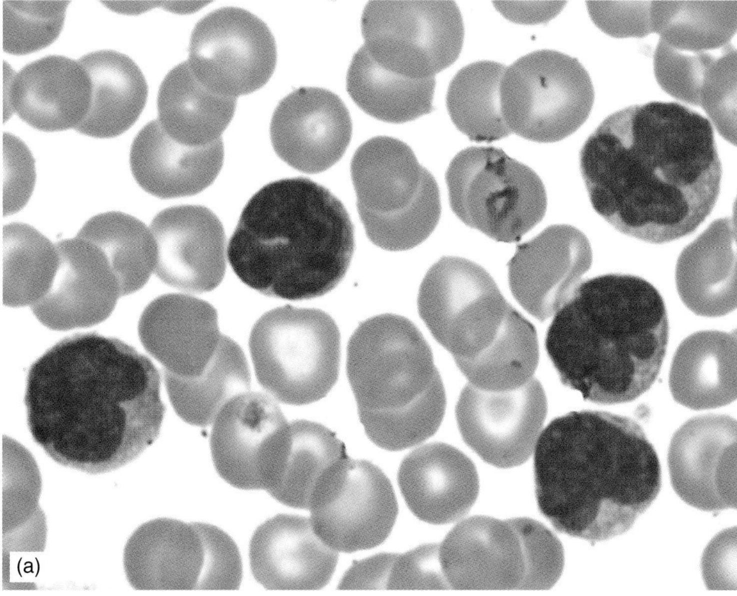

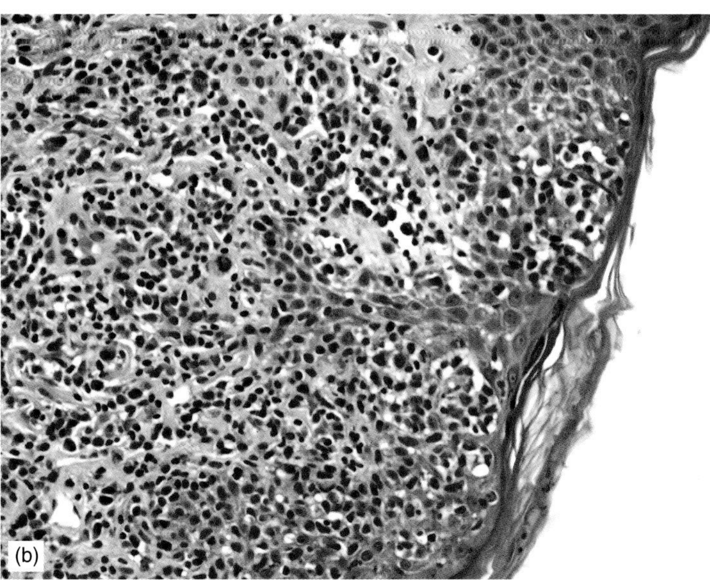

FIGURE 17.23 Adult T-cell leukemia/lymphoma. (a) Blood smear demonstrating atypical lymphocytes with lobulated nuclei. (b) Skin biopsy section showing a heavy lymphoid infiltrate in the upper dermis and epidermis.

Clinical Aspects

The geographic areas with the highest prevalence of HTLV-I include Japan, Africa, Caribbean islands, South America, and southern part of the United States [76, 95, 96]. Patients are adults with a median age of about 50 years with male: female ratio of about 3:2 [1]. Clinical symptoms may include hypercalcemia, lytic bone lesions, cutaneous lesions simulating mycosis fungoides (MF), lymphadenopathy, pulmonary lesions, and hepatosplenomegaly. Hypercalcemia is observed in over 70% of the cases during the clinical course, which appears to be due to osteoclastic proliferation and increased bone resorption. ATL cells have been shown to express receptor activator of nuclear factor-kappaB (RANK) ligand, which plays a role in the differentiation of hematopoietic precursors to osteoclasts [97]. There are four types of clinical presentation [72, 76, 85, 94, 98]:

1. Acute onset which is the most common type and occurs in approximately 60% of cases. It has an aggressive clinical course with 4-year survival rate of 5–12%.
2. The lymphomatous type representing about 20% of the cases characterized by prominent lymphadenopathy and no blood involvement but also aggressive clinical course.
3. The chronic type consisting of about 15% of the cases with skin lesions and absolute lymphocytosis, but no hypercalcemia.
4. The smoldering type representing 5% of the cases with normal blood lymphocyte counts and <5% circulating neoplastic cells and frequent skin or pulmonary lesions. There is no hypercalcemia.

Approximately 25% of the cases of chronic or smoldering types eventually progress to an acute phase. This transition is often associated with specific changes on gene expression profiling [94, 99].

The clinical outcome in majority of the cases is very poor with a median survival of <1 year despite advances in chemotherapy [100]. Combination chemotherapies such as cyclophosphamide, adriamycin, vincristine, and prednisone (CHOP), nucleoside analogs, topoisomerase inhibitors, interferon, and zudovudine are among the therapeutic chances [76].

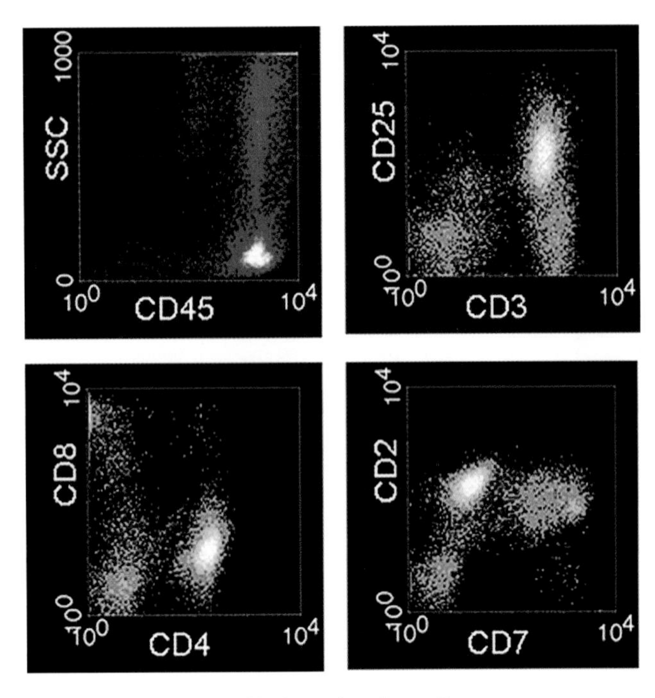

FIGURE 17.24 Adult T-cell leukemia/lymphoma. Flow cytometry demonstrating a population of helper T-cells (CD3+, CD4+) with loss of CD7 expression. The tumor cells are also CD2+ and CD25+.

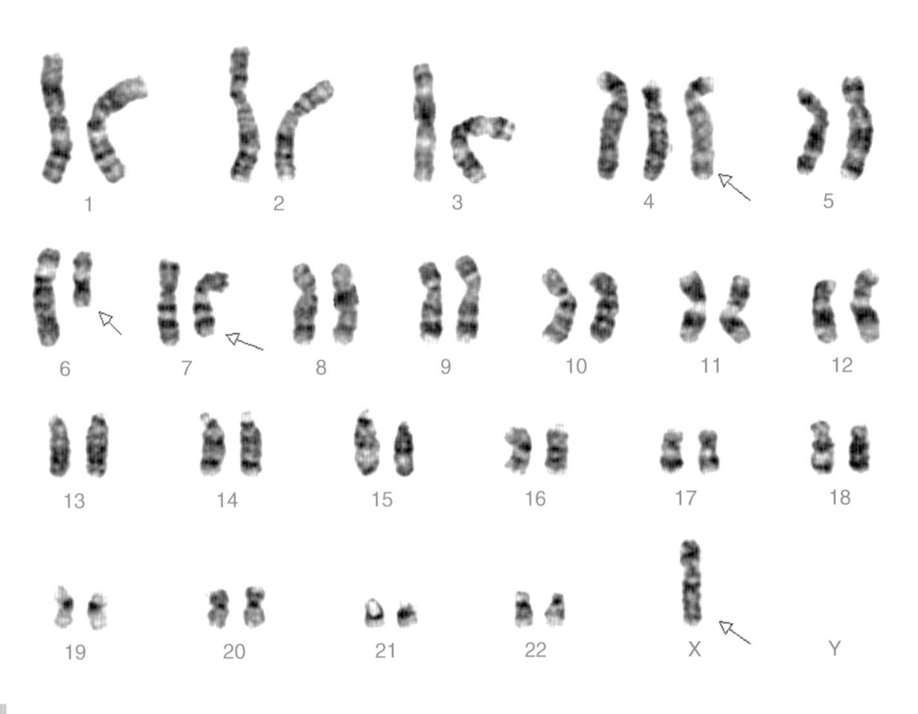

FIGURE 17.25 Karyotype of tumor cells in a patient with ATL demonstrating 46,X,−X, +4,del(6)(q15),del(7)(q35).

Differential Diagnosis

The neoplastic cells in peripheral blood and bone marrow smears may mimic Sezary cells and may share the same immunophenotypic features (expressing CD4 and pan-T-cell markers except CD7). In general, nuclear convolution in Sezary cells is more delicate and finer than that in the neoplastic ATL cells. In all ATL cases, the neoplastic cells are positive for HTLV-I, whereas the majority of the patients with SS are HTLV-I negative. Also, the median age for ATL is lower than that of MF/SS. Hypercalcemia and lytic bone lesions are frequent findings in ATL but absent in MF/SS.

The tissue infiltrates in ATL may contain anaplastic large cells and/or Reed–Sternberg-like cells mimicking anaplastic large cell lymphoma (ALCL) or Hodgkin lymphoma.

MYCOSIS FUNGOIDES AND SEZARY SYNDROME

Mycosis fungoides (MF) is a cutaneous peripheral T-cell lymphoma (CTCL) characterized by an indolent course and skin manifestations ranging from patches to plaques and tumor formation. Sezary syndrome (SS) is the leukemic and erythrodermic variant of MF with the presence of neoplastic cells in the peripheral blood.

Etiology and Pathogenesis

The etiology and pathogenesis of MF/SS are not known. An infectious etiology has been suggested [101]. For example, a *Borrelia burgdorferi*-specific sequence has been detected in 205 of the skin samples of patients with MF [102]. Coinfection of HIV-1 and HTLV-II has been reported in three patients with MF-like disorders [103, 104]. The possible association between HTLV-I and MF/SS is controversial [105–107].

Pathology

Morphology

Skin biopsies demonstrate an infiltrate of atypical lymphoid cells in the upper dermis and epidermis. These atypical cells are usually medium to large size with variable amount of cytoplasm and convoluted (cerebriform) nuclei. The intraepidermal lesions characteristically consist of small aggregates of atypical cells referred to as "Pautrier microabscesses" (Figures 17.26 and 17.27). In one study, in addition to the Pautrier abscesses, which were observed in 38% of the cases, the following morphologic features were frequently observed: haloed lymphocytes, exocytosis, epidermal lymphocytes larger than dermal lymphocytes, and lymphocytes aligned within the basal layer [1, 108–110]. In occasional cases, referred to as *pagetoid reticulosis*, the lymphoid infiltrates are strictly epidermal. Pagetoid reticulosis is usually localized and has an excellent prognosis [1, 111]. The epidermal infiltrate is patchy, diffuse, or band-like and is often associated with reactive inflammatory cells such as lymphocytes and eosinophils [1].

Lymphadenopathy is frequent in advanced cases (30–40%) and is divided into three categories as follows:

1. Dermatopathic lymphadenitis with scattered cerebriform lymphocytes, but no clusters.
2. Focal effacement of nodal architecture with clusters of atypical cerebriform lymphocytes, primarily in the paracortical regions.
3. Complete effacement of nodal architecture with diffuse infiltration of atypical cerebriform lymphocytes.

SS is considered a variant of MF associated with the presence of atypical cerebriform lymphocytes (Sezary cells) in the peripheral blood. The Sezary cells have a variable amount of non-granular cytoplasm and show the characteristic delicately convoluted, cerebriform nucleus with condensed chromatin and inconspicuous nucleoli (Figures 17.26c and 17.28). These cells may vary in size with the smaller forms referred to as Lutzner cells. The diagnosis of SS according to the International Society for Cutaneous Lymphoma (ISCL) is based on one or more of the following [112]:

A. An absolute Sezary cell count of ≥1,000 cells/μL.

B. A CD4:CD8 ratio of ≥10 caused by an increase in circulating T-lymphocytes, or an aberrant loss of pan-T-cell markers.

C. Lymphocytosis with the evidence of a T-cell clone by the Southern blot or PCR.

D. A chromosomal abnormal T-cell clone.

In the cases of B, C, and D, at least 5% of the circulating lymphocytes should demonstrate characteristic morphologic features of the Sezary cells.

Immunophenotype

The neoplastic MF/SS cells are of helper T-cell phenotype (CD4+) and express CD45 and pan-T-cell markers CD2, CD3, and CD5, and TCRβ, but are usually negative for CD7, CD8, CD26, TRA-1, and granzyme B (Figures 17.27 and 17.29) [110, 113, 114]. The MF/SS cells are usually CD158+ and may express the IL-2 receptor-α (CD25). Atypical patterns such as CD4−/CD8− or CD4−/CD8+ phenotypes are rare.

Cytogenetic and Molecular Studies

Patients with cutaneous T-cell lymphoma commonly show a wide variety of clonal or non-clonal chromosomal aberrations in their blood or skin. Standard karyotyping studies of CTCL are difficult, as the malignant cells respond poorly to mitogens needed for inducing visible, analyzable mitotic chromosomes for G-banding staining. However, no recurrent or specific abnormality has been found in CTCL, leading to a hypothesis of genetic instability.

The cytogenetic studies in MF/SS have been largely performed on blood lymphocytes and revealed a large spectrum of chromosomal abnormalities, both numerical and structural. In one study by Whang-Peng *et al.* [115] the chromosomes most often involved in structural abnormalities

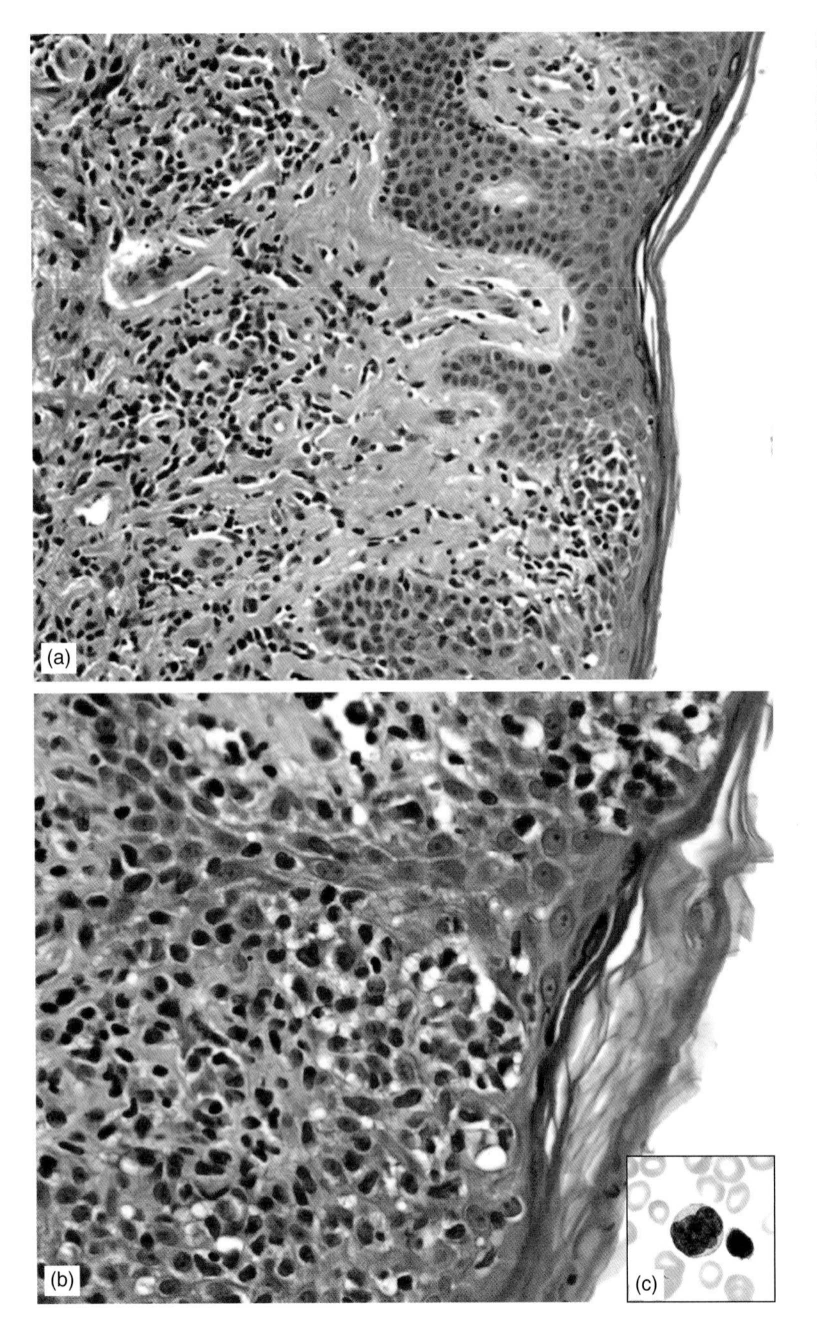

FIGURE 17.26 Mycosis fungoides/Sezary syndrome. Section of skin demonstrating an infiltrate of atypical lymphoid cells in the upper dermis and epidermis with Pautrier microabscesses: (a) low power and (b) high power views. Blood smear (c) showing a Sezary cell with convoluted nucleus.

of 1, 6, 7, 4, 9, 10, 12, 14, 15, and 17, and numerical abnormalities of chromosomes 11, 21, 22, 8, 9, 15, 16, and 17, respectively, in order of frequency. Cytogenetic abnormalities were observed prior to malignancy observed by histology and were suggested to have a significant diagnostic and prognostic value. CGH studies revealed loss of chromosomal region at bands 10q25-q26 and 13q21-q22 and gains (amplifications) in chromosomes 8 and 17q21-q25 in SS. DNA content analysis may show evidence of aneuploidy in up to one-third of the patients (Figure 17.30).

The MF/SS cells show *TCR* gene rearrangement. Additional *Her2/neu* gene copies and inactivation of *CDKN2A/p16* have been reported in MF/SS cells [116, 117]. In one study, real-time PCR studies demonstrated overexpression of five genes, *STAT4*, *GATA-3*, *PLS3*, *CD1D*, and *TRAIL*, by the circulating tumor cells in patients with SS [118].

Clinical Aspects

MF/SS comprises <1% of all non-Hodgkin lymphomas, but it is the most common primary cutaneous T-cell lymphoma accounting for about 45% of lymphomas present in the skin [119–121]. The peak age is 55–60 years, with a male:female ratio of about 2:1. MF usually presents as indolent cutaneous erythematous scaly patches or plaques, often with pruritus, mimicking common skin disorders such as eczema or psoriasis [101]. These lesions may wax and wane for many years and may eventually progress to cutaneous tumor formation, erythroderma, and the infiltration of the neoplastic cells into the circulation (SS). Extracutaneous involvement is relatively uncommon in the early stages of the disease but becomes more frequent in the advanced stages, comprising 8% in the plaque stage compared to 30–40% in the erythrodermatous stage [101, 119, 120].

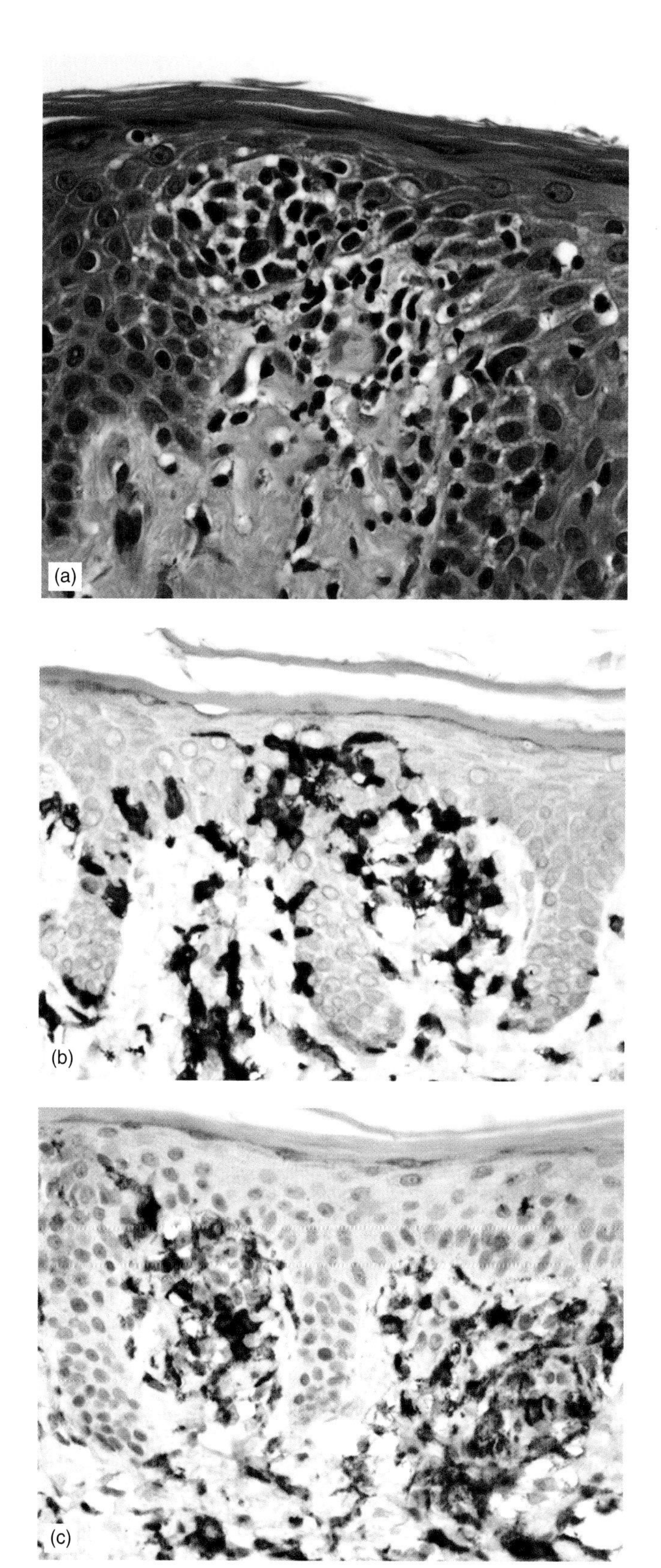

FIGURE 17.27 Mycosis fungoides. Section of skin demonstrating an infiltrate of atypical lymphoid cells in the upper dermis and epidermis (Pautrier microabscess) (a). These cells are positive for CD3 (b) and CD4 (c) by immunohistochemical stains.

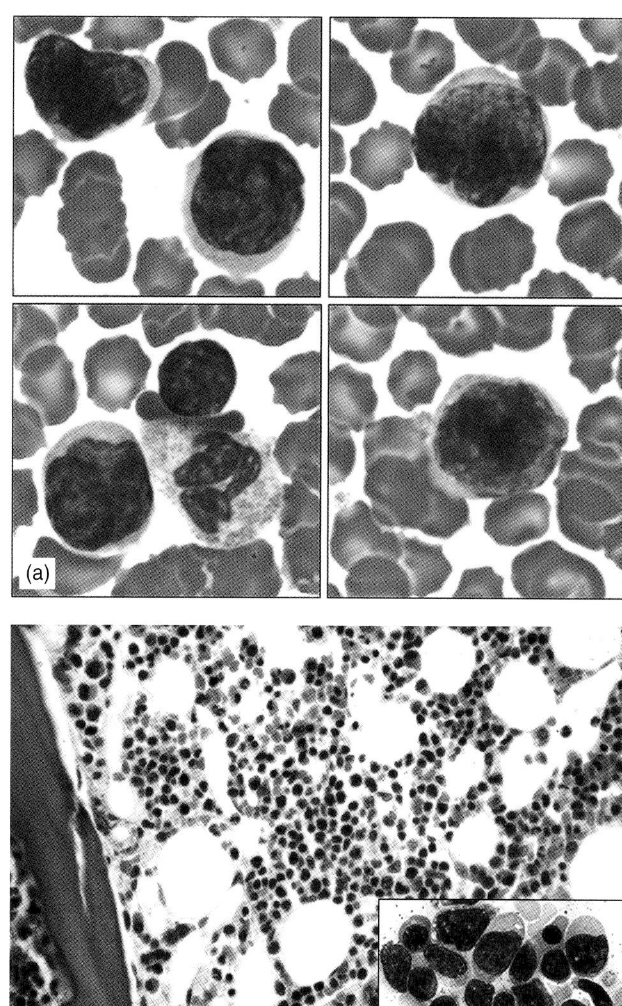

FIGURE 17.28 Mycosis fungoides/Sezary syndrome. Blood smear several Sezary cells with convoluted nuclei demonstrating (a). Bone marrow biopsy section (b) and smear (c) showing an aggregate of atypical lymphocytes with irregular nuclei. Adapted from Naeim F. (1997). *Pathology of Bone Marrow*, 2nd ed. Williams & Wilkins, Baltimore, by permission.

The regional lymph nodes are the most frequent sites of involvement followed by lungs, spleen, and liver.

The clinical staging is based on the extent of cutaneous lesions (T) and the involvement of the lymph nodes (L), viscera (M), and blood (B) (see Table 17.3) [121]. In the majority of the cases, cutaneous lesions are at stage T1 (20–25%) or T2 (35–40%). The overall 5-year survival for MF has been reported as 87% compared to 33% for SS [109, 122]. Therefore, transformation from MF to SS is an indicative of poor prognosis [122–124]. Transformation to CD30+ large cell lymphoma is relatively frequent. In two large studies of patients with MF, the cumulative probability of transformation to large cell lymphoma was 39% in 12 years, with a median time interval of 1–6.5 years after diagnosis of MF [125, 126]. The criteria for transformation were the formation of microscopic nodules of large neoplastic cells or >25% large cells in the neoplastic infiltrate [126]. The elevation of serum lactate dehydrogenase (LDH) and β2-microglobulin was predictive of transformation.

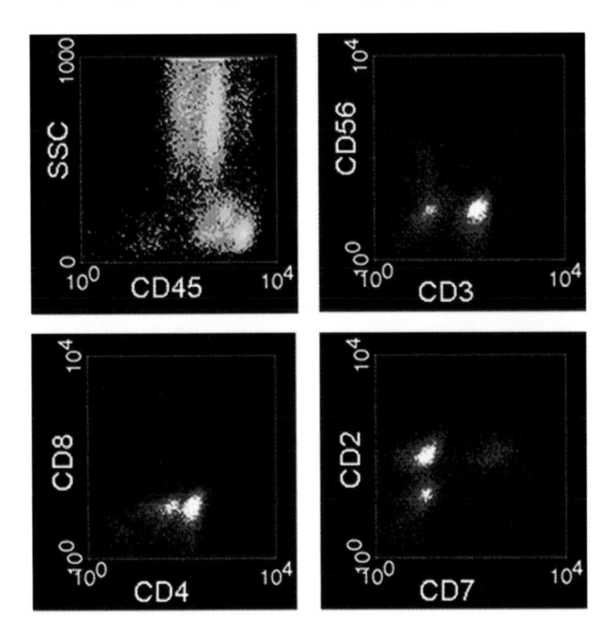

FIGURE 17.29 Mycosis fungoides/Sezary syndrome. Flow cytometry demonstrating a population of helper T-cells (CD3+, CD4+) with the loss of CD7 expression. The tumor cells are also positive for CD2.

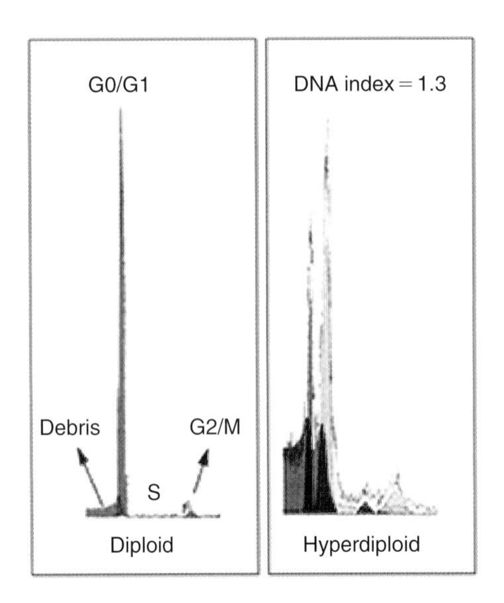

FIGURE 17.30 Mycosis fungoides/Sezary syndrome. DNA content analysis by flow cytometry demonstrating a hyperdiploid population (right yellow).

Transformation to CD30+ large cell lymphoma is associated with an aggressive clinical course. In one study, the median survival was 3 years in transformed MF patients compared to 14 years in untransformed patients [126].

Treatment includes a broad spectrum of options. The topical therapeutic measures, such as nitrogen mustard, carmustine, electron beam therapy, and phototherapy, are used for the early stages of the disease. Systemic chemotherapy, alone or in combination with topical therapy, is used in the advanced stages of the disease [101, 127]. Purine and pyrimidine analogs are the primary chemotherapy agents.

TABLE 17.3 Clinical staging for MF.*

Skin (T)
T1: Limited involvement; patches/plaques <10% of the total skin surface
T2: Generalized involvement; patches and plaques >10% of the total skin surface
T3: Cutaneous tumors
T4: Generalized erythroderma
Lymph nodes (N)
N0: No enlarged lymph node
N1: Lymphadenopathy with no evidence of histologic neoplastic involvement
N2: No lymphadenopathy but evidence of histologic involvement
N3: Lymphadenopathy and histologic involvement
Viscera (M)
M0: No visceral involvement
M1: Histologically confirmed visceral involvement
Blood (B)
B0: Circulating Sezary cells <5% of the lymphocytes
B1: Circulating Sezary cells >5% of the lymphocytes

*Adapted from the *AJCC Cancer Staging Manual*, 6th ed. Springer-Verlag, New York, 2002.

Differential Diagnosis

The differential diagnosis of MF includes a garden variety of benign reactive skin disorders, such as psoriasis, eczema, parapsoriasis, drug reactions, contact dermatitis, and photodermatitis. These distinctions can be quite challenging and often settled only by the results of *TCR* gene rearrangement studies. As the most referred skin biopsies will be paraffin-embedded, PCR analysis is the primary approach for these lesions, since formalin-fixed tissue does not yield DNA of high enough quality for Southern blot analysis to be reliable. However, it is in this setting that the potential for false-positive results due to spurious amplification of a small number of T-lymphocytes in the specimen (pseudoclonality) comes to the fore. In our laboratory, we include a disclaimer to this effect when the skin biopsy contains only scattered or scant T-lymphocytes and/or when we see an isolated clonal spike in the PCR profile in the absence of any polyclonal background signal; this is a hint that the signal may be artifactual (see Figure 17.4). MF should be distinguished from other primary cutaneous lymphomas. It shares overlapping morphologic and immunophenotypic features with ATL. ATL occurs in a younger age group and is associated with HTLV-I (Table 17.4). The neoplastic cells in a minority of T-PLL cases are Sezary-cell like (see T-PLL in this chapter). The T-PLL cells are usually CD7+ (Table 17.4).

HEPATOSPLENIC T-CELL LYMPHOMA

Hepatosplenic lymphoma is a rare extranodal peripheral T-cell lymphoma of cytotoxic type characterized by sinusoidal infiltration of liver, spleen, and bone marrow.

TABLE 17.4 Clinicopathological features of T-prolymphocytic leukemia (T-PLL), adult T-cell leukemia/lymphoma (ATL), and mycosis fungoides/Sezary syndrome (MF/SS).

Features	T-PLL	ATL	MF/SS
Median age (years)	>50	50	55–60
Male:female	?	1.5	2
Association	*ATM**	*HTLV-I*	*HTLV-I?*
Skin involvement (%)	15	15	100
Immunophenotype	CD3+, CD4+, CD7+, some CD4+/CD8+	CD3+, CD4+, CD7−	CD3+, CD4+, CD7−, CD26−
Overall prognosis	Aggressive	Aggressive	Indolent

*Ataxia-telangiectasia gene.

In most reported cases, the neoplastic cells represent the TCR$\gamma\delta$ T-cell subtype, though rare cases express TCR$\alpha\beta$ [1, 128–133].

Etiology and Pathogenesis

The etiology and pathogenesis of hepatosplenic T-cell lymphoma are not known. Frequent report of this disorder in patients with immune-compromised conditions, such as immunosuppressive therapy for solid organ transplantation, autoimmune disorders, and malaria, raises the possibility of an association between impairment of the immune system and development of hepatosplenic lymphoma. All of these conditions have been associated with the expansion of $\gamma\delta$T-cells, probably secondary to chronic antigenic stimulation [1, 134–136]. The proliferation of $\gamma\delta$T-cells in immune-compromised conditions may represent the initial step of a multistep process in the pathogenesis of hepatosplenic lymphoma.

Pathology

Morphology

Liver infiltration is sinusoidal with various degrees of portal tract involvement (Figure 17.31). Splenomegaly is a common feature. The pattern of infiltration in the spleen is diffused with the involvement of the red pulp (Figure 17.32). The neoplastic lymphoid cells are usually monomorphic, with moderate amount of cytoplasm; small to medium size, round to slightly irregular nuclei; condensed chromatin; and inconspicuous nucleoli [132, 133]. Mitotic figures are usually infrequent. Occasional cases may show a highly pleomorphic cell population. The bone marrow involvement is reported in 75–100% of the cases. The bone marrow infiltration is often subtle and sinusoidal (Figure 17.33). Occasionally, there is an increased number of histiocytes with features of hemophagocytosis [133]. Fulminant leukemic picture is rare, but scattered abnormal cells may be seen in the peripheral blood (Figure 17.33a). The lymph node

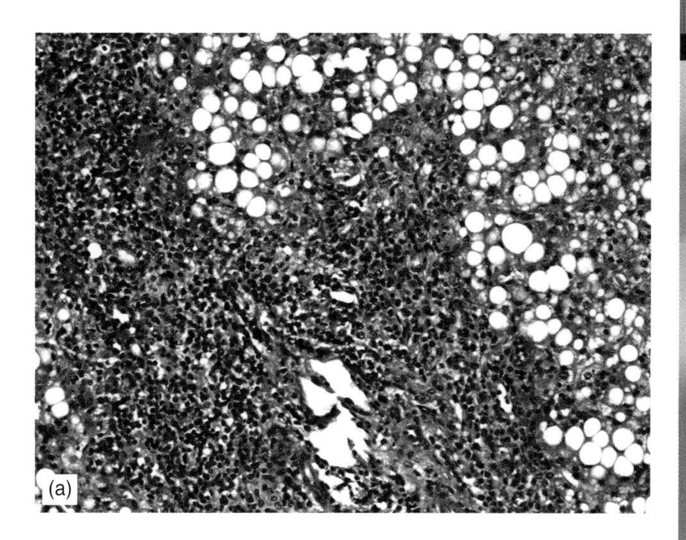

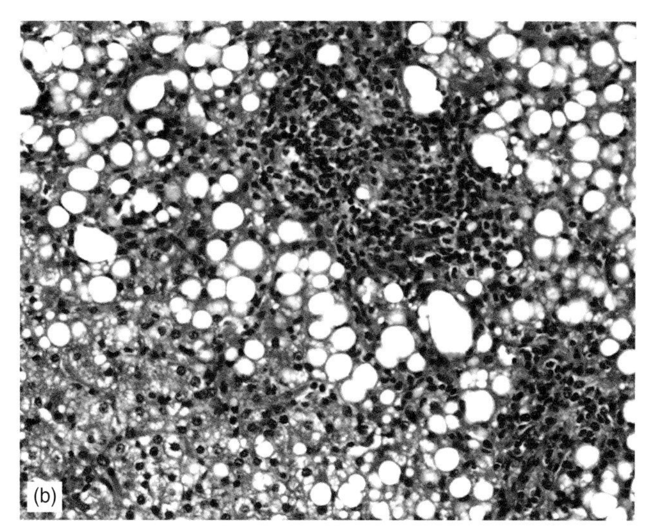

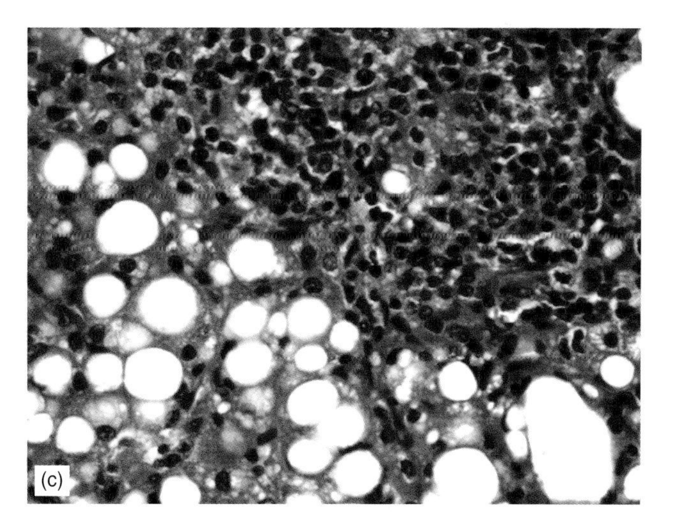

FIGURE 17.31 Hepatic involvement with hepatosplenic T-cell lymphoma. Patchy lymphoid infiltration is evident in a liver with fatty degeneration: (a) low power, (b) intermediate power, and (c) high power views.

involvement is rare. There are other variants of $\gamma\delta$T-cell lymphomas, which mostly involve extranodal tissues, such as skin, subcutaneous tissue, or intestine, which are not considered as hepatosplenic lymphoma [1, 137].

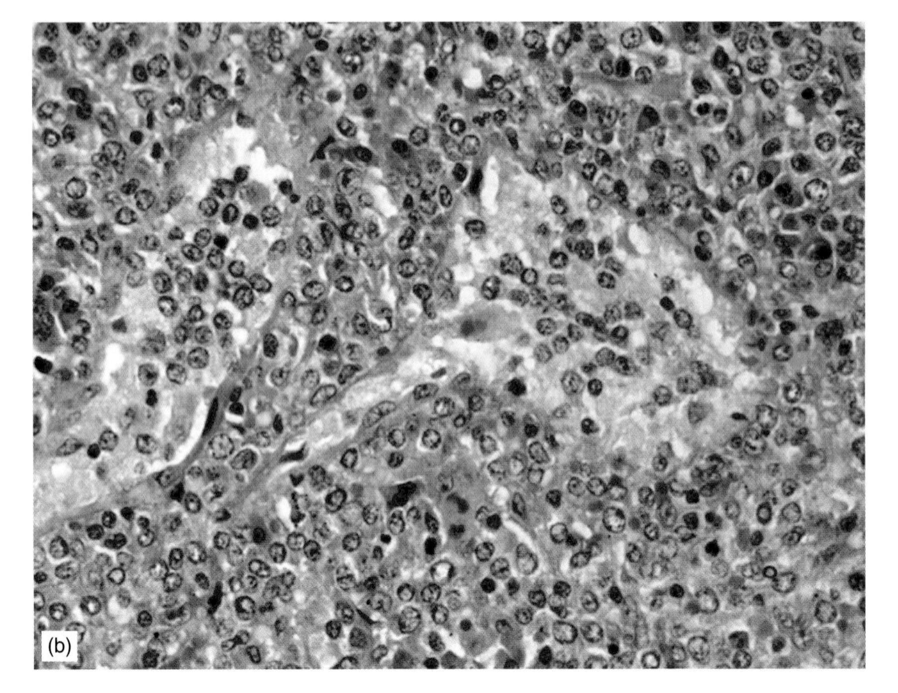

FIGURE 17.32 Hepatosplenic T-cell lymphoma with the involvement of splenic red pulp and sinuses: (a) low power and (b) high power views.

Immunophenotype

The most common immunophenotypic features of the neoplastic cells are γδTCR+(γTCR1), CD2+, CD3+, CD56+, TIA-1+, CD7±, CD4−, CD5−, CD8−, and granzyme B− (Figures 17.33 and 17.34). Rare cases may demonstrate TCRαβ or express CD8 [138, 139].

Molecular and Cytogenetic Studies

The neoplastic cells typically show *TCR* gene rearrangement by Southern blotting or PCR techniques. The *in situ* hybridization studies for EBV are negative [1]. The most frequent cytogenetic abnormalities are isochromosome 7q and trisomy 8 [133]. Isochromosome 7q is currently viewed as a pathognomonic genetic alteration in hepatosplenic T-cell lymphoma and can therefore serve as a diagnostic tool for this entity (Figure 17.35) [140, 141]. Also, del(13)(q12q14) has been reported in patients with hepatosplenic lymphoma (Figure 17.36) [131].

Clinical Aspects

Hepatosplenic T-cell lymphoma is a rare disease, accounting for about 5% of all peripheral T-cell lymphomas. The median age is about 35 years with the male:female ratio

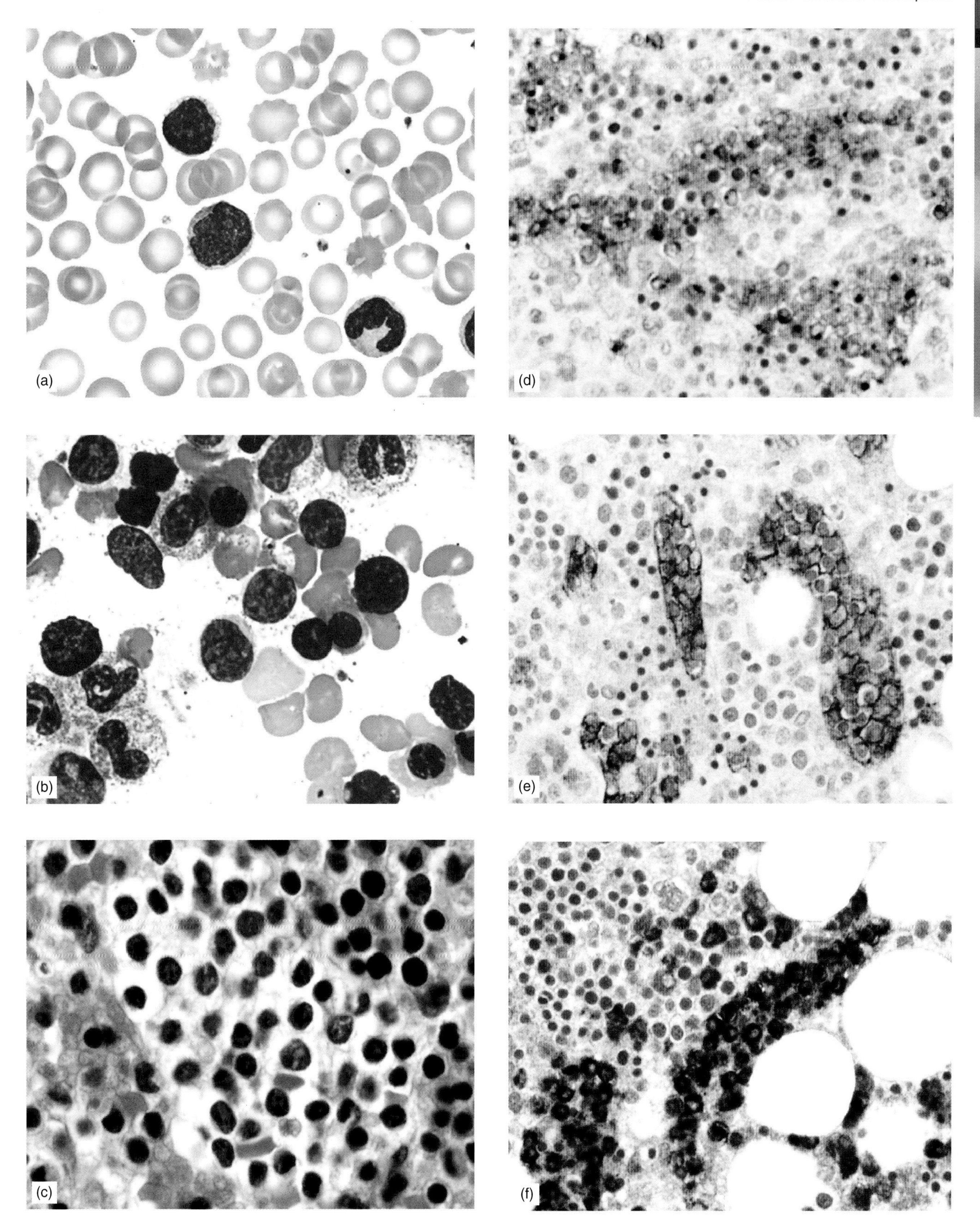

FIGURE 17.33 Hepatosplenic T-cell lymphoma. Blood (a) and bone marrow (b) smears showing atypical lymphocytes. Bone marrow involvement is often sinusoidal (c) demonstrated by accumulation of CD3+ (d), CD56+ (e), and TIA-1+ (f) cells in the sinusoids.

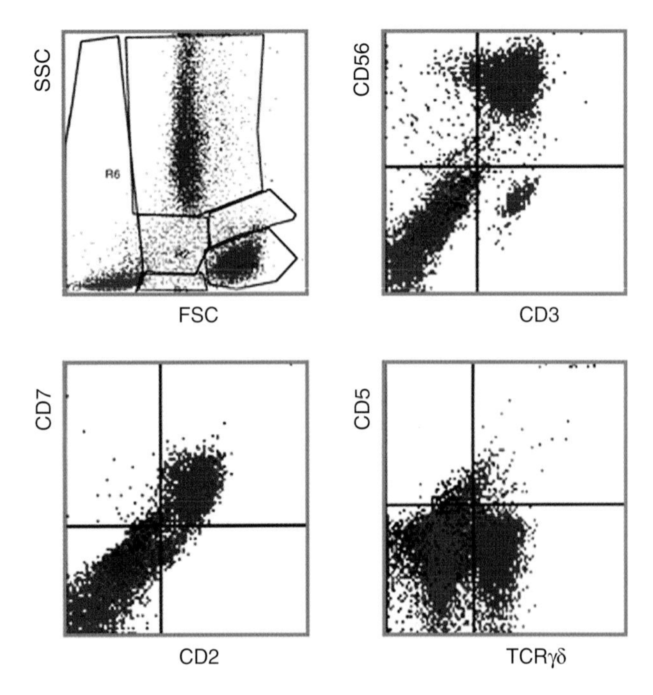

FIGURE 17.34 Flow cytometric studies demonstrating the common γδ variant of hepatosplenic T-cell lymphoma with the expression of CD2, CD3, CD7, and CD56. CD5 is negative.

of around 3:1 [133, 140]. Splenomegaly and thrombocytopenia are commonly followed by hepatomegaly, anemia, and leukopenia. Lymphadenopathy and other extranodal involvements are rare [140].

Hepatosplenic T-cell lymphoma has an aggressive clinical course with a median survival of about 16 months [133]. Combination chemotherapy is the frequent therapeutic approach, often with unsatisfactory results [133].

Differential Diagnosis

The differential diagnosis includes NK-cell lymphoproliferative disorders, because of overlapping morphologic pattern (sinusoidal involvement) and immunophenotypic features (CD56 and TIA-1 expression). However, neoplastic cells of hepatosplenic lymphoma are CD3+ and granzyme B−, and in many cases demonstrate isochromosome 7q. The T-LGL leukemia cells are CD3+ and may occasionally express *TCRγδ*, whereas lymphoid cells of hepatosplenic lymphoma are CD5−, CD56−, and CD57+ [138].

ENTEROPATHY-TYPE T-CELL LYMPHOMA

Enteropathy-type T-cell lymphoma is a rare intraepithelial T-cell intestinal lymphoma consisting of a polymorphic lymphoid infiltrate of medium to large atypical lymphocytes [1, 142–144].

Etiology and Pathogenesis

The etiology and pathogenesis of this disorder are not known. There is a strong association with celiac disease with the evidence of serologic markers such as positive antigliadin antibodies and HLA-DQA1, B1, and HLA–DRB1 types [145, 146]. Some cases in South and Central America have been associated with EBV [147].

Pathology

Morphology

The neoplasm often appears as multifocal ulcerating intestinal lesions consisting of a pleomorphic lymphoid infiltrate [1]. Jejunum and ileum are the most frequently affected sites. There is a predominance of atypical medium to large lymphoid cells with variable amount of pale cytoplasm, round or irregular vesicular nuclei, and prominent nucleoli. Anaplastic large cells or multinucleated giant cells may be present, mimicking ALCL. The infiltrate is commonly mixed with inflammatory cells such as histiocytes and eosinophils.

Immunophenotype

The neoplastic cells express CD3, CD7, TIA-1, granzyme B, and CD103. They are negative for CD4 and CD5. Tumors consisting of small- to medium-sized lymphoid cells may express CD8 and/or CD56 [1, 148]. In most cases, a various proportion of tumor cells are also CD30+ [148].

Molecular and Cytogenetic Studies

The *TCR* genes, most commonly γ and δ, are rearranged [149]. Loss of heterozygosity (LOH) at chromosome 9p21 and gains at chromosome 9q, 7q, 5q, and 1q have been reported in some cases [150–152].

Clinical Aspects

Most patients have a history of celiac disease and present with abdominal pain and sometimes evidence of intestinal perforation [1]. The clinical outcome is usually poor. Differential diagnosis includes inflammatory bowel disorders and other types of lymphomas.

ANAPLASTIC LARGE CELL LYMPHOMA

Anaplastic large cell lymphoma (ALCL) is a T-cell malignancy consisting of large anaplastic, pleomorphic CD30+ cells, often expressing ALK (large cell lymphoma kinase) and cytotoxic-associated proteins. The tumor cells have a tendency to grow cohesively in the lymph node sinuses, mimicking metastatic tumors [1, 153, 154].

Etiology and Pathogenesis

The etiology and pathogenesis of ALCL are not known. The strong association of ALCL with t(2;5)(p23;q35) suggests that this chromosomal translocation may play a role in the pathogenesis of this disease [154]. This translocation creates

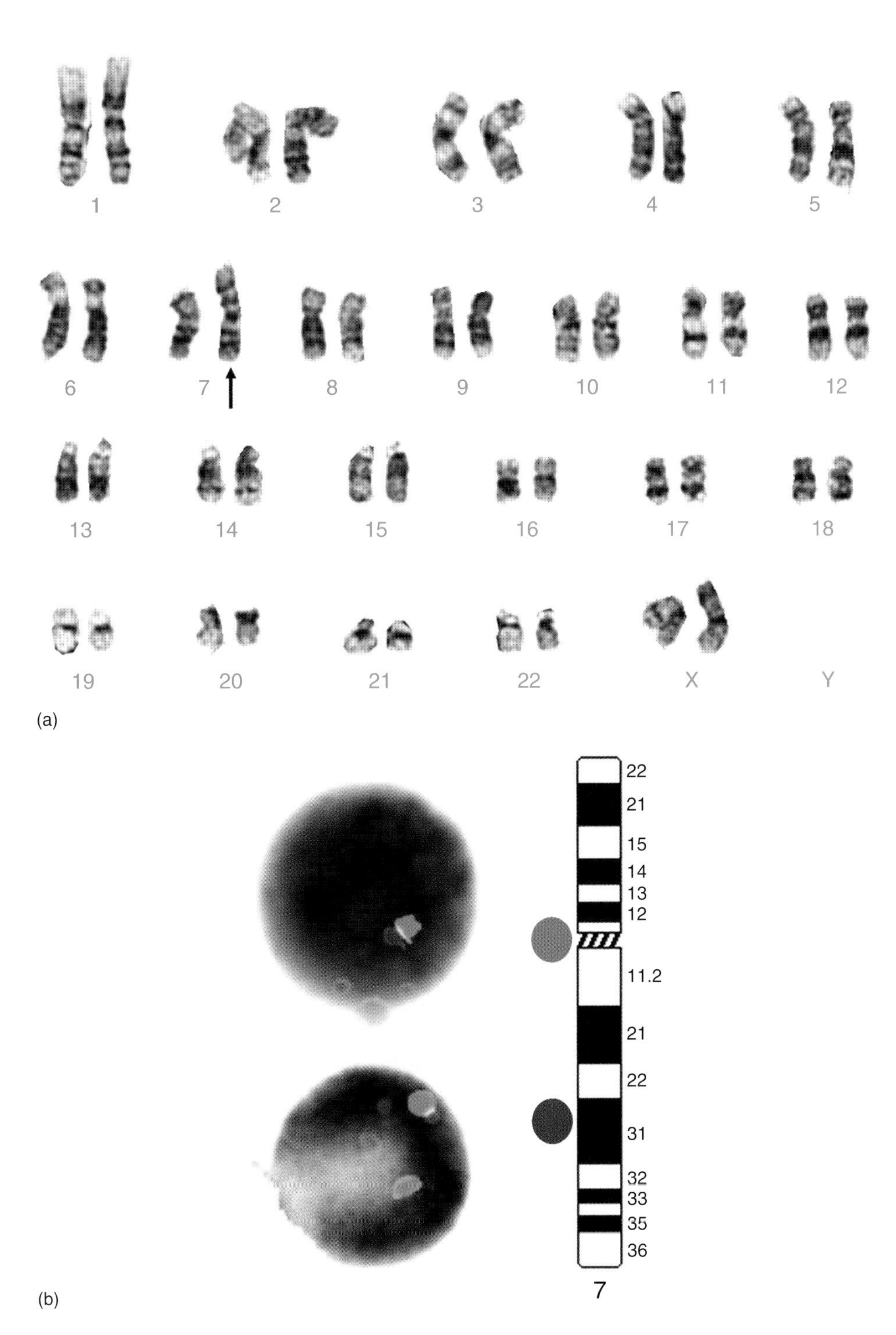

FIGURE 17.35 Karyotype (a) and FISH (b) of tumor cells in a patient with hepatosplenic lymphoma demonstrating 46,XX,iso(7)(q10).

a hybrid gene as the result of the fusion of *NPM1* (nucleophosmin) gene on chromosome 5 with *ALK* gene on chromosome 2 [153, 156]. The NPM1–ALK protein activates the anti-apoptotic PI3K–Akt pathway and a number of signal transducers and activators of transcription proteins, which are all important in cellular transformation [157, 158]. The oncogenic properties of *NMP1–ALK* have been supported by *in vivo* studies in experimental animals [158, 159].

Pathology

Morphology

The affected lymph nodes are partially or totally effaced due to the infiltration of the tumor cells that often have a tendency to grow cohesively in the sinuses, resembling metastatic tumors [1, 153]. The tumor cells are usually admixed

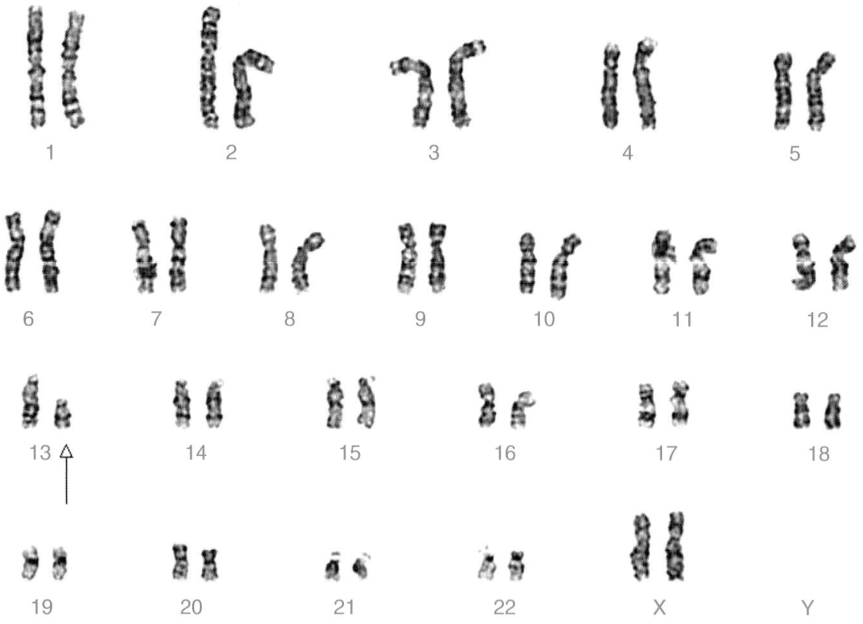

FIGURE 17.36 Karyotype of tumor cells in a patient with hepatosplenic lymphoma demonstrating 46,XX,del(13)(q12q22).

with inflammatory cells, which predominantly consist of histiocytes and plasma cells. Sclerotic thick capsule and well-formed fibrous bands are infrequent. Three major morphologic variants have been described: (1) common, (2) small cell, and (3) lymphohistiocytic variants [1, 153].

The *common variant* (70% of cases) is characterized by the predominance of large pleomorphic cells with abundant clear to light blue cytoplasm; often eccentric, horse-shoe, kidney-shaped, or multilobulated nuclei; and multiple small nucleoli (Figures 17.37 and 17.38). The large cells which have a horse-shoe or kidney-shaped nucleus next to the Golgi area are referred to as the "hallmark cells," because they are detected in all morphologic types [1, 153]. The neoplastic cells may demonstrate cytoplasmic vacuoles in touch preparations. The multilobulated cells may show pseudonuclear inclusions (doughnut cells) due to the invagination of the nuclear membrane [1, 160]. Reed–Sternberg-like multinucleated cells may be present.

The *small cell variant* (5–10% of cases) consists of a mixture of large-, medium-, and small-sized pleomorphic cells. Small- and medium-sized cells are the predominant cells depicting a clear cytoplasm and an irregular nucleus [160]. The large neoplastic cells tend to cluster around small vessels.

The *lymphohistiocytic variant* (5–10% of cases) consists of small and large neoplastic cells including "hallmark" cells as well as large numbers of reactive histiocytes. Hemophagocytic macrophages may be present. The abundance of histiocytes may mask the neoplastic cell population in the H&E stains, but immunohistochemical stains for CD30 and ALK help to identify the tumor cells.

Other morphologic variants such as sarcomatoid form with large, bizarre spindle-shaped tumor cells and giant cell-rich type with numerous multinucleated giant cells have been reported [153, 160].

Immunophenotype

The large neoplastic cells characteristically express CD30 which is confined to the cell membrane and the Golgi region. In the small cell variant, only the large anaplastic cells show strong positivity for CD30, whereas the small ones are either negative or weakly positive [160]. However, CD30 is not ALCL specific and is demonstrated on activated lymphocytes, and other types of B- and T-cell lymphoid malignancies [1, 153]. The tumor cells in the majority of the ALCL cases express epithelial membrane antigen (EMA), TCR, TIA-1, granzyme, and one or more T-cell-associated markers, particularly CD2 and CD4. The expression of CD3, CD5, and CD7 is less frequent and CD8 is usually negative [1]. Some ALCL cases may lack expression of T-cell-associated markers (previously considered "null" ALCL) [154]. CD43 and clusterin are expressed in the majority of the cases [161, 162]. Clusterin is a disulfide-linked heterodimeric protein associated with apoptosis. CD15 and EBV are usually negative in ALCL.

ALK expression is the most specific marker for the diagnosis of ALCL and is detected in 60–85% cases (Figure 17.39) [1]. ALK expression is cytoplasmic or nuclear, or both. Bcl-2 expression has been reported only in ALK-negative ALCL, whereas c-MYC nuclear expression is seen in ALK-positive pediatric cases but not ALK-negative ones [163].

Molecular and Cytogenetic Studies

The majority of the cases of ALCL demonstrate clonal *TCR* rearrangements and *NPM1-ALK* fusion gene [153, 156]. EBV sequences are not detected. The *NPM1-ALK* fusion is associated with a balanced t(2;5)(p32;q35) (Figure 17.40) [158]. This translocation is the most common ALK-related chromosomal aberration, accounting for about 75% of all ALK-positive cases of ALCL. It is most readily and sensitively detected by FISH, but PCR-based molecular testing, similar to that used to detect the *BCR–ABL* fusion in chronic myeloid leukemia (CML), is available in some centers [164–167]. The remaining 25% show various ALK-related rearrangements including t(1;2)(q21;p23)[*TPM3-ALK*], t(2;3)(p23;q21)[*ALK-TFG*]

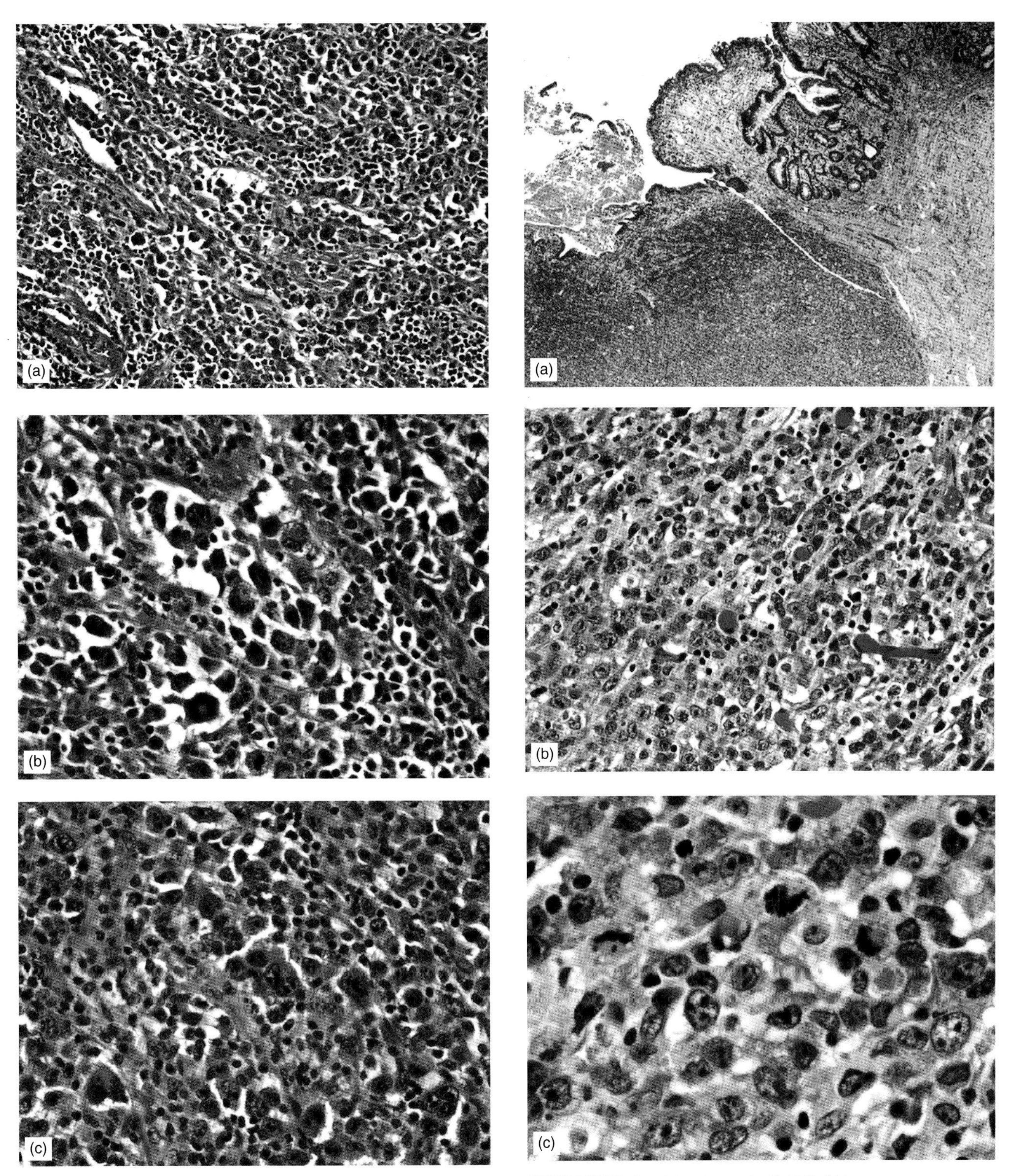

FIGURE 17.37 A lymph node section demonstrating ALCL with sinusoidal involvement. Horse-shoe and Hodgkin-like cells are present: (a) low power, (b) intermediate power, and (c) high power views.

FIGURE 17.38 Gastric involvement with ALCL: (a) low power, (b) intermediate power, and (c) high power views.

(Figure 17.41), t(2;17)(p23;q23)[*ALK-CLTC*], t(X;2)(q11-12;p23)[*MSN-ALK*], and inv(2)(p23q35) (Table 17.5) [156, 158, 168]. The expression of ALK in hematologic neoplasms is largely limited to ALCL tumors of T- or null-cell immunophenotype [169]. Because multiple chromosomal regions are involved in the translocations with the 2p23 (*ALK* locus), the rearrangements can be easily identified by a 2p23-specific dual-color "breakapart" FISH probe.

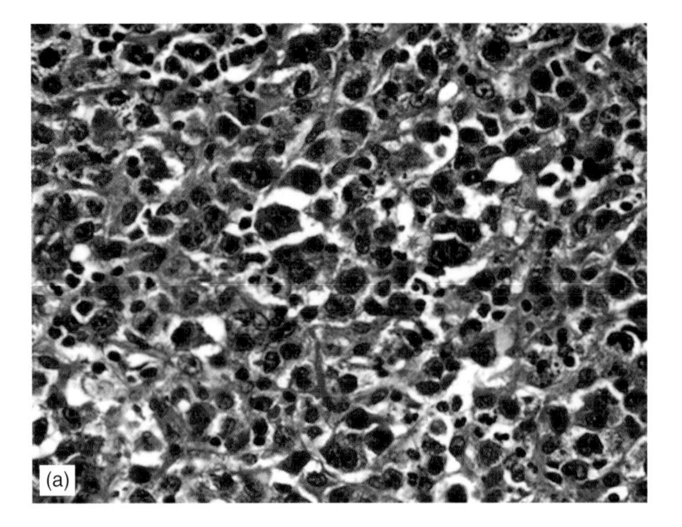

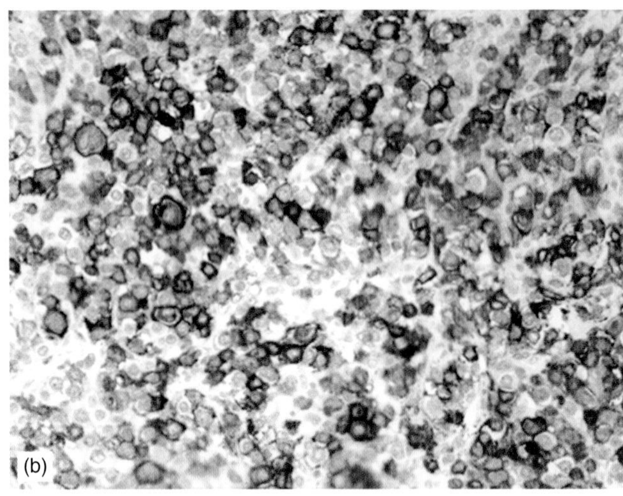

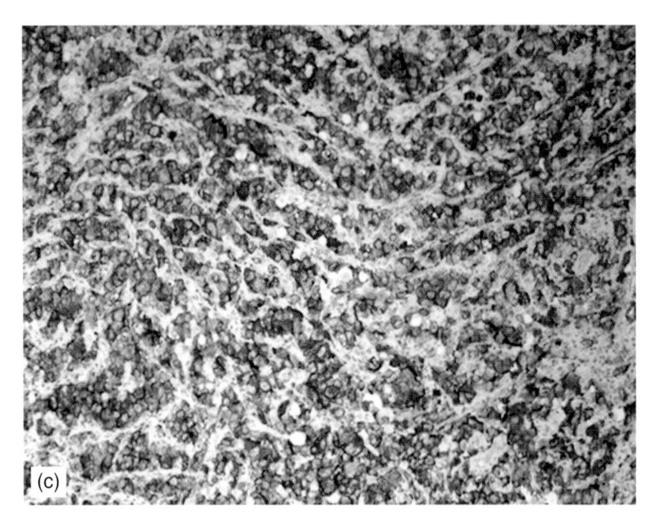

FIGURE 17.39 Anaplastic large cell lymphoma. H&E (a) and immunohistochemical stains demonstrating expression of CD30 (b) and ALK (c).

In many studies, evidence of the translocation has not only been observed in the patients, but also in the blood of healthy persons by RT-PCR studies [170].

The ALK overexpression is detected by immunohistochemical stains with anti-ALK antibodies. ALK-positive staining in the classic t(2;5) is detected in nucleolus, nucleus, and cytoplasm; whereas other translocations lead to only cytoplasmic staining [1].

Recent studies of gene expression profiling of systemic ALCL have revealed differences between the ALK-positive and the ALK-negative subgroups. The ALK-positive tumors showed overexpression of *BCL6*, *PTPN12*, *CEBPB*, and *SERPINA1* genes; whereas *CCR7*, *CNTFR*, *IL22*, and *IL21* genes were overexpressed in the ALK-negative group [171]. Study of the gene expression profiles of four ALCL and three HD cell lines showed higher levels of *BCL3* expression in ALCL than HD cell lines. *BCL3* encodes a nuclear protein which belongs to the IκB family of inhibitors of nuclear factor-κB (NF-κB) transcriptional factors [172]. Also, *JAK3* activation is significantly associated with ALK expression in ALCL [173].

Clinical Aspects

Primary systemic ALCL represents about 20–30% of the large cell lymphomas in children and 5% of all non-Hodgkin lymphomas in adults [153, 154, 160]. The male:female ratio is about 6:1 [153, 154]. Approximately 70% of the patients present with constitutional symptoms (mostly high fever and weight loss) and are in stage III/IV. Extranodal involvement is noted in about 60% of the patients, with skin (21%), bone (17%), and soft tissues (17%) being the most frequent sites [174, 175]. Approximately 10% of the patients may show liver or pulmonary involvement [153]. Leukemic presentation is uncommon.

ALCL is divided into two major groups: ALK-positive and ALK-negative. Several studies of a large series of ALCL patients have confirmed a significant prognostic difference between these two groups. The 5-year overall survival ranges from 70% to 90% in patients with ALK-positive tumors compared to 15–37% in patients with ALK-negative tumors [176–179]. The vast majority of the ALCLs in children and young adults are ALK-positive by immunohistochemical stains and show translocation of *ALK* gene [177, 180, 181]. ALCL accounts for 10–15% of childhood non-Hodgkin lymphomas.

Other factors associated with poor prognosis include mediastinal involvement, involvement of spleen, lung, or liver, CD56 expression, and small cell variant [182, 183].

Multiagent chemotherapy and autologous or allogeneic bone marrow transplantation are among the routine therapeutic approaches [184]. Anti-CD30 therapy and vaccination against ALK protein are under investigation [185].

Differential Diagnosis

ALCL may share some morphologic features with classical Hodgkin lymphoma. Sclerotic thick capsule and well-formed fibrous bands which are frequently seen in Hodgkin lymphoma are infrequent in ALCL. The Reed–Sternberg cells are CD45−, CD15+, CD30+, EMA−, and ALK−, and may express CD20 and show EBV positivity; whereas the neoplastic cells of ALCL are CD45+, CD15−, ALK±, EMA+, may express some of pan-T-cell markers, and may be positive or negative for EBV. Sinusoidal involvements in

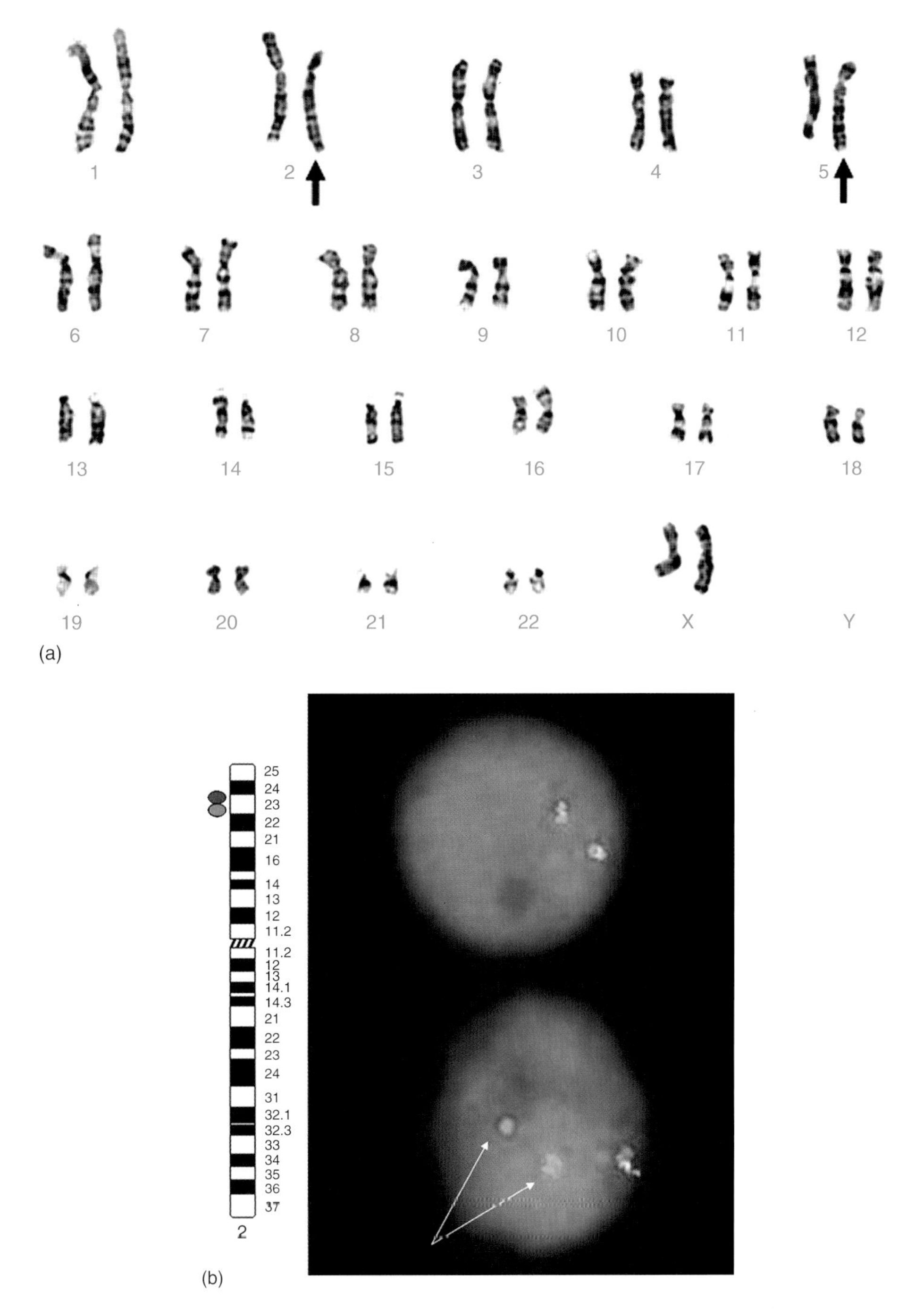

FIGURE 17.40 (a) Karyotype of tumor cells of a patient with ALCL demonstrating t(2;5)(p23;q35). (b) FISH with ALK(2p23) dual-color, breakapart rearrangement probe showing split signals.

the lymph nodes and bone marrow may mimic metastatic carcinoma.

PRIMARY CUTANEOUS ALCL

Primary cutaneous ALCL is a CD30+, ALK− lymphoma of anaplastic large cells confined to the skin, most often presenting as a solitary lesion [111, 120, 186, 187].

Etiology and Pathogenesis

The etiology and pathogenesis of this disorder are not known.

Pathology

Morphology

The morphologic features are similar to the systemic ALCL. The upper and deep dermis is diffusely infiltrated with anaplastic large cells, often with numerous Reed–Sternberg-like cells and/or multinucleated giant cells [111]. The epidermis may be infiltrated or ulcerated. A modest inflammatory background may be present.

Immunophenotype

The neoplastic cells are usually positive for CD4, CD30, TIA-1, and granzyme B. Approximately 50% of the cases are positive for HECA-452, which marks for the cutaneous lymphocyte antigen [111, 187]. Loss of expression of some of the pan-T-cell markers, such as CD2, CD3, or CD5, is frequent. Most cases of cutaneous ALCL are negative for ALK and EMA.

TABLE 17.5 Chromosomal aberrations in ALK-positive ALCL.*

Aberrations	Involved genes**	Frequency (%)
t(2;5)(p23;q35)	*ALK/NPM*	~75
t(1;2)(q21;p23)	*TPM3/ALK*	~15
t(2;3)(p23;q21)	*ALK/TFG*	~2
t(2;17)(p23;q23)	*ALK/CLTC*	~2
t(X;2)(q11-12;p23)	*MSN/ALK*	~1
inv(2)(p23;q35)	*ALK/ATIC*	~2

*Adapted from Ref. [158].

**ALK: anaplastic lymphoma kinase; CLTC: clathrin heavy chain; MSN: moesin; NPM: nucleophosmin; TFG: TRCK fusion gene; TPM: tropomyosin.

Molecular and Cytogenetic Studies

Most cases show *TCR* gene rearrangement [111] and are commonly negative for *NPM1–ALK* fusion gene or t(2;5) or other *AKL*-related translocations observed in the systemic ALCL.

Clinical Aspects

Primary cutaneous ALCL accounts for 10–25% of cutaneous lymphomas [1, 111]. It usually occurs in elderly patients with a median age of about 60 years. The male:female ratio is about 1.5–2:1 [1, 111]. It typically presents as a solitary, asymptomatic cutaneous reddish-violet tumor, sometimes superficially ulcerated. Multiple skin lesions are infrequent. Approximately 25% of the patients show partial or complete spontaneous regression. Disease-specific survival at 5 years for localized lesions is 91% compared to 50% for the generalized cutaneous ALCL [188]. The conventional treatment for localized lesions is excision with or without radiation. Combination chemotherapy is recommended for disseminated skin disease [153].

Differential Diagnosis

The differential diagnosis includes lymphomatoid papulosis, Hodgkin lymphoma, CD30+ large cell transformation of MF, secondary cutaneous involvement of ALCL, and CD30+ cutaneous NK/T-cell lymphoma [188].

Lymphomatoid papulosis is a benign, chronic recurrent skin disorder characterized by multiple spontaneously

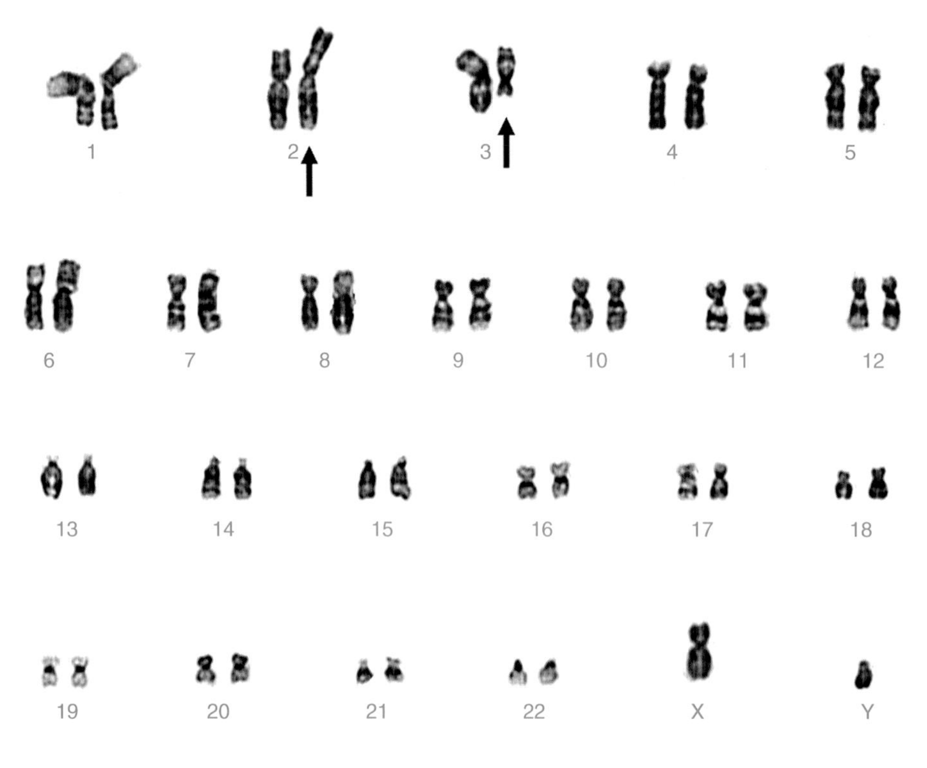

FIGURE 17.41 Karyotype of tumor cells of a patient with ALCL demonstrating 46,XY,t(2;3)(p23;q21).

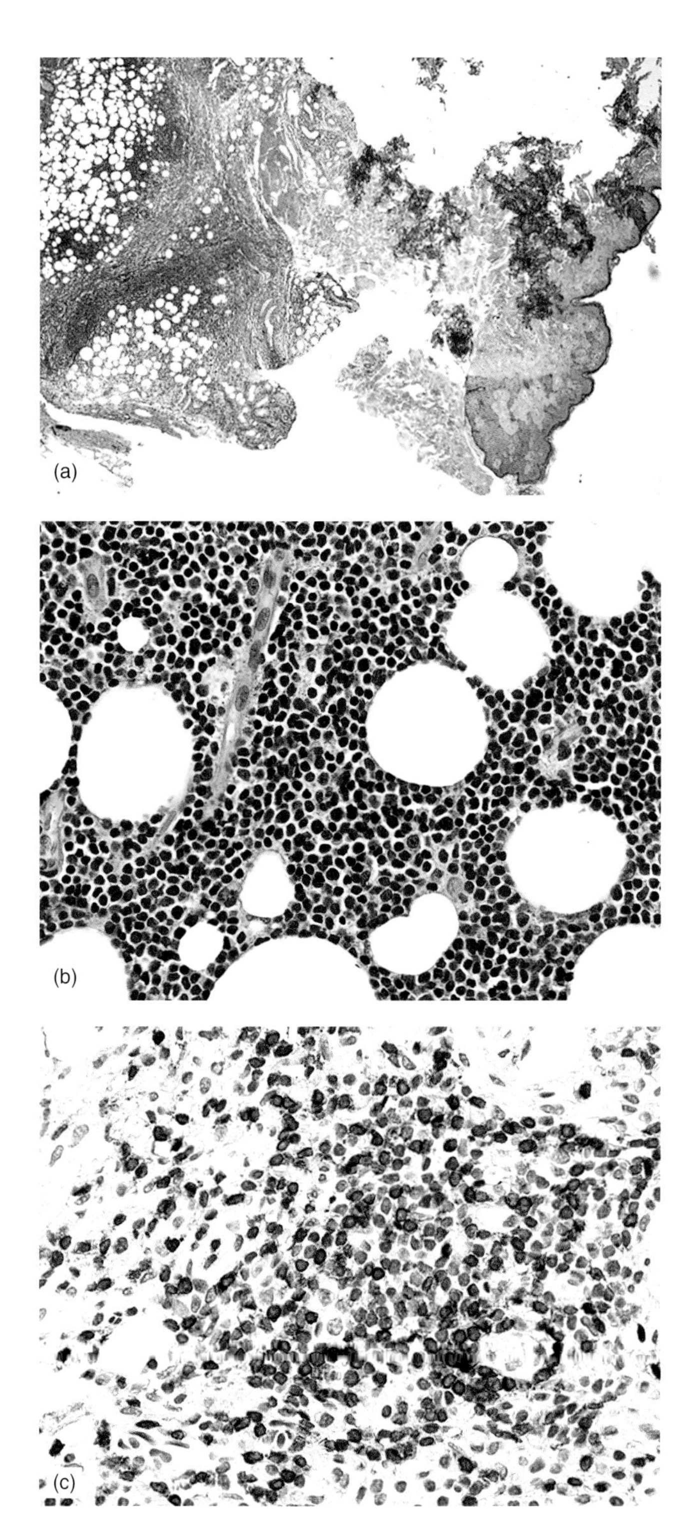

FIGURE 17.42 Subcutaneous panniculitis-like T-cell lymphoma: (a) low power and (b) high power views. (c) The infiltrating lymphocytes are CD3+.

regressing papules consisting of a polymorphic infiltrate including anaplastic large lymphocytes and cells resembling Reed–Sternberg cells. These cells represent activated CD4+ cells and coexpress CD30, but are negative for ALK (see Chapter 19).

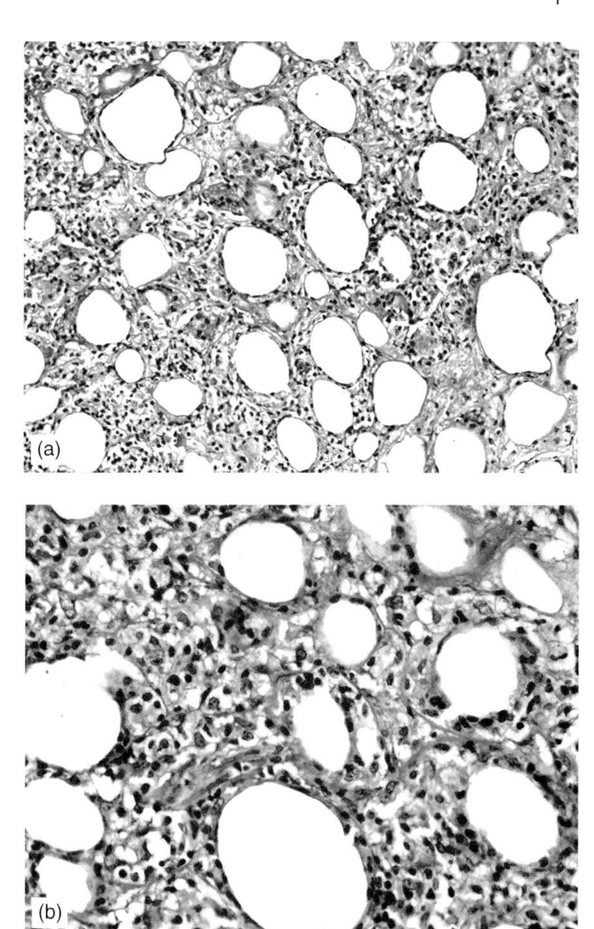

FIGURE 17.43 Subcutaneous panniculitis-like T-cell lymphoma infiltrating subcutaneous fatty tissue: (a) low power and (b) high power views.

OTHER PRIMARY CUTANEOUS T-CELL LYMPHOMAS

The classification recently proposed by the World Health Organization and European Organization for Research and Treatment of Cancer (WHO/EROTC) divides the previously known *subcutaneous panniculitis-like T-cell lymphoma* into two separate categories: (1) subcutaneous panniculitis-like T-cell lymphoma and (2) cutaneous $\gamma\delta$ T-cell lymphoma [145]. These two entities and primary cutaneous aggressive epidermotropic CD8+ cytotoxic T-cell lymphoma are briefly discussed in this section.

Subcutaneous Panniculitis-Like T-Cell Lymphoma

Subcutaneous panniculitis-like T-cell lymphoma is a rare lymphoma which involves subcutaneous fat without dermal or epidermal infiltration, leading to erythematous or violaceous nodules, plaques, or both [189, 190]. The infiltrate consists of a mixture of small and medium to large atypical cells with areas of necrosis (Figures 17.42 and 17.43).

The neoplastic T-cells have a tendency to rim around adipocytes. The infiltrate often contains reactive histiocytes, which may show hemophagocytosis [189]. The neoplastic cells represent TCRαβ class and often express CD3, CD8, granzyme B, and TIA-1. The tumor cells in most patients show clonal *TCR* gene rearrangement [189]. The EBV association is controversial [190–192].

Subcutaneous panniculitis-like T-cell lymphoma is often associated with a systemic hemophagocytosis characterized by fever, hepatosplenomegaly, lung infiltrates, liver dysfunction, coagulation abnormalities, and pancytopenia [189]. This hemophagocytic syndrome may develop before or during the manifestation of T-cell lymphoma [191].

The clinical manifestation of subcutaneous panniculitis-like T-cell lymphoma is variable, ranging from indolent course to a rapidly fatal hemophagocytic process. Local radiation therapy and/or systemic chemotherapy are used. The 5-year survival rate has been reported as 80% in a recent study [193].

Cutaneous γδ T-Cell Lymphoma

Cutaneous γδ T-cell lymphoma was previously considered a subtype of subcutaneous panniculitis-like T-cell lymphoma, accounting for 25% of the cases. Morphologic features are similar to those of subcutaneous panniculitis-like, except that dermal and epidermal involvement may be present. Cutaneous γδ T-cell lymphoma is a more aggressive disease than subcutaneous panniculitis-like T-cell lymphoma [189, 193]. The neoplastic cells are of TCRγδ type, usually negative for CD4 and CD8 and positive for CD56 [1, 194]. Note that Southern blot analysis using the $T_C\beta$ probe will be negative in these cases, but PCR analysis for TCRγ will usually be informative.

Primary Cutaneous Aggressive Epidermotropic CD8+ Cytotoxic T-Cell Lymphoma

Primary cutaneous aggressive epidermotropic CD8+ cytotoxic T-cell lymphoma is characterized by localized or disseminated eruptive skin lesions (papules, nodules and tumors) with epidermal infiltration of CD8+ cytotoxic T-cells and an aggressive clinical course [194].

ANGIOIMMUNOBLASTIC T-CELL LYMPHOMA

Angioimmunoblastic T-cell lymphoma (AITL) is a peripheral T-cell lymphoma characterized by generalized lymphadenopathy, hepatosplenomegaly, anemia, hypergammaglobulinemia, and a polymorphic infiltrate involving germinal center T-helper (GC-Th) cells and follicular dendritic cells [196, 197].

Etiology and Pathogenesis

The etiology and pathogenesis of this disorder are not clearly understood. Molecular and immunophenotypic studies suggest that the neoplastic cells of AITL derive from GC-Th cells [196, 197]. These helper T-cells are positive for CD10 and Bcl-6 and show overexpression of CXCL13, a chemokine critical for lymphocyte entry into germinal centers [198]. There have been reports of association between AITL and a number of lymphotropic viruses, such as EBV, human herpes virus (HHV), HIV, and hepatitis C virus, but their role in the pathogenesis of AITL is controversial [199].

Pathology

Morphology

The involved lymph nodes display partial or total effacement of nodal architecture by a polymorphic infiltrate, predominantly involving interfollicular areas. The lymph node sinuses are usually well preserved. There is often pericapsular infiltration. Three overlapping morphologic patterns have been described [197, 200, 201].

Pattern I represents about 20% of the cases with preservation of nodal architecture. It is characterized by follicular hyperplasia, poorly developed mantle zones, and expanded paracortex with a polymorphic infiltrate consisting of lymphocytes, eosinophils, plasma cells, macrophages, transformed large lymphoid blasts, occasional Reed–Sternberg-like cells, and vascular proliferation with abundant endothelial venules.

Pattern II accounts for about 30% of the AITL cases. In Pattern II the normal architecture is almost completely lost, except for occasional depleted follicles showing concentrically arranged follicular dendritic cells. In some cases proliferation of the follicular dendritic cells may extend beyond the follicles. A polymorphic infiltrate with numerous transformed blast cells and vascular proliferation is present.

Pattern III represents about 50% of the cases and is characterized by complete effacement of nodal architecture, prominent proliferation of follicular dendritic cells, extensive vascular proliferation, and in some cases, perivascular collections of atypical medium- to large-sized lymphoid cells with clear or pale cytoplasm (Figure 17.44).

In some cases, consecutive biopsies from the same patient have shown a transition from patterns I to III, suggesting progression from an early stage to an advanced stage [197, 200, 201].

Bone marrow, spleen, liver, skin, and lungs are the most frequent extranodal sites of involvement (Figure 17.45). The involvement of the extranodal sites is often non-specific and consists of a polymorphic infiltrate mimicking an inflammatory process [197].

Immunophenotype

The immunophenotypic hallmark of AITL is the presence of CD4+ T-cells coexpressing CD10 and Bcl-6 within the polymorphic lymphoid infiltrate along with increased numbers of follicular dendritic cells expressing CD21, CD23, or CD35 (Figures 17.46 and 17.47) [197, 200–205]. The CD4+ cells are admixed with CD8+ cells, plasma cells

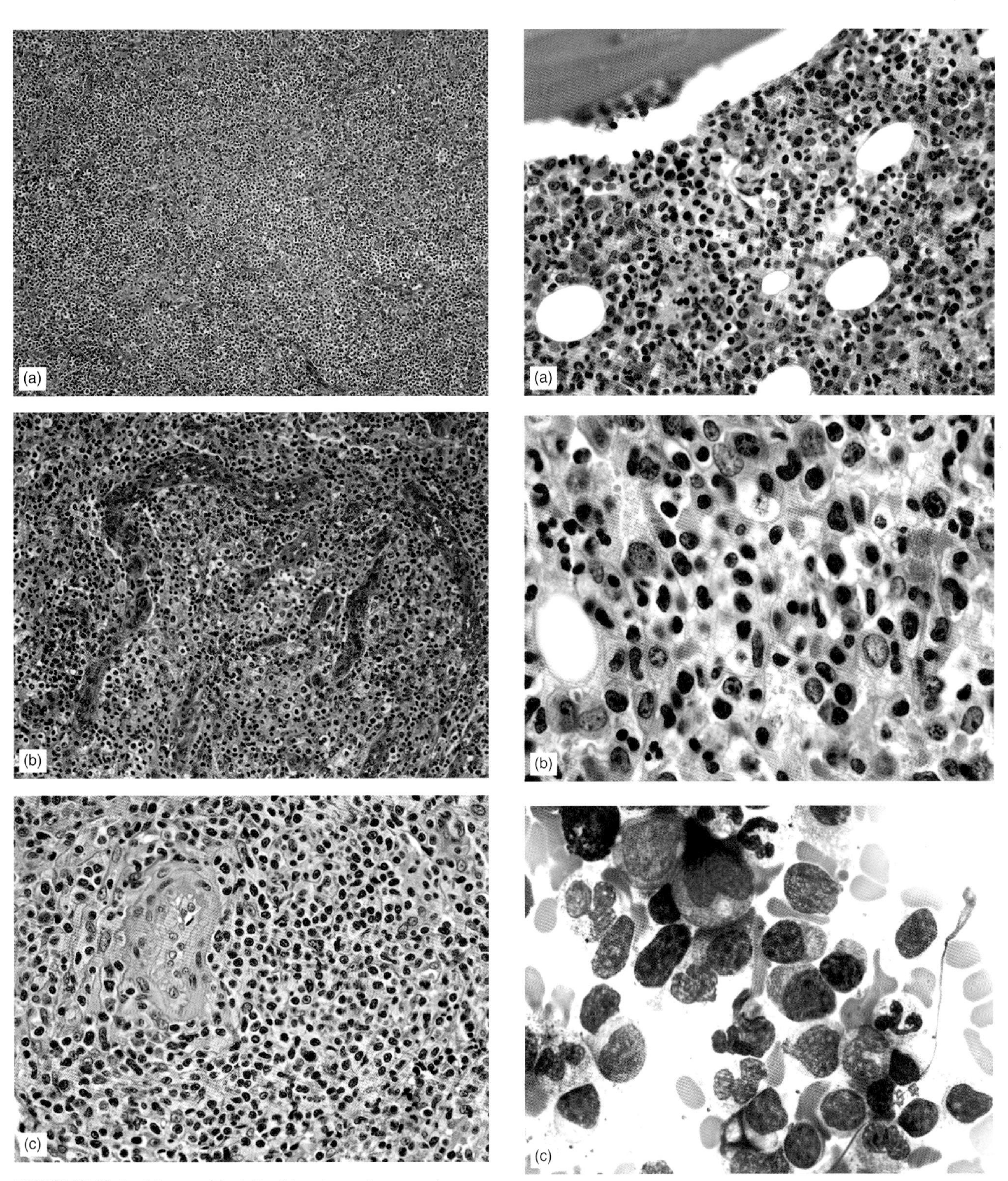

FIGURE 17.44 Angioimmunoblastic T-cell lymphoma demonstrating vascular proliferartion and predominance of atypical medium- to large-sized lymphoid cells with clear or pale cytoplasm: (a) low power, (b) intermediate power, and (c) high power views.

FIGURE 17.45 Bone marrow involvement in angioimmunoblastic T-cell lymphoma may appear non-specific with a collection of histiocytes and atypical lymphocytes: (a) low power and (b) high power views of biopsy section, and (c) bone marrow smear.

(CD138+), and histiocytes (CD68+). The residual follicles are identified by CD20 and CD79a. The CD4+/CD10+ (GC-Th) cells and follicular dendritic cells may be more prominent around high endothelial venules [206, 207].

Molecular and Cytogenetic Studies

The *TCR* rearrangement studies reveal clonal rearrangement in over 75% of the cases [200, 207]. The affected

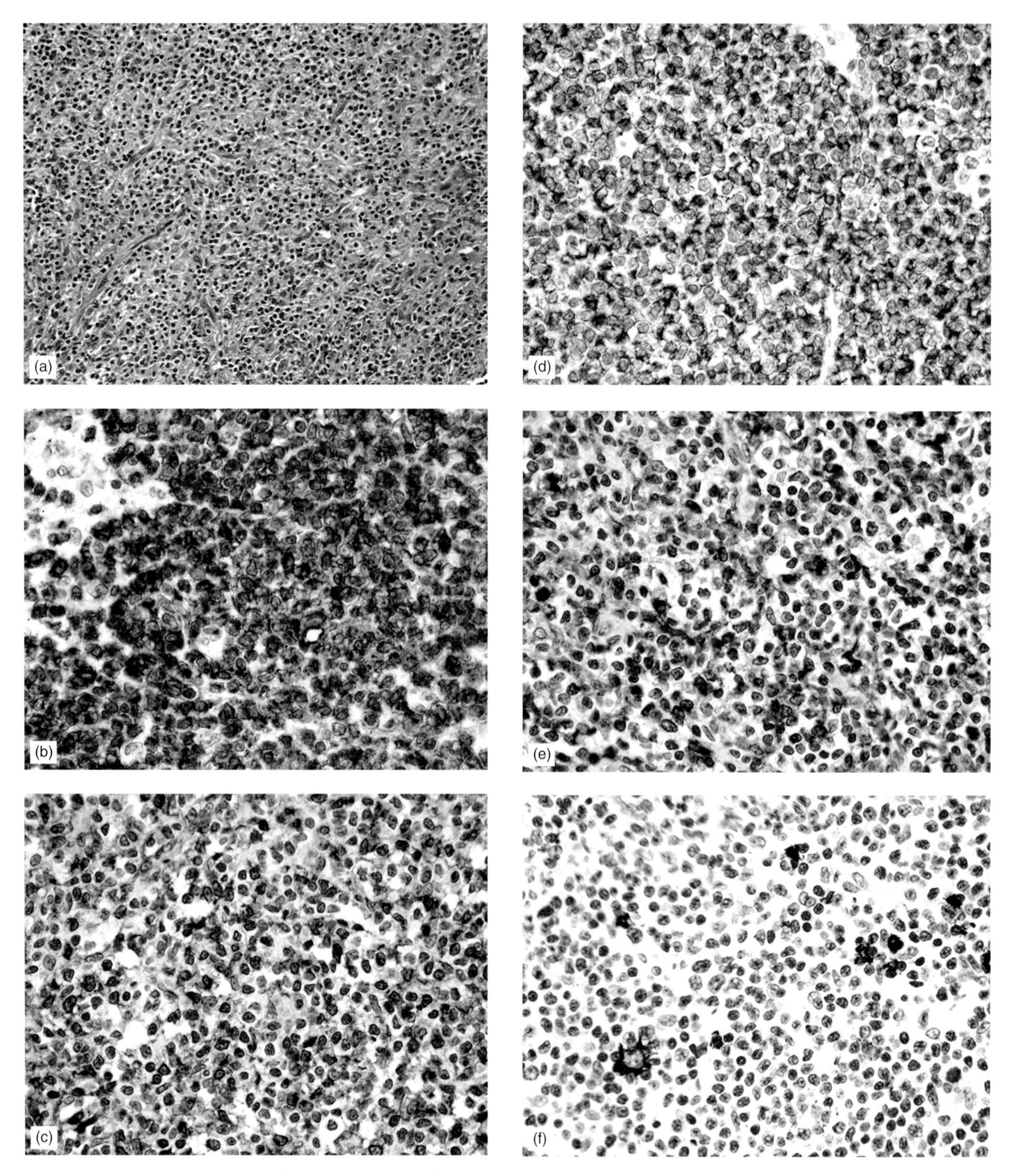

FIGURE 17.46 Angioimmunoblastic T-cell lymphoma. H&E (a) and immunohistochemical stains demonstrating expression of CD2 (b), CD3 (c) CD4 (d), CD10 (e), and CD20 (f).

lymph nodes in approximately 10% of the cases may show an expanded monoclonal B-cell population, often in association with increased number of EBV-infected large B-cells [197].

The gene expression profile of AITL cases has revealed overexpression of several gene characteristics of GC-Th cells, such as *CXCL13*, *BCL6*, *PDCD1*, *CD40L*, and *NFATC1* [208].

Approximately 70% of the AITL patients show clonal chromosomal aberrations [196, 197, 209]. The most frequent recurrent abnormalities include trisomy 3, trisomy 5, and an additional X chromosome (Table 17.6). Other

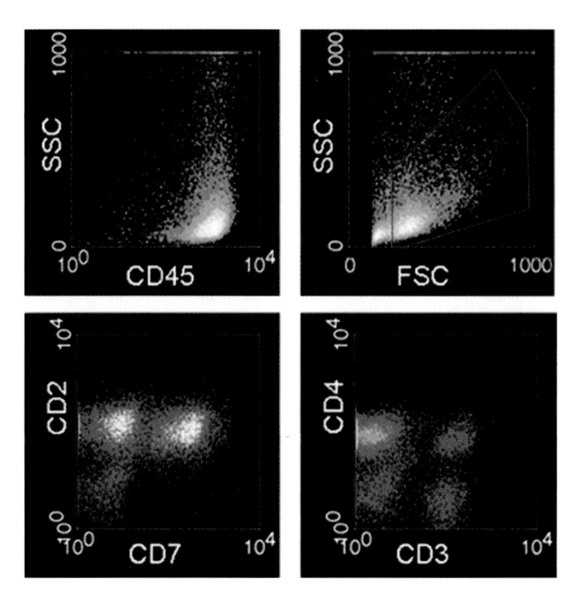

FIGURE 17.47 Angioimmunoblastic T-cell lymphoma. Flow cytometry demonstrating a large cell population (gated red) primarily consisting of CD2+, CD4+, and CD7+ cells.

TABLE 17.6 Clinical features and laboratory findings in angioimmunoblastic T-cell lymphoma.*

Symptoms and signs	Frequency (%)
B symptoms	68–85
Generalized lymphadenopathy	94–97
Splenomegaly	70–73
Hepatomegaly	52–72
Skin rash	48–58
Effusions	23–37
Laboratory findings	
Anemia	40–57
Hypergammaglobulinemia	50–83
Autoantibodies	66–77
Elevated LDH	70–74
Bone marrow involvement	61
Clonal cytogenetic aberrations such as +3, +5, and +X	70
TCR rearrangement	~100

*Adapted from Ref. [197].

frequently observed recurrent cytogenetic abnormalities are gains of 11q13, 19, and 22q [210].

Clinical Aspects

AITL accounts for about 2% of all non-Hodgkin lymphomas. The peak incidence is between the sixth and seventh decades with no significant sex predilection and wide geographical distribution [197, 211]. The clinical presentation often mimics an infectious process characterized by B symptoms and generalized lymphadenopathy [200, 212]. Hepatosplenomegaly has been reported in 50–70% of the patients. Around 50% of the patients complain of pruritus and/or show skin rashes (Table 17.6). A garden variety of autoimmune disorders have been observed in association with AITL, such as autoimmune hemolytic anemia, polyarthritis, rheumatoid arthritis, autoimmune thyroiditis, and vasculitis [212, 213]. Laboratory findings include anemia or pancytopenia, hypergammaglobulinemia, circulating autoantibodies, and elevated serum LDH (Table 17.6).

Single agent chemotherapy (steroids or methotrexate) and various combination of chemotherapeutic regiments have been tried with an overall discouraging outcome and a 5-year survival rate of 30–35% [197, 212].

Differential Diagnosis

The differential diagnosis includes various viral infections and collagen vascular disorders. Bone marrow biopsy and fine needle aspiration or needle core biopsy of the enlarged lymph node usually do not yield a definitive diagnosis. Diagnosis is achieved by the morphologic examination of the entire lymph node [197].

PERIPHERAL T-CELL LYMPHOMA, UNSPECIFIED

Peripheral T-cell lymphoma, unspecified (PTCL-u), includes all the T-cell lymphomas that are not included in the well-defined clinicopathologic entities previously described in this chapter. PTCL-u is predominantly nodal and represents the most common T-cell lymphomas in the Western countries [145, 211, 214].

Etiology and Pathogenesis

The etiology and pathogenesis of this heterogeneous group of neoplastic disorders are not known.

Pathology

Morphology

The normal nodal architecture is often effaced with a diffuse infiltration of neoplastic lymphoid cells. A garden variety of morphologic features have been described, but in most cases the predominant cells are medium- to large-sized with irregular nuclei. The nuclei are hyperchromatic or vesicular with prominent nucleoli. Mitotic figures are frequent. Large cells with clear cytoplasm and Reed–Sternberg-like cells may be present. Vascular proliferation is frequently noted [1]. The infiltrating neoplastic cells are often mixed with inflammatory cells such as small lymphocytes, eosinophils, and histiocytes. Histiocytes may appear in aggregates.

Two major subtypes have been described in the WHO classification: (1) T-zone and (2) lymphoepithelioid cell variants [1].

The *T-zone variant* is characterized by the expansion of interfollicular spaces as the result of infiltration of predominantly small- to medium-sized lymphocytes (Figure 17.48). Lymphoid follicles are usually well preserved or even hyperplastic [1]. Clusters of lymphoid cells with clear cytoplasm are often present, and scattered Reed–Sternberg-like cells may be present. There is a vascular proliferation with predominance of high endothelial venules. Inflammatory cells, such as eosinophils, plasma cells, and histiocytes, are commonly present.

The *lymphoepithelioid cell variant* (Lennert lymphoma) is characterized by the presence of numerous small aggregates of epithelioid histiocytes mixed with a lymphocytic infiltrate predominantly consisting of small lymphocytes with slightly irregular nuclei (Figure 17.49). Scattered larger lymphocytes with clear cytoplasm may be present. The pattern of lymphoid infiltration is diffuse but less frequently may be interfollicular. Vascular proliferation and presence of inflammatory cells are common features.

In a large multicenter retrospective analysis of 385 patients with PTCL-u, cases were divided into a number of morphologic subtypes such as large cell, large- and medium-sized cell, pleomorphic cell, small cell, lymphoepithelioid, T-zone, and PTCL-u not otherwise specified (Table 17.7). The most frequent morphologic subtype in this report was the large cell type accounting for 42% of the cases. Lymphoepithelioid and T-zone types presented 4.9% and 3.8% of the cases, respectively.

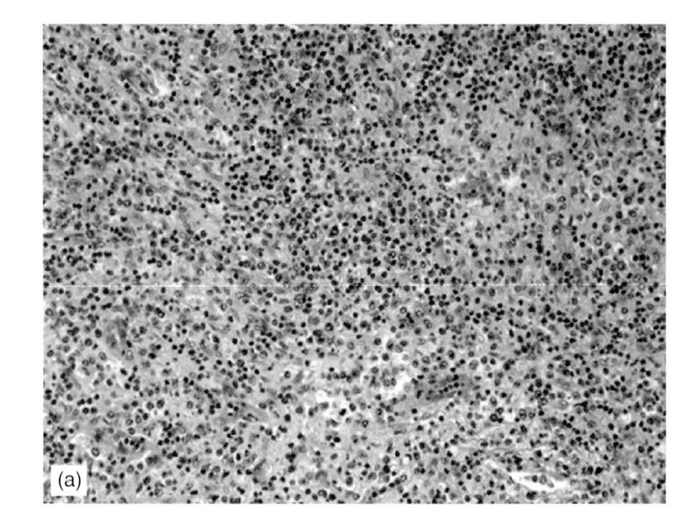

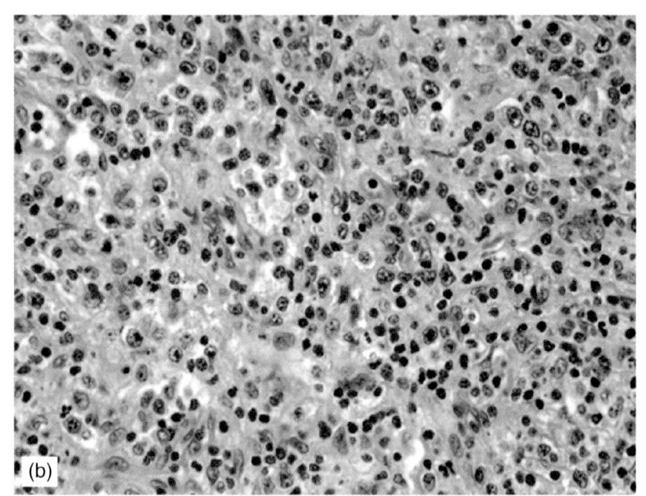

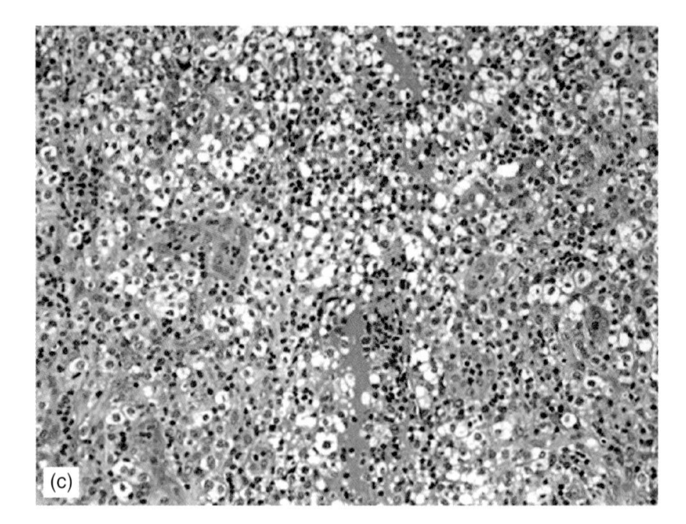

FIGURE 17.48 Peripheral T-cell lymphoma; the T-zone variant consisting of a mixture of small to large lymphocytes (a and b) and the presence of lymphoid cells with clear cytoplasm (c).

Immunophenotype

The neoplastic cells usually express pan-T-cell markers, such as CD3, CD5, and CD7, but loss of CD5 or CD7 is not infrequent. The tumor cells are more commonly CD4+ than CD8. Scattered cells may express CD30 [1, 145, 215]. Expression of CD56 or cytotoxic-associated proteins is rare, and EBV is usually negative in the tumor cells [1, 216].

Molecular and Cytogenetic Studies

Approximately 90% of the patients with PTCL-u show clonally rearranged *TCR* genes and 70–90% demonstrate cytogenetic aberrations [145]. A novel t(5;9)(q33;q22) has been reported in 17% of PTCL-u [217]. This translocation fuses *ITK* and *SYK* genes together. Complex karyotype consistent with clonal evolution is a frequent finding. Breaks involving the TCR loci are often seen.

The chromosomes most frequently altered in structural aberrations are 1, 6, 2, 4, 11, 14, and 17. Additionally, trisomies of 3 or 5 and an extra X chromosome are also common [218, 219]. CGH studies found recurrent losses on chromosomes 13q21, 6q21, 9p21, 10q23-24, 12q21-22, and 5q (Figure 17.50) [220]. Recurrent gains were found on chromosome 7q22. High-level amplifications of 12p13 were observed in a few PTCL-u cases with cytotoxic phenotype. These results suggest that certain genetic alterations may indeed exist in PTCL-u, but definitive clinicopathologic subgroups have not yet been identified, and, thus far, various molecular and cytogenetic findings do not allow a consistent model for pathogenesis to be constructed.

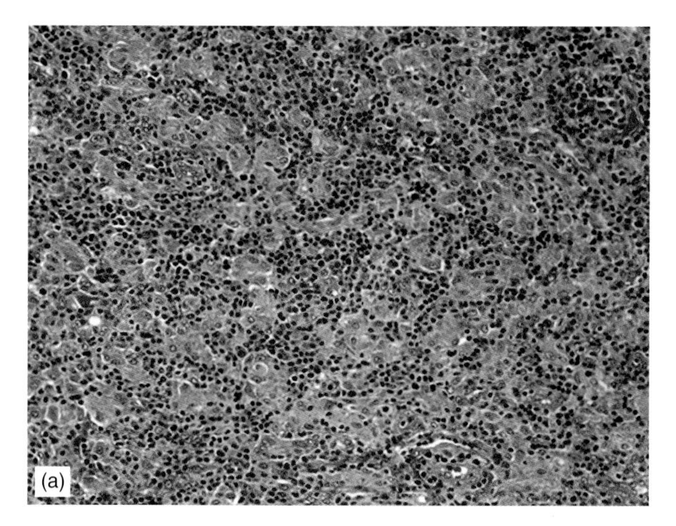

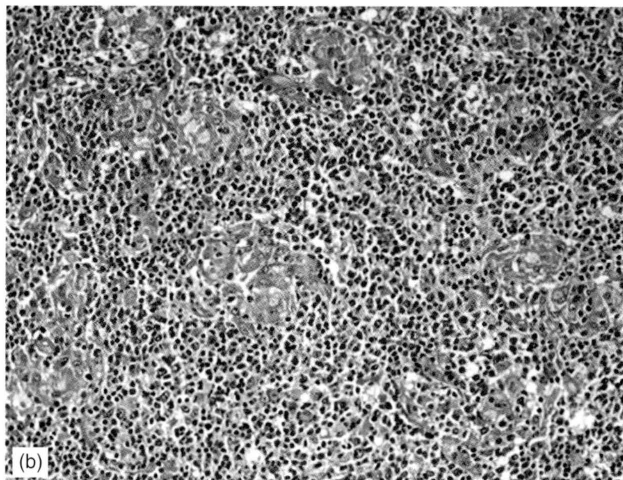

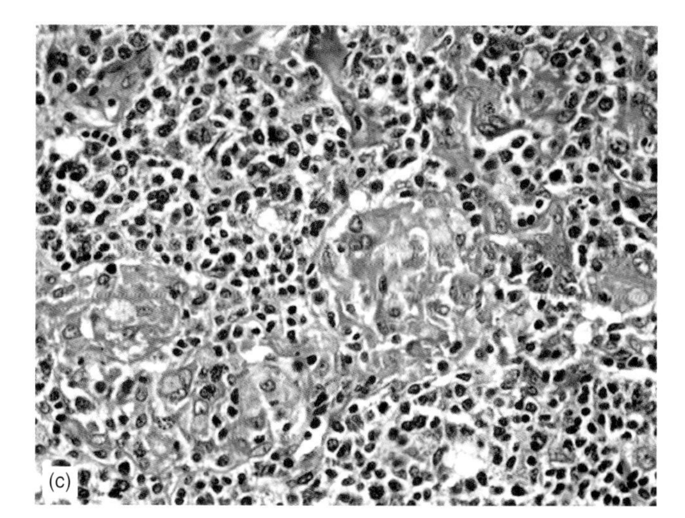

FIGURE 17.49 Peripheral T-cell lymphoma; the lymphoepithelioid cell variant (Lennert lymphoma). Epithelioid histiocytes are mixed with infiltrating lymphocytes in small (a) or large (b and c) clusters.

Clinical Aspects

PTCL-u is the most frequent peripheral T-cell lymphoma comprising approximately 60–70% of all T-cell lymphomas

TABLE 17.7 Morphologic subtypes observed in the retrospective study of 385 patients with peripheral T-cell lymphoma, unspecified.*

Morphologic subtype	% cases
T-zone	3.8
Lymphoepithelioid	4.9
Small cell	5.9
Pleomorphic cell	8.0
Large- and medium-sized cell	16.6
Large cell	18.4
PTCL-u not otherwise specified	42.0

*Adapted from Ref. [221].

[211]. It occurs in middle-aged to elderly patients with a median age of 54 years and a male:female ratio of about 1–2:1 [221]. About 65–75% of the patients are in advanced stages (stage III/IV) at diagnosis [145, 221]. Extranodal disease, B symptoms, and elevated LDH are frequent findings and are associated with unfavorable prognosis. The most frequently involved extranodal sites are bone marrow, spleen, liver, Waldeyer ring, and skin (Table 17.8).

In a large multicenter retrospective analysis of 385 patients with PTCL-u, multivariate analysis revealed four risk factors: (1) age >60 years, (2) LDH values at normal levels or above, (3) performance status (PS) ≥2, and (4) bone marrow involvement [221]. The following four prognostic groups were identified:

Group 1 with no risk factors and 5-year survival rate of 62%

Group 2 with one risk factor and 5-year survival rate of 53%

Group 3 with two risk factors and 5-year survival rate of 33%

Group 4 with three or four risk factors and 5-year survival rate of 18%.

Molecular and immunophenotypic studies have demonstrated that *p53* mutation and overexpression of p53 protein correlate with treatment failure and unfavorable prognosis [145, 222, 223].

Differential Diagnosis

The differential diagnosis includes various reactive lymphadenopathies, ALCL, AITL, and Hodgkin lymphoma. The interfollicular expansion and the presence of inflammatory cells in the T-zone variant may mimic T-zone hyperplasia. Presence of clusters of epithelioid histiocytes in the lymphoepithelioid cell (Lennert) variant may simulate

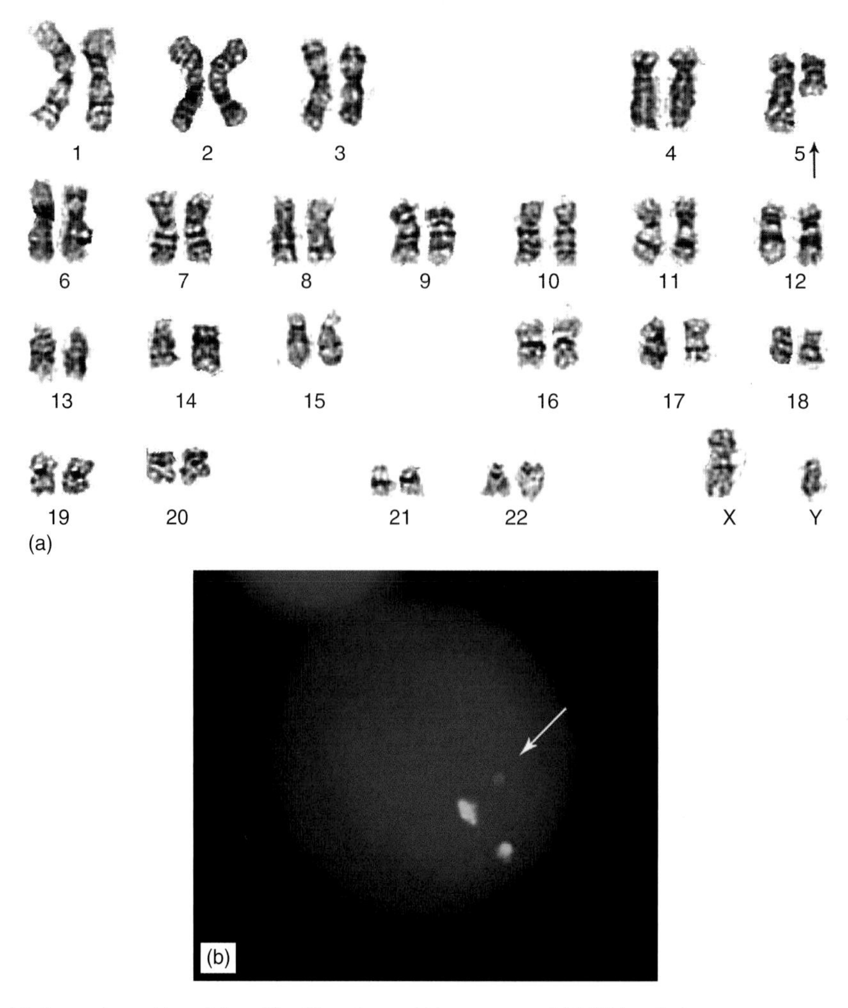

FIGURE 17.50 Deletion of 5q in a patient with peripheral T-cell lymphoma: (a) karyotype and (b) FISH analysis.

TABLE 17.8 Extranodal involvement in patients with peripheral T-cell lymphoma, unspecified.*

Site	% cases
Bone marrow	30.6
Spleen	24.6
Liver	12.9
Waldeyer ring	10.9
Skin	10.1
Lung and pleura	9.8
Bone	4.6
Soft tissue	1.2

*Adapted from Ref. [221].

toxoplasmosis, sarcoidosis, or other types of granulomatous lymphadenitis. Cases with increased proportion of anaplastic large cells or the presence of Reed–Sternberg-like cells may mimic ALCL or Hodgkin lymphoma, respectively. Vascular proliferation and polymorphous infiltrate are among the overlapping features between PTCL-u and AITL.

References

1. Jaffe ES, Harris NL, Stein H, Vardiman JW. (2001). *Pathology and Genetics. Tumors of Haematopoietic and Lymphoid Tissues.* IARC Press, Lyon.
2. Krishnan B, Matutes E, Dearden C. (2006). Prolymphocytic leukemias. *Semin Oncol* **33**, 257–63.
3. Cao TM, Coutre SE. (2003). T-cell prolymphocytic leukemia: Update and focus on alemtuzumab (Campath-1H). *Hematology* **8**, 1–6.
4. De Schouwer PJ, Dyer MJ, Brito-Babapulle VB, Matutes E, Catovsky D, Yuille MR. (2000). T-cell prolymphocytic leukaemia: Antigen receptor gene rearrangement and a novel mode of MTCP1 B1 activation. *Br J Haematol* **110**, 831–8.

5. Yamaguchi M, Yamamoto K, Miki T, Mizutani S, Miura O. (2003). T-cell prolymphocytic leukemia with der(11)t(1;11)(q21;q23) and ATM deficiency. *Cancer Genet Cytogenet* **146**, 22–6.
6. Stilgenbauer S, Schaffner C, Litterst A, Liebisch P, Gilad S, Bar-Shira A, James MR, Lichter P, Dohner H. (1997). Biallelic mutations in the ATM gene in T-prolymphocytic leukemia. *Nat Med* **3**, 1155–9.
7. Shimizu D, Nomura K, Matsumoto Y, Nishida K, Taki T, Horiike S, Inaba T, Fujita N, Taniwaki M. (2006). Small cell variant type of T-prolymphocytic leukemia with a four-year indolent course preceding acute exacerbation. *Leuk Lymphoma* **47**, 1170–2.
8. Brito-Babapulle V, Maljaie SH, Matutes E, Hedges M, Yuille M, Catovsky D. (1997). Relationship of T-leukaemias with cerebriform nuclei to T-prolymphocytic leukaemia: A cytogenetic analysis with *in situ* hybridization. *Br J Haematol* **96**, 724–32.
9. Pawson R, Matutes E, Brito-Babapulle V, Maljaie H, Hedges M, Mercieca J, Dyer M, Catovsky D. (1997). Sezary cell leukaemia: A distinct T cell disorder or a variant form of T prolymphocytic leukaemia? *Leukemia* **11**, 1009–13.
10. Matutes E, Brito-Babapulle V, Swansbury J, Ellis J, Morilla R, Dearden C, Sempere A, Catovsky D. (1991). Clinical and laboratory features of 78 cases of T-prolymphocytic leukemia. *Blood* **78**, 3269–74.
11. Maljaei SH, Brito-Babapulle V, Hiorns LR, Catovsky D. (1998). Abnormalities of chromosomes 8, 11, 14, and X in T-prolymphocytic leukemia studied by fluorescence *in situ* hybridization. *Cancer Genet Cytogenet* **103**, 110–16.
12. Soulier G, Pierron D, Vecchione R, Garand F, Brizard F, Sigaux M-H, Stern , Aurias A. (2001). A complex pattern of recurrent chromosomal losses and gains in T-cell prolymphocytic leukemia. *Genes Chromosomes Cancer* **31**, 248–54.
13. Zemlin M, Hummel M, Anagnostopoulos I, Stein H. (1998). Improved polymerase chain reaction detection of clonally rearranged T-cell receptor ß chain genes. *Diagn Mol Pathol* **7**, 138–45.
14. Arber DA. (2000). Molecular diagnostic approach to non-Hodgkin's lymphoma. *J Mol Diagn* **2**, 178–90.
15. Dearden CE. (2006). T-cell prolymphocytic leukemia. *Med Oncol* **23**, 17–22.
16. Shichishima T, Kawaguchi M, MacHii T, Matsuoka R, Ogawa K, Maruyama Y. (2000). T-prolymphocytic leukaemia with spontaneous remission. *Br J Haematol* **108**, 397–9.
17. Kruspe RC, Ashraf KK, Foran JM, Salzman DE, Reddy VV, Vaughan WP. (2007). Successful treatment of T-cell prolymphocytic leukemia with full-intensity conditioning followed by matched unrelated donor allogeneic stem cell transplantation. *Clin Adv Hematol Oncol* **5**, 882–4.
18. Dearden C. (2004). Alemtuzumab in peripheral T-cell malignancies. *Cancer Biother Radiopharm* **19**, 391–8.
19. Lamy T, Loughran Jr. TP. (2007). Natural killer (NK) cell large granular lymphocyte leukemia. *UpToDate.*
20. O'Malley DP. (2007). T-cell large granular leukemia and related proliferations. *Am J Clin Pathol* **127**, 850–9.
21. Alekshun TJ, Sokol L. (2007). Diseases of large granular lymphocytes. *Cancer Control* **14**, 141–50.
22. Loughran Jr. TP. (1993). Clonal diseases of large granular lymphocytes. *Blood* **82**, 1–14.
23. Loughran Jr. TP, Zambello R, Ashley R, Guderian J, Pellenz M, Semenzato G, Starkebaum G. (1993). Failure to detect Epstein–Barr virus DNA in peripheral blood mononuclear cells of most patients with large granular lymphocyte leukemia. *Blood* **81**, 2723–7.
24. Fouchard N, Flageul B, Bagot M, Avril MF, Hermine O, Sigaux F, Merle-Beral H, Troussard X, Delfraissy JF, deThé G. (1995). Lack of evidence of HTLV-I/II infection in T CD8 malignant or reactive lymphoproliferative disorders in France: A serological and/or molecular study of 169 cases. *Leukemia* **9**, 2087–92.
25. Jaffe ES, Krenacs L, Raffeld M. (2003). Classification of cytotoxic T-cell and natural killer cell lymphomas. *Semin Hematol* **40**, 175–84.
26. Sokol L, Loughran Jr. TP. (2006). Large granular lymphocyte leukemia. *Oncologist* **11**, 263–73.
27. Gentile TC, Loughran Jr. TP. (1995). Interleukin-12 is a costimulatory cytokine for leukemic CD3+ large granular lymphocytes. *Cell Immunol* **166**, 158–61.
28. Liu JH, Wei S, Lamy T, Li Y, Epling-Burnette PK, Djeu JY, Loughran Jr. TP. (2002). Blockade of Fas-dependent apoptosis by soluble Fas in LGL leukemia. *Blood* **100**, 1449–53.
29. Loughran Jr. TP, Hadlock KG, Perzova R, Gentile TC, Yang Q, Foung SK, Poiesz BJ. (1998). Epitope mapping of HTLV envelope seroreactivity in LGL leukaemia. *Br J Haematol* **101**, 318–24.
30. Sokol L, Agrawal D, Loughran Jr. TP. (2005). Characterization of HTLV envelope seroreactivity in large granular lymphocyte leukemia. *Leuk Res* **29**, 381–7.
31. Berliner N, Horwitz M, Loughran Jr. TP. (2004). Congenital and acquired neutropenia. *Hematol Am Soc Hematol Educ Program* **15**, 63–79.
32. Osuji N, Matutes E, Catovsky D, Lampert I, Wotherspoon A. (2005). Histopathology of the spleen in T-cell large granular lymphocyte leukemia and T-cell prolymphocytic leukemia: A comparative review. *Am J Surg Pathol* **29**, 935–41.
33. Go RS, Lust JA, Phyliky RL. (2003). Aplastic anemia and pure red cell aplasia associated with large granular lymphocyte leukemia. *Semin Hematol* **40**, 196–200.
34. Zambello R, Trentin L, Facco M, Cerutti A, Sancetta R, Milani A, Raimondi R, Tassinari C, Agostini C, Semenzato G. (1995). Analysis of the T cell receptor in the lymphoproliferative disease of granular lymphocytes: Superantigen activation of clonal CD3+ granular lymphocytes. *Cancer Res* **55**, 6140–5.
35. Makishima H, Ishida F, Ito T, Kitano K, Ueno S, Ohmine K, Yamashita Y, Ota J, Ota M, Yamauchi K, Mano H. (2006). DNA microarray analysis of T cell-type lymphoproliferative disease of granular lymphocytes. *Br J Haematol* **118**, 462–9.
36. Wlodarski MW, Schade AE, Maciejewski JP. (2006). T-large granular lymphocyte leukemia: Current molecular concepts. *Hematology* **11**, 245–56.
37. Wong KF, Zhang YM, Chan JK. (1999). Cytogenetic abnormalities in natural killer cell lymphoma/leukaemia – is there a consistent pattern? *Leuk Lymphoma* **34**, 241–50.
38. Wong KF, Chan JC, Liu HS, Man C, Kwong YL. (2002). Chromosomal abnormalities in T-cell large granular lymphocyte leukaemia: Report of two cases and review of the literature. *Br J Haematol* **116**, 598–600.
39. Epling-Burnette PK, Loughran Jr. TP. (2003). Survival signals in leukemic large granular lymphocytes. *Semin Hematol* **40**, 213–20.
40. Lamy T, Loughran Jr. TP. (2003). Clinical features of large granular lymphocyte leukemia. *Semin Hematol* **40**, 185–95.
41. Boeckx N, Uyttebroeck A, Langerak AW, Brusselmans C, Goossens W, Bossuyt X. (2004). Clonal proliferation of T-cell large granular lymphocytes. *Pediatr Blood Cancer* **42**, 275–7
42. Pandolfi F, Loughran Jr. TP, Starkebaum G, Chisesi T, Barbui T, Chan WC, Brouet JC, De Rossi G, McKenna RW, Salsano F, *et al.* (1990). Clinical course and prognosis of the lymphoproliferative disease of granular lymphocytes. A multicenter study. *Cancer* **65**, 341–8.
43. Lamy T, Loughran Jr. TP. (1998). Large granular lymphocyte leukemia. *Cancer Control* **5**, 25–33.
44. Lamy T, Loughran Jr. TP. (1999). Current concepts: Large granular lymphocyte leukemia. *Blood Rev* **13**, 230–40.
45. Dhodapkar MV, Li CY, Lust JA, Tefferi A, Phyliky RL. (1994). Clinical spectrum of clonal proliferations of T-large granular lymphocytes: A T-cell clonopathy of undetermined significance? *Blood* **84**, 1620–7.
46. Saway PA, Prasthofer EF, Barton JC. (1989). Prevalence of granular lymphocyte proliferation in patients with rheumatoid arthritis and neutropenia. *Am J Med* **86**, 303–7.
47. Cheung MM, Chan JK, Wong KF. (2003). Natural killer cell neoplasms: A distinctive group of highly aggressive lymphomas/leukemias. *Semin Hematol* **40**, 221–32.
48. Ryder J, Wang X, Bao L, Gross SA, Hua F, Ironsa RD. (2007). Aggressive natural killer cell leukemia: Report of a Chinese series and review of the literature. *Int J Hematol* **85**, 18–25.

49. Kwong YL. (2005). Natural killer-cell malignancies: Diagnosis and treatment. *Leukemia* **19**, 2186–94.
50. Kawa-Ha K, Ishihara S, Ninomiya T, Yumura-Yagi K, Hara J, Murayama F, Tawa A, Hirai K. (1989). CD3-negative lymphoproliferative disease of granular lymphocytes containing Epstein–Barr viral DNA. *J Clin Invest* **84**, 51–5.
51. Nava VE, Jaffe ES. (2005). The pathology of NK-cell lymphomas and leukemias. *Adv Anat Pathol* **12**, 27–34.
52. Chan JK. (2003). Splenic involvement by peripheral T-cell and NK-cell neoplasms. *Semin Diagn Pathol* **20**, 105–20.
53. Siu LL, Chan JK, Kwong YL. (2002). Natural killer cell malignancies: Clinicopathologic and molecular features. *Histol Histopathol* **17**, 539–54.
54. Chan WC, Hans CP. (2003). Genetic and molecular genetic studies in the diagnosis of T and NK cell neoplasia. *Hum Pathol* **34**, 314–21.
55. Oshimi K. (2003). Leukemia and lymphoma of natural killer lineage cells. *Int J Hematol* **78**, 18–23.
56. Ruskova A, Thula R, Chan G. (2004). Aggressive natural killer-cell leukemia: Report of five cases and review of the literature. *Leuk Lymphoma* **45**, 2427–38.
57. Stadlmann S, Fend F, Moser P, Obrist P, Greil R, Dirnhofer S. (2001). Epstein–Barr virus-associated extranodal NK/T-cell lymphoma, nasal type of the hypopharynx, in a renal allograft recipient: Case report and review of literature. *Hum Pathol* **32**, 1264–8.
58. Chan JK, Sin VC, Wong KF, Ng CS, Tsang WY, Chan CH, Cheung MM, Lau WH. (1997). Nonnasal lymphoma expressing the natural killer cell marker CD56: A clinicopathologic study of 49 cases of an uncommon aggressive neoplasm. *Blood* **89**, 4501–13.
59. Wong KF, Chan JK, Cheung MM, So JC. (2001). Bone marrow involvement by nasal NK cell lymphoma at diagnosis is uncommon. *Am J Clin Pathol* **115**, 266–70.
60. Zambello R, Trentin L, Ciccone E, Bulian P, Agostini C, Moretta A, Moretta L, Semenzato G. (1993). Phenotypic diversity of natural killer (NK) populations in patients with NK-type lymphoproliferative disease of granular lymphocytes. *Blood* **81**, 2381–5.
61. Zhang Y, Nagata H, Ikeuchi T, Mukai H, Oyoshi MK, Demachi A, Morio T, Wakiguchi H, Kimura N, Shimizu N, Yamamoto K. (2003). Common cytological and cytogenetic features of Epstein–Barr virus (EBV)-positive natural killer (NK) cells and cell lines derived from patients with nasal T/NK-cell lymphomas, chronic active EBV infection and hydroa vacciniforme-like eruptions. *Br J Haematol* **121**, 805–14.
62. Chim CS, Ooi GC, Shek TW, Liang R, Kwong YL. (1999). Lethal midline granuloma revisited: Nasal T/natural-killer cell lymphoma. *J Clin Oncol* **17**, 1322–5.
63. Kim GE, Cho JH, Yang WI, Chung EJ, Suh CO, Park KR, Hong WP, Park IY, Hahn JS, Roh JK, Kim BS. (2000). Angiocentric lymphoma of the head and neck: Patterns of systemic failure after radiation treatment. *J Clin Oncol* **18**, 54–63.
64. Cheung MM, Chan JK, Lau WH, Ngan RK, Foo WW. (2002). Early stage nasal NK/T-cell lymphoma: Clinical outcome, prognostic factors, and the effect of treatment modality. *Int J Radiat Oncol Biol Phys* **54**, 182–90.
65. You JY, Chi KH, Yang MH, Chen CC, Ho CH, Chau WK, Hsu HC, Gau JP, Tzeng CH, Liu JH, Chen PM, Chiou TJ. (2004). Radiation therapy versus chemotherapy as initial treatment for localized nasal natural killer (NK)/T-cell lymphoma: A single institute survey in Taiwan. *Ann Oncol* **15**, 618–25.
66. Egashira M, Kawamata N, Sugimoto K, Kaneko T, Oshimi K. (1999). P-glycoprotein expression on normal and abnormally expanded natural killer cells and inhibition of P-glycoprotein function by cyclosporin A and its analogue, PSC833. *Blood* **93**, 599–606.
67. Rabbani GR, Phyliky RL, Tefferi A. (1999). A long-term study of patients with chronic natural killer cell lymphocytosis. *Br J Haematol* **106**, 960–6.
68. Suzuki K, Ohshima K, Karube K, Suzumiya J, Ohga S, Ishihara S, Tamura K, Kikuchi M. (2004). Clinicopathological states of Epstein–Barr virus-associated T/NK-cell lymphoproliferative disorders (severe chronic active EBV infection) of children and young adults. *Int J Oncol* **24**, 1165–74.
69. Quintanilla-Martinez L, Kumar S, Fend F, Reyes E, Teruya-Feldstein J, Kingma DW, Sorbara L, Raffeld M, Straus SE, Jaffe ES. (2000). Fulminant EBV(+) T-cell lymphoproliferative disorder following acute/chronic EBV infection: A distinct clinicopathologic syndrome. *Blood* **96**, 443–51.
70. Siegel RS, Gartenhaus RB, Kuzel TM. (2001). Human T-cell lymphotropic-I-associated leukemia/lymphoma. *Curr Treat Options Oncol* **2**, 291–300.
71. Kikuchi M, Mitsui T, Takeshita M, Okamura H, Naitoh H, Eimoto T. (1986). Virus associated adult T-cell leukemia (ATL) in Japan: Clinical, histological and immunological studies. *Hematol Oncol* **4**, 67–81.
72. Carneiro-Proietti AB, Catalan-Soares BC, Castro-Costa CM, Murphy EL, Sabino EC, Hisada M, Galvao-Castro B, Alcantara LC, Remondegui C, Verdonck K, Proietti FA. (2006). HTLV in the Americas: Challenges and perspectives. *Rev Panam Salud Publica* **19**, 44–53.
73. Grassmann R, Aboud M, Jeang KT. (2005). Molecular mechanisms of cellular transformation by HTLV-1 Tax. *Oncogene* **24**, 5976–85.
74. Yasunaga J, Matsuoka M. (2003). Leukemogenesis of adult T-cell leukemia. *Int J Hematol* **78**, 312–20.
75. Mortreux F, Gabet AS, Wattel E. (2003). Molecular and cellular aspects of HTLV-1 associated leukemogenesis *in vivo*. *Leukemia* **17**, 26–38.
76. Taylor GP, Matsuoka M. (2005). Natural history of adult T-cell leukemia/lymphoma and approaches to therapy. *Oncogene* **24**, 6047–57.
77. Barnard AL, Igakura T, Tanaka Y, Taylor GP, Bangham CR. (2005). Engagement of specific T-cell surface molecules regulates cytoskeletal polarization in HTLV-1-infected lymphocytes. *Blood* **106**, 988–95.
78. Giam CZ, Jeang KT. (2007). HTLV-1 Tax and adult T-cell leukemia. *Front Biosci* **12**, 1496–507.
79. Takeda S, Maeda M, Morikawa S, Taniguchi Y, Yasunaga J, Nosaka K, Tanaka Y, Matsuoka M. (2004). Genetic and epigenetic inactivation of Tax gene in adult T-cell leukemia cells. *Int J Cancer* **109**, 559–67.
80. Yamada Y, Kamihira S. (2005). Inactivation of tumor suppressor genes and the progression of adult T-cell leukemia–lymphoma. *Leuk Lymphoma* **46**, 1553–9.
81. Sasaki H, Nishikata I, Shiraga T, Akamatsu E, Fukami T, Hidaka T, Kubuki Y, Okayama A, Hamada K, Okabe H, Murakami Y, Tsubouchi H, Morishita K. (2005). Overexpression of a cell adhesion molecule, TSLC1, as a possible molecular marker for acute-type adult T-cell leukemia. *Blood* **105**, 1204–13.
82. Yasunaga J, Taniguchi Y, Nosaka K, Yoshida M, Satou Y, Sakai T, Mitsuya H, Matsuoka M. (2004). Identification of aberrantly methylated genes in association with adult T-cell leukemia. *Cancer Res* **64**, 6002–9.
83. Yoshida M, Nosaka K, Yasunaga J, Nishikata I, Morishita K, Matsuoka M. (2004). Aberrant expression of the MEL1S gene identified in association with hypomethylation in adult T-cell leukemia cells. *Blood* **103**, 2753–60.
84. Mesnard JM, Barbeau B, Devaux C. (2006). HBZ, a new important player in the mystery of adult T-cell leukemia. *Blood* **108**, 3979–82.
85. Matutes E. (2007). Adult T-cell leukaemia/lymphoma. *J Clin Pathol* **60**, 1373–7.
86. Ohshima K. (2007). Pathological features of diseases associated with human T-cell leukemia virus I. *Cancer Sci* **98**, 772–8.
87. Dogan A, Morice WG. (2004). Bone marrow histopathology in peripheral T-cell lymphomas. *Br J Haematol* **127**, 140–54.
88. Ravandi F, O'Brien S. (2005). Chronic lymphoid leukemias other than chronic lymphocytic leukemia: Diagnosis and treatment. *Mayo Clin Proc* **80**, 1660–74.
89. Waldmann TA, Greene WC, Sarin PS, Saxinger C, Blayney DW, Blattner WA, Goldman CK, Bongiovanni K, Sharrow S, Depper JM. (1984). Functional and phenotypic comparison of human T cell leukemia/lymphoma virus positive adult T cell leukemia with human T cell

leukemia/lymphoma virus negative Sezary leukemia, and their distinction using anti-Tac. Monoclonal antibody identifying the human receptor for T cell growth factor. *J Clin Invest* **73**, 1711–18.

90. Sanada I, Tanaka R, Kumagai E, Tsuda H, Nishimura H, Yamaguchi K, Kawano F, Fujiwara H, Takatsuki K. (1985). Chromosomal aberrations in adult T cell leukemia: Relationship to the clinical severity. *Blood* **65**, 649–54.
91. Hatta Y, Yamada Y, Tomonaga M, Miyoshi I, Said JW, Koeffler HP. (1999). Detailed deletion mapping of the long arm of chromosome 6 in adult T-cell leukemia. *Blood* **93**, 613–16.
92. Chadburn A, Athan E, Wieczorek R, Knowles DM. (1991). Detection and characterization of human T-cell lymphotropic virus type I (HTLV-I) associated T-cell neoplasms in an HTLV-I nonendemic region by polymerase chain reaction. *Blood* **77**, 2419–30.
93. Taylor G. (2007). Molecular aspects of HTLV-I infection and adult T-cell leukaemia/lymphoma. *J Clin Pathol* **60**, 1392–6.
94. Oshiro A, Tagawa H, Ohshima K, Karube K, Uike N, Tashiro Y, Utsunomiya A, Masuda M, Takasu N, Nakamura S, Morishima Y, Seto M. (2006). Identification of subtype-specific genomic alterations in aggressive adult T-cell leukemia/lymphoma. *Blood* **107**, 4500–7.
95. Proietti FA, Carneiro-Proietti AB, Catalan-Soares BC, Murphy EL. (2005). Global epidemiology of HTLV-I infection and associated diseases. *Oncogene* **24**, 6058–68.
96. Bittencourt AL, da Graças Vieira M, Brites CR, Farre L, Barbosa HS. (2007). Adult T-cell leukemia/lymphoma in Bahia, Brazil: Analysis of prognostic factors in a group of 70 patients. *Am J Clin Pathol* **128**, 875–82.
97. Nosaka K, Miyamoto T, Sakai T, Mitsuya H, Suda T, Matsuoka M. (2002). Mechanism of hypercalcemia in adult T-cell leukemia: Overexpression of receptor activator of nuclear factor kappaB ligand on adult T-cell leukemia cells. *Blood* **99**, 634–40.
98. Ravandi F, Kantarjian H, Jones D, Dearden C, Keating M, O'Brien S. (2005). Mature T-cell leukemias. *Cancer* **104**, 1808–18.
99. Tsukasaki K, Tanosaki S, DeVos S, Hofmann WK, Wachsman W, Gombart AF, Krebs J, Jauch A, Bartram CR, Nagai K, Tomonaga M, Said JW, Koeffler HP. (2004). Identifying progression-associated genes in adult T-cell leukemia/lymphoma by using oligonucleotide microarrays. *Int J Cancer* **109**, 875–81.
100. Matutes E. (2007). Adult T-cell leukemia/lymphoma. *J Clin Pathol* **60**, 1373–7.
101. Hoppe RT, Kim YH. (2007). Clinical features, diagnosis, and staging of mycosis fungoides and Sezary syndrome. *UpToDate*.
102. Tothova SM, Bonin S, Trevisan G, Stanta G. (2006). Mycosis fungoides: Is it a *Borrelia burgdorferi*-associated disease? *Br J Cancer* **94**, 879–83.
103. Hall WW, Liu CR, Schneewind O, Takahashi H, Kaplan MH, Röupe G, Vahlne A. (1991). Deleted HTLV-I provirus in blood and cutaneous lesions of patients with mycosis fungoides. *Science* **253**, 317–20.
104. Poiesz B, Dube D, Dube S, Love J, Papsidero L, Uner A, Hutchinson R. (2000). HTLV-II-associated cutaneous T-cell lymphoma in a patient with HIV-1 infection. *N Engl J Med* **342**, 930–6.
105. Ghosh SK, Abrams JT, Terunuma H, Vonderheid EC, DeFreitas E. (1994). Human T-cell leukemia virus type I tax/rex DNA and RNA in cutaneous T-cell lymphoma. *Blood* **84**, 2663–71.
106. Lessin SR, Rook AH, Li G, Wood GS. (1996). HTLV-I and CTCL: The link is missing. *J Invest Dermatol* **107**, 783–4.
107. Pawlaczyk M, Filas V, Sobieska M, Gozdzicka-Jozefiak A, Wiktorowicz K, Breborowicz J. (2005). No evidence of HTLV-I infection in patients with mycosis fungoides and Sezary syndrome. *Neoplasma* **52**, 52–5.
108. Smoller BR, Bishop K, Glusac E, Kim YH, Hendrickson M. (1995). Reassessment of histologic parameters in the diagnosis of mycosis fungoides. *Am J Surg Pathol* **19**, 1423–30.
109. Fink-Puches R, Zenahlik P, Back B, Smolle J, Kerl H, Cerroni L. (2002). Primary cutaneous lymphomas: Applicability of current classification schemes (European Organization for Research and Treatment of Cancer, World Health Organization) based on clinicopathologic features observed in a large group of patients. *Blood* **99**, 800–5.
110. Kim EJ, Lin J, Junkins-Hopkins JM, Vittorio CC, Rook AH. (2006). Mycosis fungoides and sezary syndrome: An update. *Curr Oncol Rep* **8**, 376–86.
111. Willemze R, Kerl H, Sterry W, Berti E, Cerroni L, Chimenti S, Diaz-Peréz JL, Geerts ML, Goos M, Knobler R, Ralfkiaer E, Santucci M, Smith N, Wechsler J, van Vloten WA, Meijer CJ. (1997). EORTC classification for primary cutaneous lymphomas: A proposal from the Cutaneous Lymphoma Study Group of the European Organization for Research and Treatment of Cancer. *Blood* **90**, 354–71.
112. Vonderheid EC, Pena J, Nowell P. (2006). Sezary cell counts in erythrodermic cutaneous T-cell lymphoma: Implications for prognosis and staging. *Leuk Lymphoma* **47**, 1841–56.
113. Morice WG, Katzmann JA, Pittelkow MR, el-Azhary RA, Gibson LE, Hanson CA. (2006). A comparison of morphologic features, flow cytometry, TCR-Vbeta analysis, and TCR-PCR in qualitative and quantitative assessment of peripheral blood involvement by Sezary syndrome. *Am J Clin Pathol* **125**, 364–74.
114. Sokolowska-Wojdylo M, Wenzel J, Gaffal E, Steitz J, Roszkiewicz J, Bieber T, Tuting T. (2005). Absence of CD26 expression on skin-homing CLA+ CD4+ T lymphocytes in peripheral blood is a highly sensitive marker for early diagnosis and therapeutic monitoring of patients with Sezary syndrome. *Clin Exp Dermatol* **30**, 702–6.
115. Whang-Peng J, Bunn PA, Knutsen T, Matthews MJ, Schechter G, Minna JD. (1982). Clinical implications of cytogenetic studies in cutaneous T-cell lymphoma (CTCL). *Cancer* **50**, 1539–53.
116. Utikal J, Poenitz N, Gratchev A, Klemke CD, Nashan D, Tuting T, Goerdt S. (2006). Additional Her 2/neu gene copies in patients with Sezary syndrome. *Leuk Res* **30**, 755–60.
117. Navas IC, Ortiz-Romero PL, Villuendas R, Martínez P, García C, Gómez E, Rodriguez JL, García D, Vanaclocha F, Iglesias L, Piris MA, Algara P. (2000). p16(INK4a) gene alterations are frequent in lesions of mycosis fungoides. *Am J Pathol* **156**, 1565–72.
118. Nebozhyn M, Loboda A, Kari L, Rook AH, Vonderheid EC, Lessin S, Berger C, Edelson R, Nichols C, Yousef M, Gudipati L, Shang M, Showe MK, Showe LC. (2006). Quantitative PCR on 5 genes reliably identifies CTCL patients with 5% to 99% circulating tumor cells with 90% accuracy. *Blood* **107**, 3189–96.
119. Foss F. (2004). Mycosis fungoides and the Sezary syndrome. *Curr Opin Oncol* **16**, 421–8.
120. Rosen ST, Querfeld C. (2006). Primary cutaneous T-cell lymphomas, hematology. *Am Soc Hematol Educ Program*, 323–30.
121. Kashani-Sabet M, McMillan A, Zackheim HS. (2001). A modified staging classification for cutaneous T-cell lymphoma. *J Am Acad Dermatol* **45**, 700–6.
122. Tan ES, Tang MB, Tan SH. (2006). Retrospective 5-year review of 131 patients with mycosis fungoides and Sezary syndrome seen at the National Skin Centre, Singapore. *Australas J Dermatol* **47**, 248–52.
123. Ortonne N, Huet D, Gaudez C, Marie-Cardine A, Schiavon V, Bagot M, Musette P, Bensussan A. (2006). Significance of circulating T-cell clones in Sezary syndrome. *Blood* **107**, 4030–8.
124. Introcaso CE, Hess SD, Kamoun M, Ubriani R, Gelfand JM, Rook AH. (2005). Association of change in clinical status and change in the percentage of the CD4+ CD26− lymphocyte population in patients with Sezary syndrome. *J Am Acad Dermatol* **53**, 428–34.
125. Vergier B, de Muret A, Beylot-Barry M, Vaillant L, Ekouevi D, Chene G, Carlotti A, Franck N, Dechelotte P, Souteyrand P, Courville P, Joly P, Delaunay M, Bagot M, Grange F, Fraitag S, Bosq J, Petrella T, Durlach A, De Mascarel A, Merlio JP, Wechsler J. (2000). Transformation of mycosis fungoides: Clinicopathological and prognostic features of 45 cases. French Study Group of Cutaneous Lymphomas. *Blood* **95**, 2212–18.
126. Diamandidou E, Colome-Grimmer M, Fayad L, Duvic M, Kurzrock R. (1998). Transformation of mycosis fungoides/Sezary syndrome: Clinical characteristics and prognosis. *Blood* **92**, 1150–9.

127. Trautinger F, Knobler R, Willemze R, Peris K, Stadler R, Laroche L, D'Incan M, Ranki A, Pimpinelli N, Ortiz-Romero P, Dummer R, Estrach T, Whittaker S. (2006). EORTC consensus recommendations for the treatment of mycosis fungoides/Sezary syndrome. *Eur J Cancer* **42**, 1014–30.
128. Gaulard P, Belhadj K, Reyes F. (2003). Gammadelta T-cell lymphomas. *Semin Hematol* **40**, 233–43.
129. Weidmann E. (2000). Hepatosplenic T cell lymphoma. A review on 45 cases since the first report describing the disease as a distinct lymphoma entity in 1990. *Leukemia* **14**, 991–7.
130. Salmon JS, Thompson MA, Arildsen RC, Greer JP. (2006). Non-Hodgkin's lymphoma involving the liver: Clinical and therapeutic considerations. *Clin Lymphoma Myeloma* **6**, 273–80.
131. Taguchi A, Miyazaki M, Sakuragi S, Shinohara K, Kamei T, Inoue Y. (2004). Gamma/delta T cell lymphoma. *Intern Med* **43**, 120–5.
132. de Wolf-Peeters C, Achten R. (2000). Gammadelta T-cell lymphomas: A homogeneous entity? *Histopathology* **36**, 294–305.
133. Belhadj K, Reyes F, Farcet JP, Tilly H, Bastard C, Angonin R, Deconinck E, Charlotte F, Leblond V, Labouyrie E, Lederlin P, Emile JF, Delmas-Marsalet B, Arnulf B, Zafrani ES, Gaulard P. (2003). Hepatosplenic gammadelta T-cell lymphoma is a rare clinicopathologic entity with poor outcome: Report on a series of 21 patients. *Blood* **102**, 4261–9. [Epub 2003 August 7].
134. Volk HD, Reinke P, Neuhaus K, Fiebig H, von Baehr R. (1989). Expansion of a CD 3+ 4-8- TCR alpha/beta-T lymphocyte population in renal allograft recipients. *Transplantation* **47**, 556–8.
135. Kirk AD, Ibrahim S, Dawson DV, Sanfilippo F, Finn OJ. (1993). Characterization of T cells expressing the gamma/delta antigen receptor in human renal allografts. *Hum Immunol* **36**, 11–19.
136. Brennan FM, Londei M, Jackson AM, Hercend T, Brenner MB, Maini RN, Feldmann M. (1988). T cells expressing gamma delta chain receptors in rheumatoid arthritis. *J Autoimmun* **1**, 319–26.
137. Toro JR, Beaty M, Sorbara L, Turner ML, White J, Kingma DW, Raffeld M, Jaffe ES. (2000). Gamma delta T-cell lymphoma of the skin: A clinical, microscopic, and molecular study. *Arch Dermatol* **136**, 1024–32.
138. Ahmad E, Kingma DW, Jaffe ES, Schrager JA, Janik J, Wilson W, Stetler-Stevenson M. (2005). Flow cytometric immunophenotypic profiles of mature gamma delta T-cell malignancies involving peripheral blood and bone marrow. *Cytometry B Clin Cytom* **67**, 6–12.
139. Sallah S, Smith SV, Lony LC, Woodard P, Schmitz JL, Folds JD. (1997). Gamma/delta T-cell hepatosplenic lymphoma: Review of the literature, diagnosis by flow cytometry and concomitant autoimmune hemolytic anemia. *Ann Hematol* **74**, 139–42.
140. Yao M, Tien HF, Lin MT, Su IJ, Wang CT, Chen YC, Shen MC, Wang CH. (1996). Clinical and hematological characteristics of hepatosplenic T gamma/delta lymphoma with isochromosome for long arm of chromosome 7. *Leuk Lymphoma* **22**, 495–500.
141. Wang CC, Tien HF, Lin MT, Su IJ, Wang CH, Chuang SM. (1995). Consistent presence of isochromosome 7q in hepatosplenic T gamma/delta lymphoma: A new cytogenetic–clinicopathologic entity. *Genes Chromosomes Cancer* **12**, 161–4.
142. Halfdanarson TR, Litzow MR, Murray JA. (2007). Hematologic manifestations of celiac disease. *Blood* **109**, 412–21.
143. Lee MY, Tsou MH, Tan TD, Lu MC. (2005). Clinicopathological analysis of T-cell lymphoma in Taiwan according to WHO classification: High incidence of enteropathy-type intestinal T-cell lymphoma. *Eur J Haematol* **75**, 221–6.
144. Isaacson PG, Du MQ. (2005). Gastrointestinal lymphoma: Where morphology meets molecular biology. *J Pathol* **205**, 255–74.
145. Rizvi MA, Evens AM, Tallman MS, Nelson BP, Rosen ST. (2006). T-cell non-Hodgkin lymphoma. *Blood* **107**, 1255–64.
146. Howell WM, Leung ST, Jones DB, Nakshabendi I, Hall MA, Lanchbury JS, Ciclitira PJ, Wright DH. (1995). HLA-DRB, -DQA, and -DQB polymorphism in celiac disease and enteropathy-associated T-cell lymphoma. Common features and additional risk factors for malignancy. *Hum Immunol* **43**, 29–37.
147. Quintanilla-Martínez L, Lome-Maldonado C, Ott G, Gschwendtner A, Gredler E, Angeles-Angeles A, Reyes E, Fend F. (1998). Primary intestinal non-Hodgkin's lymphoma and Epstein–Barr virus: High frequency of EBV-infection in T-cell lymphomas of Mexican origin. *Leuk Lymphoma* **30**, 111–21.
148. Chott A, Haedicke W, Mosberger I, Födinger M, Winkler K, Mannhalter C, Müller-Hermelink HK. (1998). Most CD56+ intestinal lymphomas are CD8+ CD5-T-cell lymphomas of monomorphic small to medium size histology. *Am J Pathol* **153**, 1483–90.
149. Daum S, Weiss D, Hummel M, Ullrich R, Heise W, Stein H, Riecken EO, Foss HD, Intestinal Lymphoma Study Group (2001). Frequency of clonal intraepithelial T lymphocyte proliferations in enteropathy-type intestinal T cell lymphoma, coeliac disease, and refractory sprue. *Gut* **49**, 804–12.
150. Obermann EC, Diss TC, Hamoudi RA, Munson P, Wilkins BS, Camozzi ML, Isaacson PG, Du MQ, Dogan A. (2004). Loss of heterozygosity at chromosome 9p21 is a frequent finding in enteropathy-type T cell lymphoma. *J Pathol* **202**, 252 62.
151. Baumgartner AK, Zettl A, Chott A, Ott G, Muller-Hermelink HK, Starostik P. (2003). High frequency of genetic aberrations in enteropathy-type T-cell lymphoma. *Lab Invest* **83**, 1509–16.
152. Zettl A, Ott G, Makulik A, Katzenberger T, Starostik P, Eichler T, Puppe B, Bentz M, Muller-Hermelink HK, Chott A. (2002). Chromosomal gains at 9q characterize enteropathy-type T-cell lymphoma. *Am J Pathol* **161**, 1635–45.
153. Falini B. (2001). Anaplastic large cell lymphoma: Pathological, molecular and clinical features. *Br J Haematol* **114**, 741–60.
154. Jacobsen E. (2006). Anaplastic large-cell lymphoma, T-/null-cell type. *Oncologist* **11**, 831–40.
155. Kaneko Y, Frizzera G, Edamura S, Maseki N, Sakurai M, Komada Y, Sakurai M, Tanaka H, Sasaki M, Suchi T. (1989). A novel translocation, t(2;5)(p23;q35), in childhood phagocytic large T-cell lymphoma mimicking malignant histiocytosis. *Blood* **73**, 806–13.
156. Morris SW, Kirstein MN, Valentine MB, Dittmer KG, Shapiro DN, Saltman DL, Look AT. (1994). Fusion of a kinase gene, ALK, to a nucleolar protein gene, NPM, in non-Hodgkin's lymphoma. *Science* **263**, 1281–4.
157. Slupianek A, Nieborowska-Skorska M, Hoser G, Morrione A, Majewski M, Xue L, Morris SW, Wasik MA, Skorski T. (2001). Role of phosphatidylinositol 3-kinase-Akt pathway in nucleophosmin/anaplastic lymphoma kinase-mediated lymphomagenesis. *Cancer Res* **61**, 2194–9.
158. Kutok JL, Aster JC. (2002). Molecular biology of anaplastic lymphoma kinase-positive anaplastic large-cell lymphoma. *J Clin Oncol* **20**, 3691–702.
159. Kuefer MU, Look AT, Pulford K, Behm FG, Pattengale PK, Mason DY, Morris SW. (1997). Retrovirus-mediated gene transfer of NPM-ALK causes lymphoid malignancy in mice. *Blood* **90**, 2901–10.
160. Kadin ME. (1997). Anaplastic large cell lymphoma and its morphological variants. *Cancer Surv* **30**, 77–86.
161. Nascimento AF, Pinkus JL, Pinkus GS. (2004). Clusterin, a marker for anaplastic large cell lymphoma immunohistochemical profile in hematopoietic and nonhematopoietic malignant neoplasms. *Am J Clin Pathol* **121**, 709–17.
162. Lae ME, Ahmed I, Macon WR. (2002). Clusterin is widely expressed in systemic anaplastic large cell lymphoma but fails to differentiate primary from secondary cutaneous anaplastic large cell lymphoma. *Am J Clin Pathol* **118**, 773–9.
163. Rassidakis GZ, Sarris AH, Herling M, Ford RJ, Cabanillas F, McDonnell TJ, Medeiros LJ. (2001). Differential expression of BCL-2 family proteins in ALK-positive and ALK-negative anaplastic large cell lymphoma of T/null-cell lineage. *Am J Pathol* **159**, 527–35.
164. Vega F, Orduz R, Medeiros LJ. (2002). Chromosomal translocations and their role in the pathogenesis of non-Hodgkin's lymphomas. *Pathology* **34**, 397–409.
165. Amin HM, Lai R. (2007). Pathobiology of ALK+ anaplastic large-cell lymphoma. A distinct clinicopathologic entity. *Blood* **110**, 2259–67.

166. Ladanyi M, Cavalchire G, Morris SW, Downing J, Filippa DA. (1994). Reverse transcriptase polymerase chain reaction for the Ki-1 anaplastic large cell lymphoma-associated t(2;5) translocation in Hodgkin's disease. *Am J Pathol* **145**, 1296–300.
167. Cataldo KA, Jalal SM, Law ME, Ansell SM, Inwards DJ, Fine M, Arber DA, Pulford KA, Strickler JG. (1999). Detection of t(2;5) in anaplastic large cell lymphoma: Comparison of immunohistochemical studies, FISH, and RT-PCR in paraffin-embedded tissue. *Am J Surg Pathol* **23**, 1386–92.
168. Pulford K, Morris SW, Turturro F. (2004). Anaplastic lymphoma kinase proteins in growth control and cancer. *J Cell Physiol* **199**, 330–58.
169. Falini B, Bigerna B, Fizzotti M, Pulford K, Pileri SA, Delsol G, Carbone A, Paulli M, Magrini U, Menestrina F, Giardini R, Pilotti S, Mezzelani A, Ugolini B, Billi M, Pucciarini A, Pacini R, Pelicci PG, Flenghi L. (1998). ALK expression defines a distinct group of T/null lymphomas ("ALK lymphomas") with a wide morphological spectrum. *Am J Pathol* **153**, 875–86.
170. Trumper L, Pfreundschuh M, Bonin FV, Daus H. (1998). Detection of the t(2;5)-associated NPM/ALK fusion cDNA in peripheral blood cells of healthy individuals. *Br J Haematol* **103**, 1134–8.
171. Lamant L, de Reyniès A, Duplantier MM, Rickman DS, Sabourdy F, Giuriato S, Brugières L, Gaulard P, Espinos E, Delsol G. (2007). Gene-expression profiling of systemic anaplastic large-cell lymphoma reveals differences based on ALK status and two distinct morphologic ALK+ subtypes. *Blood* **109**, 2156–64.
172. Nishikori M, Maesako Y, Ueda C, Kurata M, Uchiyama T, Ohno H. (2003). High-level expression of BCL3 differentiates t(2;5)(p23;q35)-positive anaplastic large cell lymphoma from Hodgkin disease. *Blood* **101**, 2789–96.
173. Lai R, Rassidakis GZ, Lin Q, Atwell C, Medeiros LJ, Amin HM. (2005). Jak3 activation is significantly associated with ALK expression in anaplastic large cell lymphoma. *Hum Pathol* **36**, 939–44.
174. Lones MA, Sanger W, Perkins SL, Medeiros LJ. (2000). Anaplastic large cell lymphoma arising in bone: Report of a case of the monomorphic variant with the t(2;5)(p23;q35) translocation. *Arch Pathol Lab Med* **124**, 1339–43.
175. Nagasaka T, Nakamura S, Medeiros LJ, Juco J, Lai R. (2000). Anaplastic large cell lymphomas presented as bone lesions: A clinicopathologic study of six cases and review of the literature. *Mod Pathol* **13**, 1143–9.
176. Shiota M, Nakamura S, Ichinohasama R, Abe M, Akagi T, Takeshita M, Mori N, Fujimoto J, Miyauchi J, Mikata A, Nanba K, Takami T, Yamabe H, Takano Y, Izumo T, Nagatani T, Mohri N, Nasu K, Satoh H, Katano H, Fujimoto J, Yamamoto T, Mori S. (1995). Anaplastic large cell lymphomas expressing the novel chimeric protein p80NPM/ALK: A distinct clinicopathologic entity. *Blood* **86**, 1954–60.
177. Falini B, Pileri S, Zinzani PL, Carbone A, Zagonel V, Wolf-Peeters C, Verhoef G, Menestrina F, Todeschini G, Paulli M, Lazzarino M, Giardini R, Aiello A, Foss HD, Araujo I, Fizzotti M, Pelicci PG, Flenghi L, Martelli MF, Santucci A. (1999). ALK+ lymphoma: Clinico-pathological findings and outcome. *Blood* **93**, 2697–706.
178. Skinnider BF, Connors JM, Sutcliffe SB, Gascoyne RD. (1999). Anaplastic large cell lymphoma: A clinicopathologic analysis. *Hematol Oncol* **17**, 137–48.
179. ten Berge RL, de Bruin PC, Oudejans JJ, Ossenkoppele GJ, van der Valk P, Meijer CJ. (2003). ALK-negative anaplastic large-cell lymphoma demonstrates similar poor prognosis to peripheral T-cell lymphoma, unspecified. *Histopathology* **43**, 462–9.
180. Perkins SL, Pickering D, Lowe EJ, Zwick D, Abromowitch M, Davenport G, Cairo MS, Sanger WG. (2005). Childhood anaplastic large cell lymphoma has a high incidence of ALK gene rearrangement as determined by immunohistochemical staining and fluorescent *in situ* hybridisation: A genetic and pathological correlation. *Br J Haematol* **131**, 624–7.
181. Williams DM, Hobson R, Imeson J, Gerrard M, McCarthy K, Pinkerton CR, United Kingdom Children's Cancer Study Group (2002). Anaplastic large cell lymphoma in childhood: Analysis of 72 patients treated on the United Kingdom Children's Cancer Study Group chemotherapy regimens. *Br J Haematol* **117**, 812–20.
182. Suzuki R, Kagami Y, Takeuchi K, Kami M, Okamoto M, Ichinohasama R, Mori N, Kojima M, Yoshino T, Yamabe H, Shiota M, Mori S, Ogura M, Hamajima N, Seto M, Suchi T, Morishima Y, Nakamura S. (2000). Prognostic significance of CD56 expression for ALK-positive and ALK-negative anaplastic large-cell lymphoma of T/null cell phenotype. *Blood* **96**, 2993–3000.
183. Hodges KB, Collins RD, Greer JP, Kadin ME, Kinney MC. (1999). Transformation of the small cell variant Ki-1+ lymphoma to anaplastic large cell lymphoma: Pathologic and clinical features. *Am J Surg Pathol* **23**, 49–58.
184. Kim MK, Kim S, Lee SS, Sym SJ, Lee DH, Jang S, Park CJ, Chi HS, Huh J, Suh C. (2007). High-dose chemotherapy and autologous stem cell transplantation for peripheral T-cell lymphoma: Complete response at transplant predicts *survival. Ann Hematol.* January.
185. Passoni L, Gambacorti-Passerini C. (2003). ALK a novel lymphoma-associated tumor antigen for vaccination strategies. *Leuk Lymphoma* **44**, 1675–81.
186. Qyerfeld C, Kuzel TM, Guitart J, Rosen ST. (2007). Primary cutaneous CD30+ lymphoproliferative disorders: New insights into biology and therapy. *Oncology (Williston Park)* **21**, 689–96.
187. Kadin ME. (2006). Pathobiology of CD30+ cutaneous T-cell lymphomas. *J Cutan Pathol* **33**(Suppl 1), 10–17.
188. Liu HL, Hoppe RT, Kohler S, Harvell JD, Reddy S, Kim YH. (2003). CD30+ cutaneous lymphoproliferative disorders: The Stanford experience in lymphomatoid papulosis and primary cutaneous anaplastic large cell lymphoma. *J Am Acad Dermatol* **49**, 1049–58.
189. Salhany KE, Macon WR, Choi JK, Elenitsas R, Lessin SR, Felgar RE, Wilson DM, Przybylski GK, Lister J, Wasik MA, Swerdlow SH. (1998). Subcutaneous panniculitis-like T-cell lymphoma: Clinicopathologic, immunophenotypic, and genotypic analysis of alpha/beta and gamma/delta subtypes. *Am J Surg Pathol* **22**, 881–93.
190. Hoque SR, Child FJ, Whittaker SJ, Ferreira S, Orchard G, Jenner K, Spittle M, Russell-Jones R. (2003). Subcutaneous panniculitis-like T-cell lymphoma: A clinicopathological, immunophenotypic and molecular analysis of six patients. *Br J Dermatol* **148**, 516–25.
191. Marzano AV, Berti E, Paulli M, Caputo R. (2000). Cytophagic histiocytic panniculitis and subcutaneous panniculitis-like T-cell lymphoma: Report of 7 cases. *Arch Dermatol* **136**, 889–96.
192. Su IJ, Wang CH, Cheng AL, Chen RL. (1995). Hemophagocytic syndrome in Epstein–Barr virus-associated T-lymphoproliferative disorders: Disease spectrum, pathogenesis, and management. *Leuk Lymphoma* **19**, 401–6.
193. Massone C, Chott A, Metze D, Kerl K, Citarella L, Vale E, Kerl H, Cerroni L. (2004). Subcutaneous, blastic natural killer (NK), NK/T-cell, and other cytotoxic lymphomas of the skin: A morphologic, immunophenotypic, and molecular study of 50 patients. *Am J Surg Pathol* **28**, 719–35.
194. Jaffe ES, Krenacs L, Kumar S, Kingma DW, Raffeld M. (1999). Extranodal peripheral T-cell and NK-cell neoplasms. *Am J Clin Pathol* **111**(1 Suppl 1), S46–55.
195. Kim SK, Kim YC, Kang HY. (2006). Primary cutaneous aggressive epidermotropic CD8(+) cytotoxic T-cell lymphoma with atypical presentation. *J Dermatol* **33**, 632–4.
196. de Leval L, Rickman DS, Thielen C, Reynies A, Huang YL, Delsol G, Lamant L, Leroy K, Brière J, Molina T, Berger F, Gisselbrecht C, Xerri L, Gaulard P. (2007). The gene expression profile of nodal peripheral T-cell lymphoma demonstrates a molecular link between angioimmunoblastic T-cell lymphoma (AITL) and follicular helper T (TFH) cells. *Blood* **109**, 4952–63.
197. Dogan A, Attygalle AD, Kyriakou C. (2003). Angioimmunoblastic T-cell lymphoma. *Br J Haematol* **121**, 681–91.

198. Ansel KM, Ngo VN, Hyman PL, Luther SA, Förster R, Sedgwick JD, Browning JL, Lipp M, Cyster JG. (2000). A chemokine-driven positive feedback loop organizes lymphoid follicles. *Nature* **406**, 309–14.
199. Luppi M, Torelli G. (1996). The new lymphotropic herpesviruses (HHV-6, HHV-7, HHV-8) and hepatitis C virus (HCV) in human lymphoproliferative diseases: An overview. *Haematologica* **81**, 265–81.
200. Attygalle AD, Diss TC, Munson P, Isaacson PG, Du MQ, Dogan A. (2004). CD10 expression in extranodal dissemination of angioimmunoblastic T-cell lymphoma. *Am J Surg Pathol* **28**, 54–61.
201. Ree HJ, Kadin ME, Kikuchi M, Ko YH, Go JH, Suzumiya J, Kim DS. (1998). Angioimmunoblastic lymphoma (AILD-type T-cell lymphoma) with hyperplastic germinal centers. *Am J Surg Pathol* **22**, 643–55.
202. Chen W, Kesler MV, Karandikar NJ, McKenna RW, Kroft SH. (2006). Flow cytometric features of angioimmunoblastic T-cell lymphoma. *Cytometry B Clin Cytom* **70**, 142–8.
203. Warnke RA, Jones D, His ED. (2007). Morphologic and immunophenotypic variants of nodal T-cell lymphomas and T-cell lymphoma mimics. *Am J Clin Pathol* **127**, 511–27.
204. Merchant SH, Amin MB, Viswanatha DS. (2006). Morphologic and immunophenotypic analysis of angioimmunoblastic T-cell lymphoma: Emphasis on phenotypic aberrancies for early diagnosis. *Am J Clin Pathol* **126**, 29–38.
205. Yuan CM, Vergilio JA, Zhao XF, Smith TK, Harris NL, Bagg A. (2005). CD10 and BCL6 expression in the diagnosis of angioimmunoblastic T-cell lymphoma: Utility of detecting CD10+ T cells by flow cytometry. *Hum Pathol* **36**, 784–91.
206. Krenacs L, Schaerli P, Kis G, Bagdi E. (2006). Phenotype of neoplastic cells in angioimmunoblastic T-cell lymphoma is consistent with activated follicular B helper T cells. *Blood* **108**, 1110–11.
207. Grogg KL, Attygalle AD, Macon WR, Remstein ED, Kurtin PJ, Dogan A. (2005). Angioimmunoblastic T-cell lymphoma: A neoplasm of germinal-center T-helper cells? *Blood* **106**, 1501–2.
208. Weiss LM, Strickler JG, Dorfman RF, Horning SJ, Warnke RA, Sklar J. (1986). Clonal T-cell populations in angioimmunoblastic lymphadenopathy and angioimmunoblastic lymphadenopathy-like lymphoma. *Am J Pathol* **122**, 392–7.
209. Schlegelberger B, Zwingers T, Hohenadel K, Henne-Bruns D, Schmitz N, Haferlach T, Tirier C, Bartels H, Sonnen R, Kuse R. (1996). Significance of cytogenetic findings for the clinical outcome in patients with T-cell lymphoma of angioimmunoblastic lymphadenopathy type. *J Clin Oncol* **14**, 593–9.
210. Thorns C, Bastian B, Pinkel D, Roydasgupta R, Fridlyand J, Merz H, Krokowski M, Bernd HW, Feller AC. (2007). Chromosomal aberrations in angioimmunoblastic T-cell lymphoma and peripheral T-cell lymphoma unspecified: A matrix-based CGH approach. *Genes Chromosomes Cancer* **46**, 37–44.
211. Rüdiger T, Weisenburger DD, Anderson JR, Armitage JO, Diebold J, MacLennan KA, Nathwani BN, Ullrich F, Müller-Hermelink HK, Non-Hodgkin's Lymphoma Classification Project (2002). Peripheral T-cell lymphoma (excluding anaplastic large-cell lymphoma): Results from the Non-Hodgkin's Lymphoma Classification Project. *Ann Oncol* **13**, 140–9.
212. Pautier P, Devidas A, Delmer A, Dombret H, Sutton L, Zini JM, Nedelec G, Molina T, Marolleau JP, Brice P. (1999). Angioimmunoblastic-like T-cell nonHodgkin's lymphoma: Outcome after chemotherapy in 33 patients and review of the literature. *Leuk Lymphoma* **32**, 545–52.
213. Sugaya M, Nakamura K, Asahina A, Tamaki K. (2001). Leukocytoclastic vasculitis with IgA deposits in angioimmunoblastic T cell lymphoma. *J Dermatol* **28**, 32–7.
214. Au WY, Liang R. (2002). Peripheral T-cell lymphoma. *Curr Oncol Rep* **4**, 434–42.
215. Went P, Agostinelli C, Gallamini A, Piccaluga PP, Ascani S, Sabattini E, Bacci F, Falini B, Motta T, Paulli M, Artusi T, Piccioli M, Zinzani PL, Pileri SA. (2006). Marker expression in peripheral T-cell lymphoma: A proposed clinical-pathologic prognostic score. *J Clin Oncol* **24**, 2472–9.
216. Asano N, Suzuki R, Kagami Y, Ishida F, Kitamura K, Fukutani H, Morishima Y, Takeuchi K, Nakamura S. (2005). Clinicopathologic and prognostic significance of cytotoxic molecule expression in nodal peripheral T-cell lymphoma, unspecified. *Am J Surg Pathol* **29**, 1284–93.
217. Streubel B, Vinatzer U, Willheim M, Raderer M, Chott A. (2006). Novel t(5;9)(q33;q22) fuses ITK to SYK in unspecified peripheral T-cell lymphoma. *Leukemia* **20**, 313–18.
218. Lepretre S, Buchonnet G, Stamatoullas A, Lenain P, Duval C, d'Anjou J, Callat MP, Tilly H, Bastard C. (2000). Chromosome abnormalities in peripheral T-cell lymphoma. *Cancer Genet Cytogenet* **117**, 71–9.
219. Chen CY, Yao M, Tang JL, Tsay W, Wang CC, Chou WC, Su IJ, Lee FY, Liu MC, Tien HF. (2004). Chromosomal abnormalities of 200 Chinese patients with non-Hodgkin's lymphoma in Taiwan: With special reference to T-cell lymphoma. *Ann Oncol* **15**, 1091–6.
220. Zettl A, Rudiger T, Konrad MA, Chott A, Simonitsch-Klupp I, Sonnen R, Muller-Hermelink HK, Ott G. (2004). Genomic profiling of peripheral T-cell lymphoma, unspecified, and anaplastic large T-cell lymphoma delineates novel recurrent chromosomal alterations. *Am J Pathol* **164**, 1837–48.
221. Gallamini A, Stelitano C, Calvi R, Bellei M, Mattei D, Vitolo U, Morabito F, Martelli M, Brusamolino E, Iannitto E, Zaja F, Cortelazzo S, Rigacci L, Devizzi L, Todeschini G, Santini G, Brugiatelli M, Federico M, Intergruppo Italiano Linfomi (2004). Peripheral T-cell lymphoma unspecified (PTCL-U): A new prognostic model from a retrospective multicentric clinical study. *Blood* **103**, 2474–9.
222. Møller MB, Gerdes AM, Skjødt K, Mortensen LS, Pedersen NT. (1999). Disrupted p53 function as predictor of treatment failure and poor prognosis in B- and T-cell non-Hodgkin's lymphoma. *Clin Cancer Res* **5**, 1085–91.
223. Pescarmona E, Pignoloni P, Puopolo M, Martelli M, Addesso M, Guglielmi C, Baroni CD. (2001). p53 over-expression identifies a subset of nodal peripheral T-cell lymphomas with a distinctive biological profile and poor clinical outcome. *J Pathol* **195**, 361–6.

CHAPTER 18

Hodgkin Lymphoma

Sophie Song, Wayne W. Grody and Faramarz Naeim

Hodgkin lymphoma (HL) is a clonal B-cell neoplasm as demonstrated by the detection of clonal immunoglobulin (Ig) V-gene rearrangements in isolated tumor cells using microdissection and single-cell polymerase chain reaction (PCR) techniques [1–5]. It is recognized by the current WHO classification as a malignant lymphoma with unique clinicopathologic features [6]. It affects more often the young adults with a median age of 38 years at diagnosis [7] and primarily involves lymph nodes commonly found in the supradiaphragmatic areas, and the neoplastic tissue comprises a minor population of large pleomorphic neoplastic cells (Hodgkin–Reed–Sternberg cells, or HRS cells, and their variants) that are admixed with an inflammatory background (Figures 18.1 and 18.2).

The incidence of HL is about 3 per 100,000 in Western Europe and the United States and is consistently lower than that of non-Hodgkin lymphoma (NHL). It accounts for about 10–15% of all lymphomas in Europe and the United States [6–10].

Based on its clinical behaviors, as well as morphologic, immunophenotypic, and genotypic profiles, HL is divided into two entities of nodular-lymphocyte-predominant HL (NLPHL) and classical HL (CHL) with the following subtypes according to the current WHO classification [6]:

- Nodular-lymphocyte-predominant Hodgkin lymphoma (NLPHL)
- Classical Hodgkin lymphoma
  - Nodular sclerosis classical Hodgkin lymphoma (NSHL)
  - Mixed cellularity classical Hodgkin lymphoma (MCHL)
  - Lymphocyte-rich classical Hodgkin lymphoma (LRCHL)
  - Lymphocyte-depleted classical Hodgkin lymphoma (LDHL).

NODULAR-LYMPHOCYTE-PREDOMINANT HODGKIN LYMPHOMA

Nodular-lymphocyte-predominant Hodgkin lymphoma (NLPHL) is a distinct but rare subtype of HL [11]. It represents 5–7% of all HL cases in the United States and Europe with an estimated 500 new cases each year in the United States [12]. NLPHL is a monoclonal B-cell lymphoma characterized by nodular pattern, presence of sparse large pleomorphic lymphocytic and histiocytic (L&H) cells (or popcorn cells, HRS cell variants) admixed with abundant B lymphocytes that reside in an expanded meshwork of follicular dendritic cells. It most commonly presents as limited nodal disease involving peripheral lymph nodes above or below the diaphragm.

Etiology and Pathogenesis

The pathogenesis of NLPHL is poorly understood due, in part, to the rarity of the disease. Since NLPHL is virtually always Epstein–Barr-virus-negative, no EBV-related transforming event is known for the pathogenesis of NLPHL. A recent study shows aberrant somatic hypermutations (SHM) in the neoplastic cells of both CHL and NLPHL, which have been identified

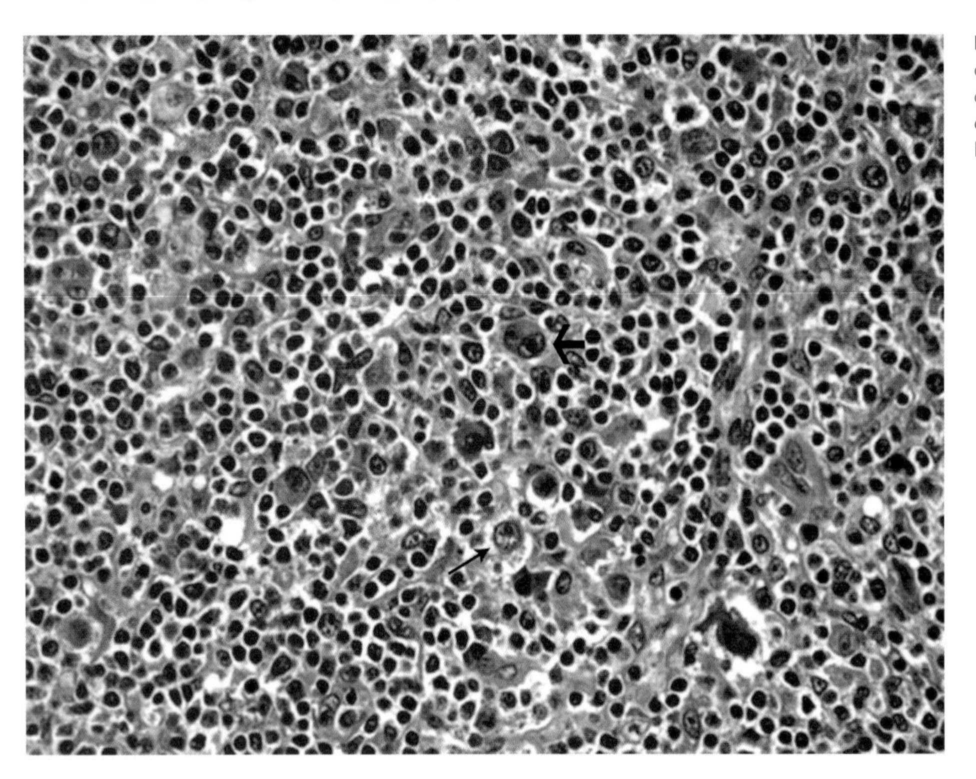

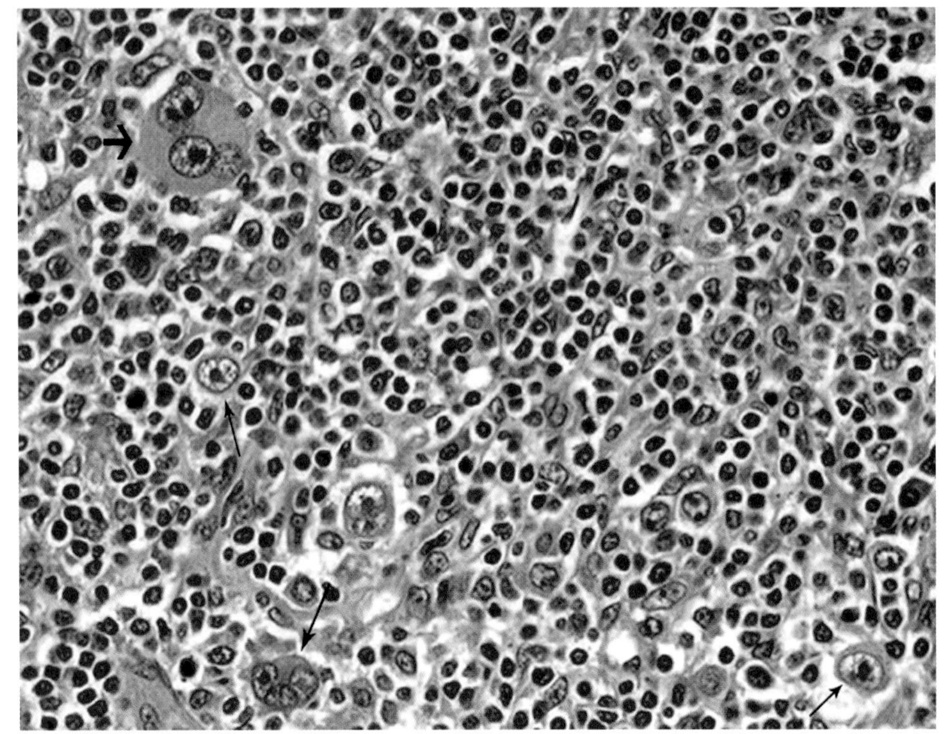

FIGURE 18.1 Reed–Sternberg cells (thick arrows) and Hodgkin cells (thin arrows) in a background of lymphocytes, plasma cells, and histiocytes.

in diffuse large B-cell lymphoma (DLBCL) as a mechanism for genome instability. This suggests common molecular pathogenetic events that are shared by these lymphomas [13]. In NLPHL, SHM of SOCS1 (suppressor of cytokine signaling 1) loci is thought to be accompanied by high expression of *JAK2* and activation of the *JAK2–STAT6* pathway in the neoplastic cells [14].

Pathology

Morphology

Normal nodal architecture is partially to completely effaced by the neoplastic infiltrate, which, by definition, demonstrates at least partially nodular growth pattern (Figure 18.3). Residual follicles, follicular hyperplasia, and progressive

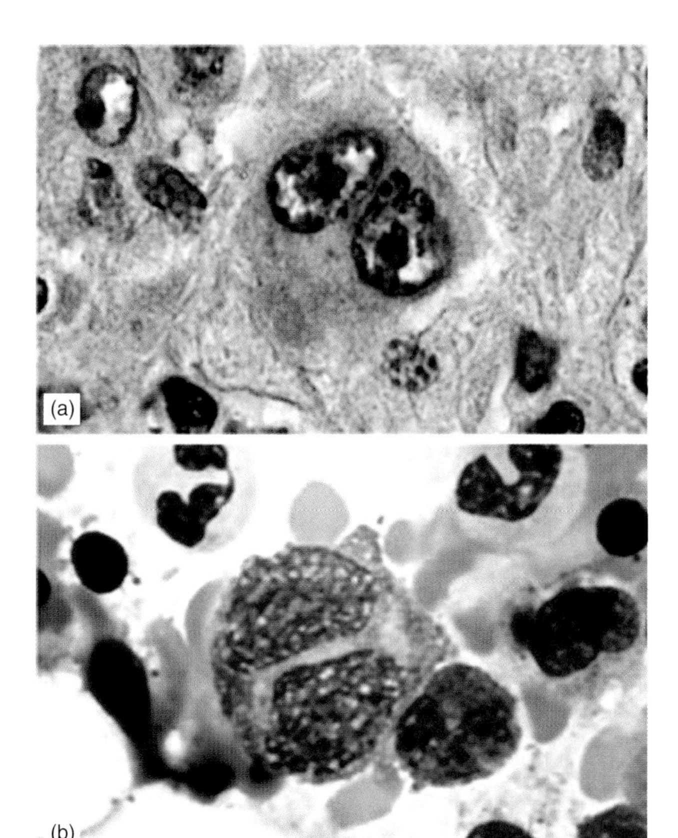

FIGURE 18.2 Classic Reed–Sternberg cells are large binucleated cells with prominent round nucleolar displaying an "owl-eye" appearance: (a) bone marrow biopsy section and (b) bone marrow smear.

transformation of germinal centers can be present simultaneously within the lymph node that is involved in NLPHL. The neoplastic nodules are usually large and poorly defined and are comprised of a minor population of scattered L&H cells, which are admixed with occasional single to clusters of histiocytes and abundant small lymphocytes that are typically B-cells (Figure 18.4). L&H cells are large and pleomorphic and often contain a single nucleus with folding and multilobation, resembling popcorn kettle and hence the name "popcorn" cells. L&H cells have vesicular chromatin, usually multiple small nucleoli with abundant basophilic cytoplasm but without distinct perinuclear halos. A follicular dendritic cell meshwork can be highlighted within the neoplastic nodules using special stains. When diffuse areas are present, they predominantly comprise T-cells as well as histiocytes with occasional L&H cells. Neutrophils, eosinophils, and plasma cells are not commonly seen. According to the current criteria, the detection of one nodule showing the typical features of NLPHL in an otherwise diffuse growth pattern is sufficient to exclude the diagnosis of T-cell/histiocyte-rich large B-cell lymphoma (T/HRBCL) [6].

Immunophenotype

The immunophenotypic profile of L&H cells is distinctly different from that of HRS cells [15, 16], and the diagnosis of NLPHL should be confirmed by appropriate immunohistochemical studies [17]. L&H cells express CD45, pan-B-cell-associated antigens including CD20, CD22, and CD79a, as well as BCL-6, J-chain, and CD75. Epithelial membrane antigen (EMA) is positive in about 50% of the cases [18]. CD15 and CD30 are negative in nearly all cases, though weak expression of CD30 is rarely detected in L&H cells [18, 19]. Overexpression of BCL-2 is not seen, and EBV infection is generally not detectable either by immunohistochemical methods for latent membrane protein (LMP) or by *in situ* hybridization studies for Epstein–Barr early RNA (EBER) [20, 21].

Immunoglobulin light chain restriction is expressed in most cases [22], and heavy chains may also be positive [23]. L&H cells of a unique subset of NLPHL are positive for IgD exhibiting distinctive clinical features including striking male predominance, younger median age, and more frequent involvement of cervical lymph node [24]. This subset of NLPHL more often involves the interfollicular region in a background that is relatively rich in T-cells.

Recently, markers for B-cell transcription factors are used in distinguishing NLPHL from CHL, which include octamer-binding transcription factor 2 (Oct2), B-cell Oct-binding protein 1 (BOB.1), B-cell-specific activator, also known as PAX-5, and PU.1. L&H cells are positive for both Oct2 and BOB.1 and PAX-5 [23, 25]. The expression of PU.1 is found in most of the NLPHL cases [26, 27].

Staining for CD21 highlights expanded meshwork of follicular dendritic cells [28, 29]. The small lymphocytes seen in NLPHL are mostly polytypic B-cells admixed with occasional T-cells, some of which ring around L&H cells forming rosettes. These rosette T-cells can be highlighted by staining for CD3 or CD57 [30, 31]. Coexpression of BCL-6 and CD57 on rosette T-cells has been reported [32].

Cytogenetic and Molecular Studies

L&H cells in NLPHL represent transformed germinal center B-cells [4, 33]. There are only few reported cytogenetic and molecular mutations in NLPHL, including common rearrangements of the *BCL-6* gene [34] and that of SHM of SOCS1 loci. Because of the scattered nature of the neoplastic cells admixed with the non-neoplastic cells in HL, standard clinical molecular diagnostic assays (gene rearrangement clonality studies, etc.) are not of great use or, if performed, are usually negative because of the dilution effect of the background cell population. As noted at the start of this chapter, single-cell and microdissection PCR techniques have been used to prove the B-cell origin of neoplastic cells [4], but this is a research modality of largely academic interest. The same approach combined with DNA sequencing has been used to demonstrate the high rate of somatic hypermutation of the variable region genes, but again this is of scientific rather than practical diagnostic impact. Rearrangements of *BCL-6* are usually detected by immunohistochemical techniques rather than molecular tests, whereas the other chromosomal changes are best detected by cytogenetic techniques.

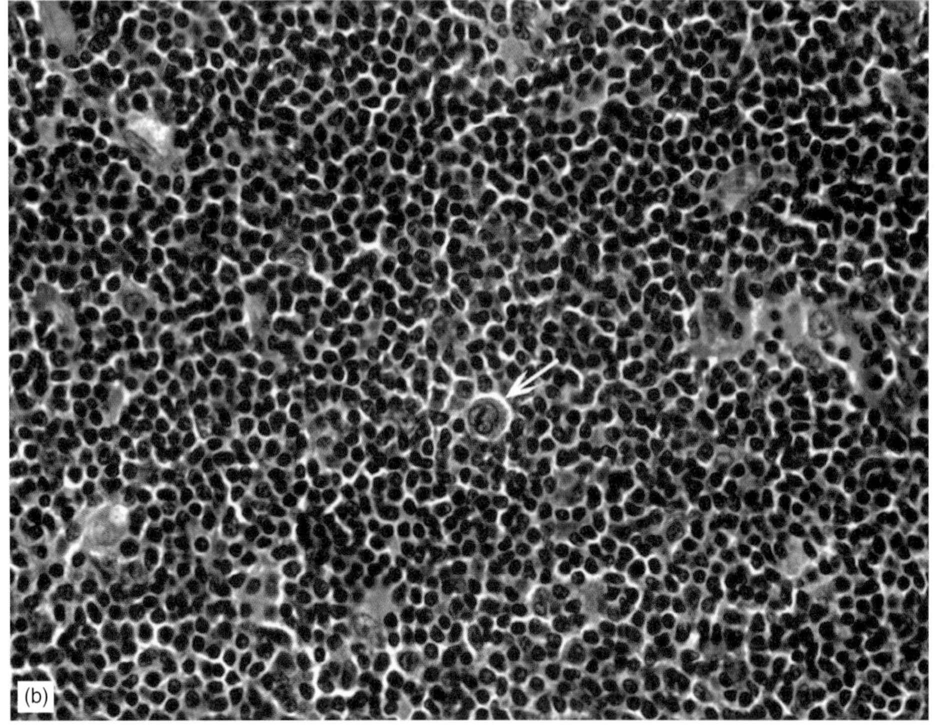

FIGURE 18.3 Nodular lymphocyte predominant Hodgkin lymphoma. (a) Low power view demonstrating a nodular pattern. (b) High power view showing an L&H cell (arrow) in a background of small lymphocytes.

Clinical Aspects

Patients with NLPHL are typically asymptomatic and present with localized peripheral lymphadenopathy frequently involving the cervical and axillary regions. Comparing with CHL, it is a more indolent disease with more frequent relapse, but good response to therapy, and overall more favorable prognosis. It is male predominant with a male:female ratio of 3–4:1 and has two peaks in age distribution, one that of children and the other that of 30- to 40-year-old adults [12]. Most of the patients present with stages I and II diseases, and <20% of the patients with NLPHL present in stage III or IV disease, whereas approximately 50% of CHL patients present in more advanced stages [35–37]. The prognostically significant factors such as bulky disease and mediastinal involvement are rare in NLPHL.

Management of early stages of NLPHL includes the use of monoclonal antibody rituximab, involved field

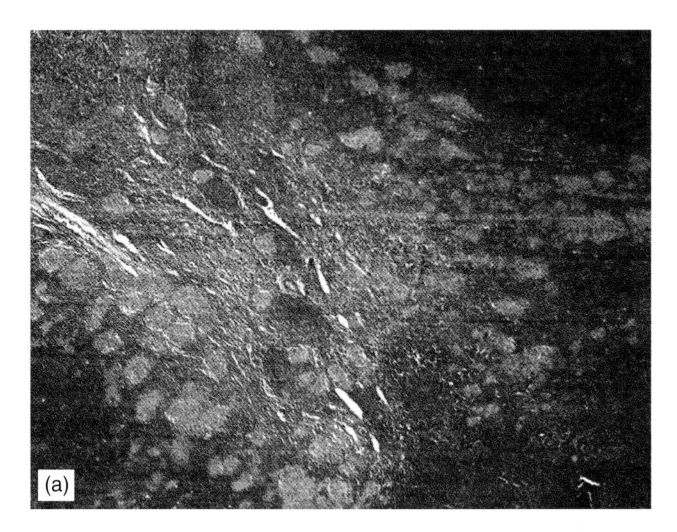

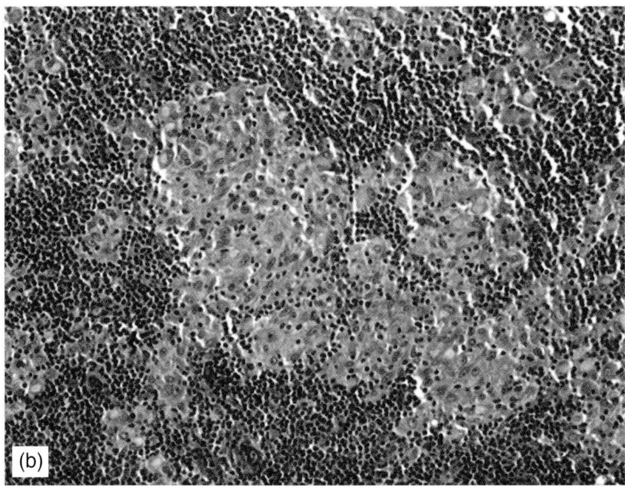

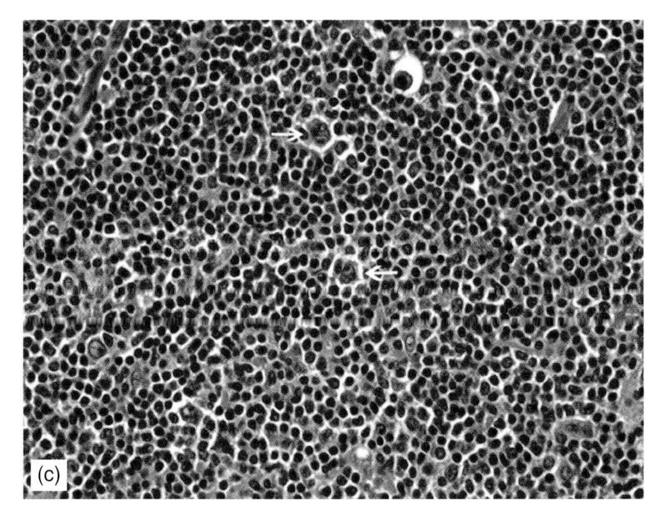

FIGURE 18.4 Nodular lymphocyte predominant Hodgkin lymphoma. (a and b) Histiocytic clusters are present mixed with small lymphocytes. (c) L&H cells (arrows) in a background of small lymphocytes.

or extended field radiotherapy, as well as "watch and wait" strategy [11, 38]. It is generally accepted that patients with early-stage NLPHL without risk factors can be treated with reduced intensity regimens resulting in less severe adverse effects, whereas patients in early stage of disease with unfavorable risk factors are treated similar to CHL patients [38]. The treatment of NLPHL with standard HL protocols leads to complete remission in >95% of the patients [38]. Prognosis is worse in advanced-stage patients than in early-stage patients.

A very small proportion (2–7%) of NLPHL transforms to DLBCL [39–41], which has a more indolent course and more favorable prognosis than *de novo* DLBCL [40, 42].

Differential Diagnosis

The differential diagnosis of NLPHL includes T/HRBCL and lymphocyte-rich classical Hodgkin lymphoma.

T-cell/histiocyte-rich diffuse large B-cell lymphoma: Although NLPHL and T/HRBCL are two distinct lymphomas requiring different clinical management, they share overlapping morphologic and immunophenotypic features. Some studies suggest a biologic continuum between NLPHL and T/HRBCL [29], which implies that they may represent different spectrums of the same disease. Both lymphomas may have concurrent nodular and diffuse growth patterns, though by definition, at least partial nodular pattern has to be present for NLPHL. The immunophenotypic profile of L&H cells cannot be clearly distinguished from that of the neoplastic cells in T/HRBCL. However, immunophenotypic studies of the background composition are helpful [29], with common findings of abundant small B-cells and prominent follicular dendritic cell meshwork along with CD3+CD4+CD57+ T-cells in NLPHL, and lack of small B-cells but abundance of CD8+ T-cells and histiocytes in T/HRBCL.

Lymphocyte-rich classical Hodgkin lymphoma: NLPHL and LRCHL cannot be distinguished morphologically or clinically [37]. LRCHL frequently presents with a nodular growth pattern and a background of abundant small lymphocytes. The neoplastic cells in LRCHL sometimes display cytologic features of L&H cells seen in NLPHL. However, the immunophenotypic profile of the neoplastic cells in LRCHL differs from that of the L&H cells. The neoplastic cells in LRCHL show immunophenotypic features of HRS cells in other subtypes of CHL, including expression of CD30, CD15, Fascin, and EBV LMP positivity in about 50% of the cases [17]. Unlike NLPHL, T-cell rosettes around the neoplastic cells generally do not express CD57 in LRCHL [18].

CLASSICAL HODGKIN LYMPHOMA

Classical Hodgkin lymphoma (CHL) represents about 95% of all HL cases in the United States and Europe. The estimated figure for the newly diagnosed cases of HL in 2007 in the United States was 8,190 [7]. CHL is a clonal B-cell lymphoma characterized by the proliferation of a minor population of mononuclear Hodgkin and multinucleated Reed–Sternberg (HRS) cells admixed with abundance of reactive infiltrate including various inflammatory cells [6]. The diagnostic (classic) Reed–Sternberg cell is a large binucleated cell with prominent round nucleoli and perinucleolar halos displaying an "owl-eye" appearance (Figures 18.1, 18.2, and 18.5) [43]. Reed–Sternberg cell variants and mononuclear Hodgkin cells may display particular morphologic features in certain subclasses, such as "lacunar" cells in nodular sclerosis HL. It most commonly presents as lymphadenopathy involving the cervical, mediastinal, axillary, and para-aortic regions. Primary extranodal involvement is rare.

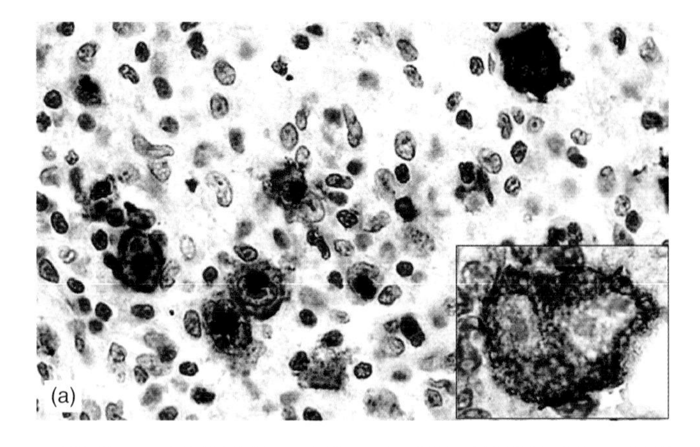

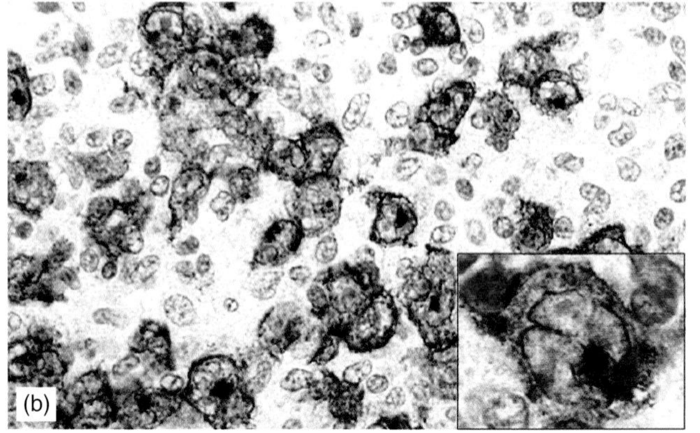

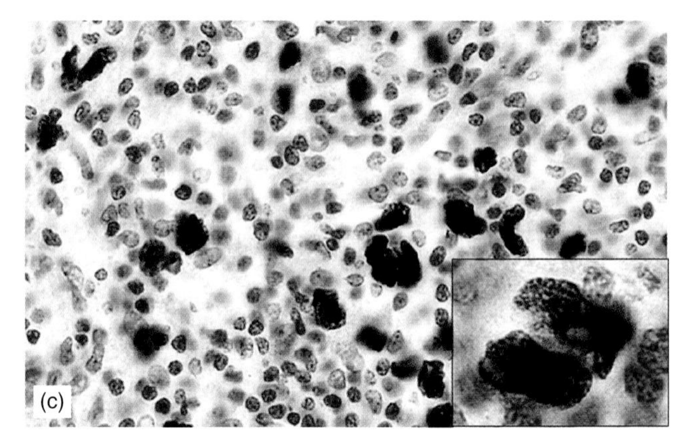

FIGURE 18.5 Reed–Sternberg and Hodgkin cells in classical Hodgkin lymphoma express: (a) CD15 and (b) CD30. They may also express EBV-encoded RNA, EBER (c and inset). Adapted from Ref. [43] by permission.

Etiology and Pathogenesis

Although the etiology is unclear, both genetic and environmental factors play a role in the pathogenesis of HL. Familial aggregation is found in approximately 4–5% of all cases of HL [44]. In addition, genetic predisposition of HL is supported by a significantly increased risk (99 times higher) of developing the disease in the identical twins of HL patients than in the dizygotic twins [45]. Furthermore, population-based data indicate ethnic variations with low incident rate of HL in Asian subgroups [46].

The genetic susceptibility of HL has been linked to human leukocyte antigens (HLA) for more than two decades [47], and many alleles and haplotypes have been implicated in either increased or reduced risks of the disease [48, 49]. More recent studies have further demonstrated the association of HLA class I with EBV+ HL, whereas HLA class III haplotypes are more associated with EBV− HL [50], and association of HLA-A*02 with reduced risk whereas HLA-A*01 with increased risk of developing EBV+ HL [51]. However, there is no consensus regarding the role of specific HLA alleles and haplotypes in HL due to the linkage disequilibrium of the HLA region [52].

HL is one of the EBV-associated lymphomas, and EBV infection appears to be the most significant environmental factor in the development of HL (Figure 18.5c). Several studies have implicated an increased risk of HL after infectious mononucleosis (IM) [53–55]. Recent studies suggest a causal association between IM-related EBV infection and the EBV+ subgroup of HL in young adults [56, 57]. Serologic analysis reveals elevated levels of IgG and IgA antibodies against EBV viral capsid antigen in patients with HL [58]. Immunohistochemical as well as fluorescence *in situ* hybridization (FISH) studies show EBV-positivity in approximately 30–40% of cases of HL [59]. Clonal EBV genome products can be detected in HRS cells in up to 50% of the CHL cases [60].

In addition to genetic predisposition and EBV infection, other factors including human immunodeficiency virus (HIV) infection, autoimmune disease, immunocompromised status, and existing lymphomas are associated with increased risks of HL [61–63].

Although the molecular pathogenesis of HL remains unclear, recent molecular studies have furthered our knowledge in understanding the pathogenetic mechanisms of HL. Aberrant activation of multiple signaling pathways in HRS cells is thought to play an important role in HRS cell survival and proliferation, which includes activation of Notch1, multiple receptor tyrosine kinases (RTKs), the PI3K, MED/ERK, NF-κB, STAT, and AP-1 [64]. Among these pathways, constitutive activation of transcription factor NF-κB is considered a hallmark of CHL [65–67]. Because EBV can be detected in up to 50% of the CHL cases, it is thought to contribute to the transforming mechanisms of HRS cells, and the pathogenesis of CHL is likely through pathways including viral-encoded gene latent membrane protein 1 (LMP1)-induced activation of NF-κB [68] and LMP2A-driven B-cell survival in the absence of normal B-cell receptor signals [69]. In addition, genomic amplification of the c-REL, JAK2, and MDM2 loci, as well as expression of Caspase 3, is implicated in the antiapoptotic phenotype and survival of HRS cells [70–72]. Furthermore, the microenvironment of HRS cells in CHL is uniquely favorable in protecting HRS cells from cell-mediated apoptosis, which is established and maintained by the unique cytokine profiles produced by HRS cells and specific regulatory T-cells [73].

Pathology

Morphology

There are four subtypes of CHL, which reveal distinct morphologic features [6], and are discussed in this section.

Nodular sclerosis classical Hodgkin lymphoma is the most frequent subtype of CHL accounting for 70–80% CHL cases in North America. It shows at least a focally nodular growth pattern with neoplastic nodules that are dissected by thick and fibroblast-poor collagen bands. Thickening of the capsule is often present. Scattered, large pleomorphic neoplastic cells including HRS cells as well as lacunar cells are admixed with abundant inflammatory cells (Figures 18.6 and 18.7). Lacunar cells, which are characteristic of NSHL, display paracellular halos as a result of retraction artifact of cytoplasmic membrane in formalin-fixed tissue. The number of HRS and lacunar cells is highly variable, as is the cellular composition of the reactive infiltrate in the background with small lymphocytes, eosinophils, neutrophils, plasma cells, histiocytes, and fibroblasts. Eosinophils and neutrophils can be abundant, and focal formation of eosinophilic or neutrophilic microabscesses can be seen.

Mixed cellularity classical Hodgkin lymphoma is the second most common subtype of CHL. It is more commonly seen in developing countries and has more frequent association with EBV infection and HIV/AIDS [61, 62]. The neoplastic infiltrate is mostly diffuse but can have a vaguely nodular pattern in focal areas. Even though interstitial fibrosis is seen, there should not be thickened fibrous capsule or broad collagen bands. The neoplastic infiltrate can present as interfollicular involvement accompanied by hyperplastic follicles in the background or regressed germinal centers. The neoplastic cells are typical HRS cells, which are admixed with highly variable reactive background (Figures 18.8 and 18.9). The reactive infiltrate in the background includes small lymphocytes, plasma cells, histiocytes, eosinophils, and neutrophils. Histiocytes can form granuloma-like aggregates.

Lymphocyte-rich classical Hodgkin lymphoma is a distinct subtype of CHL even though it has overlapping clinical and morphologic features with NLPHL. The neoplastic infiltrate displays more often a nodular but can be a diffuse growth pattern and is comprised of scattered HRS cells admixed with small lymphocytes. Eosinophils and neutrophils are absent. Occasionally, the neoplastic cells can resemble L&H cells seen in NLPHL.

Lymphocyte-depleted classical Hodgkin lymphoma is the least common subtype of CHL and is more frequently seen in developing countries and in persons with HIV/AIDS [17, 62]. Unlike the other subtypes of HL, the neoplastic infiltrate is diffuse and comprises relatively preponderant HRS cells. HRS cells can sometimes display

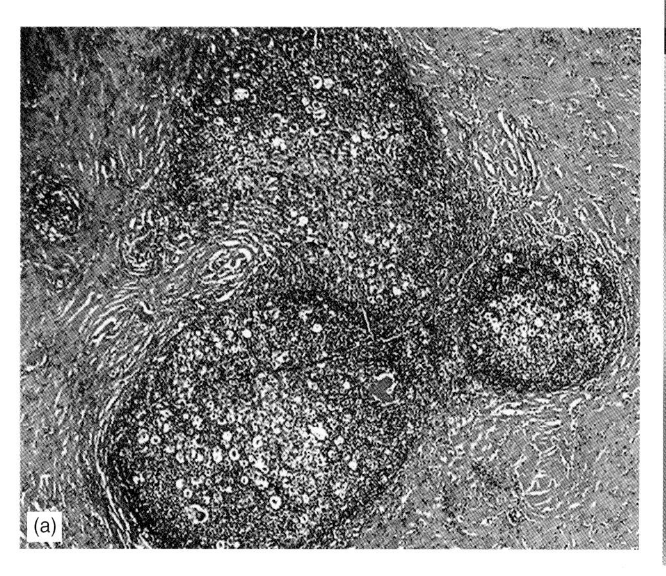

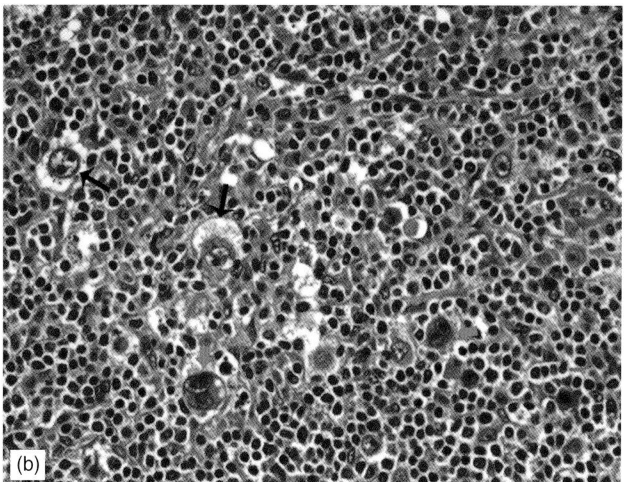

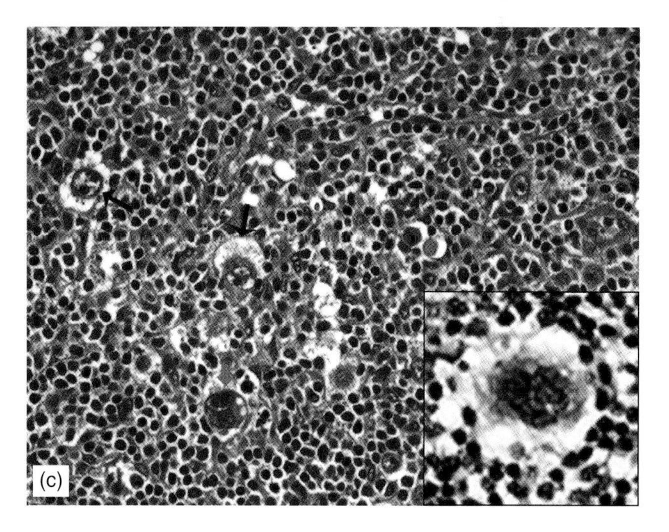

FIGURE 18.6 Nodular sclerosis Hodgkin lymphoma. (a) Low power view demonstrating a nodular pattern with numerous lacunar cells in the neoplastic nodules surrounded by thick collagen bands. (b and c) High power views showing lacunar cells (black arrows) and Reed–Sternberg cells (c, green arrows) mixed with lymphocytes and other inflammatory cells. The inset demonstrates a lacunar cell. Adapted from Ref. [43] by permission.

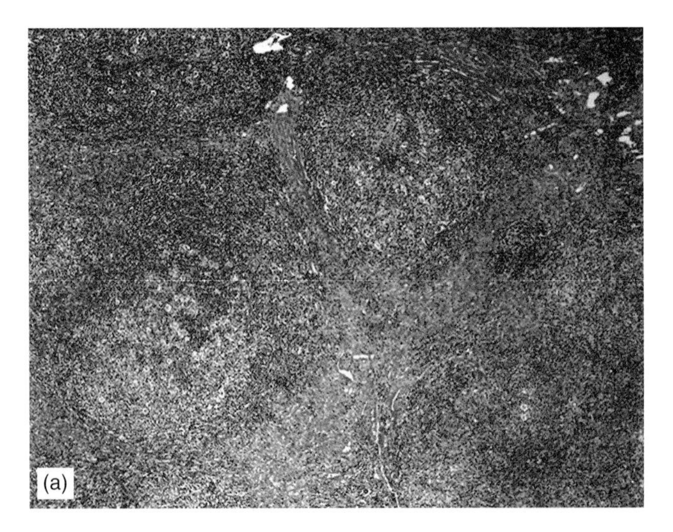

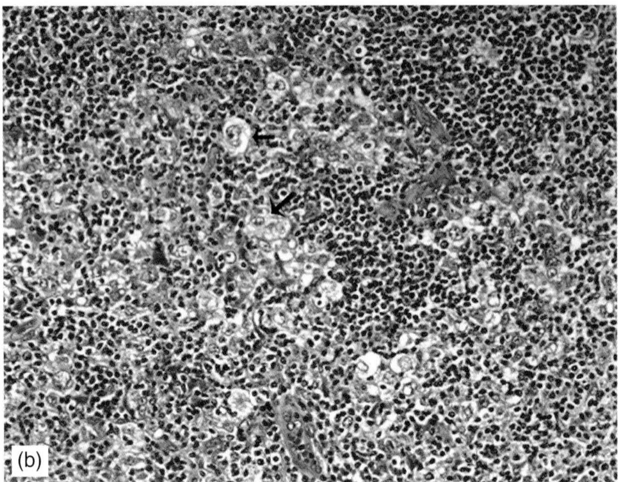

FIGURE 18.7 Nodular sclerosis Hodgkin lymphoma. (a) Low power view demonstrating a nodular pattern with the neoplastic nodules surrounded by thick collagen bands. (b) High power view showing a Reed–Sternberg cell (arrow) and several lacunar cells.

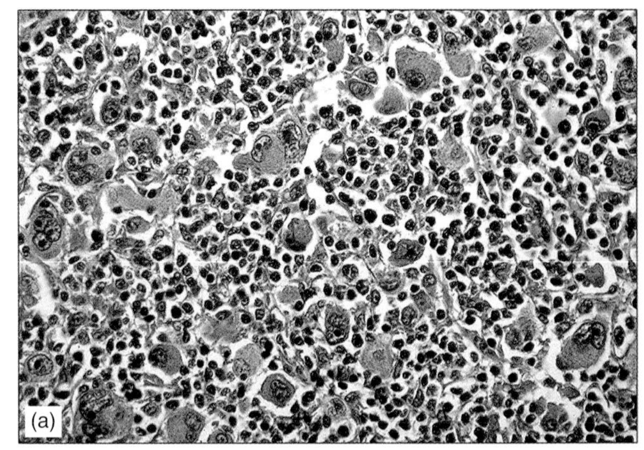

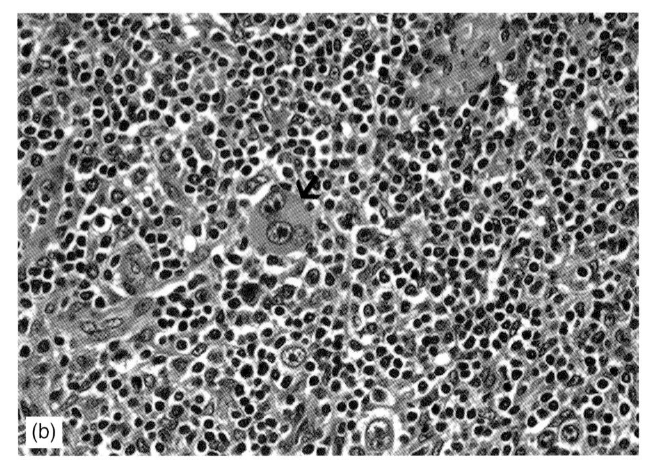

FIGURE 18.8 Mixed cellularity Hodgkin lymphoma. (a) Numerous Hodgkin and Reed–Sternberg cells are found in a background infiltrate predominantly consisting of lymphocytes. A classical Reed–Sternberg cell is demonstrated in (b, arrow).

sarcomatoid appearance. The background contains relatively fewer numbers of small lymphocytes, and fibrosis can be prominent (Figures 18.10 and 18.11). If a nodular sclerosis is present, the lymphoma should be diagnosed as NSHL.

Immunophenotype

Hodgkin–Reed–Sternberg (HRS) cells are positive for CD30 and CD15 with membrane and Golgi staining, but are negative for CD45, J-chain, and CD75 (Figure 18.12) [22, 74, 75]. CD15 may be seen in only a minority of HRS cells. Fascin, an actin-bundling protein involved in the formation of dendritic processes, is positive in all CHL cases [76, 77]. Staining for B-cell-associated markers is variable, with positivity of PAX-5 in 90%, CD20 in 30–40%, and CD79a in 10% of CHL cases [22, 78, 79]. The expression of the B-cell-associated markers may be detected in a subpopulation of HRS cells and can vary in intensity generally revealing a weaker staining than that of the background B-cells [22, 80]. Unlike L&H cells of NLPHL where both Oct2 and BOB.1 are positive, HRS cells of CHL are negative for either Oct2 or BOB.1 [25]. BCL-6 nuclear staining of HRS cells is found in 40% of CHL cases [81]. MUM1 (multiple myeloma-1), a marker for late centrocytes and plasma cells, is detected in nearly all CHL cases, whereas it is absent or weakly positive in NLPHL [82–85].

HRS cells are positive for EBV infection in about 30–40% of the CHL cases by immunohistochemical as well as FISH studies [59]. EBV positivity is higher in MCHL and LRCHL than in NSHL [17, 60].

Cytogenetic and Molecular Studies

HRS cells are characterized as preapoptotic crippled germinal center B-cells that have lost their B-cell surface receptor expression [2]. HRS cells show cytogenetic abnormalities with numerical chromosomal changes, which commonly involve gains of chromosomes 2p, 12q, 17p, 9p, and 16p [86], indicating chromosomal instability of the HRS cells [87]. Two recurrent aberrations with gains of 2p13-p16 and 9p24 are identified in CHL, but not in NLPHL using comparative genomic hybridization analysis. Microarray analysis of

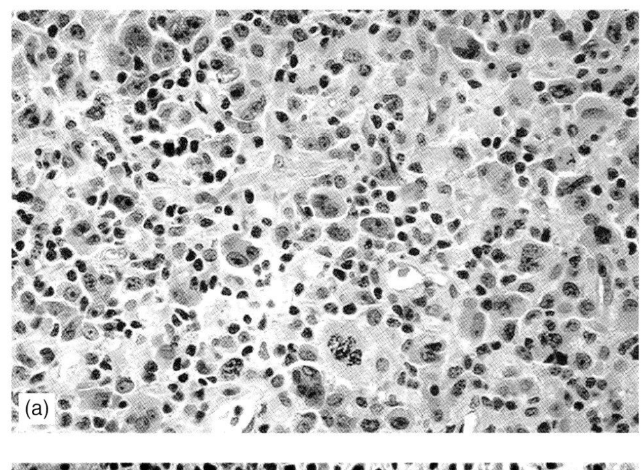
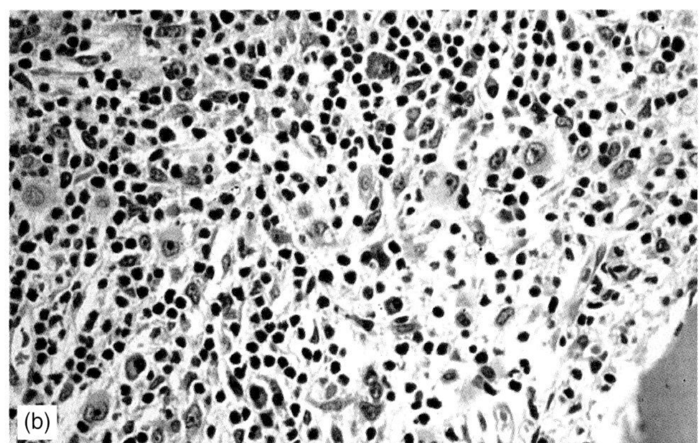

FIGURE 18.9 Mixed cellularity Hodgkin lymphoma; bone marrow biopsy section. (a) Numerous Hodgkin and Reed–Sternberg cells are present. (b) Scattered Hodgkin cells are mixed with numerous lymphocytes. Adapted from Ref. [43] by permission.

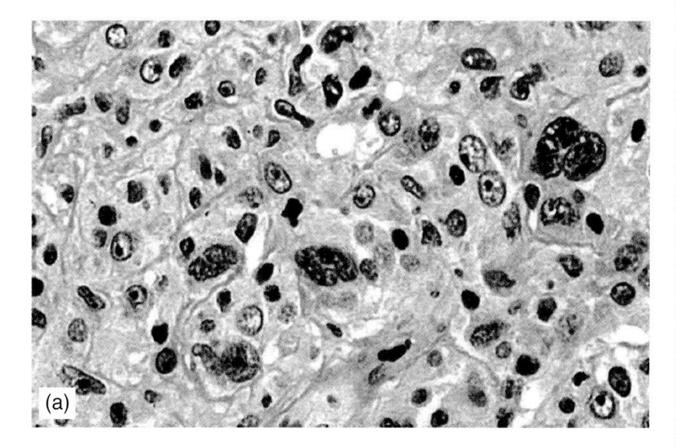
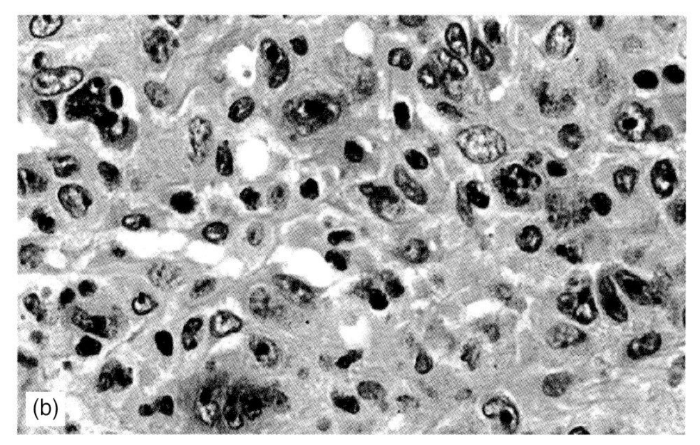
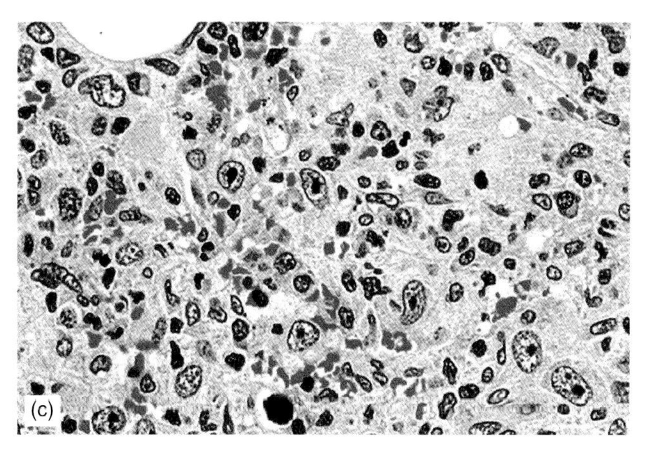

FIGURE 18.10 Lymphocyte-depleted classical Hodgkin lymphoma. Lymph node (a and b) and bone marrow biopsy (c) sections demonstrating numerous HRS cells mixed with scattered lymphocytes. Adapted from Ref. [43] by permission.

Hodgkin cell lines has identified expression of activating transcription factor 3 (ATF3), a member of the cyclic AMP response element binding protein (Figure 18.13) [88].

As in NLPHL, the molecular tests used for B- and T-cell non-Hodgkin lymphomas are not of much value in CHL. Sophisticated microdissection PCR techniques have again confirmed the neoplastic cells, and in particular, the Reed–Sternberg cells as being of B-cell origin; but these are not routinely used in the clinical setting. In contrast to NLPHL, somatic hypermutation of the immunoglobulin genes is not seen, whereas the *BCL-6* gene does show somatic mutations but no rearrangements or translocations [34, 89].

Clinical Aspects

The incidence of CHL is historically believed to have a bimodal curve with one peak in young adults (15–35 years of age) and the other at a late stage in life. According to the most recent statistics, the median age at diagnosis of HL is 38 years, and male:female ratio is slightly >1 [7]. Patients with a history of IM, HIV/AIDS, or autoimmune diseases have higher incidence of the disease [53, 62, 63]. Most of the CHL patients present with asymptomatic lymph node enlargement in the supradiaphragmatic regions. Systemic symptoms with fever, night sweats, and weight loss are reported in about 30% of the cases [10]. Approximately 50% of the patients have advanced-stage disease at diagnosis [35–37]. Splenic involvement is seen in about 20%, whereas bone marrow is involved in about 5% of CHL cases [6].

Staging of HL is based on the Ann Arbor system with the addition of the definition of bulky disease, which is currently favored by the maximum diameter of the largest

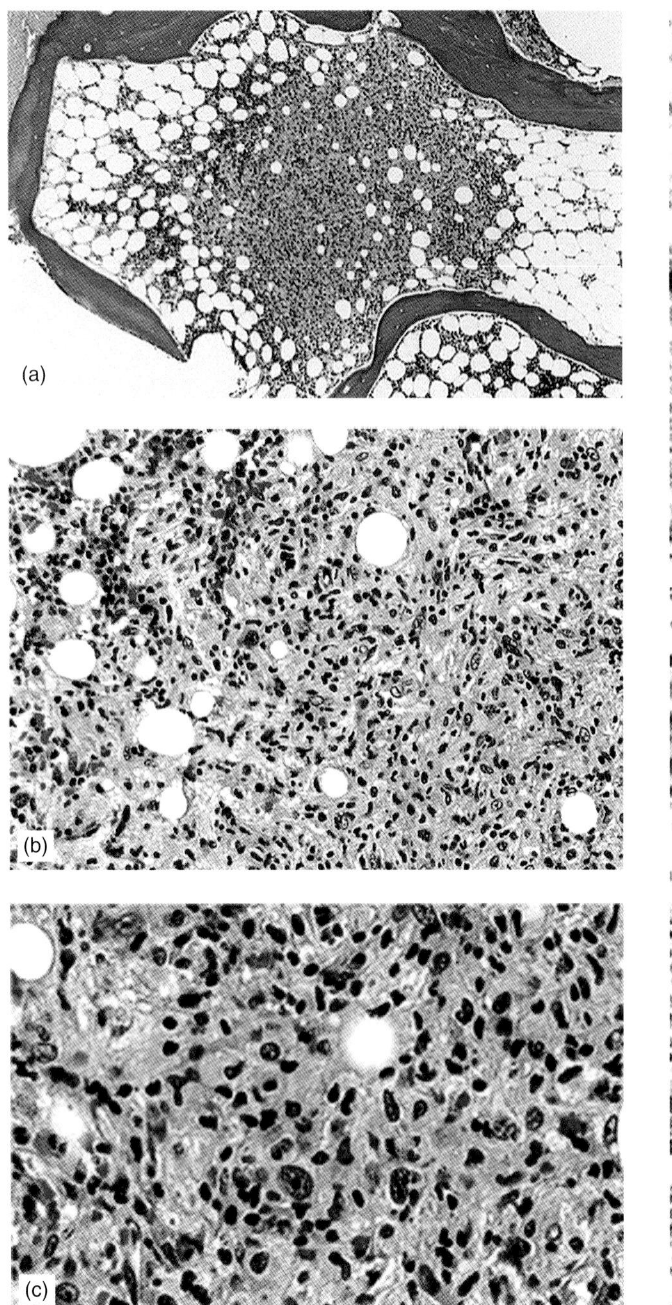

FIGURE 18.11 Bone marrow involvement in lymphocyte-depleted classical Hodgkin lymphoma demonstrating focal involvement with scattered HRS cells in a background of fibrosis and inflammatory cells: (a) low power, (b) intermediate power, and (c) high power views. Adapted from Ref. [43] by permission.

single tumor mass [90]. A limited-stage HL is usually defined by non-bulky stage IA or IIA disease, and a cure rate of >90% can be expected regardless of prognostic factor model. Seven independent prognostic factors have been identified, which are used to assess risks of primary treatment failure and possible intensified treatment in advanced-stage HL patients [91]. They include sex, age, stage, hemoglobin, WBC, lymphocyte count, and serum albumin.

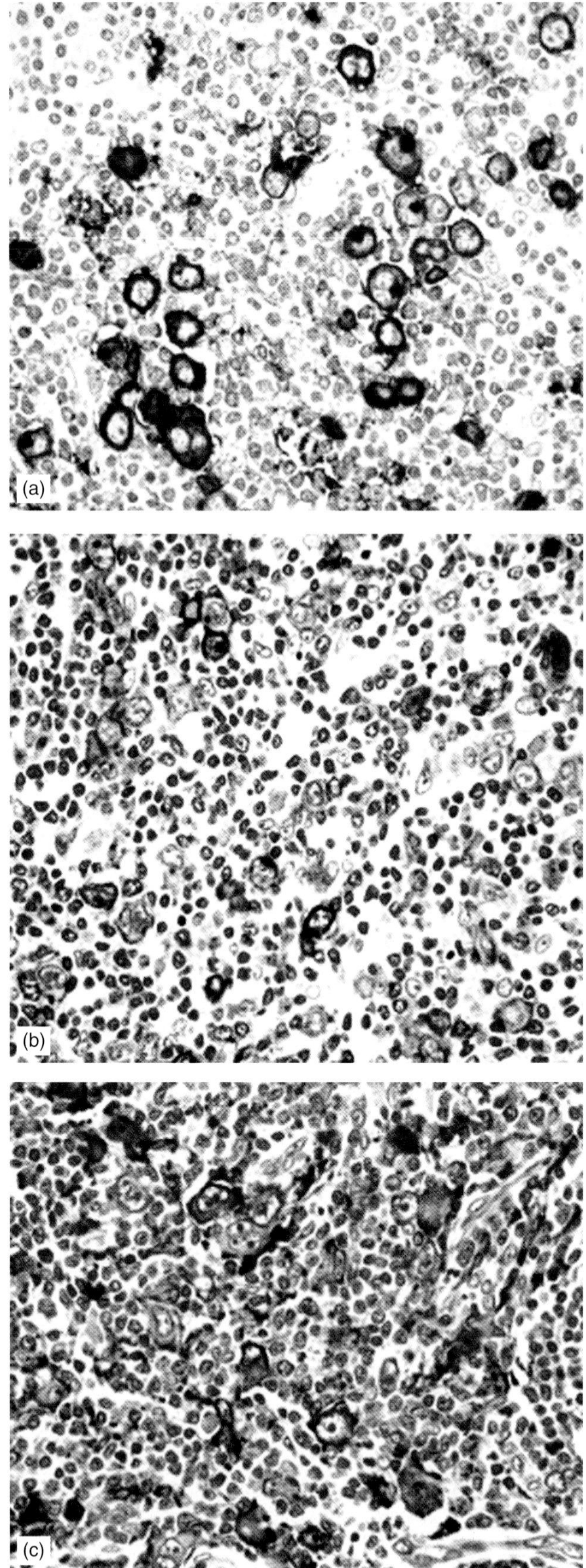

FIGURE 18.12 Immunohistochemical stains of HRS cells demonstrating positivity for (a) CD30, (b) CD15, and (c) Fascin.

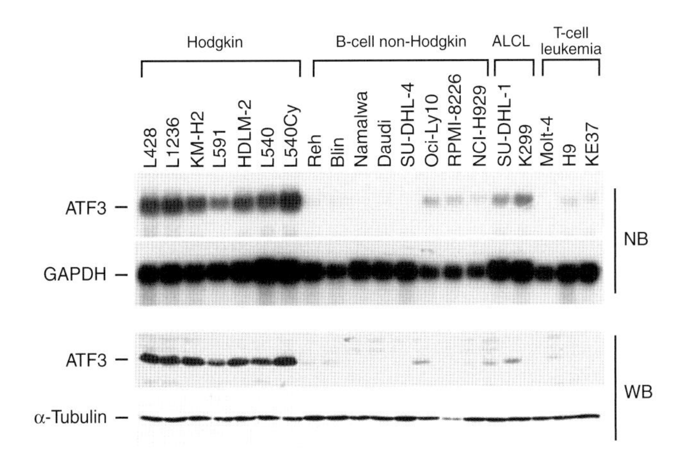

FIGURE 18.13 ATF3 mRNA and protein expression in HRS cells and Hodgkin and non-Hodgkin cell lines are determined by Northern blot and Western blot analyses. The Hodgkin cell lines and HRS tumor cells are characterized by high-constitutive ATF3 expression. Adapted from Ref. [88] by permission. This research was originally published in *Blood*.

Clinical characteristics vary among the distinct subtypes of CHL. Most patients of NSHL present with stage II disease, and mediastinal involvement is predominant [92]. In contrast to NSHL, MCHL often presents with stage III or IV disease and rare mediastinal involvement [6]. Unlike the other subtypes of CHL, LRCHL has clinicopathologic profiles resembling those of NLPHL with the exception that multiple relapse is less frequent [37]. LDHL is frequently associated with advanced-stage disease and more frequent involvement of abdominal organs, retroperitoneal lymph nodes, and bone marrow.

Treatment of HL represents one of the greatest success stories in modern medicine [93], and about 80–90% of patients in all stages achieve long-term survival [94]. The current understanding of the appropriate therapy is established on the basis of recent results of many large randomized clinical trials with focus on maximizing effectiveness while minimizing toxicity. For limited-stage HL, treatment involves brief, combined modality chemotherapy only augmented with involved field irradiation if an early CR is not achieved [93]. Patients with advanced-stage HL require extended course of chemotherapy without radiation therapy [95]. High-dose chemotherapy and irradiation and autologous hematopoietic stem cell transplantation can be an effective treatment for patients with relapsed or refractory HL [90, 96].

In addition to chemotherapy and radiotherapy, some studies suggest that immunotherapy including monoclonal antibodies play a positive role in the treatment of HL [97]. These suggestions are further supported by recent results of targeted immunotherapy using anti-CD30 chimeric receptors bound to EBV-CTLs that are persistent and can expand in HL patients for targeting the tumor and achieving complete patient response [98].

Concurrence of CHL and NHL is rare. Transformation of CHL is not well defined, even though cases have been reported where patients with CHL developed aggressive NHL including DLBCL and peripheral T-cell lymphoma [19, 99–101]. A clonal association between the concurrent CHL and NHL has been demonstrated in some cases.

Differential Diagnosis

The differential diagnosis of CHL includes diffuse large B-cell lymphoma (DLBCL), anaplastic large cell lymphoma (ALCL), and senile EBV+ B-cell lymphoproliferative disorder.

Diffuse large B-cell lymphoma: Clinical and morphologic distinction can be difficult between CHL and large B-cell lymphoma, especially T/HRBCL, primary mediastinal (thymic) large B-cell lymphoma (PMLBCL), and anaplastic variant of DLBCL. Appropriate immunophenotyping of the neoplastic cells is critical for making the distinction. The neoplastic cells of T/HRBCL are CD45+, EMA+, CD15−, and CD30−, whereas those of PMLBCL are CD45+, CD15−, and can sometimes express CD30. Staining for Fascin can assist the immunophenotypic distinction between CHL and large B-cell lymphomas as it is mostly negative in large B-cell lymphomas but 100% positive in CHL [76, 77]. However, the differential diagnosis can sometimes be extremely challenging due to, perhaps, true biologic overlap between these entities [102, 103].

Anaplastic large cell lymphoma: Immunophenotypic analysis is helpful in differentiating CHL from ALCL. Although PAX-5 is positive in about 90% of CHL cases, it is consistently negative in ALCL [6]. Negative staining for EMA and ALK in CHL is also helpful. Furthermore, ALCL commonly expresses some T-cell-associated markers with clonal T-cell receptor gene rearrangements, which are extremely rare in CHL [104].

Senile EBV+ B-cell lymphoproliferative disorder: This recently described disorder represents a spectrum of diseases ranging from polymorphous B-cell lymphoproliferative disorders to DLBCL, and occasional cases may be indistinguishable from CHL on morphologic grounds [61, 105]. Immunophenotypic analysis can be helpful, and HRS-like cells in senile EBV+ B-cell lymphoproliferative disorder are positive for CD20, CD45, and EBV. They can be variably positive for CD30, but are negative for CD15 [106].

References

1. Küppers R, Rajewsky K, Zhao M, Simons G, Laumann R, Fischer R, Hansmann ML. (1994). Hodgkin disease: Hodgkin and Reed–Sternberg cells picked from histological sections show clonal immunoglobulin gene rearrangements and appear to be derived from B cells at various stages of development. *Proc Natl Acad Sci USA* **91**, 10962–6.
2. Kanzler H, Küppers R, Hansmann ML, Rajewsky K. (1996). Hodgkin and Reed–Sternberg cells in Hodgkin's disease represent the outgrowth of a dominant tumor clone derived from (crippled) germinal center B cells. *J Exp Med* **184**, 1495–505.
3. Küppers R, Kanzler H, Hansmann ML, Rajewsky K. (1996). Single cell analysis of Hodgkin/Reed–Sternberg cells. *Ann Oncol* **7**(Suppl 4), 27–30.
4. Marafioti T, Hummel M, Anagnostopoulos I, Foss HD, Falini B, Delsol G, Isaacson PG, Pileri S, Stein H. (1997). Origin of nodular lymphocyte-predominant Hodgkin's disease from a clonal expansion of highly mutated germinal-center B cells. *N Engl J Med* **337**, 453–8.

5. Marafioti T, Hummel M, Foss HD, Laumen H, Korbjuhn P, Anagnostopoulos I, Lammert H, Demel G, Theil J, Wirth T, Stein H. (2000). Hodgkin and Reed–Sternberg cells represent an expansion of a single clone originating from a germinal center B-cell with functional immunoglobulin gene rearrangements but defective immunoglobulin transcription. *Blood* **95**, 1443–50.
6. Jaffe ES, Harris NL, Stein H, Vardiman JW. (2001). *Pathology and Genetics. Tumors of Haematopoietic and Lymphoid Tissues*. IARC Press, Lyon.
7. Ries LAG, Melbert D, Krapcho M, Mariotto A, Miller BA, Feuer EJ, Clegg L, Horner MJ, Howlader N, Eisner MP, Reichman M, Edwards BK (eds). (2007). *SEER Cancer Statistics Review*, 1975–2004, National Cancer Institute, Bethesda, MD, http://seer.cancer.gov/csr/1975_2004/, based on November 2006 SEER data submission, posted to the SEER web site.
8. Rodriguez-Abreu D, Bordoni A, Zucca E. (2007). Epidemiology of hematological malignancies. *Ann Oncol* **18**(Suppl 1), i3–ii8.
9. Nakatsuka S, Aozasa K. (2006). Epidemiology and pathologic features of Hodgkin lymphoma. *Int J Hematol* **83**, 391–7.
10. Tsang RW, Hodgson DC, Crump M. (2006). Hodgkin's lymphoma. *Curr Probl Cancer* **30**, 107–58.
11. Nogová L, Rudiger T, Engert A. (2006). Biology, clinical course and management of nodular lymphocyte-predominant Hodgkin lymphoma. *Hematology Am Soc Hematol Educ Program*, 266–72.
12. Tsai HK, Mauch PM. (2007). Nodular lymphocyte-predominant Hodgkin lymphoma. *Semin Radiat Oncol* **17**, 184–9.
13. Liso A, Capello D, Marafioti T, Tiacci E, Cerri M, Distler V, Paulli M, Carbone A, Delsol G, Campo E, Pileri S, Pasqualucci L, Gaidano G, Falini B. (2006). Aberrant somatic hypermutation in tumor cells of nodular-lymphocyte-predominant and classic Hodgkin lymphoma. *Blood* **108**, 1013–20.
14. Mottok A, Renne C, Willenbrock K, Hansmann ML, Bräuninger A. (2007). Somatic hypermutation of SOCS1 in lymphocyte-predominant Hodgkin lymphoma is accompanied by high JAK2 expression and activation of STAT6. *Blood* **Jul 25**.
15. Pinkus GS, Said JW. (1985). Hodgkin's disease, lymphocyte predominance type, nodular – a distinct entity? Unique staining profile for L&H variants of Reed–Sternberg cells defined by monoclonal antibodies to leukocyte common antigen, granulocyte-specific antigen, and B-cell-specific antigen. *Am J Pathol* **118**, 1–6.
16. Stein H, Hansmann ML, Lennert K, Brandtzaeg P, Gatter KC, Mason DY. (1986). Reed–Sternberg and Hodgkin cells in lymphocyte-predominant Hodgkin's disease of nodular subtype contain J chain. *Am J Clin Pathol* **86**, 292–7.
17. Fraga M, Forteza J. (2007). Diagnosis of Hodgkin's disease: An update on histopathological and immunophenotypical features. *Histol Histopathol* **22**, 923–35.
18. Anagnostopoulos I, Hansmann ML, Franssila K, Harris M, Harris NL, Jaffe ES, Han J, van Krieken JM, Poppema S, Marafioti T, Franklin J, Sextro M, Diehl V, Stein H. (2000). European task force on lymphoma project on lymphocyte predominance Hodgkin disease: Histologic and immunohistologic analysis of submitted cases reveals 2 types of Hodgkin disease with a nodular growth pattern and abundant lymphocytes. *Blood* **96**, 1889–99.
19. Pileri SA, Ascani S, Leoncini L, Sabattini E, Zinzani PL, Piccaluga PP, Pileri Jr. A, Giunti M, Falini B, Bolis GB, Stein H. (2002). Hodgkin's lymphoma: The pathologist's viewpoint. *J Clin Pathol* **55**, 162–76.
20. Alkan S, Ross CW, Hanson CA, Schnitzer B. (1995). Epstein–Barr virus and bcl-2 protein overexpression are not detected in the neoplastic cells of nodular lymphocyte predominance Hodgkin's disease. *Mod Pathol* **8**, 544–7.
21. Weiss LM, Chen YY, Liu XF, Shibata D. (1991). Epstein–Barr virus and Hodgkin's disease. A correlative *in situ* hybridization and polymerase chain reaction study. *Am J Pathol* **139**, 1259–65.
22. Schmid C, Sargent C, Isaacson PG. (1991). L and H cells of nodular lymphocyte predominant Hodgkin's disease show immunoglobulin light-chain restriction. *Am J Pathol* **139**, 1281–9.
23. Stein H, Marafioti T, Foss HD, Laumen H, Hummel M, Anagnostopoulos I, Wirth T, Demel G, Falini B. (2001). Down-regulation of BOB.1/OBF.1 and Oct2 in classical Hodgkin disease but not in lymphocyte predominant Hodgkin disease correlates with immunoglobulin transcription. *Blood* **97**, 496–501.
24. Prakash S, Fountaine T, Raffeld M, Jaffe ES, Pittaluga S. (2006). IgD positive L&H cells identify a unique subset of nodular lymphocyte predominant Hodgkin lymphoma. *Am J Surg Pathol* **30**, 585–92.
25. Browne P, Petrosyan K, Hernandez A, Chan JA. (2003). The B-cell transcription factors BSAP, Oct-2, and BOB.1 and the pan-B-cell markers CD20, CD22, and CD79a are useful in the differential diagnosis of classic Hodgkin lymphoma. *Am J Clin Pathol* **120**, 767–77.
26. Torlakovic E, Tierens A, Dang HD, Delabie J. (2001). The transcription factor PU.1, necessary for B-cell development is expressed in lymphocyte predominance, but not classical Hodgkin's disease. *Am J Pathol* **159**, 1807–14.
27. Marafioti T, Mancini C, Ascani S, Sabattini E, Zinzani PL, Pozzobon M, Pulford K, Falini B, Jaffe ES, Müller-Hermelink HK, Mason DY, Pileri SA. (2004). Leukocyte-specific phosphoprotein-1 and PU.1: Two useful markers for distinguishing T-cell-rich B-cell lymphoma from lymphocyte-predominant Hodgkin's disease. *Haematologica* **89**, 957–64.
28. Fan Z, Natkunam Y, Bair E, Tibshirani R, Warnke RA. (2003). Characterization of variant patterns of nodular lymphocyte predominant Hodgkin lymphoma with immunohistologic and clinical correlation. *Am J Surg Pathol* **27**, 1346–56.
29. Boudová L, Torlakovic E, Delabie J, Reimer P, Pfistner B, Wiedenmann S, Diehl V, Müller-Hermelink HK, Rüdiger T. (2003). Nodular lymphocyte-predominant Hodgkin lymphoma with nodules resembling T-cell/histiocyte-rich B-cell lymphoma: Differential diagnosis between nodular lymphocyte-predominant Hodgkin lymphoma and T-cell/histiocyte-rich B-cell lymphoma. *Blood* **102**, 3753–8.
30. Poppema S. (1989). The nature of the lymphocytes surrounding Reed–Sternberg cells in nodular lymphocyte predominance and in other types of Hodgkin's disease. *Am J Pathol* **135**, 351–7.
31. Kamel OW, Gelb AB, Shibuya RB, Warnke RA. (1993). Leu 7 (CD57) reactivity distinguishes nodular lymphocyte predominance Hodgkin's disease from nodular sclerosing Hodgkin's disease, T-cell-rich B-cell lymphoma and follicular lymphoma. *Am J Pathol* **142**, 541–6.
32. Kraus MD, Haley J. (2000). Lymphocyte predominance Hodgkin's disease: The use of bcl-6 and CD57 in diagnosis and differential diagnosis. *Am J Surg Pathol* **24**, 1068–78.
33. Braeuninger A, Küppers R, Strickler JG, Wacker HH, Rajewsky K, Hansmann ML. (1997). Hodgkin and Reed–Sternberg cells in lymphocyte predominant Hodgkin disease represent clonal populations of germinal center-derived tumor B cells. *Proc Natl Acad Sci USA* **94**, 9337–42.
34. Wlodarska I, Nooyen P, Maes B, Martin-Subero JI, Siebert R, Pauwels P, De Wolf-Peeters C, Hagemeijer A. (2003). Frequent occurrence of BCL6 rearrangements in nodular lymphocyte predominance Hodgkin lymphoma but not in classical Hodgkin lymphoma. *Blood* **101**, 706–10.
35. Mauch PM, Kalish LA, Kadin M, Coleman CN, Osteen R, Hellman S. (1993). Patterns of presentation of Hodgkin disease. Implications for etiology and pathogenesis. *Cancer* **71**, 2062–71.
36. Bodis S, Kraus MD, Pinkus G, Silver B, Kadin ME, Canellos GP, Shulman LN, Tarbell NJ, Mauch PM. (1997). Clinical presentation and outcome in lymphocyte-predominant Hodgkin's disease. *J Clin Oncol* **15**, 3060–6.
37. Diehl V, Sextro M, Franklin J, Hansmann ML, Harris N, Jaffe E, Poppema S, Harris M, Franssila K, van Krieken J, Marafioti T, Anagnostopoulos I, Stein H. (1999). Clinical presentation, course, and prognostic factors in lymphocyte-predominant Hodgkin's disease and lymphocyte-rich classical Hodgkin's disease: Report from the European Task Force on Lymphoma Project on Lymphocyte-Predominant Hodgkin's Disease. *J Clin Oncol* **17**, 776–83.

38. Nogová L, Reineke T, Josting A, Müller-Hermelink HK, Eich HT, Behringer K, Müller RP, Diehl V, Engert A. (2005). Lymphocyte-predominant and classical Hodgkin's lymphoma – comparison of outcomes. *Eur J Haematol Suppl* **75**, 106–10.
39. Miettinen M, Franssila KO, Saxén E. (1983). Hodgkin's disease, lymphocytic predominance nodular. Increased risk for subsequent non-Hodgkin's lymphomas. *Cancer* **51**, 2293–300.
40. Hansmann ML, Stein H, Fellbaum C, Hui PK, Parwaresch MR, Lennert K. (1989). Nodular paragranuloma can transform into high-grade malignant lymphoma of B type. *Hum Pathol* **20**, 1169–75.
41. Karayalcin G, Behm FG, Gieser PW, Kung F, Weiner M, Tebbi CK, Ferree C, Marcus R, Constine L, Mendenhall NP, Chauvenet A, Murphy SB. (1997). Lymphocyte predominant Hodgkin disease: Clinicopathologic features and results of treatment – the Pediatric Oncology Group experience. *Med Pediatr Oncol* **29**, 519–25.
42. Grossman DM, Hanson CA, Schnitzer B. (1991). Simultaneous lymphocyte predominant Hodgkin's disease and large-cell lymphoma. *Am J Surg Pathol* **15**, 668–76.
43. Naeim F. (1997). *Pathology of Bone Marrow*, 2nd ed. Williams & Wilkins, Baltimore.
44. Ferraris AM, Racchi O, Rapezzi D, Gaetani GF, Boffetta P. (1997). Familial Hodgkin's disease: A disease of young adulthood? *Ann Hematol* **74**, 131–4.
45. Mack TM, Cozen W, Shibata DK, Weiss LM, Nathwani BN, Hernandez AM, Taylor CR, Hamilton AS, Deapen DM, Rappaport EB. (1995). Concordance for Hodgkin's disease in identical twins suggesting genetic susceptibility to the young-adult form of the disease. *N Engl J Med* **332**, 413–18.
46. Glaser SL, Hsu JL. (2002). Hodgkin's disease in Asians: Incidence patterns and risk factors in population-based data. *Leuk Res* **26**, 261–9.
47. Hors J, Dausset J. (1983). HLA and susceptibility to Hodgkin's disease. *Immunol Rev* **70**, 167–92.
48. Taylor GM, Gokhale DA, Crowther D, Woll PJ, Harris M, Ryder D, Ayres M, Radford JA. (1999). Further investigation of the role of HLA-DPB1 in adult Hodgkin's disease (HD) suggests an influence on susceptibility to different HD subtypes. *Br J Cancer* **80**, 1405–11.
49. Harty LC, Lin AY, Goldstein AM, Jaffe ES, Carrington M, Tucker MA, Modi WS. (2002). HLA-DR, HLA-DQ, and TAP genes in familial Hodgkin disease. *Blood* **99**, 690–3.
50. Diepstra A, Niens M, Vellenga E, van Imhoff GW, Nolte IM, Schaapveld M, van der Steege G, van den Berg A, Kibbelaar RE, te Meerman GJ, Poppema S. (2005). Association with HLA class I in Epstein–Barr-virus-positive and with HLA class III in Epstein–Barr-virus-negative Hodgkin's lymphoma. *Lancet* **365**, 2216–24.
51. Niens M, Jarrett RF, Hepkema B, Nolte IM, Diepstra A, Platteel M, Kouprie N, Delury CP, Gallagher A, Visser L, Poppema S, te Meerman GJ, van den Berg A. (2007). HLA-A*02 is associated with a reduced risk and HLA-A*01 with an increased risk of developing EBV-positive Hodgkin lymphoma. *Blood* **Jul 13**.
52. Ahmad T, Neville M, Marshall SE, Armuzzi A, Mulcahy-Hawes K, Crawshaw J, Sato H, Ling KL, Barnardo M, Goldthorpe S, Walton R, Bunce M, Jewell DP, Welsh KI. (2003). Haplotype-specific linkage disequilibrium patterns define the genetic topography of the human MHC. *Hum Mol Genet* **12**, 647–56.
53. Muñoz N, Davidson RJ, Witthoff B, Ericsson JE, De-Thé G. (1978). Infectious mononucleosis and Hodgkin's disease. *Int J Cancer* **22**, 10–13.
54. Gutensohn N, Cole P. (1981). Childhood social environment and Hodgkin's disease. *N Engl J Med* **304**, 135–40.
55. Alexander FE, Jarrett RF, Lawrence D, Armstrong AA, Freeland J, Gokhale DA, Kane E, Taylor GM, Wright DH, Cartwright RA. (2000). Risk factors for Hodgkin's disease by Epstein–Barr virus (EBV) status: Prior infection by EBV and other agents. *Br J Cancer* **82**, 1117–21.
56. Hjalgrim H, Askling J, Rostgaard K, Hamilton-Dutoit S, Frisch M, Zhang JS, Madsen M, Rosdahl N, Konradsen HB, Storm HH, Melbye M. (2003). Characteristics of Hodgkin's lymphoma after infectious mononucleosis. *N Engl J Med* **349**, 1324–32.
57. Hjalgrim H, Smedby KE, Rostgaard K, Molin D, Hamilton-Dutoit S, Chang ET, Ralfkiaer E, Sundström C, Adami HO, Glimelius B, Melbye M. (2007). Infectious mononucleosis, childhood social environment, and risk of Hodgkin lymphoma. *Cancer Res* **67**, 2382–8.
58. Mueller N, Evans A, Harris NL, Comstock GW, Jellum E, Magnus K, Orentreich N, Polk BF, Vogelman J. (1989). Hodgkin's disease and Epstein–Barr virus. Altered antibody pattern before diagnosis. *N Engl J Med* **320**, 689–95.
59. Cickusić E, Mustedanagić-Mujanović J, Iljazović E, Karasalihović Z, Skaljić I. (2007). Association of Hodgkin's lymphoma with Epstein–Barr virus infection. *Bosn J Basic Med Sci* **7**, 58–65.
60. Herbst H, Pallesen G, Weiss LM, Delsol G, Jarrett RF, Steinbrecher E, Stein H, Hamilton-Dutoit S, Brousset P. (1992). Hodgkin's disease and Epstein–Barr virus. *Ann Oncol* (Suppl 4), 27–30.
61. Said JW. (2007). Immunodeficiency-related Hodgkin lymphoma and its mimics. *Adv Anat Pathol* **14**, 189–94.
62. Biggar RJ, Jaffe ES, Goedert JJ, Chaturvedi A, Pfeiffer R, Engels EA. (2006). Hodgkin lymphoma and immunodeficiency in persons with HIV/AIDS. *Blood* **108**, 3786–91.
63. Landgren O, Engels EA, Pfeiffer RM, Gridley G, Mellemkjaer L, Olsen JH, Kerstann KF, Wheeler W, Hemminki K, Linet MS, Goldin LR. (2006). Autoimmunity and susceptibility to Hodgkin lymphoma: A population-based case–control study in Scandinavia. *J Natl Cancer I* **98**, 1321–30.
64. Bräuninger A, Schmitz R, Bechtel D, Renné C, Hansmann ML, Küppers R. (2006). Molecular biology of Hodgkin's and Reed/Sternberg cells in Hodgkin's lymphoma. *Int J Cancer* **118**, 1853–61.
65. Bargou RC, Leng C, Krappmann D, Emmerich F, Mapara MY, Bommert K, Royer HD, Scheidereit C, Dörken B. (1996). High-level nuclear NF-kappa B and Oct-2 is a common feature of cultured Hodgkin/Reed–Sternberg cells. *Blood* **87**, 4340–7.
66. Bargou RC, Emmerich F, Krappmann D, Bommert K, Mapara MY, Arnold W, Royer HD, Grinstein E, Greiner A, Scheidereit C, Dörken B. (1997). Constitutive nuclear factor-kappaB-RelA activation is required for proliferation and survival of Hodgkin's disease tumor cells. *J Clin Invest* **100**, 2961–9.
67. Jost PJ, Ruland J. (2007). Aberrant NF-kappaB signaling in lymphoma: Mechanisms, consequences, and therapeutic implications. *Blood* **109**, 2700–7.
68. Eliopoulos AG, Caamano JH, Flavell J, Reynolds GM, Murray PG, Poyet JL, Young LS. (2003). Epstein–Barr virus-encoded latent infection membrane protein 1 regulates the processing of p100 NF-kappaB2 to p52 via an IKKgamma/NEMO-independent signalling pathway. *Oncogene* **22**, 7557–69.
69. Caldwell RG, Wilson JB, Anderson SJ, Longnecker R. (1998). Epstein–Barr virus LMP2A drives B cell development and survival in the absence of normal B cell receptor signals. *Immunity* **9**, 405–11.
70. Barth TF, Martin-Subero JI, Joos S, Menz CK, Hasel C, Mechtersheimer G, Parwaresch RM, Lichter P, Siebert R, Möller P. (2003). Gains of 2p involving the REL locus correlate with nuclear c-Rel protein accumulation in neoplastic cells of classical Hodgkin lymphoma. *Blood* **101**, 3681–6.
71. Joos S, Küpper M, Ohl S, von Bonin F, Mechtersheimer G, Bentz M, Marynen P, Möller P, Pfreundschuh M, Trümper L, Lichter P. (2000). Genomic imbalances including amplification of the tyrosine kinase gene JAK2 in CD30+ Hodgkin cells. *Cancer Res* **60**, 549–52.
72. Küpper M, Joos S, von Bonin F, Daus H, Pfreundschuh M, Lichter P, Trümper L. (2001). MDM2 gene amplification and lack of p53 point mutations in Hodgkin and Reed–Sternberg cells: Results from single-cell polymerase chain reaction and molecular cytogenetic studies. *Br J Haematol* **112**, 768–75.
73. Re D, Küppers R, Diehl V. (2005). Molecular pathogenesis of Hodgkin's lymphoma. *J Clin Oncol* **23**, 6379–86.
74. Stein H, Mason DY, Gerdes J, O'Connor N, Wainscoat J, Pallesen G, Gatter K, Falini B, Delsol G, Lemke H. (1985). The expression

of the Hodgkin's disease associated antigen Ki-1 in reactive and neoplastic lymphoid tissue: Evidence that Reed–Sternberg cells and histiocytic malignancies are derived from activated lymphoid cells. *Blood* **66**, 848–58.

75. Chittal SM, Caverivière P, Schwarting R, Gerdes J, Al Saati T, Rigal-Huguet F, Stein H, Delsol G. (1988). Monoclonal antibodies in the diagnosis of Hodgkin's disease. The search for a rational panel. *Am J Surg Pathol* **12**, 9–21.
76. Fan G, Kotylo P, Neiman RS, Braziel RM. (2003). Comparison of fascin expression in anaplastic large cell lymphoma and Hodgkin disease. *Am J Clin Pathol* **119**, 199–204.
77. Bakshi NA, Finn WG, Schnitzer B, Valdez R, Ross CW. (2007). Fascin expression in diffuse large B-cell lymphoma, anaplastic large cell lymphoma, and classical Hodgkin lymphoma. *Arch Pathol Lab Med* **131**, 742–7.
78. Foss HD, Reusch R, Demel G, Lenz G, Anagnostopoulos I, Hummel M, Stein H. (1999). Frequent expression of the B-cell-specific activator protein in Reed–Sternberg cells of classical Hodgkin's disease provides further evidence for its B-cell origin. *Blood* **94**, 3108–13.
79. Tzankov A, Zimpfer A, Pehrs AC, Lugli A, Went P, Maurer R, Pileri S, Dirnhofer S. (2003). Expression of B-cell markers in classical Hodgkin lymphoma: A tissue microarray analysis of 330 cases. *Mod Pathol* **16**, 1141–7.
80. Zukerberg LR, Collins AB, Ferry JA, Harris NL. (1991). Coexpression of CD15 and CD20 by Reed–Sternberg cells in Hodgkin's disease. *Am J Pathol* **139**, 475–83.
81. Carbone A, Gloghini A, Gaidano G, Franceschi S, Capello D, Drexler HG, Falini B, Dalla-Favera R. (1998). Expression status of BCL-6 and syndecan-1 identifies distinct histogenetic subtypes of Hodgkin's disease. *Blood* **92**, 2220–8.
82. Falini B, Fizzotti M, Pucciarini A, Bigerna B, Marafioti T, Gambacorta M, Pacini R, Alunni C, Natali-Tanci L, Ugolini B, Sebastiani C, Cattoretti G, Pileri S, Dalla-Favera R, Stein H. (2000). A monoclonal antibody (MUM1p) detects expression of the MUM1/IRF4 protein in a subset of germinal center B cells, plasma cells, and activated T cells. *Blood* **95**, 2084–92.
83. Carbone A, Gloghini A, Aldinucci D, Gattei V, Dalla-Favera R, Gaidano G. (2002). Expression pattern of MUM1/IRF4 in the spectrum of pathology of Hodgkin's disease. *Br J Haematol* **117**, 366–72.
84. Buettner M, Greiner A, Avramidou A, Jäck HM, Niedobitek G. (2005). Evidence of abortive plasma cell differentiation in Hodgkin and Reed–Sternberg cells of classical Hodgkin lymphoma. *Hematol Oncol* **23**, 127–32.
85. Bai M, Panoulas V, Papoudou-Bai A, Horianopoulos N, Kitsoulis P, Stefanaki K, Rontogianni D, Agnantis NJ, Kanavaros P. (2006). B-cell differentiation immunophenotypes in classical Hodgkin lymphomas. *Leuk Lymphoma* **47**, 495–501.
86. Joos S, Menz CK, Wrobel G, Siebert R, Gesk S, Ohl S, Mechtersheimer G, Trümper L, Möller P, Lichter P, Barth TF. (2002). Classical Hodgkin lymphoma is characterized by recurrent copy number gains of the short arm of chromosome 2. *Blood* **99**, 1381–7.
87. Weber-Matthiesen K, Deerberg J, Poetsch M, Grote W, Schlegelberger B. (1995). Numerical chromosome aberrations are present within the CD30+ Hodgkin and Reed–Sternberg cells in 100% of analyzed cases of Hodgkin's disease. *Blood* **86**, 1464–8.
88. Janz M, Hummel M, Truss M, Wollert-Wulf B, Mathas S, Jöhrens K, Hagemeier C, Bommert K, Stein H, Dörken B, Bargou RC. (2006). Classical Hodgkin lymphoma is characterized by high constitutive expression of activating transcription factor 3 (ATF3), which promotes viability of Hodgkin/Reed–Sternberg cells. *Blood* **107**, 2536–9.
89. Seitz V, Hummel M, Anagnostopoulos I, Stein H. (2001). Analysis of BCL-6 mutations in classic Hodgkin disease of the B- and T-cell type. *Blood* **97**, 2401–5.
90. Connors JM. (2005). State-of-the-art therapeutics: Hodgkin's lymphoma. *J Clin Oncol* **23**, 6400–8.
91. Hasenclever D, Diehl V. (1998). A prognostic score for advanced Hodgkin's disease. International Prognostic Factors Project on Advanced Hodgkin's Disease. *N Engl J Med* **339**, 1506–14.
92. Colby TV, Hoppe RT, Warnke RA. (1982). Hodgkin's disease: A clinicopathologic study of 659 cases. *Cancer* **49**, 1848–58.
93. Connors JM. (2005). Evolving approaches to primary treatment of Hodgkin lymphoma. *Hematology Am Soc Hematol Educ Program*, 239–44.
94. Fuchs M, Diehl V, Re D. (2006). Current strategies and new approaches in the treatment of Hodgkin's lymphoma. *Pathobiology* **73**, 126–40.
95. Ansell SM, Armitage JO. (2006). Management of Hodgkin lymphoma. *Mayo Clin Proc* **81**, 419–26.
96. Murphy F, Sirohi B, Cunningham D. (2007). Stem cell transplantation in Hodgkin lymphoma. *Expert Rev Anticancer Ther* **7**, 297–306.
97. Klimm B, Schnell R, Diehl V, Engert A. (2005). Current treatment and immunotherapy of Hodgkin's lymphoma. *Haematologica* **90**, 1680–92.
98. Savoldo B, Rooney CM, Di Stasi A, Abken H, Hombach A, Foster AE, Zhang L, Heslop HE, Brenner MK, Dotti G. (2007). Epstein–Barr virus-specific cytotoxic T lymphocytes expressing the anti-CD30zeta artificial chimeric T-cell receptor for immunotherapy of Hodgkin's disease. *Blood* **May 16**.
99. Bellan C, Lazzi S, Zazzi M, Lalinga AV, Palummo N, Galieni P, Marafioti T, Tonini T, Cinti C, Leoncini L, Pileri SA, Tosi P. (2002). Immunoglobulin gene rearrangement analysis in composite Hodgkin disease and large B-cell lymphoma: Evidence for receptor revision of immunoglobulin heavy chain variable region genes in Hodgkin–Reed–Sternberg cells. *Diagn Mol Pathol* **11**, 2–8.
100. Brown JR, Weng AP, Freedman AS. (2004). Hodgkin disease associated with T-cell non-Hodgkin lymphomas: Case reports and review of the literature. *Am J Clin Pathol* **121**, 701–8.
101. Prochorec-Sobieszek M, Majewski M, Sikorska A, Centkowski P, Tajer J, Lampka-Wojciechowska E, Rymkiewicz G, Konopka L, Meder J, Warzocha K, Maryniak RK. (2006). Transformation in lymphomas – morphological, immunophenotypic and molecular features. *Pol J Pathol* **57**, 63–70.
102. García JF, Mollejo M, Fraga M, Forteza J, Muniesa JA, Pérez-Guillermo M, Pérez-Seoane C, Rivera T, Ortega P, Piris MA. (2005). Large B-cell lymphoma with Hodgkin's features. *Histopathology* **47**, 101–10.
103. Traverse-Glehen A, Pittaluga S, Gaulard P, Sorbara L, Alonso MA, Raffeld M, Jaffe ES. (2005). Mediastinal gray zone lymphoma: The missing link between classic Hodgkin's lymphoma and mediastinal large B-cell lymphoma. *Am J Surg Pathol* **29**, 1411–21.
104. Seitz V, Hummel M, Marafioti T, Anagnostopoulos I, Assaf C, Stein H. (2000). Detection of clonal T-cell receptor gamma-chain gene rearrangements in Reed–Sternberg cells of classic Hodgkin disease. *Blood* **95**, 3020–4.
105. Shimoyama Y, Oyama T, Asano N, Oshiro A, Suzuki R, Kagami Y, Morishima Y, Nakamura S. (2006). Senile Epstein–Barr virus-associated B-cell lymphoproliferative disorders: A mini review. *J Clin Exp Hematop* **46**, 1–4.
106. Oyama T, Ichimura K, Suzuki R, Suzumiya J, Ohshima K, Yatabe Y, Yokoi T, Kojima M, Kamiya Y, Taji H, Kagami Y, Ogura M, Saito H, Morishima Y, Nakamura S. (2003). Senile EBV+ B-cell lymphoproliferative disorders: A clinicopathologic study of 22 patients. *Am J Surg Pathol* **27**, 16–26.

Non-neoplastic and Borderline Lymphocytic Disorders

CHAPTER 19

Faramarz Naeim, P. Nagesh Rao and Wayne W. Grody

In this chapter, the non-neoplastic lymphocytic disorders and lymphoproliferative disorders with borderline clinicopathological features (disorders of variable malignant potential) are discussed.

LYMPHOCYTOPENIA

Lymphocytopenia, or lymphopenia (absolute total blood lymphocyte count <1,500/μL), is one of the hallmarks of the primary and acquired immunodeficiency syndromes (AIDS). It also occurs in a wide variety of conditions such as aplastic anemia, tuberculosis, zinc deficiency, systemic lupus erythematous, sarcoidosis, Hodgkin lymphoma (HL), toxic shock, and renal failure [1–4]. Administration of glucocorticoids and antilymphocyte globulin, cancer chemotherapy and radiotherapy, and thoracic duct drainage are frequently associated with lymphocytopenia.

Primary Immunodeficiency Syndromes

The primary immunodeficiencies are rare congenital disorders with defective function of the immune system leading to increased susceptibility to infection, autoimmunity, and development of malignant neoplasms [5, 6]. The primary immunodeficiencies include different subtypes representing T-cell (cellular) or B-cell (humoral) defects or combined deficiencies (Table 19.1). The T-cell defects are characterized by opportunistic viral and pneumocystis infections, whereas B-cell immunodeficiencies usually lead to bacterial infections [5]. In this section, severe combined immunodeficiency (SCID), Wiskott–Aldrich syndrome (WAS), DiGeorge syndrome, and agammaglobulinemia are briefly discussed as examples of primary immunodeficiencies.

Severe combined immunodeficiency syndromes (SCID) consist of a heterogeneous group of disorders primarily arising from molecular T-cell defects leading to developmental and functional disturbances of T-cells and sometimes natural killer (NK) cells or B-cells (Table 19.2) [7–9]. There are two different modes of inheritance: X-linked and autosomal recessive. The X-linked form is associated with the mutation of the gamma chain of IL-2 receptor gene (*IL2RG*) and is characterized by reduced number of circulating T- and NK-cells and normal numbers of B-cells (Table 19.2). However, the B-cells do not mature to plasma cells, leading to a virtually non-existent serum immunoglobulin in these patients [7, 10]. Other variants of SCID are listed in Table 19.2. Of the many possible genes involved in SCID and related disorders, DNA sequencing to detect mutations is becoming available only for a few, such as *RAG1* and *RAG2*. Adenosine deaminase deficiency (ADD) is a subtype of SCID and accounts for about 50% of the autosomal recessive forms [11, 12]. ADD is characterized by a marked decline in the absolute number of both T- and B-cells in the peripheral blood and reduced number of hematogones and plasma cells in the bone marrow.

The typical symptoms of SCID are recurrent viral and bacterial infections, chronic diarrhea, and failure to thrive in newborns. These findings are associated with the lack of thymic shadow on chest X-ray studies. In some cases, clinical symptoms may be delayed by several months due to the protective effects of the maternally derived antibodies.

Wiskott–Aldrich syndrome (WAS) is a rare congenital disorder characterized by the triad of immunodeficiency, eczema, and abnormal

TABLE 19.1 Primary immunodeficiency syndromes associated with defective lymphocytes.*

Type	Inheritance	Molecular defect
T-cell defects		
SCID		
Reticular dysgenesis	AR	?
Alymphocytosis	AR	*RAG1; RAG2*
Deficit of T- and NK-cells	X-linked; AR	γ-chain-*IL-2R; JAK3*
PNP deficiency	AR	*PNP*
Omenn syndrome	AR	5′-nucleotidase
T-cell activation defects (CID)		
HLA deficiency class-I	AR	*TAP2*
HLA deficiency class-II	AR	*CIITA-RFX5*
CD3 deficiency	AR	γ,ε CD3
Zap-70 deficiency	AR	*ZAP-70*
Calcium influx deficiency	AR	?
Defect in IL-2 synthesis	AR	?
Defects in DNA repair		
Ataxia telangiectasia	AR	*ATM*
Bloom syndrome	AR	BLM
Nijmegen syndrome	AR	NBSI
Xeroderma pigmentosum	AR	Complementation groups XPA to G
Others		
Wiskott–Aldrich syndrome	X-linked; AR	*WASP*; ?
DiGeorge syndrome	Sporadic	del(22q11.2)/*TBX1*
Hyper IgM syndrome	X-linked; AR	CD40 L; ?
B-cell defects		
Lymphoproliferative syndrome	X-linked; AD	?; fas
Bruton agammaglobulinemia	X-linked	BTK
Common variable immunodeficiency	Sporadic	?
IgA deficiency	AD; AR; sporadic	?
IgG subclass deficiency	AR	?
Hyper IgE syndrome	AR	?

*Adapted from Ref. [5].

AD: autosomal dominant, AR: autosomal recessive, BTK: Bruton tyrosine kinase, CID: combined immunodeficiency, PNP: purine nucleoside phosphorylase, SCID: severe combined immunodeficiency, WASP: Wiskott–Aldrich syndrome protein.

platelets (thrombocytopenia, small platelets, and platelet dysfunction) (see Chapter 24). Mutation of the *WASP* gene located on the chromosome Xp11.22-23 is the primary molecular defect. Numerous mutations have been reported, mostly found in exons 1 and 2 [13–15]. These mutations lead to the absence or aberrant expression of the WAS protein in lymphocytes and megakaryocytes. The WAS protein, a 53-kD cytoplasmic protein, appears to be involved in the regulation of actin in the cytoskeletal structures of T-cells and platelets. Lack of WAS protein expression may lead to decreased platelet size and defective T-cell and platelet function [5, 13].

WAS is an X-linked recessive disorder primarily involving males with an average age of about 21 months at diagnosis [16]. An autosomal dominant mode of inheritance has also been reported in some families [17], and complete sequencing of the *WAS* gene is available in a number of specialized laboratories to detect the mutations.

Clinical manifestations include bleeding secondary to thrombocytopenia and abnormal platelet function, recurrent infections, autoimmune manifestations, and eczema. There is an increased risk of hematologic malignancies,

TABLE 19.2 Gene defects and blood lymphocyte alterations in certain subtypes of severe combined immunodeficiency.*

Subtype	Gene	CD4	CD8	B	NK
X-linked, common γ-chain	*IL2RG*	↓	↓	N	
Janus kinase 3	*JAK3*	↓	↓	N	
IL-2 receptor-α (CD25)	*IL2RA*	↓	↓	N	N
IL-7 receptor-α (CD127)	*IL7RA*	↓	↓	N	N
CD3 complex	*CD3D*	↓	↓	N	
Recombinase activating genes	*RAG1, RAG2*	↓	↓	↓	N
Adenosine deaminase	*ADA*	↓	↓	↓	±
MHC class-II	*MHCIID*	↓	N	N	N
Zeta-associated protein	*ZAP70*	N	↓	N	N
Protein tyrosine phosphatase (CD45)	*PTPRC*	↓	↓	N	

*Adapted from Ref. [9].

particularly non-HL [18]. The platelet counts are often <50,000/μL and platelet size is usually half the normal size [19]. There is an inverse correlation between the severity of the clinical manifestations and the detectable levels of the WAS protein in the peripheral blood [20, 21]. The differential diagnosis includes immune-associated thrombocytopenia, other primary immunodeficiencies, myelodysplastic syndromes, and hematopoietic malignancies [22].

DiGeorge syndrome is a rare congenital immunodeficiency disorder characterized by the deletion of 22q11.2, developmental abnormalities of the third and fourth pharyngeal pouches, including thymic hypoplasia and hypoparathyroidism, and midline cardiac defects. The suggested criteria for definitive diagnosis of DiGeorge syndrome consist of reduced numbers of total CD3+ T-cells plus the presence of at least two of the following three findings [23]:

1. Deletion of chromosome 22q11.2 (Figures 19.1 and 19.2)
2. Hypocalcemia
3. Congenital cardiac defect.

Up to 90% of patients with DiGeorge syndrome show microdeletion of 22q11.2, and about 75% of the patients demonstrate congenital heart disease including tetralogy of Fallot, ventricular septal defect, or interrupted aortic arch [24–30]. As testing for the 22q11.2 deletion by fluorescence *in situ* hybridization (FISH) has become more widespread, the range of cardiac defects associated with the syndrome has expanded. Some centers have begun checking for the deletion in almost every patient with even an isolated congenital heart defect, and the diagnostic yield has been significant. Discovery of the deletion can then raise alertness to the possibility of other immunologic and endocrine complications (which are not present in all cases). Observation of the characteristic facial dysmorphism can also raise suspicion (the synonymous velocardiofacial syndrome). The deletion can also be detected in the context of a whole-genome scan for copy number variants (CNVs) using array comparative genomic hybridization (aCGH). The parents should also be tested to distinguish *de novo* from inherited deletions, and to distinguish benign CNVs from pathologic deletions/insertions. Despite much work aimed at identifying the causative gene(s) in the deleted critical region, targeted molecular testing for mutations in the *TXB1* gene is not yet routine.

Agammaglobulinemia is one of the primary humoral immunodeficiencies and consists of two congenital types: X-linked and autosomal recessive [31–34].

The *X-linked* variant (Bruton agammaglobulinemia) affects boys with clinical manifestations between 6 and 18 months of age [32]. This disorder is caused by mutation in a tyrosine kinase gene called *BTK* (Bruton tyrosine kinase) mapped at Xq21 [35]. Complete sequencing of the *BTK* gene to detect these variants is available in several reference laboratories. The BTK product is a signal transduction molecule expressed in B-cells and other hematopoietic cells. The affected patients show profound hypogammaglobulinemia, reduced number of peripheral blood B-cells (<1% CD19+ or CD20+ cells), and recurrent bacterial and viral infections [32, 36]. An atypical form of X-linked agammaglobulinemia has been reported with less severe lymphopenia and hypogammaglobulinemia [37].

The clinical presentations of the *autosomal recessive* variant are similar to those of the X-linked variant except for the lack of *BTK* mutation and manifestation of the disease in both genders. The autosomal recessive type is caused by mutations in several genes that regulate B-cell development, such as *IGHM, CD179B, CD79A, BLNK*, and *LRRCB* [9, 38, 39].

Human Immunodeficiency Virus/Acquired Immunodeficiency Syndrome

The human immunodeficiency virus/acquired immunodeficiency syndrome (HIV/AIDS) has spread throughout the world. AIDS is associated with lymphocytopenia and defective-cell-mediated immunity leading to recurrent

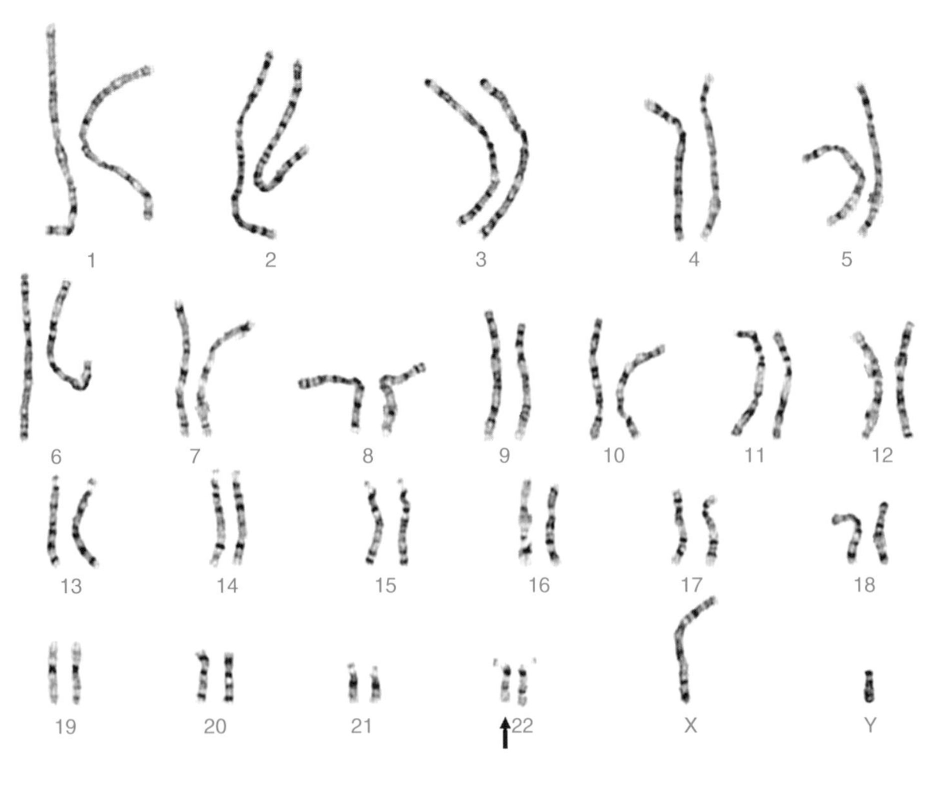

FIGURE 19.1 The characteristic karyotypic feature of DiGeorge syndrome is 22q11.2 (arrow).

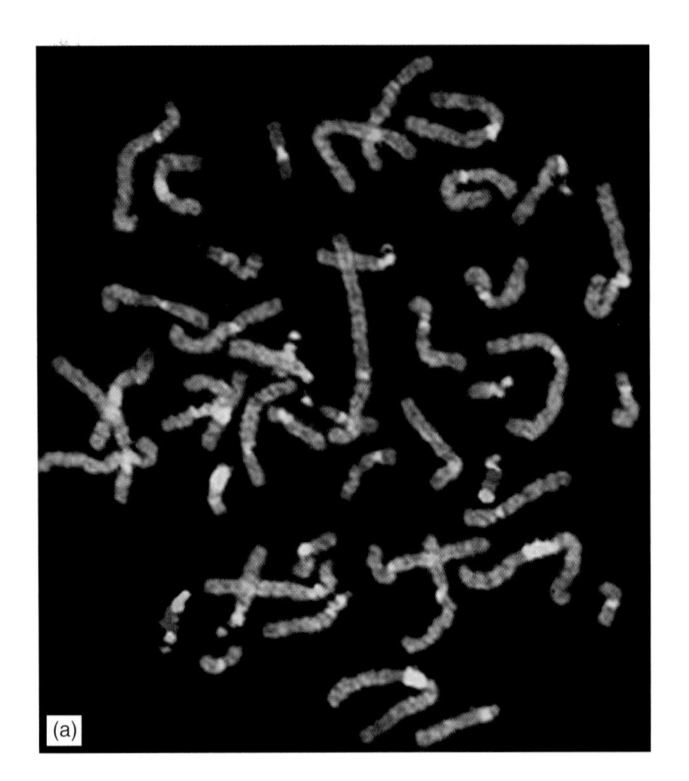

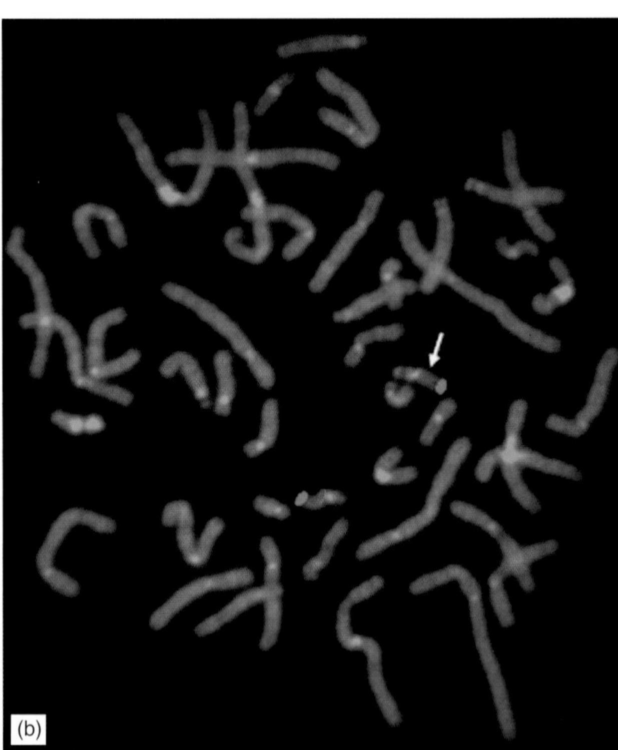

FIGURE 19.2 FISH analysis demonstrating a normal metaphase (a) and an abnormal metaphase with a subtle deletion of the 22q11.2 band (arrow) in a patient with DiGeorge syndrome (b).

infections, Kaposi sarcoma, lymphoma, and cervical cancer (Table 19.3) [40]. All HIV-positive patients with a CD4 cell count of $<200/\mu L$ are considered to have AIDS regardless of the presence or absence of clinical symptoms.

Etiology and Pathogenesis

AIDS is an HIV-induced illness with a broad spectrum of clinical manifestations. There are two strains of HIV: HIV-1 and HIV-2. HIV-1 was discovered first and

TABLE 19.3 Clinical conditions included in the 1993 AIDS surveillance.

Candidiasis (respiratory system, esophageal)
Cervical cancer, invasive*
Coccidiomycosis, disseminated or extrapulmonary
Cryptococcosis, extrapulmonary
Cryptosporidiosis, intestinal
Cytomegalovirus disease (excluding liver, spleen, and lymph node involvements)
Herpes simplex infections, chronic
Histoplasmosis, disseminated or extrapulmonary
HIV encephalopathy
Isosporiasis
Kaposi sarcoma
Lymphoma (Burkitt, immunoblastic, primary brain)
Mycobacterium avium complex or kansasii, disseminated or extrapulmonary
*Mycobacterium tuberculosis**
Mycobacterium, other species
Pneumocystis carinii pneumonia
Pneumonia, recurrent*
Progressive multifocal leukoencephalopathy
Salmonella septicemia, recurrent
Toxoplasmosis, brain
Wasting syndrome

Adapted from CDC, www.cdc.gov/mmwr/preview/mmwrhtml/00018871.htm
*Added in 1993.

is widespread in the United States, Europe, and other parts of the world, whereas HIV-2 is virtually homologous to the simian immunodeficiency virus (SIV) and is primarily detected in West Africa [41–43]. The general consensus is that HIV is the result of the cross-species transmission of SIV from African primates to humans [43].

HIV infection usually spreads through three major routes: sexual intercourse (70–80%), exposure to contaminated blood (5–10%), and prenatal transmission (5–10%) [44]. Viral penetration of mucosal epithelium is followed by infection of CD4+ T-cells, dendritic cells, and macrophages with subsequent spread to the lymph nodes and blood (viremia) [40, 45]. The viral envelope protein, GP-120, binds to CD4 expressed on T-cells, dendritic cells, and macrophages. The entry to the CD4+ cells is mediated by the chemokine receptor CCR5 [46]. Patients homozygous for a 32 base pair deletion in CCR5 are resistant to HIV infection [47], whereas heterozygous-infected individuals may have a

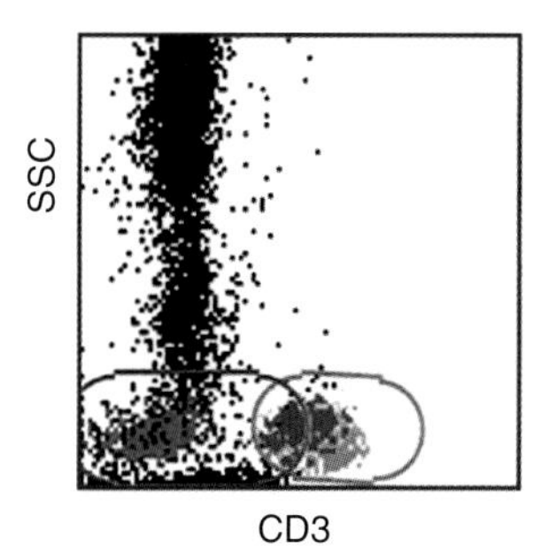

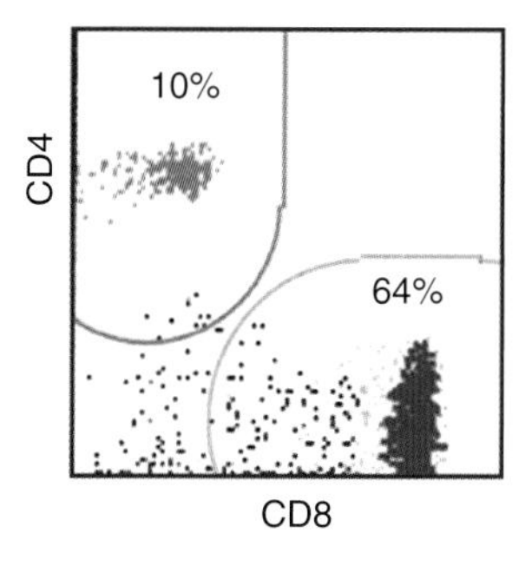

FIGURE 19.3 Flow cytometry of peripheral blood demonstrating a population of CD3+ lymphocytes (absolute count = 980/μL) with reversed CD4:CD8 ratio (CD4 orange and CD8 green).

less aggressive clinical course. The heterozygote frequency in the general population is about 10% [48]. Although the molecular test for this deletion is straightforward, it has not found a place in routine HIV diagnosis or treatment.

The infected CD4+ T-cells are predominantly of the CD29+/CD45+ subset (memory cells). Several mechanisms have been suggested for the depletion and dysfunction of the infected cells, such as formation of syncytia (multinucleated giant cells from fusion of infected T-cells), cytotoxic effects of antiviral antibodies, release of cytotoxic cytokines, and inappropriate induction of apoptosis [49].

Pathology

The peripheral blood reveals lymphopenia (often <500/μL) with a reversed CD4:CD8 ratio (Figure 19.3) (CDC (1987). *MMWR* **36**(suppl), 35). Scattered reactive lymphocytes may be present. Pancytopenia is observed in over 50% of the cases. Anemia is common and is usually normocytic normochromic, with decreased reticulocyte count and elevated serum iron and ferritin levels. Mild to moderate neutropenia is a common feature and approximately 30% of the patients demonstrate monocytopenia. Thrombocytopenia is frequent and in some cases it is autoimmune associated (ITP-like syndrome) [50].

HIV antibodies are raised against various viral components, such as envelope glycoproteins GP-120 and GP-41 and core protein p24. These antibodies are detected by a variety of techniques such as enzyme-linked immunosorbent assay (ELISA), Western blot, radioimmunoassay, and immunofluorescence [51]. Many patients show positive HIV serology (seroconversion) within 4–10 weeks after exposure to the virus, and >95% are seroconvert within 6 months [52, 53].

HIV RNA levels are detected in early stages of viremia by a sensitive reverse transcriptase-polymerase chain reaction (RT-PCR) assay [54, 55]. The detection of HIV RNA usually coincides with seroconversion [54, 55]. The quantitative RT-PCR assay is especially useful for monitoring viral load in patients on anti-retroviral therapy.

The bone marrow specimens from AIDS patients reveal a normocellular to hypercellular marrow, often with myeloid preponderance and left shift, and mild to severe dysplastic changes in one or more hematopoietic lines [56, 57]. An increased frequency of naked megakaryocyte nuclei has been observed in the bone marrow samples (biopsy sections

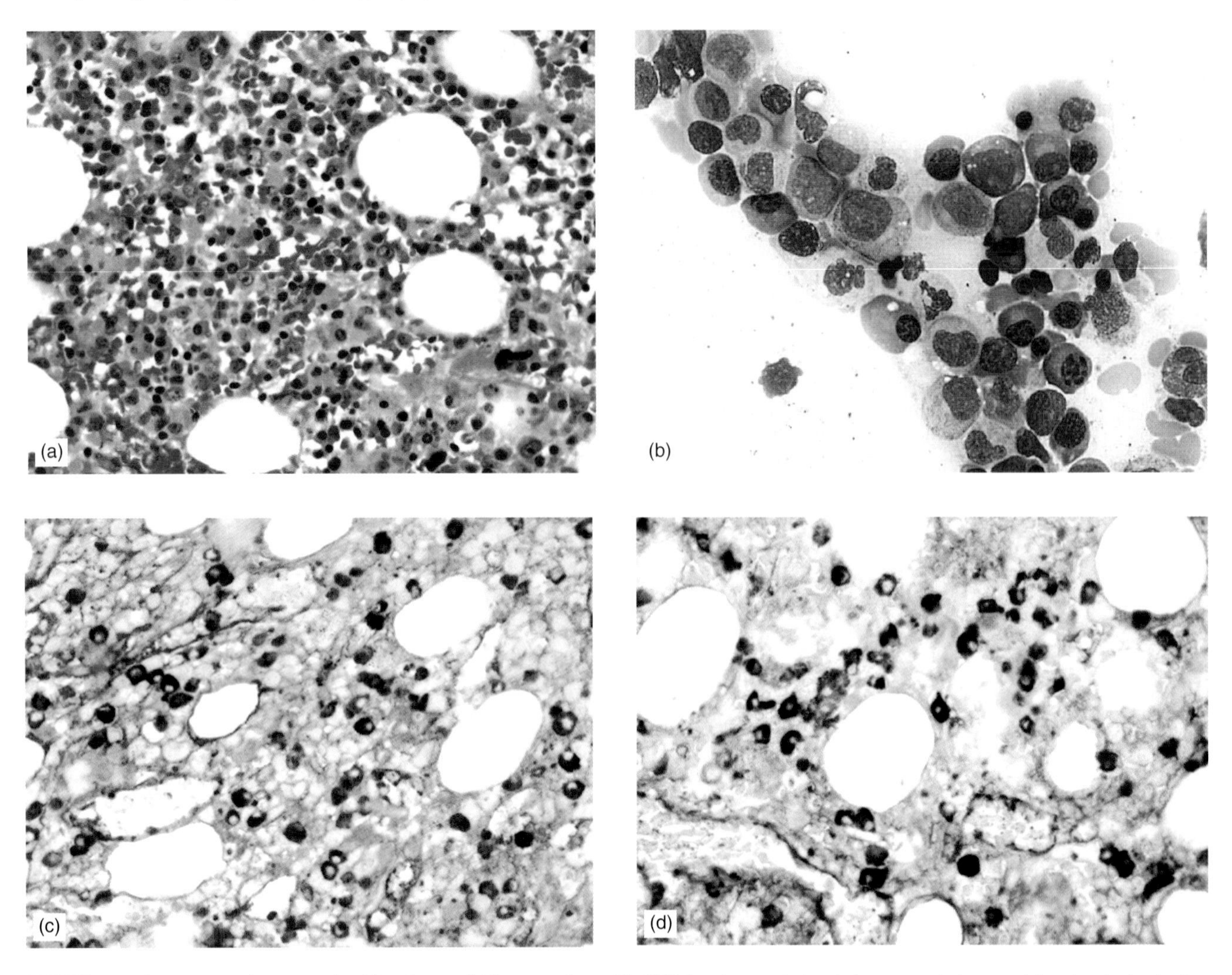

FIGURE 19.4 Bone marrow biopsy section (a) and smear (b) from a patient with AIDS showing a polyclonal plasmacytosis. Immunohistochemical stains for kappa light chain (c) and lambda light chain (d) demonstrate a polyclonal plasma cell population.

and marrow smears) of patients with AIDS [58]. A polyclonal lymphoplasmacytosis is common, sometimes with reactive lymphocytes and immature plasma cells (Figure 19.4). CD4+ cells are reduced, particularly those of the CD45RA-negative phenotype, whereas the proportion of CD8+ cells is increased, especially the HLA-DR+/CD45RA+ subtype [59]. Reticulin fibrosis and gelatinous transformation of bone marrow may be present.

The affected lymph nodes show various morphologic changes. Florid follicular hyperplasia is the most frequent pattern, particularly at the early stages of the disease. Numerous, large, irregular follicles with expanded germinal centers (GC), numerous mitotic figures, and tingible body macrophages are present [60]. Mantle zones are often ill-defined or effaced. The interfollicular areas show prominent vascularity and a mixed cellular component consisting of plasma cells, immunoblasts, histiocytes, lymphocytes, and monocytoid B-cells. The CD4:CD8 ratio is reversed. Sinus histiocytosis is a frequent feature and may be associated with hemophagocytosis [60].

In the later stages of the disease, there is evidence of lymphoid depletion by reduced number of the lymphoid follicles, decreased number of lymphocytes in the interfollicular areas, and the presence of amorphous eosinophilic deposits and/or fibrosis [60]. The affected lymph nodes may show signs of opportunistic infection – Kaposi sarcoma or lymphoma.

Clinical Aspects

The estimated number of people living with HIV in 2005 was 38.6 million with about 4.1 million newly infected ones and an estimated 2.8 million deaths due to AIDS (www.unaids.org, 2007). The major categories for HIV-1 infection include homosexual men, injection drug users, blood product recipients, and healthcare workers exposed to needle sticks [Trends in HIV/AIDS diagnoses: 33 states – 2001–2004 (225). *MMWR* **54**, 1149–1153, http://www.unaids.org, 2004]. The major risk factors for the transmission of HIV infection include viral load, sexual intercourse, ulcerated sexually transmitted diseases, nitrate inhalant use, and the host genetic background [44, 61, 62]. HIV infection is divided into several stages advancing from viral transmission to primary (acute) HIV infection, seroconversion, clinical latent period, early symptomatic stage, and finally to AIDS. The primary HIV infection is characterized by acute non-specific flu-like symptoms such as fever, headache, sore throat, myalgia/arthralgia, diarrhea, nausea/vomiting, and lymphadenopathy. Clinical latent period refers to the 6-month asymptomatic period (except lymphadenopathy) after seroconversion. In this period, HIV is trapped by the follicular dendritic cells in the lymphoid tissues [63]. The early symptomatic stage (Class B), previously called "AIDS-related complex," is associated with a number of clinical symptoms and opportunistic infections such as fever, chronic diarrhea, oral leukoplakia, peripheral neuropathy, thrombocytopenia, herpes zoster infection, vaginal candidiasis, and cervical dysplasia [44].

The revised 1993 definition of AIDS includes several clinical conditions (Table 19.3) plus a CD4+ lymphocyte count of <500/μL in HIV-infected patients. All HIV-positive patients with a CD4+ cell count of <200/μL are considered to have AIDS regardless of the presence or absence of clinical symptoms (www.unaids.org, 2007, www.cdc.gov/mmwr/preview/mmwrhtml/00018871.html). The most frequent clinical presentation associated with AIDS is *Pneumocystis carinii* pneumonia (42.6%) followed by esophageal candidiasis (15%), wasting (10.7%), and Kaposi sarcoma (10.7%) [44]. The spectrum and natural course of these complications have changed dramatically with the advent of anti-retroviral therapy.

Numerous drugs have been approved by the Food and Drug Administration (FDA) for the treatment of HIV infection including nucleoside analogs, protease inhibitors, and fusion inhibitors (www.niaid.nih.gov/factsheets/hivinf.html). Nucleoside analogs are nucleoside reverse transcriptase inhibitors, such as azidothymidine (AZT), zalcitabine (ddC), and stavudine (d4T). Examples for protease inhibitors include ritonavir (Norvir), indinavir (Crixivan), and saquinivir (Invirase). Fuzeon (Enfuvirtide) is the first approved fusion inhibitor drug.

Idiopathic CD4+ T-Lymphocytopenia

Idiopathic CD4+ T-lymphocytopenia is a condition characterized by acquired immunodeficiency, CD4+ lymphocyte depletion, and opportunistic infections *without* evidence of HIV infection [64–66]. These patients have an overall better prognosis than patients with AIDS, and a proportion of them may show a spontaneous regression [67]. The peripheral blood absolute CD4+ count is usually <300/μL.

LYMPHOCYTOSIS

Lymphocytosis is referred to the increased number of peripheral blood lymphocytes, >4,500/μL in individuals older than 12 years of age. The absolute lymphocyte count in children ≤12 years may reach to as high as 8,000/μL in normal conditions [68]. Lymphocytosis is a common finding in most viral infections, certain bacterial infections, X-linked lymphoproliferative (XLP) disease, post-splenectomy, thyrotoxicosis, certain lymphoid malignancies, and a number of other disorders (Table 19.4). Reactive lymphocytosis, particularly in viral infections, is associated with the presence of large, activated, or atypical lymphocytes. Atypical lymphocytosis is one of the characteristic features of infectious mononucleosis (IM) but has also been observed in other viral infections such as cytomegalovirus (CMV), varicella-zoster, rubella, and hepatitis.

TABLE 19.4 Major types of lymphocytosis.

Lymphoid malignancies
Virus-associated lymphocytosis
EBV
CMV
HIV-1
Hepatitis
Influenza
Measles
Mumps
Rubella
Varicella
Lymphocytosis associated with other infectious agents
Babesia microti
Bartonella henselae (cat-scratch fever)
Bordetella pertussis (Whooping cough)
Brucella
Mycobacterium tuberculosis
Toxoplasma gondii
Treponema pallidum (syphilis)
X-linked lymphoproliferative disease
Chronic polyclonal B-cell lymphocytosis
Polyclonal immunoblastic proliferation
Idiopathic lymphocytosis
Lymphocytosis associated with hypersensitivity reactions
Acute serum sickness
Drug-induced (e.g. ceftriaxone or carbamazepine)
Stress-induced lymphocytosis
Cardiac emergencies
Sickle cell anemia
Status epilepticus
Trauma
Other conditions associated with lymphocytosis
Autoimmune disorders
Cigarette smoking
Post-splenectomy
Thyrotoxicosis

Infectious Mononucleosis

Infectious mononucleosis (IM) is the clinical manifestation of Epstein–Barr virus (EBV) infection characterized by fever, oropharyngitis, lymphadenopathy, and lymphocytosis with the presence of atypical lymphocytes in the peripheral blood [69, 70].

Etiology and Pathogenesis

EBV primarily spreads through saliva (kissing), infecting the epithelial cells of the oropharynx. The virus is replicated in the epithelial cells and released in the lymphoid-enriched surrounding environment, infecting B-cells through the EBV receptor CD21 (CR2, C3d receptor) [71]. The entry of EBV into the B-lymphocytes causes polyclonal B-cell proliferation. The EBV-transformed B-cells are able to induce a massive T-cell proliferation, primarily CD8+ and CD45RO+ cytotoxic T-cells [72, 73]. This rapid proliferation is associated with a short survival of CD8+ T-cells in the circulation [73]. Recent studies suggest that at least some degree of CD8+ cell proliferation is associated with the upregulation of Bim, a proapoptotic member of the Bcl-2 protein family and a major regulator of T-cell deletion [74].

Pathology

Lymphocytosis (>4500/μL) with the presence of more than 10% atypical lymphocytes is the characteristic morphologic feature of EBV infection. Atypical lymphocytes are large, pleomorphic cells with abundant shady gray-blue cytoplasm with or without vacuoles (Figures 19.5 and 19.6). They may show scalloping of the cytoplasmic membrane around red blood cells. The nucleus is round, oval, or irregular; the chromatin is clumped; and the nucleoli are often small or inconspicuous [75]. Anemia, granulocytopenia, and thrombocytopenia may occur, and leukocyte alkaline phosphatase (LAP) activity tends to be low [1].

Bone marrow examination reveals lymphocytosis with the presence of atypical lymphocytes. Small granulomas may be present, and there may be evidence of hemophagocytosis. The affected lymph nodes show expansion of paracortical regions with a population of polymorphous lymphoid cells ranging from small lymphocytes to large immunoblasts, mixed with tingible body macrophages and plasma cells [60]. Reed–Sternberg-like cells may be present. Other morphologic findings include small areas of necrosis, reactive follicles, dilated sinuses containing polymorphous lymphoid cells, and the presence of a polymorphous lymphoid infiltrate in the perinodal tissue.

The atypical lymphocytes represent activated lymphocytes that are predominantly of CD8+, CD45RO+, and HLA-DR+ phenotypes [76, 77]. The EBV-transformed immunoblasts are also present in the paracortical regions. The Reed–Sternberg-like cells may express CD30 but are negative for CD15 and EMA and positive for CD45.

The heterophil antibody test is positive, indicating the presence of cross-reacting antibodies to antigens from phylogenetically unrelated species, such as sheep (Paul-Bunnell test), equine (Monospot test), ox, and goat red blood cells [78, 79]. Specific antibodies against EBV antigens such as viral capsid antigen (VCA), EBV nuclear antigen (EBNA), and early antigen (EA) are detected [79]. EBV-encoded RNA (EBER) can be detected in the transformed cells by molecular techniques such as PCR assays and *in situ*

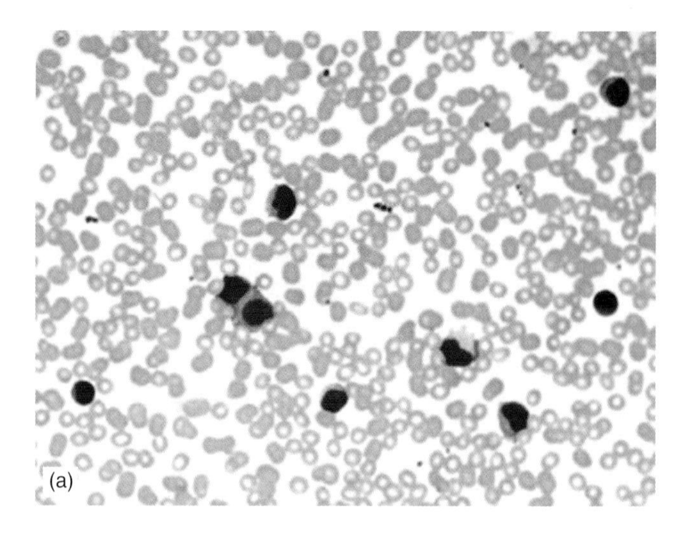

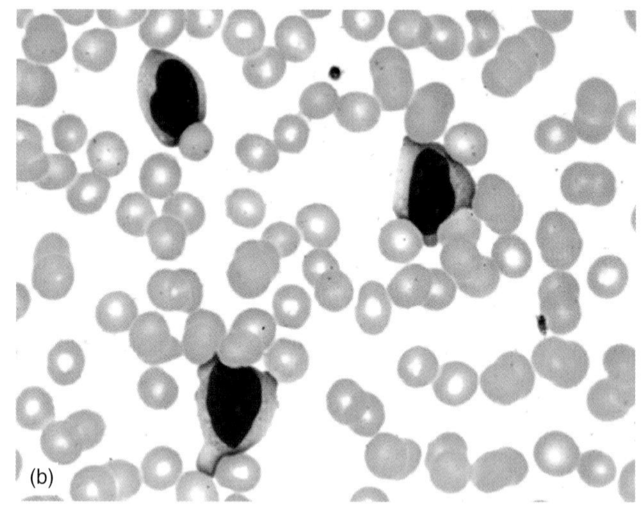

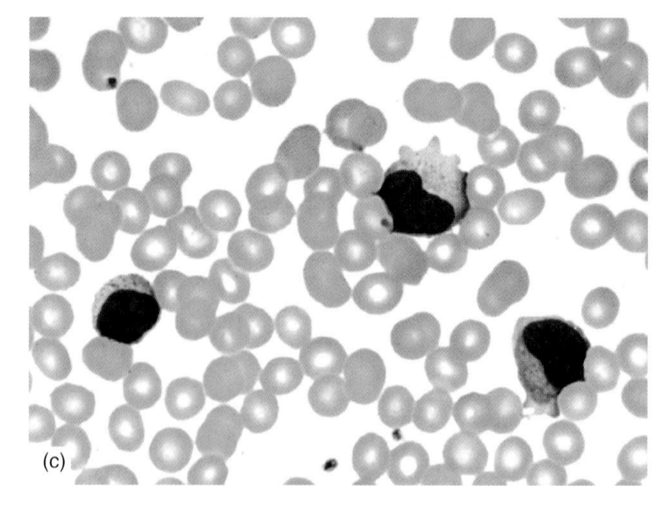

FIGURE 19.5 Blood smear from a patient with IM demonstrates large, pleomorphic atypical (activated) lymphocytes with abundant cytoplasm, round, oval, or irregular nucleus, and dense chromatin. The cytoplasm demonstrates some degree of basophilia and scalloping of the cytoplasmic membrane around erythrocytes. Some cells may show cytoplasmic granules: (a) low power, (b and c) high power views.

hybridization in the atypical lymphocytes or by immunohistochemical stains in the affected lymph node sections [80].

Clinical Aspects

Although EBV infects over 90% of the human population worldwide, the clinical manifestations of IM are uncommon in children and account for <3% of pharyngitis in adults [81]. The vast majority of primary EBV infections are not clinically detected [69].

Symptoms often begin with malaise, headache, and fever followed by lymphadenopathy and pharyngitis [82, 83]. Lymphadenopathy usually involves posterior cervical chains but could become systematic. It peaks in the first week and then gradually disappears within 2–3 weeks [69]. Other clinical findings include splenomegaly, neurologic symptoms such as facial palsies or meningoencephalitis, hepatitis, acute renal failure, and hemophagocytic lymphohistiocytosis [84–87]. IM is usually a self-limited disease and clinical symptoms disappear within 3–4 weeks. However, EBV infection in patients with X-linked lymphoproliferative syndrome may be fatal or lead to non-Hodgkin lymphoma. Also, rare cases of fatal T-cell lymphoproliferative disorders have been reported in association with EBV infection [88].

Supportive therapy is the recommended approach in treating patients with IM. Administration of antiviral drugs, such as Acyclovir, helps to protect people from EBV infection but has no effect on curing the infection [89].

Differential Diagnosis

Lymphocytosis with the presence of atypical lymphocytes is found in various conditions. Approximately 10% of the patients with the clinical symptoms of IM (atypical lymphocytes, fever, pharyngitis, and lymphadenopathy) are EBV- negative, and the condition is caused by other infectious agents, such as toxoplasmosis, CMV, human herpesvirus 6 (HHV-6), and hepatitis B [69, 90–92]. Some drugs such as phenytoin, carbamazepine, isoniazid, and minocycline may also cause atypical lymphocytosis [93, 94]. The diagnosis of EBV infection is confirmed by heterophil and/or specific EBV antibody tests or the identification of EBV by molecular studies, such as PCR assays [78–80]. However, one must exercise caution in interpretation of PCR results, given the high proportion of healthy EBV carriers in the population.

X-Linked Lymphoproliferative Syndrome

XLP syndrome is a rare inherited immunodeficiency characterized by lymphocytosis, dysgammaglobulinemia, fatal IM, or lymphoma usually developing in response to infection with EBV.

Etiology and Pathogenesis

Mutations in two X-linked genes, *SAP* and *XIAP*, have been reported in XLP patients [95–97]. *SAP (SH2D1A)*, the most commonly affected gene, is located in the Xq25 region and encodes a signaling lymphocyte activation molecule (SLAM)-associated protein. SLAM has a number of functions including regulation of T-cell cytotoxicity, T-cell/B-cell co-stimulation, and induction of interferon-γ in the Th_1 cells [98, 99]. SAP protein binds to the cytoplasmic tail of SLAM and also binds to other IG superfamily members, such as 2B4 expressed on NK-cells and cytotoxic T-cells [100, 101]. *SAP* mutations probably result in the defective T- and NK-cell responses and dysregulated cytokine release [98].

Mutation in *XIAP* (or *BIRC4*) gene has been recently reported in some patients with XLP who showed no evidence of *SAP* mutation. *XIAP* encodes the X-linked inhibitor of apoptosis. The *XIAP*-deficient patients with XLP have low numbers of NK-cells, suggesting that *XIAP* is required for the survival and/or differentiation of NK-cells [97].

Pathology

The most common pathologic finding is a massive, systemic lymphohistiocytosis associated with a clinical picture of fatal IM. The proliferating cells are EBV-infected B-cells and cytotoxic T-cells along with histiocytes. This lymphohistiocytic proliferation is accompanied by hemophagocytosis and dysregulated cytokine release resulting in extensive tissue damage, such as hepatic necrosis and profound bone marrow hypoplasia. The lymphohistiocytic infiltration has been described in other organs such as bone marrow, brain, heart, and kidney [98–102].

Approximately 35% of the affected children develop lymphoma, usually of B-cell type. Lymphoma is often extranodal and involves ileocaecal, central nervous system, liver, and kidney [103]. Burkitt lymphoma is the most frequent subtype (53%) followed by immunoblastic (12%) and follicular lymphomas (12%) [103]. HL is rare [98].

Abnormal production of serum immunoglobulin is a common finding and is often associated with a defective cellular immune function, presenting a picture of common variable immunodeficiency [104]. The degree of hypogammaglobulinemia ranges from moderately decreased levels of IgG to severe panhypogammaglobulinemia [105].

Clinical Aspects

XLP is a rare X-linked inherited disorder affecting boys with the age of onset ranging from 2 to 19 years [106]. The most common clinical presentation is a fatal IM following EBV infection, which has a very high mortality rate and a survival rate of <5% [98]. However, in approximately one-third of the affected patients, EBV infection is not fatal. These patients eventually develop dysgammaglobulinemia and/or lymphoma. The definitive diagnostic criteria include a male patient with lymphoma, fatal EBV infection, immunodeficiency, aplastic anemia or lymphohistiocytic disorder, and *SAP* mutation [102] (Table 19.5).

The treatment of choice is allogeneic hematopoietic stem cell transplantation [102]. Antiviral agents such as acyclovir or foscarnet, high dose immunoglobulin, immunosuppressive drugs, interferons α and γ, and HLH 94 have been tried with debatable outcomes [107, 108]. The HLH 94 (an antihistiocytic regimen) has been shown to induce long-term remissions.

TABLE 19.5 Diagnostic criteria for X-linked lymphoproliferative syndrome.*

Definitive Male patient with lymphoma, immunodeficiency, aplastic anemia, lymphohistiocytic disorder, or fatal EBV infection and mutation in *SAP* gene.
Probable Male patient with lymphoma, immunodeficiency, aplastic anemia, lymphohistiocytic disorder, or fatal EBV infection. Patient has maternal cousins, uncles, or nephews with a history of similar disorder.
Possible Male patient with lymphoma, aplastic anemia, or lymphohistiocytic disorder, resulting in death, following EBV infection.

*Adapted from Refs. [98, 102].

Stress-Induced Lymphocytosis

A transient atypical absolute lymphocytosis with lymphocyte counts of up to 13,000/μL has been observed in adult patients with cardiac emergencies, trauma, status epilepticus, or sickle cell anemia crisis [68, 109–111]. The absolute lymphocytosis in these cases is usually the result of the increased numbers of B-, T-, and NK-cells [111].

Persistent Polyclonal B-Cell Lymphocytosis

Persistent (chronic) polyclonal B-cell lymphocytosis (PPBL) is a rare condition that has been reported in young to middle-aged women [112, 113]. An association with heavy smoking and HLA-DR 7 has been reported, suggesting that both environmental and genetic factors are involved [114, 115]. Reports of familial occurrences further support underlying genetic defects in this disorder [116, 117]. These individuals have absolute lymphocytosis ranging from 5,000 to 15,000/μL, with the presence of binucleated and/or atypical lymphocytes (Figure 19.7). There is a polyclonal increase in serum IgM levels with no lymphadenopathy or splenomegaly. The polyclonal B-cells express pan-B-cell markers such as CD19, CD20, and CD22 and show lack of or dim expression of CD5, CD10, and CD23 [114]. They may also express FMC7, CD11c, and CD25. In spite of its polyclonal nature and benign clinical behavior, PPBL, in some cases, has been associated with multiple bcl-2/Ig gene rearrangements and chromosomal abnormalities such as +i(3q), del (6q), and +8, respectively [115, 117, 118].

Polyclonal Immunoblastic Proliferation

Polyclonal immunoblastic proliferation is a rare, transient condition characterized by proliferation of B-immunoblasts and plasma cells in the lymph nodes, bone marrow, and peripheral blood [119, 120]. Hepatosplenomegaly and generalized lymphadenopathy are frequent clinical presentations. The etiology and pathogenesis are not clearly understood,

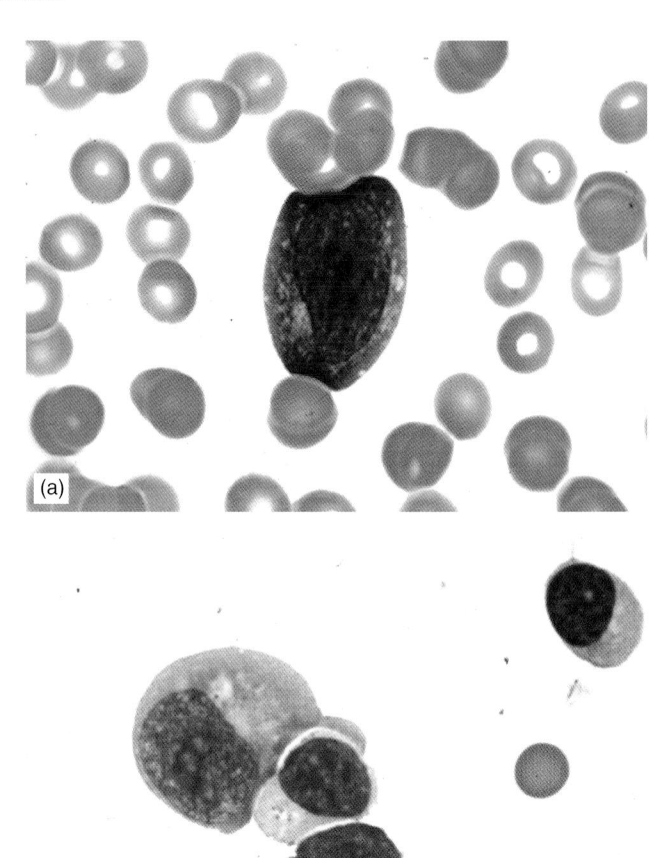

FIGURE 19.6 Blood smear (a) and pleural effusion (b) from a patient with IM showing atypical lymphocytes.

though an underlying abnormal immune response or viral infection has been suggested [119, 121]. Similar changes have also been associated with patients who receive methotrexate.

Bone Marrow Benign Lymphoid Aggregates

Benign lymphoid aggregates (lymphoid nodules, lymphoid follicles) are relatively frequent in bone marrow sections, ranging from <5% to >45% of the cases in various reports [122, 123]. They appear to be more frequent in older individuals and in women [124]. The presence of lymphoid aggregates in younger individuals usually indicates an underlying cause, such as autoimmune disorder, drug reaction, or viral infection. Benign lymphoid aggregates have also been reported in association with aplastic anemia, myeloproliferative disorders, myelodysplastic syndromes, mastocytosis, and HL and non-HL [125–128].

Lymphoid aggregates consist of small, well-defined clusters of mature lymphocytes that are sometimes mixed with scattered plasma cells, eosinophils, mast cells, or histiocytes (Figures 19.8 and 19.9). Lymphoid aggregates are

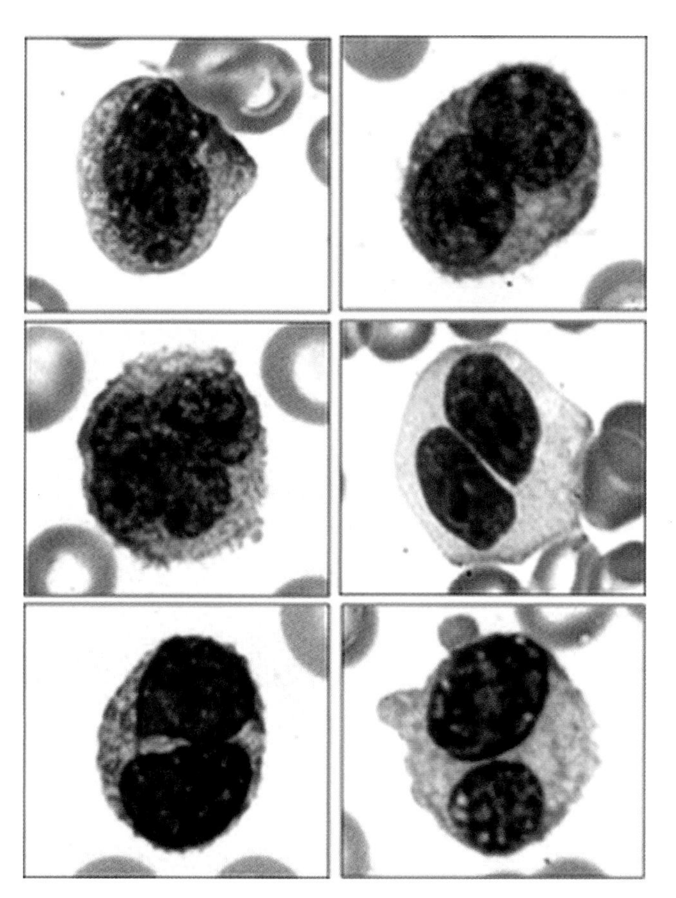

FIGURE 19.7 Binucleated lymphocytes and lymphocytes with lobated nuclei are frequently seen in patients with chronic PPBL. From Naeim F. (2001). *Atlas of Bone Marrow and Blood Pathology*. WB Saunders, Philadelphia, by permission.

usually interstitial, surrounded by fat or hematopoietic cells. They are usually distant from bone trabeculae. Approximately 5% of lymphoid aggregates may show germinal centers. Their presence may indicate reaction to a marked or prolonged immunologic stimulation. The term *lymphoid nodular hyperplasia* is used when four or more lymphoid aggregates are seen in a low power microscopic field, or if an aggregate exceeds 0.6 mm in its greatest dimension [124]. *Reactive polymorphous lymphohistiocytic lesion* refers to aggregates consisting of a mixture of lymphocytes, histiocytes, and other inflammatory cells [122]. These lesions may be large, ill-defined, and paratrabecular.

Differentiation of benign lymphoid aggregates from lymphomatous involvement in the bone marrow is sometimes problematic [129] (Table 19.6). Benign lymphoid aggregates are often well defined, lack an infiltrative pattern, are not paratrabecular, and are primarily composed of small, mature lymphocytes with round nuclei and condensed chromatin. They consist of a mixture of B- and T-cells with no evidence of monoclonality based on immunophenotypic, molecular, and/or cytogenetic studies. The B-cell component of the lymphoid aggregates is negative for bcl-2 and usually lacks the expression of CD5, CD10, and CD23 (Table 19.7) [130–132].

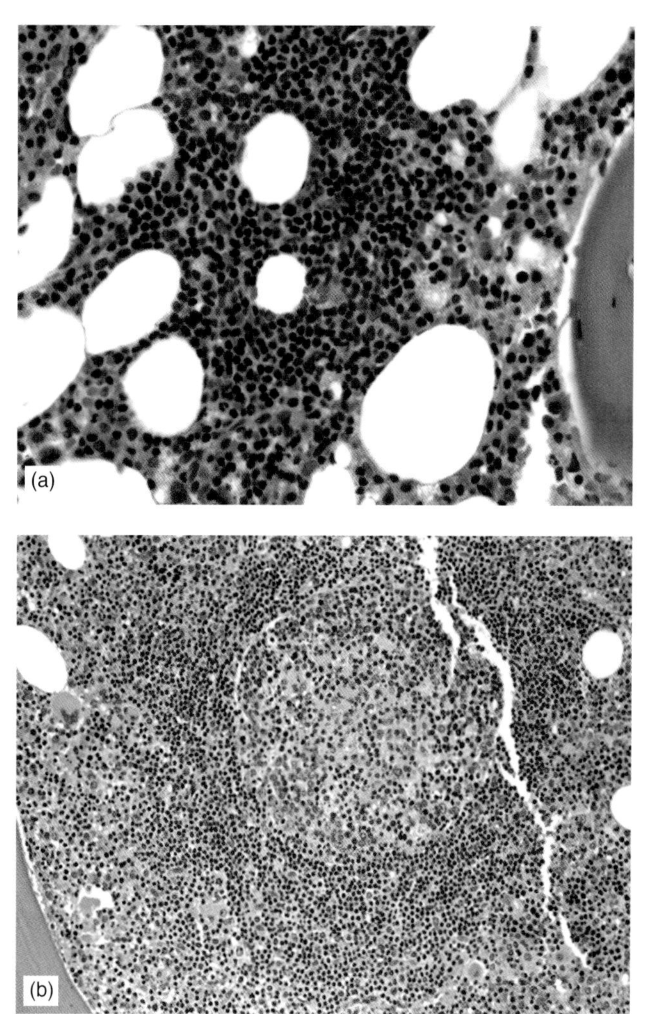

FIGURE 19.8 Benign lymphoid aggregates in bone marrow. Biopsy sections demonstrating lymphoid aggregates (a and b), one with germinal center (b).

LYMPHOPROLIFERATIVE DISORDERS OF VARIABLE MALIGNANT POTENTIAL

Post-transplant Lymphoproliferative Disorders

Post-transplant lymphoproliferative disorders (PTLD) are benign or malignant lymphoid disorders which develop after solid organ or bone marrow allogeneic transplantation [133, 134]. They represent a complex group with a wide spectrum of clinicopathological features ranging from lymphoid hyperplasia to full-blown lymphoma. The World Health Organization (WHO) classification defines four major categories: (1) early lesions, including reactive plasmacytic hyperplasia and IM-like PTLD, (2) polymorphic PTLD, (3) monomorphic PTLD (lymphoma), and (4) Hodgkin lymphoma and Hodgkin lymphoma-like PTLD [133] (Table 19.8).

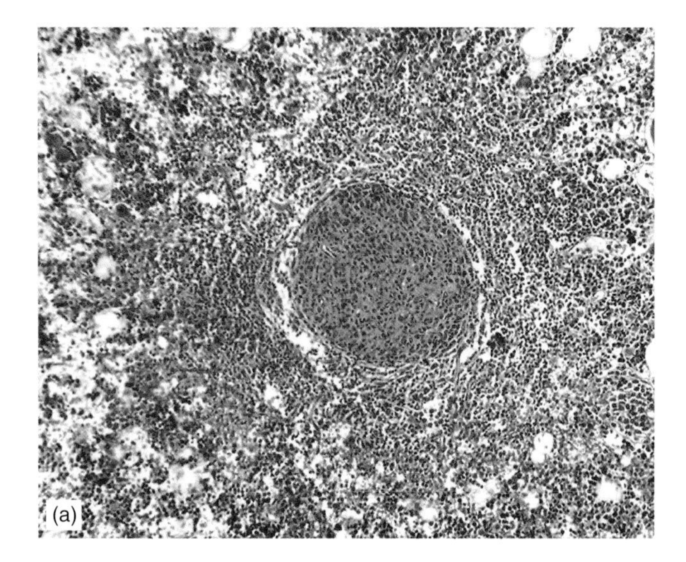

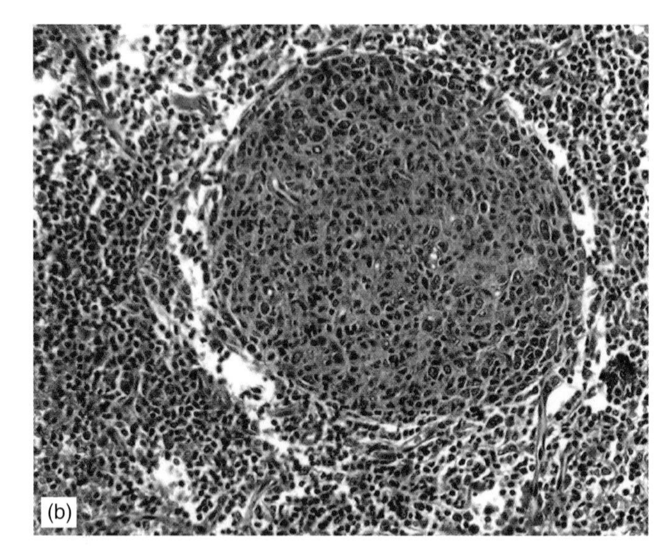

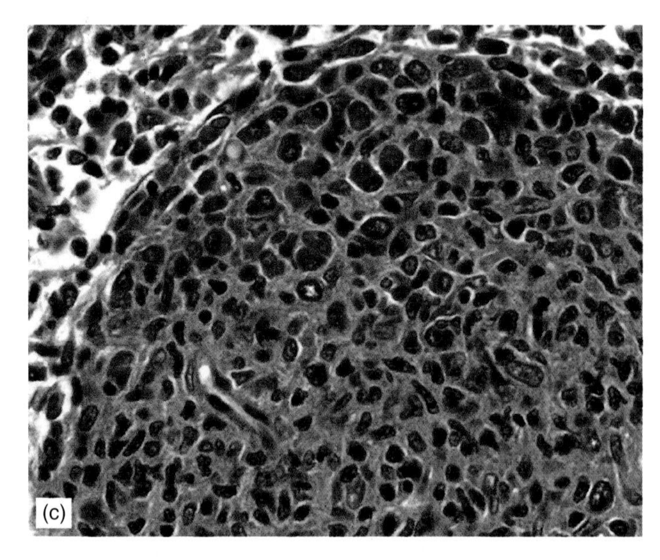

FIGURE 19.9 Bone marrow biopsy section from a patient with AIDS demonstrating an atypical lymphoid aggregate with germinal center: (a) low power, (b) intermediate power, and (c) high power views.

TABLE 19.6 Features of benign and malignant lymphoid aggregates in bone marrow sections.

Benign lymphoid aggregates	Malignant lymphoid aggregates
Often well defined and circumscribed	Usually irregular and infiltrating into the adjacent marrow
Usually interstitial	Frequently paratrabecular
Infrequent cellular atypia	Common cellular atypia
Germinal centers may be present	Germinal centers are not present
Polymorphous lesions lack Reed–Sternberg cells and Reed–Sternberg variants	Presence of Reed–Sternberg cells and variants in Hodgkin lymphoma
Lack of significant fibrosis	May be associated with significant fibrosis
B-cells in aggregates are usually negative for bcl-2, CD5, CD10, and CD23	Malignant B-cells often express bcl-2 and may also express CD5, CD10, or CD23
No evidence of monoclonality by immunophenotypic, molecular, and/or cytogenetic studies	Non-Hodgkin lymphomas often show evidence of monoclonality

TABLE 19.7 Conditions associated with benign lymphoid aggregates in the bone marrow.

Autoimmune disorders Rheumatoid arthritis Systemic lupus erythematosus Autoimmune hemolytic anemia Idiopathic thrombocytopenia Hashimoto thyroiditis
Myelodysplastic syndromes
Myeloproliferative disorders
Mastocytosis
Aplastic anemia
Lymphoid malignancies
Viral infections
Drugs
Unknown

Etiology and Pathogenesis

The primary predisposition conditions associated with PTLD are immunosuppression and EBV infection [135–137]. EBV infection is documented in 50–80% of patients with PTLD cases. The EBV-negative cases are mostly renal allograft recipients which tend to occur later than the EBV-positive cases (>5 years after transplantation) [138, 139]. The vast majority of the PTLD cases (>90%) in the solid

TABLE 19.8 WHO categories of post-transplant lymphoproliferative disorders.*

Type	Major characteristic features
Early lesions	
Reactive plasmacytic hyperplasia	Preserved nodal architecture, increased plasma cells, and rare immunoblasts
IM-like lesions	Preserved nodal architecture, paracortical expansion with numerous immunoblasts.
Polymorphic PTLD	Destructive infiltrates consisting of a mixture of small and large lymphoid cells and plasma cells. Scattered atypical cells, areas of necrosis, and frequent mitotic figures may be present. Rearrangement of Ig or presence of EBV genome. Lack of mutations of *MYC*, *RAS*, and *TP53* genes.
Monomorphic PTLD	Destructive infiltrates of monomorphic atypical lymphoid cells consistent with lymphoma. Most frequent B-cell types are diffuse large B-cell, Burkitt lymphoma, and plasma cell myeloma. Rearrangement of Ig or presence of EBV genome. Mutations of *MYC*, *RAS*, and *TP53* genes may be present. Most frequent T-cell type is peripheral T-cell lymphoma, not otherwise specified. TCR gene rearrangement and up to 25% clonal EBV genome.
Hodgkin lymphoma (HL) and HL-like PTLD	Presence of classical Reed–Sternberg cells expressing CD15 and CD30. Presence of Reed–Sternberg-like cells and morphology consistent with lymphoma.

*Adapted from Ref. [133].

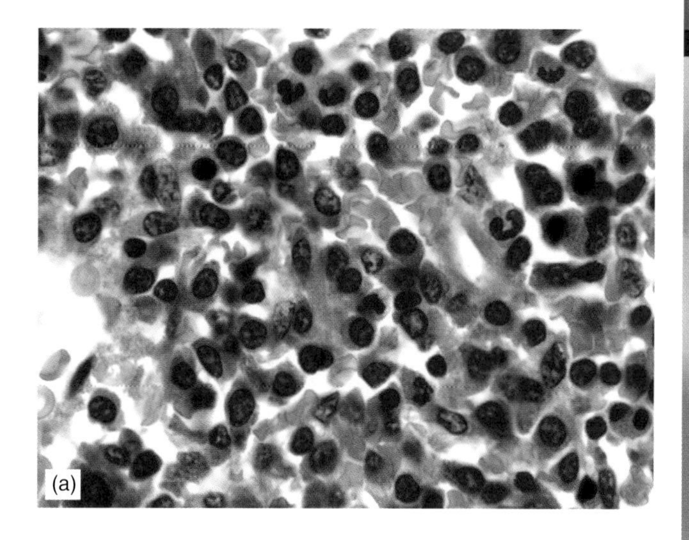

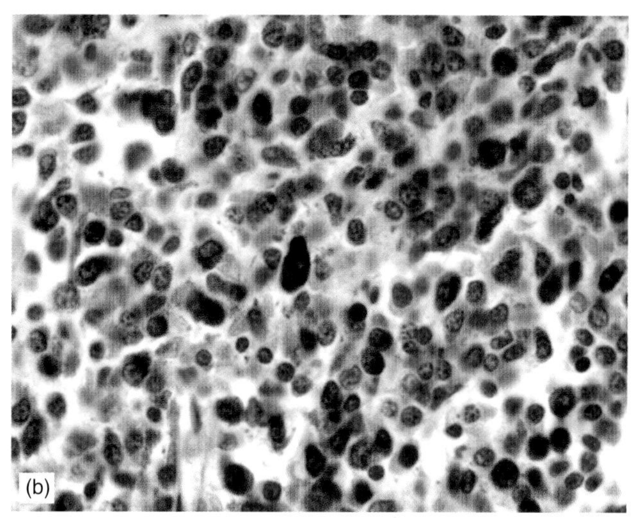

FIGURE 19.10 An early lesion of post-transplant lymphoproliferative disorder showing plasmacytic hyperplasia: (a) H&E and (b) CD138 stains.

organ transplant population are of recipient origin [140]. This may suggest the escape of host EBV-infected cells from the regulatory mechanisms of the immune system. An EBV-associated protein, LMP-1, is engaged in signaling host proteins from the tumor necrosis factor (TNF) receptor family, leading to cell growth and transformation [141]. Unlike PTLD in solid organ transplants, the overwhelming majority of the PTLD cases in the bone marrow allograft recipients are of donor origin [142].

The proliferating B-cells in PTLD are BCL-6− and MUM1+, consistent with a post-germinal center stage of B-cell differentiation [143]. Molecular studies have demonstrated amplification of *PAX5* at chromosome 9p13 region suggesting that this gene may play a role in the pathogenesis of PTLD [144].

Pathology

Early lesions consist of plasmacytic hyperplasia and IM-like disorders (Figure 19.10). These lesions mainly occur in oropharynx and lymph nodes and are characterized by preservation of the nodal sinuses and residual reactive follicles with diffuse interfollicular proliferation of plasma cells and B-immunoblasts mixed with T-cells [133, 138]. Immunoblasts are commonly positive for EBV-LMP [145].

Polymorphic PTLD leads to the effacement of nodal architecture and/or destructive extranodal tissues [133]. The lymphoid infiltrates are polymorphic and consist of a mixture of small to large cells, including plasma cells and immunoblasts, mimicking mixed small and large cell lymphoma (Figure 19.11a) [133]. Scattered atypical large cells and necrotic areas may be present [133]. Some cases may show frequent mitosis. Immunophenotypic studies reveal a mixture of B- and T-cells, but B-cells may express monotypic or polytypic Ig light chains. The immunoblasts often express EBV-LMP and EBNA [133]. Molecular studies often show Ig gene rearrangement and/or EBV genome, frequently of type A [133, 146]. These lesions usually lack mutations in oncogenes and tumor suppressor genes such as *MYC*, *RAS*, and *TP53* [147].

Monomorphic PTLD demonstrates significant architectural alteration, monomorphic features, and cellular atypia consistent with the diagnosis of lymphoma (Figure 19.11b). These lesions are mostly of B-cell origin, but the

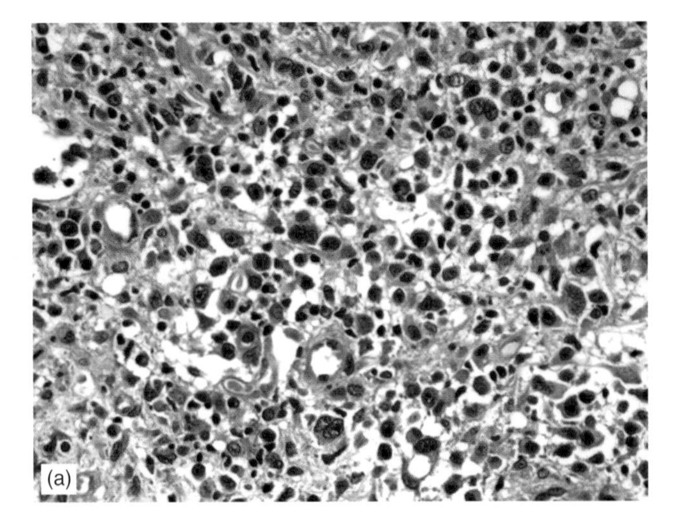

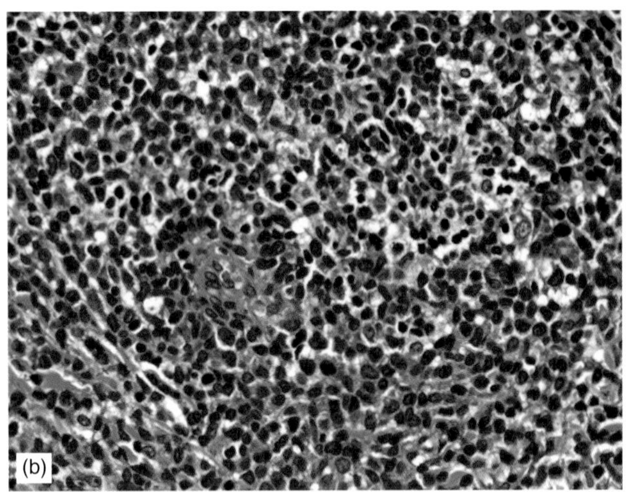

FIGURE 19.11 Post-transplant lymphoproliferative disorder. Lymph node biopsy sections demonstrating an example of polymorphic (a) and monomorphic (b) PTLD.

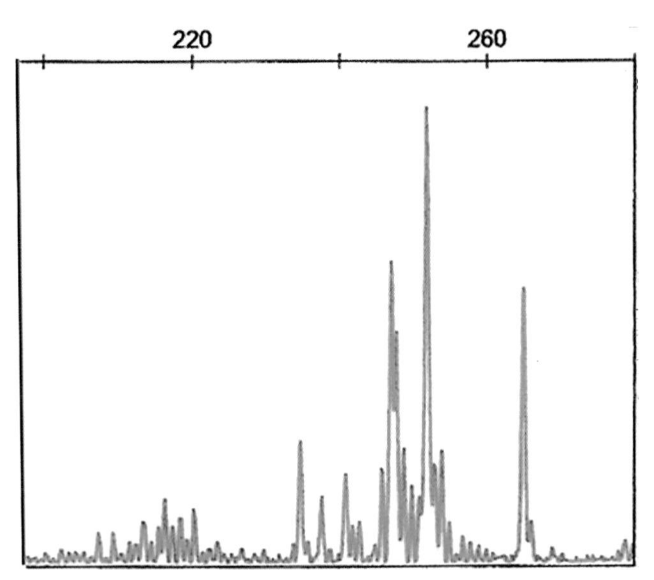

FIGURE 19.12 An example of oligoclonal post-transplant lymphoproliferative disorder by PCR analysis. The number of clonal peaks (IgM heavy chain) is too great to be considered monoclonal, but they are discrete enough to represent several clonal subsets (oligoclonal proliferation).

T-cell variants have also been reported in 4–14% of the cases [139, 148]. Monomorphic PTLD should be classified according to the WHO guidelines for the classification of B- and T-cell lymphomas (see Chapters 15–18). In some cases, the affected organs may show both polymorphic and monomorphic infiltrates in the same tissue section, and at the molecular level, they often evolve from polyclonal through oligoclonal and finally to monoclonal when a particular clone achieves predominance (Figure 19.12) [147].

Hodgkin lymphoma (HL) and HL-like PTLD are characterized by the presence of Hodgkin and Reed–Sternberg (HRS) cells or HRS-like cells. The classical HRS cells typically express CD15 and CD30 and lack CD45 expression, whereas HRS-like cells show atypical immunophenotypic features [149, 150]. The HRS-like cells are commonly EBV-positive.

Clinical Aspects

The major risk factors for the development of PTLD are immunosuppression and EBV infection. The more severe the immunosuppression, the greater the risk for PTLD [151–153]. The incidence of PTLD is significantly higher in the EBV-seronegative patients than in the EBV-seropositive ones, because of the higher risk of EBV infection in recipients who have no pre-transplant immunity to EBV [154, 155].

Other risk factors include age of recipient under the age of 25 years, fewer HLA matches, and history of pre-transplant malignancy [156, 157]. The risk of PTLD is highest in the first year of the post-transplant period [158].

The incidence of PTLD in solid organ transplants is high in intestinal and multiorgan transplants, ranging from 11% to 33% followed by 2% to 9% for lung, 2% to 6% for heart, 1% to 3% for kidney, and 1% to 2% for liver transplantations [152–154, 159]. The incidence of PTLD in bone marrow allograft recipients is about 1%, except for those who receive HLA-mismatched or T-cell-depleted bone marrow or are treated by immunosuppressive drugs for graft versus host disease. In this group of patients, the risk of PTLD is up to 20% [158].

Patients typically demonstrate. local or generalized lymphadenopathy, sometimes with graft dysfunction or other organ failures due to extranodal lymphoid infiltrate [160]. EBV-negative and late-occurring cases have a higher tendency to be monomorphic [139, 148].

The therapeutic approaches include reduction in immunosuppression, antiviral therapy, and chemotherapy and/or radiation therapy (for the treatment of monotypic PTLD and HL) [161, 162]. A significant proportion of early PTLD lesions and polymorphic forms may regress by the reduction of immunosuppression [163]. The overall prognosis is poor, particularly in monomorphic PTLD [157, 164].

Differential Diagnosis

The differential diagnosis between different categories of PTLD, particularly between polymorphic and monomorphic variants, is often difficult. The major morphologic, immunophenotypic, and molecular genetic differences between the PTLD subtypes are presented in Table 19.8.

Methotrexate-Associated Lymphoproliferative Disorder

Methotrexate-associated lymphoproliferative disorders are rare conditions seen in patients with autoimmune disorders (such as rheumatoid arthritis, psoriasis) treated with methotrexate. The lymphoproliferation may mimic polymorphous PTLD (about 14% of the cases) or a garden variety of lymphomas [133, 165, 166] (see Chapters 15–17). Partial or total regression of the lesion is seen in up to 60% of the cases in response to withdrawal of methotrexate therapy [133, 167, 168].

Lymphomatoid Granulomatosis

Lymphomatoid granulomatosis is an extranodal, angiocentric lymphoproliferative disorder consisting of large, EBV-positive B-cells in a background of polymorphous reactive cells, predominantly T-lymphocytes. It appears to be an EBV-induced disorder in an immunodeficiency setting, such as HIV infection, XLP syndrome, methotrexate therapy, allogeneic organ transplantations, or WAS [133, 169–171].

Morphology: Lymphomatoid granulomatosis is an angiocentric and angiodestructive polymorphous lymphoproliferative process characterized by the presence of a small number of large B-cells in a background of inflammatory cells. The inflammatory component is polymorphic consisting of a mixture of lymphocytes, plasma cells, immunoblasts, and histiocytes with scattered eosinophils and rare neutrophils (Figure 19.13). The large EBV-positive cells resemble immunoblasts, but bizarre large cells with multilobated nucleus or Reed–Sternberg-like cells may be present. This lymphoproliferative disorder characteristically infiltrates into the vascular structures and may cause vascular damage, fibrinoid necrosis, and ischemic changes in the surrounding tissues. There are three histological grades [172–174].

Grade I	Polymorphous lymphoid infiltrate with absent or rare large, immunoblastic, EBV-positive lymphocytes (>5/hpf).
Grade II	Polymorphous lymphoid infiltrate with moderate numbers of large, immunoblastic, EBV-positive lymphocytes (5–20/hpf). Necrosis is often present.
Grade III	Is considered a variant of DLBCL and consists of numerous large, immunoblastic, and EBV-positive lymphocytes (5–20/hpf). In some areas, these cells may appear in clusters or small sheets. Necrosis is often extensive. Large bizarre cells with multilobed nucleus and Reed–Sternberg-like cells are often present.

Immunophenotype: The large immunoblast-like cells are usually positive for CD20, CD45, and LMP1 [133]. The expression of CD79a and CD30 is variable, and CD15 is negative. The background lymphocyte population consists primarily of CD3+ T-cells. The CD4:CD8 ratio is often elevated.

Cytogenetic and molecular studies: No recurrent chromosomal abnormalities have been reported. There is evidence of clonal Ig gene rearrangement in most grade II and III cases. The clonal population may not be identical in different sites. Establishment of clonality in the grade I lesions may not be conclusive. Most of the cases show EBV infection demonstrated by EBER or other molecular markers. The EBV infection in some cases is clonal.

Clinical aspects: Lymphomatoid granulomatosis is a rare condition observed in immunodeficiency states, more often in adults and males [133, 172–174]. The most common sites of involvement in order of frequency are lung, skin, kidney, liver, and brain. Other sites such as upper respiratory and gastrointestinal tracts, spleen, and lymph nodes are occasionally affected. Clinical symptoms are related to the involved organ, with respiratory symptoms being the most common presentation. The clinical course, particularly for the grades II and III, is often aggressive. However, some patients may show a fluctuating clinical course with occasional spontaneous remission.

Senile EBV-Associated B-Cell Lymphoproliferative Disorder

Senile EBV-associated B-cell lymphoproliferative disorder is a recently described entity with a high incidence in elderly people (>60 years old) and without underlying immunodeficiencies [175, 176]. It is characterized by an EBV-positive B-cell proliferation and a polymorphic cellular composition consisting of varying numbers of centroblasts, immunoblasts, and HRS-like cells [175, 177]. Angiocentric growth pattern and areas of necrosis are often present. This condition involves both lymph nodes and extranodal tissues. Two morphologic subtypes have been described: (1) polymorphic type with a relatively favorable clinical course and (2) large cell lymphoma type with aggressive clinical course. The EBV+ HRS-like cells in this disorder are positive for CD20 and CD45. They may variably express CD30 but are negative for CD15 [176].

Lymphomatoid Papulosis

Lymphomatoid papulosis is a benign, chronic recurrent skin disorder which occurs at all ages, but the peak incidence is around 50 years of age [178, 179]. The male:female ratio is about 2:1. It is characterized by multiple spontaneously regressing papules consisting of a polymorphic infiltrate including anaplastic large lymphocytes and cells resembling HRS cells (Figure 19.14). These cells represent activated CD4+ cells and coexpress CD30 but are negative for anaplastic lymphoma kinase. In some cases, lymphomatoid papulosis after a long period (~15 years) may evolve into anaplastic large cell lymphoma, Hodgkin lymphoma, mycosis fungoides, or other T-cell malignancies [180–183].

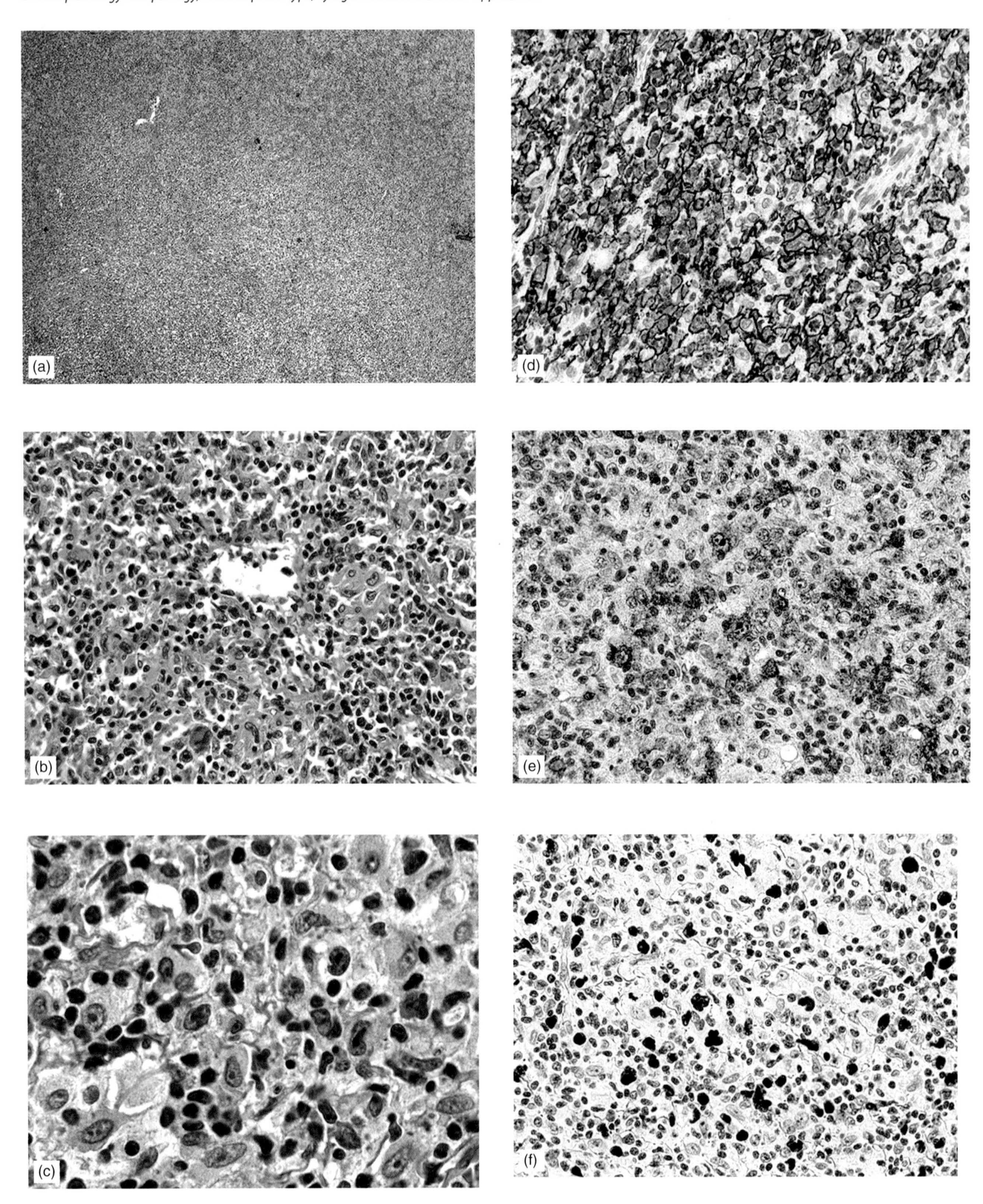

FIGURE 19.13 Lymphomatoid granulomatosis. Lung biopsy section demonstrates a dense polymorphic, angiocentric lymphoid infiltrate: (a) low power, (b) intermediate power, and (c) high power views. Numerous large cells are CD20-positive (d) and some express CD30 (e) and/or EBV-EBER (f). Courtesy of Sophie Song, MD, PhD, Department of Pathology and Laboratory Medicine, UCLA Medical Center.

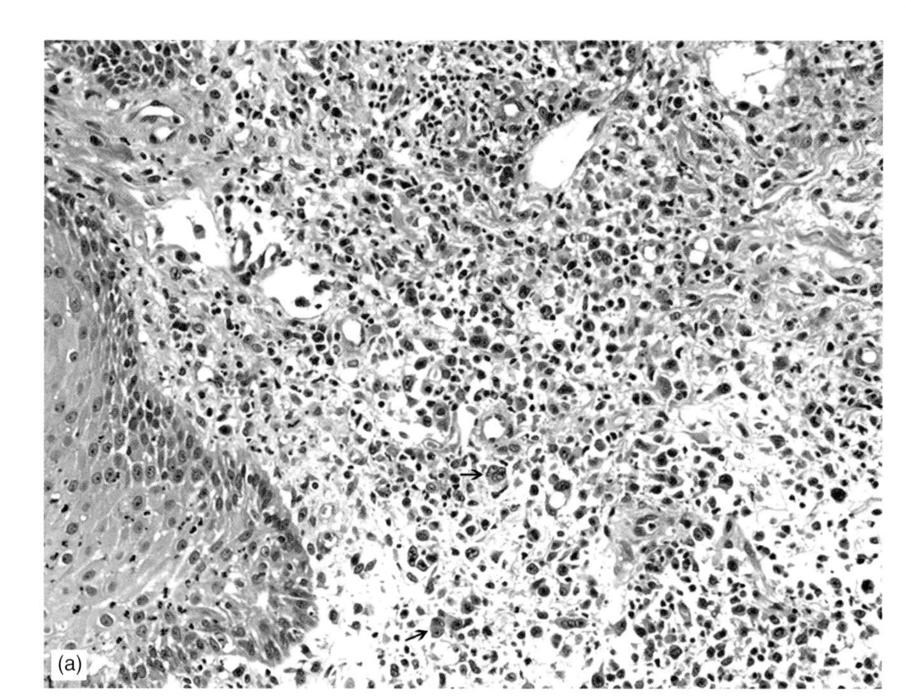

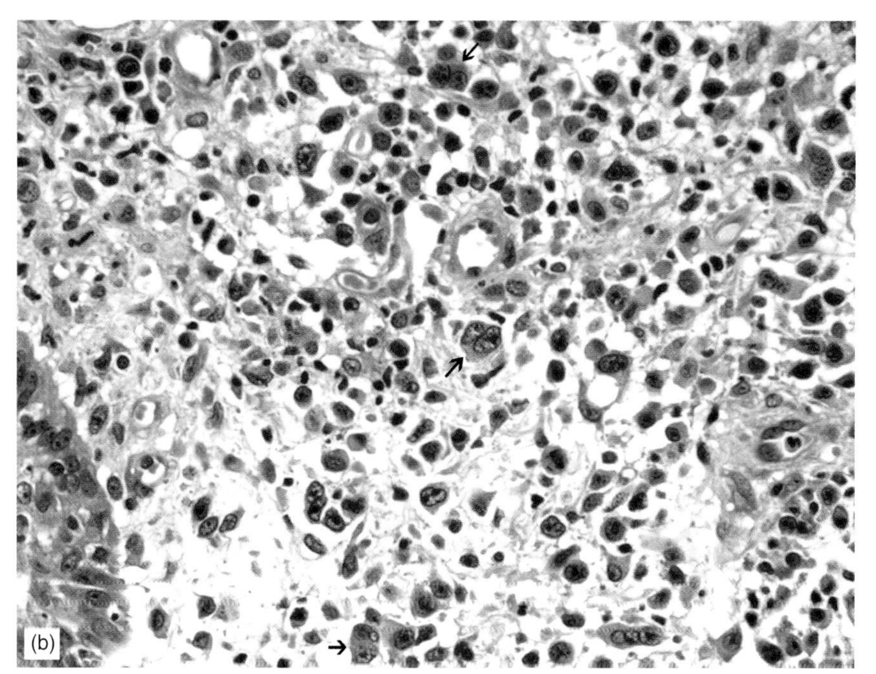

FIGURE 19.14 Lymphomatoid papulosis. Skin biopsy demonstrates a polymorphous infiltrate with atypical large lymphocytes and Reed–Sternberg-like cells: (a) low power and (b) high power views.

References

1. Naeim F. (1997). *Pathology of Bone Marrow*, 2nd ed. Williams & Wilkins, Baltimore.
2. Biggar RJ, Jaffe ES, Goedert JJ, Chaturvedi A, Pfeiffer R, Engels EA. (2006). Hodgkin lymphoma and immunodeficiency in persons with HIV/AIDS. *Blood* **108**, 3786–91.
3. Galus MA, Stern J. (2004). Extreme lymphocytopenia associated with toxic shock syndrome. *J Intern Med* **244**, 351–4.
4. Wenzel J, Montag S, Wilsmann-Theis D, Gaffal E, Bieber T, Tuting T. (2004). Successful treatment of recalcitrant Wegener's granulomatosis of the skin with tacrolimus (Prograf). *Br J Dermatol* **151**, 927–8.
5. Ten RM. (1998). Primary immunodeficiencies. *Mayo Clin Proc* **73**, 865–72.
6. Ming JE, Stiehm ER, Graham Jr. JM. (1999). Syndromes associated with immunodeficiency. *Adv Pediatr* **46**, 271–351.
7. Buckley RH, Schiff RI, Schiff SE, Markert ML, Williams LW, Harville TO, Roberts JL, Puck JM. (1997). Human severe combined immunodeficiency: Genetic, phenotypic, and functional diversity in one hundred eight infants. *J Pediatr* **130**, 378–87.
8. Puck JM, Pepper AE, Henthorn PS, Candotti F, Isakov J, Whitwam T, Conley ME, Fischer RE, Rosenblatt HM, Small TN, Buckley RH. (1997). Mutation analysis of IL2RG in human X-linked severe combined immunodeficiency. *Blood* **89**, 1968–77.
9. Bonilla F. (2007). Combined immune deficiencies. *UpToDate*
10. Noguchi M, Yi H, Rosenblatt HM, Filipovich AH, Adelstein S, Modi WS, McBride OW, Leonard WJ. (1993). Interleukin-2 receptor gamma chain mutation results in X-linked severe combined immunodeficiency in humans. *Cell* **73**(1), 147–57.

11. Hirschhorn R. (1990). Adenosine deaminase deficiency. *Immunodefic Rev* **2**, 175–98.
12. Carson DA, Carrera CJ. (1990). Immunodeficiency secondary to adenosine deaminase deficiency and purine nucleoside phosphorylation deficiency. *Semin Hematol* **27**, 260–9.
13. Derry JM, Ochs HD, Francke U. (1994). Isolation of a novel gene mutated in Wiskott–Aldrich syndrome. *Cell* **78**, 635–44.
14. Kwan SP, Hagemann TL, Radtke BE, Blaese RM, Rosen FS. (1995). Identification of mutations in the Wiskott–Aldrich syndrome gene and characterization of a polymorphic dinucleotide repeat at DXS6940, adjacent to the disease gene. *Proc Natl Acad Sci USA* **92**, 4706–10.
15. Binder V, Albert MH, Kabus M, Bertone M, Meindl A, Belohradsky BH. (2006). The genotype of the original Wiskott phenotype. *N Engl J Med* **355**, 1790–3.
16. Sullivan KE, Mullen CA, Blaese RM, Winkelstein JA. (1994). A multiinstitutional survey of the Wiskott–Aldrich syndrome. *J Pediatr* **125**(6 Pt 1), 876–85.
17. Rocca B, Bellacosa A, De Cristofaro R, Neri G, Della Ventura M, Maggiano N, Rumi C, Landolfi R. (1996). Wiskott–Aldrich syndrome: Report of an autosomal dominant variant. *Blood* **87**, 4538–43.
18. Cotelingam JD, Witebsky FG, Hsu SM, Blaese RM, Jaffe ES. (1985). Malignant lymphoma in patients with the Wiskott–Aldrich syndrome. *Cancer Invest* **3**, 515–22.
19. Ochs HD, Slichter SJ, Harker LA, Von Behrens WE, Clark RA, Wedgwood RJ. (1980). The Wiskott–Aldrich syndrome: Studies of lymphocytes, granulocytes, and platelets. *Blood* **55**, 243–52.
20. Imai K, Morio T, Zhu Y, Jin Y, Itoh S, Kajiwara M, Yata J, Mizutani S, Ochs HD, Nonoyama S. (2004). Clinical course of patients with WASP gene mutations. *Blood* **103**, 456–64.
21. MacCarthy-Morrogh L, Gaspar HB, Wang YC, Katz F, Thompson L, Layton M, Jones AM, Kinnon C. (1998). Absence of expression of the Wiskott–Aldrich syndrome protein in peripheral blood cells of Wiskott–Aldrich syndrome patients. *Clin Immunol Immunopathol* **88**, 22–7.
22. Shcherbina A. (2007). Wiskott–Aldrich syndrome. *UpToDate.*
23. Conley ME. (1999). Diagnostic guidelines – An International Consensus document. *Clin Immunol* **93**, 189.
24. Staple L, Andrews T, McDonald-McGinn D, Zackai E, Sullivan KE. (2005). Allergies in patients with chromosome 22q11.2 deletion syndrome (DiGeorge syndrome/velocardiofacial syndrome) and patients with chronic granulomatous disease. *Pediatric Allergy and Immunol* **16**, 226–30.
25. Sullivan KE. (2004). The clinical, immunological, and molecular spectrum of chromosome 22q11.2 deletion syndrome and DiGeorge syndrome. *Curr Opin Allergy and Clinical Immunol* **4**, 505–12.
26. Baldini A. (2004). DiGeorge syndrome: An update. *Curr Opin Cardiol* **19**, 201–4.
27. Baldini A. (2002). DiGeorge syndrome: The use of model organisms to dissect complex genetics. *Hum Mol Genet Special Rev Issue* **11**, 2363–9.
28. Seroogy CM. (2007). Pathogenesis, epidemiology, and clinical manifestations of DiGeorge syndrome. *UpToDate.*
29. Greenberg F, Elder FF, Haffner P, Northrup H, Ledbetter DH. (1988). Cytogenetic findings in a prospective series of patients with DiGeorge anomaly. *Am J Hum Genet* **43**, 605–11.
30. McDonald-McGinn DM, Kirschner R, Goldmuntz E, Sullivan K, Eicher P, Gerdes M, Moss E, Solot C, Wang P, Jacobs I, Handler S, Knightly C, Heher K, Wilson M, Ming JE, Grace K, Driscoll D, Pasquariello P, Randall P, Larossa D, Emanuel BS, Zackai EH. (1999). The Philadelphia story: The 22q11.2 deletion: Report on 250 patients. *Genet Couns* **10**, 11–24.
31. Ochs HD, Smith CI. (1996). X-linked agammaglobulinemia. A clinical and molecular analysis. *Medicine (Baltimore)* **75**, 287–99.
32. Winkelstein JA, Marino MC, Lederman HM, Jones SM, Sullivan K, Burks AW, Conley ME, Cunningham-Rundles C, Ochs HD. (2006). X-linked agammaglobulinemia: Report on a United States registry of 201 patients. *Medicine (Baltimore)* **85**, 193–202.
33. Minegishi Y, Coustan-Smith E, Wang YH, Cooper MD, Campana D, Conley ME. (1998). Mutations in the human lambda5/14.1 gene result in B cell deficiency and agammaglobulinemia. *J Exp Med* **187**, 71–7.
34. Yel L, Minegishi Y, Coustan-Smith E, Buckley RH, Trubel H, Pachman LM, Kitchingman GR, Campana D, Rohrer J, Conley ME. (1996). Mutations in the mu heavy-chain gene in patients with agammaglobulinemia. *N Engl J Med* **335**, 1486–93.
35. Kwan SP, Terwilliger J, Parmley R, Raghu G, Sandkuyl LA, Ott J, Ochs H, Wedgwood R, Rosen F. (1990). Identification of a closely linked DNA marker, DXS178, to further refine the X-linked agammaglobulinemia locus. *Genomics* **6**, 238–42.
36. Lee AH, Levinson AI, Schumacher Jr. HR. (1993). Hypogammaglobulinemia and rheumatic disease. *Semin Arthritis Rheum* **22**, 252–64.
37. Saffran DC, Parolini O, Fitch-Hilgenberg ME, Rawlings DJ, Afar DE, Witte ON, Conley ME. (1994). Brief report: A point mutation in the SH2 domain of Bruton's tyrosine kinase in atypical X-linked agammaglobulinemia. *N Engl J Med* **330**, 1488–91.
38. Minegishi Y, Coustan-Smith E, Rapalus L, Ersoy F, Campana D, Conley ME. (1999). Mutations in Igalpha (CD79a) result in a complete block in B-cell development. *J Clin Invest* **104**, 1115–21.
39. Minegishi Y, Rohrer J, Coustan-Smith E, Lederman HM, Pappu R, Campana D, Chan AC, Conley ME. (1999). An essential role for BLNK in human B cell development. *Science* **286**, 1954–7.
40. Colson AE, Sax PE. (2007). Primary HIV-1 infection: Diagnosis and treatment. *UpToDate.*
41. Essex M. (1998). State of the HIV pandemic. *J Hum Virol* **1**, 427–9.
42. Sankale JL, Mboup S, Essex ME, Kanki PJ. (1998). Genetic characterization of viral quasispecies in blood and cervical secretions of HIV-1- and HIV-2-infected women. *AIDS Res Hum Retroviruses* **14**, 1473–81.
43. Heeney JL, Dalgleish AG, Weiss RA. (2006). Origins of HIV and the evolution of resistance to AIDS. *Science* **313**, 462–6.
44. Bartlett JG. (2007). The stages and natural history of HIV infection. *UpToDate.*
45. Nowak MA, Lloyd AL, Vasquez GM, Wiltrout TA, Wahl LM, Bischofberger N, Williams J, Kinter A, Fauci AS, Hirsch VM, Lifson JD. (1997). Viral dynamics of primary viremia and antiretroviral therapy in simian immunodeficiency virus infection. *J Virol* **71**, 7518–25.
46. Trkola A, Dragic T, Arthos J, Binley JM, Olson WC, Allaway GP, Cheng-Mayer C, Robinson J, Maddon PJ, Moore JP. (1996). CD4-dependent, antibody-sensitive interactions between HIV-1 and its coreceptor CCR-5. *Nature* **384**, 184–7.
47. Liu R, Paxton WA, Choe S, Ceradini D, Martin SR, Horuk R, MacDonald ME, Stuhlmann H, Koup RA, Landau NR. (1996). Homozygous defect in HIV-1 coreceptor accounts for resistance of some multiply-exposed individuals to HIV-1 infection. *Cell* **86**, 367–77.
48. Iyer RK, Bando JM, Lu KV, Kim PS, Gregg JP, Grody WW. (2001). A multiethnic study of Δ32ccr5 and ccr2b-V64I allele distribution in four Los Angeles populations. *Diagn Molec Pathol* **10**, 105–10.
49. Ameisen JC, Capron A. (1991). Cell dysfunction and depletion in AIDS: The programmed cell death hypothesis. *Immunol Today* **12**, 102–5.
50. Ratner L. (1989). Human immunodeficiency virus-associated autoimmune thrombocytopenic purpura: A review. *Am J Med* **86**, 194–8.
51. Jackson JB, Kwok SY, Sninsky JJ, Hopsicker JS, Sannerud KJ, Rhame FS, Henry K, Simpson M, Balfour Jr. HH. (1990). Human immunodeficiency virus type 1 detected in all seropositive symptomatic and asymptomatic individuals. *J Clin Microbiol* **28**, 16–19.
52. Simmonds P, Lainson FA, Cuthbert R, Steel CM, Peutherer JF, Ludlam CA. (1998). HIV antigen and antibody detection: Variable responses to infection in the Edinburgh haemophiliac cohort. *Br Med J (Clin Res Ed)* **296**, 593–8.
53. Sheppard HW, Busch MP, Louie PH, Madej R, Rodgers GC. (1993). HIV-1 PCR and isolation in seroconverting and seronegative

homosexual men: Absence of long-term immunosilent infection. *J Acquir Immune Defic Syndr* **6**, 1339–46.
54. Little SJ, McLean AR, Spina CA, Richman DD, Havlir DV. (1999). Viral dynamics of acute HIV-1 infection. *J Exp Med* **190**, 841–50.
55. Fiebig EW, Heldebrant CM, Smith RI, Conrad AJ, Delwart EL, Busch MP. (2005). Intermittent low-level viremia in very early primary HIV-1 infection. *J Acquir Immune Defic Syndr* **39**, 133–7.
56. Mehta KU, Gascon P, Tannir N, Lombardo J, Robboy SJ. (1989). Impaired bone marrow in AIDS. *N J Med* **86**, 623–7.
57. Sun NC, Shapshak P, Lachant NA, Hsu MY, Sieger L, Schmid P, Beall G, Imagawa DT. (1989). Bone marrow examination in patients with AIDS and AIDS-related complex (ARC), morphologic and *in situ* hybridization studies. *Am J Clin Pathol* **92**, 589–94.
58. Gordon S, Lee S. (1994). Naked megakaryocyte nuclei in bone marrows of patients with acquired immunodeficiency syndrome: A somewhat specific finding. *Mod Pathol* **7**, 166–8.
59. von Laer D, Modrau B, Dietrich M, Kern P. (1994). Leukocyte differentiation antigens in the bone marrow of patients with HIV-related disease. *Acta Haematol* **92**, 197–203.
60. Ferry JA, Harris NL. (1997). *Atlas of Lymphoid Hyperplasia and Lymphoma*. W.B. Saunders, Philadelphia.
61. Quinn TC, Wawer MJ, Sewankambo N, Serwadda D, Li C, Wabwire-Mangen F, Meehan MO, Lutalo T, Gray RH. (2000). Viral load and heterosexual transmission of human immunodeficiency virus type 1, Rakai Project Study Group. *N Engl J Med* **342**, 921–9.
62. Dorak MT, Tang J, Penman-Aguilar A, Westfall AO, Zulu I, Lobashevsky ES, Kancheya NG, Schaen MM, Allen SA, Kaslow RA. (2004). Transmission of HIV-1 and HLA-B allele-sharing within serodiscordant heterosexual Zambian couples. *Lancet* **363**, 2137–9.
63. Pantaleo G, Graziosi C, Demarest JF, Butini L, Montroni M, Fox CH, Orenstein JM, Kotler DP, Fauci AS. (1993). HIV infection is active and progressive in lymphoid tissue during the clinically latent stage of disease. *Nature* **362**, 355–8.
64. Smith DK, Neal JJ, Holmberg SD. (1993). Unexplained opportunistic infections and CD4+ T-lymphocytopenia without HIV infection. An investigation of cases in the United States. The Centers for Disease Control Idiopathic CD4+ T-lymphocytopenia Task Force. *N Engl J Med* **328**, 373–9.
65. Moore JP, Ho DD. (1992). HIV-negative AIDS. *Lancet* **340**, 475.
66. Spira TJ, Jones BM, Nicholson JK, Lal RB, Rowe T, Mawle AC, Lauter CB, Shulman JA, Monson RA. (1993). Idiopathic CD4+ T-lymphocytopenia – an analysis of five patients with unexplained opportunistic infections. *N Engl J Med* **328**(6), 386–92.
67. Smith DK, Neal JJ, Holmberg SD. (1993). Idiopathic CD4+ T-lymphocytopenia – immunodeficiency without evidence of HIV infection *N Engl J Med* **328**, 380–5.
68. Baehner RL. (2007). Approach to the patient with lymphocytosis. *UpToDate*.
69. Aronson MD, Auwaerter PG. (2007). Infectious mononucleosis in adults and adolescents. *UpToDate*.
70. Evans AS. (1974). The history of infectious mononucleosis. *Am J Med Sci* **267**, 189–95.
71. Jabs WJ, Paulsen M, Wagner HJ, Kirchner H, Kluter H. (1999). Analysis of Epstein–Barr virus (EBV) receptor CD21 on peripheral B lymphocytes of long-term EBV− adults. *Clin Exp Immunol* **116**, 468–73.
72. Issekutz TB, Chin W, Hay JB. (1982). The characterization of lymphocytes migrating through chronically inflamed tissues. *Immunology* **46**, 59–66.
73. Macallan DC, Wallace DL, Irvine AJ, Asquith B, Worth A, Ghattas H, Zhang Y, Griffin GE, Tough DF, Beverley PC. (2003). Measurement and modeling of human T cell kinetics. *Eur J Immunol* **33**, 2316–26.
74. Sandalova E, Wei CH, Masucci MG, Levitsky V. (2006). T-cell receptor triggering differentially regulates bim expression in human lymphocytes from healthy individuals and patients with infectious mononucleosis. *Hum Immunol* **67**, 958–65.
75. Tomkinson BE, Wagner DK, Nelson DL, Sullivan JL. (1987). Activated lymphocytes during acute Epstein-Barr virus infection. *J Immunol* **139**, 3802–7.
76. Giuliano VJ, Jasin HE, Ziff M. (1974). The nature of the atypical lymphocyte in infectious mononucleosis. *Clin Immunol Immunopathol* **3**, 90–8.
77. Williams H, Crawford DH. (2006). Epstein–Barr virus: The impact of scientific advances on clinical practice. *Blood* **107**, 862–9.
78. Linderholm M, Boman J, Juto P, Linde A. (1994). Comparative evaluation of nine kits for rapid diagnosis of infectious mononucleosis and Epstein–Barr virus-specific serology. *J Clin Microbiol* **32**, 259–61.
79. Bruu AL, Hjetland R, Holter E, Mortensen L, Natas O, Petterson W, Skar AG, Skarpaas T, Tjade T, Asjo B. (2000). Evaluation of 12 commercial tests for detection of Epstein–Barr virus-specific and heterophile antibodies. *Clin Diagn Lab Immunol* **7**, 451–6.
80. Weinberger B, Plentz A, Weinberger KM, Hahn J, Holler E, Jilg W. (2004). Quantitation of Epstein–Barr virus mRNA using reverse transcription and real-time PCR. *J Med Virol* **74**, 612–18.
81. Aronson MD, Komaroff AL, Pass TM, Ervin CT, Branch WT. (1982). Heterophil antibody in adults with sore throat: Frequency and clinical presentation. *Ann Intern Med* **96**, 505–8.
82. Hoagland RJ. (1975). Infectious mononucleosis. *Prim Care* **2**, 295–307.
83. Rea TD, Russo JE, Katon W, Ashley RL, Buchwald DS. (2001). Prospective study of the natural history of infectious mononucleosis caused by Epstein–Barr virus. *J Am Board Fam Pract* **14**, 234–42.
84. Long CM, Kerschner JE. (2001). Parotid mass: Epstein–Barr virus and facial paralysis. *Int J Pediatr Otorhinolaryngol* **59**, 143–6.
85. Schellinger PD, Sommer C, Leithauser F, Schwab S, Storch-Hagenlocher B, Hacke W, Kiessling M. (1999). Epstein–Barr virus meningoencephalitis with a lymphoma-like response in an immunocompetent host. *Ann Neurol* **45**, 659–62.
86. Devereaux CE, Bemiller T, Brann O. (1999). Ascites and severe hepatitis complicating Epstein–Barr infection. *Am J Gastroenterol* **94**, 236–40.
87. Lindemann TL, Greene JS. (2005). Persistent cervical lymphadenopathy in an adolescent with Epstein–Barr induced hemophagocytic syndrome: Manifestations of a rare but often fatal disease. *Int J Pediatr Otorhinolaryngol* **69**, 1011–14.
88. Hauptmann S, Meru N, Schewe C, Jung A, Hiepe F, Burmester G, Niedobitek G, Buttgereit F. (2001). Fatal atypical T-cell proliferation associated with Epstein–Barr virus infection. *Br J Haematol* **112**, 377–80.
89. Tynell E, Aurelius E, Brandell A, Julander I, Wood M, Yao QY, Rickinson A, Akerlund B, Andersson J. (1996). Acyclovir and prednisolone treatment of acute infectious mononucleosis: A multicenter, double-blind, placebo-controlled study. *J Infect Dis* **174**, 324–31.
90. Klemola E, Von Essen R, Henle G, Henle W. (1970). Infectious-mononucleosis-like disease with negative heterophil agglutination test. Clinical features in relation to Epstein–Barr virus and cytomegalovirus antibodies. *J Infect Dis* **121**, 608–14.
91. Horwitz CA, Henle W, Henle G, Polesky H, Balfour Jr. HH, Siem RA, Borken S, Ward PC. (1997). Heterophil-negative infectious mononucleosis and mononucleosis-like illnesses. Laboratory confirmation of 43 cases. *Am J Med* **63**, 947–57.
92. Steeper TA, Horwitz CA, Ablashi DV, Salahuddin SZ, Saxinger C, Saltzman R, Schwartz B. (1990). The spectrum of clinical and laboratory findings resulting from human herpesvirus-6 (HHV-6) in patients with mononucleosis-like illnesses not resulting from Epstein–Barr virus or cytomegalovirus. *Am J Clin Pathol* **93**, 776–83.
93. Lupton JR, Figueroa P, Tamjidi P, Berberian BJ, Sulica VI. (1999). An infectious mononucleosis-like syndrome induced by minocycline: A third pattern of adverse drug reaction. *Cutis* **64**, 91–6.
94. Brown M, Schubert T. (1986). Phenytoin hypersensitivity hepatitis and mononucleosis syndrome. *J Clin Gastroenterol* **8**, 469–77.
95. Sayos J, Wu C, Morra M, Wang N, Zhang X, Allen D, van Schaik S, Notarangelo L, Geha R, Roncarolo MG, Oettgen H, De Vries JE, Aversa G, Terhorst C. (1998). The X-linked lymphoproliferative-disease gene product SAP regulates signals induced through the co-receptor SLAM. *Nature* **395**, 462–9.

96. Nichols KE, Harkin DP, Levitz S, Krainer M, Kolquist KA, Genovese C, Bernard A, Ferguson M, Zuo L, Snyder E, Buckler AJ, Wise C, Ashley J, Lovett M, Valentine MB, Look AT, Gerald W, Housman DE, Haber DA. (1998). Inactivating mutations in an SH2 domain-encoding gene in X-linked lymphoproliferative syndrome. *Proc Natl Acad Sci USA* **95**, 13765–70.
97. Rigaud S, Fondaneche MC, Lambert N, Pasquier B, Mateo V, Soulas P, Galicier L, Le Deist F, Rieux-Laucat F, Revy P, Fischer A, de Saint Basile G, Latour S. (2006). XIAP deficiency in humans causes an X-linked lymphoproliferative syndrome. *Nature* **444**, 110–14.
98. Gaspar HB, Sharifi R, Gilmour KC, Thrasher AJ. (2002). X-linked lymphoproliferative disease: Clinical, diagnostic and molecular perspective. *Br J Haematol* **119**, 585–95.
99. Henning G, Kraft MS, Derfuss T, Pirzer R, de Saint-Basile G, Aversa G, Fleckenstein B, Meinl E. (2001). Signaling lymphocytic activation molecule (SLAM) regulates T cellular cytotoxicity. *Eur J Immunol* **31**, 2741–50.
100. Poy F, Yaffe MB, Sayos J, Saxena K, Morra M, Sumegi J, Cantley LC, Terhorst C, Eck MJ. (1999). Crystal structures of the XLP protein SAP reveal a class of SH2 domains with extended, phosphotyrosine-independent sequence recognition. *Mol Cell* **4**, 555–61.
101. Tangye SG, Cherwinski H, Lanier LL, Phillips JH. (2003). Functional requirements for interactions between CD84 and Src homology 2 domain-containing proteins and their contribution to human T cell activation. *J Immunol* **171**, 2485–95.
102. Seemayer TA, Gross TG, Egeler RM, Pirruccello SJ, Davis JR, Kelly CM, Okano M, Lanyi A, Sumegi J. (1995). X-linked lymphoproliferative disease: Twenty-five years after the discovery. *Pediatr Res* **38**, 471–8.
103. Harrington DS, Weisenburger DD, Purtilo DT. (1987). Malignant lymphoma in the X-linked lymphoproliferative syndrome. *Cancer* **59**, 1419–29.
104. Nistala K, Gilmour KC, Cranston T, Davies EG, Goldblatt D, Gaspar HB, Jones AM. (2001). X-linked lymphoproliferative disease: Three atypical cases. *Clin Exp Immunol* **126**, 126–30.
105. Purtilo DT, Sakamoto K, Barnabei V, Seeley J, Bechtold T, Rogers G, Yetz J, Harada S. (1982). Epstein–Barr virus-induced diseases in boys with the X-linked lymphoproliferative syndrome (XLP): Update on studies of the registry. *Am J Med* **73**, 49–56.
106. Purtilo DT, Grierson HL, Davis JR, Okano M. (1991). The X-linked lymphoproliferative disease: From autopsy toward cloning the gene 1975–1990. *Pediatr Pathol* **11**, 685–710.
107. Okano M, Bashir RM, Davis JR, Purtilo DT. (1991). Detection of primary Epstein–Barr virus infection in a patient with X-linked lymphoproliferative disease receiving immunoglobulin prophylaxis. *Am J Hematol* **36**, 294–6.
108. Henter JI, Arico M, Egeler RM, Elinder G, Favara BE, Filipovich AH, Gadner H, Imashuku S, Janka-Schaub G, Komp D, Ladisch S, Webb D. (1997). HLH-94: A treatment protocol for hemophagocytic lymphohistiocytosis. HLH study Group of the Histiocyte Society. *Med Pediatr Oncol* **28**, 342–7.
109. Teggatz JR, Parkin J, Peterson L. (1987). Transient atypical lymphocytosis in patients with emergency medical conditions. *Arch Pathol Lab Med* **111**, 712–14.
110. Pinkerton PH, McLellan BA, Quantz MC, Robinson JB. (1989). Acute lymphocytosis after trauma – early recognition of the high-risk patient? *J Trauma* **29**, 749–51.
111. Groom DA, Kunkel LA, Brynes RK, Parker JW, Johnson CS, Endres D. (1990). Transient stress lymphocytosis during crisis of sickle cell anemia and emergency trauma and medical conditions. An immunophenotyping study. *Arch Pathol Lab Med* **114**, 570–6.
112. Gordon DS, Jones BM, Browning SW, Spira TJ, Lawrence DN. (1982). Persistent polyclonal lymphocytosis of B lymphocytes. *N Engl J Med* **307**, 232–6.
113. Machii T, Yamaguchi M, Inoue R, Tokumine Y, Kuratsune H, Nagai H, Fukuda S, Furuyama K, Yamada O, Yahata Y, Kitani T. (1997). Polyclonal B-cell lymphocytosis with features resembling hairy cell leukemia-Japanese variant. *Blood* **89**, 2008–14.
114. Salcedo I, Campos-Caro A, Sampalo A, Reales E, Brieva JA. (2002). Persistent polyclonal B lymphocytosis: An expansion of cells showing IgVH gene mutations and phenotypic features of normal lymphocytes from the CD27+ marginal zone B-cell compartment. *Br J Haematol* **116**, 662–6.
115. Mossafa H, Tapia S, Flandrin G, Troussard X. (2004). Groupe Francais d'Hematologie Cellulaire (GFHC). Chromosomal instability and ATR amplification gene in patients with persistent and polyclonal B-cell lymphocytosis (PPBL). *Leuk Lymphoma* **45**, 1401–6.
116. Carr R, Fishlock K, Matutes E. (1997). Persistent polyclonal B-cell lymphocytosis in identical twins. *Br J Haematol* **96**, 272–4.
117. Delage R, Jacques L, Massinga-Loembe M, Poulin J, Bilodeau D, Mignault C, Leblond PF, Darveau A. (2001). Persistent polyclonal B-cell lymphocytosis: Further evidence for a genetic disorder associated with B-cell abnormalities. *Br J Haematol* **114**, 666–70.
118. Delage R, Roy J, Jacques L, Darveau A. (1998). All patients with persistent polyclonal B cell lymphocytosis present Bcl-2/Ig gene rearrangements. *Leuk Lymphoma* **31**, 567–74.
119. Peterson LC, Kueck B, Arthur DC, Dedeker K, Brunning RD. (1988). Systemic polyclonal immunoblastic proliferations. *Cancer* **61**, 1350–8.
120. Poje EJ, Soori GS, Weisenburger DD. (1992). Systemic polyclonal B-immunoblastic proliferation with marked peripheral blood and bone marrow plasmacytosis. *Am J Clin Pathol* **98**, 222–6.
121. Kojima M, Motoori T, Hosomura Y, Tanaka H, Sakata N, Masawa N. (2006). Atypical lymphoplasmacytic and immunoblastic proliferation from rheumatoid arthritis: A case report. *Pathol Res Pract* **202**, 51–4.
122. Navone R, Valpreda M, Pich A. (1985). Lymphoid nodules and nodular lymphoid hyperplasia in bone marrow biopsies. *Acta Haematol* **74**, 19–22.
123. Maeda K, Hyun BH, Rebuck JW. (1977). Lymphoid follicles in bone marrow aspirates. *Am J Clin Pathol* **67**, 41–8.
124. Rywlin AM, Ortega RS, Dominguez CJ. (1974). Lymphoid nodules of bone marrow: Normal and abnormal. *Blood* **43**, 389–400.
125. Douglas VK, Gordon LI, Goolsby CL, White CA, Peterson LC. (1999). Lymphoid aggregates in bone marrow mimic residual lymphoma after rituximab therapy for non-Hodgkin lymphoma. *Am J Clin Pathol* **112**, 844–53.
126. Engels K, Oeschger S, Hansmann ML, Hillebrand M, Kriener S. (2001). Bone marrow trephines containing lymphoid aggregates from patients with rheumatoid and other autoimmune disorders frequently show clonal B-cell infiltrates. *Hum Pathol* **Jun 7**.
127. Magalhaes SM, Duarte FB, Vassallo J, Costa SC, Lorand-Metze I. (2001). Multiple lymphoid nodules in bone marrow biopsy in immunocompetent patient with cytomegalovirus infection: An immunohistochemical analysis. *Rev Soc Bras Med Trop* **34**, 365–8.
128. Magalhaes SM, Filho FD, Vassallo J, Pinheiro MP, Metze K, Lorand-Metze I. (2002). Bone marrow lymphoid aggregates in myelodysplastic syndromes: Incidence, immunomorphological characteristics and correlation with clinical features and survival. *Leuk Res* **26**, 525–30.
129. Thiele J, Zirbes TK, Kvasnicka HM, Fischer R. (1999). Focal lymphoid aggregates (nodules) in bone marrow biopsies: Differentiation between benign hyperplasia and malignant lymphoma – a practical guideline. *J Clin Pathol* **52**, 294–300.
130. West RB, Warnke RA, Natkunam Y. (2002). The usefulness of immunohistochemistry in the diagnosis of follicular lymphoma in bone marrow biopsy specimens. *Am J Clin Pathol* **117**, 636–43.
131. Gandhi AM, Ben-Ezra JM. (2004). Do Bcl-2 and survivin help distinguish benign from malignant B-cell lymphoid aggregates in bone marrow biopsies? *J Clin Lab Anal* **18**, 285–8.
132. Fakan F, Skalova A, Kuntscherova J. (1996). Expression of bcl-2 protein in distinguishing benign from malignant lymphoid aggregates in bone marrow biopsies. *Gen Diagn Pathol* **141**, 359–63.

133. Jaffe ES, Harris NL, Stein H, Vardiman JW. (2001). *Pathology and Genetics. Tumors of Haematopoietic and Lymphoid Tissues.* IARC Press, Lyon.
134. Chadburn A, Suciu-Foca N, Cesarman E, Reed E, Michler RE, Knowles DM. (1995). Post-transplantation lymphoproliferative disorders arising in solid organ transplant recipients are usually of recipient origin. *Am J Pathol* **147**, 1862–70.
135. Hanto DW. (1995). Classification of Epstein–Barr virus-associated posttransplant lymphoproliferative diseases: Implications for understanding their pathogenesis and developing rational treatment strategies. *Annu Rev Med* **46**, 381–94.
136. Randhawa PS, Jaffe R, Demetris AJ, Nalesnik M, Starzl TE, Chen YY, Weiss LM. (1992). Expression of Epstein–Barr virus-encoded small RNA (by the EBER-1 gene) in liver specimens from transplant recipients with post-transplantation lymphoproliferative disease. *N Engl J Med* **327**, 1710–14.
137. Kotton CN, Fishman JA. (2005). Viral infection in the renal transplant recipient. *J Am Soc Nephrol* **16**, 1758–74.
138. Ferry JA, Jacobson JO, Conti D, Delmonico F, Harris NL. (1989). Lymphoproliferative disorders and hematologic malignancies following organ transplantation. *Mod Pathol* **2**, 583–92.
139. Leblond V, Davi F, Charlotte F, Dorent R, Bitker MO, Sutton L, Gandjbakhch I, Binet JL, Raphael M. (1998). Posttransplant lymphoproliferative disorders not associated with Epstein–Barr virus: A distinct entity? *J Clin Oncol* **16**, 2052–9.
140. Weissmann DJ, Ferry JA, Harris NL, Louis DN, Delmonico F, Spiro I. (1995). Posttransplantation lymphoproliferative disorders in solid organ recipients are predominantly aggressive tumors of host origin. *Am J Clin Pathol* **103**, 748–55.
141. Mosialos G, Birkenbach M, Yalamanchili R, VanArsdale T, Ware C, Kieff E. (1995). The Epstein–Barr virus transforming protein LMP1 engages signaling proteins for the tumor necrosis factor receptor family. *Cell* **80**, 389–99.
142. Zutter MM, Martin PJ, Sale GE, Shulman HM, Fisher L, Thomas ED, Durnam DM. (1988). Epstein–Barr virus lymphoproliferation after bone marrow transplantation. *Blood* **72**, 520–9.
143. Capello D, Cerri M, Muti G, Berra E, Oreste P, Deambrogi C, Rossi D, Dotti G, Conconi A, Vigano M, Magrini U, Ippoliti G, Morra E, Gloghini A, Rambaldi A, Paulli M, Carbone A, Gaidano G. (2003). Molecular histogenesis of posttransplantation lymphoproliferative disorders. *Blood* **102**, 3775–85.
144. Rinaldi A, Kwee I, Poretti G, Mensah A, Pruneri G, Capello D, Rossi D, Zucca E, Ponzoni M, Catapano C, Tibiletti MG, Paulli M, Gaidano G, Bertoni F. (2006). Comparative genome-wide profiling of post-transplant lymphoproliferative disorders and diffuse large B-cell lymphomas. *Br J Haematol* **134**, 27–36.
145. Lones MA, Mishalani S, Shintaku IP, Weiss LM, Nichols WS, Said JW. (1995). Changes in tonsils and adenoids in children with posttransplant lymphoproliferative disorder: Report of three cases with early involvement of Waldeyer's ring. *Hum Pathol* **26**, 525–30.
146. Frank D, Cesarman E, Liu YF, Michler RE, Knowles DM. (1995). *Am J Pathol* **147**, 1862–70.
147. Knowles DM, Cesarman E, Chadburn A, Frizzera G, Chen J, Rose EA, Michler RE. (1995). Correlative morphologic and molecular genetic analysis demonstrates three distinct categories of posttransplantation lymphoproliferative disorders. *Blood* **85**, 552–65.
148. Nelson BP, Nalesnik MA, Bahler DW, Locker J, Fung JJ, Swerdlow SH. (2000). Epstein–Barr virus-negative post-transplant lymphoproliferative disorders: A distinct entity? *Am J Surg Pathol* **24**, 375–85.
149. Rowlings PA, Curtis RE, Passweg JR, Deeg HJ, Socie G, Travis LB, Kingma DW, Jaffe ES, Sobocinski KA, Horowitz MM. (1999). Increased incidence of Hodgkin's disease after allogeneic bone marrow transplantation. *J Clin Oncol* **17**, 3122–7.
150. Nalesnik MA, Randhawa P, Demetris AJ, Casavilla A, Fung JJ, Locker J. (1993). Lymphoma resembling Hodgkin disease after posttransplant lymphoproliferative disorder in a liver transplant recipient. *Cancer* **72**, 2568–73.
151. Penn I. (1990). Cancers complicating organ transplantation. *N Engl J Med* **323**, 1767–9.
152. Opelz G, Henderson R. (1993). Incidence of non-Hodgkin lymphoma in kidney and heart transplant recipients. *Lancet* **34**, 1514–16.
153. Armitage JM, Kormos RL, Stuart RS, Fricker FJ, Griffith BP, Nalesnik M, Hardesty RL, Dummer JS. (1991). Posttransplant lymphoproliferative disease in thoracic organ transplant patients: Ten years of cyclosporine-based immunosuppression. *J Heart Lung Transplant* **10**, 877–86.
154. Aris RM, Maia DM, Neuringer IP, Gott K, Kiley S, Gertis K, Handy J. (1996). Post transplantation lymphoproliferative disorder in the Epstein–Barr virus-naive lung transplant recipient. *Am J Respir Crit Care Med* **154**(6 Pt 1), 1712–17.
155. Shahinian VB, Muirhead N, Jevnikar AM, Leckie SH, Khakhar AK, Luke PP, Rizkalla KS, Hollomby DJ, House AA. (2003). Epstein–Barr virus seronegativity is a risk factor for late-onset posttransplant lymphoroliferative disorder in adult renal allograft recipients. *Transplantation* **75**, 851–6.
156. Caillard S, Dharnidharka V, Agodoa L, Bohen E, Abbott K. (2005). Posttransplant lymphoproliferative disorders after renal transplantation in the United States in era of modern immunosuppression. *Transplantation* **80**, 1233–43.
157. Friedberg JW, Jessup, Brennan, DC. (2007). Lymphoproliferative disorders following solid organ transplantation. *UpToDate*
158. Curtis RE, Travis LB, Rowlings PA, Socie G, Kingma DW, Banks PM, Jaffe ES, Sale GE, Horowitz MM, Witherspoon RP, Shriner DA, Weisdorf DJ, Kolb HJ, Sullivan KM, Sobocinski KA, Gale RP, Hoover RN, Fraumeni Jr. JF, Deeg HJ. (1999). Risk of lymphoproliferative disorders after bone marrow transplantation: A multi-institutional study. *Blood* **94**, 2208–16.
159. Cockfield SM. (2001). Identifying the patient at risk for post-transplant lymphoproliferative disorder. *Transpl Infect Dis* **3**, 70–8.
160. Kasiske BL, Vazquez MA, Harmon WE, Brown RS, Danovitch GM, Gaston RS, Roth D, Scandling JD, Singer GG. (2000). Recommendations for the outpatient surveillance of renal transplant recipients, American Society of Transplantation. *J Am Soc Nephrol* **15**(Suppl), S1–86.
161. Allen U, Hebert D, Moore D, Dror Y, Wasfy S. (2001). Canadian PTLD Survey Group – 1998. Epstein–Barr virus-related post-transplant lymphoproliferative disease in solid organ transplant recipients, 1988–97: A Canadian multi-centre experience. *Pediatr Transplant* **5**, 198–203.
162. Tsai DE, Hardy CL, Tomaszewski JE, Kotloff RM, Oltoff KM, Somer BG, Schuster SJ, Porter DL, Montone KT, Stadtmauer EA. (2001). Reduction in immunosuppression as initial therapy for post-transplant lymphoproliferative disorder: Analysis of prognostic variables and long-term follow-up of 42 adult patients. *Transplantation* **71**, 1076–88.
163. Starzl TE, Nalesnik MA, Porter KA, Ho M, Iwatsuki S, Griffith BP, Rosenthal JT, Hakala TR, Shaw Jr. BW, Hardesty RL. (1984). Reversibility of lymphomas and lymphoproliferative lesions developing under cyclosporin-steroid therapy. *Lancet* **1**, 583–7.
164. Savage P, Waxman J. (1997). Post-transplantation lymphoproliferative disease. *Q J Med* **90**, 497–503.
165. Fahey JB, DiMaggio C. (2007). High-dose methotrexate and primary central nervous system lymphoma. *J Neurosci Nurs* **39**, 83–8.
166. Said JW. (2007). Immunodeficiency-related Hodgkin lymphoma and its mimics. *Adv Anat Pathol* **14**, 189–94.
167. Salloum E, Cooper DL, Howe G, Lacy J, Tallini G, Crouch J, Schultz M, Murren J. (1996). Spontaneous regression of lymphoproliferative disorders in patients treated with methotrexate for rheumatoid arthritis and other rheumatic diseases. *J Clin Oncol* **14**, 1943–9.
168. Thonhofer R, Gaugg M, Kriessmayr M, Neumann HJ, Erlacher L. (2005). Spontaneous remission of marginal zone B cell lymphoma in

a patient with seropositive rheumatoid arthritis after discontinuation of infliximab–methotrexate treatment. *Ann Rheum Dis* **64**, 1098–9.

169. Kwon EJ, Katz KA, Draft KS, Seykora JT, Rook AH, Wasik MA, Junkins-Hopkins JM. (2006). Posttransplantation lymphoproliferative disease with features of lymphomatoid granulomatosis in a lung transplant patient. *J Am Acad Dermatol* **54**, 657–63.
170. Johnston A, Coyle L, Nevell D. (2006). Prolonged remission of refractory lymphomatoid granulomatosis after autologous hemopoietic stem cell transplantation with post-transplantation maintenance interferon. *Leuk Lymphoma* **47**, 323–8.
171. Kameda H, Okuyama A, Tamaru JI, Itoyama S, Iizuka A, Takeuchi T. (2007). Lymphomatoid granulomatosis and diffuse alveolar damage associated with methotrexate therapy in a patient with rheumatoid arthritis. *Clin Rheumatol* **Jan 3**.
172. Wu SM, Min Y, Ostrzega N, Clements PJ, Wong AL. (2005). Lymphomatoid granulomatosis: A rare mimicker of vasculitis. *J Rheumatol* **3**, 2242–5.
173. Percik R, Serr J, Segal G, Stienlauf S, Trau H, Shalmon B, Shimoni A, Sidi Y. (2005). Lymphomatoid granulomatosis: A diagnostic challenge. *Isr Med Assoc J* **7**, 198–9.
174. Lipford EH, Margolick JB, Longo DL, Fauci AS, Jaffe ES. (1988). Angiocentric immunoproliferative lesions: A clinicopathologic spectrum of post-thymic T-cell proliferations. *Blood* **72**, 1674–81.
175. Shiozawa E, Saito B, Yamochi-Onizuka T, Makino R, Takimoto M, Nakamaki T, Tomoyasu S, Ota H. (2007). Senile EBV-associated B-cell lymphoproliferative disorder of indolent clinical phenotype with recurrence as aggressive lymphoma. *Pathol Int* **57**, 688–93.
176. Oyama T, Yamamoto K, Asano N, Oshiro A, Suzuki R, Kagami Y, Morishima Y, Takeuchi K, Izumo T, Mori S, Ohshima K, Suzumiya J, Nakamura N, Abe M, Ichimura K, Sato Y, Yoshino T, Naoe T, Shimoyama Y, Kamiya Y, Kinoshita T, Nakamura S. (2007). Age-related EBV-associated B-cell lymphoproliferative disorders constitute a distinct clinicopathologic group: A study of 96 patients. *Clin Cancer* **13**, 5124–32.
177. Shimoyama Y, Oyama T, Asano N, Oshiro A, Suzuki R, Kagami Y, Morishima Y, Nakamura S. (2006). Senile Epstein–Barr virus-associated B-cell lymphoproliferative disorders: A mini review. *J Clin Exp Hematop* **46**, 1–4.
178. Cabanillas F, Armitage J, Pugh WC, Weisenburger D, Duvic M. (1995). Lymphomatoid papulosis: A T-cell dyscrasia with a propensity to transform into malignant lymphoma. *Ann Intern Med* **122**, 210–17.
179. Shabrawi-Caelen L, Kerl H, Cerroni L. (2004). Lymphomatoid papulosis: Reappraisal of clinicopathologic presentation and classification into subtypes A, B, and C. *Arch Dermatol* **140**, 441–7.
180. Kadin ME. (2006). Pathobiology of CD30+ cutaneous T-cell lymphomas. *J Cutan Pathol* **33**(Suppl 1), 10–17.
181. Drews R, Samel A, Kadin ME. (2000). Lymphomatoid papulosis and anaplastic large cell lymphomas of the skin. *Semin Cutan Med Surg* **19**, 109–17.
182. Jaffe ES, Wilson WH. (1997). Lymphomatoid granulomatosis: Pathogenesis, pathology and clinical implications. *Cancer Surv* **30**, 233–48.
183. Zackheim HS, Jones C, Leboit PE, Kashani-Sabet M, McCalmont TH, Zehnder J. (2003). Lymphomatoid papulosis associated with mycosis fungoides: A study of 21 patients including analyses for clonality. *J Am Acad Dermatol* **49**, 620–3.

20 CHAPTER

Mastocytosis

Faramarz Naeim

Mastocytosis refers to an abnormal proliferation of mast cells in one or multiple organs. It covers a wide spectrum of clinicopathologic disorders from localized to systemic and from indolent to aggressive forms. Most systemic variants of mastocytosis are the result of clonal expansion of mast cells [1–3].

Mast cells are derived from the hematopoietic stem cells. The committed mast cell progenitors express CD33, CD34, and CD117 (*c-kit*) and are detectable in the bone marrow and peripheral blood [4, 5]. Mast cells are distinguished from other granulocytic cells by their unique phenotypic and functional properties (Tables 20.1 and 20.2). Mast cells produce a substantial amount of histamine and heparin and express surface IgE receptor [6–8]. In contrast to basophils and other granulocytic cells, mast cells have a significantly longer *in vivo* life span ranging from several months to years [7, 8].

ETIOLOGY AND PATHOGENESIS

The etiology and pathogenesis of mastocytosis are not clearly understood. The *c-kit* protooncogene encodes KIT, a tyrosine kinase receptor, which binds stem cell factor (SCF) [5, 9]. Systemic mastocytosis (SM) has been associated with somatic *c-kit* mutation at codon 816, substituting valine to aspartate (Figure 20.1) [2, 3, 10, 11]. Activated mutated *c-kit* along with increased production of SCF may play a role in the pathogenesis of SM [12, 13]. Genetic factors appear to play an important role in the childhood mastocytosis based on the report of 25% of congenital mastocytosis in pediatric cases and concordant symptoms of mastocytosis observed in 10 pairs of homozygotic twins [14, 15].

PATHOLOGY

Morphology

Mastocytosis is demonstrated as multifocal clusters or diffuse infiltration of mast cells in the skin, bone marrow, spleen, liver, gastrointestinal tract, and other tissues [16–18]. Mast cells in smears stained with Wright's or Giemsa stains are very distinct and appear as medium- to large-sized cells with abundant cytoplasm loaded with small deeply basophilic granules, often masking the nucleus (Figure 20.2). The nuclei are round, oval, or spindle-shaped and show a dense chromatin [16]. Some mast cells may be hypogranular or appear immature. Binucleated or multinucleated mast cells may be present [16].

Mast cells in the H&E-stained sections appear as medium-sized cells with variable amounts of granular cytoplasm and round, oval, or spindle-shaped nucleus with condensed chromatin (Figure 20.3). The cytoplasmic granules are faintly eosinophilic in the H&E sections and variable in amount. The granules in some mast cells are sparse and difficult to detect. The hypogranular mast cells may demonstrate an abundant pale cytoplasm resembling histiocytes, monocytoid B-cells, or hairy cells [1]. The spindle-shaped mast cells may mimic fibroblasts. Because of these overlapping morphologic features, it is highly recommended to perform additional accessory studies such as cytochemical stains, immunophenotyping, and molecular analysis to establish the diagnosis of mastocytosis (Table 20.3). Mast cell infiltration is often associated with various degrees of fibrosis and the presence of inflammatory cells, such as lymphocytes and eosinophils (Figures 20.4 and 20.5).

TABLE 20.1 Immunophenotypic features of mast cells.*

	Mast cells			
CD	**Normal**	**Neoplastic**	**Basophils**	**Monocytes**
CD2	−	+	−	−
CD13	−	±	+	+
CD14	−	−	−	+
CD15	−	−	−	+
CD25	−	+	+	±
CD33	+	+	+	+
CD34	−	−	−	−
CD45	+	+	+	+
CD117	+	+	−	−

*Adapted from Ref. [17].

TABLE 20.2 Major mast-cell-derived mediators and their effects in systemic mastocytosis.*

Mediators	Clinicopathological effects
Histamine	Vascular instability, urticaria, headache, edema, flushing, gastric hypersecretion, abdominal pain, bronchial constriction, diarrhea
Heparin	Coagulation abnormalities, bleeding
Tryptases	Fibrosis, angiogenesis, tissue remodeling, degradation of matrix molecules, bone resorption
tPA	Fibrinolysis
VEGF	Increased angiogenesis, edema
PGD2	Edema, urticaria, flushing, bronchial constriction
bFGF	Fibrosis, osteosclerosis, angiogenesis
TNF-α	Activation of endothelial cells, vascular instability
TNF-β	Fibrosis, abnormal bone remodeling, osteopenia
Interleukins (IL-1, -2, -3, -5, -6, -9, -10, -11, -13, GM-CSF)	Eosinophilia, bone marrow lymphocytosis, myeloid hyperplasia, activation of stromal cells, fibrosis
Chemokines (MCP-1, MIP-1α, others)	Leukocyte activation, accumulation of lymphocytes, monocytes, and eosinophils

*Adapted from Ref. [17].

The WHO criteria for the diagnosis of mastocytosis are presented in Table 20.3. For the diagnosis of SM, the major criteria are the presence of multifocal infiltrate of mast cells in bone marrow biopsy sections or other extracutaneous sites, confirmed by tryptase immunohistochemistry or other special stains [1]. Minor criteria include (1) the presence of 25% of mast cells being spindle-shaped, atypical, or immature, (2) detection of *KIT* mutation, (3) co-expression of CD117, CD2, and CD25 by the infiltrating mast cells, and (4) serum tryptase levels of >20 ng/mL. Diagnosis of SM could be made when one major and one minor criteria or three minor criteria are present.

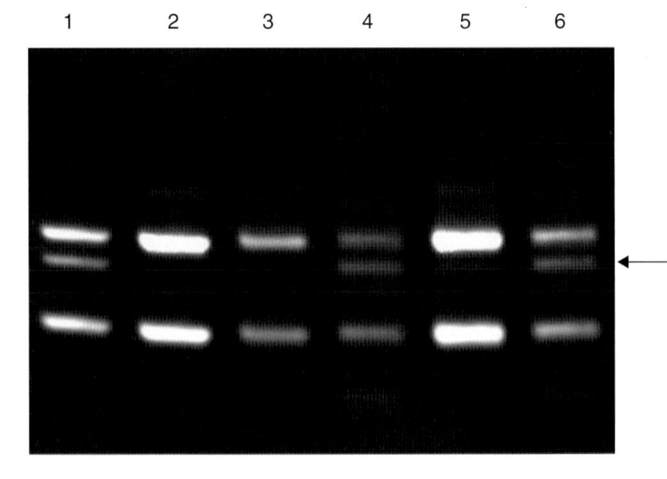

FIGURE 20.1 Demonstration of D816V *C-KIT* mutation by restriction fragment length polymorphism. Lane 1 is a positive control. Patient samples in lanes 4 and 6 show the presence of an additional band (arrow) as a result of a new restriction site by an A→T nucleotide change in *c-kit* codon 816. From Ref. [10] by permission.

Skin and bone marrow biopsies are the most frequent tissue samples obtained for the diagnosis of cutaneous and systemic mastocytosis, respectively. In cutaneous mastocytosis (CM), aggregates or sheets of elongated or spindle-shaped mast cells are present in the papillary and/or reticular dermis. These infiltrates are often around vascular structures [1, 16].

The bone marrow biopsy sections often show multiple mast cell aggregates. These aggregates may be either paratrabecular or interstitial (Figures 20.3–20.5). Paratrabecular infiltrates may show extensive fibrosis and osteosclerosis with scattered or aggregates of spindle-shaped mast cells which are identified by the Giemsa stain or immunohistochemical stains for tryptase or CD117 (Figures 20.6 and 20.7). Mastocytosis may be a part of a clonal primary hematopoietic disorder, such as myelodysplastic syndrome, chronic myeloproliferative disorder, or acute myelogenous leukemia.

In mast cell leukemia, mast cells account for ≥20% of the bone marrow nucleated cells, often with the presence of circulating mast cells in the peripheral blood. The bone marrow differential count on the smears should be performed in the areas away from bone marrow tissue particles [17].

Splenic involvement in mastocytosis usually consists of infiltrating mast cells randomly distributed in the red pulp or appearing in aggregates adjacent to the white pulps or trabeculae [1, 19]. The mast cell clusters are often associated with variable amounts of fibrosis and are mixed or surrounded by lymphocytes, plasma cells, and/or eosinophils. In the liver, mast cells are found in the sinuses and/or portal areas with focal areas of fibrosis (Figure 20.8). The infiltrating mast cells are detected by special stains, such as Giemsa, or immunophenotypic studies (discussed later).

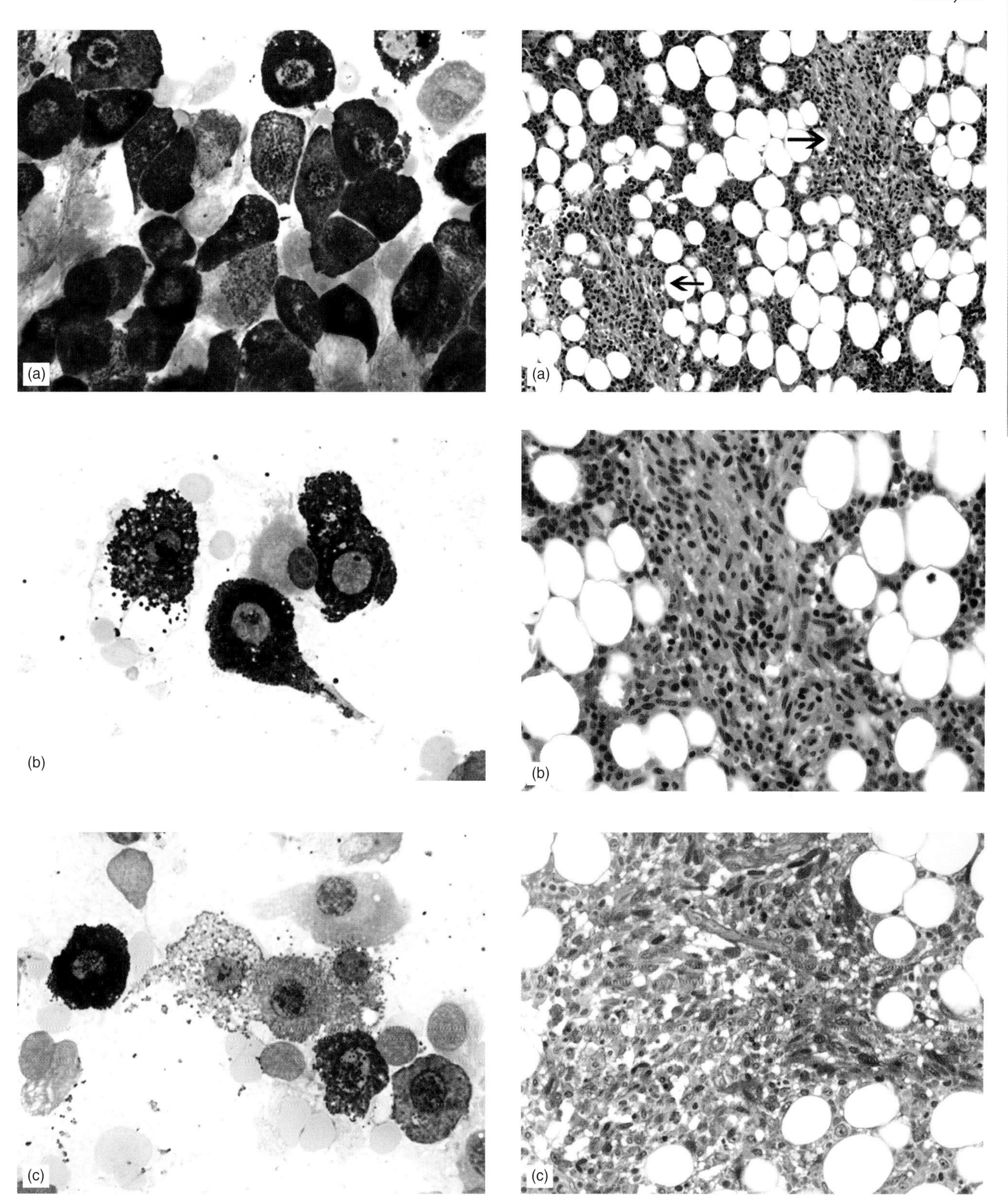

FIGURE 20.2 Mast cells. Wright-stained bone marrow smears demonstrating mast cells with abundant cytoplasm containing variable amounts of small, deeply basophilic granules (a, b, and c).

FIGURE 20.3 Systemic mastocytosis. Bone marrow biopsy section demonstrating cluster of mast cells resembling histiocytes or fibroblasts: (a) low power view, arrows, and (b) high power view. Mast cells are often easily identified by Giemsa stain (c).

TABLE 20.3 WHO criteria for the diagnosis of cutaneous and systemic mastocytosis.*

Cutaneous mastocytosis
Mast cell infiltrates in a multifocal or diffuse pattern in the skin biopsies with typical clinical findings.
Systemic mastocytosis
Major criterion
Multifocal infiltrates of mast cells (≥15 mast cells in each aggregate) in one or more extracutaneous sites confirmed by tryptase immunohistochemistry or other special stains.
Minor criteria
1. More than 25% of mast cells are spindle-shaped or atypical in the extracutaneous infiltrates in the biopsy sections and/or smear preparations.
2. Detection of *KIT* mutation at codon 816.
3. Co-expression of CD117, CD2, and/or CD25 by the infiltrating mast cells.
4. Persistent total serum tryptase levels >20 ng/mL in cases not associated with clonal myeloid disorders.

*Adapted from Ref. [1].

Immunophenotype

Normal mast cells express CD33, CD45, CD68, CD117, and tryptase and are negative for CD14, CD15, CD16, and MPO (Figures 20.4, 20.5, and 20.7). The neoplastic mast cells show aberrant expression of CD2 and CD25 [1, 17, 20–24].

Molecular and Cytogenetic Studies

SM has been associated with somatic *c-kit* mutations. The most common mutation is reported at codon 816, substituting valine to aspartate. This can be detected by RFLP or DNA sequence analysis (Figures 20.1 and 20.9) [2, 3, 10, 11]. A report in gene expression analysis in mastocytosis demonstrated significant upregulation of genes for α-tryptase, the activating transcription factor type 3, and the muscle aponeurotic fibrosarcoma type F as surrogate markers, strongly correlated with serum tryptase levels [25].

CLINICAL ASPECTS

Mast cell diseases are divided into two major clinical entities: (1) cutaneous mastocytosis and (2) systemic mastocytosis [1, 8, 26].

Cutaneous Mastocytosis

Cutaneous mastocytosis consists of mast cell disorders limited to skin without evidence of systemic involvement, such as elevated levels of serum tryptase, bone marrow

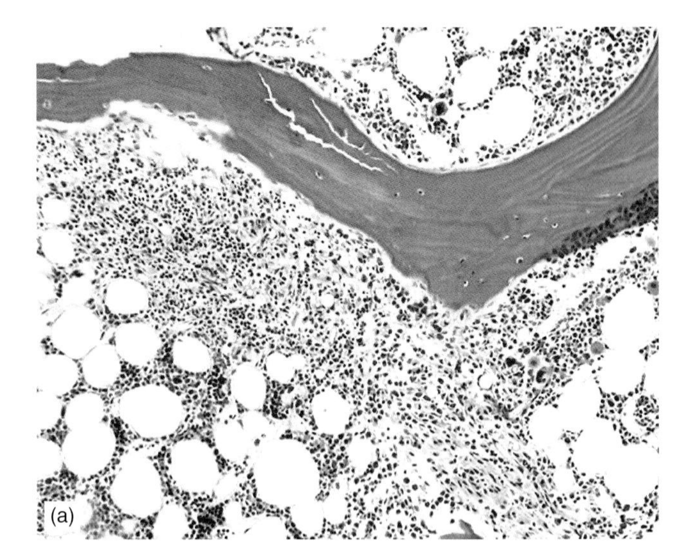

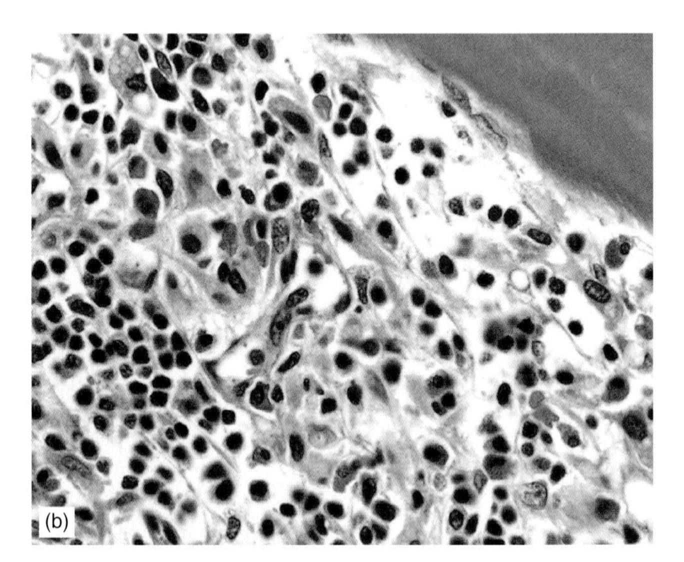

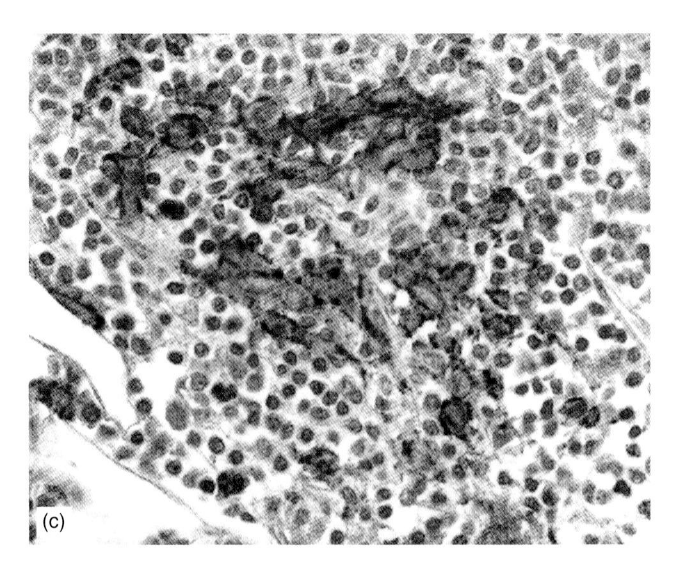

FIGURE 20.4 A cluster of mast cells mixed with lymphocytes and histiocytes is present next to bone: (a) low power and (b) high power views. Mast cells show expression of CD117 by immunohistochemical technique (c).

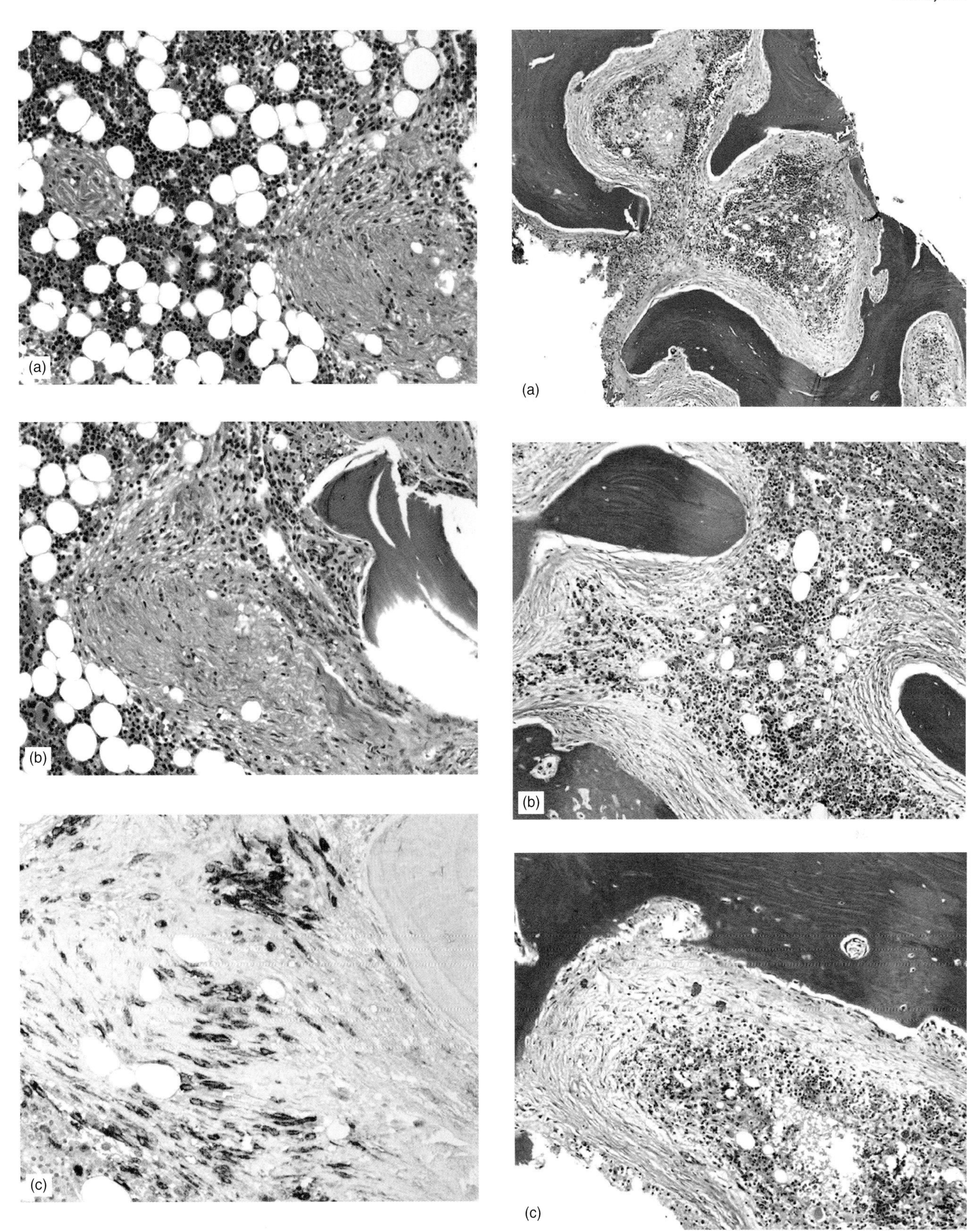

FIGURE 20.5 Bone marrow biopsy section demonstrating areas of dense fibrosis: (a) low power and (b) high power views. Clusters of CD117-positive-elongated cells representing mast cells are present (c).

FIGURE 20.6 Bone marrow involvement in systemic mastocytosis. Extensive paratrabecular fibrosis and some osteosclerosis are present: (a) low power, (b) intermediate power, and (c) high power views.

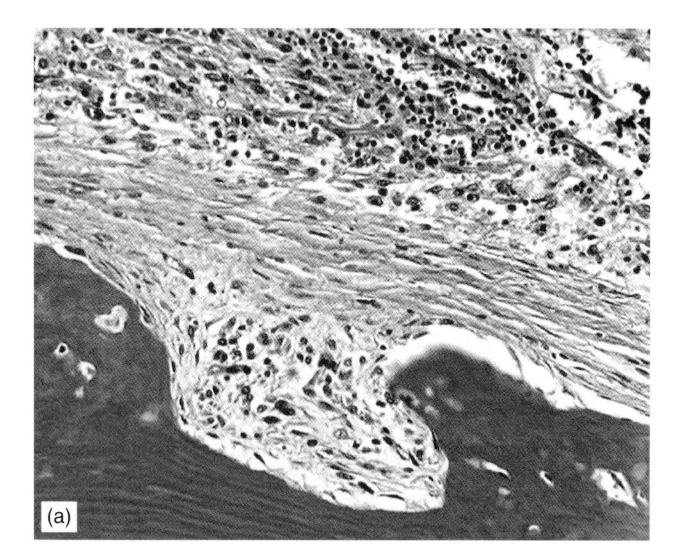

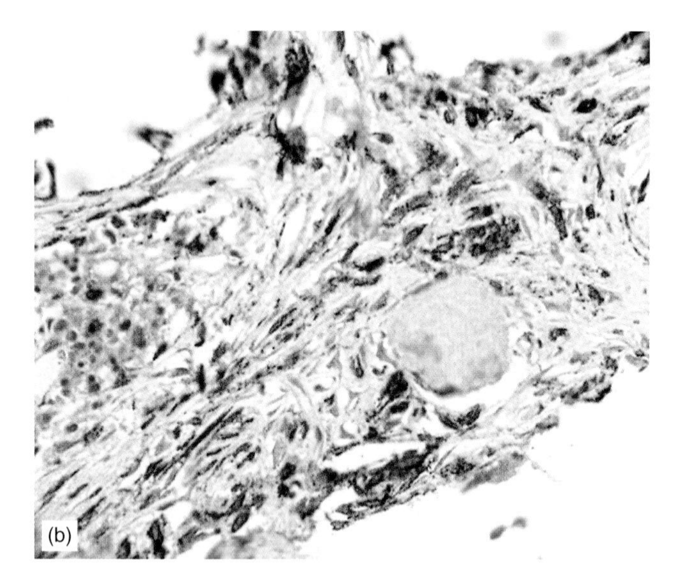

FIGURE 20.7 Systemic mastocytosis. Bone marrow biopsy section showing paratrabecluar fibrosis (a) with the presence of numerous tryptase-positive cells in the fibrotic area: (b) low power and (c) high power views.

infiltration, or organomegaly. There are four major clinicopathologic subtypes: (1) urticaria pigmentosa, (2) cutaneous mastocytoma, (3) diffuse CM, and (4) telangiectasia macularis eruptiva perstans. The diagnosis is made based on the clinical presentation and by a skin biopsy demonstrating significant increase in mast cells (usually ≥20 mast cells per high power field), which are particularly found around vascular structures [9, 27, 28].

Urticaria pigmentosa (UP) is the most common mast cell disorder in children and adults. The *KIT* point mutation in the pediatric and adult UP appears to be different from the mutation of codon 816 observed in SM. There are reports of mutations in codon 839 and codon 516 in pediatric and adult UP, respectively [3, 29].

The cutaneous lesions are usually small yellow-tan to reddish-brown macules or papules. Plaque-like lesions may also occur [9]. The upper and lower extremities are the most frequently affected sites. The face, palms, and soles are not involved. Most of the affected children are under the age of 1 and rarely show systemic involvement. The UP-associated pruritis is exacerbated by a variety of stimulants, such as change in temperature, spicy food, or local friction [9, 29].

Cutaneous mastocytoma is typically a solitary lesion occurring in early childhood, usually before 6 months of age [9, 30, 31]. The trunk and wrist are frequent sites of involvement [1]. Large clusters or sheets of mast cells are present in the papillary and reticular dermis.

Diffuse cutaneous mastocytosis is a childhood disorder usually occurring before the age of 3. The skin is diffusely infiltrated but relatively smooth. It may show increased thickness and/or a yellowish-brown color. The maculopapular lesions are usually absent [9, 27, 32].

Telangiectasia macularis eruptiva perstans is a rare cutaneous mast cell disorder mainly occurring in adults. It is characterized by tan-brown macules with telangiectasia but no blisters or pruritis [9, 33].

Systemic Mastocytosis

Systemic mastocytosis represents mast cell disease beyond skin. The increased mast cells are found in extracutaneous sites with or without skin involvements. The frequently involved extracutaneous sites include bone marrow, liver, spleen, lymph nodes, and gastrointestinal tract [1, 9, 34, 35a]. The WHO requirements for the diagnosis and classification of SM are demonstrated in Tables 20.3 and 20.4 [1].

Clinical manifestations of mastocytosis are the result of two different mechanisms: (1) mediator release from mast cells and (2) growth and infiltration of the mast cells in various organs.

A wide variety of mediators are released from mast cells resulting in clinical symptoms, such as headache, flushing, pruritis, hypotension, and diarrhea (Table 20.2). The growth and infiltration of mast cells in various organs may lead to organomegaly as well as organ dysfunction, leading to ascites, cytopenia, malabsorption, and pathologic fractures. The organopathy-related clinical symptoms are referred to as *C-findings*, whereas organomegalies without any evidence of organopathy are termed *B-findings* [17, 18].

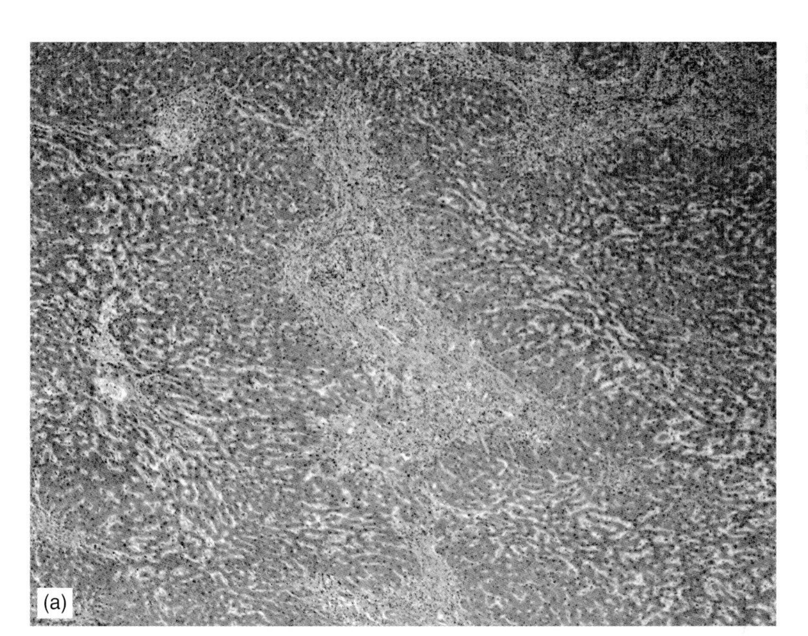

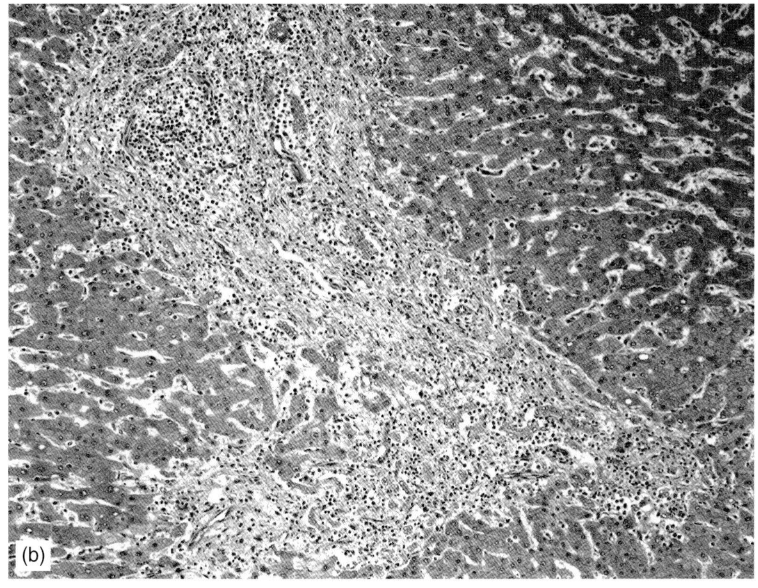

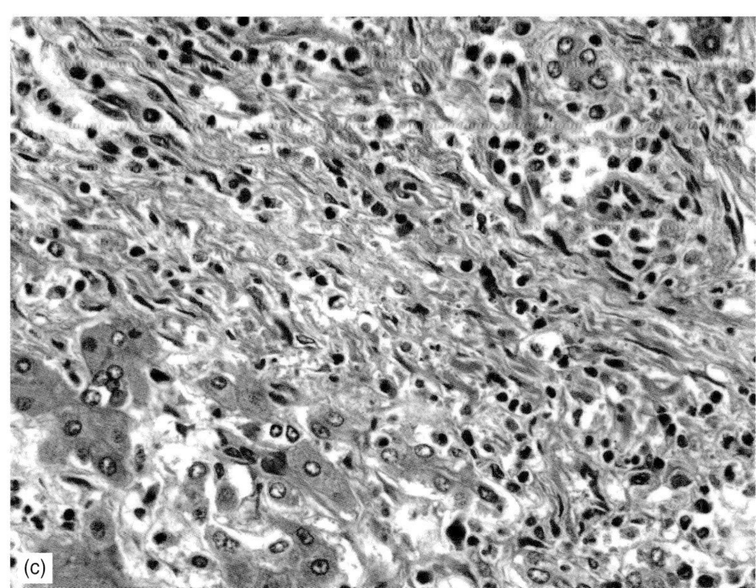

FIGURE 20.8 Mastocytosis involving the liver. Fibrosis and infiltration of the inflammatory cells are noted in the expanded portal areas: (a) low power and (b) intermediate power views. Cells with elongated or spindle-shaped nuclei represent mast cells. (c) high power view.

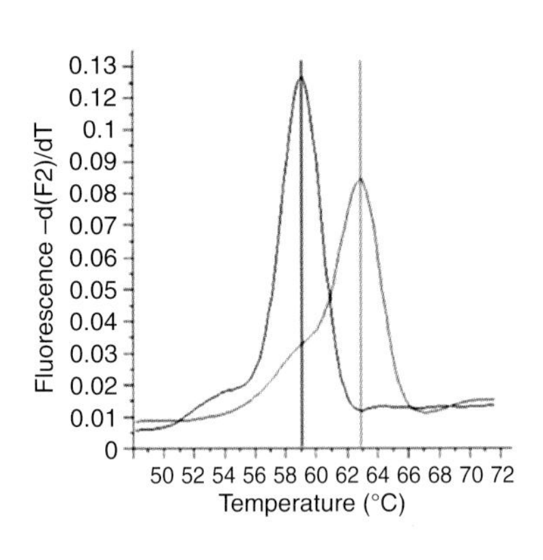

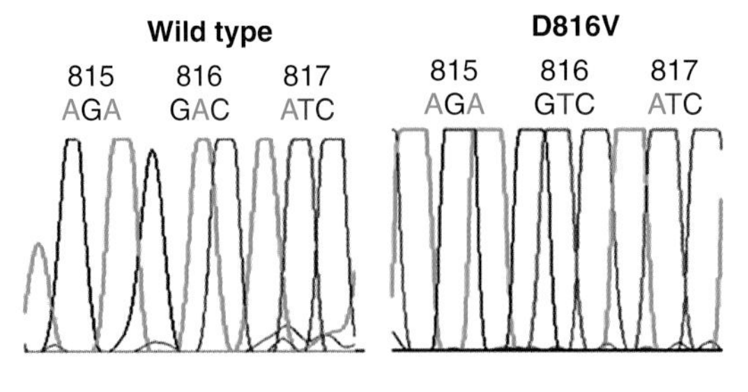

FIGURE 20.9 D816V *c-KIT* mutation in a patient with systemic mastocytosis. From Ref. [11] by permission. This research was originally published in *Blood*.

TABLE 20.4 WHO classification of systemic mastocytosis.*

Indolent systemic mastocytosis
Systemic mastocytosis with associated clonal, hematological non-mast cell lineage disease
Aggressive systemic mastocytosis
Extracutaneous mastocytoma
Mast cell sarcoma

*Adapted from Ref. [1].

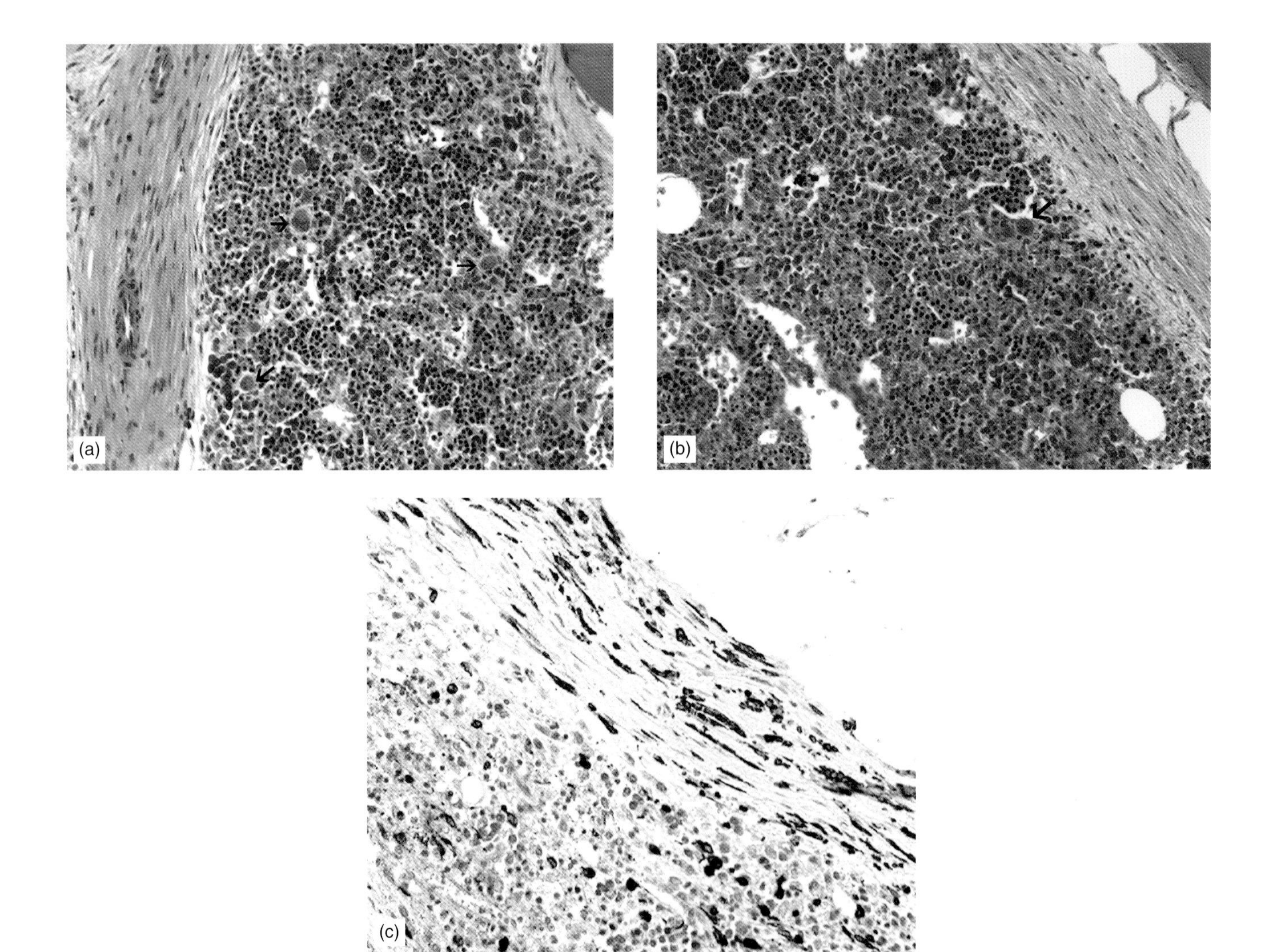

FIGURE 20.10 Mastocytosis in a patient with myelodysplastic syndrome. A biopsy section showing paratrabecular fibrosis. The bone marrow is hypercellular with the presence of numerous micromegakaryocytes (a and b; arrows). Numerous spindle cells are positive for CD117 in the fibrotic area (c).

TABLE 20.5 Comparison of B- and C-findings in subtypes of systemic mastocytosis.*

Findings	ISM			SM-AHNMD	ASM	MCL
	Typical ISM	BMM	SSM			
B-findings						
Hepatomegaly	–	–	±	±	±	±
Splenomegaly	–	–	+	±	+	±
Lymphadenopathy	–	–	±	±	±	±
Tryptase >200 ng/mL	–	–	+	±	±	±
C-findings						
Anemia (Hb <10 g/dL)	–	–	–	±	+	+
Thrombocytopenia (<100 × 10^9/L)	–	–	–	±	+	+
Neutrophil count <1 × 10^9/L	–	–	–	±	+	+
Ascites or portal hypertension	–	–	–	–	+	+
Hypersplenism	–	–	–	±	+	±
Malabsorption with weight loss	–	–	–	–	±	±
Osteolysis	–	–	–	–	+	+
Others						
Urticaria pigmentosa-like lesions	+	–	±	±	±	–
Elevated serum LDH	–	–	–	±	±	+
Abnormal coagulation	–	–	±	±	±	+

*Adapted from Ref. [17].

ISM: indolent systemic mastocytosis, BMM: isolated bone marrow mastocytosis, SSM: smoldering systemic mastocytosis, SM-AHNMD: systemic mastocytosis with associated clonal hematological non-mast cell lineage disease, ASM: aggressive systemic mastocytosis, MCL: mast cell leukemia.

SM has been divided into the following categories (Table 20.4).

Indolent systemic mastocytosis (ISM) is referred to cases with relatively low burden of mast cells and therefore to an indolent clinical course and good prognosis. The majority of the patients with ISM have UP and show evidence of systemic involvement but lack C-findings. ISM accounts for >80% of all cases of SM [9]. Two subtypes of ISM have been described: smoldering systemic mastocytosis (SSM) and isolated bone marrow mastocytosis (BMM) [17]. In SSM, B-findings are present, and in BMM, there is lack of skin involvement (Table 20.5).

Systemic mastocytosis with associated clonal hematological non-mast cell lineage disease (SM-AHNMD) is mastocytosis associated with acute myeloid leukemias, acute lymphoid leukemias, myelodysplastic syndromes, chronic myeloproliferative disorders, or lymphoma (Figures 20.10 and 20.11).

While the FIP1L1/PDGFRA fusion gene is mainly known for its association with hypereosinophilia (Chapter 9, p. 174 and Figure 9.20), this abnormal gene is also identified in patients with systemic mastocytosis. Eosinophilia is also frequently observed in systemic mastocytosis, and up to half of these cases are also positive for the *FIP1L1/PDGFRA* fusion gene [35b].

Aggressive systemic mastocytosis (ASM) is characterized by the presence of organ-function impairment and C-findings, leading to an aggressive clinical course [24, 36]. C-findings include (1) anemia, thrombocytopenia, and/or leukopenia, (2) hepatomegaly with ascites or portal hypertension, (3) splenomegaly with hypersplenism, (4) malabsorption and weight loss, and (5) osteolysis and pathologic fractures (Table 20.5).

Less than 50% of patients in this category show UP lesions. Mast cells account for <20% of the bone marrow nucleated cells. The *KIT* mutation in codon 816 is the typical molecular finding [36].

Mast cell leukemia is a rare condition characterized by diffuse infiltration of the bone marrow by atypical and/or immature mast cells. The pattern of bone marrow infiltration is usually interstitial. Mast cells comprise ≥20% of the nucleated cells in the bone marrow smears and ≥10% of the leukocyte differential counts in the peripheral blood (Figure 20.12) [1, 37–39]. Prognosis is extremely poor with an estimated survival of 6–12 months [9].

Extracutaneous mastocytoma is an extremely rare lesion consisting of an accumulation of mature mast cells in extracutaneous sites, such as the lung [40].

Mast cell sarcoma is another extremely rare lesion consisting of an infiltrating growth of atypical and/or immature mast cells with a potential of distant metastasis or progression to a leukemic phase [41, 42].

In general, the therapeutic approaches depend on the clinical symptoms and extent of the disease. Mediator-related

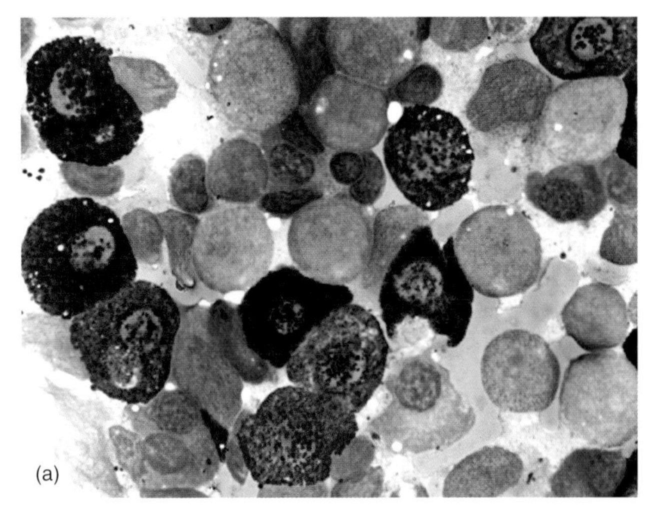

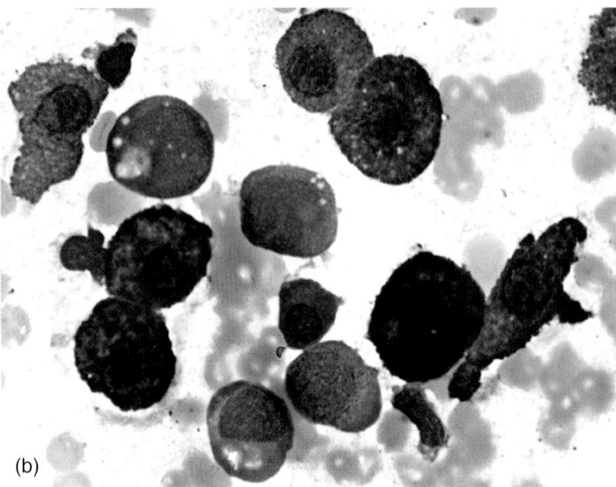

FIGURE 20.11 Mastocytosis in a patient with acute myelogenous leukemia. Bone marrow smear consisting of a mixture of myeloblasts and mast cells (a and b).

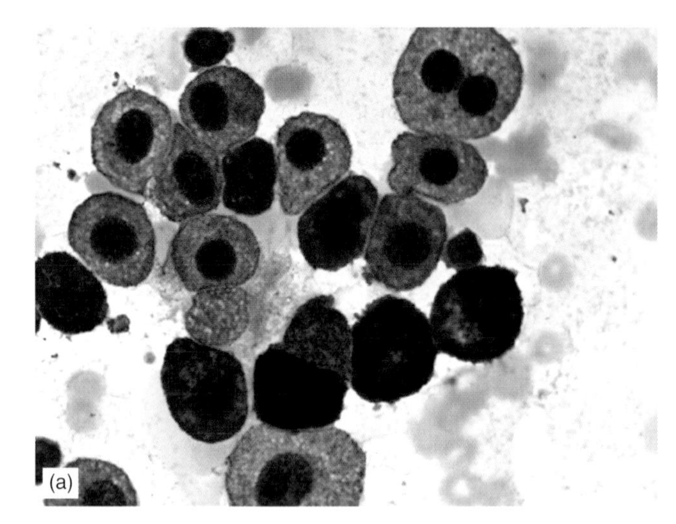

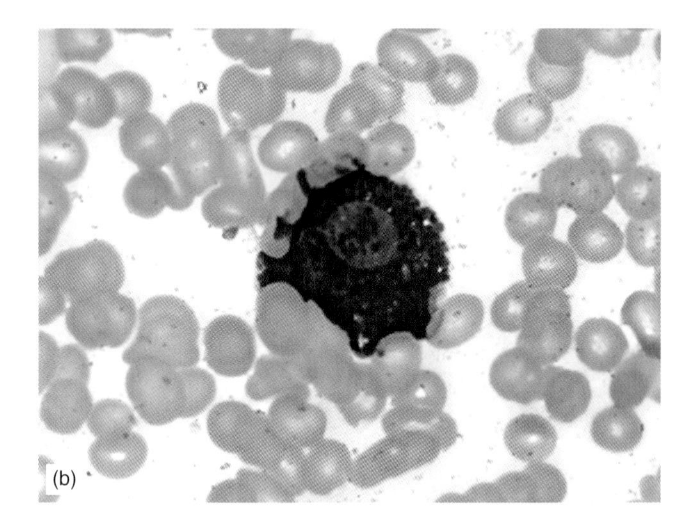

FIGURE 20.12 Bone marrow smear from a patient with mast cell leukemia demonstrating numerous mast cells (a). A mast cell is shown in peripheral blood smear (b).

symptoms are treated with drugs that interfere with mediator production/release or mediator functions, such as histamine and leukotriene antagonists, glucocorticoids, cromylin sodium, and aspirin [17, 26]. In addition to the anti-mediator drugs, patients with CM may receive psoralen and ultraviolet-A [43]. Cytoreductive drugs such as interferon-α, cytosine arabinoside, cladribine, vincristine, and doxorubicin are preserved for patients who have clear signs of aggressive disease [17, 44].

DIFFERENTIAL DIAGNOSIS

The differential diagnosis includes two major categories: (1) disorders with similar clinical manifestation but lack of histologic evidence of CM or SM and (2) disorders associated with increased mast cells or elevated serum tryptase (Table 20.6).

Disorders with similar clinical manifestation but lack of histologic evidence of CM or SM include anaphylaxis (may show elevated serum tryptase during, but not in the period between, the acute events), angioedema, carcinoid

TABLE 20.6 Differential diagnosis of mastocytosis.

Categories	Examples
Similar clinical manifestations but lack of histologic evidence of cutaneous or systemic mastocytosis	Anaphylaxis, angioderma, carcinoid syndrome, pheochromocytoma, Zollinger–Ellison syndrome
Associated with increased mast cells	Basal cell carcinoma, melanoma, lymphoma, helminth infection
Associated with elevated serum tryptase but lack of mastocytosis	Acute myelogenous leukemia, myelodysplastic syndrome, chronic myeloproliferative disorders
Morphologic overlap	Basophilic leukemia, histiocytic disorders, bone marrow metastasis, disorders associated with paratrabecular bone marrow fibrosis, such as chronic renal failure

syndrome, pheochromocytoma, and Zollinger–Ellison syndrome. All these disorders except acute episodes of anaphylaxis lack elevated serum tryptase or urinary histamine [9].

Disorders associated with increased mast cells or elevated serum tryptase consist of reactive conditions associated with mastocytosis or hematopoietic malignancies associated with elevated serum tryptase but lack of mastocytosis [44]. Reactive mast cell hyperplasia has been observed in various conditions, such as basal cell carcinoma, melanoma, helminth infection, and lymphomas. Elevated serum tryptase levels have been reported in a variety of clonal myeloid disorders, such as acute myelogenous leukemia, myelodysplastic syndromes, and chronic myeloproliferative disorders [8, 45]. Also, some patients with myelodysplastic syndrome may show *c-kit* mutation [46].

References

1. Jaffe ES, Harris NL, Stein H, Vardiman JW. (2001). *Pathology and Genetics. Tumors of Haematopoietic and Lymphoid Tissues.* IARC Press, Lyon.
2. Nagata H, Worobec AS, Oh CK, Chowdhury BA, Tannenbaum S, Suzuki Y, Metcalfe DD. (1995). Identification of a point mutation in the catalytic domain of the protooncogene c-kit in peripheral blood mononuclear cells of patients who have mastocytosis with an associated hematologic disorder. *Proc Natl Acad Sci USA* **92**, 10560–4.
3. Longley Jr. BJ, Metcalfe DD, Tharp M, Wang X, Tyrrell L, Lu SZ, Heitjan D, Ma Y. (1999). Activating and dominant inactivating c-KIT catalytic domain mutations in distinct clinical forms of human mastocytosis. *Proc Natl Acad Sci USA* **96**, 1609–14.
4. Kirshenbaum AS, Goff JP, Semere T, Foster B, Scott LM, Metcalfe DD. (1999). Demonstration that human mast cells arise from a progenitor cell population that is CD34(+), c-kit(+), and expresses aminopeptidase N (CD13). *Blood* **94**, 2333–42.
5. Valent P. (1994). The riddle of the mast cell: kit(CD117)-ligand as the missing link? *Immunol Today* **15**, 111–14.
6. Metcalfe DD. (2008). Mast cells and mastocytosis. *Blood* **112**, 946–56.
7. Galli SJ. (1990). New insights into "the riddle of the mast cells": Microenvironmental regulation of mast cell development and phenotypic heterogeneity. *Lab Invest* **62**, 5–33.
8. Valent P, Akin C, Sperr WR, Horny HP, Metcalfe DD. (2003). Mast cell proliferative disorders: Current view on variants recognized by the World Health Organization. *Hematol Oncol Clin North Am* **17**, 1227–41.
9. Castells MC. (2007). Clinical manifestations and pathogenesis of mastocytosis (cutaneous and systemic). *UpToDate.*
10. Akin C. (2006). Molecular diagnosis of mast cell disorders: A paper from the 2005 William Beaumont Hospital Symposium on Molecular Pathology. *J Mol Diagn*, **8**, 412–19.
11. Garcia-Montero AC, Jara-Acevedo M, Teodosio C, Sanchez ML, Nunez R, Prados A, Aldanondo I, Sanchez L, Dominguez M, Botana LM, Sanchez-Jimenez F, Sotlar K, Almeida J, Escribano L, Orfao A. (2006). KIT mutation in mast cells and other bone marrow hematopoietic cell lineages in systemic mast cell disorders: A prospective study of the Spanish Network on Mastocytosis (REMA) in a series of 113 patients. *Blood* **108**, 2366–72.
12. Valent P, Akin C, Sperr WR, Mayerhofer M, Fodinger M, Fritsche-Polanz R, Sotlar K, Escribano L, Arock M, Horny HP, Metcalfe DD. (2005). Mastocytosis: Pathology, genetics, and current options for therapy. *Leuk Lymphoma* **46**, 35–48.
13. Castells MC, Friend DS, Bunnell CA, Hu X, Kraus M, Osteen RT, Austen KF. (1996). The presence of membrane-bound stem cell factor on highly immature nonmetachromatic mast cells in the peripheral blood of a patient with aggressive systemic mastocytosis. *J Allergy Clin Immunol* **98**, 831–40.
14. Kettelhut BV, Metcalfe DD. (1991). Pediatric mastocytosis. *J Invest Dermatol* **96**, 15S–18S.
15. Hartmann K, Metcalfe DD. (2000). Pediatric mastocytosis. *Hematol Oncol Clin North Am* **14**, 625–40.
16. Valent P, Horny HP, Escribano L, Longley BJ, Li CY, Schwartz LB, Marone G, Nunez R, Akin C, Sotlar K, Sperr WR, Wolff K, Brunning RD, Parwaresch RM, Austen KF, Lennert K, Metcalfe DD, Vardiman JW, Bennett JM. (2001). Diagnostic criteria and classification of mastocytosis: A consensus proposal. *Leuk Res* **25**, 603–25.
17. Valent P, Akin C, Sperr WR, Horny HP, Arock M, Lechner K, Bennett JM, Metcalfe DD. (2003). Diagnosis and treatment of systemic mastocytosis: State of the art. *Br J Haematol* **122**, 695–717.
18. Valent P, Sperr WR, Schwartz LB, Horny HP. (2004). Diagnosis and classification of mast cell proliferative disorders: Delineation from immunologic diseases and non-mast cell hematopoietic neoplasms. *J Allergy Clin Immunol* **114**, 3–11.
19. Wimazal F, Schwarzmeier J, Sotlar K, Simonitsch I, Sperr WR, Fritsche-Polanz R, Fodinger M, Schubert J, Horny HP, Valent P. (2004). Splenic mastocytosis: Report of two cases and detection of the transforming somatic C-KIT mutation D816V. *Leuk Lymphoma* **45**, 723–9.
20. Jordan JH, Walchshofer S, Jurecka W, Mosberger I, Sperr WR, Wolff K, Chott A, Buhring HJ, Lechner K, Horny HP, Valent P. (2001). Immunohistochemical properties of bone marrow mast cells in systemic mastocytosis: Evidence for expression of CD2, CD117/Kit, and bcl-x(L). *Hum Pathol* **32**, 545–52.
21. Horny HP, Sotlar K, Sperr WR, Valent P. (2004). Systemic mastocytosis with associated clonal haematological non-mast cell lineage diseases: A histopathological challenge. *J Clin Pathol* **57**, 604–8.
22. Sotlar K, Horny HP, Simonitsch I, Krokowski M, Aichberger KJ, Mayerhofer M, Printz D, Fritsch G, Valent P. (2004). CD25 indicates the neoplastic phenotype of mast cells: A novel immunohistochemical marker for the diagnosis of systemic mastocytosis (SM) in routinely processed bone marrow biopsy specimens. *Am J Surg Pathol* **28**, 1319–25.
23. Escribano L, Diaz-Agustin B, Bellas C, Navalon R, Nunez R, Sperr WR, Schernthaner GH, Valent P, Orfao A. (2001). Utility of flow cytometric analysis of mast cells in the diagnosis and classification of adult mastocytosis. *Leuk Res* **25**, 563–70.
24. Schernthaner GH, Jordan JH, Ghannadan M, Agis H, Bevec D, Nunez R, Escribano L, Majdic O, Willheim M, Worda C, Printz D, Fritsch G, Lechner K, Valent P. (2001). Expression, epitope analysis, and functional role of the LFA-2 antigen detectable on neoplastic mast cells. *Blood* **98**, 3784–92.
25. D'ambrosio C, Akin C, Wu Y, Magnusson MK, Metcalfe DD. (2003). Gene expression analysis in mastocytosis reveals a highly consistent profile with candidate molecular markers. *J Allergy Clin Immunol* **112**, 1162–70.
26. Horny HP, Sotlar K, Valent P. (2007). Mastocytosis: State of the art. *Pathobiology* **74**, 121–32.
27. Ben-Amitai D, Metzker A, Cohen HA. (2005). Pediatric cutaneous mastocytosis: A review of 180 patients. *Isr Med Assoc J* **7**, 320–2.
28. Kinsler VA, Hawk JL, Atherton DJ. (2005). Diffuse cutaneous mastocytosis treated with psoralen photochemotherapy: Case report and review of the literature. *Br J Dermatol* **152**, 179–80.
29. Novembre E, Cianferoni A, Mori F, Calogero C, Bernardini R, Di Grande L, Pucci N, Azzari C, Vierucci A. (2007). Urticaria and urticaria related skin condition/disease in children. *Eur Ann Allergy Clin Immunol* **39**, 253–8.
30. Scheck O, Horny HP, Ruck P, Schmelzle R, Kaiserling E. (1987). Solitary mastocytoma of the eyelid. A case report with special reference to the immunocytology of human tissue mast cells, and a review of the literature. *Virchows Arch A Pathol Anat Histopathol* **412**, 31–6.
31. Kacker A, Huo J, Huang R, Hoda RS. (2000). Solitary mastocytoma in an infant-case report with review of literature. *Int J Pediatr Otorhinolaryngol* **52**, 93–5.

32. Hannaford R, Rogers M. (2001). Presentation of cutaneous mastocytosis in 173 children. *Australas J Dermatol* **42**, 15–21.
33. Rishpon A, Matz H, Gat A, Brenner S. (2006). Telangiectasia macularis eruptiva perstans: Unusual presentation and treatment. *Skinmed* **5**, 300–2.
34. Patnaik MM, Rindos M, Kouides PA, Tefferi A, Pardanani A. (2007). Systemic mastocytosis: A concise clinical and laboratory review. *Arch Pathol Lab Med* **131**, 784–91.
35a. Barbie DA, Deangelo DJ. (2006). Systemic mastocytosis: Current classification and novel therapeutic options. *Clin Adv Hematol Oncol* **4**, 768–75.
35b. Pardanani A, Ketterling RP, Brockman SR, Flynn HC, Paternoster SF, Shearer BM, Reeder TL, Li CY, Cross NC, Cools J, Gilliland DG, Dewald GW, Tefferi A. (2003). CHIC2 deletion, a surrogate for FIP1L1-PDGFRA fusion, occurs in systemic mastocytosis associated with eosinophilia and predicts response to imatinib mesylate therapy. *Blood* **102**, 3093–6.
36. Kupfer SS, Hart J, Mohanty SR. (2007). Aggressive systemic mastocytosis presenting with hepatic cholestasis. *Eur J Gastroenterol Hepatol* **19**, 1–5.
37. Hashimoto K, Tsujimura T, Moriyama Y, Yamatodani A, Kimura M, Tohya K, Morimoto M, Kitayama H, Kanakura Y, Kitamura Y. (1996). Transforming and differentiation-inducing potential of constitutively activated c-kit mutant genes in the IC-2 murine interleukin-3-dependent mast cell line. *Am J Pathol* **148**, 189–200.
38. Valentini CG, Rondoni M, Pogliani EM, Van Lint MT, Cattaneo C, Marbello L, Pulsoni A, Giona F, Martinelli G, Leone G, Pagano L. (2008). Mast cell leukemia: A report of ten cases. *Ann Hematol* **Jan 3**.
39. Noack F, Sotlar K, Notter M, Thiel E, Valent P, Horny HP. (2004). Aleukemic mast cell leukemia with abnormal immunophenotype and c-kit mutation D816V. *Leuk Lymphoma* **45**, 2295–302.
40. Charrette EE, Mariano AV, Laforet EG. (1966). Solitary mast cell "tumor" of lung. Its place in the spectrum of mast cell disease. *Arch Intern Med* **118**, 358–62.
41. Horny HP, Parwaresch MR, Kaiserling E, Muller K, Olbermann M, Mainzer K, Lennert K. (1986). Mast cell sarcoma of the larynx. *J Clin Pathol* **39**, 596–602.
42. Kojima M, Nakamura S, Itoh H, Ohno Y, Masawa N, Joshita T, Suchi T. (1999). Mast cell sarcoma with tissue eosinophilia arising in the ascending colon. *Mod Pathol* **12**, 739–43.
43. Hartmann K, Bruns SB, Henz BM. (2001). Mastocytosis: Review of clinical and experimental aspects. *J Investig Dermatol Symp Proc* **6**, 143–7.
44. Tefferi A, Li CY, Butterfield JH, Hoagland HC. (2001). Treatment of systemic mast-cell disease with cladribine. *N Engl J Med* **344**, 307–9.
45. Sperr WR, Jordan JH, Baghestanian M, Kiener HP, Samorapoompichit P, Semper H, Hauswirth A, Schernthaner GH, Chott A, Natter S, Kraft D, Valenta R, Schwartz LB, Geissler K, Lechner K, Valent P. (2001). Expression of mast cell tryptase by myeloblasts in a group of patients with acute myeloid leukemia. *Blood* **98**, 2200–9.
46. Fritsche-Polanz R, Jordan JH, Feix A, Sperr WR, Sunder-Plassmann G, Valent P, Fodinger M. (2001). Mutation analysis of C-KIT in patients with myelodysplastic syndromes without mastocytosis and cases of systemic mastocytosis. *Br J Haematol* **113**, 357–64.

Histiocytic and Dendritic Cell Disorders

21

CHAPTER

Faramarz Naeim

Histiocytes/macrophages and most subtypes of dendritic cells are derived from the hematopoietic stem cells and play an important role in the regulation of immune functions. Histiocytes/macrophages are derived from monocytes and are involved in different aspects of host defense and tissue repair, such as phagocytosis, cytotoxic activities, regulation of inflammatory and immune responses, and wound healing. Dendritic cells (DC) are primarily involved in antigen processing and antigen presentation to the B- and T-lymphocytes [1, 2]. There are three major subclasses of dendritic cells: Langerhans cells (LC), interdigitating dendritic cells (IDC), and follicular dendritic cells (FDC). LC and IDC, similar to histiocytes/macrophages, are derived from hematopoietic stem cells, whereas FDC are derived from mesenchymal cells in the follicular structures of the lymph nodes [3–5]. The immunophenotypic features of histiocytes/macrophages and subclasses of DC are presented in Table 21.1.

MONOCYTIC AND HISTIOCYTIC DISORDERS

Disorders of monocytes and histiocytes/macrophages are divided into three major categories: (1) functional defects, (2) reactive responses, and (3) neoplastic proliferations. Functional defects are mostly hereditary, such as lysosomal storage diseases. Reactive disorders are non-neoplastic conditions associated with hypoplasia, hyperplasia, or hyperactivation of the monocytic/histiocytic system. Neoplastic disorders are the result of clonal proliferation of monocytic/histiocytic cells (see Chapters 8–11). A summary of monocytic and histiocytic disorders is demonstrated in Table 21.2.

In this section, the following entities are discussed as examples of monocytic and histiocytic disorders:

Functional disorders

- Gaucher disease (GD)
- Niemann–Pick disease (NPD)
- Chediak–Higashi syndrome (CHS)

Reactive responses

- Monocytopenia and monocytosis
- Histiocytic proliferations
- Hemophagocytic histiocytosis

Neoplastic disorders

- Histiocytic sarcoma
- Dendritic cell tumors

Gaucher Disease

Gaucher disease (GD) is an inherited autosomal recessive inborn error of metabolism resulting in the accumulation of glucocerebroside (glucosylceramide) in macrophages. GD is the most common lysosomal storage disease occurring in about 1 in 75,000 births worldwide, mostly affecting the Ashkenazi Jews [6–9]. Lysosomal storage diseases are a group of disorders caused by deficiencies of lysosomal enzymes necessary for the degradation of glycolipids and glycoproteins (Table 21.3).

TABLE 21.1 Immunophenotypic features of macrophages/histiocytes and subclass of dendritic cells.

Markers	MP	LC	IDC	FDC
CD1a	−	+	−	−
CD4	+	+	−	−
CD21	±	−	−	+
CD35	±	−	−	+
CD45	+	+	±	−
CD68	+	±	±	−
CD207 (Langerin)	−	+	−	−
HLA-DR	+	+	+	−
S-100	−	+	+	−
FcR	+	+	−	−
Lysozyme	+	±	±	−
NSE	+	−	−	−

MP: macrophage/histiocyte, LC: Langerhans cells, IDC: interdigitating dendritic cells, FDC: follicular dendritic cell, FcR: Fc IgG receptors, NSE: non-specific esterase.

TABLE 21.2 Disorders of monocytes and histiocytes.

Functional disorders
1. Lysosomal disorders
2. Chronic granulomatous disease
3. Defective monocyte chemotaxis
4. Others

Reactive disorders
1. Monocytopenia
2. Monocytosis
3. Granulomatous disorders
4. Hemophagocytosis
5. Sinus histiocytosis with massive lymphadenopathy
6. Others

Neoplastic disorders
1. Chronic myelomonocytic leukemia
2. Acute myelomonocytic leukemia
3. Acute monocytic leukemia
4. Histiocytic sarcoma

TABLE 21.3 Lysosomal storage disease.*

Sphingolipidoses
- Gaucher disease
- Niemann–Pick disease
- Farber disease
- Fabry disease
- Krabbe disease
- Metachromatic leukodystrophy
- GM1 gangliosidosis
- GM2 gangliosidosis (Tay–Sachs disease)

Mucopolysaccharidoses
- Mucopolysaccharidosis I (Hurler disease)
- Hunter disease
- Sanfilippo disease
- Morquio disease
- Maroteaux–Lamy syndrome
- Multiple sulfatase deficiency

Glycoproteinoses
- Sialidosis
- Fucosidosis
- Mannosidosis
- Aspartylglycosaminuria

Mucolipidoses
- Mucolipidosis II
- Pseudo-Hurler polydystrophy
- Mucolipodosis IV

Others
- Cystinosis
- Pompe disease
- Wolman disease

*Adapted from Ref. [20].

Pathogenesis and Molecular Genetics

GD results from deficiency of the lysosomal enzyme glucocerebrosidase (or acid beta-glucosidase) [6, 10, 11]. The deficiency is secondary to mutations in the glucocerebrosidase gene located on chromosome 1q21 [12]. More than 180 distinct mutations are listed in the Human Mutation Database for glucocerebrosidase gene, which are mostly point mutations. Three major mutant alleles are identified in affected patients: *N370S*, *L444P*, and *84GG* [6, 13–15]. These three mutations account for >90% of alleles in Ashkenazi patients.

The glucocerebrosidase deficiency leads to accumulation of glucocerebroside and other glycolipids in the macrophages leading to organomegaly, osteopenia, and cytopenia.

Pathology

Morphology

The morphologic features are characterized by the accumulation of glycolipid-laden macrophages, known as *Gaucher cells*, in the spleen, liver, bone marrow, and other tissues (Figures 21.1 and 21.2). Gaucher cells have abundant cytoplasm containing a large amount of hydrophobic glucocerebroside molecules in bilayered membranous sheets [16]. For this reason, the cytoplasm appears striated, like a wrinkled tissue paper. By electron microscopy, Gaucher cells reveal spindle- or rod-shaped, membrane-bound cytoplasmic inclusions consisting of numerous small, tubular structures measuring 13–75 nm in diameter [17].

The bone marrow biopsy sections demonstrate focal, interstitial, or diffuse accumulation of the Gaucher

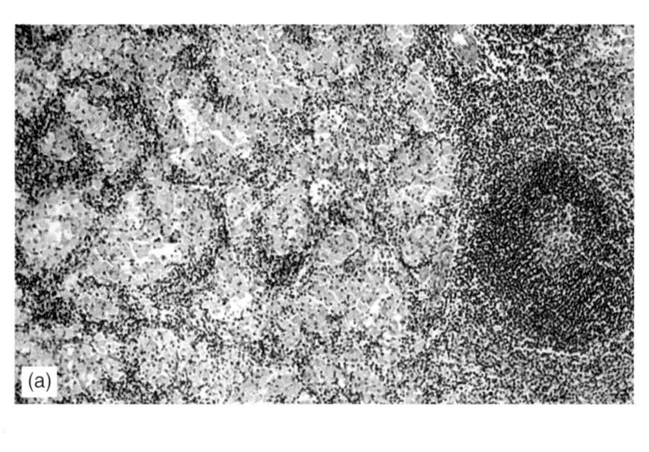

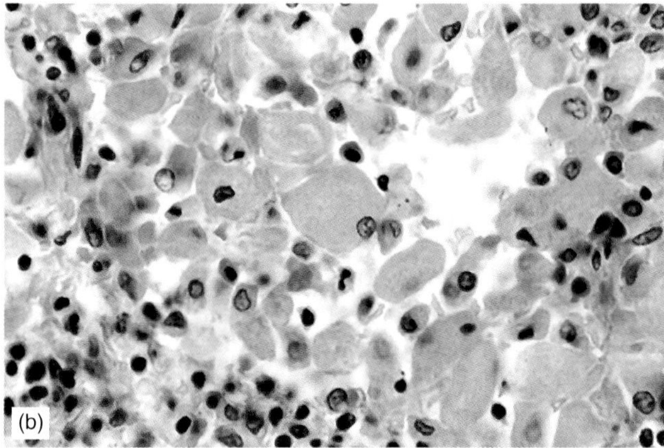

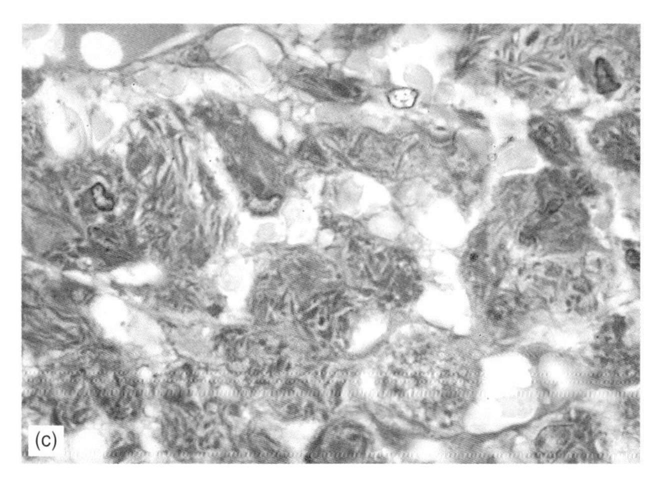

FIGURE 21.1 Gaucher disease. Section of spleen demonstrates diffuse infiltration of the red pulp by Gaucher cells: (a) low power and (b) high power. PAS stain shows accumulation of glucocerebroside molecules in membrane sheets creating cytoplasmic striation (c).

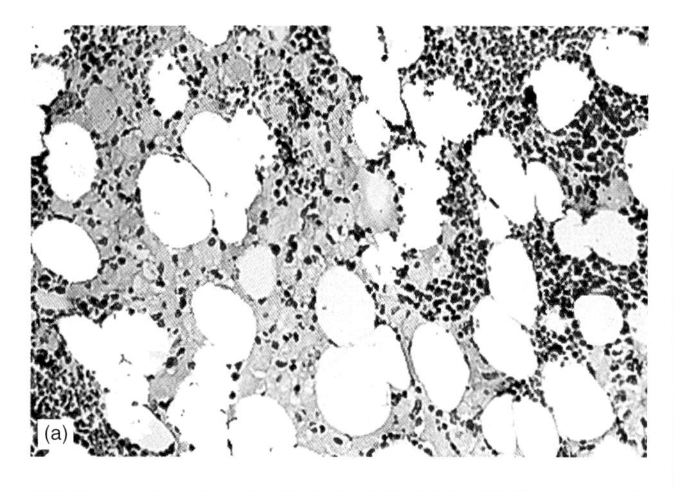

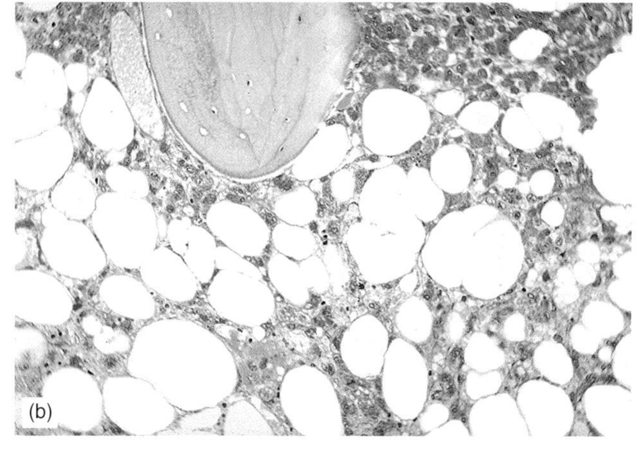

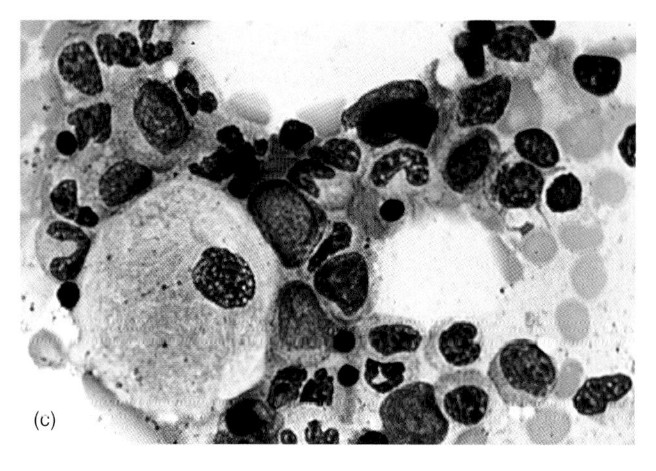

FIGURE 21.2 Gaucher disease. Bone marrow section demonstrates interstitial infiltration by Gaucher cells: (a) H&E, and (b) PAS stains. Bone marrow smear (c) shows one Gaucher cell characterized by abundant cytoplasm with striation.

cells, sometimes associated with fibrosis (Figure 21.2). Accumulation of Gaucher cells is also present in the splenic red pulp in virtually all patients [18]. Gaucher cells are also found in the centrilobular and portal areas of the liver, sometimes in association with hepatic fibrosis [19].

Immunophenotype and Special Stains

The Gaucher cells are glucocerebroside-laden macrophages and, therefore, demonstrate expression of monocyte-associated markers such as CD45, lysozyme, and CD68 by immunohistochemical stains (Table 21.1). They are typically negative for CD1a and S-100 protein. The Gaucher cells are PAS-positive and demonstrate a strong acid phosphatase activity, which is usually resistant to tartaric acid inhibition. They are also Sudan Black B and non-specific esterase (NSE) positive [20].

Clinical Aspects

There are three types of GD. All three types share splenomegaly, hepatomegaly, bone marrow involvement, anemia, and skeletal changes. They differ in the presence or lack of neurologic manifestations, type of mutation, ethnic predilection, and clinical outcome (Table 21.4).

Type 1 GD represents about 90% of the patients with GD and is predominantly seen in the Ashkenazi Jewish population. The disease may occur at any age, but about 70% of the patients are diagnosed by the age of 20 years [21, 22]. Hepatosplenomegaly is common and there is no neurologic manifestation. Mild-to-moderate degree of anemia and thrombocytopenia is present, mainly due to hypersplenism. Liver enzymes, serum angiotensin converting enzyme, and acid phosphatase levels may be elevated. The most common mutation is *N370S*. The clinical course is variable with an overall more rapid progression in children than in adults [23].

Type 2 GD is the rarest form and is also known as acute neuronopathic GD or infantile cerebral GD. It usually occurs in the first year of life and is characterized by extensive visceral involvement and rapidly progressive neurologic deterioration, including oculomotor dysfunction and bulbar palsy [24]. The average survival is <12 months [25].

Type 3 GD is a subacute or chronic neuropathic form with a later onset and more variable course than type 2 GD. A subtype of this category (type 1a) which is associated with *L444P* mutation, progressive dementia, and ataxia has been described in Norrbottnian region of Sweden (Table 21.3) [26].

Diagnosis is confirmed by the demonstration of reduced glucocerebrosidase activity in peripheral leukocytes and/or the presence of mutation by molecular genetic studies [27, 28].

Enzyme replacement therapy with recombinant glucocerebrosidase (imiglucerase) is the treatment of choice for most symptomatic type 1 patients [29]. Substrate reduction therapy, such as treatment with miglustat, is used for those patients who cannot afford the high expense of enzyme replacement therapy. Miglustat is an inhibitor of glycosylceramide synthesis and reduces glycolipid accumulation [9]. Splenectomy and bone marrow transplantations are other alternative therapeutic approaches [30, 31].

TABLE 21.4 Classification of Gaucher disease.

Features	Type 1	Type 2	Type 3
Onset	Variable	First year	Childhood
Anemia	+	+	+
Thrombocytopenia	+	–	–
Splenomegaly	+	+	+
Hepatomegaly	+	+	+
Bone marrow involvement	+	+	+
Skeletal changes	+	±	+
Neurologic manifestations	–	+	+
Mutations	*N370S*	Diverse	*L444P*
Ethnic predilection	Ashkenazi Jews	None	Norrbottnian, Sweden
Progression	Slow	Rapid	Variable

Differential Diagnosis

The differential diagnosis includes those conditions that are associated with splenomegaly and cytopenia, such as leukemia, lymphoma, collagen vascular diseases, and other lysosomal storage diseases. The diagnosis of GD is usually made by the detection of Gaucher cells in bone marrow and is confirmed by the reduced leukocyte glucocerebrosidase activity and molecular genetic studies.

Niemann–Pick Disease

Niemann–Pick disease represents a group of autosomal recessive disorders that are associated with tissue accumulation of sphingomyelin, splenomegaly, and manifestation of variable degrees of neurologic defects [32, 33].

Pathogenesis and Molecular Genetics

This group of disorders has been associated with mutations in three different genes: (1) sphingomyelin phosphodiesterase 1 gene (*SMPD1*) mapped at chromosome 11p15, (2) Niemann–Pick C1 gene (*NPC1*) on chromosome 18q11-q12, and (3) Niemann–Pick C2 gene (*NPC2*) on chromosome 14q24.3 (Table 21.5). Mutations of *SMPD1* are most prevalent in the Ashkenazi Jews. Three *SMPD1* mutations, R496L, L302P, and fsP330, account for >90% of the type 1 NPD [34, 35]. These mutations lead to the deficiency of SMPD enzyme and accumulation of lysosphingomyelin in the macrophages and the neural tissues. Lysosphingomyelin is believed to be toxic to the nervous system [35].

The *NPC1* gene product is localized in vesicles that transiently interact with cholesterol-laden lysosomes to facilitate sterol and probably glycolipid relocation [36, 37]. The *NPC2* gene encodes a small soluble lysosomal cholesterol-binding protein [33, 38]. The *NPC1* and *NPC2* mutations, therefore, play a role in the accumulation of cholesterol and glycolipids in neurons and other cells [35].

Pathology

Morphology

The characteristic morphologic feature of NPD is the presence of foamy cells in various tissues secondary to the duplication and expansion of the sphingomyelin-laden lysosomal system. The expansion of lysosomal structure may eventually lead to the total occupation of the cytoplasmic space, creating a foamy appearance of the affected cells, known as "Niemann–Pick cells" (NP cells) (Figures 21.3 and 21.4). By electron microscopy, the NP cells are loaded with lysosomes, which may appear as membrane-bound wavy concentric structures or homogenous (washed out or lucent) deposits [20, 39]. NP cells consist of macrophages in the spleen, bone marrow and lymph nodes, endothelial cells, neurons, Schwann cells, and retinal cells [33].

TABLE 21.5 Major classes of Niemann–Pick disease.

NPD	Gene	Chromosome
Type 1A	Sphingomyelin phosphodiesterase 1 gene (*SMPD1*)	11p15
Type 1S	Sphingomyelin phosphodiesterase 1 gene (*SMPD1*)	11p15
Type 2S	Niemann–Pick C1 gene (*NPC1*) or	18q11-q12
	Niemann–Pick C2 gene (*NPC2*)	14q24.3

The bone marrow biopsy sections reveal focal, interstitial, or diffuse accumulation of the NP cells (Figures 21.3 and 21.4). The accumulation of NP cells is also present in the splenic red pulp in virtually all patients and may be demonstrated in the centrilobular and portal areas of the liver and lymph node sinuses [18, 20].

Immunophenotype and Special Stains

The NP cell accumulation in the hematopoietic tissues such as bone marrow, spleen, and lymph nodes are sphingomyelin-laden macrophages and, therefore, demonstrate expression of monocyte-associated markers such as CD45, lysozyme, and CD68 by immunohistochemical stains. They are typically negative for CD1a and S-100 protein (Table 21.1). The NP cells are Sudan Black B and NSE positive and contain lipopigment that stains intensively with iron hematoxylin [20, 40]. Cresyl violet and PAS stains are either negative or weakly positive [20, 40]. Filipin and BC theta stains are helpful in demonstrating unesterified lysosomal cholesterol droplets in the fibroblast cell culture or neurons by fluorescence or confocal microscopy [41, 42].

Clinical Aspects

Currently, NPD is divided into three major types: 1A, 1S, and 2S [33, 35].

Type 1A is the acute neuronopathic form (formerly known as type A) and the most common type of NPD with the highest incidence among Ashkenazi Jews. The affected children, usually >1 year of age, present with hepatosplenomegaly, feeding difficulties, and loss of motor skills with a progressive clinical downhill and death within 2–3 years [43]. Type 1A is caused by mutations of *SMPD1* gene (Table 21.6). Decreased serum HDL, increased serum LDL, and hypertriglyceridemia are frequent laboratory findings [43].

Type 1S is a chronic non-neuronopathic form of NPD which occurs during infancy or childhood (formerly called type B). It is characterized by hepatosplenomegaly, short stature, delayed skeletal maturation, and ocular abnormalities. The affected children usually survive into their adulthood [44]. In type 1S, the SMPD1 enzyme activity is partially preserved, and decreased serum HDL, increased serum LDL, and hypertriglyceridemia may be present. Type 1C is a subtype of this category representing the adult non-neuronopathic form associated with hepatosplenomegaly.

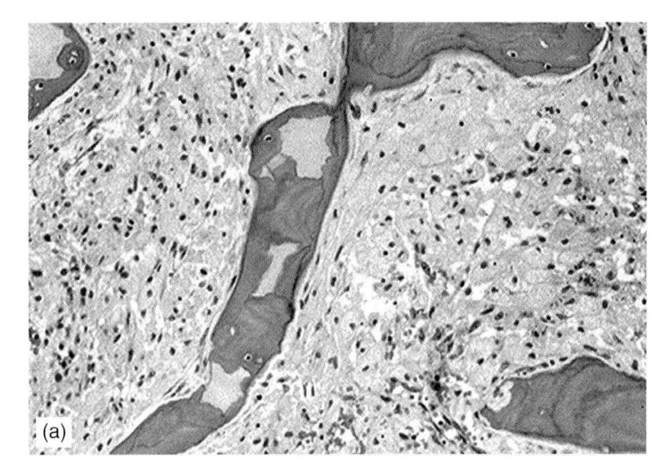

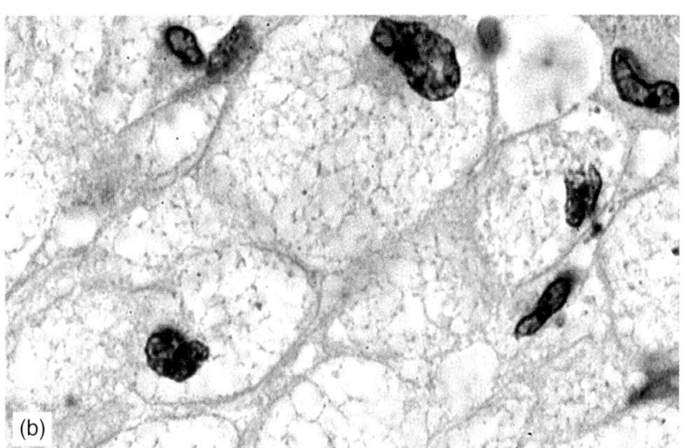

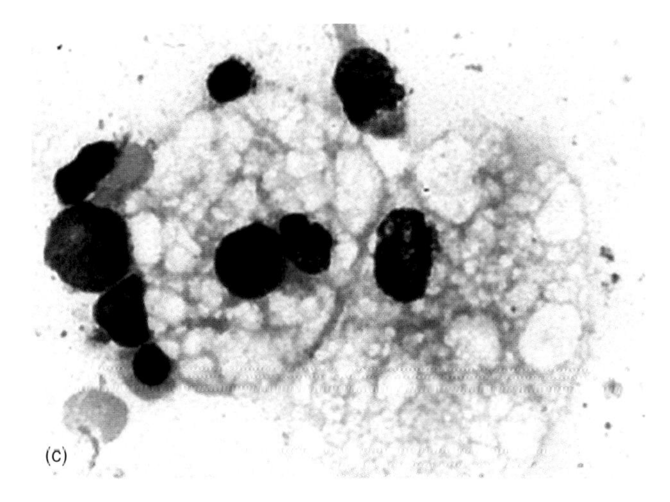

FIGURE 21.3 Niemann–Pick disease. Bone marrow biopsy section demonstrates diffuse infiltration of large vacuolated histiocytes: (a) low power and (b) high power views. Two large vacuolated histiocytes are shown in bone marrow smear (c).

Type 2S represents the abnormalities of intracellular transport of cholesterol and its sequestration in lysosomes [35]. This functional defect is the result of mutations in *NPC1* and *NPC2* genes (Table 21.6). Most patients demonstrate neurologic disease with a late infantile or juvenile onset. Cerebellar involvement, dystonia, ophthalmoplegia, and seizures are among neurologic manifestations [33, 35]. Type 2S consists of two subtypes: type C (more frequent, younger age

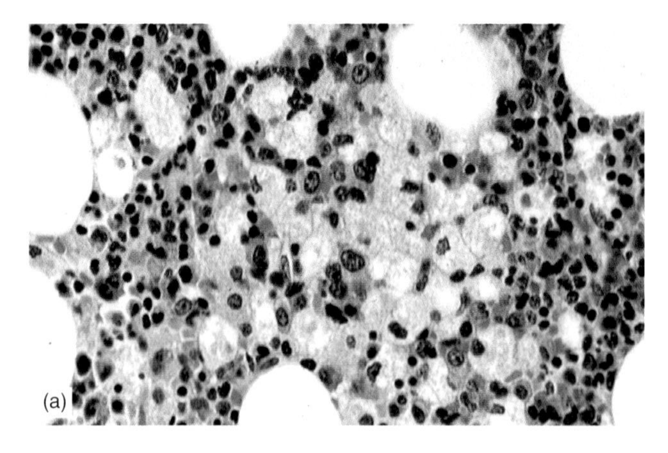

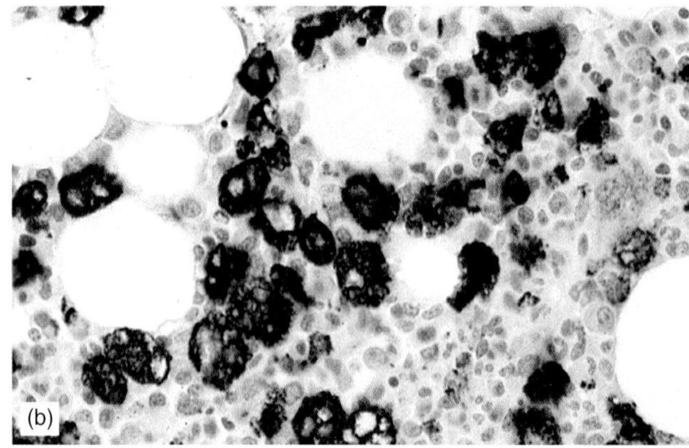

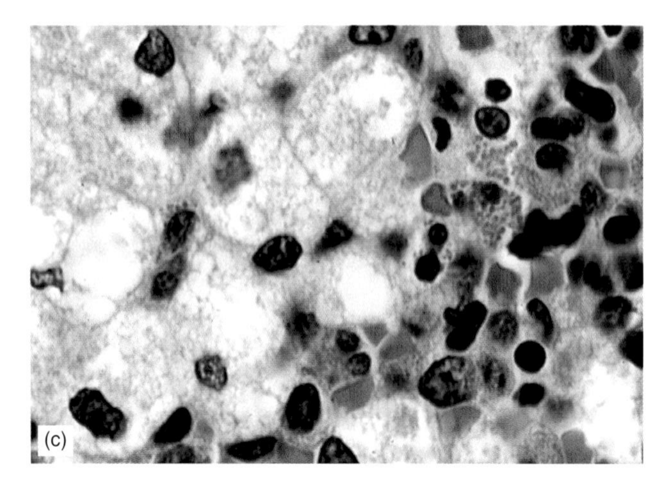

FIGURE 21.4 Niemann–Pick disease. Bone marrow biopsy section demonstrates clusters of finely vacuolated histiocytes (a) that are CD68-positive (b). A high power view is shown in (c).

of onset, and more aggressive) and type D (later age of onset and slower progress).

There is no effective treatment for NPD. Cholesterol-lowering drugs reduce the hepatic-free cholesterol levels but do not change the clinical course. Bone marrow transplantation does not reverse the neurologic symptoms [45, 46].

Differential Diagnosis

The differential diagnosis includes neurologic disorders and conditions that are associated with hepatosplenomegaly.

TABLE 21.6 Conditions associated with monocytosis.*

Infections
1. Tuberculosis
2. Syphilis
3. Subacute bacterial endocarditis
4. Cytomegalovirus infection
5. Disseminated candidiasis
6. Others
Inflammatory and immune-associated disorders
1. Myositis
2. Temporal arteritis
3. Polyarteritis nodosa
4. Rheumatoid arthritis
5. Inflammatory bowel diseases
6. Alcoholic liver disease
7. Others
Hematologic disorders
1. Neoplastic disorders
(a) Chronic myelomonocytic leukemia
(b) Acute myelomonocytic and acute monocytic leukemias
(c) Lymphoma
(d) Plasma cell myeloma
2. Non-neoplastic conditions
(a) Hemolytic anemia
(b) Chronic neutropenia
(c) Postsplenectomy
(d) Idiopathic thrombocytopenic purpura
(e) Others
Non-hematopoietic malignancies
Drug-induced
1. Chlorpromazine
2. Ampicillin
3. Glucocorticoids
4. Others

*Adapted from Ref. [20].

Diagnosis is established by morphologic findings in biopsy sections, use of special stains (such as Filipin and BC theta stains), and molecular genetic studies demonstrating mutations of *SMPD1*, *NPC1*, or *NPC2*.

Chediak–Higashi Syndrome

Chediak–Higashi syndrome (CHS) is a rare autosomal recessive disorder characterized by severe immune deficiency, partial albinism, bleeding tendencies, and recurrent bacterial infections [47, 48]. The defective gene, *LYST*, is a lysozyme-trafficking regulator mapped at 1q42 [49, 50]. *LYST* is a large, highly conserved gene encoding a protein with 3,801 amino acids and a predicted molecular weight of 429 kDa [51]. Multiple different mutations in CHS patients have been described, but molecular genetic testing is not yet readily available. The *LYST* gene is involved with intracellular protein trafficking, and its mutation may lead to impairment in the fusion of cytoplasmic vesicles [47, 48].

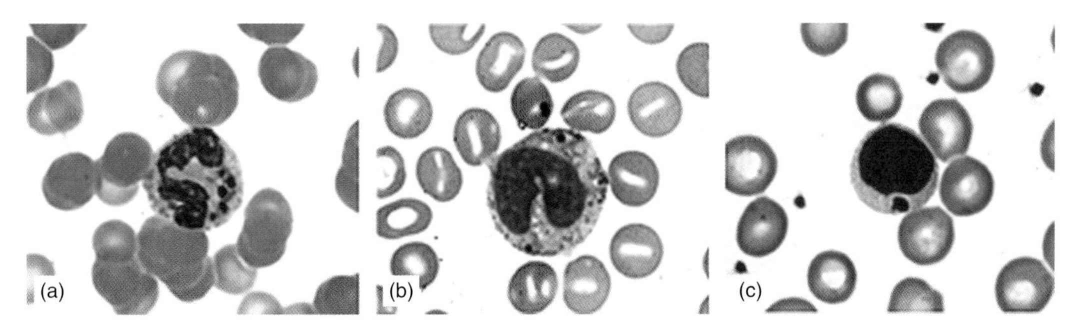

FIGURE 21.5 Chediak–Higashi syndrome. Cytoplasmic granules of variable sizes are present in neutrophils (a), monocytes (b) and lymphocytes (c).

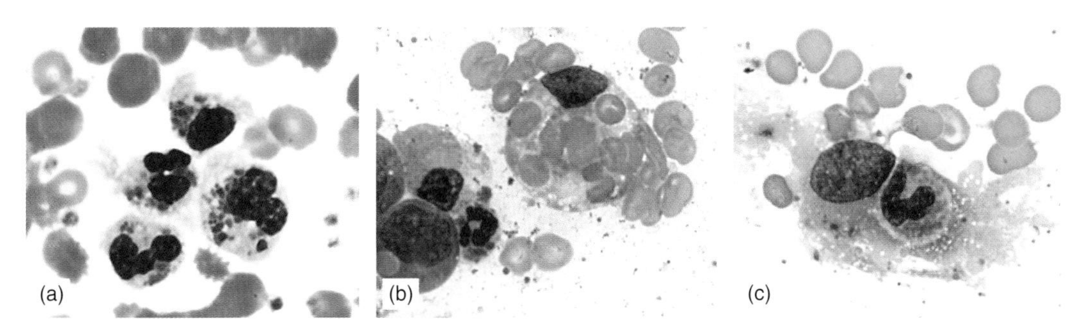

FIGURE 21.6 Chediak–Higashi syndrome in accelerated phase. Blood smear demonstrates several white cells with cytoplasmic granules (a). Bone marrow smear shows an erythrophagocytic histiocytes (b) and a histiocyte containing a neutrophilic band (c).

The morphologic hallmark of CHS is the presence of giant cytoplasmic granules in various cells. These granules include lysosomes (such as monocytes, granulocytes, cytotoxic T-, and NK-cells), melanosomes (melanocytes), and cytoplasmic granules in Schwann cells (Figures 21.5 and 21.6) [47, 52]. The platelets contain abnormal dense bodies [47].

Clinical features are presented in early childhood and reflect the functional defects in leukocytes, melanocytes, platelets, and Schwann cells and consist of recurrent pyogenic infection, partial oculocutaneous hypopigmentation, coagulation defect with petechiae, bruising and mucosal bleeding, and neurologic disturbances such as peripheral neuropathy and dysfunction of the spinal tract and the cerebellum [53]. The evolution to an "accelerated phase" had been reported in some cases characterized by T-cell lymphocytosis and hemophagocytic histiocytosis (Figure 21.6) [47, 52]. An association between EBV infection and accelerated phase has been observed, suggesting that the hemophagocytic lymphohistiocytosis (HLH) is EBV-induced [54, 55]. The treatment of choice is bone marrow transplantation, which improves leukocyte and platelet defects and immunologic problems [47].

Monocytopenia and Monocytosis

Monocytopenia is less frequent than monocytosis and occurs in aplastic anemia, hairy cell leukemia, severe thermal injuries, and treatment with corticosteroids [20, 56–58]. Decreased monocyte blood count has also been reported in patients with rheumatoid arthritis (RA), systemic lupus erythematosus (SLE), and AIDS [59, 60]. Cyclic neutropenia may be associated with intermittent monocytopenia [20].

Monocytosis is observed in a wide variety of conditions such as chronic and subacute infections, collagen vascular disorders, hematologic and non-hematologic malignancies, hemolytic anemia, and idiopathic thrombocytopenic purpura (Table 21.6). Certain drugs such as chlorpromazine, ampicillin, and tetrachloroethane may also induce monocytosis.

Reactive Histiocytic Proliferations

Reactive histiocytic proliferation is often associated with other inflammatory cells such as lymphocytes, plasma cells, eosinophils, and neutrophils. The reactive histiocytes may be diffusely intermixed with other inflammatory cells or may appear as aggregates of cohesive (epithelioid) histiocytes standing alone or accompanied by other inflammatory cells, making granulomas (see Chapter 5). Reactive histiocytic proliferations may be secondary to infectious diseases, autoimmune disorders, or malignancies, or may be idiopathic.

Infections

A wide variety of infections cause histiocytic proliferation predominantly in epithelioid clusters or granulomatous formation (see Chapter 5). For example, mycobacterial infections, syphilis, leprosy, Q fever, cat-scratch fever, and fungal infections typically cause granulomatous formation in the bone marrow, lymph nodes, and other tissues (see Figures 5.5 and 5.6). Small clusters of epithelioid histiocytes are present in the lymph nodes infected with *Toxoplasma gondii* and occasionally may contain microorganism. Protozoan-laden epithelioid histiocytes are identified in the bone marrow, lymph nodes, and other tissues of patients with leishmaniasis (see Figure 5.8).

Autoimmune Disorders

Autoimmune disorders such as SLE and RA are sometimes associated with lymphadenopathy and increased histiocytic proliferations. In SLE, affected lymph nodes show follicular hyperplasia and areas of necrosis in paracortical areas (see Chapter 6). Necrotic areas are surrounded by histiocytes and other inflammatory cells [61, 62]. The enlarged lymph nodes in RA show follicular hyperplasia, sinus histiocytosis, interfollicular plasmacytosis, and deposition of PAS-positive, Congo-red-negative hyaline material (see Figure 6.1) [61, 63, 64].

Tumor-Associated

Histiocytes are one of the prominent background inflammatory cells in Hodgkin lymphoma. They may also appear as discrete epithelioid aggregates or granulomas. Histiocytic proliferation is also frequently observed in T-cell malignancies, sometimes associated with hemophagocytosis (discussed later). In the peripheral T-cell lymphoma of Lennert type, neoplastic T-cells are mixed with sheets or aggregates of epithelioid histiocytes (see Chapter 17). Metastatic carcinomas to lymph nodes may be associated with sinus histiocytosis.

Idiopathic

Sarcoidosis, Kikuchi disease, Erdheim–Chester disease, and sinus histiocytosis with massive lymphadenopathy are examples of reactive histiocytic proliferations with no known etiology (see Chapter 6).

Sarcoidosis is a rare granulomatous disorder, more prevalent in Blacks than in Caucasians and in women than in men [61]. Mediastinal and pulmonary hilar lymph nodes are most frequently affected, but other tissues and organs such as the bone marrow, liver, spleen, and lungs may also be affected. The affected tissues show multiple well-defined granulomas consisting of epithelioid histiocytes and multinucleated giant cells without significant necrosis (see Figure 6.9) [65].

Kikuchi disease is a rare histiocytic necrotizing lymphadenitis, more frequently seen in Asia than in Western countries [61]. It is more prevalent in young adult women. The affected lymph nodes show discrete or confluent eosinophilic areas consisting of phagocytic and non-phagocytic histiocytes, lymphocytes, immunoblasts, eosinophils, and plasma cells with rare or absent neutrophils (see Figure 6.14).

Erdheim–Chester disease is a rare condition characterized by a symmetrical sclerosis of lower extremities and involvement of many organs and tissues, including the lung, kidney, orbit, skin, pericardium, and retroperitoneum [66, 67]. The affected tissues show aggregates or sheets of lipid-laden histiocytes and scattered multinucleated Touton giant cells (lipid-laden histiocytes in which multiple nuclei are grouped around a small island of cytoplasm) (see Figure 5.9).

Sinus histiocytosis with massive lymphadenopathy (Rosai–Dorfman disease) is a rare condition affecting children and young adults. It is characterized by massive, bilateral, painless, cervical lymphadenopathies, sometimes with extranodal involvement such as the skin, bone, upper respiratory tract, or central nervous system [61]. The cortical and medullary sinuses of the affected lymph nodes are markedly dilated and filled with large histiocytes with abundant, pale, vacuolated cytoplasm. These histiocytes show evidence of emperipolesis by containing intact lymphocytes, and less frequently, plasma cells, neutrophils, or erythrocytes (see Figure 6.7) [68].

Iatrogenic

Lymphangiogram-associated lipogranuloma (Figure 21.7), prosthetic- or implanted-induced foreign body reactions are among the examples of iatrogenic histiocytic proliferations [61, 69, 70].

Hemophagocytic Lymphohistiocytosis

Hemophagocytic lymphohistiocytosis (HLH) is characterized by proliferation of reactive non-dendritic histiocytes with evidence of hemophagocytic activities [71–74]. The hemophagocytosis is either associated with an underlying genetic disorder or secondary to an underlying infection, autoimmune or neoplastic process [75].

Etiology and Pathogenesis

The primary underlying pathophysiologic mechanism in HLH appears to be related to cytokine abnormalities, leading to uncontrolled accumulation of activated T-lymphocytes and histiocytes [76, 77]. High plasma levels of interferon-gamma (IFN-γ), tumor necrosis factor-alpha (TNF-α), interleukin (IL)-1, IL-2, IL-6, IL-10, IL-12, and IL-18 have been reported in patients with HLH [77–79]. Mutation of perforin gene (*PRF1*) plays an important role in the pathogenesis of HLH in a significant proportion of patients. Perforin protein is stored in the cytoplasmic granules of large granular lymphocytes, monocytes, and other hematopoietic precursors, and plays an important role in the formation of pores in the membrane of target cells. Once the target cell membrane is perforated, granzymes and other cytolytic components are able to enter the target cells. A defective perforin protein results in the failure of killing the target cells and removal of the antigenic stimulation by viruses or other infectious agents. The persistent stimulation of T-cells results in the production of large amounts of cytokines and activation of macrophages.

There are four major subtypes of hemophagocytic lymphohistiocytosis: (1) familial HLH, (2) infection-associated HLH, (3) hemophagocytosis associated with autoimmune and immunodeficiency disorders, and (4) cancer-associated HLH.

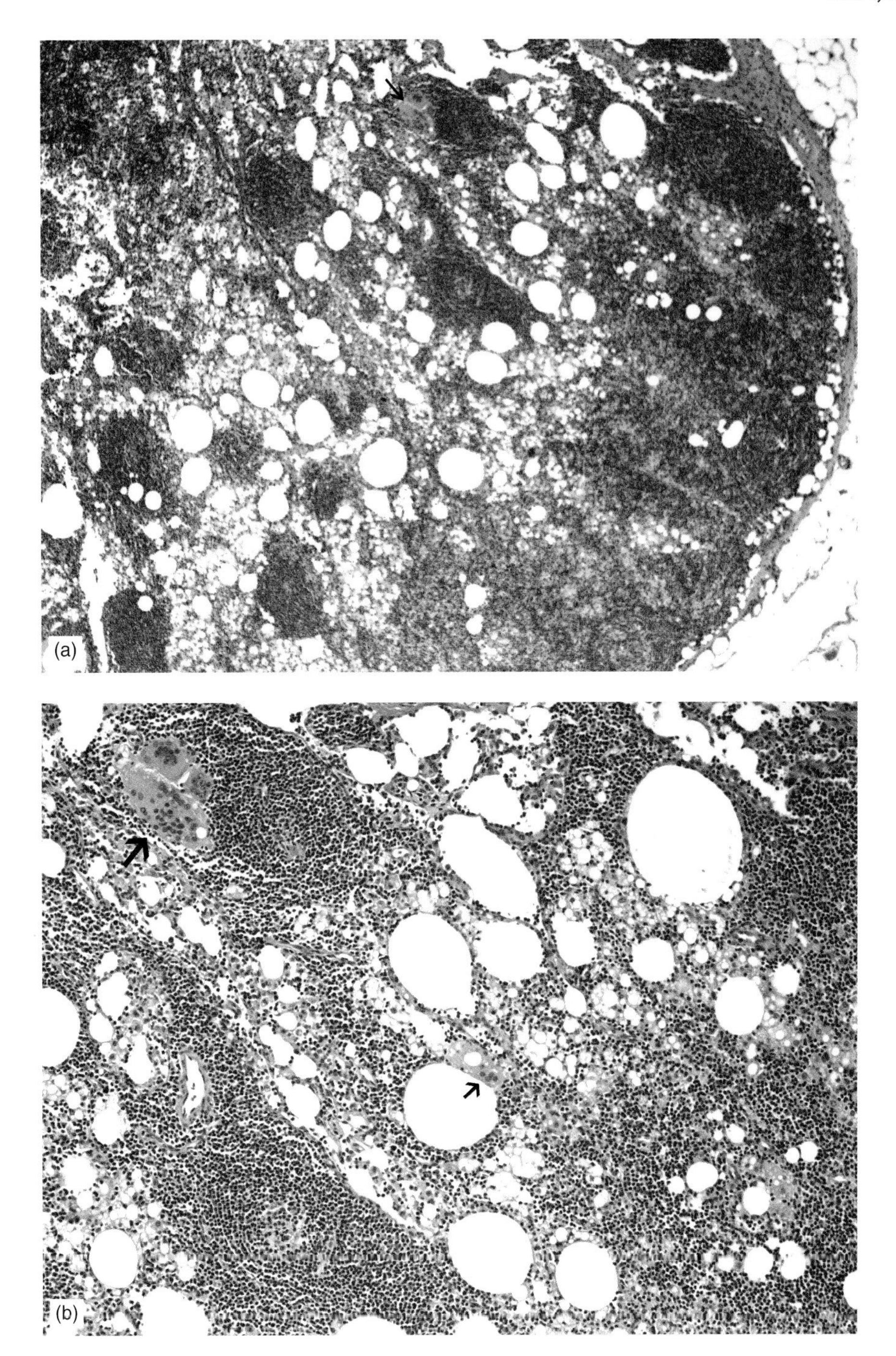

FIGURE 21.7 Lymphangiogram-associated lipogranuloma. Lymph node section reveals sinus histiocytosis with numerous foamy, lipid-containing histiocytes and the presence of multinucleated giant cells (arrows): (a) low power, (b) intermediate power, and (c) high power views. Courtesy of G. Pezeshkpour, M.D., Department of Pathology, VA Greater Los Angeles Healthcare System.

Familial HLH is an autosomal recessive disorder affecting infants from birth to 18 months of age [80, 81]. Mutation of *PRF1, UNC13D*, and *STX11* genes is most frequently reported in the familial type [82–84].

Infection-associated HLH has been reported in a wide variety of viral and bacterial infections including EBV, CMV, parvovirus, herpes simplex, varicella zoster, measles, HIV, tuberculosis, gram-negative bacteria, as well as fungal and parasitic infections [72, 75, 85–88].

Hemophagocytosis associated with autoimmune and immunodeficiency disorders have been observed in SLE, RA, polyarteritis nodosa, pulmonary sarcoidosis, Sjogren's syndrome, and a wide variety of immunodeficiency conditions such as X-linked lymphoproliferative syndrome, Kawasaki disease, and Chediak–Higashi syndrome [75, 89–92].

Cancer-associated HLH has been reported primarily in T-cell lymphoid malignancies as well as NK-cell leukemia and B-cell lymphoma [93–96].

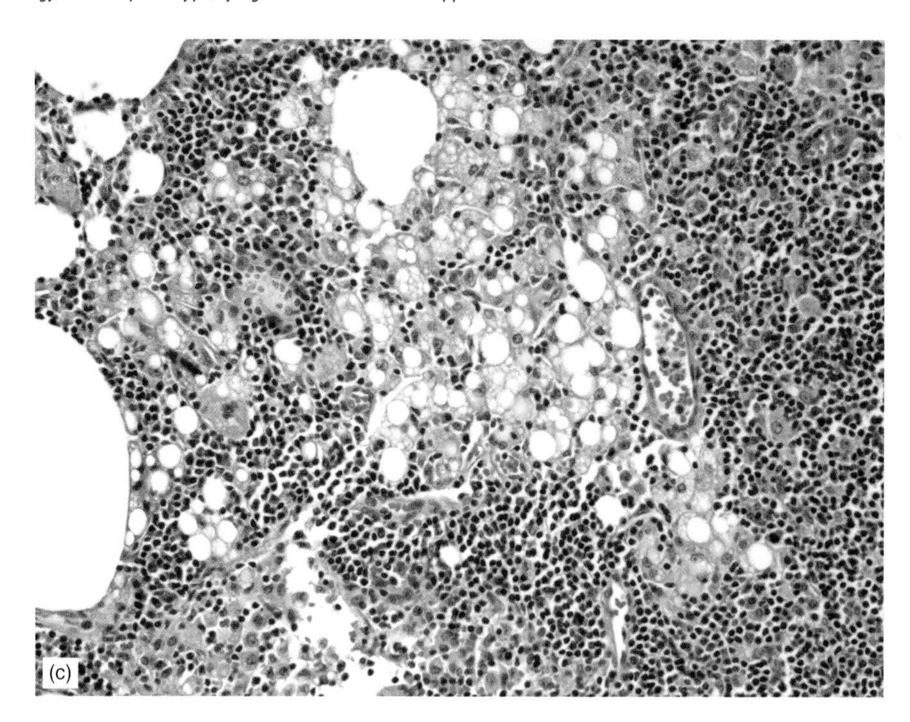

FIGURE 21.7 (Continued)

Pathology

Morphology and Laboratory Findings

HLH is usually a systemic disorder involving various organs with predilection of bone marrow, spleen, liver, and lymph nodes (Figures 21.8–21.10). Other sites of involvement include meninges, lungs, gastrointestinal tract, thymus gland, and genitourinary system. Cutaneous manifestations are rare [97]. The involved tissues show increased number of a mixture of lymphocytes and histiocytes in focal, interstitial, sinusoidal, or diffuse patterns. The histiocytes show abundant finely granular or vacuolated cytoplasm. The nucleus is bland, round, oval, or cleaved, and the nucleoli are inconspicuous. Numerous histiocytes show hemophagocytosis, which is predominantly erythrophagocytosis, and also includes phagocytosis of other hematopoietic cells such as platelets, neutrophils, and lymphocytes. There is a lack of significant cytologic atypia, and mitotic figures are absent or rare. Bone marrow appears to be the best tissue resource for the establishment of the diagnosis. However, in some cases, repeated bone marrow samples are required to document hemophagocytosis.

Other findings include cytopenia, hypertriglyceridemia and/or hypofibrinogenemia, low or absent NK-cell activity, and elevated serum ferritin concentration and soluble CD25 (Table 21.7) [98].

Immunophenotype and Cytochemical Stains

Histiocytes are of the non-dendritic cell type and therefore express CD68 and lysozyme. They are negative for CD1a and S-100 by immunohistochemical stains (Table 21.1). Histiocytes show positive reactions for NSE, acid phosphatase, and alpha-1-antitrypsin.

Molecular and Cytogenetic Studies

Mutation of *PRF1*, *UNC13D*, and *STX11* genes is most frequently reported in the familial type [82, 83]. In one study, 30% of the German patients and 80% of patients from Turkish origin with familial HLH showed *PRF1*, *UNC13D*, or *STX11* mutation [82]. The *PRF1* gene is mapped to the long arm of chromosome 10 in the 10q21-22 region [77].

Clinical Aspects

The initial signs and symptoms of HLH may simulate systemic infection, hepatitis, or encephalitis [75]. The most common clinical signs include fever and hepatomegaly (~90%), splenomegaly (~80%), neurologic symptoms (~45%), and lymphadenopathy (~40%) [75]. The clinical outcome is poor, and delay in therapy may lead to irreversible multiorgan failure and death. The recommended treatment is based on the guidelines provided by HLH-94 protocol [99], which includes induction therapy by dexamethasone and etoposide, followed by cyclosporine and dexamethasone. With this protocol, the reported overall 3-year survival is about 55%. Hematopoietic stem cell transplantation is the treatment of choice [99].

Differential Diagnosis

HLH clinically may mimic multiple organ failure syndrome including respiratory, cardiovascular, hepatic, and renal failures. The CNS involvement may simulate encephalitis, and pancytopenia may suggest bone marrow failure or leukemia. The majority of the HLH cases are associated with underlying causes such as infection and autoimmune disorder of lymphoid malignancies. HLH is distinguished from LCH by the presence of numerous hemophagocytic cells, expression

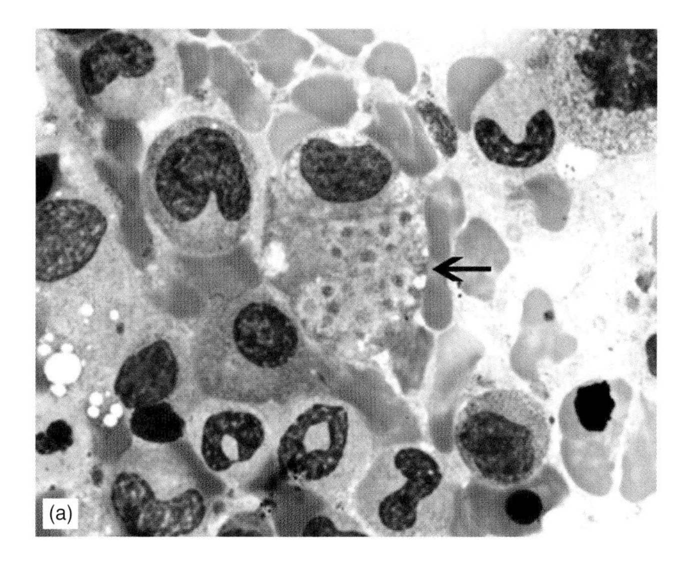

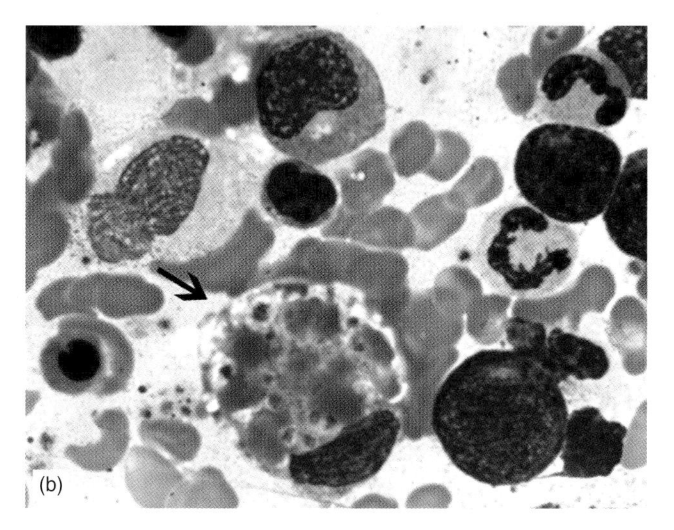

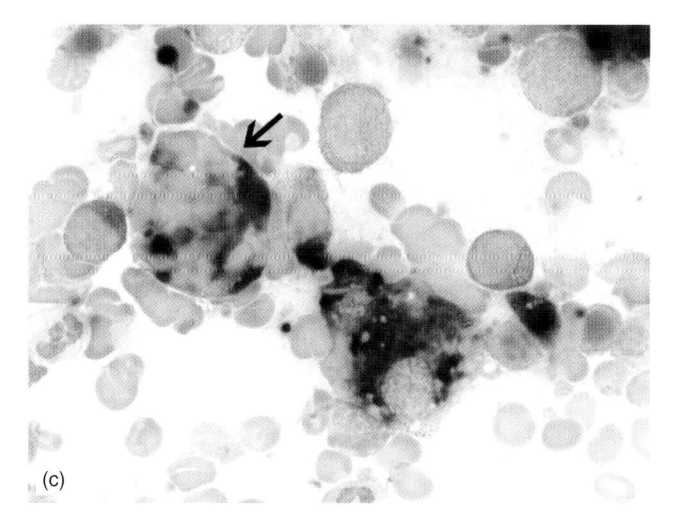

FIGURE 21.8 Hemophagocytosis. Bone marrow smear demonstrates histiocytes containing platelets (a, arrow) and erythrocytes (b, arrow). Immunohistochemical stain for CD68 shows a positive hemophagocytic histiocyte (c, arrow).

of NSE, lysozyme, and CD68, and lack of expression of CD1a and S-100. No Birbeck granules are demonstrated by electron microscopy (discussed later). Clinical manifestations and the presence of numerous hemophagocytic histiocytes separate HLH from other histiocytic disorders such as lysosomal storage diseases.

Histiocytic Sarcoma

Histiocytic sarcoma is a rare extramedullary malignant neoplasm demonstrating morphologic and immunophenotypic features of mature tissue histiocytes with lack of expression of lymphoid and dendritic-cell-associated markers. Acute myelogenous leukemias with monocytic differentiation are excluded [5, 100–102].

The etiology and pathogenesis are not known. The extranodal involvement is frequent. The infiltrating tumor cells show a diffuse growth pattern composed of large cells with abundant eosinophilic cytoplasm, round or irregular nuclei with vesicular chromatin, and one or more distinct nucleoli. Some cases may show significant pleomorphism with focal areas of spindle cells or the presence of multinucleated giant cells, and/or evidence of hemophagocytosis [100]. Histiocytic sarcoma may resemble diffuse large B-cell lymphoma, anaplastic large cell lymphoma, carcinoma, or melanoma [100–102].

Immunohistochemical stains show strong positivity for CD68 and CD163. Lysozyme stain may be weakly positive and there may be focal positivity for S-100 protein. The flow cytometric studies reveal a group of cells expressing CD45, CD4, CD11c, CD14, CD64, and HLA-DR. CD33 and CD34 are negative [5, 100, 102].

Histiocytic sarcoma is an aggressive disease with a poor response to chemotherapy.

DENDRITIC CELL DISORDERS

Langerhans Cell Histiocytosis

Langerhans cell histiocytosis (LCH), previously referred to as histiocytosis X, is a neoplasm of Langerhans dendritic cells characterized by the expression of CD1a and S-100 and the presence of ultrastructural cytoplasmic Birbeck granules [5]. LCH represents a wide variety of clinical manifestations with their specific terminology such as *eosinophilic granuloma* (solitary bone or extraosseous lesions), *HandSchuller–Christian* (multifocal, unisystem) disease, *Letterer–Siwe* (multiple organ system) disease, and *Hashimoto–Pritzker* (spontaneously resolving) syndrome (Table 21.8) [5, 102–108].

Etiology and Pathogenesis

The etiology and pathogenesis of LCH are not known. The LC proliferation appears to be clonal, but it is still not clear whether this clonal proliferation is induced by environmental conditions (viruses, cytokines) or genetic predispositions,

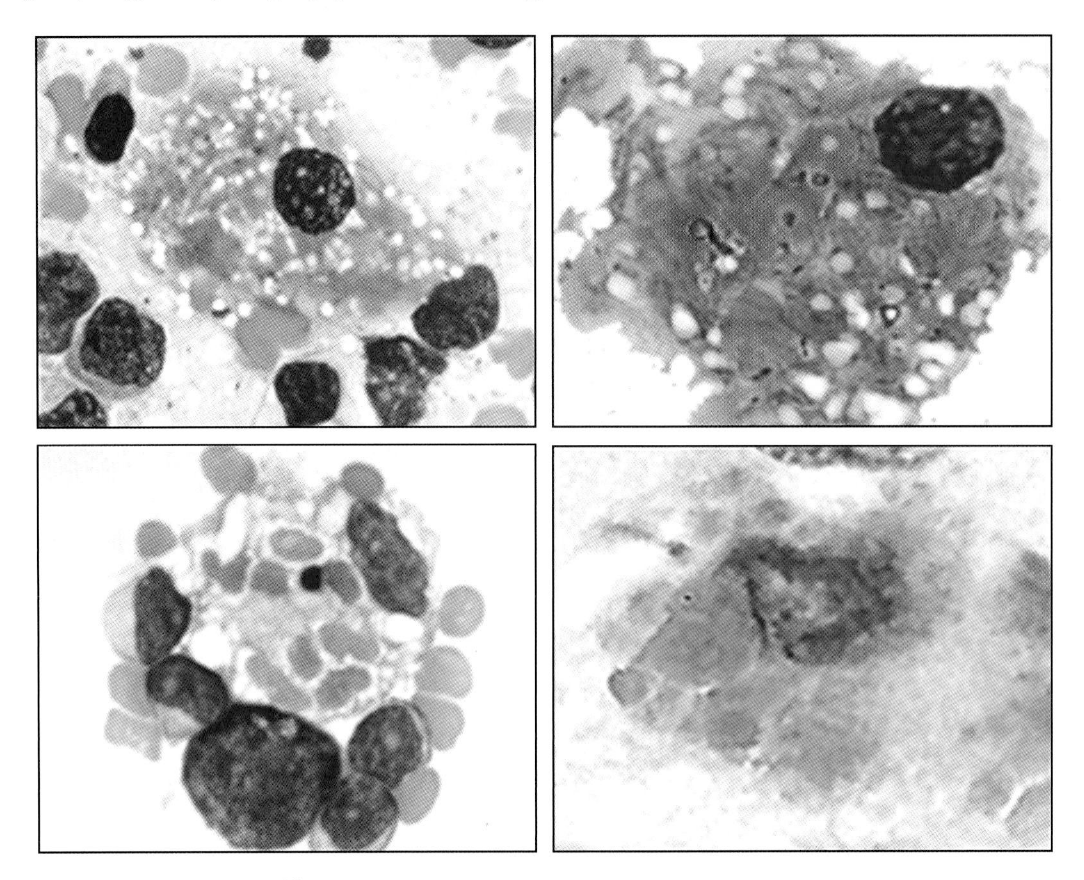

FIGURE 21.9 Erythrophagocytic histiocytes in bone marrow.

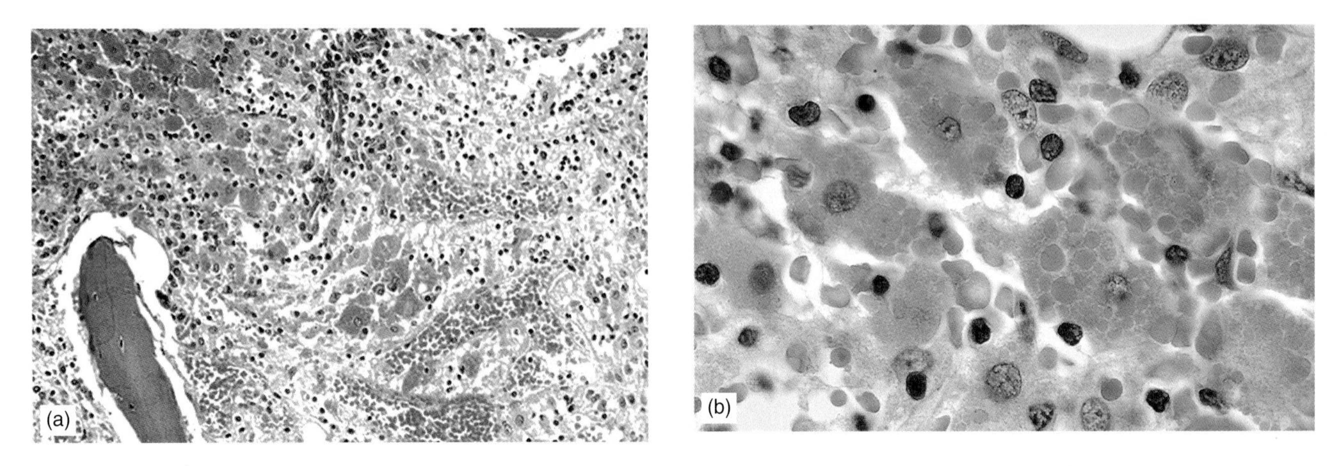

FIGURE 21.10 Bone marrow sections after chemotherapy and/or irradiation may show erythrophagocytic histiocytes: (a) low power and (b) high power views.

or both [108–113]. The significance of human herpes virus-6 (HHV-6) or EBV infections in the pathogenesis of LCH is still not clear [114–116].

Some investigators believe that development of LCH is primarily the result of an immunologic dysfunction and elevated levels of numerous cytokines such as GM-CSF, IFN-γ, IL-1, IL-4, and IL-10 [117]. It has been suggested that aberrant expression of certain chemokine receptors such as CCR6, CRR7, and CCL20/MIP-3α may play a role in pathogenesis of LCH [109, 118].

An association between HLA and LCH has been reported. For example, HLA-DRB1 was frequently found in the Nordic patients with unisystem LCH, whereas HLA-Cw7 and HLA-DR4 were found more prevalent in Caucasian LCH patients with solitary bone lesions [119, 120]. Several molecular cytogenetic abnormalities have been reported in patients with LCH, such as gains of DNA copy number on chromosomes 2q, 4q, and 12 and losses of DNA sequences on chromosomes 1p, 5, 6, 16, and 22q [121].

The cells of LCH are immature dendritic cells with reduced or absent antigen presenting capability and loss of ability to migrate from the involved tissue [122]. Altered cellular biology in LCH is also evident by the overexpression of Bcl-2, Ki-67, TGF-bR1, p53, RB, p16, and p21 [123, 124].

Pathology

Morphology

In the H&E sections, LC appear as large cells with abundant eosinophilic or pale cytoplasm, grooved, folded, indented, or convoluted nuclei with fine chromatin, and inconspicuous nucleoli (Figures 21.11 and 21.12). Mitotic figures are rare and hemophagocytosis is infrequent. The involved tissues are focally or diffusely infiltrated by the LC, often with increased eosinophils. Multinucleated giant cells and areas of necrosis may be present (Figure 21.12a). In more chronic lesions, cellular elements are replaced by fibrosis. Bone marrow smears or tissue touch preparations may show the presence of large histiocytic cells with abundant gray-blue vacuolated or finely granular cytoplasm (Wright's stain), sometimes with dendritic cytoplasmic projections (Figure 21.12c).

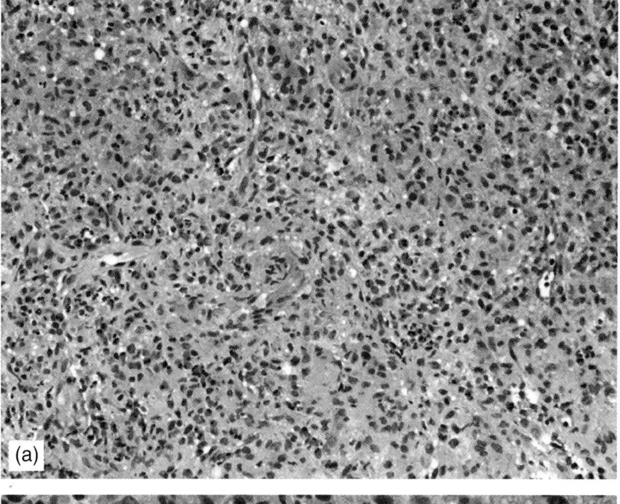

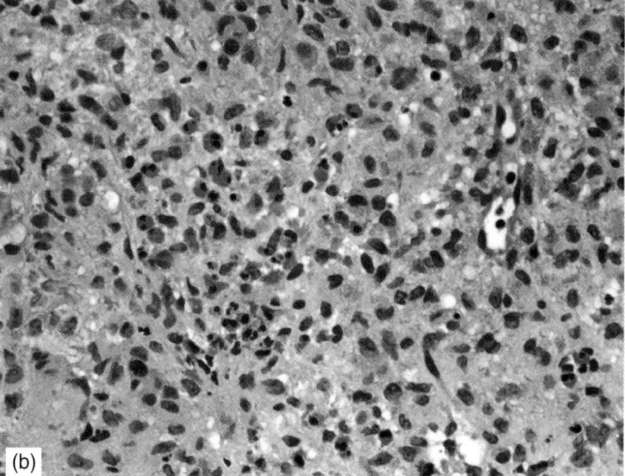

FIGURE 21.11 Langerhans cell histiocytosis. Bone marrow biopsy section demonstrates diffuse infiltration of large mononuclear cells with abundant eosinophilic cytoplasm. Scattered eosinophils are present: (a) low power and (b) high power views.

TABLE 21.7 Diagnostic criteria for hemophagocytic lymphohistiocytosis.*

Major criteria **1.** High fever for ≥7days **2.** Splenomegaly **3.** Cytopenia **4.** Hypertriglyceridemia or hypofibrinogenemia **5.** Hemophagocytosis
Minor criteria A. Low or absent NK-cell activity B. Serum ferritin level >500 μg/L C. Soluble CD25 >2400 U/mL
Diagnosis of HLH requires all 5 major criteria or 4 major criteria plus (A), or 4 major criteria plus (B) and (C).

*Adapted from Ref. [98].

TABLE 21.8 Clinical classification of Langerhans cell histiocytosis.*

Type	Clinical features
Unifocal	Single site of involvement, most commonly bone Older children and adults Good prognosis
Multifocal single system	Multiple sites of involvement, most commonly bone Young children Intermediate prognosis
Multifocal multisystem	Multiple involved sites in more than one organ system Most commonly bone, skin, liver, spleen, and lymph nodes Children <2 years of age Poor prognosis
Congenital self-healing	Multiple skin lesions involving neonates and infants Self-healing involution
Pulmonary LCH	Young adult smokers Indolent, progression to pulmonary fibrosis

*Proposed by the Histiocyte Society (Favara B, *Med Pediatr Oncol* 1997) and adapted from Ref. [122].

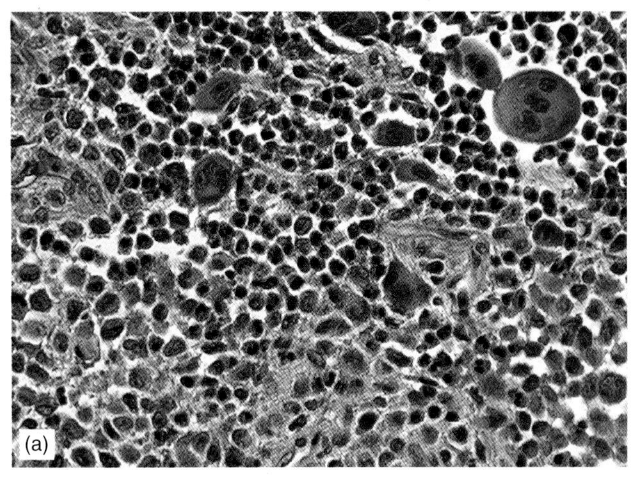

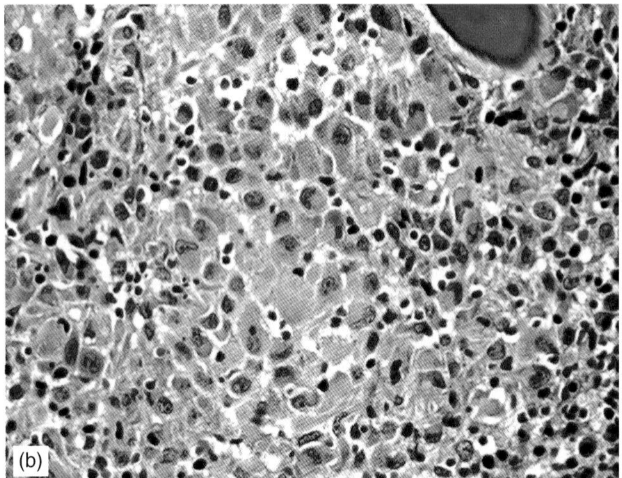

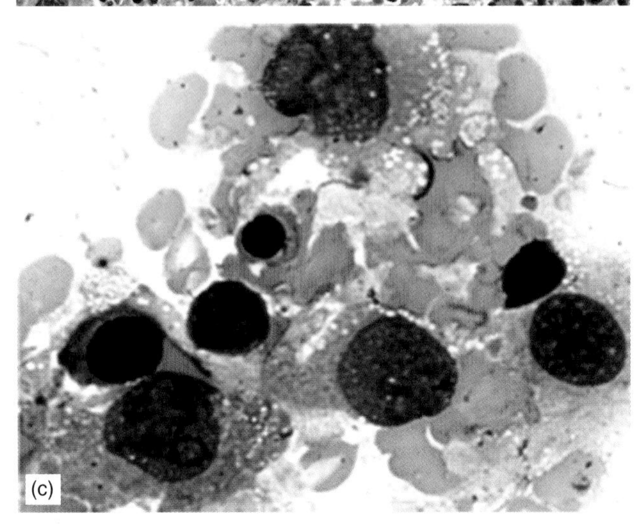

FIGURE 21.12 Langerhans cell histiocytosis. Bone marrow biopsy section (a and b) demonstrates diffuse infiltration of large mononuclear cells with abundant eosinophilic cytoplasm. Numerous eosinophils and scattered multinucleated giant cells are present (a). Bone marrow smear (c) shows several large cells with abundant finely vacuolated cytoplasm.

Electron microscopy demonstrates the characteristic cytoplasmic Birbeck granules. These granules are rod-shaped, often with the expanded end, resembling a tennis racket (Figure 21.13) [20, 125]. These structures consist of superimposed and zippered membranes. A protein known as "Langerin" (CD207) (see Table 3.1) is constitutively associated with Birbeck granules [125].

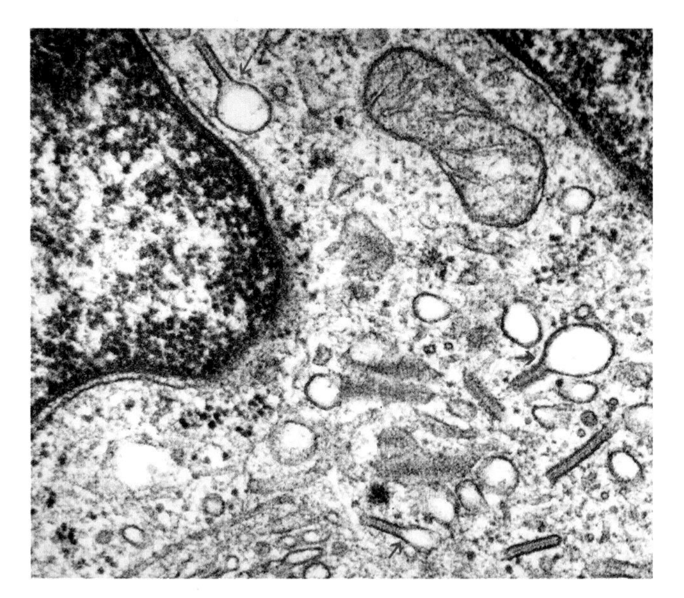

FIGURE 21.13 Electron micrograph of a Langerhans cell with Birbeck granules appearing like tennis rackets (arrows). Courtesy of Sunita Bhuta, M.D., Department of Pathology and Laboratory Medicine, UCLA Medical Center.

The most common single sites of involvement include the bone, skin, or lymph nodes. Multisystem disease presentation may include the liver, spleen, bone marrow, lung, and the endocrine, gastrointestinal, and central nervous systems [123].

Immunohistochemistry

The major immunophenotypic characteristics of LCH are similar to the normal LC and are demonstrated in Table 21.1. The expression of CD1a and CD207 (Langerin) is considered the immunophenotypic hallmark of the LC [109, 125]. These cells are positive for CD4, S-100 protein, and HLA-DR and are negative for CD21 and CD35 (Figure 21.14). The LCH may aberrantly express CD52 [126]. The extracutaneous lesions appear to be consisting of less mature LCH cells, and in addition to CD1a and Langerin, express CD14 and CD68, whereas the skin lesions consist of more mature LCH and lack the expression of CD14 and CD68 [109]. CD45 and lysozyme are weakly positive and CD15, CD30, CD33, CD34, and MPO are negative [108, 109, 122, 123].

Molecular and Cytogenetic Studies

Monoclonal nature of LCH has been suggested by using X-linked polymorphic DNA probes [112]. There is an association between LCH and monosomy 7 [121, 122]. Loss of heterozygosity at 9p21, 17p, and 22q has been reported in patients with LCH [111, 127]. As mentioned earlier, several molecular cytogenetic abnormalities have been reported in patients with LCH, such as gains of DNA copy number on chromosomes 2q, 4q, and 12 and loss of DNA sequences on chromosomes 1p, 5, 6, 16, and 22q [121].

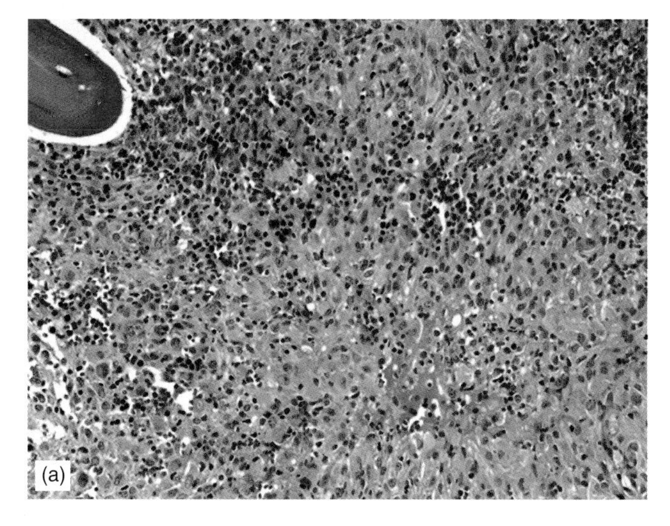

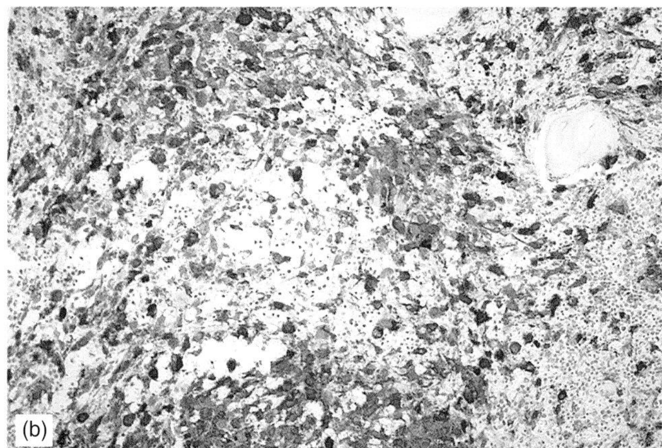

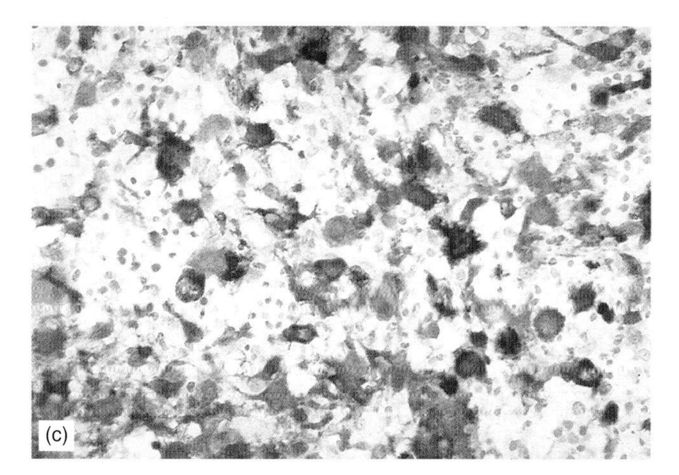

FIGURE 21.14 Langerhans cell histiocytosis. Bone marrow biopsy section demonstrates a diffuse infiltration by the Langerhans cells (a) H&E. These cells express S-100 (b) low power and (c) high power views, immunohistochemical stain.

Clinical Aspects

LCH is primarily a disease of childhood and early adulthood with a male:female ratio of 3–4:1. People from the northern Europe are more commonly affected [123]. The disease is rare in Afro-Americans. A wide spectrum of clinical manifestations have been described with their own specific terminology, ranging from congenital self-healing form (Hashimoto–Pritzker disease) to solitary disease (eosinophilic granuloma) with good prognosis to multifocal multisystem disease with poor prognosis (Table 21.8).

Bone is the most frequent site of involvement, usually presenting as a lytic lesion of the skull, which may be painful or asymptomatic [128]. Other frequent sites of bone involvement include femur, ribs, vertebra, and humerus. The skin lesions in infants appear as brown to purplish papules and may mimic congenital neuroblastoma or leukemia [123, 129]. Hepatosplenomegaly and lymphadenopathy are the major clinical presentations of the multisystem LCH. Lung lesions are associated with heavy smoking and are frequently observed in young adults. There is an association between multisystem LCH and acute lymphoblastic leukemia [5].

Different therapeutic approaches have been proposed. Recommendations for unifocal osseous lesions include surgical curettage or excision, radiation or single agent chemotherapy, or combination of all. For pediatric multisystem LCH, combination chemotherapy, such as LCH-1 protocol (vinblastine versus etoposide in combination with intravenous steroids), LCH-2 protocol (vinblastine, oral prednisone, and mercaptopurine with or without etoposide), and LCH-3 protocol (vinblastine and prednisone with or without methotrexate), is recommended [130–132].

Differential Diagnosis

The differential diagnosis includes a garden variety of histiocytic disorders such as lysosomal storage diseases, HLH, and granulomatous disorders. Hepatosplenomegaly and lymphadenopathy may raise the possibility of leukemia/lymphoma. Expression of CD1a and Langerin and the presence of ultrastructural Birbeck granules are diagnostic features for LCH. LCH is distinguished from LC sarcoma by the lack of atypical cytologic features such as pleomorphic, hyperchromatic nuclei, prominent nucleoli, or numerous mitotic figures (see the following section).

Langerhans Cell Sarcoma

Langerhans cell sarcoma is a rare neoplasm of LC characterized by explicit malignant cytologic features (Figure 21.15) [5, 133, 134]. These features include hyperchromatic and pleomorphic nuclei, prominent nucleoli, and high mitotic figures (usually >5/hpf). Associated inflammatory cells, such as eosinophils, are lacking or minimal [5, 134]. The neoplastic cells in most cases, similar to the LCH, express CD1a, Langerin, and S-100 protein and demonstrate the ultrastructural Birbeck granules. They may also aberrantly express CD31 or CD56 [125, 135]. The Ki-67 index is usually high. Langerhans cell sarcoma has been reported in both children and adults. It is characterized by multisystem involvement and poor prognosis. The involved organs include bone, skin, spleen, liver, lymph nodes, and lung [135].

Agranular CD4+, CD56+ Hematodermic Neoplasms (Blastic NK-Cell Lymphoma)

This entity is called "Blastic NK-cell lymphoma" in the WHO classification and is defined as a lymphoid malignancy

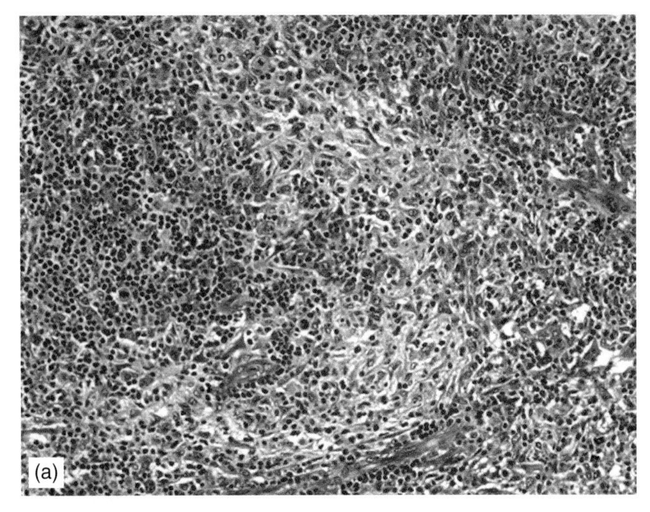

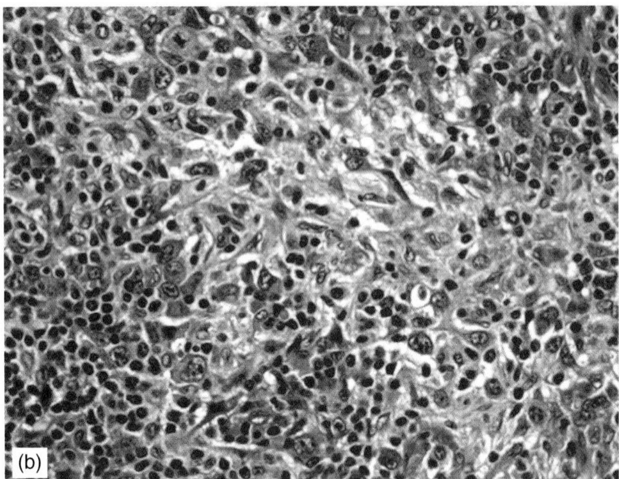

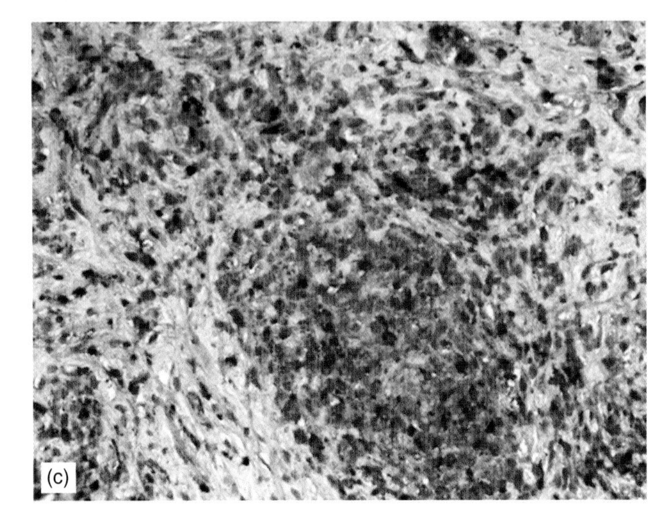

FIGURE 21.15 Lymph node section demonstrating clusters of large, atypical pleomorphic cells that were positive for CD1c and S-100. The lesion may represent Langerhans cell sarcoma. (a) and (b) are low and high power views, (c) represents immunohistochemical stain for S-100.

consisting of blast cells expressing NK-cell-associated marker, CD56 [5]. However, more recent studies suggest that these tumors originate from plasmacytoid dendritic cells [136–139].

Etiology and Pathogenesis

The etiology and pathogenesis are not known. There is no evidence of a viral etiology.

Pathology

Morphology

The skin involvement is usually multifocal, often with formation of nodules (Figure 21.16). Large clusters of monomorphous cells are present in the dermis and the hypodermis with infiltration of the cutaneous appendages. A single-file infiltration may be present in some areas. The epidermis is spared and there is no evidence of angiocentrism or angiodestruction [137].

The neoplastic cells are blastic with variable amount of weakly basophilic, non-granular cytoplasm with round, oval, or cleaved nuclei, fine chromatin, and multiple prominent nucleoli (Figure 21.17). The cytoplasm may show peripheral microvacuoles or pseudopods. The tumor cells may vary in size from small to large and in some cases consist of a mixture of small and large blastic cells. The blood, bone marrow, and lymph nodes are other frequent sites of involvement.

The bone marrow is involved in >80% of the cases showing focal or diffuse infiltration by the blastic tumor cells. Circulating blast cells are detected in about 60% of the patients ranging from 1% to >90% of the leukocyte counts [140]. The affected lymph nodes usually show a leukemic pattern of infiltration with the involvement of medulla, sinuses, and interfollicular areas.

Immunophenotype

The major immunophenotypic features of the hematodermic neoplasms are presented in Table 21.9. These cells characteristically express CD4, CD56, CD43, CD45(dim to strong), TCL1, HLA-DR, and CD123. CD68 is positive in about 50% and TdT is positive in about 25% of the cases (Figure 21.18). The overwhelming majority of cases are negative for other T- and NK-cell markers and B-cell- and myeloid-associated antigens. Most of these immunophenotypic features (coexpression of CD4, CD43, CD45, TCL1, HLA-DR, CD123, and CD68) along with the exception of CD56 are shared with the plasmacytoid dendritic cells, suggesting a lineage relationship [136–139].

Molecular and Cytogenetic Studies

Molecular studies have not shown TCR rearrangements in these neoplasms. Approximately 65% of cases are associated with cytogenetic abnormalities, which are often complex and with an aneuploid karyotype [137, 141]. Six major recurring abnormalities have been reported: deletions of 12p13 and 6q23, monosomy 9, monosomy 15, and abnormalities of chromosomes 5 (5q21 and 5q34) and 13 (between 13q13 and 13q21) (Figure 21.19) [141].

Clinical Aspects

Hematodermic neoplasm is a rare disease affecting patients at any age, ranging from 8 to 96 years, with a median age of 65 years [137, 140, 142]. The most frequent

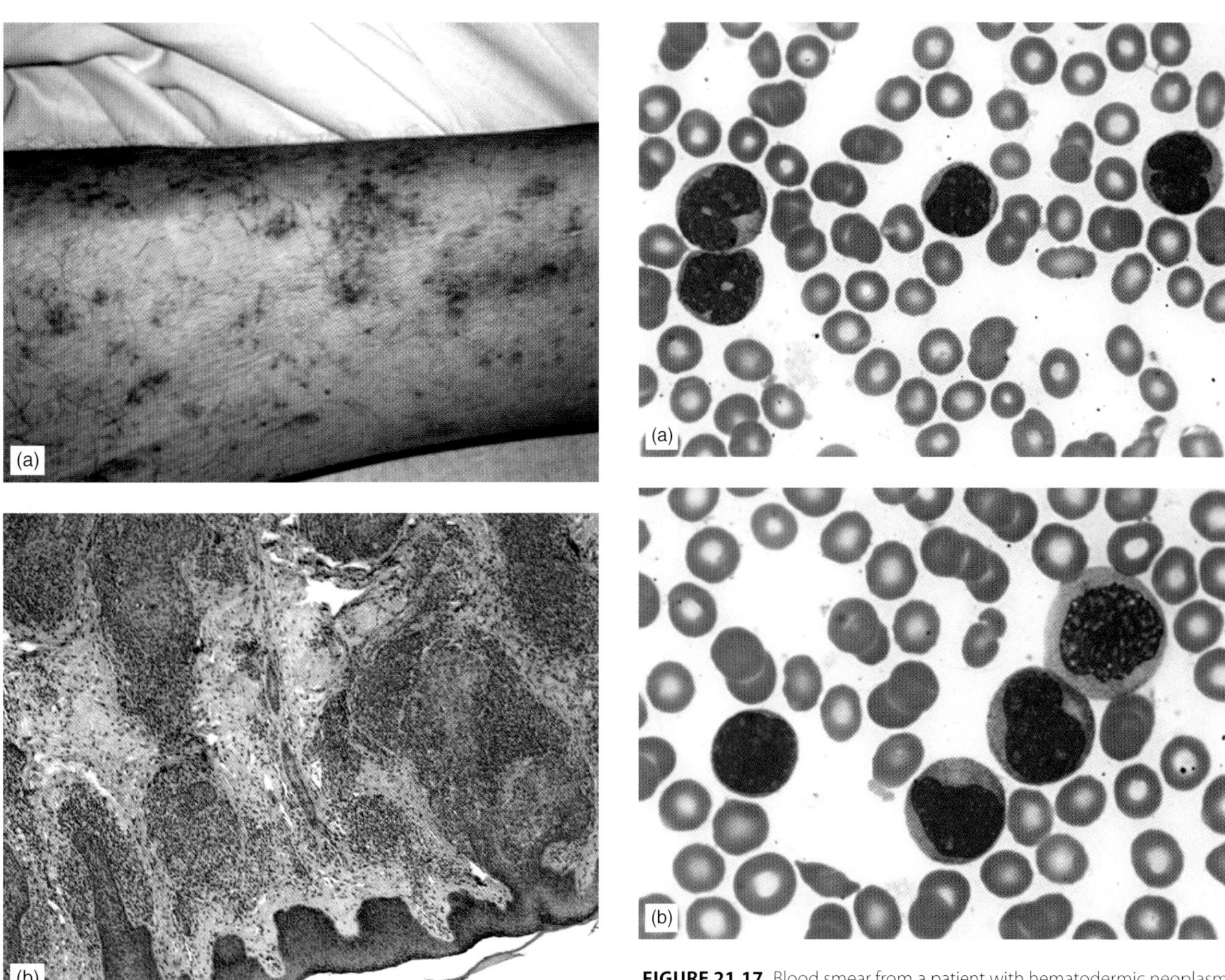

FIGURE 21.16 Agranular CD4+, CD56+ hematodermic neoplasm. Skin demonstrates multifocal lesions in various sizes and shapes (a) with heavy infiltration of the neoplastic cells in the dermis and surrounding skin appendages (b). High power view shows immature mononuclear cells with oval or irregular nuclei, fine chromatin, and prominent nucleoli (c).

FIGURE 21.17 Blood smear from a patient with hematodermic neoplasm demonstrates numerous immature mononuclear cells in various sizes and shapes (a and b). Some of the immature cells mimic monoblasts.

site of involvement is skin that is affected in >90% of the patients [137, 143–145]. The cutaneous lesions are usually purple and solitary or localized at the beginning but become multiple with time. They may appear as nodules or patches. A leukemic presentation with circulating blasts is reported in about 60%, and splenomegaly and/or lymphadenopathy are reported in about 60% of the cases [140, 141]. Other sites of involvement include the liver, lung, and central nervous system [140].

Combination chemotherapy may achieve complete remission, but in most patients, the disease relapses between 3 and 18 months. The overall reported 2-year survival rate is about 25% [140].

Differential Diagnosis

The differential diagnosis includes primary and secondary cutaneous lymphomas/leukemias such as nasal type T/NK-cell lymphoma, aggressive NK-cell leukemia/lymphoma, and mycosis fungoides. The main distinguishing features between the hematodermic neoplasm and the cutaneous nasal type T/NK lymphoma are the lack of angiocentric and angiodestructive lesions, TIA-1 expression, and EBV genome

TABLE 21.9 Immunophenotypic features of hematodermic neoplasms*.

Markers	Results
CD4	+
CD56	+
CD123	+
CD43	+
CD45	+
HLA-DR	+
TCL1	+
CD68	±
CD7	±
CD2	±
CD36	±
CD38	±
TdT	±
CD1a	−
CD3	−
CD5	−
CD8	−
CD16	−
CD57	−
TIA-1	−
CD10	−
CD19	−
CD20	−
CD79a	
CD13	−
CD14	−
CD15	−
CD33	−
MOP	−
CD117	−
CD34	−

*Adapted from Ref. [137].

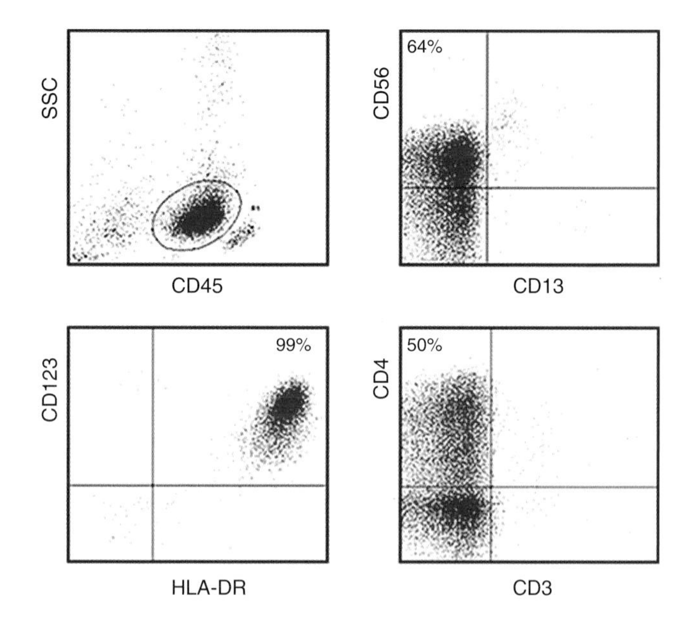

FIGURE 21.18 Flow cytometry of blood sample of a patient with agranular CD4+, CD56+ hematodermic leukemia. Neoplastic cells express $CD45^{dim}$, CD123, HLA-DR, CD56 (partial), and CD4 (partial).

in the hematodermic neoplasms. The NK-cell tumors often show cytoplasmic granules and are mostly CD4−, CD8+, and TIA-1+, whereas hematodermic neoplasms are CD4+, CD8−, and negative for TIA-1. Mycosis fungoides is distinguished from hematodermic neoplasm by epidermal involvement (Pautrier abscesses) and CD3+, CD56− immunophenotype.

Also, cutaneous infiltrations of myelomonocytic leukemias may show significant overlapping of immunophenotypic features with hematodermic neoplasms by expressing CD4, CD56, CD68, and HLA-DR. However, TCL1 is expressed in 90% of the hematodermic neoplasms and <20% of the acute myeloid leukemias [137], and myeloid-associated markers, such as CD13, CD14, CD15, CD33, and CD117, are negative in hematodermic neoplasms (Table 21.9).

Interdigitating Dendritic Cell Tumor (Sarcoma)

Interdigitating dendritic cell tumor (sarcoma) is an extremely rare neoplasm consisting of cells with spindle-shaped nuclei and immunophenotypic features similar to the IDC [5, 146]. Lymphadenopathy is the most frequent clinical findings which may present as localized or generalized. Cases of extranodal involvement such as the skin, intestine, spleen, and soft tissues have been observed [5, 147–149]. This disorder has been reported in ages from 8 to 77 years, but most patients are adults with a median age of over 50 years [146, 150]. Clinical outcome is variable, and therefore, IDC *tumor* may be a more suitable term.

The involved lymph nodes often show paracortical infiltration of elongated cells with ovoid or spindle-shaped nuclei in a storiform pattern (Figure 21.20). These cells have a variable amount of cytoplasm with indistinct cell borders,

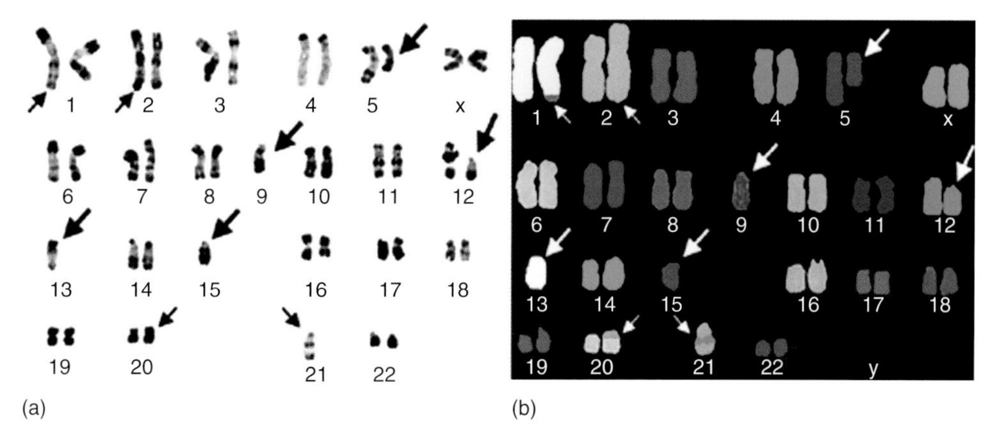

FIGURE 21.19 Agranular CD4+, CD56+ hematodermic leukemia. (a) Representative R-banded and (b) M-FISH-Metasystems probe karyotypes. Arrows indicate the presence of chromosomal abnormalities; large arrows identify recurrent anomalies. From Ref. [140] by permission. This research was originally published in *Blood*.

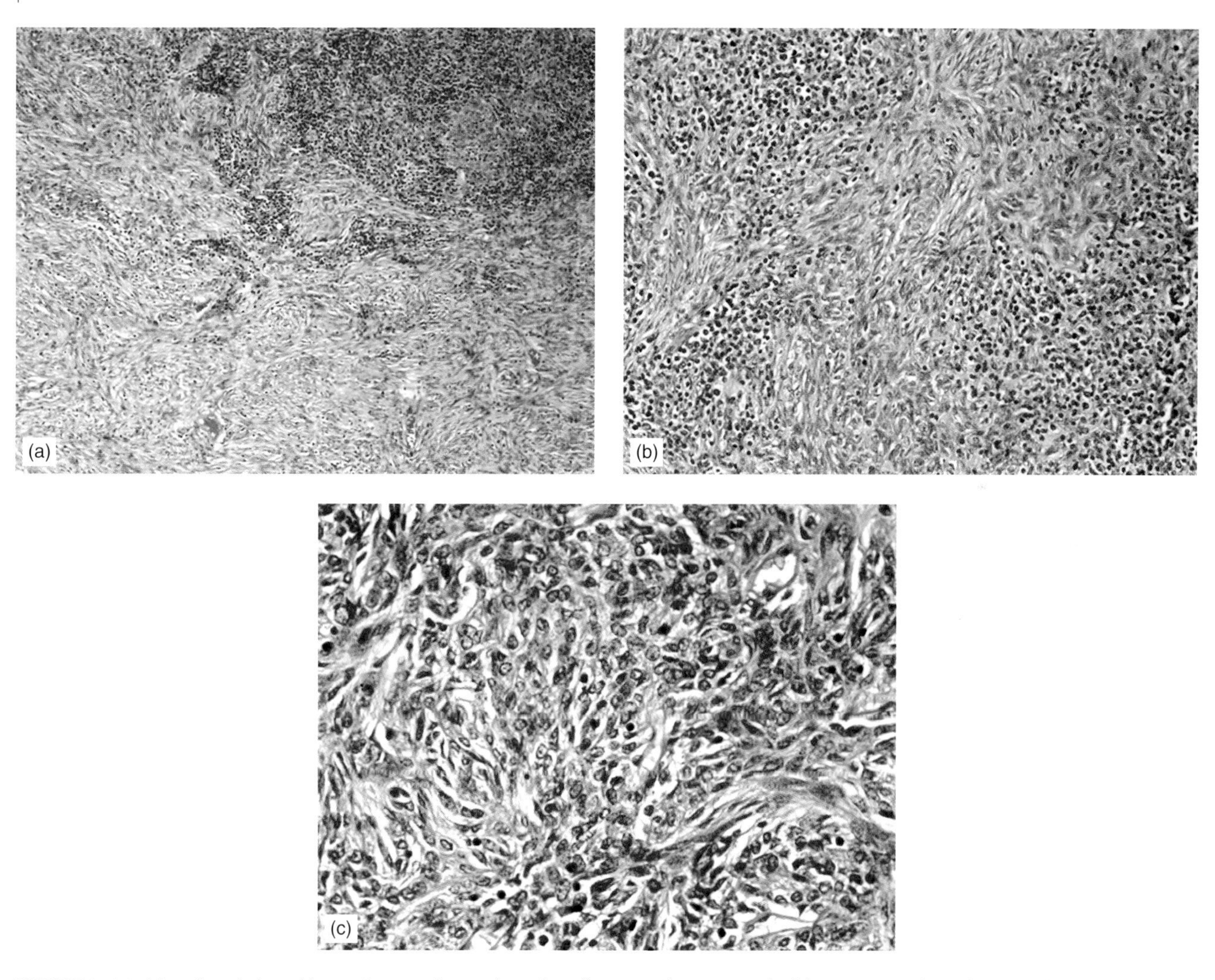

FIGURE 21.20 A lymph node-based lesion showing sheets of spindle cells in a storiform pattern, highly suggestive of interdigitating dendritic cell tumor at (a) low power, (b) intermediate power, and (c) high power views. Tissue blocks were not available for immunohistochemical stains.

a vesicular nuclear chromatin, and a distinct nucleus [5, 146]. A variable degree of cytologic atypia may be present. Mitotic figures are usually low (<5/hpf) and there is lack of necrosis. Residual follicular structures are often present [5].

Immunophenotypic characteristics are the expression of S-100 protein, Vimentin, and HLA-DR and the lack of expression of CD1a, CD21, CD35, and pan-B- and pan-T-cell markers (Table 21.1). The neoplastic cells may show weak positive reactions for CD45, CD68, or lysozyme [5, 147–149].

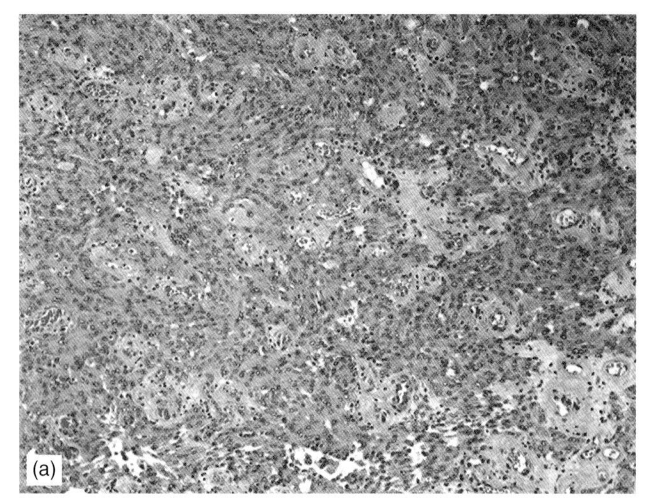

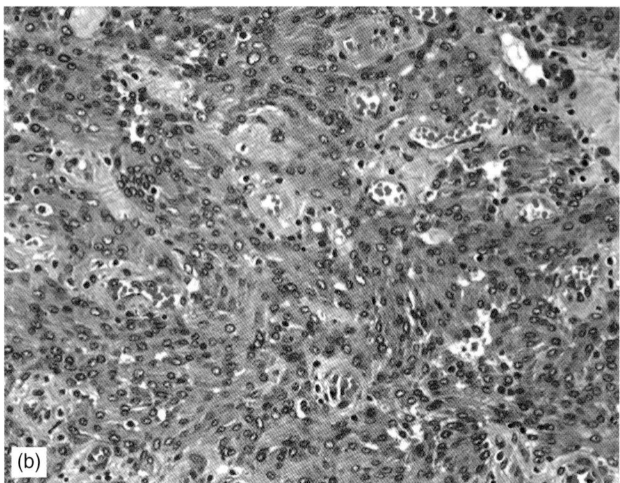

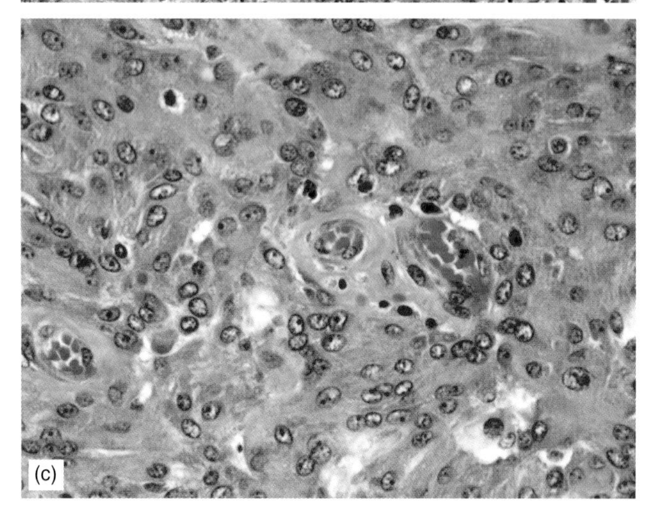

FIGURE 21.21 Follicular dendritic cell tumor. Sheets of tumor cells with spindle to oval nuclei and relatively abundant cytoplasm in a vascular stroma: (a) low power, (b) intermediate power, and (c) high power views. Slide from CAP, PIP-D 2007, Case # 2007-33. Neoplastic cells were reported to be positive for CD21, CD35, and clusterin.

They are negative for CD30, CD34, epithelial membrane antigen (EMA), and myeloperoxidase (MPO). Ki-67 is expressed in about 10–20% of the tumor cells. The immunophenotypic features are helpful in distinguishing these tumors from the follicular dendritic neoplasms and other sarcomas.

Follicular Dendritic Cell Tumor (Sarcoma)

Follicular dendritic cell tumor (sarcoma) is an extremely rare neoplasm consisting of cells with spindle-shaped nuclei and immunophenotypic features similar to the FDC [5, 151–153]. Lymph nodes are the most frequent sites of involvement [152, 153]. Extranodal involvement has been reported in about 25% of the cases, primarily affecting intra-abdominal organs [153–155]. It is a disease of young and middle-aged adults, usually with an indolent clinical course, high recurrent rate, and low risk of metastasis [5, 155]. For these reasons, FDC *tumor* appears to be a more suitable term.

The neoplastic cells have a variable amount of cytoplasm with indistinct cell borders, a vesicular nuclear chromatin, and a distinct nucleus (Figure 21.21) [5, 155]. A variable degree of cytologic atypia and occasional multinucleated giant cell may be present. Mitotic figures are usually ≤10/hpf, but occasional cases with >30/hpf have been reported [5]. The neoplastic cells form fascicles and whorls and/or demonstrate a storiform pattern. Residual follicular structures or lymphoid tissues are often present [5].

Immunophenotypic features include the expression of CD21, CD23, and CD35 as well as HLA-DR and Vimentin. The tumor cells may variably express S-100 protein, EMA, and CD68 [152, 154]. Ki-67 staining is usually ≤25% [5]. CD1a, CD30, CD34, cytokeratin, and lysozyme are negative. The immunophenotypic features are helpful in distinguishing these tumors from the interdigitating dendritic neoplasms and other sarcomas.

References

1. Shepard JL, Zon LI. (2000). Developmental derivation of embryonic and adult macrophages. *Curr Opin Hematol* **7**, 3–8.
2. Wu L, Dakic A. (2004). Development of dendritic cell system. *Cell Mol Immunol* **1**, 112–18.
3. Allen CD, Cyster JG. (2008). Follicular dendritic cell networks of primary follicles and germinal centers: Phenotype and function. *Semin Immunol* **20**, 14–25 [Epub 2008 Feb 7].
4. Park CS, Choi YS. (2005). How do follicular dendritic cells interact intimately with B cells in the germinal centre? *Immunology* **114**, 2–10.
5. Jaffe ES, Harris NL, Stein H, Vardiman JW. (2001). *Pathology and Genetics. Tumors of Haematopoietic and Lymphoid Tissues.* IARC Press, Lyon.
6. Butters TD. (2007). Gaucher disease. *Curr Opin Chem Biol* **11**, 412–18.
7. Meikle PJ, Hopwood JJ, Clague AE, Carey WF. (1999). Prevalence of lysosomal storage disorders. *JAMA* **281**, 249–54.
8. Zhao H, Grabowski GA. (2002). Gaucher disease: Perspectives on a prototype lysosomal disease. *Cell Mol Life Sci* **59**, 694–707.
9. Eng CM. (2007). Genetics; clinical manifestations; and diagnosis of Gaucher disease. *UpToDate.*
10. Sun Y, Quinn B, Witte DP, Grabowski GA. (2005). Gaucher disease mouse models: Point mutations at the acid beta-glucosidase locus combined with low-level prosaposin expression lead to disease variants. *J Lipid Res* **46**, 2102–13.
11. Futerman AH, van Meer G. (2004). The cell biology of lysosomal storage disorders. *Nat Rev Mol Cell Biol* **5**, 554–65.
12. Cormand B, Montfort M, Chabás A, Vilageliu L, Grinberg D. (1997). Genetic fine localization of the beta-glucocerebrosidase (GBA), and

prosaposin (PSAP) genes: Implications for Gaucher disease. *Hum Genet* **100**, 75–9.
13. Germain DP. (2004). Gaucher's disease: A paradigm for interventional genetics. *Clin Genet* **65**, 77–86.
14. Beutler E. (1993). Modern diagnosis and treatment of Gaucher's disease. *Am J Dis Child* **147**, 1175–83.
15. Beutler E, Nguyen NJ, Henneberger MW, Smolec JM, McPherson RA, West C, Gelbart T. (1993). Gaucher disease: Gene frequencies in the Ashkenazi Jewish population. *Am J Hum Genet* **52**, 85–8.
16. Glew RH, Basu A, LaMarco KL, Prence EM. (1988). Mammalian glucocerebrosidase: Implications for Gaucher's disease. *Lab Invest* **58**, 5–25.
17. Brady RO, King FM. (1973). Gaucher's disease. In *Lysosomes and Storage Diseases* (Hers HG, van Hoof F, eds), p. 381. Academic Press, New York.
18. Enriquez P, Neiman RS. (1976). *The Pathology of the Spleen. A Functional Approach*. ASCP Press, Chicago.
19. James SP, Stromeyer FW, Stowens DW, Barranger JA. (1982). Gaucher disease: Hepatic abnormalities in 25 patients. *Prog Clin Biol Res* **95**, 131–42.
20. Naeim F. (1998). *Pathology of Bone Marrow*, 2nd ed. Williams & Wilkins, Baltimore.
21. Charrow J, Andersson HC, Kaplan P, Kolodny EH, Mistry P, Pastores G, Rosenbloom BE, Scott CR, Wappner RS, Weinreb NJ, Zimran A. (2000). The Gaucher registry: Demographics and disease characteristics of 1698 patients with Gaucher disease. *Arch Intern Med* **160**, 2835–43.
22. Grabowski GA. (2005). Recent clinical progress in Gaucher disease. *Curr Opin Pediatr* **17**, 519–24.
23. Weinreb NJ, Aggio MC, Andersson HC, Andria G, Charrow J, Clarke JT, Erikson A, Giraldo P, Goldblatt J, Hollak C, Ida H, Kaplan P, Kolodny EH, Mistry P, Pastores GM, Pires R, Prakash-Cheng A, Rosenbloom BE, Scott CR, Sobreira E, Tylki-Szymańska A, Vellodi A, vom Dahl S, Wappner RS, Zimran A. (2004). International Collaborative Gaucher Group (ICGG). Gaucher disease type 1: Revised recommendations on evaluations and monitoring for adult patients. *Semin Hematol* **41**(4 Suppl 5), 15–22.
24. Harris CM, Campbell P. (2007). The importance of correct phenotyping in Gaucher disease. *J Child Neurol* **22**, 1056–7.
25. Mignot C, Doummar D, Maire I, De Villemeur TB. (2006). French Type 2 Gaucher Disease Study Group. Type 2 Gaucher disease: 15 new cases and review of the literature. *Brain Dev* **28**, 39–48.
26. Blom S, Erikson A. (1983). Gaucher disease – Norrbottnian type. Neurodevelopmental, neurological, and neurophysiological aspects. *Eur J Pediatr* **140**, 316–22.
27. Wenger DA, Clark C, Sattler M, Wharton C. (1978). Synthetic substrate beta-glucosidase activity in leukocytes: A reproducible method for the identification of patients and carriers of Gaucher's disease. *Clin Genet* **13**, 145–53.
28. Sidransky E. (2004). Gaucher disease: Complexity in a "simple" disorder. *Mol Genet Metab* **83**, 6–15.
29. Pastores GM, Weinreb NJ, Aerts H, Andria G, Cox TM, Giralt M, Grabowski GA, Mistry PK, Tylki-Szymańska A. (2004). Therapeutic goals in the treatment of Gaucher disease. *Semin Hematol* **41** (4 Suppl 5), 4–14.
30. Grabowski GA. (2008). Treatment perspectives for the lysosomal storage diseases. *Expert Opin Emerg Drugs* **13**, 197–211.
31. Steward CG, Jarisch A. (2005). Haemopoietic stem cell transplantation for genetic disorders. *Arch Dis Child* **90**, 1259–63.
32. Kolodny EH. (2000). Niemann–Pick disease. *Curr Opin Hematol* **7**, 48–52.
33. Cruse RP. (2007). Overview of Niemann–Pick disease. *UpToDate*.
34. Pavlů H, Elleder M. (1997). Two novel mutations in patients with atypical phenotypes of acid sphingomyelinase deficiency. *J Inherit Metab Dis* **20**, 615–16.
35. Loftus SK, Morris JA, Carstea ED, Gu JZ, Cummings C, Brown A, Ellison J, Ohno K, Rosenfeld MA, Tagle DA, Pentchev PG, Pavan WJ. (1997). Murine model of Niemann–Pick C disease: Mutation in a cholesterol homeostasis gene. *Science* **277**, 232–5.
36. Neufeld EB, Wastney M, Patel S, Suresh S, Cooney AM, Dwyer NK, Roff CF, Ohno K, Morris JA, Carstea ED, Incardona JP, Strauss III JF, Vanier MT, Patterson MC, Brady RO, Pentchev PG, Blanchette-Mackie EJ. (1999). The Niemann–Pick C1 protein resides in a vesicular compartment linked to retrograde transport of multiple lysosomal cargo. *J Biol Chem* **274**, 9627–35.
37. Vanier MT. (2002). Prenatal diagnosis of Niemann–Pick diseases types A, B and C. *Prenat Diagn* **22**, 630–2.
38. Landrieu P, Saïd G. (1984). Peripheral neuropathy in type A Niemann–Pick disease. A morphological study. *Acta Neuropathol* **63**, 66–71.
39. Elleder M, Hrodek J, Cihula J. (1983). Niemann–Pick disease: Lipid storage in bone marrow macrophages. *Histochem J* **15**, 1065–77.
40. Lefevre M. (1988). Localization of lipoprotein unesterified cholesterol in nondenaturing gradient gels with filipin. *J Lipid Res* **29**, 815–18.
41. Reid PC, Sakashita N, Sugii S, Ohno-Iwashita Y, Shimada Y, Hickey WF, Chang TY. (2004). A novel cholesterol stain reveals early neuronal cholesterol accumulation in the Niemann–Pick type C1 mouse brain. *J Lipid Res* **45**, 582–91.
42. McGovern MM, Pohl-Worgall T, Deckelbaum RJ, Simpson W, Mendelson D, Desnick RJ, Schuchman EH, Wasserstein MP. (2004). Lipid abnormalities in children with types A and B Niemann Pick disease. *J Pediatr* **145**, 77–81.
43. Wasserstein MP, Larkin AE, Glass RB, Schuchman EH, Desnick RJ, McGovern MM. (2003). Growth restriction in children with type B Niemann–Pick disease. *J Pediatr* **142**, 424–8.
44. Schuchman EH. (2007). The pathogenesis and treatment of acid sphingomyelinase-deficient Niemann–Pick disease. *J Inherit Metab Dis* **30**, 654–63.
45. Victor S, Coulter JB, Besley GT, Ellis I, Desnick RJ, Schuchman EH, Vellodi A. (2003). Niemann–Pick disease: Sixteen-year follow-up of allogeneic bone marrow transplantation in a type B variant. *J Inherit Metab Dis* **26**, 775–85.
46. Shiflett SL, Kaplan J, Ward DM. (2002). Chediak–Higashi syndrome: A rare disorder of lysosomes and lysosome related organelles. *Pigment Cell Res* **15**, 251–7.
47. Ward DM, Shiflett SL, Kaplan J. (2002). Chediak–Higashi syndrome: A clinical and molecular view of a rare lysosomal storage disorder. *Curr Mol Med* **2**, 469–77.
48. Certain S, Barrat F, Pastural E, Deist Le F, Goyo-Rivas J, Jabado N, Benkerrou M, Seger R, Vilmer E, Beullier G, Schwarz K, Fischer A, de Saint Basile G. (2000). Protein truncation test of LYST reveals heterogenous mutations in patients with Chediak–Higashi syndrome. *Blood* **95**, 979–83.
49. Fukai K, Oh J, Karim MA, Moore KJ, Kandil HH, Ito H, Bürger J, Spritz RA. (1996). Homozygosity mapping of the gene for Chediak–Higashi syndrome to chromosome 1q42-q44 in a segment of conserved synteny that includes the mouse beige locus (bg). *Am J Hum Genet* **59**, 620–4.
50. Perou CM, Leslie JD, Green W, Li L, Ward DM, Kaplan J. (1997). The Beige/Chediak–Higashi syndrome gene encodes a widely expressed cytosolic protein. *J Biol Chem* **272**, 29790–4.
51. Ahluwalia J, Pattari S, Trehan A, Marwaha RK, Garewal G. (2003). Accelerated phase at initial presentation: An uncommon occurrence in Chediak–Higashi syndrome. *Pediatr Hematol Oncol* **20**, 563–7.
52. Roberts RL, Bonilla FA. (2007). Primary disorders of phagocytic function: An overview. *UpToDate*.
53. Merino F, Henle W, Ramírez-Duque P. (1986). Chronic active Epstein–Barr virus infection in patients with Chediak–Higashi syndrome. *J Clin Immunol* **6**, 299–305.
54. Okano M, Gross TG. (2000). A review of Epstein–Barr virus infection in patients with immunodeficiency disorders. *Am J Med Sci* **319**, 392–6.
55. Twomey JJ, Douglass CC, Sharkey Jr. O. (1973). The monocytopenia of aplastic anemia. *Blood* **41**, 187–95.
56. Janckila AJ, Wallace JH, Yam LT. (1982). Generalized monocyte deficiency in leukaemic reticuloendotheliosis. *Scand J Haematol* **29**, 153–60.

57. Peterson V, Hansbrough J, Buerk C, Rundus C, Wallner S, Smith H, Robinson WA. (1983). Regulation of granulopoiesis following severe thermal injury. *J Trauma* **23**, 19–24.
58. Treacy M, Lai L, Costello C, Clark A. (1987). Peripheral blood and bone marrow abnormalities in patients with HIV related disease. *Br J Haematol* **65**, 289–94.
59. Isenberg DA, Martin P, Hajirousou V, Todd-Pokropek A, Goldstone AH, Snaith ML. (1986). Haematological reassessment of rheumatoid arthritis using an automated method. *Br J Rheumatol* **25**, 152–7.
60. Ferry JA, Harris NL. (1997). *Atlas of Lymphoid Hyperplasia and Lymphoma*. W.B. Saunders, Philadelphia.
61. Medeiros LJ, Kaynor B, Harris NL. (1989). Lupus lymphadenitis: Report of a case with immunohistologic studies on frozen sections. *Hum Pathol* **20**, 295–9.
62. McCluggage WG, Bharucha H. (1994). Lymph node hyalinisation in rheumatoid arthritis and systemic sclerosis. *J Clin Pathol* **47**, 138–42.
63. Kondratowicz GM, Symmons DP, Bacon PA, Mageed RA, Jones EL. (1990). Rheumatoid lymphadenopathy: A morphological and immunohistochemical study. *J Clin Pathol* **43**, 106–13.
64. Roncalli M, Servida E. (1989). Granulomatous and nongranulomatous lymphadenitis in sarcoidosis. An immunophenotypic study of seven cases. *Pathol Res Pract* **185**, 351–7.
65. Khamseh ME, Mollanai S, Hashemi F, Rezaizadeh A, Azizi F. (2002). Erdheim–Chester syndrome, presenting as hypogonadotropic hypogonadism and diabetes insipidus. *J Endocrinol Invest* **25**, 727–9.
66. Al-Quran S, Reith J, Bradley J, Rimsza L. (2002). Erdheim–Chester disease: Case report, PCR-based analysis of clonality, and review of literature. *Mod Pathol* **15**, 666–72.
67. Maric I, Pittaluga S, Dale JK, Niemela JE, Delsol G, Diment J, Rosai J, Raffeld M, Puck JM, Straus SE, Jaffe ES. (2005). Histologic features of sinus histiocytosis with massive lymphadenopathy in patients with autoimmune lymphoproliferative syndrome. *Am J Surg Pathol* **29**, 903–11.
68 O'Connell JX, Rosenberg AE. (1993). Histiocytic lymphadenitis associated with a large joint prosthesis. *Am J Clin Pathol* **99**, 314–16.
69. Woda BA, Sullivan JL. (1993). Reactive histiocytic disorders. *Am J Clin Pathol* **99**, 459–63.
70. Albores-Saavedra J, Vuitch F, Delgado R, Wiley E, Hagler H. (1994). Sinus histiocytosis of pelvic lymph nodes after hip replacement. A histiocytic proliferation induced by cobalt-chromium and titanium. *Am J Surg Pathol* **18**, 83–90.
71. Rouphael NG, Talati NJ, Vaughan C, Cunningham K, Moreira R, Gould C. (2007). Infections associated with haemophagocytic syndrome. *Lancet Infect Dis* **7**, 814–22.
72. Filipovich AH. (2006). Hemophagocytic lymphohistiocytosis and related disorders. *Curr Opin Allergy Clin Immunol* **6**, 410–15.
73. Ishii E, Ohga S, Imashuku S, Kimura N, Ueda I, Morimoto A, Yamamoto K, Yasukawa M. (2005). Review of hemophagocytic lymphohistiocytosis (HLH) in children with focus on Japanese experiences. *Crit Rev Oncol Hematol* **53**, 209–23.
74. McClain KL. (2007). Hemophagocytic lymphohistiocytosis. *UpToDate*.
75. Osugi Y, Hara J, Tagawa S, Takai K, Hosoi G, Matsuda Y, Ohta H, Fujisaki H, Kobayashi M, Sakata N, Kawa-Ha K, Okada S, Tawa A. (1997). Cytokine production regulating Th1 and Th2 cytokines in hemophagocytic lymphohistiocytosis. *Blood* **89**, 4100–3.
76. Aricò M, Danesino C, Pende D, Moretta L. (2001). Pathogenesis of haemophagocytic lymphohistiocytosis. *Br J Haematol* **114**, 761–9.
77. Komp DM, McNamara J, Buckley P. (1989). Elevated soluble interleukin-2 receptor in childhood hemophagocytic histiocytic syndromes. *Blood* **73**, 2128–32.
78. Mazodier K, Marin V, Novick D, Farnarier C, Robitail S, Schleinitz N, Veit V, Paul P, Rubinstein M, Dinarello CA, Harlé JR, Kaplanski G. (2005). Severe imbalance of IL-18/IL-18BP in patients with secondary hemophagocytic syndrome. *Blood* **106**, 3483–9.
79. Henter JI, Arico M, Elinder G, Imashuku S, Janka G. (1998). Familial hemophagocytic lymphohistiocytosis. Primary hemophagocytic lymphohistiocytosis. *Hematol Oncol Clin North Am* **12**, 417–33.
80. Fadeel B, Orrenius S, Henter JI. (2001). Familial hemophagocytic lymphohistiocytosis: Too little cell death can seriously damage your health. *Leuk Lymphoma* **42**, 13–20.
81. Zur Stadt U, Beutel K, Kolberg S, Schneppenheim R, Kabisch H, Janka G, Hennies HC. (2006). Mutation spectrum in children with primary hemophagocytic lymphohistiocytosis: Molecular and functional analyses of PRF1, UNC13D, STX11, and RAB27A. *Hum Mutat* **27**, 62–8.
82. Ueda I, Ishii E, Morimoto A, Ohga S, Sako M, Imashuku S. (2006). Correlation between phenotypic heterogeneity and gene mutational characteristics in familial hemophagocytic lymphohistiocytosis (FHL). *Pediatr Blood Cancer* **46**, 482–8.
83. Yamamoto K, Ishii E, Horiuchi H, Ueda I, Ohga S, Nishi M, Ogata Y, Zaitsu M, Morimoto A, Hara T, Imashuku S, Sasazuki T, Yasukawa M. (2005). Mutations of syntaxin 11 and SNAP23 genes as causes of familial hemophagocytic lymphohistiocytosis were not found in Japanese people. *J Hum Genet* **50**, 600–3.
84. Imashuku S. (2002). Clinical features and treatment strategies of Epstein–Barr virus-associated hemophagocytic lymphohistiocytosis. *Crit Rev Oncol Hematol* **44**, 259–72.
85. Chen TL, Wong WW, Chiou TJ. (2003). Hemophagocytic syndrome: An unusual manifestation of acute human immunodeficiency virus infection. *Int J Hematol* **78**, 450–2.
86. Fardet L, Blum L, Kerob D, Agbalika F, Galicier L, Dupuy A, Lafaurie M, Meignin V, Morel P, Lebbé C. (2003). Human herpesvirus 8-associated hemophagocytic lymphohistiocytosis in human immunodeficiency virus-infected patients. *Clin Infect Dis* **37**, 285–91.
87. Brastianos PK, Swanson JW, Torbenson M, Sperati J, Karakousis PC. (2006). Tuberculosis-associated haemophagocytic syndrome. *Lancet Infect Dis* **6**, 447–54.
88. Pringe A, Trail L, Ruperto N, Buoncompagni A, Loy A, Breda L, Martini A, Ravelli A. (2007). Macrophage activation syndrome in juvenile systemic lupus erythematosus: An under-recognized complication? *Lupus* **16**, 587–92.
89. Wong KF, Hui PK, Chan JK, Chan YW, Ha SY. (1991). The acute lupus hemophagocytic syndrome. *Ann Intern Med* **114**, 387–90.
90. Dhote R, Simon J, Papo T, Detournay B, Sailler L, Andre MH, Dupond JL, Larroche C, Piette AM, Mechenstock D, Ziza JM, Arlaud J, Labussiere AS, Desvaux A, Baty V, Blanche P, Schaeffer A, Piette JC, Guillevin L, Boissonnas A, Christoforov B. (2003). Reactive hemophagocytic syndrome in adult systemic disease: Report of twenty-six cases and literature review. *Arthritis Rheum* **49**, 633–9.
91. Palazzi DL, McClain KL, Kaplan SL. (2003). Hemophagocytic syndrome in children: An important diagnostic consideration in fever of unknown origin. *Clin Infect Dis* **36**, 306–12.
92. Falini B, Pileri S, De Solas I, Martelli MF, Mason DY, Delsol G, Gatter KC, Fagioli M. (1990). Peripheral T-cell lymphoma associated with hemophagocytic syndrome. *Blood* **75**, 434–44.
93. Okuda T, Sakamoto S, Deguchi T, Misawa S, Kashima K, Yoshihara T, Ikushima S, Hibi S, Imashuku S. (1991). Hemophagocytic syndrome associated with aggressive natural killer cell leukemia. *Am J Hematol* **38**, 321–3.
94. Miyahara M, Sano M, Shibata K, Matsuzaki M, Ibaraki K, Shimamoto Y, Tokunaga O. (2000). B-cell lymphoma-associated hemophagocytic syndrome: Clinicopathological characteristics. *Ann Hematol* **79**, 378–88.
95. Shimazaki C, Inaba T, Nakagawa M. (2000). B-cell lymphoma-associated hemophagocytic syndrome. *Leuk Lymphoma* **38**, 121–30.
96. Morrell DS, Pepping MA, Scott JP, Esterly NB, Drolet BA. (2002). Cutaneous manifestations of hemophagocytic lymphohistiocytosis. *Arch Dermatol* **138**, 1208–12.
97. Henter JI, Elinder G, Ost A. (1991). Diagnostic guidelines for hemophagocytic lymphohistiocytosis. The FHL Study Group of the Histiocyte Society. *Semin Oncol* **18**, 29–33.

98. Horne A, Janka G, Maarten Egeler R, Gadner H, Imashuku S, Ladisch S, Locatelli F, Montgomery SM, Webb D, Winiarski J, Filipovich AH, Henter JI, Histiocyte Society (2005). Haematopoietic stem cell transplantation in haemophagocytic lymphohistiocytosis. *Br J Haematol* **129**, 622–30.
99. Vos JA, Abbondanzo SL, Barekman CL, Andriko JW, Miettinen M, Aguilera NS. (2005). Histiocytic sarcoma: A study of five cases including the histiocyte marker CD163. *Mod Pathol* **18**, 693–704.
100. Kobayashi S, Kimura F, Hama Y, Ogura K, Torikai H, Kobayashi A, Ikeda T, Sato K, Aida S, Kosuda S, Motoyoshi K. (2008). Histiocytic sarcoma of the spleen: Case report of asymptomatic onset of thrombocytopenia and complex imaging features. *Int J Hematol* **87**, 83–7.
101. Yoshida C, Takeuchi M. (2008). Histiocytic sarcoma: Identification of its histiocytic origin using immunohistochemistry. *Intern Med* **47**, 165–9.
102. Weitzman S, Egeler RM. (2008). Langerhans cell histiocytosis: Update for the pediatrician. *Curr Opin Pediatr* **20**, 23–9.
103. Allen CE, McClain KL. (2007). Langerhans cell histiocytosis: A review of past, current and future therapies. *Drugs Today (Barc)* **43**, 627–43.
104. Hoover KB, Rosenthal DI, Mankin H. (2007). Langerhans cell histiocytosis. *Skeletal Radiol* **36**, 95–104.
105. Barton III CP, Horlbeck D. (2007). Eosinophilic granuloma: Bilateral temporal bone involvement. *Ear Nose Throat J* **86**, 342–3.
106. Greenlee JD, Fenoy AJ, Donovan KA, Menezes AH. (2007). Eosinophilic granuloma in the pediatric spine. *Pediatr Neurosurg* **43**, 285–92.
107. Herzog KM, Tubbs RR. (1998). Langerhans cell histiocytosis. *Adv Anat Pathol* **5**, 347–58.
108. Geissmann F, Lepelletier Y, Fraitag S, Valladeau J, Bodemer C, Debré M, Leborgne M, Saeland S, Brousse N. (2001). Differentiation of Langerhans cells in Langerhans cell histiocytosis. *Blood* **97**, 1241–8.
109. Annels NE, Da Costa CE, Prins FA, Willemze A, Hogendoorn PC, Egeler RM. (2003). Aberrant chemokine receptor expression and chemokine production by Langerhans cells underlies the pathogenesis of Langerhans cell histiocytosis. *J Exp Med* **197**, 1385–90.
110. Scappaticci S, Danesino C, Rossi E, Klersy C, Fiori GM, Clementi R, Russotto VS, Bossi G, Aricò M. (2000). Cytogenetic abnormalities in PHA-stimulated lymphocytes from patients with Langerhans cell histocytosis. AIEOP-Istiocitosi Group. *Br J Haematol* **111**, 258–62.
111. Willman CL, Busque L, Griffith BB, Favara BE, McClain KL, Duncan MH, Gilliland DG. (1994). Langerhans'-cell histiocytosis (histiocytosis X) – a clonal proliferative disease. *N Engl J Med* **331**, 154–60.
112. McClain KL, Cai YH, Hicks J, Peterson LE, Yan XT, Che S, Ginsberg SD. (2005). Expression profiling using human tissues in combination with RNA amplification and microarray analysis: Assessment of Langerhans cell histiocytosis. *Amino Acids* **28**, 279–90.
113. Jenson HB, McClain KL, Leach CT, Deng JH, Gao SJ. (2000). Evaluation of human herpesvirus type 8 infection in childhood Langerhans cell histiocytosis. *Am J Hematol* **64**, 237–41.
114. Leahy MA, Krejci SM, Friednash M, Stockert SS, Wilson H, Huff JC, Weston WL, Brice SL. (1993). Human herpesvirus 6 is present in lesions of Langerhans cell histiocytosis. *J Invest Dermatol* **101**, 642–5.
115. Shimakage M, Sasagawa T, Kimura M, Shimakage T, Seto S, Kodama K, Sakamoto H. (2004). Expression of Epstein–Barr virus in Langerhans' cell histiocytosis. *Hum Pathol* **35**, 862–8.
116. Egeler RM, Favara BE, van Meurs M, Laman JD, Claassen E. (1999). Differential *in situ* cytokine profiles of Langerhans-like cells and T cells in Langerhans cell histiocytosis: Abundant expression of cytokines relevant to disease and treatment. *Blood* **94**, 4195–201.
117. Fleming MD, Pinkus JL, Fournier MV, Alexander SW, Tam C, Loda M, Sallan SE, Nichols KE, Carpentieri DF, Pinkus GS, Rollins BJ. (2003). Coincident expression of the chemokine receptors CCR6 and CCR7 by pathologic Langerhans cells in Langerhans cell histiocytosis. *Blood* **101**, 2473–5.
118. Bernstrand C, Carstensen H, Jakobsen B, Svejgaard A, Henter JI, Olerup O. (2003). Immunogenetic heterogeneity in single-system and multisystem langerhans cell histiocytosis. *Pediatr Res* **54**, 30–6.
119. McClain KL, Laud P, Wu WS, Pollack MS. (2003). Langerhans cell histiocytosis patients have HLA Cw7 and DR4 types associated with specific clinical presentations and no increased frequency in polymorphisms of the tumor necrosis factor alpha promoter. *Med Pediatr Oncol* **41**, 502–7.
120. Murakami I, Gogusev J, Fournet JC, Glorion C, Jaubert F. (2002). Detection of molecular cytogenetic aberrations in Langerhans cell histiocytosis of bone. *Hum Pathol* **33**, 555–60.
121. Hicks J, Flaitz CM. (2005). Langerhans cell histiocytosis: Current insights in a molecular age with emphasis on clinical oral and maxillofacial pathology practice. *Oral Surg Oral Med Oral Pathol Oral Radiol Endod* **100**(2 Suppl), S42–66.
122. McClain KL. (2007). Langerhans cell histiocytosis (histiocytosis X, eosinophilic granuloma). *UpToDate*.
123. Dina A, Zahava V, Iness M. (2005). The role of vascular endothelial growth factor in Langerhans cell histiocytosis. *J Pediatr Hematol Oncol* **27**, 62–6.
124. Valladeau J, Ravel O, Dezutter-Dambuyant C, Moore K, Kleijmeer M, Liu Y, Duvert-Frances V, Vincent C, Schmitt D, Davoust J, Caux C, Lebecque S, Saeland S. (2000). Langerin, a novel C-type lectin specific to Langerhans cells, is an endocytic receptor that induces the formation of Birbeck granules. *Immunity* **12**, 71–81.
125. Jordan MB, McClain KL, Yan X, Hicks J, Jaffe R. (2005). Anti-CD52 antibody, alemtuzumab, binds to Langerhans cells in Langerhans cell histiocytosis. *Pediatr Blood Cancer* **44**, 251–4.
126. Betts DR, Leibundgut KE, Feldges A, Plüss HJ, Niggli FK. (1998). Cytogenetic abnormalities in Langerhans cell histiocytosis. *Br J Cancer* **77**, 552–5.
127. Slater JM, Swarm OJ. (1980). Eosinophilic granuloma of bone. *Med Pediatr Oncol* **8**, 151–64.
128. Stein SL, Paller AS, Haut PR, Mancini AJ. (2001). Langerhans cell histiocytosis presenting in the neonatal period: A retrospective case series. *Arch Pediatr Adolesc Med* **155**, 778–83.
129. Broadbent V, Gadner H. (1998). Current therapy for Langerhans cell histiocytosis. *Hematol Oncol Clin North Am* **12**, 327–38.
130. Haupt R, Nanduri V, Calevo MG, Bernstrand C, Braier JL, Broadbent V, Rey G, McClain KL, Janka-Schaub G, Egeler RM. (2004). Permanent consequences in Langerhans cell histiocytosis patients: A pilot study from the Histiocyte Society-Late Effects Study Group. *Pediatr Blood Cancer* **42**, 438–44.
131. Minkov M, Grois N, Heitger A, Pötschger U, Westermeier T, Gadner H. (2001). Treatment of multisystem Langerhans cell histiocytosis. Results of the DAL-HX 83 and DAL-HX 90 studies. DAL-HX Study Group. *Klin Padiatr* **212**, 139–44.
132. Bohn OL, Ruiz-Arguelles G, Navarro L, Saldivar J, Sanchez-Sosa S. (2007). Cutaneous langerhans cell sarcoma: A case report and review of the literature. *Int J Hematol* **85**, 116–20.
133. Ferringer T, Banks PM, Metcalf JS. (2006). Langerhans cell sarcoma. *Am J Dermatopathol* **28**, 36–9.
134. Kawase T, Hamazaki M, Ogura M, Kawase Y, Murayama T, Mori Y, Nagai H, Tateno M, Oyama T, Kamiya Y, Taji H, Kagami Y, Naoe T, Takahashi T, Morishima Y, Nakamura S. (2005). CD56/NCAM-positive Langerhans cell sarcoma: A clinicopathologic study of 4 cases. *Int J Hematol* **81**, 323–9.
135. Herling M, Teitell MA, Shen RR, Medeiros LJ, Jones D. (2003). TCL1 expression in plasmacytoid dendritic cells (DC2s) and the related CD4+ CD56+ blastic tumors of skin. *Blood* **101**, 5007–9.
136. Petrella T, Bagot M, Willemze R, Beylot-Barry M, Vergier B, Delaunay M, Meijer CJ, Courville P, Joly P, Grange F, De Muret A, Machet L, Dompmartin A, Bosq J, Durlach A, Bernard P, Dalac S, Dechelotte P, D'Incan M, Wechsler J, Teitell MA. (2005). Blastic NK-cell lymphomas (agranular CD4+ CD56+ hematodermic neoplasms): A review. *Am J Clin Pathol* **123**, 662–75.

137. Petrella T, Comeau MR, Maynadié M, Couillault G, De Muret A, Maliszewski CR, Dalac S, Durlach A, Galibert L. (2002). Agranular CD4+ CD56+ hematodermic neoplasm (blastic NK-cell lymphoma) originates from a population of CD56+ precursor cells related to plasmacytoid monocytes. *Am J Surg Pathol* **26**, 852–62.
138. Jaye DL, Geigerman CM, Herling M, Eastburn K, Waller EK, Jones D. (2006). Expression of the plasmacytoid dendritic cell marker BDCA-2 supports a spectrum of maturation among CD4+ CD56+ hematodermic neoplasms. *Mod Pathol* **19**, 1555–62.
139. Feuillard J, Jacob MC, Valensi F, Maynadie M, Gressin R, Chaperot L, Arnoulet C, Brignole-Baudouin F, Drenou B, Duchayne E, Falkenrodt A, Garand R, Homolle E, Husson B, Kuhlein E, Le Calvez G, Sainty D, Sotto MF, Trimoreau F, Bene MC. (2002). Clinical and biologic features of CD4(+)CD56(+) malignancies. *Blood* **99**, 1556–63.
140. Leroux D, Mugneret F, Callanan M, Radford-Weiss I, Dastugue N, Feuillard J, Le Mee F, Plessis G, Talmant P, Gachard N, Uettwiller F, Pages MP, Mozziconacci MJ, Eclache V, Sibille C, Avet-Loiseau H, Lafage-Pochitaloff M. (2002). CD4(+), CD56(+) DC2 acute leukemia is characterized by recurrent clonal chromosomal changes affecting 6 major targets: A study of 21 cases by the Groupe Francais de Cytogenetique Hematologique. *Blood* **99**, 4154–9.
141. Niakosari F, Sur M. (2007). Agranular CD4 + /CD56+ hematodermic neoplasm: A distinct entity described in the recent World Health Organization-European Organization for Research and Treatment of Cancer classification for cutaneous lymphomas. *Arch Pathol Lab Med* **131**, 149–51.
142. Martin JM, Nicolau MJ, Galan A, Ferrandez-Izquierdo A, Ferrer AM, Jorda E. (2006). CD4+/CD56+ haematodermic neoplasm: A preculsor haematological neoplasm that frequently first presents in the skin. *J Eur Acad Dermatol Venereol* **20**, 1129–32.
143. Ng AP, Lade S, Rutherford T, McCormack C, Prince HM, Westerman DA. (2006). Primary cutaneous CD4+/CD56+ hematodermic neoplasm (blastic NK-cell lymphoma): A report of five cases. *Haematologica* **91**, 143–4.
144. Touahri T, Belaouni H, Mossafa H, Pulik M, Bourguigna L, Ibbora C, Dal Soglio D, Davi F, Fourcade C. (2002). Agranular CD4+/ CD56+ cutaneous neoplasm. *Leuk Lymphoma* **43**, 1475–9.
145. Jo S, Babb MJ, Hilsinger Jr. RL. (2006). Interdigitating dendritic cell sarcoma of cervical lymph nodes. *Arch Otolaryngol Head Neck Surg* **132**, 1257–9.
146. Kawachi K, Nakatani Y, Inayama Y, Kawano N, Toda N, Misugi K. (2002). Interdigitating dendritic cell sarcoma of the spleen: Report of a case with a review of the literature. *Am J Surg Pathol* **26**, 530–7.
147. Barwell N, Howatson R, Jackson R, Johnson A, Jarrett RF, Cook G. (2004). Interdigitating dendritic cell sarcoma of salivary gland associated lymphoid tissue not associated with HHV-8 or EBV infection. *J Clin Pathol* **57**, 87–9.
148. Olnes MJ, Nicol T, Duncan M, Bohlman M, Erlich R. (2002). Interdigitating dendritic cell sarcoma: A rare malignancy responsive to ABVD chemotherapy. *Leuk Lymphoma* **43**, 817–21.
149. Pillay K, Solomon R, Daubenton JD, Sinclair-Smith CC. (2004). Interdigitating dendritic cell sarcoma: A report of four paediatric cases and review of the literature. *Histopathology* **44**, 283–91.
150. Monda L, Warnke R, Rosai J. (1986). A primary lymph node malignancy with features suggestive of dendritic reticulum cell differentiation. A report of 4 cases. *Am J Pathol* **122**, 562–72.
151. Soriano AO, Thompson MA, Admirand JH, Fayad LE, Rodriguez AM, Romaguera JE, Hagemeister FB, Pro B. (2007). Follicular dendritic cell sarcoma: A report of 14 cases and a review of the literature. *Am J Hematol* **Mar 20**.
152. McDuffie C, Lian TS, Thibodeaux J. (2007). Follicular dendritic cell sarcoma of the tonsil: A case report and literature review. *Ear Nose Throat J* **86**, 234–5.
153. Chang KC, Jin YT, Chen FF, Su IJ. (2001). Follicular dendritic cell sarcoma of the colon mimicking stromal tumour. *Histopathology* **38**, 25–9.
154. Araújo VC, Martins MT, Salmen FS, Araújo NS. (1999). Extranodal follicular dendritic cell sarcoma of the palate. *Oral Surg Oral Med Oral Pathol Oral Radiol Endod* **87**, 209–14.

Granulocytic Disorders

Faramarz Naeim

This chapter is devoted to the qualitative and quantitative non-neoplastic granulocytic disorders or granulocytic abnormalities associated with primary bone marrow disorders. It includes topics such as morphologic and functional abnormalities, neutropenia, neutrophilia, eosinophilia, and basophilia.

MORPHOLOGIC ABNORMALITIES

Toxic Granulation

This term is used to describe the presence of purple to dark-blue granules, resembling primary granules in segmented neutrophils, bands, and metamyelocytes (Figure 22.1). They represent lysosomal granules, contain myeloperoxidase (MPO), and show increased alkaline phosphatase activity. They indicate a shortened maturation time and activation of post-mitotic neutrophilic pool. They are found in infections and acute inflammations. They are often associated with Dohle bodies.

Dohle Inclusion Bodies

Dohle bodies are round, oval, elongated, or triangular, blue, gray-blue, or gray-green cytoplasmic inclusions (Figure 22.1b). They are RNA-contacting structures apparently derived from rough endoplasmic reticulum [1, 2]. Dohle bodies are found in activated post-mitotic neutrophilic series, including metamyelocytes, bands, and segmented neutrophils. They are frequently associated with toxic granulation and are found in infections, burns, traumas, and drug toxicity [1, 3, 4].

May–Hegglin Anomaly

May–Hegglin anomaly is a hereditary condition in which Dohle body–like inclusions are found in neutrophils, eosinophils, basophils, and monocytes (Figure 22.2). May–Hegglin anomaly is a member of a group of rare, autosomal dominant disorders characterized by thrombocytopenia, giant platelets, and Dohle body–like inclusions. It is associated with mutations in the *MYH9* gene mapped to chromosome 22q12.3-q13.2 [5–9]. It represents variable clinical manifestations including sensorineural deafness, cataracts, and nephritis.

Alder–Reilly Anomaly

Alder–Reilly anomaly is characterized by the presence of dense azurophilic granules (resembling toxic granules) in neutrophils, eosinophils, basophils, and, sometimes, lymphocytes and monocytes (Figure 22.3). This anomaly is seen in patients with mucopolysaccharidosis (Hurler's and Hunter's syndromes) and myelodysplastic syndromes [10–12]. Alder–Reilly anomaly has been reported in association with a mutation of the MPO structural gene [12].

Chediak–Higashi Granules

Chediak–Higashi syndrome (CHS) is a rare autosomal recessive disorder characterized by severe immune deficiency, partial albinism, bleeding tendencies, and recurrent bacterial

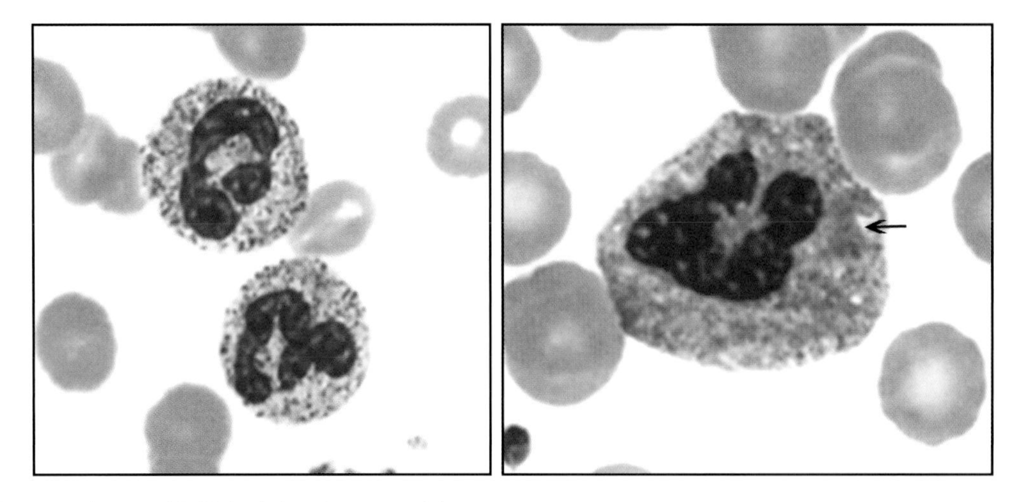

FIGURE 22.1 Toxic granulation and Dohle body (arrow) in neutrophils.

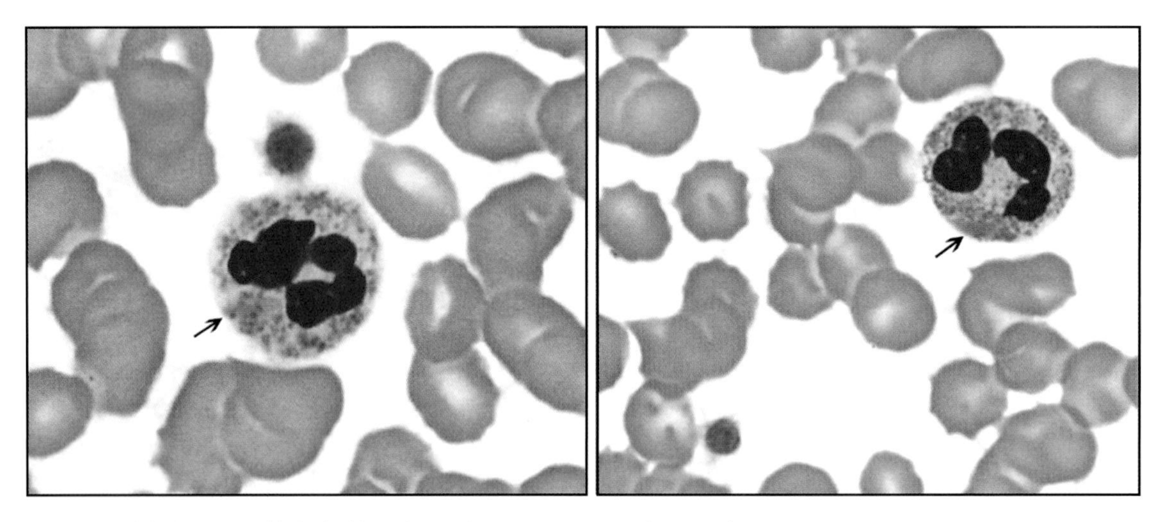

FIGURE 22.2 Neutrophils showing Dohle body–like inclusions (arrows) in May–Hegglin anomaly.

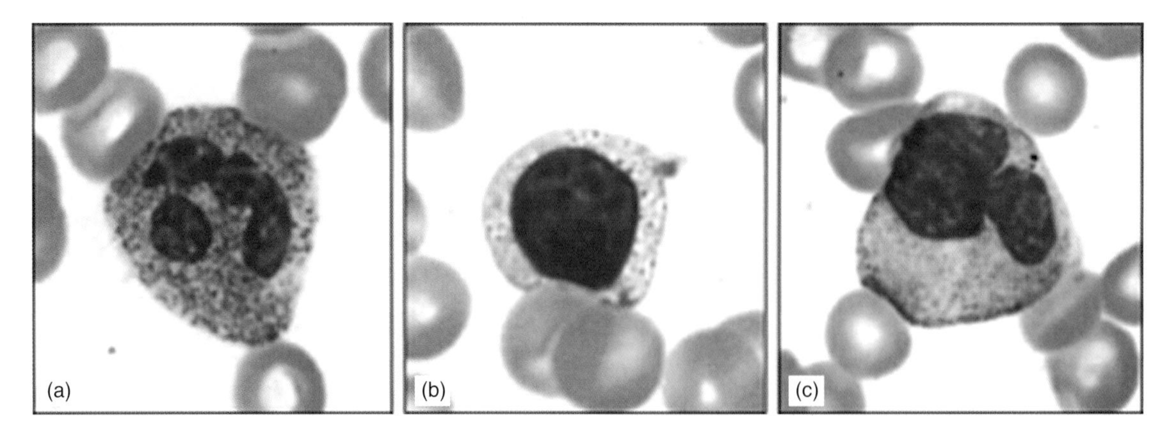

FIGURE 22.3 Alder–Reilly anomaly is characterized by the presence of dense azurophilic granules (resembling toxic granules) in neutrophils (a), lymphocytes (b), and monocytes (c).

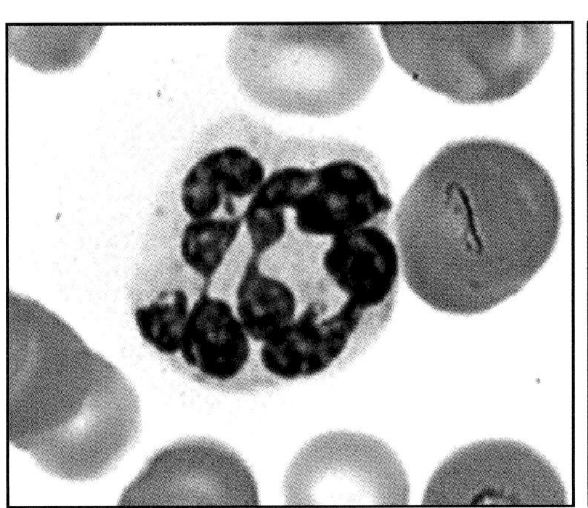
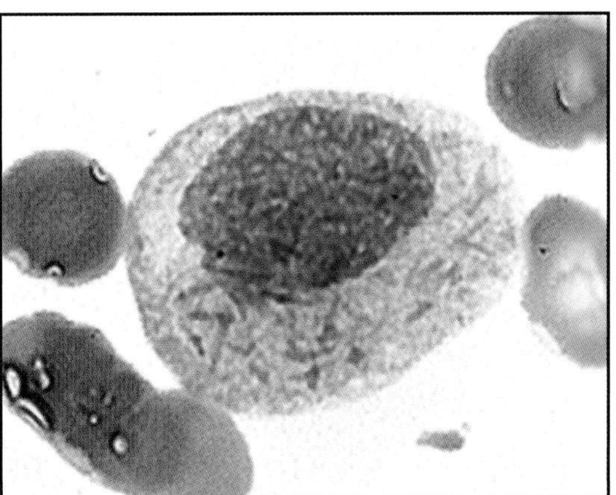

FIGURE 22.4
A hypogranular, hypersegmented neutrophil is shown on the left and a late promyelocyte loaded with Auer rods is depicted on the right.

infections [13, 14]. The morphologic hallmark of CHS is the presence of giant, blue or greenish gray cytoplasmic granules in various cells, including monocytes, granulocytes, cytotoxic T-cells, NK-cells, melonocytes, and Schwann cells [13, 14] (see Chapter 21).

Auer Rods

Auer rods are rod-like cytoplasmic inclusions found in myeloid precursors in patients with acute myelogenous leukemia (Figure 22.4). They are formed from the fusion of primary granules and, therefore, are MPO-positive.

Absence or Reduction of Cytoplasmic Neutrophilic Granules

Absence or reduced cytoplasmic granules are characteristic features of granulocytic cells in myelodysplastic syndrome (Figure 22.4). Selective defect in lactoferrin gene expression has been reported in association with neutrophil-specific granule deficiency [15].

Atypical Eosinophils and Eosinophilic Granules

In certain conditions, such as a subtype of acute myelomonocytic leukemia with inversion of chromosome 16(p13q32) and rare cases of chronic myelogenous leukemia, eosinophils contain atypical basophilic granules (Figure 22.5) [16, 17]. In hypereosinophilic syndrome, eosinophils may show hypersegmentation (Figure 22.5, inset).

Abnormal Nuclear Morphology

Pelger–Huet anomaly is an autosomal dominant disorder characterized by defective nuclear segmentation (hyposegmentation) in neutrophils (Figure 22.6) [18–20]. An association has been reported between this entity and *LBR* (lamin B-receptor) gene mutation located at chromosome 1q41-43 [21, 22]. The acquired hyposegmentation of the neutrophil nucleus, also called pseudo-Pelger–Huet anomaly, has been observed in sepsis, myelodysplastic syndromes, myeloproliferative disorders, some leukemias/lymphomas, and patients with solid organ transplantation [23–25].

Neutrophil nuclear hypersegmentation (more than five lobes) is observed in megaloblastic anemia, chronic infection, myelodysplastic syndrome, iron deficiency, hypogonadism, and chronic myelogenous leukemia (Figure 22.7) [3, 26, 27]. Neutrophilic hypersegmentation has been reported in two siblings, suggesting an autosomal recessive trait [28].

FUNCTIONAL ABNORMALITIES

Functional abnormalities of granulocytes are of two major types: (1) intrinsic defects, such as chemotactic disorders, CHS, adhesion defects, and MPO deficiency, and (2) extrinsic disorders, such as abnormalities of opsonizing systems due to abnormal and/or complement defects. CHS demonstrates both morphologic and functional abnormalities in a variety of hematopoietic and non-hematopoietic cells (see Chapter 21). In this section, the following intrinsic defects are briefly discussed: MPO deficiency, chronic granulomatous disease (CGD), and leukocyte adhesion deficiency (LAD).

MPO Deficiency

MPO deficiency is the most common intrinsic granulocytic functional deficiency [29–33]. It is an autosomal recessive inherited disorder involving the *MPO* gene located at 17q23 [34, 45]. MPO plays a critical role in the microbicidal activity of neutrophils. Activated neutrophils release MPO into the phagolysosomes or the extracellular spaces [34]. MPO, hydrogen peroxide, and chloride ion make up the cytocidal toxic mediators in infections [31, 32].

A secondary type of MPO deficiency has been observed in patients with clonal hematopoietic disorders such as

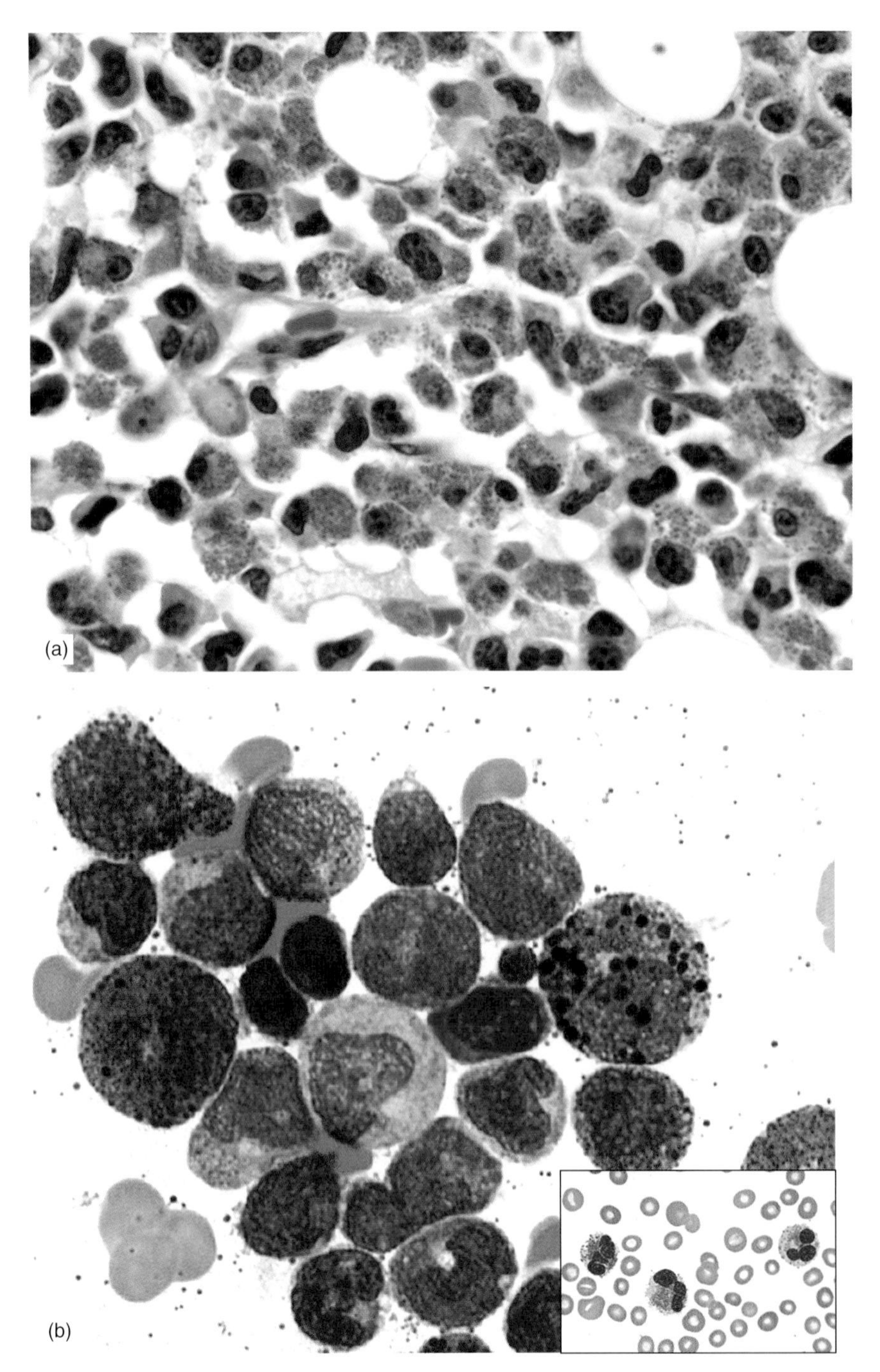

FIGURE 22.5 Bone marrow biopsy section (a) shows sheets of eosinophils. Bone marrow smear (b) demonstrates eosinophilic precursors with basophilic granules and blood smear (inset) shows eosinophils with hypersegmented nuclei.

myelodysplastic syndrome, acute myeloid leukemia, and lymphoma. The MPO deficiency in these patients is often due to discrete chromosomal aberrations involving the *MPO* gene [35].

Chronic Granulomatous Disease

CGD is characterized by the deficiency of respiratory burst oxidase and inability to manufacture superoxide (O_2^-) for the formation of microbicidal oxidants [36, 37]. CGD is an inherited disorder. It is either X-linked or autosomal recessive [38, 39]. The X-linked CGD almost exclusively involves men, except in rare homozygosity conditions or when there is a coexistent inactivated normal gene. The autosomal-recessive CGD occurs with equal frequency in male and female.

Neutrophils and other phagocytes (such as eosinophils and monocytes) use an NADPH oxidase to generate superoxide. The NADPH oxidase consists of five subunits. Two subunits, gp91-PHOX and p22-PHOX, form the heavy and light chains of the cytochrome b_{558} (CYBB) and are membrane-associated (membranes of secretory vesicles and granules) [40]. The three remaining subunits, p40-PHOX, p47-PHOX, and p67-PHOX, are cytosolic components. The gene encoding gp91-PHOX is located on X-chromosome. Mutation of this gene occurs in approximately 70% of the patients with CGD [40]. The remaining 30% of the cases

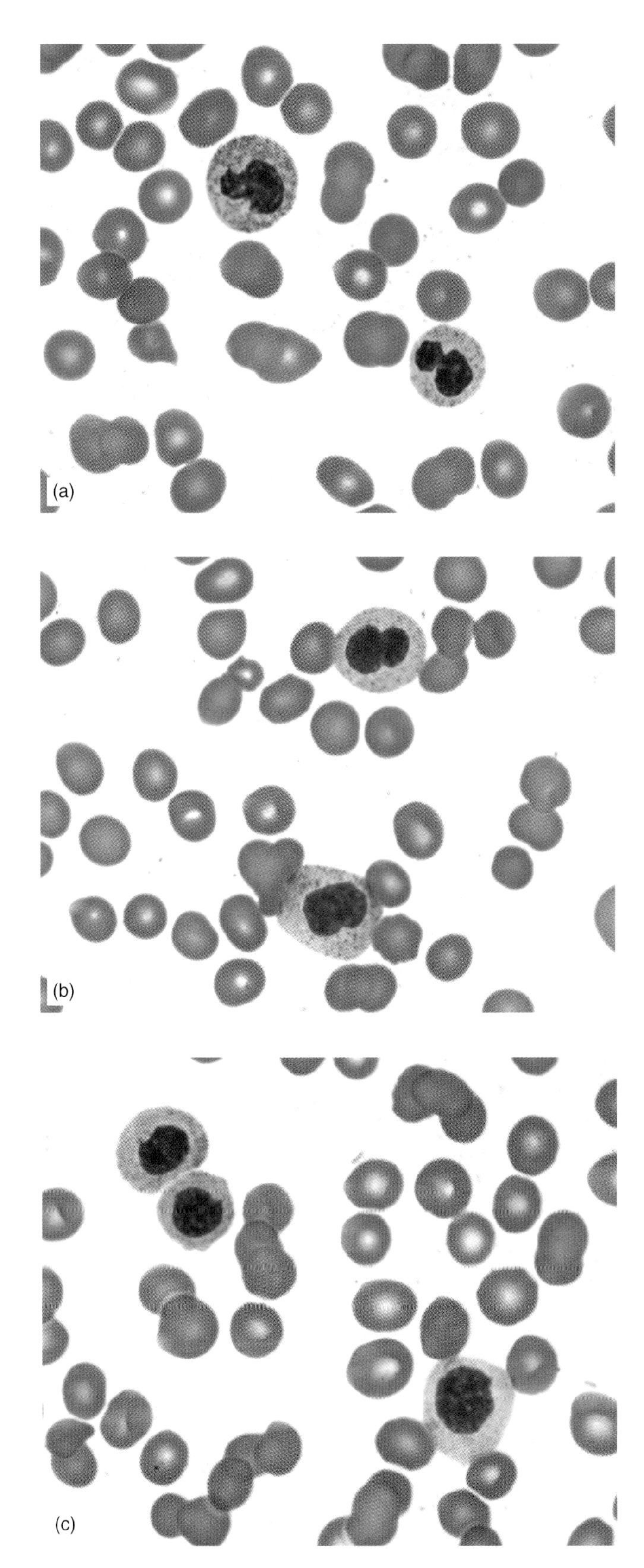

FIGURE 22.6 Blood smears (a, b, and c) demonstrate Pelger–Huet anomaly characterized by defective nuclear segmentation (hyposegmentation) in neutrophils.

are caused by mutations of autosomal genes encoding p22-, p47-, and p67-PHOX subunits.

Patients with CGD suffer recurrent bacterial and fungal infections often presenting as pneumonia, abscesses (skin, soft tissues, and organs), suppurative lymphadenitis, osteomyelitis, bacteremia, and fungemia. The severity of infection and the recurrence rate are variable, and the onset of clinical manifestations may range from early infancy to adulthood. The X-linked variant is seen in the younger population with a mean age of 3 years compared to about 8 years for the autosomal recessive forms [34]. The diagnosis of CGD is based on the demonstration of defective respiratory burst oxidase in neutrophils. The conventional laboratory test is the nitroblue tetrazolium (NBT) test, in which the oxygen produced in the course of a respiratory burst reduces the yellow, water-soluble tetrazolium dye to an insoluble deep-blue pigment. In normal conditions, at least 95% of neutrophils show positive reaction. The NBT test should be confirmed by DNA sequencing of the patient's *PHOX* genes [40].

Leukocyte Adhesion Deficiency

LAD syndromes are autosomal recessive disorders characterized by defective adhesion, binding and/or rolling of the leukocytes on the sites of microbial invasion [41]. These adhesion/rolling defects are of three major types: LAD I, LAD II, and LAD III [42, 43] (Figure 22.8).

LAD I syndromes are caused by impairment of expression of the leukocyte adhesion molecules (integrins). Integrins represent a family of glycoproteins, each composed of α- and β-subunits (CD11/CD18). The α-subunit, CD11, represents three different types of glycoprotein, CD11a, CD11b, and CD11c, respectively, for LFA-1 (expressed on all leukocytes), glycoprotein Mac-1, and p150/90 (both are expressed on monocytes, neutrophils, and NK-cells). The β-subunit, CD18, is shared in all integrins.

The LAD I is relatively rare. Its clinical outcome is often severe and includes recurrent or unresolved localized or systemic infections. The frequently reported clinical manifestations are delayed separation of the umbilical cord, recurrent bacterial infections, periodontitis, absence of pus formation, and impaired wound healing [41, 44]. Due to the lack of adhesion molecules and impaired mobilization of the leukocytes into the extravascular sites, these patients often have marked peripheral blood leukocytosis (5–20 times normal values) [43]. The diagnosis is established by flow cytometry demonstrating lack or marked reduction of the expression of integrins (CD11/CD18). DNA sequencing will identify relevant gene mutations. Mild to moderate cases are treated by antibiotics, and severe cases require bone marrow or stem cell transplantation.

LAD II syndromes are extremely rare and are caused by defective fucosylated carbohydrate ligands for p-selectin glycoprotein-1 (PSGL-1) (Figure 22.9) [42, 45]. The primary defect is in fucosylation of macromolecules, particularly at the stage of transport of fucose to the Golgi apparatus. This defect results in the lack of expression of certain glycans, such as CD15a (SLex) and H-antigen (Bombay blood group) in these patients [43].

The LAD II patients have less severe and fewer infection episodes than the LAD I patients. This syndrome has been

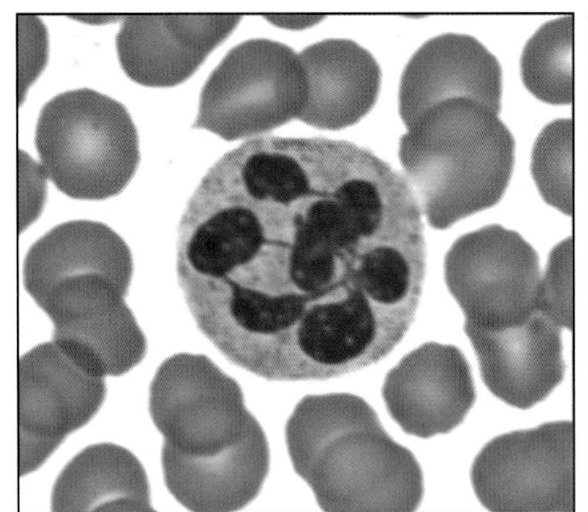

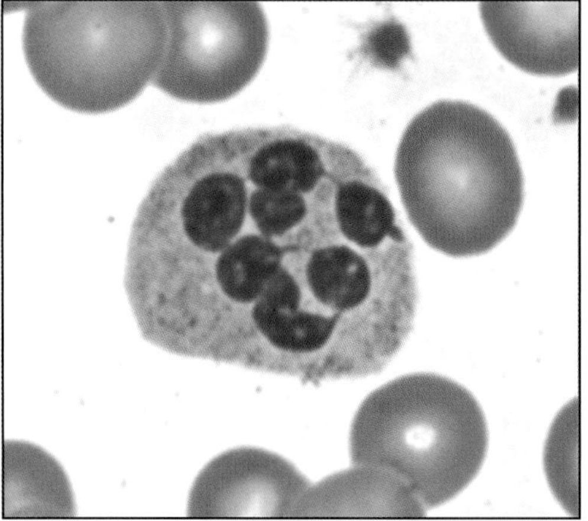

FIGURE 22.7 Neutrophil nuclear hypersegmentation (more than five lobes) is observed in megaloblastic anemia and other conditions such as chronic infection, myelodysplastic syndrome, hypogonadism, and chronic myelogenous leukemia.

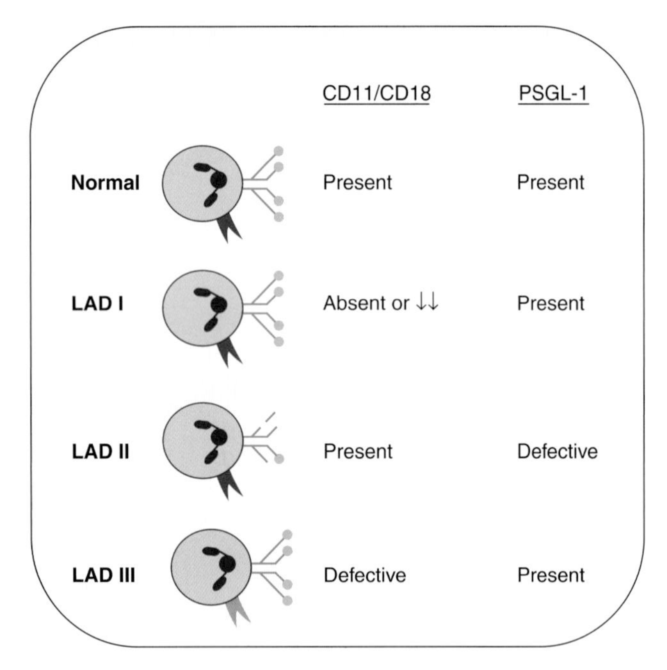

FIGURE 22.8 Altered surface phenotype in leukocyte adhesion deficiency. Expression of CD11/CD18 integrins and p-selectin glycoprotein 1 in various types of leukocyte adhesion deficiencies. Adapted from Ref. [41].

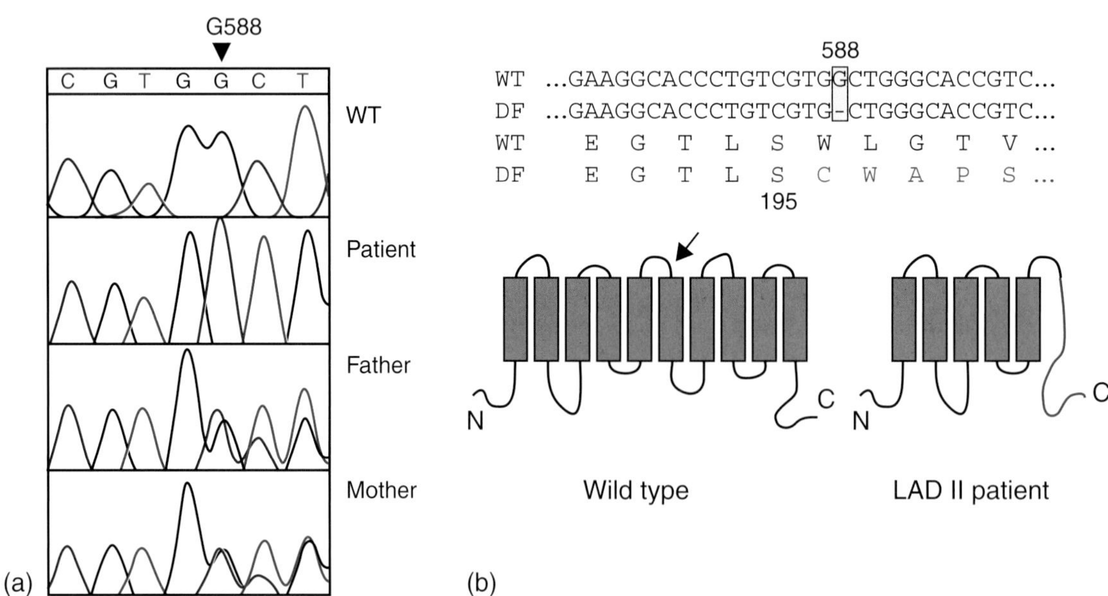

FIGURE 22.9 A single nucleotide deletion in position 588 (G588) in the *GDP-fucose transporter* gene of a patient with LAD II. (a) Chromatograms show a partial sequence for the *GDP-fucose transporter* gene from a healthy donor, the LAD II patient, and both parents. Note that G588 is absent in the patient, while both parents are heterozygotes in this position (overlaid sequence). (b) Partial primary sequence of the gene and the predicted protein region in which the deletion is found. A shift in the open-reading frame alters the protein sequence after Ser195 (arrow), as shown in the schematic representation of the predicted structure of the transporter. Adapted from Ref. [46] by permission. This research was originally published in *Blood*.

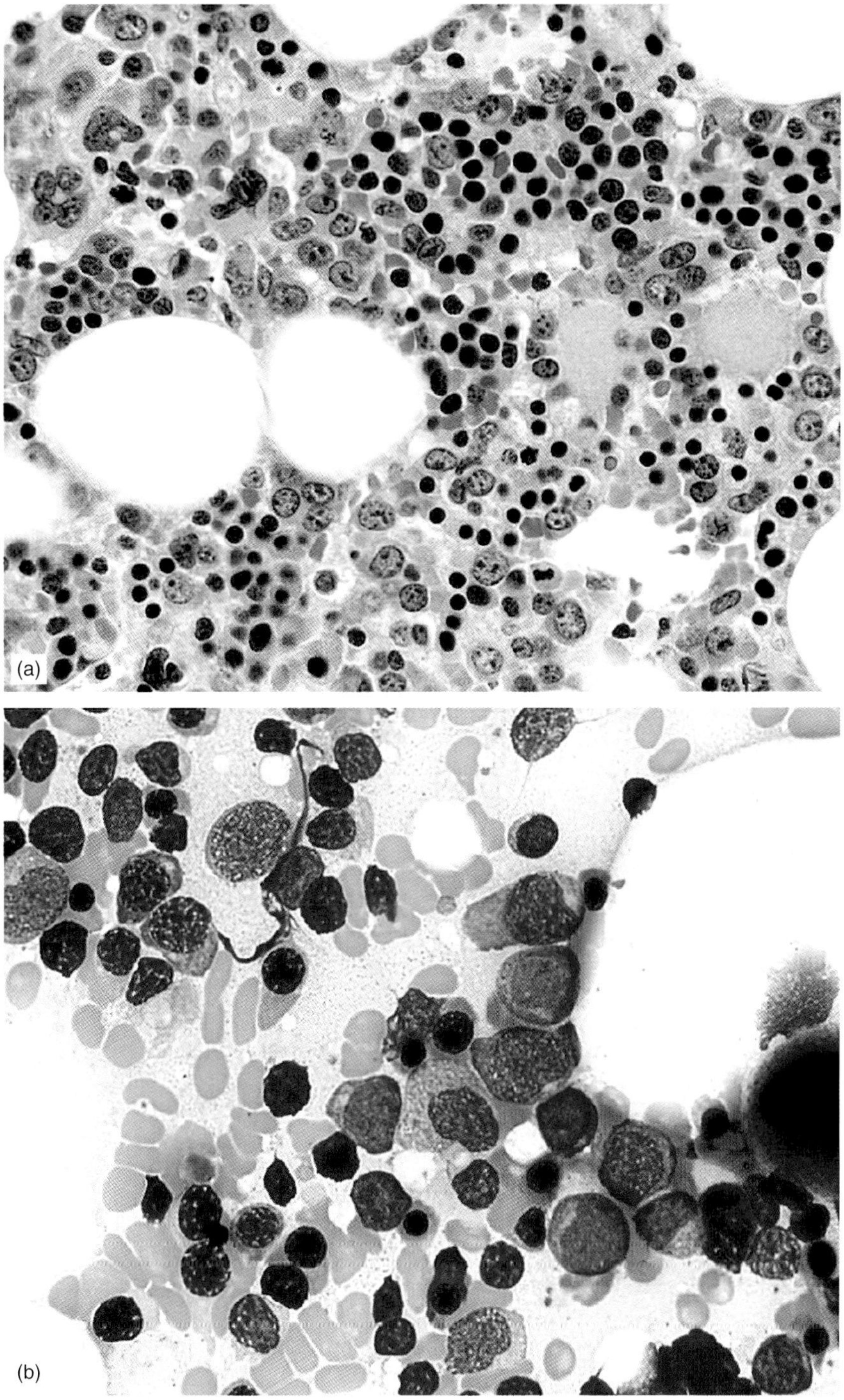

FIGURE 22.10 Agranulocytosis. Bone marrow biopsy section (a) and smear (b) demonstrate absence of late-stage granulocytic cells and sparse immature myeloid forms.

associated with mental retardation, short stature, microcephaly, depressed nasal bridge, and delayed motor development in affected children [43, 45, 46]. Other findings are leukocytosis (mainly neutrophilia) and Bombay blood group. The CD15a expression is lacking on the leukocytes by flow cytometry.

LAD III (previously called LAD I variant) is extremely rare and is characterized by a defect in integrin activation. The structure and expression of CD11 and CD18 appear to be intact [47]. Clinical symptoms are similar to those of LAD I. The prognosis is poor, and bone marrow transplantation is the treatment of choice.

NEUTROPENIA

Neutropenia refers to the peripheral blood absolute neutrophil count (ANC) of <1,500/μL. Neutropenia is considered mild when ANC is between 1,000 and 1,500/μL, moderate when ANC is between 500 and 1,000/μL, and severe when ANC is <500/μL. Severe neutropenia is also referred to as *agranulocytosis* (Figure 22.10). Three major mechanisms are involved in the development of neutropenia: (1) reduced production or ineffective granulopoiesis,

TABLE 22.1 Major causes of neutropenia.

Infections-associated
Drug- induced
Immune-associated
Isoimmune neonatal neutropenia
Chronic autoimmune neutropenia
Transfusion neutropenia
Transfusion reactions
Associated with hematologic disorders
Myelodysplastic syndromes
Aplastic anemia
Leukemia
Bone marrow replacement by fibrosis or metastasis
Hypersplenism
Congenital neutropenia
Kostmann syndrome
Congenital cyclic neutropenia
Associated with other congenital anomalies
Schwachman–Diamond syndrome
Cartilage-hair hypoplasia
Associated with functional abnormalities
Chediak–Higashi syndrome
Myelokathexis
Other types of neutropenia
Acquired cyclic neutropenia
Neutropenia due to nutritional deficiencies
Neutropenia associated with endocrine disorders
Lazy leukocyte syndrome
Chronic idiopathic neutropenia

(2) increased destruction or utilization of neutrophils, and (3) a shift from the circulating to the marginal pool [48–50]. Neutropenia is either acquired or congenital [48]. Infections and drugs are the most common etiologic factors in acquired neutropenia. Other contributing causes include primary immune deficiencies, bone marrow disorders, and congenital factors (Table 22.1).

Post-Infectious Neutropenia

Post-infectious neutropenia is the most common type of neutropenia caused by relocation and accumulation of granulocytes from circulation into the infected sites and/or destruction by circulating antibodies [48, 49]. Severe forms of neutropenia have been reported in hepatitis B virus and EBV and HIV infections [48].

Drug-Induced Neutropenia

Drug-induced neutropenia is the second most common type of neutropenia caused by two possible mechanisms: (1) drug-induced cytotoxicity and (2) immune-mediated. The drug-induced cytotoxicity affects protein synthesis or cell replication in the granulocytic precursors and is usually dose-dependent. Drugs or their metabolites may be the source of haptens or antigens, causing antibody production and drug–antibody interactions with neutrophils and their destruction [51]. Antithyroid drugs, clozapine, and sulfasalazine are among the drugs that could cause severe neutropenia [48]. A list of the drugs which may induce neutropenia is presented in Table 22.2. Drug-induced neutropenia is often associated with severe infectious complications and has a mortality rate of about 10% [52, 53].

Neutropenia Associated with Primary Immune Disorders

Neutropenia associated with primary immune disorders is caused by antineutrophil antibodies which are either autoimmune or alloimmune [51]. The mechanism of neutrophil destruction is either through complement-mediated neutrophil lysis or by splenic sequestration of opsonized neutrophils [49, 54]. The immune-associated neutropenia includes isoimmune neonatal neutropenia secondary to transplacental IgG antineutrophil antibodies, chronic autoimmune neutropenia primarily occurring in children younger than 4 years, transfusion reactions, complement activation, pure white cell aplasia due to antibody-mediated GM-CFU inhibitory activity, or antibodies to G-CSF [49].

Non-immune Chronic Idiopathic Neutropenia

Non-immune chronic idiopathic neutropenia is an acquired syndrome with no underlying autoimmune disease, nutritional deficiency, drug-association, or clonal bone marrow disorders [55]. The clinical course is usually benign and neutropenia is an incidental laboratory finding, with no history of infection or other symptoms [51, 56].

Neutropenia Associated with Bone Marrow Disorders

Neutropenia associated with bone marrow disorders include aplastic anemia, myelodysplastic syndromes, leukemias, and post-chemotherapy neutropenia.

Congenital Neutropenias

Congenital neutropenias consist of two major forms: cyclic neutropenia and severe congenital neutropenia.

Cyclic neutropenia is characterized by oscillation of peripheral neutrophil counts, from $<500/\mu L$ to near normal range, with approximately 3 week intervals. The congenital form is autosomal dominant and is associated with mutations in the neutrophil elastase gene, *ELA2* [57]. The acquired forms of cyclin neutropenia have been observed in various conditions, such as chronic myelogenous leukemia, large granular lymphocytosis, and hypereosinophilic syndromes [48].

Severe congenital neutropenia refers to various congenital syndromes including infantile agranulocytosis

TABLE 22.2 Major drugs with potential risk of severe neutropenia.

Drug class / drug
Antidepressant and psychotropic drugs
Clozapine
Phenothiazines
Meprobamate
Tricyclic and tetracyclic antidepressants
Antithyroid drugs
Methimazole
Carbimazole
Propylthiouracil
Antiinflammatory drugs
Sulfasalazine
Gold salts
Penicillamine
Phenylbutazone
Dipyrone
Phenacetin
Antihistamines: H2-receptor blockers
Cimetidine
Ranitidine
Antibacterial and antifungal drugs
Chloramphenicol
Sulfonamides
Vancomycin
Cephalosporin
Amphotricin B
Flucytosine
Sulfametoxazole
Antimalarial
Amodiaquine
Chloroquine
Quinine
Cardiovascular drugs
Propranolol
Dipyridamole
Digoxin
Antiarrhythmic drugs
ACE inhibitors
Anticonvulsants
Carbamazepine
Phenytoin
Ethosuximide
Valpoate
Diuretics
Thiazides
Acetazolamide
Furosemide
Spironolactone
Others
Aminoglutethimide
Chlorpropamide
Tolbotamide
Isotretinoin
Dapsone

(Kostmann syndrome), Shwachman–Diamond–Oski syndrome, myelokathexis, CHS, and congenital dysgranulopoietic neutropenia [48, 49].

Kostmann syndrome is an autosomal recessive agranulocytosis of infancy characterized by frequent, severe infections, mostly due to staphylococci and streptococci [58]. Bone marrow often reveals myeloid hypoplasia and maturation arrest at the promyelocyte stage [59].

Shwachman–Diamond–Oski syndrome represents a triad of neutropenia, pancreatic insufficiency, and metaphyseal dysplasia [50]. Patients are usually under 10 years of age with a history of recurrent infection and steatorrhea. Neutropenia appears to be secondary to increased apoptosis in the bone marrow [60].

Myelokathexis is an extremely rare form of chronic, childhood neutropenia with recurrent infections [61–65]. The bone marrow is often hypercellular and shows dysmyelopoietic features [63]. The dysplastic changes include nuclear hypersegmentation, cytoplasmic vacuolization, and hypogranularity [61]. One possible mechanism for neutropenia is defective release and prolonged retention of neutrophils in the bone marrow [61, 63]. Myelokathexis has been reported in association with accelerated apoptosis and defective expression of bcl-x in granulocytic precursors. It is considered a part of WHIM (warts, hypogammaglobulinemia, infections, myelokathexis) syndrome [65–67].

Chediak–Higashi syndrome is described in Chapter 21.

Congenital dysgranulopoietic neutropenia is a rare autosomal recessive disorder characterized by repeated severe infections and dysmyelopoietic features in the bone marrow, including defects in the synthesis of primary and specific granules and premature cell lysis [68, 69].

Neutropenia has also been associated with lazy leukocyte syndrome, bone marrow stem cell disorders (such as aplastic anemia, leukemias, and myelodysplastic syndrome, in cobalamin deficiency), and hereditary disorders, such as dyskeratosis congenital, reticular dysgenesis, and glycogen storage disease type 1 [48, 49]. Lazy leukocyte syndrome is an extremely rare condition characterized by severe neutropenia, defective neutrophil chemotaxis, and impaired random mobility of granulocytes [70, 71].

NEUTROPHILIA

Neutrophilia refers to an increase in the absolute number of neutrophils in the peripheral blood (absolute neutrophil count >7,700/μL in adults). Since neutrophils account for the majority of the circulating leukocytes (about 60%), in most instances, white blood cell (WBC) counts of over 11,000/μL represent neutrophilia [72]. There are two major causes for neutrophilia: (1) a reactive response to ongoing processes, such as infection, inflammation, smoking, stress, medication, or malignancy and (2) due to primary abnormalities in the regulation of bone marrow neutrophil production (Table 22.3). Leukemoid reaction is referred to leukocytosis in excess of 50,000/μL with a left shift, caused by conditions other than a leukemic process [72].

TABLE 22.3 Major causes of neutrophilia.

Reactive
Acute infections
Chronic inflammations
Cigarette smoking
Exercise
Stress
Drugs
Bone marrow stimulation
Non-hematopoietic malignancies
Heatstroke
Others
Primary
Hereditary neutrophilia
Down syndrome
Muckle–Wells syndrome
Leukocyte adhesion deficiency
Chronic myeloproliferative disorders
Others

Reactive Neutrophilia

Acute infections commonly cause various degrees of neutrophilia due to the release of segmented neutrophils and bands from the bone marrow and marginating pool (Figure 22.11). Acute bacterial infections, such as pneumococcal, staphylococcal, or leptospiral infections, are the most frequent causes of infection-induced neutrophilia. Certain viral infections, such as herpes complex, varicella, and EBV infections, may also cause neutrophilia. Reactive neutrophilia is often associated with the presence of toxic granulation, Dohle bodies, cytoplasmic vacuoles, and elevated levels of leukocyte (neutrophil) alkaline phosphatase (LAP) (Figure 22.11c).

Chronic inflammations, such as rheumatoid arthritis, Kawasaki disease, and inflammatory bowel disorders are often associated with neutrophilia. Neutrophilia in these conditions are in part due to the release of cytokines, such as TNF-α, G-CSF, GM-CSF, IL-6, and IL-8 [72–74].

Cigarette smoking is usually associated with elevated leukocyte count and neutrophilia (up to 25%), which may last as long as 5 years after quitting smoking [74–77]. The exact mechanism of smoking-induced leukocytosis is not known.

Stress-associated neutrophilia appears to be related to the redistribution of neutrophils from the marginating pool into the circulating pool, probably due to the reduced neutrophil adhesion by the release of epinephrine [78]. Post-operative neutrophilia and leukocytosis during acute myocardial infarction are considered stress-related [79, 80].

Exercise-induced neutrophilia is probably due to a combination of increased plasma epinephrine levels and a change in the cardiac output, leading to the redistribution of neutrophils from the marginating pool into the circulating pool [81, 82]. Delayed leukocytosis in exercise is probably due to the release of leukocytes from the bone marrow [83].

Medications such as beta agnostics (e.g. epinephrine), glucocorticosteroids, lithium, recombinant colony stimulating factors, and all-*trans* retinoic acid (ATRA) are associated with neutrophilia (Figure 22.12) [84–86].

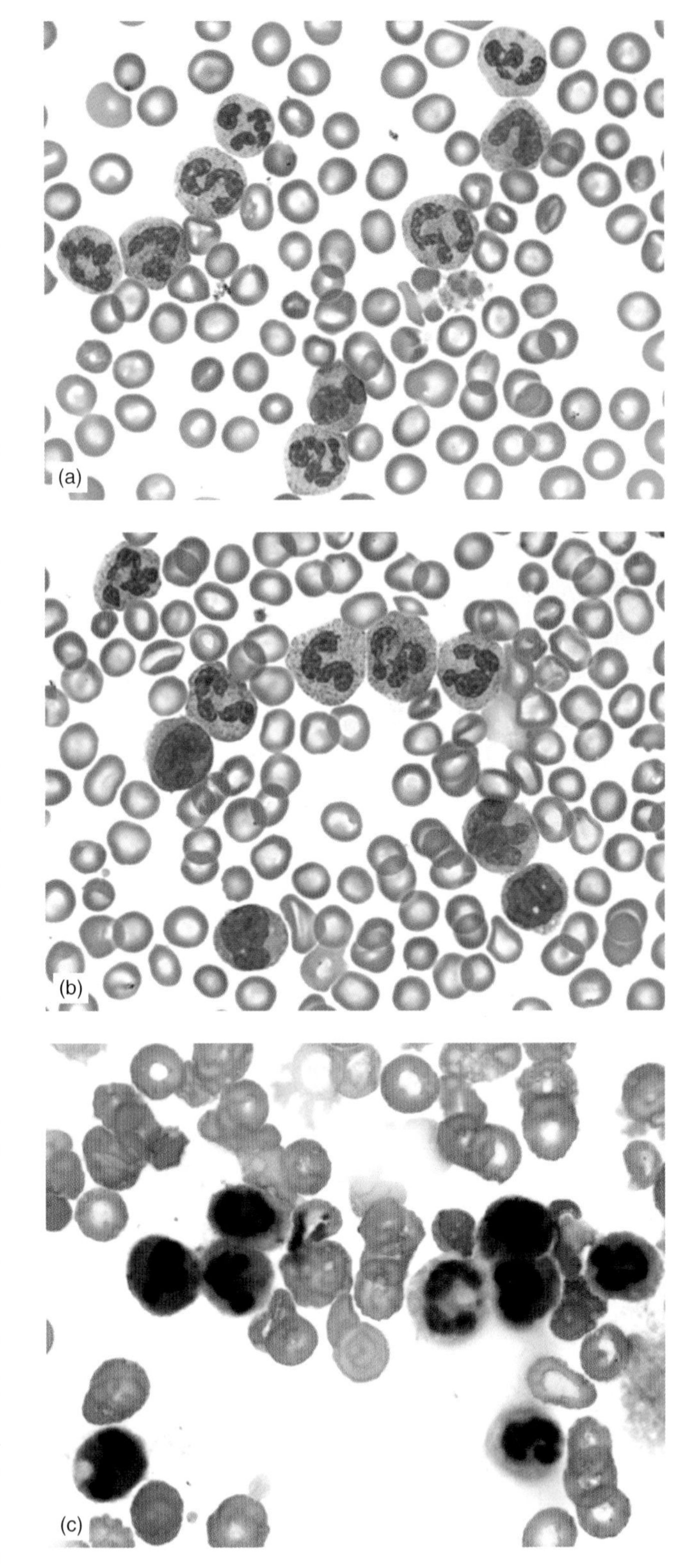

FIGURE 22.11 Leukemoid reaction. Blood smears demonstrate numerous neutrophils and bands (a), neutrophils and monocytes (b), and alkaline phosphatase-positive neutrophils (c).

Other causes of neutrophilia include bone marrow stimulation (such as hemolytic anemia or immune thrombocytopenia), non-hematologic malignancies, heatstroke, and post-splenectomy.

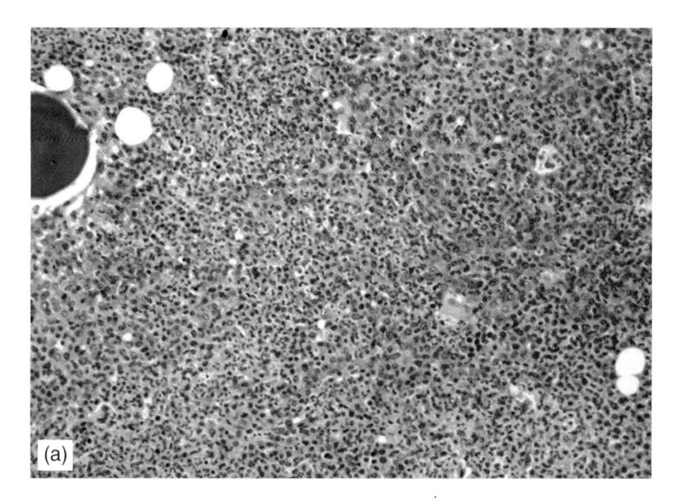

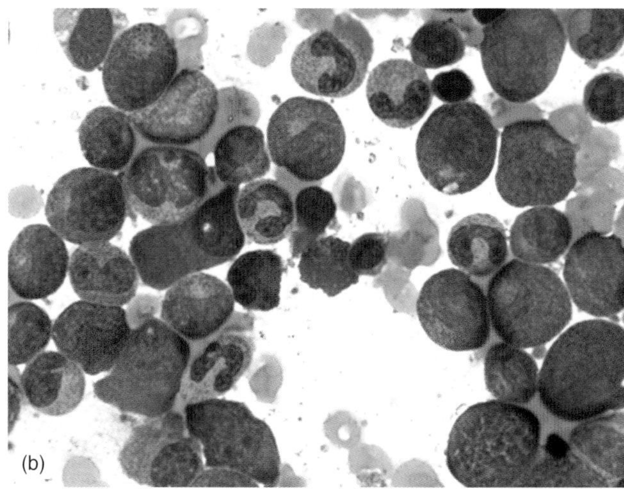

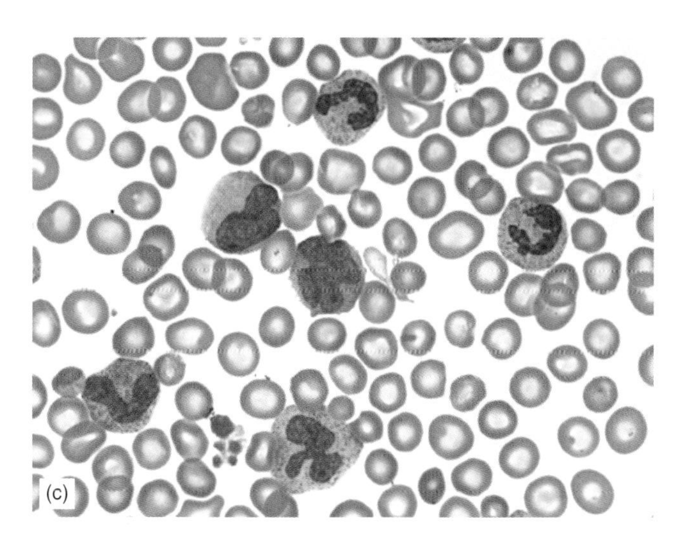

FIGURE 22.12 Effect of G-CSF therapy. Bone marrow biopsy section (a) and smear (b) show myeloid preponderance and left shift. Blood smear (c) reveals leukocytosis with the presence of neutrophils and monocytes.

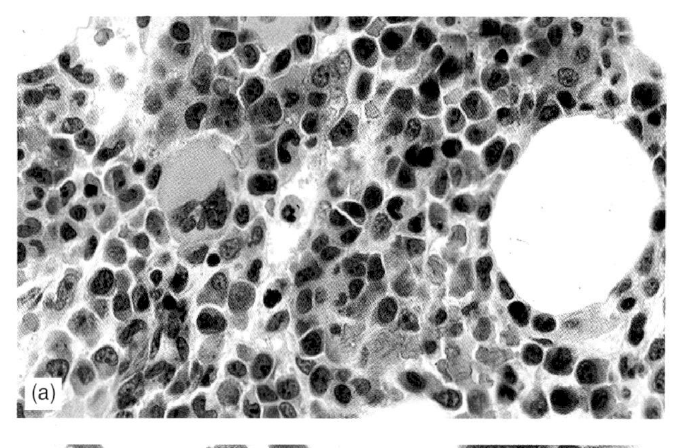

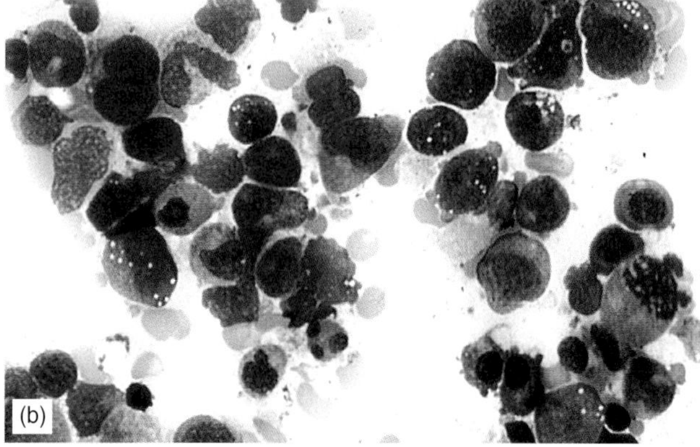

FIGURE 22.13 Transient myeloproliferative disorder in Down syndrome. Bone marrow biopsy section (a) and smear (b) show myeloid preponderance and left shift.

Primary Neutrophilia

Primary neutrophilia is due to primary abnormalities in the regulation of bone marrow neutrophil production, such as hereditary neutrophilia, chronic myeloproliferative disorders (see Chapter 9), myeloproliferative/myelodysplastic syndromes (see Chapter 10), transient myeloproliferative disorder in Down syndrome, and leukocyte adhesion deficiency. In this section, hereditary neutrophilia and transient myeloproliferative disorder in Down syndrome are briefly discussed.

Hereditary neutrophilia is a rare autosomal disorder characterized by chronic neutrophilia (ranging from 20,000 to over 100,000/μL), splenomegaly, elevated LAP, and widened dipole of the skull [87]. Neutrophil function is normal, but affected individuals may demonstrate bleeding complications due to platelet dysfunction [87].

Transient myeloproliferative disorder (TMD) in Down syndrome or transient abnormal myelopoiesis is a leukemoid reaction occasionally observed in some neonates with Down syndrome (trisomy of chromosome 21) [80–90]. The affected neonates are usually under 1 month old and demonstrate peripheral blood leukocytosis with a left shift and the presence of blast cells (Figures 22.13 and 22.14). Blasts cells are predominantly of megakaryocytic and erythroid origin. Mutation of *GATA-1* gene encoding for erythroid/megakaryocytic transcription factor GATA-1 has been reported in Down syndrome patients with TMD [91]. TMD in Down syndrome usually disappears spontaneously after 4–6 weeks [91, 92]. A small proportion of patients may eventually develop acute myelogenous leukemia [93, 94].

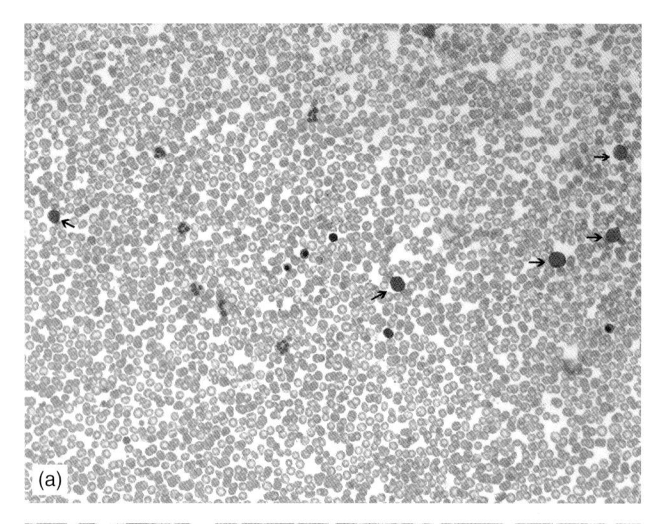

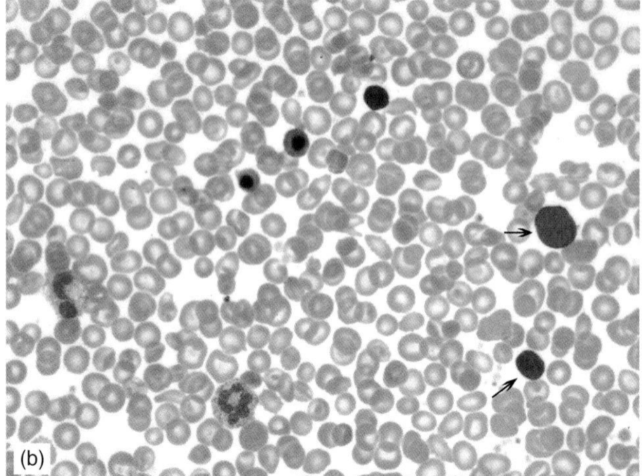

FIGURE 22.14 Transient myeloproliferative disorder in Down syndrome. Blood smear demonstrate several blast cells: (a) low power and (b) high power views.

Spurious Neutrophilia

Spurious neutrophilia is a falsely elevated peripheral blood neutrophil count resulting from various reasons. Formation of precipitated cryoglobulin particles may result in erroneous elevated WBC and/or platelet counts [95]. Also, platelet clumps may be counted as leukocytes by automated cell counters.

EOSINOPHILIA

Eosinophilia is defined as an absolute peripheral blood eosinophil count exceeding 600/μL [96]. It is observed in a wide variety of conditions including allergic and inflammatory processes, protozoan and metazoan infections, immunodeficiencies and autoimmune disorders, chronic myeloproliferative disorders, leukemias and lymphomas, and certain non-hematopoietic malignancies (Table 22.4). Idiopathic hypereosinophilic syndrome and chronic eosinophilic leukemia are classified under chronic myeloproliferative disorders and are discussed in Chapter 9.

TABLE 22.4 Conditions associated with eosinophilia.*

Protozoan infections
Pneumocystic, toxoplasmosis, amebiasis, malaria
Metazoan infections
Nematodes, trematodes, cestodes, arthropods
Allergic and autoimmune disorders
Hay fever, asthma, angioneurotic edema, urticaria, serum sickness, allergic vasculitis, pemphigus vulgaris, dermatitis herpetiformis, ulcerative colitis, regional enteritis
Hematopoietic disorders
Hodgkin and non-Hodgkin lymphomas, mycosis fungoides, acute myelogenous leukemia, chronic myeloproliferative disorders, plasma cell myeloma, Langerhans cell histiocytosis, familial hemophagocytic lymphohistiocytosis
Solid tumors
Carcinomas, brain tumors, melanoma
Others
Immunodeficiency syndromes, sarcoidosis, chronic renal disease, peritoneal dialysis, pleural effusion, radiotherapy, splenectomy

*Adapted from Ref. [3].

A hypereosinophilic syndrome has been reported in association with the use of tryptophan derivatives [97, 98]. It is characterized by peripheral blood eosinophilia and scleroderma-like features, including muscle tenderness, fatigue, edema, arthralgia, nephropathy, rash, cough, and dyspnea. Tryptophan derivatives have been used for the treatment of insomnia and depression.

Eosinophilia is reported in association with a variety of solid tumors such as bronchogenic carcinoma, medullary carcinoma of the thyroid gland, and transitional cell carcinoma of the bladder. A garden variety of hematopoietic malignancies, such as chronic myelogenous leukemia, acute lymphoblastic leukemia, acute myelogenous leukemia, and Hodgkin and non-Hodgkin lymphomas, may demonstrate eosinophilia in the involved tissues and peripheral blood. A subtype of acute myelomonocytic leukemia is associated with atypical eosinophilia and chromosomal aberrations involving 16q22 (see Chapter 10). A subtype of precursor B-cell acute lymphoblastic leukemia with t(5;14) has been associated with eosinophilia [99, 100]. Similarly, eosinophilia has been reported in precursor T-cell lymphoblastic leukemias with t(8;13) [101].

BASOPHILIA

Basophilia is defined as an absolute peripheral blood basophil count exceeding 200/μL [72]. It may occur in a wide variety of infections or inflammatory and autoimmune conditions, such as tuberculosis, chickenpox, smallpox, influenza, ulcerative colitis, and rheumatoid arthritis [72, 102, 103]. Other conditions associated with basophilia include irradiation, iron deficiency, hypothyroidism, diabetes mellitus, chronic myeloproliferative disorders, and myelodysplastic syndromes [104]. Certain subtypes of acute myelogenous leukemia, such as acute promyelocytic leukemia and acute myeloid leukemias with t(6;9), t(3;6), and aberrations of chromosome 16 may demonstrate bone marrow basophilia [105].

MASTOCYTOSIS

Mastocytosis is discussed in Chapter 20.

Reference

1. Cawley JC, Hayhoe FG. (1972). The inclusions of the May–Hegglin anomaly and Dohle bodies of infection: An ultrastructural comparison. *Br J Haematol* **22**, 491–6.
2. Jordan SW, Larsen WE. (1965). Ultrastructural studies of the May–Hegglin anomaly. *Blood* **25**, 921–32.
3. Naeim F. (1998). *Pathology of Bone Marrow*, 2nd ed. Williams & Wilkins, Baltimore.
4. Al-Gwaiz LA, Babay HH. (2007). The diagnostic value of absolute neutrophil count, band count and morphologic changes of neutrophils in predicting bacterial infections. *Med Prin Pract* **16**(5), 344–7.
5. Kunishima S, Kojima T, Tanaka T, Kamiya T, Ozawa K, Nakamura Y, Saito H. (1999). Mapping of a gene for May–Hegglin anomaly to chromosome 22q. *Hum Genet* **105**, 379–83.
6. Deutsch S, Rideau A, Bochaton-Piallat ML, Merla G, Geinoz A, Gabbiani G, Schwede T, Matthes T, Antonarakis SE, Beris P. (2003). Asp1424Asn MYH9 mutation results in an unstable protein responsible for the phenotypes in May–Hegglin anomaly/Fechtner syndrome. *Blood* **102**, 529–34.
7. Kunishima S, Yoshinari M, Nishio H, Ida K, Miura T, Matsushita T, Hamaguchi M, Saito H. (2007). Haematological characteristics of MYH9 disorders due to MYH9 R702 mutations. *Eur J Haematol* **78**, 220–6.
8. Kelley MJ, Jawien W, Lin A, Hoffmeister K, Pugh EW, Doheny KF, Korczak JF. (2000). Autosomal dominant macrothrombocytopenia with leukocyte inclusions (May–Hegglin anomaly) is linked to chromosome 22q12-13. *Hum Genet* **106**, 557–64.
9. Toren A, Rozenfeld-Granot G, Rocca CJ, Epstein B, Amariglio N, Laghi F, Landolfi R, Brok-Simoni F, Carlsson LE, Rechavi G, Greinacher A. (2000). Autosomal-dominant giant platelet syndromes: A hint of the same genetic defect as in Fechtner syndrome owing to a similar genetic linkage to chromosome 22q11-13. *Blood* **96**, 3447–51.
10. Peterson L, Parkin J, Nelson A. (1982). Mucopolysaccharidosis type VII. A morphologic, cytochemical, and ultrastructural study of the blood and bone marrow. *Am J Clin Pathol* **78**, 544–8.
11. Ghandi MK, Howard MR, Hamilton PJ. (1996). The Alder–Reilly anomaly in association with the myelodysplastic syndrome. *Clin Lab Haematol* **18**, 39–40.
12. Presentey B. (1986). Alder anomaly accompanied by a mutation of the myeloperoxidase structural gene. *Acta Haematol* **75**, 157–9.
13. Ward DM, Griffiths GM, Stinchcombe JC, Kaplan J. (2000). Analysis of the lysosomal storage disease Chediak–Higashi syndrome. *Traffic* **1**(11), 816–22.
14. Shiflett SL, Kaplan J, Ward DM. (2002). Chediak–Higashi syndrome: A rare disorder of lysosomes, lysosome related organelles. *Pigm Cell Res* **15**, 251–7.
15. Lomax KJ, Gallin JI, Rotrosen D, Raphael GD, Kaliner MA, Benz Jr. EJ, Boxer LA, Malech HL. (1989). Selective defect in myeloid cell lactoferrin gene expression in neutrophil specific granule deficiency. *J Clin Invest* **83**, 514–19.
16. Le Beau MM, Larson RA, Bitter MA, Vardiman JW, Golomb HM, Rowley JD. (1983). Association of an inversion of chromosome 16 with abnormal marrow eosinophils in acute myelomonocytic leukemia. A unique cytogenetic–clinicopathological association. *N Engl J Med* **309**, 630–6.
17. Takemori N, Saito N, Tachibana N, Hayashishita N, Miyazaki T. (1988). Hybrid eosinophilic–basophilic granulocytes in chronic myeloid leukemia. *Am J Clin Pathol* **89**, 702–3.
18. Ware R, Kurtzberg J, Brazy J, Falletta JM. (1988). Congenital Pelger–Huet anomaly in triplets. *Am J Hematol* **27**, 226–7.
19. Kalfa TA, Zimmerman SA, Goodman BK, McDonald MT, Ware RE. (2006). Pelger–Huet anomaly in a child with 1q42.3-44 deletion. *Pediatr Blood Cancer* **46**, 645–8.
20. Oosterwijk JC, Mansour S, van Noort G, Waterham HR, Hall CM, Hennekam RC. (2003). Congenital abnormalities reported in Pelger–Huet homozygosity as compared to Greenberg/HEM dysplasia: Highly variable expression of allelic phenotypes. *J Med Genet* **40**, 937–41.
21. Hoffmann K, Dreger CK, Olins AL, Olins DE, Shultz LD, Lucke B, Karl H, Kaps R, Müller D, Vayá A, Aznar J, Ware RE, Sotelo Cruz N, Lindner TH, Herrmann H, Reis A, Sperling K. (2002). Mutations in the gene encoding the lamin B receptor produce an altered nuclear morphology in granulocytes (Pelger–Huët anomaly). *Nat Genet* **31**, 410–14.
22. Best S, Salvati F, Kallo J, Garner C, Height S, Thein SL, Rees DC. (2003). Lamin B-receptor mutations in Pelger–Huet anomaly. *Br J Haematol* **123**, 542–4.
23. Dusse LM, Silva E, Morais RM, Freitas VM, Medeiros de Paula GM, Vieira LM, Carvalho MG. (2006). Pseudo-Pelger–Huet in kidney-transplanted patients. *Acta Haematol* **116**, 272–4.
24. Etzell JE, Wang E. (2006). Acquired Pelger–Huet anomaly in association with concomitant tacrolimus and mycophenolate mofetil in a liver transplant patient: A case report and review of the literature. *Arch Pathol Lab Med* **130**, 93–6.
25. Liesveld J, Smith BD. (1988). Acquired Pelger–Huet anomaly in a case of non-Hodgkin's lymphoma. *Acta Haematol* **79**, 46–9.
26. Jbour AK, Mubaidin AF, Till M, El-Shanti H, Hadidi A, Ajlouni KM. (2003). Hypogonadotrophic hypogonadism, short stature, cerebellar ataxia, rod-cone retinal dystrophy, and hypersegmented neutrophils: A novel disorder or a new variant of Boucher–Neuhauser syndrome? *J Med Genet* **40**(1), e2.
27. Zittoun J, Zittoun R. (1999). Modern clinical testing strategies in cobalamin and folate deficiency. *Semin Hematol* **36**, 35–46.
28. Bohinjec J. (1981). Myelokathexis: Chronic neutropenia with hyperplastic bone marrow and hypersegmented neutrophils in two siblings. *Blut* **42**, 191–6.
29. Kitahara M, Eyre HJ, Simonian Y, Atkin CL, Hasstedt SJ. (1981). Hereditary myeloperoxidase deficiency. *Blood* **57**, 888–93.
30. Nauseef WM. (1988). Myeloperoxidase deficiency. *Hematol Oncol Clin North Am* **2**, 135–58.
31. Nguyen C, Katner HP. (1997). Myeloperoxidase deficiency manifesting as pustular candidal dermatitis. *Clin Infect Dis* **24**, 258–60.
32. Lehrer RI, Hanifin J, Cline MJ. (1969). Leukocyte myeloperoxidase deficiency and disseminated candidiasis: The role of myeloperoxidase in resistance to Candida infection. *J Clin Invest* **48**, 1478–88.
33. Cech P, Papathanassiou A, Boreux G, Roth P, Miescher PA. (1979). Hereditary myeloperoxidase deficiency. *Blood* **53**, 403–11.
34. Winterbourn CC, Vissers MC, Kettle AJ. (2000). Myeloperoxidase. *Curr Opin Hematol* **7**, 53–8.
35. Rosenzweig SD, Holland SM. (2007). Myeloperoxidase deficiency and other enzymatic WBC defects causing immunodeficiency. *UpToDate.*
36. Teimourian S, Rezvani Z, Badalzadeh M, Kannengiesser C, Mansouri D, Movahedi M, Zomorodian E, Parvaneh N, Mamishi S, Pourpak Z, Moin M. (2008). Molecular diagnosis of X-linked chronic granulomatous disease in Iran. *Int J Hematol* **Mar 7**.
37. Soler-Palacín P, Margareto C, Llobet P, Asensio O, Hernández M, Caragol I, Español T. (2007). Chronic granulomatous disease in pediatric patients: 25 years of experience. *Allergol Immunopathol (Madr)* **35**, 83–9.

38. Assari T. (2006). Chronic granulomatous disease: Fundamental stages in our understanding of CGD. *Med Immunol* **5**, 4.
39. Heyworth PG, Cross AR, Curnutte JT. (2003). Chronic granulomatous disease. *Curr Opin Immunol* **15**, 578–84.
40. Fraser IP. (2007). Chronic granulomatous disease. *UpToDate.*
41. Bunting M, Harris ES, McIntyre TM, Prescott SM, Zimmerman GA. (2002). Leukocyte adhesion deficiency syndromes: Adhesion and tethering defects involving beta 2 integrins and selectin ligands. *Curr Opin Hematol* **9**, 30–5.
42. Etzioni A, Tonetti M. (2000). Leukocyte adhesion deficiency II – from A to almost Z. *Immunol Rev* **178**, 138–47.
43. Etzioni A. (2007). Leukocyte adhesion deficiency. *UpToDate.*
44. Lekstrom-Himes JA, Gallin JI. (2000). Immunodeficiency diseases caused by defects in phagocytes. *N Engl J Med* **343**, 1703–14.
45. Wild MK, Luhn K, Marquardt T, Vestweber D. (2002). Leukocyte adhesion deficiency II: Therapy and genetic defect. *Cells Tissues Organs* **172**, 161–73.
46. Hidalgo A, Ma S, Peired AJ, Weiss LA, Cunningham-Rundles C, Frenette PS. (2003). Insights into leukocyte adhesion deficiency type 2 from a novel mutation in the GDP-fucose transporter gene. *Blood* **101**, 1705–12.
47. McDowall A, Inwald D, Leitinger B, Jones A, Liesner R, Klein N, Hogg N. (2003). A novel form of integrin dysfunction involving beta1, beta2, and beta3 integrins. *J Clin Invest* **111**, 51–60.
48. Berliner N, Horwitz M, Loughran Jr. TP. (2004). Congenital and acquired neutropenia. *Hematol Am Soc Hematol Educ Prog*, 63–79.
49. Baehner RL. (2007). Overview of neutropenia. *UpToDate.*
50. Bishop CR, Rothstein G, Ashenbrucker HE, Athens JW. (1971). Leukokinetic studies. XIV. Blood neutrophil kinetics in chronic, steady-state neutropenia. *J Clin Invest* **50**, 1678–89.
51. Palmblad J, Papadaki HA, Eliopoulos G. (2001). Acute and chronic neutropenias. What is new? *J Intern Med* **250**, 476–91.
52. Andres E, Kurtz JE, Maloisel F. (2002). Nonchemotherapy drug-induced agranulocytosis: Interest of haematopoietic growth factors. *J Intern Med* **251**, 533–4.
53. van der Klauw MM, Goudsmit R, Halie MR, van't Veer MB, Herings RM, Wilson JH, Stricker BH. (1999). A population-based case–cohort study of drug-associated agranulocytosis. *Arch Intern Med* **159**, 369–74.
54. Bux J. (2002). Molecular nature of antigens implicated in immune neutropenias. *Int J Hematol* **76**(Suppl 1), 399–403.
55. Papadaki HA, Palmblad J, Eliopoulos GD. (2001). Non-immune chronic idiopathic neutropenia of adult: An overview. *Eur J Haematol* **67**, 35–44.
56. Kyle RA. (1980). Natural history of chronic idiopathic neutropenia. *N Engl J Med* **302**, 908–9.
57. Horwitz M, Benson KF, Person RE, Aprikyan AG, Dale DC. (1999). Mutations in ELA2, encoding neutrophil elastase, define a 21-day biological clock in cyclic haematopoiesis. *Nat Genet* **23**, 433–6.
58. Kostmann R. (1975). Infantile genetic agranulocytosis: A review with presentation of 10 cases. *Acta Pediatr Scand* **64**, 362.
59. Amato D, Freedman MH, Saunders EF. (1976). Granulopoiesis in severe congenital neutropenia. *Blood* **47**, 531–8.
60. Dror Y, Freedman MH. (2001). Shwachman–Diamond syndrome marrow cells show abnormally increased apoptosis mediated through the Fas pathway. *Blood* **97**, 3011–16.
61. Aprikyan AA, Liles WC, Park JR, Jonas M, Chi EY, Dale DC. (2000). Myelokathexis, a congenital disorder of severe neutropenia characterized by accelerated apoptosis and defective expression of bcl-x in neutrophil precursors. *Blood* **95**, 320–7.
62. Zuelzer WW. (1964). "Myelokathexis" – a new form of chronic granulocytopenia. Report of a case. *N Engl J Med* **270**, 699–704.
63. O'Regan S, Newman AJ, Graham RC. (1977). "Myelokathexis". Neutropenia with marrow hyperplasia. *Am J Dis Child* **131**, 655–8.
64. Rassam SM, Roderick P, al-Hakim I, Hoffbrand AV. (1989). A myelokathexis-like variant of myelodysplasia. *Eur J Haematol* **42**, 99–102.
65. Kawai T, Choi U, Cardwell L, DeRavin SS, Naumann N, Whiting-Theobald NL, Linton GF, Moon J, Murphy PM, Malech HL. (2007). WHIM syndrome myelokathexis reproduced in the NOD/SCID mouse xenotransplant model engrafted with healthy human stem cells transduced with C-terminus-truncated CXCR4. *Blood* **109**, 78–84.
66. Balabanian K, Lagane B, Pablos JL, Laurent L, Planchenault T, Verola O, Lebbe C, Kerob D, Dupuy A, Hermine O, Nicolas JF, Latger-Cannard V, Bensoussan D, Bordigoni P, Baleux F, Le Deist F, Virelizier JL, Arenzana-Seisdedos F, Bachelerie F. (2005). WHIM syndromes with different genetic anomalies are accounted for by impaired CXCR4 desensitization to CXCL12. *Blood* **105**, 2449–57.
67. Latger-Cannard V, Bensoussan D, Bordigoni P. (2006). The WHIM syndrome shows a peculiar dysgranulopoiesis: Myelokathexis. *Br J Haematol* **132**, 669.
68. Parmley RT, Crist WM, Ragab AH, Boxer LA, Malluh A, Lui VK, Darby CP. (1980). Congenital dysgranulopoietic neutropenia: Clinical, serologic, ultrastructural, and *in vitro* proliferative characteristics. *Blood* **56**, 465–75.
69. Lightsey AL, Parmley RT, Marsh WL, Garg AK, Thomas WJ, Wolach B, Boxer LA. (1985). Severe congenital neutropenia with unique features of dysgranulopoiesis. *Am J Hematol* **18**, 59–71.
70. Pinkerton PH, Robinson JB, Senn JS. (1978). Lazy leukocyte syndrome – disorder of the granulocyte membrane? *J Clin Pathol* **31**, 300–8.
71. Komiyama A, Kawai H, Yamada S, Aoyama K, Yamazaki M, Saitoh H, Miyagawa Y, Akabane T, Uehara Y. (1985). Impaired natural killer cell recycling in childhood chronic neutropenia with morphological abnormalities and defective chemotaxis of neutrophils. *Blood* **66**, 99–105.
72. Coates TD, Baehner RL. (2007). Causes of neutrophilia. *UpToDate.*
73. Xing Z, Gauldie J, Cox G, Baumann H, Jordana M, Lei XF, Achong MK. (1998). IL-6 is an antiinflammatory cytokine required for controlling local or systemic acute inflammatory responses. *J Clin Invest* **101**, 311–20.
74. Dale DC, Liles WC, Llewellyn C, Price TH. (1998). Effects of granulocyte-macrophage colony-stimulating factor (GM-CSF) on neutrophil kinetics and function in normal human volunteers. *Am J Hematol* **57**, 7–15.
75. Parry H, Cohen S, Schlarb JE, Tyrrell DA, Fisher A, Russell MA, Jarvis MJ. (1997). Smoking, alcohol consumption, and leukocyte counts. *Am J Clin Pathol* **107**, 64–7.
76. Schwartz J, Weiss ST. (1994). Cigarette smoking and peripheral blood leukocyte differentials. *Ann Epidemiol* **4**, 236–42.
77. Van Tiel E, Peeters PH, Smit HA, Nagelkerke NJ, Van Loon AJ, Grobbee DE, Bueno-de-Mesquita HB. (2002). Quitting smoking may restore hematological characteristics within five years. *Ann Epidemiol* **12**, 378–88.
78. Boxer LA, Allen JM, Baehner RL. (1980). Diminished polymorphonuclear leukocyte adherence. Function dependent on release of cyclic AMP by endothelial cells after stimulation of beta-receptors by epinephrine. *J Clin Invest* **66**, 268–74.
79. Zalokar JB, Richard JL, Claude JR. (1981). Leukocyte count, smoking, and myocardial infarction. *N Engl J Med* **304**, 465–8.
80. Jakobsen BW, Pedersen J, Egeberg BB. (1986). Postoperative lymphocytopenia and leucocytosis after epidural and general anaesthesia. *Acta Anaesthesiol Scand* **30**, 668–71.
81. Christensen RD, Hill HR. (1987). Exercise-induced changes in the blood concentration of leukocyte populations in teenage athletes. *Am J Pediatr Hematol Oncol* **9**, 140–2.
82. Foster NK, Martyn JB, Rangno RE, Hogg JC, Pardy RL. (1986). Leukocytosis of exercise: Role of cardiac output and catecholamines. *J Appl Physiol* **61**, 2218–23.
83. McCarthy DA, Perry JD, Melsom RD, Dale MM. (1987). Leucocytosis induced by exercise. *Br Med J (Clin Res Ed)* **295**, 636.
84. Dale DC, Fauci IV AS, Guerry D, Wolff SM. (1975). Comparison of agents producing a neutrophilic leukocytosis in man. Hydrocortisone, prednisone, endotoxin, and etiocholanolone. *J Clin Invest* **56**, 808–13.

85. Boggs DR, Joyce RA. (1983). The hematopoietic effects of lithium. *Semin Hematol* **20**, 129–38.
86. Vahdat L, Wong ET, Wile MJ, Rosenblum M, Foley KM, Warrell Jr. RP. (1994). Early mortality and the retinoic acid syndrome in acute promyelocytic leukemia: Impact of leukocytosis, low-dose chemotherapy, PMN/RAR-alpha isoform, and CD13 expression in patients treated with all-trans retinoic acid. *Blood* **84**, 3843–9.
87. Herring WB, Smith LG, Walker RI, Herion JC. (1974). Hereditary neutrophilia. *Am J Med* **56**, 729–34.
88. Brodeur GM, O'Neill PJ, Willimas JA. (1980). Transient leukemoid reaction and trisomy 21 mosaicism in a phenotypically normal newborn. *Blood* **55**, 691–3.
89. Weinstein HJ. (1978). Congenital leukaemia and the neonatal myeloproliferative disorders associated with Down's syndrome. *Clin Haematol* **7**, 147–54.
90. Cantu-Rajnoldi A, Cattoretti G, Caccamo ML, Biasini A, Bagnato L, Schiro R, Polli N. (1988). Leukaemoid reaction with megakaryocytic features in newborns with Down's syndrome. *Eur J Haematol* **40**, 403–9.
91. Xu G, Nagano M, Kanezaki R, Toki T, Hayashi Y, Taketani T, Taki T, Mitui T, Koike K, Kato K, Imaizumi M, Sekine I, Ikeda Y, Hanada R, Sako M, Kudo K, Kojima S, Ohneda O, Yamamoto M, Ito E. (2003). Frequent mutations in the GATA-1 gene in the transient myeloproliferative disorder of Down syndrome. *Blood* **102**, 2960–8.
92. Schwab M, Niemeyer C, Schwarzer U. (1998). Down syndrome, transient myeloproliferative disorder, and infantile liver fibrosis. *Med Pediatr Oncol* **31**, 159–65.
93. Liang DC, Ma SW, Lu TH, Lin ST. (1993). Transient myeloproliferative disorder and acute myeloid leukemia: Study of six neonatal cases with long-term follow-up. *Leukemia* **7**, 1521–4.
94. Yokoyama S, Nito T, Irimada K, Yoshimura Y, Hayashi T. (1984). Acute megakaryoblastic leukemia preceded by refractory anemia with an excess of blast cells: Leukemic transformation in an infant of Down syndrome recovering from transient abnormal myelopoiesis. *Nippon Ketsueki Gakk Zasshi* **47**, 62–70.
95. Patel KJ, Hughes CG, Parapia LA. (1987). Pseudoleucocytosis and pseudothrombocytosis due to cryoglobulinaemia. *J Clin Pathol* **40**, 120–1.
96. Tefferi A. (2005). Blood eosinophilia: A new paradigm in disease classification, diagnosis, and treatment. *Mayo Clin Proc* **80**, 75–83.
97. Hertzman PA, Blevins WL, Mayer J, Greenfield B, Ting M, Gleich GJ. (1990). Association of the eosinophilia–myalgia syndrome with the ingestion of tryptophan. *N Engl J Med* **322**, 869–73.
98. Sindransky H. (1994). Eosinophilia–myalgia syndrome: A recent syndrome serving as an alert to new diseases ahead . *Mod Pathol* **7**, 806–810.
99. Hogan TF, Koss W, Murgo AJ, Amato RS, Fontana JA, VanScoy FL. (1987). Acute lymphoblastic leukemia with chromosomal 5;14 translocation and hypereosinophilia: Case report and literature review. *J Clin Oncol* **5**, 382–90.
100. Baumgarten E, Wegner RD, Fengler R, Ludwig WD, Schulte-Overberg U, Domeyer C, Schüürmann J, Henze G. (1989). Calla-positive acute leukaemia with t(5q;14q) translocation and hypereosinophilia – a unique entity? *Acta Haematol* **82**, 85–90.
101. Fagan K, Hyde S, Harrison P. (1993). Translocation (8;13) and T-cell lymphoma. A case report. *Cancer Genet Cytogenet* **65**, 71–3.
102. May ME, Waddell CC. (1984). Basophils in peripheral blood and bone marrow. A retrospective review. *Am J Med* **76**, 509–11.
103. Mitre E, Nutman TB. (2006). Basophils, basophilia and helminth infections. *Chem Immunol Allergy* **90**, 141–56.
104. Matsushima T, Handa H, Yokohama A, Nagasaki J, Koiso H, Kin Y, Tanaka Y, Sakura T, Tsukamoto N, Karasawa M, Itoh K, Hirabayashi H, Sawamura M, Shinonome S, Shimano S, Miyawaki S, Nojima Y, Murakami H. (2003). Prevalence and clinical characteristics of myelodysplastic syndrome with bone marrow eosinophilia or basophilia. *Blood* **101**, 3386–90.
105. Lillington DM, MacCallum PK, Lister TA, Gibbons B. (1993). Translocation t(6;9)(p23;q34) in acute myeloid leukemia without myelodysplasia or basophilia: Two cases and a review of the literature. *Leukemia* **7**, 527–31.

Disorder of Red Blood Cells: Anemias

Faramarz Naeim

Anemia, a decline in blood hemoglobin (Hb) level, is caused by three major mechanisms: (1) blood loss, (2) inefficient erythropoiesis, and (3) increased red blood cell (RBC) destruction. These mechanisms are often associated with certain morphologic features reflecting the size (normocytic, microcytic, or macrocytic) or the Hb content (normochromic, hypochromic, or hyperchromic) of the erythrocytes.

Four RBC indices are measured by automated hematology instruments in clinical laboratories. These indices correlate with the size, hemoglobin content, and degree of anisocytosis in the RBCs. They include (Table 23.1):

1. Mean corpuscular volume (MCV) calculated as: HCT (%) × 10/RBC count (million/μL). MCV indicates average RBC volume.
2. Mean corpuscular hemoglobin (MCH) calculated as: Hb (g/dL) × 10/RBC count (million/μL). MCH indicates average amount of Hb per RBC.
3. Mean corpuscular hemoglobin concentration (MCHC) calculated as: Hb (g/dL) × 100/HCT (%). MCHC indicates the average concentration of Hb per RBC.
4. Red cell distribution width (RDW) indicates the degree of anisocytosis.

Biochemical analyses such as measurement of serum iron, iron-binding capacity, ferritin, folate, and vitamin B_{12} levels, RBC enzyme assays, and Hb electrophoresis provide valuable information regarding the cause of anemia. Molecular genetic studies add additional dimensions to the understanding, classification, and treatment of certain anemias, particularly the hereditary variants.

Routine examination of the peripheral blood plays a key role in the diagnosis and classification of anemias. It provides basic information regarding RBC counts, morphology and indices, Hb and hematocrit (HCT, percentage of

TABLE 23.1 Red blood cell (RBC) values and related parameters in healthy adults.*

Parameters	Men	Women	Both
RBC count, million/μL	5.2 ± 0.7**	4.6 ± 0.5	
Hemoglobin (Hb), g/dL	15.7 ± 1.7	13.8 ± 1.5	
Hematocrit, %	46.0 ± 4.0	40.0 ± 4.0	
Reticulocytes, %	1.6 ± 0.5	1.4 ± 0.5	
Mean corpuscular volume, fl (MCV)			88.0 ± 8.0
Mean cell hemoglobin, pg/RBC (MCH)			30.4 ± 2.8
Mean cell Hb concentration, g/dL of RBC (MCHC)			34.4 ± 1.1
Red cell volume distribution width (RDW), %			13.1 ± 1.4

*Adapted from Butler E, Lichtman MA, Coller BS, *et al.* (2001). *Williams' Hematology*, 6th ed., McGraw-Hill, New York.
**Values = mean + standard deviation.

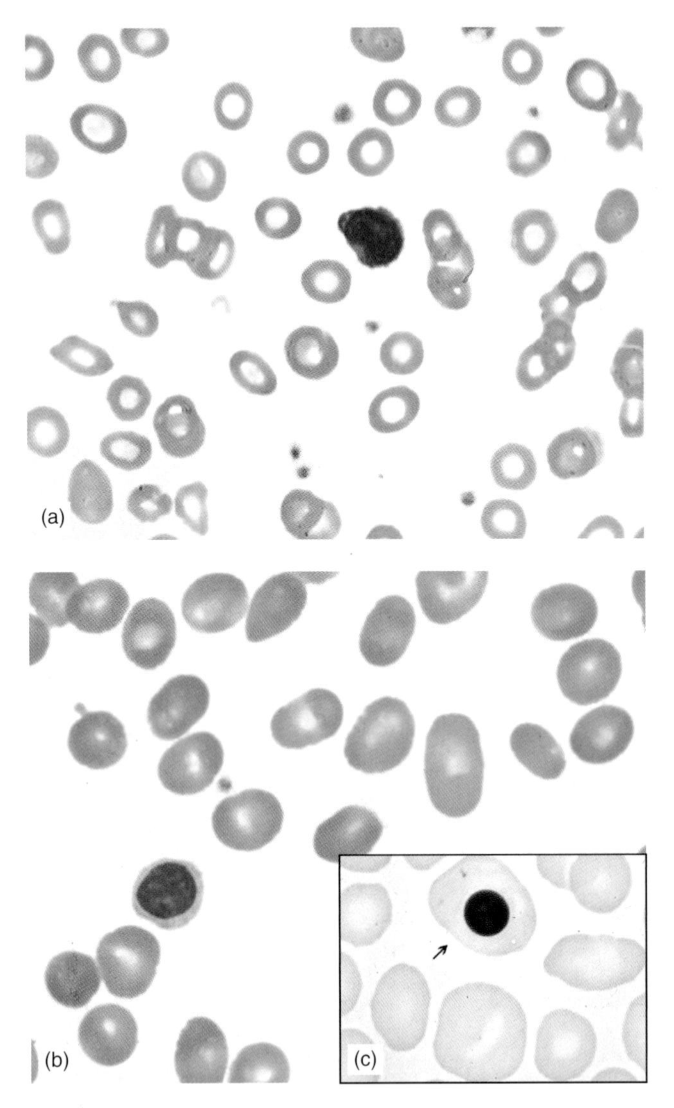

FIGURE 23.1 (a) Blood smear demonstrating microcytic hypochromic erythrocytes in a patient with iron deficiency anemia. (b and c) Macro-ovalocytes in a patient with megaloblastic anemia. A late stage nucleated red cell with abundant cytoplasm is present (c, arrow).

the packed RBC volume in blood) levels (Table 23.1). The WHO criteria for anemia are <13 and <12 g/dL of Hb for men and women, respectively [1, 2], or more than two standard deviations (SD) below the mean of normal range in an age- and sex-matched representative population [3]. In general, the range of Hb, HCT, and RBC counts are higher in smokers and in people who live in an air-polluted environment or at high altitudes, and are lower in women, Afro-Americans, and the elderly [3–7].

Morphologic variations are often associated with certain categories of anemias. For example, microcytic, hypochromic RBCs are seen in iron deficiency anemia (IDA), thalassemia, and lead poisoning (Figure 23.1a); macrocytic RBCs are observed in folate or vitamin B_{12} deficiencies, liver disease, hypothyroidism, and newborns (Figure 23.1b and c); target cells are often associated with thalassemia and IDA (Figure 23.2a); stomatocytes are seen in hereditary conditions, liver diseases, and electrolyte imbalances (Figure 23.2b); burr cells (echinocytes) are found in uremia, pyruvate kinase (PK) deficiency, and acute blood loss (Figure 23.3a); acanthocytes may be present in a β-lipoproteinemia, liver diseases, and anorexia nervosa (Figure 23.3b); and fragmented RBCs (schistocytes) are seen in microangiopathic hemolytic anemias such as disseminated intravascular coagulopathies, thrombotic thrombocytopenic purpura, and severe burns (Figure 23.4a). Teardrop-shaped erythrocytes (dacrocytes) are seen in bone marrow fibrosis, thalassemia syndromes, and hemolytic anemias (Figure 23.4b). Reticulocytes (polychromatophilic RBCs) account for about 0.5–1.5% of the RBCs and are primarily increased in conditions associated with elevated production of bone marrow erythropoiesis, such as hemolytic anemias (Figure 23.5).

Erythrocytes may demonstrate a variety of cytoplasmic inclusions. For example, basophilic stippling is seen in lead poisoning, impaired Hb synthesis, alcoholism, and megaloblastic anemias (Figure 23.6a), and iron particles (Pappenheimer bodies) are noted in myelodysplastic syndrome (MDS), congenital dyserythropoietic anemia, and post-splenectomy (Figure 23.6b). Erythrocytes may contain remnants of DNA (Howell-Jolly bodies), such as in megaloblastic anemia or after splenectomy. Cabot rings are inclusions observed in pernicious anemia or lead poisoning (Figure 23.7). Heinz bodies (precipitated abnormal Hb structures) and Hb H inclusions are visible by supravital stains (Figure 23.8). RBCs may also contain a variety of microorganisms (Figure 23.9).

Although peripheral blood is the most informative sample in the diagnosis and classification of anemias, bone marrow examination, in certain conditions, provides additional valuable information such as bone marrow cellularity, myeloid:erythroid (M:E) ratio, estimation of stored iron and presence or lack of bone marrow replacement by fibrosis, inflammatory processes, or primary or secondary malignancies. In general, anemia caused by blood loss or RBC destruction is associated with bone marrow erythroid hyperplasia and reticulocytosis, whereas anemia due to ineffective erythropoiesis is characterized by reticulocytopenia and a bone marrow which may be hypo-, normo-, or hypercellular.

In this chapter, most anemias, especially those associated with significant bone marrow changes, are discussed. Anemias secondary to the pluripotent hematopoietic stem cell disorders such as aplastic anemia, chronic myeloproliferative disorders, and myelodysplastic syndromes are discussed in previous chapters.

PURE RED CELL APLASIA

Pure red cell aplasia (PRCA) is a rare condition characterized by severe anemia, lack of reticulocytes, and marked reduction or virtual absence of erythroid precursors in the bone marrow.

Etiology and Pathogenesis

The primary defect in PRCA appears to be the inability of the committed erythroid progenitor cells, BFU-E and

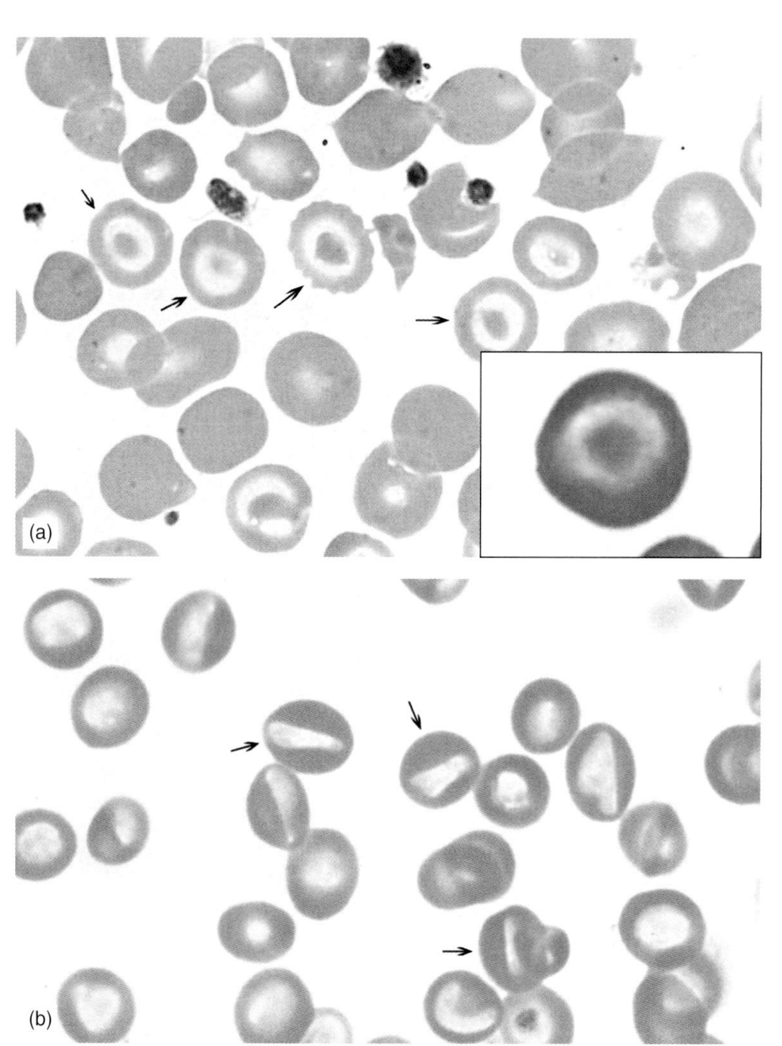

FIGURE 23.2 Blood smears demonstrating target cells (a, arrows and inset) and stomatocytes (b).

CFU-E, to differentiate into pronormoblasts [8]. There are two forms of PRCA: congenital and acquired.

The congenital form or the Diamond–Blackfan anemia apparently represents a heterogenous group and has been observed in both autosomal recessive and dominant forms [9–11]. A significant proportion of these patients show mutations on chromosomes 19q or 8p [10, 12]. The 19q mutation involves *RPS19* gene, encoding the ribosomal protein S19, suggesting that a defect in ribosome biogenesis plays a role in the pathogenesis of Diamond–Blackfan anemia [13, 14].

Acquired PRCA may be drug-induced or associated with a wide variety of clinical conditions such as viral infections, autoimmune disorders, and lymphoid malignancies (Table 23.2) [3, 8, 14, 15]. Acquired PRCA may also precede myelodysplastic syndrome [14, 15].

The B19 parvovirus-induced PRCA can occur in various conditions such as sickle cell (SC) anemia, post-organ transplantation, congenital immunodeficiency syndromes, and lymphoproliferative disorders [16–21]. The erythroid progenitor cells are targeted by B19 parvovirus via the red cell receptor globoside (blood group P antigens) [19].

The majority of immunologically mediated PRCAs are caused by antibodies which either inhibit erythropoiesis and Hb synthesis or complement-binding and have direct cytotoxic effects on erythroblasts [22, 23].

Pathology and Laboratory Findings

Patients usually have a profound anemia that is often macrocytic. White blood cell (WBC) and platelet counts are unaffected. Bone marrow specimens of patients with PRCA show variable cellularity with markedly elevated M:E ratio due to severe erythroid depletion and rare late erythroid progenitor cells (Figure 23.10). The erythroid cells may appear dysplastic. Parvovirus-associated aplasia is often characterized by prominent megaloblastic changes and the presence of giant rubriblasts (gigantoblasts) (Figures 23.11–23.13). These giant cells are characterized by dark basophilic cytoplasm and large vesicular nuclei with fine chromatin and prominent nucleoli [24]. They may demonstrate viral inclusions. The granulocytic and megakaryocytic lines are unremarkable and show progressive maturation. No significant changes are noted in the lymphocytic and plasmacytic population, except when the underlying cause is a lymphoproliferative process.

Anemia is severe and macrocytic [3, 16, 18]. The majority of the patients show an elevated red cell adenosine deaminase activity [25] and may also demonstrate an increased proportion of fetal Hb for age and an increased expression of the I antigen [26].

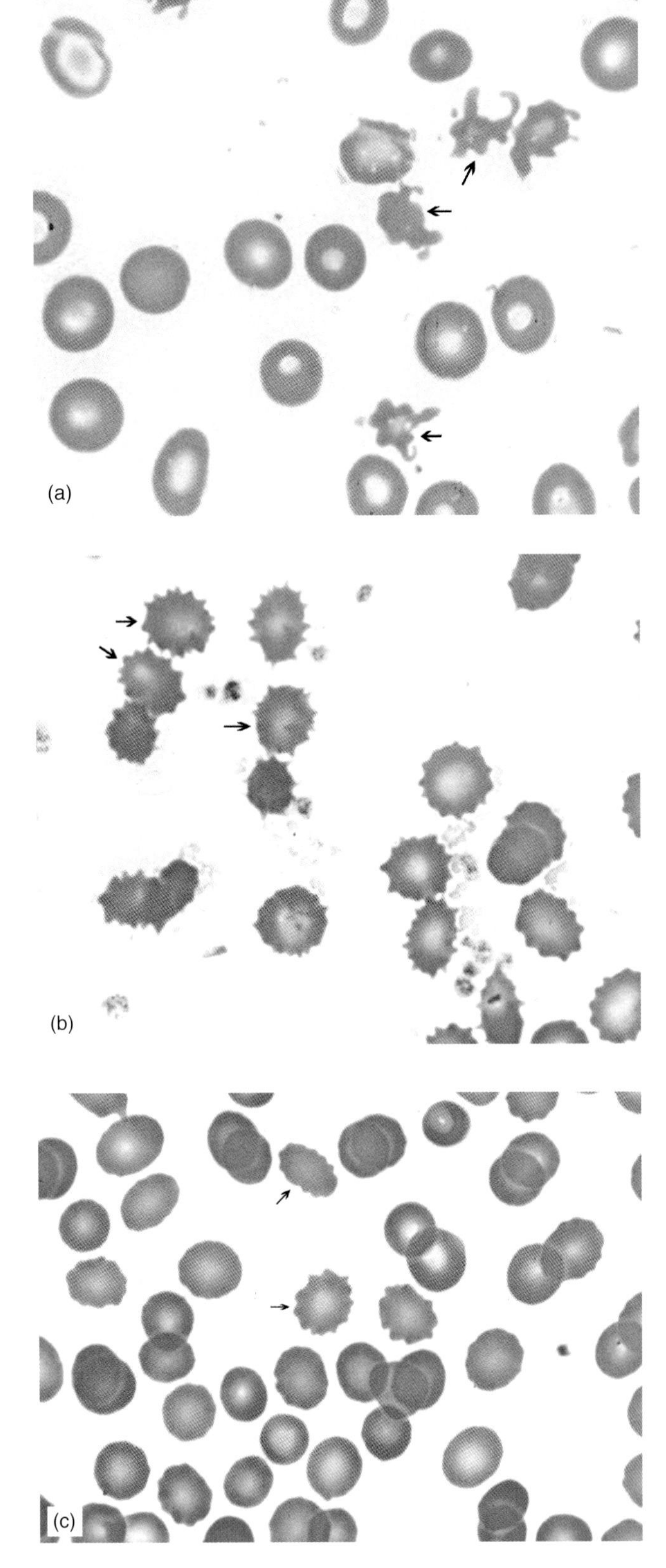

FIGURE 23.3 Blood smears demonstrate acanthocytes (a) with irregularly spaced, thorn-like projections, echinocytes (b) with evenly spaced pointed projections, and crenated red cells (c) with blunt projections evenly distributed over the surface. Crenated red cells are usually caused by faulty drying of the blood smear, change in pH, or excess EDTA (shrinkage of the erythrocytes).

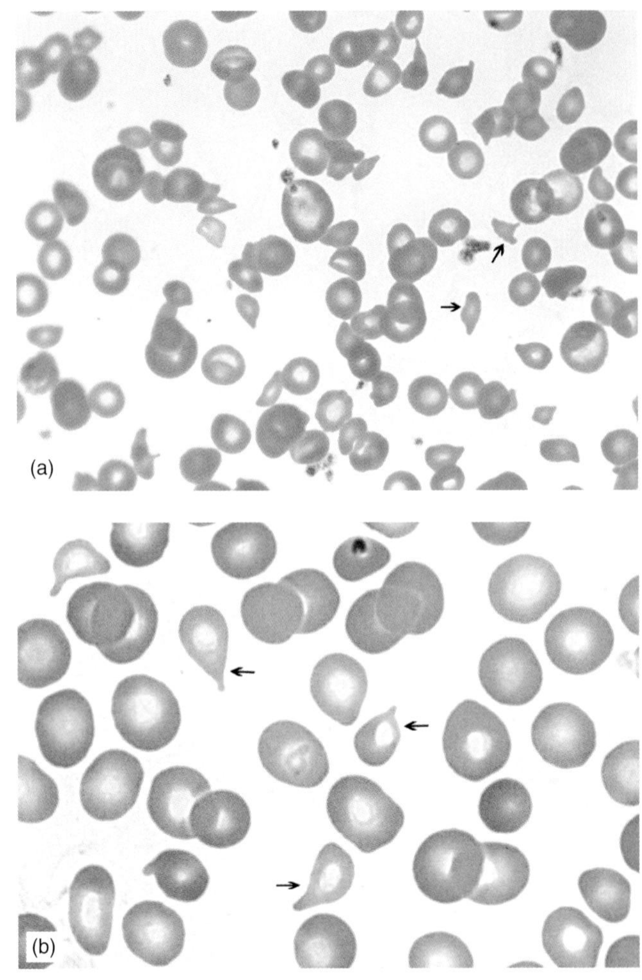

FIGURE 23.4 Blood smears demonstrate fragmented red cells (schistocytes) (a) and teardrop red cells (dacrocytes) (b, arrows).

Clinical Aspects

Diamond–Blackfan anemia is a disease of early infancy with over 90% of cases diagnosed within the first year of life [9, 16–18, 26]. Males and females are equally affected. The disease is sporadic in about 35% of the cases, but autosomal recessive and dominant forms have been reported in several families [16]. Approximately one-third of the patients with Diamond–Blackfan anemia demonstrate physical abnormalities such as short stature, microcephali, cleft palate, atrial or ventricular septal defect, and/or urogenital abnormalities [27]. Spontaneous remissions have been reported in up to 25% of the patients [28]. Therapeutic approaches include corticosteroid and metoclopramide therapy, blood transfusion, and bone marrow transplantation [29, 30].

The *transient erythroblastoma of childhood* is a temporary acquired anemia with reticulocytopenia and bone marrow erythroblastopenia usually observed in children <1 year old [28, 31]. It is mostly idiopathic or in some cases autoimmune related.

Major underlying causes of non-congenital forms of PRCA are large granular lymphocytic leukemia, thymoma, hypoplastic MDS, parvovirus B19, HIV, drugs, autoimmune

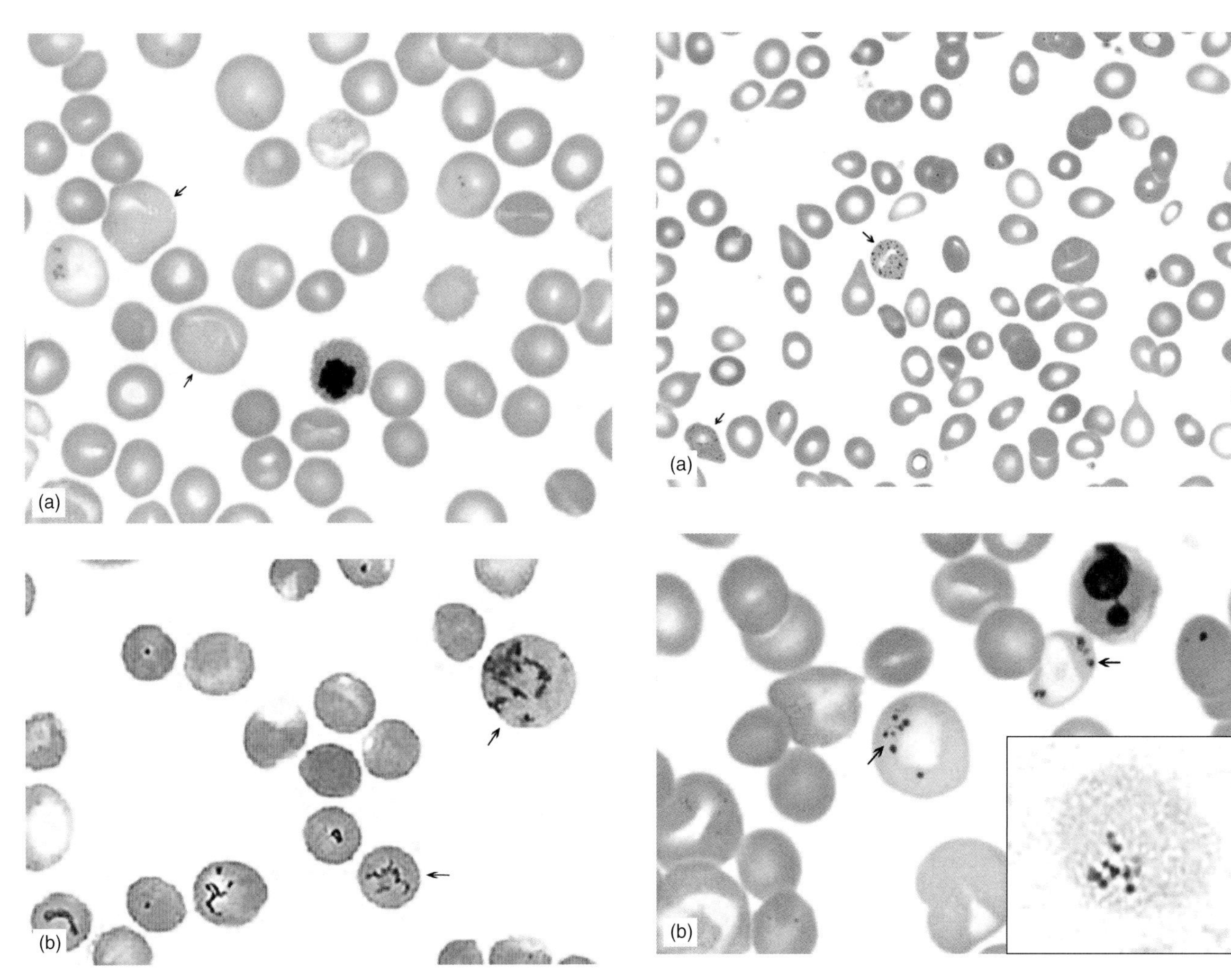

FIGURE 23.5 Polychromatophilic red cells (reticulocytes) (a, arrows) contain ribosomes, which are demonstrated by supravital stains (b, arrows).

FIGURE 23.6 Blood smear showing basophilic stippling (a, arrows), polychromatophilic red cells, and teardrops. Dark, angular bodies of various sizes on Wright-stained smear (b) represent Pappenheimer bodies (iron particles). The inset is iron stain.

disorders, and idiopathic (Table 23.2) [18–20, 32–34]. Therapeutic approaches include treatment of the underlying disorder and supportive therapy.

Differential Diagnosis

The major distinguishing features of Diamond–Blackfan anemia are family history and the manifestation of the disease in early infancy. Patients may also show mutations on chromosomes 19q or 8p. Anemia secondary to acquired PRCA is often transient and is mostly seen in individuals older than 1 year.

CONGENITAL DYSERYTHROPOIETIC ANEMIAS

Congenital dyserythropoietic anemias (CDAs) are a group of rare congenital anemias characterized by ineffective erythropoiesis and dysplastic changes in erythroid precursors [35–38]. Originally, there were three well-established CDA types, I, II, and III (Table 23.3), but several new variants such as types IV, V, VI, and VII have been recognized in recent years [36, 37].

Congenital Dyserythropoietic Anemia Type I

Congenital dyserythropoietic anemia type I is a rare autosomal recessive disorder manifested in various ages, ranging from infancy to adulthood [39, 40]. It is characterized by variable degrees of anemia with mild to distinct macrocytosis, moderate hyperbilirubinemia, iron overload, and often splenomegaly [41, 42]. Congenital malformations such as presence of sixth toe, ventricular septal defect, short stature, and hip dysplasia have been reported in up to one-third of the cases [43]. The majority of the patients demonstrate mutations of *CDAN1* gene mapped to chromosome 15q15.1-15.3 [44–46].

The bone marrow is hypercellular with marked erythroid hyperplasia and dysplasia. Dysplastic changes are mostly confined to the middle and late stages of erythroid maturation and include nuclear irregularity, double segmented nuclei, and binucleation with nuclei of different sizes, textures, and stainability (Figures 23.14 and 23.15). Pairs of

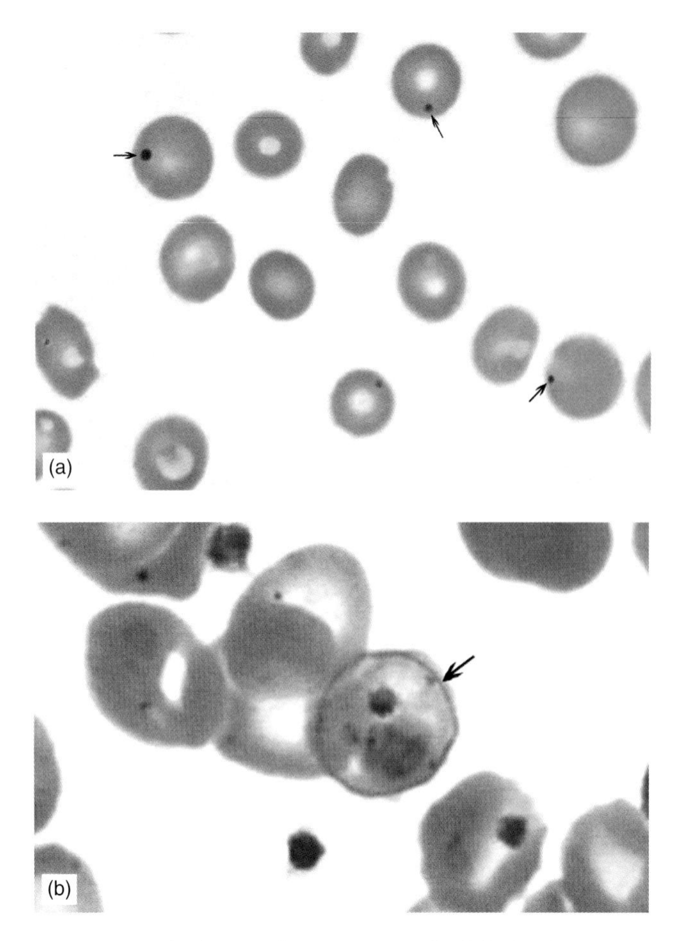

FIGURE 23.7 Blood smears demonstrate Howell-Jolly bodies (a, arrows) and a Cabot ring (b, arrow). Courtesy of Diana Tanaka-Mukai, Clinical Laboratories, UCLA Medical Center.

normoblasts attached by a chromatin bridge may be present. There are also mild to moderate megaloblastic changes. Iron stores are increased [37, 41, 45]. Electron microscopy reveals widening of the nuclear membrane pores with vacuolization and disintegration of nuclear chromatin ("Swiss cheese" appearance [47, 48]).

The peripheral blood shows anemia with Hb levels ranging from 6.5 to 13 g/dL. Anisocytosis, poikilocytosis, and macrocytosis are common features with a normal to slightly elevated reticulocyte count (1–5%) [45]. Basophilic stippling, Howell-Jolly bodies, and Cabot rings may be present. Some cases may show elevated Hb A_2 levels and/or unbalanced globin chain synthesis with increased α:β ratio [37]. The serum levels of haptoglobin are low, bilirubin is high, and iron is normal to elevated [37, 41, 45]. Patients with CDA type I show a significant positive response to recombinant interferon-α_2 treatments [44].

Congenital Dyserythropoietic Anemia Type II

Congenital dyserythropoietic anemia type II is an autosomal recessive disorder also known as *hereditary erythroblastic multinuclearity with a positive acidified serum lysis test* (HEMPAS) (Table 23.3). CDA type II is the most common type of CDA with >300 cases reported. The extent of anemia varies from mild to severe. Hyperbilirubinemia and consequently gallstone formation are frequent findings and over two-thirds of the patients show splenomegaly. There is a progressive elevation of serum ferritin levels with about 20% chance of development of hepatic cirrhosis [49, 50]. Dysmorphic features are less common than type I [49]. Linkage studies have localized a gene (*CDAN2*) in the region of chromosome 20q11.2 [51].

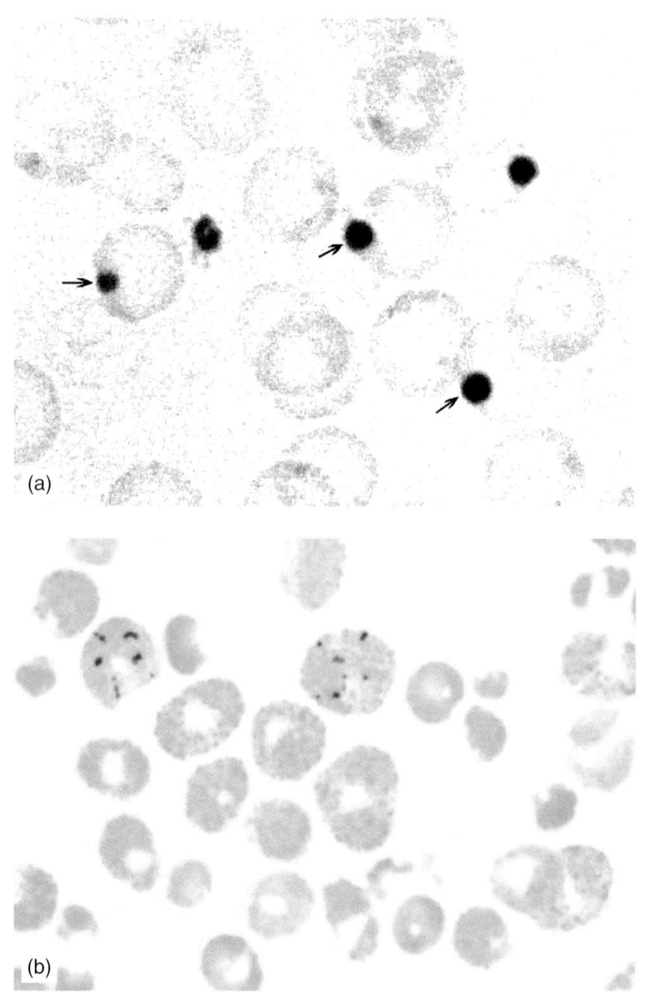

FIGURE 23.8 Heinz bodies are unstable Hb precipitates attached to red cell membrane visible by supravital stains (a, arrows). Hemoglobin H represented as small evenly dispersed deposits (like golf ball), visible by supravital stains (b). Courtesy of Diana Tanaka-Mukai, Clinical Laboratories, UCLA Medical Center.

The bone marrow is hypercellular and shows erythroid hyperplasia with dysplastic changes in middle to late stage normoblasts. A significant proportion of these cells (10–30%) show bi- or multilobated nuclei and many are binucleated. Overdestruction of erythroid precursors and increased cell debris often lead to the accumulation of macrophages and the presence of pseudo-Gaucher cells or sea-blue histiocytes. Electron microscopy reveals an excess of smooth endoplasmic reticulum parallel to the cell membrane of the normoblasts ("double membrane" appearance) [52].

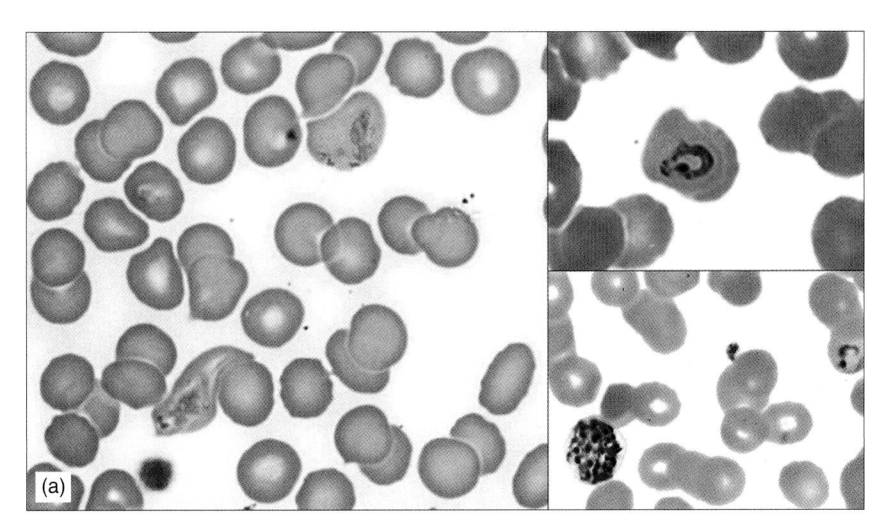

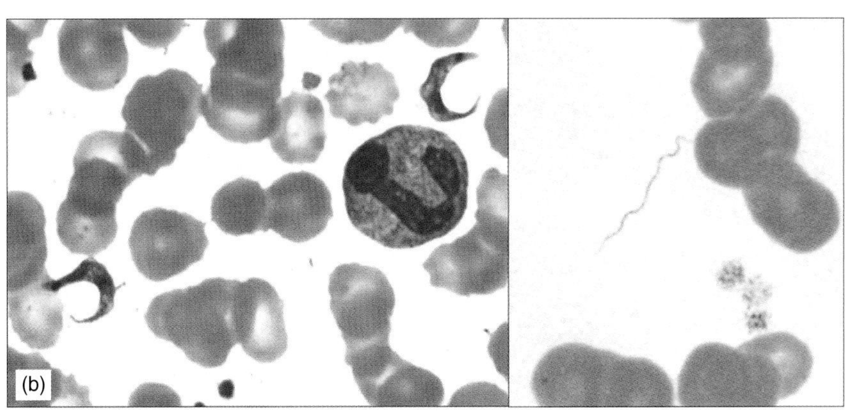

FIGURE 23.9 (a) Blood smears demonstrate ring and schizont forms of malarial parasites in the erythrocytes and (b) blood smears showing *Trypanosoma brucei* (left) and *Borrelia recurrentis* (right). Courtesy of Diana Tanaka-Mukai, Clinical Laboratories, UCLA Medical Center.

TABLE 23.2 Conditions associated with acquired pure red cell aplasia.*

Lymphoproliferative disorders
  Large granular lymphocyte-mediated
  B-cell lymphoproliferative disorders

Myelodysplastic syndromes
Myeloproliferative disorders
Viral infections (parvovirus B19, viral hepatitis, HIV)
Drugs (phenytoin, chloramphenicol, zidovudine)

Immunologically mediated
  Autoimmune hemolytic anemia
  Rheumatoid arthritis
  Systemic lupus erythematosus
  ABO-incompatible stem cell transplantation
  Antibodies against erythropoietin

Thymoma and other cancers
Pregnancy
Idiopathic

*Adapted from Ref. [16].

The peripheral blood smear shows varying degrees of anisopoikilocytosis, often with a normal MCV. Occasional basophilic stippling or teardrops may be present. The RBCs of most patients are lysed in the acidified serum lysis test (Ham test) by normal sera but not by the patient's own serum [43]. Band 3 and 4.5 glycoproteins are underglycosylated, although there may be overglycosylation of glycolipids [49, 53, 54]. Therapeutic approaches include supportive therapy, such as RBC transfusion, and splenectomy if severity of anemia compromises patients' performance [49].

Congenital Dyserythropoietic Anemia Type III

Congenital dyserythropoietic anemia type III is an extremely rare condition and has been observed in both familial and sporadic forms (Table 23.3) [55]. The familial form is autosomal dominant. CDA type III is characterized by a mild to moderate macrocytic anemia, bone marrow erythroid hyperplasia, and the presence of dysplastic, giant erythroid precursor cells with one or multiple nuclei [56, 57]. The erythrocytes may react with anti-I and/or anti-i sera, but serum acid test is negative. Linkage analysis and recombination studies in a Swedish family have suggested a gene (*CDAN3*) located in the region of chromosome 15q22 [55].

Other Types of Congenital Dyserythropoietic Anemia

Several forms of CDA other than types I, II, and III have been reported. These briefly include the following.

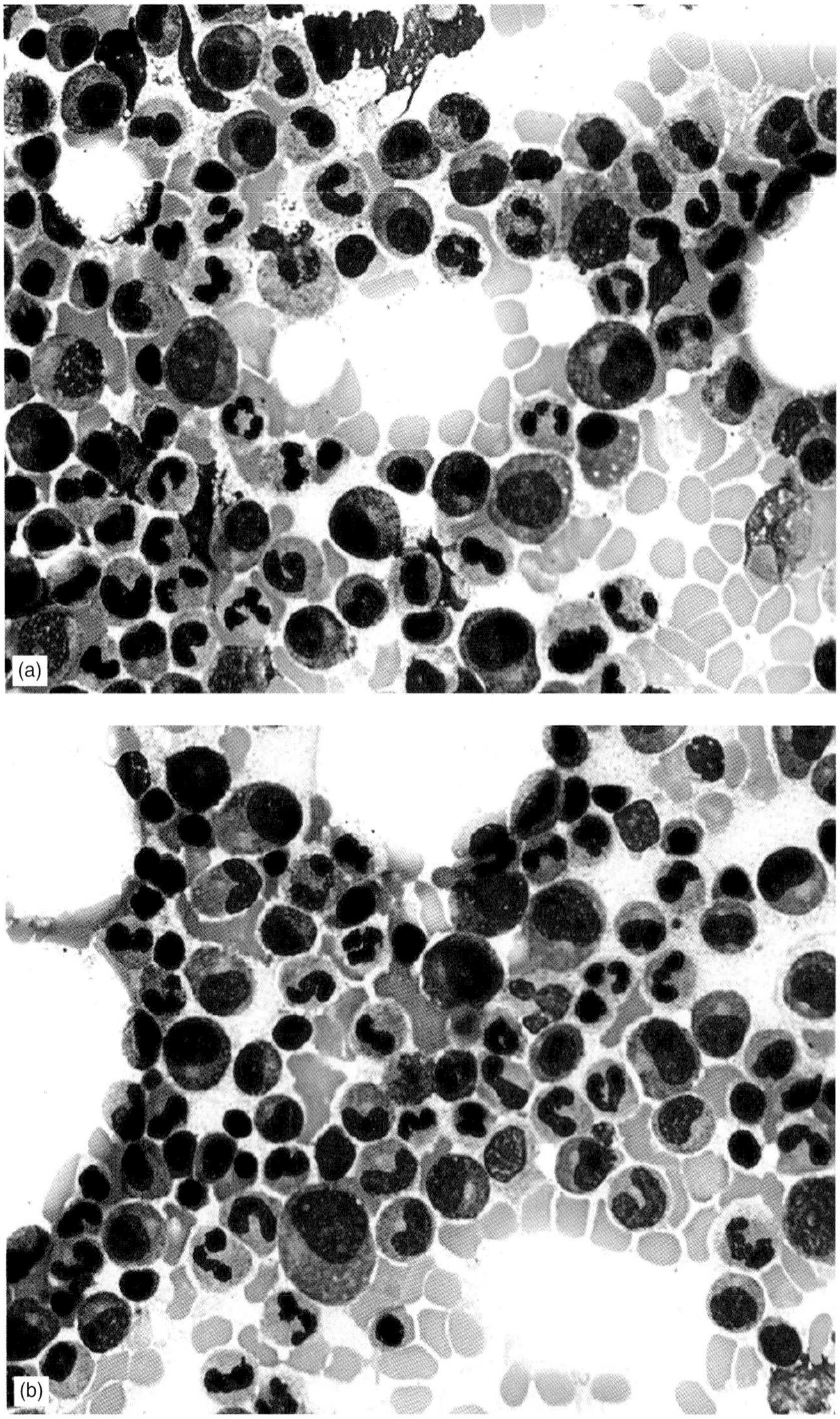

(a)

(b)

FIGURE 23.10 Bone marrow smears of a patient with pure red cell aplasia (a and b) showing marked myeloid preponderance and occasional erythroid precursors.

CDA group IV is characterized by severe transfusion-dependent anemia since birth with absence of precipitated protein in erythroblasts [37].

CDA group V represents patients with congenital ineffective erythropoiesis with insignificant dysplasia, mild anemia, and elevated unconjugated bilirubin [58].

CDA group VI is characterized by marked congenital macrocytosis and folate- and vitamin B_{12}-independent megaloblastic erythropoiesis with mild anemia [37].

Differential Diagnosis

The differential diagnosis includes all congenital and acquired anemias known to be associated with dyserythropoiesis, such as β-thalassemia, hereditary sideroblastic anemias, PK deficiency anemia, myelodysplastic syndromes, megaloblastic anemia, parvovirus B19 infection, arsenic poisoning, and severe IDA [37, 59]. Diagnosis of CDA should

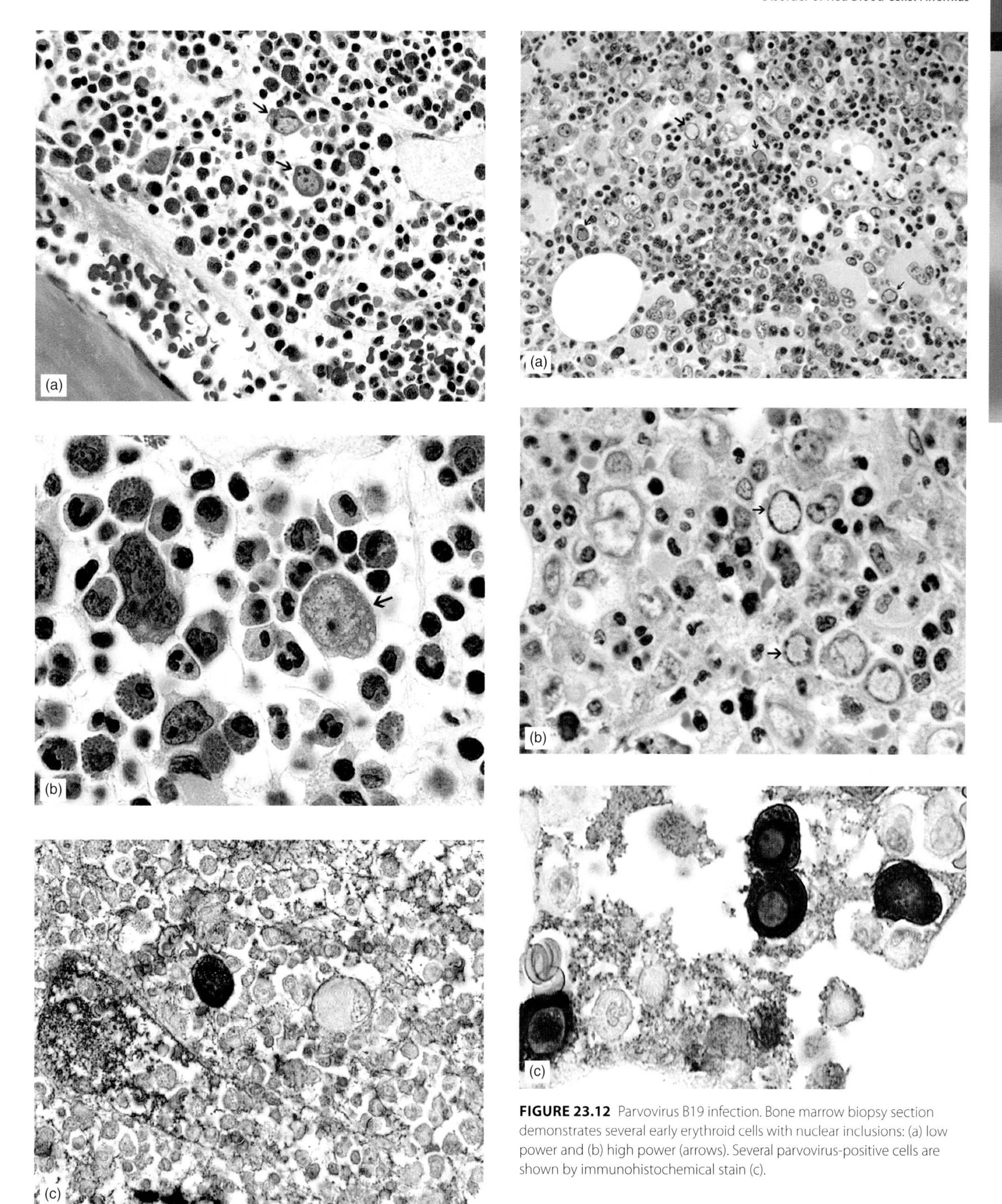

FIGURE 23.11 Pure red cell aplasia caused by parvovirus B19 infection. Bone marrow biopsy section demonstrates myeloid preponderance, eosinophilia, and scattered giant early erythroid precursors: (a) low power and (b) high power (arrows). A parvovirus-positive cell is shown by immunohistochemical stain (c).

FIGURE 23.12 Parvovirus B19 infection. Bone marrow biopsy section demonstrates several early erythroid cells with nuclear inclusions: (a) low power and (b) high power (arrows). Several parvovirus-positive cells are shown by immunohistochemical stain (c).

be considered when the reticulocyte count does not correlate with the degree of anemia in a patient with erythroid hyperplasia or when there is unexplained hyperbilirubinemia or iron overload [37].

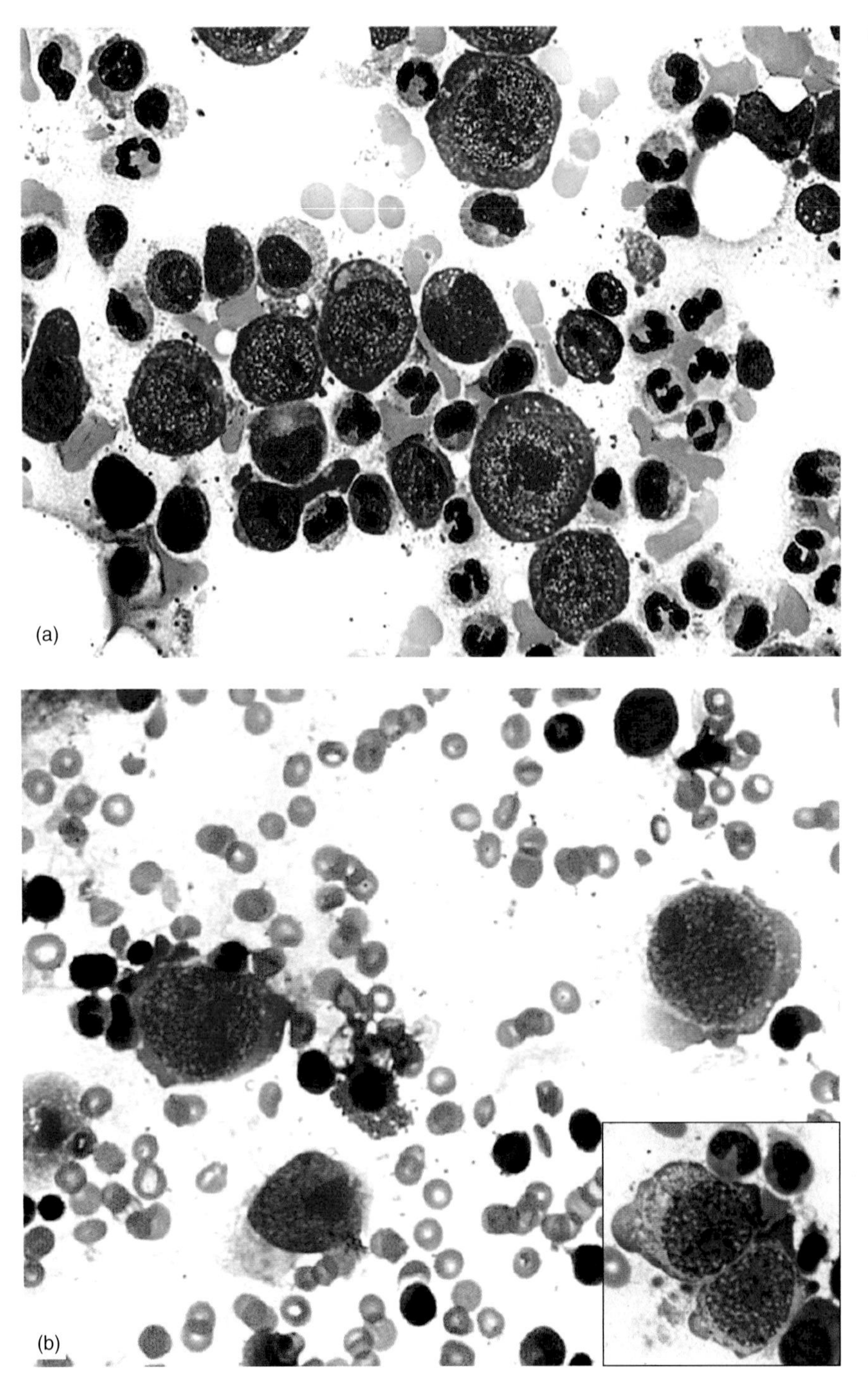

FIGURE 23.13 Bone marrow smear (a, b, and inset) demonstrates several large erythroblasts in a patient infected with parvovirus B19.

MACROCYTIC ANEMIAS

Macrocytic anemias are a group of anemias in which the RBCs are larger than normal (MCV >100 fl). Macrocytic anemia is caused by various disorders such as folate and vitamin B_{12} deficiencies, alcoholism, liver disease, hypothyroidism, and myelodysplastic syndromes (Table 23.4). In general, macrocytic anemia can be divided into two major categories: (1) megaloblastic anemias and (2) non-megaloblastic anemias. In this chapter, megaloblastic anemias are discussed.

Megaloblastic Anemias

Megaloblastic anemias are a group of anemias characterized by megaloblastic erythropoiesis and macrocytosis.

Etiology and Pathogenesis

The underlying defect in megaloblastic anemia is the decline in the rate of DNA synthesis leading to a delay in cell division in all proliferating cells. This defect is due to abnormal purine or pyrimidine metabolism, or inhibition of DNA

TABLE 23.3 Features of the major types of CDA.*

Features	Type I	Type II	Type III
Inheritance	Autosomal recessive	Autosomal recessive	Autosomal dominant or recessive
Gene; chromosome	*CDAN1*; 15q15.1-15.3	*CDAN2*; 20q11.2	*CDAN3*; 15q22
Red cells	Macrocytic	Normocytic	Large macrocytic
Erythroblasts	Megaloblastic; nuclear chromatin bridges	Normoblastic; binucleated cells	Megaloblastic; giant mono- or multinucleated cells
Ham test	Negative	Usually positive	Negative
Glycosylation	Some abnormality	Markedly abnormal	Some abnormality

*Adapted from Ref. [37].

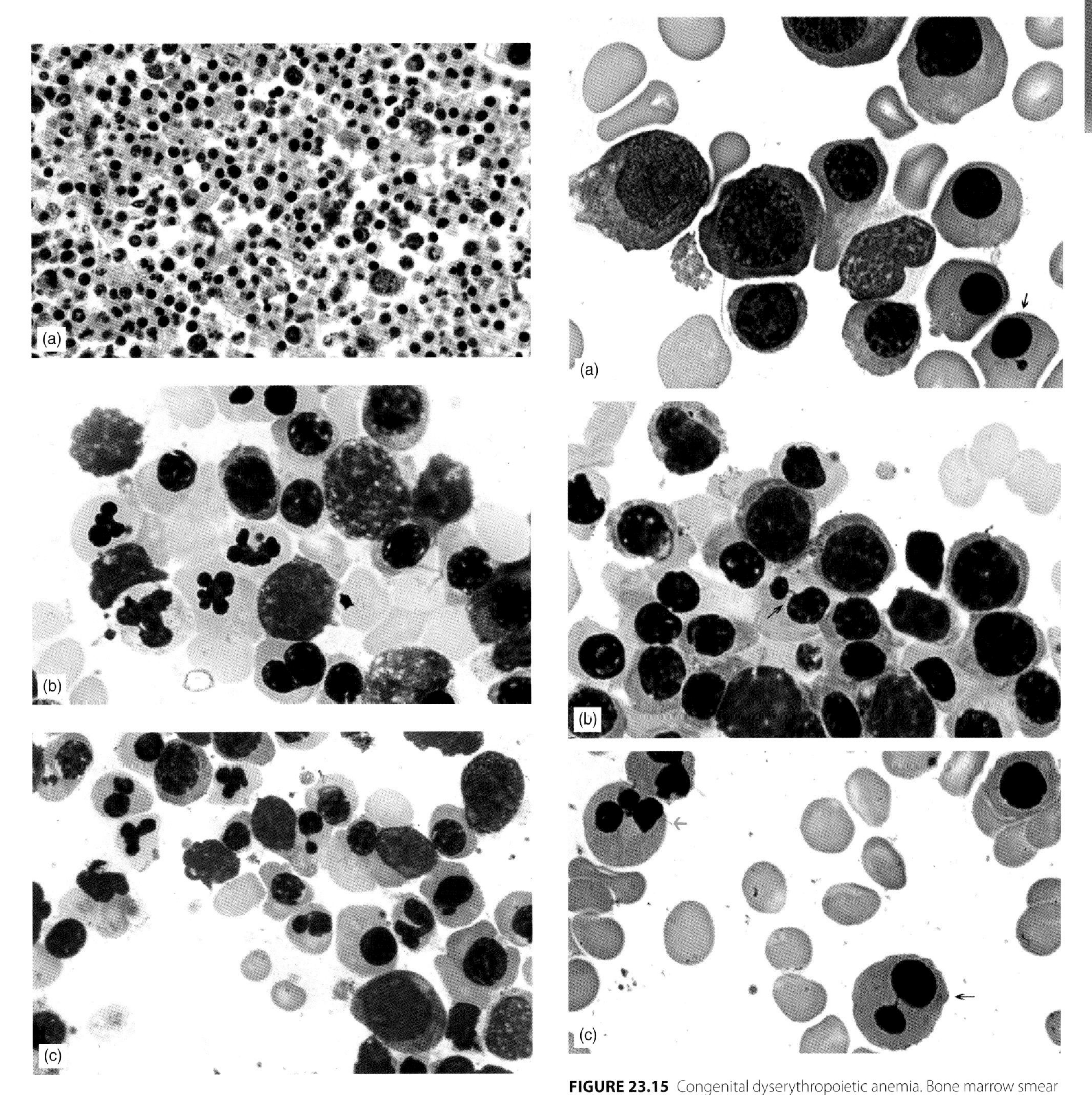

FIGURE 23.14 Congenital dyserythropoietic anemia. Bone marrow biopsy (a) and smear (b and c, arrows) demonstrate numerous intermediate and late-stage erythroid precursors with lobulated nuclei.

FIGURE 23.15 Congenital dyserythropoietic anemia. Bone marrow smear (a, b, and c) shows double segmented nuclei, each segment in different size (black arrows). A dysplastic, multilobated erythroid precursor is shown (c, green arrow).

TABLE 23.4 Common causes of macrocytosis.*

Newborn
Alcoholism
Reticulocytosis
Folate deficiency
Cobalamin (vitamin B_{12}) deficiency
Liver disease
Hypothyroidism
Myelodysplastic syndromes
Aplastic anemia
Hairy cell leukemia
Acute leukemia
Drugs Chemotherapeutic (e.g. cyclophosphamide, methotrexate) Antiviral (e.g. zidovudine, stavudine) Hypoglycemic (metformin) Antimicrobial (e.g. sulfamethoxazole, valacyclovir) Diuretics (triamterene) Anticonvulsant (e.g. phenytoin, primidone) Anti-inflammatory (sulfasalazine)

*Adapted from Ref. [62].

TABLE 23.5 Major causes of folate deficiency.*

1. Decreased intake a. Malnutrition (e.g. poverty, old age, alcoholism) b. Special diet (e.g. goat's milk) c. Hyperalimentation
2. Malabsorption a. Gluten-induced b. Tropical sprue c. Other diseases of small intestine
3. Excess loss of folate a. Dialysis b. Congestive heart failure
4. Increased requirement a. Pregnancy b. Premature infants c. Excessive marrow turnover (e.g. hemolytic anemia) d. Inflammatory diseases (e.g. rheumatoid arthritis, exfoliative dermatitis) e. Cancers
5. Drugs a. Folate antagonists (e.g. methotrexate) b. Anticonvulsants c. Barbiturates
6. Congenital defects a. Congenital folate malabsorption b. Dihydrofolate reductase deficiency c. Homocysteine methyltransferase deficiency

*Adapted from Naeim F. (1998). *Pathology of Bone Marrow*, 2nd ed. Williams & Wilkins, Baltimore.

polymerization [60–64]. The major etiologic factors in megaloblastic anemia are (1) dietary insufficiency and acquired or congenital conditions which lead to folate (folic acid) or cobalamin (vitamin B_{12}) deficiencies and (2) congenital or acquired defects of purine or pyrimidine metabolism.

Folate Deficiency

Folate is primarily absorbed through the upper third of the small intestine and is transported to the cells mainly as 5-methyl tetrahydrofolate (5M-THF). THF is required in the methylation of deoxyuridine monophosphate (dUMP) to deoxythymidine monophosphate (dTMP) in the presence of thymidylate synthetase [61, 65, 66]. Folate deficiency leads to decreased synthesis of dTMP and increased levels of dUMP, and this imbalance apparently slows down DNA synthesis and delays cell proliferation [67, 68]. The end result of this retardation in DNA synthesis is inappropriate enlargement of the cells (megaloblasts), which contain more than the normal amount of DNA but not enough for cell division [60].

Folate deficiency is caused by decreased intake, malabsorption, excess loss, increased requirements, drug effects, and metabolic defects (Table 23.5). The major cause of folate deficiency is dietary. Folate is heat labile and is present in various green vegetables, yeast, mushrooms, kidney, and liver.

Vitamin B_{12} Deficiency

Vitamin B_{12} is involved in the conversion of 5M-THF to THF, which is required for the methylation of dUMP to dTMP. Vitamin B_{12} deficiency, similar to folate deficiency, leads to disturbance of DNA synthesis, and consequently, megaloblastic anemia [60, 69–71]. Vitamin B_{12} is heat resistant and is abundant in animal proteins. The most common cause of vitamin B_{12} deficiency is impaired intestinal absorption. Other less frequent causes include inadequate intake and metabolic defects (Table 23.6).

In the digestive system, vitamin B_{12} attaches to salivary and gastric vitamin B_{12} binders (R binders; haptocorrins). The R-vitamin B_{12} complexes are broken when exposed to pancreatic enzymes and the R binders are digested. The released vitamin B_{12} binds to the intrinsic factor (IF), a glycoprotein secreted by the gastric parietal cells [72]. The IF–vitamin B_{12} complex is then carried to the ileum, where it binds to specific receptors. The IF is then separated from vitamin B_{12}, and the free vitamin B_{12} is absorbed. The absorbed vitamin B_{12} enters the portal circulation and binds to transcobalamins, mainly transcobalamin II. IF deficiency leads to magaloblastic anemia. Megaloblastic anemia secondary to IF deficiency is known as *pernicious anemia* [73].

Pernicious anemia is either congenital or acquired. The congenital form is a rare disorder characterized by mutations in the *IF* gene localized on chromosome 11q13 and lack of IF production. In congenital IF deficiency, gastric acid secretion and mucosal cytology are normal [74–76]. The acquired pernicious anemia, in most cases, is an autoimmune disorder characterized by chronic atrophic gastritis and reduced IF production due to the presence of autoantibodies against gastric parietal cells [73, 77].

TABLE 23.6 Major causes of vitamin B_{12} deficiency.*

1. Impaired absorption a. Gastric origin i. Pernicious anemia ii. Zollinger–Ellison syndrome iii. Gastrectomy, partial or total b. Intestinal origin i. Iliac disease or resection ii. Blind loop syndrome iii. Chronic tropical sprue iv. Fish tapeworm v. Drugs (e.g. metformin)
2. Inadequate diet
3. Metabolic defects a. Congenital i. Transcobalamin II deficiency ii. Homocystinuria and methylmalonic aciduria iii. Hereditary orotic aciduria iv. Lesch–Nyhan syndrome b. Acquired i. Anesthesia with nitrous oxide ii. Drug-induced
4. Others (e.g. congenital R-binding deficiency)

*Adapted from Naeim F. (1998). *Pathology of Bone Marrow*, 2nd ed. Williams & Wilkins, Baltimore.

Disturbance of Purine and Pyrimidine Metabolism

The defects of purine and pyrimidine metabolism are either acquired or congenital. The major causes of acquired defects are antimetabolite drugs (purine and pyrimidine analogs). *Hereditary orotic aciduria* and *Lesch–Nyhan syndrome* represent the congenital deficiencies. Hereditary orotic aciduria is an autosomal recessive disorder of pyrimidine metabolism characterized by orotidylic decarboxylase deficiency [63, 78, 79]. Lesch–Nyhan syndrome is an X-linked disorder caused by a deficiency of hypoxanthine-guanine phosphoribosyltransferase [80].

Pathology and Laboratory Studies

The impaired DNA synthesis in megaloblastic anemia slows nuclear replication and cell division and leads to ineffective erythropoiesis and premature destruction of the RBCs. Bone marrow sections and smears are hypercellular and reveal erythroid hyperplasia. There is often a shift to the left with numerous megaloblasts. Megaloblastic erythroid precursors are larger than their normal counterparts, show asynchronous nuclear and cytoplasmic maturation, display coarser nuclear chromatin, and have more cytoplasm relative to the size of the nucleus (Figures 23.16–23.18). These megaloblastic features are more striking in the intermediate and late stages of erythroid maturation, demonstrated as unevenly speckled nuclear chromatin and abundant Hb-loaded cytoplasm [81]. Final condensation of chromatin (pyknosis), which is seen in late orthochromatic normoblasts, is either delayed or fails to occur. Dysplastic changes such as nuclear fragmentation and nuclear irregularity or lobulation are common. The megaloblastic changes and the

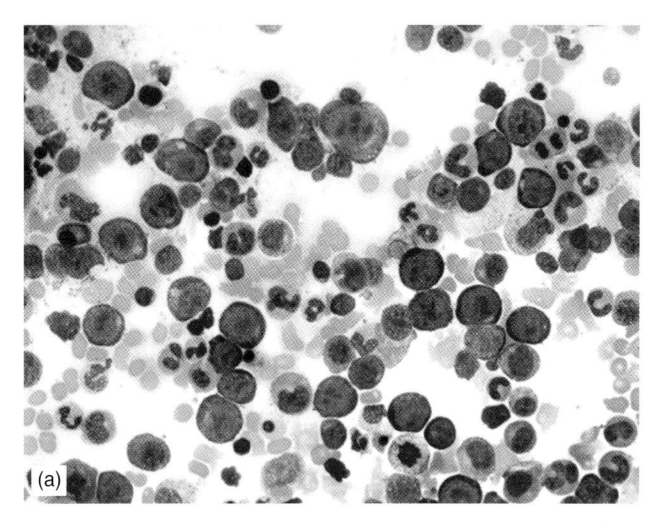

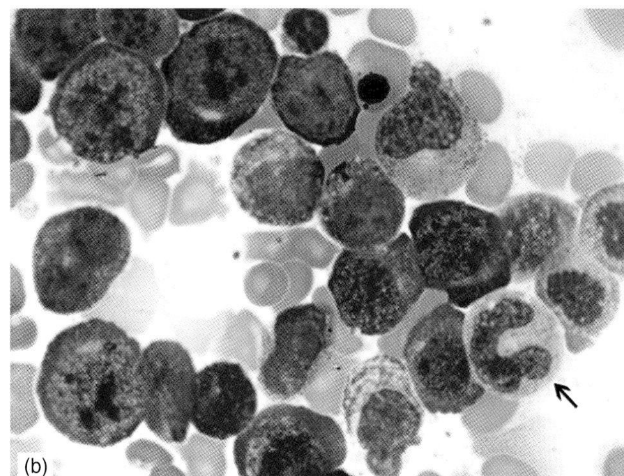

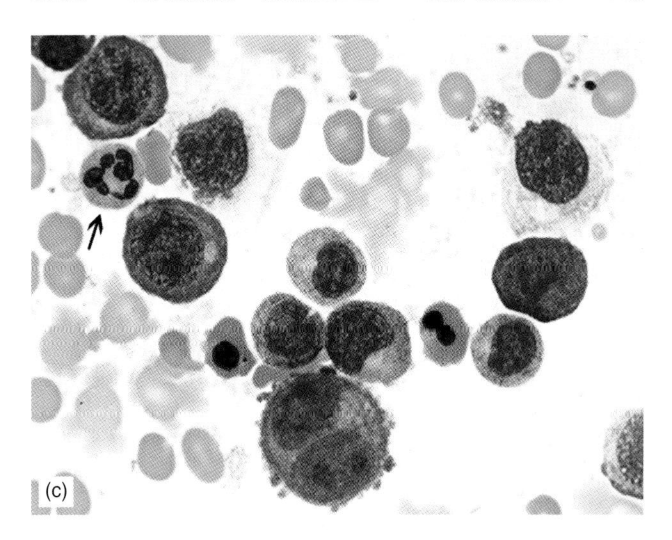

FIGURE 23.16 Megaloblastic anemia. Bone marrow smear demonstrating megaloblastic erythroid precursors (a and b) with a giant band (b, arrow); (c) shows a hypersegmented neutrophil (arrow), two binucleated erythroid precursors, and one with a small nuclear fragment.

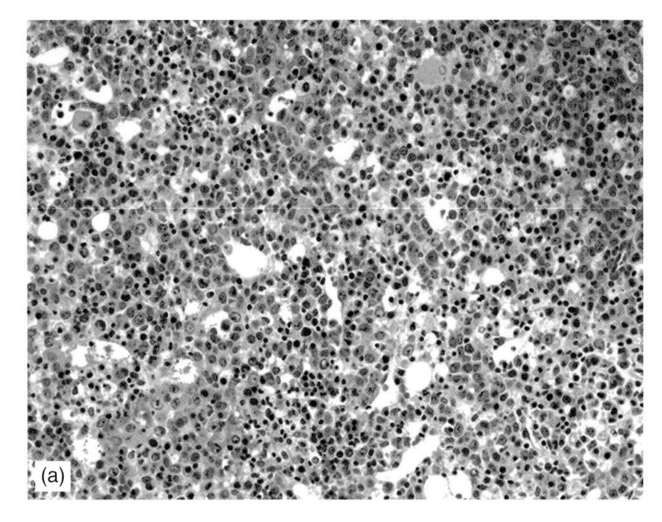

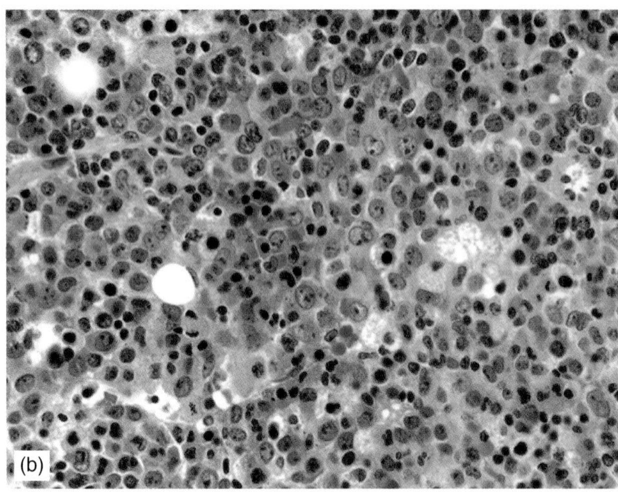

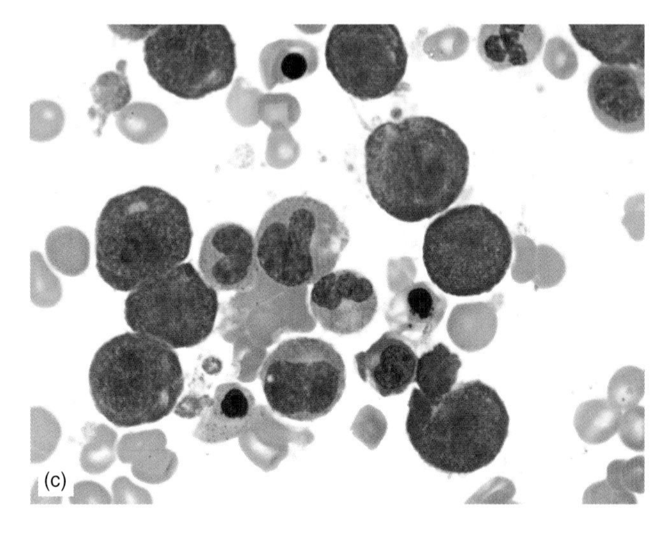

FIGURE 23.17 Megaloblastic anemia. Biopsy section demonstrates a hypercellular marrow with a high proportion of large early erythroid precursors: (a) low power and (b) high power views. Bone marrow smear shows numerous megaloblasts and scattered late erythroid precursors with irregular nuclei (c).

erythroid left shift may simulate an acute leukemic process, particularly in H&E sections, where clusters of immature erythroid cells with open nuclear chromatin and prominent nucleoli are found.

The granulocytic series are also affected and show nuclear-cytoplasmic asynchrony with the presence of giant metamyelocytes and bands, and hypersegmented neutrophils. Mild to moderate myeloid left shift is a common feature. Megakaryocytic lineage may display nuclear hypersegmentation.

Peripheral blood examination reveals pancytopenia with increased MCV of usually >115 fl. Smears show anisopoikilocytosis, macro-ovalocytes, and hypersegmented neutrophils (five nuclear segments in >5% of neutrophils or more than six nuclear segments in >1% of neutrophils) [82]. Other features include basophilic stippling, Howell-Jolly bodies, and occasionally, Cabot rings. In severe cases, numerous nucleated RBCs are present. Coincident iron deficiency or thalassemia can mask macrocytosis [83].

Laboratory studies reveal reduced concentrations of serum and RBC folate in patients with megaloblastic anemia due to folate deficiency. Pregnancy, alcohol intake, and certain anticonvulsant drugs may cause a decrease in serum levels despite adequate tissue stores [62, 84]. Hemolysis may falsely elevate serum folate levels due to high concentration of folate in RBCs [62, 85].

Different laboratory methods are used for the measurement of serum vitamin B_{12} levels. Serum vitamin B_{12} levels may be falsely low in pregnancy, use of oral contraceptives, congenital deficiency of serum haptocorrins (R binders), and plasma cell myeloma [62, 85]. The Schilling test is performed to evaluate vitamin B_{12} absorption and to distinguish various causes of its reduced absorption, such as pernicious anemia/gastrectomy, iliac disease, or intestinal bacterial over growth [86].

Folate and vitamin B_{12} are both required for the conversion of homocysteine to methionine. Therefore, their deficiencies may lead to the elevation of serum homocysteine concentration and increased risk of atherosclerosis and venous thromboembolism [83, 87]. Serum methylmalonic acid levels are elevated only in patients with vitamin B_{12} deficiency [62, 88, 89].

Clinical Aspects

Folate and vitamin B_{12} deficiencies are the second and third most common causes of nutritional anemia in the world, respectively (iron deficiency is the first) [83, 90]. The major difference in the clinical manifestation between these two deficiencies is that only B_{12} deficiency manifests neurologic symptoms. The vitamin B_{12} stores in body are so large relative to the daily requirements that it takes years of inadequate supply before clinical symptoms develop. On the contrary, symptoms of megaloblastic anemia due to folate deficiency can occur within a few months after its supply is diminished [83].

The most common cause of folate deficiency is nutritional due to poor diet or alcoholism. The normal daily requirement of folate is about 200–400 μg/day, but it increases to 500–800 μg/day during pregnancy and lactation. The most frequent cause of vitamin B_{12} deficiency is

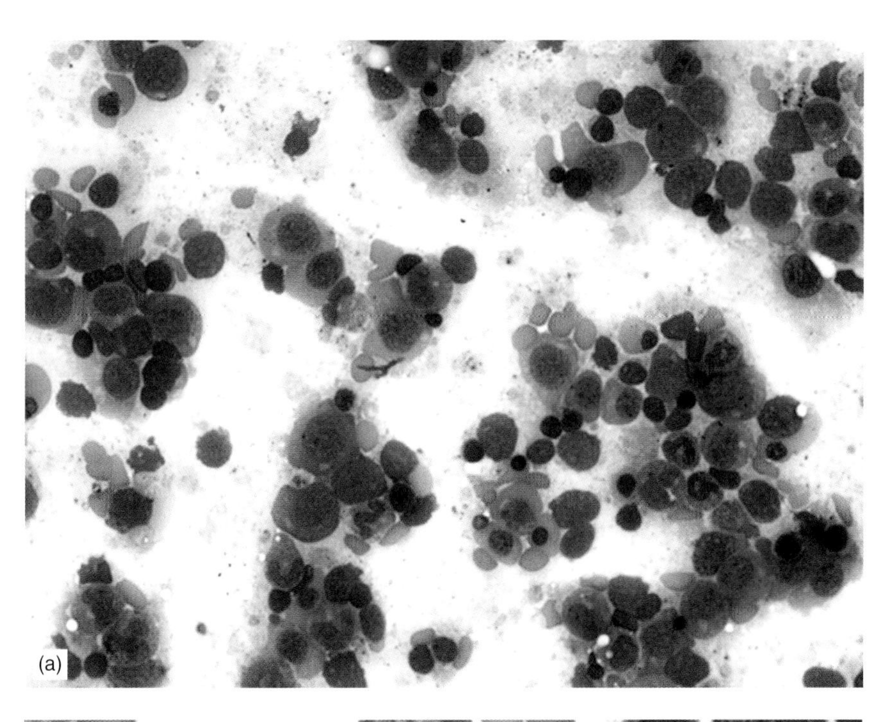

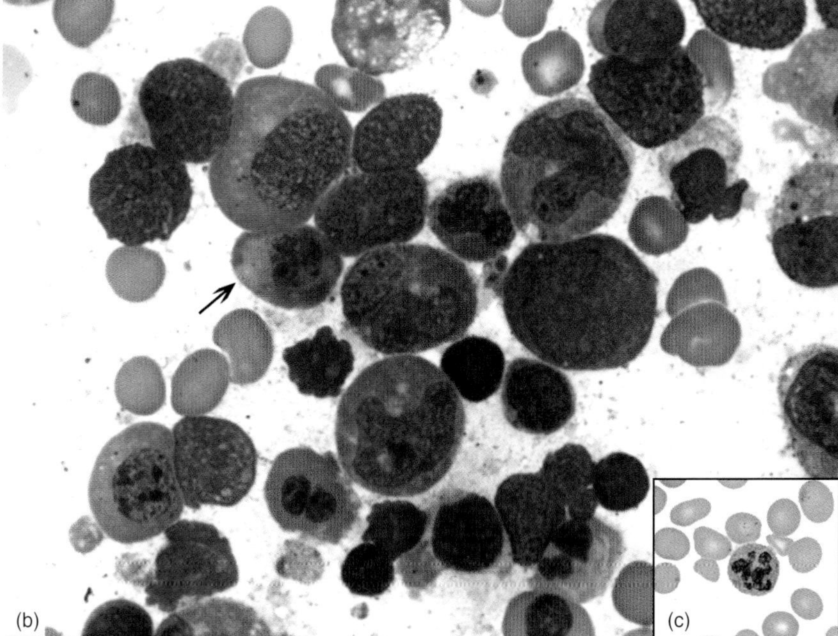

FIGURE 23.18 Megaloblastic anemia. Bone marrow smear demonstrates erythroid preponderance with a high proportion of megaloblasts: (a) low power and (b) high power views. An unevenly hemoglobinized erythroid cell is shown by an arrow (b). Blood smear shows macro-ovalocytes and a hypersegmented neutrophil (c).

inadequate absorption due to IF deficiency (pernicious anemia), gastric disease, or gastrectomy. The minimum dietary requirement of vitamin B_{12} is 6–9 µg/day. Unlike folate deficiency, vitamin B_{12} deficiency may be associated with neurologic symptoms. These symptoms result from degenerative changes of the dorsal and lateral columns of the spinal cord (subacute combined system disease) and have been attributed to the impairment of myelin synthesis due to the accumulation of methylmalonyl CoA or impairment of methyl group metabolism [91, 92].

A potentially fatal megaloblastic anemia due to a rapid tissue folate depletion has been observed in association with nitrous oxide (N_2O) anesthesia [93].

With the administration of folate or vitamin B_{12}, reticulocyte count will be increased in 3–4 days, followed by a fall in MCV and a rise in Hb levels within 10 days.

Hypersegmented neutrophils disappear in 10–14 days and Hb levels return to normal within 8 weeks [81].

Differential Diagnosis

Megaloblastic changes are observed in various conditions such as myelodysplastic syndromes, erythroleukemia, CDA, parvovirus infection, and following chemotherapy. In these conditions, serum folate and vitamin B_{12} levels are elevated or within normal limits. Increased number of early erythroid precursors and the presence of myeloid left shift in bone marrow biopsy sections may sometimes resemble acute leukemia. Immunophenotypic studies and review of the bone marrow smears will help to identify the erythroid nature of the immature cells. Coexistence of megaloblastic anemia with microcytic anemias, such as IDA or thalassemia, may mask macrocytosis, but hypersegmented neutrophils and giant bands are still present.

MICROCYTIC ANEMIAS

Microcytic anemias are anemias with red cells smaller than normal (MCV <80 fl). The most common cause of microcytic anemia is iron deficiency followed by thalassemias. Anemia of chronic disease, sideroblastic anemia, copper deficiency, and zinc intoxication may also be associated with microcytic anemia.

Iron Deficiency Anemia

Iron deficiency anemia (IDA) is the most common anemia worldwide [94]. The highest frequency of IDA is seen in women of reproductive age, pregnant women, and premature infants.

Etiology and Pathogenesis

Iron is usually lost through blood loss or loss of cells as they slough [94]. Major causes of IDA are inadequate dietary intake, blood loss, hemoglobinuria, iron malabsorption, renal dialysis, or inability of erythroid precursors to utilize iron [94–98].

Inadequate dietary intake is seen in infants fed milk without supplementary iron [99, 100]. Iron is low in milk and milk products, and prolonged (>6 months) breast or bottle feeding without iron supplementation may lead to IDA. Premature infants may become iron deficient as early as 10–12 weeks after birth, if their diet is not supplemented with iron. Iron dietary requirements increase during pregnancy and lactation.

Blood loss, particularly chronic bleeding, is one of the major causes of iron deficiency in adults [101, 102]. Chronic gastrointestinal bleeding due to peptic ulcer, gastritis, ulcerative colitis, amebiasis, hiatus hernia, esophageal and gastric varices, and cancers are the most common causes of IDA in men and post-menstrual women [94, 101, 103]. Bleeding of genitourinary tract, such as hematuria, hemoglobinuria (e.g. paroxysmal nocturnal hemoglobinuria, PNH), menstruation, and menorrhage often leads to the depletion of iron store. Similarly, IDA is observed in respiratory tract blood loss secondary to infection, epistaxis, idiopathic pulmonary hemosiderosis, and Goodpasture syndrome [103–105].

Iron deficiency is associated with the depletion and disappearance of hemosiderin and ferritin from bone marrow and other storage sites, and decreased concentration of several iron-containing proteins such as Hb, myoglobin, cytochrome c, cytochrome oxidase, and xanthine oxidase [106]. Approximately 1–2 mg/day iron is absorbed through the digestive system. Iron absorption is retarded by phosphate and phytates from cereals and is enhanced by vitamin C [107]. Meat, eggs, and liver are rich in iron. All vegetables except legumes and all fruits are either poor in iron or contain unabsorbable iron chelates.

Pathology and Laboratory Studies

Bone marrow examination often reveals mild to moderate hypercellularity with erythroid preponderance. Erythropoid precursors, particularly intermediate and late normoblasts, are small and show scanty, ragged rims of poorly hemoglobinized cytoplasm (Figure 23.19). These morphologic findings are not consistent and do not correlate with the severity of anemia. Bone marrow hemosiderin (demonstrated by Prussian blue stain) is reduced or absent. However, lack of stainable iron has been noted in patients with chronic myelogenous leukemia and myelofibrosis. These patients usually lack other evidences of iron deficiency.

Peripheral blood examination in the early stages of IDA reveals mild anisocytosis with slightly elevated red cell distribution width (RDW) and decreased MCV (<80 fl) (Figure 23.19). In more severe forms of IDA, erythrocytes are clearly microcytic and hypochromic and show pronounced anisocytosis with the presence of target cells.

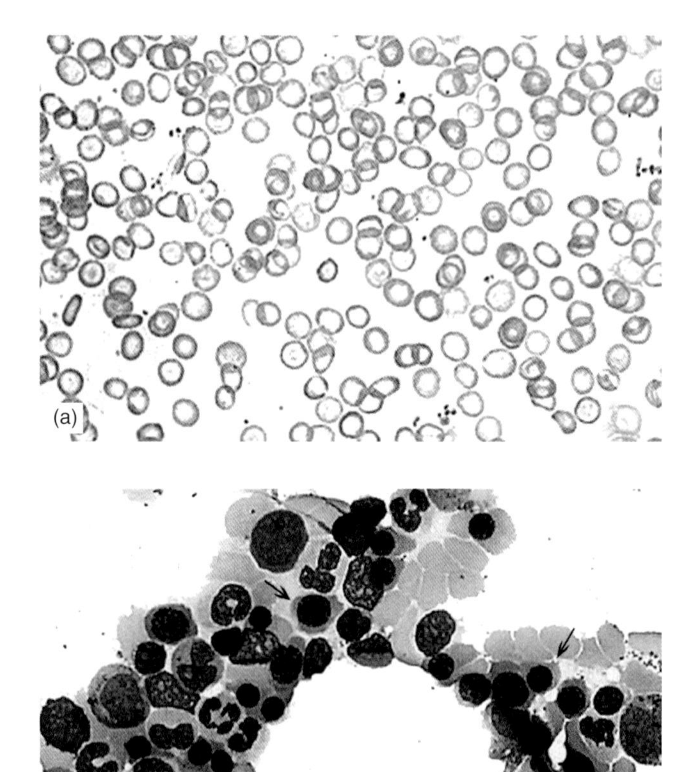

FIGURE 23.19 Iron deficiency anemia. Blood smear demonstrates hypochromic microcytic red cells (a) and bone marrow smear shows erythroid precursors with ragged rims (b, arrows).

The RDW is markedly elevated, and the MCV, MCH, and Hb levels are significantly lower than the normal range. Serum iron and ferritin levels are low and serum transferrin or total iron-binding capacity (TIBC) is elevated, as is the concentration of erythrocyte protoporphyrin (Table 23.7).

Serum ferritin is invariably low in IDA, but it is elevated in anemia of chronic disease, sideroblastic anemia, and thalassemia (Table 23.8) [108–110].

Clinical Aspects

Symptoms of anemia such as fatigue, headache, paresthesia, and burning sensation of the oropharyngeal mucosa are often preceded by the depletion of iron stores [94]. There is a poor correlation between the severity of the symptoms and the blood Hb level, suggesting that some of the symptoms are caused by a deficiency of iron-containing enzymes or proteins rather than by a low concentration of Hb. *Pica* (appetite for substances not considered as food, such as clay or dirt) and *pagophagia* (pica for ice) may be early clinical symptoms [111, 112].

Impaired learning ability and growth in children, defects in cell-mediated immunity and bactericidal function of leukocytes, increased frequency of premature contractions during pregnancy, and possibly an increased rate of premature births have been reported in patients with iron deficiency [113–115]. Most of the symptoms subside a few days after initiation of iron therapy. The reticulocyte count reaches its peak at about 1–2 weeks and then gradually levels off. The Hb level begins to improve after 2 weeks and usually comes back to normal levels after 2 months of adequate iron therapy.

TABLE 23.7 Laboratory findings in iron deficiency anemia.

Parameter	Results
Hemoglobin	<13 g/dL in men <12 g/dL in women
MCV	<80 fl
Serum iron	<75 μg/dL in men <65 μg/dL in women
Transferrin (TIBC)	<450 μg/dL
Transferrin saturation	<16%
Serum ferritin	<10 μg/L
Erythrocyte protoporphyrin	>7 μg/dL RBC
Bone marrow iron stain	Markedly reduced or negative

Thalassemia Syndromes

Thalassemia syndromes are a group of disorders caused by inherited defects in the synthesis of one or more of the globin chains and are characterized by a microcytic anemia [116–118]. The defect in globin chain production leads to a change in the proportion of the Hb classes, which consists of Hb A ($\alpha_2\beta_2$), Hb A2 ($\alpha_2\delta_2$), and Hb F ($\alpha_2\gamma_2$), constituting >95%, >3%, and <2% of the Hb molecules in normal adults, respectively.

The main β- and α-globin gene clusters are located on chromosomes 11 and 16, respectively (Figure 23.20), and encode globin subunits specific for the embryonic, fetal, and adult developmental stages [119–122]. In normal conditions, a balance is maintained between α- and β-cluster gene expression so that functional Hb tetramers are assembled. In thalassemia this balance is lost [120, 121]. The reduction in the synthesis of certain globin chain(s) leads to relative excess of the non-affected globin chain(s), causing Hb instability and hemolysis. Thalassemia syndromes are divided into β, δ, βδ, and α. δ-Thalassemia has no clinical significance.

β-Thalassemia

β-Thalassemia is the result of impaired production of β-chains of Hb A, leading to excess α-globin chains [117, 123–126]. Excess α-globin chains are unstable and precipitate in the erythrocytes (Heinz bodies) leading to the disruption of cell membrane and hemolysis. The excess α-chain deposition also accelerates apoptosis, leading to shortened red cell survival and ineffective erythropoiesis [127]. The exact mechanism of apoptosis is not clear, but a death-receptor-mediated pathway with Fas–Fas interactions has been suggested [128]. The abnormal β-globin gene expression is primarily the result of point mutations, but it may also be due to deletion of long stretches of nucleotides, or substitution of a small number of nucleotides within or close to the β-globin gene [119]. More than 200 point mutations have been reported [123], but ethnic-specific mutations allow for practical testing using more limited panels. The principal techniques used are DNA sequencing and allele-specific probe hybridization. Other genetic changes are less frequent. β-Thalassemia has two major genotypic types: homozygous with both β-genes affected (β_0) and heterozygous with one β-gene involved (β_+). There is a considerable phenotypic variation, depending on the nature of the mutations and the extent of the affected genes.

TABLE 23.8 Clinicopathologic features of common β-thalassemia syndromes.*

Features	Thalassemia trait	Thalassemia intermedia	Thalassemia major
Genetic	Heterozygous with one mutated β-globin gene	Heterozygous with more than one mutation in β-globin genes, one of which being partially defective	Homozygous, two β-globin genes affected
Clinical	Mild microcytic anemia; no other clinical symptoms	Moderate anemia, splenomegaly, and iron overload	Severe transfusion-dependent anemia, splenomegaly, iron overload, bone deformities

*Adapted from Ref. [121].

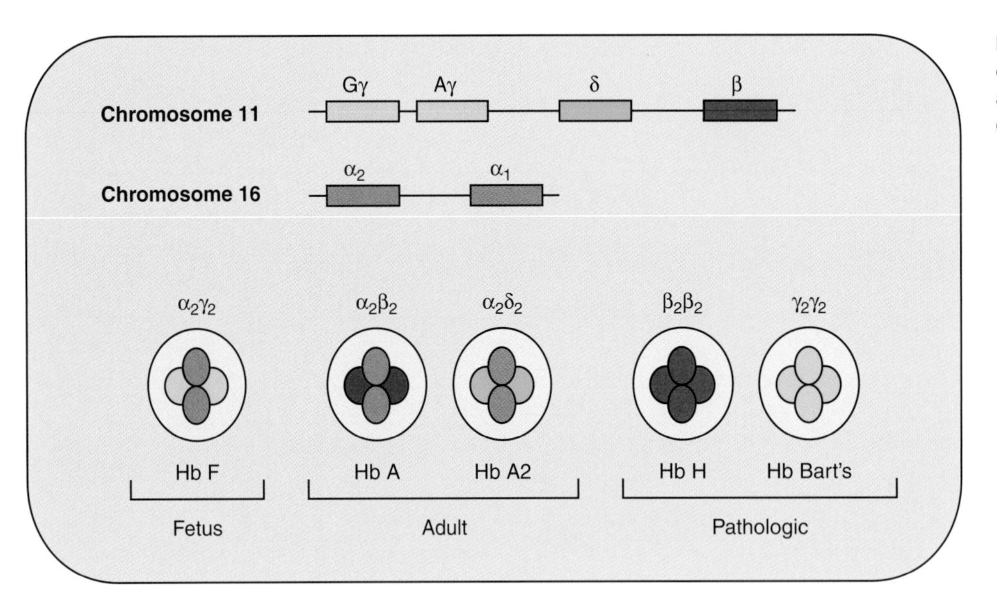

FIGURE 23.20 Hb genes are located on chromosomes 11 and 16. Normal and abnormal Hb structures are demonstrated.

Pathology and Laboratory Studies

Pathologic findings correlate with the severity of the clinical manifestations. In general, symptomatic patients show a hypercellular bone marrow with marked erythroid hyperplasia and left shift with some megaloblastic features (Figure 23.21). The iron stores are increased with abundant iron-laden macrophages. Scattered macrophages may contain phagocytosed normoblasts. The polychromatophilic normoblasts are poorly hemoglobinized, show ragged cytoplasm, and contain precipitated α-globin chains. The severely affected patients may show extramedullary erythropoiesis with masses of erythroid colonies in the thoracic or pelvic bone marrow spaces, causing spinal cord compression or other symptoms [126, 129].

The peripheral blood shows marked microcytic hypochromic anemia with low MCV and MCH, and elevated RDW and reticulocyte count (Figure 23.21). Anisocytosis is prominent, and there is a variable degree of poikilocytosis, target cell formation, and basophilic stippling. Heinz bodies may be detected in the affected RBCs using supravital stains such as methyl violet. Nucleated RBCs are often present. The WBC count is usually elevated but the platelet count is within normal limits. The serum iron and ferritin levels are elevated. The transferring saturation rate (ratio of serum iron to transferrin) is also high. Sera of the affected patients usually demonstrate increased concentration of indirect (unconjugated) bilirubin, elevated levels of lactate dehydrogenase (LDH), and reduced haptoglobin concentration, all indicative of hemolysis.

One of the characteristic features of thalassemia major (β_0-thalassemia) is elevated levels of fetal Hb. The proportion of Hb A2 to Hb A is also elevated. Hb electrophoresis (alkaline cellulose acetate and acidic citrate agar) is the most widely used method for diagnosis of hemoglobinopathies. However, newer techniques such as high-performance liquid chromatography (HPLC) provide higher sensitivity and specificity (Figure 23.22) [130, 131]. Isoelectric focusing provides excellent resolution but lacks quantitative accuracy and is labor intensive [130].

Clinical Features

β-Thalassemia syndromes are widespread in the Mediterranean basin, Middle East, and Southeast Asia with a spectrum of clinical manifestations ranging from a very mild asymptomatic anemia to severe transfusion-dependent anemia [121, 129, 132]. These syndromes are divided into three main categories: (1) β-thalassemia minor, (2) β-thalassemia major, and (3) β-thalassemia intermedia (Table 23.8).

β-Thalassemia minor, also known as β-thalassemia trait, is referred to heterozygous status when only one β-globin gene is affected. Patients with β-thalassemia trait are asymptomatic and may show mild anemia, usually detected as an incidental finding by a routine blood examination. Anemia may get worse during stressful conditions such as severe infection or pregnancy [133, 134]. The Hb levels are about 9–11 g/dL, but the MCV and MCH are markedly reduced.

β-Thalassemia major is referred to the patients who are homozygote and have no ability for effective production of β-globin. The disease, also known as Cooley's anemia, usually starts during the first year of life with profound transfusion-dependent anemia (Hb as low as 2.5–6 g/dL), hepatosplenomegaly, complications of iron overload, and skeletal deformities due to bone marrow expansion [123, 132]. Endocrinopathies, particularly hypogonadism, are frequent complications as the result of chronic anemia and iron overload [123]. Children with β-thalassemia major are at risk for parvovirus B19-induced aplastic crisis [124, 126].

β-Thalassemia intermedia represents patients with intermediate severity [135]. These patients usually have a compound heterozygous thalassemia consisting of more than one mutation.

The therapeutic approaches are based on the severity of the clinical manifestations and the thalassemia subtype. Patients with β-thalassemia minor usually do not need any specific treatment. Most patients with β-thalassemia intermediate need transfusion therapy, which can be delayed by splenectomy in certain patients [126]. Iron chelation therapy may be indicated [136]. Therapy for β-thalassemia major includes chronic hypertransfusion, splenectomy, iron

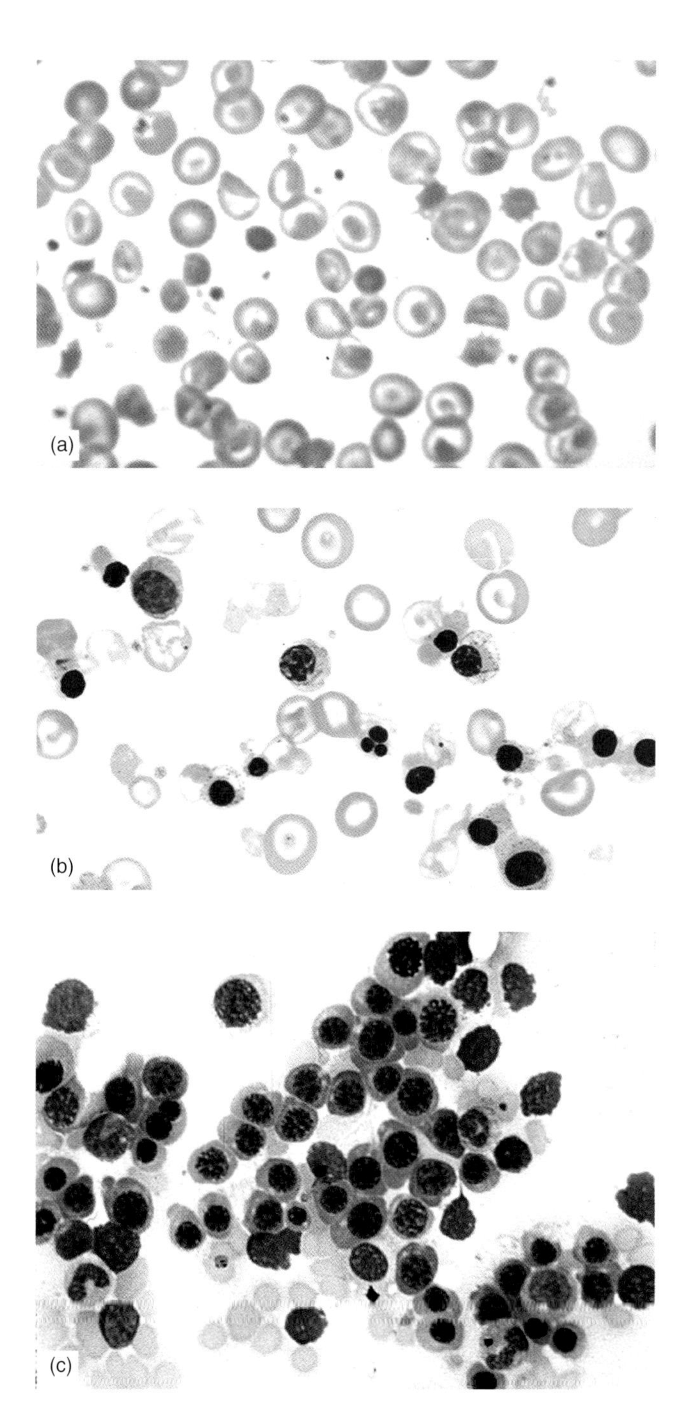

FIGURE 23.21 β-Thalassemia major. Blood smears showing target cells and abnormal nucleated red cells (a and b). Bone marrow smear shows erythroid hyperplasia (c).

chelation, hematopoietic stem cell transplantation, and molecular therapy [123, 137–140].

α-Thalassemia

α-Thalassemia results from the deletion of one or more α-globin genes (Figure 23.20) [141–143]. The single α-globin gene deletion appears to be the most common mutation worldwide [141–143]. The α-globin gene deletion, similar to other types of hemoglobinopathies, is believed to be protective against malaria infection [144, 145]. The non-deletion mutations generally fall into three major categories: RNA processing defect, RNA translation defect, and post-translational instability [146, 147].

The deficiency in α-chain production results in the accumulation of excessive γ- and/or β-globin chains in α-thalassemia which may lead to the formation of unstable homotetramers β_4 (Hb H) and γ_4 (Hb Bart's) [148–150]. The severity of clinical manifestation and anemia in α-thalassemia correlates with several factors including the affected gene (α_1 or α_2), the total number of the affected α-globin genes, and the extent of functional loss resulting from specific mutations. For example, the α_2-globin gene is expressed at 2.6-fold higher levels than the α_1-globin gene and therefore plays a more significant clinical role [121].

Pathology and Laboratory Studies

Pathologic features are similar to β-thalassemia and correlate with the severity of the clinical manifestations. In general, symptomatic patients show a hypercellular bone marrow with marked erythroid hyperplasia and left shift. The polychromatophilic normoblasts are poorly hemoglobinized and show ragged cytoplasm. The iron stores are increased with abundant iron-laden macrophages.

The peripheral blood shows marked microcytic hypochromic anemia with low MCV and MCH and elevated RDW and reticulocyte count. Anisocytosis is prominent, and there is a variable degree of poikilocytosis, target cell formation, and basophilic stippling. Red cells may contain Hb H (tetramers of β-globin chains), which are precipitated when exposed to oxidants such as supravital stains, methylene blue and brilliant cresyl blue. Hb H precipitates are small, evenly dispersed deposits creating a golf ball appearance. Nucleated RBCs are often present. The WBC count may be elevated but the platelet count is within normal limits. The serum iron and ferritin levels are elevated. The transferring saturation rate (ratio of serum iron to transferrin) is also high.

Hb electrophoresis, HPLC, and molecular genetic studies are used for the diagnosis of thalassemia syndromes and other hemoglobinopathies (see Figure 23.22) [130]. The number of deleted α-globin genes can be determined by Southern blot analysis or dosage-dependent quantitative PCR.

Clinical Features

There are four types of α-thalassemia: (1) a silent carrier with three functional α-genes, (2) α-thalassemia trait with two functional α-genes, (3) Hb H disease with only one functional α-gene, and (4) Hb Bart's (hydrops fetalis) or α_0-thalassemia with no functional α-gene [121, 141, 146, 148].

The silent carrier form is prevalent in the Mediterranean basin, Middle East, India, Southeast Asia, and Indonesia and affects approximately 30% of Afro-Americans [121, 148, 151]. The affected individuals are asymptomatic and display unremarkable hematologic parameters except for a borderline low MCV and MCH. The Hb A2 concentration is normal. The diagnosis is usually made by molecular genetic studies.

Patients with α-thalassemia trait have no clinical symptoms but display mild anemia. The MCV and MCH are low and the Hb A2 concentration is normal. It affects approximately 3% of Afro-Americans [152].

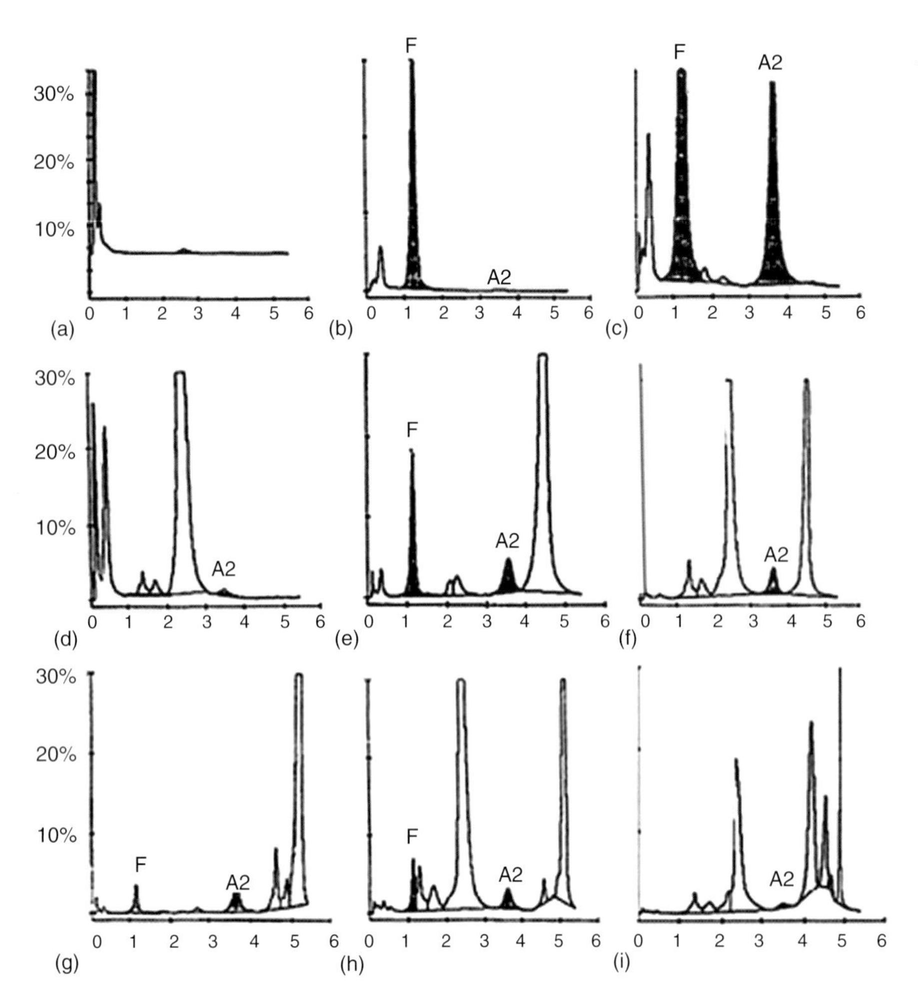

FIGURE 23.22 HPLC on the Bio-Rad Variant β Thal short program for Hb Bart's (a), β_0 thalassemia major (b), β_+ thalassemia homozygous E (c), Hb H (d), homozygous S (e), S trait (f), homozygous C (g), C trait (h), and Hb S-Hb $G_{Philadelphia}$ (i). From Ref. [131] by permission.

The Hb H disease is associated with the precipitation of β-globin chain tetramers, leading to red cell membrane damage and hemolysis. Patients suffer from moderate to severe anemia and often demonstrate hepatosplenomegaly. Diagnosis is made by demonstration of Hb Bart's in 15–30% of the cord red cells at birth or by detection of Hb H (see Figure 23.8b) [148].

Hb Bart's leads to hydrops fetalis, the most severe form of α-thalassemia [148, 150]. Hb Bart's binds oxygen with high affinity and cannot release it to tissues. The affected fetuses have severe anemia leading to congestive heart failure with anasarca and capillary leak (hydrops) [150]. This condition is incompatible with extrauterine life.

The therapeutic approaches for α-thalassemias are similar to those discussed for β-thalassemias.

Other Thalassemia Variants

In addition to the classical hereditary α- and β-thalassemias, there are reports of acquired Hb H disease associated with myelodysplastic syndromes and myeloproliferative disorders [153, 154]. There are also variants of β-thalassemia associated with other structural abnormalities of β-globin chain such as Hb S, Hb C, and Hb E.

δβ-Thalassemia is the result of deletions of the δ- and β-globin genes or crossover between part of the δ locus on one chromosome and part of the β locus on the complementary chromosome. The δβ-gene crossovers are apparently caused by misalignment of chromosome pairing during meiosis, resulting in a δβ-fusion gene. The gene product is an abnormal hemoglobin called *Hb Lepore* [155, 156].

Acquired α-thalassemia (Hb H disease) has been reported in myelodysplastic and myeloproliferative disorders. Two mechanisms have been proposed for this acquired process: (1) acquired deletion of the α-globin gene cluster limited to the abnormal hematopoietic clone and (2) inactivation of somatic mutations of the transacting chromatin-associated factor ATRX leading to downregulation of α-globin gene expression [153, 154].

Differential Diagnosis

The differential diagnosis of microcytic anemias, in addition to IDA and thalassemia syndromes, includes ACD, sideroblastic anemia, and anemias due to copper deficiency and zinc toxicity. The patient's clinical history and presentation help to distinguish these different categories. IDA is the only microcytic anemia with reduced serum ferritin and reduced or absence of the bone marrow iron stores (Table 23.9). The proportion of Hb A2 and fetal Hb is elevated, and Heinz bodies are present in β-thalassemias, and the presence of Hb H is the hallmark of α-thalassemia.

TABLE 23.9 Laboratory findings in microcytic anemias.*

	Serum Fe	TIBC*	% Saturation	Serum ferritin	Marrow Fe stain
Iron deficiency	↓	↑ (<16%)	↓	↓	Absent
Chronic disease	↓	↓	↓ or N	↑	↑
Sideroblastic	↑ or N	↑	↑	↑	↑ + RS**
Thalassemia	↑ or N	↑ or N	↑	↑	↑

*Total iron binding capacity (transferrin).
**Ringed sideroblasts.

SICKLE CELL DISEASE

Sickle cell (SC) disease consists of a family of hereditary hemoglobinopathies caused by mutations in the β-globin chain gene [157–159]. It includes Hb S (sickle β-globin), Hb C, and Hb E disorders. Hb D disease is the result of either β- or α-chain gene mutation. Sickle cell (SC) anemia is a homozygous state which represents the most severe form of the disease in this group, resulting from the inheritance of Hb S from both parents [160, 161].

Etiology and Pathogenesis

The Hb S mutation is a single base change in the DNA (GAG to GTG) in codon 6, which results in the substitution of the amino acid valine for glutamic acid at the sixth amino acid position in the β-globin chain. Deoxygenated Hb S molecules align in liquid crystals (tactoids) and distort the erythrocytes into rigid sickle shapes [159, 162, 163]. Reoxygenation disassembles the tactoids, and the erythrocytes become discoid and flexible again. Repeated sickling and unsickling is associated with red cell membrane changes such as changes in the membrane phospholipid composition and perturbation of the interaction between membrane phospholipids and cytoskeletal proteins. These changes eventually lead to the inability of the erythrocytes to switch from sickle shape to discoid shape [164, 165]. The irreversible SCs have a short intravascular life span and, because of their rigid shape, may block small vascular structures, leading to ischemia and endothelial cell damage [162, 166].

In Hb C, glutamic acid in the sixth position of the β-globin chain is replaced by lysine [167, 168]. The Hb C-containing erythrocytes are more rigid than normal red cells and may form rod-like crystals in hypoxic conditions. Erythrocyte damage and fragmentation may lead to the formation of microspherocytes.

Hb D involves either the β-globin chain or the α-globin chain. In the β-globin chain variant, glutamate is substituted for lysine at the 121th position, and in the α-globin variant, also known as Hb $G_{Philadelphia}$, asparagine is replaced by lysine at the 68th position [169, 170]. Several other variants of Hb D have been reported.

Hb E is the result of a β-chain mutation in which glutamine is substituted for lysine at the 26th position [171, 172].

Pathology and Laboratory Studies

Pathologic features correlate with the severity of anemia. In general, patients show a hypercellular bone marrow with marked erythroid hyperplasia and left shift. The extensive erythroid response and bone marrow overgrowth may lead to skeletal deformities [173]. Bone deformities may be secondary to bone infarcts, such as irregularities in the size of fingers and toes, partial vertebral collapse, or avascular necrosis of the pelvis. The femoral shafts, which in normal conditions are not actively involved in hematopoiesis, may become active. Extramedullary hematopoiesis may occur.

Blood examination in patients with homozygous SC anemia reveals anemia with a normal MCV, an Hb level ranging from 5 to 11 g/dL, and reticulocytosis (5–15%) (Figure 23.23). There is anisopoikilocytosis with the presence of SCs, target cells, and nucleated RBCs. Serum indirect bilirubin and lactate dehydrogenase levels are elevated, and serum haptoglobin and creatinine concentrations are low [174]. Hb C disease is often associated with the presence of rod-like crystals (Figure 23.24). Fetal Hb is slightly to moderately elevated.

Diagnosis of SC disease is established by the demonstration of Hb S on hemoglobin electrophoresis, HPLC, and by molecular genetic studies (Figure 23.25). In alkaline Hb electrophoresis, Hb S moves slower than Hb A and occupies a position between Hb A and Hb A2. Hb C and Hb E move slower than Hb S, whereas Hb D has a mobility similar to that of Hb S. These variations are further delineated by an agar gel electrophoresis at an acid pH [175, 176]. A simple inexpensive test, SC test, has been traditionally used for the screening of patients suspicious of SC disease. This test is based on the principle that the reducing agents, such as sodium metabisulfite, are able to induce precipitation of Hb S in erythrocytes and cause sickle cell formation.

Molecular studies such as Southern blot analysis and polymerase chain reaction (PCR) have been used for the diagnosis of SC anemia and related disorders [176, 177]. For prenatal testing, fetal DNA samples are obtained from chorionic villi at 8–10 weeks gestation or from amniocetesis at 18 weeks. Although isolation of fetal SCs from maternal circulation has been reported, such approaches are not yet routinely utilized [178, 179].

Clinical Aspects

The prevalence of Hb S is greatest in Africa with a heterozygote frequency of about 20%. In Afro-Americans, SC

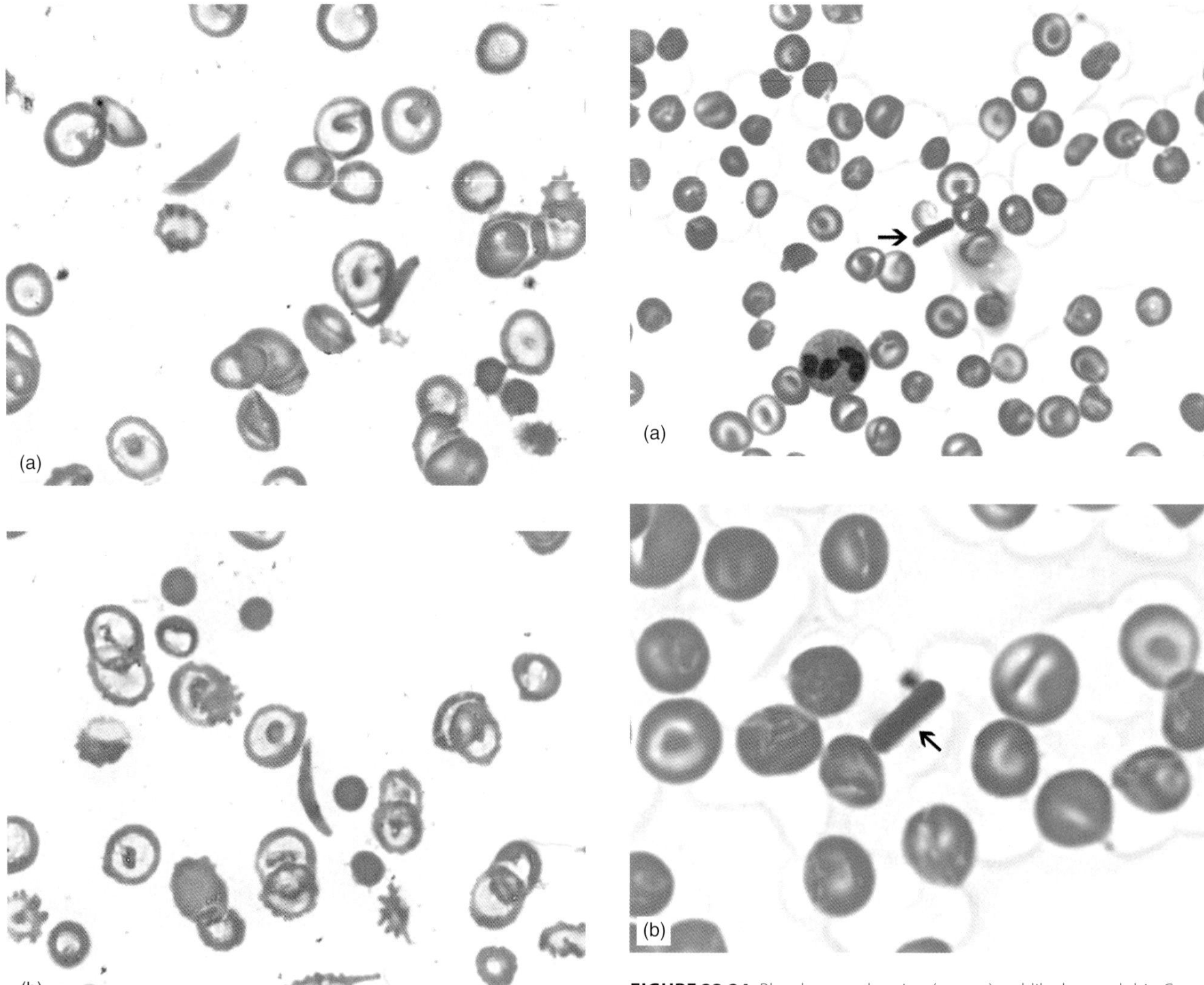

FIGURE 23.23 Blood smear (a and b) demonstrating numerous sickle cells and target cells.

FIGURE 23.24 Blood smear showing (arrows) rod-like hemoglobin C crystals: (a) low power and (b) high power.

trait occurs in approximately 8% of the population [157–159, 180]. There is a high prevalence of the Hb S in the areas of the world where malaria is epidemic. The heterozygous SC patients appear to be more resistant to malaria infection, particularly *Plasmodium falciparum*, than normal persons [181–183].

The clinical symptoms of SC anemia in affected infants are usually manifested 8–10 weeks after birth. Prior to this period, the newborn is protected by the high concentration of Hb F in the erythrocytes. Patients have a steady-state course that is intermittently interrupted by a sudden onset of severe clinical course (crisis). The most common cause of SC crisis is vaso-occlusive events leading to ischemia and subsequent infarction of the affected tissues (frequently bone and spleen) [166, 184]. This type of crisis is usually associated with severe pain in the bones, chest, and abdomen. Other crisis-related clinical complications include the development of aplastic, megaloblastic, or hemolytic anemia.

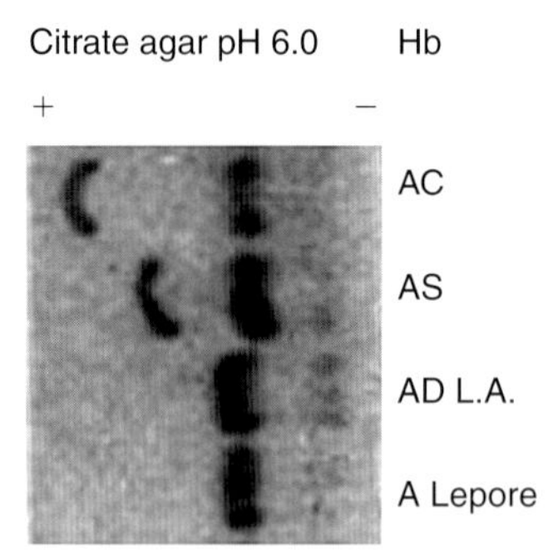

FIGURE 23.25 Citrate agar hemoglobin electrophoresis (pH 6.0) demonstrating bands of various types of hemoglobin (A: Hb A; C: Hb C; S: Hb S; D: Hb D). From the International Committee for Standardization in Hematology. (1978). Simple electrophoretic system for presumptive identification of abnormal Hbs. *Blood* **52**, 1058–63, by permission. This research was originally published in *Blood*.

Affected children often demonstrate growth retardation and skeletal abnormalities. Retinal damage, cerebrovascular accidents, leg ulcers, renal papillary necrosis, and recurrent infections are common complications [184–186].

Dactylitis (acute pain in fingers and toes) before age 1, Hb concentration <7 g/dL, and leukocytosis in the absence of infections are considered predictors of an adverse outcome [187–190]. The estimated survival rate at 18 years in one large study was reported at about 88% [188, 189].

Other forms of SC disease and related disorders such as SC trait, Hb SC, Hb C, Hb D, Hb E, Hb S-β-thalassemia, and Hb C-β-thalassemia cause either no symptoms or mild symptoms with slight to moderate anemia.

The therapeutic approaches include treatment with hydroxyurea to increase Hb F levels, blood transfusion, iron chelation therapy, administration of antibiotics for prevention and treatment of the infections, and anticoagulation therapy to prevent or treat vaso-occlusive events [191–193]. Bone marrow transplantation may be effective in selected patients [194].

Differential Diagnosis

Family history, clinical manifestations, and evidence of Hb S establish the diagnosis. Bone marrow samples during SC crisis may be markedly hypoplastic or may show evidence of parvovirus B19 infection.

OTHER HEMOGLOBINOPATHIES

So far, over 800 different mutations have been reported in human Hb genes [195]. The majority of Hb mutations are clinically asymptomatic and have been discovered in conjunction with large population studies. The most prevalent mutations, thalassemia syndromes, and Hb S and related disorders were discussed earlier in this chapter. Unstable Hbs are infrequent Hb mutations that are briefly discussed in the following section.

Unstable Hemoglobins

Mutations of the globin chains may result in the formation of Hb molecules that are less soluble and have increased tendency of precipitation in the erythrocytes, leading to red cell membrane damage and hemolysis. The intracellular Hb precipitates are detected by supravital stains as dark globular aggregates called Heinz bodies. Over 250 unstable Hb molecules have been identified [196–198]. Patients with unstable Hb demonstrate a broad spectrum of clinical manifestations ranging from normal Hb levels to severe anemia. Jaundice and splenomegaly may be present. In some cases, high oxygen affinity may cause polycythemia. Bone marrow is usually hypercellular and shows erythroid preponderance. Anisopoikilocytosis and reticulocytosis are frequent peripheral blood morphologic findings. The most frequent type of unstable Hb in the West is Hb Koln which is characterized by mild anemia, reticulocytosis, splenomegaly, and pigmenturia [198, 199].

ERYTHROCYTE MEMBRANE SKELETON DEFECTS

A group of hereditary hemolytic anemias are caused by defects in erythrocyte membrane skeleton. These disorders are relatively common and are characterized by abnormal shape and decreased deformability of the red cells [200, 201].

Etiology and Pathogenesis

The red cell membrane consists of a lipid bilayer, integral membrane proteins, and a complex skeletal protein comprising α- and β-spectrin, ankyrin, protein 4.1, protein 4.2 (pallidin), and actin (Figure 23.26) [200, 202–204]. The membrane skeleton proteins interact with the integral membrane protein band 3 (anion exchanger, AE1) and glycophorin C to provide erythrocyte integrity and deformability. The α- and β-spectrin subunits bind side-to-side and form flexible rod-like heterodimers which self-associate head-to-head and make tetramers [200]. The spectrin tetramers are linked by ankyrin to protein 4.2 and protein band 3. Protein 4.1 attaches to the integral membrane glycoprotein C and interacts with β-spectrin at the actin-binding domain to increase the affinity of the spectrin–actin binding [200]. The spectrin tetramers at their tail ends interact with actin protofilaments, tropomyosin, tropomodulin, and adducin to form junctional complexes [200]. The genes of the major erythrocyte membrane proteins have been mapped and cloned, and structural and functional domains in each protein have been characterized [200, 203, 205]. Mutations of these genes are associated with erythrocyte membrane skeleton defects.

Defects in spectrin and other membrane-associated skeletal proteins lead to membrane lipid loss and surface area deficiency, alteration in cation content and membrane permeability, and decreased deformability of the erythrocytes. The affected red cells are not able to pass through the Billroth cords to the splenic sinuses and stagnate in an environment that has a lower pH and a decreased level of glucose. The erythrocytes are eventually destroyed or removed by the splenic macrophages [205–207].

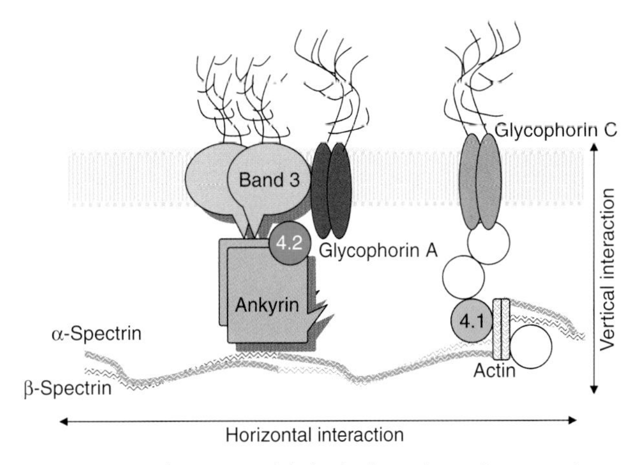

FIGURE 23.26 Schematic model of red cell membrane demonstrating vertical and horizontal interactions of various cytoskeletal components. Adapted from Ref. [200].

Clinical and Pathologic Features

Hereditary hemolytic anemias due to defective erythrocyte membrane cytoskeleton include clinicopathologic entities such as spherocytosis, elliptocytosis, acanthocytosis, and stomatocytosis.

Hereditary Spherocytosis

Hereditary spherocytosis (HS) is the most common cause of hemolytic anemia of non-immune nature characterized by the presence of numerous spherocytes in the peripheral blood (Figure 23.27a). The incidence of HS is significantly higher in northern European countries than in other parts of the world [207, 208]. The autosomal dominant form is the most frequent type accounting for approximately 75% of the cases. The disease is caused by defective erythrocyte membrane skeletal proteins in vertical interactions (see Figure 23.26), demonstrated by reduced β-spectrin (dominant) or α-spectrin (recessive) production, impaired binding to protein 4.1, and reduced or unstable ankyrin production (Table 23.10) [200, 207–209].

The clinical manifestation of HS may range from no symptoms to severe anemia. Most patients show mild to moderate anemia, mild jaundice, and a palpable spleen with a family history of anemia involving one or more siblings or parents. The peripheral blood smears reveal spherocytosis with a variable degree of reticulocytosis. MCHC and RDW are elevated [210]. Red blood cells show increased osmotic fragility. Occasionally, HS is complicated by aplastic crisis caused by parvovirus B19 infection [211, 212]. Patients with severe HS are treated with folic acid supplementation, blood transfusion, and splenomegaly.

Hereditary Elliptocytosis

Hereditary elliptocytosis (HE) represents a diverse group of hemolytic anemias characterized by the presence of numerous oval, elliptical, or elongated erythrocytes in the peripheral blood (Figure 23.27b). The prevalence is relatively low in the United States and Europe and high in the regions where malaria is endemic, such as Africa and Southeast Asia [213, 214]. The disorder is associated with defective erythrocyte membrane skeletal proteins in horizontal interactions, demonstrated by defective self-association of subunits or dimmer–dimmer association of spectrin and/or deficiencies of glycoprotein C, protein 4.1, or protein 4.2 (Table 23.10) [215–218]. HE has been divided into three major clinical groups: (1) common type, (2) spherocytic type, and (3) stomatocytic type.

Common HE is mostly reported as autosomal dominant disorder with no clinical symptoms or mild to moderate anemia. Jaundice, splenomegaly, and reticulocytosis may be present in some cases. An autosomal recessive variant of HE, *hereditary pyropoikilocytosis* (HPP), is characterized by severe hemolytic anemia, jaundice, splenomegaly, and marked spherocytosis and poikilocytosis [219]. HPP is found predominantly in the black population. Elliptocytosis may provide resistance against malaria [214, 220]. Similar to HS, patients with severe HE are treated with folic acid supplementation, blood transfusion, and splenomegaly.

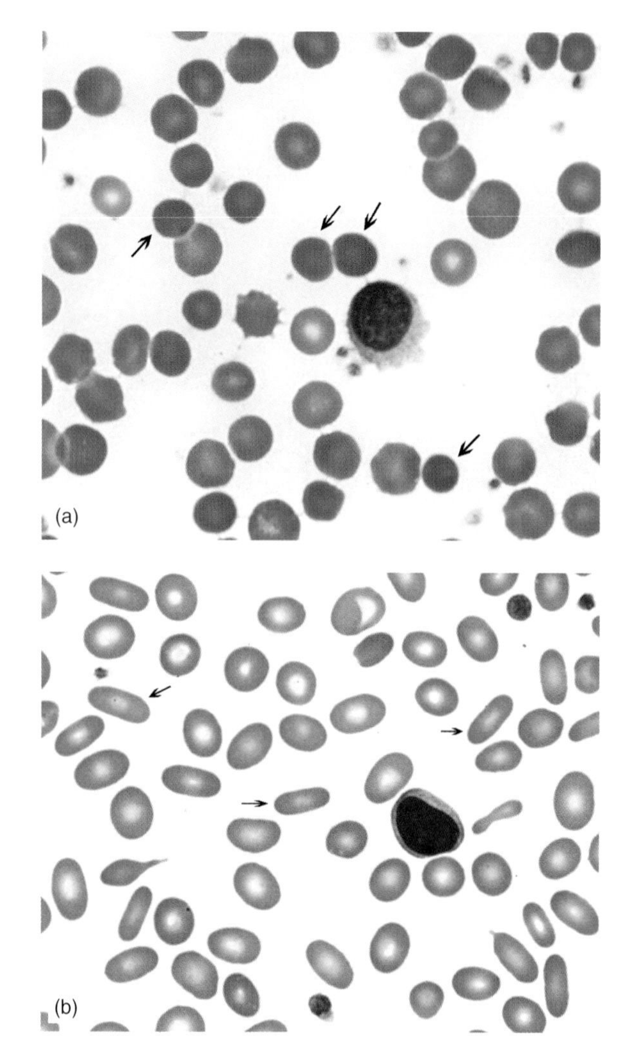

FIGURE 23.27 Blood smears demonstrate spherocytes (a) and elliptocytes (b).

Spherocytic HE (hereditary ovalocytosis) is associated with rounder erythrocytes that show increased osmotic fragility. It is an autosomal dominant disorder observed only in Caucasians. There is mild to moderate hemolysis with splenomegaly.

Stomatocytic HE has been reported in up to 30% of the population of Malayan aborigines in Southeast Asia and it is also known as Melanesian ovalocytosis [221]. It is characterized by spoon-shaped erythrocytes with absent or mild clinical manifestations.

Differential Diagnosis

The differential diagnosis of HS includes autoimmune hemolytic anemia (AIHA) and unstable hemoglobinopathies. The HS patients are negative for Coobs' test and lack Heinz bodies. Elliptocytosis and poikilocytosis have been associated with a garden variety of hematologic disorders such as IDA, thalassemia syndromes, megaloblastic anemia,

TABLE 23.10 Major abnormalities of skeletal proteins of the red blood cell membrane.*

Gene product	Chromosome	Phenotypic features	Defect
α-Spectrin	1q21	HS (recessive)	Spectrin 30–75% normal; abnormal proteolytic domain
		HE_C/HPP	Defective α-subunit self-association; abnormal proteolytic domain
β-Spectrin	14q22-q23.2	HS (dominant)	Spectrin 60–85% normal; impaired binding to protein 4.1
		HE_C	Diminished spectrin tetramer self-association; impaired binding to ankyrin
Ankyrin	8p11.2	HS	Spectrin 50% normal; unstable ankyrin
Protein 3	17q21-q22	Acanthocytosis	Reduced ankyrin binding
Protein 4.1	1p34-p36.2	HE_C	Partial or total 4.1 deficiency
Protein 4.2	15q15	HE_S	4.2 deficiency; decreased ankyrin stability
Glycophorin C	2q14-q21	HE_C	Weakened 4.1 association with membrane

*Adapted from Davies KA, Lux SE, McGuire M, Agre P. (1988). Clinical disorders of the erythrocyte membrane skeleton. *Hematol Pathol* **2**, 1–14.
HS: hereditary spherocytosis; HE_C: hereditary elliptocytosis, common type; HPP: hereditary pyropoikilocytosis; HE_S: hereditary elliptocytosis, stomatocytic type.

myelofibrosis, and myelodysplastic syndromes. Elliptocytes in these conditions, however, are generally below 60% of the red cells. Definitive diagnosis of HS or HE is established by identification of the underlying molecular defects of the erythrocyte membrane skeletal proteins.

Other Membrane-Associated Erythrocyte Abnormalities

Acanthocytosis

Acanthocytes, or spur cells, are red cells with multiple irregularly shaped and randomly distributed cytoplasmic projections (see Figure 23.3a). These abnormal erythrocytes are seen in a β-lipoproteinemia, amyotrophic chorea (chorea-acanthocytosis), McLeod syndrome (Kell antigen defects), cystic fibrosis, anorexia nervosa, severe liver disease, and hypothyroidism [222–226]. These disorders appear to cause overexpansion of the outer half of the membrane bilayer and formation of the irregular projections [225–227].

Acanthocytes should be distinguished from echinocytes (burr cells). Echinocytes are erythrocytes with evenly distributed pointed projections (see Figure 23.3b). Echinocytosis is observed in uremia, liver disease, hypomagnesemia, hypophosphatemia, post-chemotherapy, and in athletes after heavy physical exercise [228, 229]. Crenated red cells are cells with evenly distributed blunt projections, usually a common laboratory artifact caused by blood storage, contact with glass, or elevated pH (see Figure 23.3c).

Stomatocytosis

Stomatocytes are cup- or bowel-shaped erythrocytes which in blood smears appear as cells with a wide slit or stoma (mouth-like) area of central pallor (see Figure 23.2b). The stomatocyte shape is the result of the decreased ratio of the surface area to the volume in the erythrocytes. The increased red cell volume in almost all cases is due to increased permeability. Stomatocytes are trapped and consequently hemolyzed in the microvasculature of spleen and other organs. Stomatocytosis is either hereditary or acquired.

Hereditary stomatocytosis is an autosomal dominant genetic disorder leading to the increased permeability of the red cells to sodium [216, 230–232]. The increased permeability to sodium in some studies was associated with a deficiency of the erythrocyte membrane protein stomatin (band 7.2b) [232]. Hemolytic anemia with stomatocytosis (up to 40–60%), elevated reticulocyte count, elevated serum bilirubin levels, and reduced serum haptoglobin concentration in children or adolescents are characteristic features of hereditary stomatocytosis.

Acquired stomatocytosis is usually manifested in adults. It has been observed in alcoholism, chronic liver disease, malignancies, and cardiovascular disorders [233, 234].

HEMOLYTIC ANEMIA SECONDARY TO ERYTHROCYTE ENZYME DEFICIENCIES

Numerous enzymes are involved in the RBC metabolic activities such as Embden–Meyerhof glycolytic pathway, the Rapoport-Leubering shunt, and the pentose phosphate pathway. These metabolic activities maintain the structural and functional activities of the erythrocytes. Erythrocyte enzyme deficiencies are associated with a wide variety of clinical syndromes, some of which demonstrate hemolytic anemia of non-spherocytic type [235–239]. The most prominent erythrocyte enzyme deficiencies, glucose-6-phosphate dehydrogenase (G6PD) deficiency and pyruvate kinase (PK) deficiency, are briefly discussed here.

Glucose-6-Phosphate Dehydrogenase Deficiency

Glucose-6-phosphate dehydrogenase deficiency results from a wide variety of mutations in the *G6PD* gene located on X chromosome (Xq28) [235, 236, 240–242]. Over 350 mutations of *G6PD* gene have been defined. G6PD is an enzyme essential for basic cellular functions, including protection of red cell proteins from oxidative damage [235, 236]. The G6PD activity is necessary for the generation of NADPH that is utilized for glutathione reduction [235, 236]. Reduced glutathione restores soluble Hb. The G6PD deficiency increases Hb vulnerability to oxidative damage, leading to Hb instability and precipitation as Heinz bodies [237, 243]. There are three clinical variants of G6PD deficiency associated with (1) acute intermittent hemolytic anemia, (2) chronic hemolysis, and (3) no obvious risk of hemolysis [235, 236]. The A-G6PD deficiency with acute intermittent hemolysis is the most common clinical presentation and is observed in Africans. The Mediterranean type of G6PD deficiency is more severe than the type A-G6PD deficiency. The G6PD deficiency affects about 10% of Afro-Americans and West Africans, 5–15% of Kurds, and 5–35% of Sardinians [240–242]. The distribution of populations with high frequency of G6PD deficiency geographically overlaps closely with the prevalence of malaria, suggesting that G6PD deficiency may play a protective role against malaria [244].

The vast majority of G6PD-deficient patients do not demonstrate hemolysis if they are not suffering from infection or are not exposed to oxidants [240, 245, 246]. Drugs, fava beans (particularly in patients with the Mediterranean type), and viral or bacterial infections are the most common inducers of hemolysis in the G6PD-deficient patients (Table 23.11) [235, 236].

TABLE 23.11 Examples of substances known to induce hemolysis in patients with glucose-6-phosphate dehydrogenase deficiency.

Acetanilid
Aspirin, high dose
Dimercaprol
Fava beans
Isobutyl nitrate
Methylene blue
Naphthalene
Nitrofurantoin (Furadantin)
Pamaquine
Pentaquine
Phenacetin
Phenylhydrazine
Primaquine
Sulfacetamide
Sulfonamide
Sulfapyridine
Toluidine blue
Vitamin K (water-soluble analogs)

During the hemolysis, red cells are normocytic normochromic or may show some degree of anisocytosis and poikilocytosis. Spherocytosis is not evident. Supravital stains may show Heinz bodies. The hemolytic episode stimulates erythropoiesis and leads to bone marrow erythroid hyperplasia and peripheral blood reticulocytosis. Similar to other hemolytic anemias, serum bilirubin level is elevated and serum haptoglobin concentration is reduced. G6PD activity is assessed by a fluorescent screening test or by quantitative spectrophotometric analysis. Alternatively, the actual gene mutation can be detected by DNA sequencing.

Patients with G6PD deficiency are instructed to avoid drugs and substances that may induce hemolysis. Patients who develop severe hemolysis may require blood transfusion. Folate supplementation is provided in those patients with chronic hemolysis [235, 236].

Pyruvate Kinase Deficiency

Pyruvate kinase (PK) deficiency is an autosomal recessive disorder with clinical manifestations of non-spherocytic hemolytic anemia in homozygotes or double heterozygotes [235, 236, 247–249]. PK is involved in the conversion of phosphoenolpyruvate to pyruvate and the generation of ATP in the Embden–Meyerhof pathway. Erythrocyte PK is synthesized under the control of the *PK-LR* gene located on chromosome 1 (1q21) [247, 250–253]. So far, over 150 *PK-LR* mutations associated with non-spherocytic hemolytic anemia have been reported [250]. PK deficiency is distributed worldwide, but it is more prevalent in people of northern Europe, China, and the Amish community in Pennsylvania [235, 236].

The severity of hemolysis in the PK-deficient patients is highly variable and may range from a mild fully compensated hemolysis with no anemia to chronic, severe, life-threatening hemolytic anemia [235, 236, 247]. The red cell morphology is unremarkable or may show variable degrees of anisopoikilocytosis and reticulocytosis. Splenomegaly is a frequent finding. Severe cases may develop clinical complications, such as chronic jaundice, development of gallstones, folate deficiency, and transient aplastic anemia crisis, often due to parvovirus B19 infection. Quantitative PK enzyme assays establish the diagnosis. Molecular diagnostic methods, due to the large number of mutations and their low prevalence, are not in routine laboratory use [235, 236].

IMMUNE-MEDIATED HEMOLYTIC ANEMIA

Immune-mediated hemolytic anemia is the most common form of acquired hemolytic anemias. It is caused by a variety of antibodies against antigens expressed or attached to the patient's red cell membrane. These include autoantibodies, antibodies against transfused red cells (ABO-incompatibility), maternal antibodies against fetal erythrocytes, and drug–antibody complexes [254, 255].

Autoimmune Hemolytic Anemia

Autoimmune hemolytic anemia (AIHA) is the most common type of acquired hemolytic anemias and is caused by the destruction of RBCs by autoantibodies directed against erythrocytes [254–256].

Etiology and Pathogenesis

Autoimmune hemolytic anemia is most often idiopathic. However, several etiologic factors have been proposed for

TABLE 23.12 Classification of autoimmune hemolytic anemia.

1. Serologic classification a. Warm autoantibody (agglutinin): antibody with enhanced activity at 37°C b. Cold autoantibody (agglutinin): antibody with enhanced activity <37°C c. Mixed warm and cold autoantibodies
2. Classification based on etiology a. Idiopathic b. In association with i. Lymphoproliferative disorders ii. Other malignancies (e.g. ovarian carcinoma) iii. Autoimmune disorders (e.g. SLE, ulcerative colitis) iv. Infections (e.g. mycoplasma, EBV) v. Drugs (e.g. α-methyldopa)

AIHA, such as altered self-antigen, abnormalities in antigen presentation, B-cell hyperactivity, and abnormalities in suppressor T-cell number and/or function [254–258]. One or a combination of these immunologic aberrations, in association with environmental and genetic factors (drugs, infections, malignancies), lead to an autoimmune hemolytic disorder.

Autoimmune hemolytic anemia is divided into two major categories based on the nature of the antibodies (agglutinins): (1) warm-reacting antibodies and (2) cold-reacting antibodies (Table 23.12).

Warm-reacting antibodies (warm agglutinins) primarily represent IgG antibodies that react with red cell membrane proteins at body temperature. The IgG-coated red cells adhere to the Fc receptors of the tissue macrophages and are damaged either by erythrophagocytosis or by lysosomal enzymes released during antibody-dependent cellular cytotoxicity. Erythrophagocytosis is either complete or incomplete. Incomplete erythrophagocytosis leads to the formation of spherocytes. The red cell destruction occurs in tissues and therefore is called *extravascular hemolysis*. Approximately 2–3% of patients with severe AIHA demonstrate erythrocyte-bound IgA [259].

The efficacy of red cell–macrophage interactions depends on a number of factors such as the IgG subclass, the antibody concentration, and the concentration and subclass of Fc receptors on the macrophages [255, 257]. IgG1 and IgG3 are the most predominant Ig subtypes in macrophage-induced AIHA. Splenic macrophages are the most efficient effector cells and play a major role in the extravascular red cell destruction in autoimmune conditions [260].

Various conditions may initiate the autoantibody production and development of AIHA, such as viral infections (usually in children), collagen vascular diseases (especially systemic lupus erythematosus), lymphoid malignancies, prior allogeneic blood transfusion or stem cell transplantation, and certain drugs, particularly those used in the treatment of lymphoid malignancies (Table 23.12) [256, 261, 262].

Cold-reacting antibodies (cold agglutinins) in general consist of IgM antibodies that react with red cell membrane polysaccharides (such as I and i antigens) at temperatures below body temperature [263]. Cold agglutinins are not very effective in extravascular hemolysis, because macrophages and other cytotoxic cells of the immune system do not carry IgM receptors. The red cell destruction is primarily due to the complement fixation and happens within the blood vessels (*intravascular hemolysis*) (Figure 23.28). The IgM autoantibodies at low temperature activate the complement cascade and bind the erythrocyte membrane, causing small holes in the red cell membrane and consequently hemolysis [264, 265].

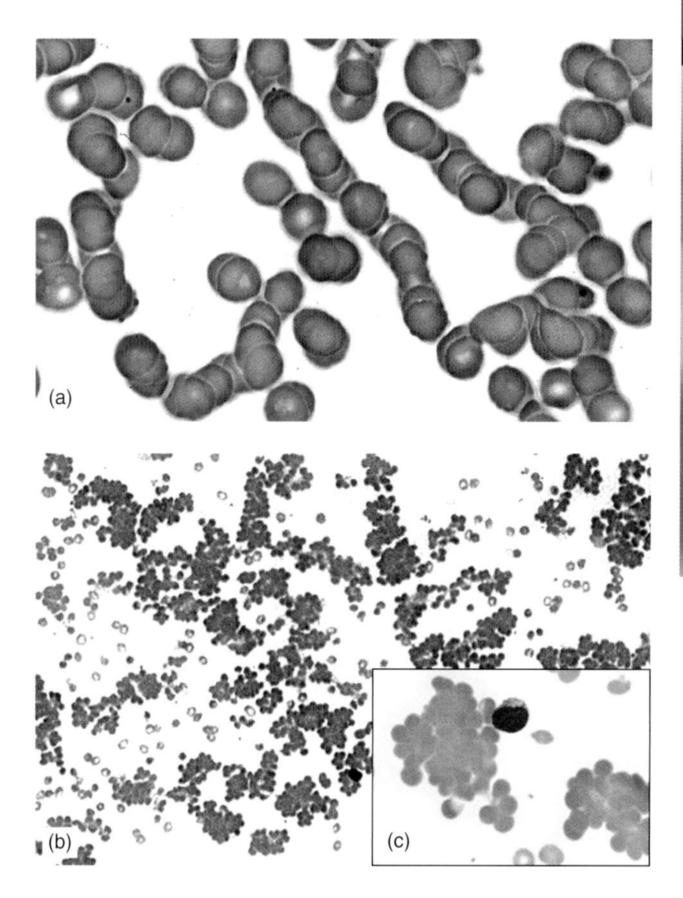

FIGURE 23.28 Blood smears demonstrate red cell rouleaux formation (a) and agglutinated red cells (b and c) for comparison.

Two major causes of cold agglutinin are infections and lymphoid neoplasms (Table 23.12). The most frequent infections associated with cold agglutinin production are *Mycoplasma pneumoniae* and EBV [266, 267]. Chronic lymphocytic leukemia, Waldenstrom macroglobulinemia, and high-grade B-cell lymphomas are among the neoplasms that are frequently associated with cold agglutinins [268, 269].

Clinical and Pathologic Features

Autoimmune hemolytic anemia with warm-reacting antibodies occurs at any age, but the majority of the patients are over age 40 with a peak incidence at around age 70. Clinical features are highly variable, ranging from a very mild chronic anemia with no clinical symptoms to a severe acute form with jaundice, splenomegaly, and hepatomegaly. Most cases have a mild clinical course.

Peripheral blood examination shows various degrees of anisopoikilocytosis, spherocytosis, reticulocytosis, and the presence of nucleated red cells. In severe cases, peripheral blood monocytes may occasionally show erythrophagocytosis. Granulocytosis and thrombocytosis may occur, and some severe cases may demonstrate a leukoerythroblastic blood picture. The bone marrow shows erythroid hyperplasia.

Demonstration of IgG and/or complement bound to the patient's erythrocytes is a diagnostic test for AIHA. The use of polyclonal antiglobulin reagent (Coombs' test), which contains antibodies against IgG and complement components, is a routine screening procedure. In the direct antiglobulin test (DAGT), or direct Coombs' test, the patient's red cells are examined for the presence of bound Ig or complement [256, 262]. In the indirect antiglobulin test, the patient's serum is screened for the presence of autoantibodies. The Coombs' test can be quantitated by methods such as enzyme-linked immunoabsorbent assay (ELISA) or other immunoassay techniques [262, 270]. Monospecific antisera are used for further characterization of the autoantibody and the nature of the hemolytic process.

Autoimmune hemolytic anemia with cold-reacting antibodies (cryopathic hemolytic syndromes) is caused by autoantibodies which have enhanced activities below 37°C and usually below 20°C. These syndromes account for about 30% of AIHA. Two major syndromes are recognized in this group: cold hemagglutinin disease (CHAD) and paroxysmal cold hemoglobinuria (PCH).

Cold hemagglutinin disease is characterized by a positive DAGT due to the presence of complement (C3d and C4d) and high-titer anti-erythrocyte antibodies with maximum erythrocyte agglutination at 0–15°C. In cold weather, cold-reacting antibodies bind to erythrocytes and mediate complement fixation with cooler peripheral circulation, leading to red cell aggregation and intravascular hemolysis. Intravascular hemolysis is associated with anemia, hemoglobinuria, and hemosiderinuria. Acrocyanosis and splenomegaly may be present. Occasionally, CHAD develops as complication of EBV or *Mycoplasma pneumonia*, lasting for 1–3 weeks [271]. In such cases, the cold-reacting antibodies are polyclonal IgM, and often demonstrate anti-I or anti-I activities. A chronic form of CHAD exists, which is associated with lymphoproliferative disorders and is characterized by the presence of monoclonal IgM kappa with anti-I specificity in the serum.

Paroxysmal cold hemoglobinuria is a rare form of AIHA which has been characterized by episodes of massive intravascular hemolysis and hemoglobinuria in children when exposed to cold. The disease appears to be secondary to a number of viral infections (such as measles, chickenpox, mumps, and influenza) or congenital and tertiary syphilis [263, 264, 272]. Clinical features include fever, muscle pains, headache, vomiting, diarrhea, urticaria, and acrocyanosis. The antibody responsible for the hemolysis is of the IgG class, usually with anti-P specificity, able to bind to complement components at low temperature.

Therapeutic approaches in AIHA include blood transfusion, administration of corticosteroids, splenectomy, and the use of cytotoxic drugs, singly or in combination based on the severity of the anemia and its duration. Plasma exchange or plasmapheresis has been used in some patients with warm-reacting antibodies.

Differential Diagnosis

Spherocytosis, reticulocytosis, and positive antiglobulin (Coombs') tests are characteristic laboratory features of AIHA. AIHA is distinguished from hereditary spherocytosis by lack of a family history and positive antiglobulin tests. Alloantibody hemolytic anemia, sometimes observed in recipients of organ transplants, may mimic AIHA. Patients with PNH, similar to patients with CHAD, demonstrate hemoglobinuria. However, in PNH patients, flow cytometric studies for CD55 and CD59 show reduced or lack of expression of these antigens in blood cells, as well as positive acidified serum test and sucrose hemolysis test, whereas all these studies are negative in CHAD.

OTHER IMMUNE-MEDIATED HEMOLYTIC ANEMIAS

Hemolytic Transfusion Reactions

Hemolytic transfusion reactions occur when the recipient's plasma contains antibody against the transfused red cells [273–276]. In rare occasions, it also happens when plasma with high titer of antibody is transfused to a patient whose erythrocytes carry the relevant antigen.

Transfusion reactions may take place during or immediately after transfusion or may be delayed 6–8 days after transfusion [273, 274, 276]. Immediate transfusion reactions are typically the result of ABO incompatibilities. Rapid intravascular hemolysis may lead to hemoglobinuria and jaundice, and may be complicated by disseminated intravascular coagulopathy (DIC) and renal failure [273]. Fever, chill, chest pain, dyspnea, hypotension, rigors, vomiting, and diarrhea are among the initial clinical features.

Delayed transfusion reactions occur in previously immunized patients. Rh incompatibility is a classical example of this kind of reaction [277, 278]. The red cell destruction in delayed reactions is predominantly extravascular with a positive DAGT. Common features are anemia, fever, jaundice, and spherocytosis.

Hemolytic Disease of the Newborn

Hemolytic disease of the newborn is the result of interaction of maternal IgG antibodies (crossed placenta) and incompatible fetal red cells in fetal circulation [279, 280]. ABO and Rh incompatibilities are the first and second most common causes, respectively. The ABO type is less severe and occurs predominantly in group A or B infants and group O mothers. Rh-related hemolysis is more severe and may lead to intrauterine death or hydrops fetalis [281].

Drug-Induced Immune Hemolytic Anemia

Certain drugs may initiate an immune-mediated red cell destruction [282–284]. Three possible mechanisms have been suggested for drug-induced immune hemolysis [257, 283].

1. Drug (or one of its metabolites) acts as a hapten and binds to the red cell membrane to generate antibodies. The antibody-coated red cells are then subject to extravascular clearance by the phagocytic system (Figure 23.29a). Penicillin and other cephalosporins are examples.
2. A drug–antibody complex formed in plasma leads to complement activation at the erythrocyte surface and, consequently, red cell hemolysis (Figure 23.29b). Examples are quinine, phenacetin, *para*-aminosalicylic acid, and sulfonamides. This mechanism has been challenged [285].
3. Certain drugs, such as α-methyldopa, initiate the formation of autoantibodies against red cells (Figure 23.29c). Approximately 10–25% of patients on long-term α-methyldopa therapy develop IgG anti-red cell antibodies, though <1% may develop overt hemolytic anemia. Possible mechanisms include inhibition of suppressor T-cell activation by methyldopa or interaction of methyldopa metabolites with red cell peptides to form antigens [286, 287].

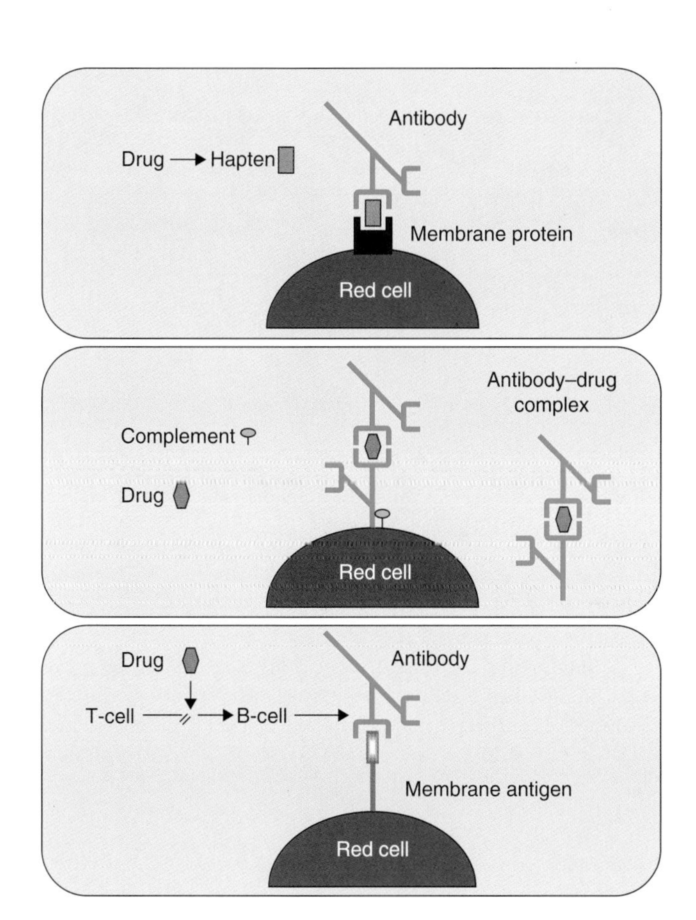

FIGURE 23.29 Schematic demonstration of possible mechanisms of drug-induced hemolytic anemia. Adapted from Naeim F. (1998). *Pathology of Bone Marrow*, 2nd ed. Williams & Wilkins, Baltimore.

ACQUIRED NON-IMMUNE HEMOLYTIC ANEMIAS

A wide variety of conditions may lead to an acquired non-immune hemolytic anemia. Examples include mechanical trauma and heat, drugs and other chemicals, infections, and hypersplenism.

Hemolysis Induced by Mechanical Trauma and Heat

Traumatic injury to the erythrocytes may cause red cell fragmentation and hemolysis. Erythrocyte fragmentation has been observed in patients with cardiac valve prosthesis, thrombotic thrombocytopenic purpura, malignant hypertension, generalized vasculitis, and carcinomatosis (Figures 23.30 and 23.31) [288–290]. Traumatic hemolysis and hemoglobinuria have also been reported soon after walking or running long distances, bongo-drumming, or karate exercise. Thermal damage is often associated with spherocytosis and generation of microvesicles [289].

Hemolysis Caused by Drugs and Other Chemicals

Drug-induced hemolysis unrelated to enzyme deficiencies is reported in various conditions such as exposure to arsenic hydride or nitrobenzene derivatives, copper, certain nitrites, and naphthalene [284, 291, 292].

Acute alcoholism may be associated with stomatocytosis and a transient hemolytic episode. Patients with alcohol-induced fatty liver may develop a syndrome (Zeive syndrome) characterized by episodes of hypercholesterolemia, hypertriglyceridemia, and hemolysis [293]. Venoms of certain species of snakes, bees, spiders, and wasps may cause hemolysis.

Hemolysis Caused by Infections

As discussed earlier, infection may initiate or provoke hemolytic episodes in patients suffering from red cell enzyme deficiencies or hemoglobinopathies, presumably due to the generation of oxidant substances. Certain microorganisms may damage erythrocyte membrane and cause hemolysis by the release of substances such as phospholipases (*Clostridium perfringens*) and hemolysins (*Streptococcus pyogenes*). Activation of macrophages in some viral or bacterial infections may lead to extensive erythrophagocytosis and anemia [294]. Malaria, particularly *Plasmodium falciparum*, may cause hemolysis.

Hypersplenism

One of the main functional roles of the spleen is to serve as a filter by removing foreign materials, cell debris, and defective blood cells [295]. This is accomplished by the passage

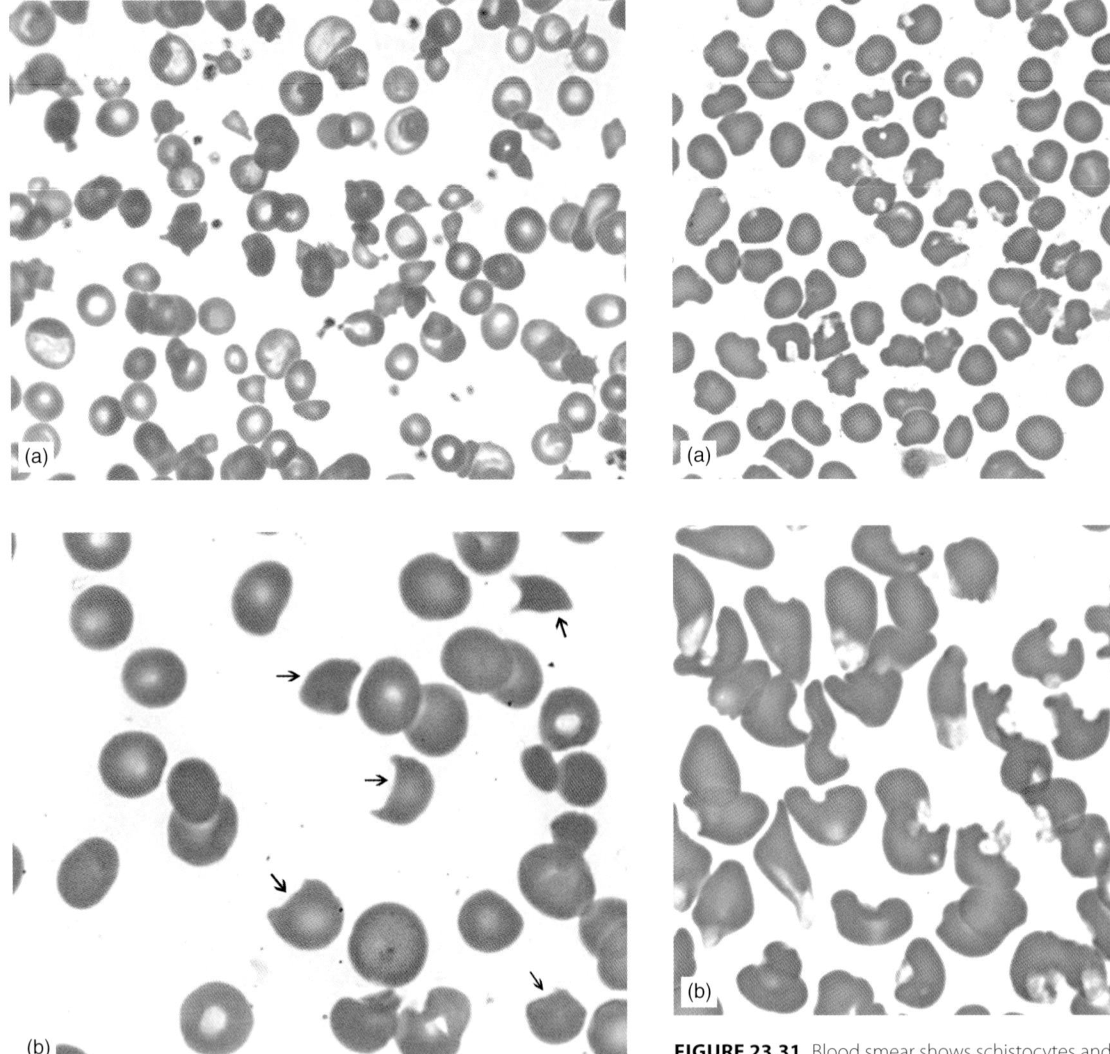

FIGURE 23.30 Microangiopathic changes. Blood smears demonstrate numerous schistocytes (a) and several helmet cells (b, arrows).

FIGURE 23.31 Blood smear shows schistocytes and punched out red cells in a patient with severe burn: (a) low power and (b) high power.

of a proportion of splenic blood supply through the non-endothelialized, macrophage-containing spaces present in the red pulp, functioning as a filter bed. In splenomegaly, the proportion of the blood channeled through the red pulp and the filter bed increases, causing an inappropriate sequestration of both normal and abnormal blood cells, particularly red cells, because of their limited, self-sufficient metabolic resources [296, 297]. The increase in splenic sequestration is more pronounced when the splenomegaly is caused by congestion than by an infiltrative process.

OTHER TYPES OF ANEMIA

Anemia of Chronic Disease

Anemia of chronic disease (ACD) is an anemia associated with chronic infections, chronic inflammations, and malignancies, as well as a number of other conditions such as severe trauma, cardiovascular disorders, and diabetes mellitus [298–301].

Hepcidin appears to play an important role in iron metabolism and pathogenesis of ACD [302]. Hepcidin is an acute phase protein, a regulator of iron absorption in the small intestine and iron release from macrophages [303]. Increased hepcidin production has been observed in patients with ACD [304]. Recent studies suggest that IL-6 induces production of hepcidin [303].

ACD is characterized by a mild to moderate anemia, reduced absolute reticulocyte count, modest shortening of the erythrocyte life span, a low serum iron-binding capacity, an increased bone marrow iron store, and relatively ineffective erythropoiesis. Anemia is usually normocytic and normochromic, but it may be microcytic and hypochromic. Elevated levels of serum IL-6, fibrinogen, and C-reactive protein, as well as increased erythrocyte sedimentation rates, are frequently noted [305, 306].

An acute variant of ACD has been reported in patients with major trauma, myocardial infarction, and sepsis [300, 307].

The differential diagnosis includes microcytic and refractory anemias (see Table 23.9). ACD, unlike IDA, shows reduced serum TIBC and increased bone marrow iron stores. ACD is distinguished from refractory and sideroblastic anemias by reduced serum TIBC, lack of significant dysplastic changes, and absent ringed sideroblasts.

Deficiency Anemias Other than Vitamin B_{12} and Folate

Copper Deficiency

Anemia of copper deficiency has been described in malnourished children and in patients receiving parenteral alimentation [308, 309]. Excessive and chronic administration of zinc may also cause copper deficiency. Copper deficiency is associated with microcytic anemia and neutropenia. The erythroid precursors in the bone marrow are often vacuolated.

Vitamin Deficiencies

Vitamin A and Vitamin B_6 deficiencies may lead to microcytic, hypochromic anemia. Vitamin B_2 (riboflavin) may cause pure red cell aplasia [310]. Anemia associated with vitamin C deficiency is usually normochromic and normocytic. Vitamin E deficiency, a common deficiency in patients with cystic fibrosis, is usually associated with abnormal red cell morphology and hemolysis [311].

Anemia of Chronic Renal Failure

A complex process is involved in the anemia of chronic renal failure consisting of a decline in erythropoietin production, the suppressive effects of uremia on erythropoiesis, and plasma inhibitors of heme synthesis [312, 313]. Anemia is usually normochromic and normocytic with a normal or reduced reticulocyte counts. Echinocytosis is often present. The bone marrow is normo- to hypocellular and may show severe erythroid hypoplasia.

Anemia Associated with Marrow Infiltration (Myelophthisic Anemia)

Metastatic carcinomas, particularly carcinoma of the lung, breast, and prostate, are the most common causes of bone marrow infiltration [314, 315]. Other causes of bone marrow infiltration are hematopoietic malignancies, marrow fibrosis, lysosomal storage diseases, and inflammatory processes such as granulomas. The infiltrative process disrupts the bone marrow's microenvironment and reduces hematopoietic activities. Myelophthisic anemia is often associated with anisocytosis and poikilocytosis, presence of teardrop red cells and leukoerythroblastosis [316]. Bone marrow biopsy sections are the most reliable source for the diagnosis of marrow infiltration.

References

1. World Health Organization. (1994). Indicators and Strategies for Iron Deficiency and Anemia Programmes. Report of the WHO/UNICEF/UNU Consultation. Geneva, Switzerland, December 6–10, 1993.
2. Gibson R. (1993). *Nutritional Assessment: A Laboratory Manual.* Oxford University Press, New York.
3. Schrier S. (2007). Approach to the adult patient with anemia. *UpToDate.*
4. Stewart RD, Baretta ED, Platte LR, Stewart EB, Kalbfleisch JH, Van Yserloo B, Rimm AA. (1974). Carboxyhemoglobin levels in American blood donors. *JAMA* **229**, 1187–95.
5. Ruíz-Argüelles GJ. (2006). Altitude above sea level as a variable for definition of anemia. *Blood* **108**, 2131.
6. Garn SM, Keating MT, Falkner F. (1981). Hematological status and pregnancy outcomes. *Am J Clin Nutr* **34**, 115–7.
7. Reed WW, Diehl LF. (1991). Leukopenia, neutropenia, and reduced hemoglobin levels in healthy American blacks. *Arch Intern Med* **151**, 501.
8. Dessypris EN. (1991). The biology of pure red cell aplasia. *Semin Hematol* **28**, 275–84.
9. Halperin DS, Freedman MH. (1989). Diamond–Blackfan anemia: Etiology, pathophysiology, and treatment. *Am J Pediatr Hematol Oncol* **11**, 380–94.
10. Orfali KA, Ohene-Abuakwa Y, Ball SE. (2004). Diamond–Blackfan anaemia in the UK: Clinical and genetic heterogeneity. *Br J Haematol* **125**, 243–52.
11. Gordon-Smith EC. (2005). Congenital bone marrow failure involving the red blood cells. *Hematology* **10**(Suppl 1), 312–15.
12. Gazda HT, Sieff CA. (2006). Recent insights into the pathogenesis of Diamond–Blackfan anaemia. *Br J Haematol* **135**, 149–57.
13. Flygare J, Karlsson S. (2007). Diamond–Blackfan anemia: Erythropoiesis lost in translation. *Blood* **109**, 3152–4.
14. Choesmel V, Bacqueville D, Rouquette J, Noaillac-Depeyre J, Fribourg S, Crétien A, Leblanc T, Tchernia G, Da Costa L, Gleizes PE. (2007). Impaired ribosome biogenesis in Diamond–Blackfan anemia. *Blood* **109**, 1275–83.
15. Lim LC. (2005). Acquired red cell aplasia in association with the use of recombinant erythropoietin in chronic renal failure. *Hematology* **10**, 255–9.
16. Fisch P, Handgretinger R, Schaefer HE. (2000). Pure red cell aplasia. *Br J Haematol* **111**, 1010–22.
17. Djaldetti M, Blay A, Bergman M, Salman H, Bessler H. (2003). Pure red cell aplasia – a rare disease with multiple causes. *Biomed Pharmacother* **57**, 326–32.
18. Perkins SL. (2004). Pediatric red cell disorders and pure red cell aplasia. *Am J Clin Pathol* **122**(Suppl), S70–86.
19. Brown KE, Young NS. (1995). Parvovirus B19 infection and hematopoiesis. *Blood Rev* **9**, 176–82.
20. Florea AV, Ionescu DN, Melhem MF. (2007). Parvovirus B19 infection in the immunocompromised host. *Arch Pathol Lab Med* **131**, 799–804.
21. Frickhofen N, Chen ZJ, Young NS, Cohen BJ, Heimpel H, Abkowitz JL. (1994). Parvovirus B19 as a cause of acquired chronic pure red cell aplasia. *Br J Haematol* **87**, 818–24.
22. Krantz SB, Kao V. (1969). Studies on red cell aplasia. II. Report of a second patient with an antibody to erythroblast nuclei and a remission after immunosuppressive therapy. *Blood* **34**, 1–13.
23. Zaentz SD, Krantz SB. (1971). Studies on pure red cell aplasia. VI. Development of two-stage erythroblast cytotoxicity method and role of complement. *J Lab Clin Med* **82**, 31–43.
24. Schaefer HE. (1992). Aplastic crisis in haemolytic anaemia due to infection parvovirus B19. *Pathol Res Pract* **188**, 817–23.
25. Glader BE, Backer K, Diamond LK. (1983). Elevated erythrocyte adenosine deaminase activity in congenital hypoplastic anemia. *N Engl J Med* **309**, 1486–90.
26. Bomgaars L. (2007). Approach to the child with anemia. *UpToDate.*
27. Diamond LK. (1978). Congenital hypoplastic anemia: Diamond–Blackfan syndrome. Historical and clinical aspects. *Blood Cells* **4**, 209–13.
28. Janov AJ, Leong T, Nathan DG, Guinan EC. (1996). Diamond–Blackfan anemia. Natural history and sequelae of treatment. *Medicine (Baltimore)* **75**, 77–8.

29. Nawel G, Fethi M, Kouki R, Mohamed B. (2007). Successful treatment of Diamond Blackfan anemia with metoclopramide. *J Pediatr Hematol Oncol* **29**, 728.
30. Mugishima H, Ohga S, Ohara A, Kojima S, Fujisawa K, Tsukimoto I. For the Aplastic Anemia Committee of the Japanese Society of Pediatric Hematology (2007). Hematopoietic stem cell transplantation for Diamond–Blackfan anemia: A report from the Aplastic Anemia Committee of the Japanese Society of Pediatric Hematology. *Pediatr Transplant* **11**, 601–7.
31. Miller R, Berman B. (1994). Transient erythroblastopenia of childhood in infants <6 months of age. *Am J Pediatr Hematol Oncol* **16**, 246–8.
32. Majluf-Cruz A, Luna-Castaños G, Nieto-Cisneros L. (1996). AIDS-related pure red cell aplasia. *Am J Hematol* **51**, 171.
33. Stricker RB, Goldberg B. (1997). AIDS and pure red cell aplasia. *Am J Hematol* **54**, 264.
34. García Vela JA, Pérez V, Monteserin MC, Oña F. (1998). Pure red cell aplasia associated with large granular lymphocytic leukemia: A rare association in Western countries. *Haematologica* **83**, 664–5.
35. Wickramasinghe SN. (1998). Congenital dyserythropoietic anaemias: Clinical features, haematological morphology and new biochemical data. *Blood Rev* **12**, 178–200.
36. Wickramasinghe SN. (2000). Congenital dyserythropoietic anemias. *Curr Opin Hematol* **7**, 71–8.
37. Wickramasinghe SN, Wood WG. (2005). Advances in the understanding of the congenital dyserythropoietic anaemias. *Br J Haematol* **131**, 431–46.
38. Tchernia G, Bader-Meunier B, Beauchamp-Nicoud A, Cynober T, Feneant-Thibault M, Delaunay J. (2004). Congenital dyserythropoietic anemias. *Hematol J* **5**(Suppl 3), S191–4.
39. Shalev H, Moser A, Kapelushnik J, Karplus M, Zucker N, Yaniv I, Tamary H. (2000). Congenital dyserythropoietic anemia type I presenting as persistent pulmonary hypertension of the newborn. *J Pediatr* **136**, 553–5.
40. Shalev H, Kapleushnik Y, Haeskelzon L, Degani O, Kransnov T, Sphilberg O, Moser A, Yaniv I, Tamary H. (2002). Clinical and laboratory manifestations of congenital dyserythropoietic anemia type I in young adults. *Eur J Haematol* **68**, 170–4.
41. Tamary H, Shalev H, Luria D, Shaft D, Zoldan M, Shalmon L, Gruinspan A, Stark B, Chaison M, Shinar E, Resnitzky P, Zaizov R. (1996). Clinical features and studies of erythropoiesis in Israeli Bedouins with congenital dyserythropoietic anemia type I. *Blood* **87**, 1763–70.
42. Kuribayashi T, Uchida S, Kuroume T, Umegae S, Omine M, Maekawa T. (1979). Congenital dyserythropoietic anemia type I: Report of a pair of siblings in Japan. *Blut* **39**, 201–9.
43. Heimpel H. (2004). Congenital dyserythropoietic anemias: Epidemiology, clinical significance, and progress in understanding their pathogenesis. *Ann Hematol* **83**, 613–21.
44. Goede JS, Benz R, Fehr J, Schwarz K, Heimpel H. (2006). Congenital dyserythropoietic anemia type I with bone abnormalities, mutations of the CDAN I gene, and significant responsiveness to alpha-interferon therapy. *Ann Hematol* **85**, 591–5.
45. Heimpel H, Schwarz K, Ebnöther M, Goede JS, Heydrich D, Kamp T, Plaumann L, Rath B, Roessler J, Schildknecht O, Schmid M, Wuillemin W, Einsiedler B, Leichtle R, Tamary H, Kohne E. (2006). Congenital dyserythropoietic anemia type I (CDA I): Molecular genetics, clinical appearance, and prognosis based on long-term observation. *Blood* **107**, 334–40.
46. Tamary H, Dgany O, Proust A, Krasnov T, Avidan N, Eidelitz-Markus T, Tchernia G, Geneviève D, Cormier-Daire V, Bader-Meunier B, Ferrero-Vacher C, Munzer M, Gruppo R, Fibach E, Konen O, Yaniv I, Delaunay J. (2005). Clinical and molecular variability in congenital dyserythropoietic anaemia type I. *Br J Haematol* **130**, 628–34.
47. Heimpel H, Forteza-Vila J, Queisser W, Spiertz E. (1971). Electron and light microscopic study of the erythroblasts of patients with congenital dyserythropoietic anemia. *Blood* **37**, 299–310.
48. Lewis SM, Nelson DA, Pitcher CS. (1972). Clinical and ultrastructural aspects of congenital dyserythropoietic anaemia type I. *Br J Haematol* **23**, 113–19.
49. Heimpel H, Anselstetter V, Chrobak L, Denecke J, Einsiedler B, Gallmeier K, Griesshammer A, Marquardt T, Janka-Schaub G, Kron M, Kohne E. (2003). Congenital dyserythropoietic anemia type II: Epidemiology, clinical appearance, and prognosis based on long-term observation. *Blood* **102**, 4576–81.
50. Heimpel H, Wilts H, Hirschmann WD, Hofmann WK, Siciliano RD, Steinke B, Wechsler JG. (2007). Aplastic crisis as a complication of congenital dyserythropoietic anemia type II. *Acta Haematol* **117**, 115–18.
51. Iolascon A, Servedio V, Carbone R, Totaro A, Carella M, Perrotta S, Wickramasinghe SN, Delaunay J, Heimpel H, Gasparini P. (2000). Geographic distribution of CDA-II: Did a founder effect operate in Southern Italy? *Haematologica* **85**, 470–4.
52. Wong KY, Hug G, Lampkin BC. (1972). Congenital dyserythropoietic anemia type II: Ultrastructural and radioautographic studies of blood and bone marrow. *Blood* **39**, 23–30.
53. Fukuda MN, Papayannopoulou T, Gordon-Smith EC, Rochant H, Testa U. (1984). Defect in glycosylation of erythrocyte membrane proteins in congenital dyserythropoietic anaemia type II (HEMPAS). *Br J Haematol* **56**, 55–68.
54. Zdebska E, Anselstetter V, Pacuszka T, Krauze R, Chełstowska A, Heimpel H, Kościelak J. (1987). Glycolipids and glycopeptides of red cell membranes in congenital dyserythropoietic anaemia type II (CDA II). *Br J Haematol* **66**, 385–91.
55. Sandstrom H, Wahlin A. (2000). Congenital dyserythropoietic anemia type III. *Haematologica* **85**, 753–7.
56. Wickramasinghe SN, Wahlin A, Anstee D, Parsons SF, Stopps G, Bergstrom I, Eriksson M, Sandstrom H, Shiels S. (1993). Observations on two members of the Swedish family with congenital dyserythropoietic anaemia, type III. *Eur J Haematol* **50**, 213–21.
57. Goudsmit R, Beckers D, De Bruijne JI, Engelfriet CP, James J, Morselt AF, Reynierse E. (1972). Congenital dyserythropoietic anaemia, type 3. *Br J Haematol* **23**, 97–105.
58. Bird AR, Knottenbelt E, Jacobs P, Maigrot J. (1991). Primary shunt hyperbilirubinaemia: A variant of the congenital dyserythropoietic anaemias. *Postgrad Med J* **67**, 396–8.
59. Carpenter SL, Zimmerman SA, Ware RE. (2004). Acute parvovirus B19 infection mimicking congenital dyserythropoietic anemia. *J Pediatr Hematol Oncol* **26**, 133–5.
60. Herbert V. (1985). Megaloblastic anemias. *Lab Invest* **52**, 3–19.
61. Chanarin I. (1987). Megaloblastic anaemia, cobalamin, and folate. *J Clin Pathol* **40**, 978–84.
62. Aslinia F, Mazza JJ, Yale SH. (2006). Megaloblastic anemia and other causes of macrocytosis. *Clin Med Res* **4**, 236–41.
63. Whitehead VM. (2006). Acquired and inherited disorders of cobalamin and folate in children. *Br J Haematol* **134**, 125–36.
64. Grasbeck R. (2005). Megaloblastic anaemia (MA). *Hematology* **10**(Suppl 1), 227–8.
65. Donnelly JG. (2001). Folic acid. *Crit Rev Clin Lab Sci* **38**, 183–223.
66. James SJ, Basnakian AG, Miller BJ. (1994). *In vitro* folate deficiency induces deoxynucleotide pool imbalance, apoptosis, and mutagenesis in Chinese hamster ovary cells. *Cancer Res* **54**, 5075–80.
67. Wickremasinghe RG, Hoffbrand AV. (1980). Conversion of partially single-stranded replicating DNA to double-stranded DNA is delayed in megaloblastic anaemia. *Biochim Biophys Acta* **607**, 411–19.
68. Killmann SA. (1964). Effect of deoxyuridine on incorporation of tritiated thymidine: Difference between normoblasts and megaloblasts. *Acta Med Scand* **175**, 483–8.
69. Ramsahoye BH, Burnett AK, Taylor C. (1996). Nucleic acid composition of bone marrow mononuclear cells in cobalamin deficiency. *Blood* **87**, 2065–70.
70. Ramsahoye BH, Davies CS, Mills KI. (1996). DNA methylation: Biology and significance. *Blood Rev* **10**, 249–61.
71. Wickramasinghe SN. (2006). Diagnosis of megaloblastic anaemias. *Blood Rev* **20**, 299–318.

72. Herzlich B, Herbert V. (1984). The role of the pancreas in cobalamin (vitamin B12) absorption. *Am J Gastroenterol* **79**, 489–93.
73. Toh BH, Alderuccio F. (2004). Pernicious anaemia. *Autoimmunity* **37**, 357–61.
74. Hewitt JE, Gordon MM, Taggart RT, Mohandas TK, Alpers DH. (1991). Human gastric intrinsic factor: Characterization of cDNA and genomic clones and localization to human chromosome 11. *Genomics* **10**, 432–40.
75. Yassin F, Rothenberg SP, Rao S, Gordon MM, Alpers DH, Quadros EV. (2004). Identification of a 4-base deletion in the gene in inherited intrinsic factor deficiency. *Blood* **103**, 1515–17.
76. Gordon MM, Brada N, Remacha A, Badell I, del Rio E, Baiget M, Santer R, Quadros EV, Rothenberg SP, Alpers DH. (2004). A genetic polymorphism in the coding region of the gastric intrinsic factor gene (GIF) is associated with congenital intrinsic factor deficiency. *Hum Mutat* **23**, 85–91.
77. De Aizpurua HJ, Cosgrove LJ, Ungar B, Toh BH. (1983). Autoantibodies cytotoxic to gastric parietal cells in serum of patients with pernicious anemia. *N Engl J Med* **309**, 625–9.
78. Fox RM, O'Sullivan WJ, Firkin BG. (1969). Orotic aciduria. Differing enzyme patterns. *Am J Med* **47**, 332–6.
79. Fox RM, Wood MH, Royse-Smith D, O'Sullivan WJ. (1973). Hereditary orotic aciduria: Types I and II. *Am J Med* **55**, 791–8.
80. van der Zee SP, Schretlen ED, Monnens LA. (1968). Megaloblastic anaemia in the Lesch–Nyhan syndrome. *Lancet* **1**, 1427.
81. Krouse JR. (1988). The bone marrow in nutritional deficiencies. *Hematol Oncol Clin North Am* **2**, 557–66.
82. Harkins LS, Sirel JM, McKay PJ, Wylie RC, Titterington DM, Rowan RM. (1994). Discriminant analysis of macrocytic red cells. *Clin Lab Haematol* **16**, 225–34.
83. Schrier SL. (2007). Macrocytosis. *UpToDate*.
84. Barney-Stallings RA, Heslop SD. (2001). What is the clinical utility of obtaining a folate level in patients with macrocytosis or anemia? *J Fam Pract* **50**, 544.
85. Snow CF. (1999). Laboratory diagnosis of vitamin B12 and folate deficiency: A guide for the primary care physician. *Arch Intern Med* **159**, 1289–98.
86. Brugge WR, Goff JS, Allen NC, Podell ER, Allen RH. (1980). Development of a dual label Schilling test for pancreatic exocrine function based on the differential absorption of cobalamin bound to intrinsic factor and R protein. *Gastroenterology* **78**, 937–49.
87. Sumner AE, Chin MM, Abrahm JL, Berry GT, Gracely EJ, Allen RH, Stabler SP. (1996). Elevated methylmalonic acid and total homocysteine levels show high prevalence of vitamin B12 deficiency after gastric surgery. *Ann Intern Med* **124**, 469–76.
88. Allen RH, Stabler SP, Savage DG, Lindenbaum J. (1990). Diagnosis of cobalamin deficiency: I. Usefulness of serum methylmalonic acid and total homocysteine concentrations. *Am J Hematol* **34**, 90–8.
89. Lindenbaum J, Savage DG, Stabler SP, Allen RH. (1990). Diagnosis of cobalamin deficiency: II. Relative sensitivities of serum cobalamin, methylmalonic acid, and total homocysteine concentrations. *Am J Hematol* **34**, 99–107.
90. Chanarin I, Deacon R, Lumb M, Perry J. (1992). Cobalamin and folate: Recent developments. *J Clin Pathol* **45**, 277–83.
91. Stabler SP, Allen RH, Savage DG, Lindenbaum J. (1990). Clinical spectrum and diagnosis of cobalamin deficiency. *Blood* **76**, 871–81.
92. Beck WS. (1988). Cobalamin and the nervous system. *N Engl J Med* **318**, 1752–4.
93. Kondo H, Osborne ML, Kolhouse JF, Binder MJ, Podell ER, Utley CS, Abrams RS, Allen RH. (1981). Nitrous oxide has multiple deleterious effects on cobalamin metabolism and causes decreases in activities of both mammalian cobalamin-dependent enzymes in rats. *J Clin Invest* **67**, 1270–83.
94. Killip S, Bennett JM, Chambers MD. (2007). Iron deficiency anemia. *Am Fam Physician* **75**, 671–8.
95. Krantz SB. (1994). Pathogenesis and treatment of the anemia of chronic disease. *Am J Med Sci* **307**, 353–9.
96. Massey AC. (1992). Microcytic anemia. Differential diagnosis and management of iron deficiency anemia. *Med Clin North Am* **76**, 549–66.
97. Means RT. (2004). Hepcidin and cytokines in anaemia. *Hematology* **9**, 357–62.
98. Looker AC, Dallman PR, Carroll MD, Gunter EW, Johnson CL. (1997). Prevalence of iron deficiency in the United States. *JAMA* **277**, 973–6.
99. Irwin JJ, Kirchner JT. (2001). Anemia in children. *Am Fam Physician* **64**, 1379–86.
100. Richardson M. (2007). Microcytic anemia. *Pediatr Rev* **28**, 5–14.
101. Tefferi A. (2003). Anemia in adults: A contemporary approach to diagnosis. *Mayo Clin Proc* **78**, 1274–80.
102. Yip R, Limburg PJ, Ahlquist DA, Carpenter HA, O'Neill A, Kruse D, Stitham S, Gold BD, Gunter EW, Looker AC, Parkinson AJ, Nobmann ED, Petersen KM, Ellefson M, Schwartz S. (1997). Pervasive occult gastrointestinal bleeding in an Alaska native population with prevalent iron deficiency. Role of *Helicobacter pylori* gastritis. *JAMA* **277**, 1135–9.
103. Schrier SL. (2007). Causes and diagnosis of anemia due to iron deficiency. *UpToDate*.
104. Niles JL, Böttinger EP, Saurina GR, Kelly KJ, Pan G, Collins AB, McCluskey RT. (1996). The syndrome of lung hemorrhage and nephritis is usually an ANCA-associated condition. *Arch Intern Med* **156**, 440–5.
105. Ioachimescu OC, Sieber S, Kotch A. (2004). Idiopathic pulmonary haemosiderosis revisited. *Eur Respir J* **24**, 162–70.
106. Fillet G, Beguin Y, Baldelli L. (1989). Model of reticuloendothelial iron metabolism in humans: Abnormal behavior in idiopathic hemochromatosis and in inflammation. *Blood* **74**, 844–51.
107. Monsen ER. (1988). Iron nutrition and absorption: Dietary factors which impact iron bioavailability. *J Am Diet Assoc* **88**, 786–90.
108. Cook JD, Skikne BS. (1989). Iron deficiency: Definition and diagnosis. *J Intern Med* **226**, 349–55.
109. Farley PC, Foland J. (1990). Iron deficiency anemia. How to diagnose and correct. *Postgrad Med* **87**, 89–93, 96, 101.
110. Lipschitz DA, Cook JD, Finch CA. (1974). A clinical evaluation of serum ferritin as an index of iron stores. *N Engl J Med* **290**, 1213–16.
111. Kettaneh A, Eclache V, Fain O, Sontag C, Uzan M, Carbillon L, Stirnemann J, Thomas M. (2005). Pica and food craving in patients with iron-deficiency anemia: A case–control study in France. *Am J Med* **118**, 185–8.
112. Reynolds RD, Binder HJ, Miller MB, Chang WW, Horan S. (1968). Pagophagia and iron deficiency anemia. *Ann Intern Med* **69**, 435–40.
113. Dallman PR. (1987). Iron deficiency and the immune response. *Am J Clin Nutr* **46**, 329–34.
114. Goepel E, Ulmer HU, Neth RD. (1988). Premature labor contractions and the value of serum ferritin during pregnancy. *Gynecol Obstet Invest* **26**, 265–73.
115. Lieberman E, Ryan KJ, Monson RR, Schoenbaum SC. (1988). Association of maternal hematocrit with premature labor. *Am J Obstet Gynecol* **159**, 107–14.
116. Kutlar F. (2007). Diagnostic approach to hemoglobinopathies. *Hemoglobin* **31**, 243–50.
117. Birgens H, Ljung R. (2007). The thalassaemia syndromes. *Scand J Clin Lab Invest* **67**, 11–25.
118. Weatherall DJ, Clegg JB. (1996). Thalassemia – a global public health problem. *Nat Med* **2**, 847–9.
119. Rund D, Rachmilewitz E. (1995). Advances in the pathophysiology and treatment of thalassemia. *Crit Rev Oncol Hematol* **20**, 237–54.
120. Schwartz E, Cohen A, Surrey S. (1988). Overview of the beta thalassemias: Genetic and clinical aspects. *Hemoglobin* **12**, 551–64.
121. Liebhaber SA. (1989). Alpha thalassemia. *Hemoglobin* **13**, 685–731.
122. Jones RW, Old JM, Trent RJ, Clegg JB, Weatherall DJ. (1981). Major rearrangement in the human beta-globin gene cluster. *Nature* **291**, 39–44.

123. Rund D, Rachmilewitz E. (2005). Beta-thalassemia. *N Engl J Med* **353**, 1135–46.
124. Olivieri NF. (1999). The beta-thalassemias. *N Engl J Med* **341**, 99–109.
125. Efremov GD. (2007). Dominantly inherited beta-thalassemia. *Hemoglobin* **31**, 193–207.
126. Benz. (2007). Molecular pathology of the thalassemic syndromes. *UpToDate.*
127. Pootrakul P, Sirankapracha P, Hemsorach S, Moungsub W, Kumbunlue R, Piangitjagum A, Wasi P, Ma L, Schrier SL. (2000). A correlation of erythrokinetics, ineffective erythropoiesis, and erythroid precursor apoptosis in thai patients with thalassemia. *Blood* **96**, 2606–12.
128. Stassi G, Todaro M, Bucchieri F, Stoppacciaro A, Farina F, Zummo G, Testi R, De Maria R. (1999). Apoptotic role of fas/fas ligand system in the regulation of erythropoiesis. *Blood* **93**, 796–803.
129. Schrier SL. (2007). Pathophysiology of beta thalassemia. *UpToDate.*
130. Ou CN, Rognerud CL. (2001). Diagnosis of hemoglobinopathies: Electrophoresis vs. HPLC. *Clin Chim Acta* **313**, 187–94.
131. Clarke GM, Higgins TN. (2000). Laboratory investigation of hemoglobinopathies and thalassemias: Review and update. *Clin Chem* **46**, 1284–90.
132. Angastiniotis M, Modell B. (1998). Global epidemiology of hemoglobin disorders. *Ann NY Acad Sci* **850**, 251–69.
133. Sheiner E, Levy A, Yerushalmi R, Katz M. (2004). Beta-thalassemia minor during pregnancy. *Obstet Gynecol* **103**, 1273–7.
134. Telfer P, Constantinidou G, Andreou P, Christou S, Modell B, Angastiniotis M. (2005). Quality of life in thalassemia. *Ann NY Acad Sci* **1054**, 273–82.
135. Taher A, Isma'eel H, Cappellini MD. (2006). Thalassemia intermedia: Revisited. *Blood Cells Mol Dis* **37**, 12–20.
136. Pootrakul P, Sirankapracha P, Sankote J, Kachintorn U, Maungsub W, Sriphen K, Thakernpol K, Atisuk K, Fucharoen S, Chantraluksri U, Shalev O, Hoffbrand AV. (2003). Clinical trial of deferiprone iron chelation therapy in beta-thalassaemia/haemoglobin E patients in Thailand. *Br J Haematol* **122**, 305–10.
137. Benz EJ. (2007). Treatment of beta thalassemia. *UpToDate.*
138. Quek L, Thein SL. (2007). Molecular therapies in beta-thalassaemia. *Br J Haematol* **136**, 353–65.
139. Sadelain M. (2006). Recent advances in globin gene transfer for the treatment of beta-thalassemia and sickle cell anemia. *Curr Opin Hematol* **13**, 142–8.
140. Borgna-Pignatti C, Cappellini MD, De Stefano P, Del Vecchio GC, Forni GL, Gamberini MR, Ghilardi R, Origa R, Piga A, Romeo MA, Zhao H, Cnaan A. (2005). Survival and complications in thalassemia. *Ann NY Acad Sci* **1054**, 40–7.
141. Higgs DR, Weatherall DJ. (1983). Alpha-thalassemia. *Curr Top Hematol* **4**, 37–97.
142. Forget BG. (2006). *De novo* and acquired forms of alpha thalassemia. *Curr Hematol Rep* **5**, 11–14.
143. Yenchitsomanus PT, Summers KM, Bhatia KK, Cattani J, Board PG. (1985). Extremely high frequencies of alpha-globin gene deletion in Madang and on Kar Kar Island, Papua New Guinea. *Am J Hum Genet* **37**, 778–84.
144. Clegg JB, Weatherall DJ. (1999). Thalassemia and malaria: New insights into an old problem. *Proc Assoc Am Physicians* **111**, 278–82.
145. Flint J, Hill AV, Bowden DK, Oppenheimer SJ, Sill PR, Serjeantson SW, Bana-Koiri J, Bhatia K, Alpers MP, Boyce AJ. (1986). High frequencies of alpha-thalassaemia are the result of natural selection by malaria. *Nature* **321**, 744–50.
146. Kazazian Jr. HH. (1990). The thalassemia syndromes: Molecular basis and prenatal diagnosis in 1990. *Semin Hematol* **27**, 209–28.
147. Adams III JG, Coleman MB. (1990). Structural hemoglobin variants that produce the phenotype of thalassemia. *Semin Hematol* **27**, 229–38.
148. Chui DH. (2005). Alpha-thalassemia: Hb H disease and Hb Barts hydrops fetalis. *Ann NY Acad Sci* **1054**, 25–32.
149. Chui DH. (2005). Alpha-thalassaemia and population health in Southeast Asia. *Ann Hum Biol* **32**, 123–30.
150. Abrams ME, Meredith KS, Kinnard P, Clark RH. (2007). Hydrops fetalis: A retrospective review of cases reported to a large national database and identification of risk factors associated with death. *Pediatrics* **120**, 84–9.
151. Kan YW, Dozy AM, Stamatoyannopoulos G, Hadjiminas MG, Zachariades Z, Furbetta M, Cao A. (1979). Molecular basis of hemoglobin-H disease in the Mediterranean population. *Blood* **54**, 1434–8.
152. Dozy AM, Kan YW, Embury SH, Mentzer WC, Wang WC, Lubin B, Davis Jr. JR, Koenig HM. (1979). Alpha-globin gene organisation in blacks precludes the severe form of alpha-thalassaemia. *Nature* **280**, 605–7.
153. Alli NA. (2005). Acquired haemoglobin H disease. *Hematology* **10**(5), 413–18.
154. Steensma DP, Gibbons RJ, Higgs DR. (2005). Acquired alpha-thalassemia in association with myelodysplastic syndrome and other hematologic malignancies. *Blood* **105**, 443–52.
155. Sloane-Stanley J, Roberts NA, Olivieri N, Weatherall DJ, Wood WG. (2006). Globin gene expression in Hb Lepore-BAC transgenic mice. *Br J Haematol* **135**, 735–7.
156. Chan V, Au P, Yip B, Chan TK. (2004). A new cross-over region for hemogloboin-Lepore-Hollandia. *Haematologica* **89**, 610–11.
157. Kutlar A. (2007). Sickle cell disease: A multigenic perspective of a single gene disorder. *Hemoglobin* **31**, 209–24.
158. Driscoll MC. (2007). Sickle cell disease. *Pediatr Rev* **28**, 259–68.
159. Meremikwu M. (2006). Sickle cell disease. *Clin Evid* **15**, 45–59.
160. Kutlar A. (2005). Sickle cell disease: A multigenic perspective of a single-gene disorder. *Med Princ Pract* **14**(Suppl 1), 15–19.
161. Frenette PS, Atweh GF. (2007). Sickle cell disease: Old discoveries, new concepts, and future promise. *J Clin Invest* **117**, 850–8.
162. Raphael RI. (2005). Pathophysiology and treatment of sickle cell disease. *Clin Adv Hematol Oncol* **3**, 492–505.
163. Redding-Lallinger R, Knoll C. (2006). Sickle cell disease – pathophysiology and treatment. *Curr Probl Pediatr Adolesc Health Care* **36**, 346–76.
164. Eaton JW, Jacob HS, White JG. (1979). Membrane abnormalities of irreversibly sickled cells. *Semin Hematol* **16**, 52–64.
165. Lubin B, Kuypers F, Chiu D. (1989). Lipid alterations and cellular properties of sickle red cells. *Ann NY Acad Sci* **565**, 86–95.
166. Alli NA, Wainwright RD, Mackinnon D, Poyiadjis S, Naidu G. (2007). Skull bone infarctive crisis and deep vein thrombosis in homozygous sickle cell disease – case report and review of the literature. *Hematology* **12**, 169–74.
167. Hunt JA, Ingram VM. (1958). Allelomorphism and the chemical differences of the human haemoglobins A, S and C. *Nature* **181**, 1062–3.
168. Hirsch RE, Samuel RE, Fataliev NA, Pollack MJ, Galkin O, Vekilov PG, Nagel RL. (2001). Differential pathways in oxy and deoxy HbC aggregation/crystallization. *Proteins* **42**, 99–107.
169. Lawrence C, Hirsch RE, Fataliev NA, Patel S, Fabry ME, Nagel RL. (1997). Molecular interactions between Hb alpha-G Philadelphia, HbC, and HbS: Phenotypic implications for SC alpha-G Philadelphia disease. *Blood* **90**, 2819–25.
170. Refaldi C, Mocellin MC, Cappellini MD. (2007). Gene symbol: HBB. *Hum Genet* **121**, 298.
171. Hunt JA, Ingram VM. (1961). Abnormal human haemoglobins. VI. The chemical difference between haemoglobins A and E. *Biochim Biophys Acta* **49**, 520–36.
172. Kishore B, Khare P, Gupta RJ, Bisht S, Majumdar K. (2007). Hemoglobin E disease in North Indian population: A report of 11 cases. *Hematology* **12**, 343–7.
173. Aguilar C, Vichinsky E, Neumayr L. (2005). Bone and joint disease in sickle cell disease. *Hematol Oncol Clin North Am* **19**, 929–41.
174. West MS, Wethers D, Smith J, Steinberg M. (1992). Laboratory profile of sickle cell disease: A cross-sectional analysis. The Cooperative Study of Sickle Cell Disease. *J Clin Epidemiol* **45**, 893–909.
175. Gupta R, Jarvis M, Yardumian A. (2000). Compound heterozygosity for haemoglobin S and haemoglobin E. *Br J Haematol* **108**, 463.

176. Vrettou C, Traeger-Synodinos J, Tzetis M, Palmer G, Sofocleous C, Kanavakis E. (2004). Real-time PCR for single-cell genotyping in sickle cell and thalassemia syndromes as a rapid, accurate, reliable, and widely applicable protocol for preimplantation genetic diagnosis. *Hum Mutat* **23**, 513–21.
177. Bermudez MG, Piyamongkol W, Tomaz S, Dudman E, Sherlock JK, Wells D. (2003). Single-cell sequencing and mini-sequencing for preimplantation genetic diagnosis. *Prenat Diagn* **23**, 669–77.
178. Xu K, Shi ZM, Veeck LL, Hughes MR, Rosenwaks Z. (1999). First unaffected pregnancy using preimplantation genetic diagnosis for sickle cell anemia. *JAMA* **281**, 1701–6.
179. Lamvu G, Kuller JA. (1997). Prenatal diagnosis using fetal cells from the maternal circulation. *Obstet Gynecol Surv* **52**, 433–7.
180. Lane PA. (1996). Sickle cell disease. *Pediatr Clin North Am* **43**, 639–64.
181. Hebbel RP. (2003). Sickle hemoglobin instability: A mechanism for malarial protection. *Redox Rep* **8**, 238–40.
182. Williams TN, Mwangi TW, Wambua S, Peto TE, Weatherall DJ, Gupta S, Recker M, Penman BS, Uyoga S, Macharia A, Mwacharo JK, Snow RW, Marsh K. (2005). Negative epistasis between the malaria-protective effects of alpha+-thalassemia and the sickle cell trait. *Nat Genet* **37**, 1253–7.
183. Stiehm ER. (2006). Disease versus disease: How one disease may ameliorate another. *Pediatrics* **117**, 184–91.
184. Bunn HF. (1997). Pathogenesis and treatment of sickle cell disease. *N Engl J Med* **337**, 762–9.
185. Bainbridge R, Higgs DR, Maude GH, Serjeant GR. (1985). Clinical presentation of homozygous sickle cell disease. *J Pediatr* **106**, 881–5.
186. Almeida A, Roberts I. (2005). Bone involvement in sickle cell disease. *Br J Haematol* **129**, 482–90.
187. Miller ST, Sleeper LA, Pegelow CH, Enos LE, Wang WC, Weiner SJ, Wethers DL, Smith J, Kinney TR. (2000). Prediction of adverse outcomes in children with sickle cell disease. *N Engl J Med* **342**, 83–9.
188. Quinn CT, Lee NJ, Shull EP, Ahmad N, Rogers ZR, Buchanan GR. (2008). Prediction of adverse outcomes in children with sickle cell anemia: A study of the Dallas Newborn Cohort. *Blood* **111**, 544–8.
189. Quinn CT, Rogers ZR, Buchanan GR. (2004). Survival of children with sickle cell disease. *Blood* **103**, 4023–7.
190. Steinberg MH. (2005). Predicting clinical severity in sickle cell anaemia. *Br J Haematol* **129**, 465–81.
191. Okpala IE. (2005). New therapies for sickle cell disease. *Hematol Oncol Clin North Am* **19**, 975–87.
192. Cokic VP, Smith RD, Beleslin-Cokic BB, Njoroge JM, Miller JL, Gladwin MT, Schechter AN. (2003). Hydroxyurea induces fetal hemoglobin by the nitric oxide-dependent activation of soluble guanylyl cyclase. *J Clin Invest* **111**, 231–9.
193. Rodgers GP. (2007). Specific therapies for sickle cell disease. *UpToDate*.
194. Eggleston B, Patience M, Edwards S, Adamkiewicz T, Buchanan GR, Davies SC, Dickerhoff R, Donfield S, Feig SA, Giller RH, Haight A, Horan J, Hsu LL, Kamani N, Lane P, Levine JE, Margolis D, Moore TB, Ohene-Frempong K, Redding-Lallinger R, Roberts IA, Rogers ZR, Sanders JE, Scott JP, Sleight B, Thompson AA, Sullivan KM, Walters MC. Multicenter Study of HCT for SCA (2007). Effect of myeloablative bone marrow transplantation on growth in children with sickle cell anaemia: Results of the multicenter study of haematopoietic cell transplantation for sickle cell anaemia. *Br J Haematol* **136**, 673–6.
195. Hardison RC, Chui DH, Riemer C, Giardine B, Lehväslaiho H, Wajcman H, Miller W. (2001). Databases of human hemoglobin variants and other resources at the globin gene server. *Hemoglobin* **25**, 183–93.
196. Nagel. (2007). Unstable hemoglobin variants. *UpToDate*.
197. Williamson D. (1993). The unstable haemoglobins. *Blood Rev* **7**, 146–63.
198. Sogami M, Uyesaka N, Era S, Kato K. (2003). Saturation transfer in human red blood cells with normal and unstable hemoglobin. *NMR Biomed* **16**, 19–28.
199. Miller DR, Weed RI, Stamatoyannopoulos G, Yoshida A. (1971). Hemoglobin Köln disease occurring as a fresh mutation: Erythrocyte metabolism and survival. *Blood* **38**, 715–29.
200. Tse WT, Lux SE. (1999). Red blood cell membrane disorders. *Br J Haematol* **104**, 2–13.
201. McGuire M, Agre P. (1988). Clinical disorders of the erythrocyte membrane skeleton. *Hematol Pathol* **2**, 1–14.
202. Delaunay J. (2007). The molecular basis of hereditary red cell membrane disorders. *Blood Rev* **21**, 1–20.
203. Delaunay J. (2002). Molecular basis of red cell membrane disorders. *Acta Haematol* **108**, 210–18.
204. Iolascon A, Perrotta S, Stewart GW. (2003). Red blood cell membrane defects. *Rev Clin Exp Hematol* **7**, 22–56.
205. Palek J, Lambert S. (1990). Genetics of the red cell membrane skeleton. *Semin Hematol* **27**, 290–332.
206. Kunze D, Rüstow B. (1993). Pathobiochemical aspects of cytoskeleton components. *Eur J Clin Chem Clin Biochem* **31**, 477–89.
207. Eber S, Lux SE. (2004). Hereditary spherocytosis – defects in proteins that connect the membrane skeleton to the lipid bilayer. *Semin Hematol* **41**, 118–41.
208. Iolascon A, Miraglia del Giudice E, Perrotta S, Alloisio N, Morle L, Delaunay J. (1998). Hereditary spherocytosis: From clinical to molecular defects. *Haematologica* **83**, 240–57.
209. Agre P, Asimos A, Casella JF, McMillan C. (1986). Inheritance pattern and clinical response to splenectomy as a reflection of erythrocyte spectrin deficiency in hereditary spherocytosis. *N Engl J Med* **315**, 1579–83.
210. Cynober T, Mohandas N, Tchernia G. (1996). Red cell abnormalities in hereditary spherocytosis: Relevance to diagnosis and understanding of the variable expression of clinical severity. *J Lab Clin Med* **128**, 259–69.
211. Rappaport ES, Quick G, Ransom D, Helbert B, Frankel LS. (1989). Aplastic crisis in occult hereditary spherocytosis caused by human parvovirus (HPV B19). *South Med J* **82**, 247–51.
212. Green DH, Bellingham AJ, Anderson MJ. (1984). Parvovirus infection in a family associated with aplastic crisis in an affected sibling pair with hereditary spherocytosis. *J Clin Pathol* **37**, 1144–6.
213. Glele-Kakai C, Garbarz M, Lecomte MC, Leborgne S, Galand C, Bournier O, Devaux I, Gautero H, Zohoun I, Gallagher PG, Forget BG, Dhermy D. (1996). Epidemiological studies of spectrin mutations related to hereditary elliptocytosis and spectrin polymorphisms in Benin. *Br J Haematol* **95**, 57–66.
214. Nagel RL. (1990). Red-cell cytoskeletal abnormalities – implications for malaria. *N Engl J Med* **323**, 1558–60.
215. Gallagher PG. (2004). Hereditary elliptocytosis: Spectrin and protein 4.1R. *Semin Hematol* **41**, 142–64.
216. An X, Mohandas N. (2008). Disorders of red cell membrane. *Br J Haematol* **Mar 12**.
217. Liu SC, Palek J, Prchal JT. (1982). Defective spectrin dimer-dimer association with hereditary elliptocytosis. *Proc Natl Acad Sci USA* **79**, 2072–6.
218. Marchesi SL, Letsinger JT, Speicher DW, Marchesi VT, Agre P, Hyun B, Gulati G. (1987). Mutant forms of spectrin alpha-subunits in hereditary elliptocytosis. *J Clin Invest* **80**, 191–8.
219. Palek J. (1987). Hereditary elliptocytosis, spherocytosis and related disorders: Consequences of a deficiency or a mutation of membrane skeletal proteins. *Blood Rev* **1**, 147–68.
220. Mentzer WC. (2007). Lubin. Hereditary elliptocytosis: Clinical features and diagnosis. *UpToDate*.
221. Jarolim P, Wichterle H, Hanspal M, Murray J, Rubin HL, Palek J. (1995). Beta spectrin PRAGUE: A truncated beta spectrin producing spectrin deficiency, defective spectrin heterodimer self-association and a phenotype of spherocytic elliptocytosis. *Br J Haematol* **91**, 502–10.

222. Rampoldi L, Danek A, Monaco AP. (2002). Clinical features and molecular bases of neuroacanthocytosis. *J Mol Med* **80**, 475–91.
223. Hooper AJ, van Bockxmeer FM, Burnett JR. (2005). Monogenic hypocholesterolaemic lipid disorders and apolipoprotein B metabolism. *Crit Rev Clin Lab Sci* **42**, 515–45.
224. Redman CM, Marsh WL. (1993). The Kell blood group system and the McLeod phenotype. *Semin Hematol* **30**, 209–18.
225. Cooper RA. (1980). Hemolytic syndromes and red cell membrane abnormalities in liver disease. *Semin Hematol* **17**, 103–12.
226. Flamm M, Schachter D. (1982). Acanthocytosis and cholesterol enrichment decrease lipid fluidity of only the outer human erythrocyte membrane leaflet. *Nature* **298**, 290–2.
227. Lange Y, Cutler HB, Steck TL. (1980). The effect of cholesterol and other intercalated amphipaths on the contour and stability of the isolated red cell membrane. *J Biol Chem* **255**, 9331–7.
228. Morse EE. (1990). Mechanisms of hemolysis in liver disease. *Ann Clin Lab Sci* **20**, 169–74.
229. Dumez H, Reinhart WH, Guetens G, de Bruijn EA. (2004). Human red blood cells: Rheological aspects, uptake, and release of cytotoxic drugs. *Crit Rev Clin Lab Sci* **41**, 159–88.
230. Syfuss PY, Ciupea A, Brahimi S, Cynober T, Stewart GW, Grandchamp B, Beaumont C, Tchernia G, Delaunay J, Wagner JC. (2006). Mild dehydrated hereditary stomatocytosis revealed by marked hepatosiderosis. *Clin Lab Haematol* **28**, 270–4.
231. Delaunay J. (2004). The hereditary stomatocytoses: Genetic disorders of the red cell membrane permeability to monovalent cations. *Semin Hematol* **41**, 165–72.
232. Fricke B, Argent AC, Chetty MC, Pizzey AR, Turner EJ, Ho MM, Iolascon A, von Düring M, Stewart GW. (2003). The "stomatin" gene and protein in overhydrated hereditary stomatocytosis. *Blood* **102**, 2268–77.
233. Wislöff F, Boman D. (1979). Acquired stomatocytosis in alcoholic liver disease. *Scand J Haematol* **23**, 43–50.
234. Douglass CC, Twomey JJ. (1970). Transient stomatocytosis with hemolysis: A previously unrecognized complication of alcoholism. *Ann Intern Med* **72**, 159–64.
235. Prchal JT, Gregg XT. (2005). Red cell enzymes. *Hematology Am Soc Hematol Educ Program*, 19–23.
236. Stewart GW, Hepworth-Jones BE, Keen JN, Dash BC, Argent AC, Casimir CM. (1992). Isolation of cDNA coding for an ubiquitous membrane protein deficient in high Na+, low K+ stomatocytic erythrocytes. *Blood* **79**, 1593–601.
237. Ronquist G, Theodorsson E. (2007). Inherited, non-spherocytic haemolysis due to deficiency of glucose-6-phosphate dehydrogenase. *Scand J Clin Lab Invest* **67**, 105–11.
238. Jacobasch G. (2000). Biochemical and genetic basis of red cell enzyme deficiencies. *Baillieres Best Pract Res Clin Haematol* **13**, 1–20.
239. Mason PJ. (1998). Red cell enzyme deficiencies: From genetic basis to gene transfer. *Semin Hematol* **35**, 126–35.
240. Beutler E. (1994). G6PD deficiency. *Blood* **84**, 3613–36.
241. Kugler W, Lakomek M. (2000). Glucose-6-phosphate isomerase deficiency. *Baillieres Best Pract Res Clin Haematol* **13**, 89–101.
242. Beutler E. (2008). Glucose-6-phosphate dehydrogenase deficiency: A historical perspective. *Blood* **111**, 16–24.
243. Thompson SF, Fraser IM, Strother A, Bull BS. (1989). Change of deformability and Heinz body formation in G6PD-deficient erythrocytes treated with 5-hydroxy-6-desmethylprimaquine. *Blood Cells* **15**, 443–52.
244. Ruwende C, Khoo SC, Snow RW, Yates SN, Kwiatkowski D, Gupta S, Warn P, Allsopp CE, Gilbert SC, Peschu N. (1995). Natural selection of hemi- and heterozygotes for G6PD deficiency in Africa by resistance to severe malaria. *Nature* **376**, 246–9.
245. Doll DC. (1983). Oxidative haemolysis after administration of doxorubicin. *Br Med J (Clin Res Ed)* **287**, 180–1.
246. Tishler M, Abramov A. (1983). Phenazopyridine-induced hemolytic anemia in a patient with G6PD deficiency. *Acta Haematol* **70**, 208–9.
247. Zanella A, Bianchi P, Fermo E. (2007). Pyruvate kinase deficiency. *Haematologica* **92**, 721–3.
248. Zanella A, Bianchi P. (2000). Red cell pyruvate kinase deficiency: From genetics to clinical manifestations. *Baillieres Best Pract Res Clin Haematol* **13**, 57–81.
249. Beutler E. (1990). Red cell enzyme defects. *Hematol Pathol* **4**, 103–14.
250. Zanella A, Fermo E, Bianchi P, Valentini G. (2005). Red cell pyruvate kinase deficiency: Molecular and clinical aspects. *Br J Haematol* **130**, 11–25.
251. Demina A, Varughese KI, Barbot J, Forman L, Beutler E. (1998). Six previously undescribed pyruvate kinase mutations causing enzyme deficiency. *Blood* **92**, 647–52.
252. Pissard S, Max-Audit I, Skopinski L, Vasson A, Vivien P, Bimet C, Goossens M, Galacteros F, Wajcman H. (2006). Pyruvate kinase deficiency in France: A 3-year study reveals 27 new mutations. *Br J Haematol* **133**, 683–9.
253. Fermo E, Bianchi P, Chiarelli LR, Cotton F, Vercellati C, Writzl K, Baker K, Hann I, Rodwell R, Valentini G, Zanella A. (2005). Red cell pyruvate kinase deficiency: 17 new mutations of the PK-LR gene. *Br J Haematol* **129**, 839–46.
254. Hoffman PC. (2006). Immune hemolytic anemia – selected topics. *Hematology Am Soc Hematol Educ Program*, 13–18.
255. Rosse WF, Hillmen P, Schreiber AD. (2004). Immune-mediated hemolytic anemia. *Hematology Am Soc Hematol Educ Program*, 48–62.
256. Gehrs BC, Friedberg RC. (2002). Autoimmune hemolytic anemia. *Am J Hematol* **69**, 258–71.
257. Gibson J. (1988). Autoimmune hemolytic anemia: Current concepts. *Aust NZ J Med* **18**, 625–37.
258. Beutler E, Miwa S, Palek J. (1994). Hemolytic anemias. *Rev Invest Clin* (Suppl), 162–8.
259. Clark DA, Dessypris EN, Jenkins Jr. DE, Krantz SB. (1984). Acquired immune hemolytic anemia associated with IgA erythrocyte coating: Investigation of hemolytic mechanisms. *Blood* **64**, 1000–5.
260. Sokol RJ, Hewitt S. (1985). Autoimmune hemolysis: A critical review. *Crit Rev Oncol Hematol* **4**, 125–54.
261. Young PP, Uzieblo A, Trulock E, Lublin DM, Goodnough LT. (2004). Autoantibody formation after alloimmunization: Are blood transfusions a risk factor for autoimmune hemolytic anemia? *Transfusion* **44**, 67–72.
262. Rosse WF, Schrier, SL. (2007). Clinical features and diagnosis of autoimmune hemolytic anemia: Warm agglutinins. *UpToDate*.
263. Rosse WF, Schrier SL. (2007). Clinical features and diagnosis of autoimmune hemolytic anemia: Cold agglutinin disease. *UpToDate*.
264. Gertz MA. (2006). Cold hemolytic syndrome. *Hematology Am Soc Hematol Educ Program*, 19–23.
265. Schreiber AD, Herskovitz BS, Goldwein M. (1977). Low-titer cold-hemagglutinin disease. Mechanism of hemolysis and response to corticosteroids. *N Engl J Med* **296**, 1490–4.
266. Feizi T, Taylor-Robinson D. (1967). Cold agglutinin anti-I and *Mycoplasma pneumoniae*. *Immunology* **13**, 405–9.
267. Horwitz CA, Moulds J, Henle W, Henle G, Polesky H, Balfour Jr. HH, Schwartz B, Hoff T. (1997). Cold agglutinins in infectious mononucleosis and heterophil-antibody-negative mononucleosis-like syndromes. *Blood* **50**, 195–202.
268. Crisp D, Pruzanski W. (1982). B-cell neoplasms with homogeneous cold-reacting antibodies (cold agglutinins). *Am J Med* **72**, 915–22.
269. Isbister JP, Cooper DA, Blake HM, Biggs JC, Dixon RA, Penny R. (1978). Lymphoproliferative disease with IgM lambda monoclonal protein and autoimmune hemolytic anemia. A report of four cases and a review of the literature. *Am J Med* **64**, 434–40.
270. Bencomo AA, Diaz M, Alfonso Y, Valdés O, Alfonso ME. (2003). Quantitation of red cell-bound IgG, IgA, and IgM in patients with autoimmune hemolytic anemia and blood donors by enzyme-linked immunosorbent assay. *Immunohematol* **19**, 47–53.
271. Petz LD, Garratty G. (1980). *Acquired Immune Hemolytic Anemia*. Churchill Livingstone, New York.

272. Nordhagen R, Stensvold K, Winsnes A, Skyberg D, Støren A. (1984). Paroxysmal cold haemoglobinuria. The most frequent acute autoimmune haemolytic anaemia in children? *Acta Paediatr Scand* **73**, 258–62.
273. Kopko PM, Holland PV. (2001). Mechanisms of severe transfusion reactions. *Transfus Clin Biol* **8**, 278–81.
274. Talano JA, Hillery CA, Gottschall JL, Baylerian DM, Scott JP. (2003). Delayed hemolytic transfusion reaction/hyperhemolysis syndrome in children with sickle cell disease. *Pediatrics* **111**, e661–5.
275. Davenport RD. (2005). Pathophysiology of hemolytic transfusion reactions. *Semin Hematol* **42**, 165–8.
276. Kim HH, Park TS, Oh SH, Chang CL, Lee EY, Son HC. (2004). Delayed hemolytic transfusion reaction due to anti-Fyb caused by a primary immune response: A case study and a review of the literature. *Immunohematol* **20**, 184–6.
277. Urbaniak SJ. (2006). Alloimmunity to RhD in humans. *Transfus Clin Biol* **13**, 19–22.
278. Mochizuki K, Ohto H, Hirai S, Ujiie N, Amanuma F, Kikuta A, Miura S, Yasuda H, Ishijima A, Suzuki H. (2006). Hemolytic disease of the newborn due to anti-Di: A case study and review of the literature. *Transfusion* **46**, 454–60.
279. Murray NA, Roberts IA. (2007). Haemolytic disease of the newborn. *Arch Dis Child Fetal Neonatal Ed* **92**, F83–8.
280. May-Wewers J, Kaiser JR, Moore EK, Blackall DP. (2006). Severe neonatal hemolysis due to a maternal antibody to the low-frequency Rh antigen C(w). *Am J Perinatol* **23**, 213–17.
281. Joshi DD, Nickerson HJ, McManus MJ. (2004). Hydrops fetalis caused by homozygous alpha-thalassemia and Rh antigen alloimmunization: Report of a survivor and literature review. *Clin Med Res* **2**, 228–32.
282. Arndt PA, Garratty G. (2005). The changing spectrum of drug-induced immune hemolytic anemia. *Semin Hematol* **42**, 137–44.
283. Garratty G. (2004). Review: Drug-induced immune hemolytic anemia – the last decade. *Immunohematol* **20**, 138–46.
284. Wright MS. (1999). Drug-induced hemolytic anemias: Increasing complications to therapeutic interventions. *Clin Lab Sci* **12**, 115–18.
285. Salama A, Mueller-Eckhardt C. (1987). On the mechanisms of sensitization and attachment of antibodies to RBC in drug-induced immune hemolytic anemia. *Blood* **69**, 1006–10.
286. Kirtland III HH, Mohler DN, Horwitz DA. (1980). Methyldopa inhibition of suppressor-lymphocyte function: A proposed cause of autoimmune hemolytic anemia. *N Engl J Med* **302**, 825–32.
287. Owens NA, Hui HL, Green FA. (1982). Induction of direct Coombs positivity with alpha-methyldopa in chimpanzees. *J Med* **13**, 473–7.
288. Maraj R, Jacobs LE, Ioli A, Kotler MN. (1998). Evaluation of hemolysis in patients with prosthetic heart valves. *Clin Cardiol* **21**, 387–92.
289. Wallner SF, Vautrin RM, Buerk C, Robinson WA, Peterson VM. (1982). The anemia of thermal injury: Studies of erythropoiesis *in vitro*. *J Trauma* **22**, 774–80.
290. Issa SA, Qasem Q. (2007). Central retinal vein occlusion associated with thrombotic thrombocytopenic purpura/hemolytic uremic syndrome: Complete resolution is possible. *J Postgrad Med* **53**, 183–4.
291. Saif MW, McGee PJ. (2005). Hemolytic-uremic syndrome associated with gemcitabine: A case report and review of literature. *JOP* **6**, 369–74.
292. Petroianu A. (2007). Drug-induced splenic enlargement. *Expert Opin Drug Saf* **6**, 199–206.
293. Zieve L. (1958). Jaundice, hyperlipemia and hemolytic anemia: A heretofore unrecognized syndrome associated with alcoholic fatty liver and cirrhosis. *Ann Intern Med* **48**, 471–96.
294. Eakle JF, Bressoud PF. (2000). Hemophagocytic syndromes and infection. *Emerg Infect Dis* **6**, 601–8.
295. Rosse WF. (1987). The spleen as a filter. *N Engl J Med* **317**, 704–6.
296. Peck-Radosavljevic M. (2001). Hypersplenism. *Eur J Gastroenterol Hepatol* **13**, 317–23.
297. Kraus MD. (2003). Splenic histology and histopathology: An update. *Semin Diagn Pathol* **20**, 84–93.
298. Thomas C, Thomas L. (2005). Anemia of chronic disease: Pathophysiology and laboratory diagnosis. *Lab Hematol* **11**, 14–23.
299. Weiss G, Goodnough LT. (2005). Anemia of chronic disease. *N Engl J Med* **352**, 1011–23.
300. Schrier SL. (2007). Anemia of chronic disease (anemia of chronic inflammation). *UpToDate*.
301. Cavill I, Auerbach M, Bailie GR, Barrett-Lee P, Beguin Y, Kaltwasser P, Littlewood T, Macdougall IC, Wilson K. (2006). Iron and the anaemia of chronic disease: A review and strategic recommendations. *Curr Med Res Opin* **22**, 731–7.
302. Hugman A. (2006). Hepcidin: An important new regulator of iron homeostasis. *Clin Lab Haematol* **28**, 75–83.
303. Ganz T. (2006). Molecular pathogenesis of anemia of chronic disease. *Pediatr Blood Cancer* **46**, 554–7.
304. Nemeth E, Valore EV, Territo M, Schiller G, Lichtenstein A, Ganz T. (2003). Hepcidin, a putative mediator of anemia of inflammation, is a type II acute-phase protein. *Blood* **101**, 2461–3.
305. Macciò A, Madeddu C, Massa D, Mudu MC, Lusso MR, Gramignano G, Serpe R, Melis GB, Mantovani G. (2005). Hemoglobin levels correlate with interleukin-6 levels in patients with advanced untreated epithelial ovarian cancer: Role of inflammation in cancer-related anemia. *Blood* **106**, 362–7.
306. Vreugdenhil G, Löwenberg B, van Eijk HG, Swaak AJ. (1990). Anaemia of chronic disease in rheumatoid arthritis. Raised serum interleukin-6 (IL-6) levels and effects of IL-6 and anti-IL-6 on *in vitro* erythropoiesis. *Rheumatol Int* **10**, 127–30.
307. van Iperen CE, van de Wiel A, Marx JJ. (2001). Acute event-related anaemia. *Br J Haematol* **115**, 739–43.
308. Kumar N. (2006). Copper deficiency myelopathy (human swayback). *Mayo Clin Proc* **81**, 1371–84.
309. Todd LM, Godber IM, Gunn IR. (2004). Iatrogenic copper deficiency causing anaemia and neutropenia. *Ann Clin Biochem* **41**, 414–16.
310. Foy H, Kondi A, Macdougall L. (1961). Pure red-cell aplasia in marasmus and kwashiorkor treated with riboflavine. *Br Med J* **1**, 937–41.
311. Oski FA, Barness LA. (1968). Hemolytic anemia in vitamin E deficiency. *Am J Clin Nutr* **21**, 45–50.
312. Fisher JW. (1980). Mechanism of the anemia of chronic renal failure. *Nephron* **25**, 106–11.
313. Al-Khoury S, Afzali B, Shah N, Thomas S, Gusbeth-Tatomir P, Goldsmith D, Covic A. (2007). Diabetes, kidney disease and anaemia: Time to tackle a troublesome triad? *Int J Clin Pract* **61**, 281–9.
314. Makoni SN, Laber DA. (2004). Clinical spectrum of myelophthisis in cancer patients. *Am J Hematol* **76**, 92–3.
315. Lin YC, Chang HK, Sun CF, Shih LY. (1995). Microangiopathic hemolytic anemia as an initial presentation of metastatic cancer of unknown primary origin. *South Med J* **88**, 683–7.
316. Shamdas GJ, Ahmann FR, Matzner MB, Ritchie JM. (1993). Leukoerythroblastic anemia in metastatic prostate cancer. Clinical and prognostic significance in patients with hormone-refractory disease. *Cancer* **71**, 3594–600.

Disorders of Megakaryocytes and Platelets

24

CHAPTER

Tom E. Howard
and
Faramarz Naeim

MEGAKARYOCYTIC HYPOPLASIA

Megakaryocytic hypoplasia is one of the features of congenital or acquired aplastic anemia. A lack or markedly reduced production of megakaryocytes without bone marrow aplasia is rare and has been associated with prolonged administration of prednisone, estrogens, interferon, and chlorothiazide [1–3]. Chronic alcoholism and certain infectious diseases, such as measles, varicella, infectious mononucleosis, and cytomegalovirus, have been associated with megakaryocytic hypoplasia [2–4]. Amegakaryocytosis has also been reported in patients with thymic aplasia [5].

Congenital Megakaryocytic Hypoplasias

Congenital megakaryocytic hypoplasias are rare conditions characterized by thrombocytopenic purpura [6–10]. Two major subtypes have been described: (1) congenital amegakaryocytic thrombocytopenia (CAMT) and (2) congenital thrombocytopenia with absent radii (TAR).

CAMT is associated with an impaired response to thrombopoietin (TPO) due to the presence of mutations in the TPO receptor, c-Mpl [9, 10]. Serum TPO levels are elevated. Thrombocytopenia may be persistent with fast progression into pancytopenia (CAMT I) or transient and less severe CAMT II [10].

The TAR syndrome is associated with bilateral aplasia of the radii, cardiac and/or renal malformations, and dysmegakaryocytopoiesis characterized by cells blocked at early stages of megakaryocytic differentiation [7, 8]. Thrombocytopenia is usually severe and is sometimes associated with a transient leukemoid reaction. Platelets may demonstrate abnormal function [11]. Bone marrow examination reveals a marked decrease in the number of megakaryocytes (Figure 24.1). TAR syndrome platelets demonstrate impaired Mpl expression compared to platelets from adult controls (Figure 24.2).

MEGAKARYOCYTOSIS

Megakaryocytosis and thrombocytosis are associated with acute infections, iron deficiency, diabetes mellitus, and postsplenectomy status [3, 11–13]. Postsplenectomy thrombocytosis is typically a transient event. Peripheral destruction of platelets in conditions such as autoimmune-associated thrombocytopenic purpura and drug-induced thrombocytopenia leads to megakaryocytosis secondary to bone marrow compensation.

Bone marrow metastasis is a frequent cause of reactive megakaryocytosis due to either the release of TPO-like substances from tumor cells or the increased expression of extracellular matrix (ECM) proteins such as tenascin [14, 15].

Megakaryocytosis is associated with a garden variety of primary bone marrow disorders such as myeloproliferative disorders, myelodysplastic syndromes, and acute megakaryoblastic leukemia (Figures 24.3 and 24.4). Unlike reactive megakaryocytosis, these conditions are often associated with significant dysplastic changes and arrangement of megakaryocytes in clusters or sheets.

Hereditary thrombocythaemia (HT) is an inherited autosomal dominant disorder caused by mutations within the genes encoding TPO or c-Mpl (TPO receptor) genes [16, 17].

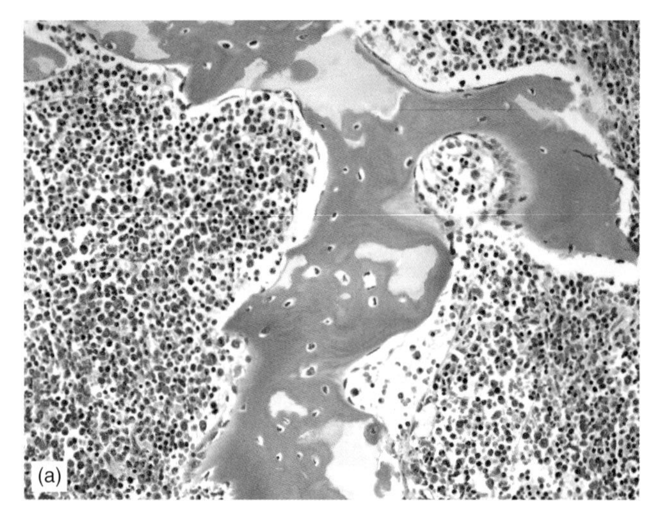

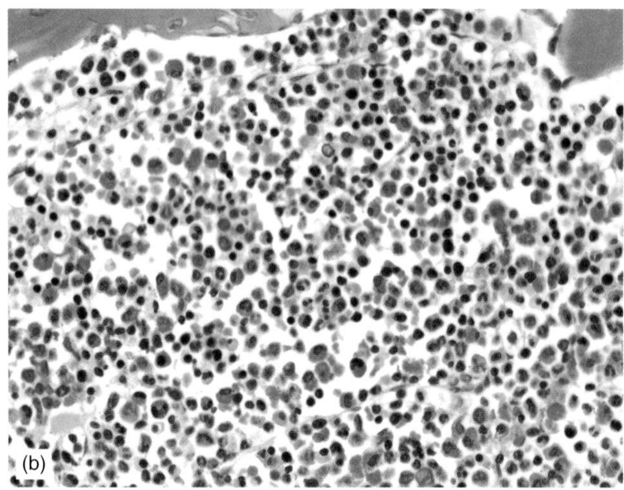

FIGURE 24.1 Bone marrow biopsy section of a child with TAR syndrome demonstrating markedly decreased megakaryocytes (a and b).

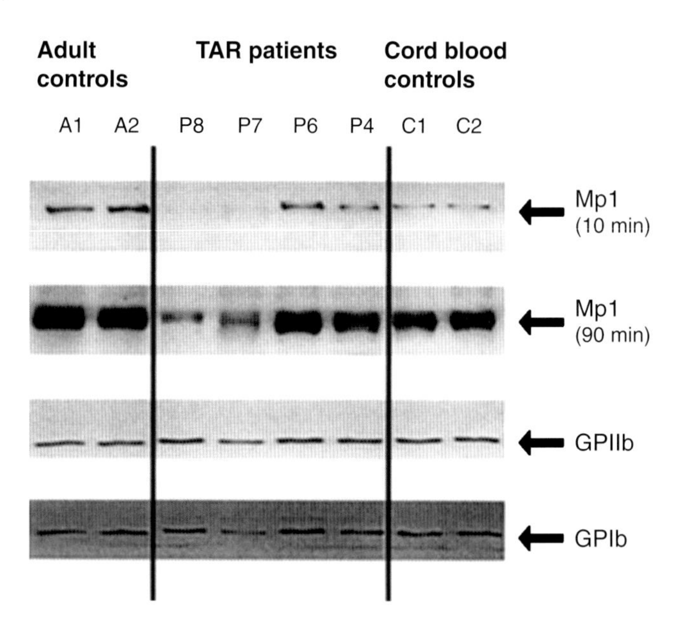

FIGURE 24.2 Western blot analysis of Mp1, GPIIb, and GPIb expression in platelets from normal adult controls, newborns, and TAR syndrome patients. Platelet lysates (20 μg) were separated by SDS-PAGE and probed with polyclonal antibodies against Mpl, GPIIb, and GPIbα. From Ref. [8] by permission.

THROMBOCYTOPENIA

Thrombocytopenia is due to either impaired platelet production or increased platelet destruction. Impaired platelet production is caused by congenital and acquired aplastic anemias, paroxysmal nocturnal hemoglobinuria, congenital amegakaryocytosis, bone marrow metastasis, myelofibrosis, drugs, radiation, certain infections, and malnutrition. Increased platelet destruction may occur by two major mechanisms: (1) immunologic and (2) nonimmunologic.

Autoimmune Thrombocytopenic Purpura

Etiology and Pathogenesis

Autoimmune thrombocytopenic purpura (AITP), also known as idiopathic thrombocytopenic purpura (ITP), is an antibody-mediated thrombocytopenia [18–23]. In a vast majority of these cases, antiplatelet antibodies are against platelet membrane glycoproteins. Glycoprotein (GP) IIb-IIIa (CD41/CD61) is the most frequent target [19, 21–24]. Autoantibodies against GPIb-IX (CD42a) and GPV (CD42d) have also been reported [19, 25]. The autoantibodies are typically of the IgG class and can cross placenta, causing fetal thrombocytopenia in pregnant patients. A T-cell-mediated cytotoxicity has been suggested as an alternative mechanism in patients who do not demonstrate detectable autoantibodies [22, 26]. Molecular parody between GPIIb-IIIa and HIV proteins may play an important role in the pathogenesis of thrombocytopenia in patients with AIDS [22].

AITP is often associated with other autoimmune disorders such as systemic lupus erythematosus, rheumatoid arthritis, systemic sclerosis, Hashimoto thyroiditis, ulcerative colitis, Crohn disease, biliary cirrhosis, myasthenia gravis, and pernicious anemia [27–31]. Other diseases such as sarcoidosis, lymphoproliferative disorders, Gaucher disease, IgA deficiency, and panhypogammaglobulinemia may demonstrate AITP [32–35].

Clinicopathologic Features

AITP has two clinical presentations: acute and chronic. Acute AITP is predominantly observed in children between the ages of 2 and 9 years. It is one of the most common causes of thrombocytopenia in children and is often preceded by a history of viral infection such as chickenpox, rubella, or rubeola [18, 19, 21]. Vaccination for measles, mumps, or chickenpox viruses may occasionally lead to AITP. It is characterized by a severe onset of thrombocytopenia leading to petechiae, purpura, as well as gastrointestinal and/or genitourinary tract hemorrhages. Intracranial bleeding is

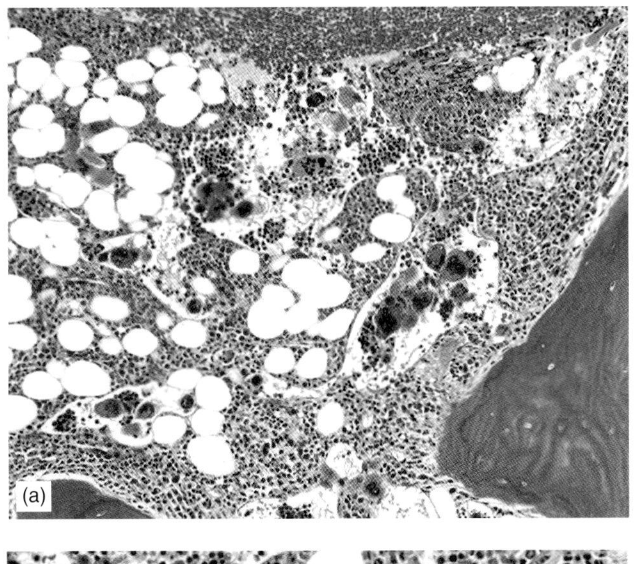

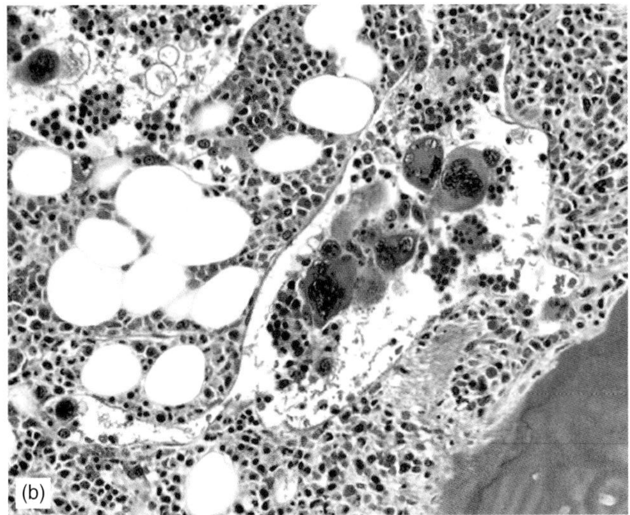

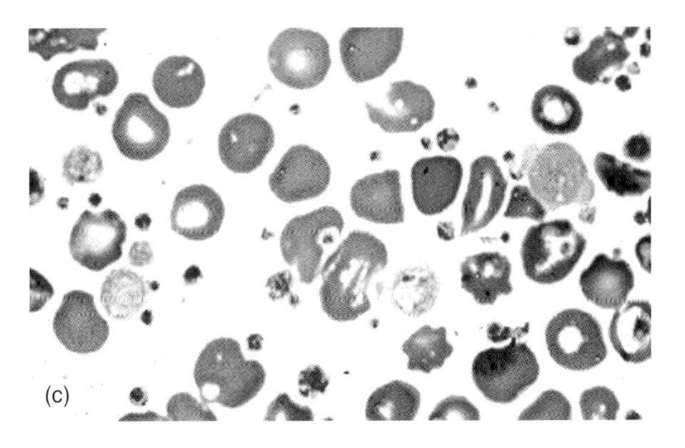

FIGURE 24.3 Bone marrow biopsy section from a patient in early stage of idiopathic myelofibrosis demonstrating megakaryocytosis. Aggregates of megakayocytes are present in dilated sinuses: (a) low power and (b) high power views. Blood smear (c) shows thrombocytosis with several giant platelets.

rare. In approximately 80% of affected children, the thrombocytopenia resolves spontaneously within 6–12 months.

Chronic AITP occurs in adults, usually between 20 and 50 years of age [18, 19, 22]. The female:male ratio is

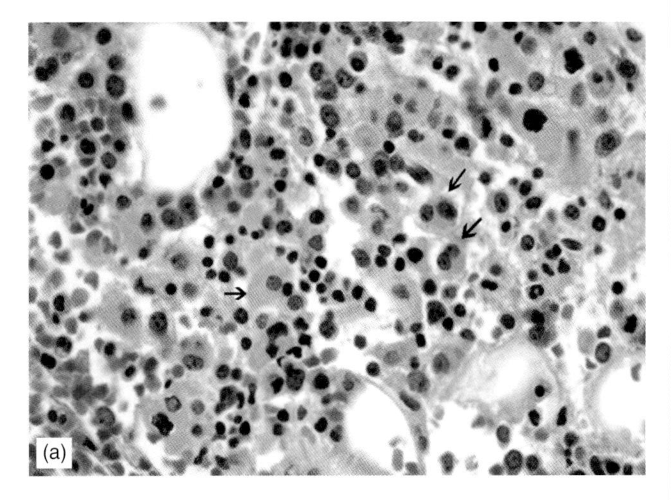

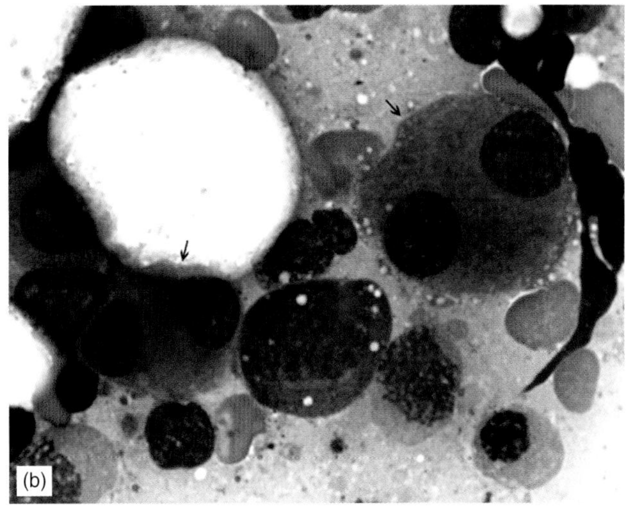

FIGURE 24.4 Biopsy section (a) of a patient with myelodysplastic syndrome showing numerous micromegakaryocytes. Bone marrow smear (b) demonstrates a binucleated micromegakaryocyte.

about 3:1. Overall, thrombocytopenia is less severe, and the chance of spontaneous remission is less frequent than the acute form of AITP. Clinical manifestations may vary from scattered petechiae to purpuric and ecchymotic lesions, epistaxis, and even intracranial bleeding.

Examination of the peripheral blood reveals thrombocytopenia with marked variation in the shape and size of the platelets. The platelet count may reach $<30,000/\mu L$ with prolongation of bleeding time. Rarely patients may also demonstrate an autoimmune hemolytic anemia (Evans syndrome). Iron deficiency anemia may occur due to chronic or excessive bleeding.

The clinical use of antiplatelet antibody studies for the diagnosis of AITP is controversial [22, 36]. In several studies, the rate of positive results for antiplatelet antibody tests ranged from 49% to 66% in patients with AITP and 7% to 28% in patients with nonimmune thrombocytopenia (e.g. congenital thrombocytopenia, myelodysplastic syndrome) [22, 37, 38].

Bone marrow examination reveals megakaryocytosis, diffusely dispersed between other hematopoietic cells. Micromegakaryocytes and giant megakaryocytes are often present. Some megakaryocytes may show more basophilic and less granular cytoplasm with no platelet budding [3].

AITP in children is usually self-limiting with spontaneous remission. In adults, <10% of the patients may achieve spontaneous remission. In life-threatening forms, treatment includes platelet transfusion and intravenous immunoglobulin. Patients with mild to moderate thrombocytopenia are treated with corticosteroids, followed by splenectomy in nonresponders or in patients who relapse [21, 22, 39–41].

Differential Diagnosis

Differential diagnosis includes all conditions associated with thrombocytopenia, such as congenital thrombocytopenia, infections, primary bone marrow disorders, drugs, and disseminated intravascular coagulation (DIC). EDTA-dependent agglutinins may also cause platelet clumping and platelet satellitism around leukocytes and therefore a reduction in the platelet count (pseudo-thrombocytopenia).

Drug-Induced Immune Thrombocytopenia

The most common cause of drug-induced thrombocytopenia is immune mediated with two major proposed mechanisms:

1. The drug or its metabolites bind to a plasma protein, and antibodies are generated against the drug–protein complexes.
2. The drug binds to one or more components of the platelet membrane and induces a structural change. This drug–platelet complex provokes antibody production.

Platelet membrane glycoproteins such as GP1b-IX (CD42a) and GPIIb-IIIa (CD41/CD61) are the primary targets [42–46]. Quinine, quinidine, sulfonamides, organic arsenicals, sulfisoxazole, ranitidine, rifampin, alpha methyl-dopa, para-aminosalicylate, heparin, and gold salts are among the drugs that have caused immune-mediated thrombocytopenia [42–48].

Heparin-induced thrombocytopenia (HIT) accounts for 10–20% of patients receiving unfractionated heparin [49–51]. HIT is frequently associated with thromboembolism. Two different mechanisms have been proposed for heparin-associated thrombocytopenia: (1) a direct heparin–platelet interaction leading to platelet activation, aggregation, and clearance (HIT type I) and (2) an immune-mediated thrombocytopenia (HIT type II) [49, 50]. Approximately 2% of HIT is immune mediated. Antibodies raised against heparin–platelet surface membrane complexes trigger platelet activation and aggregation and eventually immune clearance of the affected platelets. Platelet aggregates may cause emboli in both the arterial and the venous circulation.

Neonatal Immune-Mediated Thrombocytopenia

Neonatal immune-mediated thrombocytopenia is caused when maternal antiplatelet antibodies cross the placenta and bind to fetal platelets [52, 53]. Antiplatelet antibody production in the mother is due to either feto-maternal incompatibility in platelet antigens (neonatal alloimmune thrombocytopenia) or maternal AITP. Alloimmune thrombocytopenia is caused by maternal IgG antibodies raised against fetal platelets. The most common cause of neonatal alloimmune thrombocytopenia is feto-maternal incompatibility in human platelet alloantigens (HPA)-1a, accounting for 50–90% of the cases [54, 55]. The risk of maternal sensitization correlates with the maternal human leukocyte antigen (HLA) type and appears to be confined to HLA-B8 and HLA-DR3 mothers. Affected children present with scattered or generalized petechial and/or purpuric hemorrhages at birth or soon after delivery. Intracranial hemorrhage may occur in up to 25% of the affected neonates [56]. Thrombocytopenia may persist for 2–3 weeks with a platelet count of around 30,000/μL or lower.

Thrombocytopenia associated with maternal AITP is observed in up to 50% of neonates born to mothers with AITP. The maternal IgG autoantibody crosses the placenta and binds to fetal platelets. Thrombocytopenia may last for 1–6 months. The incidence of spontaneous abortion is increased in mothers with AITP.

Posttransfusion Purpura

Posttransfusion purpura is caused by platelet alloantigenic incompatibilities between the recipient and the donor [57, 58]. It is usually manifested 7–10 days after blood or platelet transfusion. Alloantibodies are usually against HPA-1a. It is most commonly observed in recipients with specific HLA class II phenotypes, particularly in those who are HLA-DR3-positive [59, 60].

Thrombotic Thrombocytopenic Purpura/ Hemolytic Uremic Syndrome

Thrombotic thrombocytopenic purpura (TTP) and hemolytic uremic syndrome (HUS) are overlapping acute syndromes with abnormalites in multiple organ systems [61–65]. Neurologic abnormalities are the dominant clinical features in TTP, and acute renal failure is the main clinical manifestation in HUS. The peak incidence of TTP is in the third decade, while HUS is observed primarily in infancy and early childhood.

Several etiologic factors have been postulated, which are primarily related to endothelial cell damage. Abnormalities of endothelial cell function, such as defects in prostacyclin metabolism and deficiency of plasminogen activator, have been demonstrated in some patients with TTP [66–69]. Recent studies suggest that TTP is caused most commonly by an autoimmune-mediated deficiency of the circulating von Willebrand factor (VWF) cleaving metalloprotease, ADAMTS13 [70]. There is a close association between TTP and collagen vascular disorders such as systemic lupus erythematosus, rheumatoid arthritis, Sjogren syndrome, and polyarteritis nodosa [71–73]. Certain viruses and bacteria such as HIV, coxsackie A and B, *Mycoplasma pneumoniae*, and *Legionella pneumophila* have been associated with TTP.

HUS is strongly associated with *Escherichia coli* O157:H7, which produces the Shiga-like toxins Stx-1 and Stx-2.

These toxins damage endothelial cells and cause increased release of inflammatory mediators by these cells [64, 74]. Other microorganisms reported in association with HUS are *Pneumococcus pneumoniae*, *Yersinia pseudotuberculosis*, *Salmonella typhi*, varicella, coxsackie enteroviruses, and ECHO virus [75–81].

Reports of the hereditary forms of TTP/HUS suggest a genetic predisposition in pathogenesis of this disorder in some patients [82, 83].

The characteristic clinicopathologic features of TTP/HUS are fluctuating neurologic symptoms (more in TTP), acute renal failure, and diarrhea (more in HUS), fever, microangiopathic hemolytic anemia, and thrombocytopenia with purpura, but usually not severe bleeding [62, 63, 84, 85].

Blood smears demonstrate moderate to severe anemia with anisopoikilocytosis, schistocytosis, and reticulocytosis. Nucleated red cells are often present. Thrombocytopenia is marked and in some patients it is <20,000/μL. The most prominent histologic finding is disseminated platelet/fibrin thrombi in the capillaries and arterioles of various tissues such as brain, kidney, pancreas, adrenal glands, and heart (Figure 24.5). Microthrombi are rarely detected in bone marrow biopsy sections.

Plasma exchange is the treatment of choice for all but minimally affected patients. Other therapeutic approaches include administration of corticosteroids, intravenous immunoglobulin, and antiplatelet aggregation agents [86].

Differential diagnosis includes conditions that are associated with hemolysis and thrombocytopenia, such as Evans syndrome (autoimmune hemolytic anemia and thrombocytopenia), DIC, systemic lupus erythematosus, microangiopathic hemolytic anemia in patients with disseminated cancers, and paroxysmal nocturnal hemoglobinuria.

Other Conditions Associated with Thrombocytopenia

A wide spectrum of clinical conditions are associated with thrombocytopenia, such as infections, glomerulonephritis, renal transplant rejection, renal vein thrombosis, DIC, giant cavernous hemangioma, burns, snake bites, aortic valvular disease, and primary pulmonary hypertension. Certain drugs such as hematin, protamine sulfate, bleomycin, and heparin can have direct damaging effects on platelets and cause thrombocytopenia. Administration of antithymocyte globulin may cause thrombocytopenia possibly due to the formation of immune complexes. Splenomegaly may cause thrombocytopenia due to sequestration of the platelets.

QUALITATIVE DISORDERS OF PLATELETS AND MEGAKARYOCYTES

Introduction

Analogous to the quantitative disorders just discussed, qualitative defects of platelet function result in a hemorrhagic

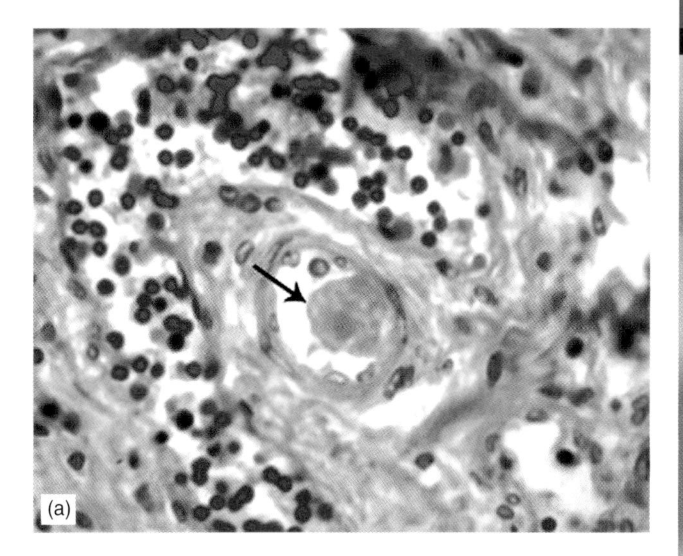

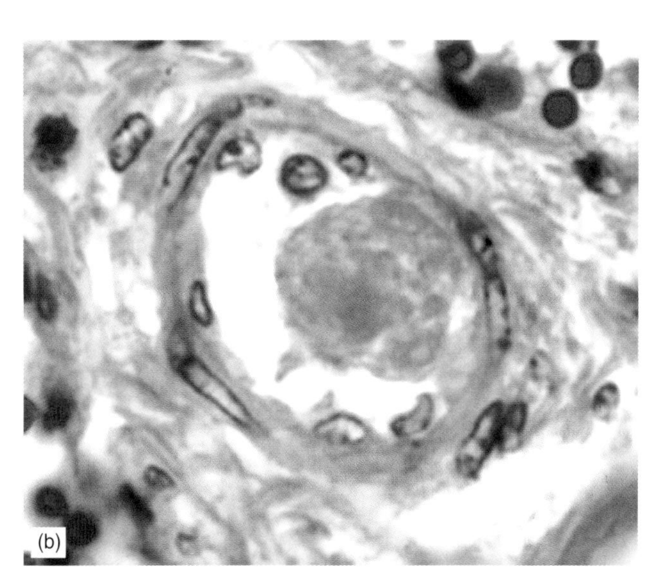

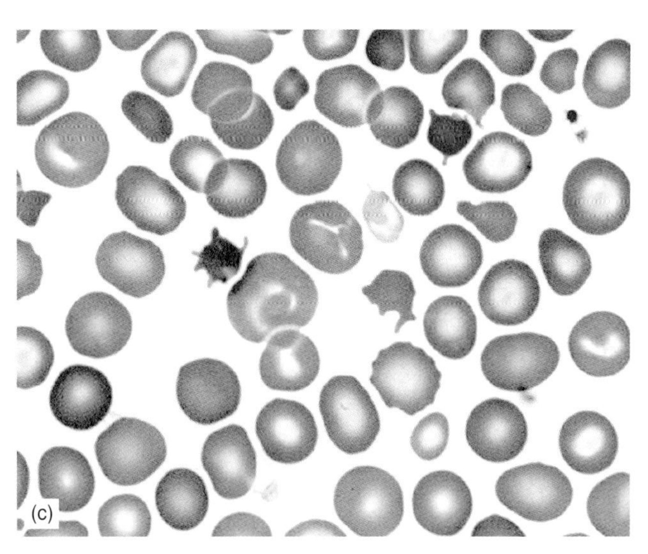

FIGURE 24.5 Platelet/fibrin thrombus in a patient with TTP: (a) low power and (b) high power views; (c) blood smear showing fragmented red cells.

diathesis that most commonly manifests as mucocutaneous bleeding. The main purpose of this section is to provide important diagnostic information useful for evaluating the cause of bleeding in patients suspected of having platelet function disorders. As for the patient with bleeding due to thrombocytopenia, a thorough history and physical examination are important components of the overall diagnostic evaluation to identify the specific abnormality underlying hemorrhage in patients with dysfunctional platelets [87]. The qualitative disorders of platelet function can be divided and discussed within the context of two broad categories based on whether they are inherited and present at birth or are acquired after birth and manifest at some point later in life. Although acquired defects are by far more common, we first discuss the inherited disorders since they cause specific biochemical abnormalities and typically predominantly impair only one of the major aspects of platelet function including adhesion, activation, secretion, aggregation, and formation of a surface for concentrating and facilitating interactions between proteins of the coagulation system [88, 89]. Because for most acquired disorders the specific biochemical abnormality is poorly understood and more than one aspect of platelet function is usually impaired, these disorders are discussed here in the context of their associated underlying medical condition, which is usually evident [90]. Herein, we provide a brief overview of the most common platelet function disorders.

CONGENITAL PLATELET DISORDERS

Bernard–Soulier Syndrome

The Bernard–Soulier syndrome (BSS), previously known as hemorrhagic thrombocytopathic dystrophy, was first reported in 1948 by Jean Bernard and Jean-Pierre Soulier, who described a child from consanguineous parents that had mucocutaneous bleeding out of proportion to what would be expected based solely on the mildly decreased platelet count [88, 91]. Characterized by a prolonged bleeding time, macrothrombocytes (i.e. giant platelets) and thrombocytopenia, with counts ranging between 20,000 and 100,000 platelets/μL, the BSS is an inherited autosomal recessive hemorrhagic disorder that is characterized by a deficient and/or dysfunctional platelet membrane glycoprotein (GP) complex known as GPIb-IX-V, which is the main receptor for VWF. Although BSS is extremely rare, with only about 100 cases reported in the literature thus far [91], the discovery of this syndrome, together with the finding – from numerous subsequent studies performed in many laboratories – that platelets from these patients display abnormalities in adhesion due to an inability to interact with VWF, ultimately led to the discovery of ristocetin (and botrocetin) and the development of new *in vitro* assays useful in the diagnosis and management of patients with von Willebrand's disease (VWD) and platelet function disorders.

Etiology and Pathogenesis

GPIb-IX-V plays a major role in primary hemostasis as it represents the main platelet surface receptor for VWF. Because platelets from BSS patients have a defective and/or deficient GPIb-IX-V complex, they are unable to bind the VWF that is present at a high concentration in the subendothelial ECM and therefore cannot adhere adequately to sites of vascular injury. GPIb-IX-V is a macromolecular complex that comprises four distinct membrane-spanning glycoproteins, all belonging to the leucine-rich motif containing protein family [92]. Each of these glycoproteins, referred to as GPIbα, GPIbβ, GPIX, and GPV, is synthesized and assembled in mature megakaryocytes, the precursors of mature circulating platelets, to form the functional VWF receptor [92–96]. All four subunits of this receptor are encoded by single copy genes (*GPIBA*, *GPIBB*, *GPIX*, and *GPV*) that are exclusively expressed in the platelet lineage [97–100]. GPIb consists of two disulfide linked subunits, GPIbα (135 kDa) and GPIbβ (26 kDa), both of which must undergo normal posttranslational processing for the entire GPIb-IX-V complex to be expressed at wild-type levels on the platelet surface [101, 102]. Although GPIX (20 kDa) and GPV (82 kDa) both bind GPIb noncovalently, only the interaction with GPIX is essential for normal expression (and function) of this VWF receptor. Since the associations between GPIbα, GPIbβ, and GPIX are required for efficient transport of the entire GPIb-IX-V complex to the platelet membrane, and the absence of even a single subunit can abolish its surface expression, their biosynthesis must be coordinated and tightly regulated [88]. As described later, this is consistent with the fact that loss-of-function mutations have been identified in the genes encoding GPIBA, GPIBB, and GPIX of patients with the BSS. Because, in contrast, no mutations have, as of yet, been identified in the gene encoding GPV in BSS patients, GPV is likely to be less tightly associated with the other three proteins comprising the GPIb-IX-V complex [103]. The fact that knock-out mice, which have had a targeted disruption of their GPV gene, do not develop a BSS-like disorder is consistent with this notion [104, 105]. Although these mice were found to be hyper-responsive to thrombin activation, they demonstrated that a congenital absence of GPV does not prevent the surface expression of the heterotrimeric complex GPIb-IX, which itself is capable of VWF binding.

The GPIb-IX-V complex is the major link between the platelet plasma membrane and the underlying actin cytoskeleton, through an intermediate interaction with filamin. Because these interactions restrict the mobility of the spectrin-based membrane skeleton, the loss of an actin filament attachment to the plasma membrane explains the large size of BSS platelets as well as their fragility, both of which likely contribute to the thrombocytopenia observed in patients with this disorder.

Morphology and Immunophenotype

Platelet counts typically are only mildly decreased (80,000–100,000 platelets/μL) but can vary substantially and even be normal. Because electronic cell counters frequently underestimate the actual platelet count in patients with macrothrombocytes and therefore yield an artificially low mean platelet volume (MPV), a manual platelet count should be performed in patients suspected of having the BSS or

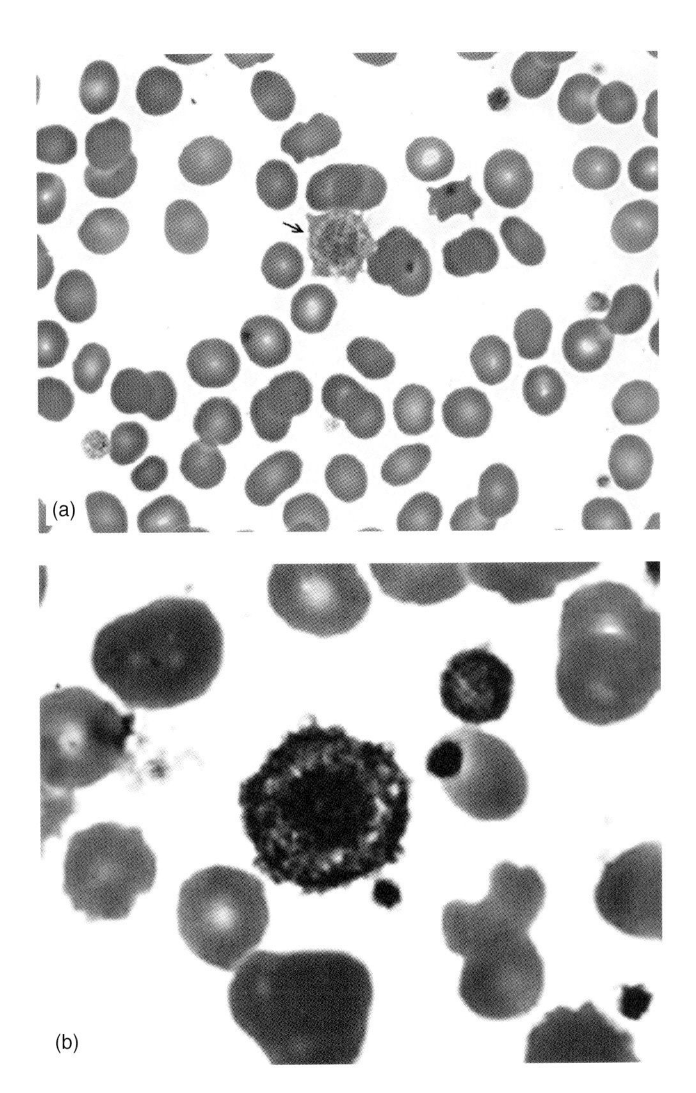

FIGURE 24.6 Blood smear demonstrates giant platelet: (a) low power (arrow) and (b) high power views.

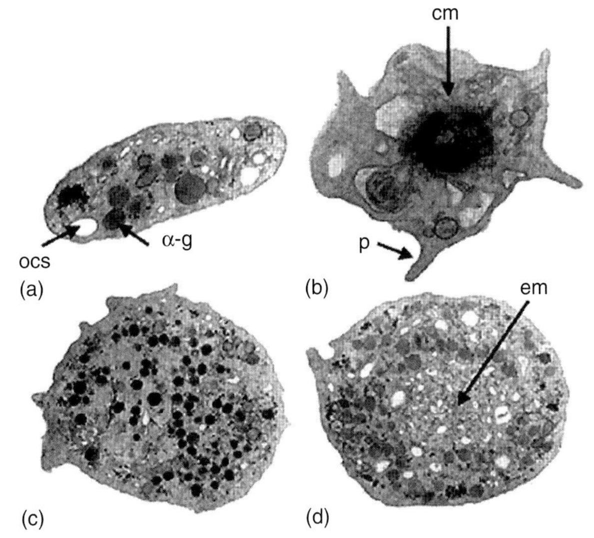

FIGURE 24.7 Electron microscopy of platelets in the BSS. (a) A normal discoid platelet showing both α-granules (α-g) and the open canalicular system (ocs). (b) A normal platelet after thrombin stimulation showing its rounded form, dark central mass (cm) of contractile protein, protruding pseudopods, but no granules. (c) A giant platelet from a BSS patient that is rounded in shape and has abundant granules distributed throughout the cytoplasm. (d) A giant platelet from a May–Hegglin anomaly patient that is similarly appearing except for the cytoplasmic zones rich in entangled membranes (em). From Alan D. Michelson (2007). *Platelets*, 2nd ed. (Figure 57.1), Academic Press, Amsterdam, by permission.

one of the other disorders accompanied by giant platelets (Figure 24.6). During light microscopic examination of peripheral blood smears from patients with the BSS, giant platelets up to 20 μm in diameter are frequently observed and can comprise up to 80% of the entire platelet population. In ultrastructural examinations by electron microscopy, platelets are observed to have rounded shapes (instead of the normal discoid structure) and an abundant supply of granules distributed throughout the cytoplasm except in zones enriched for vacuoles and entangled membrane complexes (Figure 24.7). Precursor structural correlates of these ultrastructural abnormalities have been observed in megakaryocytes, with a dystrophic demarcation membrane system being one of the most common. Ultrastructural abnormalities of the nuclease, such as an increased ploidy in mature megakaryocytes, are also present [106].

Because the BSS is the consequence of an inherited deficiency of the GPIb-IX-V complex, flow cytometric analysis of platelets from suspected patients, using monoclonal antibodies specific for the component proteins GPIb, GPIX, and GPV, represents a rapid diagnostic strategy to identify individuals who are homozygous and heterozygous for this disorder. With flow cytometry, giant BSS platelets can be analyzed directly in whole blood without having to be separated from the similar-sized red and white blood cells. However, light scatter (LS) gates may require adjustment when evaluating patients with macrothrombocyte syndromes such as the BSS, since forward-LS (F-LS) correlates with platelet size. Because this adjustment can result in some overlap of the F-LS from giant platelets with that from a subpopulation of both red and white blood cells, it is essential to include a platelet-specific monoclonal antibody in the assay as a platelet identifier. The identifier antibody used will depend on the specific platelet disorder that is suspected; for BSS platelets, this antibody cannot be GPIb-, GPIX-, or GPV-specific.

Molecular and Cytogenetic Studies

GPIbα, GPIbβ, GPIX, and GPV are encoded by four distinct genes – *GP1BA*, *GP1BB*, *GP9*, and *GP5*, respectively – which are expressed solely within the megakaryocyte/platelet lineage. Almost 45 distinct loss-of-function genetic defects associated with the BSS have been reported; most of these can be reviewed online at the web site for the database of BSS mutations (Figure 24.8) [107, 108]. Because 19 different mutant alleles of *GP1BA*, which is located on chromosome (chr)17p12, have been identified thus far, defects in this gene represent the most common cause of the BSS. This is likely because *GP1BA* encodes GPIbα, which is the largest protein of the complex and bears the VWF-binding site. Although together 25 different mutations have been identified in *GP1BB* ($N = 15$), located on chr22q11.2, and *GP9* ($N = 10$), located on chr3q29, no loss-of-function alleles of

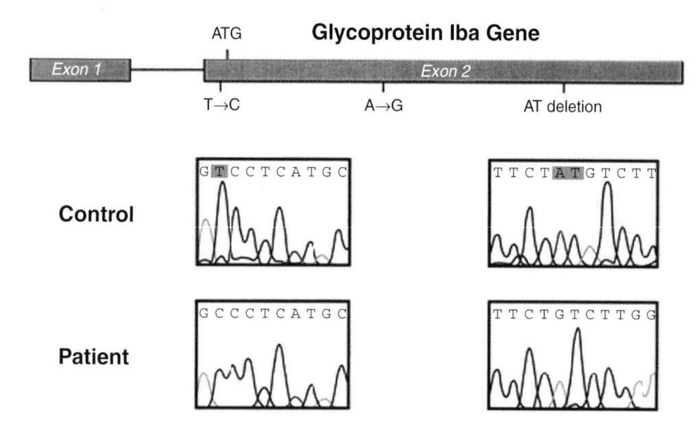

FIGURE 24.8 Mutations found in the GPIbα gene of a BSS patient. Sequencing results from the patient and from a control subject are compared in the bottom panels, with the mutated nucleotides shown enclosed in gray boxes. From Afshar-Kharghan V, López JA. (1997). Bernard–Soulier syndrome caused by a dinucleotide deletion and reading frameshift in the region encoding the glycoprotein Ib alpha transmembrane domain. *Blood* **90**, 2634–43, by permission.

GP5 (chr3q21) have been found in patients with the BSS. As described earlier, this is consistent with the fact that the coordinate expression of GPIbα, GPIbβ, and GPIX, and their subsequent association, is essential for efficient transport of the overall complex to the platelet surface, whereas GPV biosynthesis is known to be expendable for the expression and function of the VWF receptor.

That the BSS usually exhibits an autosomal recessive pattern of inheritance typically means that affected children are born to consanguineous parents, both of which are carriers for (and "passed on") the same loss-of-function allele of the *GP1BA*, *GP1BB*, or *GP9* gene. There are a number of rare exceptions to this, however, as the disease can occur in association with some syndromes associated with chromosomal abnormalities or can occur in an autosomal dominant manner. This means that rarely the disease can be inherited from just one parent who carries the gene for the abnormality.

Clinical Aspects

The BSS is extremely rare with a prevalence estimated at less than one patient per one million people. Consistent with its autosomal recessive pattern of inheritance, BSS patients can be either male or female. Most of the approximately 100 cases reported since 1948 have presented with an identical disorder to that found in the original patients described by Bernard and Soulier: namely, a severe hemorrhagic defect characterized by a prolonged bleeding time and a low platelet count with very large platelets (i.e. a macrothrombocytopenia). In most of these patients, the bleeding symptoms manifested shortly after birth or during early childhood. Clinical manifestations usually include easy bruisability, petechiae, epistaxis, and gingival bleeding, and less frequently gastrointestinal bleeding and hematuria. In older female patients, menorrhagia is common and can be severe. Episodes of severe bleeding are usually associated with trauma and surgical procedures such as tonsillectomy, appendectomy, and splenectomy, but also occur during dental extractions and menses. Although the frequency and severity of hemorrhagic episodes can vary from patient to patient, even within the same family, bleeding mainly affects mucocutaneous tissues, with joint bleeds, major hematomas, and intracranial hemorrhages only very rarely being observed.

Quality care for BSS patients involves both prophylactic and therapeutic measures. Effective prophylaxis requires good dental hygiene, educating patients how to locally control nosebleeds, and frequent follow-up for premenopausal women. Although platelet transfusion is the only broadly and consistently effective therapy for severe bleeding, it must be reserved only for those patients with uncontrollable or life-threatening hemorrhage so as to avoid unnecessary exposures to foreign antigens and the possible development of refractoriness to future platelet transfusions due to alloimmunization. However, most BSS patients require transfusions at multiple points in their lives. Alternative therapy with DDAVP has been employed successfully for quite some time. Infusions of recombinant factor VIIa (rFVIIa), also known as NovoSeven®, have been found to be effective in many platelet function disorders and are therefore increasingly being employed. In women of childbearing age, menorrhagia is quite responsive to hormonal regulation therapy in most cases but occasionally requires supplementation with one of the strategies described earlier.

Differential Diagnosis

Functional analysis of platelets by aggregometry, using a patient's platelet-rich plasma (PRP), is necessary to differentiate the BSS from other rare, inherited platelet function disorders accompanied by macrothrombocytopenia that include (1) *MYH9*-related diseases (i.e. May–Hegglin anomaly, Sebastian syndrome, Fechtner syndrome, and Epstein syndrome); (2) DiGeorge syndrome (i.e. velocardiofacial syndrome [VCFS]); (3) benign Mediterranean macrothrombocytopenia; (4) platelet-type VWD; (5) gray platelet syndrome; (6) Montreal platelet syndrome; (7) macrothrombocytopenia with platelet expression of glycophorin-A; and (8) macrothrombocytopenia with platelet β1-tubulin Q43P polymorphism. In addition to macrothrombocytopenia, therefore, a diagnosis of the BSS is based on a prolonged bleeding time, defective ristocetin-induced platelet agglutination (RIPA), and low or absent platelet levels of the GPIb-IX-V complex (CD42a–d) by flow cytometry.

Glanzmann Thrombasthenia

Glanzmann thrombasthenia (GT) was first described in 1918 by Eduard Glanzmann in a patient with recurrent mucocutaneous bleeding, a prolonged bleeding time, and an isolated, rather than clumped, appearance of platelets on a peripheral blood smear. Nearly 40 years later, Braunsteiner and Pakesch described GT as an inherited bleeding disorder characterized by normal size platelets that neither spread on surfaces nor support clot retraction [109]. Caen *et al.* subsequently established the absence of platelet aggregation as the hallmark diagnostic feature of GT in a report on 15 French patients [110]. To date, impaired or absent aggregation to all known physiologic agonists remains the defining characteristic of GT, the most common platelet function disorder. GT exhibits autosomal recessive inheritance and is caused by a deficiency and/or dysfunction of the αIIbβ3 integrin, a heterodimeric platelet membrane protein also known as GPIIb-IIIa. Platelet-to-platelet interactions essential for thrombus propagation at sites of vascular injury are not possible in GT patients because wild-type αIIbβ3 functions by binding the aggregative plasma proteins fibrinogen and VWF, which in turn converts the initial layer of adherent platelets into a reactive surface that supports continued platelet deposition. Although GT is a rare Mendelian disorder, studies of GT platelets have helped elucidate important structure/function properties of αIIbβ3 and have led to the development of a widely used class of antithrombotic agents including Abciximab, Eptifibatide, and Tirofiban, which function by binding and inhibiting the conformationally active form of this integrin [111].

Etiology and Pathogenesis

GPIIb-IIIa is expressed exclusively by megakaryocytes [26], and under normal conditions, it is the most abundant integral platelet plasma membrane protein, with up to 80,000 copies per cell [112–114]. Such a high surface density is important because αIIbβ3 is a multifaceted receptor that is essential for both initiating and propagating thrombus formation. Despite the fact that GPIb-IX-V mediates the first adhesive interaction of platelets with the subendothelial matrix by binding VWF molecules immobilized within its collagen fibers, these interactions are transient and break if additional contacts between other platelet membrane proteins and the ECM are not made to stabilize them. GPIIb-IIIa is required for successful initiation of thrombus formation as these additional contacts are mediated largely by the activated form of αIIbβ3, which binds tightly to VWF through a high-affinity interaction with the peptide sequence Arg–Gly–Asp (RGD). Activated αIIbβ3 is essential for thrombus growth as well, because it is also the principal receptor for plasma fibrinogen and VWF, the soluble proteins required for continued deposition of platelets through aggregation, the *in vivo* process by which nonadherent platelets are tethered to adherent platelets in the initial monolayer. Wild-type αIIbβ3 molecules are also essential for platelets, suspended in plasma, to aggregate in response to the various agonists that are used *in vitro* to test platelet functioning by aggregometry. Central for normal αIIbβ3 function is its ability to undergo an inducible allosteric switch from a bent, low-affinity resting form to an extended, active conformation that has a high affinity for the adhesive and aggregative ligands mentioned earlier. This affinity modulation is initiated when various platelet adhesion and/or G-protein-coupled receptors are stimulated, through interactions with one or more of several agonists, to generate second messengers that bind the cytoplasmic tails of αIIbβ3 and, in a process referred to as "inside-out" signaling, [115–118] trigger a switch of its extracellular domain to a conformation competent for binding multivalent VWF and divalent fibrinogen molecules. Binding of αIIbβ3 to either molecule induces "outside-in" signaling, a process that leads to the assembly of signaling and actin/myosin complexes on the cytoplasmic domains of activated integrins [119].

Because αIIbβ3 is deficient and/or defective in GT, platelets from these patients are unable to stably adhere to or aggregate at sites of vascular injury. Additionally, aggregation is absent or markedly diminished in response to all agonists used in aggregometry except ristocetin, since this bacterial glycopeptide induces agglutination (not aggregation) through an integrin-independent mechanism that promotes interaction between VWF and GPIb-IX-V, both of which are wild-type in GT. The inability of GT platelets to stably adhere to or aggregate on exposed subendothelial tissues is consistent with the fact that they do not spread on these surfaces or support clot retraction.

Morphology and Immunophenotype

Platelet counts in patients with GT are typically normal, whether performed with electronic cell counters or manually. Routinely measured parameters of platelet structure, such as MPV, are also generally within normal limits. In light microscopic examination of peripheral blood smears and ultrastructural examinations by electron microscopy, platelets from GT patients are morphologically normal. Because GT is caused by a deficiency of GPIIb-IIIa, also designated CD41/CD61, flow cytometric analysis of platelets from suspected patients, using monoclonal antibodies specific for the component proteins αIIb and β3, represents a rapid diagnostic strategy to identify homozygotes and heterozygotes for this disorder [120, 121]. Once platelet function studies demonstrate absent aggregation to all agonists, flow cytometry is the method of choice to analyze platelet surface glycoproteins and screen for GT because it can be performed on small blood samples and allows diagnoses to be made in children.

Molecular and Cytogenetic Features

Like all integrins, αIIbβ3 comprises an alpha (αIIb) and beta (β3) heterodimer with the two subunits associated noncovalently in a 1:1 complex. Since association between αIIb and β3 – which are encoded by the single copy genes *ITGA2B* and *ITGB3*, respectively, located adjacently on chr17q21-23 – is required for efficient transport of this integrin to the platelet membrane, absence of either subunit can abolish its surface expression. This is consistent with the fact that (1) non-complexed or incorrectly folded subunits are not processed in the Golgi apparatus and undergo intracellular degradation and (2) loss-of-function mutations have been identified in both *ITGA2B* and *ITGB3* in unrelated GT patients. A database is available on the Internet that currently lists more than 100 different GT-causing mutations. Although large deletions are rare, single-base-substitution abnormalities encoding missense, nonsense, and splice-site mutations are common, as are small deletions and insertions. Missense mutants of both *ITGA2B* and *ITGB3*, which encode aberrant αIIb and β3 proteins, respectively, that differ from the wild-type subunits by a single amino acid, represent the most common overall cause of GT.

Consistent with its autosomal recessive inheritance, GT patients are most frequently born to consanguineous parents and are homozygous for a mutation in either *ITGA2B* or *ITGB3*. However, a large number of compound heterozygous patients with two distinct loss-of-function *ITGA2B* or *ITGB3* alleles have been reported. *ITGA2B* mutations, which have been found in all functional regions of this 30 exon containing approximately 17 kb gene, are somewhat more common than mutations in the approximately

65 kb *ITGB3* locus, which have been identified in most of its 15 exons and flanking intronic sequences. Although most patients have *ITGA2B* or *ITGB3* defects that result in either the absence of (type I) or a severely decreased level of (type II) αIIbβ3 on the surface of their platelets, some carry one (or two) of a number of missense *ITGA2B* or *ITGB3* alleles that encode a dysfunctional form of the protein. Such patients comprise a category of this disorder referred to as variant GT because they express immunologically recognizable αIIbβ3 molecules that are unable to bind fibrinogen. Studies performed on subjects with variant GT, which differ from those with type I or type II disease and comprise a heterogeneous collection of patients with a broad range of bleeding diatheses from severe to mild, have been very useful in elucidating integrin function.

Although αIIbβ3 is normally found solely in platelets and because αIIb is expressed only by megakaryocytes, β3 is expressed by a number of other cell types including endothelial cells, osteoblasts, chondrocytes, fibroblasts, smooth muscle cells, monocytes, and certain lymphocytes. Despite this and the fact that platelets express a second, less abundant (i.e. ~50 copies/cell) β3-integrin, the vitronectin receptor αVβ3, patients with *ITGB3* defects do not appear to have a more severe form of the disease for unknown reasons.

Clinical Aspects

As an autosomal recessive disorder, GT affects both males and females. Patients typically manifest moderate to severe mucocutaneous bleeding, with purpura, epistaxis, gingival hemorrhage, and menorrhagia being the most common at presentation [122, 143]. However, there is considerable variability in the associated bleeding diatheses, even in patients with the same gene abnormality, with some manifesting only easy bruisability and others severe, life-threatening hemorrhages. Although genitourinary and gastrointestinal hemorrhages also occur, deep-visceral, intra-articular, and intramuscular bleeding is rare. Although all patients share the hallmark diagnostic feature of GT, that is, the absence of platelet aggregation in response to all known agonists, they differ with respect to whether their platelets support residual clot retraction or express immunologically detectable surface αIIbβ3 molecules. Patients whose platelets lack clot retraction and express <5% of the mean normal αIIbβ3 level on their surface are defined as having type I GT. Platelets from patients with type II disease demonstrate residual clot retraction and express up to 20% of the normal αIIbβ3 level. In contrast, platelets from variant GT patients have nearly normal to normal surface levels of a dysfunctional αIIbβ3 molecule. Consequently, although platelets from all patients with variant disease lack agonist-induced aggregation, they have a variable ability to support clot retraction ranging from normal to absent, which likely underlies their heterogeneity in bleeding symptoms that range from mild to severe.

Platelet transfusion is the treatment of choice for severe bleeding episodes, but some patients become refractory due to alloimmunization, presumably because they lack preexisting tolerance to the wild-type allele of either the αIIb or the β3 subunit of GPIIb-IIIa, the mismatched antigen(s) of the HLA system, or both. Although rFVIIa infusion may represent a therapeutic alternative to platelet transfusion, especially in those with antiplatelet antibodies, it evidently is not effective in all patients. However, invasive procedures or surgery can be performed in rFVIIa-unresponsive alloimmunized patients if their platelet antibody titers are first transiently reduced using protein-A immunoabsorption. As a last resort, bone marrow transplantation can be used successfully to treat and cure severe hemorrhagic, alloimmunized GT patients.

Differential Diagnosis

GT is the only disease in which platelet aggregation is absent or markedly impaired in response to all physiologic agonists used in aggregometry, including adenosine diphosphate (ADP), collagen, epinephrine, arachidonic acid, and thrombin. Because platelets from GT patients are also normal in size and agglutinate normally in the RIPA assay, one can readily exclude the BSS, as platelets from these patients are unusually large and despite aggregating normally in response to all agonists do not agglutinate in the RIPA assay. Furthermore, additional inherited disorders can be excluded including the nonmuscle myosin heavy chain IIA-associated syndromes (i.e., May–Hegglin anomaly, Fechtner syndrome, Sebastian syndrome, and Epstein syndrome) and the Montreal platelet syndrome, which are all associated with abnormally large platelets, and the Wiskott–Aldrich syndrome (WAS), in which abnormally small platelets are observed.

Although congenital GT patients typically present with hemorrhage in early childhood, acquired causes must be ruled out when there is no family history. Acquired GT has been reported in patients with acute promyelocytic leukemia (APL) [123]. Analogous to congenital GT, platelets from these patients were deficient in αIIbβ3 and failed to aggregate in response to all agonists. Although not definitively demonstrated at the DNA level, a disruption of either *ITGA2B* or *ITGB3* is the likely etiology of acquired GT in this setting as chr17 breakpoints underlying the chr15-17 translocations in most APL patients are heterogeneous and known to occur occasionally at 17q21, the location of these genes [124]. Other possible etiologies in the differential diagnosis for congenital bleeding disorders can be excluded on this basis. For instance, impaired platelet aggregation, specifically in response to ADP or collagen, implies defects in either the primary receptors for these agonists or the signaling pathways they induce.

Wiskott–Aldrich Syndrome

Wiskott–Aldrich syndrome (WAS) is an X-linked congenital disorder characterized by severe thrombocytopenia, immunodeficiency, recurrent infections, and eczema [125–129]. The Wiskott–Aldrich syndrome protein (WASP) structural gene (*WASP*) resides at Xp11.22-23. About 300 *WASP* mutations have been discovered, including single-base substitutions, deletions, insertions, and splice-site mutations [128, 130–133]. The WASP appears to play critical roles in signal transduction and in regulating cytoskeletal reorganization [130, 131].

Approximately 40% of the patients with WAS demonstrate an associated autoimmune disorder. Hemolytic anemia, vasculitis, nephritis, and inflammatory bowel disorders are the most frequently encountered [128, 134]. Thrombocytopenia is usually severe and <50,000/μL. Platelets are small (about half the average size of normal platelets) and are associated with functional defects and shortened survival [129, 135]. A variant of WAS with milder clinical presentation exists, which is referred to as X-linked thrombocytopenia (XLT) [130].

Disorders of Platelet Secretion and Granule Deficiencies

These disorders fall into two major categories: (1) deficiencies in the platelet granules or their contents and (2) defects in the ability of platelets to release their contents [136, 137]. Platelet dense granules contain serotonin, pyrophosphate, calcium, ADP, and ATP [138, 139]. Platelet α-granules contain numerous proteins such as fibrinogen, platelet factor 4, platelet-derived growth factor, β-thromboglobulin, factor V, VWF, and high-molecular-weight kininogen [137, 140]. In addition to dense and α-granules, platelets contain vesicles with acid hydrolases. These vesicles are involved in the arachidonic acid pathways and thromboxane A_2 production.

Dense Granule Deficiency

Dense granule deficiency (or δ-storage pool disease) is characterized by marked decrease or absence of platelet dense granules [136–138, 141]. Platelets are normal in size and show unremarkable ultrastructural features, except for a marked decline or absence of dense granules [136, 138, 141].

Patients demonstrate a mild to moderate bleeding diathesis such as easy bruising, epistaxis, gingival bleeding, and menorrhagia. The bleeding time is prolonged and *in vitro* platelet function studies reveal absence of the second wave of aggregation with either ADP or epinephrine and an impaired aggregation response to collagen [142]. Dense granule deficiency has been associated with a number of other congenital abnormalities such as Hermansky–Pudlak syndrome (oculocutaneous albinism and increased ceroid in the monocytic/histiocytic system), Chediak–Higashi syndrome, WAS, and congenital thrombocytopenia with absent radii (TAR) syndrome [125, 143, 144].

Alpha-Granule Deficiency

Platelet α-granule deficiency, *gray platelet syndrome* or *α-storage pool disease*, is caused by the reduction or absence of platelet α-granules (Figure 24.9) [136, 145–147]. The affected patients have a history of a bleeding diathesis and demonstrate a mild thrombocytopenia. Platelets appear grayish in blood smears stained with Wright's stain. Ultrastructural studies reveal absence or markedly reduced numbers of α-granules in the affected platelets. The affected megakaryocytes also show absent or decreased numbers of α-granules [148]. The basic defect seems to be the inability of megakaryocytes to transfer endogenously synthesized proteins into α-granule precursors [145]. Affected patients show a prolonged bleeding time and variable responses in platelet aggregation studies. The most consistent finding is impairment of thrombin-induced aggregation [142].

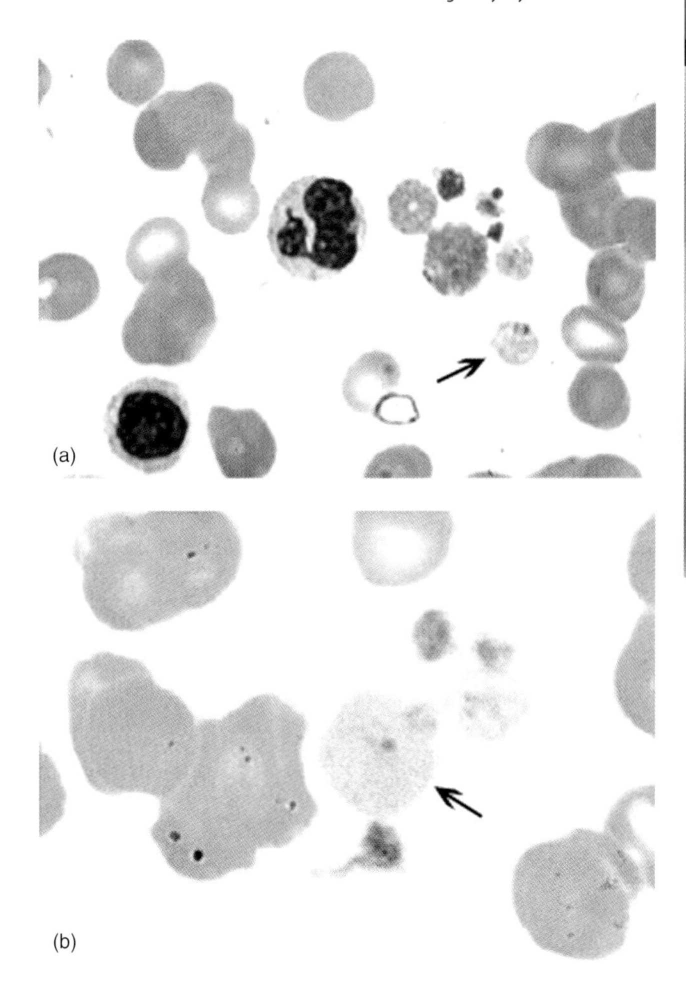

FIGURE 24.9 Blood smears demonstrate markedly hypogranular platelets (a and b, arrows).

Abnormalities in Platelet Arachidonic Acid Pathways

Abnormalities of arachidonic acid pathways are extremely rare and are of two major types: (1) defect in the release of arachidonic acid from phospholipids and (2) deficiencies of cyclooxygenase or thromboxane synthetase [142, 149]. Affected patients are usually adults and often demonstrate mild to moderate hemorrhages [142]. Severe bleeding is rare [150].

Formation of thromboxane A_2 is one of the major responses of platelets during activation. Thromboxane A_2 is necessary for platelet secretion during the stimulation of platelets with ADP, epinephrine, and low concentration of collagen and thrombin. Thromboxane A_2 is also a potent vasoconstrictor [142].

ACQUIRED PLATELET DISORDERS

Drug-Induced Disorders

Many commonly used drugs are known to affect platelet function. Although some of these agents were developed mainly for treating patients at risk for thromboembolism because of their ability to inhibit specifically one (or more) of the several distinct molecular events required for normal platelet function, and therefore to impair primary hemostasis, most were developed for clinical indications unrelated to their hemostatic effects and were found subsequently to nonspecifically inhibit platelet function. Because the mechanism(s) by which the agents in the former group impair platelet function have been studied extensively and are well established, they will only be discussed briefly here. However, because much of our knowledge in this area on drugs from the latter, much larger group comes from *in vitro* studies performed on platelets exposed to one pharmacologic agent at a time, the overall impact on the hemostasis system *in vivo* has not been established for most of them. Furthermore, the clinical relevance of this knowledge is not clear since most patients are administered more than one drug simultaneously. Inhibitors of platelet cyclooxygenase-1 (COX-1), including aspirin and other non-steroidal anti-inflammatory drugs (NSAIDs), are among the most commonly used medications.

Aspirin

Aspirin irreversibly inactivates cyclooxygenase and thereby inhibits production of thromboxane A_2 from arachidonic acid and impairs platelet secretion [151]. The end result is defective platelet aggregation and prolonged bleeding time. Prolongation of the bleeding time may last up to 4 days after administration of aspirin is stopped [152, 153]. Ethanol ingestion may enhance prolongation of the bleeding time in patients who take aspirin [154]. Other NSAIDs also inhibit COX-1 but do so reversibly.

Beta-Lactam Antibiotics

Beta-lactam antibiotics, such as penicillin and cephalosporin derivatives, may prolong bleeding time and induce abnormal platelet aggregation [155–157]. These antibiotics seem to interfere with the function of platelet membrane integrins such as GPIIb-IIIa and GPIa-IIa. The effect is dose- and duration-dependent.

Others

Excessive garlic ingestion may induce platelet dysfunction and inhibits cyclooxygenase activity [158, 159]. Long-term dietary supplementation with marine oils reduces the platelet content of arachidonic acid and may cause abnormal platelet aggregation and slight prolongation of the bleeding time [160]. Dextran may slightly prolong the bleeding time without increasing operative or postoperative bleeding. Therefore, it has been used for the prevention of postsurgical thromboembolic complications [161, 162].

PLATELET DYSFUNCTION ASSOCIATED WITH PATHOLOGIC CONDITIONS

Cardiopulmonary Bypass

Prolonged bleeding time, abnormal platelet aggregation, and thrombocytopenia are some of the common features of cardiopulmonary bypass [163, 164]. During bypass surgery, platelets adhere to fibrinogen absorbed by the bypass circuit. Bypass procedures also enhance thrombin and ADP generation and complement activation [165]. Mechanical trauma from the bypass pump may also degranulate platelets.

Chronic Renal Failure

Uremia may lead to platelet dysfunction and abnormal aggregation [166, 167]. The bleeding time is often prolonged, and there may be bleeding manifestations such as purpura, epistaxis, menorrhagia, gastrointestinal bleeding, and hematuria.

Hematologic Disorders

Abnormal platelet function and morphology may occur in association with myelodysplastic syndromes, myeloproliferative disorders, and acute myelogenous leukemia (see Chapters 9–12). Abnormal platelet functions include decreased platelet aggregation and secretion in response to ADP, epinephrine, and collagen, and reduced platelet procoagulant activity. Morphologic changes include abnormal shapes, giant forms, and hypogranularity. A case of hairy-cell leukemia with abnormal platelet morphology and severe platelet dysfunction has been reported [168].

References

1. Burkhardt R. (1988). Bone marrow in megakaryocytic disorders. *Hematol Oncol Clin North Am* **2**, 695–733.
2. Samsygina GA. (1991). Thrombocytopenic states in the neonatal period. *Gematol Transfuziol* **36**, 14–17.
3. Naeim F. (1998). *Pathology of Bone Marrow*, 2nd ed. Williams & Wilkins, Baltimore.
4. Slater LM, Katz J, Walter B, Armentrout SA. (1985). Aplastic anemia occurring as amegakaryocytic thrombocytopenia with and without an inhibitor of granulopoiesis. *Am J Hematol* **18**, 251–4.
5. Emberger JM, Izarn P, Gueraud L. (1973). Thymic aplasia and amegakaryocytosis. *Lancet* **1**, 159.
6. Alter BP. (2002). Bone marrow failure syndromes in children. *Pediatr Clin North Am* **49**, 973–88.
7. Geddis AE. (2006). Inherited thrombocytopenia, congenital amegakaryocytic thrombocytopenia and thrombocytopenia with absent radii. *Semin Hematol* **43**, 196–203.
8. Letestu R, Vitrat N, Massé A, Le Couedic JP, Lazar V, Rameau P, Wendling F, Vuillier J, Boutard P, Plouvier E, Plasse M, Favier R, Vainchenker W, Debili N. (2000). Existence of a differentiation blockage at the stage of a megakaryocyte precursor in the thrombocytopenia and absent radii (TAR) syndrome. *Blood* **95**, 1633–41.
9. Ballmaier M, Germeshausen M, Schulze H, Cherkaoui K, Lang S, Gaudig A, Krukemeier S, Eilers M, Strauss G, Welte K. (2001). c-mpl mutations are the cause of congenital amegakaryocytic thrombocytopenia. *Blood* **97**, 139–46.

10. Germeshausen M, Ballmaier M, Welte K. (2006). MPL mutations in 23 patients suffering from congenital amegakaryocytic thrombocytopenia, the type of mutation predicts the course of the disease. *Hum Mutat* **27**, 296.
11. Day HJ, Holmsen H. (1972). Platelet adenine nucleotide "storage pool deficiency" in thrombocytopenic absent radii syndrome. *JAMA* **221**, 1053–4.
12. Tschoepe D. (1995). The activated megakaryocyte-platelet-system in vascular disease, focus on diabetes. *Semin Thromb Hemost* **21**, 152–60.
13. Wolach B, Morag H, Drucker M, Sadan N. (1990). Thrombocytosis after pneumonia with empyema and other bacterial infections in children. *Pediatr Infect Dis J* **9**, 718–21.
14. Levin J, Conley CL. (1964). Thrombocytosis associated with malignant disease. *Arch Intern Med* **114**, 497–500.
15. Soini Y, Kamel D, Apaja-Sarkkinen M, Virtanen I, Lehto VP. (1993). Tenascin immunoreactivity in normal and pathological bone marrow. *J Clin Pathol* **46**, 218–21.
16. Kikuchi M, Tayama T, Hayakawa H, Takahashi I, Hoshino H, Ohsaka A. (1995). Familial thrombocytosis. *Br J Haematol* **89**, 900–2.
17. Tecuceanu N, Dardik R, Rabizadeh E, Raanani P, Inbal A. (2006). A family with hereditary thrombocythaemia and normal genes for thrombopoietin and c-Mpl. *Br J Haematol* **135**, 348–51.
18. Fogarty PF, Segal JB. (2007). The epidemiology of immune thrombocytopenic purpura. *Curr Opin Hematol* **14**, 515–19.
19. Psaila B, Bussel JB. (2007). Immune thrombocytopenic purpura. *Hematol Oncol Clin North Am* **21**, 743–59.
20. McMillan R. (2007). The pathogenesis of chronic immune thrombocytopenic purpura. *Semin Hematol* **44**(4 Suppl 5), S3–11.
21. Steuber CP. (2007). Clinical manifestations and diagnosis of immune (idiopathic) thrombocytopenic purpura in children. *UpToDate*.
22. George JN. (2007). Clinical manifestations and diagnosis of immune (idiopathic) thrombocytopenic purpura in adults. *UpToDate*.
23. Godeau B, Provan D, Bussel J. (2007). Immune thrombocytopenic purpura in adults. *Curr Opin Hematol* **14**, 535–56.
24. Van Leeuwen EF, van der Ven JT, Engelfriet CP, von dem Borne AE. (1982). Specificity of autoantibodies in autoimmune thrombocytopenia. *Blood* **59**, 23–6.
25. Beardsley DS, Spiegel JE, Jacobs MM, Handin RI, Lux IV SE. (1984). Platelet membrane glycoprotein IIIa contains target antigens that bind anti-platelet antibodies in immune thrombocytopenias. *J Clin Invest* **74**, 1701–7.
26. Sukati H, Watson HG, Urbaniak SJ, Barker RN. (2007). Mapping helper T-cell epitopes on platelet membrane glycoprotein IIIa in chronic autoimmune thrombocytopenic purpura. *Blood* **109**, 4528.
27. Harris EN, Gharavi AE, Hegde U, Derue G, Morgan SH, Englert H, Chan JK, Asherson RA, Hughes GR. (1985). Anticardiolipin antibodies in autoimmune thrombocytopenic purpura. *Br J Haematol* **59**, 231–4.
28. Kurata Y, Tsubakio T, Nishioeda Y, Kitani T. (1981). Seven cases of idiopathic thrombocytopenic purpura associated with Graves' disease. *Nippon Ketsueki Gakkai Zasshi* **44**, 951–6.
29. Gupta S, Saverymuttu SH, Marsh JC, Hodgson HJ, Chadwick VS. (1986). Immune thrombocytopenic purpura, neutropenia and sclerosing cholangitis associated with ulcerative colitis in an adult. *Clin Lab Haematol* **8**, 67–9.
30. Kosmo MA, Bordin G, Tani P, McMillan R. (1986). Immune thrombocytopenia and Crohn's disease. *Ann Intern Med* **104**, 136.
31. Selinger S, Tsai J, Pulini M, Saperstein A, Taylor S. (1987). Autoimmune thrombocytopenia and primary biliary cirrhosis with hypoglycemia and insulin receptor autoantibodies. A case report. *Ann Intern Med* **107**, 686–8.
32. Field SK, Poon MC. (1987). Sarcoidosis presenting as chronic thrombocytopenia. *West J Med* **146**, 481–2.
33. Waddell CC, Cimo PL. (1979). Idiopathic thrombocytopenic purpura occurring in Hodgkin disease after splenectomy, report of two cases and review of the literature. *Am J Hematol* **7**, 381–7.
34. Lester TJ, Grabowski GA, Goldblatt J, Leiderman IZ, Zaroulis CG. (1984). Immune thrombocytopenia and Gaucher's disease. *Am J Med* **77**, 569–71.
35. Aghai E, Quitt M, Lurie M, Antal S, Cohen L, Bitterman H, Froom P. (1987). Primary hepatic lymphoma presenting as symptomatic immune thrombocytopenic purpura. *Cancer* **60**, 2308–11.
36. George JN, Woolf SH, Raskob GE. (1996). Idiopathic thrombocytopenic purpura: A practice guideline developed by explicit methods for the American Society of Hematology. *Blood* **88**, 3.
37. McMillan R. (2005). The role of antiplatelet autoantibody assays in the diagnosis of immune thrombocytopenic purpura. *Curr Hematol Rep* **4**, 160–5.
38. Davoren A, Bussel J, Curtis BR, Moghaddam M, Aster RH, McFarland JG. (2005). Prospective evaluation of a new platelet glycoprotein (GP)-specific assay (PakAuto) in the diagnosis of autoimmune thrombocytopenia (AITP). *Am J Hematol* **78**, 193–7.
39. Medeiros D, Buchanan GR. (1996). Current controversies in the management of idiopathic thrombocytopenic purpura during childhood. *Pediatr Clin North Am* **43**, 757.
40. Arnold DM, Kelton JG. (2007). Current options for the treatment of idiopathic thrombocytopenic purpura. *Semin Hematol* **44**(4 Suppl 5), S12–23.
41. Stasi R, Evangelista ML, Stipa E, Buccisano F, Venditti A, Amadori S. (2008). Idiopathic thrombocytopenic purpura, current concepts in pathophysiology and management. *Thromb Haemost* **99**, 4–13.
42. Christie DJ, Mullen PC, Aster RH. (1985). Fab-mediated binding of drug-dependent antibodies to platelets in quinidine- and quinine-induced thrombocytopenia. *J Clin Invest* **75**, 310–14.
43. Smith ME, Reid DM, Jones CE, Jordan JV, Kautz CA, Shulman NR. (1987). Binding of quinine- and quinidine-dependent drug antibodies to platelets is mediated by the Fab domain of the immunoglobulin G and is not Fc dependent. *J Clin Invest* **79**, 912–17.
44. George JN. (2007). Drug-induced thrombocytopenia. *UpToDate*.
45. Coutre S. (2007). Heparin-induced thrombocytopenia. *UpToDate*.
46. Aster RH, Bougie DW. (2007). Drug-induced immune thrombocytopenia. *N Engl J Med* **357**, 580–7.
47. Bougie DW, Wilker PR, Aster RH. (2006). Patients with quinine-induced immune thrombocytopenia have both "drug-dependent" and "drug-specific" antibodies. *Blood* **108**, 922.
48. Kojouri K, Vesely SK, George JN. (2001). Quinine-associated thrombotic thrombocytopenic purpura-hemolytic uremic syndrome: Frequency, clinical features, and long-term outcomes. *Ann Intern Med* **135**, 1047.
49. Swanson JM. (2007). Heparin-induced thrombocytopenia: A general review. *J Infus Nurs* **30**, 232–40.
50. Coutre S, Leung LK, Landaw SA. (2008). Heparin-induced thrombocytopenia. *UpToDate*.
51. Warkentin TE, Greinacher A. (2004). Heparin-induced thrombocytopenia: Recognition, treatment, and prevention: The Seventh ACCP Conference on Antithrombotic and Thrombolytic Therapy. *Chest* **126**, 311S.
52. Roberts IA, Murray NA. (2006). Neonatal thrombocytopenia. *Curr Hematol Rep* **5**, 55–63.
53. Roberts IA, Murray NA. (2003). Thrombocytopenia in the newborn. *Curr Opin Pediatr* **15**, 17–23.
54. Kaplan C, Forestier F, Dreyfus M, Morel-Kopp MC, Tchernia G. (1995). Maternal thrombocytopenia during pregnancy, diagnosis and etiology. *Semin Thromb Hemost* **21**, 85–94.
55. Sternbach MS, Malette M, Nadon F, Guévin RM. (1986). Thrombocytopenia due to specific HLA antibodies. *Curr Stud Hematol Blood Transfus* **52**, 97–103.
56. Morales WJ, Stroup M. (1985). Intracranial hemorrhage *in utero* due to isoimmune neonatal thrombocytopenia. *Obstet Gynecol* **65**(3 Suppl), 20S–21S.
57. McCrae KR, Herman JH. (1996). Posttransfusion purpura, two unusual cases and a literature review. *Am J Hematol* **52**, 205–11.
58. Rahav G, Wollner A, Shalev O. (1988). Posttransfusion thrombocytopenia. *Isr J Med Sci* **24**, 271–3.
59. Mueller-Eckhardt C. (1986). Post-transfusion purpura. *Br J Haematol* **64**, 419–24.

60. de Waal LP, van Dalen CM, Engelfriet CP, von dem Borne AE. (1986). Alloimmunization against the platelet-specific Zwa antigen, resulting in neonatal alloimmune thrombocytopenia or posttransfusion purpura, is associated with the supertypic DRw52 antigen including DR3 and DRw6. *Hum Immunol* **17**, 45–53.
61. Fakhouri F, Frémeaux-Bacchi V. (2007). Does hemolytic uremic syndrome differ from thrombotic thrombocytopenic purpura? *Nat Clin Pract Nephrol* **3**, 679–87.
62. Mannucci PM. (2007). Thrombotic thrombocytopenic purpura and the hemolytic uremic syndrome, much progress and many remaining issues. *Haematologica* **92**, 878–80.
63. Schneider M. (2007). Thrombotic microangiopathy (TTP and HUS), advances in differentiation and diagnosis. *Clin Lab Sci* **20**, 216–20.
64. Nangaku M, Nishi H, Fujita T. (2007). Pathogenesis and prognosis of thrombotic microangiopathy. *Clin Exp Nephrol* **11**, 107–14.
65. Sadler JE. (2006). Thrombotic thrombocytopenic purpura: A moving target. *Hematology Am Soc Hematol Educ Program*, 415–20.
66. Chen YC, McLeod B, Hall ER, Wu KK. (1981). Accelerated prostacyclin degradation in the thrombotic thrombocytopenic purpura. *Lancet* **2**, 267–9.
67. Wu KK, Hall ER, Rossi EC, Papp AC. (1985). Serum prostacyclin binding defects in thrombotic thrombocytopenic purpura. *J Clin Invest* **75**, 168–74.
68. Glas-Greenwalt P, Hall JM, Panke TW, Kant KS, Allen CM, Pollak VE. (1986). Fibrinolysis in health and disease, abnormal levels of plasminogen activator, plasminogen activator inhibitor, and protein C in thrombotic thrombocytopenic purpura. *J Lab Clin Med* **108**, 415–22.
69. Kwaan HC. (1987). Clinicopathologic features of thrombotic thrombocytopenic purpura. *Semin Hematol* **24**, 71–81.
70. Tsai HM. (2007). Thrombotic thrombocytopenic purpura: A thrombotic disorder caused by ADAMTS13 deficiency. *Hematol Oncol Clin North Am* **21**, 609–32.
71. Dekker A, O'Brien ME, Cammarata RJ. (1974). The association of thrombotic thrombocytopenic purpura with systemic lupus erythematosus: A report of two cases with successful treatment of one. *Am J Med Sci* **267**, 243–9.
72. Steinberg AD, Green Jr. WT, Talal N. (1971). Thrombotic thrombocytopenic purpura complicating Sjögren's syndrome. *JAMA* **215**, 757–61.
73. Benitez L, Mathews M, Mallory GK. (1964). Platelet thrombosis with polyarteritis nodosa: Report of a case. *Arch Pathol* **77**, 116–25.
74. Karmali MA, Petric M, Lim C, Fleming PC, Arbus GS, Lior H. (1985). The association between idiopathic hemolytic uremic syndrome and infection by verotoxin-producing *Escherichia coli*. *J Infect Dis* **151**, 775–82.
75. Moorthy B, Makker SP. (1979). Hemolytic-uremic syndrome associated with pneumococcal sepsis. *J Pediatr* **95**, 558–9.
76. Prober CG, Tune B, Hoder L. (1979). *Yersinia pseudotuberculosis* septicemia. *Am J Dis Child* **133**, 623–4.
77. Delans RJ, Biuso JD, Saba SR, Ramirez G. (1984). Hemolytic uremic syndrome after Campylobacter-induced diarrhea in an adult. *Arch Intern Med* **144**, 1074–6.
78. Baker NM, Mills AE, Rachman I, Thomas JE. (1974). Haemolytic-uraemic syndrome in typhoid fever. *Br Med J* **2**, 84–7.
79. Sharman VL, Goodwin FJ. (1980). Hemolytic uremic syndrome following chicken pox. *Clin Nephrol* **14**, 49–51.
80. Austin TW, Ray CG. (1973). Coxsackie virus group B infections and the hemolytic-uremic syndrome. *J Infect Dis* **127**, 698–701.
81. O'Regan S, Robitaille P, Mongeau JG, McLaughlin B. (1980). The hemolytic uremic syndrome associated with ECHO 22 infection. *Clin Pediatr (Phila)* **19**, 125–7.
82. Rodrigues RG. (2006). Two generations with familial thrombotic thrombocytopenic purpura. *Int J Clin Pract* **60**, 95–8.
83. Caprioli J, Brioschi S, Remuzzi G. (2003). Molecular basis of familial thrombotic thrombocytopenic purpura and hemolytic uremic syndrome. *Saudi J Kidney Dis Transpl* **14**, 342–50.
84. Rose BD, George JN. (2007). Causes of thrombotic thrombocytopenic purpura-hemolytic uremic syndrome in adults. *UpToDate*.
85. Rose BD, George JN. (2007). Diagnosis of thrombotic thrombocytopenic purpura-hemolytic uremic syndrome in adults. *UpToDate*.
86. Rose BD, Kaplan AA, George JN. (2007). Treatment of thrombotic thrombocytopenic purpura-hemolytic uremic syndrome in adults. *UpToDate*.
87. Michelson AD. (2007). The clinical approach to disorders of platelet number and function. In *Platelets* (Michelson AD, ed.), 2nd ed. Academic Press, Elsevier.
88. Nurden AT, Nurden P. (2007). Inherited disorders of platelet function. In *Platelets* (Michelson AD, ed.), 2nd ed. Elsevier, Academic Press.
89. Weiss HJ. (1994). Scott syndrome: A disorder of platelet coagulant activity. *Semin Hematol* **31**(4), 312–19.
90. Rao AK. (2007). Acquired disorders of platelet function. In *Platelets* (Michelson AD, ed.), 2nd ed. Academic Press, Elsevier.
91. Lanza F. (2006). Bernard–Soulier syndrome hemorrhagiparous thrombocytic dystrophy. *Orphanet J Rare Dis* **1**, 46.
92. Pham A, Wang J. (2007). Bernard–Soulier syndrome: An inherited platelet disorder. *Arch Pathol Lab Med* **131**, 1834–6.
93. Kobe B, Deisenhofer J. (1994). The leucine-rich repeat: A versatile binding motif. *Trends Biochem Sci* **19**, 415–8.
94. Ruggeri ZM. (1991). The platelet glycoprotein Ib-IX complex. *Prog Hemost Thromb* **10**, 35–68.
95. Berndt MC, Shen Y, Dopheide SM, Gardiner EE, Andrews RK. (2001). The vascular biology of the glycoprotein Ib-IX-V complex. *Thromb Haemost* **86**, 178–88.
96. Clemetson KJ, Clemetson JM. (2004). Platelet receptor signalling. *Hematol J* **3**(5 Suppl), S159–163.
97. Andrews RK, Shen Y, Gardiner EE, Dong JF, Lopez JA, Berndt MC. (1999). The glycoprotein Ib-IX-V complex in platelet adhesion and signaling. *Thromb Haemost* **82**, 357–64.
98. Wenger RH, Wicki AN, Kieffer N, Adolph S, Hameister H, Clemetson KJ. (1989). The 5′flanking region and chromosomal localization of the gene encoding human platelet membrane glycoprotein Ib alpha. *Gene* **85**, 517–24.
99. Yagi M, Edelhoff S, Disteche CM, Roth GJ. (1994). Structural characterization and chromosomal location of the gene encoding human platelet glycoprotein Ib beta. *J Biol Chem* **269**, 17424–7.
100. Hickey MJ, Deaven LL, Roth GJ. (1990). Human platelet glycoprotein IX. Characterization of cDNA and localization of the gene to chromosome 3. *FEBS Lett* **274**, 189–92.
101. Lanza F, Morales M, de La Salle C, *et al.* (1993). Cloning and characterization of the gene encoding the human platelet glycoprotein V. A member of the leucine-rich glycoprotein family cleaved during thrombin-induced platelet activation. *J Biol Chem* **268**(28), 20801–7.
102. Lopez JA, Chung DW, Fujikawa K, Hagen FS, Davie EW, Roth GJ. (1988). The alpha and beta chains of human platelet glycoprotein Ib are both transmembrane proteins containing a leucine-rich amino acid sequence. *Proc Natl Acad Sci USA* **85**(7), 2135–9.
103. Lopez JA. (1994). The platelet glycoprotein Ib-IX complex. *Blood Coagul Fibrinolysis* **5**(1), 97–119.
104. Modderman PW, Admiraal LG, Sonnenberg A, von dem Borne AE. (1992). Glycoproteins V and Ib-IX form a noncovalent complex in the platelet membrane. *J Biol Chem* **267**, 364–9.
105. Lanza F. (2007). Murine models of platelet diseases. *Transfus Clin Biol* **14**, 35–40.
106. Tomer A, Scharf RE, McMillan R, Ruggeri ZM, Harker, LA. (1994). Bernard–Soulier syndrome, quantitative characterization of megakaryocytes and platelets by flow cytometric and platelet kinetic measurements. *Eur J Haematol* **52**, 193–200.
107. Afrasiabi A, Lecchi A, Artoni A. (2007). Genetic characterization of patients with Bernard–Soulier syndrome and their relatives from Southern Iran. *Platelets* **18**, 409–13.
108. The Bernard–Soulier Syndrome Database (http://www.bernardsoulier.org/index.html). Royal College of Surgeons in Ireland, Dublin.
109. Braunsteiner H, Pakesch F. (1956). Thrombocytoasthenia and thrombocytopathia – old names and new diseases. *Blood* **11**, 965–76.

110. Caen JP, Castaldi PA, Lecrec JC. (1966). Glanzmann's thrombasthenia. I. Congenital bleeding disorders with long bleeding time and normal platelet count (report of fifteen patients). *Am J Med* **44**, 4–26.
111. Xiao T, Takagi J, Coller BS, Wang JH, Springer TA. (2004). Structural basis for allostery in integrins and binding to fibrinogen-mimetic therapeutics. *Nature* **432**, 59–67.
112. Uzan G, Prenant M, Prandini MH, Martin F, Marguerie G. (1991). Tissue-specific expression of the platelet GPIIb gene. *J Biol Chem* **266**, 8932–9.
113. Wagner CL, Mascelli MA, Neblock DS, Weisman HF, Coller BS, Jordan RE. (1996). Analysis of GPIIb/IIIa receptor number by quantification of 7E3 binding to human platelets. *Blood* **88**, 907–14.
114. Phillips DR, Charo IF, Parise LV, Fitzgerald LA. (1988). The platelet membrane glycoprotein IIb-IIIa complex. *Blood* **71**, 831–43.
115. Ginsberg MH, Du X, Plow EF. (1992). Inside-out integrin signalling. *Curr Opin Cell Biol* **4**, 766–71.
116. O'Toole TE, Katagiri Y, Faull RJ, Peter K, Tamura R, Quaranta V, Loftus JC, Shattil SJ, Ginsberg MH. (1994). Integrin cytoplasmic domains mediate inside-out signal transduction. *J Cell Biol* **124**, 1047–59.
117. Schwartz MA, Schaller MD, Ginsberg MH. (1995). Integrins: Emerging paradigms of signal transduction. *Annu Rev Cell Dev Biol* **11**, 549–999.
118. Clark EA, Brugge JS. (1995). Integrins and signal transduction pathways: The road taken. *Science* **268**, 233–9.
119. Shattil SJ, Newman PJ. (2004). Integrins: Dynamic scaffolds for adhesion and signaling in platelets. *Blood* **104**, 1606–15.
120. Michelson AD. (1987). Flow cytometric analysis of platelet surface glycoproteins, phenotypically distinct subpopulations of platelets in children with chronic myeloid leukemia. *J Lab Clin Med* **110**, 346–54.
121. Jennings LK, Ashmun RA, Wang WC, Dockter ME. (1986). Analysis of human platelet glycoproteins IIb-IIIa and Glanzmann's thrombasthenia in whole blood by flow cytometry. *Blood* **68**, 173–9.
122. George JN, Caen JP, Nurden AT. (1990). Glanzmann's thrombasthenia: The spectrum of clinical disease. *Blood* **75**, 1383–95.
123. Nurden AT. (2006). Glanzmann thrombasthenia. *Orphanet J Rare Dis* **1**, 10.
124. Wilhide CC, Jin Y, Guo Q, Li L, Li SX, Rubin E, Bray PF. (1997). The human integrin beta3 gene is 63 kb and contains a 5′-UTR sequence regulating expression. *Blood* **90**, 3951–61.
125. Notarangelo LD, Miao CH, Ochs HD. (2008). Wiskott–Aldrich syndrome. *Curr Opin Hematol* **15**, 30–6.
126. Ochs HD, Thrasher AJ. (2006). The Wiskott–Aldrich syndrome. *J Allergy Clin Immunol* **117**, 725–38.
127. Notarangelo LD, Mori L. (2005). Wiskott–Aldrich syndrome: Another piece in the puzzle. *Clin Exp Immunol* **139**, 173–5.
128. Shcherbina A, Stiehm ER, Feldweg AM. (2008). Wiskott–Aldrich syndrome. *UpToDate*.
129. Baldini MG. (1972). Nature of the platelet defect in the Wiskott Aldrich syndrome. *Ann NY Acad Sci* **201**, 437–9.
130. Jin Y, Mazza C, Christie JR, Giliani S, Fiorini M, Mella P, Gandellini F, Stewart DM, Zhu Q, Nelson DL, Notarangelo LD, Ochs HD. (2004). Mutations of the Wiskott–Aldrich Syndrome Protein (WASP), hotspots, effect on transcription, and translation and phenotype/genotype correlation. *Blood* **104**, 4010–19.
131. Derry JM, Ochs HD, Francke U. (1994). Isolation of a novel gene mutated in Wiskott–Aldrich syndrome. *Cell* **78**, 635–44.
132. Ochs HD. (1998). The Wiskott–Aldrich syndrome. *Springer Semin Immunopathol* **19**, 435–43.
133. Imai K, Nonoyama S, Ochs HD. (2003). WASP (Wiskott–Aldrich syndrome protein) gene mutations and phenotype. *Curr Opin Allergy Clin Immunol* **3**, 427–36.
134. Dupuis-Girod S, Medioni J, Haddad E, Quartier P, Cavazzana-Calvo M, Le Deist F, de Saint Basile G, Delaunay J, Schwarz K, Casanova JL, Blanche S, Fischer A. (2003). Autoimmunity in Wiskott–Aldrich syndrome, risk factors, clinical features, and outcome in a single-center cohort of 55 patients. *Pediatrics* **111**(5 Pt 1), e622–7.
135. Grottum KA, Hovig T, Holmsen H. (1969). Wiskott–Aldrich syndrome: Qualitative platelet defects and short platelet survival. *Br J Haematol* **17**, 373–6.
136. White JG. (1986). Platelet granule disorders. *Crit Rev Oncol Hematol* **4**, 337–77.
137. Fukami MH, Salganicoff L. (1977). Human platelet storage organelles: A review. *Thromb Haemost* **38**, 963–70.
138. Gunay-Aygun M, Huizing M, Gahl WA. (2004). Molecular defects that affect platelet dense granules. *Semin Thromb Hemost* **30**, 537–47.
139. McNicol A, Israels SJ. (1999). Platelet dense granules: Structure, function and implications for haemostasis. *Thromb Res* **95**, 1–18.
140. Maynard DM, Heijnen HF, Horne MK, White JG, Gahl WA. (2007). Proteomic analysis of platelet alpha-granules using mass spectrometry. *J Thromb Haemost* **5**, 1945–55.
141. De Munnynck K, Van Geet C, De Vos R, Van de Voorde W. (2007). Delta-storage pool disease: A pitfall in the forensic investigation of sudden anal blood loss in children: A case report. *Int J Legal Med* **121**, 44–7.
142. Rao AK. (1990). Congenital disorders of platelet function. *Hematol Oncol Clin North Am* **4**, 65–86.
143. Wei ML. (2006). Hermansky–Pudlak syndrome: A disease of protein trafficking and organelle function. *Pigment Cell Res* **19**, 19–42.
144. Kaplan J, De Domenico I, Ward DM. (2008). Chediak–Higashi syndrome. *Curr Opin Hematol* **15**, 22–9.
145. Nurden AT, Nurden P. (2007). The gray platelet syndrome: Clinical spectrum of the disease. *Blood Rev* **21**, 21–36.
146. Nurden AT, Nurden P. (2001). Inherited defects of platelet function. *Rev Clin Exp Hematol* **5**, 314–34.
147. Lages B, Sussman II, Levine SP, Coletti D, Weiss HJ. (1997). Platelet alpha granule deficiency associated with decreased P-selectin and selective impairment of thrombin-induced activation in a new patient with gray platelet syndrome (alpha-storage pool deficiency). *J Lab Clin Med* **129**, 364–75.
148. Breton-Gorius J, Vainchenker W, Nurden A, Levy-Toledano S, Caen J. (1981). Defective alpha-granule production in megakaryocytes from gray platelet syndrome: Ultrastructural studies of bone marrow cells and megakaryocytes growing in culture from blood precursors. *Am J Pathol* **102**, 10–19.
149. Rao AK, Koike K, Willis J, Daniel JL, Beckett C, Hassel B, Day HJ, Smith JB, Holmsen H. (1984). Platelet secretion defect associated with impaired liberation of arachidonic acid and normal myosin light chain phosphorylation. *Blood* **64**, 914–21.
150. Mestel F, Oetliker O, Beck E, Felix R, Imbach P, Wagner HP. (1980). Severe bleeding associated with defective thromboxane synthetase. *Lancet* **1**, 157.
151. Patrono C, Rocca B. (2008). Aspirin: Promise and resistance in the new millennium. *Arterioscler Thromb Vasc Biol* **28**, s25–32.
152. Mielke Jr. CH, Kaneshiro MM, Maher IA, Weiner JM, Rapaport SI. (1969). The standardized normal Ivy bleeding time and its prolongation by aspirin. *Blood* **34**, 204–15.
153. Kaneshiro MM, Mielke Jr. CH, Kasper CK, Rapaport SI. (1969). Bleeding time after aspirin in disorders of intrinsic clotting. *N Engl J Med* **281**, 1039–42.
154. Deykin D, Janson P, McMahon L. (1982). Ethanol potentiation of aspirin-induced prolongation of the bleeding time. *N Engl J Med* **306**, 852–4.
155. Fass RJ, Copelan EA, Brandt JT, Moeschberger ML, Ashton JJ. (1987). Platelet-mediated bleeding caused by broad-spectrum penicillins. *J Infect Dis* **155**, 1242–8.
156. Brown III CH, Bradshaw MJ, Natelson EA, Alfrey Jr. CP, Williams TW Jr. (1976). Defective platelet function following the administration of penicillin compounds. *Blood* **47**, 949–56.
157. Natelson EA, Brown III CH, Bradshaw MW, Alfrey Jr. CP, Williams TW Jr. (1979). Influence of cephalosporin antibiotics on blood coagulation and platelet function. *Antimicrob Agents Chemother* **9**, 91–3.

158. Rose KD, Croissant PD, Parliament CF, Levin MB. (1990). Spontaneous spinal epidural hematoma with associated platelet dysfunction from excessive garlic ingestion: A case report. *Neurosurgery* **26**, 880–2.
159. Rahman K. (2007). Effects of garlic on platelet biochemistry and physiology. *Mol Nutr Food Res* **51**, 1335–44.
160. von Schacky C, Fischer S, Weber PC. (1985). Long-term effects of dietary marine omega-3 fatty acids upon plasma and cellular lipids, platelet function, and eicosanoid formation in humans. *J Clin Invest* **76**, 1626–31.
161. Just M, Tripier D, Seiffge D. (1991). Studies of the antithrombotic effects of dextran 40 following microarterial trauma. *Br J Plast Surg* **44**, 15–22.
162. Matthiasson SE, Lindblad B, Mätzsch T, Molin J, Qvarford P, Bergqvist D. (1994). Study of the interaction of dextran and enoxaparin on haemostasis in humans. *Thromb Haemost* **72**, 722–7.
163. Murphy GJ, Angelini GD. (2004). Side effects of cardiopulmonary bypass: What is the reality? *J Card Surg* **19**, 481–8.
164. Holloway DS, Summaria L, Sandesara J, Vagher JP, Alexander JC, Caprini JA. (1988). Decreased platelet number and function and increased fibrinolysis contribute to postoperative bleeding in cardiopulmonary bypass patients. *Thromb Haemost* **59**, 62–7.
165. George JN, Shattil SJ. (1991). The clinical importance of acquired abnormalities of platelet function. *N Engl J Med* **324**, 27–39.
166. Di Minno G, Cerbone A, Usberti M, Cianciaruso B, Cortese A, Farace MJ, Martinez J, Murphy S. (1986). Platelet dysfunction in uremia II. Correction by arachidonic acid of the impaired exposure of fibrinogen receptors by adenosine diphosphate or collagen. *J Lab Clin Med* **108**, 246–52.
167. Escolar G, Cases A, Bastida E, Garrido M, López J, Revert L, Castillo R, Ordinas A. (1990). Uremic platelets have a functional defect affecting the interaction of von Willebrand factor with glycoprotein IIb-IIIa. *Blood* **76**, 1336–40.
168. Rosove MH, Naeim F, Harwig S, Zighelboim J. (1980). Severe platelet dysfunction in hairy cell leukemia with improvement after splenectomy. *Blood* **55**, 903–6.

Index

Page numbers in *italics* refer to illustrations or tables

A

ABL1 gene, 61, 167
  chronic myeloproliferative diseases, 155–6
    chronic myelogenous leukemia, 158–9, 160, 163–7
  See also BCR-ABL1 fusion gene
Abnormal localization of immature precursors (ALIP), 131, *132*, 144
Absence of cytoplasmic neutrophilic granules, 515
Acanthocytes, 530, *532*, 553
Acanthocytosis, 553
Acquired immunodeficiency syndrome (AIDS), *86*, *88*, 447, 455, 457–61
  clinical aspects, *459*, 460–1
  etiology and pathogenesis, 458–9
  pathology, 459–60, *460*, *466*
Acute basophilic leukemia (ABL), 246
Acute erythroid leukemias (AML-M6), 239–43
  clinical aspects, 242
  differential diagnosis, 242–3
  etiology and pathogenesis, 239
  immunophenotype, 242, *243*
  molecular and cytogenetic studies, 242
  pathology, 240–1, *240–3*
Acute leukemias of ambiguous lineage, 279–85, *279*
  clinical aspects, 283–4
  differential diagnosis, 284–5
  etiology and pathogenesis, 279–80
  immunophenotype, 281–3
    acute leukemias with expression of aberrant markers, 281, *281*, *282*
    bilineal acute leukemia, 282–3, *284*, *285*
    biphenotypic acute leukemia, 281, *281*, 282, 283–4, *283*, *285*
  molecular and cytogenetic studies, 283, *285*
  pathology, 280–1, *280*
Acute lymphoblastic leukemia (ALL), 257
  Burkitt type, 274–5
  cytogenetics, 58
  with cytoplasmic granules, 273
  with eosinophilia, 273–4, *274*
  with hypoplastic marrow, 274, *275*
  See also Precursor B-lymphoblastic leukemia/lymphoma; Precursor T-lymphoblastic leukemia/lymphoblastic lymphoma
Acute megakaryoblastic leukemia (AMKL), 243–6
  clinical aspects, 245
  differential diagnosis, 245–6
  etiology and pathogenesis, 243
  immunophenotype, 245
  molecular and cytogenetic studies, 245, *245*
  pathology, 243–4, *244*
Acute monoblastic leukemia (AML-M5), 234–9
  clinical aspects, 236–7
  differential diagnosis, 238–9
  etiology and pathogenesis, 235
  immunophenotype, 235–6
  molecular and cytogenetic studies, 236
  pathology, 235, *237*
Acute monocytic leukemia (AML-M5b), 234–9
  clinical aspects, 236–7
  differential diagnosis, *227*, 238–9
  etiology and pathogenesis, 235
  immunophenotype, 235–6, *238*, *239*
  molecular and cytogenetic studies, 236, *239*
  pathology, 235, *238*, *239*
Acute myeloid leukemia (AML), 207–49
  classification, 207, *208*
  clinical aspects, 212
  cytogenetics, 208, 211, 212
  differential diagnosis, *124*, 151–2, *151*, 222, 226, 229, 231, 232
  etiology and pathogenesis, 207–8
  Fanconi anemia and, 116
  hypocellular, 151–2
  immunohistochemistry, 209–11, *209*, *210*, *211*
  mastocytosis, *486*
  minimally differentiated (AML-M0), 228–9, *228*, *230*
  molecular studies, 211
  molecular techniques, 72
  myelodysplastic syndromes and, 142, *150*
  pathology, 208–11
  therapy-related (t-AML), 207, *225*, 226–8
    alkylating agent/radiation-related AML, 227

Acute myeloid leukemia (AML) (*Continued*)
topoisomerase II inhibitor-related AML, 227–8
with 11q23 (*MLL*) abnormalities, 219–22, *224*
with aberrant expression of CD19, *282*
with chromosome 3 aberrations and thrombocytosis, 248
with inv(16)(p13q22) or t(16;16)(p13;q22);(*CBFβ/MYH11*), 217–19, *221*, *222*, *223*
with maturation, 231–2, *233*
with multilineage dysplasia, 223–6, *225*, *226*
with t(8;21)(q22;q22);(*RUNX1/RUNXT1*), 212–14, *213*, *214*
without maturation (AML-M1), 229–31, *231*
Acute myelomonocytic leukemia (AML-M4), 232–4
clinical aspects, 233–4
differential diagnosis, 234
etiology and pathogenesis, 232
immunophenotype, 233, *236*
molecular and cytogenetic studies, 233
pathology, 232–3, *234*, *235*
Acute panmyelosis with myelofibrosis (APMF), 246–7
Acute promyelocytic leukemia (APL), 214–17
clinical aspects, 217
cytogenetics, 57, 216–17, *220*
differential diagnosis, 222
etiology and pathogenesis, 214–15
immunophenotype, 215–16, *219*
pathology, 215, *215–18*
Adenocarcinoma, metastatic, *50*
Adhesion molecules, *2*
Adipocytes, 15–16
Adipose tissue, *17*
Adult T-cell leukemia/lymphoma (ATL), 409–13
clinical aspects, 412
cytogenetics, 411, *412*
differential diagnosis, 413, *417*
etiology and pathogenesis, 410
immunophenotype, 410–11, *412*
pathology, 410, *411*
AF4/MLL fusion gene, 261
Agammaglobulinemia, 457
Agglutinins, 555–6
Aggressive NK-cell leukemia, 404–5, *406*, *407*, *408*, *410*
Agranulocytosis, 519, *519*
AIDS, *See* Acquired immunodeficiency syndrome (AIDS)
Alcoholism, 557
Alder–Reilly anomaly, 513, *514*
ALK gene, 293
See also Anaplastic lymphoma kinase (ALK)
Alkylating agent/radiation-related AML, 227
α-granule deficiency, 577
α-naphthyl butyrate esterase marker, 211
α-storage pool disease, 577
α-thalassemia, 547–8
acquired, 548
Amegakaryocytosis, 118
AML1 gene, 136, 212, 229, 243
See also TEL-AML1 fusion gene
Amyloidosis, 83–5, *84*, 390
categories, 390, *392*
primary (AL), 390–2, *392*, *393*
Analyte Specific Reagents (ASRs), 73
Anaplastic large cell lymphoma (ALCL), 420–5
clinical aspects, 424
cytogenetics, 422–4, *425*, *426*
differential diagnosis, 424–5, 426–7, 451
etiology and pathogenesis, 420–1
immunophenotype, 422, *424*
pathology, 421–2, *423*
primary cutaneous, 425–7
variants, 422
Anaplastic lymphoma kinase (ALK), 422–4
ALK positive DLBCL, 358
Anemia, 529
diagnostic criteria, 530
of chronic disease (ACD), 558–9
See also Specific types of anemia
Aneuploidy, 61–3
plasma cell myeloma, 380, 382–3
See also Cytogenetics; *Specific monosomies and trisomies*
Angiofollicular hyperplasia, 103
Angioimmunoblastic T-cell lymphoma, 428–31
clinical aspects, 431, *431*
cytogenetics, 429–31, *431*
differential diagnosis, 431
etiology and pathogenesis, 428
immunophenotype, 428–9, *430*, *431*
pathology, 428, *429*
Ann Arbor staging system, 294, *294*
Anorexia nervosa, 81
Anti-hemoglobin antibodies, 42
Aplastic anemia (AA), *96*, 115
acquired, 118–20, *120*
clinical aspects, 120
etiology and pathogenesis, 118–19
genetics, 119–20
pathology, 119
associated drugs, 118–19, *119*
differential diagnosis, *124*, 151, *151*
See also Bone marrow aplasia
Apoptosis, in myelodysplastic syndrome, 129–30
Arachidonic acid pathway abnormalities, 577
Ashkenazi Jews, 116, 489, 492, 493
Aspirin, 578
Asymptomatic (smoldering) myeloma, 386, *387*
Ataxia telangiectasia, *456*
ATM gene, 347–8, 397, 399
ATRA (all-*trans* retinoic acid), 217
Atypical chronic myeloid leukemia (aCML), 197–9
clinical aspects, 199
cytogenetics, 199
diagnostic criteria, *199*
etiology and pathogenesis, 197
pathology, 199
See also Chronic myelogenous leukemia (CML)
Auer rods, *209*, 212, *216*, *217*, 515, *515*
Autoimmune disorders, *96*, 496, 497
See also Specific disorders
Autoimmune hemolytic anemia (AIHA), 554–6
classification, *555*
differential diagnosis, 556
drug-induced, 556–7, *557*
etiology and pathogenesis, 554–5
pathology, 555–6
with cold-reacting antibodies, 555, 556
with warm-reacting antibodies, 555–6
Autoimmune thrombocytopenic purpura (AITP), 568–70
differential diagnosis, 570
etiology and pathogenesis, 568
pathology, 568–70
Azurophilic granules, 401, *402*

B

B19 parvovirus-induced RBC aplasia, 531, *537*, *538*
B progenitor cells, 5
B-cell lineage-specific activator protein (BSAP), 44, 316
B-cell lymphoma, *49*, *298*
diffuse large B-cell lymphoma, 349–53
intravascular large B-cell lymphoma, 356–7, *356*, *357*
MALT lymphoma, 324–8
mediastinal large B-cell lymphoma, 353–6
See also Specific types of lymphoma
B-cell prolymphocytic leukemia (B-PLL), 313–16
clinical aspects, 315
cytogenetics, 314–15
differential diagnosis, 315–16
immunophenotype, *292*, 314
pathology, 314, *314*, *315*
B-cells, 12, *14*
associated markers, 27–39
differentiation, *288*, *298*
lymph nodes, 22–3, 24, *49*
monoclonal expansion in the elderly, 312
monocytoid B-cell hyperplasia, 106, *106*
spleen, 18, 19, 20, *48*
See also Mature B-cell neoplasms
Bands, 8, *8*, 17, *19*
Bartonella henselae, 107
Basket cells, 302, *302*
Basophilia, 524–5
chronic myelomonocytic leukemia, 193
chronic myeloproliferative diseases, 159, 186, 203
Basophils, 8, *8*, *9*, *18*, *19*
acute basophilic leukemia, 246
maturation, 5
BCL-1 gene, 293
BCL-2 gene, 293, 336–7, 351
BCL-2 protein, 130, 336–7
BCL-6 gene, 293, 337, 351, 360
BCR gene, 61, 167
chronic myeloproliferative diseases, 155
chronic myelogenous leukemia, 158–9, 160, 163–7
BCR-ABL1 fusion gene, 158–9
acute erythroid leukemia, 242
B-cell neoplasms, 293
chronic myelogenous leukemia, 158–9, 160, 163–7, 169, 186
chronic neutrophilic leukemia, 170
precursor B-lymphoblastic leukemia/lymphoma, 260–1, *261*, *262*
Bence-Jones protein, 388
Bernard–Soulier syndrome (BSS), 572–4

clinical aspects, 574
differential diagnosis, 574
etiology and pathogenesis, 572
molecular and cytogenetic studies, 573–4
pathology, 572–3, *573*
Beta-lactam antibiotics, 578
ß-thalassemia, 545–7, *545*, *547*, *548*
ß-thalassemia intermedia, 546
ß-thalassemia major, 546
ß-thalassemia minor, 546
Bilineal acute leukemia, 282–3, *284*, *285*
Biopsy, 5, *6*, *7*
bone marrow recovery, 94, *95*
Biphenotypic acute leukemia, 281, *281*, 282, 283–4, *283*, *285*
Birbeck granules, 9, 502, *502*
BIRC3 gene, 324
Blastic NK-cell lymphoma, 409, 503–6
clinical aspects, 504–5
cytogenetics, 504
differential diagnosis, 505–6
immunophenotype, 504, 506, *506*
pathology, 504, *505*
Blood loss, 544
Blood smears, 16–17
Bloom syndrome, *456*
Blotting techniques, 76–7
BOB.1/OBF.1, 44
Bone changes, 93
Bone marrow, 1–5
aplasia, *See* Bone marrow aplasia
benign lymphoid aggregates, 464–5, *465*, *466*
cellularity, 5, *7*
ranges of differential counts, *8*
cytogenetic studies, 58
examination, 5, *6*
extracellular matrix, 1–2, *2*
lymphoid aggregates, *466*
microvascular circulation, 2, *2*
morphological abnormalities, 81–94
amyloidosis, 83–5, *84*
fibrosis, 89, *90*, 94, *96*, *97*
gelatinous transformation, 81, *82*
granulomas, 85–9, *85–9*
metastasis, 89–91, *90*, *91*, *92*
necrosis, 81–3, *83*, *93*
post-therapeutic changes, 91–3, *93*
previous biopsy site, 94, *95*
vascular changes, 94, *97*
Bone marrow aplasia, 115–24
acquired aplastic anemia, 118–20, *120*
amegakaryocytosis, 118
classification, *115*
Diamond–Blackfan anemia, 117, *118*
differential diagnosis, 124, *124*
dyskeratosis congenita, 117
Fanconi anemia, 115–16
paroxysmal nocturnal hemoglobinuria, 120–4
Shwachman–Diamond syndrome, 117, *117*
Break-apart probe, 68, *70*
Brucellosis, *88*
Bruton agammaglobulinemia, *456*, 457
BTK gene, 457
Burkitt leukemia/lymphoma (BL), 359–64
acute lymphoblastic leukemia, Burkitt type, 274–5
atypical, 360
BL with plasmacytoid differentiation, 360
clinical aspects, 360–4
cytogenetics, 360, *363*
differential diagnosis, 364
endemic, 360–4
etiology and pathogenesis, 359
immunodeficiency-associated, 364
immunophenotype, *292*, 360, *362–3*
pathology, 359–60, *361*, *362*
sporadic, 364
Burst-forming units (BFUs):
BFU-K, 5
erythroid (BFU-E), 4

C

C-bands, 59
c-kit mutation, 477, *478*, 480
C-MYC gene, 130, 347, 359
amplification, 63
overexpression, 359, *360*
C/EBPα (CCAAT/enhancer-binding protein alpha), 248
Cabot rings, 530, *534*
Cancer, 57
chromosome abnormalities, 57
See also Specific types of cancer
Carcinoma, metastatic, 89
prostatic carcinoma, 50, 91
Cardiopulmonary bypass, 578
Castleman's disease, 103
hyaline vascular type, 103, *104*, *105*
plasma cell type, 103
Cat-scratch disease, 107, *109*
CBFß gene, 212
CBFß/MYH11 fusion gene, 217, 219, 232
CCND genes, 347, 382
CD molecules, 27, *28–37*
acute basophilic leukemia, 246
acute erythroid leukemia, 242, *243*
acute leukemias of ambiguous lineage, 281
aberrant expression, 281, *282*
acute megakaryoblastic leukemia, 245
acute monoblastic/monocytic leukemias, 235–6, *238*, *239*
acute myeloid leukemia, 209–10, *209*, *210*
minimally differentiated, 229
therapy-related, 227
with 11q23 abnormalities, 219–21
with inv(16)(p13q22) or t(16;16)(p13;q22), 218–19
with maturation, 232
with multilineage dysplasia, 223
with t(8;21), 212
without maturation, 229
acute myelomonocytic leukemia, 233, *236*
acute panmyelosis with myelofibrosis, 246
acute promyelocytic leukemia, 215–16, *219*
adult T-cell leukemia/lymphoma, 410–11, *412*
anaplastic large cell lymphoma, 422
primary cutaneous, 426
angioimmunoblastic T-cell lymphoma, 428–9, *430*, *431*
B-cell associated, 27–38
B-cell prolymphocytic leukemia, 314
blastic NK-cell lymphoma, 504, 506, *506*
Burkitt lymphoma, 360, *362*
chronic eosinophilic leukemia, 172–3
chronic lymphocytic leukemia, 303–7, *307*, *308*, 309
chronic myelogenous leukemia, 161–3
chronic myelomonocytic leukemia, 193–5, *197*
dendritic cell tumors, 507, 508
diffuse large B-cell lymphoma, 349, *354*
enteropathy-type T-cell lymphoma, 420
erythroid-associated, 42
extranodal NK/T-cell lymphoma, nasal type, 406–7
follicular lymphoma, 336, *338*, *339*
granulocytic/monocytic-associated, 41–2
hairy cell leukemia, 331
hepatosplenic T-cell lymphoma, 418
histiocytic sarcoma, 499
HIV/AIDS, 459
Hodgkin lymphoma, 443, 448
hypereosinophilic syndrome, 172–3
intravascular large B-cell lymphoma, 357, *357*
juvenile myelomonocytic leukemia, 202
Langerhans cell histiocytosis, 502
large granular lymphocyte-associated, 40–1
leukocyte adhesion deficiency, 517–19
lymphomatoid granulomatosis, 469
lymphoplasmacytic lymphoma, 316, 319
MALT lymphoma, 327
mantle cell lymphoma, 342–7, *345*
mastocytosis, 478, 480
mature B-cell neoplasms, 292, *292*
mature T-cell neoplasms, 292
mediastinal large B-cell lymphoma, 355
megakaryocyte/platelet-associated, 42
mycosis fungoides, 413
myelodysplastic syndromes, 133–4
NK-cell leukemia, 405
NK-cell neoplasms, 292
peripheral T-cell lymphoma, 432
persistent polyclonal B-cell lymphocytosis, 464
plasma cell myeloma, 376–7, *381*
plasmablastic lymphoma, 358
precursor B-lymphoblastic leukemia/lymphoma, 259–60, *260*
precursor T-lymphoblastic leukemia/lymphoblastic lymphoma, 269–70, *270*
precursor-associated, 43
primary effusion lymphoma, 357
Sezary syndrome, 413
splenic marginal zone lymphoma, 321
T-cell large granular lymphocytic leukemia, 403
T-cell prolymphocytic leukemia, 398, *399*
T-cell-associated, 39–40
CD1, 39, 107
CD1a, 9, *28*, 39, 269, 281, 499, 502, 503
CD1b, *28*, 39
CD1c, *28*, 39
CD2, *2*, *28*, 39, 163, 215, 219, 229, 233, 269, 273, 281, 292, 307, 398, 405, 407, 410, 413, 418, 422, 478, *478*
CD3, 12, *28*, 39, 163, 229, 232, 269, 281, 292, 357, 398, 401, 403, 405, 409, 410, 413, 418, 420, 422, 428, 432, 443
deficiency, *456*

CD4, 9, 23, *28*, 39, 193, 202, 219, 223, 227, 234, 235, 269, 292, 398, 403, 409, 422, 426, 428, 459–61, 499, 502, 504
use in flow cytometry, 45, *46*
CD5, 23, *28*, 38, 39–40, 163, 292, 301, 302, 303, 305, 312, 316, 324, 333, 336, 342, 349, 355, 357, 398, 410, 413, 422, 432
CD6, *28*
CD7, *28*, 40, 134, 162, 163, 223, 227, 229, 233, 245, 269, 273, 281, 292, 307, 398, 401, 405, 407, 416, 420, 422, 432
CD8, 12, 23, *28*, 40, 118, 269, 292, 398, 401, 403, 405, 409, 420, 428, 462
use in flow cytometry, 45, *46*
CD9, *28*, 172, 215, 246
CD10, 12, 16, 23, 27, *28*, 134, 163, 193, 202, 223, 227, 229, 260, 262, 267, 270, 292, 307, 312, 324, 335, 339, 349, 352, 357, 359, 360, 377, 388, 428
CD11, 43, 517
CD11a, *2*, *28*, 43, 134, 223, 227
CD11b, *2*, 9, *28*, 43, 193
CD11c, *2*, 8, 9, *28*, 43, 163, 193, 218, 219, 229, 232, 233, 235, 303, 331, 336, 464, 499
CDw12, *28*
CD13, 7, 8, 9, *28*, 41, 134, 162, 163, 193, 209, 212, 215–16, 218, 219, 229, 231, 232, 233, 235, 242, 246, 249, 260, 268, 270, 273, 281, 307, 477, *478*, 480
CD14, 9, *28*, 41, 120, 123, 193, 202, 218, 219, 229, 232, 233, 234, 235, 281, *478*, 480, 499, 502
CD15, 7, 8, 9, *28*, 41, 123, 134, 163, 193, 215, 218, 233, 235, 246, 260, 261, 281, 448, 468, *478*, 480
CD16, 8, 12, *28*, 40, 120, 123, 193, 202, 260, 273, 292, 403, 405, 480
CDw17, *28*
CD18, *2*, *28*, 517
CD19, 12, 13, 23, 27–37, *28*, 134, 163, 212, 229, 232, 259, 262, 267, 281, *282*, 292, 302, 303, 305, 312, 314, 316, 321, 327, 331, 336, 342, 349, 355, 357, 358, 360, 386, 464
CD20, 23, *28*, 37, 163, 229, 232, 260, 261, 262, 267, 292, 302, 303, 305, 316, 321, 327, 331, 336, 342, 349, 355, 357, 358, 360, 429, 443, 448, 464, 469
CD21, 10, 23, *28*, 38, 102, 335, 336, 428, 443, 462, 508
CD22, 23, *28*, 38, 162, 229, 232, 259, 281, 303, 305, 316, 331, 336, 342, 349, 355, 357, 360, 443, 464
CD23, *28*, 38, 292, 301, 302, 303, 305, 311, 312, 316, 324, 335, 336, 342, 355, 428, 508
CD24, *28*, 38, 120, 281
CD25, *28*, 246, 303, 324, 331, 333, 411, 464, 478, *478*, 498
CD26, *28*, 40, 403
CD27, *28*
CD28, *28*
CD29, *28*
CD30, *28*, 43, 349, 355, 357, 411, 422, 426, 432, 443, 448, 468, 469
CD31, 11, 16, *28*, 135, 163, 503
CD32, *29*, 172
CD33, 3, 7, 9, *29*, 41, 123, 134, 162, 163, 193, 209, 212, 215, 218, 219, 229, 231, 232, 233, 235, 242, 246, 249, 260, 268, 270, 273, 281, *282*, *478*, 480
CD34, 1, *2*, 7, 12, 16, *29*, 43, 134, 148, 151, 162, 163, 173, 209, 212, 215, 218–19, 221, 223, 227, 229, 232, 233, 235, 242, 246, 260, 267, 307, 364, 477, *478*
CD35, 10, 23, *29*, 428, 508
CD36, *29*, 133, 163, 218, 219, 233, 235, 245
CD37, *29*
CD38, 1, 13, *29*, 43, 229, 269, 303, 305, 308, 309, 316, 357, 376
CD39, *29*
CD40, *29*
CD41, 11, *29*, 42, 123, 133, 134, 245, 246, 570
CD42, 11, 42, 245, 246
CD42a, *29*, 570
CD42b, *29*
CD42c, *29*
CD42d, *29*
CD43, 23, *29*, 43–4, 303, 336, 342, 360, 422, 504
CD44, *29*, 269
CD45, *2*, 13, *29*, 40, 162, 193, 212, 259, 267, 269, 273, 292, 357, 358, 413, 443, 468, 469, *478*, 480, 491, 493, 499, 504
CD45RA, *29*, 40
CD45RB, *29*
CD45RO, *29*, 40
use in flow cytometry, 45
CD46, *29*
CD47, *29*
CD48, *29*, 120, 123
CD49a, *2*, *29*
CD49b, *2*, *29*
CD49c, *2*, *29*
CD49d, *2*, *29*
CD49e, *2*, *29*
CD49f, *2*, *29*
CD50, *2*, *29*
CD51, *29*
CD52, *29*, 120, 411, 502
CD53, *29*
CD54, *2*, *29*
CD55, *29*, 44, 120, 122–3, 151
CD56, 12, *30*, 40, 134, 163, 193, 202, 212, 215–16, 223, 227, 229, 232, 233, 234, 235, 260, 273, 281, 292, 377, 386, 388, 401, 403, 405, 418, 420, 432, 503, 504
CD57, 12, *30*, 40–1, 292, 401, 403, 405, 407, 443
CD58, *2*, *30*, 120
CD59, *30*, 44, 120, 122–3, 151
CD60, *30*
CD61, 11, *30*, 42, 123, 133, 134, 135, 163, 246, 570
CD62, 245
CD62E, *2*, *30*
CD62L, *2*, *30*
CD62P, *2*, *30*
CD63, *30*
CD64, 9, *30*, 41, 193, 202, 218, 219, 233, 235, 281, 499
CD65, *30*, 281
CD66, 120, 123, 134, 223, 227
CD66a, *30*
CD66b, *30*
CD66c, *30*, 281
CD66d, *30*
CD66e, *30*
CD66f, *30*
CD68, 9, *30*, 41–2, 105, 107, 135, 163, 202, 218, 221, 229, 233, 236, 260, 480, 491, 493, 498, 499, 502, 504, 508
CD69, *30*
CD70, *30*
CD71, 10, *30*, 42, 134, 242
CD72, *30*
CD73, *30*, 120
CD74, *30*, 38
CD75, *30*, 443
CD75s, *30*
CD77, *30*
CD79, 38
CD79a, 23, *30*, 38, 163, 229, 232, 259, 270, 281, 292, 303, 305, 316, 321, 327, 331, 336, 342, 349, 355, 357, 358, 360, 376, 429, 443, 448
CD79b, 13, *30*, 38, 303
CD80, *30*
CD81, *30*
CD82, *30*
CD83, *30*
CD84, *31*
CD85, *31*
CD86, *31*
CD87, *31*
CD88, *31*, 42
CD89, *31*
CD90, *31*, 43
CDw92, *31*
CDw93, *31*
CD94, *31*
CD95, *31*
CD96, *31*
CD97, *31*
CD98, *31*
CD99, *31*, 43
CD100, *31*, 107
CD101, *31*
CD102, *2*, *32*
CD103, *32*, 38, 316, 324, 331, 333, 420
CD104, *31*
CD105, *31*, 134
CD106, *2*, *31*
CD107a, *31*
CD107b, *31*
CD108, *31*
CD109, *31*
CD110, *31*, 42, 118
CD111, *31*
CD112, *31*
CDw113, *31*
CD114, *31*, 42
CD115, *31*, 42
CD116, *31*, 42, 134, 223, 227
CD117, 7, 8, 13, *32*, 43, 134, 148, 151, 163, 173, 209, 212, 215, 218–19, 221, 223, 227, 229, 231, 232, 233, 235, 242, 246, 249, 268, 270, 273, 377, 477, 478, *478*, 480
CD118, *32*
CD119, *32*
CD120a, *32*
CD120b, *32*
CD121a, *32*
CD121b, *32*
CD122, *32*, 403
CD123, *32*, 504
CD124, *32*

CDw125, *32*, 172
CD126, *32*
CD127, *32*
CD129, *32*
CD130, *32*
CD131, *32*
CD132, *32*
CD133, *32*, 134
CD134, *32*
CD135, *32*
CDw136, *32*
CDw137, *32*
CD138, 13, *32*, 38, 316, 319, 349, 357, 358, 376
CD139, *32*
CD140a, *32*
CD140b, *32*
CD141, *32*
CD142, *32*
CD143, *32*
CD144, *32*
CDw145, *32*
CD146, 16, *32*
CD147, *32*
CD148, *32*
CD150, *33*
CD151, *33*
CD152, *33*
CD153, *33*
CD154, *33*
CD155, *33*
CD156a, *33*
CD156b, *33*
CD157, *33*
CD158, *33*, 41, 413
CD159a, *33*
CD160, *33*
CD161, *33*, 41
CD162, *2*, *33*
CD163, *33*, 499
CD164, *33*
CD165, *33*
CD166, *33*
CD167, *33*
CD168, *33*
CD169, *33*
CD170, *33*
CD171, *33*
CD172, *33*
CD173, *33*
CD174, *33*
CD175, *33*
CD176, *33*
CD177, *33*
CD178, *33*
CD179a, *33*
CD179b, *33*
CD180, *33*
CD181, *33*
CD182, *33*
CD183, *34*
CD184, *34*
CD185, *34*
CDw186, *34*
CD191, *34*
CD192, *34*
CD193, *34*, 172
CD195, *34*
CD196, *34*
CD197, *34*
CDw198, *34*
CDw199, *34*
CD200, *34*
CD201, *34*
CD202b, *34*
CD203c, *34*, 246
CD204, *34*
CD205, *34*
CD206, *34*
CD207, 9, *34*, 107, 502
CD208, *34*
CD209, *34*
CDw210, *34*
CD212, *34*
CD213a1, *34*
CD213a2, *34*
CD217, *34*
CDw218, *34*
CD220, *34*
CD221, *34*
CD222, *34*
CD223, *34*
CD224, *34*
CD225, *34*
CD226, *34*
CD227, *35*
CD228, *35*
CD229, *35*
CD230, *35*
CD231, *35*
CD232, *35*
CD233, *35*
CD234, *35*
CD235, 10, 42
CD235a, *35*
CD235b, *35*
CD236, *35*
CD236R, *35*
CD238, 10, *35*, 42
CD239, *35*
CD240, 10, 42
CD240CE, *35*
CD240D, *35*
CD240DCE, *35*
CD241, *35*
CD242, 10, *35*, 42
CD243, *35*
CD244, *35*
CD245, *35*
CD246, *35*, 40
CD247, *35*, 40
CD248, *35*
CD249, *35*
CD252, *35*
CD253, *35*
CD254, *35*
CD256, *35*
CD257, *35*
CD258, *35*
CD261, *35*
CD262, *35*
CD263, *35*
CD264, *36*
CD265, *36*
CD266, *36*
CD267, *36*
CD268, *36*
CD269, *36*
CD271, *36*
CD272, *36*
CD273, *36*
CD274, *36*
CD275, *36*
CD276, *36*
CD277, *36*
CD278, *36*
CD279, *36*
CD280, *36*
CD281, *36*
CD282, *36*
CD283, *36*
CD284, *36*
CD289, *36*
CD292, *36*
CDw293, *36*
CD294, *36*
CD295, *36*
CD296, *36*
CD297, *36*
CD298, *36*
CD299, *36*
CD300, *36*
CD301, *36*
CD302, *36*
CD303, *36*
CD304, *36*
CD305, *36*
CD306, *37*
CD307, *37*
CD309, *37*
CD312, *37*
CD314, *37*
CD315, *37*
CD316, *37*
CD317, *37*
CD318, *37*
CD319, *37*
CD320, *37*
CD321, *37*
CD322, *37*
CD324, *37*
CDw325, *37*
CD326, *37*
CDw327, *37*
CDw328, *37*
CDw329, *37*
CD331, *37*
CD332, *37*
CD333, *37*
CD334, *37*
CD335, *37*, 41
CD336, *37*, 41
CD337, *37*, 41
CDw338, *37*
CD339, *37*
CDAN1 gene, 533
CDAN2 gene, 534
CDAN3 gene, 535
CDK4 gene, 347, 348
CDK6 gene, 321
Centroblasts, 23
  follicular lymphoma, 334–5, *334*, *335*
Centrocytes, 23
  follicular lymphoma, 334, *334*, *335*
Centromeric probes, 66, *66*

Charcot-Leyden crystals, 171–2
CHC1L gene, 373–5
Chediak–Higashi granules, 513–15
Chediak–Higashi syndrome (CHS), 494–5, *495*, 513–15
Chemotherapy, 226
  acute myeloid leukemia and, 207
  bone marrow morphological changes, 91–3, *93*
  myelodysplastic syndromes and, 148, 149
Children:
  acute myeloid leukemia with maturation, 232
  myelodysplastic syndromes, 148
  precursor B-lymphoblastic leukemia/ lymphoma, 257, 261, 262, 266–7
  precursor T-lymphoblastic leukemia/ lymphoblastic lymphoma, 272–3
  See also Juvenile myelomonocytic leukemia (JMML)
Chloroma, 247
Chromosome abnormalities, 57
  aneuploidy, 61–3
  balanced rearrangements, 60–1
  loss of heterozygosity (LOH), 63
  See also Cytogenetics; Deletions; Inversions; Translocations
Chronic eosinophilic leukemia (CEL), 171–4
  clinical aspects, 174
  cytogenetics, 173, *173*
  diagnostic criteria, *173*
  etiology and pathogenesis, 171
  immunophenotypic studies, 172–3
  molecular studies, 173
  pathology, 171–2
Chronic granulomatous disease (CGD), 516–17
Chronic idiopathic myelofibrosis (CIMF), 177–81, *178–82*
  clinical aspects, 180–1
  cytogenetics, 180
  etiology and pathogenesis, 177–8
  molecular studies, 179–80
  pathology, 178–80
Chronic lymphocytic leukemia (CLL), *291*, 301–13
  clinical aspects, 309–11
  cytogenetics, 58, 307–9, *308–11*, 311
  diagnostic criteria, 302
  differential diagnosis, 312–13, *314*, 315–16, 349
  etiology and pathogenesis, 301
  immunophenotype, *292*, 303–7, *307*, *308*
  pathology, 301–3, *302–6*
  staging, 309, *312*
  transformation to more aggressive disease, 311–12
Chronic myelogenous leukemia (CML), 157–69
  blast crisis, 160–1, 162–3, *164*, *165*, 167
  chromosome abnormalities, 57, 61
  clinical aspects, 169
  cytogenetics, 58, 167–9, *167*, *168*
  differential diagnosis, *202*, 203
  etiology and pathogenesis, 157–9
  evolution of, *163*
  flow cytometry, 161–3
  immunohistochemistry, 163
  molecular studies, 163–7, *166*
  molecular techniques, 67, 70, 73–4, 75
  pathology, 159–61
  See also Atypical chronic myeloid leukemia (aCML)
Chronic myelomonocytic leukemia (CMML), 191–7
  clinical aspects, 196–7
  cytogenetics, 195–6, *198*
  diagnostic criteria, *191*
  etiology and pathogenesis, 191–2
  immunophenotype studies, 193–5, *197*
  molecular studies, 195, *197*
  pathology, 192–3, *192–6*
  subcategories, 193
Chronic myeloproliferative diseases (CMPD), *96*, 151, 155–86
  classification, 155
  cytogenetics, 156–7, *156–9*, 186
  differential diagnosis, *151*, 184–6, *185*
  morphological features, 155, *156*
  See also Myelodysplastic/myeloproliferative diseases (MDS/MPD); *Specific diseases*
Chronic neutrophilic leukemia (CNL), 169–71
  clinical aspects, 171
  cytogenetics, 170–1, *171*
  diagnostic criteria, *169*
  etiology and pathogenesis, 169
  immunophenotypic studies, 170
  molecular studies, 170
  pathology, 169–70
Chronic polyclonal B-cell lymphocytosis, 312
Chronic renal failure, *96*, 559, 578
Clonal succession hypothesis, 1
Clot sections, 5, *6*
Clusters of differentiation (CD), 27, *28–37*
  See also CD molecules
Coagulation necrosis, 81–3, *83*
Cold hemagglutin disease (CHAD), 556
Collagen, *2*, 179
Colony-forming units (CFUs), 1, 4–5
  CFU-Baso, 4
  CFU-E, 4
  CFU-Eo, 4
  CFU-G, 4
  CFU-GEMM, 4, 5
  CFU-GM, 4
  CFU-M, 4
  CFU-Meg, 5
  HPP-CFU-MK, 5
Common acute lymphoblastic leukemia antigen (CALLA), 27
  See also CD10
Comparative genomic hybridization (CGH), 70–2
  array CGH (aCGH), 71, *72*
Complete blood count (CBC), 16
Congenital amegakaryocytic thrombocytopenia (CAMT), 567
Congenital dyserythropoietic anemias (CDAs), 150, 533–7, *539*
  differential diagnosis, *151*, 536–7
  type I, 533–4, *539*
  type II, 534–5, *539*
  type III, 535, *539*
  type IV, 536
  type V, 536
  type VI, 536
Congenital dysgranulopoietic neutropenia, 521
Congenital megakaryocytic hypoplasias, 567
Coombs' test, 556
Copper deficiency, 559
Core-binding factor beta (CBFß), 212, 217
Cryptococcosis, *88*
Cryptococcus neoformans, 87
Cutaneous mastocytoma, 482
Cutaneous mastocytosis (CM), 478, 480–2
  diffuse, 482
Cutaneous T-cell lymphoma (CTCL), 413, 414, 428
  anaplastic large cell lymphoma, 425–7
Cyclin D, 347, 375, 382
Cyclooxygenase-1 (COX-1) inhibitors, 578
Cytogenetics, 57–8
  acute basophilic leukemia, 246
  acute erythroid leukemia, 242
  acute leukemias of ambiguous lineage, 283, *285*
  acute lymphoblastic leukemia with eosinophilia, *274*
  acute megakaryoblastic leukemia, *245*
  acute monoblastic/monocytic leukemias, 236, *239*
  acute myeloid leukemia, 208, 211, 212
    minimally differentiated, 229, *230*
    therapy-related, 227, 228
    with 11q23 abnormalities, 221
    with inv(16)(p13q22) or t(16;16)(p13;q22), 219, *222*, *223*
    with maturation, 232, *233*
    with multilineage dysplasia, 223–5, *226*
    with t(8;21), 212–13, *214*
    without maturation, 229
  acute myelomonocytic leukemia, 233
  acute panmyelosis with myelofibrosis, 246
  acute promyelocytic leukemia, 216–17, *220*
  adult T-cell leukemia/lymphoma, 411, *412*
  analysis, 59–63
  anaplastic large cell lymphoma, 422–4, *425*, *426*
    primary cutaneous, 426
  aplastic anemia, 120, *121*
  atypical chronic myeloid leukemia, 199
  B-cell prolymphocytic anemia, 314–15
  banding techniques, 59
  blastic NK-cell lymphoma, 504
  Burkitt lymphoma, 360, *363*
  cell preparation, 58
  chronic lymphocytic leukemia, 58, 307–9, *308–11*, 311
  chronic myelomonocytic leukemia, 195–6, *198*
  chronic myeloproliferative diseases, 156–7, *156–9*, 186
    chronic eosinophilic leukemia, 173, *173*
    chronic idiopathic myelofibrosis, 180
    chronic myelogenous leukemia, 58, 167–9, *167*, *168*
    chronic neutrophilic leukemia, 170–1, *171*
  diffuse large b-cell lymphoma, 349–51
  enteropathy-type T-cell lymphoma, 420
  essential thrombocythemia, 184
  extranodal NK/T-cell lymphoma, nasal type, 407–8, *410*
  follicular lymphoma, 336–7, *340*, *341*
  hairy cell leukemia, 332
  hemophagocytic lymphohistiocytosis, 498
  hepatosplenic T-cell lymphoma, 418, *421*, *422*

Hodgkin lymphoma, 443, 448–9
hypereosinophilic syndrome, 173–4, *174*
juvenile myelomonocytic leukemia, *201*, 202
Langerhans cell histiocytosis, 502
lymphomatoid granulomatosis, 469
lymphoplasmacytic lymphoma, 316, *319*
MALT lymphoma, 327, *328*
mantle cell lymphoma, 347–8, *348*
mastocytosis, 480
mature B-cell neoplasms, 293, *293*
mature T-cell neoplasms, 293
mediastinal large B-cell lymphoma, 355
monoclonal gammopathy of undetermined significance, 377, 382–3
multiple myeloma, 380–2, 383–4
mycosis fungoides, 413–14
myelodysplastic syndromes, 137–42, *137*, *138–42*, *143*, *148*
NK-cell leukemia, 405, *407*
non-Hodgkin lymphoma, 288
paroxysmal nocturnal hemoglobinuria, 123
peripheral T-cell lymphoma, 432–3, *434*
plasma cell myeloma, 377–86, *383*, *384*, *385*
polycythemia vera, 176, *177*
precursor B-lymphoblastic leukemia/ lymphoma, 257–8, 260–6, *261–6*
precursor T-lymphoblastic leukemia/ lymphoblastic lymphoma, 270–2, *271*, *272*
primary effusion lymphoma, 357
Sezary syndrome, 413–14
splenic marginal zone lymphoma, 321–2, *324*, *325*
T-cell large granular lymphocytic leukemia, 403, *403*, *404*, *405*
T-cell prolymphocytic leukemia, 398–401, *399*, *400*, *401*
Cytokeratin, *50*
Cytokines, *3*, *4*
in hematopoieses, 4–5, *4*
in Langerhans cell histiocytosis, 500
Cytomegalovirus, *88*

D

Dacrocytes, 186, 530
Decay accelerating factor (DAF), 44, 120
Deletions, *61*, *66*, *69*
See also Cytogenetics
Dendritic cells (DC), 9–10, 489
disorders, 499–508
follicular dendritic cell tumor, 508, *508*
immunophenotypic features, *490*
interdigitating dendritic cell tumor, 506–8, *507*
spleen, 20–1
Dense granule deficiency, 577
Dermatopathic lymphadenitis, 107, *109*, *110*
Diamond–Blackfan anemia (DBA), 117, *118*, 531, 532, 533
Diffuse follicle center lymphoma, 339
Diffuse large B-cell lymphoma (DLBCL), 349–53
anaplastic lymphoma kinase (ALK) positive, 358
anaplastic variant, 349, *353*
centroblastic variant, 349
clinical aspects, 351–2
cytogenetics, 349–51
differential diagnosis, 352–3, 364, 451
etiology and pathogenesis, 349
immunoblastic variant, 349, *352*
immunophenotype, 349, *354*
pathology, 349, *350*, *351*
primary central nervous system, 358
primary cutaneous, 359
T-cell/histiocyte-rich variant, 349, *353*
with chronic inflammation, 358
DiGeorge syndrome, *456*, 457, *458*
Direct antiglobulin test (DAGT), 556
Disseminated intravascular coagulopathy (DIC), 217
DKC1 gene, 117
DNA chip, 77
DNA sequencing, 77–9
chemical methods, 78
limitations, 79
sequence detection and analysis, 78, *78*
Dohle inclusion bodies, 513, *514*
Dot blot, 76–7
Double minutes, 61, 62
Down syndrome, 151, 208, 243, 245, 523, *523*, *524*
See also Trisomy 21
Dutcher bodies, 13, *15*, 316, 376, *379*
Dyskeratosis congenita (DC), 117
Dysmyelopoietic syndromes, *See* Myelodysplastic syndromes (MDS)

E

Echinocytes, 530, 553
Echinocytosis, 553
ELA2 gene, 520
Electrophoresis, 78, *78*
Elliptocytes, 552–3, *552*
Embden–Meyerhof glycolytic pathway, 553, 554
Emperipolesis, 105, 177
Endothelial cells, 16, *17*
Enteropathy-type T-cell lymphoma, 420
Enumerating probe, 67, *69*
Eosinophilia, 524
acute lymphoblastic leukemia with eosinophilia, 273–4, *274*
acute myeloid leukemia with inv(16)(p13q22), 217–18, *221*, *222*
acute myeloid leukemia with maturation, 232
associated conditions, *524*
chronic myelomonocytic leukemia, 193
chronic myeloproliferative diseases, 159, 171–2, 186
Eosinophilic granules, 515
Eosinophilic hyperplastic lymphogranuloma, 109
Eosinophils, 8, *8*, *9*, *14*, *18*, *19*
atypical, 515, *516*
chronic eosinophilic leukemia, 171–2
idiopathic hypereosinophilic syndrome, 171–2
maturation, 5
Epstein–Barr virus (EBV), 287–8, 359
extranodal NK/T-cell lymphoma, 407–8
fulminant EBV-positive T-cell lymphoproliferative disorder, 409
Hodgkin lymphoma, 446
infectious mononucleosis, 461–3
NK-cell lymphoma, 405–6
post-transplant lymphoproliferative disease, 466–7, 468
senile EBV-associated B-cell lymphoproliferative disorder, 451, 469
X-linked lymphoproliferative syndrome, 463
Erdheim–Chester disease, *87*, 496
Erythroblastic islands, 4
Erythrocytes, 16
anemia and, 530
membrane skeleton defects, 551–3, *551*, *553*
differential diagnosis, 552–3
pathologic features, 552, *552*
myelodysplastic syndromes, 132, *133*
See also Red blood cells (RBCs)
Erythrocytosis, polycythemia vera, 174
Erythroid associated CD markers, 42
Erythroid burst-forming unit (BFU-E), 4
Erythroid precursors, 10–11, *11*
myelodysplastic syndromes, 132, 134
Erythroleukemia (AML-M6a), 151, 239, 240–1, *240*, *241*
differential diagnosis, *151*
See also Acute erythroid leukemias (AML-M6)
Erythropoiesis, 4
myelodysplastic syndromes, 132, *133*, *151*
Erythropoietin (EPO), *3*, 4
polycythemia vera and, 174
Essential thrombocythemia (ET), 181–4
clinical aspects, 184
cytogenetics, 184
diagnostic criteria, *182*
etiology and pathogenesis, 182–3
molecular studies, 183–4
pathology, 183, *183*, *184*, *185*
versus reactive thrombocytosis, *182*
ETV6 gene, *264*
ETV6-RUNX1 fusion gene, 262
European Group for the Immunologic Classification of Leukemia (EGIL), 281
Extramedullary plasmacytoma, 390, *390*
Extranodal marginal zone B-cell lymphoma of mucosa-associated lymphoid tissue (MALT lymphoma), 324–8
clinical aspects, 327
cytogenetics, 327, *328*
differential diagnosis, 327–8, 339
etiology and pathogenesis, 324
immunophenotype, 325–7, *327*
pathology, 324–5, *326*
Extranodal NK/T-cell lymphoma, nasal type, 406–8
clinical aspects, 408
cytogenetics, 407–8, *410*
differential diagnosis, 408–9, *410*
immunophenotype, 406–7
pathology, 406, *408*, *409*

F

FA-A gene, 115–16
Factor VII, 217
Factor VIII, 11, 43, 133, 135, 245, 246
Factor X, 217
FAD1 gene, 116

Familial Mediterranean fever, 391–2, *393*
Fanconi anemia (FA), 115–16, *116*
clinical aspects, 116
etiology and pathogenesis, 115–16
genetics, 116
pathology, 116
Ferritin, 545
FGFR3 gene, 382
Fibrinoid necrosis, 83
Fibroblast-like cells, 16
Fibronectin, *2*
Fibrosis:
bone marrow, 89, *90*, 94, *96*, *97*
chronic myeloproliferative diseases, 177–8, 185
precursor B-lymphoblastic leukemia/ lymphoma, 258
mastocytosis, 478, *481*, *483*, *484*
myelofibrosis, 177
acute panmyelosis with myelofibrosis, 246–7
See also Chronic idiopathic myelofibrosis (CIMF)
FIP1L1 gene, 173
FIP1L1-PDGFRA fusion gene, 171, 173–4
FISH, *See* Fluorescence *in situ* hybridization (FISH)
Flow cytometry (FCM), 44–7, *44*
aneuploidy detection, 61
compensation, 45, *45*, *46*
data analysis, 45–6, *46*, *47*
forward scatter (FSC), 45
gating, 45
quality control, 47
side scatter (SSC), 45
See also Immunohistochemistry
FLT3 gene, 136
Fluorescein isothiocyanide (FITC), 45, *45*
Fluorescence *in situ* hybridization (FISH), 61, *62*, 65–74
FFPE sections, 73
limitations, 68–70
multi-color FISH, 67
nomenclature, 73–4
probe strategies, 67–74
break-apart probe, 68, *70*
enumerating probe, 67, *69*
fusion probe, 67, *70*, *71*
probe types, 66–7, *66*, *67*, *68*
FMC7, 23, 38–9, 303, 305, 315, 316, 321, 331, 342, 349
Folate deficiency, 540, *540*, 542
Follicular dendritic cell tumor, *508*
Follicular dendritic cells (FDCs), 9, 10, 489
lymph nodes, 23
Follicular hyperplasia, 101–3, *101*, *102*
Castleman's disease, 103, *104*, *105*
differential diagnosis, 102, *103*
rheumatoid arthritis, 102–3, *104*
Follicular lymphoma (FL), 298, 333–9
clinical aspects, 337
cytogenetics, 336–7, *340*, *341*
differential diagnosis, 102, *103*, 339, *342*
etiology and pathogenesis, 333
grading, *336*
immunophenotype, *292*, 336, *338*, *339*
pathology, 334–6, *334–8*
variants, 337–9
diffuse follicle center lymphoma, 339
primary cutaneous follicular center cell lymphoma, 339
Foreign-body granulomas, 85, *88*
Franklin disease, 393
Fulminant EBV-positive T-cell lymphoproliferative disorder, 409
Fusion probe, 67, *70*, *71*

G

G-banding, 59, *69*
See also Karyotype
GATA-1 gene, 177, 523
Gaucher cells, 490, *491*
Gaucher disease (GD), 489–92
classification, 492, *492*
clinical aspects, 492
differential diagnosis, 492
immunophenotype, 491
molecular genetics, 490
pathogenesis, 490
pathology, 490–1, *491*
Gelatinous transformation, bone marrow, 81, *82*
Gene-specific probes, 66, *66*
Germinal center, 22–3, *22*, *23*
in follicular hyperplasia, 101–2, *102*
84GG gene, 490
Giant cell arteritis, *97*
Giant lymph node hyperplasia, 103
Giemsa stain, 5, 59
Glanzmann thrombasthenia (GT), 574–6
clinical aspects, 576
differential diagnosis, 576
etiology and pathogenesis, 575
molecular and cytogenetic studies, 575–6
pathology, 575
Glucocerebrosidase deficiency, 490, 492
Glucose-6-phosphate dehydrogenase (G6PD) deficiency, 553–4, *554*
Glycophorin A, 42
Glycophorin B, 42
Glycoproteinoses, *490*
Glycoproteins (GP), 570, 572
GPIb-IX-V complex, 572, 573, 575
GPIIb-IIIa complex, 42, 575
Glycosyl phosphatidylinositol (GPI) anchor defect, paroxysmal nocturnal hemoglobinuria, 120–1, *122*, 123
GP, *See* Glycoproteins
GPI genes, 572, 573
GPV gene, 572
Granular megakaryocytes, 11
Granulocyte colony-stimulating factor (G-CSF) receptor, *3*, 42
Granulocyte-macrophage colony stimulating factor (GM-CSF), *3*, 4, 5
receptor, 42
Granulocytes, 5–10
associated markers, 41–2
maturation, 6–7
paroxysmal nocturnal hemoglobinuria, 123
Granulocytic sarcoma, 247–8, *247*
Granulocytosis:
chronic myelomonocytic leukemia, 193
chronic myeloproliferative diseases, 186
juvenile myelomonocytic leukemia, 201
Granulomas, 85–9, *86*, *96*
detection, 89
foreign-body, 85, *88*
viral-associated, 85–9, *86*, *89*
Granulomatous lymphadenitis, *101*, 107
Granulopoiesis, 2
myelodysplastic syndromes, 132, *134*, *135*
Gray platelet syndrome, *96*, 577

H

Hairy cell leukemia (HCL), *291*, 329–33
clinical aspects, 332, *332*
cytogenetics, 332
differential diagnosis, *124*, 316, *320*, 324, 333
etiology and pathogenesis, 329
immunophenotype, *292*, 331, *332*
pathology, 329–30, *329*, *330*, *331*
variants, *330*, 333, *333*
Hallmark cells, 422
Hand-mirror cells, 258–9, *259*
Heavy chain deposition disease (HCDD), 392
Heavy chain diseases, 392–3
α heavy chain disease, 392–3
μ heavy chain disease, 392–3
τ heavy chain disease, 392–3
Heinz bodies, 530, *534*, 551
Helicobacter pyloris, 324
Helmet cells, *558*
Hematogones, 5, 12–13, *13*, *14*, 273
precursor B-lymphoblastic leukemia/ lymphoma, 267, *267*, *268*
Hematopoiesis, 4–5, *4*
extramedullary, 177, 179
Hematopoietic stem cells (HSCs), 1, 2
Hematoxylin and eosin (H&E) staining, 5
Hemoglobin (Hb):
Hb Bart's, 547, 548, *548*
sickle cell disease, 549
thalassemia, 545–8, *546*, *548*
unstable, 551
Hemolytic anemia:
drug-induced, 556–7, *557*
immune-mediated, 554–7
infection-induced, 557
secondary to enzyme deficiencies, 553–4
substances inducing hemolysis, *554*
trauma-induced, 557, *558*
Hemolytic disease of the newborn, 556
Hemolytic transfusion reactions, 556
Hemolytic uremic syndrome (HUS), 570–1
Hemonectin, *2*
Hemophagocytic lymphohistiocytosis (HLH), 496–9
cancer-associated, 497
clinical aspects, 498
cytogenetics, 498
diagnostic criteria, *501*
differential diagnosis, 498–9
etiology and pathogenesis, 496–7
familial, 496–7
immunophenotype, 498
infection-associated, 497
pathology, 498, *499*, *500*
Hemophagocytic syndromes, 107

Hemosiderin, 9
Heparin, *478*
Heparin-induced thrombocytopenia (HIT), 570
Hepatitis, aplastic anemia and, 119
Hepatosplenic T-cell lymphoma, 416–20
  clinical aspects, 418–20
  cytogenetics, 418, *421*, *422*
  differential diagnosis, 420
  etiology and pathogenesis, 417
  immunophenotype, 418, *420*
  pathology, 417, *417*, *418*, *419*
Hepcidin, 558
Hereditary elliptocytosis (HE), 552, *552*
Hereditary erythroblastic multinuclearity with a positive acidified serum lysis test (HEMPAS), 534
Hereditary orotic aciduria, 541
Hereditary spherocytosis (HS), 552, *552*
Hereditary thrombocythaemia (HT), 567
Herpes zoster, *88*
Histamine, *478*
Histiocytes, 85–9, *87*, *89*, 489
  immunophenotypic features, *490*
Histiocytic disorders, 489, *490*
Histiocytic sarcoma, 499
Histiocytosis:
  Langerhans cell (LCH), 107, 499–503, *501*, *502*, *503*
  sea-blue, 159
  sinus, 103–5, *105*
    with massive lymphadenopathy, 105, *106*, 496
  See also Hemophagocytic lymphohistiocytosis (HLH)
Histiocytosis X, *See* Langerhans cell histiocytosis (LCH)
Histograms, 46, *47*
Histoplasma capsulatum, *87*
Histoplasmosis, *88*
HLA (human leukocyte antigen), 500, 570
  deficiency, *456*
HLA-B14, 119
HLA-DR, 1, 7, 9, 23, 134, 193, 202, 209, 218, 219, 229, 232, 233, 235, 242, 246, 303, 499, 502, 504, 508
  HLA-DR2, 119
Hodgkin lymphoma (HL), 287, 290–1, 441–51, *442*, *443*
  classical Hodgkin lymphoma (CHL), 445–51
    clinical aspects, 449–51
    cytogenetics, 448–9
    differential diagnosis, 451
    etiology and pathogenesis, 446–7
    immunophenotype, 448, *450*
    lymphocyte-depleted CHL, 447–8, *449*, *450*, 451
    lymphocyte-rich (LRCHL), 445, 447, 451
    mixed cellularity CHL, 447, *448*, *449*
    nodular sclerosis CHL, 447, *447*, *448*, 451
    pathology, *446*, 447–8
    staging, 449–50
  classification, 291, *292*, 441
  morphological diversity, 290
  nodular lymphocyte predominant Hodgkin lymphoma (NLPHL), 441–5
    clinical aspects, 444–5
    cytogenetics, 443
    differential diagnosis, 445
    etiology and pathogenesis, 441–2
    immunophenotype, 443
    pathology, 442–3, *444*, *445*
  post-transplant, 465, *467*, 468
  therapy, 451
  See also Lymphoid malignancies; Lymphoma
Hodgkin–Reed–Sternberg (HRS) cells, 441, 445–6, *446*, 448, *449*, *450*
Homologous staining regions (HSRs), 61, 62–3, *63*
Howell-Jolly body, *19*, 530, *534*
HPA (human platelet alloantigen), 570
Human cell differentiation molecules (HCDM), 27, *28–37*
  See also CD molecules
Human Genome Project, 77
Human herpes virus 8 (HHV-8), 357, 358
Human immunodeficiency virus (HIV), 457, 459
  See also Acquired immunodeficiency syndrome (AIDS)
Human T-cell lymphotrophic virus (HTLV-1), 288, 409–12
Hyper IgM syndrome, *456*
Hyperdiploid karyotype, *62*
  plasma cell myeloma, 380–2, 383–4, *383*, 388
  precursor B-lymphoblastic leukemia/lymphoma, 265, *265*
Hypereosinophilic syndrome (HES), 171–4
  clinical aspects, 174
  cytogenetics, 173–4, *174*
  diagnostic criteria, *173*
  etiology and pathogenesis, 171
  immunophenotypic studies, 172–3
  molecular studies, 173
  pathology, 171–2, *172*
Hyperparathyroidism, primary, *96*
Hypersplenism, 557–8
Hypodiploid karyotype:
  plasma cell myeloma, 380, 382, *383*
  precursor B-lymphoblastic leukemia/lymphoma, 265, *266*
Hypoplastic acute leukemia, 248–9, *248*
  lymphoblastic, 274, *275*

I

ICAM-1, *2*
ICAM-2, *2*
ICAM-3, *2*
ICAM-4, *2*
Idiopathic CD4+ T-lymphocytopenia, 461
Idiopathic thrombocytopenic purpura (ITP), 568
IGH gene, 316, 337, 382, 386
IGK gene, 337
IGL gene, 337
IGVH gene, 309, 320
IL2RG gene, 455
IL-1β gene, 403
Image cytometry (ICM), 61
Imatinib mesylate (IM), 169, 173
Immunoglobulin gene rearrangement clonality:
  mature B-cell neoplasms, 297–9, *299*, *300*, *301*
  precursor B-lymphoblastic leukemia/lymphoma, 265–6, *266*
Immunoglobulins, *2*
  monoclonal immunoglobulin deposition diseases, 390–3
  transcription factors, 44
Immunohistochemistry, 47–50, *47*, *48*, *49*
  acute basophilic leukemia, 246
  acute erythroid leukemia, 242, *243*
  acute leukemias of ambiguous lineage, 281–3, *282*, *283*
    bilineal acute leukemia, 282–3, *285*
    biphenotypic acute leukemia, *281*, 282, *283*
    with expression of aberrant markers, 281, *281*, *282*
  acute megakaryoblastic leukemia, 245
  acute monoblastic/monocytic leukemias, 235–6, *238*, *239*
  acute myeloid leukemia, 209–11, *209*, *210*, *211*
    minimally differentiated, 229
    therapy-related, 227
    with 11q23 abnormalities, 219–21
    with inv(16)(p13q22) or t(16;16)(p13;q22), 218–19
    with maturation, 232
    with multilineage dysplasia, 223
    with t(8;21), 212
    without maturation, 229
  acute myelomonocytic leukemia, 233, *236*
  acute panmyelosis with myelofibrosis, 246
  acute promyelocytic leukemia, 215–16, *219*
  adult T-cell leukemia/lymphoma, 410–11, *412*
  anaplastic large cell lymphoma (ALCL), 422, *424*
    primary cutaneous, 426
  angioimmunoblastic T-cell lymphoma, 428–9, *430*, *431*
  B-cell prolymphocytic leukemia, 314
  blastic NK-cell lymphoma, 504, 506, *506*
  Burkitt lymphoma, 360, *362–3*
  chronic eosinophilic leukemia, 172–3
  chronic lymphocytic leukemia, 303–7, *307*, *308*
  chronic myelogenous leukemia, 163
  chronic myelomonocytic leukemia, 193–5, *197*
  chronic neutrophilic leukemia, 170
  dendritic cell tumors, 507, 508
  diffuse large B-cell lymphoma, 349, *354*
  enteropathy-type T-cell lymphoma, 420
  extranodal NK/T-cell lymphoma, nasal type, 406–7
  follicular lymphoma, 336, *338*, *339*
  Gaucher disease, 491
  hairy cell leukemia, 331, *332*
  hematodermic neoplasms, *506*
  hemophagocytic lymphohistiocytosis, 498
  hepatosplenic T-cell lymphoma, 418, *420*
  histiocytic sarcoma, 499
  Hodgkin lymphoma, 443, 448, *450*
  hypereosinophilic syndrome, 172–3
  intravascular large B-cell lymphoma, 357, *357*
  juvenile myelomonocytic leukemia, 202
  Langerhans cell histiocytosis, 502
  lymphomatoid granulomatosis, 469
  lymphoplasmacytic lymphoma, 316
  MALT lymphoma, 325–7, *327*
  mantle cell lymphoma, 342–7, *345*
  mastocytosis, 480
  mature B-cell neoplasms, 292, *292*
  mature T-cell neoplasms, 292

Immunohistochemistry (*Continued*)
mediastinal large B-cell lymphoma, 355
mycosis fungoides, 413, *416*
myelodysplastic syndromes, 134–5
Niemann–Pick disease, 493
NK-cell leukemia, 405, *406*
peripheral T-cell lymphoma, 432
plasma cell myeloma, 376–7, *381*, *382*
plasmablastic lymphoma, 358
precursor B-lymphoblastic leukemia/ lymphoma, 259–60, *260*
precursor T-lymphoblastic leukemia/ lymphoblastic lymphoma, 269–70, *270*
primary effusion lymphoma, 357
Sezary syndrome, 413, *416*
splenic marginal zone lymphoma, 321
T-cell large granular lymphocytic leukemia, *403*
T-cell prolymphocytic leukemia, 398, *399*
Indolent myeloma, 386
Infantile (childhood) monosomy 7 syndrome, 203
Infections, 495–6, 497
See also Specific infections
Infectious mononucleosis (IM), *88*, 446, 461–3
clinical aspects, 463
differential diagnosis, 463
etiology and pathogenesis, 462
pathology, 462–3, *462*, *464*
Integrins, 517
Interdigitating dendritic cell tumor, 506–8, *507*
Interdigitating dendritic cells (IDCs), 9, 10, 489
lymph nodes, 23
Interferons, *3*
Interleukins (ILs), *3*, 4–5, *478*
IL-2, *3*
IL-3, *3*
IL-4, *3*
IL-5, *3*
IL-6, *3*
IL-7, *3*
IL-8, *3*
IL-9, *3*
IL-10, *3*
IL-11, *3*
IL-12, *3*
IL-13, *3*
IL-15, *3*
IL-16, *3*
IL-17, *3*
IL-18, *3*
IL-19, *3*
IL-20, *3*
IL-21, *3*
IL-22, *3*
IL-23, *3*
IL-24, *3*
IL-25, *3*
IL-26, *3*
in hematopoiesis, 4–5
Intravascular large B-cell lymphoma, 356–7, *356*, *357*
Intrinsic factor (IF), 540
Inversions, 60–1, 67
paracentric, *60*
See also Cytogenetics
Ionizing radiation, *See* Radiation
Iron deficiency anemia (IDA), *530*, 544–5, *544*, *545*
Irradiation, *See* Radiation
ITGA2B gene, 575–6
ITGB3 gene, 575–6

J

JAK2 gene, 171, 183, 186
chronic idiopathic myelofibrosis, 177, 179
essential thrombocythemia, 183–4
polycythemia vera, 174, 175–6, *176*
Juvenile myelomonocytic leukemia (JMML), 200–3
clinical aspects, 202–3
cytogenetics, *201*, 202
diagnostic criteria, *200*
etiology and pathogenesis, 200
immunophenotypic studies, 202
pathology, 200–2, *200*

K

Karyotype, 59, *59–63*, 65, *69*
acute leukemias of ambiguous lineage, *285*
acute lymphoblastic leukemia with eosinophilia, *274*
acute megakaryoblastic leukemia, *245*
acute monocytic leukemia, *239*
acute myeloid leukemia:
minimally differentiated, *230*
with inv(16)(p13q22) or t(16;16)(p13;q22), *22*, *223*
with maturation, *233*
with multilineage dysplasia, *226*
with t(8:21), *214*
acute promyelocytic leukemia, *220*
adult T-cell leukemia/lymphoma, 411, *412*
anaplastic large cell lymphoma, *425*, *426*
aplastic anemia, *121*
Burkitt lymphoma, *363*
chronic lymphocytic leukemia, *308–11*
chronic myelomonocytic leukemia, *188*
chronic myeloproliferative diseases, *157–9*
chronic eosinophilic leukemia, *173*
chronic myelogenous leukemia, *167*, *168*
chronic neutrophilic leukemia, *171*
DiGeorge syndrome, *458*
extranodal NK/T-cell lymphoma, nasal type, *410*
follicular lymphoma, *340*
hepatosplenic T-cell lymphoma, *421*, *422*
juvenile myelomonocytic leukemia, *201*
lymphoplasmacytic lymphoma, *319*
MALT lymphoma, *328*
mantle cell lymphoma, *348*
myelodysplastic syndromes, *138–42*
peripheral T-cell lymphoma, *434*
plasma cell myeloma, 377, 380, *383*, *384*, *385*
polycythemia vera, *177*
precursor B-lymphoblastic leukemia/ lymphoma, *263–6*
precursor T-lymphoblastic leukemia/ lymphoblastic lymphoma, *271*, *272*
splenic marginal zone lymphoma, *324*, *325*
T-cell prolymphocytic leukemia, *399*, *400*, *401*
See also Cytogenetics
Kell blood group transmembrane protein, 42
Ki-67, 44
Kikuchi's disease, 107–9, *111*, 496
differential diagnosis, 109
Kimura's disease, 109
KIT, 477
Kostmann syndrome, 521

L

L444P gene, 490, 492
Lactoferrin, *3*
Landsteiner–Wiener (LW) blood group antigen, 42
Langerhans cell histiocytosis (LCH), 107, 499–503
classification, *501*
clinical aspects, 503
differential diagnosis, 503
etiology and pathogenesis, 499–500
immunohistochemistry, 502
molecular and cytogenetic studies, 502
pathology, 501–2, *501*, *502*, *503*
Langerhans cell sarcoma, 503, *504*
Langerhans cells (LCs), 9–10, *10*, 489
Langerin, 107, 502, 503
Large granular lymphocytes (LGL), 12, *13*, *18*, 289, *289*, 401, *402*
associated markers, 40–1
Large granular lymphocytic leukemias, 401
differential diagnosis, 408–9, *410*
Lazy leukocyte syndrome, 521
Legionnaire's disease, *88*
Leishmania donovani, *86*
Lennert lymphoma, 432, *433*
Leprosy, *88*
Lesch–Nyhan syndrome, 541
Leukemia, *See Specific types of leukemia*
Leukemic reticuloendotheliosis, *See* Hairy cell leukemia (HCL)
Leukemogenesis, 208, 217
Leukocyte adhesion deficiency (LAD), 517–19, *518*
Leukocyte cell adhesion molecules (Leu CAM), *2*
Leukocyte morphology, 17
Leukocytosis:
chronic myeloproliferative leukemia, 159, *162*
Leukoerythroblastosis:
chronic myeloproliferative diseases, 179, 186
Light chain deposition disease (LCDD), 392, *393*
Lipogranulomas, 89
lymphangiogram-associated, 496, *497–8*
Loss of heterozygosity (LOH), 63
Lutzner cells, 413
Lymph nodes, 22–4, *22*
chronic lymphocytic leukemia, 303, *306*
cortex, 22, *22*
follicular structures, 22–3, *22*
germinal center, 22–3, *22*, *23*
mantle zone, *22*, 23
See also Follicular hyperplasia
lymphoma, *289*
mantle cell lymphoma, 342, *343*, *345*
medulla, *22*, 24

paracortex, 22, *22*, 23
vascular and lymphatic structures, 24
Lymphadenitis:
dermatopathic, 107, *109*, *110*
granulomatous, *101*, 107
mixed pattern, *101*, 107–9
paracortical, *101*, 103
sinus pattern, *101*, 103–7
Lymphangiogram-associated lipogranuloma, 496, *497–8*
Lymphoblastic lymphoma (LBL), 257
Lymphoblasts, 11
Lymphocytes, *8*, 11–13, *13*, *14*, *16*, *18*, *19*
activated, 12
lymph nodes, 23
paroxysmal nocturnal hemoglobinuria, 123
See also B-cells; T-cells
Lymphocytic and histiocytic (L&H) cells, 443, *444*, *445*
Lymphocytopenia, 455–61
HIV/AIDS, 457–61
idiopathic CD4+ T-lymphocytopenia, 461
primary immunodeficiency syndromes, 455–7
Lymphocytosis, 461–5, *461*
bone marrow benign lymphoid aggregates, 464–5, *465*, *466*
infectious mononucleosis, 461–3
persistent polyclonal B-cell lymphocytosis (PPBL), 464, *465*
polyclonal immunoblastic proliferation, 464
stress-induced, 464
X-linked lymphoproliferative syndrome (XLP), 463, *464*
Lymphoid aggregates, 13, *16*
benign, 464–5, *465*, *466*
Lymphoid lineage, 11–13
Lymphoid malignancies:
classification, 287, 291, *292*
clinical aspects, 293–4
cytogenetics, 293, *293*
differential diagnosis, 294–5
etiology and pathogenesis, 287–8
immunophenotype, 292–3, *292*
pathology, 288–91
See also Specific malignancies
Lymphoid nodular hyperplasia, 465
Lymphoma, *96*, *289*
cytogenetics, 58, 288
etiology and pathogenesis, 287–8
staging, 293, *294*
See also Specific types of lymphoma
Lymphomagenesis, 288
Lymphomatoid granulomatosis, 469, *470*
Lymphomatoid papulosis, 426–7, 469–71, *471*
Lymphomatous polyposis, 342
Lymphoplasmacytic lymphoma (LPL), 316–20
clinical aspects, 316–19
cytogenetics, 316, *319*
differential diagnosis, 312–13, 319–20, *320*, 324
etiology and pathogenesis, 316
immunophenotype, *292*, 316
pathology, 316, *317*, *318*
Lymphopoiesis, 5
Lysosomal storage disease, *96*, *490*
LYST gene, 494

M

M-component, 373
M-protein, 387–8
Macrocytic anemias, 538–44
Macrocytosis:
causes, *540*
juvenile myelomonocytic leukemia, 201
Macrophage colony-stimulating factor (M-CSF), *3*
receptor, 42
Macrophages, 8–9, *10*, 489
immunophenotypic features, *490*
spleen, 19, 20, 21
MAF genes, 382
MAL gene, 355
Malaria infection, *535*, 550
MALT lymphoma, *See* Extranodal marginal zone B-cell lymphoma of mucosa-associated lymphoid tissue (MALT lymphoma)
MALT1 gene, 324
Mantle cell lymphoma (MCL), 103, 339–49
clinical aspects, 348
cytogenetics, 347–8, *348*
differential diagnosis, 312–13, *314*, 324, 348–9
etiology and pathogenesis, 342
immunophenotype, *292*, 342–7, *345*
pathology, 342, *343–4*
variants, *347*
Mast cell leukemia, 478, 485, *486*
Mast cell sarcoma, 485–6
Mast cells, 8, *10*, 477, *479*
immunophenotypic features, *478*
mediators, *478*
Mastocytoma:
cutaneous, 482
extracutaneous, 485
Mastocytosis, *94*, *96*, 477–87
clinical aspects, 480–6
cutaneous (CM), 478, 480–2
diagnostic criteria, 478, *480*
differential diagnosis, 486–7, *486*
etiology and pathogenesis, 477
immunophenotype, 480
pathology, 477–8, *478–83*
systemic, 482–6, *482–5*
aggressive (ASM), 485
indolent (ISM), 485
with associated clonal hematological non-mast cell lineage disease (SM-AHNMD), 485
Mature B-cell neoplasms, 287, 297–364
classification, *292*
clinical aspects, 293–4
differential diagnosis, 294–5
immunophenotype, 292, *292*
molecular and cytogenetic studies, 293, *293*
immunoglobulin gene rearrangement clonality, 297–9, *300*, *301*
minimal residual disease (MRD) detection, 299–300
See also Lymphoid malignancies; *Specific neoplasms*
Mature T-cell neoplasms, 287, 289, 397
classification, *292*, *397*
cytogenetics, 293
differential diagnosis, 294–5
immunophenotype, 292
See also Lymphoid malignancies; *Specific neoplasms*
May–Hegglin anomaly, 513
Mean corpuscular hemoglobin concentration (MCHC), 529
Mean corpuscular hemoglobin (MCH), 529
Mean corpuscular volume (MCV), 529
Mediastinal large B-cell lymphoma (MLBCL), 353–6
clinical aspects, 355
cytogenetics, 355
differential diagnosis, 355–6
etiology and pathogenesis, 354
immunophenotype, 355
pathology, 354, *355*
Mediterranean lymphoma, 392
MEFV gene, 392
Megakaryoblasts, 11, *12*
See also Acute megakaryoblastic leukemia (AMKL)
Megakaryocytes, 4, *8*, *12*
associated markers, 42–3
chronic myeloproliferative diseases, 185–6
myelodysplastic syndromes, 132–3, *136*
Megakaryocytic hypoplasia, 567
Megakaryocytosis, 567, *569*
Megaloblastic anemia, 538–44
clinical aspects, 542–3
differential diagnosis, 150, *151*, 242–3, 543–4
etiology and pathogenesis, 538–41
pathology, 541–2, *542*, *543*
Melanoma, metastatic, *91*
Metamyelocytes, 7–8, *8*
Metarubricytes, *8*, 11
Metastatic lesions, 50, *50*, *96*
bone marrow, 89–91, *90*, *91*, *92*
Methotrexate-associated lymphoproliferative disorder, 469
Microarray techniques, 77, *77*
Microcytic anemias, 544–8, *549*
Mixed lineage leukemia (MLL), 62
Mixed pattern reactive lymphadenitis, *101*, 107–9
dermatopathic lymphadenitis, 107, *109*, *110*
Kikuchi's disease, 107–9, *111*
Kimura's disease, 109
toxoplasmosis, 107, *110*
MLL gene, 62, 68, 72, 219, 221, 235
acute leukemias of ambiguous lineage, 279–80
acute myeloid leukemia with 11q23 (*MLL*) abnormalities, 219–22
MLL/AF4 fusion gene, 261
MMSET gene, 382
Monoclonal B-cell expansion in the elderly, 312
Monoclonal gammopathy of undetermined significance (MGUS), 377, 386, *387*
clinical aspects, *386*
cytogenetics, 377, 382–3
Monoclonal immunoglobulin deposition diseases, 390–3
heavy chain deposition disease (HCDD), 392
heavy chain diseases, 392–3
light chain deposition disease (LCDD), 392, *393*
primary amyloidosis, 390–2

Monocytes, 8–9, *8*, *18*, *19*
associated markers, 41–2
differentiation , acute myeloid leukemia with 11q23 abnormalities, 219
maturation, 5, *10*
myelodysplastic syndromes, *134*, *135*
paroxysmal nocturnal hemoglobinuria, 123
Monocytic disorders, 489, *490*
Monocytoid B-cell hyperplasia, 106, *106*
Monocytopenia, 495
Monocytosis, 495
associated conditions, *494*
chronic myelomonocytic leukemia, 191, 193, *194*, 203
juvenile myelomonocytic leukemia, *200*, 201, 203
See also Acute monocytic leukemia
Monosomy, 3, *60*
Monosomy, 7, 61, *121*
infantile (childhood) monosomy 7 syndrome, 203
juvenile myelomonocytic leukemia, *201*, 202, 203
myelodysplastic syndromes, 138, *140*, *142*
Monosomy, 18, 391
Morula cells, 13, 376
Mott cells, 13, *15*, 376
MPL gene, 176
c-MPL, 183
Mucolipidoses, *490*
Mucopolysaccharidoses, *490*
Mucosa-associated lymphoid tissues (MALT), 288
See also Extranodal marginal zone B-cell lymphoma of mucosa-associated lymphoid tissue (MALT lymphoma)
Multiple myeloma (MM), 380–2, 383–4
MYC gene, 293, 314, 360
See also C-MYC gene
MYCN gene amplification, 62
Mycobacterium avium-intercellulare, *86*, *88*
Mycoplasma, *88*
Mycosis fungoides (MF), 107, 413–16
clinical aspects, 414–16
cytogenetics, 413–14
differential diagnosis, 416, *417*
etiology and pathogenesis, 413
immunophenotype, 413, *416*
pathology, 413, *414*, *415*
staging, 415, *416*
Myeloblasts, 7, *8*, *9*
Myelocytes, 7, *8*
Myelodysplasia:
autoimmune, 148–9
heavy-metal associated, 149
HIV-associated, 149
hypocellular, 148, *149*
non-clonal, 148–9, *151*, 152
paraneoplastic, 149
Myelodysplastic syndromes (MDS), *96*, 129–52, *484*
classification, 129, *130*, 143–8
5q– syndrome, 145–6, *147*
MDS, unclassifiable, 146–8
refractory anemia, 143, *144*
refractory anemia with excess blasts, 144–5, *146*
refractory anemia with ringed sideroblasts, 143, *145*
refractory cytopenia with multilineage dysplasia, 143–4, *145*
clinical aspects, 149–50
cytogenetics, 137–42, *137*, *138–42*, *143*, *148*
differential diagnosis, 150–2, *151*
etiology and pathogenesis, 129–31, *131*
Fanconi anemia and, 116
flow cytometry, 133–4, *136*
hypocellular, *124*, 148, *149*
immunohistochemistry, 134–5
International Prognostic Scoring System (IPSS), 149–50, *150*
molecular studies, 136–43
morphological features, 131–3
dyserythropoiesis, 132, *133*
dysgranulopoiesis, 132, *134*, *135*
megakaryocyte abnormalities, 132–3, *569*
platelet abnormalities, 133
non-clonal, 148–9, *151*, 152
pediatric, 148
therapy-related (t-MDS), 148, 226
transformation to acute leukemia, *150*
See also Myelodysplastic/myeloproliferative diseases (MDS/MPD)
Myelodysplastic/myeloproliferative diseases (MDS/MPD), 191
classification, 191
differential diagnosis, *202*, 203
See also Chronic myeloproliferative diseases (CMPD); Myelodysplastic syndromes (MDS); *Specific diseases*
Myelofibrosis, 177, *569*
acute panmyelosis with myelofibrosis, 246–7
See also Chronic idiopathic myelofibrosis (CIMF)
Myelokathexis, 521
Myelomatosis, 387
Myeloperoxidase (MPO), 7, 8
deficiency, 515–16
stain, 210, *210*
Myelophthisic anemia, 559
Myelopoiesis, 4–5, *9*
Myeloproliferative disorder (MPD), 156–7
See also Chronic myeloproliferative diseases (CMPD)
MYH9 gene, 513

N

N370S gene, 490, 492
NADPH oxidase, 516
Naphthol AS-D acetate esterase marker, 211
Naphthol AS-D chloroacetate marker, 211
Natural killer (NK) cells, 5, 12, *289*
associated markers, 41
NK-cell neoplasms, 287, 289, 397
aggressive NK-cell leukemia, 404–5, *406*, *407*
blastic NK-cell lymphoma, 409, 503–6
classification, *292*, *397*
immunophenotype, 292
NK-cell lymphoblastic leukemia/ lymphoma, 273
Necrosis, bone marrow, 81–3, *83*, *93*
Neonatal immune-mediated thrombocytopenia, 570
Neuroblastoma, metastatic, 89, *92*
Neutropenia, 519–21
associated with bone marrow disorders, 520
associated with primary immune disorders, 520
causes, *520*
congenital, 520–1
drug-induced, 520, *521*
non-immune chronic idiopathic, 520
post-infectious, 520
Neutrophil nuclear hypersegmentation, 515, *518*
Neutrophilia, 521–4
causes, *522*
chronic neutrophilic leukemia, 170, *170*
primary, 523, *523*, *524*
reactive, 522, *522*, *523*
spurious, 524
Neutrophilic granulocytosis, 17
Neutrophils, *8*
maturation, 4–5
myelodysplastic syndromes, *134*, *135*
segmented, 8
NF1 gene, 200, 202
NFKB-2 gene, 293
Niemann–Pick disease (NPD), 492–4
classification, 493–4, *493*
clinical aspects, 493–4
differential diagnosis, 494
immunophenotype, 493
molecular genetics, 492
pathology, 492–3, *493*, *494*
Niemann–Pick (NP) cells, 492–3
Nijmegen syndrome, *456*
Nitroblue tetrazolium (NBT) test, 517
Nodal marginal zone B-cell lymphoma (nodal MZL), 328
Nodular lymphocyte predominant Hodgkin lymphoma (NLPHL), *See* Hodgkin lymphoma
Non-Hodgkin lymphoma, 287
classification, 291, *292*
clinical aspects, 293, *294*
cytogenetics, 288
international prognostic index (IPI), 293–4, *294*
Richter syndrome, 311–12, *313*
staging, 293, *294*
See also Lymphoid malignancies; Lymphoma
Non-secretory myeloma, 388–9
Non-specific esterase (NSE) marker, 211, *211*
Northern blot, 76
NPC1 gene, 492, 493
NPC2 gene, 492, 493
NPM1-ALK fusion gene, 421, 422
NPM gene, 217, 248
Nucleolus organizer regions (NORs):
visualization, 59
NuMA gene, 217

O

Oct1, 44
Oct2, 44
Omenn syndrome, *456*

Osteoblasts, 4, 13, *16*
Osteoclasts, 13–14, *17*
Osteomalacia, *96*
Osteomyelitis, *96*
Osteopenia, 93
Osteopetrosis, *96*
Osteoporosis, 93
Osteosclerosis, 93, *94*, 179
Osteosclerotic myeloma, 390

P

p15 gene, 136
p53 gene, 314, *315*, 321, 433
  myelodysplastic syndromes, 136–7
Pagetoid reticulosis, 413
Paget's disease, *96*
Pancytopenia, 223
Pappenheimer bodies, 530, *533*
Paracortical hyperplasia, *101*, 103
Paratrabecular disease, 290, *290*
Paroxysmal cold hemoglobinuria (PCH), 556
Paroxysmal nocturnal hemoglobinuria (PNH), *96*, 120–4
  clinical aspects, 123–4
  differential diagnosis, *124*
  etiology and pathogenesis, 120–1
  flow cytometry, 122–3, *122*
  genetics, 123
  pathology, 121–2, *122*
  protein deficiencies, *122*
Parvovirus-induced RBC aplasia, 531, *537*, *538*
Pautrier microabscesses, 413, *414*, *415*
PAX5, 44, 321
PAX5 gene, 44, 293, 316
PBX1/E2A fusion gene, 262
PCR, *See* Polymerase chain reaction (PCR)
PDGFRA gene, 171, 173
  See also FIP1L1-PDGFRA fusion gene
Pelger–Huet anomaly, 515, *517*
Perforin, 401, 496
Periodic acid Schiff (PAS) staining, 5, 211
Peripheral T-cell lymphoma, unspecified, 431–4
  clinical aspects, 433
  cytogenetics, 432, *434*
  differential diagnosis, 433–4
  extranodal involvement, 433, *434*
  immunophenotype, 432
  pathology, 431–2, *432*, *433*
  variants, 432, *433*
Pernicious anemia, 540
Persistent polyclonal B-cell lymphocytosis (PPBL), 464, *465*
Phagocytosis, spleen, 21
Philadelphia (*Ph*) chromosome, 57
  acute leukemias of ambiguous lineage, 283
  chronic myelogenous leukemia, 155–9, 167, *167*, 168, *168*
  precursor B-lymphoblastic leukemia/lymphoma, 260–1
Phycoerythrin (PE), 45, *45*
PIG-A gene, 120–1, 123
PK-LR gene, 554
Plasma cell leukemia, 389, *389*
Plasma cell myeloma (PCM), *96*, 373–86, *374*, 386–9
  asymptomatic (smoldering) myeloma, 386, *387*
  cytogenetics, 377–86, *383*, *384*, *385*
  diagnostic criteria, *387*, *388*
  etiology and pathogenesis, 373–5
  immunophenotype, 376–7, *381*, *382*
  indolent myeloma, 386
  non-secretory myeloma, 388–9
  pathology, 375–6, *375–81*
  prognostic factors, *389*
  staging, 388
  symptomatic, 387–8, *387*
Plasma cells, *8*, 13, *15*, *49*
  See also Plasma cell myeloma (PCM)
Plasmablastic lymphoma, 357–8, *359*
Plasmacytoma, 389–90, *391*
  extramedullary, 390, *390*, *391*
  solitary, of bone, 389–90, *390*
Platelet glycoprotein IIb, 42
Platelet glycoprotein IIIa, 42
Platelets, 11, *18*
  arachidonic acid pathway abnormalities, 577
  associated markers, 42–3
  congenital platelet disorders, 572–8
  drug-induced disorders, 578
  morphology, 17
  myelodysplastic syndromes, 133
  paroxysmal nocturnal hemoglobinuria, 123
  precursors, 11
PMC1 gene, 171
PML-RARα fusion gene, *220*, *221*
PML-RARα fusion protein, 214, 217
POEMS syndrome, 390
Polyclonal immunoblastic proliferation, 464
Polycythemia vera (PV), 174–7
  category A, 174
  category B, 174
  clinical aspects, 176–7
  cytogenetics, 176, *177*
  molecular studies, 175–6, *176*
  pathology, 174–6
Polymerase chain reaction (PCR), 65, 74–6
  basic technique, 74, *75*
  chronic myelogenous leukemia, 166, *166*
  primer design, 74–5
  product analysis, 75
  quality control, 75
  real-time PCR, 75–6
  reverse transcriptase PCR (RT-PCR), 74, 75
Polymorphisms, 79
Post-transfusion purpura, 570
Post-transplant lymphoproliferative disorders (PTLD), 465–9
  classification, 465, *467*
  clinical aspects, 468
  differential diagnosis, 468–9
  early lesions, 467, *467*
  etiology and pathogenesis, 466–7
  Hodgkin lymphoma, *467*, 468
  monomorphic, 467–8, *467*, *468*
  pathology, 467–8
  polymorphic, 467, *467*, *468*
PRAD1 gene, 347
Precursor B-lymphoblastic leukemia/lymphoma (B-ALL/B-LBL), 257–68
  clinical aspects, 266–7
  cytogenetics, 260–6, *260*
    hyperdiploidy, 265, *265*
    hypodiploidy, 265, *266*
    immunoglobulin gene rearrangement clonality, 265–6, *266*
    Philadelphia chromosome, 260–1, *261*, *262*
    t(1;19)(q23;p13.3), 261–2, *264*
    t(4;11)(q21;q23), 261, *262*, *263*
    t(12;21)(p13;q22), 262–4, *264*
  differential diagnosis, 267–8, *267*
  etiology and pathogenesis, 257–8
  immunophenotype, 259–60, *260*
  pathology, 258–9, *258*, *259*
Precursor T-lymphoblastic leukemia/lymphoblastic lymphoma (T-ALL/T-LBL), 268–73
  clinical aspects, 272–3
  cytogenetics, 270–2, *271*, *272*
  differential diagnosis, 273
  etiology and pathogenesis, 268
  immunophenotype, 269–70, *270*
  pathology, 268, *269*
PRF1 gene, 496, 497, 498
Primary amyloidosis (AL), 390–2, *392*, *393*
Primary central nervous system DLBCL, 358
Primary cutaneous aggressive epidermotropic CD8+ cytotoxic T-cell lymphoma, 428
Primary cutaneous follicle center cell lymphoma, 339
Primary cutaneous DLBCL, 359
Primary effusion lymphoma, 357, *358*
Primary immunodeficiency syndromes, 455–7, *456*
  agammaglobulinemia, 457
  DiGeorge syndrome, 457, *458*
  severe combined immunodeficiency syndrome (SCID), 455, *456*
  Wiskott–Aldrich syndrome (WAS), 455–7
Prolymphocytes, 12, *13*, 313, *314*
  See also B-cell prolymphocytic leukemia
Promegakaryocytes, 11
Promonocytes, *234*, 235, *235*, *238*, *239*
Promyelocytes, 7, *8*
  hypergranular, 215
Prorubricytes, *8*, 11
Prostaglandin E (PGE), *3*
Prostatic carcinoma:
  metastatic, 50, 91
Proteoglycans, *2*
Prussian blue stain, 9
Pseudo-Chediak–Higashi granules, 132
Pseudo-Gaucher cells, 159, *160*
Pseudo-Pelger–Huet anomaly, 132, 515
Pseudodiploidy, plasma cell myeloma, 380, 382
PTPN11 gene, 200, 202
Pure erythroid leukemia (AML-M6b), 239, 241, 242–3, *242*, *243*
  See also Acute erythroid leukemias (AML-M6)
Pure red cell aplasia (PRCA), 530–3
  associated conditions, *535*
  clinical aspects, 532–3
  differential diagnosis, 533
  etiology and pathogenesis, 530–1
  pathology, 531, *536*
Purine metabolism defects, 541
Pyrimidine metabolism defects, 541
Pyruvate kinase (PK) deficiency, 554

Q

Q fever, 85–9, *88*
5q− syndrome, *136*, 137, 145–6, *147*
Q-bands, 59
Quality control (QC):
  flow cytometry, 47
  PCR, 75

R

R-bands, 59
Radiation, 207, 226
  acute myeloid leukemia and, 207, 227
  aplastic anemia and, 119
  bone marrow morphological changes, 91–3
  chronic myelogenous leukemia and, 157–8
  myelodysplastic syndromes and, 148, 149
Radiotherapy, *See* Radiation
RAG genes, 455
RAN gene, 373
Rapoport-Leubering shunt, 553
RARα gene, 214, 216–17
  See also PML-RARα fusion gene
RAS genes, 177, 199, 200, 202
  K-RAS, 192, 195
  N-*ras*, 136, 192, 195
RB1 gene, 179
Reactive follicular hyperplasia (RFH), differential diagnosis, 339, *342*
Reactive histiocytic proliferations, 495–6
Reactive lymphadenopathies, 101, *101*
  See also Specific conditions
Reactive neutrophilia, 522, *522*, *523*
Reactive polymorphous lymphohistiocytic lesion, 465
Reactive thrombocytosis (RT):
  versus essential thrombocythemia, *182*
Real-time PCR, 75–6
Receptor activator of nuclear factor-kappaB (RANK), 93, 412
Red blood cells (RBCs), 10, 529–30
  anemia and, 530, *530–5*
  indices, 529, *529*
  microcytic hypochromic, 530, *530*
  morphology, 16–17
  paroxysmal nocturnal hemoglobinuria, 122–3
  pure red cell aplasia (PRCA), 530–3
  See also Erythrocytes; Erythropoiesis
Red cell distribution width (RDW), 529
Reduction of cytoplasmic neutrophilic granules, 515
Reed–Sternberg cells, 292–3, *442*, *443*, 446, *446*
Refractory anemia (RA), *130*, 143, *144*, 241
  with excess blasts (RAEB), *130*, 144–5, *146*, 226, 241
  with ringed sideroblasts (RARS), *130*, 143, *145*
  See also Myelodysplastic syndromes (MDS)
Refractory cytopenia with multilineage dysplasia (RCMD), *130*, 143–4, *145*
  with ringed sideroblasts (RCMD-RS), *130*, 144
Repetitive sequence probes, 66, *66*
Reticulin, 179
Reticulocytes, 11, 16, *19*, 530, *533*
Retroviruses, 208
Reverse transcriptase PCR (RT-PCR), 74, 75
Rhabdomyosarcoma, metastatic, 91
Rheumatoid arthritis (RA), 102–3, *104*, 495, 496
Richter syndrome, 311–12, *313*
Ringed sideroblasts, 143, 150
Rocky Mountain spotted fever, *88*
Rosai–Dorfman disease, 105, 496
RPS19 gene, 117, 531
Rubriblasts, *8*, 10–11
Rubricytes, *8*, 11
RUNX1 gene, 229, 243
  see also ETV6-RUNX1 fusion gene
RUNX1/RUNXT1 fusion gene, 208, 211, *214*
  acute myeloid leukemia with maturation, 232
  acute myeloid leukemia with t(8;21), 212–14
Russell bodies, 13, *15*, 316, 376, *379*

S

SAP genes, 463
Sarcoidosis, 85, *85*, 107, *108*, 496
Sarcoma:
  follicular dendritic cell, 508, *508*
  granulocytic, 247–8, *247*
  histiocytic, 499
  interdigitating dendritic cell, 506–8, *507*
  Langerhans cell, 503, *504*
  mast cell, 485–6
Schistocytes, 530, *532*, *558*
Sclerosis:
  bone marrow, *90*
  osteosclerosis, 93, *94*
Segmented cells (Segs), 8, *8*, 17, *18*, *19*
Selectins, *2*
Senile EBV-associated B-cell lymphoproliferative disorder, 451, 469
Serum amyloid P component (SAP), 391
Severe combined immunodeficiency syndrome (SCID), 455, *456*
Sezary cells, 413, *414*
Sezary syndrome (SS), 413–16
  clinical aspects, 414–16
  cytogenetics, 413–14
  differential diagnosis, 416, *417*
  etiology and pathogenesis, 413
  immunophenotype, 413, *416*
  pathology, 413, *414*, *415*
  staging, 415
Shwachman–Bodian–Diamond syndrome (SBDS), 117
Shwachman–Diamond syndrome, 117, *117*
Shwachman–Diamond–Oski syndrome, 521
Sialomucins, *2*
Sickle cell (SC) disease, 549–51
  clinical aspects, 549–51
  differential diagnosis, 551
  etiology and pathogenesis, 549
  pathology, 549, *550*
Sinus histiocytosis with massive lymphadenopathy, 105, *106*, 496
Sinus pattern reactive lymphadenitis, *101*, 103–7
  hemophagocytic syndromes, 107
  Langerhans cell histiocytosis, 107
  monocytoid B-cell hyperplasia, 106, *106*
  sinus histiocytosis, 103–5, *105*
  with massive lymphadenopathy, 105, *106*
  Whipple's disease, 106–7
Sinuses:
  lymph nodes, 24
  spleen, 21, *21*
Small lymphocytic lymphoma (SLL), *292*, 301, 309
  staging, 309, *312*
Smears:
  blood, 16–17
  bone marrow, 5, *6*
Smoldering myeloma, 386, *387*
SMPD1 gene, 492, 493
Smudge cells, 302, *302*
Southern blot, 76, *76*
  chronic myelogenous leukemia, 165–6
Spectral karyotyping (SKY), 67
Spectrin, 551
Spherocytes, 552, *552*
Sphingolipidoses, *490*
Spleen, 17–21, *19*
  extramedullary hematopoiesis, 179
  hepatosplenic T-cell lymphoma, *418*
  hypersplenism, 557–8
  marginal zone, 18–21
  red pulp, *20*, 21, *21*, *48*
  sinuses, 21, *21*
  white pulp, 18, *20*, *21*, *48*
  See also Splenomegaly
Splenic artery, 18
Splenic cords, 21
Splenic marginal zone lymphoma (SMZL), 320–4
  clinical aspects, 322–3
  cytogenetics, 321–2, *324*, *325*
  differential diagnosis, *320*, 323–4
  etiology and pathogenesis, 320
  immunophenotype, 321
  pathology, 320–1, *321*, *322*, *323*
Splenomegaly:
  chronic idiopathic myelofibrosis, 180, 181
  chronic myelomonocytic leukemia, 193
  chronic myeloproliferative leukemia, 159, *161*
  chronic neutrophilic leukemia, 170
  See also Spleen
Stabs, 8, 17
Steel factor (SF), *3*, 4, 5
Stem cells, 1
Still's disease, 103
Stomatocytes, 530, *531*, 553
Stomatocytosis, 553
Stromal cells, 1–2, *17*
STX11 gene, 497, 498
Subcutaneous panniculitis-like T-cell lymphoma, 427–8, *427*
Subtelomeric probes, 66–7, *67*
Sudan Black B stain, 210, *211*
Supernumerary marker chromosomes (SMCs), 61–2
survivin, 195
Systemic lupus erythematosus (SLE), 495, 496

T

T-cell large granular lymphocytic leukemia (T-LGL), 402–4

clinical aspects, 404
cytogenetics, 403, *403*, *404*, *405*
differential diagnosis, 408, *410*
etiology and pathogenesis, 402–3
immunophenotype, 403, *403*
pathology, 403
T-cell lymphoma, *See Specific types of lymphoma*
T-cell prolymphocytic leukemia (T-PLL), 397–401
clinical aspects, 401
cytogenetics, 398–401, *399*, *400*, *401*
differential diagnosis, 401, *417*
etiology and pathogenesis, 397
immunophenotype, 398, *399*
pathology, 397–8, *398*
T-cell receptor (TCR), 40, 270–2, 288
T-cell large granular lymphocytic leukemia, 403
T-cell prolymphocytic leukemia, 398–400
T-cell receptor (TCR) complex, 39, *39*, 40
T-cell/histiocyte-rich diffuse large B-cell lymphoma, 445
T-cells, 5, 12, *289*
associated markers, 39–40
cytotoxic, 12
helper, 23
lymph nodes, 23, 24, *49*
spleen, 18, *48*
suppressor, 23
See also Mature T-cell neoplasms
Tartrate resistant acid phosphatase (TRAP), 316, *329*, 331
TAX genes, 410
TCR genes, 270–2, 283, 414, 418, 420, 432
T-cell prolymphocytic leukemia, 398–400
TCRA gene, 397
TdT (terminal deoxynucleotidyl transferase), 12, 43, 259, 267, 269, 281, 364
TEL-AML1 fusion gene, 262
Telangiectasia macularis eruptiva perstans, 482
Tetrahydrofolate (THF), 540
Thalassemia syndromes, 545–8
α-thalassemia, 547–8
acquired, 548
ß-thalassemia, 545–7, *545*, *547*, *548*
δß-thalassemia, 548
differential diagnosis, 548
Thrombocytopenia, 568–71
amegakaryocytic, 118
congenital (CAMT), 567
autoimmune thrombocytopenic purpura, 568–70
drug-induced, 570
neonatal, 570
with absent radius (TAR syndrome), 118, 567, *568*
Thrombocytosis:
acute myeloid leukemia with chromosome 3 aberrations and thrombocytosis, 248
chronic myeloproliferative disorders, 186
See also Essential thrombocythemia (ET); Reactive thrombocytosis (RT)
Thrombopoiesis, 5
Thrombopoietin (TPO), *3*, 567
Thrombopoietin receptor (TPO-R), 42
Thrombospondin, *2*
Thrombotic thrombocytopenic purpura (TTP), 570–1, *571*
Thromboxane A, 577
Topoisomerase II inhibitor-related AML, 227–8
Touch preparations, 5, *6*
Toxic granulation, 513, *514*
Toxoplasmosis, 107, *110*
TPO gene, 183
Transferrin receptor, 42
Transforming growth factor-beta (TGF-ß), *3*, 177
Transient erythroblastoma of childhood, 532
Transient myeloproliferative disorder (TMD), *151*, 523
Translocations, 60–1, *60*, *61*, 67, *68*
1;7, *177*
1;14, 327
1;19, 261–2, *264*
1;22, 245, *245*, 248
4;11, 261, *262*, *263*
5;9, 432
5;11, *233*
5;12, 173, *173*, *198*
5;14, *142*
5;17, 217
7;14, *272*
8;16, *239*
8;21, 212–14, 232
9;22, 61, 260–1, *261*
11;14, *59*, 314, 347, *348*
11;17, 217, *220*
11;18, 327
12;21, *71*, 257, 262–4, *264*
14;18, 336
15;17, 57, 214, 217, *220*
16;16, *222*
See also Cytogenetics
TRAP (tartrate resistant acid phosphatase), 316, *329*, 331
Trisomy, 3, 327, 328
Trisomy, 8, 61, *66*, 70
acute myeloid leukemia, 208
aplastic anemia, *121*
chronic myelomonocytic leukemia, *198*
myelodysplastic syndromes, *141*, 142
Trisomy, 9, *157*
Trisomy, 12, 59, *60*
chronic lymphocytic leukemia, 307–8, *309*
Trisomy, 13, *141*
Trisomy, 21, 61, *69*
acute myeloid leukemia, 208
chronic neutrophilic leukemia, *171*
See also Down syndrome
Trisomy, 22, *223*
Tropheryma whippeli, 106–7
Tryptases, *478*
Tuberculosis, *85*, *88*
Tumor necrosis factor (TNF), *3*, *478*
Typhoid fever, *88*
Tyrosine kinase, in chronic myeloproliferative diseases, 155–6, 171

U

UNC13D gene, 497, 498
Urticaria pigmentosa, 482

V

Very late activation antigens (VLA), *2*
Vitamin deficiencies, 559
vitamin B12, 540, *541*, 542–3
von Willebrand factor (VWF), 572, 575

W

Waldenstrom macroglobulinemia (WM), *See* Lymphoplasmacytic lymphoma (LPL)
Whipple's disease, 106–7
White blood cells (WBC) counts, 17, *19*
Whole-chromosome painting (WCP), 67, *68*
Wiskott–Aldrich syndrome (WAS), 455–7, *456*, 576–7
Wrights stain, 5, 16

X

X-linked lymphoproliferative syndrome (XLP), 463, *464*
X-linked thrombocytopenia, 577
Xeroderma pigmentosum, *456*
XLAP genes, 463

Z

ZAP-70, 44, 303, 305, *308*, 309–10
deficiency, *456*
Zeive syndrome, 557
ZHX-2 gene, 375